AF527632

Network Analysis

Third Edition

Network Analysis

Third Edition

Dr. N.C. Jagan
B.E., M.E., Ph.D., MISTE, FIE

Professor in Electrical Engineering (Retd.)

University College of Engineering
Osmania University
Hyderabad.

Dr. C. Lakshminarayana
B.E., M.S., Ph.D.

Professor in Electrical Engineering (Retd.)

University College of Engineering
Osmania University
Hyderabad.

4-4-309/316, Giriraj Lane, Sultan Bazar,
Hyderabad - 500 095
Phone : 040 - 23445605, 23445688

Network Analysis, ***Third Edition***

Published by :

4-4-309/316, Giriraj Lane, Sult an Bazar,
Hyderabad - 500 095
Phone : 040 - 23445605, 23445688
e-mail : info@bspbooks.net

ISBN : 978-93-85433-75-7 (HB)

Preface to Third Edition

We are glad to bring the third edition of the book. It is felt that the students will be benefited by a detailed study of Laplace Transform method of Analysis of Networks. Hence a separate chapter is included in which Laplace Transforms are defined, their properties explained and then their application to networks is discussed in detail.

The chapter, 'Two Port Networks' is shifted suitably so that the RLC two ports can be analysed using Laplace Transforms. The Equalizers portion in filters chapter is deleted as this is not considered in syllabi of many Universities. A new chapter on Fourier series and Fourier transforms is added. Many universities are including this topic in Network Analysis course.

Multiple Choice Questions (MCQs) are included chapter wise for the new chapters also. Qucstions on topics in Nctwork Analysis which have occurred in Gate Question papers from Years 2004 to 2014 with answers, yearwise, are added at the end.

We earnestly hope that this edition will satisfy the students and teachers of this course as before.

We wish to thank M/s BS Publications for bringing out this new edition of the Book.

- Authors

Preface to Second Edition

Network Analysis or Electrical Circuits Analysis is a foundation course for many Engineering disciplines like Electronics and Communications, Electrical and Electronic Engineering, Computer Science Engineering etc. The authors have vast experience in teaching this course over several years and it is felt that this experience can be shared with the new generation of teachers and students.

In this Second Edition, the chapters are reorganised to present the contents in proper sequence and make it systematic.

The first two chapters introduce the basic elements of the network and present simple anaysis techniques for resistive networks. Steady state sinusoidal analysis is presented in chapters 3 and 4. Topological properties of networks and the analysis of networks based on these properties are discussed in chapter 5. Chapter 6 describes some additional useful techniques for solving networks. Properties and analysis of 2-port networks are discussed in chapter 7. Behaviour of RLC networks for suddenly applied excitations, is described next in chapters 8 and 9. This behaviour of the network is known as transient behaviour and plays a very important role in netwrok analysis.

Chapter 10 describes the analysis and design of frequency selective two port networks called filters. Design of attenuators and equalizers also are discussed in this chapter.

Number of examples are worked out to illustrate the concepts developed and also number of problems are given for the student to solve, at the end of each chapter. Answers are given at the end of the book, to enable the students to verify their answers, which will boost their confidence.

Over 500 Multiple Choice Questions (MCQs) are given chapter-wise at the end. These questions are selected from various competetive examinations like UPSC, GATE etc. This will also help the students of JNTU and other universities who have online internal tests based on MCQs. Answers for all these questions are also provided at the end.

We are extremely thankful to all the students and teachers who have used this book and made valuable suggestions.

We are also thankful to M/S BS Publications for bringing out this new edition in record time. We are also thankful to all the members of our families for their encouragement and support throughout the preparation of the material for the book.

-Authors

CONTENTS

Preface to Third Edition .... (v)
Preface to Second Edition .... (vii)

CHAPTER 1

Network Variables and Elements

1.1 Introduction 1
1.2 Network Model 1
1.3 Network Variables 2
1.4 Network Elements 6
1.4.1 Active elements 6
1.4.2 Passive elements 9
1.5 The Resistor 10
1.6 The Inductor 13
1.6.1 Faraday's law of electromagnetic induction 13
1.6.2 Self inductance 14
1.6.3 Mutual inductance 19
1.6.4 Dot convention 20
1.6.5 Energy consideration in the magnetically coupled coils 22
1.7 The Capacitor 26
Problems 32

CHAPTER 2

Kirchhoff's Laws and Resistive Networks

2.1 Introduction 37
2.2 Kirchhoff's Laws 38
2.2.1 Kirchhoff 's current law 38
2.2.2 Kirchhoff 's voltage law 39

2.3 Resistive Networks 40
2.3.1 Resistors in series 41
2.3.2 Resistors in parallel 44
2.3.3 The voltage divider 46
2.3.4 The current divider 47
2.3.5 Ladder network 49
2.3.6 Source transformations 51
2.3.7 Voltage source with no series resistance 52
2.3.8 Current source with no parallel resistor 53
2.3.9 Voltage sources with parallel resistors or current sources 54
2.3.10 Current sources in series with resistors or voltage sources 55
2.3.11 Star – delta or Y – Δ transformation 60

2.4 Node Voltage Method 66

2.5 Mesh Current Method 75
2.5.1 Dummy mesh current method 79

2.6 Magnetic Circuits 84
2.6.1 Magnetic fields 84
2.6.2 Some definitions 85

2.7 Properties of Magnetic Materials 86
2.7.1 B-H curves 86
2.7.2 Classes of magnetic materials 87

2.8 Simple Magnetic Circuits 87
2.8.1 Leakage flux and leakage coefficient 88
2.8.2 Composite magnetic circuits 89

Problems 92

CHAPTER 3

Single Phase Circuits

3.1 Introduction 101

3.2 Characteristics of Periodic Functions 101

3.3 Average Value, R.M.S Value, form Factor 103
3.3.1 Average value 104
3.3.2 Effective value or RMS value 104
3.3.3 Form factor 107

3.4 Representation of a Sine Function 108
3.4.1 Trignometric representation 108
3.4.2 Phasor representation 109
3.4.3 Complex algebra representation 110

3.5 Steady State Response of R, L and C Elements to Sinusoidal Excitations 113
3.5.1 Response of pure resistance 113
3.5.2 Response of pure inductance 116
3.5.3 Response of a pure capacitor 119
3.5.4 Series R – L circuit 123
3.5.5 Series RC circuit 128
3.5.6 Series RLC circuit 130
3.5.7 Parallel RL circuit 132
3.5.8 Parallel R C circuit 133

3.6 Series Resonance 141
3.6.1 Characteristics of a series resonant circuit 141
3.6.2 Q – factor 145
3.6.3 Selectivity and bandwidth 147
3.6.4 Voltage across the capacitor 149
3.6.5 Voltage across the inductor 150
3.6.6 Resonance by varying capacitance 151

3.7 Parallel Resonance 153
3.7.1 Maximum impedance by adjusting frequency 155
3.7.2 Selectivity and bandwidth 157

3.8 Locus Diagrams 159
3.8.1 Impedance locus of series connected elements 160
3.8.2 Admittance locus of series connected elements 160
3.8.3 Admittance locus of parallel connected elements 162
3.8.4 Impedance locii of parallel connected elements 163
3.8.5 Current locus of a parallel circuit with one variable branch . 163
3.8.6 Inversion of circle 165

3.9 Magnetically Coupled Circuits 168
3.9.1 Sinusoidal steady state analysis of magnetically coupled circuits 169
3.9.2 The linear transformer 172
3.9.3 The ideal transformer 176

Problems 181

Chapter 4

Three Phase Circuits

4.1 Introduction 189

4.2 Advantages of Three Phase Over Single Phase 189

4.3 Generation of Three Phase Voltages 190

4.4 Interconnection of Three Phase Windings 192

4.4.1 The star connection 192

4.4.2 The delta connection 194

4.5 Analysis of 3 Phase Networks 196

4.5.1 Balanced star connected load supplied from balanced 3 phase 4 wire supply 197

4.5.2 Balanced star connected load supplied from a balanced 3 phase 3 wire supply 198

4.5.3 Power in a star connected system 202

4.5.4 Balanced Δ – connected load connected to a 3 – phase balanced supply 203

4.5.5 Power in a Δ – connected balanced system 205

4.6 Unbalanced Three Phase Network 208

4.6.1 Unbalanced star connected load supplied from a 3 – phase balanced 4 – wire supply 208

4.6.2 Unbalanced Δ – connected load 209

4.6.3 Unbalanced star connected load supplied from a 3 phase balanced 3 wire supply 211

4.7 Measurement of Three Phase Power 216

4.7.1 Two wattmeter method 216

4.7.2 Wattmeter readings and power factor in a balanced load 218

Problems 226

Chapter 5

Network Topology

5.1 Introduction 229

5.2 The Concept of a Network Graph 229

5.3 Chords or Links 235
5.3.1 Incidence matrix (A) 235
5.3.2 Circuit matrix or tie set matrix B 238
5.3.3 Cutset matrix Q 241
5.4 Formulation of Network Equations 243
5.4.1 A typical branch of a network 244
5.4.2 Node voltage equations 245
5.4.3 Loop current equations 249
5.4.4 Cutset equations 255
Problems 262

CHAPTER 6

Network Theorems

6.1 Introduction 265
6.2 General Network Equations 265
6.3 Superposition Theorem 268
6.4 Thevenin's Theorem 274
6.5 Norton's Theorem 276
6.6 Millman's Theorem 286
6.7 Reciprocity Theorem 289
6.8 Compensation Theorem 291
6.9 Tellegen's Theorem 296
6.10 Maximum Power Transfer Theorem 301
6.10.1 When source has internal resistance and the load is purely resistive 301
6.10.2 When source has an impedance and the load has variable impedance 302
6.10.3 When source has impedance and the load is purely resistive 303
Problems 309

Chapter 7

Differential Equations and Initial Conditions in RLC Networks

7.1 Introduction 316
7.2 Formulation of Network Equations 316
7.3 Duality 320
7.4 Initial Conditions in Networks 326
7.4.1 Behaviour of R, L and C at t = O 327
7.4.2 Behaviour of R, L and C at t = ∞ 329
7.4.3 Special cases 339
Problems 348

Chapter 8

Time Response of R L C Networks

8.0 Response of R L C Networks 356
8.1 Types of Energy Sources 356
8.2 Response of RL Networks 359
8.2.1 Internal energy excitation 359
8.2.2 Unit impulse excitation 361
8.2.3 Unit step excitation 362
8.2.4 Sinusoidal excitation 366
8.3 Response of RC Networks 368
8.3.1 Internal energy excitation 369
8.3.2 Impulse excitation 370
8.3.3 Unit step excitation 371
8.3.4 Sinusoidal excitation 373
8.4 Response of RLC Networks 381
8.4.1 Internal energy excitation 381
8.4.2 Impulse excitation 390
8.4.3 Unit step excitation 391
8.4.4 Sinusoidal excitation 394
Problems 399

CHAPTER 9

Laplace Transform Method of Analysis of Networks

9.1 Introduction 411

9.2 The Laplace Transformation 412

9.3 Laplace Transform of Some Important Time Functions 413

9.4 Some Theorems of Laplace Transformation 415

9.5 Waveform Synthesis 434

9.6 Gate Function 438

9.7 Periodic Function 441

9.8 Application to Solution of Differential Equations 444

9.9 Heaviside's Partial Fraction Expansion Technique 445

9.10 Application to Networks 449

9.11 Transform Networks 452
- 9.11.1 Resistance 452
- 9.11.2 Inductance 452
- 9.11.3 Capacitance 454
- 9.11.4 Solution of transform networks 455

9.12 Transfer Function 461
- 9.12.1 Poles and zeroes of transfer function 464
- 9.12.2 Significance of poles and zeros 466

9.13 Network Functions 469
- 9.13.1 Restrictions on poles and zeros of driving point functions 470
- 9.13.2 Restrictions on locations of poles and zeros of transfer functions 473

Problems 478

CHAPTER 10

Two Port Networks

10.1 Introduction 487

10.2 Two Port Networks 488

10.3 Open Circuit Impedance Parameters 489

10.4 Short Circuit Admittance Parameters 490

10.5 Transmission Parameters 495

10.6 Hybrid Parameters or h – Parameters 496

10.7 Relationships Between Different Sets of Parameters 499

10.7.1 Relation between y and z – parameters 499

10.7.2 Transmission parameters to z – parameters 500

10.7.3 h – parameters to z – parameters 500

10.7.4 Inverse hybrid parameters to z – parameters 501

10.8 Reciprocity and Symmetry 502

10.9 Interconnection of Two Port Networks 510

10.9.1 Cascade connection of networks 510

10.9.2 Series connection networks 514

10.9.3 Parallel connection of networks 516

10.9.4 Series parallel connections 519

10.9.5 Parallel – series connection 521

Problems 530

CHAPTER 11

Network Filters and Attenuators

11.1 Introduction 538

11.2 Symmetrical Networks 538

11.2.1 Image impedances of networks 539

11.2.2 Characteristic impedance of symmetrical networks 541

11.3 Measure of Attenuation : Neper and Decibel 543

11.4 Propagation Constant 545

11.5 Propagation Constant of Symmetrical Networks 546

11.6 Filter Characteristics 548

11.7 Nature of Characteristic Impedance in Pass and Stop Bands 551

11.8 Constant K – Filters 552

11.8.1 Constant – K low pass filter 552

11.8.2 Phase shift of a constant – K LP filter 554

11.8.3 Attenuation of a constant–K LP filter 554

11.8.4 Characteristic impedance of a constant – K LP filter 555

11.8.5 Constant – K high pass filter 555

11.9 m–Derived Filters 557
11.9.1 m–Derived T–filter 558
11.9.2 m–Derived π–filter 559
11.9.3 m–Derived low pass filter 560
11.9.4 m–Derived high pass filter 564
11.9.5 Terminating half or L – section 567

11.10 Bandpass Filters 571
11.10.1 Bandpass constant K – filter 571
11.10.2 Bandpass m–derived filter 574

11.11 Bandstop or Band Elimination Filter 577
11.11.1 Constant K bandstop filter 577
11.11.2 m–Derived bandstop filter 579

11.12 Attenuators 592
11.12.1 Symmetrical attenuators 592
11.12.2 Symmetrical T – attenuator 592
11.12.3 Symmetrical π – attenuator 593
11.12.4 Symmetrical bridged T – attenuator 594
11.12.5 Symmetrical lattice attenuator 597

Problems 600

CHAPTER 12

Fourier Series and Fourier Transforms

12.1 Introduction 602

12.2 Fourier Series 602

12.3 Convergence of Fourier Series 607

12.4 Symmetry Conditions in Fourier Series 608
12.4.1 Even function symmetry 609
12.4.2 Odd function symmetry 612
12.4.3 Halfwave or rotation symmetry 613
12.4.4 Quarterwave symmetry 616
12.4.5 Concealed symmetry 618

12.5 Complex Fourier Series 620
12.5.1 Some comments about complex fourier series 623

12.6 Fourier Transforms 629
12.6.1 Existence of the fourier transform 631
12.7 Fourier Transform Properties 639
12.8 Application to Networks 654
12.8.1 Periodic input functions 654
12.8.2 Pulse input functions 655
Problems 656

Answers to Problems 663
Multiple Choice Questions (MCQs) for Competitive Examinations 687
Answers to MCQs for Competitive Examinations 805
MCQs from Gate Examination, Yearwise from 2004 to 2014 811

1

NETWORK VARIABLES AND ELEMENTS

1.1 INTRODUCTION :

Since the first discovery of electricity in the year 600 BC by Thales of Miletus, the knowledge and use of electricity has multiplied by enormous counts. All the modern day conveniences are dependent to the most extent on electricity. Without electricity the life of the modern man will come to a standstill.

Network theory is a basic discipline of electric science. It describes how the energy is transferred from one device to another. Thorough understanding of this subject is essential for every type of engineer and physicist. The objective of this text is to provide a thorough understanding and proficiency in the subject of engineering circuit analysis.

1.2 NETWORK MODEL :

Just as in every branch of science and engineering, where the behaviour of a complex physical device is studied by constructing its mathematical model, in network theory also, an idealised model for the device is constructed and its behaviour is analysed. Network theory deals with the determination of response of this network model to a stimulus or excitation as shown in Fig. 1.1. In electrical engineering it is often possible to separate the excitation and response.

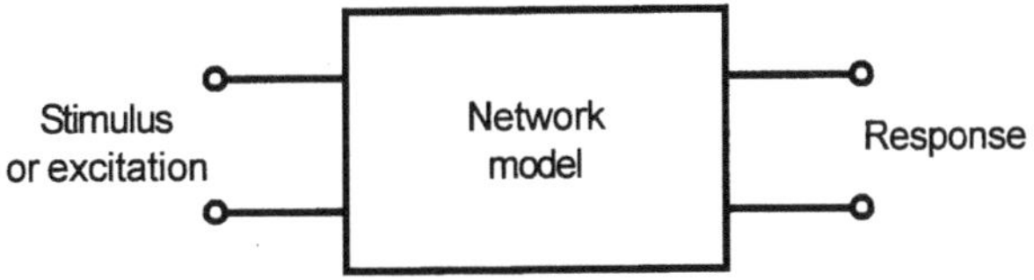

Fig. 1.1 Excitation and response

The network model itself is an interconnection of various idealised models of physical devices. A basic characteristic of many physical networks encountered in engineering practice is that the response is proportional to the excitation. If the input is doubled, the output is also doubled. This property of the network is called *linearity*. In this text we will be concerned with the study of networks which obey this property.

We are often concerned with the flow of electrical energy from one device to another and not in the internal distribution of the energy in any particular device. Thus we will treat the basic building blocks of our network models to be '*lumped*' elements rather than distributed elements. Thus their behaviour can be analysed by the effects produced at the terminals of the device, through which the energy leaves or enters the device. A simple model of an electric device is represented in Fig. 1.2.

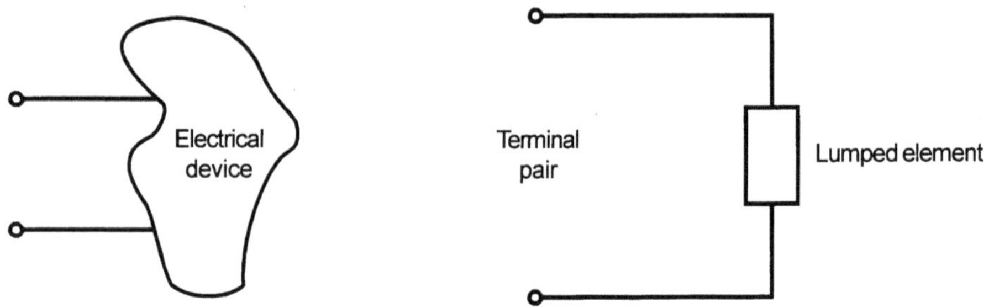

Fig. 1.2 An electrical device and its lumped model

The networks are constructed with these ideal elements interconnected by ideal connectors which absorb zero energy.

1.3 NETWORK VARIABLES :

As we are concerned with the flow of energy from one device to another in a network, we must know what are the variables that are used in the study of the flow of energy. Energy by itself is not a measurable quantity and is therefore measured by association with other measurable quantities. For example, the mechanical energy is measured in terms of force and distance.

The electrical energy can be measured by defining a measurable quantity called *electric charge*. Charges were first observed by rubbing dry substances together. Suppose a light material such as pith ball is suspended by a string and a hard rubber comb, rubbed with a woollen cloth, is brought near it, it is observed that the pith ball tends to swing away from it. If the pith ball is approached with the woollen cloth now, the pith ball is found to be attracted towards it. This phenomenon is explained by saying that there are forces acting due to electric charges present on the pith ball, the comb and the woollen cloth. Since these forces can be either attraction or repulsion as seen in the above experiment, it is postulated that there are two kinds of electrical charges : positive and negative and like charges repel and unlike charges attract. Benjamin Franklin first named the charges that are present on the comb to be negative and that on the woollen cloth, positive.

According to atomic physicists, the entire matter in the universe is made up of fundamental building blocks called *atoms* and these atoms are composed of different kinds of fundamental particles. The three most important particles are the electron, proton and the neutron. The ultimate unit of electric charge is the electron, with a mass of 9.107×10^{-31} kg. The fundamental unit of charge, called *the Coulomb*, named after Charles Coulomb, is 6.24×10^{18} electrons. These electrons move in orbits around the nucleus of an atom and many of them break loose and move at random in the interatomic spaces. An excess of electrons in a body represents negative charge and deficit of electrons represents a positive charge.

Hans Christian Oersted, in the year 1819, observed that a flow of electrical charge produced a force on a nearby magnetic compass needle and he found that the force was proportional to the rate of flow of charge. Since the measurement of this force was much easier than the measurement of forces between static charges, a new variable, current, was defined as the rate of flow of electric charge and its unit, ampere = 1 Coulomb/sec. Thus the current, rather than the charge, has come to be the basic variable of circuit analysis. Charges are represented by the symbols q or Q and the currents are represented by i or I depending on whether they are varying with respect to time or constant respectively. Thus

$$i = \frac{dq}{dt} \qquad .....(1.1)$$

We observe two important aspects of the flow of current. First, the current has direction. The number of coulombs per second has a direction and the current is said to be flowing in a circuit in a particular direction. Secondly, the current flow may be due to the flow of electrons in a particular direction or due to motion of positive charge, or due to motion of positive charge in one direction and negative charge in the other direction. Conventionally, the direction of current is taken to be the flow of positive charge in the direction of a reference arrow placed for convenience.

In Fig. 1.3, two equivalent methods are shown to represent the current in a conductor. 1A of current in Fig. 1.3(a) means a net positive charge of 1 C/sec is flowing in the direction of arrow, whereas, in Fig. 1.3(b) a net negative charge of –1 C/sec

Fig. 1.3 Refrence direction for currents

is flowing in the direction of arrow. Since the magnetic effect produced by a flow of positive charge in one direction and that due to a flow of negative charge in the opposite direction is identical, the currents are equivalent in both the cases. Hence to avoid ambiguity, the current is always considered to be an equivalent flow of positive charge. Thus a current of positive charge flowing in the direction of an arbitrary reference direction is taken to be positive. Again coming back to our lumped model of the elements, the current is measured at the terminals and no consideration is given to the finite speed of propogation of electricity through the material. Thus the current entering one terminal of the element is taken to be equal to the current leaving the other terminal and currents are measured and defined only at the terminals of the lumped element as shown in Fig. 1.4.

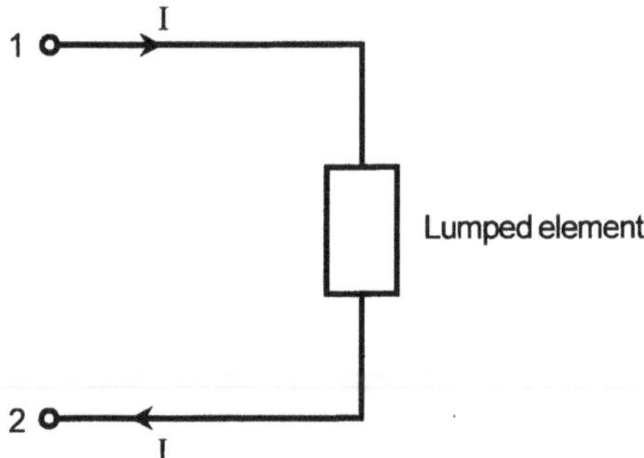

Fig. 1.4 Current in a lumped element

Coming back to the measurement of electrical energy, in addition to the concepts of charge and current, we need to define an additional variable to measure the amount of energy lost or gained by each charge as it passes through the element. This variable is called *voltage* or *potential* and is denoted by the letters e, E or v, V.

The voltage across a terminal pair of a device is a measure of the work required to move the electric charge through the device. Thus

$$v = \frac{dW}{dQ} \qquad .....(1.2)$$

where v = voltage in Volts
W = energy in Joules
Q = charge in Coulombs

From the definitions of current and voltage given in eqs. (1.1) and (1.2), it is clear that the product of voltage and current is the power associated with the device. Power is measured in Joules/sec or watts and is represented by the symbols p or P. Thus

$$p = v \cdot i \qquad .....(1.3)$$

Just as a current has direction, voltage has a polarity depending on whether energy is supplied to the device or received from it. A plus sign at the terminal where the current enters the device indicates that the energy is absorbed by the device and negative sign at that terminal indicates that the device is supplying energy. We therefore use a pair of signs, plus and minus, at the two terminals of an electric device to give a reference polarity for the voltage. Thus, placing a +ve sign at the terminal 1 and –ve sign at the terminal 2 indicates that the terminal 1 is at a higher potential than the terminal 2. As for the current, the reference marks for voltage are for algebraic reference only. If the actual voltage has the same polarity as the reference marks, the voltage is positive and it is negative if it has opposite polarity. Thus the two figures 1.5(a) and 1.5(b) are equivalent.

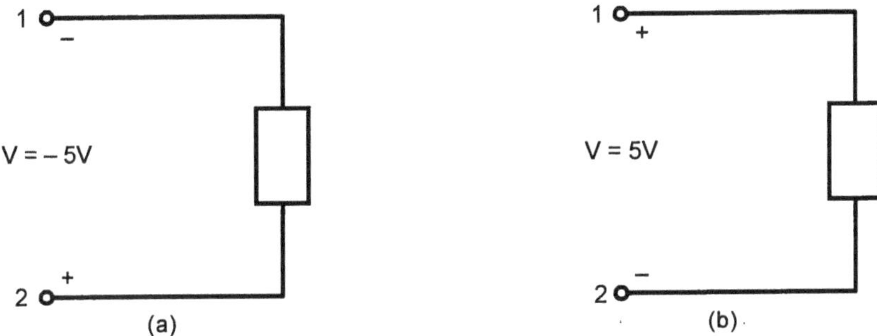

Fig. 1.5 Reference directions for voltage

Consistent signs must be assigned to the current and voltage in a device if the power is absorbed by the device or supplied by the device. First an arbitrary direction is assigned for the current and the sign of voltage is made to agree. A plus sign is assigned to the terminal at which the current enters the device. With this convention, a positive sign for power indicates that the device is absorbing power or it is a sink, whereas a negative sign for the power indicates that the device is supplying power and hence it is a source.

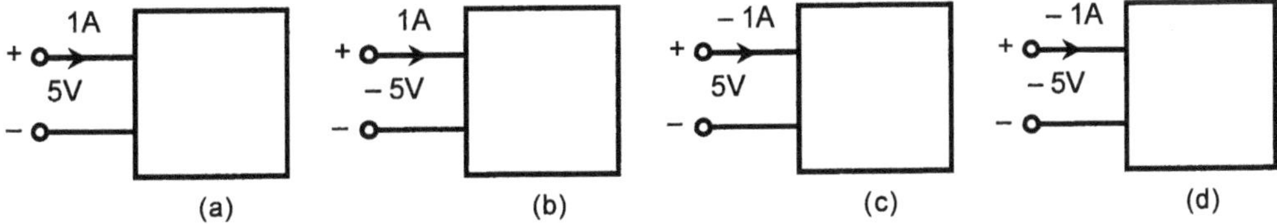

Fig. 1.6 Reference directions for voltage and current

In Fig. 1.6 four different combinations of the polarities are shown for an electric device. In Fig. 1.6(a) the power is P = 5 × 1 = 5 watts and hence it is a sink. In Fig. 1.6(b) the power is P = – 5 × 1 = – 5 W and hence it is a source. In a similar manner Fig. 1.6(c) is a source and Fig. 1.6(d) is a sink.

1.4 NETWORK ELEMENTS :

The network elements are the mathematical models of two terminal electric devices which can be characterised by their voltage and current relationship at the terminals. These network elements are interconnected to form a network and can be divided into two major groups as shown in Fig. 1.7.

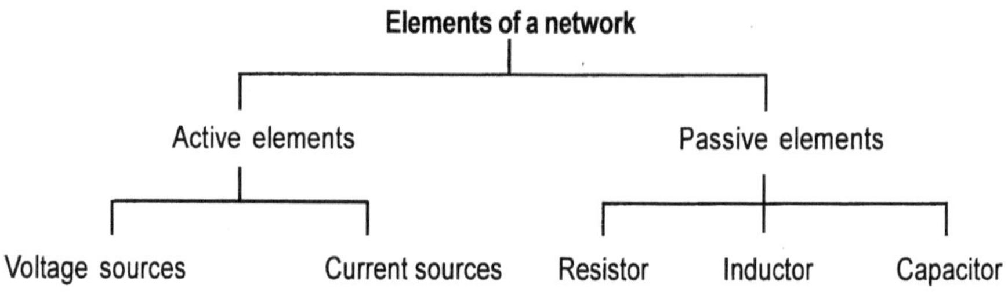

Fig. 1.7 Classification of elements of a network

1.4.1 Active elements :

Active elements are those elements which are capable of delivering energy to the networks or devices which are connected across them. There are essentially two kinds of active elements in a network.

Independent voltage source :

An independent voltage source is characterised by the property that the voltage across its terminals is independent of current passing through it. The symbolic representation and the v – i relationship of the element is given in Fig. 1.8.

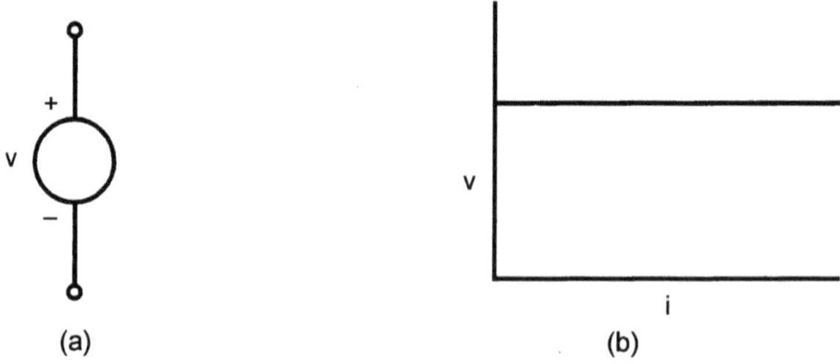

Fig. 1.8 Independent voltage source (a) its circuit representation and (b) v – i characteristic

If the voltage source has a constant voltage, it is termed as a d.c voltage source and is represented by either of the symbols shown in Fig. 1.9(a) or (b).

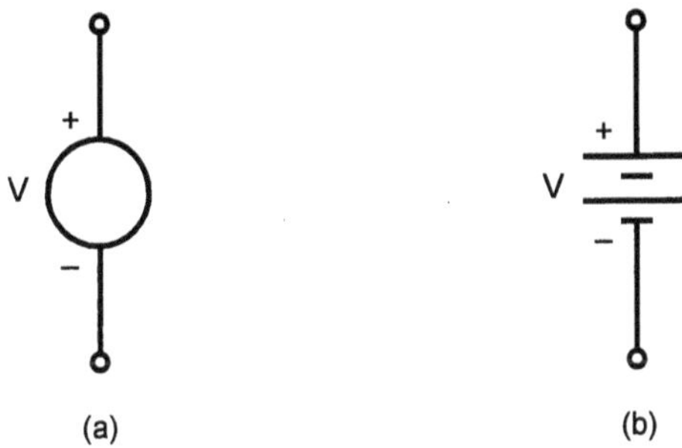

Fig. 1.9. Symbolic representation of d.c. sources

The sources described above are '*ideal*' sources and do not represent exactly any practical device encountered in physical world, because, theoretically it can deliver infinite amount of power to the circuit to which it is connected. This device is supposed to deliver unlimited number of coulombs per second, each receiving 'v' Joules as it passes through the source. This, clearly, is impossible, but there are several practical devices which can be modelled very closely with these elements. The v – i characteristic of a practical voltage source is characterised by a drooping characteristic as given in Fig. 1.10.

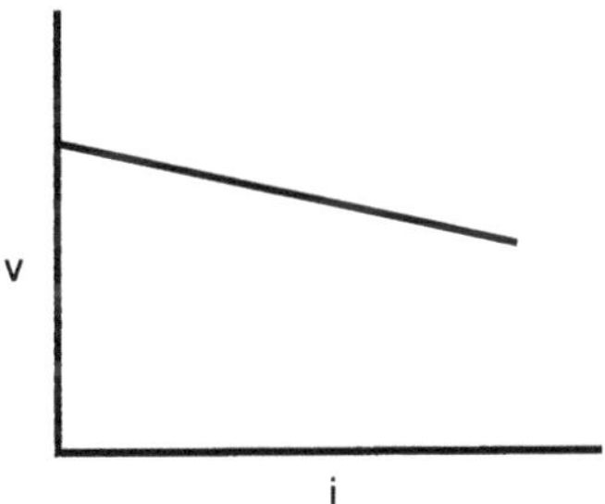

Fig. 1.10 v – i characteristic of a practical voltage source

One point worth emphasising here is about the polarity signs placed in Fig. 1.8(a). The '*plus*' sign placed at the upper terminal does not mean that it is always positive with respect to the lower terminal. In fact, it only means that the upper terminal is 'v' volts positive with respect to the other terminal at a particular instant of time. If 'v' is negative at any instant the upper terminal is actually 'v' volts negative with respect to the lower terminal at that instant.

Independent current source :

The second ideal source is an independent current source which delivers a constant current independent of the voltage across its terminals. The symbol and the v – i characteristic are given in Fig. 1.11.

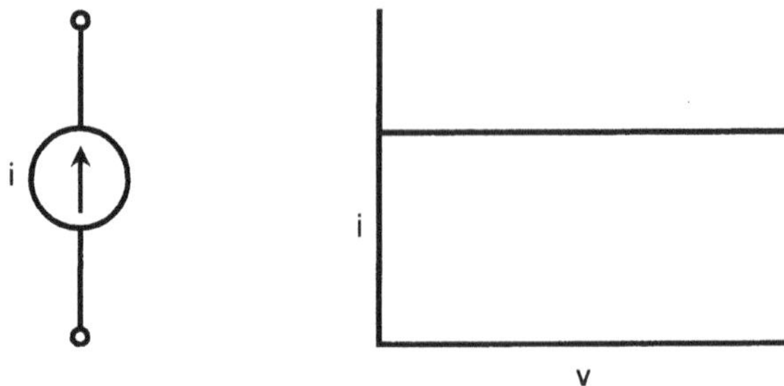

Fig. 1.11 The symbol and the v – i characteristic of an independent current source

This, again, can be considered as a close approximation to a practical device. This too, theoretically, is capable of supplying infinite power from its terminals since it will be able to deliver the same current even if the voltage across it is infinite ! However, this element is very useful in representing the behaviour of some electronic devices. The v – i characteristic of a practical device is given in Fig. 1.12 which is a drooping characteristic.

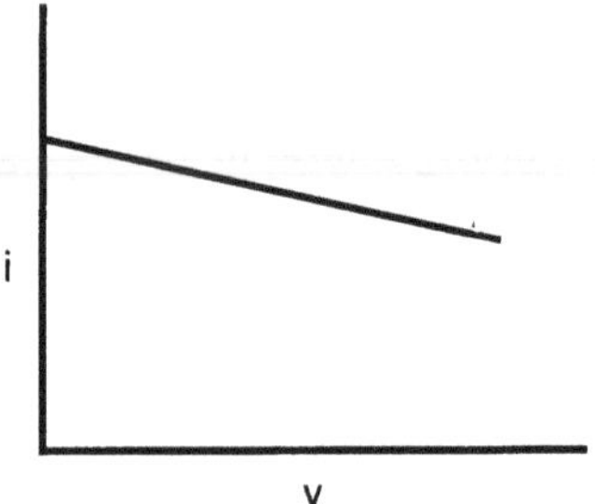

Fig. 1.12 v – i characteristic of a practical current source

The two active elements described above are called *independent sources* because their values are not affected by the currents or voltages in other elements of the network to which these elements are connected. There are other types of sources which are very useful in describing many electronic devices that are represented by their equivalent circuits. These are known as *dependent* or *controlled sources* in which the voltage or the current associated with the element is dependent on either the current or voltage in some other element in the network. In order to distinguish these sources from the independent sources, a diamond shaped symbol is used to represent them as shown in Fig. 1.13.

Fig. 1.13 Symbols for dependent sources

There are essentially four types of controlled sources depending on whether it is a current source or a voltage source and whether it is controlled by a current or voltage in some other element of the network. Thus we have

a) Voltage controlled voltage source
b) Current controlled voltage source
c) Voltage controlled current source
and d) Current controlled current source.

1.4.2 Passive elements :

The passive elements are those elements which are capable of only receiving power. They cannot deliver power. However there are some passive elements which can store finite energy and then return this energy to external elements. Since these elements cannot deliver unlimited energy over an infinite time interval, they are treated as passive elements only. There are essentially three kinds of passive elements called *the resistor, the inductor* and *the capacitor* as shown in Fig. 1.7.

As described earlier, a network is an interconnection of two or more elements. If this interconnection involves atleast one closed circuit, the network is also called as *an electrical circuit.* Thus every circuit is a network but every network need not be a circuit. Also, if the network contains at least one independent source, the network is called *an active network.* If the network does not contain any active element it is called *a passive network.*

Example 1. 1

Find the power absorbed by each element in the circuit of Fig. 1.14.

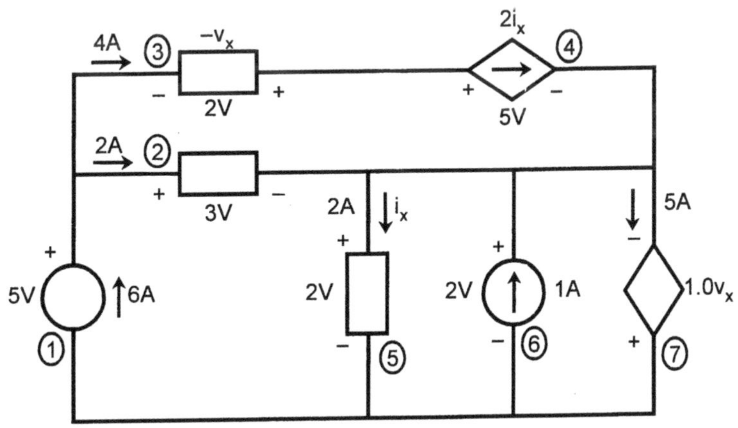

Fig. 1.14 Network for example 1.1

Solution :

The elements are identified by the encircled numbers. For any element if the current is entering the +ve terminal, the current is taken as +ve and hence the power absorbed will be positive.

Thus for element (1)

Power absorbed is

$P_1 = 5 \times (-6) = -30$ W

and for the other elements

$P_2 = 3 \times 2 \times = 6$ W

$P_3 = 2 \times (-4) = -8$ W

$P_4 = 5 \times 4 = 20$ W

$P_5 = 2 \times 2 = 4$ W

$P_6 = 2 \times (-1) = -2$ W

$P_7 = (1.0 \times -2) \times (-5) = +10$ W

1.5 THE RESISTOR :

The passive elements introduced in section 1.4 will now be described in detail. First, the idealised passive element called *linear resistor*, will be considered. The mathematical model for a linear resistor is described by the famous Ohm's law for most of the conducting materials. Ohm's law states that :

"*The voltage across any conducting material is directly proportional to the current flowing through the conductor*" or

$$v = R\,i \qquad \text{......(1.4)}$$

Where 'R', the constant of proportionality, is called *the resistance of the material.* The resistance is measured in ohms. The v – i characteristic of a resistor is shown in Fig. 1.15.

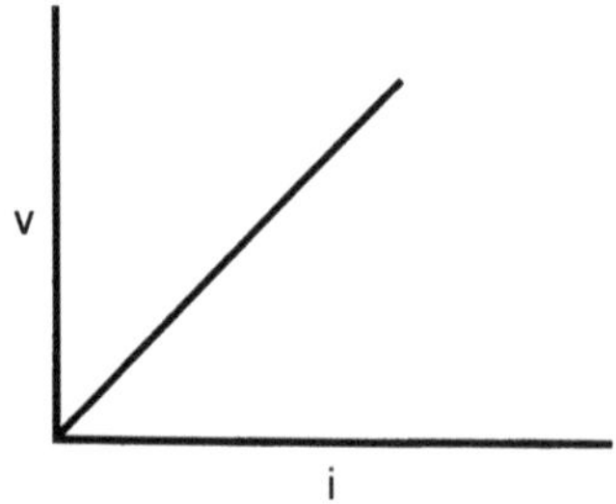

Fig. 1.15 v – i characteristic of a resistor

Since the characteristic is a linear relationship, the resistor is a linear element. The network symbol and the polarities of voltage and current for absorbing power in a resistor are shown in Fig. 1.16.

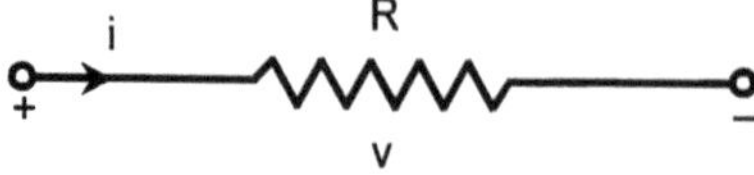

Fig. 1.16 Network model of a resistor and sign convention for voltage and current

The product v . i gives the power absorbed by the resistor and this power manifests itself as heat in the resistor. As resistor is a passive element it cannot deliver or store power. The expressions for power are

$$P = vi = i^2R = \frac{v^2}{R} \qquad .....(1.5)$$

It is pertinent to emphasise that the linear resistor is an idealised model of a physical device and the relationship of eq. (1.4) holds good for a certain range of currents only. If it exceeds a particular value, excessive heat is produced and the value of the resistance is found to change with temperature. It is no longer linear and hence the eq. (1.4) cannot be used. However, in this text, we will be concerned with the linear resistors only. The resistance of a wire of length 'l' meters, and a cross sectional area of A Sq. m. is given by

$$R = \frac{\rho l}{A}$$

where is ρ the resistivity of the wire in ohm – meters

The reciprocal of resistance is defined as conductance. Thus, conductance is the ratio of current to voltage.

$$G = \frac{1}{R} = \frac{i}{v} \qquad .....(1.6)$$

The unit of conductance is Siemens (S).

Example 1.2

Consider the resistor in Fig. 1.17(a). A voltage v(t) of waveform given in Fig. 1.17(b) is applied at its terminals. Obtain the waveform of current through it.

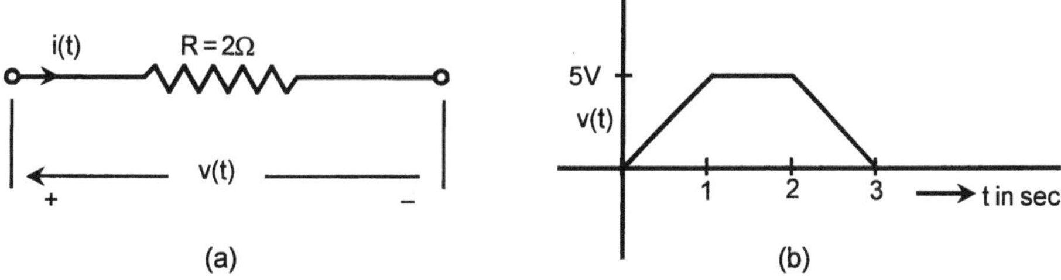

Fig. 1.17 The resistance and voltage waveform for example 1.2

Solution :

The v – i relationship for a resistor is

$$v = i R$$

or $$i = \frac{v}{R} = vG$$

$$G = \frac{1}{R} = \frac{1}{2} = 0.5S$$

For $0 \le t \le 1$ sec

$$v(t) = 5\ t$$

$$i(t) = 5t(0.5) = 2.5t\ A$$

For $1 \le t \le 2$ sec

$$v(t) = 5\ V$$

$$\therefore \quad i(t) = 5(0.5) = 2.5\ A$$

For $2 \le t \le 3$ sec

$$v(t) = -\ 5t + 15$$

$$i(t) = (-\ 5t + 15)0.5$$

$$= -\ 2.5t + 7.5\ A$$

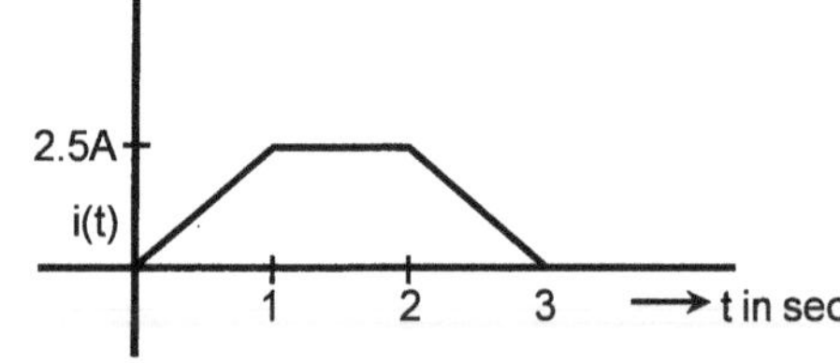

Fig. 1.17(c) Current waveform for example 1.2

The current waveform is given in Fig. 1.17(c).

Example 1.3

Through a resistor of value 2Ω a current of i(t) = 2 sin50t A is passed. What is the voltage across its terminals and what is the power consumed by it ?

Solution :

$$R = 2\ \Omega$$

$$i(t) = 2\ \text{Sin}\ 50t\ \ A$$

$$v(t) = i(t)\ .\ R = (2\text{Sin}\ 50t) \times 2$$

$$= 4\text{Sin}\ 50t\ \ V$$

$$p(t) = v(t)\ .\ i(t)$$

$$= 4\ \text{Sin}\ 50t\ .\ 2\ \text{Sin}\ 50t$$

$$= 8\ \text{Sin}^2\ 50t\ \ \text{Watts}$$

1.6 THE INDUCTOR :

This is another important passive element which does not consume energy but is capable of storing energy. Once it stores energy, it is then capable of supplying this energy to external devices. But unlike active source, it is not capable of supplying unlimited energy or a finite average power over an infinite interval of time.

The mathematical model for an inductor will be given purely in a network point of view, as a v – i relationship at the terminals. But it will be worth reviewing how this element was conceived as a result of magnetic effects of current. A scientist Danish Oersted showed that a current carrying conductor produced a magnetic field in the air surrounding it. Michael Faraday in the year 1831 observed a magnetic effect in the air surrounding a wire carrying current, and a magnetic needle always pointed in the direction of a circle around the wire. Faraday explained the effect in terms of magnetic lines of force in the 'aether' surrounding the wire. He postulated that the current in the wire was responsible in stretching the lines of force, like elastic bands, around the wire, and the energy thus stored in the form of magnetic energy was returned to the wire in the form of a voltage generated when the current decayed to zero.

1.6.1 Faraday's law of electromagnetic induction :

Faraday's law of electromagnetic induction states that whenever a current in a conductor changes, it produces a changing magnetic flux, and this in turn induces a voltage in the conductor which opposes the current. The magnitude of this induced voltage is proportional to the rate of change of flux. If instead of a single conductor, we have a coil of N turns, then

$$e = -\frac{d(N\phi)}{dt} \text{ volts} \qquad \text{.....(1.7)}$$

where e is the induced voltage, $\Psi = N\phi$ is the flux linkage in weber turns. The negative sign is taken to indicate that the induced voltage opposes the current producing the flux ϕ. Since N is a constant, we can write eq. (1.7) as,

$$e = -N\frac{d\phi}{dt}.$$

If a voltage v is applied to a coil of N turns, it produces a current i in the coil which in turn produces a flux ϕ. The induced voltage e, opposes the applied voltage and hence,

$$v = -e = N\frac{d\phi}{dt}.$$

1.6.2 Self inductance :

The flux linkage $\Psi = N\phi$ can be shown to be proportional to the current in the coil and the proportionality constant is termed as the self inductance, L, of the coil.

Thus,

$$v = \frac{d\psi}{dt} = \frac{d(Li)}{dt} = L\frac{di}{dt} \qquad .....(1.8)$$

Eq. (1.8) is the mathematical model of an inductor. L is measured in henries.

Practical inductors are usually made of many turns of fine wire wound in a coil or helix to increase the magnetic effects without increasing the size of the element. The inductance of a coil of a long helix of a small pitch is given by

$$L = \frac{\mu N^2 A}{l}$$

where A = cross sectional area of the coil in Sq. m.

N = number of turns of the coil

l = axial length of the helix in m

and μ = constant of the material inside the helix called *permeability* and for free space or air $\mu = \mu_o = 4\pi \times 10^{-7}$ H/m.

The network symbol and the reference directions for voltage and current are shown in Fig. 1.18 for passivity.

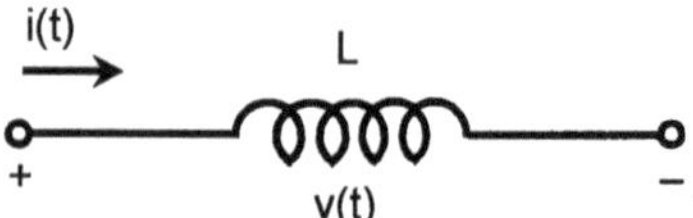

Fig. 1.18 Network symbol for inductor

From eq. (1.8) it is clear that if the current is a constant, the voltage across the inductor is zero, which means that the two terminals of the inductor are connected by a wire of zero resistance. This is called a short circuit between the two terminals. Hence we can say that the inductance behaves like a short circuit to d.c. Another characteristic we observe from eq. (1.8) is that if there is an instantaneous or sudden change in the current, the rate of change of current is infinite and the voltage across the inductor is also infinite. Thus a sudden change in the current in an inductor can occur only if the voltage across it is infinite. Hence the inductor opposes abrupt changes in the current. Further if an inductor carrying current is open circuited, for example, by opening a switch, an arc appears across the switch due to the same reason.

An alternate form of eq. (1.8) describes the relation between i and v.

Rewriting eq. (1.8) as

$$di = \frac{1}{L} v\,dt \qquad .....(1.9)$$

Integrating on both sides, and assuming that the current in the inductor was zero at the time $t = -\infty$. (The time $t = -\infty$ is a conceptual time to ensure that the current was zero at $t = -\infty$. This time could be the time at which the inductance coil was wound and obviously the current at that time was zero !)

$$\int_0^i di = \frac{1}{L}\int_{-\infty}^t vdt$$

$$i = \frac{1}{L}\int_{-\infty}^t vdt \qquad ....(1.10)$$

We observe that both the equations, eq. (1.8) and eq. (1.10), are linear and hence the inductor is a linear element.

The power entering the inductor at any instant is given by

$$p = v\,.\,i = Li\frac{di}{dt} \qquad .....(1.11)$$

When the current is constant, $\frac{di}{dt} = 0$ and $p = 0$ and no additional energy is stored in the inductor. The energy associated with the lines of flux will be fixed. If the current increases, the derivative of the current is positive and the power is positive. Thus additional energy is stored in the inductor. The total energy stored in the inductor at any time is given by

$$w_L = \int_{-\infty}^t vidt = \int_{-\infty}^t iL\frac{di}{dt}\,dt$$

$$= \int_0^i Li\,di = \frac{1}{2}Li^2 \qquad .....(1.12)$$

Observe that the limits of integration are chosen appropriately in eq. (1.12) to suit the variable of integration. Note also that the energy stored in the inductor at any instant depends on the value of the current at that instant only and not on its past history.

Example 1.4

An inductor shown in Fig. 1.19(a) is supplied with a current waveform given in Fig. 1.19(b). Draw the waveforms for the voltage and energy in the inductor.

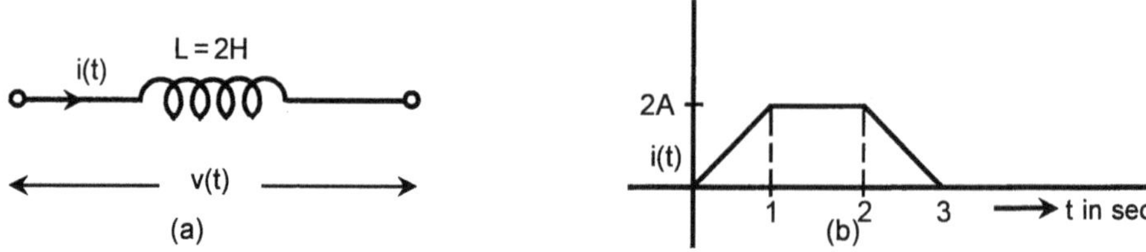

Fig. 1.19 An inductor and its associated current waveform for example 1.4

Solution :

For $0 \leq t \leq 1$ sec

$$i(t) = 2t \text{ A}$$

$$\therefore \quad v(t) = L\frac{di}{dt} = 2 \times 2 = 4V \qquad w_L(t) = \frac{1}{2} L i^2 = \frac{1}{2} \times 2 \times (2t)^2 = 4t^2 \text{ J}$$

For $1 \leq t \leq 2$ sec

$$i(t) = 2 \text{ A} \qquad w_L(t) = \frac{1}{2} \times 2 \times 2^2 = 4 \text{ J}$$

$$v(t) = 0$$

For $2 \leq t \leq 3$ sec

$$i(t) = -2t + 6 \text{ A} \qquad w_L(t) = \frac{1}{2} \times 2 \times (-2t + 6)^2 = 4t^2 - 24t + 36 \text{ J}$$

$$v(t) = 2 \times -2 = -4V$$

The resulting waveform of voltage and energy are given in Fig. 1.19(c).

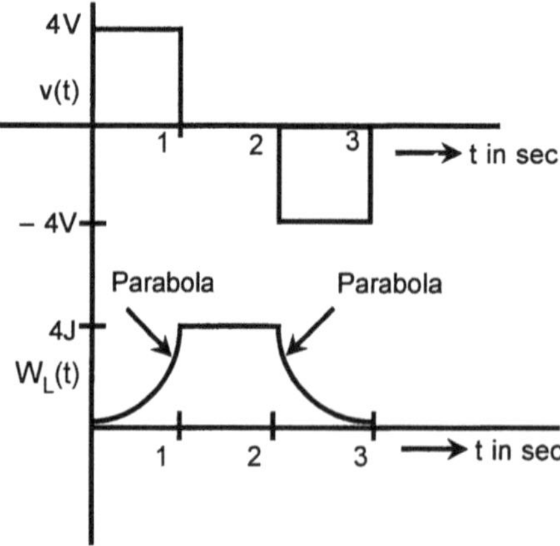

Fig. 1.19(c) Solution of example 1.4

Example 1.5

A waveform of current shown in the figure is applied to an inductor of value 0.5 H. Obtain the waveform of voltage across it.

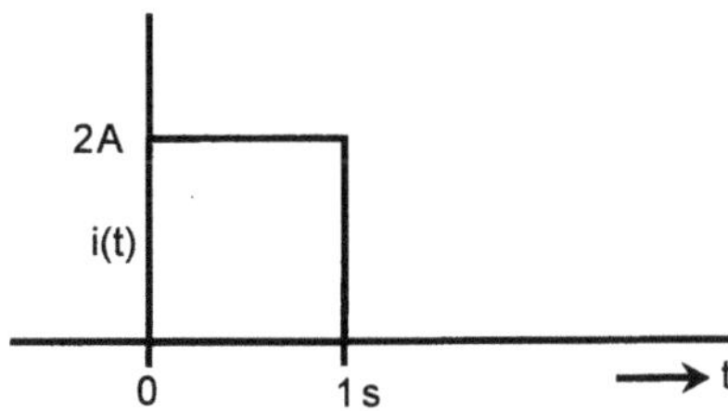

Current wave for example 1.5

Solution :

At t = 0 and t = 1s there is an abrupt change in the current. Here $\frac{di}{dt} = \infty$.

Hence at t = 0 and t = 1s we have infinite voltage spikes appearing across the inductor terminals. As discussed earlier, these infinite spikes are required to allow the current to change suddenly. These spikes are called *impulses*, which are mathematically defined later. These spikes are physically not possible as, a finite time, however small it may be, is required for the current to rise to a given value. However these spikes can be of very large magnitude.

For $0 < t < 1$ (observe that $t \neq 0$ or 1s)

$$i(t) = 2A \quad \text{and} \quad v(t) = L\,\frac{di}{di} = 0v$$

Thus the waveform of voltage across inductor consists of two infinite spikes called *impulses* at t = 0 and 1 sec and is zero for $0 < t < 1s$. The waveform is shown in the figure.

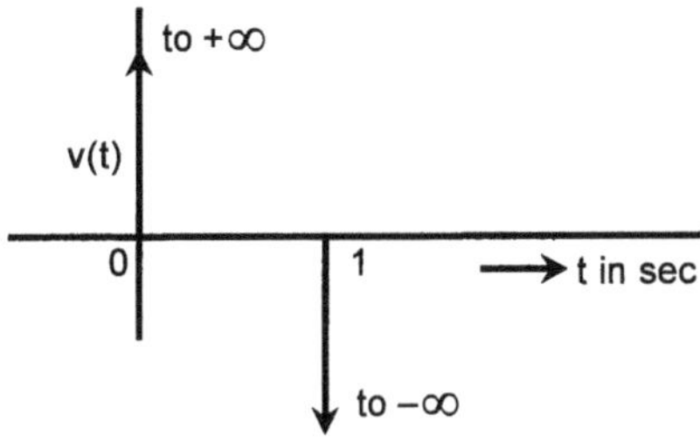

Waveform of v(t) for example 1.5

Example 1.6

Consider a waveform of voltage given in Fig. 1.20(a) applied to an inductor of value 2 mH. Obtain the waveforms of current and energy in the inductor. Assume that at t = 0 the energy and thus current in it to be zero.

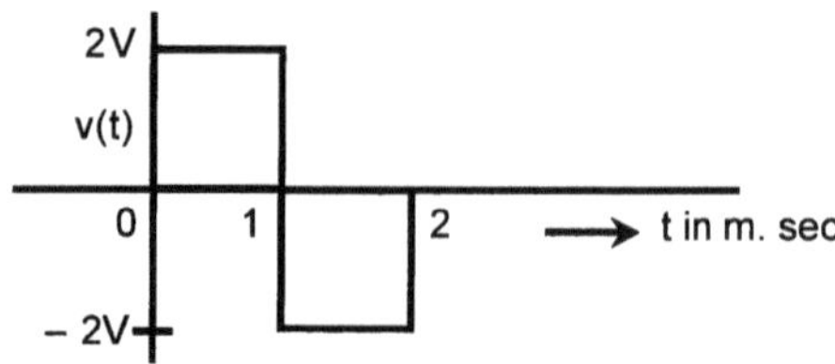

Fig. 1.20(a) waveform of voltage for example 1.6

Solution :

For $0 \le t \le 1$ ms

$$v(t) = 2\text{ V}$$

$$i(t) = \frac{1}{L}\int_{-\infty}^{t} v(t)dt = \frac{1}{L}\int_{-\infty}^{0} v(t)dt + \frac{1}{L}\int_{0}^{t} v(t)dt \qquad .....(1.13)$$

The first term in eq. (1.13) represents the current in the inductor at t = 0. Thus

$$i(t) = i(0) + \frac{1}{L}\int_{0}^{t} 2dt = 0 + \frac{1}{2\times10^{-3}}2t\Big|_0^t = 10^3 t \text{ A} \qquad .....(1.14)$$

$$w_L(t) = \frac{1}{2}Li^2 = \frac{1}{2} \times 2 \times 10^{-3} \times 10^6 t^2 = 10^3 t^2 \text{ J}$$

For $1 \le t \le 2$ ms

$$v(t) = -2\text{V}$$

Again, $$i(t) = \frac{1}{L}\int_{-\infty}^{t} vdt$$

$$= \frac{1}{L}\int_{-\infty}^{1\times10^{-3}} vdt + \int_{1\times10^{-3}}^{t}(-2)dt$$

$$= i(1\times10^{-3}) + \frac{1}{2\times10^{-3}}[-2t]_{1\times10^{-3}}^{t}$$

From eq. (1.14) at t = 1 m sec.

$$i(1\times10^{-3}) = 10^3 \times 1 \times 10^{-3} = 1\text{A}$$

$$i(t) = 1 + 10^3[-t + 10^{-3}] = -10^3 t + 2 \text{ A}$$

$$w_L(t) = \frac{1}{2} \times 2 \times 10^{-3}(-10^3 t + 2)^2 = 10^3 t^2 - 4t + 4 \times 10^{-3} \text{ J}$$

The waveform of current and energy are shown in Fig. 1.20(b).

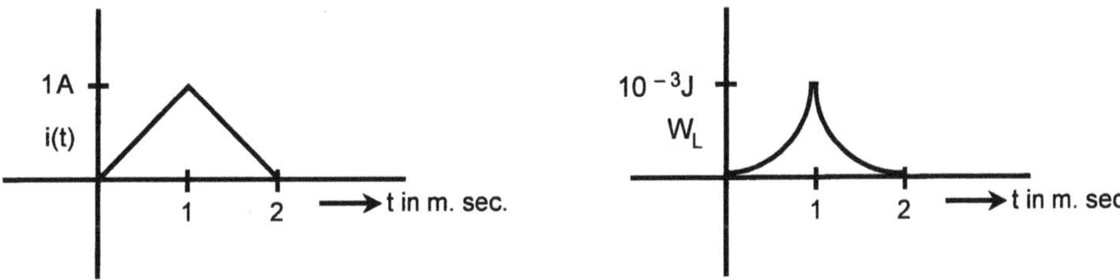

Fig. 1.20(b) Waveform of current and energy for example 1.6

1.6.3 Mutual inductance :

The inductor was defined in terms of the magnetic field produced by a current carried in a coil. The voltage produced in the coil due to the change in the magnetic flux surrounding the coil is essentially due to flux caused by the change in current in itself. Thus this inductance is appropriately called as *self inductance*. The time varying magnetic flux which is produced by a changing current in one coil, may also cause a voltage in the vicinity of a second coil. In 1831, Michael Faraday discovered this property and termed the voltage thus induced as mutually induced voltage and the property of the coils which causes this voltage as mutual inductance. The voltage induced in the second coil was found to be proportional to the time rate of change of current in the first coil. The circuit symbol for mutual inductor is shown in Fig. 1.21.

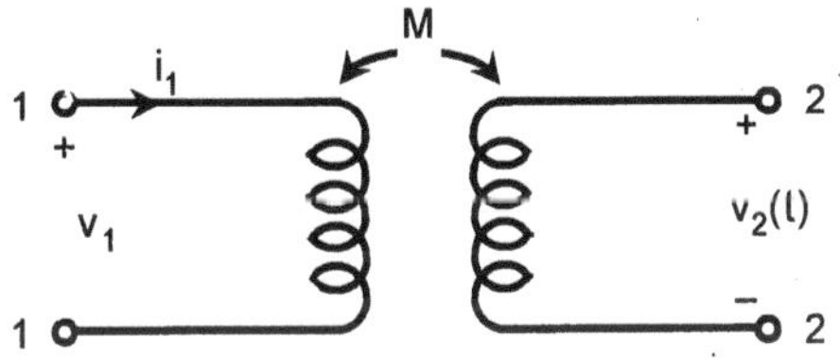

Fig. 1.21 Circuit symbol for mutual inductor

The v – i relationship is given by eq. (1.15).

$$v_2 = M_{21} \frac{di_1}{dt} \qquad .....(1.15)$$

Unlike self inductance where the voltage and current signs are given considering the passivity, in mutual inductor, the signs of current and voltage have to be given on different considerations as they are associated with different terminal pairs 1, 1 and 2, 2. The symbol M_{21} indicates that the current is applied at terminals of coil 1 and the voltage is measured at coil 2. If current is applied at coil 2 and voltage is measured at coil 1, we can write

$$v_1 = M_{12} \frac{di_2}{dt} \qquad .....(1.16)$$

On considerations of energy in the two cases one can easily prove that $M_{21} = M_{12}$. Hence we will use only one coefficient of mutual inductance $M_{12} = M_{21} = M$. The two coils with mutual inductance between them are said to be magnetically coupled coils. Mutual inductance is also measured in *henrys*.

1.6.4 Dot convention :

As explained earlier, the sign of the voltage induced in the second coil can not be decided by passivity considerations. For a given direction of current in one coil, the voltage induced may be positive or negative in the second coil depending on the sense of direction of the winding of the coil. Often, to increase the mutual inductance of the coils, they are wound on iron core so that the flux produced by the coil will be concentrated in the core and it links with the other coil wound on the same core. To indicate the sense of winding of these coils on the circuit symbol of mutual inductor, a dot convention is developed.

Consider two coils wound on a magnetic core as shown in Fig. 1.22.

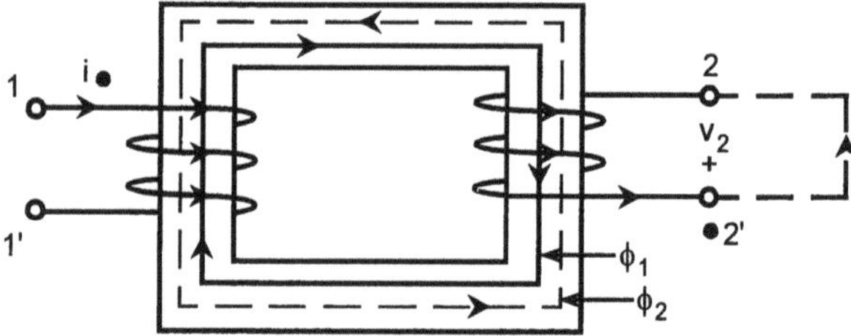

Fig. 1.22 Dot convention for mutual inductor

Place a dot at the terminal of coil 1 where the current i is entering, that is the current is increasing positively. This current produces a magnetic flux ϕ_1 in the core in a direction indicated by the arrow. This flux links with the second coil and induces a voltage at the terminals 2, 2' as per Faraday's laws of electromagnetic induction. The polarity of this voltage is given by Lenz's law. According to this, the voltage produced at terminals 2, 2' should have a polarity so as to produce a flux opposing the flux produced by coil 1, ϕ_1, if coil 2 were short circuited. This means that when coil 2 is short circuited, a current flows in the coil and the current in turn produces a flux ϕ_2 opposing the flux ϕ_1. The flux ϕ_2 should have a direction as shown in Fig. 1.21. In order to produce a flux in the indicated direction the current in the second coil must flow from left to right, or from terminal 2' to 2 in the short circuit across them. Thus the terminal 2' must be positive with respect to terminal 2 and a dot is placed at that terminal as indicated in Fig. 1.22. If the sense of winding in coil 2 is reversed as shown in Fig. 1.23, the dot must be placed at the terminal 2.

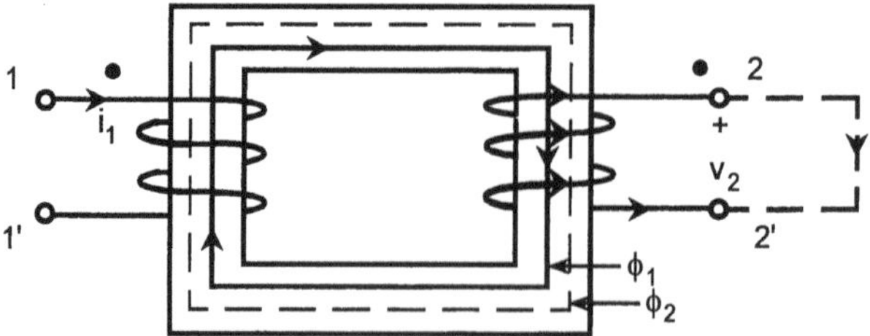

Fig. 1.23 Mutual inductor with sense of winding of coil 2 changed

It is very inconvenient to show the sense of winding for all the coils with mutual inductance in a network representation. Hence the dot convention helps to overcome this problem. The two coils are shown with a dot placed at one terminal of each coil to indicate the sense of winding and hence the polarity of the voltage induced. The following rule establishes the polarity of the mutually induced voltage.

" *If the current enters the dotted terminal of one coil, the voltage induced in the second coil will have a reference direction such that its dotted terminal is +ve* ".

In the same way, the current entering at the undotted terminal (or equivalently current leaving the dotted terminal) of first coil, produces a voltage in the second coil such that the undotted terminal is positive.

The two rules are illustrated in Fig. 1.24.

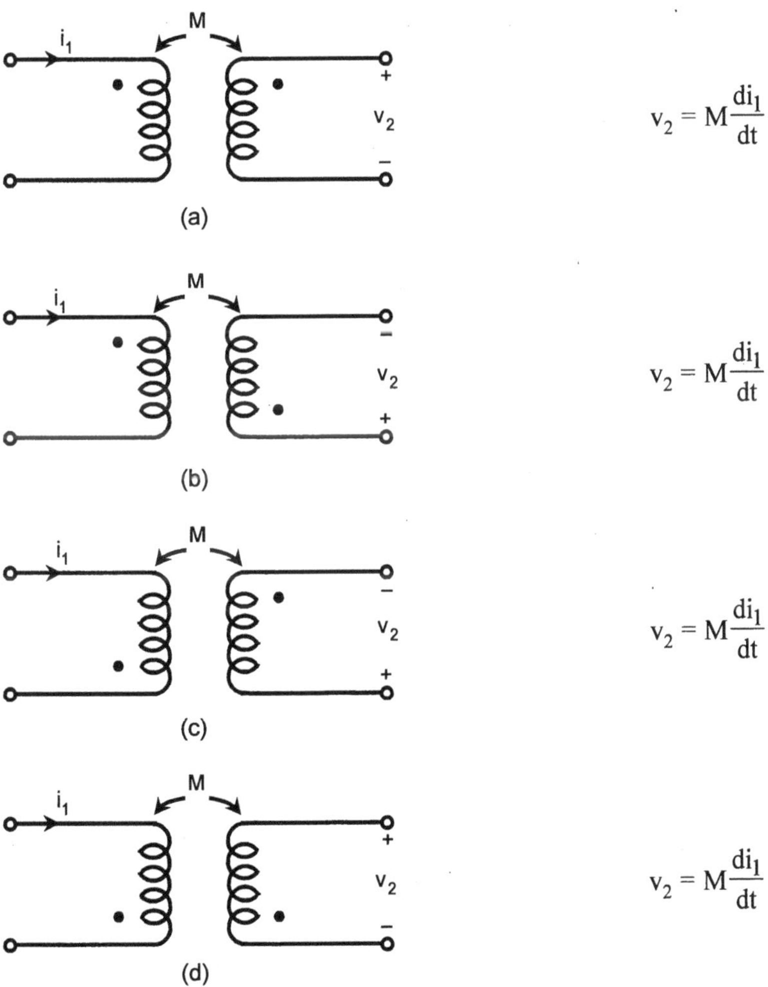

Fig. 1.24 Different possibilities of applying dot convention rules

Thus far, we had considered only the voltage produced in a open circuited coil, due to a change of current in another coil. If currents flow in each of these coil in addition to mutually induced voltages, self induced voltages will also be present. The mutual voltage induced in the coil is independent of the self induced voltage and will be in addition to it. Hence for the coils shown in Fig. 1.25, with reference direction of current and voltages as indicted, the v – i equation are

$$v_1 = L_1 \frac{di_1}{dt} + M \frac{di_2}{dt} \qquad .....(1.17)$$

$$v_2 = L_2 \frac{di_2}{dt} + M \frac{di_1}{dt} \qquad .....(1.18)$$

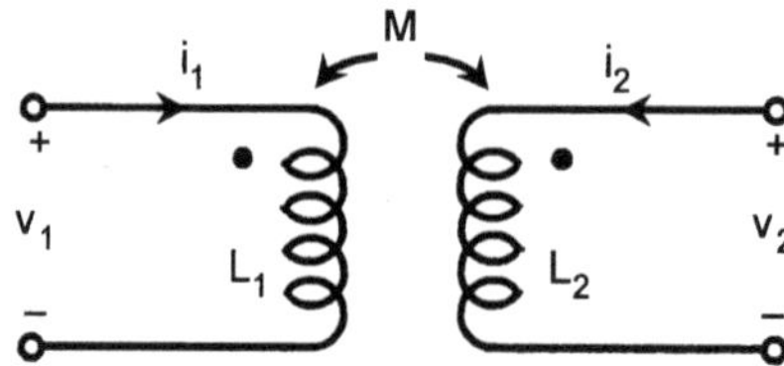

Fig. 1.25 Circuit with both self and mutual induction voltages

If the reference direction for i_2 is reversed in Fig. 1.25, the voltage due to self induction and the voltage due to mutual induction will have opposite sign and thus

$$v_1 = L_1 \frac{di_1}{dt} - M \frac{di_2}{dt} \qquad .....(1.19)$$

$$v_2 = L_2 \frac{di_2}{dt} - M \frac{di_1}{dt} \qquad .....(1.20)$$

1.6.5 Energy consideration in the magnetically coupled coils :

Consider two mutually coupled coil as shown in Fig. 1.26. Let us change the current i_1 from 0 to some value i_1 (t_1), keeping the second coil open. Also assuming

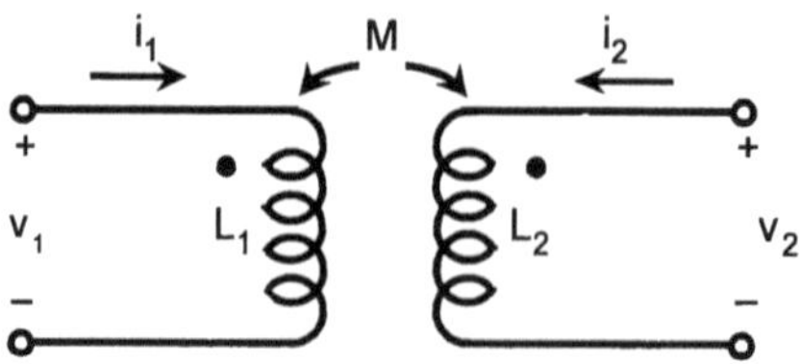

Fig. 1.26 Energy in coupled coils

zero energy storage in the coils to start with, the power delivered to the circuit is

$$P = v_1 i_1 + v_2 i_2 = L_1 i_1 \frac{di_1}{dt} + 0$$

The energy stored in the coils when $i_1 = i_1(t_1)$ is

$$\int_0^{t_1} v_1 i_1 dt = \int_0^{i_1(t_1)} L_1 i_1 di_1 = \frac{1}{2} L_1 [i_1(t_1)]^2$$

Now, if the current i_1 is held constant at $i_1(t_1)$ and the current i_2 is changed from 0 to $i_2(t_2)$, then the power in the two coils will be

$$p = i_2 L_2 \frac{di_2}{dt} + M_{12} \frac{di_2}{dt} \cdot i_1$$

and the energy stored between the instants t_1 and t_2 is

$$\int_{t_1}^{t_2} (v_2 i_2 + v_1 i_1) dt = \int_{t_1}^{t_2} L_2 \frac{di_2}{dt} . i_2 dt + \int_{t_1}^{t_2} M_{12} \frac{di_2}{dt} . i_1 dt$$

$$= L_2 \int_0^{i_2(t_2)} i_2 di_2 + M_{12} i_1 \int_0^{i_2(t_2)} di_2$$

$$= \frac{1}{2} L_2 [i_2(t_2)]^2 + M_{12} i_2(t_2) i_1(t_1)$$

The total energy in the coils at $t = t_2$ is given by

$$w_{Total} = \frac{1}{2} L_1 [i_1(t_1)]^2 + \frac{1}{2} L_2 [i_2(t_2)]^2 + M_{12}\, i_1(t_1)\, i_2(t_2) \quad .....(1.21)$$

If the process is repeated by first changing the current i_2 form 0 to $i(t_2)$ with $i_1 = 0$ and then changing i_1 from 0 to $i(t_1)$, we get

$$w_{Total} = \frac{1}{2} L_1 [i_1(t_1)]^2 + \frac{1}{2} L_2 [i_2(t_2)]^2 + M_{21}\, i_1(t_1)\, i_2(t_2) \quad .....(1.22)$$

Note that the only difference in eq. (1.21) and eq. (1.22) is the mutual inductance term, namely, instead of M_{12} we have M_{21}.

The two values of energy given by eq. (1.21) and eq. (1.22) must be same since the initial and final conditions are same in both cases. Hence we can conclude that

$$M_{12} = M_{21} = M \quad .....(1.23)$$

Thus the energy stored in the coil is given by

$$W = \frac{1}{2} L_1 i_1^2 + \frac{1}{2} L_2 i_2^2 + M i_1 i_2. \qquad .....(1.24)$$

If one current enters the dot and the other leaves the dot then

$$W = \frac{1}{2} L_1 i_1^2 + \frac{1}{2} L_2 i_2^2 - M i_1 i_2. \qquad .....(1.25)$$

Thus the energy stored in a pair of mutually coupled coils is given by

$$W = \frac{1}{2} L_1 i_1^2 + \frac{1}{2} L_2 i_2^2 \pm M i_1 i_2. \qquad .....(1.26)$$

Since the energy is stored in a passive network, it must be positive. The only way it can be negative is if the energy due to the mutual inductance is negative i.e..

$$W = \frac{1}{2} L_1 i_1^2 + \frac{1}{2} L_2 i_2^2 - M i_1 i_2. \qquad .....(1.27)$$

This may be written as

$$W = \frac{1}{2} \left(\sqrt{L_1}\, i_1 - \sqrt{L_2}\, i_2\right)^2 + \sqrt{L_1 L_2}\, i_1 i_2 - M i_1 i_2 \qquad .....(1.28)$$

Since this can not be – ve. It follows that

$$M \leq \sqrt{L_1 L_2}$$

or

$$M = K \sqrt{L_1 L_2} \qquad .....(1.29)$$

Where K lies between 0 and 1. It has a zero value when the two coils have no magnetic link between them and is equal to 1 when the two coils are perfectly coupled. Thus the maximum value of mutual inductance is equal to the geometric mean of the self inductances of the two coils. The degree of coupling is given by the factor K, which is known as the *coefficient of coupling.*

Example 1. 7

Place the dots at the appropriate terminals of the three coils shown in Fig. 1.27.

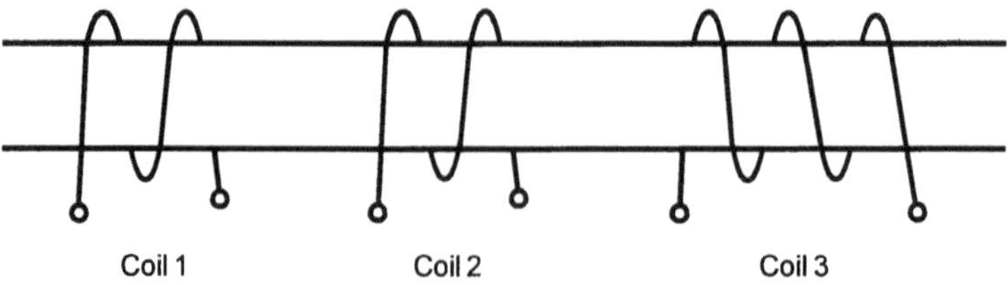

Fig. 1.27(a) The three coils on a common magnetic core

Solution :

First place a dot at the left terminal of coil 1 and assume that a current i_1 enters the dot.

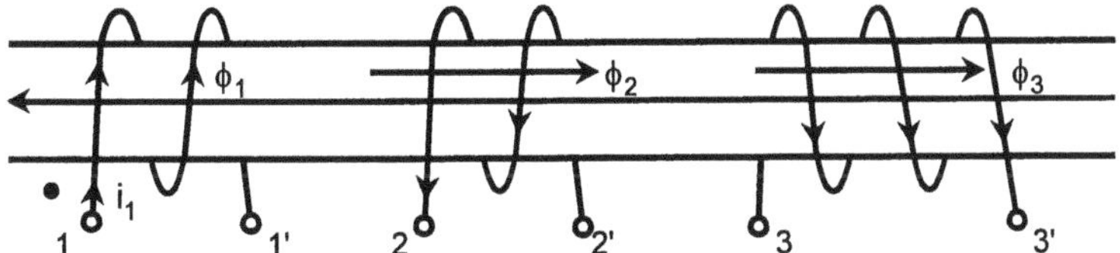

Fig. 1.27(b) Coils showing directions of currents and fluxes

The flux ϕ_1 will have a direction from right to left. The flux produced by coil 2 must oppose this flux. Thus the current direction in coil 2 must be as shown and hence this current must leave the terminal 2. Hence a dot must be placed at terminal 2 for coil 2.

Similarly, for coil 3, a current in the direction as shown in Fig. 1.27(b) must flow to produce a flux ϕ_3, opposing the flux ϕ_1. This current must leave the terminal 3' and hence a dot must be placed at 3' of the coil 3. Hence the circuit representation of these coils is given in Fig. 1.27(c).

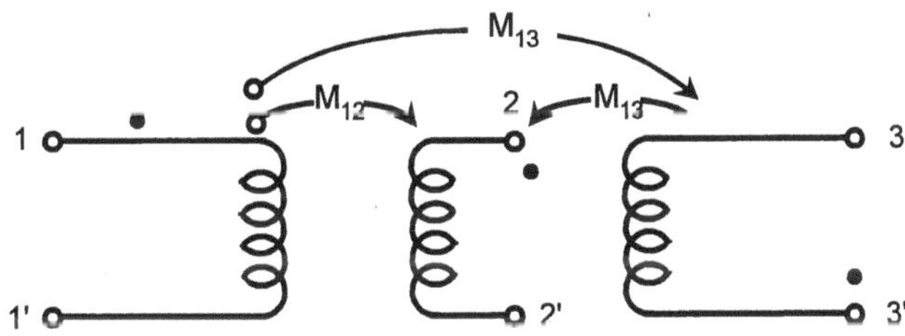

Fig. 1.27(c) Coils with dots placed to indicate polarity

When more than two coils are magnetically coupled, different types of dots (• ▲ ■) are used to indicate the polarities between various coils. In the above examples we can use • for coils 1 and 2, ▲ for coils 1 and 3 and ■ for coils 2 and 3. The circuit symbol with dot convention is shown in Fig. 1.27(d).

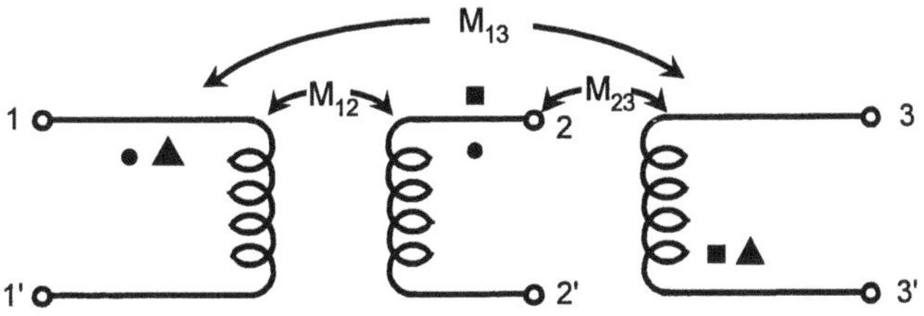

Fig. 1.27(d) Representation of coils with mutual inductance indicating polarities

1.7 THE CAPACITOR :

The capacitor is the third network element which is a passive element, but, like inductor, it can also store energy. Once it stores energy, it is capable of supplying this energy to external devices. It cannot provide unlimited energy or a finite average power over an infinite time interval. Hence it is not an active element like an ideal voltage or current source.

The inductor stores energy because of the current carried by it. In a similar manner, the capacitor is a device which stores energy by virtue of the voltage across it. Historically the capacitor was the first element to be discovered. In 1745 Van Mussenbrock devised an experiment to store static electricity by placing on insulator between two metal sheets, and then charged it by rubbing. Cunaeus, one of his friends, touched this device and received a violent shock ! Thus a device to store electricity was discovered.

Later in the year 1812 Simeon Poisson, gave a mathematical explanation for the energy storage on a capacitor. He compared the forces between charge on the plates to the gravitational force between a mass and earth and the energy associated with these stored charges to the potential energy of a mass at rest above the earth. He also showed that the energy is proportional to the area of the plates and inversely proportional to the spacing between them. Thus, capacitance, which is the property of a capacitor by which it can store energy, is given by

$$C = \frac{\in A}{d}$$

where A is the area of the plates in sq. m.
d is the distance between the plates in m.
and $\in$ is the permittivity of the insulating material between the two conductors

For air or vacuum,

$$\in = \in_0 = 8.854 \text{ pF/m.}$$

Coming back to the circuit model of a capacitor, it is described by the v – i relationship

$$i = C\frac{dv}{dt} \qquad \text{.....(1.30)}$$

The two commonly used symbols and the polarities of voltage and current for passivity requirement are shown in Fig. 1.28.

Fig. 1.28 circuit models of capacitor

An alternate form of eq. (1.30) can be derived as follows :

$$dv = \frac{1}{C}\ i\,dt \qquad .....(1.31)$$

Integrating on both sides of eq. (1.31) and assuming that the voltage across the capacitor was zero at t = − ∞, a conceptual time when the capacitor was manufactured and hence the voltage across it was zero, we get

$$\int_0^v dv = \int_{-\infty}^{t} \frac{1}{C} i\,dt$$

$$v = \frac{1}{C}\int_{-\infty}^{t} i\,dt \qquad .....(1.32)$$

From eq. (1.30), it is clear that, if the voltage across the capacitor is a constant, the current through it would be zero. But if the voltage changes suddenly, the current would by infinite. Thus a capacitor opposes any sudden change in voltage across it. An infinite current is required to be passed through it if the voltage has to change suddenly by a finite value. If the two terminals of a charged capacitor are shorted, a spark is produced because of this infinite current.

The two eqs. (1.30) and (1.32) describing the behaviour of a capacitor, are linear and hence the capacitor is a linear element. The power entering the capacitor at any instant is given by

$$P = v \cdot i$$

$$= v \cdot c\frac{dv}{dt} \qquad .....(1.33)$$

When $\frac{dv}{dt} = 0$ or v(t) is a constant, p = 0 and no additional energy is stored in the capacitor. If the voltage is increasing, the power is positive and additional energy is stored in the capacitor. The total energy in the capacitor is given by

$$w_C = \int_{-\infty}^{t} vi\,dt = \int_{-\infty}^{t} C\,v\frac{dv}{dt}\,dt$$

$$= C\int_0^v v\,dv = \frac{1}{2}Cv^2 \qquad .....(1.34)$$

Thus the energy stored in the capacitor at any instant depends on the value of the voltage at that instant only and not on the past history.

Example 1.8

The capacitor in Fig 1.29(a) is supplied with a voltage waveform shown in Fig. 1.29(b). Obtain the current and energy waveforms in the capacitor.

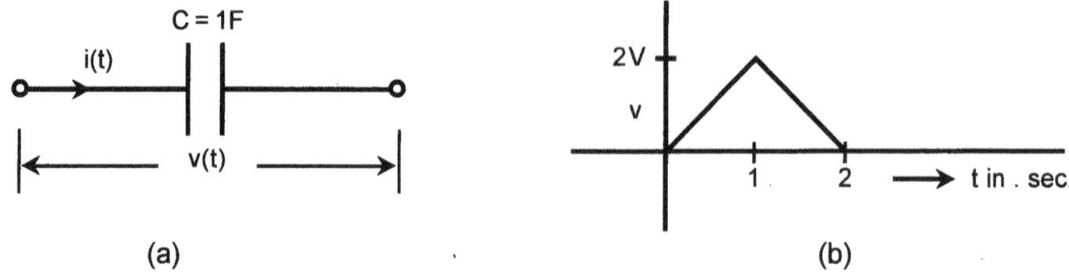

Fig. 1.29 The capacitor and the voltage waveform for example 1.8

Solution :

For $0 \leq t \leq 1$

$$v(t) = 2t \ \text{V}$$

$$i = C\frac{dv}{dt} = 1.2 = 2 \ \text{A}$$

$$w_C = \frac{1}{2}Cv^2 = \frac{1}{2} \cdot 1 \times 4t^2 = 2t^2 \ \text{J}$$

Thus the current is a constant during the interval $0 \leq t \leq 1$.

For $1 \leq t \leq 2$

$$v(t) = -2t + 4 \ \text{V}$$

and $$i(t) = 1 \, . \, (-2) = -2 \ \text{A}$$

The energy stored in the capacitor is

$$w_C = \frac{1}{2}Cv^2 = \frac{1}{2} \times 1 \times (-2t + 4)^2 = 2t^2 - 8t + 8 \ \text{J}$$

The waveforms of current and energy are given in Fig. (1.30).

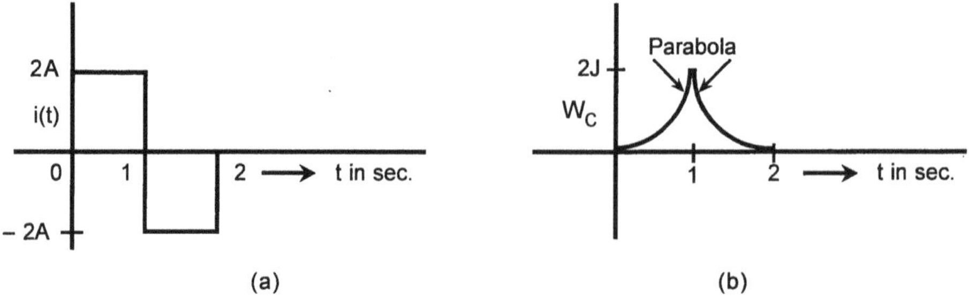

Fig. 1.30 Waveforms of i and W_C for example 1.8

Example 1.9

Consider the capacitor in Fig. 1.31(a) and a waveform of voltage shown in Fig. 1.31(b) applied across it. Find the waveform of current and energy in the capacitor.

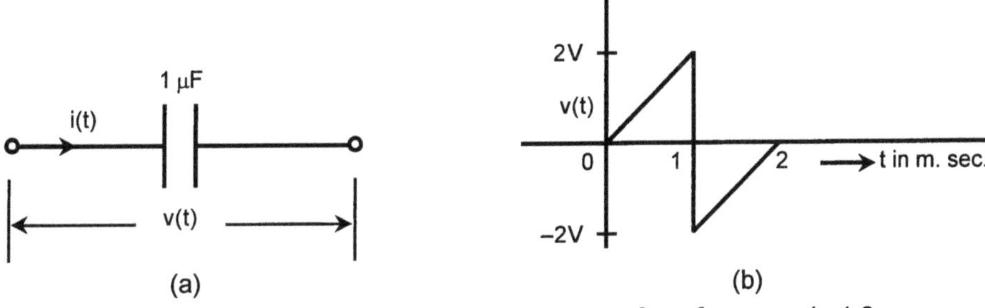

Fig. 1.31 The capacitor and the voltage waveform for example 1.9

Solution :

For $0 \le t < 1$ m.sec.

$$v(t) = \frac{2}{1\times10^{-3}}\, t = 2 \times 10^3 t$$

$$i(t) = C\frac{dv}{dt} = 1 \times 10^{-6} \times 2 \times 10^3 = 2\ \text{mA}$$

$$w_C = \frac{1}{2}Cv^2 = \frac{1}{2} \times 1 \times 10^{-6} \times 4 \times 10^6 t^2 = 2t^2\ \text{J}$$

At t = 1 m.sec the voltage across the capacitor instantaneously changes from +2V to –2V thus making $\frac{dv}{dt}$ infinite. Hence an impulse of current will be produced at t = 1 m.sec.

For $1 < t \le 2$ m. sec.

$$v(t) = 2 \times 10^3\, t - 4$$

$$i(t) = C\frac{dv}{dt} = 1 \times 10^{-6} \times 2 \times 10^3 = 2\ \text{mA}.$$

The energy during this interval is

$$W_C(t) = \frac{1}{2}\, C\, v^2 = \frac{1}{2} \times 1 \times 10^{-6} \times (2 \times 10^3 t - 4)^2$$

$$= 2\, t^2 - 8 \times 10^{-3}\, t + 8 \times 10^{-6}\ \text{J}$$

The waveform of current and energy are given in Fig. 1.32.

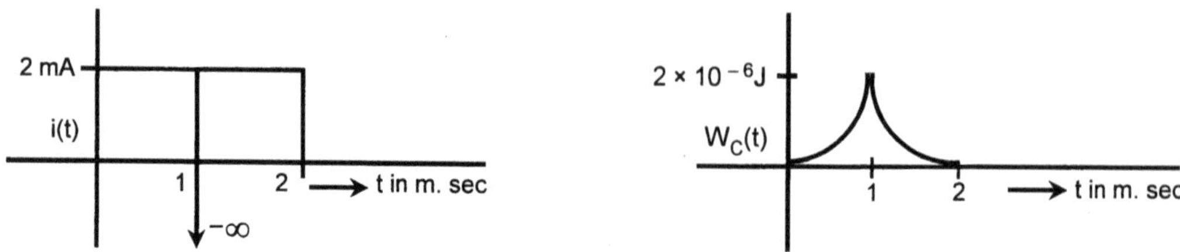

Fig. 1.32 The current and energy waveforms for example 1.8

Example 1.10

A current of waveform shown in Fig. 1.33(a) is applied to a capacitor of value 2 μ F. Find the voltage waveform.

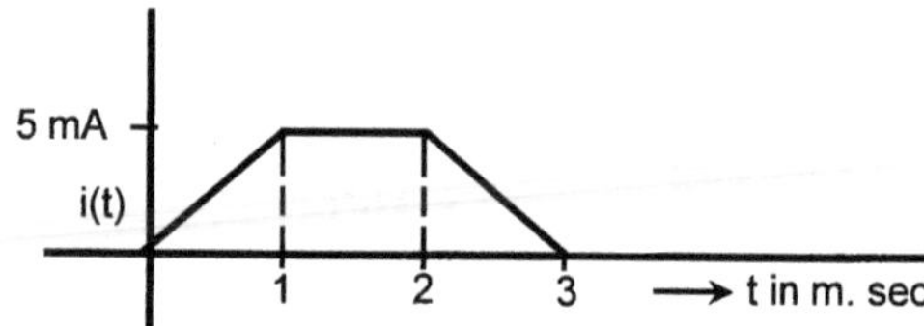

Fig. 1.33(a) Current waveform for example 1.10

Solution :

For $0 \leq t \leq 1$ m sec.

$$i(t) = \frac{5\times10^{-3}}{1\times10^{-3}}t = 5t \text{ A}$$

$$v(t) = \frac{1}{C}\int_{-\infty}^{0} i(t)dt$$

Assuming no voltage to be present across the capacitor at t = 0

$$v(t) = \frac{1}{C}\int_{-\infty}^{0} i(t)dt + \frac{1}{C}\int_{0}^{t} i(t)dt$$

$$= v(0) + \frac{1}{2\times10^{-6}}\int_{0}^{t} 5t\,dt$$

$$= 0 + \frac{5}{2\times10^{-6}}\left.\frac{t^2}{2}\right|_0^t$$

$$= 1.25 \times 10^6 t^2 \text{ V} \qquad .....(1.35)$$

For $1 \le t \le 2$ m sec

$$i(t) = 5 \times 10^{-3}\ A$$

$$v(t) = \frac{1}{C}\int_{-\infty}^{t} i(t)\,dt$$

$$= \frac{1}{C}\int_{-\infty}^{1\times10^{-3}} i(t)\,dt + \frac{1}{C}\int_{1\times10^{-3}}^{t} 5\times10^{-3}\,dt \qquad .....(1.36)$$

The first integral in the above expression is the voltage across the capacitor at t = 1 m sec. From eq. (1.35).

$$v(1 \times 10^{-3}) = 1.25 \times 10^{6} \times 1 \times 10^{-6} = 1.25\ V$$

Eq. (1.36) becomes

$$v(t) = 1.25 + \frac{1}{2\times10^{-6}}\int_{1\times10^{-3}}^{t} 5\times10^{-3}\,dt$$

$$= 1.25 + 2.5 \times 10^{3} t \Big|_{1\times10^{-3}}^{t}$$

$$= 2.5 \times 10^{3}\,t - 1.25 \qquad .....(1.37)$$

For $2 \le t \le 3$

$$i(t) = -5\,t + 15 \times 10^{-3} A$$

$$v(t) = \frac{1}{C}\int_{-\infty}^{2\times10^{-3}} i(t)dt + \frac{1}{2\times10^{-6}}\int_{2\times10^{-3}}^{t}\left(-5t+15\times10^{-3}\right)dt$$

$$= v(2\times10^{-3}) + \frac{1}{2\times10^{-6}}\left[\frac{-5t^{2}}{2}+15\times10^{-3}t\right]_{2\times10^{-3}}^{t}$$

From eq. (1.37),

$$v(2 \times 10^{-3}) = 3.75\ V$$

$$v(t) = 3.75 - 1.25 \times 10^{6}\,t^{2} + 7.5 \times 10^{3}t - 15 + 5$$

$$= -1.25 \times 10^{6}t^{2} + 7.5 \times 10^{3}t - 6.25\ V$$

The waveform of voltage is shown in Fig. 1.33(b).

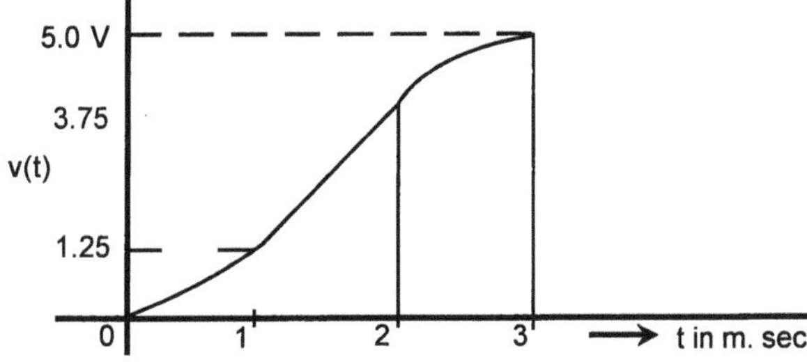

Fig. 1.33(b) Waveform of voltage across the capacitor for example 1.10

Problems

1.1 Ten Coulombs of positive charge per second are passing through a wire in a direction from 1 to 2.

(a) What is the current if the assumed reference direction is from 1 to 2 ?

(b) What is the current if the assumed reference direction is from 2 to 1 ?

(c) What would be the answer for (a) and (b) if the charge is negative instead of positive ?

1.2 A Coulomb of charge changes its energy by 30 Joules in moving from point 1 to 2. What is the voltage of point 1 with respect to point 2 if

(a) the charge is positive and the energy is lost.

(b) the charge is negative and the energy is lost.

(c) the charge is positive and the energy is gained

and (d) the charge is negative and the energy is gained.

1.3 A 2 terminal device has a positive current of 15 A entering at terminal 1 and leaving at terminal 2. What is the power absorbed in the device ? When

(a) the voltage at 1 is 5 V positive with respect to 2.

(b) the voltage at 1 is 5 V negative with respect to 2

and (c) the voltage at 2 is – 10 V with respect to 1.

1.4 Identify the source in the devices shown in Fig. P.1.1.

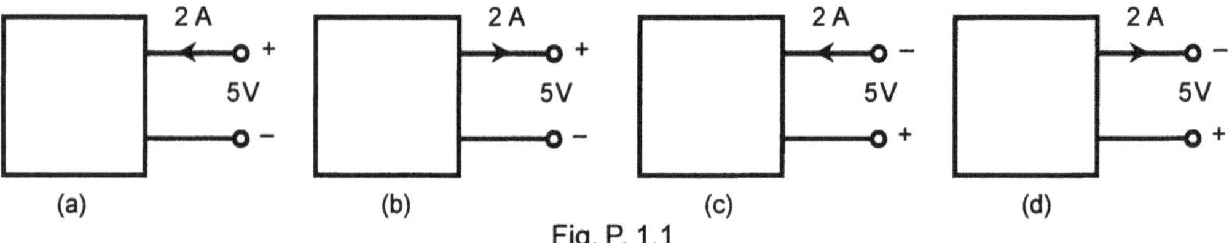

Fig. P. 1.1

1.5 A resistor of value 2 Ω is connected across a voltage whose waveform is shown in Fig. P. 1.2. Draw the waveforms of current and power.

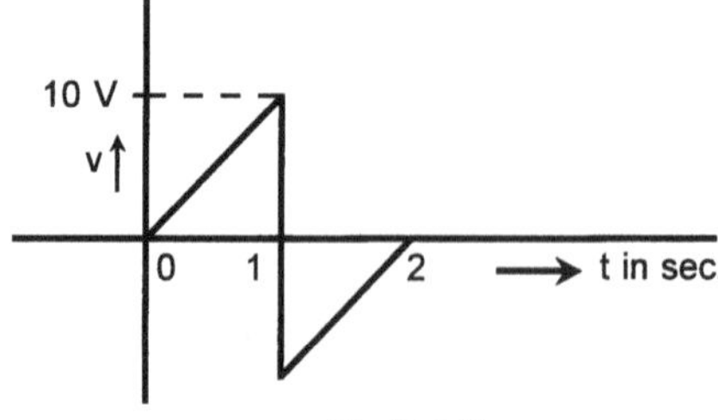

Fig. P. 1.2

1.6 The volt – ampere characteristic of a device is shown in Fig. P. 1.3. What type of device is it and what is its value ?

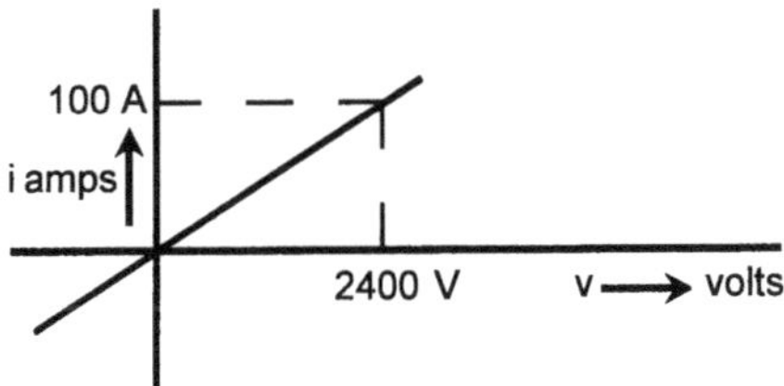

Fig. P. 1.3

1.7 A current $i = 10\ e^{-t}$ is applied to

(a) a 3Ω resistor (b) a 2H inductor and (c) a 0.1 F capacitor.

What are the respective voltages ? Write down the expressions for power in each case.

1.8 A number of waveforms are shown in Fig. P. 1.4(a). Similarly a number elements are shown in Fig. P. 1.4(b). Write down the waveforms of the response for each of these elements for each of the waveforms in Fig. P. 1.4(a) where

(a) f(t) is a current and the response is the voltage and

(b) f(t) is voltage and the response is the current.

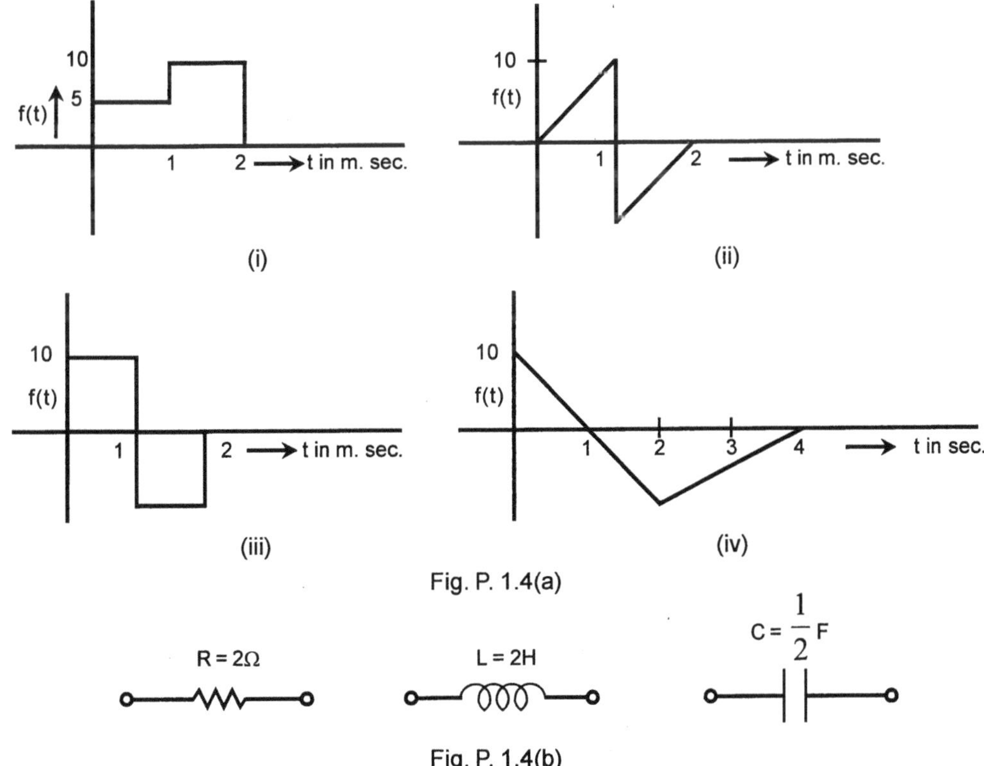

Fig. P. 1.4(a)

Fig. P. 1.4(b)

1.9 A voltage v(t) = 200 Sin 100 πt volts is applied across a capacitor of value C = 0.05 farads. Find the expression for the current i(t) and the energy stored in the capacitor.

1.10 A voltage v(t) = 25 Sin 1000 t volts is applied across a 5 mH inductor at t = 0 when the current in it is 2 A. Find the expression for the current i(t) for t ≥ 0.

1.11 The current in a coil is changing at a rate of 2 A/sec. The voltage across another coil placed very near to it was found to be 10 m V. What is the value of the mutual inductance between the two coils ?

1.12 Two coils with self inductances 0.3 mH and 0.2 mH are coupled together so that the mutual inductance between them is 0.22 mH. A current waveform shown in Fig. P. 1.5 is applied to the first inductor. Draw the waveform of the self induced emf's in the first coil and the mutually induced emf in the second coil.

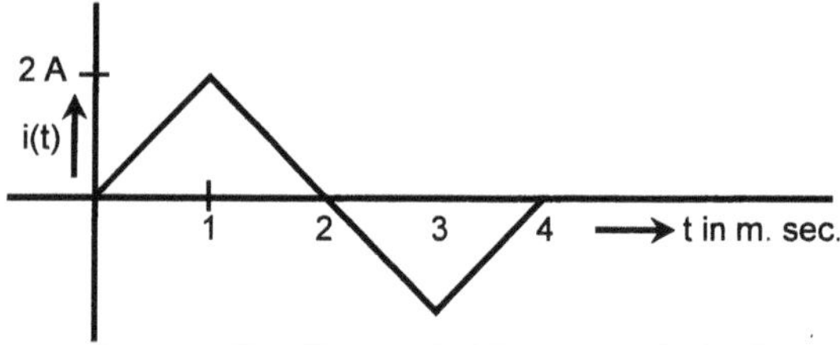

Fig. P. 1.5

1.13 For the magnetically coupled coils shown in Fig. P.1.6 establish the polarity markings using different kinds of dots.

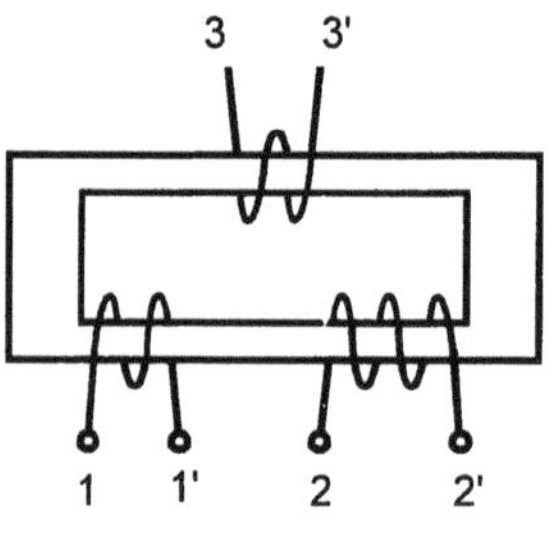

Fig. P. 1.6

1.14 Three coils have terminals 1, 1' ; 2, 2' and 3, 3'. Place these coils on a common core with proper winding sense such that

(a) The terminals 1 and 2, 2 and 3 and 3 and 1 have same polarity mark.

(b) 1 and 2' ; 2 and 3' ; 3 and 1 have same polarity mark.

1.15 Two inductors have self inductance of 0.1 mH and 0.4 mH and a mutual inductance of 0.15 mH. What is the value of the coefficient of coupling between them ? If a current

$$i(t) = 3 \text{ Sin } t + 1.5 \text{ Sin } 2t$$

is passed through the first inductor what is the expression for the voltage induced in the second coil ?

1.16 If a voltage v is applied to a coil, the quantity $\int_{-\infty}^{t} v\,dt$ is called the *flux linkage*, ψ, in the coil at time 't'. A current given by $i(t) = (1 - e^{-2t})$ A, $t > 0$ is flowing through a coil of inductance 0.5 H. The current has a value 0.865 A at a certain time. At this time

(a) What is the rate of change of current ?

(b) What is the value of flux linkages ?

(c) What is the rate of change of flux linkages ?

(d) What is the value of voltage across the coil ?

and (e) What is the energy stored in the inductor ?

1.17 The quantity $\int_{-\infty}^{t} i\,dt$ in a capacitor is called the *charge* 'q'. In a capacitor of value $\frac{1}{2}$ F and charged to a voltage of 1 V, a current of the from $i(t) = e^{-2t}$, $t > 0$ is flowing. What is the value of the current at t = 1 sec. ? At t = 1 sec.

(a) What is the total charge accumulated in the capacitor ?

(b) What is the rate of change of voltage across the capacitor ?

(c) What is the voltage across the capacitor ?

and (d) At what rate is energy being taken from the electric field of the capacitor ?

1.18 The voltage induced in the inductor is given by

$$v = \frac{d}{dt}(Li) = \frac{d\psi}{dt}$$

where ψ is the flux linkages. A voltage is induced in a coil if L is constant and i is changing with respect to time. If the inductance value changes with time a constant current also produces an emf in the coil. When a dc current of 1 A is passed through a time varying inductor whose inductance changes as shown in Fig. P. 1.7 what is the waveform of voltage across the inductor ?

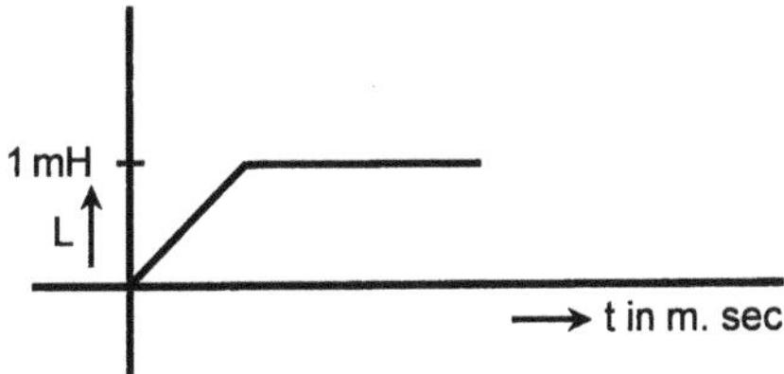

Fig. P. 1.7

1.19 A nonlinear capacitor has its v Vs q characteristic shown in Fig. P. 1.8. If a voltage v(t) = 10 Sin 2t is applied to such a capacitor what is the expression for the current through it ?

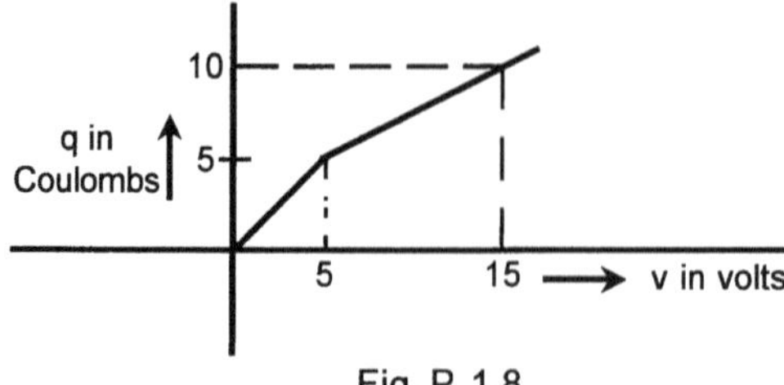

Fig. P. 1.8

1.20 Fig. P. 1.8 represents characteristic of a nonlinear inductor if the x and y axis variables are current and flux linkages respectively. If the current in the inductance is given by the waveform shown in Fig. P. 1.9 sketch the waveform of voltage across the inductor.

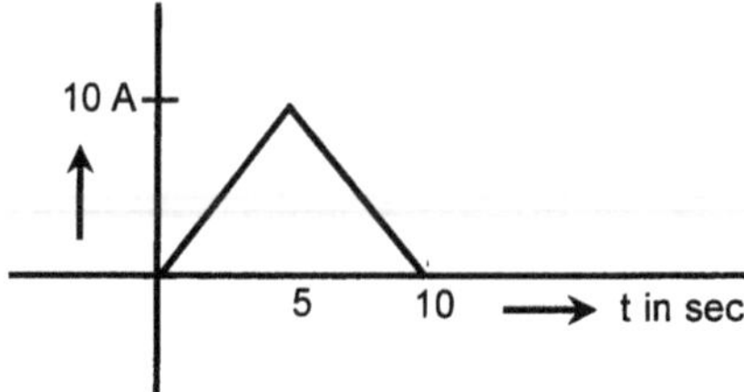

Fig. P. 1.9

2

KIRCHHOFF'S LAWS AND RESISTIVE NETWORKS

2.1 INTRODUCTION :

An electrical network is an interconnection of passive and active devices which are represented by their network models. As discussed in chapter 1, the network models are two terminal lumped elements and we are concerned with the current through these elements and voltages across them. The elements are interconnected with wires having zero resistance and absorbing zero power. These wires may have different lengths and shapes and the elements may be oriented in different positions. The same network may be drawn in different ways to look completely dissimilar and an engineer must be able to recognise the equivalence of these networks.

The four networks shown in Fig. 2.1 appear to be different but they are equivalent. One must be able to redraw the network in its simplest form.

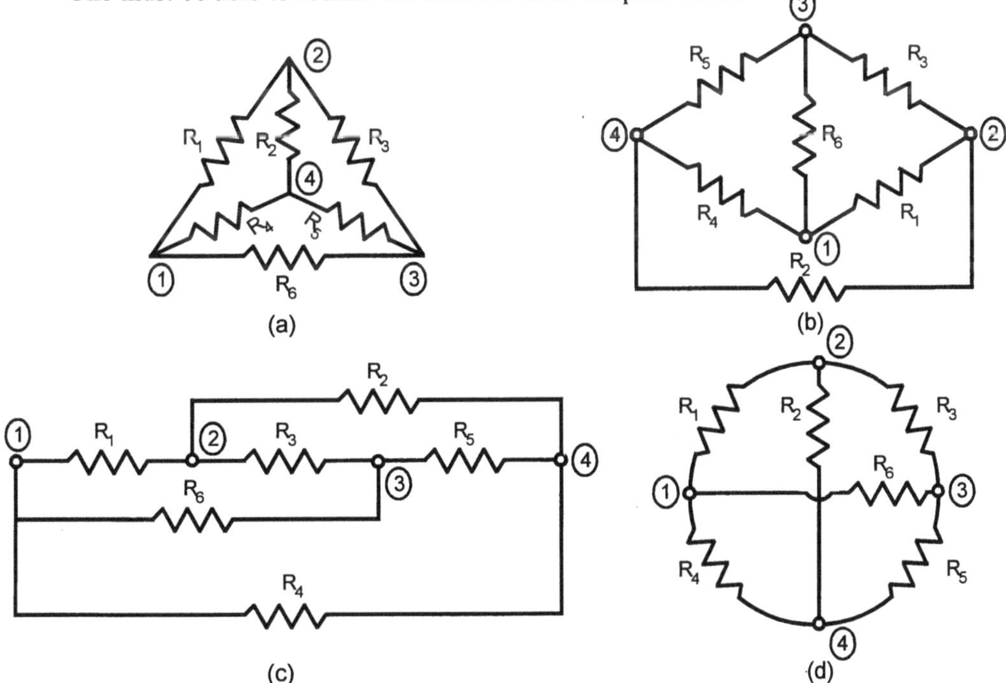

Fig. 2.1 A network represented in 4 different configurations

The junction points where two or more elements are connected together are known as *nodes*. One single element with two nodes at its each end is called as a *branch*. Thus in Fig. 2.1(a) there are four nodes and six branches. Suppose we start at a node, say node 1, and travel along the branches to the other nodes and come back to the starting node without going through any node twice, we call this path *a circuit*.

For example, in Fig. 2.1(a) the branches with elements R_1, R_2, R_4 constitute a circuit. In a similar manner we can recognise more circuits in Fig. 2.1 (a). Elements R_1, R_3, R_5, R_2, R_1 do not constitute a circuit because the node 2 is encountered twice in travelling from node 1 to itself. Having defined a node and a circuit we are now ready to consider two basic laws, which form the foundation for the study of network theory. These are the two laws enunciated by a German Professor, Gustav Robert Kirchhoff, in 1848.

2.2 KIRCHHOFF'S LAWS :

Kirchhoff 's laws are the consequence of the law of conservation of energy and conservation of charge.

2.2.1 Kirchhoff 's current law :

The first law, which is known as Kirchhoff's current law, or simply KCL, states that

" *The algebraic sum of the currents at a node is equal to zero* ".

Thus at a node

$$\Sigma i = 0 \quad .....(2.1)$$

This law is based on the fact that there cannot be accumulation of charge at a node and thus store energy, since node is not a circuit element. Any charge which enters a node at any instant must therefore leave the node immediately. In terms of current, the total current entering the node must be equal to the total current leaving the node. If currents leaving the node are taken to be positive and currents entering are taken to be negative, this choice being arbitrary, the algebraic sum of the currents at the node will be zero. This is illustrated in Fig. 2.2.

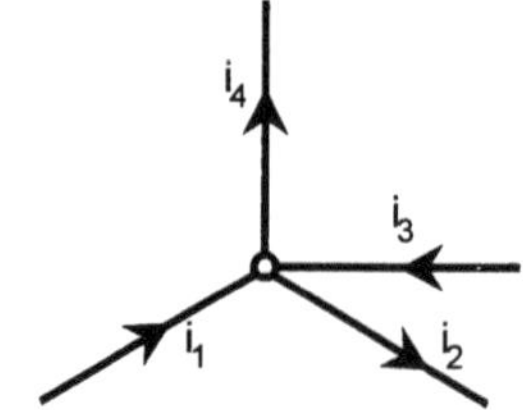

Fig. 2.2 An example of KCL

$$-i_1 + i_2 - i_3 + i_4 = 0 \quad ....(2.2)$$

or

$$i_1 + i_3 = i_2 + i_4 \quad .....(2.3)$$

The directions of currents shown in Fig. 2.2 are the assumed reference directions. The actual currents may have negative values. For example all the reference directions of currents at a junction may be pointing away form the node as in Fig. 2.3 and according to KCL.

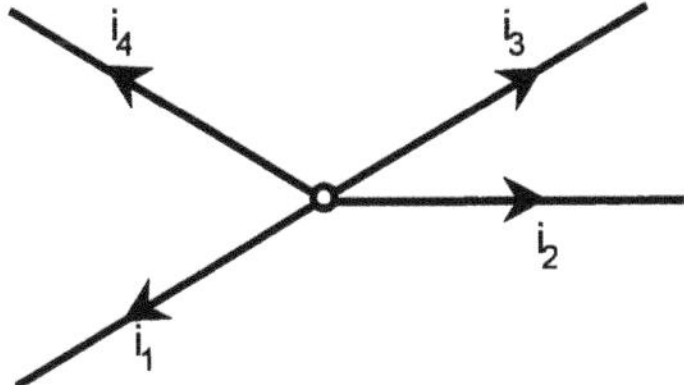

Fig. 2.3 All currents leaving the node

$$i_1 + i_2 + i_3 + i_4 = 0 \qquad .....(2.4)$$

This may appear to be absurd, but recognising the fact that these are reference directions only and some of the currents may have negative values, eq. (2.4) is quite appropriate.

2.2.2 Kirchhoff 's voltage law :

The second law of Kirchhoff, known as Kirchhoff 's voltage law, or in short KVL, is stated as

" The algebraic sum of voltages around any closed path in a network is equal to zero ".

This, again, is based on law of conservation of energy. This law can be best understood in terms of its gravitational analogy. The energy level of a point A with respect to another point B can be found by moving a unit mass from point A to point B and finding the energy given up by the mass in this process. If a mass is moved from point A to B and then from point B to A, the net work done on the mass must be equal to zero. In a similar manner, the energy level of a point A with respect to a point B can be obtained by moving 1 coulomb of charge from A to B and noting the energy gained or lost by the charge. The energy gained or lost by the charge is independent of the path taken in travelling form A to B, as otherwise, a charge can be made to gain energy continuously by taking the charge repeatedly form A to B along the high energy path and back to A along the low energy path. In Fig. 2.4, if 1 coulomb of charge moves form A to B along the element 1, it loses v_1 Joules of energy. If the same charge is moved through elements 2 and 3 to reach the point B, it loses $(v_2 - v_3)$ Joules of energy. Since these two energies must be the same, we have.

$$v_1 = v_2 - v_3 \quad .....(2.5)$$

or

$$v_2 - v_3 - v_1 = 0 \quad .....(2.6)$$

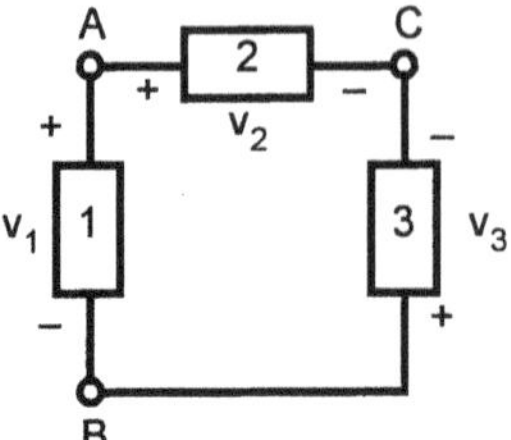

Fig. 2.4 Illustration of Kirchhoff 's Voltage Law

Thus if we move 1 coulomb of charge, starting from any node of a network, along any closed path or a circuit, the net energy gained or lost must be equal to zero.

In applying the KVL, one may adopt the following procedure to minimise errors in writing the equations. Starting from any node in the closed path, move along a clockwise direction (if you chose, you may follow anti – clockwise direction also !) and assign positive sign for the voltage across the element, if you are entering it at positive reference terminal and assign negative sign for the voltage, if you are entering a negative, terminal. The algebraic sum of all these voltages must be zero. Following this procedure, starting from node A, the KVL for the closed path of network in Fig. 2.4, will be,

$$v_2 - v_3 - v_1 = 0 \quad .....(2.7)$$

Eq. (2.7) is same as eq. (2.6). It is to be emphasised here that the polarities of voltages shown in Fig. 2.4 are only reference directions and the voltages themselves may have positive or negative values.

The Ohm's law, combined with the two laws of Kirchhoff, enables us to analyse a network consisting of resistors and sources. By the analysis of a network, we mean, determining the currents and voltages in all the elements forming the network.

2.3 RESISTIVE NETWORKS :

A resistive network consists of only resistors and sources. In this chapter we will consider sources which have a constant value, i.e., d.c sources only. The method of solution we would follow, is to define certain unknown variables, usually some currents or voltages, write equations in terms of these variables and then solve these equations using rules of algebra. The equations are based on the volt ampere relations of the elements and the two Kirchhoff 's laws. This procedure is a mechanical process and can be applied to solve any network easily. First, we will consider simple networks consisting of several resistors connected in series or parallel, and then analysis more complicated networks.

2.3.1 Resistors in series :

By definition, all the elements that carry the same current are said to be in series. It is to be noted here that the elements must carry 'same' current and not equal current to be considered as connected in series. In a complex network two or more resistor in different parts of the network may carry equal currents but they are not connected in series. The elements must form a single chain between two terminals. It is often convenient to replace this chain of resistors connected in series with a single equivalent resistor between the two terminals. The two networks have same volt ampere relation at the two terminals, as shown in Fig. 2.5.

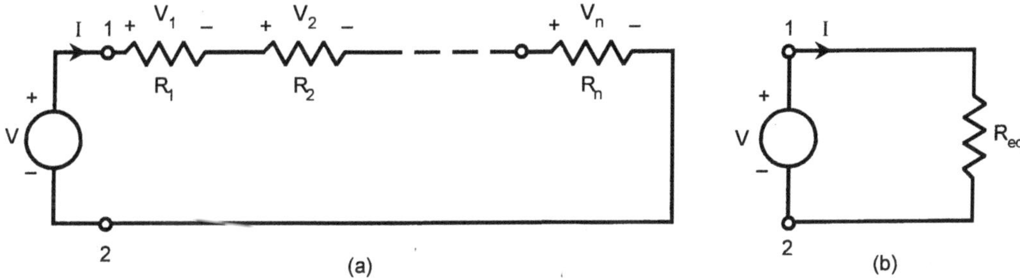

Fig. 2.5 Resistors connected in series and its equivalent

From a trivial application of KCL it can be seen that the same current 'I' flows through each of the resistors in the circuit of Fig. 2.5(a). Assuming a reference direction for the current to be directed out from the positive terminal of the source and using passivity condition the voltages across the resistors are taken to have positive signs at the terminal where the current enters the resistor. Using Kirchhoff 's law around the circuit travelling in a clockwise direction starting from terminal 1,

$$V_1 + V_2 + ... + V_n - V = 0 \qquad .....(2.8)$$

But $$V_1 = IR_1,\ V_2 = IR_2\ .....V_n = IR_n \qquad .....(2.9)$$

Therefore $$IR_1 + IR_2 + ... + IR_n - V = 0$$

or $$I(R_1 + R_2 + ... + R_n) = V \qquad .....(2.10)$$

$$I = \frac{V}{R_1 + R_2 + ... + R_n} \qquad .....(2.11)$$

For equivalent network in Fig. 2.5(b), if we assume that the same current is produced when the applied source is same, we have

$$I = \frac{V}{R_{eq}} \qquad .....(2.12)$$

From eq. (2.11) and (2.12)

$$R_{eq} = R_1 + R_2 + \text{- - -} + R_n \qquad \text{....(2.13)}$$

Thus the two networks of Fig. 2.5(a) and (b) are equivalent at the two terminals 1 and 2 if they satisfy the relation given by eq. (2.13). If the two networks to the right of the terminals 1 and 2 in Fig. 2.5(a) and (b) are enclosed in a 'black box', it is impossible to tell one from the other.

Example 2.1

Solve for the current I in the circuit of Fig. 2.6(a).

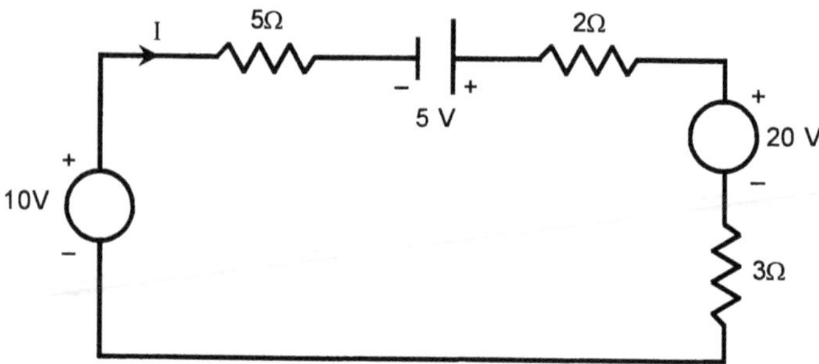

Fig. 2.6(a) Circuit for example 2.1

Solution :

As the order in which the elements are connected is immaterial in a series circuit, the network may be redrawn as shown in Fig. 2.6(b).

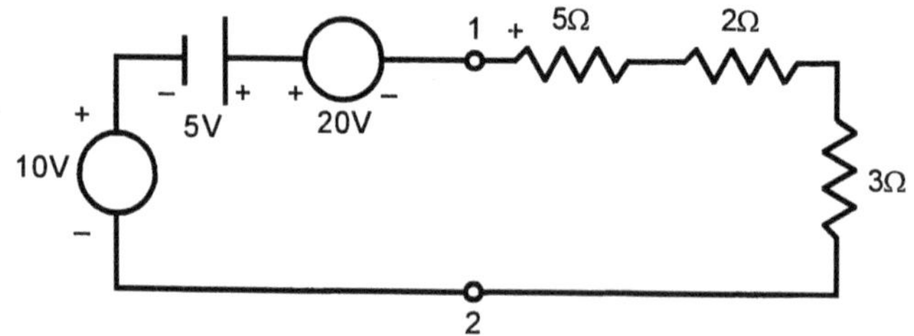

Fig. 2.6(b) Network of Fig. 2.6(a) redrawn

Now the sources connected between the terminals 1 and 2 can be replaced by a single voltage source V_{eq} and the resistors connected in series between the same terminals can be replaced with a single resistor R_{eq}. as shown in Fig. 2.6(c).

$$V_{eq} = 10 + 5 - 20 = -5 \text{ V}$$

$$R_{eq} = 5 + 2 + 3 = 10 \ \Omega$$

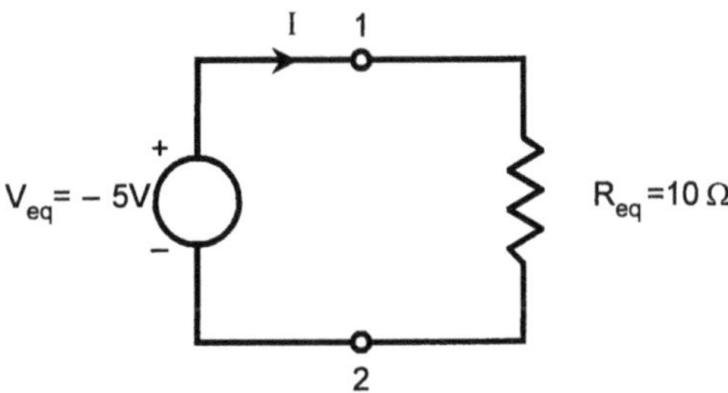

Fig. 2.6(c) Equivalent network for circuit in Fig. 2.6(a)

Observe the polarity for V_{eq}. Since the reference polarity for voltage source is taken such that terminal 1 is at a higher potential compared to the terminal 2, the actual voltage is – 5 V.

Thus $$I = \frac{V_{eq}}{R_{eq}} = \frac{-5}{10} = -0.5 \text{ A}$$

or equivalently a current of 0.5 A is flowing in the resistor form terminal 2 to 1. If the reference polarity is reversed the circuit may be written as shown in Fig. 2.6(d).

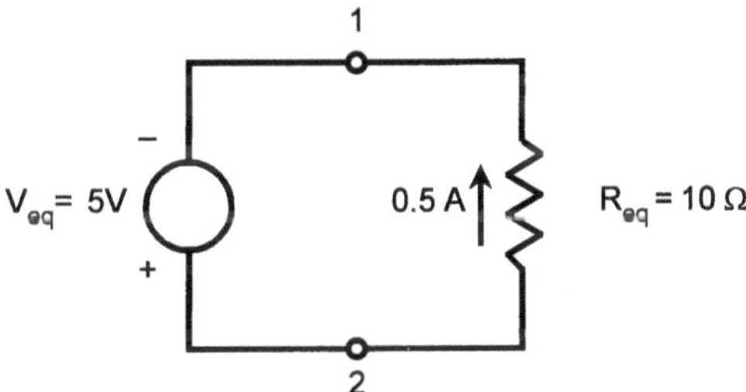

Fig. 2.6(d) Alternate form for circuit in Fig. 2.6(c)

The example 2.1 illustrates the following important points :

1. The order in which the elements are connected in series can be changed as desired.
2. The reference polarities for voltages at any set of terminals is arbitrary.
3. Once the reference polarities are chosen, the actual voltages are either positive or negative.
4. All the voltage sources connected in series between two terminals can be replaced by a single equivalent voltage source.
5. All the resistors connected in series between two terminals can also be replaced by a single equivalent resistor.

2.3.2 Resistors in parallel :

A set of resistors are in parallel if the same voltage is present across each of the resistors. For the network in Fig. 2.7(a).

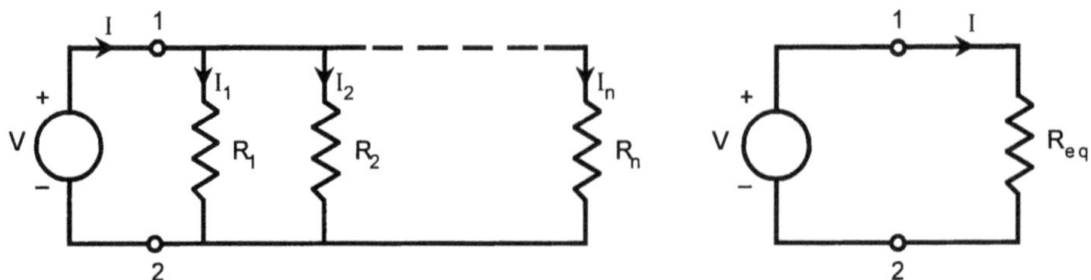

Fig. 2.7(a) Resistors in parallel

By applying KCL at node 1

$$I = I_1 + I_2 + ... + I_n \qquad .....(2.14)$$

But
$$I_1 = \frac{V}{R_1}, \; I_2 = \frac{V}{R_2} \; ..... \; I_n = \frac{V}{R_n} \qquad ....(2.15)$$

Therefore
$$I = \frac{V}{R_1} + \frac{V}{R_2} + ... + \frac{V}{R_n}$$

$$= V\left(\frac{1}{R_1} + \frac{1}{R_2} + ... + \frac{1}{R_n}\right) \qquad .....(2.16)$$

Similarly for network in Fig. 2.7(b), for the same voltage and current

$$I = \frac{V}{R_{eq}} \qquad .....(2.17)$$

Since both the networks have same V – I equations, we have from eq. (2.16) and eq. (2.17).

$$\frac{1}{R_{eq}} = \frac{1}{R_1} + \frac{1}{R_2} + ... + \frac{1}{R_n} \qquad .....(2.18)$$

From the definition of conductances of the resistors, eq (2.18) may be written as

$$G_{eq} = G_1 + G_2 + ... + G_n \qquad .....(2.19)$$

As in the case of series connected resistors, the order in which these resistors are connected in parallel between two terminals is not important. Any number of independent current sources connected across these terminals may also be combined into a single equivalent current source with a value of current equivalent to the algebraic sum of the individual currents.

Example 2.2

Consider the network in Fig. 2.7(b). Reduce it to a single current source and single resistor network at the terminals 1 and 2 and find the voltage across them.

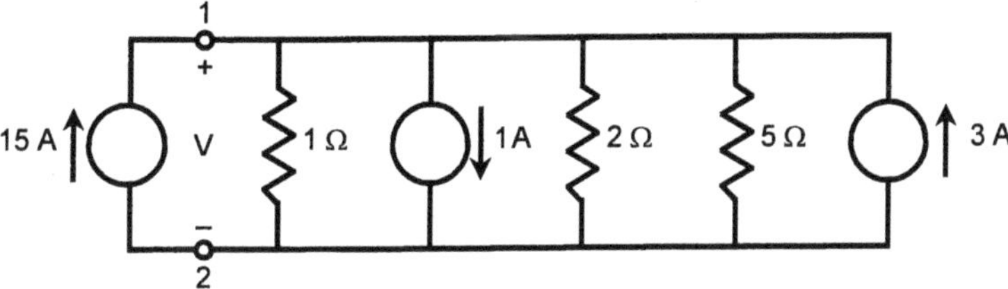

Fig. 2.7(b) Network for example 2.2

Solution :

The network is redrawn as shown in Fig. 2.7(c).

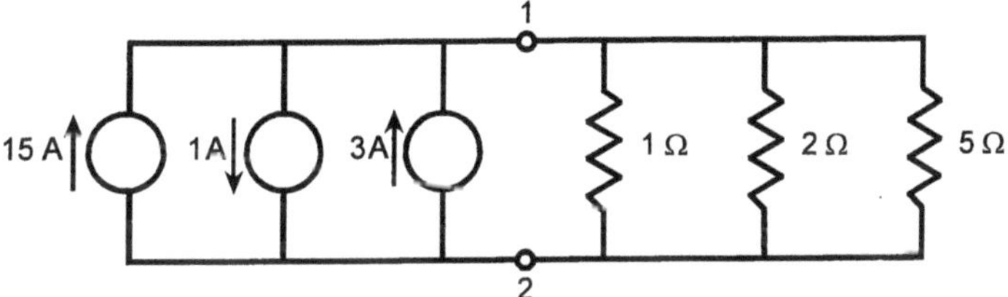

Fig. 2.7(c) Network of Fig. 2.7(b) redrawn

The equivalent network between terminals 1 and 2 is given in Fig. 2.7(d) with

$$I_{eq} = 15 - 1 + 3 = 17 \text{ A}$$

$$\frac{1}{R_{eq}} = \frac{1}{1} + \frac{1}{2} + \frac{1}{5} = 1.7$$

or $$R_{eq} = \frac{1}{1.7} \ \Omega$$

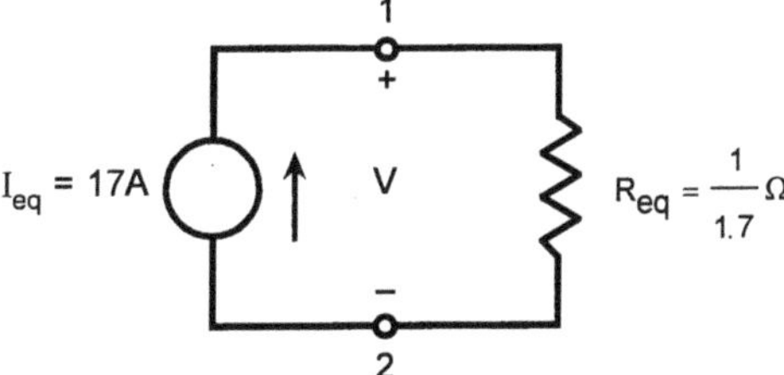

Fig. 2.7(d) Equivalent network of circuit in Fig. 2.7(b)

$$V = I_{eq} \times R_{eq}$$

$$= 17 \times \frac{1}{1.7}$$

$$= 10 \text{ V}$$

In many networks, two resistors connected in parallel is very common and thus it merits special attention. For two resistors in parallel,

$$\frac{1}{R_{eq}} = \frac{1}{R_1} + \frac{1}{R_2} \qquad ...(2.20)$$

$$R_{eq} = \frac{R_1 R_2}{R_1 + R_2} \qquad .....(2.21)$$

It is easier to use relation (2.21) rather than eq. (2.20) to find R_{eq} when two resistors are connected in parallel.

2.3.3 The voltage divider :

A practical application of resistors connected in series across a voltage source is the voltage divider. When a large voltage source is available and we desire to have a smaller voltage, we use a voltage divider circuit shown in Fig. 2.8.

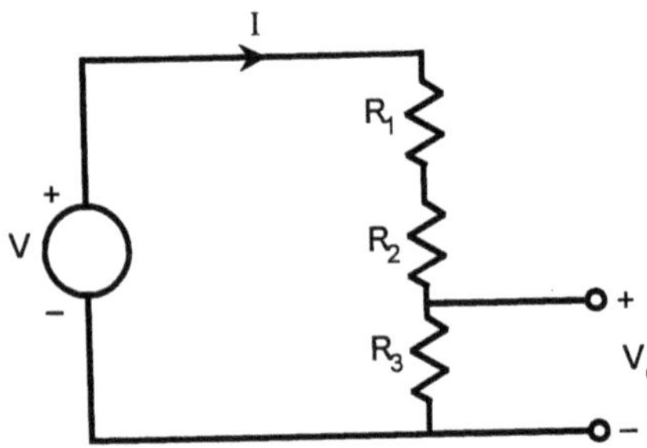

Fig. 2.8 A voltage divider circuit

In the circuit of Fig. (2.8)

$$R_{eq} = R_1 + R_2 + R_3$$

and $$I = \frac{V}{R_1 + R_2 + R_3}$$

The voltage across the resistor R_3 is

$$V_0 = I\, R_3$$

$$= \frac{R_3}{R_1 + R_2 + R_3} \cdot V \qquad .....(2.22)$$

Eq. (2.22) implies that the voltage across any resistor in a voltage divider circuit will have the same proportion to the total voltage as that resistance to the total resistance. Thus for a set of series resistors, voltages across the resistors will be in proportion to the corresponding resistances.

2.3.4 The current divider :

When several resistors are connected in parallel across a current source, we have a current divider problem as shown in Fig. 2.9.

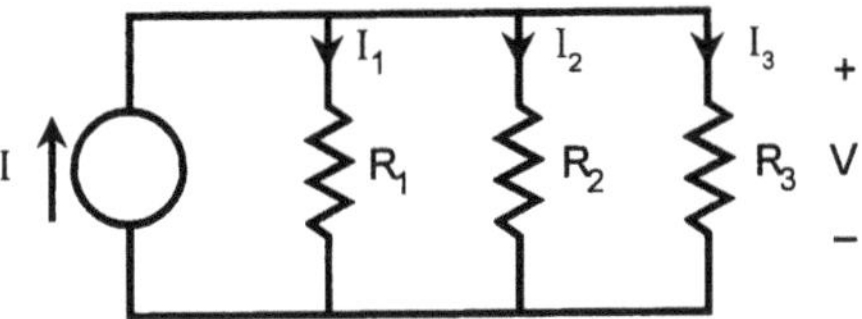

Fig. 2.9. A Current divider circuit

The equivalent resistance is given by

$$R_{eq} = \frac{1}{\frac{1}{R_1} + \frac{1}{R_2} + \frac{1}{R_3}}$$

and

$$V = I\,R_{eq}$$

$$= I\frac{1}{\frac{1}{R_1} + \frac{1}{R_2} + \frac{1}{R_3}}$$

Once the voltage is known across the resistor, the individual current can be obtained as

$$I_1 = \frac{V}{R_1} = \frac{\frac{1}{R_1}}{\frac{1}{R_1} + \frac{1}{R_2} + \frac{1}{R_3}} I \quad .....(2.23)$$

Eq. (2.23) shows that the current divides as the inverse of the resistance or directly to the conductance.

The case of two resistors connected in parallel occurs frequently and the current in individual resistors can be calculated easily. In Fig. 2.10.

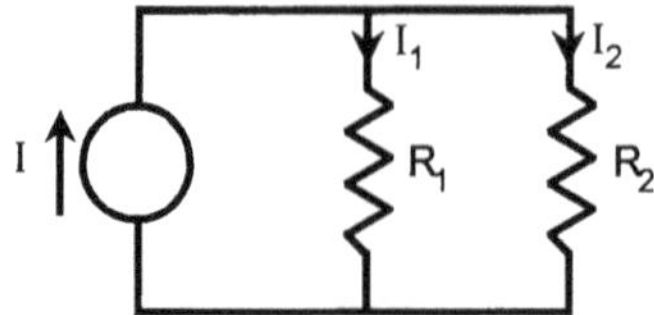

Fig. 2.10 Two resistors is parallel

the currents I_1 and I_2 are given by

$$I_1 = \frac{\frac{1}{R_1}}{\frac{1}{\frac{1}{R_1}+\frac{1}{R_2}}} I$$

$$= \frac{1}{R_1} \times \frac{R_1 R_2}{R_1 + R_2} I = \frac{R_2 I}{R_1 + R_2} \qquad .....(2.24)$$

Similarly $$I_2 = \frac{R_1 I}{R_1 + R_2} \qquad .....(2.25)$$

Eqs. (2.24) and (2.25) show that the current in any branch is equal to the total current multiplied by the resistance in the other branch divided by the sum of the resistances.

Example 2.3

Find the current in the 2 Ω resistor in Fig. 2.11

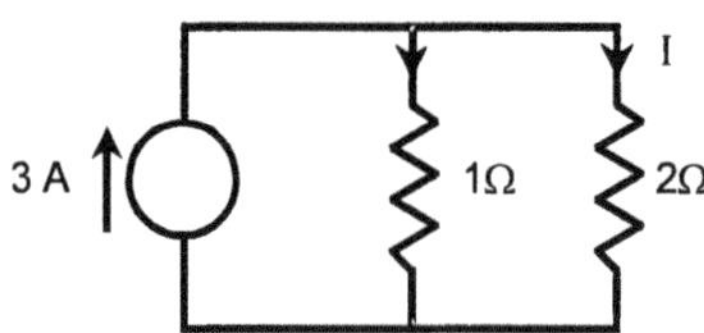

Fig. 2.11 Network for example 2.3

From current division principle

$$I = \frac{1}{1+2} \cdot 3$$

$$= 1 \text{ A}$$

2.3.5 Ladder network :

A set of resistors connected alternately in series and parallel is known as a *ladder network.* By combining the resistors in series and in parallel, a ladder network between two terminals can be replaced by an equivalent resistor at these terminals.

Example 2.4

Find the currents in all the branches of the network in Fig. 2.12.

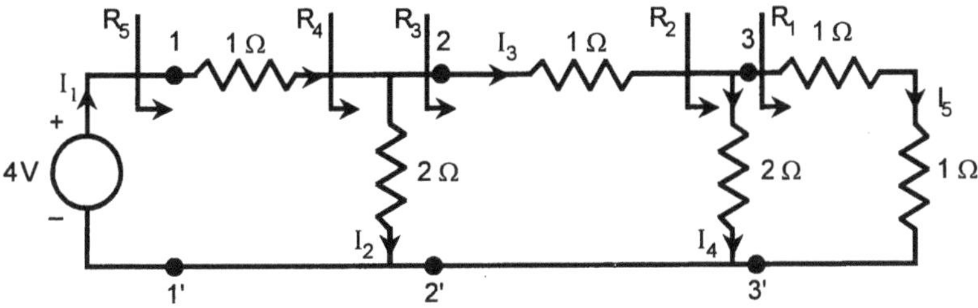

Fig. 2.12 (a) Ladder network for example 2.4

Solution :

Reducing the network successively starting from right at terminals 3 and 3', the resulting networks at different stages of combination are shown in Fig. 2.12.

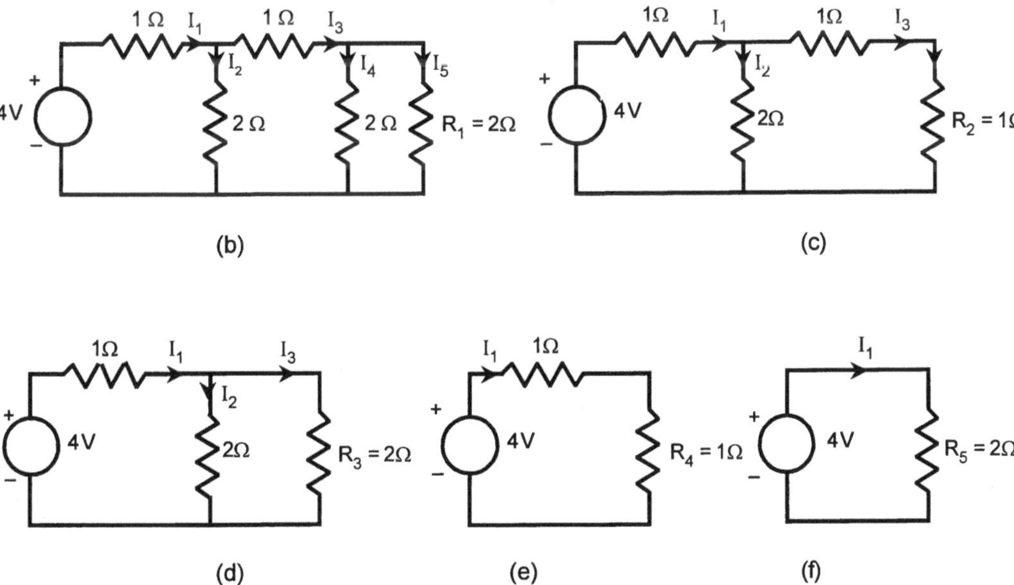

Fig. 2.12 Various intermediate networks obtained by series parallel combination of resistors

In Fig. 2.12(b), R_1 is the series combination of the two right most resistances.

$$R_1 = 1 + 1 = 2\ \Omega$$

In Fig. 2.12(c) R_2 is the parallel combination of 2 Ω resistor and R_1 i.e.,

$$R_2 = \frac{2\times 2}{2+2} = 1\ \Omega$$

Proceeding in a similar manner

$$R_3 = 2\ \Omega \qquad R_4 = 1\ \Omega \quad \text{and finally} \quad R_5 = 2\ \Omega$$

Thus the current supplied by the source is, from Fig. 2.12(f)

$$I_1 = \frac{4}{2} = 2\ \text{A}$$

To find the currents in all the branches, we proceed backwards and apply the principle of current division. Thus the current in 1 Ω resistor and R_4 is same as current I_1 in Fig. 2.2(f). The currents I_2 and I_3 in Fig. 2.12(d) are given by

$$I_2 = \frac{I_1 \times 2}{2+2} = \frac{2\times 2}{4} = 1\ \text{A}$$

$$I_3 = \frac{I_1 \times 2}{2+2} = \frac{2\times 2}{4} = 1\ \text{A}$$

The currents I_1, I_2, I_3 are shown in Fig. 2.12(c). Again currents I_4, I_5 are given by current division as

$$I_4 = I_3 \times \frac{2}{2+2} = 1 \times \frac{2}{4} = 0.5\ \text{A}$$

$$I_5 = I_3 \times \frac{2}{2+2} = 1 \times \frac{2}{4} = 0.5\ \text{A}$$

Thus the currents in all the resistors are found.

If the network to the right of terminals 1 and 1' is to be replaced by its equivalent resistance, the network appears as in Fig. 2.12(g).

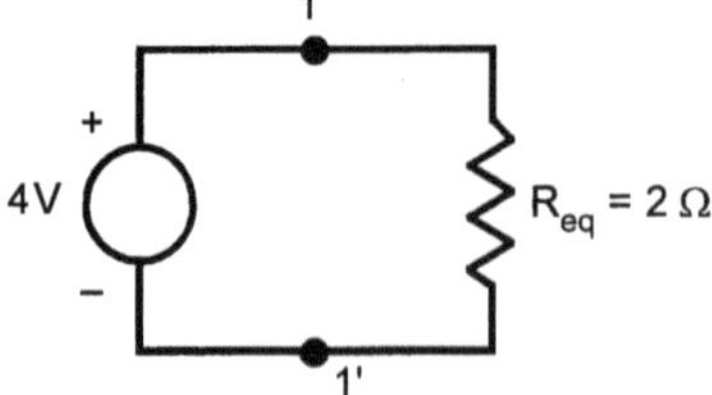

Fig. 2.12(g) Equivalent network

2.3.6 Source transformations :

We had defined ideal voltage and current sources in Chapter 1. But real world sources are seldom ideal. The practical sources have a drooping v – i characteristic rather than a constant value as shown in Fig. 2.13(a) and (b).

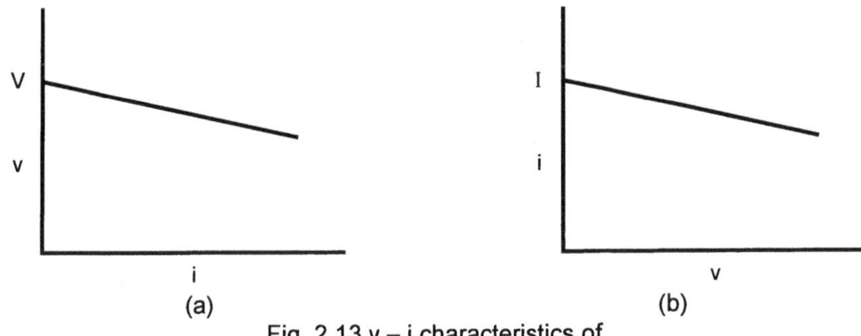

Fig. 2.13 v – i characteristics of
(a) Practical voltage source (b) Practical current source

These real physical sources can be represented by an ideal voltage source in series with a resistance or an ideal current source in parallel with a resistance as shown in Fig. 2.14(a) and (b).

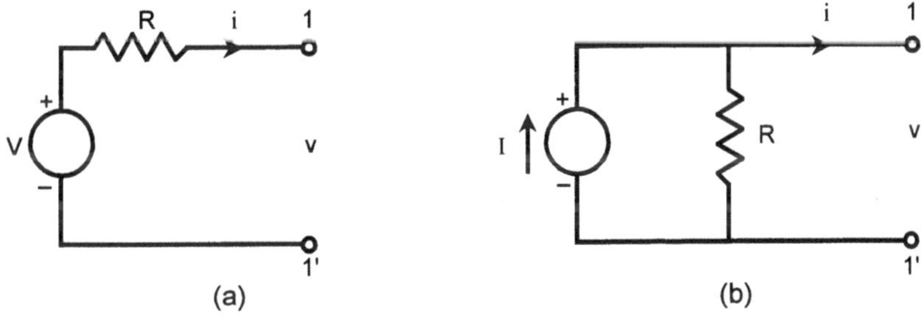

Fig. 2.14. Representation of (a) practical voltage source (b) practical current source

The volt – ampere relation for the circuit in Fig. 2.14(a) is

$$v = V - iR \qquad .....(2.26)$$

This relationship gives rise to the characteristic shown in Fig. 2.13(a)

Similarly the volt – ampere equation for Fig. 2.14(b) is given by

$$i = I - \frac{v}{R} \qquad .....(2.27)$$

This relationship gives rise to the characteristic shown in Fig. 2.13(b).

It is often convenient to transform a branch containing an ideal voltage source in series with a resistor to a branch containing an ideal current source in parallel with a resistor or vice versa.

Towards this end, let us solve eq. (2.27) for v to get,

$$v = IR - iR \qquad .....(2.28)$$

Comparing eq. (2.28) with eq. (2.26), we get

$$V = IR \qquad .....(2.29)$$

Similarly solving for i in eq. (2.26)

$$i = \frac{V}{R} - \frac{v}{R} \qquad .....(2.30)$$

Comparing with eq. (2.27), we get

$$I = \frac{V}{R} \qquad .....(2.31)$$

From eq. (2.26), eq. (2.27), eq. (2.29) and eq. (2.31) we can develop the equivalent circuits shown in Fig. 2.15(a) and (b).

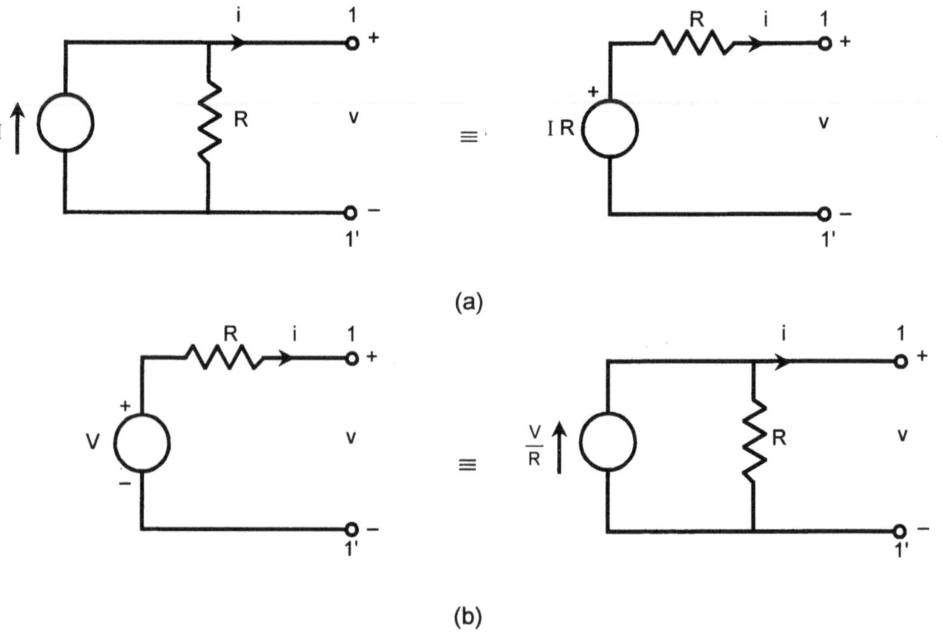

Fig. 2.15 Source transformations

2.3.7 Voltage source with no series resistance :

Practical voltage sources will always have a series resistance along with an ideal voltage source. But sometimes these resistances are so small that they can be neglected. But such branches with only ideal voltage sources in a network cause problems. It is not possible to transform this branch into an equivalent current source branch. To avoid this problem, a technique by which the ideal voltage source with no series resistance is removed and replaced with a real voltage source with a series resistance is developed.

Consider the network in Fig. 2.16(a).

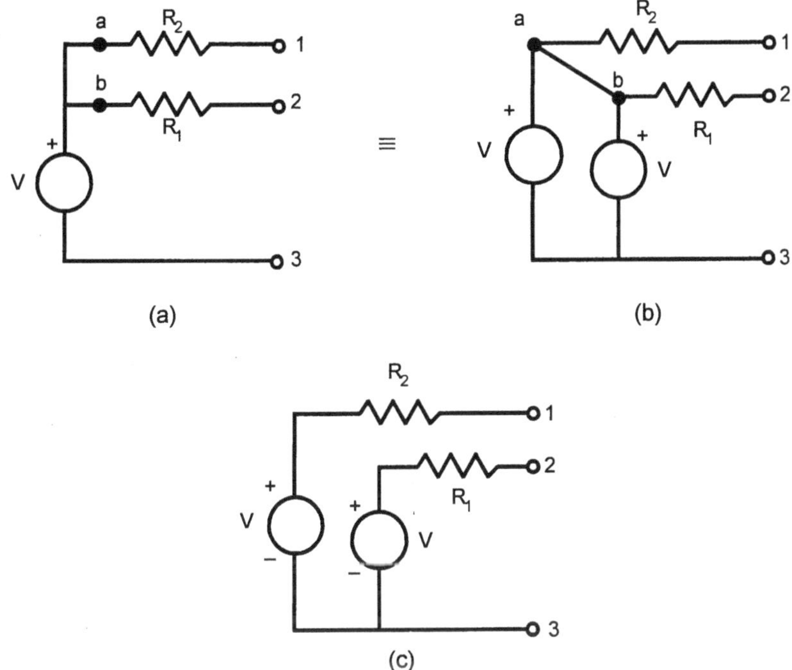

Fig. 2.16 A transformation for voltage sources without resistance in series

The voltage source V is replaced by two voltage sources of value V in parallel as shown in Fig. 2.16 (b). The points a and b, are at the same potential V, whether the short circuit between a and b is present or not. Hence removing the short circuit we get the equivalent circuit shown in Fig. 2.16(c).

2.3.8 Current source with no parallel resistor :

A real current source can be represented by an ideal current source in parallel with a resistance. But sometimes this resistance is so large in comparison with other resistances in the network, that it can be neglected. Hence we are left with an ideal current source with no resistance in parallel. This also poses problem in that, it cannot be converted to a branch with a voltage source. This problem can be overcome by considering the following circuit in Fig. 2.17(a) and replacing the current source between terminals 1 and 2 by three current sources of same value as shown in Fig. 2.17(b).

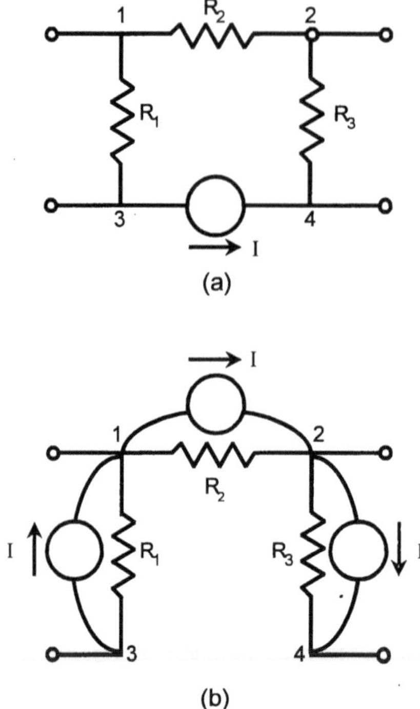

Fig. 2.17 A transformation for current source with no parallel resistance

The effect of current source is to deliver 1A from node 3 to node 4. The same effect is produced by the 3 sources of 1A connected across the three resistors without affecting the currents or voltages at the other nodes. 1 Amp of current enters the node 1 from node 3 and the same current immediately leaves the node 1 and enters the node 2. Again the same current leaves the node 2 and enters the node 4. This in effect sends a current from node 3 to 4 via nodes 1 and 2 instead of directly, without affecting the currents at other nodes.

2.3.9 Voltage sources with parallel resistors or current sources :

When a voltage source is connected in parallel with a set of resistors or current sources at two terminals in a network, the combination can be replaced by the voltage source only. Two examples are given in Fig. 2.18.

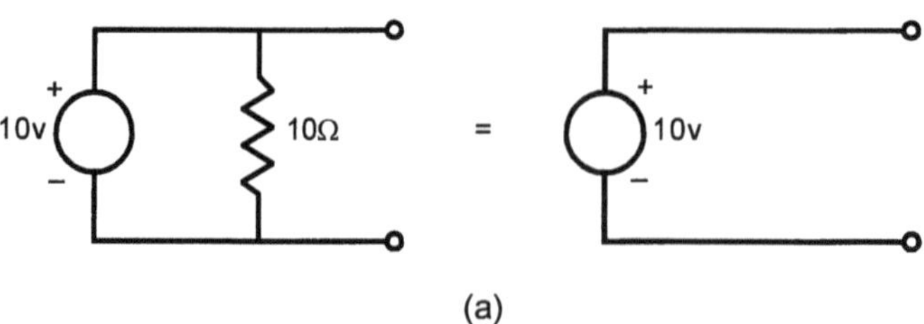

(a)

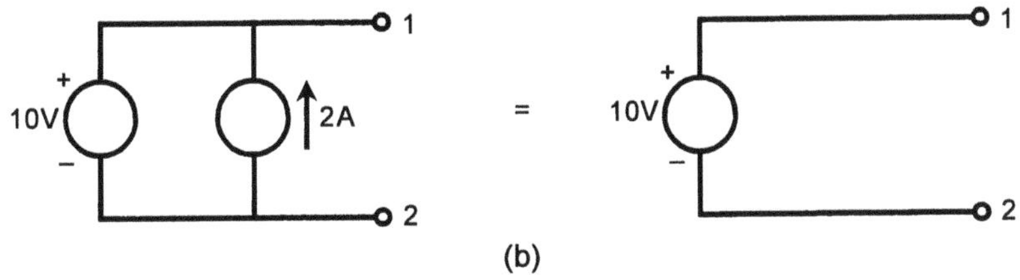

(b)

Fig. 2.18 Example of voltage source in parallel with (a) resistor (b) current source

More complex example is given in Fig. 2.19.

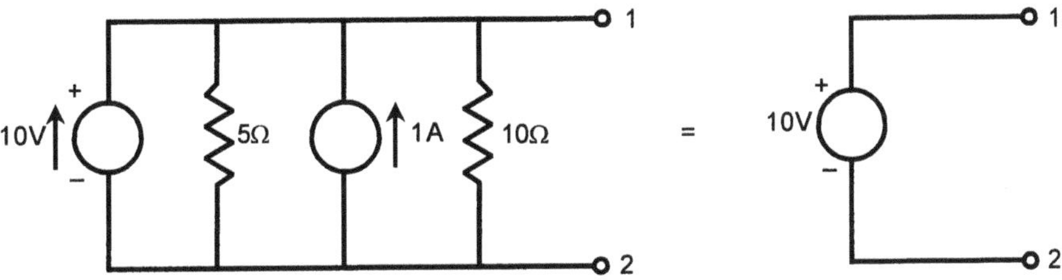

Fig. 2.19 Another example of voltage source in parallel with other elements

2.3.10 Current sources in series with resistors or voltage sources :

When a current source is in series with resistors or voltage sources or any combination of these at a set of two terminals, the combination can be replaced by a single current source at these terminals. Three examples are given in Fig. 2.20 illustrating this equivalence.

(a)

(b)

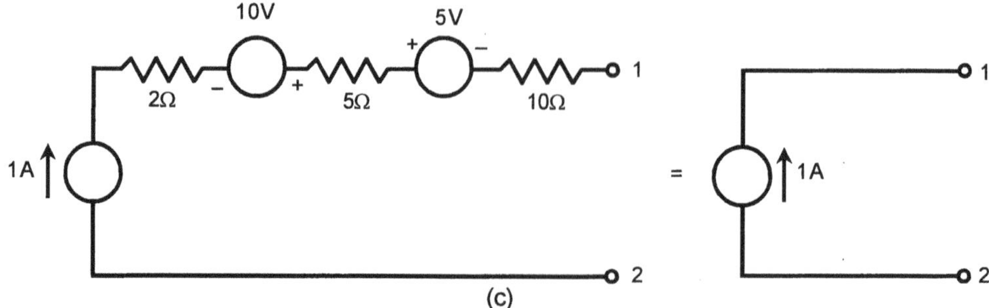

(c)

Fig. 2.20 Examples of current source in series with other elements

Example 2.5

Use source transformation to simplify the network in Fig. 2.21(a) and find the equivalent network containing only one voltage source and a resistance at the terminals 1,2.

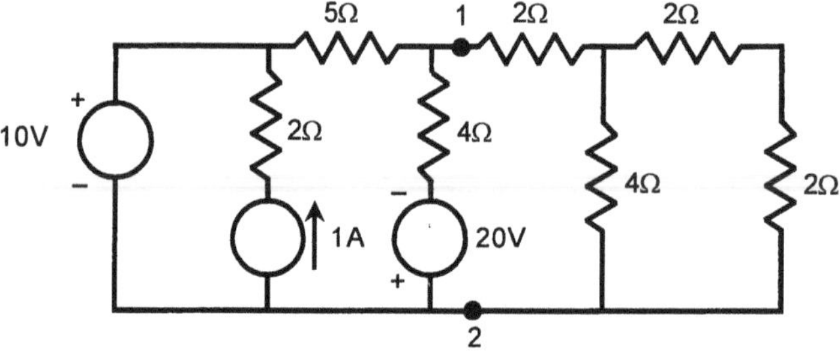

Fig. 2.21(a) Network for example 2.5

Solution :

The network to the right of terminals 1 and 2 is a ladder network and can be simplified to get its equivalent resistance

$$R_{eq} = 2 + \frac{4(2+2)}{4+2+2} = 4\ \Omega$$

Ignoring the branch containing the current source and a series resistance which is connected in parallel with a 10 V voltage source, the network can be written in simplified form as :

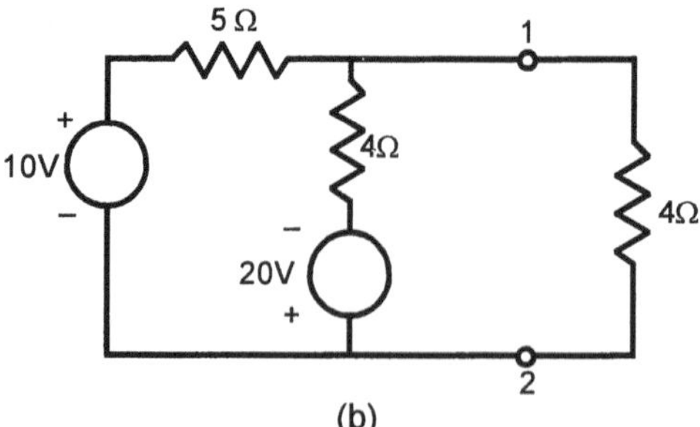

(b)

Converting the two branches containing voltage source and series resistance to branches containing current source in parallel with a resistance, using source transformation, we have

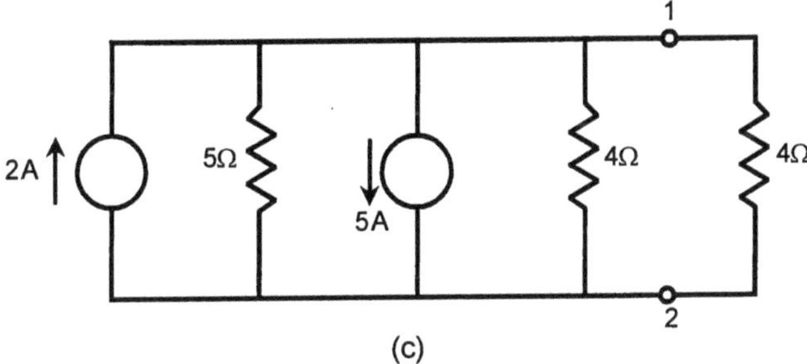

(c)

Combining the current sources which are in parallel, and resistances in parallel, across terminals 1 and 2, we get

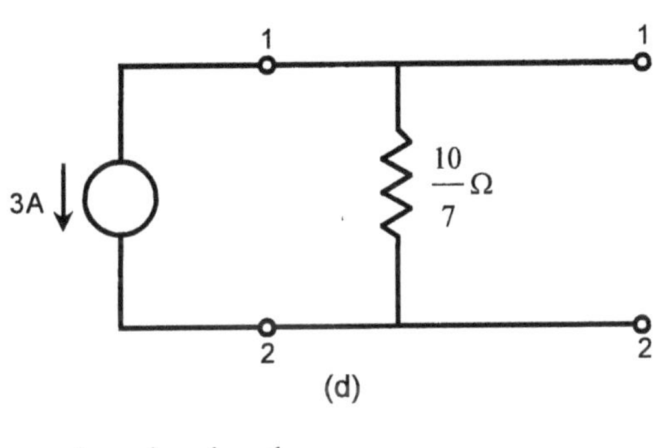

(d)

$$\frac{1}{R} = \frac{1}{5} + \frac{1}{4} + \frac{1}{4} = 0.7$$

$$R = \frac{1}{0.7} = \frac{10}{7}\ \Omega$$

Finally applying source transformation to the current source in parallel with the resistance, we get the equivalent network shown in Fig. 2.21(e) which is the desired result.

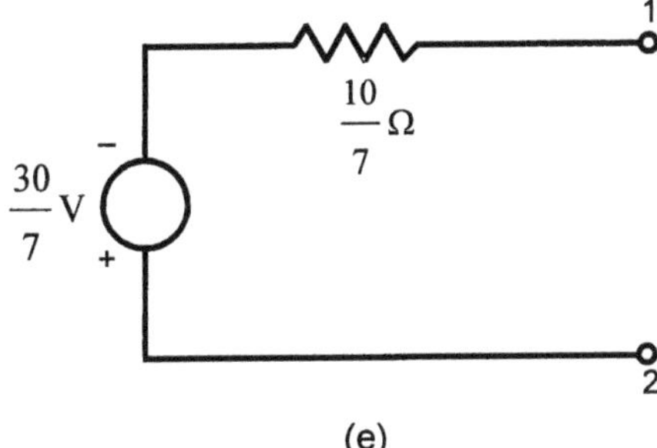

(e)

Example 2.6

Find the power delivered by the 5 V voltage source in the network of Fig. 2.22.

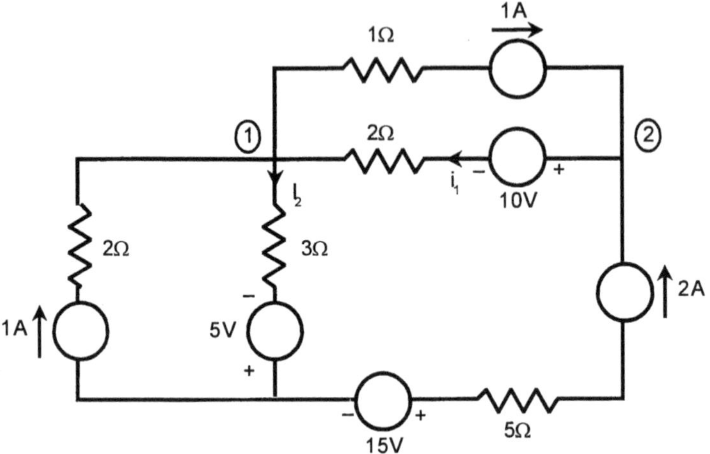

Fig. 2.22(a) Network for example 2.6

Solution :

Applying KCL at node 2, we have

$$I_1 = 2 + 1 = 3 \text{ A}$$

Applying KCL at node 1, we have

$$I_2 = 1 - 1 + 3 = 3 \text{ A}$$

Since the current in the 5 V source branch is known and the direction of current is directed out of positive terminal of the 5 V voltage source, the power delivered by the source is given by

$$P_{5V} = 5 \times 3 = 15 \text{ W}$$

If desired, the power delivered by each of the other sources in the network can also be calculated.

Power delivered by the 15 V voltage source is

$$P_{15V} = 15 \times 2 = 30 \text{ W}$$

Similarly

$$P_{10V} = 10 \times (-3) = -30 \text{ W}$$

To calculate the power delivered by the current sources, we need to know the voltages across them. These voltages can be calculated by applying KVL to a closed loop containing these sources.

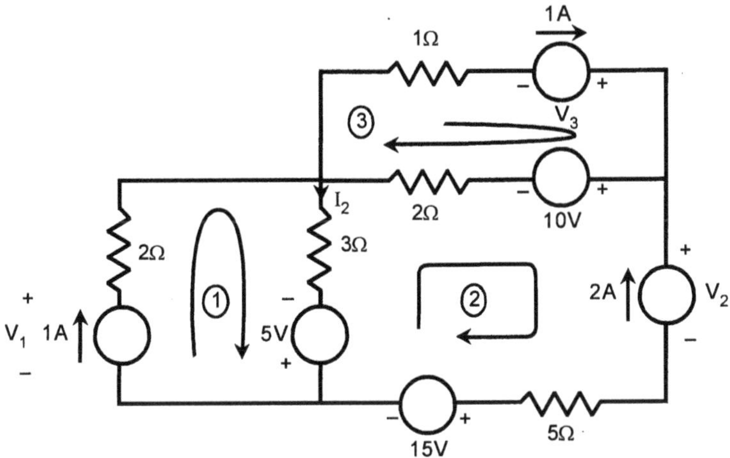

Fig. 2.22(b) Network of example 2.6 with three closed paths identified

For closed path 1, applying KVL

$$-V_1 + 1 \times 2 + 3 \times 3 - 5 = 0$$

which gives,

$$V_1 = 6\text{ V}$$

$$\therefore \quad P_{1A} = 6 \times 1 = 6\text{ W}$$

For closed path 2

$$5 - 3 \times 3 - 3 \times 2 - 10 + V_2 - 5 \times 2 + 15 = 0$$

$$V_2 = +15\text{V}$$

$$P_{2A} = 15 \times 2 = 30\text{ W}$$

For closed path 3

$$1 \times 1 - V_3 + 10 + 3 \times 2 = 0$$

$$V_3 = 17\text{ V}$$

$$P_{1A} = 17 \times 1 = 17\text{ W}$$

The power delivered to the resistors can be calculated by summing the power consumed by each resistor.

Power delivered to the resistors

$$P_R = 1^2 \times 2 + 3^2 \times 3 + 2^2 \times 5 + 3^2 \times 2 + 1^2 \times 1$$

$$= 2 + 27 + 20 + 18 + 1$$

$$= 68\text{ W}$$

Total power delivered by the sources is the algebraic sum of powers delivered by the sources,

$$P_S = 15 + 30 - 30 + 6 + 30 + 17$$
$$= 68 \text{ W}$$

It is obvious that the total power delivered by the sources is consumed by the resistors in the network.

2.3.11 Star – delta or Y – Δ transformation :

In many of the networks, the resistances are neither connected in series nor in parallel and hence the network cannot be simplified using the series or parallel combination of resistors. In such cases star – delta or wye – delta (Y – Δ) transformation, which will be developed in this section, is useful.

Consider a 3 – terminal network in which 3 resistors are connected at a common point O as shown in Fig. 2.23(a). Similarly consider another 3 – terminal network with 3 resistors connected to form a triangle as shown in Fig. 2.23(b).

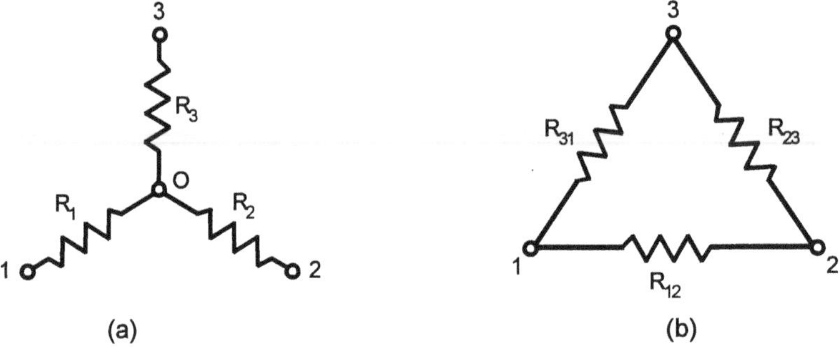

Fig. 2.23 Star – Δ or Y – Δ networks

If we are interested only in the effects produced by these networks when they are connected to other networks at these three terminals or when they form a part of another network, we can establish equivalence of these two networks and replace one with the other at the three terminals.

The conditions for the equivalence are shown in Fig. 2.23(c).

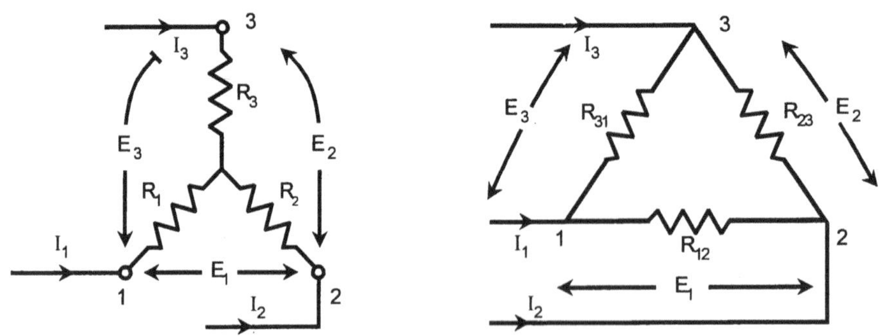

Fig. 2.23(c) Conditions for equivalence of the two networks

For similar conditions existing at the three terminals of these two networks, i.e. if the currents entering the three terminals I_1, I_2 and I_3 and the three voltages across terminal pairs, E_1, E_2 and E_3 are equal to the corresponding currents and voltages of the other network, then these two networks are equivalent.

Finding the equivalent resistance between the terminals 1, 2 with terminal 3 open, for both the networks in Fig. 2.23(c), and equating them, we get

$$R_1 + R_2 = \frac{R_{12}(R_{23} + R_{31})}{R_{12} + R_{23} + R_{31}} \quad .....(2.32)$$

($\therefore R_{31}$ and R_{23} are in series and together they are in parallel with R_{12} in the Δ network)

Similarly, repeating this with each of the other pair of terminals, we have

$$R_2 + R_3 = \frac{R_{23}(R_{31} + R_{12})}{R_{12} + R_{23} + R_{31}} \quad .....(2.33)$$

and

$$R_3 + R_1 = \frac{R_{31}(R_{23} + R_{12})}{R_{12} + R_{23} + R_{31}} \quad .....(2.34)$$

Solving the set of equations (2.32), (2.33) and (2.34)

we get

$$R_1 = \frac{R_{12}R_{31}}{R_{12} + R_{23} + R_{31}} \quad .....(2.35)$$

$$R_2 = \frac{R_{23}R_{12}}{R_{12} + R_{23} + R_{31}} \quad .....(2.36)$$

and

$$R_3 = \frac{R_{23}R_{31}}{R_{12} + R_{23} + R_{31}} \quad .....(2.37)$$

A general expression for a resistance in Y – network can be written in terms of the resistances of the Δ – network with reference to Fig. 2.23(a), we have

$$\text{Resistance in } Y = \frac{\text{Product of the resistances of adjacent branches of } \Delta}{\text{Sum of the } \Delta \text{ resistances}} \quad .....(2.38)$$

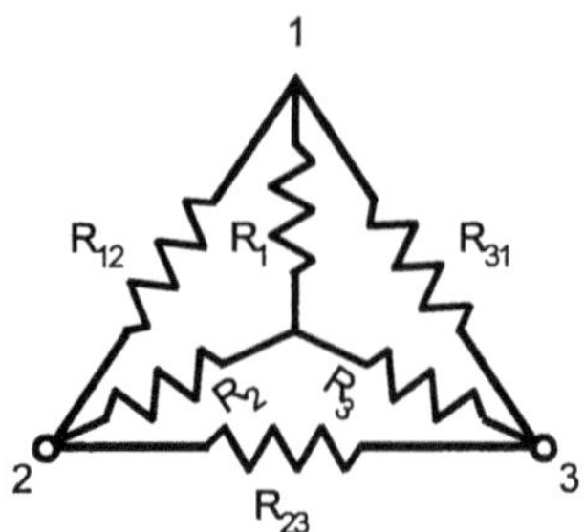

Fig. 2.23(d) Δ to star conversion

For example, in Fig. 2.23(d) R_1 can be obtained as the product of the two adjacent resistances R_{12} and R_{31} divided by the sum of the Δ – resistances.

In a similar manner, the equivalent Δ resistances can be obtained in terms of resistances in star, by shorting a pair of terminals, say, 1 and 2 and finding the equivalent resistance between 1 and 3. Using conductances rather than resistances, the equivalent conductances of the two networks can be equated. Thus

$$G_{31} + G_{32} = \frac{G_3(G_1 + G_2)}{G_1 + G_2 + G_3} \qquad \text{.....(2.39)}$$

Similarly for other pairs of terminals, we have

$$G_{12} + G_{31} = \frac{G_1(G_2 + G_3)}{G_1 + G_2 + G_3} \qquad \text{.....(2.40)}$$

and

$$G_{12} + G_{32} = \frac{G_2(G_1 + G_3)}{G_1 + G_2 + G_3} \qquad \text{.....(2.41)}$$

Solving the set of equation (2.39) to (2.41), we get

$$G_{12} = \frac{G_1 G_2}{G_1 + G_2 + G_3} \qquad \text{.....(2.42)}$$

$$G_{23} = \frac{G_2 G_3}{G_1 + G_2 + G_3} \qquad \text{.....(2.43)}$$

$$G_{31} = \frac{G_3 G_1}{G_1 + G_2 + G_3} \qquad \text{.....(2.44)}$$

A general expression for the conductances of Δ network in terms of the Y – conductances is given by

$$\text{Conductance of } \Delta \text{ branch} = \frac{\text{Product of conductances of adjacent branches of Y}}{\text{Sum of Y conductances}} \quad .....(2.45)$$

For example in Fig. 2.23(e) G_{12} can be obtained as the product of conductances the two adjacent branches G_1 at G_2, divided by sum of the Y conductances.

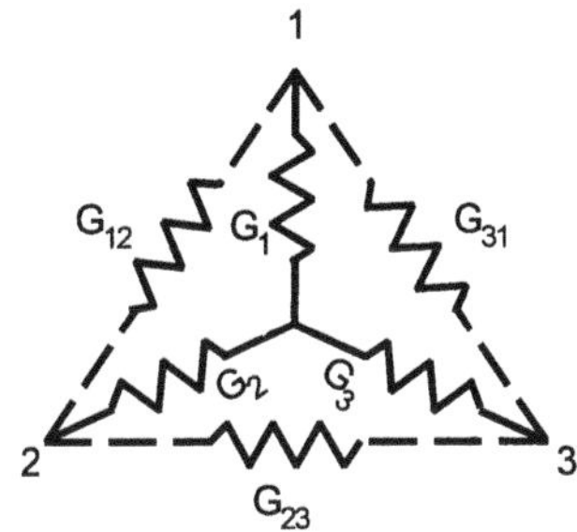

Fig. 2.23(e) Y – Δ transformation

It is often more convenient to deal with the resistances than conductances. Therefore, by taking the reciprocals of the respective conductances in equations (2.42), (2.43) and (2.44) and simplifying, we get

$$R_{12} = \frac{R_1R_2 + R_2R_3 + R_3R_1}{R_3} \quad .....(2.46)$$

$$R_{23} = \frac{R_1R_2 + R_2R_3 + R_3R_1}{R_1} \quad .....(2.47)$$

$$R_{31} = \frac{R_1R_2 + R_2R_3 + R_3R_1}{R_2} \quad .....(2.48)$$

A general expression for Δ – resistances can be written as

$$\text{Resistances in } \Delta = \frac{\text{sum of products of resistances of Y taken two at a time}}{\text{The resistance of Y not connected to this } \Delta - \text{resistance}}$$

For example in Fig. 2.23(e), R_{12} is obtained as the sum of product of resistances of Y – network taken two at a time and divided by the resistances R_3 which is not connected to R_{12}.

Using the Δ – Y or Y to Δ transformation and series parallel combinations of resistances any passive resistive network can be reduced to its equivalent resistance at a given pair of terminals.

Example 2.7

Find the equivalent resistance between the terminals 1 and 2 of the network shown in Fig. 2.24.

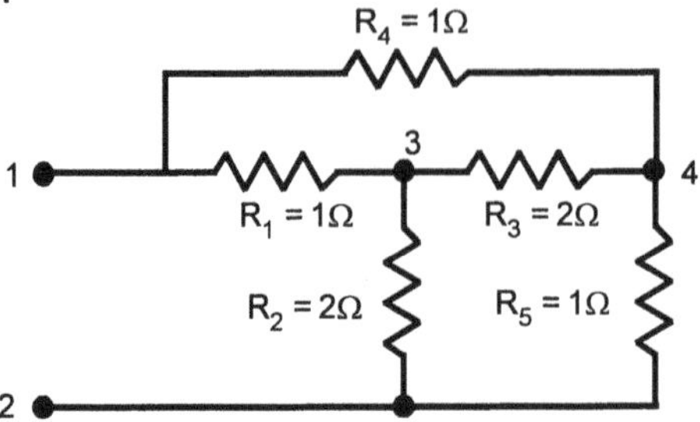

Fig. 2.24(a) Network for example 2.7

Solution :

We observe that, in this network, no two resistances are either in series or in parallel. Hence no simplification is possible by series or parallel combinations. However, one can recognise, 3 resistances connected in Δ or Y and convert them to Y or Δ. There are 3 resistor R_1, R_3, and R_4 connected in Δ between terminals 1, 3 and 4. Similarly there are 3 resistor R_2, R_3, and R_5 connected in Δ between terminals 3, 4 and 2. Any one of them can be converted into an equivalent Y . Alternatively there is a set of Y connected resistors R_1, R_2, and R_3 between terminals 1, 4 and 2 and another set of star connected resistor R_4, R_3 and R_5 between terminals 1, 3 and 2. Any one of these Y's can also be converted to a corresponding Δ. Converting the Δ – between the terminals 1, 3 and 4, we get the network in, Fig. 2.24(c).

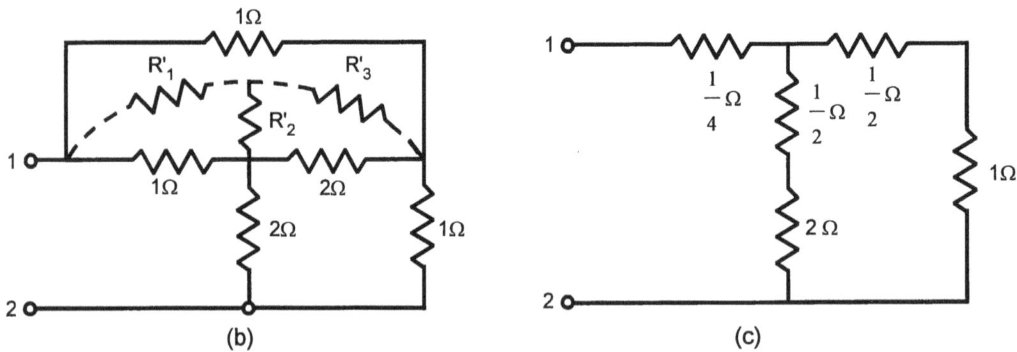

Fig. 2.24 Δ of network 2.24(b) is converted to Y and redrawn in (c)

$$R'_1 = \frac{1 \times 1}{1+1+2} = \frac{1}{4}\ \Omega$$

$$R'_2 = \frac{1 \times 2}{1+1+2} = \frac{2}{4} = \frac{1}{2}\ \Omega$$

$$R'_3 = \frac{2 \times 1}{1+1+2} = \frac{2}{4} = \frac{1}{2}\ \Omega$$

Now the resistance in Fig. 2.24(c) can be combined using series parallel combinations. The resulting networks are shown in Fig. 2.24(d), (e) and (f).

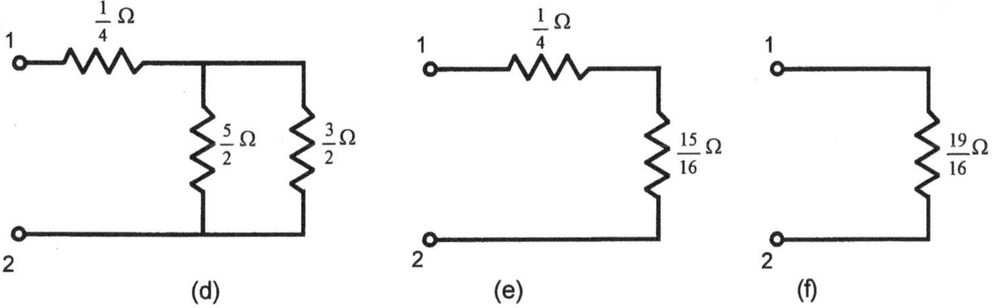

Fig. 2.24 Reduction of network in Fig. 2.24 to its equivalent resistance at terminals 1 and 2

Thus the equivalent resistances between the terminals 1 and 2 is

$$R_{eq} = \frac{19}{16}\Omega$$

Alternatively :

The Y – between the terminals 1, 3 and 2 of Fig. 2.24 can be converted to its Δ equivalent and the steps in the reduction are shown in Fig. 2.24 (g), (h), (i), (j) and (k).

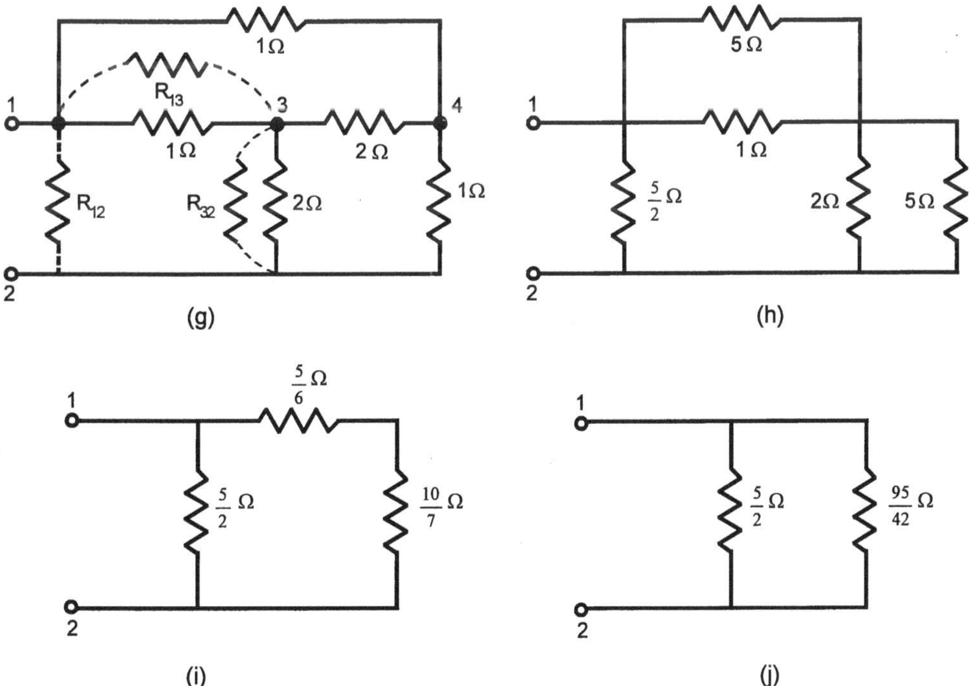

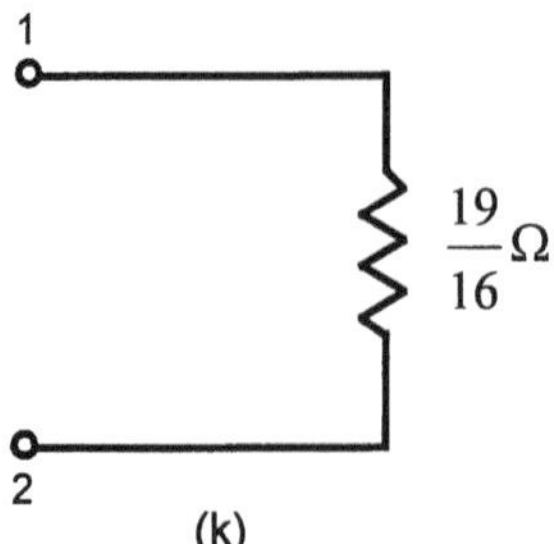

Fig. 2.24 Steps in the simplification of network of Fig. 2.24(a) by converting a star to delta

In Fig. 2.24(g)

$$R_{13} = \frac{2.1+2.1+1.1}{1} = 5\ \Omega$$

$$R_{32} = \frac{2.1+2.1+1.1}{1} = 5\ \Omega$$

$$R_{12} = \frac{2.1+2.1+1.1}{1} = \frac{5}{2}\Omega$$

The resulting network is shown in Fig. 2.24(h). Using series parallel combinations as shown in Fig. 2.24(i) and (j) the equivalent resistance between terminals 1 and 2 is again obtained as

$$R_{eq} = \frac{19}{16}\ \Omega$$

In this section we have discussed some techniques by which some simple networks can be solved. We shall now consider more complex circuits and develop methods to analyse these circuits. We shall first consider a method based on Kirchhoff 's current law.

2.4 NODE VOLTAGE METHOD :

Consider the network shown in Fig. 2.25. It is easy to identify the nodes of the

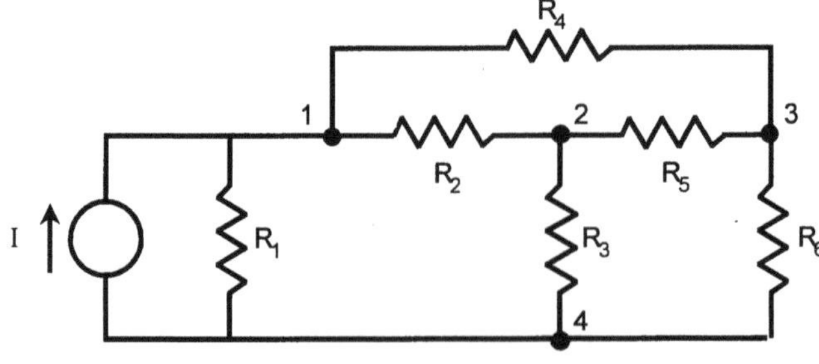

Fig. 2.25 A 4 node network

network. There are four nodes in this network. If we know the voltages across the branches of the network, we can find the currents in them using ohm's law. If instead of knowing voltage across each branch, if the voltage of every node, called as the node voltage, is known with respect to one of the nodes, called *datum node*, or *reference node*, the branch voltages can be easily calculated using Kirchhoff 's voltage law. For example, if node 4 is taken as reference in Fig. 2.25, and if the node voltages V_1, V_2 and V_3 are known with respect to node 4, the branch voltages can be easily calculated as shown in Fig. 2.26.

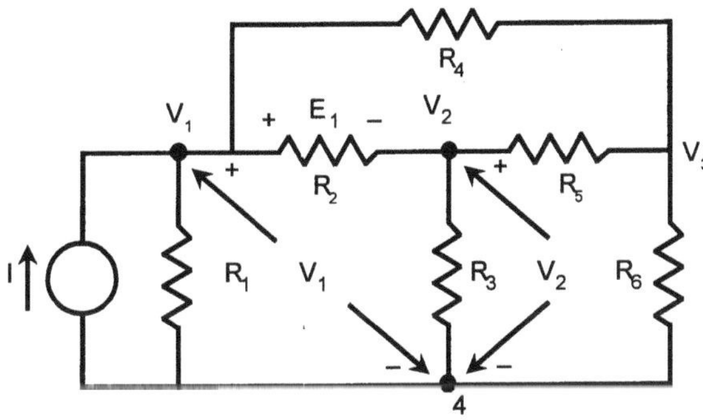

Fig. 2.26 Network with branch voltages and node voltages indicated

For simplicity, one branch voltage E_1 and two node voltages V_1 and V_2 only are shown in Fig. 2.26. Applying Kirchhoff's voltage law around the closed path indicated we have

$$V_1 - E_1 - V_2 = 0$$

or $$E_1 = V_1 - V_2 \qquad .....(2.49)$$

Thus the branch voltage can be obtained as the difference of the two node voltage V_1 and V_2. In a similar manner other branch voltages can be obtained easily by knowing the node to datum voltages of all the nodes. If there are N nodes, there will be N – 1 nodes, whose voltages have to be found with reference to the datum node. Thus to determine these N – 1 node voltages we have to write N – 1 equations. This can be accomplished by writing Kirchhoff 's Current Law equations for every node other than the reference node. In Fig. 2.26, the node voltages of nodes 1, 2 and 3 are designated as V_1, V_2, and V_3. These are the voltages with respect to the datum node arbitrarily chosen to be node 4.

Applying Kirchhoff 's current law at node 1, we get

$$\frac{V_1}{R_1} + \frac{V_1 - V_2}{R_2} + \frac{V_1 - V_3}{R_4} - I = 0 \qquad .....(2.50)$$

It is necessary to reiterate that V_1, $V_1 - V_2$, $V_1 - V_3$ are the voltages across the branches containing resistors R_1, R_2 and R_4 respectively, and each of the terms in eq. (2.50) represents the current leaving the node in that branch. Without any loss of generality it is assumed that the voltage at the node at which KCL is written, is higher than the other nodes, so that the currents in all the branches connected to it leave the node. If, for example, the actual current R_2 is entering the node instead of leaving it, we would have written the equation as

$$\frac{V_1}{R_1} - \frac{V_2 - V_1}{R_2} + \frac{V_1 - V_3}{R_4} - I = 0 \qquad .....(2.51)$$

It is easy to observe the equivalence of eqs. (2.50) and (2.51). So it is customary to assume that the node at which we are writing the Kirchhoff 's current law equation, is at a higher potential compared to other nodes to which it is connected. Thus the other two equations at nodes 2 and 3 are,

$$\frac{V_2 - V_1}{R_2} + \frac{V_2}{R_3} + \frac{V_2 - V_3}{R_5} = 0 \qquad .....(2.52)$$

$$\frac{V_3 - V_1}{R_4} + \frac{V_3}{R_6} + \frac{V_3 - V_2}{R_5} = 0 \qquad .....(2.53)$$

The unknown voltages V_1, V_2 and V_3 can be obtained by solving eqs. (2.51), (2.52) and (2.53). Once the node voltages are known, the currents in various branches can be easily calculated.

Example 2.8

Find the node voltages in the circuit of Fig. 2.27.

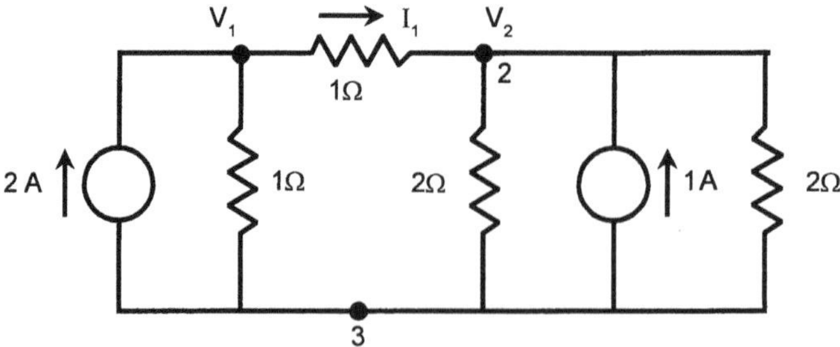

Fig. 2.27 Network for example 2.8

Solution :

There are 3 nodes and the node 3 is taken as reference. The node voltages are marked as V_1 and V_2. Writing the Kirchhoff 's current law equation at each of the nodes 1 and 2 we have

$$\frac{V_1}{1}+\frac{V_1-V_2}{1}=2 \qquad .....(2.54)$$

$$\frac{V_2-V_1}{1}+\frac{V_2}{2}+\frac{V_2}{2}=1 \qquad .....(2.55)$$

Simplifying eqs. (2.54) and (2.55), we have.

$$2V_1 - V_2 = 2 \qquad .....(2.56)$$

$$-V_1 + 2V_2 = 1 \qquad .....(2.57)$$

Solving these two equations, we get

$$V_1 = \frac{5}{3} \text{ V}$$

$$V_2 = \frac{4}{3} \text{ V}$$

If in the above example, the current in 1 ohm resistor between nodes 1 and 2 is desired, then the voltage across this resistance is first calculated using the node voltages V_1 and V_2 and then the current can be easily calculated.

Thus the current through 1 Ω resistor connected between nodes 1 at 2, I_1, is given by

$$I_1 = \frac{V_1 - V_2}{1}$$

$$= \frac{\frac{5}{3}-\frac{4}{3}}{1} = \frac{1}{3} \text{ A}$$

The currents in other branches can be similarly calculated.

If a network consists of only current sources the application of node voltage method is direct and straight forward since KCL equation is used at each node. However if a voltage source is connected at a node, the current through the source is not directly available. If a resistance is present in series with this source, this branch can be transformed into a current source in parallel with the resistor.

This is shown in Fig. 2.28.

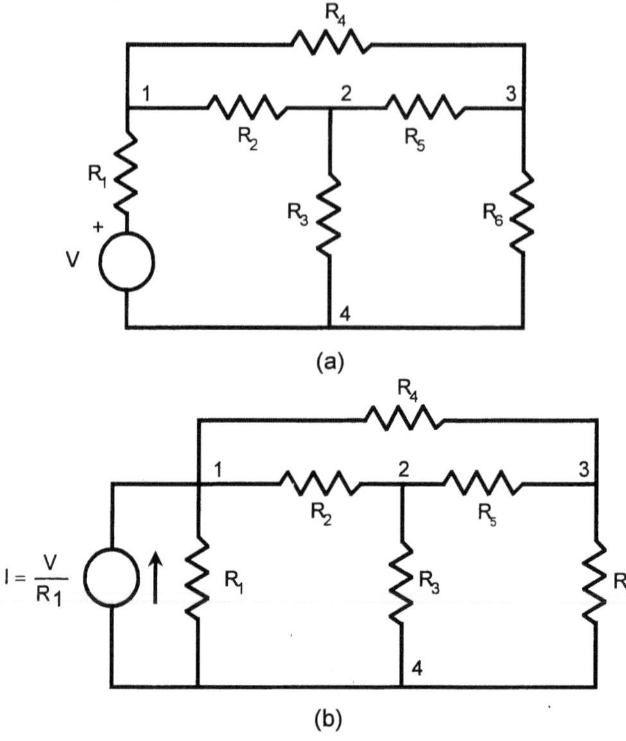

Fig. 2.28(a) A network with voltage source and (b) the same network after source transformation

Now it is a straight forward procedure to apply the node voltage method.

It is often cumbersome to convert all voltage sources to current sources and hence it will be convenient to apply the node voltage method directly to the given network. Consider the network of Fig. 2.28(a). Let us designate the junction between R_1 and V as node 5 and its voltage with respect to reference node 4 as V_5. Now there is an additional node at which the KCL has to be applied. But as the voltage of this node is known and is equal to V, it is not necessary to write KCL equation at this node and

$$V_5 = V$$

Now applying the KCL at nodes 1, 2 and 3, we get

$$\frac{V_1 - V}{R_1} + \frac{V_1 - V_2}{R_2} + \frac{V_1 - V_3}{R_4} = 0 \qquad .....(2.58)$$

$$\frac{V_2 - V_1}{R_2} + \frac{V_2}{R_3} + \frac{V_2 - V_3}{R_4} = 0 \qquad .....(2.59)$$

$$\frac{V_3 - V_2}{R_5} + \frac{V_3}{R_6} + \frac{V_3 - V_1}{R_4} = 0 \qquad .....(2.60)$$

The unknown voltages V_1, V_2 and V_3 can now be calculated using eqs. (2.58) to (2.60).

Example 2.9

Find the node voltages V_1, V_2 and V_3 in the network of Fig. 2.29 and find the current I_x.

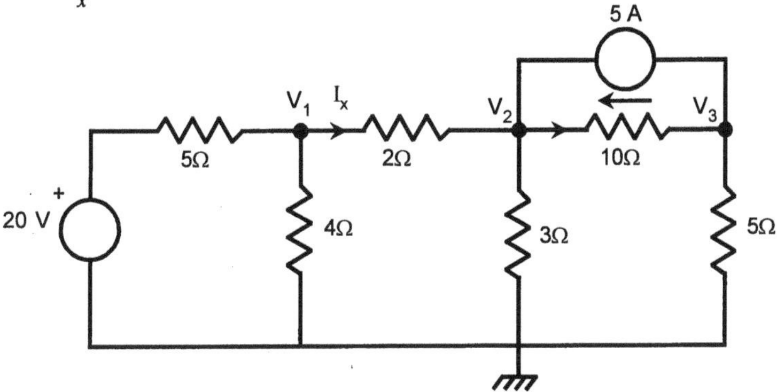

Fig. 2.29 Network for example 2.9

Solution :

Designating the node voltage as V_1, V_2 and V_3 with respect to the datum node, and writing the node voltage equations,

At node 1

$$\frac{V_1 - 20}{5} + \frac{V_1}{4} + \frac{V_1 - V_2}{2} = 0 \qquad .....(2.61)$$

$$V_1\left(\frac{1}{5} + \frac{1}{4} + \frac{1}{2}\right) - \frac{V_2}{2} = 4$$

$$19\ V_1 - 10\ V_2 = 80 \qquad .....(2.62)$$

At node 2

$$\frac{V_2 - V_1}{2} + \frac{V_2}{3} + \frac{V_2 - V_3}{10} - 5 = 0 \qquad .....(2.63)$$

$$-\frac{V_1}{2} + V_2\left(\frac{1}{2} + \frac{1}{3} + \frac{1}{10}\right) - \frac{V_3}{10} = 5$$

$$-15\ V_1 + 28\ V_2 - 3\ V_3 = 150 \qquad .....(2.64)$$

At node 3

$$\frac{V_3 - V_2}{10} + \frac{V_3}{5} + 5 = 0 \qquad .....(2.65)$$

$$\frac{-V_2}{10} + V_3\left(\frac{1}{10} + \frac{1}{5}\right) = -5$$

$$-V_2 + 3V_3 = -50 \qquad .....(2.66)$$

Solving for V_1 from eqs. (2.62), (2.64) and (2.66)

$$V_1 = \frac{\begin{vmatrix} 80 & -10 & 0 \\ 150 & 28 & -3 \\ -50 & -1 & 3 \end{vmatrix}}{\begin{vmatrix} 19 & -10 & 0 \\ -15 & 28 & -3 \\ 0 & -1 & 3 \end{vmatrix}} = \frac{9480}{1089} = 8.7 \text{ V}$$

$$V_2 = \frac{\begin{vmatrix} 19 & 80 & 0 \\ -15 & 150 & -3 \\ 0 & -50 & 3 \end{vmatrix}}{1089} = \frac{9300}{1089} = 8.54\text{V}$$

and

$$V_3 = \frac{\begin{vmatrix} 19 & -10 & 80 \\ -15 & 28 & 150 \\ 0 & -1 & -50 \end{vmatrix}}{1089} = \frac{-15050}{1089} = -13.82\text{V}$$

The current I_x can now be calculated

$$I_x = \frac{V_1 - V_2}{2} = \frac{8.7 - 8.54}{2}$$

$$= 0.08 \text{ A}$$

A special case occurs when a voltage source is connected at a node without any series resistance. An example is considered to illustrate this case.

Example 2.10

Find the node voltages V_1, V_2 and V_3 in Fig. 2.30.

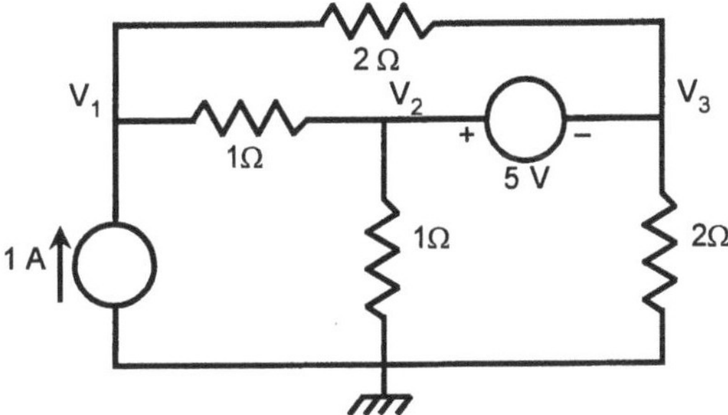

Fig. 2.30(a) Network for example 2.10

Solution :

Applying KCL at node 1

$$\frac{V_1 - V_2}{1} + \frac{V_1 - V_3}{2} = 1 \qquad \text{.....(2.67)}$$

$$\frac{3V_1}{2} - V_2 - \frac{V_3}{2} = 1$$

$$3\,V_1 - 2\,V_2 - V_3 = 2 \qquad \text{.....(2.68)}$$

If we try to apply KCL at node 2 or 3 we face a difficulty. The current in the branch containing the voltage source cannot be found and hence the KCL cannot be written at node 2 or node 3.

This difficulty can be overcome by considering the nodes 2 and 3 together and considering the combination as a '*super node*' as shown in Fig. 2.30(b). Writing KCL for this super node.

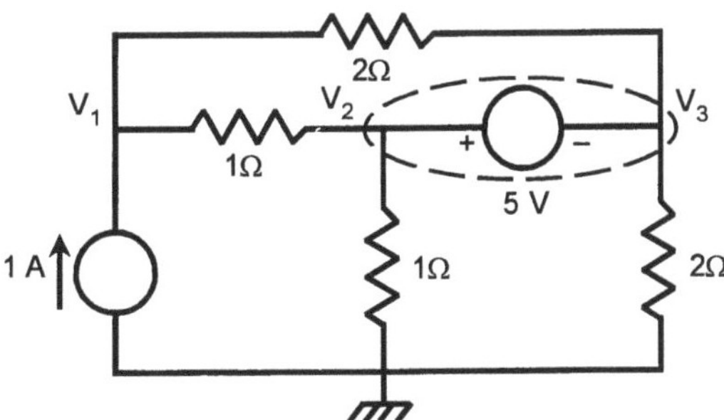

Fig. 2.30(b) Network of example 2.10 showing the supernode

$$\frac{V_2 - V_1}{1} + \frac{V_2}{1} + \frac{V_3}{2} + \frac{V_3 - V_1}{2} = 0 \qquad \text{.....(2.69)}$$

$$-V_1\left[1+\frac{1}{2}\right]+V_2[1+1]+V_3\left[\frac{1}{2}+\frac{1}{2}\right]=0$$

$$-3V_1+4V_2+2V_3=0 \qquad .....(2.70)$$

The third equation is obtained as

$$V_2-V_3=5 \qquad .....(2.71)$$

Solving the eqs. (2.68), (2.70), and (2.71), we get

$$V_1=\frac{4}{3}\text{ V} \qquad V_2=\frac{7}{3}\text{ V} \qquad V_3=\frac{-8}{3}\text{ V}$$

This example illustrates the method to be used when voltage sources are connected between nodes without any series resistances.

Let us now consider an example consisting of dependent sources.

Example 2.11

Obtain the node voltages in the network of Fig. 2.31.

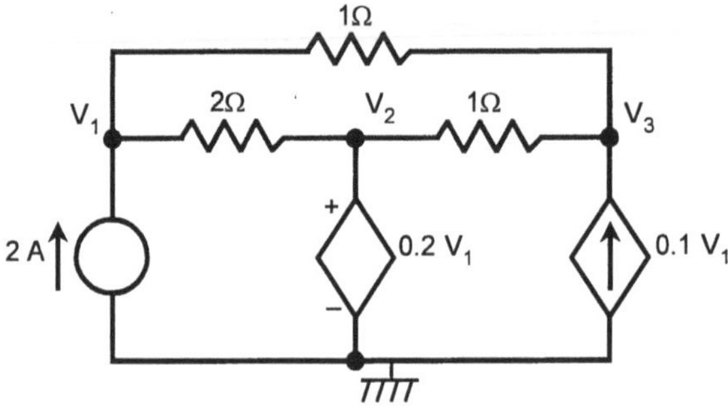

Fig. 2.31 Network or example 2.11

Solution :

Designating the node voltages as V_1, V_2 and V_3 as shown in Fig. 2.31 and writing node voltage equations :

$$\frac{V_1-V_2}{2}+\frac{V_1-V_3}{1}=2 \qquad .....(2.72)$$

$$\frac{3}{2}V_1-\frac{V_2}{2}-V_3=2$$

$$3\,V_1-V_2-2\,V_3=4 \qquad .....(2.73)$$

$$V_2=0.2\,V_1 \qquad .....(2.74)$$

$$\frac{V_3-V_1}{1}+\frac{V_3-V_2}{1}-0.1\,V_1=0$$

$$-\,1.1\,V_1-V_2+2\,V_3=0 \qquad .....(2.75)$$

Solving eqs. (2.73), (2.74) and (2.75), we get

$$V_1 = \frac{8}{3}\ V \quad V_2 = \frac{8}{15}\ V \quad V_3 = \frac{26}{15}\ V$$

Now we will consider the second method based on K V L.

2.5 MESH CURRENT METHOD :

In a network if there are 'b' branches, there are 'b' currents and b voltages to be determined, i.e., a total of '2b' unknowns. Thus if we have to determine all the currents and voltages, we need to write '2b' linearly independent equations. By linear independence we mean, no equation, out of a set of 'n' equations written to determine 'n' unknowns, be a linear combination of other equations. For example, consider the set of three equations.

$$x + 2y + 3z = 4$$
$$2x - y - z = 6$$
$$8x + y + 3z = 26$$

The third equation is obtained by multiplying the first by 2 and the second by 3 and then adding the two equations. This set of equations have no unique solution.

Now coming back to the '2b' equations for the network, the 'b' voltages are related to the 'b' currents by means of the Ohm's law for the elements. Thus we need to know either b currents or b voltages only. Once the currents are known, voltages can be found or vice versa. We have seen in section 2.4 that if we choose node voltages instead of branch voltages as variables, we had to write N – 1 node voltage equations for a N – node network to solve for all the branch voltages and currents. N – 1 for a network is always less than 'b' as will be shown in Chapter 5.

Another method, by which we write fewer equations than 'b', is known as *'mesh current method'*. To understand the concept of mesh current method let us consider a network shown in Fig. 2.32.

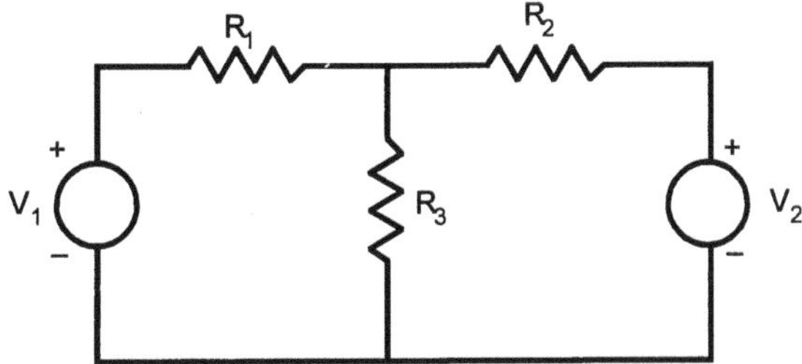

Fig. 2.32 Development of mesh current method

Let us consider the closed paths in the network in Fig. 2.33. One can identify 3 closed paths.

The closed paths containing the elements are :

1. V_1, R_1 and R_3
2. R_2, R_3 and V_2

and 3. V_1, R_1, R_2 and V_2

The first two closed paths are called either *loops* or *meshes* and the third is called only a *loop*. The mesh is defined as a loop which doesnot contain any other loop. The third closed path contains the first two loops and hence is not a mesh thus we can say every mesh is a loop but every loop is not a mesh.

Mesh is a property of a set of networks which are known as planar networks. Planar networks are those networks which can be drawn on paper without crossing of lines. Some planar networks may appear to be non planar if they are not drawn properly.

Thus Fig. 2.33(a) is an example of a planar network, Fig. 2.33(b) is a non planar network, Fig. 2.33(c) is a planar network drawn to appear as a non planar network and Fig. 2.33(d) is the network of Fig. 2.33(c) drawn, avoiding the crossing of lines.

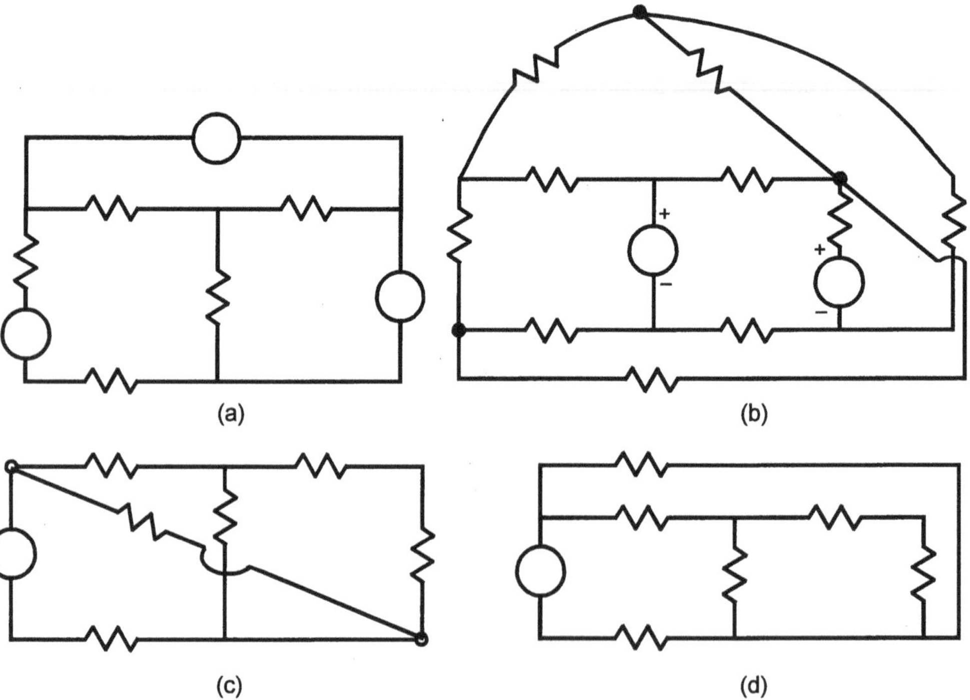

Fig. 2.33 Examples of (a) a planar network (b) non planar network (c) planar network drawn to look like non planar network (d) network in (c) without crossing of lines

The mesh current method, to be developed in this section is applicable only to planar networks.

Returning to Fig. 2.32, redrawn in Fig. 2.34 for ready reference, let us identify the branch currents I_1, I_2, and I_3.

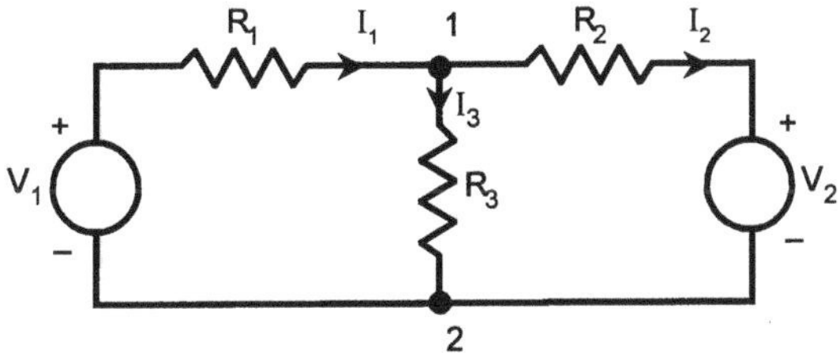

Fig. 2.34 Network in Fig. 2.32 redrawn

There are three branches and thus there are three unknown currents. But by applying Kirchhoff 's Current Law at node 1, we observe that

$$I_3 = I_1 - I_2$$

Hence if I_1 and I_2 are known, I_3 could be easily found out.

Consider imaginary currents I_1 and I_2 circulating in meshes 1 and 2 as shown in Fig. 2.35.

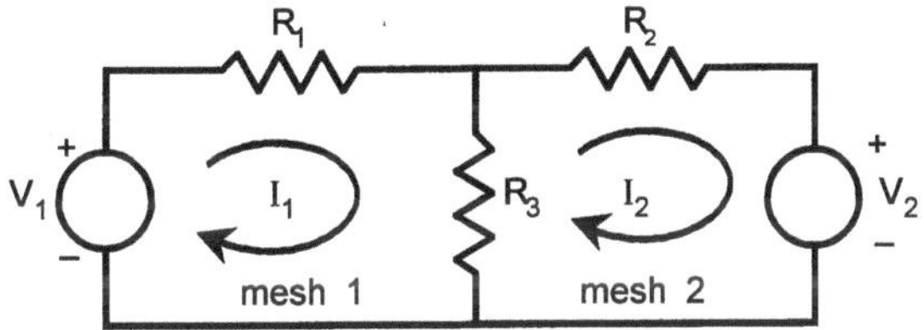

Fig. 2.35. Network showing mesh currents

These currents are customarily taken to flow in the meshes in a clockwise direction. If the branch currents are compared for the two networks in Fig. 2.34 and 2.35, they are exactly same in both the networks.

Branch current in R_1 is same as mesh current I_1
Branch current in R_2 is same as mesh current I_2
Branch current in R_3 is the difference of mesh currents I_1 and I_2.

Applying K V L to the two loops.

$$I_1 R_1 + (I_1 - I_2) R_3 = V_1 \quad .....(2.76)$$

$$I_2 R_2 + (I_2 - I_1) R_3 = - V_2 \quad .....(2.77)$$

Observe that the current in R_3 in eq. (2.77) is taken as $(I_2 - I_1)$ flowing from bottom to top, so that in the direction of mesh current, voltage across R_3 is a drop and hence a positive sign can be taken in that equation for the term $(I_2 - I_1) R_3$.

If the current in R_3 is taken as $(I_1 - I_2)$ as in eq. (2.76), the term $(I_1 - I_2)\ R_3$ will be taken with a – ve sign and is same as $(I_2 - I_1)\ R_3$.

The two equations (2.76) and (2.77) are linearly independent and therefore they have a unique solution.

Thus if we take mesh currents as unknown variables rather than branch currents, we will have a set of linearly independent equations and less number of equations too. It will be shown later in chapter 5 that the number of independent equations to be written is equal to $b - n + 1$ where 'b' is the total number of branches and 'n' is the number of nodes in the network.

How does one recognise the meshes in a network ? A planar network looks like a window with number of panes of glass and a mesh can be identified with each single pane of glass in the window, as shown in Fig. 2.36.

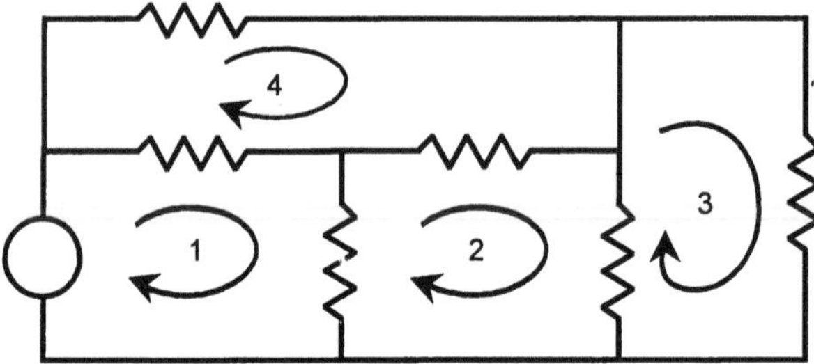

Fig. 2.36 Network with meshes identified as panes is a window

Now assign a mesh current for each of the meshes, usually taken in clockwise direction, and write KVL for each mesh. This rule is known as 'window pane rule'. Since we are writing Kirchhoff's voltage law equations, voltage sources pose no problem as can be seen from the example 2.12.

Example 2.12

Find the current in 2 Ω resistor using mesh current method in the network of Fig. 2.37.

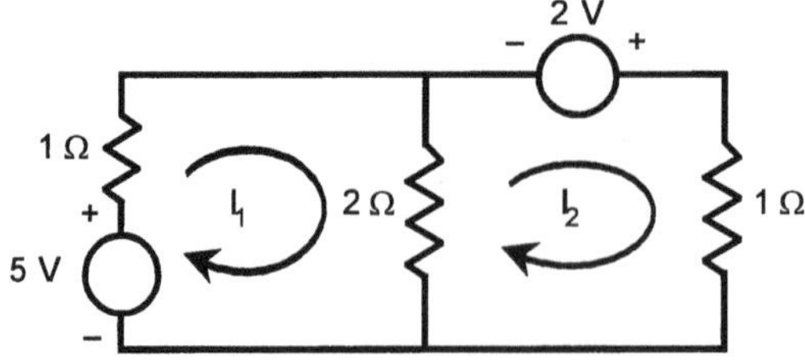

Fig. 2.37 Network for example 2.12

Solution :

There are two meshes in the network. Assigning two mesh currents I_1 and I_2 in clockwise directions and writing KVL equations for the two meshes, we have

For mesh 1

$$I_1 (1) + (I_1 - I_2)2 = 5$$

$$3 I_1 - 2 I_2 = 5 \quad .....(2.78)$$

For mesh 2

$$I_2 (1) + (I_2 - I_1)2 = 2$$

$$-2 I_1 + 3 I_2 = 2 \quad .....(2.79)$$

Solving for I_1 and I_2 from eqs. (2.78) and (2.79) we get

$$I_1 = \frac{19}{5} \text{ A} \quad \text{and} \quad I_2 = \frac{16}{5} \text{ A}$$

The current in 2Ω resistor is

$$I_1 - I_2 = \frac{19}{5} - \frac{16}{5}$$

$$= \frac{3}{5} \text{ A}$$

As can be seen from this example, voltage sources do not cause any problems. But if the network contains current sources, writing KVL in the loop containing the current source is not possible since the voltage across the current source is not known.

2.5.1 Dummy mesh current method :

In order to overcome the problem of current sources, the mesh currents are assigned for the network making the current source equal to zero, or open circuiting the current source. Then the current source branch is reintroduced and an additional dummy mesh is formed with the current source branch and other branches. This dummy mesh current is equal to the current of the current source itself. The procedure will be clear from the example 2.13.

Example 2.13

Obtain the current I_x in the network of Fig. 2.38 using mesh current method.

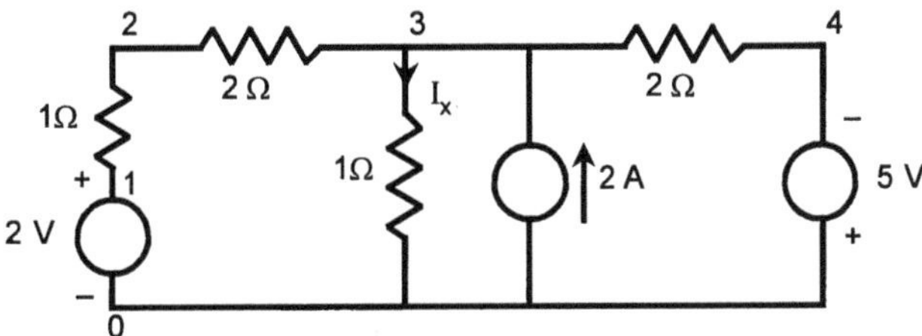

Fig. 2.38 Network for example 2.13

Solution :

Open circuiting the current source and indicating the mesh currents I_1 and I_2 as shown in Fig. 2.39, we have

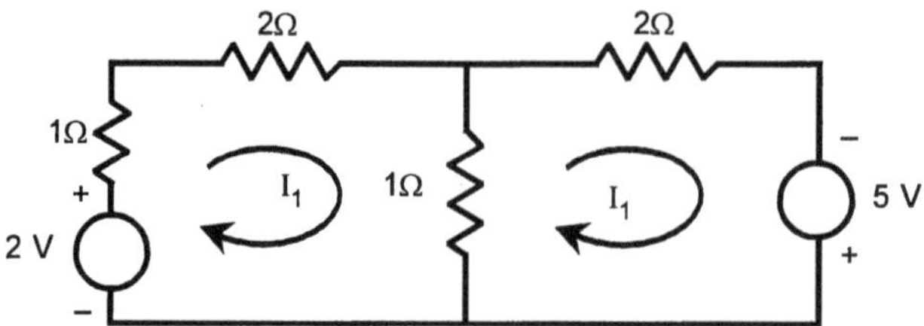

Fig. 2.39 Network with current source open circuited

Reintroducing the current source a dummy mesh is formed with mesh current equal to 2A in the direction indicated in Fig. 2.40.

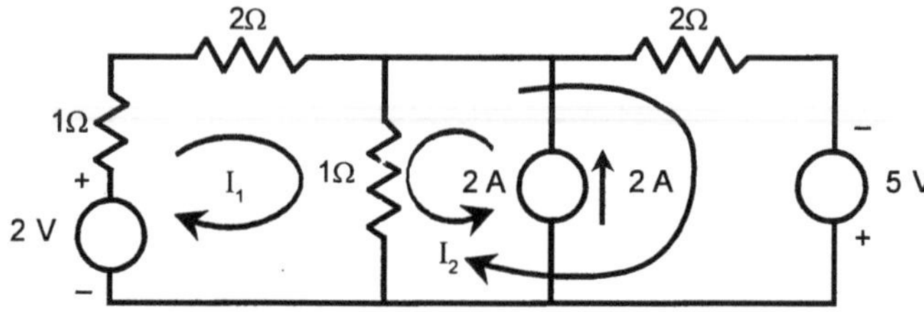

Fig. 2.40 Network with dummy mesh current

Now, writing the mesh equations for the two meshes, we have

For the mesh 1

$$3\,I_1 + (I_1 - I_2 + 2)1 = 2 \qquad \text{.....(2.80)}$$

We note here that the current in the branch common to the 2 meshes and dummy mesh is given by $(I_1 - I_2 + 2)$.

Simplifying eq. (2.80)

$$4\,I_1 - I_2 = 0 \qquad \text{.....(2.81)}$$

For mesh 2

$$2\,I_2 + (I_2 - I_1 - 2)1 = 5$$

$$-I_1 + 3\,I_2 = 7 \qquad \text{.....(2.82)}$$

Solving 2.81 and 2.82 for I_1 and I_2 we have

$$I_1 = \frac{7}{11}\text{ A} \qquad I_2 = \frac{28}{11}\text{ A}$$

and $$I_x = I_1 - I_2 + 2$$

$$= \frac{7}{11} - \frac{28}{11} + 2$$

$$= \frac{1}{11} \text{ A.}$$

Important note :

A clear understanding of the difference between the branch currents and mesh currents is in order here. Branch currents are the actual currents that are flowing in the branches. Mesh currents are imaginary circulating currents in the meshes. If in a branch only one mesh current is flowing, then the branch current is equal to the mesh current. If more than one mesh currents are flowing in a branch which is common to two or more meshes, the branch current is equal to algebraic sum of the mesh currents flowing in that branch. In Fig. 2.40 the branch current in 1Ω and 2Ω resistors in series with 2V source is equal to the mesh current I_1 and the branch current in 2Ω resistor in series with the 5V source is equal to the mesh current I_2 where as the current in 1Ω resistor common to the three meshes is the algebraic sum of the currents I_1, I_2 and 2A. Further, only one current, i.e., 2A is flowing through the branch containing the current source and therefore the current in the dummy mesh is assigned a current of 2A.

Now let us consider an example which contains dependent sources.

Example 2.14

Find the power supplied by the 6 V voltage source in the network of Fig. 2.41.

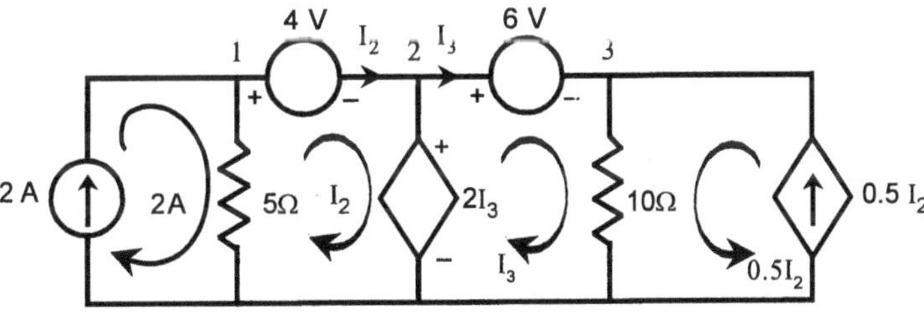

Fig. 2.41 Network for example 2.14

Solution :

There are four meshes in which two are dummy meshes. The dummy mesh currents are 2 A and 0.5 I_2 as indicated in Fig. 2.41. The mesh currents are taken as I_2 and I_3 as, for the indicated meshes, the branches containing 4V and 6V sources are not common to any other meshes and the currents in them are given to be I_2 and I_3 respectively. Writing the mesh equations for these two meshes we have

$$(I_2 - 2)5 + 4 + 2\, I_3 = 0 \qquad .....(2.83)$$

and $$(I_3 + 0.5\, I_2)10 - 2\, I_3 + 6 = 0 \qquad .....(2.84)$$

Simplifying eqs. (2.83) and (2.84), we get

$$5\,I_2 + 2\,I_3 = 6 \qquad .....(2.85)$$

$$5\,I_2 + 8\,I_3 = -\,6 \qquad .....(2.86)$$

Solving eqs. (2.85) and (2.86), we get

$$I_2 = 2A$$

$$I_3 = -\,2\text{ A.}$$

Power supplied by the 6V source is

$$P_6 = 6 \times (2) = 12\text{ watts}$$

One more example will illustrate clearly the use of mesh current method when dependent sources are present in a network.

Example 2.15

Find the current I_A in the network of Fig. 2.42(a).

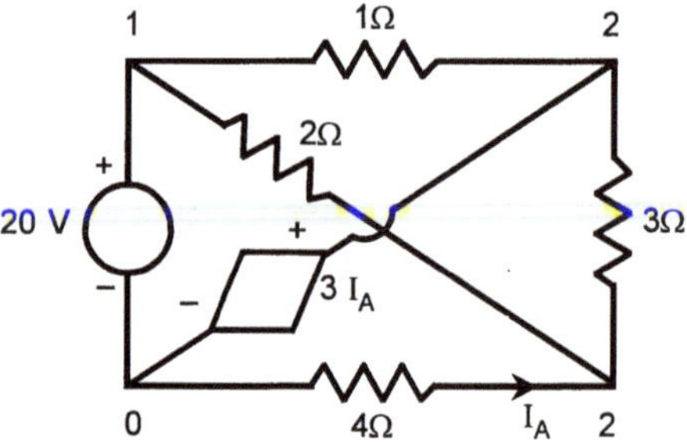

Fig. 2.42(a) Network for example 2.15

Solution :

The network, as it is drawn appears to be a non – planar network. It is actually a planar network and can be redrawn without crossing of lines as shown in Fig. 2.42(b).

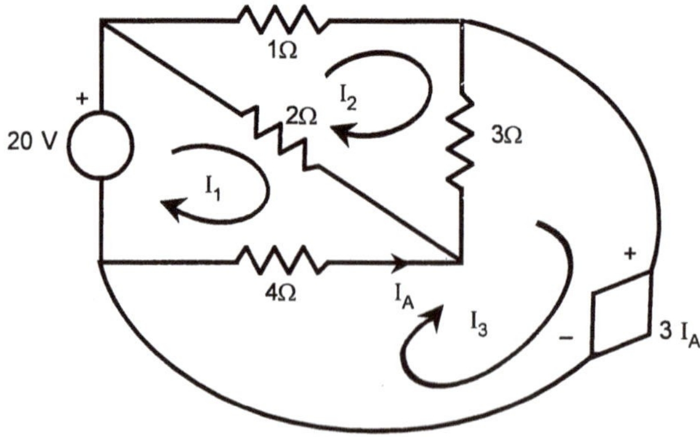

Fig. 2.42(b) The network in Fig. 2.42(a) is redrawn as a planar network

There are three 'window panes' in the network and assigning 3 mesh currents for the 3 meshes and writing mesh equations, we have

$$(I_1 - I_2)2 + (I_1 - I_3)4 = 20 \qquad .....(2.87)$$

$$I_2\,(1) + (I_2 - I_3)3 + (I_2 - I_1)2 = 0 \qquad .....(2.88)$$

$$(I_3 - I_1)4 + (I_3 - I_2)3 + 3I_A = 0 \qquad .....(2.89)$$

But the branch current I_A can be expressed in terms of the loop currents I_1 and I_3 as

$$I_A = I_3 - I_1 \qquad .....(2.90)$$

Using eq. (2.90) in eq. (2.89) and simplifying eqs. (2.87), (2.88) and (2.89) we get

$$6\,I_1 - 2\,I_2 - 4\,I_3 = 20 \qquad .....(2.91)$$

$$-2\,I_1 + 6\,I_2 - 3\,I_3 = 0 \qquad .....(2.92)$$

$$-7\,I_1 - 3\,I_2 + 10\,I_3 = 0 \qquad .....(2.93)$$

Solving eqs. (2.91) to (2.93), we get

$$I_1 = \frac{255}{8}\,A \qquad I_2 = \frac{205}{8}\,A \qquad I_3 = 30A$$

and

$$I_A = I_3 - I_1$$

$$= 30 - \frac{255}{8}$$

$$= -\frac{15}{8}A$$

Given any network, the analysis can be done by either node voltage method or mesh current method. In node voltage method the number equations to be written is given by N – 1. In mesh current method the number of equations to be written is given by b – N + 1 where b is the number of branches in the network. It is always prudent to solve by the method, which requires the least number of equations, to conserve our effort.

2.6 MAGNETIC CIRCUITS :

Magnetic circuits play a very important role in the performance of a number of industrial devices. The design of devices like electric generators, motors, transformers, control system components is based on the behaviour of magnetic circuits and in turn on the type of magnetic materials used for the construction of such devices. Relationships analogous to those in electric circuits exist in magnetic circuits. It is, therefore, necessary to study the properties of magnetic circuits for a complete understanding of any electrical device.

2.6.1 Magnetic fields :

Oersted discovered that a relationship exists between electricity and magnetism by placing a magnetic needle near a current carrying conductor. The magnetic needle pointed in a definite direction around the conductor suggesting the presence of a magnetic field similar to the field around a magnet. The rule to find the direction of this field was clearly stated by Ampere through "Ampere rule". The rule states thus :

> *"Imagine an observer in an electric circuit with the current flowing from his feet to his head. Then, on facing the compass needle, the north pole will be deflected towards his left."*

The direction of the magnetic field can also be determined using "Right hand thumb rule" which states :

> *"When the thumb of the right hand points in the direction of the current in the conductor, the direction in which the other four fingers tend to close gives the direction of the magnetic field."*

The nature of the magnetic field produced also depends on the type of the conductor structure. If the current flows in a long straight conductor, the magnetic field around it is in the form of concentric circles as shown in Fig. 2.43(a). Similarly, in the case of other shapes into which the conductor is bent, the fields vary as shown in Fig. 2.43(b) and (c).

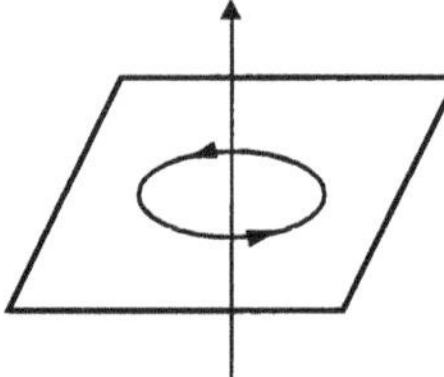

Fig. 2.43(a) Field due to current in a long straight conductor

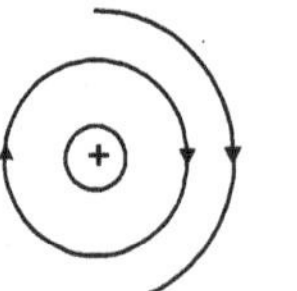

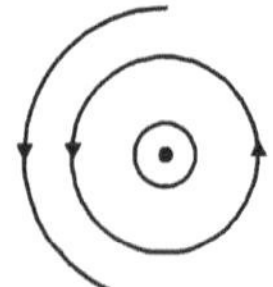

Fig. 2.43(b) Field due to current in a circular loop

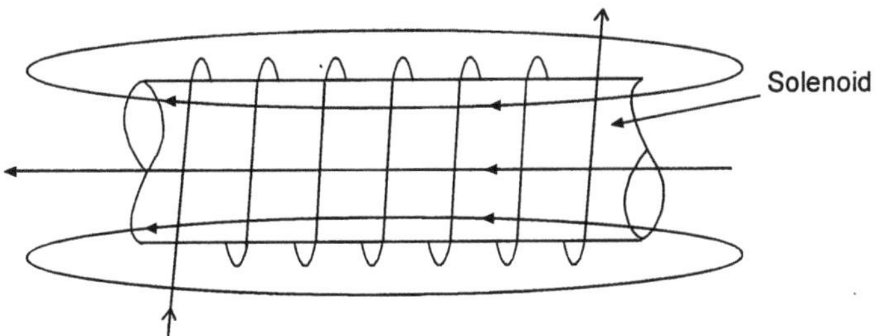

Fig. 2.43(c) Field due to current in a solenoid

In all the above cases, each line called a "line of magnetic flux", has a closed path. Having established the existence of a magnetic field due to current in a conductor bent in some shape, the following definitions are useful in the study of magnetic circuits.

2.6.2 Some definitions :

(a) *Magnetic circuit* : The path taken by the magnetic flux is called the magnetic circuit. As mentioned earlier, the magnetic flux follows a closed path. It may traverse through air or any other magnetic medium, exactly similar to the magnetic field of a magnet.

(b) *Magnetomotive force* : It is defined as the force that drives or tends to drive a magnetic flux through a magnetic circuit. It is very similar to the electromotive force in an electric circuit.

To illustrate the concept of magnetomotive force, consider an iron ring wound with a coil of wire of 'N' turns as shown in Fig. 2.44.

Let a current of 'i' amperes flow in the wire when a voltage of 'v' volts is applies to it. Consequently, a magnetic field is produced in the ring as shown by the dotted line. It can be shown that the magnitude of the flux 'ϕ' is directly proportional to the current 'i' and also to 'N'. Thus

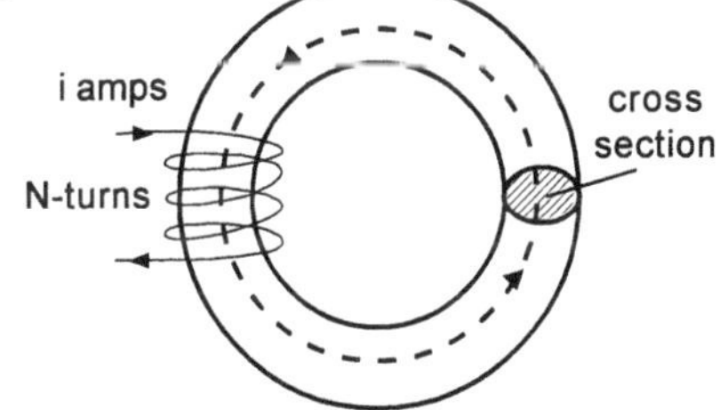

Fig. 2.44 Iron ring wound with a coil

$$\Phi \propto Ni$$

or $$\Phi = kNi \qquad .....(2.94)$$

The relationship may be called "Ohms's Law" in magnetic circuits because of its similarity with "Ohm's law" in electric circuits given by

$$i = v/R \qquad .....(2.95)$$

where 'R' is the resistance of the coil carrying the current 'i'. Since Ni is similar to the emf 'v', it is called magnetomotive force required to drive the flux 'ϕ' in the magnetic circuit. The unit of mmf is "ampere-turn".

In terms of the work done, mmf may be defined as the work done in carrying a unit magnetic pole around the closed magnetic path.

(c) *Reluctance* : Equation (2.94) may be rewritten as

$$\phi = Ni/(l/k) = Ni/R \qquad .....(2.96)$$

and is similar to R in eq. (2.95). R is called '*Reluctance*'. It is the property of the magnetic circuit that opposes the passage of magnetic flux through it. It's unit is ampere-turn per weber.

(d) *Magnetic field intensity (H)* : It is the magnetomotive force per unit length of the magnetic path. The unit of magnetic intensity is ampere-turns per meter.

(e) *Permeance* : It is the reciprocal of the reluctance and is similar to conductance in electric circuits. Its unit is webers per ampere-turn.

(f) *Magnetic flux density (B)* : The magnetic flux crossing a unit section of the magnetic circuit perpendicularly is defined as magnetic flux density. The unit of flux density is webers/sq. meter or Tesla.

(g) *Permeability* (μ) : Permeability is a property of the magnetic medium in which a magnetic field exists. It relates the magnetic flux density and the magnetic filed intensity according to the relation

$$\mathbf{B} = \mu\,\mathbf{H} \qquad .....(2.97)$$

Its unit is henries/meter.

(h) *Relative permeability* μ_r : It is permeability of any magnetic material with respect to the permeability of free air (μ_0). In the m.k.s. system of units, the permeability of free air is $4\pi \times 10^{-7}$. Relative permeability is dimensionless. Thus μ, μ_r and μ_0 are related as

$$\mu = \mu_r \mu_0 \qquad .....(2.98)$$

2.7 PROPERTIES OF MAGNETIC MATERIALS :

2.7.1 B-H curves :

All magnetic materials possess magnetic properties to a greater or lesser extent. These properties are exhibited in terms of the relative permeability the material has. The permeability of any magnetic material, is not a constant and is a function of the flux density. The value of permeability at any flux density can be obtained from the B-H curve or the hysteresis loop. A typical B-H curve is shown in Fig. 2.45. This curve describes the cycling of a magnetic material in closed circuit as it is brought to a fully magnetized state called "saturation", demagnetized, saturated in the opposite direction and then demagnetized again in the opposite direction under the influence of external magnetic field. The part OAB of the curve is of importance and has the following three distinct regions as shown in Fig. 2.46. In the region OD, the flux density increase slowly but increases rapidly and linearly in the region DE. After the point E, the increase in flux density is not appreciable and remains almost constant after the point F. The specimen is then said to be saturated. The portion OG in Fig. 2.45 is called residual magnetism or remnance. It is called so, as there exists some flux in the material even when there is no magnetic intensity (H).

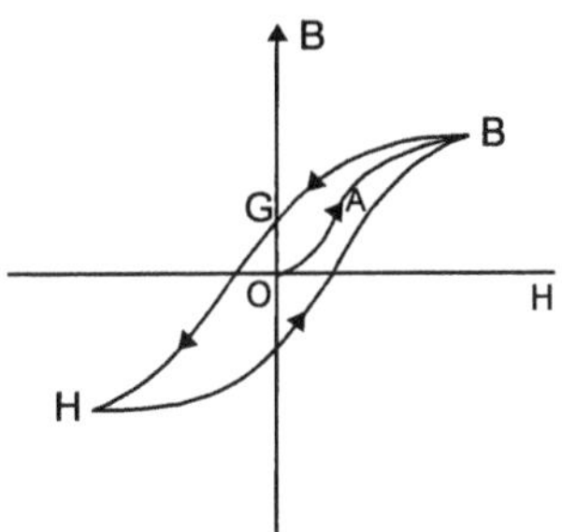

Fig. 2.45 A typical B-H curve

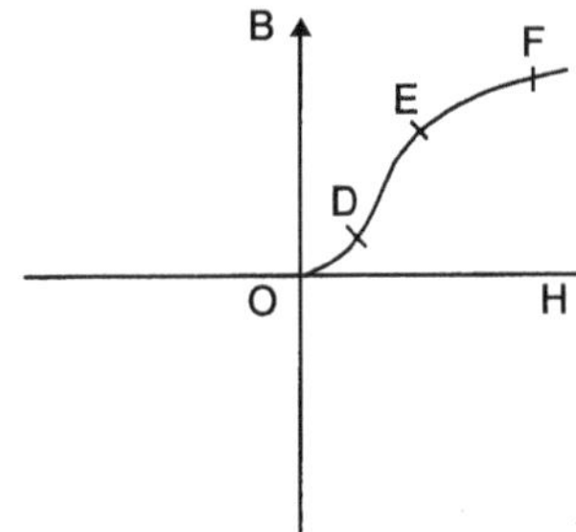

Fig. 2.46 First part of B-H curve

2.7.2 Classes of magnetic materials :

Based on the values of the permeability of the material over a certain range, all magnetic materials are classified into three broad categories :

(a) Ferromagnetic materials
(b) Paramagnetic materials
(c) Diamagnetic materials.

Ferromagnetic materials have relative permeability far above unity and these values of the permeability are dependent upon the magnetizing force. Paramagnetic materials have their relative permeability only slightly greater than unity. Diamagnetic materials have permeability slightly less than unity. In both paramagnetic and diamagnetic materials, the permeability is independent of the magnetizing force. Iron, Cobalt, Nickel and many of their alloys are examples of ferromagnetic materials.

2.8 SIMPLE MAGNETIC CIRCUITS :

A magnetic circuit in its simplest form consists of an annular ring of cross section 'A' sq.m. made of magnetic material of permeability μ with a coil of wire of N turns wound round it as shown in Fig. 2.47. If the current flowing in the coil is 'i' amps, then the total ampere turns produced is Ni.

$$\text{The magnetising force H inside the ring} = \frac{\text{Ni}}{l} \text{ amp.-turns/meter} \qquad \text{.....(2.99)}$$

where 'l' is the mean length of the magnetic path.

The flux density B inside the material is then given by

$$B = \mu_0 \mu_r H \quad .....(2.100)$$

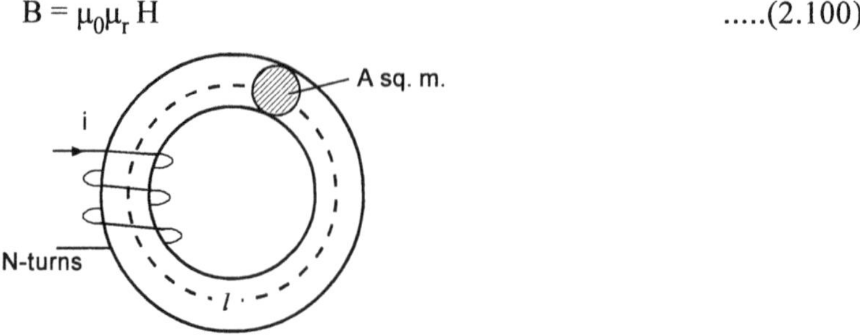

Fig. 2.47 Simple magnetic circuit

Since $\phi = BA$,

$$\phi = \frac{\mu_0 \mu_r NiA}{l} = \frac{\frac{Ni}{l}}{\mu_0 \mu_r A} \quad .....(2.101)$$

In eq. (2.101), $\frac{l}{\mu_0 \mu_r A}$ is the reluctance of the magnetic circuit and has the same structure as the resistance. Similar to a resistance, it is directly proportional to the length of the magnetic path and inversely proportional to the area of cross section A.

The quantity $\frac{1}{\mu_0 \mu_r}$ is similar to the specific resistance of a conductor.

2.8.1 Leakage flux and leakage coefficient :

Leakage flux is that part of the flux that does not follow the desired path. It is generally assumed that all the flux produced by the current 'i' in Fig. 2.47 will be confined to the core of the magnetic material. However, in practice some lines of flux jump from the material and force themselves through air as shown in Fig. 2.48. Such of these lines are called leakage fluxes (ϕ_l). All flux through iron is considered as useful flux (ϕ). Thus , the total flux (ϕ_t) produced is the sum of the fluxes $\phi_l + \phi$. The extent of leakage is estimated through a factor called "Leakage Coefficient". It is defined as :

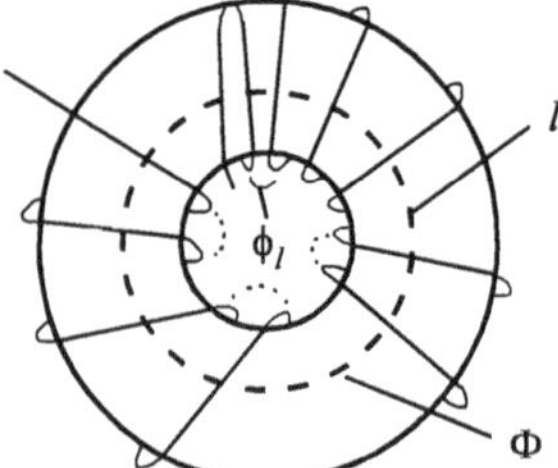

Fig. 2.48 Leakage fluxes

Leakage coefficient = Total flux/Useful flux

$$= \phi_t/\phi \quad .....(2.102)$$

The leakage coefficient is usually in the range of 1.1 to 1.2.

Example 2.16

An iron ring of mean diameter 25 cm and of cross sectional area 15 sq. cm is wound uniformly with 500 turns of wire. What is the current required to be passed through the wire to produce a flux of 5×10^{-4} webers in the ring ? Take permeability of iron as 800.

Solution :

Area A of the ring = 15 sq.cm. = 15×10^{-4} sq. m.
Mean length of the magnetic path = $25\ \pi$ cm = $0.25\ \pi$ meters
Permeability of iron = 800
Hence Reluctance of the magnetic circuit $\boldsymbol{R} = 0.052 \times 10^7$ At/wb
Flux required to be produced = 5×10^{-4} wb
So, Ampere turns Ni = Reluctance x flux = $5 \times 0.052 \times 10^3$
Current i = Ni/turns = 0.52 amps.

2.8.2 Composite magnetic circuits :

In any practical situations, magnetic circuits are developed using several regions having different magnetic properties (i.e., magnetic permeability). The situation is similar to electric circuits having a number of resistances connected in different ways depending on the requirement. We have seen that any electric circuit may be formed by a series, parallel or series-parallel combination of resistances. Similarly, it is possible to develop magnetic circuits with series, parallel or series-parallel combination of materials with different reluctances. Such circuits are called composite magnetic circuits. Figs. 2.49 (a), (b) and (c) are examples of such composite circuits. In Fig. 2.49 (a), a line of flux traverses through media of different reluctances $\boldsymbol{R}_1$, $\boldsymbol{R}_2$ and $\boldsymbol{R}_3$. This is an example of a series circuit. The equivalent reluctance is the sum of the three reluctances. Thus,

$$\boldsymbol{R} = \boldsymbol{R}_1 + \boldsymbol{R}_2 + \boldsymbol{R}_3 \qquad .....(2.103)$$

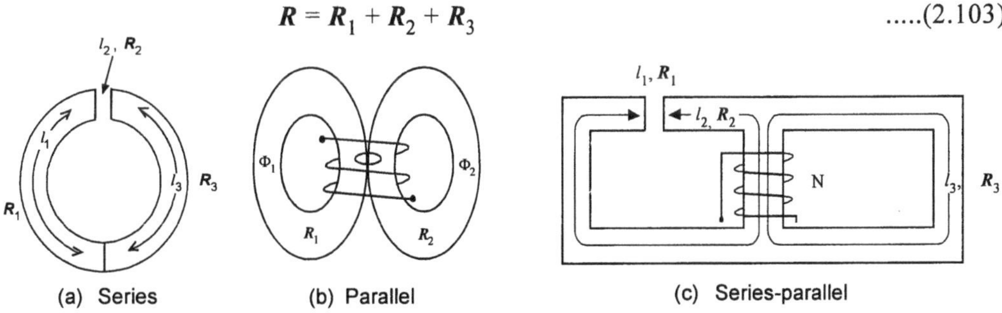

(a) Series (b) Parallel (c) Series-parallel

Fig. 2.49 Composite magnetic circuits

In Fig. 2.49 (b), a total flux ϕ produced in the central limb divides into two ϕ_1 and ϕ_2 respectively along $\boldsymbol{R}_1$ and $\boldsymbol{R}_2$ as a current divides in a circuit consisting of two resistances in parallel. In Fig. 2.49 (c), one can observe that a combination of series and parallel circuits exists. It should, however, be emphasized that reluctance of a magnetic circuit cannot be taken as a constant as in the case of resistances. Reluctance varies with the magnetic flux and has to be calculated each time there is a change in the magnetic flux.

Example 2.17

An iron ring of circular section of 2 cm diameter, has a mean diameter of 40 cm and is wound with 200 turns of wire. An air gap of 0.1 cm is cut across the ring as shown in Fig. 2.50. The useful flux in the air gap is to be 2.0×10^{-4} wb. Calculate the current required in the wire. Assume relative permeability of iron to be 900.

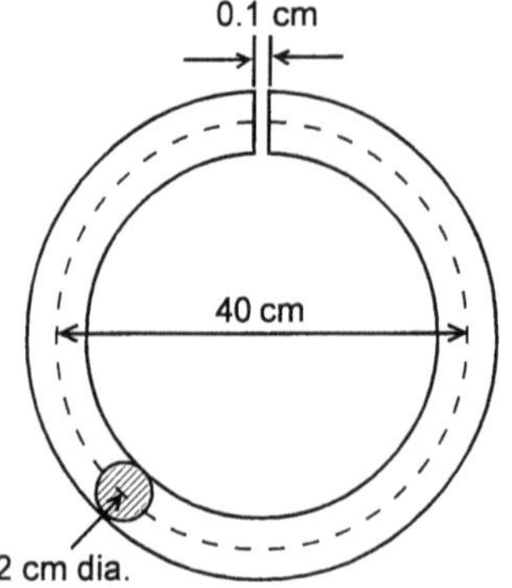

Fig. 2.50 Magnetic circuit for example 2.17

Solution :

Total reluctance = reluctance of air gap + reluctance of iron

Length of the air gap = 0.001m

Area of cross section = $\pi \times (0.01)^2 = \pi \times 10^{-4}$ sq.m.

$$\text{Reluctance of air gap} = \frac{l}{\mu_o \mu_r A} = \frac{0.001}{4\pi \times 10^{-7} \times 1 \times \pi \times 10^{-4}}$$

$$= 25.33 \times 10^5 \ \text{AT/wb}$$

Length of the iron path = $\pi d - 0.001 = \pi \times 0.4 - 0.001 = 1.256$ m

$$\text{Reluctance of iron path} = \frac{l}{\mu_o \mu_r A} = \frac{1.256}{4\pi \times 10^{-7} \times 900 \times \pi \times 10^{-4}}$$

$$= 35.35 \times 10^5 \ \text{AT/wb}$$

Total reluctance = $25.33 \times 10^5 + 35.35 \times 10^5 = 60.68 \times 10^5$ AT/wb

M mf required = flux × reluctance = $2 \times 10^{-4} \times 60.68 \times 10^5 = 1213.6$ AT

$$\therefore \ \text{current required} = \frac{\text{M mf}}{\text{No. of turns}} = \frac{1213.6}{200} = 6.068 \ \text{A}$$

Example 2.18 :

A cast steel ring has a mean diameter of 30 cm and has a circular section of 20 sq. cm area. An iron bar of circular section of 10 sq. cm. is fixed in the ring centrally as shown in Fig. 2.51 without any air gap. A winding of 800 turns is provided around this iron bar and caries a current of 1.0 amp. Assuming the relative permeabilities of the ring and bar respectively as 1800 and 900, calculate the flux in the ring.

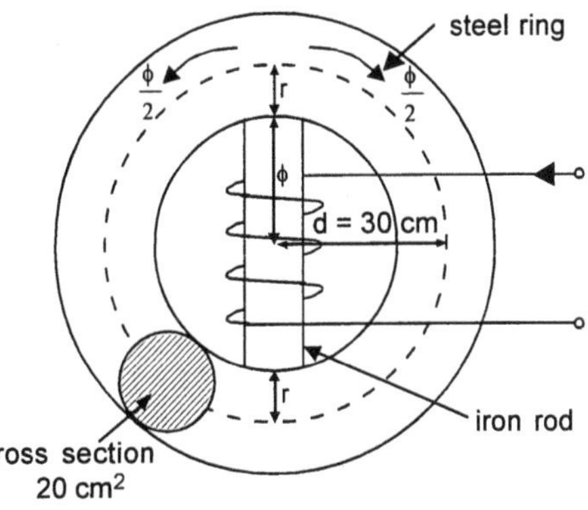

Fig. 2.51 For example 2.18

Solution :

Mean dia of steel ring, d = 30 cm

Area of cross section of steel ring, A_s = 20 cm^2

∴ radius of the ring, r = 2.52 cm

Length of the iron bar, l_i = Mean dia of steel ring – 2r = 30 – 2 × 2.52 = 24.96 cm

$$\text{Reluctance of the iron path, } R_i = \frac{l_i}{\mu_o \mu_{ri} A_i} = \frac{24.96 \times 10^{-2}}{4\pi \times 10^{-7} \times 900 \times 10 \times 10^{-4}}$$

$$= 2.207 \times 10^5 \ \text{AT}/\text{wb}$$

Half of the flux produced in the iron bar passes through the right half of the steel ring and the other half passes through the left half of the steel ring.

∴ The mean length of the flux path in steel ring in each half, $l_s = \dfrac{\pi d}{2} = 0.15\ \pi$ m

∴ Reluctance of each half of steel ring, $R_s = \dfrac{l_s}{\mu_o \mu_{rs} A_s}$

$$= \frac{0.15\pi}{4\pi \times 10^{-7} \times 1800 \times 20 \times 10^{-4}}$$

$$= 1.042 \times 10^5 \ \text{AT}/\text{wb}$$

Since the two halves of the steel ring are in parallel, the equivalent reluctance,

$$R_{es} = \frac{R_s}{2} = 0.521 \times 10^5 \ \text{AT}/\text{wb}$$

Total reluctance of the flux path, $R_t = R_i + R_{es} = 2.728 \times 10^5$ AT/wb

M mf produced = NI = 800 × 1 = 800 AT

$$\therefore \text{ Flux produced in the iron rod, } \phi = \frac{\text{Mmf}}{R_t} = \frac{800}{2.728 \times 10^5} = 2.93 \text{ m wb}$$

$$\text{Hence flux in the steel ring} = \frac{\phi}{2} = 1.465 \text{ mwb.}$$

Problems

2.1 By using series parallel combinations, find the equivalent resistance in Fig.P. 2.1 at the terminals a, b.

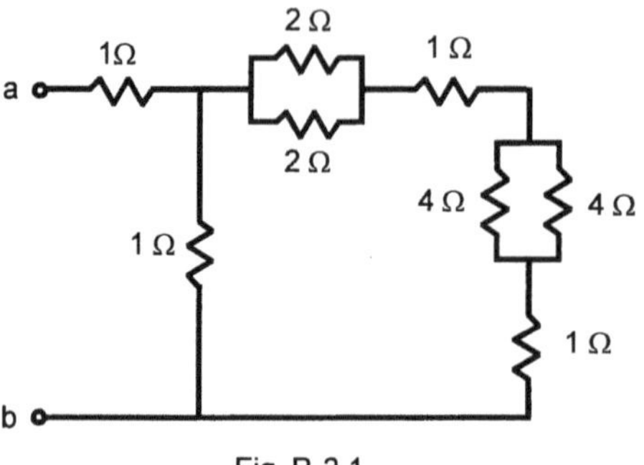

Fig. P. 2.1

2.2 Find the input resistance at terminals a, b of the network of Fig. P. 2.2.

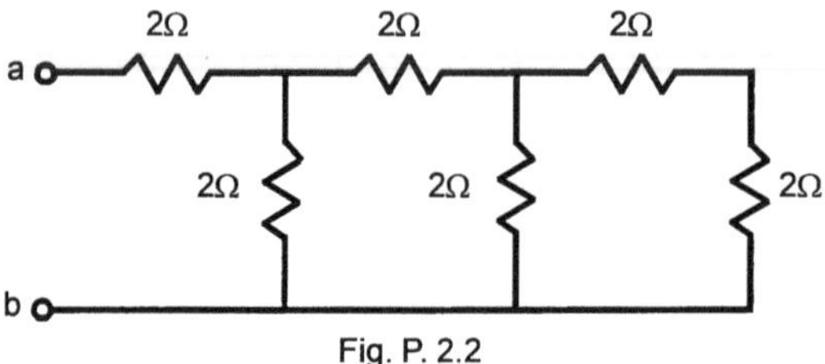

Fig. P. 2.2

2.3 Write the v – i equation of the branches shown in Fig. P. 2.3.

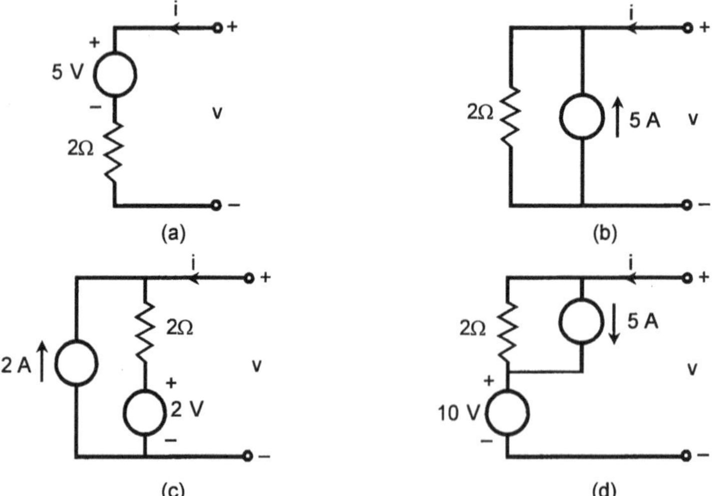

Fig. P. 2.3

2.4 Reduce the branches in Fig. P. 2.4 to single sources.

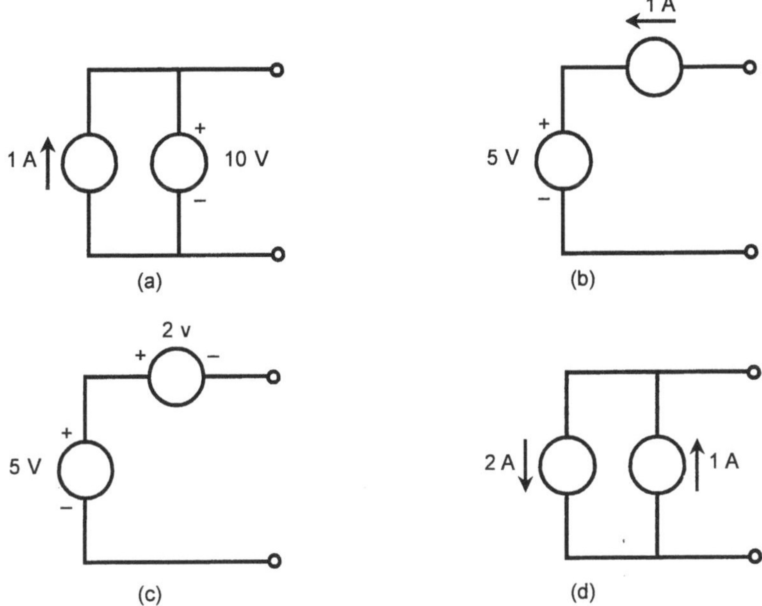

Fig. P. 2.4

2.5 Convert the voltage source branches in Fig. P. 2.5 to equivalent current source branches.

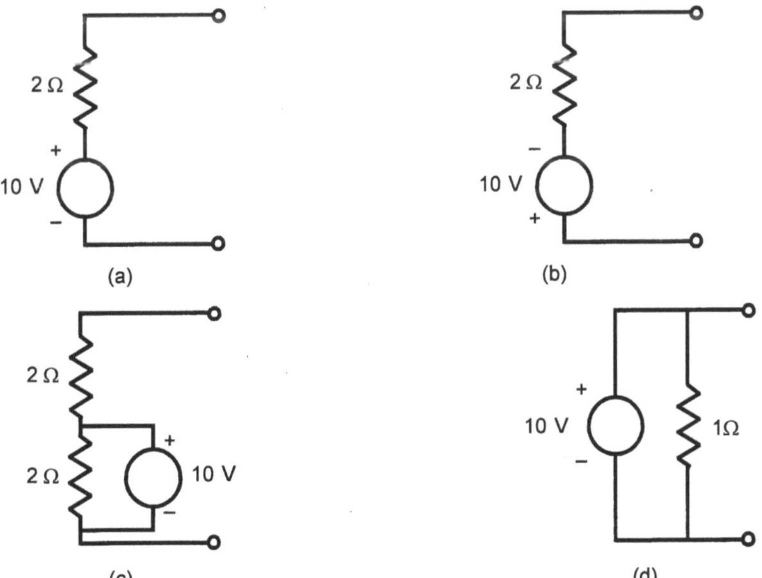

Fig. P. 2.5

2.6 Convert the current source branches in Fig. P. 2.6 to equivalent voltage source branches.

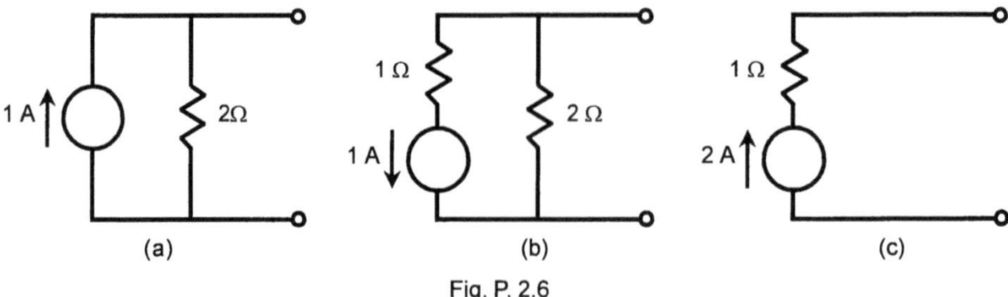

Fig. P. 2.6

2.7 Convert the branches in Fig. P. 2.7 to single voltage sources in series with single resistors.

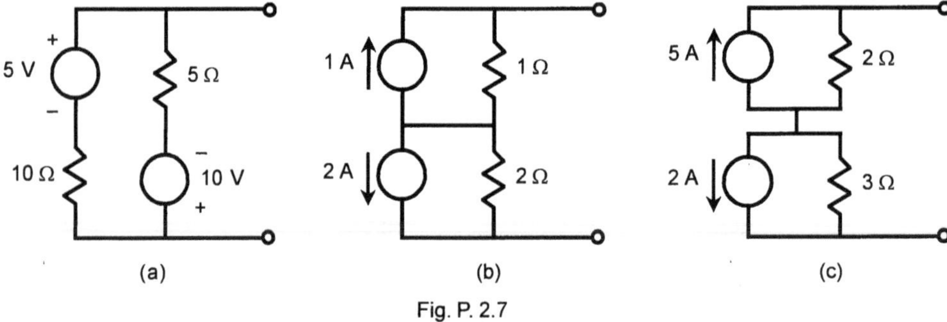

Fig. P. 2.7

2.8 Using network reduction, find the voltage V in Fig. P. 2.8.

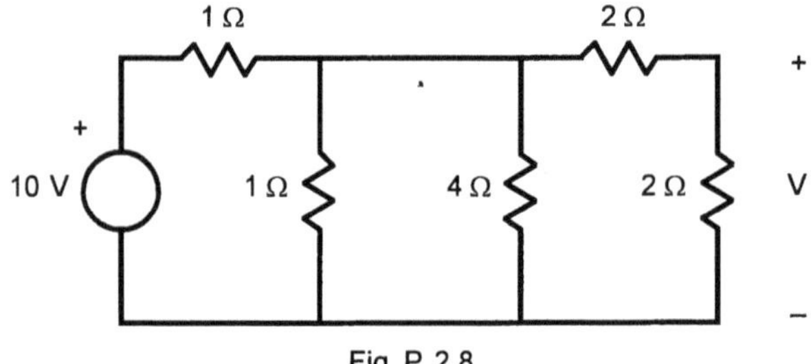

Fig. P. 2.8

2.9 Find V in Fig. P. 2.9 using the network reduction.

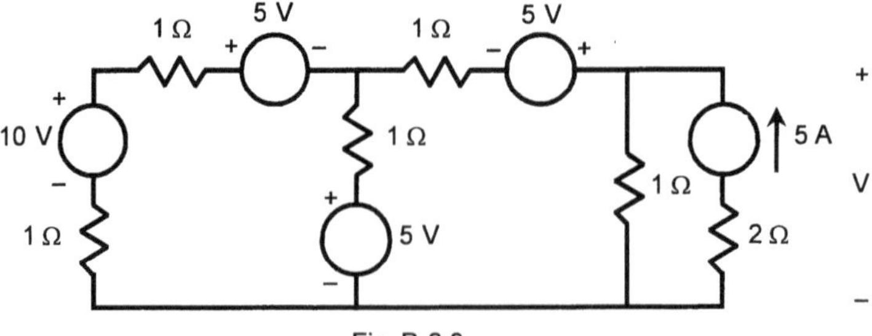

Fig. P. 2.9

2.10 Find the equivalent resistance between the terminals 1, 2 of network shown in Fig. P. 2.10. Resistor values are in ohms.

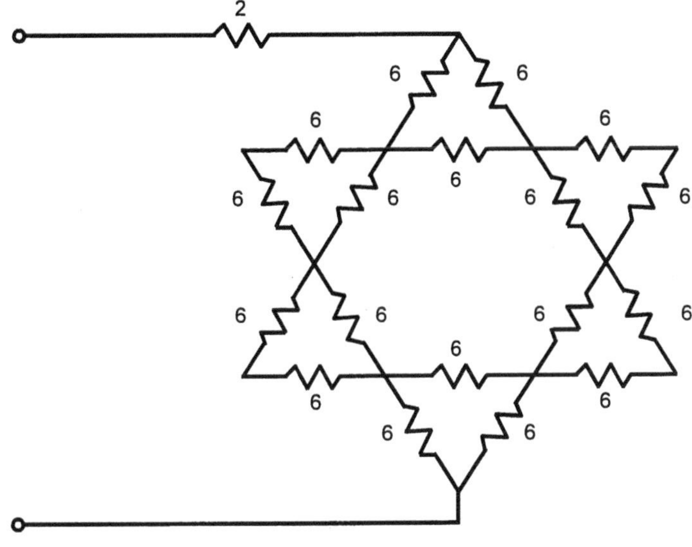

Fig. P. 2.10

2.11 Find equivalent resistance R_{ab} in Fig. P. 2.11. Resistor values are in ohms.

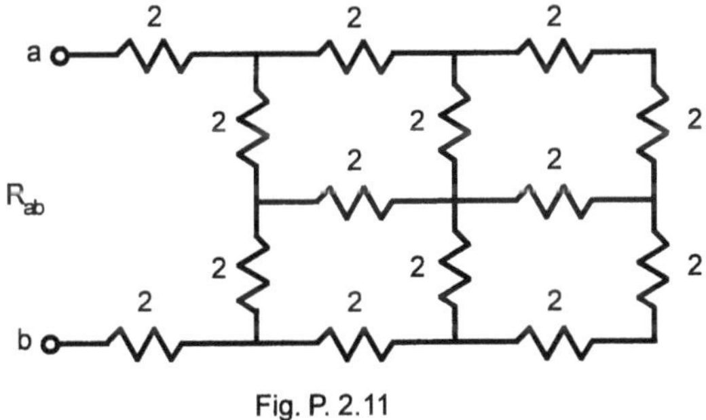

Fig. P. 2.11

2.12 Find R_{eq} in Fig. P. 2.12. Resistor values are in ohms.

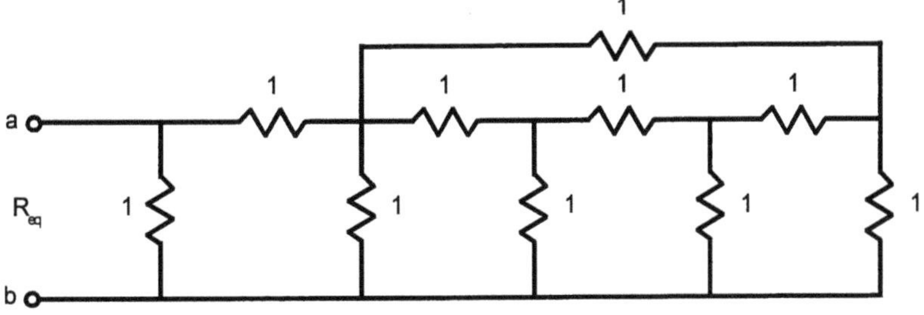

Fig. P. 2.12

2.13 1 set of twelve 1 Ω resistors are connected to form a cube as shown in Fig. P. 2.13. Find R_{eq}.

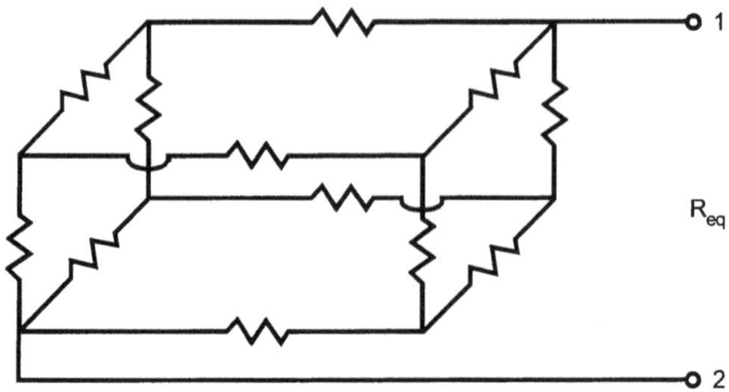

Fig. P. 2.13

2.14 The network in Fig. P. 2.13 can be written in the form shown in Fig. P. 2.14. Find the R_{eq}. Resistor values are in ohms.

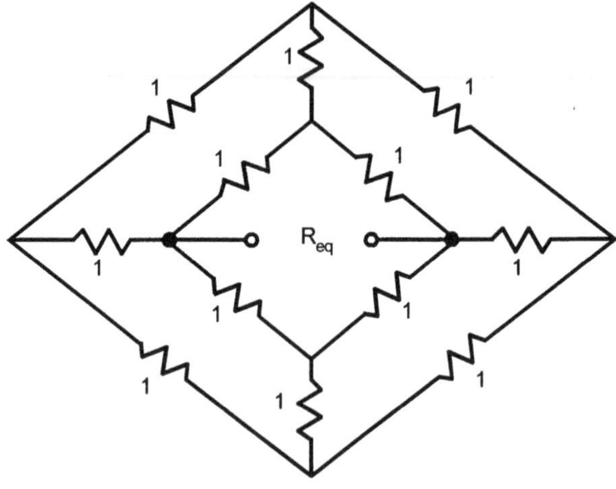

Fig. P. 2.14

2.15 Find the input resistance of the circuit in Fig. P. 2.15. (Hint; Assume V = 2V and I_x = x amps and solve for x.)

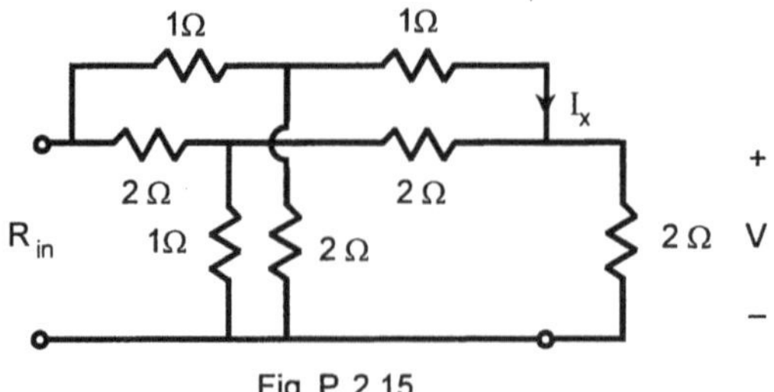

Fig. P. 2.15

2.16 Find the voltage V in Fig. P. 2.16. Using node voltage method. Redraw the figure so that 2 Ω branch connected between the two current sources is drawn above the T network consisting of 1 Ω, 1 Ω and 4 Ω resistors. Find 'V' again by inspection.

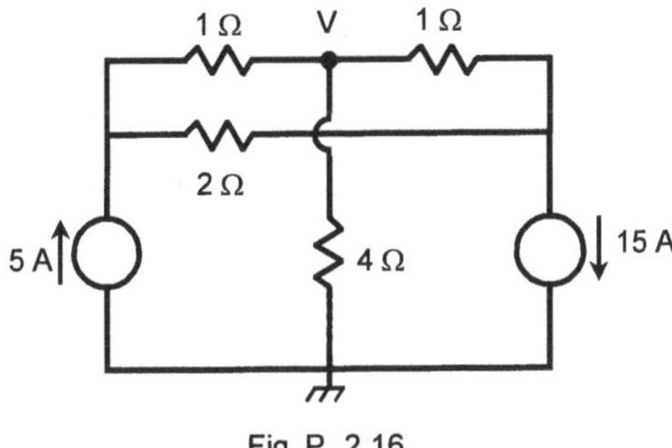

Fig. P. 2.16

2.17 Use nodal analysis to find the voltages V_1 and V_2 in Fig. P. 2.17.

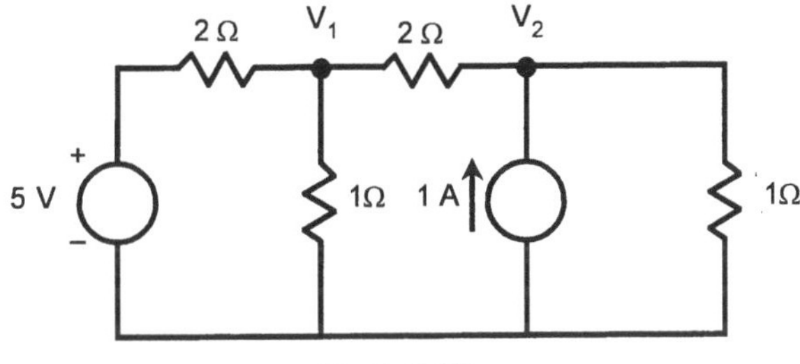

Fig. P. 2.17

2.18 Use nodal analysis to find the current 'I' in Fig. P. 2.18. What is the power delivered by the 1 A current source.

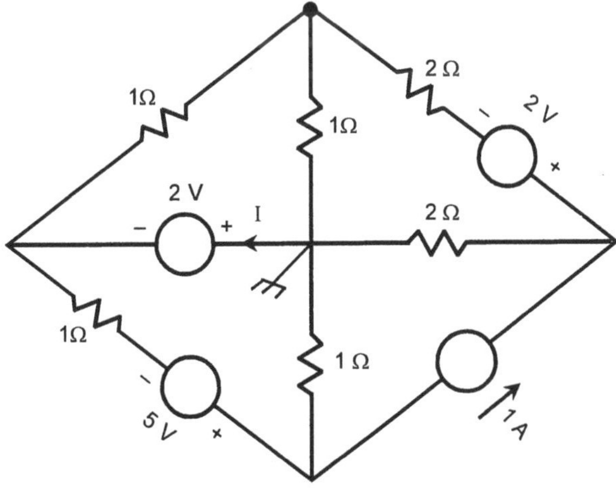

Fig. P. 2.18

2.19 Use nodal analysis to find the power delivered by the 2 V source in Fig. P. 2.19.

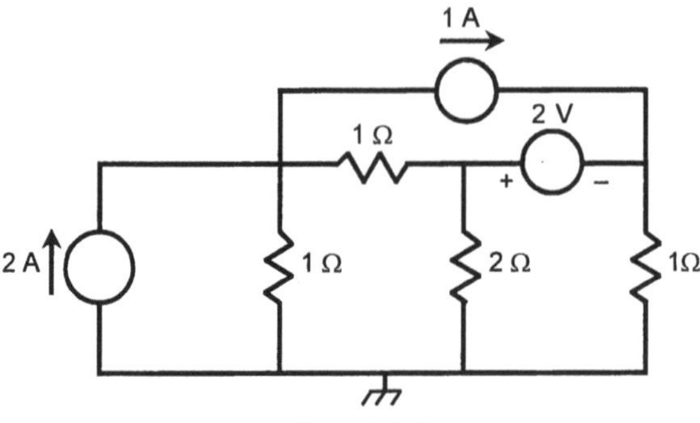

Fig. P. 2.19

2.20 Use nodal analysis to find the power delivered by the 4 A current source in Fig. P. 2.20.

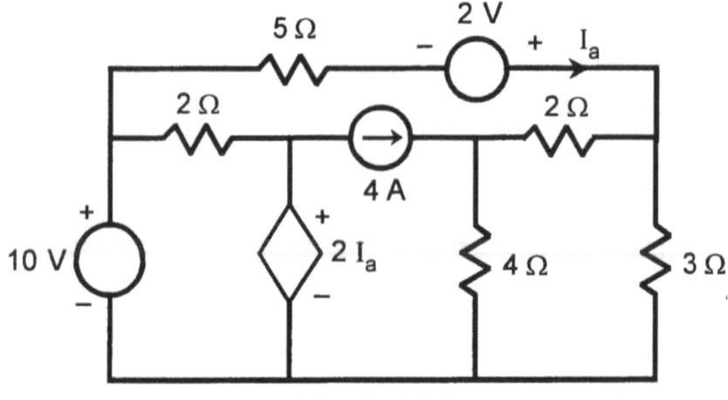

Fig. P. 2.20

2.21 Use node voltage method to find the current I_A in Fig. P. 2.21.

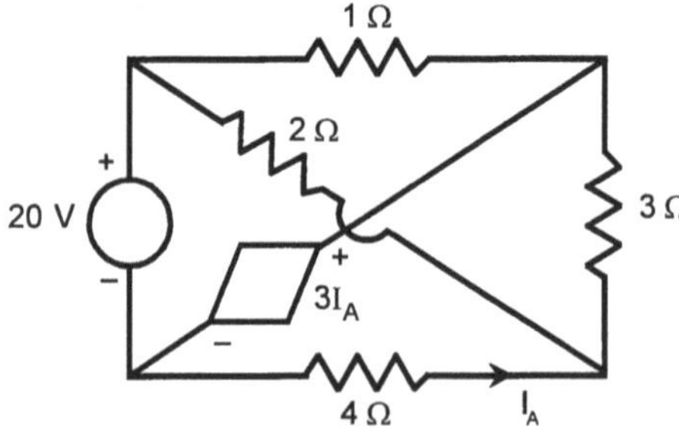

Fig. P. 2.21

2.22 Use mesh analysis to find the current I_a in Fig. P. 2.22.

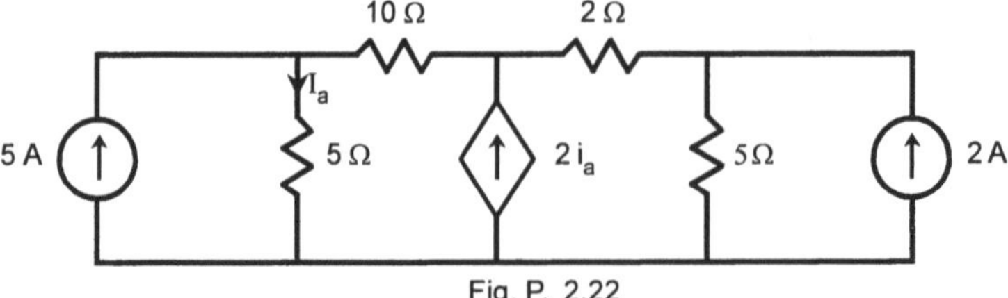

Fig. P. 2.22

2.23 Use mesh analysis to solve problem 2.18.

2.24 Use mesh analysis to solve problem 2.19.

2.25 Use mesh analysis to solve problem 2.20.

2.26 An iron ring of mean length of magnetic path of 50cm has a cross sectional area of 6 sq.cm. An air gap of 1 mm is cut in the ring as shown in the Fig. P. 2.23. A coil of 300 turns is wound round the ring. Taking the relative permeability of iron as 600, calculate the flux in the ring if a current of 2 amps flows in the coil.

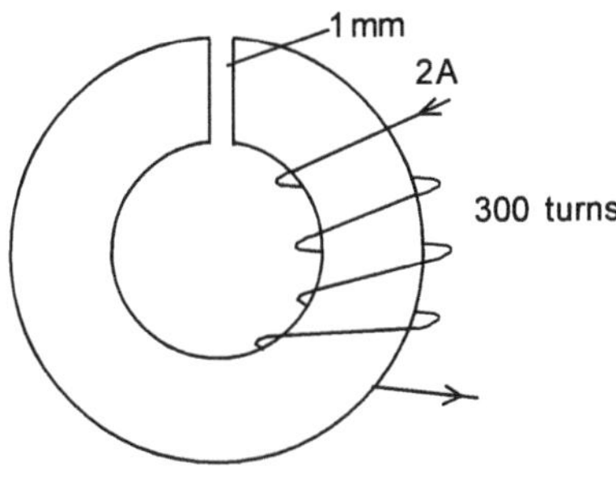

Fig. P 2.23

2.27 An iron ring of mean length of magnetic path. 60 cm has an air gap of 1 mm and a winding of 250 turns. The relative permeability of iron is 300. What is the flux density in the iron ring when a current of 1.2 amps flows through the ring ?

2.28 A cast steel structure is made of a rod of square section 2.5 cm × 2.5 cm. as shown in Fig. P. 2.24. What is the current that should be passed in a 500 turn coil on the left limb,

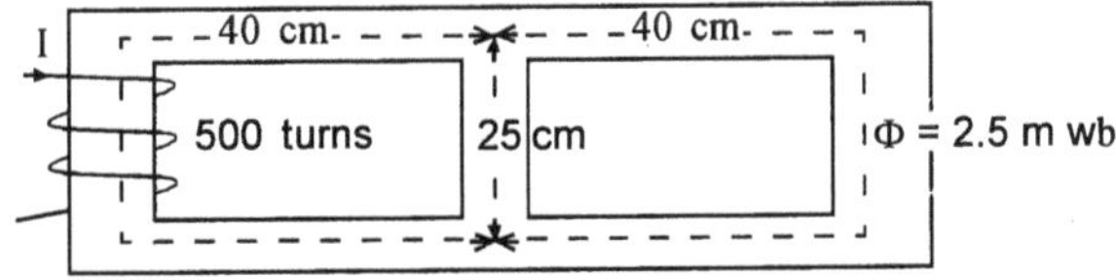

Fig. P 2.24

so that a flux of 2.5 mwb is made to pass in the right limb. Assume permeability as 750 and neglect leakage.

2.29 An iron ring of 0.20 meter diameter and 1.5×10^{-3} sq.m. in cross section with a saw cut 1 mm wide, is wound with 200 turns of copper wire. The gap density is 0.8 Tesla. The relative permeability of iron is 600. Calculate the exciting current. Ignore leakage.

2.30 The flux in the air gap of the cast steel frame shown in Fig. P.2.24 is 1.6 mWb. The exciting coil has 680 turns wound on the middle limb. All dimensions given are in mm. Assuming a uniform gap flux density and neglecting leakage, calculate the current in the coil given the following data :

B Tesla	0.2	0.4	0.5	0.6	0.8	1.0	1.2
H (AT/m)	300	480	520	600	720	900	1230

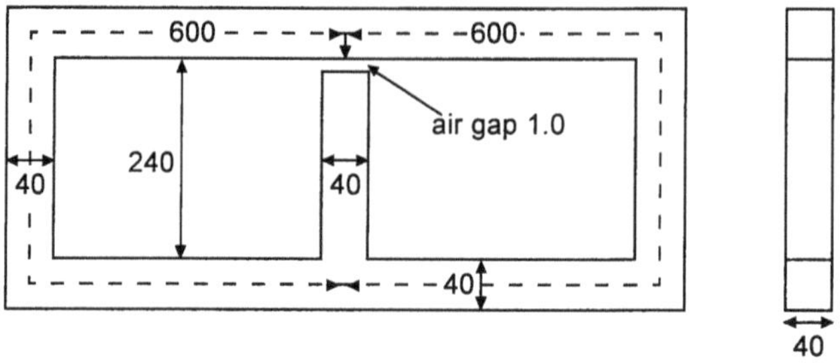

Fig. P 2.25

2.31 A mild steel magnetic circuit has a uniform cross sectional area of 5 sq.cm. and a length of 25 cm. A coil of 180 turns is wound round the steel core uniformly. When the current is 1.5 amps in the coil, the total flux is 0.6 mWb and when the current is 5 amps the flux is 1.0 mWb. For each value of the current calculate (a) the magnetizing field strength and (b) the relative permeability of steel.

2.32 A steel ring having a mean circumference of 750 mm and a cross sectional area of 500 sq. mm. is wound with a magnetising coil of 120 turns. Using the following data, calculate the current required to set up a magnetic flux of 600 μWb in the ring.

Flux density (B) Tesla	0.9	1.1	1.2	1.3
Magnetic field Strength (AT/m)	260	450	600	720

3

SINGLE PHASE CIRCUITS

3.1 INTRODUCTION :

We have seen in earlier chapters that the total response of any electric circuit is the sum of the natural and forced responses. Whereas the natural response of the circuit is a characteristic of the circuit, the forced response depends on the type of external source applied to the circuit. Accordingly, responses to different types of input were considered in the earlier chapters. In this chapter we confine our attention to the steady state response of a network to sinusoidal excitations. Here we assume that sufficient time has elapsed after the input is applied to the circuit and transient response has died down. The selection of the sinusoid as a forcing function among all periodic functions is due to the following reasons :

1. The natural response of a lossless underdamped second order system is a pure sinusoid.
2. Many natural phenomena like the motion of a simple pendulum, vibrations of musical strings etc., exhibit sinusoidal character.
3. Any periodic wave can be shown to be a combination of a number of sinusoidal functions.
4. Derivatives and integrals of sine functions are again sinusoidal functions.
5. It can be generated easily in the laboratory.

3.2 CHARACTERISTICS OF PERIODIC FUNCTIONS :

Periodic functions are those which repeat regularly after a certain time called ***'Period'***. Mathematically we can say that a periodic function is one which satisfies the relation

$$f(t) = f(t + T) \qquad \text{for} \quad -\infty < t < \infty.$$

A common example of a periodic function is a sine function. Some examples are shown in Fig. 3.1, in which T is called the period.

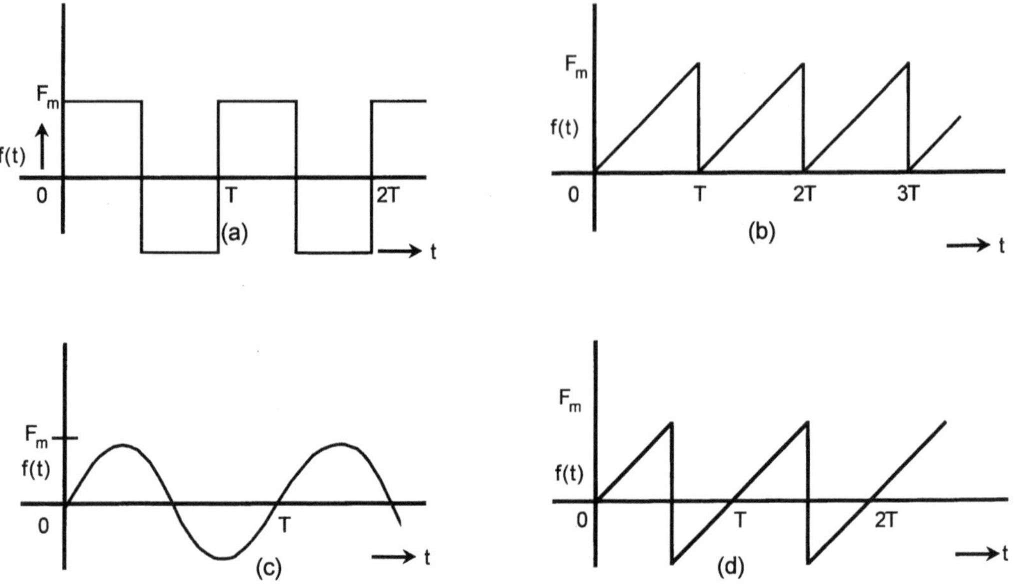

Fig. 3.1 Some examples of periodic functions

That part of the waveform which lies in one period is known as one ***'cycle'***. Number of such cycles in one second is called as ***'frequency'***. Its unit is *Hertz*.

Thus, the frequency f is given by

$$f = \frac{1}{T} \text{ Hz} \qquad .....(3.1)$$

The quantity 2π f is known as angular velocity or radian frequency 'ω'. Thus

$$\omega = 2\pi f = \frac{2\pi}{T} \text{ rad/ sec.}$$

Sinusoidal voltages can be generated by rotating a conductor in a constant magnetic field. If a conductor of length 'ℓ' is rotated with an angular velocity ω in a magnetic field of density B tesla, the voltage induced in the conductor at any instant 't' is given by

$$E = B\,\ell\,v \sin\omega t \qquad .....(3.2)$$

where v is the linear velocity of the conductor. In general this can be written as

$$E = E_m \sin\omega t \qquad .....(3.3)$$

where E_m is the maximum value of E in eq. (3.3).

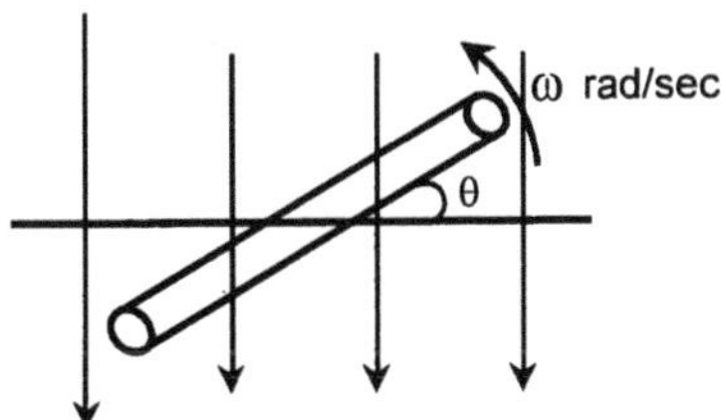

Fig. 3.2 Production of sinusoidal voltage

A more general form of the sinusoidal waveform is shown in Fig. 3.3 and is represented mathematically as

$$e(t) = E_m \sin(\omega t + \theta)$$

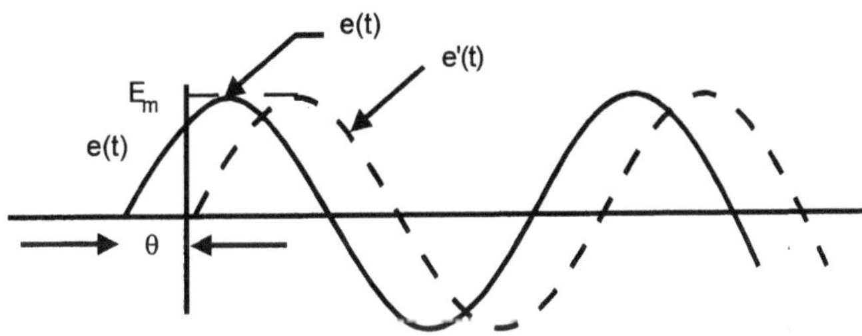

Fig. 3.3 More general sinusoidal functions

'θ' is called the phase angle of the waveform with respect to a reference waveform $e'(t) = E_m \sin\omega t$. If θ is positive, e(t) has zero value, $\frac{\theta}{\omega}$ sec earlier than e'(t) and is said to be leading e(t). If θ is negative, e(t) has zero value, $\frac{\theta}{\omega}$ sec later than e'(t) and is said to be lagging e(t).

3.3 AVERAGE VALUE, R.M.S VALUE, FORM FACTOR :

Since the periodic functions are varying with respect to time, it is difficult to specify its value as in the case of D.C. In the case of D.C we can specify the voltage as 'V'. But same is not the case with perodic functions, which may have values which are changing with respect to time and may have both positive and negative values. Hence, in order to specify a periodic function it is compared with D.C in producing the same effect in a particular device or circuit. For example let a certain amount of charge be transferred to a circuit during a given time by a d.c source of value 'V'. If a periodic voltage transfers the same charge to the circuit during the same time, the value of the periodic voltage is said to be 'V' volts. This is usually called as the average value of the periodic waveform. Similarly if a periodic current produces the same heating in a given resistance 'R' in a given time as a d.c current of I amperes in the same resistance 'R' in the same time 't', the periodic current is said to be of value I amperes. This value is usually known as effective value or root mean square value or in short RMS value.

3.3.1 Average value :

The average or mean value 'F_{av}' of a periodic waveform f(t) is defined mathematically as

$$F_{av} = \frac{1}{T}\int_0^T f(t)dt \qquad .....(3.4)$$

For periodic waveforms which have equal positive and negative half cycles over a period, this value would be zero. Hence it is customary in all such cases to define the average value over half the period. Thus for a sinusoid shown in Fig.3.1(c), the average value is given by,

$$F_{av} = \frac{2}{T}\int_0^{\frac{T}{2}} F_m \sin\omega t\, dt$$

$$= \frac{-2F_m}{\omega T}\cdot\cos\omega t\Big|_0^{\frac{T}{2}}$$

$$= \frac{-2F_m}{\omega T}[\cos\pi - \cos o] = \frac{2F_m}{\pi} = 0.637\ F_m$$

For a square wave of Fig. 3.1(a)

$$F_{av} = \frac{2}{T}\int_0^{\frac{T}{2}} F_m dt$$

$$= \frac{2F_m}{T}[t]_0^{\frac{T}{2}}$$

$$= F_m \qquad .....(3.5)$$

For the triangular waveform of Fig. 3.1(b) and (d) it can be easily shown that

$$F_{av} = \frac{F_m}{2} \qquad .....(3.6)$$

3.3.2 Effective value or RMS value :

Effective value of a periodic waveform is obtained by comparing the heat produced by a current of given waveform and an equivalent d.c current in a given resistance over a given time T. Since the heating effect is proportional to the square of the current, negative values, as well as positive values of the current produce the same heating effect. Hence consider a general waveform consisting of only positive values as shown in Fig. 3.4.

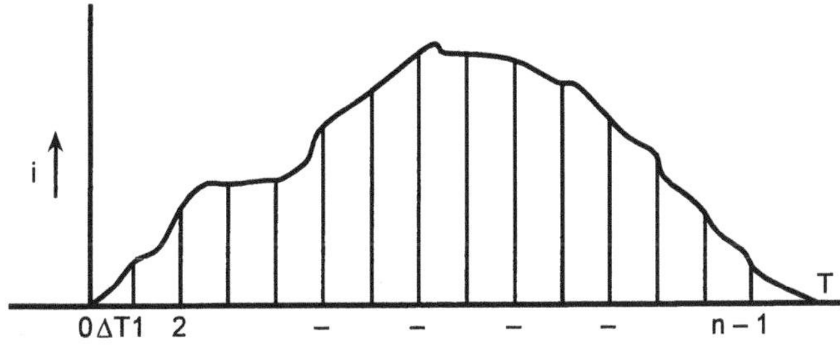

Fig. 3.4 A current waveform

Dividing the period T into n equal intervals, for sufficiently large 'n', the current can be assumed to be a constant over the small interval. Hence the heat produced in a resistance R can be obtained as the sum of the heats produced in each interval.

$$\text{Heat produced} \propto \sum_{j=0}^{n} i^2(j\Delta T)R\Delta T \qquad .....(3.7)$$

Where $\Delta T = T/n$ and $i(j\Delta t)$ is the current in j^{th} interval.

Consider a Direct Current of value I amps in the same resistance R over the same time 'T'.

$$\text{Heat produced by D.C} \propto I^2 RT \qquad .....(3.8)$$

If the two quantities in eq. (3.7) and (3.8) are equal, we have

$$I^2 R T = \sum_{j=0}^{n} I^2(j\Delta T)R\Delta T$$

$$I^2 = \frac{1}{T}\sum_{j=0}^{n} I^2(j\Delta T)\Delta T$$

If n is made very large, $\Delta T \rightarrow dt$ and summation can be replaced by an integral. Thus

$$I^2 = \frac{1}{T}\int_{O}^{T} i^2(t)dt$$

or

$$I = \sqrt{\frac{1}{T}\int_{O}^{T} i^2(t)dt} \qquad .....(3.9)$$

This value given by eq. (3.9) is known as the effective value or root mean square value since it is obtained by taking the root of the mean of the squared values of the current over a period T.

In general eq. (3.9) is true for periodic waveform of any quantity, not necessarily a current waveform.

The RMS value F of a sinusoidal waveform is given by

$$F^2 = \frac{1}{T}\int_O^T (F_m \sin \omega t)^2 dt$$

$$= \frac{F_m^2}{T}\left[\int_O^T \left(\frac{1-\cos 2\omega t}{2}\right) dt\right]$$

$$= \frac{F_m^2}{2T}\left[t - \frac{\sin 2\omega t}{2\omega}\right]_O^T$$

$$= \frac{F_m^2}{2T}\left[T - \frac{\sin 2\omega T}{2\omega}\right]$$

$$= \frac{F_m^2}{2T}[T] \quad (\because \omega T = 2\pi, \sin 2\omega T = 0)$$

$$= \frac{F_m^2}{2}$$

$$F = \frac{F_m}{\sqrt{2}} \qquad .....(3.10)$$

Similarly the RMS values of other periodic waveforms can be easily found out as shown in the following examples.

Example 3.1

Find the effective value of a square wave shown in Fig. 3.1(a).

Solution :

Since the waveform is symmetrical about x – axis, the positive half cycle produces the same heating effect as that of the negative half cycle. Hence it is sufficient to find the effective value over one half cycle.

$$F^2 = \frac{2}{T}\int_0^{\frac{T}{2}} F_m{}^2 dt = \frac{2F_m{}^2}{T}\left[\frac{T}{2}\right] = F_m{}^2$$

Thus $F = F_m$

Example 3.2

Find the effective value of the triangular waveform shown in Fig. 3.5.

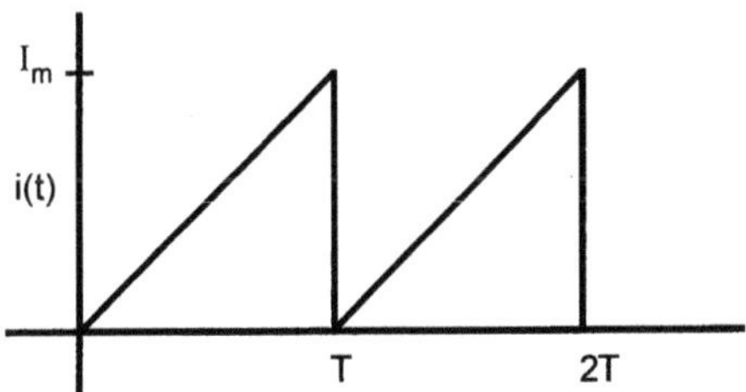

Fig. 3.5 Waveform for example 3.2

Solution :

$$I^2 = \frac{1}{T}\int_0^T \left(\frac{I_m t}{T}\right)^2 dt$$

$$= \frac{1}{T}\left[\frac{I_m^2}{T^2}\cdot\frac{t^3}{3}\Bigg|_0^T\right]$$

$$= \frac{I_m^2}{3}$$

$$I = \sqrt{\frac{I_m^2}{3}} = \frac{I_m}{\sqrt{3}}$$

The RMS values of periodic waveforms are usually denoted by the capital letters and the average values by the capital letters with a suffix 'av'. Thus V, I, F indicate effective values where as V_{av}, I_{av} and F_{av} indicate average values.

3.3.3 Form factor :

The ratio of RMS value to the average value of a periodic waveform is defined as the ***'form factor'***. This gives an indication to the shape of the waveform. If the waveform is flat topped, the form factor approaches unity. The form factor of a sinusoidal function is

$$\text{Form factor} = \frac{I_{rms}}{I_{av}} = \frac{I_m}{\sqrt{2}} \times \frac{\pi}{2I_m}$$

$$= \frac{\pi}{2\sqrt{2}} = 1.11$$

Similarly the form factor of a triangular waveform is given by

$$\text{Form factor} = \frac{I_{max}}{I_{av}}$$

$$= \frac{I_m}{\sqrt{3}} \times \frac{2}{I_m}$$

$$= 1.1547$$

The form factor of a square waveform is

$$\text{Form factor} = \frac{I_{rms}}{I_{av}}$$

$$= \frac{I_m}{I_m}$$

$$= 1$$

3.4 REPRESENTATION OF A SINE FUNCTION :

If a sinusoidal source is applied to a network consisting of R, L and C elements the currents and voltages in the network will also be sinusoidal in nature as will be shown later in the chapter. Intuitively this is to be expected because, the time derivative and integral of a sine function results in sine functions of same angular frequency ω only. Hence in order to solve these networks by using Kirchhoff's laws we have to add and subtract sinusoidal quantities. In this section we would like to develop a convenient representation of the sinusoidal function so that the algebraic manipulations become easy.

3.4.1 Trignometric representation :

One commonly used representation is by the trigonometric representation,

$$v(t) = V_m \sin(\omega t + \theta) \quad \text{....(3.11)}$$

This representation is cumbersome as it is difficult to add these quantities.

For example if

$$v_1(t) = V_{m1} \sin(\omega t + \theta_1)$$

and $$v_2(t) = V_{m2} \sin(\omega t + \theta_2),$$

it is difficult to find $v_1(t) + v_2(t)$, and express it in the same general form as in eq. (3.11).

Hence a representation called phasor representation is developed in the next section.

3.4.2 Phasor representation :

Consider an alternating current quantity i(t) represented by

$$i(t) = I_{max} \sin(\omega t + \theta) \qquad .....(3.12)$$

The instantaneous value of the quantity at $t = 0$ is $I_{max} \sin\theta$ and assumes different positive and negative values for $t > 0$. This alternating variation of the quantity at a constant angular velocity ω (or constant frequency f) can also be described using the two parameters I_{max} and θ. An oriented line OP whose length is I_{max} and positioned at an angle θ with respect to x – axis at time $t = 0$, is assumed to rotate in the counter clockwise direction at an angular velocity 'ω', as shown in Fig. 3.6.

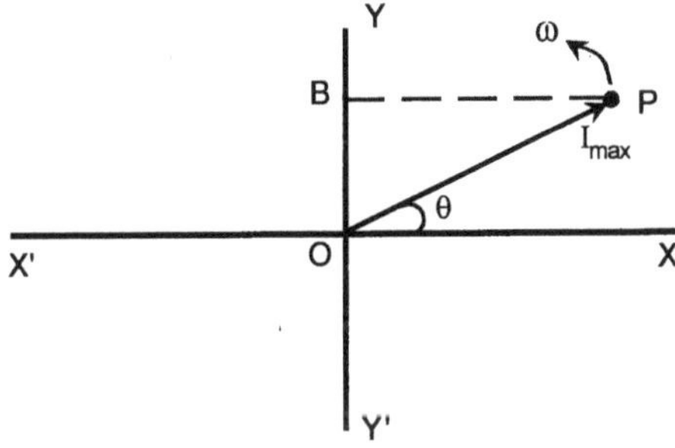

Fig. 3.6 The concept of a phasor

The instantaneous value at time 't' will be the projection OB of the line OP on the y – axis.

Thus a time varying sinusoidal quantity is represented by a rotating vector OP. Such a rotating vector is called a phasor. Though the most appropriate term is *'phasor'*, the term *'vector'* is used more commonly.

Since we would like to specify the sinusoidal quantities by their effective values, as these can be measured easily, and there is a fixed relationship between the maximum values and RMS values, the length of the vector is taken to be I_{rms} rather than I_m. The arrow indicating the direction of rotation and angular velocity ω is often omitted in the vector diagrams.

Let two currents be given by

$$i_1(t) = I_{m1} \sin(\omega t + \theta_1)$$

$$i_2(t) = I_{m2} \sin(\omega t + \theta_2)$$

These currents can be represented by two phasors in the phasor diagram (or vector diagram) shown in Fig. 3.7.

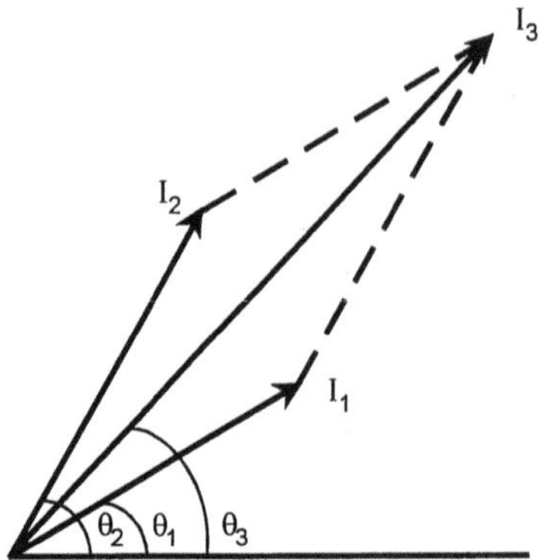

Fig. 3.7 Vector diagram showing addition of two vectors

When we wish to add these currents it may appear difficult because these vectors are rotating and hence are not fixed. But considering the fact that all the phasors are rotating with the same angular velocity 'ω', they appear to be stationary to a person rotating with the same angular velocity. Thus their relative positions do not change with time. We can consider the positions of the vectors at a reference time $t = 0$ and manipulate these vectors. Thus using the rules of vector addition, we can get the sum of the two vectors I_1 and I_2 as shown in Fig. 3.7. The magnitude I_3 and angle θ_3 can be easily found by graphical construction.

The sum of the two currents is therefore given by

$$i_3(t) = I_{m3}\sin(\omega t + \theta_3)$$

where $$I_{m3} = \sqrt{2}\, I_3$$

Phase difference :

It can be seen that phase angle θ_1 and θ_2 of the vectors I_1 and I_2 in Fig. 3.7 are measured with respect to the horizontal reference axis. But, one can measure the phase angle of I_2 with respect to I_1. This is $(\theta_2 - \theta_1)$ and is called phase difference between I_1 and I_2 with I_1 lagging I_2.

3.4.3 Complex algebra representation :

The vector representation is also not very convenient because, the additions and subtractions have to be done by geometry. Since a vector can be represented by its magnitude and angle, the sinusoidal quantity can also be represented as

$$\mathbf{I} = I\angle\theta \qquad .....(3.13)$$

Eq. (3.13) is called polar representation of the vector. Any vector OP, as in Fig. 3.6, can be resolved into two components : one, along the x-axis given by OP cosθ and second, along the y-axis, given by OP Sinθ. In terms of these components, the sum of two vector can be obtained by adding the x-components and y-components respectively of the two vectors.

In Fig. 3.7, x component of I_3 is given by :

$$I_3 \cos\theta_3 = I_1 \cos\theta_1 + I_2 \cos \theta_2$$

Similarly y component of I_3 is given by :

$$I_3 \sin\theta_3 = I_1 \sin\theta_1 + I_2 \sin\theta_2$$

Thus the components of the resultant current I can be obtained without actually constructing the parallelogram as shown in Fig. 3.7. But a concise way of representing the vector in terms of its x and y components has to be developed. This is achieved by distinguishing the y – component from x – component by multiplying it with an operator called 'j' and adding it to the x – component. Thus, we write

$$\mathbf{I_1} = I_1 \cos\theta_1 + jI_1 \sin\theta_1 \quad .....(3.14)$$

$$\mathbf{I_2} = I_2 \cos\theta_2 + jI_2 \sin\theta_2 \quad .....(3.15)$$

We get

$$\mathbf{I_3} = (I_1 \cos\theta_1 + I_2 \cos\theta_2) + j(I_1 \sin\theta_1 + I_2 \sin\theta_2)$$

Thus the addition can be accomplished very easily.

Interpretation of j – operator :

Consider a vector OA along the x – axis as shown in Fig. 3.8.

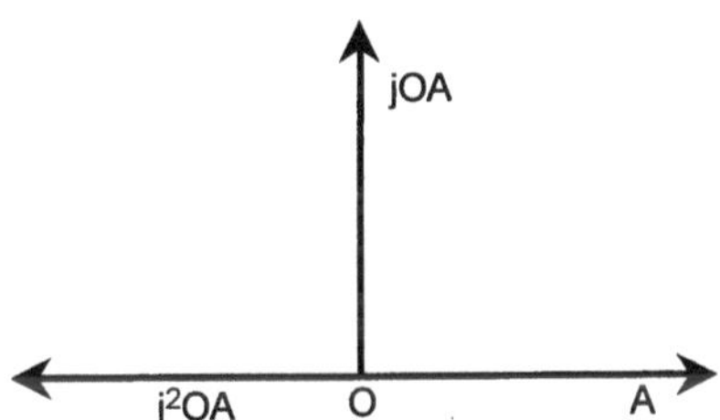

Fig. 3.8 Interpretation of j – operator

Let the vector OA be rotated by 90^0 in the counter clockwise direction. We shall denote this operation in terms of an operator called j. Whenever a vector is multiplied by the operator 'j', it results in the rotation of the vector 90^0 in counter clockwise direction. Thus in Fig. 3.8 the vector in the y – direction is j OA. If we further rotate it by 90^0 in anticlockwise direction, we get j^2 OA and the resulting vector is equal in magnitude to OA and opposite in sign. Thus, we have

$$j^2\, OA = -\, OA$$

or $$j^2 = -1$$

$$\therefore \quad j = \sqrt{-1} \quad .....(3.16)$$

Hence the 'j' operator is none other than the 'i' operator used in representing complex numbers. Hence complex algebra rules can be used to add, subtract, multiply or divide the sinusoidal quantities. The representation given in eq. (3.13) or (3.14) is called complex algebra representation. Eq. (3.13) is called polar form and eq. (3.14) is known as rectangular form. It is to be noted that 'j' is preferred here instead of 'i' used in complex algebra, since 'i' is used for current in electrical engineering. We also note that the rectangular form is more suitable when we want to add or subtract vector quantities. The multiplication or division can be done easily by using the polar form representation of the sinusoidal quantities.

Further, any complex number can also be represented in exponential form, given by

$$\mathbf{I}\angle\theta = Ie^{j\theta}$$
$$= I\cos\theta + j\, I\sin\theta$$

Thus any sinusoidal quantity can also be represented by an exponential function.

In summary, a sinusoidal quantity given by

$$v(t) = V_m \sin(\omega t + \theta)$$

can be represented by any of the following forms :

(a) *Vector representation :*

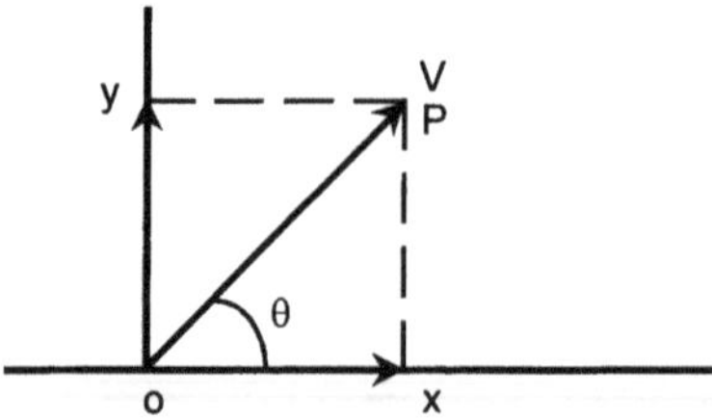

Fig. 3.9 Different representations of sine function

(b) *Complex algebra representation :*

(i) $\mathbf{V}\angle\theta$ (polar form)

(ii) $x + jy$ (rectangular form)

(iii) $\mathbf{V}e^{j\theta}$ (exponential form)

Example 3.3

Add two currents given by

$$i_1(t) = 10\sqrt{2}\,\sin(2t + 45)$$
$$i_2(t) = 5\sqrt{2}\,\sin(2t - 45)$$

Solution :

The two currents can be represented in complex algebra representation as (using RMS values)

$$\mathbf{I}_1 = 10\angle 45^0$$
$$= 7.07 + j\,7.07 \text{ A}$$

and $$\mathbf{I}_2 = 5\angle -45^0 \text{ A}$$
$$= 3.535 - j\,3.535 \text{ A}$$

$$\mathbf{I} = \mathbf{I}_1 + \mathbf{I}_2$$
$$= 7.07 + j\,7.07 + 3.535 - j\,3.535$$
$$= 10.605 + j\,3.535$$
$$= 11.179\angle 18.44^0 \text{ A}$$

The current i(t) is given by

$$i(t) = 11.179\sqrt{2}\ \sin(2t + 18.44^0)$$

Example 3.4

(a) Multiply two phasors $X_1 = 10\angle 30^0$ and $X_2 = 15\angle 45^0$

Solution :

$$X_1\, X_2 = 10\angle 30^0 \times 15\angle 45^0$$

$$= 150\angle 75^0$$

Though multiplication can be performed using rectangular form also, it is more tedious.

(b) Divide $X_1 = 4 + j\,3$ by $X_2 = 1 - j\,1$

Solution :

Using rectangular coordinates

$$\frac{X_1}{X_2} = \frac{4+j3}{1-j1} = \frac{(4+j3)(1+j1)}{2} = 0.5 + j\,3.5$$

The process in which the numerator and denominator are multiplied by 1 + j1 i.e., conjugate of 1 – j1 is called *rationalization.*

3.5 STEADY STATE RESPONSE OF R, L AND C ELEMENTS TO SINUSOIDAL EXCITATIONS :

The steady state response of networks containing R, L and C elements and sinusoidal sources can be studied by first understanding the nature of the response of individual elements R, L and C to sinusoidal inputs. The complex algebra representation of the sinusoidal functions will be used to simplify the analysis of these networks.

3.5.1 Response of pure resistance :

Consider a pure resistance supplied with a sinusoidal voltage as shown in Fig. 3.10.

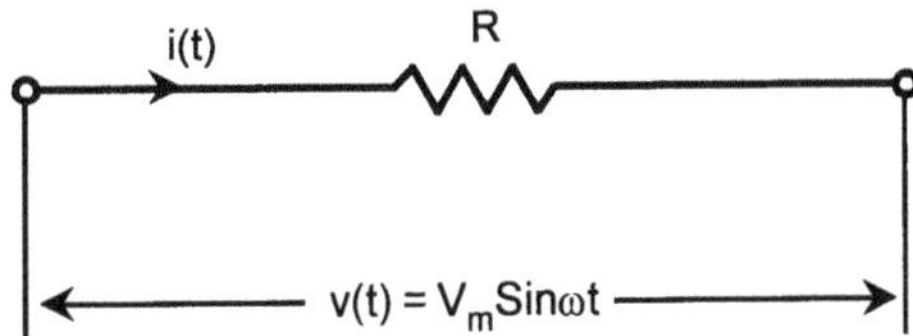

Fig. 3.10 Pure resistance

$$v(t) = V_m \sin\omega t \qquad .....(3.17)$$

By Ohm's law

$$i(t) = \frac{v(t)}{R} = \frac{V_m}{R} \sin\omega t \qquad .....(3.18)$$

Let $\frac{V_m}{R} = I_m$, then

$$i(t) = I_m \sin\omega t \qquad .....(3.19)$$

Since $I_m = \frac{V_m}{R}$

$$\frac{I_m}{\sqrt{2}} = \frac{V_m}{\sqrt{2}R}$$

$$\therefore \quad I = \frac{V}{R} \qquad .....(3.20)$$

Eq. (3.20) shows that the R.M.S values of voltage and current also obey Ohms law. In phasor representation.

$$\mathbf{V} = V \angle 0^0$$

$$\mathbf{I} = \frac{V}{R} \angle 0^0$$

Thus the resistance in Fig. 3.10 can be redrawn using phasor representation as shown in Fig. 3.11.

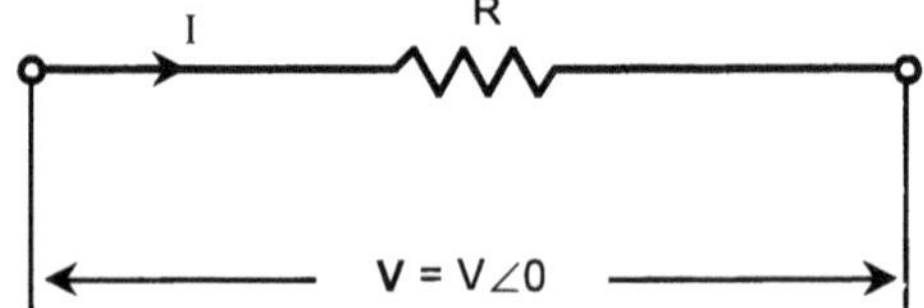

Fig. 3.11 The resistance with phasor representation of current and voltage

To understand the behaviour of any a.c circuit, it will be useful to draw a phasor (vector) diagram. A phasor diagram is one in which all the currents and voltages are shown in a single diagram. Usually one of the quantities is taken as reference and is drawn along the x – axis. In Fig. 3.10, since

$$v(t) = V_m \sin\omega t$$

and $\mathbf{V} = V \angle 0^0$

this vector is taken along the x – axis and is called the reference vector. Since current in the resistance also is given by

$$i(t) = I_m \sin\omega t$$

or $$\mathbf{I} = I\angle 0,$$

this is also shown along the x – axis only.

Fig. 3.12. Vector diagram for resistance

It is to be understood that it is not necessary to draw the diagram to scale and relative magnitudes of voltages and currents are unrelated. More often, when two or more quantities are in the same direction, instead of drawing two separate lines, different arrow marks are shown on the same line as in Fig. 3.13.

Fig. 3.13 Alternate method of showing V and I

Power : Power consumed by the resistor also varies with time when an alternating current is applied to a resistance. Since power is a scalar quantity we measure the power in terms of its average value over a period. If we denote the instantaneous power by p(t), the average power is denoted by 'P'. In a resistance,

$$\begin{aligned} p(t) &= (V_m \sin\omega t)(I_m \sin\omega t) \\ &= V_m I_m \sin^2 \omega t \\ &= \frac{V_m I_m}{2}(1 - \cos 2\,\omega t) \\ &= VI - VI\cos 2\,\omega t \end{aligned} \qquad .....(3.21)$$

The average value of p(t) over a period is equal to VI since average value of cos2 ωt is equal to zero. Thus

$$P = VI \qquad .....(3.22)$$

Using eq. (3.20) we get

$$P = \frac{V^2}{R} = I^2R \qquad .....(3.23)$$

Thus the power in a resistance is given by the same expression as in D.C circuits if we use RMS values of current and voltage. This power is actually consumed by the resistance and is called the active power. The voltage, current and power waveforms are given in Fig. 3.14.

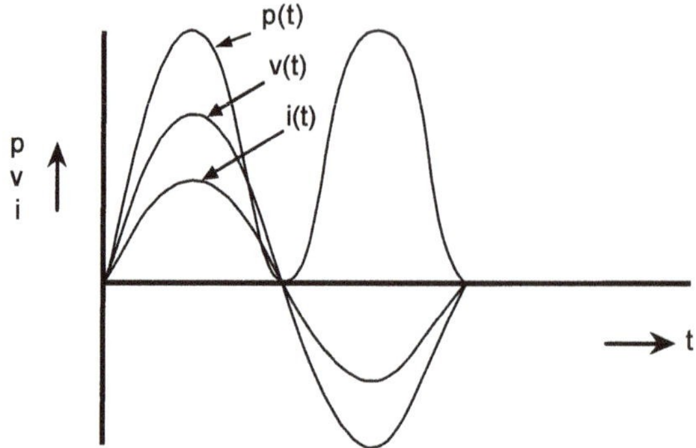

Fig. 3.14 Voltage, Current and Power waveforms in a resistance

It is to be observed that the frequency of power waveform is twice that of either current or voltage. This is obvious from eq. (3.21) and Fig. 3.14.

In summary, we can say that the current and voltage in a resistance are in phase with each other and the RMS values satisfy Ohm's law.

3.5.2 Response of pure inductance :

Consider a pure inductance of L henries supplied with a current source $i(t) = I_m \sin\omega t$ as shown in Fig. 3.15.

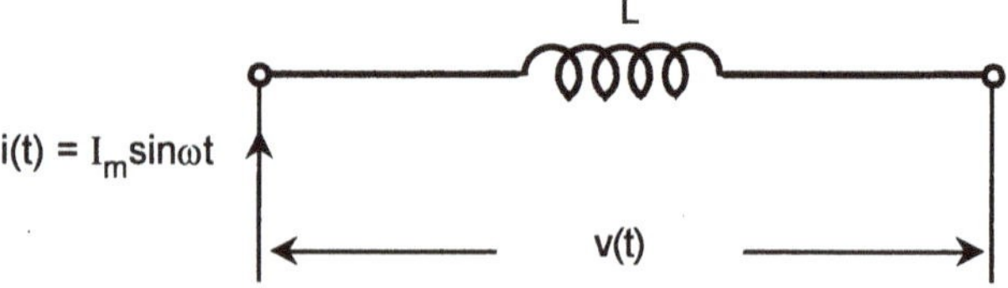

Fig. 3.15 Inductance supplied from a sinusoidal current source

Since $\qquad i(t) = I_m \sin\omega t \qquad$ .....(3.24)

$$v(t) = L\frac{di}{dt}$$

$$= LI_m\omega \cos\omega t \qquad .....(3.25)$$

Writing $\omega L I_m$ as V_m and expressing $\cos\omega t$ in terms of $\sin\omega t$, we have

$$v(t) = V_m \sin(\omega t + 90^0) \qquad .....(3.26)$$

Since $V_m = \omega L \, I_m$

$$V = \omega L \, I \qquad .....(3.27)$$

Comparing eq. (3.27) with the Ohm's law equation

$$V = I R$$

We observe that when we multiply L with ω, we get a quantity similar to the resistance. This quantity ωL is called inductive reactance and is represented by X_L. Its unit is ohm. Thus,

$$V = I X_L \qquad .....(3.28)$$

Further, from eq. (3.24) and eq. (3.26), we observe that the voltage leads the current by 90^0. Using the phasor notation,

$$I = I \angle 0^0$$

and

$$V = I X_L \angle 90^0 \qquad .....(3.29)$$

Since $1 \angle 90^0$ represents a unit vector rotated by 90^0 in the counter clockwise direction, it can be replaced by the operator 'j'. Using 'j' operator in eq. (3.29), we have

$$V = j X_L I \qquad .....(3.30)$$

Eq. (3.30) resembles the familiar ohms law equation and, the quantity $j X_L$ or $X \angle 90^0$, which is similar to a resistance, is termed as impedance, denoted by Z.

Thus

$$Z = j X_L = X_L \angle 90^0 \qquad .....(3.30)$$

Z is also measured in ohms.

In the light of the above observations we can write eq. (3.30) as

$$V = ZI \qquad .....(3.31)$$

Eq. (3.31) suggests that if we replace the inductance value by $j\omega L$ and use phasors V and I, we can apply ohms law as in D.C circuits. Fig. 3.15 can be modified as shown in Fig. 3.16 which is called as frequency domain representation of the inductance.

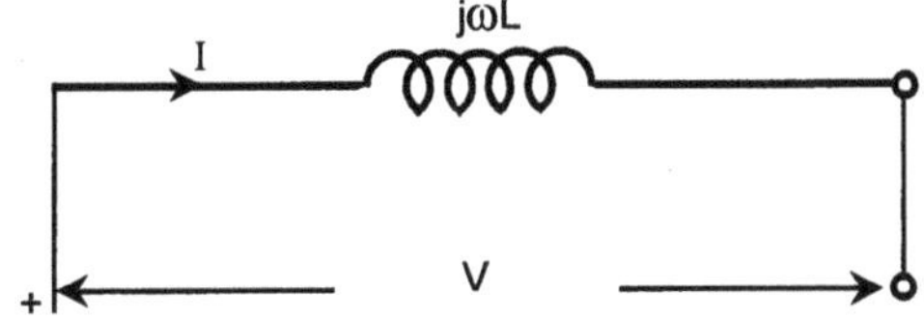

Fig. 3.16 Frequency domain representation of inductance

The vector diagram for inductance is given in Fig. 3.17.

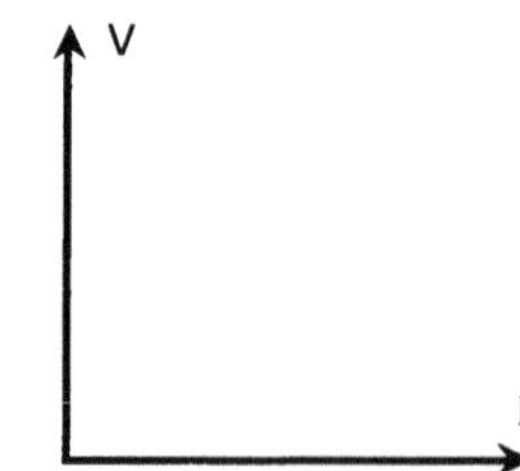

Fig. 3.17 Vector diagram for inductance

If, instead of taking current as $i = I_m \sin\omega t$, had we taken $v = V_m \sin\omega t$, we would have had

$$V = V\angle 0^0$$

$$I = \frac{V\angle 0^0}{j\omega L} = \frac{V}{\omega L}\angle -90^0$$

The vector diagram would be as given in Fig. 3.18. It is customary to say that current in an inductance lags behind the voltage by 90^0.

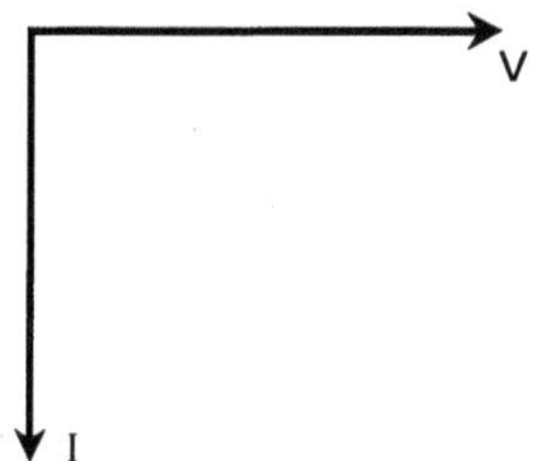

Fig. 3.18. Vector diagram with V as reference

Since we are interested only in relative positions of the vectors at a given instant of time, it is immaterial which vector we choose as reference. It will be convenient to choose current as reference in series connected elements as current is common for all the elements. Similarly if the elements are connected in parallel, it is convenient to choose voltage as reference.

Power : Now coming to power consumed by an inductance, we have

$$p = iv$$
$$= I_m \sin\omega t \cdot V_m \cos\omega t$$
$$= \frac{V_m I_m}{2} \sin 2\omega t$$
$$= VI \sin 2\omega t \qquad .....(3.32)$$

The average power P is zero because the average value of sin2 ωt is zero over a period. Thus inductance doesnot consume any power. The voltage, current and power waveforms are shown in Fig. 3.19. The frequency of power waveform is twice that of current or voltage, as in the case of a pure resistance.

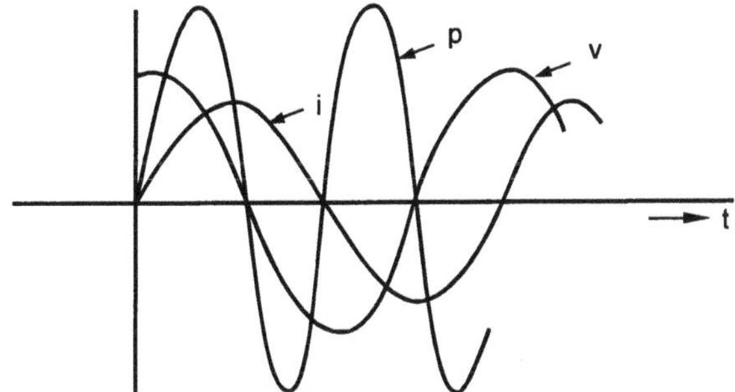

Fig. 3.19 Voltage, Current and Power waveforms in a pure inductor

When the current is increasing, the magnetic field is built up in the inductance and energy is taken from the source and stored in the magnetic field, making the power positive. When the current is decreasing, the magnetic field collapses and the energy is given back to the source, making the power negative. The average over one cycle is thus zero. Whatever power is drawn from the source is given back to the source. The amplitude VI of power waveform is similar to the power consumed by a resistance and is called reactive power and is measured in volt ampere reactive or in short VAR (or Vars). It is represented by Q and is given by

$$Q = VI$$

$$= \frac{V^2}{X_L} = I^2 X_L \text{ Vars} \qquad .....(3.33)$$

3.5.3 Response of a pure capacitor :

Let a voltage v (t) = V_m sinωt be applied to a pure capacitor of 'C' farads as shown in Fig. 3.20.

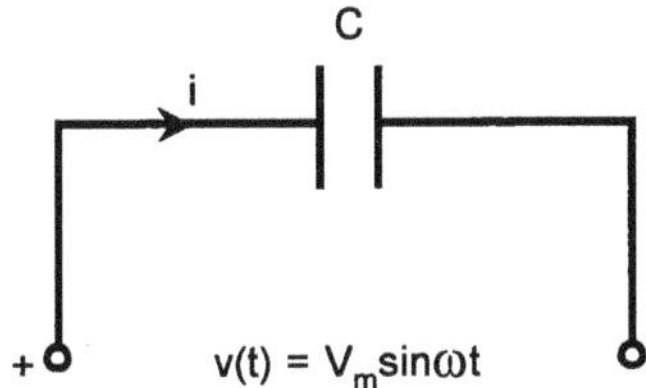

Fig. 3.20 Response of a pure capacitor

The current i(t) is given by

$$i(t) = C\frac{dv}{dt} \qquad .....(3.34)$$

$$i(t) = CV_m\omega \cos\omega t$$

$$= \frac{V_m}{\frac{1}{\omega C}} \sin(\omega t + 90^0) = I_m \sin(\omega t + 90^0) \qquad .....(3.35)$$

where $$I_m = \frac{V_m}{\frac{1}{\omega C}}$$

or $$I = \frac{V}{\frac{1}{\omega C}} \qquad .....(3.36)$$

Since this is similar to ohms law equation, the quantity $\frac{1}{\omega C}$ is similar to a resistance. $\frac{1}{\omega C}$ is called reactance and is also measured in ohms. It is denoted by X_C. Thus,

$$I = \frac{V}{X_C} \qquad .....(3.37)$$

From eq. (3.35), it can be observed that the current through a capacitor leads the voltage by 90^0. Using phasor representation

$$V = V \angle 0^0$$

$$I = \frac{V}{\frac{1}{\omega C}} \angle 90^0$$

$$= \frac{jV}{\frac{1}{\omega C}} = \frac{V}{\left(\frac{1}{j\omega C}\right)} \qquad .....(3.38)$$

As in the case of inductance, the quantity $\frac{1}{j\omega C}$ is called the inpedance and is denoted by Z. Thus the impedance Z is given by

$$Z = \frac{1}{j\omega C} = -j X_C \qquad .....(3.39)$$

In terms of impedance, eq. (3.38) can be written as

$$I = \frac{V}{Z} \qquad .....(3.40)$$

This is similar to Ohm's law and therefore in frequency domain, the capacitor can be represented as shown in Fig. 3.21.

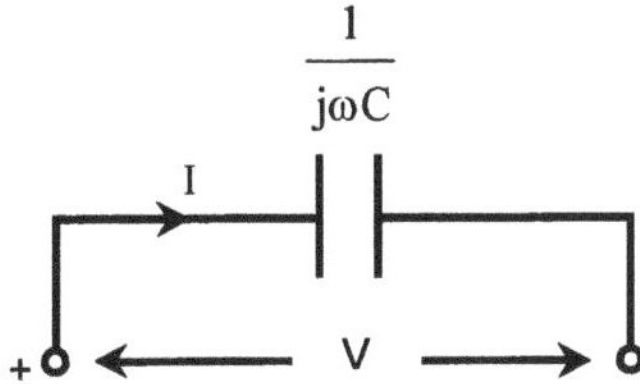

Fig. 3.21 Frequency domain representation of a capacitor

The vector diagram is shown in Fig. 3.22.

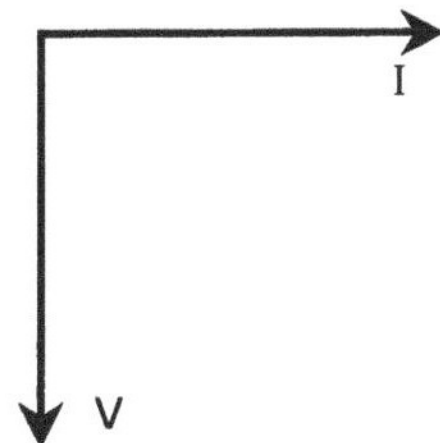

Fig. 3.22 Vector diagram for a capacitor

Power : The instantaneous power is given by

$$p = V_m \sin\omega t \cdot I_m \cos\omega t$$

$$= \frac{V_m I_m}{2} \sin 2\omega t$$

$$= VI \sin 2\omega t \qquad .....(3.41)$$

The average power P is equal to zero. The instantaneous power has double the frequency as that of voltage or current. The waveforms of voltage, current and power are given in Fig. 3.23.

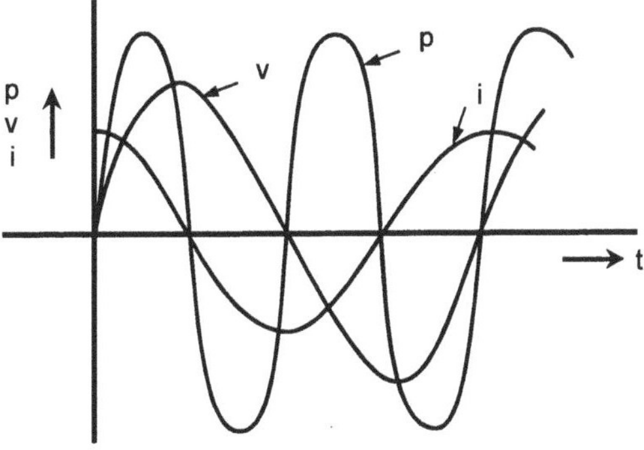

Fig. 3.23 v, i and p waveforms in a capacitor

From Fig. (3.23) we observe that the power is positive when the voltage 'v' is increasing in the first quarter cycle. The capacitor takes power from the source and stores this energy in electrostatic field. When the voltage decreases in the next quarter cycle, the electrostatic field collapses and the energy is pumped back to the source. Thus the average power is zero. The amplitude of the power waveform is VI and is called the reactive power. Thus

$$Q = VI = I^2 X_C = \frac{V^2}{X_C} \quad \text{VARS} \qquad .....(3.42)$$

Thus the inductance and capacitance are only energy storing elements and do not consume any active power. The elements L and C are called reactive elements. When a purely reactive circuit is supplied with a sinusoidal source, no useful power is given by the source, but it must have the capacity to supply power so that energy can be stored and retrieved from the reactive elements.

If the power generating companies charge only for the power actually consumed by a consumer, it will be a loss making proposition for the company. If a consumer has a purely reactive load, the company has to still maintain its infrastructure, but will not be getting any returns for its investment! Thus reactive power is detrimental to the system.

Example 3.5

Calculate the inductive reactance of a 100 mH inductor connected to a 230 V, 50 Hz sinusoidal supply. If the frequency is increased to 5 KHz what will be its reactance?

Solution :

When the supply frequency is 50 Hz

$$X_L = 2\pi f L$$
$$= 2\pi \times 50 \times 100 \times 10^{-3}$$
$$= 31.4 \text{ ohms}$$

When the frequency is 5 KHz,

$$X_L = 2\pi \times 5000 \times 100 \times 10^{-3}$$
$$= 3140 \text{ ohms}$$

Example 3.6

A voltage $v(t) = 100\sqrt{2}\ \sin 377\ t$ is applied to a pure capacitor of 530 μF. Write an expression for i(t).

Solution :

The voltage applied

$$v(t) = 100\sqrt{2}\ \sin 377\ t$$

In phasor form,

$$V = 100\angle 0$$

The current is given by

$$I = \frac{V}{Z}$$

where $$Z = \frac{1}{j\omega C}$$

$$= \frac{-j}{377 \times 530 \times 10^{-6}}$$

$$= -j5 \text{ ohms}$$

$\therefore$ $$I = \frac{100\angle 0^0}{5\angle -90^0}$$

$$= 20\angle 90^0$$

Therefore the current i(t) is given by

$$i(t) = 20\sqrt{2}\ \sin(\omega t + 90^0)$$

3.5.4 Series R – L circuit :

Consider a series R – L circuit shown in Fig. 3.24.

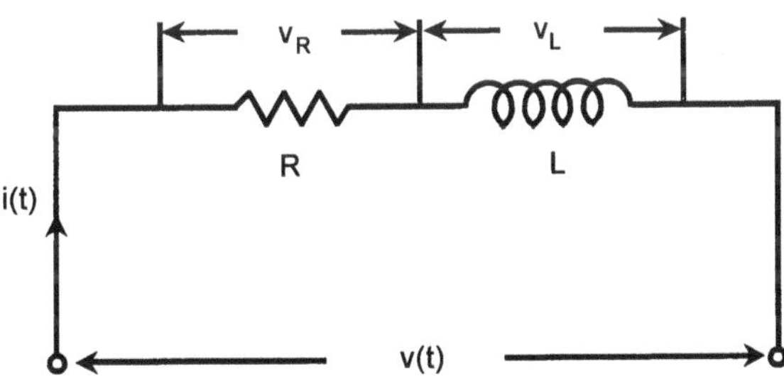

Fig. 3.24 An R – L series circuit

Let the current be

$$i(t) = I_m \sin\omega t$$

or $$\mathbf{I} = I\angle 0^0 \qquad .....(3.43)$$

Since this current is flowing through the resistance R, we have

$$\mathbf{V_R} = IR\angle 0^0 \qquad .....(3.44)$$

The voltage across L is given by

$$\mathbf{V_L} = I\,X_L\ \angle 90^0 \qquad .....(3.45)$$

$$= j\,X_L\,I$$

By Kirchhoff's voltage law,

$$v(t) = v_R(t) + v_L(t)$$

Since, these quantities are all sinusoidal quantities they can be represented by vectors and complex addition can be performed.

Thus
$$\begin{aligned} V &= IR\angle 0 + IX\angle 90^0 \\ &= IR + jX_L I \\ &= I(R + jX_L) \\ &= \mathbf{IZ} \end{aligned} \qquad .....(3.46)$$

where **Z** is called the impedance and is a complex quantity. The real part of the impedance is the resistance and the imaginary part is the reactance.

Therefore

$$\begin{aligned} \mathbf{Z} &= R + jX_L \\ &= \sqrt{R^2 + X_L^2}\angle \tan^{-1}\frac{X_L}{R} \\ &= |Z|\angle\theta^0 \end{aligned}$$

The magnitude of the impedance is given by

$$|Z| = \sqrt{R^2 + X_L^2} = \sqrt{R^2 + \omega^2 L^2}$$

and the impedance angle $\theta = \tan^{-1}\dfrac{X}{R} = \tan^{-1}\dfrac{\omega L}{R}$.

Thus $\qquad \mathbf{V} = I|Z|\angle\theta^0$

The voltage in time domain can be written as

$$\begin{aligned} v(t) &= I|Z|\sqrt{2}\,\sin(\omega t + \theta) \\ &= V_m \sin(\omega t + \theta) \end{aligned}$$

The phasor diagram is shown in Fig. 3.25.

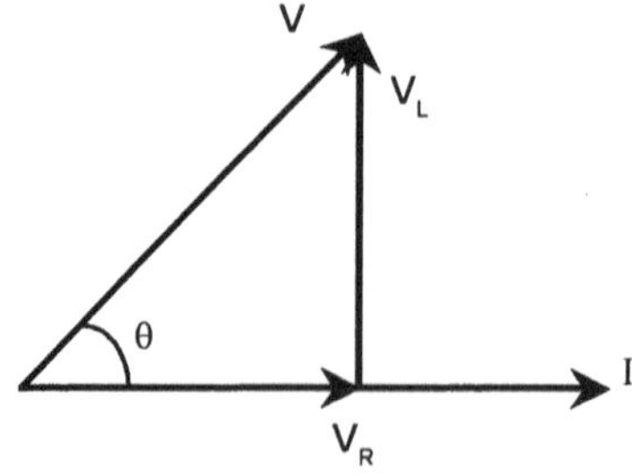

Fig. 3.25 Phasor diagram of RL circuit

The frequency domain equivalent of R – L circuit is given in Fig. 3.26.

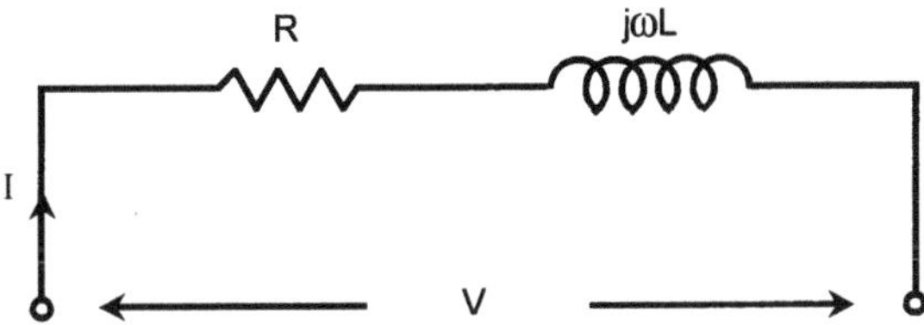

Fig. 3.26 Frequency domain equivalent of a series RL circuit

When the elements are represented by their frequency domain equivalents and voltages and currents are given by phasors, the circuit behaves like a resistive circuit and the currents and voltages can be obtained by ohms law. Thus in Fig. (3.26) if voltage V is taken as reference i.e.,

$$v(t) = V_m \sin\omega t$$

Then $$I = \frac{V\angle 0^0}{R + j\omega L} = \frac{V}{\sqrt{R^2 + \omega^2 L^2}} \angle -\tan^{-1}\frac{\omega L}{R}$$

In time domain I can be written as

$$i(t) = \frac{V\sqrt{2}}{\sqrt{R^2 + \omega^2 L^2}} \sin\left(\omega t - \tan^{-1}\frac{\omega L}{R}\right)$$

From now onwards we will use only phasors to represent voltages and currents and replace the elements R, L and C by their impedances. As there is a one to one correspondence between time domain quantities and frequency domain quantities, it is always possible to get one form from the other.

Power :

The instantaneous power is given by

$$P = v \cdot i$$

$$= V_m \sin\omega t \cdot I_m \sin(\omega t - \theta)$$

$$= \frac{V_m I_m}{2}[\cos\theta - \cos(2\omega t - \theta)]$$

$$= VI\cos\theta - VI\cos(2\,\omega t - \theta) \qquad .....(3.47)$$

The average value of eq. (3.47) is equal to VI cosθ as this term is a constant and the second term in eq. (3.47) contributes zero value for the average of P.

Thus $$P = VI\cos\theta \qquad .....(3.48)$$

where V and I are the rms values of the voltage and current and θ is the angle between the voltage and current vectors. θ is also the impedance angle.

This is a very important expression. The power is not simply given by the product of R M S values of voltage and current, but is multiplied by the factor cosθ. This factor is known as *'power factor'* of the circuit. If a d .c. voltage of V volts produces a d .c. current of I amps in a circuit, the power would be VI. But same values of voltage and current in an a .c. circuit produce less power and is given by VI cosθ expressed in watts. We can get alternate expression for the average power in terms of the resistance in the circuit. We have from eq. (3.48).

$$P = VI\cos\theta$$

But $$|V| = |I||Z|$$

$\therefore$ $$P = |I||Z||I|\cos\theta$$

$$= |I^2||Z|\cos\theta$$

But $$|Z|\cos\theta = R$$

$\therefore$ $$P = I^2R \qquad .....(3.49)$$

Since power is consumed only in a resistance and no power is consumed by the inductance, power can be directly calculated using eq. (3.49).

From the vector diagram of Fig. 3.25 we observe the vectors V_R, V_L and V form a triangle, which is shown in Fig. 3.27(a).

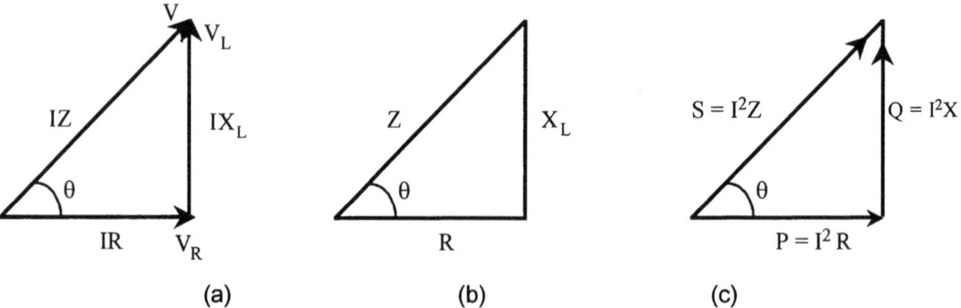

Fig. 3.27 (a) Voltage triangle (b) Impedance triangle and (c) Power triangle

The triangle formed by the voltage vectors V_R, V_L and V is known as a voltage triangle. Each side of this triangle is proportional to the current I. Hence if we divide each side of this triangle by I, we get a similar triangle known as impedance triangle as shown in Fig. 3.27 (b). The base of the right angled triangle represents the resistance, the altitude represents the reactance and the hypotenuse represents the impedance. The angle 'θ' is known as the impedance angle and is given by

$$\theta = \tan^{-1} \frac{\omega L}{R}$$

and $$\cos\theta = \frac{R}{Z} \qquad .....(3.50)$$

Further if the sides of the voltage triangle are multiplied by I, we get another similar triangle called power triangle. The base of this triangle is the real power $P = I^2R$ and the vertical side is the reactive power $Q = I^2X$. The hypotenuse is called apparent power and is denoted by S. Its unit is volt amperes. Thus

$$S = I^2Z$$

$$= I \cdot \frac{V}{Z} \cdot Z = VI \qquad .....(3.51)$$

Hence apparent power is the power that would have been consumed in the circuit if D.C voltage and current of same values are considered in the circuit.

The angle between the voltage applied and current produced is the same as the impedance angle. The apparent power is also called as the vector power and is denoted as

$$S = P + jQ \qquad .....(3.52)$$

If voltage and current are given in complex form as $\mathbf{V} = V\angle 0^0$ and $\mathbf{I} = I\angle -\theta^0$; the complex or vector power S is also given by

$$S = VI^*$$

where * stands for complex conjugate. Thus

$$S = VI\cos\theta + j\,VI\sin\theta.$$

The quantity VI sinθ is the reactive power and is denoted be 'Q'. Further the reactive power 'Q' is given by

$$Q = I^2X = VI\sin\theta \qquad .....(3.53)$$

The power factor can be expressed by any of the following expressions

(i) $$\cos\theta = \frac{R}{Z} \qquad .....(3.54)$$

(ii) $$\cos\theta = \frac{V_R}{V} \qquad .....(3.55)$$

(iii) $$\cos\theta = \frac{P}{S} = \frac{\text{Active power}}{\text{Apparant power}} \qquad .....(3.56)$$

3.5.5 Series RC circuit :

A series RC circuit is shown in Fig. 3.28.

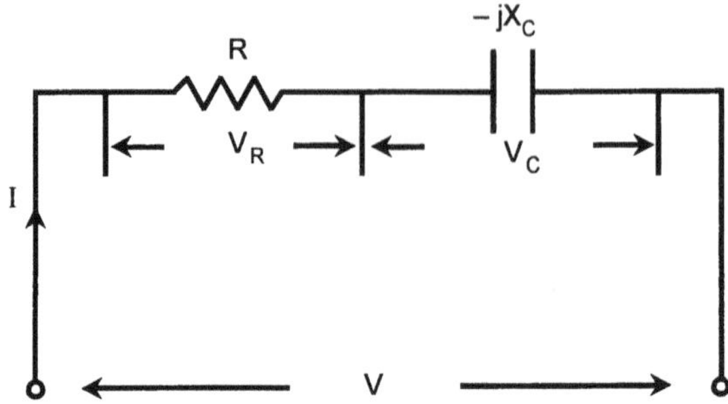

Fig. 3.28 RC circuit

Let $\mathbf{I} = I\angle 0$

$\mathbf{V}_R = IR\angle 0$

$\mathbf{V}_C = -I\,jX_C$

$\mathbf{V} = \mathbf{V}_R + \mathbf{V}_C$

$= IR - jIX_C$

$= I(R - jX_C)$

$= I\mathbf{Z}$ .....(3.57)

where $Z = R - jX_C$

$$|Z| = \sqrt{R^2 + X_C^2}\angle -\tan^{-1}\frac{X_C}{R}$$

$\therefore$ $$V = I\sqrt{R^2 + X_C^2}\angle -\tan^{-1}\frac{X_C}{R}$$ .....(3.58)

$= I|Z|\angle -\theta$

The vector diagram is shown in Fig. 3.29.

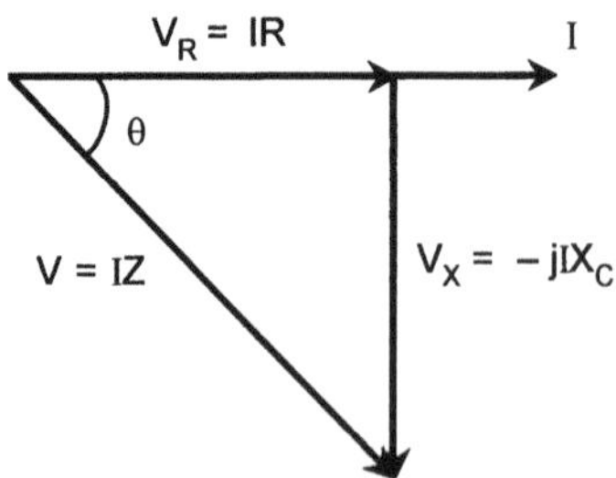

Fig. 3.29 Vector diagram for RC circuit

The voltage triangle, Impedance triangle and the power triangle are given in Fig. 3.30. (a), (b) and (c).

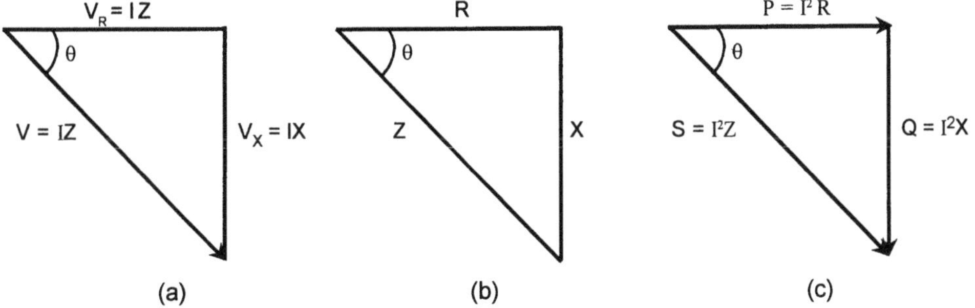

Fig. 3.30 (a) Voltage triangle (b) Impedance triangle and (c) Power triangle for RC circuit

The active power is given by

$$P = VI\cos\theta$$

$$= I^2Z\cos\theta$$

$$= I^2R \text{ watts}$$

$$Q = VI\sin\theta$$

$$= I^2 X \quad \text{VARS} \qquad .....(3.59)$$

and $$S = VI = I^2|Z| \quad \text{volt amperes} \qquad .....(3.60)$$

The complex impedance Z is given by

$$Z = R - \frac{j}{\omega C}$$

$$= \sqrt{R^2 + \frac{1}{\omega^2 C^2}} \angle -\tan^{-1}\frac{1}{\omega CR}$$

The complex power is given by

$$S = P - jQ \qquad .....(3.61)$$

If voltage and current are given in complex form, the complex power is obtained from the equation

$$S = VI^* \qquad .....(3.62)$$

where * stands for complex conjugate. Thus if

$$V = V\angle 0$$

and $$I = I\angle\theta$$

$\therefore$ The complex power S is given by

$$S = VI^*$$

or $S = V\angle 0 \cdot I\angle -\theta_2$

$= VI\ \angle -\theta$

$= VI\cos(\theta) + jVI\sin(-\theta)$

$= VI\cos\theta - j\ VI\sin\theta.$

It can be seen that the reactive power in a capacitive circuit is opposite to that of an inductive circuit.

3.5.6 Series RLC circuit :

A series RLC circuit is shown in Fig. 3.31.

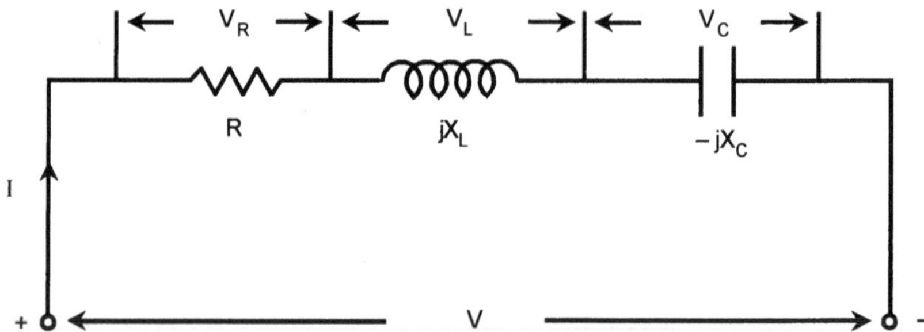

Fig. 3.31 A series RLC circuit

Taking current as reference,

$$\mathbf{I} = I\angle 0^0$$

$$\mathbf{V_R} = IR\angle 0^0 = IR + jO$$

$$\mathbf{V_L} = I\,j\,X_L$$

$$\mathbf{V_C} = -I\,j\,X_C$$

and $\mathbf{V} = \mathbf{V_R} + \mathbf{V_L} + \mathbf{V_C}$

$= IR + j\,I\,X_L - j\,I\,X_C$

$= I(R + j\,(X_L - X_C))$

$= I\mathbf{Z}$

where $\mathbf{Z} = R + j\,(X_L - X_C)$ .....(3.63)

Depending on the values of X_L and X_C we have 3 different cases.

(i) $X_L > X_C$

The inductive reactance is greater than the capacitive reactance. Hence the net reactance is inductive and the circuit behaves like an RL circuit Thus

$$Z = R + jX^1$$

where $X^1 = X_L - X_C$

The vector diagram is shown in Fig. 3.31 (a)

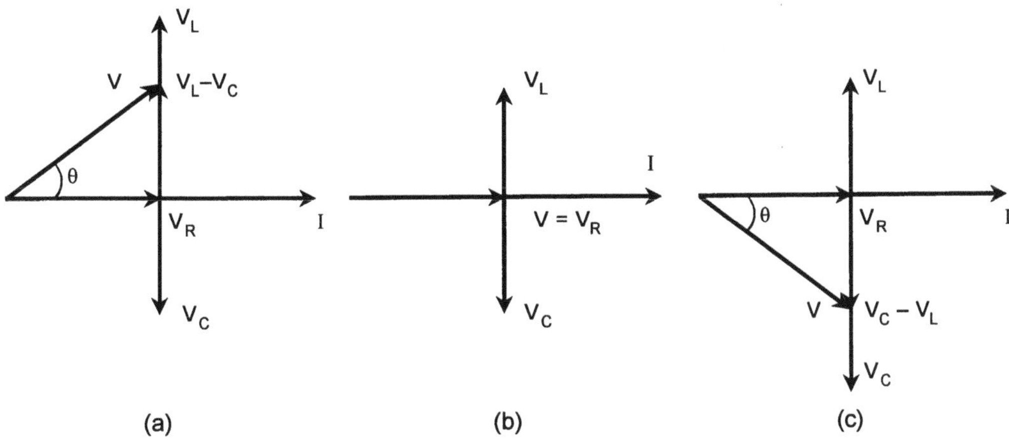

Fig. 3.31 Three different cases in a series RLC circuit (a) $X_L > X_C$ (b) $X_L = X_C$ (c) $X_L < X_C$

The impedance Z is given by

$$Z = \sqrt{R^2 + \left(\omega L - \frac{1}{\omega C}\right)^2} \angle \tan^{-1} \frac{\omega L - \dfrac{1}{\omega C}}{R} \qquad .....(3.64)$$

(ii) $X_L = X_C$

The inductive reactance is cancelled by the capacitive reactance. Hence the impedance is purely resistive and

$$Z = R \qquad .....(3.65)$$

The current will be in phase with the voltage as shown in Fig. 3.31(b). This condition is known as resonance.

The power factor under this condition is cosθ = 1. This is called unity power factor condition. Resonance will be discussed in detail later in this chapter.

(iii) $X_L < X_C$

The circuit behaves like a capacitive circuit and the impedance Z is

$$Z = R - jX^1$$

where $$X^1 = X_C - X_L = \sqrt{R^2 + \left(\omega L - \frac{1}{\omega C}\right)^2} \angle -\tan^{-1} \frac{\left(\dfrac{1}{\omega C} - \omega L\right)}{R} \qquad .....(3.66)$$

The current leads the voltage by an angle

$$\theta = \tan^{-1} \frac{\left(\dfrac{1}{\omega C} - \omega L\right)}{R}, \qquad .....(3.67)$$

as shown in Fig. 3.31 (c).

3.5.7 Parallel RL circuit :

A parallel RL circuit is shown in Fig. 3.32.

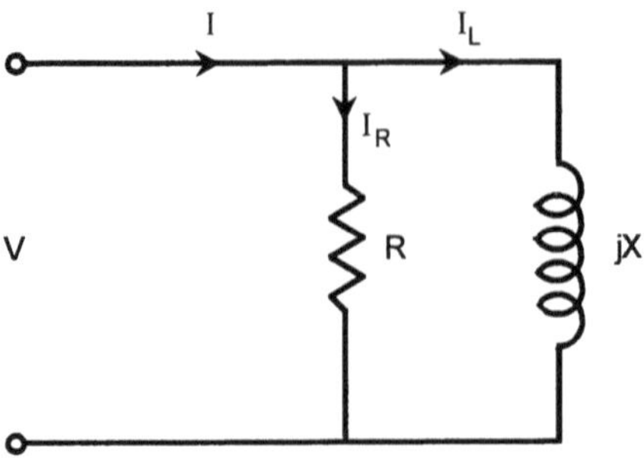

Fig. 3.32 A parallel RL circuit

Since voltage applied is common to the two elements in parallel, we take voltage as reference

$$\mathbf{V} = V\angle 0$$

$$\mathbf{I_R} = \frac{V}{R}\angle 0$$

and
$$\mathbf{I_L} = \frac{V\angle 0}{jX}$$

The total current I is given by

$$\mathbf{I} = \mathbf{I_R} + \mathbf{I_L}$$

$$= \frac{V\angle 0}{R} + \frac{V\angle 0}{jX}$$

$$= V\left(\frac{1}{R} + \frac{1}{jX}\right) \qquad .....(3.68)$$

The reciprocal of resistance is called the conductance and is represented by G. Similarly we can define the reciprocal of reactance as susceptance, B and the reciprocal of impedance as admittance, Y. Eq. (3.68) can be written as

$$I = V\,(G - jB) = VY$$

where Y is the admittance given by

$$Y = G - jB$$

for a parallel RL circuit. The unit of Y is Seimens and is denoted by S.

The vector diagram is given in Fig. 3.33.

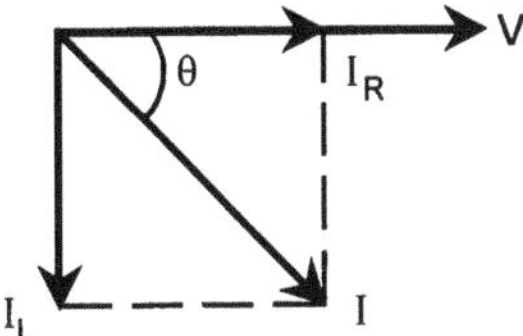

Fig. 3.33 Vector diagram for parallel RL circuit

3.5.8 Parallel R C circuit :

The parallel RC circuit can be dealt with in a similar manner. The circuit is shown in Fig. 3.34(a)

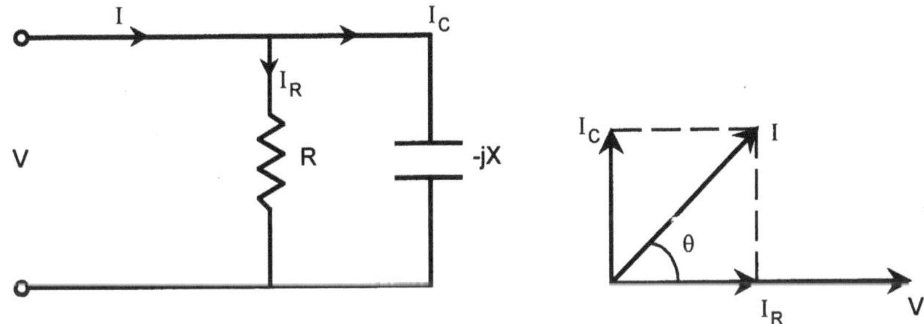

Fig. 3.34 (a) Parallel RC circuit and (b) Its vector diagram

$$\mathbf{V} = V\angle 0$$

$$\mathbf{I}_R = \frac{V}{R}\angle 0$$

$$\mathbf{I}_C = \frac{V\angle 0}{-jX}$$

$$\mathbf{I} = \mathbf{I}_R + \mathbf{I}_C$$

$$= V\left(\frac{1}{R} - \frac{1}{jX}\right)$$

$$= V(G + jB)$$

$$= V\mathbf{Y} \qquad \text{.....(3.69)}$$

where $\quad Y = G + jB \qquad$ .....(3.70)

is the admittance of a parallel RC circuit.

A parallel RLC circuit can be analysed in a similar manner.

Example 3.7

A coil takes a current of 1 A at 0.6 lagging power factor from a 220 volt 60 Hz single phase source. If the coil is modelled by a series RL circuit find (a) the complex power in the coil and (b) the values of R and L.

Solution :

(a) *The complex power* :

$$S = P + jQ$$

where $P = VI\cos\theta$

and $Q = VI\sin\theta$

$\therefore$ $P = 220 \times 1 \times 0.6 = 132$ watts

$Q = 220 \times 1 \times 0.8 = 176$ vars

$\therefore$ $S = 132 + j\,176$ voltamps

(b) *The impedance :*

$$\mathbf{Z} = \frac{V\angle 0}{1\angle -\cos^{-1}0.6} = \frac{220\angle 0}{1\angle -53.13^0} = 220\ \angle 53.13^0$$

$\therefore$ $R = 220\cos 53.13^0 = 132$ ohms

$X = 220\sin 53.13^0 = 176$ ohms

Example 3.8

A series circuit consisting of a 10 ohm resistor, a 100 μf capacitance and a 10 mH inductance is driven by a 50 Hz a.c voltage source of maximum value 100 volts . Calculate the equivalent impedance, current in the circuit, the power factor and power dissipated in the circuit.

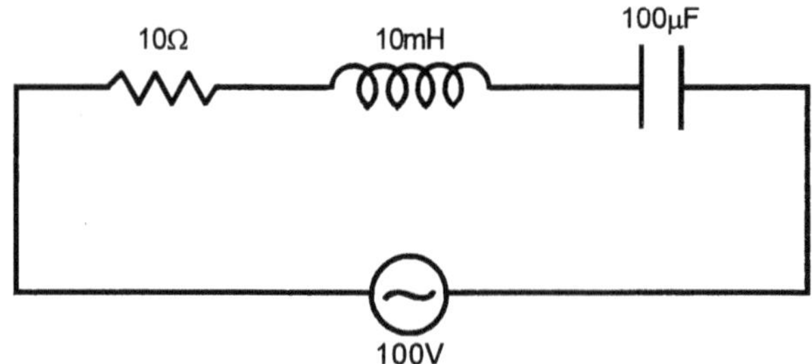

Fig. 3.35 Circuit for example 3.8

Solution :

The inductive reactance

$$X_L = 2\pi\ fL = 2 \times \pi \times 50 \times 10^{-3} \times 10$$
$$= 3.142 \text{ ohms}$$

$$\text{The Capacitive reactance} = \frac{1}{2\pi fC} = \frac{10^6}{2\pi \times 50 \times 100}$$

$$= 31.83 \text{ ohms}$$

$$\text{Resistance} = 10 \text{ ohms}$$

$$\therefore \quad \text{Impedance} = 10 + j(3.142 - 31.83) = 10 - j28.69 \text{ ohms}$$

$$= 30.38 \angle -70.78^0 \text{ ohms}$$

$$\text{Current} = \frac{100\angle 0^0}{30.38\angle -70.78^0}$$

$$= 3.292 \angle 70.78^0 \text{ amps}$$

Power factor of the circuit, cosθ, is given by

$$\text{P.F.} = \cos 70.78^0$$

$$= 0.329 \text{ leading}$$

Example 3.9

A series circuit to which 100 volts is applied, consists of a 10 ohm resistance, a 5 ohm condenser and a resistor R in which 50 watts are lost and a reactance X_L which absorbed a reactive power of 100 vars. Calculate the values of R and X_L that satisfy the stated conditions.

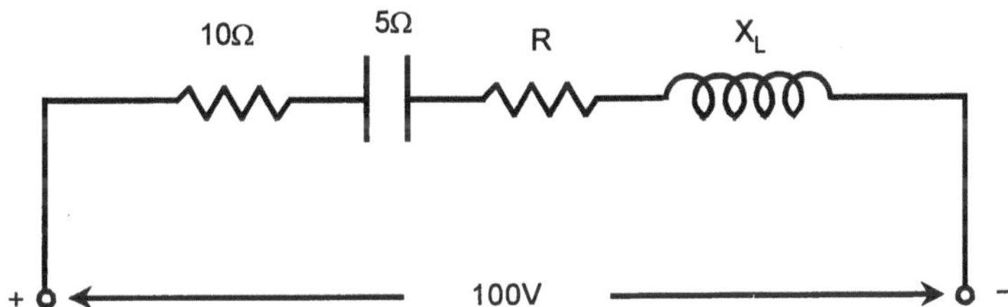

Fig. 3.36 Circuit for example 3.9

Solution :

The total impedance Z of the circuit

$$= (10 + R) + j(X_L - 5) \text{ ohms}$$

$$\therefore \quad |Z| = \sqrt{(10+R)^2 + (X_L - 5)^2} \;\angle \tan^{-1}\frac{X_L - 5}{10+R} \text{ ohms}$$

Voltage applied = 100 volts

$$\therefore \quad \text{Power in R} = \left[\frac{100}{\sqrt{(10+R)^2 + (X_L - 5)^2}}\right]^2 \cdot R$$

But the power in R is given to be 50 W and reactive power in L is given as 100 vars.

$$\therefore \qquad \left[\frac{100}{\sqrt{(10+R^2)+(X_L-5)^2}}\right]^2 \cdot R = 50 \qquad .....(3.71)$$

and $\qquad I^2 \times X_L = 100$

$$\therefore \qquad \left[\frac{100}{\sqrt{(10+R)^2+(X_L-5)^2}}\right]^2 X_L = 100 \qquad .....(3.72)$$

From eqs. (3.71) and (3.72), we have

$$\frac{X_L}{R} = \frac{100}{50} = 2$$

or $\qquad X_L = 2R$

Substituting the value of X_L in eq. (3.71) and solving for R we get

$$R - 39.365 \ \Omega \text{ or } 0.635 \ \Omega$$

and $\qquad X = 78.73 \ \Omega$ or $1.27 \ \Omega$

Example 3.10

A single phase load takes 300 watts and draws 5 amps at a lagging power from a 120 V,1ϕ supply. Determine the reactance of a pure capacitor required to be placed in series with this load so that it takes the same current when connected to a 240 volt supply.

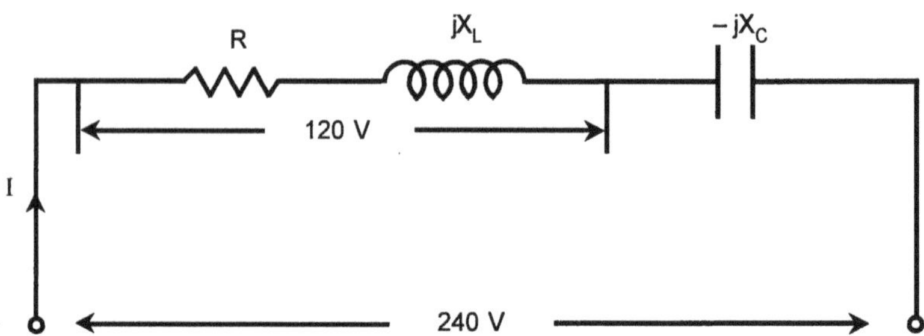

Fig. 3.37 Circuit for example 3.10

Solution :

Case (a) : With no capacitance

power drawn P = 300 watts

current input I = 5 amps

Voltage across the load V = 120 volt

∴ Impedance of the load

$$Z = \frac{V}{I} = \frac{120}{5} = 24\ \Omega$$

Also, resistance of load

$$R = \frac{P}{I^2} = \frac{300}{5^2} = 12\ \Omega$$

But $Z = \sqrt{R^2 + X_L^{\,2}}$

or $Z^2 = R^2 + X_L^{\,2}$

∴ $X_L = \sqrt{Z^2 - R^2} = \sqrt{24^2 - 12^2} = 20.78\ \Omega$

Case (b) : With a series capacitor

When 240 V is applied across the series combination of the load and capacitance as shown in Fig. 3.37, the load should take 5 A of current. Thus total impedance should be

$$Z = \frac{240}{5} = 48\ \Omega$$

But $Z = \sqrt{R^2 + X^2}$

where $X = X_L - X_C$

and $X = \pm\sqrt{Z^2 - R^2}$

Since the impedance of the combination in case (b) is more than that of case (a) $X_C > X_L$ and hence X should be negative.

Thus
$$X = -\sqrt{Z^2 - R^2}$$
$$= -\sqrt{48^2 - 12^2}$$
$$= -46.48\ \Omega$$

Since $X_L = 20.78\ \Omega$

$$X_C = X_L - X$$
$$= 20.78 - (-46.48)$$
$$= 20.78 + 46.48$$
$$= 67.26\ \Omega$$

Example 3.11

For the circuit shown in Fig. 3.38, calculate

(a) conductance and susceptance of each branch

(b) the resultant admittance

(c) the current in each branch

(d) total current input

(e) the power delivered to the circuit.

Draw the phasor diagram.

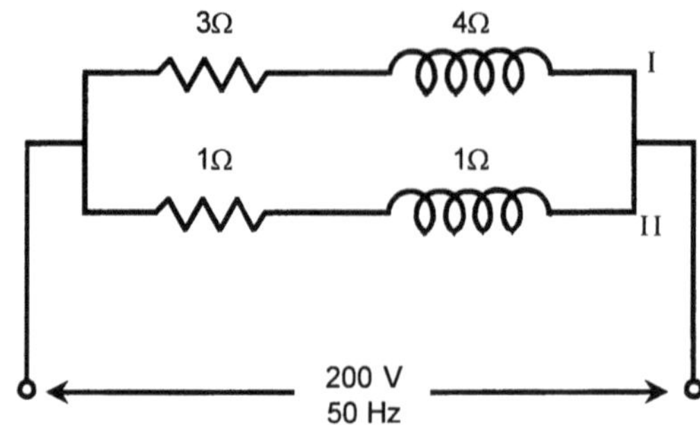

Fig. 3.38 The circuit for example 3.11

Solution :

Impedance of branch I = 3 + j4 Ω

Impedance of branch II = 1 + j1 Ω

$$\text{Admittance } Y_1 \text{ of branch I} = \frac{1}{3+j4}$$

$$= 0.12 - j0.16 \text{ S}$$

$$\text{Admittance } Y_2 \text{ of branch II} = \frac{1}{1+j1}$$

$$= 0.5 - j0.5 \text{ S}$$

Total admittance $Y = Y_1 + Y_2 = (0.62 - j0.66)$ S

(a) ∴ Conductance of branch I = 0.12 S

Conductance of branch II = 0.5 S

Susceptance of branch I = 0.16 S

Susceptance of branch II = 0.5 S

(b) Resultant admittance = (0.62 – j0.66) S

(c) Current I_1 in branch I = $V\,Y_1 = 200\ \angle 0^0\ (0.12 - j0.16)$

$$= 24 - j32$$

$$= 40\ \angle -53.1^0 \text{ amps}$$

Current in branch II = 200(0.5 – j0.5)

$$= 100 - j100 = 141.4\angle -45^0 \text{ amps}$$

(d) Total current I = (24 – j32) + (100 – j100)

$$= 124 - j132 \text{ amps}$$

$$= 181.01\ \angle -46.79^0 \text{ amps}$$

(e) Power delivered to the circuit = VI cosϕ

$$= 200 \times 181.1 \cos 46.79$$
$$= 24799 \text{ watts}$$

The phasor diagram is drawn in Fig. 3.39.

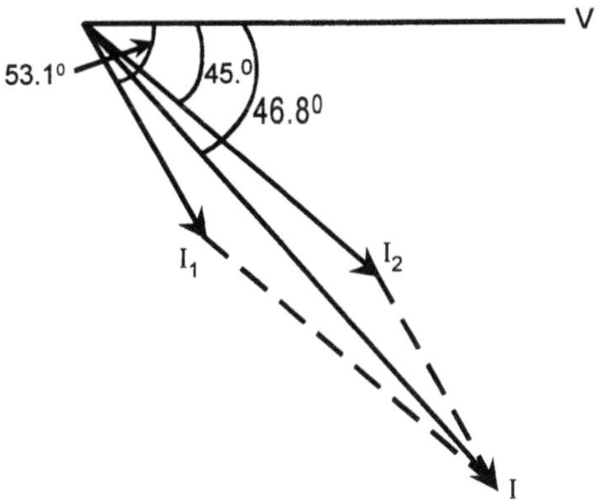

Fig. 3.39 The phasor diagram

In some circuits the elements are neither connected in series, nor in parallel. Analysis of these circuits can be performed using loop current method or node voltage method in the same way as in d.c circuits. Network reduction techniques also can be used as illustrated in the following example.

Example 3.12

The parameters of the circuit of Fig. 3.40 are

$$Z_1 = 10 + j30 \ \Omega$$
$$Z_2 = 5 + j10 \ \Omega$$
$$Z_3 = 4 - j16 \ \Omega$$

If a voltage of 100 V, 1ϕ, 50 Hz is applied to the circuit find

(i) I_1, I_2 and I_3 (ii) voltage V_1 and V_2.

Draw the complete phasor diagram.

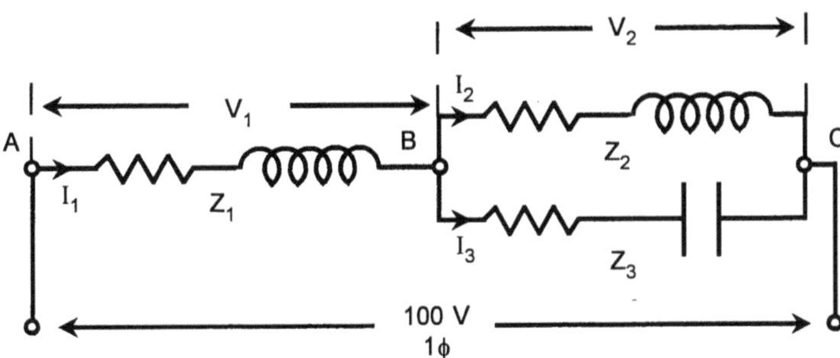

Fig. 3.40 Circuit for example 3.12

Solution :

$$Z_2 = 5 + j10\ \Omega$$
$$Z_3 = 4 - j16\ \Omega$$

Equivalent impedance between B and C

$$Z_C = \frac{Z_2 Z_3}{Z_2 + Z_3} = \frac{11.18\angle 63.43 \times 16.5\angle -75.96}{10.82\angle -33.69}$$
$$= 17.05\angle 21.16^0$$
$$= 15.9 + j6.15$$

$\therefore$ Total impedance between A and C = $Z_1 + Z_c = 10 + j30 + 15.9 + j6.15$
$= 25.9 + j36.15$

$$\text{Admittance Y between A and C} = \frac{1}{25.9 + j36.15}$$
$$= \frac{1}{44.47\angle 54.38} = 0.0225\angle -54.38^0$$

Taking V as reference, we have $V = 100\angle 0^0$

(i) Current input

$$I_1 = V\,Y = 2.25\ \angle -54.38^0$$
$$I_2 = \frac{Z_3}{Z_2 + Z_3} \cdot I = \frac{4 - j16}{9 - j6} \cdot 2.25\angle -54.38^0$$
$$= 3.416\angle -96.66^0$$
$$= -0.396 - j3.40\ \text{A}$$
$$I_3 = I_1 - I_2$$
$$= 2.25\angle -54.38 - 3.416\angle -96.66^0$$
$$= 1.706 + j1.57$$
$$= 2.318\angle 42.62^0\ \text{A}$$

(ii) Voltage across AB

$$V_1 = I_1 Z_1$$
$$= 2.25\angle -54.38 \times (10 + j30)$$
$$= 71.15\angle 17.18^0\ \text{V}$$
$$= 67.97 + j21\ \text{V}$$
$$V_2 = V - V_1$$
$$= 100 + j0 - 67.97 - j21$$
$$= 32.03 - j\,21$$
$$= 38.31\angle -33.27^0\ \text{V}$$

The complete phasor diagram is given in Fig. 3.41.

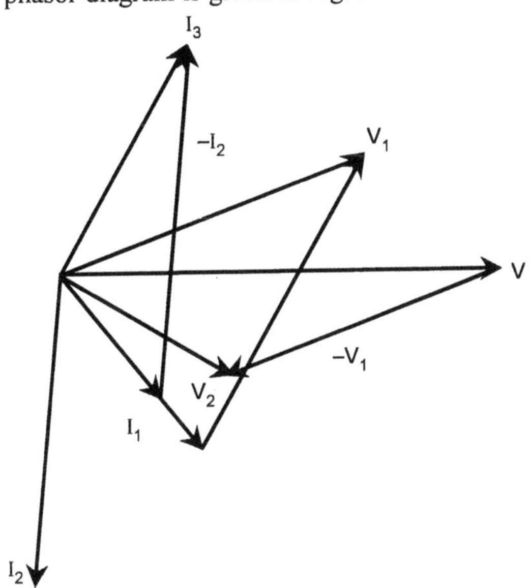

Fig. 3.41 Phasor diagram for example 3.12

3.6 SERIES RESONANCE :

Some networks consisting of R, L and C elements have an output of power, current or voltage that is greater at one frequency or over a small range of frequencies than at all other frequencies. Such circuits are called frequency selective circuits or resonant circuits. Depending on whether the R, L and C elements are connected in series or in parallel, we have series resonance or parallel resonance respectively.

3.6.1 Characteristics of a series resonant circuit :

A series R, L, C circuit with a constant voltage, variable frequency supply is shown in Fig. 3.42.

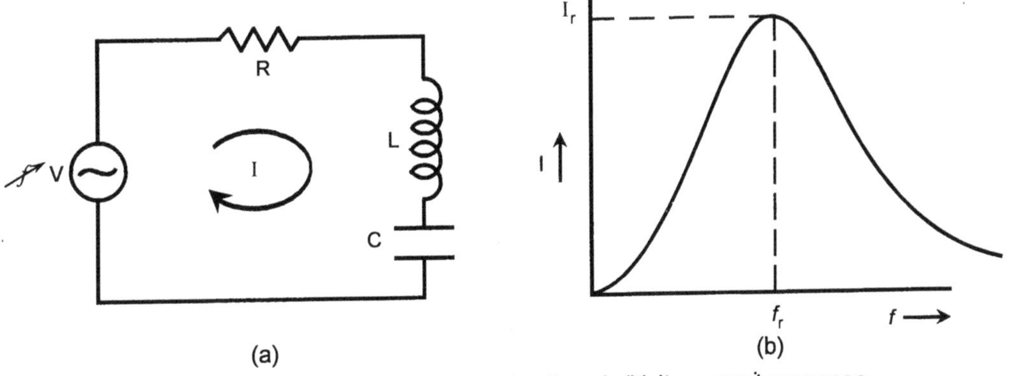

Fig. 3.42 (a) A series resonant circuit and (b) its current response

When the frequency of the source is varied keeping the voltage constant, the inductive reactance $X_L = 2\pi fL$ varies directly proportional to the frequency and the capacitive reactance $X_c = \dfrac{1}{2\pi fC}$ varies inversely with the frequency as shown in Fig. 3.43.

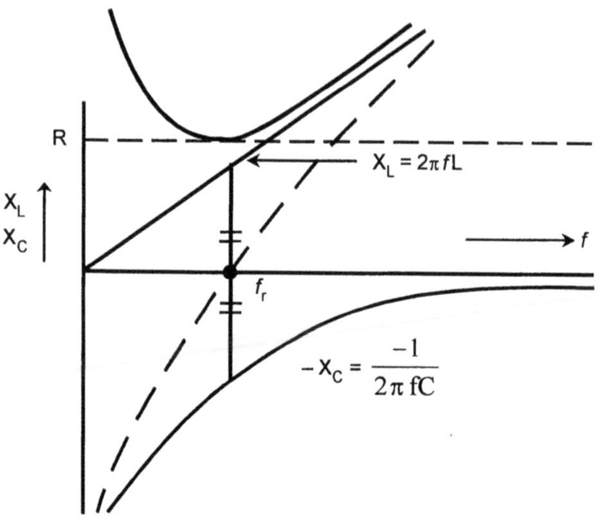

Fig. 3.43 Reactance curves of a series resonant circuit

At a particular frequency, f_r, the inductive reactance X_L and capacitive reactance X_C are equal and since they are opposite in sign, they cancel each other. The resultant impedance at this frequency is purely resistive and is equal to R. The current at this frequency is in phase with the source voltage. The variation of the magnitude of impedance, Z, is also shown in Fig. 3.43. For frequencies $f < f_r$, the capacitive reactance dominates over inductive reactance and hence the circuit behaves like an RC circuit. For frequencies $f > f_r$, the inductive reactance dominates over capacitive reactance and hence the circuit behaves like an RL circuit. At $f = f_r$ the circuit behaves like a pure resistance. The phasor diagram for these three cases are given in Fig. 3.44.

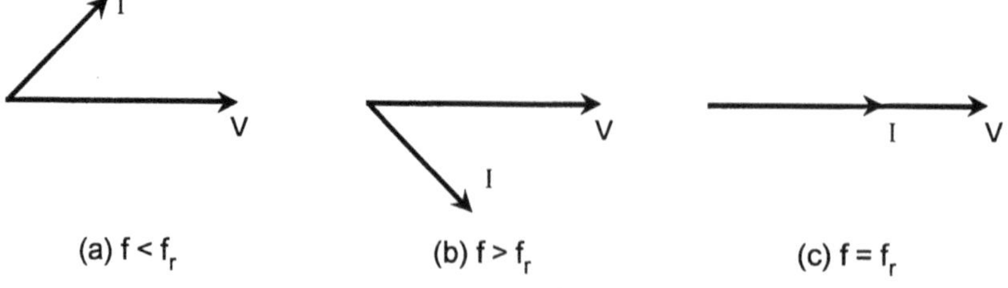

Fig. 3.44 The phasor diagram for (a) $f < f_r$ (b) $f > f_r$ and (c) $f = f_r$

Since the impedance of the series RLC circuit varies with the frequency as indicated in Fig. 3.43, the current in the circuit changes as shown in Fig. 3.42(b). As frequency is varied from 0 to infinity the current increases from 0 to a maximum value given by I_r at $f = f_r$ and then decreases to zero as $f \to \infty$.

At any frequency, the current in the circuit is given by,

$$I = \frac{V}{R + j\left(\omega L - \frac{1}{\omega C}\right)} \quad .....(3.73)$$

When $\omega = \omega_r = 2\pi f_r$

$$\omega_r L = \frac{1}{\omega_r C}$$

$$\therefore \quad \omega_r^2 = \frac{1}{LC}$$

or $$\omega_r = \frac{1}{\sqrt{LC}}$$

$$f_r = \frac{1}{2\pi\sqrt{LC}} \quad .....(3.74)$$

Hence the frequency at which a series RLC circuit resonates is dependent on the values of L and C.

At $\omega = \omega_r$

$$I_r = \frac{V}{R} \quad .....(3.75)$$

The condition of resonance can be obtained by either varying the frequency for fixed L and C values or in a circuit with a given frequency, by varying either L or C. At this frequency the current is a maximum and is in phase with the voltage.

Eqs. (3.73) can be put in a more useful form as,

$$I = \frac{\frac{V}{R}}{1 + j\frac{\omega L}{R}\left(1 - \frac{1}{\omega^2 LC}\right)}$$

Multiplying the imaginary term in the denominator with $\frac{\omega_r}{\omega_r}$ and replacing $\frac{1}{LC}$ by ω_r^2 and noting that $I_r = \frac{V}{R}$, we have

$$I = \frac{I_r}{1 + j\frac{\omega_r}{\omega_r}\frac{\omega L}{R}\left(1 - \frac{\omega_r^2}{\omega^2}\right)} \quad .....(3.76)$$

We now define an important quantity called quality factor, Q_r.

Let $$Q_r = \frac{\omega_r L}{R} \qquad \text{.....(3.77)}$$

The Q factor of the circuit is defined as the ratio of the resonant reactance of the inductor to the total resistance in the circuit. The use of Q – factor is justified because, the quality factor practically remains constant over a wide range of frequencies as shown in Fig. 3.45.

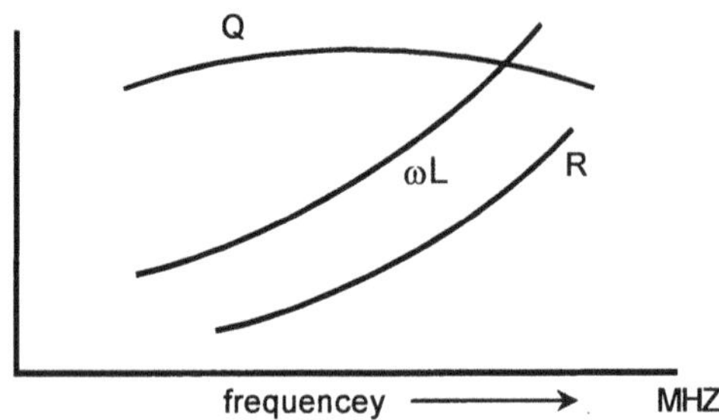

Fig. 3.45 The variation of R, X_L and Q factors

At high frequencies, the resistance of the inductive coil usually changes in the same manner as its reactance due to skin effect. Thus the ratio of $\frac{\omega L}{R}$ remains practically constant. Hence it is more convenient to express the variation of current in term of Q factor rather than the individual element values. In terms of Q factor, eq. (3.76) can be written as

$$I = \frac{I_r}{1 + jQ_r \frac{\omega}{\omega_r}\left(1 - \frac{\omega_r^2}{\omega^2}\right)} \qquad \text{.....(3.78)}$$

Circuits with same Q_r factor will give the same percent variation of current with $\frac{f}{f_r}$ regardless of the individual parameter values of R, L and C. Hence it is convenient to draw general resonance curves by plotting the variation of $\frac{I}{I_r}$ with $\frac{f}{f_r}$ instead of showing the variation of I with f for an individual circuit. The magnitude of $\frac{I}{I_r}$ is given by

$$\left(\frac{I}{I_r}\right) = \frac{1}{\sqrt{1 + Q_r^2 \frac{\omega^2}{\omega_r^2}\left(1 - \frac{\omega_r^2}{\omega^2}\right)}} \qquad \text{.....(3.79)}$$

It is evident that the rate at which $\left(\frac{I}{I_r}\right)$ changes with frequency 'f' depends on the second term of eq. (3.79) and the magnitude of this term at any frequency depends on Q_r. If Q_r is large $\left(\frac{I}{I_r}\right)$ is small and vice versa. Consequently, for small Q_r, the ratio $\frac{I}{I_r}$ changes slowly with frequency, where as it changes rapidly with frequency for large Q_r. It is also apparent that the change near the resonant frequency f_r is more rapid compared to frequencies away from it. The power factor angle of the circuit is given by

$$\theta = \tan^{-1}\left[-\frac{f}{f_r}Q_r\left(1-\frac{f_r^2}{f^2}\right)\right] \qquad .....(3.80)$$

θ approaches $+90^0$ for $f <<< f_r$ and -90^0 for $f >>> f_r$. However at $f = f_r$, $\theta = 0$.

The variation of $\frac{I}{I_r}$ and θ are shown in Fig. 3.46.

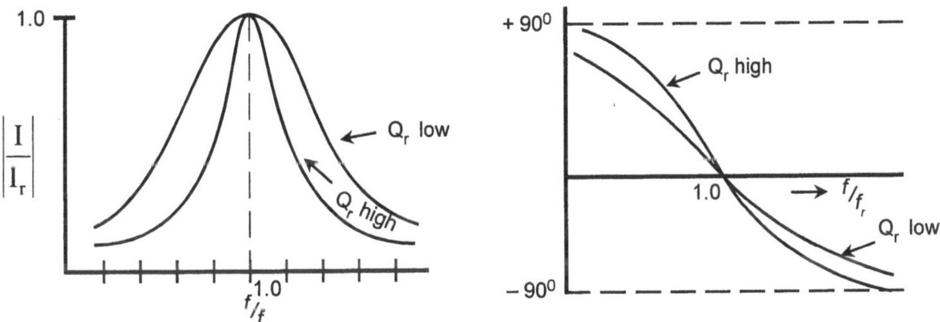

Fig. 3.46 (a) Variation of normalised current with normalised frequency for different Q – factors
(b) Variation of phase angle for different Q – factors with normalised frequency

3.6.2 Q - factor :

We had introduced the concept of Q – factor in the previous section. Let us elaborate this further in terms of the energy stored and the power dissipated in the circuit. Let us first understand the nature of the energy stored in the inductance and capacitance. At resonance, if

$$v = V_m \sin \omega_r t$$

$$i = \frac{V_m}{R} \sin \omega_r t$$

and

$$v_C(t) = \frac{1}{C}\int i dt$$

$$= \frac{-V_m}{\omega_r CR}\cos \omega_r t$$

The total energy stored in the circuit is given by

$$W(t) = W_L(t) + W_C(t)$$

$$= \frac{1}{2}Li^2 + \frac{1}{2}Cv_C^{\,2}$$

$$= \frac{1}{2}L \cdot \frac{V_m^{\,2}}{R^2}\sin^2\omega_r t + \frac{1}{2}C\frac{V_m^{\,2}}{\omega_r^{\,2}C^2R^2}\cos^2\omega_r t$$

But $\omega_r^{\,2} = \dfrac{1}{LC}$

$$\therefore \quad W(t) = \frac{1}{2}\frac{V_m^{\,2}L}{R^2}\sin^2\omega_r t + \frac{1}{2}\frac{LV_m^{\,2}}{R^2}\cos^2\omega_r t$$

$$= \frac{1}{2}L\frac{V_m^{\,2}}{R^2}(\sin^2\omega_r t + \cos^2\omega_r t)$$

$$= \frac{1}{2}L\frac{V_m^{\,2}}{R^2} \qquad .....(3.81)$$

Thus the energy stored in the circuit at any instant is equal to the maximum energy stored in the inductor. It can also be shown to be equal to the maximum energy stored in the capacitor. Thus the total energy stored in the circuit at any instant is a constant. Though the energy stored by the inductor in its magnetic field and the energy stored by the capacitor in its electrostatic field, are changing with respect to time, their sum at any instant is a constant and is equal to the maximum energy stored by either the capacitor or the inductor. This is analogous to the case of a simple pendulum when the sum of Kinetic energy and Potential energy remain constant at any instant of time. The variation of W, W_L and W_C are shown in Fig. 3.47.

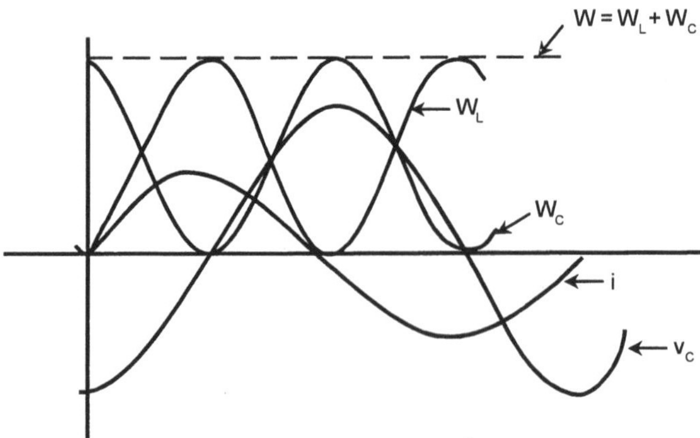

Fig. 3.47 Energy waveforms

Now we are ready to define the Q – factor. The Q – factor is defined as :

$$Q = \omega_r \left[\frac{\text{Max. energy stored}}{\text{Power dissipated}} \right]$$

$$= \omega_r \cdot \frac{W}{P}$$

$$= \omega_r \left(\frac{1}{2} L \frac{V_m^2}{R^2} \right) \left(\frac{R}{\left(\frac{V_m}{\sqrt{2}} \right)^2} \right)$$

$$= \frac{\omega_r L}{R} = \frac{1}{\omega_r CR} = \frac{1}{R}\sqrt{\frac{L}{C}} \qquad .....(3.82)$$

3.6.3 Selectivity and bandwidth :

The property of resonance is useful in discriminating a range of frequencies from the others. Since the circuit gives maximum response around the resonant frequency, we would like to classify this range of frequencies as useful band of frequencies for which the response is maximum. Since the current at resonce is a maximum, the power dissipated in the resistance is a maximum at resonance. If we are satisfied with a power of atleast half this value, for frequencies around the resonance frequency, we can define this range of frequencies as bandwidth of the circuit. It means that, at two frequencies f_1 and f_2 around f_r, at which the current falls to $\frac{I_r}{\sqrt{2}}$, the power dissipated will be equal to $\frac{1}{2}\left(I_r^2 R\right)$ or half the power at resonance as shown in Fig. 3.48.

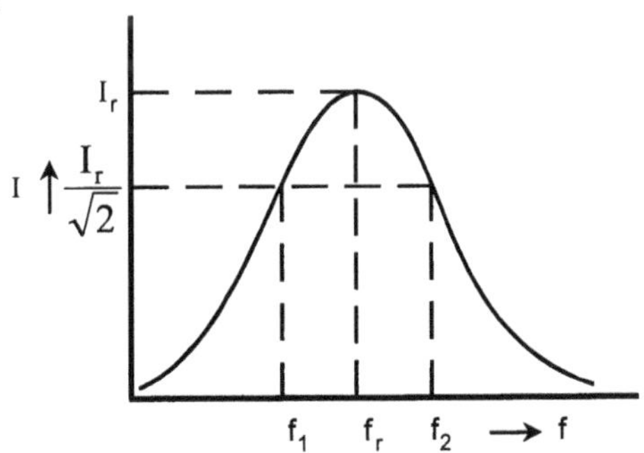

Fig. 3.48 Bandwidth of a series resonant circuit

Since the current at f_1 and f_2 is $\frac{I_r}{\sqrt{2}}$, the impedance should be $\sqrt{2}$ times the impedance at resonance. But the impedance at resonance is equal to R and hence, the impedance at f_1 or f_2 is given by

$$Z = R \pm jR$$

Thus the total reactance at these two frequencies must be equal to $\pm$ R

$$\therefore \qquad \omega_1 L - \frac{1}{\omega_1 C} = -R \qquad .....(3.83)$$

and

$$\omega_2 L - \frac{1}{\omega_2 C} = R \qquad .....(3.84)$$

Adding eq. (3.83) and (3.84), we have

$$L(\omega_1 + \omega_2) - \frac{1}{C}\left(\frac{1}{\omega_1} + \frac{1}{\omega_2}\right) = 0$$

$$(\omega_1 + \omega_2)\left[L - \frac{1}{\omega_1 \omega_2 C}\right] = 0 \qquad .....(3.85)$$

Eq. (3.85) yields

$$\omega_1 \omega_2 = \frac{1}{LC} = \omega_r^2 \qquad .....(3.86)$$

Eq. (3.86) implies that the resonance frequency is the geometric mean of the frequencies at which the power dissipated is half of the power dissipated at resonance. These two frequencies ω_1 and ω_2 are known as lower and upper half power frequencies respectively. The range of frequencies lying between ω_1 and ω_2, i.e., $\omega_2 - \omega_1$ is known as the useful bandwidth or simply bandwidth of the series resonant circuit. Let us now obtain an expression for the bandwidth of the series resonant circuit.

Subtracting eq. (3.83) from eq. (3.84), we get

$$L(\omega_2 - \omega_1) + \frac{1}{C}\left(\frac{1}{\omega_1} - \frac{1}{\omega_2}\right) = 2R$$

$$(\omega_2 - \omega_1)\left[L + \frac{1}{C.\omega_1 \omega_2}\right] = 2R$$

$$(\omega_2 - \omega_1)\left[L + \frac{1}{\omega_r^2 C}\right] = 2R$$

$$\omega_2 - \omega_1 = \frac{2\omega_r^2 RC}{\omega_r^2 LC + 1} = \omega_r^2 RC \qquad (\therefore\ \omega_r^2 LC = 1)$$

$$= \omega_r \cdot \frac{R}{\omega_r L} \qquad \left[\because \omega_r C = \frac{1}{\omega_r L}\right]$$

$$\therefore \qquad \omega_2 - \omega_2 = \frac{\omega_r}{Q_r} \qquad .....(3.87)$$

$$\text{Thus} \qquad BW = \frac{\text{Resonance frequency}}{\text{Quality factor}} \qquad .....(3.88)$$

This is an important relation. If the Q – factor is high, the bandwidth will be small for a given resonant frequency. This means that if one wants to select a particular frequency only and reject all other frequencies from a signal, the circuit Q – factor must be chosen to be a very high value. Selectivity is defined as the ratio of Bandwidth to resonance frequency and is given by

$$\text{Selectivity} = \frac{BW}{\omega_r} = \frac{1}{Q_r}$$

If Q_r is large, the circuit is highly selective and vice versa.

3.6.4 Voltage across the capacitor :

In many application of series resonant circuits, more direct use is made of the voltage across the capacitor rather than the current through the circuit . Let us obtain the nature of variation of the capacitor voltage as the frequency is varied. The voltage across the capacitor is given by

$$V_C = \frac{-jV}{\omega C\left[R + j\left(\omega L - \frac{1}{\omega C}\right)\right]} \qquad .....(3.89)$$

Since we are interested only in the magnitude of the voltage,

$$\left|\frac{V_C}{V}\right|^2 = \frac{1}{\omega^2 C^2\left[R^2 + \left(\omega L - \frac{1}{\omega C}\right)^2\right]} \qquad .....(3.90)$$

This ratio is a maximum when the denominator of eq. (3.90) is a minimum.

$$\text{Let} \qquad D = \omega^2 C^2 \left[R^2 + \left(\omega L - \frac{1}{\omega C}\right)^2\right]$$

$$= \omega^2 C^2 R^2 + (\omega^2 LC - 1)^2 \qquad ......(3.91)$$

$$\frac{dD}{d\omega} = 2\omega C^2 R^2 + 2(\omega^2 LC - 1) \cdot 2\omega LC = 0$$

Solving for ω, and denoting it by ω_C, we get

$$\omega_C = \frac{1}{\sqrt{LC}}\sqrt{1-\frac{R^2C}{2L}}$$

$$= \omega_r\sqrt{1-\frac{R^2C}{2L}} = \omega_r\sqrt{1-\frac{1}{2Q_r^2}} \qquad .....(3.92)$$

Eq. (3.92) indicates that the frequency at which the capacitor voltage becomes a maximum is not the same as the frequency at which the current becomes a maximum. It occurs at a frequency less than the resonance frequency. But for circuits with large Q_r factor, usually greater than 10, $\omega_C \simeq \omega_r$. From eq. (3.90) we have, at resonance,

$$\left(\frac{V_C}{V}\right) = \frac{1}{\omega_r CR} \qquad \left(\because \omega_r L = \frac{1}{\omega_r C}\right)$$

$$= Q_r \qquad .....(3.93)$$

The voltage across the capacitor at resonance is Q_r times the supply voltage and therefore the circuit produces a high magnification of voltage. Thus Q_r is also known as a voltage magnification factor in a series resonant circuit. The voltage across the capacitor may assume dangerous proportions for a high Q – resonant circuit !

3.6.5 Voltage across the inductor :

The magnitude of the voltage across the inductor is given by

$$(V_L) = \frac{V\omega L}{\left[R^2+\left(\omega L-\frac{1}{\omega C}\right)^2\right]^{\frac{1}{2}}}$$

$$\left(\frac{V_L}{V}\right)^2 = \frac{1}{\left[\frac{R^2}{\omega^2 L^2}+\left(1-\frac{1}{\omega^2 LC}\right)^2\right]}$$

The maximum value of voltage across the inductor, as frequency is changed, is obtained in the usual way. Denoting this frequency by ω_L, we get

$$\omega_L = \omega_r \cdot \frac{1}{\sqrt{1-\frac{1}{2Q_r^2}}} \qquad .....(3.94)$$

Thus it is observed that the voltage across the inductance is a maximum at a frequency greater than the resonance frequency. But, if the Q_r factor of the circuit is high, $\omega_L \simeq \omega_r$. The variation of V_C, V_L and I in a series resonant circuit are shown in Fig. 3.49.

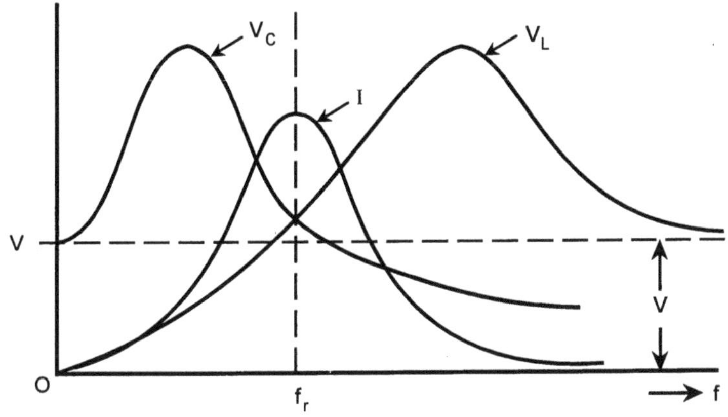

Fig. 3.49 Variation of V_C, V_L and I with frequency

3.6.6 Resonance by varying capacitance :

In practical application of the series resonant circuits, it is aften required to tune the circuit for a particular frequency by changing the value of the capacitance. Differentiating the expression in eq. (3.91) with respect to C, rather than frequency , we get

$$\frac{dD}{dC} = 2\omega^2 CR^2 + 2(\omega^2 LC - 1)\omega^2 L = 0$$

$$C(R^2 + \omega^2 L^2) = L$$

or
$$C = \frac{L}{R^2 + \omega^2 L^2}$$

$$= \frac{1}{\omega^2 L\left(1 + \frac{1}{Q^2}\right)} \qquad .....(3.95)$$

Substituting the value of C in eq. (3.90), we get

$$\left|\frac{V_C}{V}\right|^2 = \frac{1}{\frac{\omega^2 R^2 . L^2}{\left(R^2 + \omega^2 L^2\right)^2} + \left(\frac{\omega^2 L^2}{R^2 + \omega^2 L^2} - 1\right)^2}$$

$$= \frac{\left(R^2 + \omega^2 L^2\right)^2}{\omega^2 R^2 L^2 + R^4}$$

$$= \frac{R^2 + \omega^2 L^2}{R^2}$$

$$= 1 + Q^2$$

$$\therefore \qquad V_C = V\sqrt{1+Q^2}$$

$$= VQ\sqrt{1+\frac{1}{Q^2}}$$

If the Q – factor is high, the value of capacitance for maximum voltage across the capacitance is same as that required for obtaining maximum current in the circuit, which is

$$C = \frac{1}{\omega^2 L} \qquad .....(3.96)$$

Example 3.13

An inductance of 0.5 H, a resistance of 5 ohm, and a capacitance of 8 μf are in series across a 220 V a.c supply. Calculate the frequency at which the circuit resonates. Find the current at resonance, Bandwidth, half power frequencies and the voltage across capacitance at resonance.

Solution :

The resonant frequency is given by

$$f_r = \frac{1}{2\pi\sqrt{LC}}$$

$$= \frac{1}{2\pi\sqrt{0.5 \times 8 \times 10^{-6}}}$$

$$= 79.58 \text{ Hz}$$

$$I_r = \frac{V}{R}$$

$$= \frac{220}{5}$$

$$= 44 \text{ A}$$

The quality factor is given by

$$Q_r = \frac{\omega_r L}{R}$$

$$= \frac{2\pi \times 79.58 \times 0.5}{5}$$

$$= 50$$

The bandwidth is given by

$$BW = \frac{f_r}{Q_r}$$

$$f_2 - f_1 = \frac{79.58}{50}$$

$$f_2 - f_1 = 1.59 \text{ Hz}$$

But $\omega_1\omega_r = \omega_r^2$ or $f_1 f_2 = f_r^2$

$$\therefore \quad f_1 = \frac{(79.58)^2}{f_2} = \frac{6.333\times10^3}{f_2}$$

$$\therefore \quad f_2 - \frac{6.333\times10^3}{f_2} = 1.59$$

$$f_2^2 - 1.59f_2 - 6.333\times10^3 = 0$$

Solving for f_2

$$f_2 = 80.38 \text{ Hz}$$

and $f_1 = 78.79$ Hz

The voltage across the capacitor at resonance is

$$V_C = Q_r V = 50 \times 220 = 11 \text{ KV}$$

3.7 PARALLEL RESONANCE :

A parallel circuit consisting of resistance, inductance and capacitance is shown in Fig. 3.50(a). The current through the inductive branch lags the voltage by an angle less than 90^0 and the current through the capacitive branch leads the voltage by an angle less than 90^0. The vector diagram is shown in Fig.3.50(b).

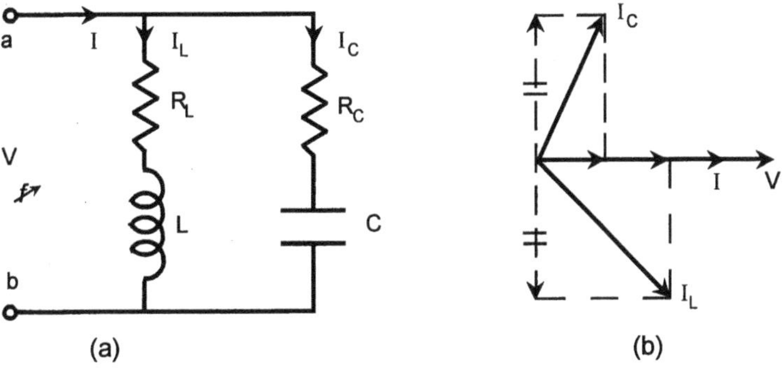

Fig. 3.50 (a) Parallel resonant circuit (b) Phasor diagram

If the reactive components of I_L and I_C are equal, the resultant current I will be in phase with the voltage. This is the condition of resonance in the parallel circuit. The line current is nearly a minimum under this condition. The parallel circuit is best analysed in terms of admittances rather than impedances. The admittance of the parallel circuit is,

$$Y = Y_L + Y_C = \frac{1}{R_L + j\omega L} + \frac{1}{R_C - \frac{j}{\omega C}}$$

Rationalising and collecting the real and imaginary terms,

$$Y = \frac{R_L}{R_L^2 + \omega^2 L^2} + \frac{R_C}{R_C^2 + \frac{1}{\omega^2 C^2}} - j\left(\frac{\omega L}{R_L^2 + \omega^2 L^2} - \frac{\frac{1}{\omega C}}{R_C^2 + \frac{1}{\omega^2 C^2}}\right) \qquad .....(3.97)$$

For resonance, the imaginary part of Y must be zero.

$$\frac{\omega L}{R_L^2 + \omega^2 L^2} = \frac{\omega C}{R_C^2 \omega^2 C^2 + 1}$$

Solving for ω

$$\omega = \frac{1}{\sqrt{LC}} \sqrt{\frac{CR_L^2 - L}{CR_C^2 - L}} \qquad .....(3.98)$$

This is the frequency at which the parallel circuit resonates. In a more practical circuit, the resistance in the capacitive branch is negligible. Thus, making $R_C = 0$ we have,

$$\omega = \frac{1}{\sqrt{LC}} \sqrt{1 - \frac{CR_L^2}{L}} \qquad .....(3.99)$$

Defining, $\omega_r = \frac{1}{\sqrt{LC}}$, and the frequency of resonance in parallel circuit as ω_{ar},

$$\omega_{ar} = \omega_r \sqrt{1 - \frac{CR_L^2}{L}} \qquad .....(3.100)$$

Expressing $\frac{CR_L^2}{L}$ in terms of Q_r factor,

$$Q_r^2 = \frac{\omega_r^2 L^2}{R_L^2} = \frac{1}{LC} \cdot \frac{L^2}{R_L^2} = \frac{L}{CR_L^2}$$

$$\therefore \qquad \omega_{ar} = \omega_r \sqrt{1 - \frac{1}{Q_r^2}} \qquad .....(3.101)$$

Here we had defined $\omega_r = \frac{1}{\sqrt{LC}}$, which is the frequency at which the circuit would resonate if the elements were connected in series. For circuits with large Q_r factors,

$$\omega_{ar} = \omega_r$$

Or, in other words, the condition of resonance occurs when the capacitive reactance is equal to the reactance in the inductive branch. The current in the circuit is nearly a minimum and hence the resonant frequency in parallel circuit is known as anti - resonant frequency. The resonance frequency is appropriately denoted by ω_{ar}. The admittance at resonance is

$$Y_{ar} = \frac{R_L}{R_L^2 + \omega_{ar}^2 L^2}$$

From eq. (3.100)

$$\omega_{ar}^2 = \frac{1}{LC}\left(1 - \frac{CR_L^2}{L}\right)$$

$$\omega_{ar}^2 L^2 C = L - CR_L^2$$

$$\omega_{ar}^2 L^2 + R_L^2 = \frac{L}{C}$$

Substituting in expression for Y, we get

$$Y_{ar} = \frac{R_L C}{L}$$

or $$Z_{ar} = \frac{L}{R_L C}$$

This impedance, which is purely resistive is known as the dynamic resistance of the parallel resonant circuit.

3.7.1 Maximum impedance by adjusting frequency :

Let us consider the impedance of the parallel circuit and obtain the condition under which it is a maximum.

$$Z_{ab} = \frac{(R + j\omega L)\left(-\frac{j}{\omega C}\right)}{R + j\left(\omega L - \frac{1}{\omega C}\right)}$$

The magnitude of Z_{ab} is

$$|Z_{ab}|^2 = \frac{\left(R^2 + \omega^2 L^2\right)\dfrac{1}{\omega^2 C^2}}{R^2 + \left(\omega L - \dfrac{1}{\omega C}\right)^2}$$

$$= \frac{R^2 + \omega^2 L^2}{\omega^2 C^2 R^2 + \omega^4 L^2 C^2 - 2\omega^2 LC + 1}$$

Differentiating w.r.t ω and equating to zero

$$\frac{(\omega^2 C^2 R^2 + \omega^4 L^2 C^2 - 2\omega^2 LC + 1)2\omega L^2 - (R^2 + \omega^2 L^2)(2\omega C^2 R^2 + 4\omega^3 L^2 C^2 - 4\omega LC)}{(\omega^2 C^2 R^2 + \omega^4 L^2 C^2 - 2\omega^2 LC + 1)^2} = 0$$

Simplifying, we get

$$\omega^4 L^4 C^2 + 2\omega^2 R^2 L^2 C^2 + R^4 C^2 - 2R^2 LC - L^2 = 0$$

This is a quadratic in ω^2 and hence

$$\omega^2 = \frac{-2R^2 L^2 C^2 \pm \sqrt{4R^4 L^4 C^4 - 4L^4 C^2\left(R^4 C^2 - 2R^2 LC - L^2\right)}}{2L^4 C^2}$$

$$= -\frac{R^2}{L^2} \pm \sqrt{\frac{2R^2}{L^3 C} + \frac{1}{L^2 C^2}} \qquad \text{.....(3.102)}$$

Expressin eq. (3.102) in terms of

$$\omega_r^2 = \frac{1}{LC} \quad \text{and} \quad Q_r = \frac{\omega_r L}{R},$$

and considering the positive value of ω^2 only, we get

$$\omega^2 = -\frac{\omega_r^2}{Q_r^2} + \sqrt{\frac{2\omega_r^4}{Q_r^2} + \omega_r^4} = \omega_r^2\left[\sqrt{1 + \frac{2}{Q_r^2}} - \frac{1}{Q_r^2}\right]$$

$$\omega = \omega_r \sqrt{\sqrt{1 + \frac{2}{Q_r^2}} - \frac{1}{Q_r^2}} \qquad \text{.....(3.103)}$$

Thus the frequency at which maximum impedance occurs is not the same as the resonance frequency. However if the Q – factor is high, the frequencies at which the impedance becomes a maximum is the same as ω_{ar} or ω_r.

The variation of impedance and current in the circuit is shown in Fig. 3.51.

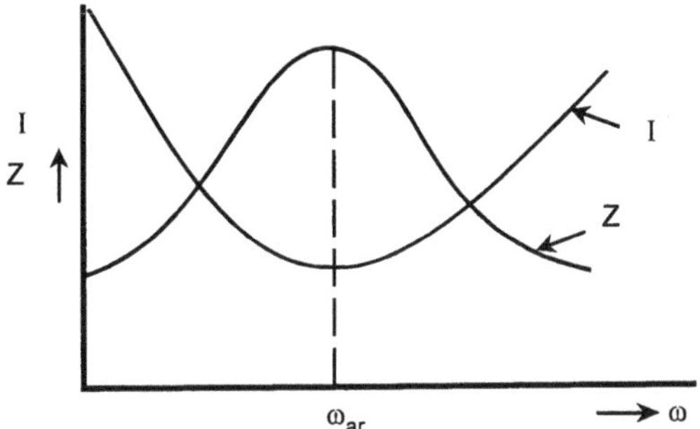

Fig. 3.51 Variation of impedance and current in a parallel circuit

3.7.2 Selectivity and bandwidth :

In the parallel resonant circuit the bandwidth is defined as the band of frequencies for which the impedance is greater than $\frac{1}{\sqrt{2}}$ times the impedance at resonance. Consider the impedance characteristic given in Fig. 3.52.

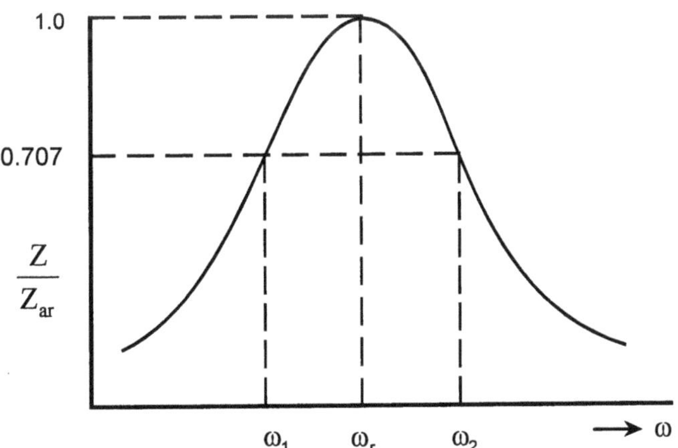

Fig. 3.52 Bandwidth of a parallel resonant circuit

The impedance of the parallel resonant circuit is given by

$$Z = \frac{(R + j\omega L)\left(\frac{-j}{\omega C}\right)}{R + j\left(\omega L - \frac{1}{\omega C}\right)} = \frac{\frac{L}{RC}\left(1 - j\frac{R}{\omega L}\right)}{1 + j\frac{\omega L}{R}\left(1 - \frac{1}{\omega^2 LC}\right)}$$

Let $$\frac{\omega - \omega_r}{\omega_r} = \delta \qquad .....(3.104)$$

$$\frac{\omega}{\omega_r} = 1 + \delta \qquad .....(3.105)$$

Thus $$\frac{\omega L}{R} = \frac{\omega_r L}{R} \cdot \frac{\omega}{\omega_r}$$

$$= Q_r (1 + \delta) \qquad .....(3.106)$$

and $$\frac{1}{\omega^2 LC} = \frac{\omega_r^2}{\omega^2} = \frac{1}{(1+\delta)^2}$$

Substituting these values in Z

$$Z = \frac{Z_{ar}\left[1 - \frac{j}{Q_r(1+\delta)}\right]}{1 + jQ_r(1+\delta)\left[1 - \frac{1}{(1+\delta)^2}\right]}$$

for large values of Q_r and small δ

$$Z = \frac{Z_{ar}}{1 + j2Q_r\delta} \qquad .....(3.107)$$

But at half power points, $Z = \frac{Z_{ar}}{\sqrt{2}}$

$$\therefore \quad 2Q_r\delta = \pm 1$$

$$\delta = \pm \frac{1}{2Q_r}$$

$$\therefore \quad \frac{\omega - \omega_r}{\omega_r} = \pm \frac{1}{2Q_r} \qquad .....(3.108)$$

$$\frac{\omega_2 - \omega_r}{\omega_r} = \frac{1}{2Q_r} \qquad .....(3.109)$$

and
$$\frac{\omega_1 - \omega_r}{\omega_r} = -\frac{1}{2Q_r} \qquad .....(3.110)$$

Solving for $\omega_2 - \omega_1$ from eqs. (3.109) and (3.110) we have

$$\omega_2 - \omega_1 = BW = \frac{\omega_r}{Q_r} \qquad .....(3.111)$$

Again, defining the selectivity as the ratio of B.W to resonant frequency, we have

$$\text{Selectivity} = \frac{BW}{\omega_r} = \frac{1}{Q_r} \qquad .....(3.112)$$

If the voltage across the capacitor is used for obtaining frequency selectivity, a source with high impedance in series must be used as shown in Fig. 3.53.

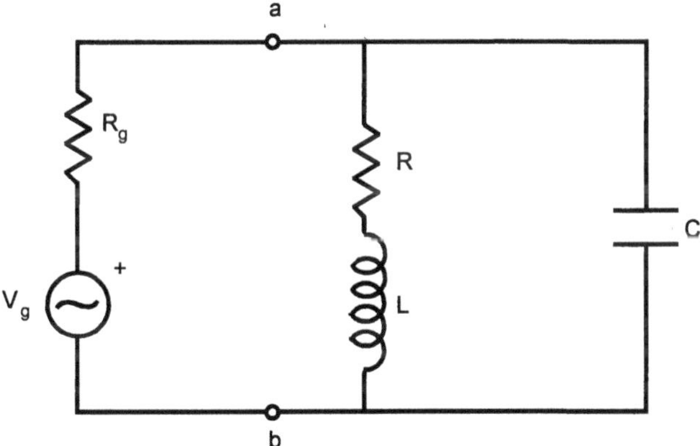

Fig. 3.53 Parallel resonant circuit supplied from a voltage source of finite resistance

It can be shown that if R_g is made large, the circuit becomes highly selective, but the voltage across the capacitor falls.

3.8 LOCUS DIAGRAMS :

In some applications it is necessary to know how the current varies as one of the elements is varied with fixed voltage applied to the circuit. As the element value is changed the impedance, admittance and hence the current change and we would like to obtain the locus of these quantities. As the voltage applied to the circuit is a constant, the current locus is same as the admittance locus since,

$$I = VY \qquad .....(3.113)$$

3.8.1 Impedance locus of series connected elements :

The locus of impedance of series connected element for different cases are shown in Fig. 3.54.

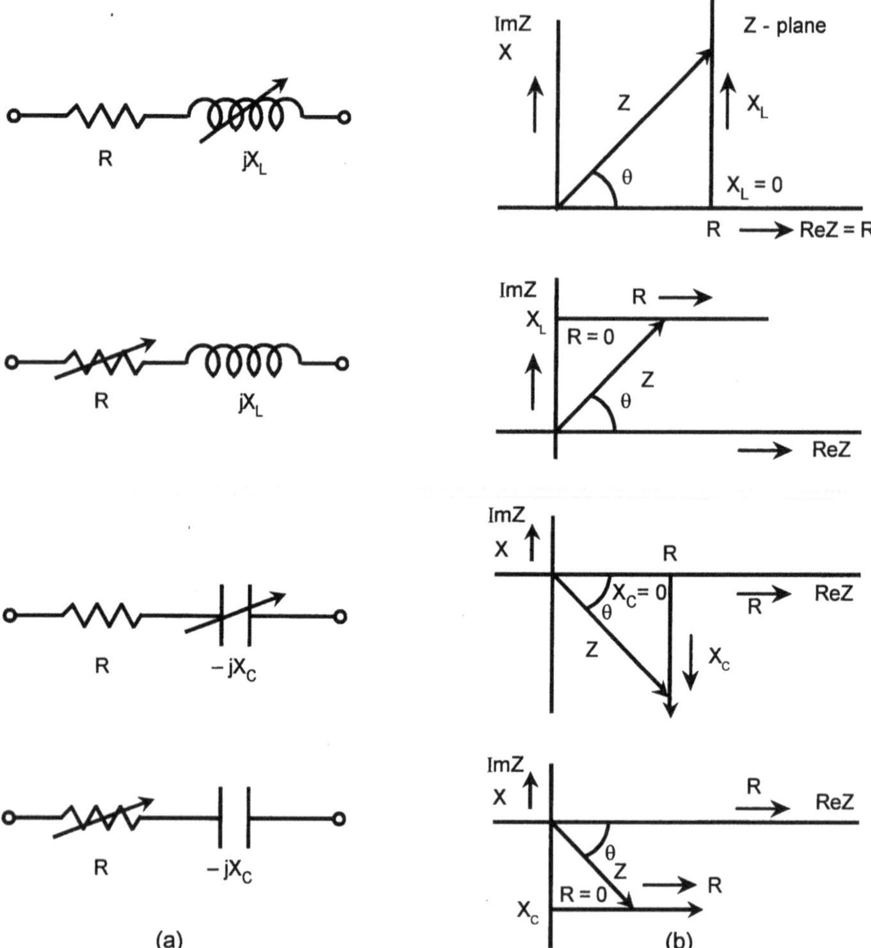

Fig. 3.54 (a) Series connected elements (b) Impedance locus

3.8.2 Admittance locus of series connected elements :

Since admittance is the reciprocal of the impedance, the locus of admittance is the inverse locus of the impedance. It is well known that the inverse of a circle is another circle and the straight line is considered as a special case of a circle with infinite radius. Thus the locus of admittance is a circle as can be seen from the following for an RL circuit with L varying, we have

$$Z = R + j\,X_L = Z\angle\phi$$

and $$Y = \frac{1}{Z} = \frac{1}{R + jX_L} = \frac{1}{Z}\angle -\phi$$

$$= \frac{R}{R^2 + X_L{}^2} - j\frac{X_L}{R^2 + X_L{}^2} = G - jB$$

$$\therefore \quad G^2 + B^2 = \frac{R^2}{\left(R^2 + X_L^2\right)^2} + \frac{X_L^2}{\left(R^2 + X_L^2\right)^2}$$

$$= \frac{1}{\left(R^2 + X_L^2\right)}$$

$$= \frac{1}{R} \cdot \frac{R}{R^2 + X_L^2} = \frac{1}{R} \cdot G$$

$$\left(G - \frac{1}{2R}\right)^2 + B^2 = \left(\frac{1}{2R}\right)^2 \qquad .....(3.114)$$

Thus it is seen that the locus of admittance, as X is varied, is a circle with radius $\frac{1}{2R}$ and centre at $\left(\frac{1}{2R}, 0\right)$ as shown in Fig. 3.55(a). Similarly the admittance loci for corresponding impedance loci of Fig. 3.54 are shown in Fig. 3.55(b), (c) and (d).

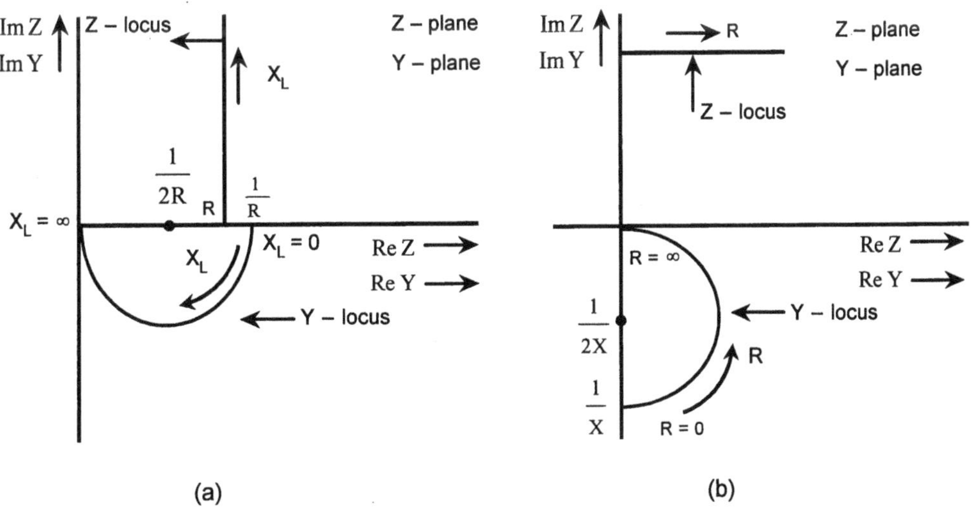

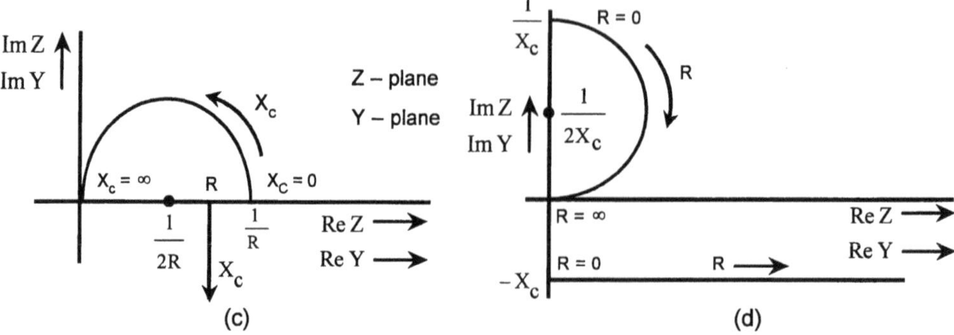

Fig. 3.55 Admittance loci for different cases of Fig. 3.27(a), (b), (c) and (d)

As discussed earlier, the current locus is same as admittance locus and therefore Fig. 3.55 represents the current locus also, if the scale to which the admittance locus is drawn, is multiplied by V, the applied voltage.

3.8.3 Admittance locus of parallel connected elements :

The admittance of parallel connected elements is given by

$$Y = G \pm jB$$

Hence in Y plane, the loci of admittances for four different cases shown in Fig. 3.56(a), are shown in Fig. 3.56(b).

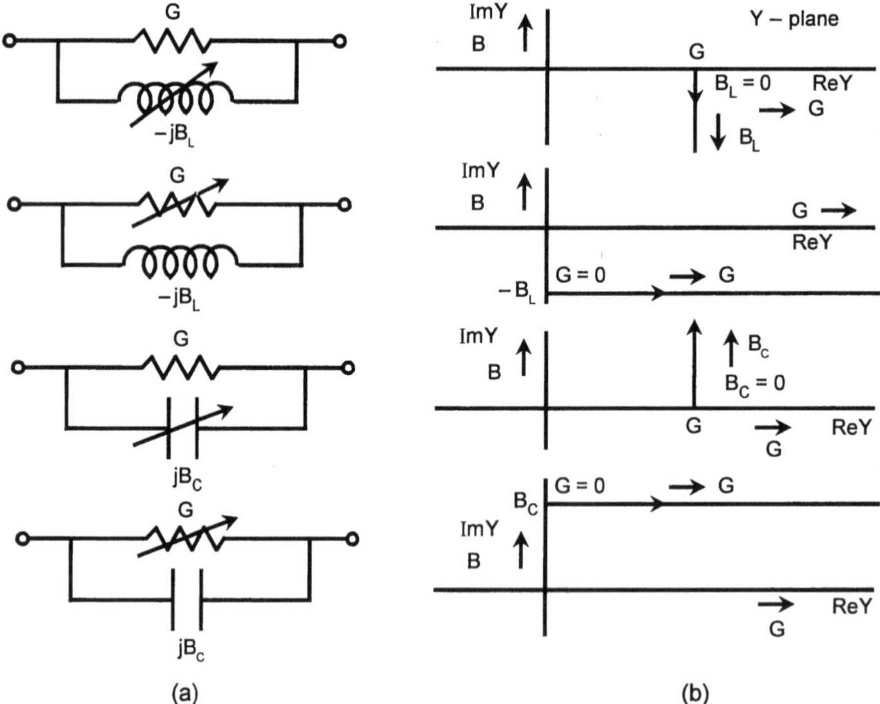

Fig. 3.56 (a) Parallel connected elements (b) Their admittance loci

3.8.4 Impedance locii of parallel connected elements :

Since the impedance Z is the reciprocal of the admittance as given in section 3.8.2. the impedance locus can be shown to be a circle. The impedance loci for the different cases shown in Fig. 3.56 are shown in Fig. 3.57.

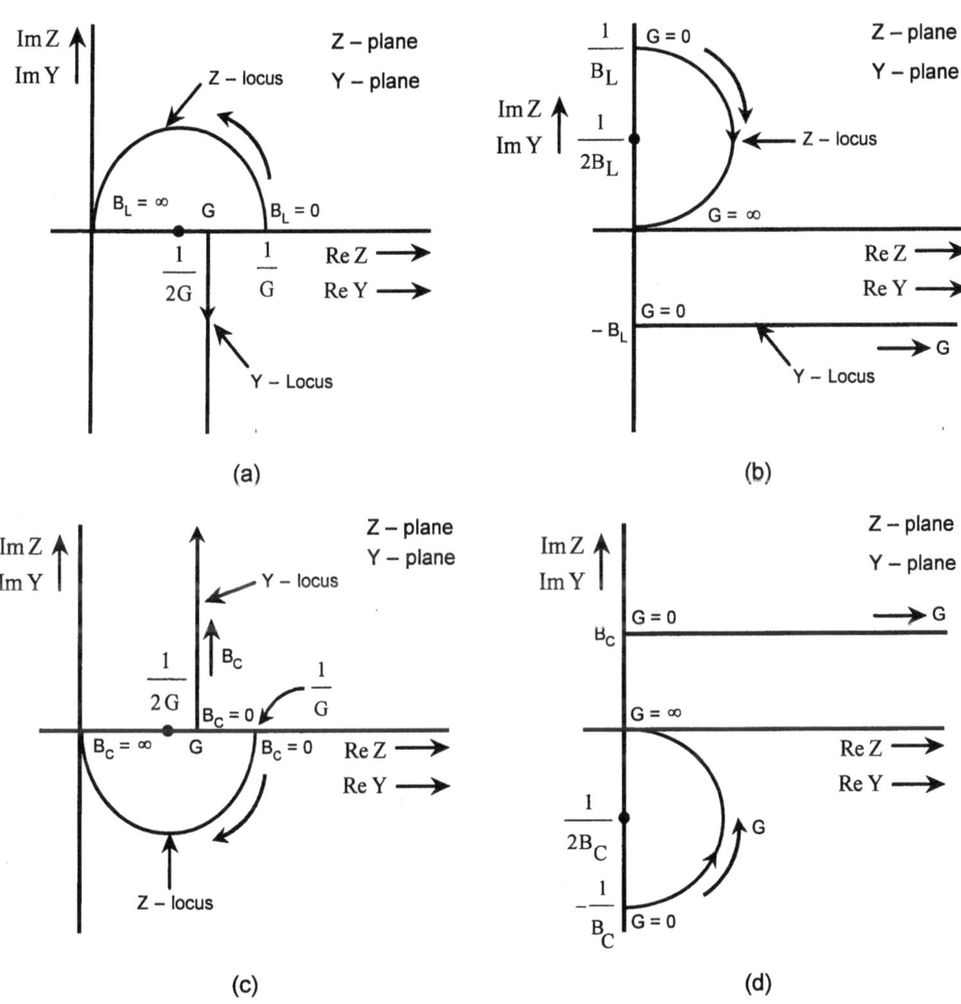

Fig. 3.57 The impedance loci of parallel connected elements

3.8.5 Current locus of a parallel circuit with one variable branch :

Consider a two branch parallel circuit in which one element of a branch is variable as shown in Fig. 3.58.

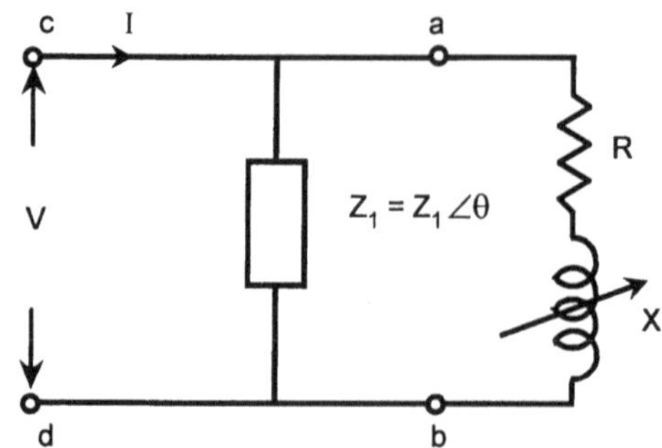

Fig. 3.58 Parallel circuit with one variable branch

The impedance and admittance loci of branch a, b are shown in Fig. 3.59(a).

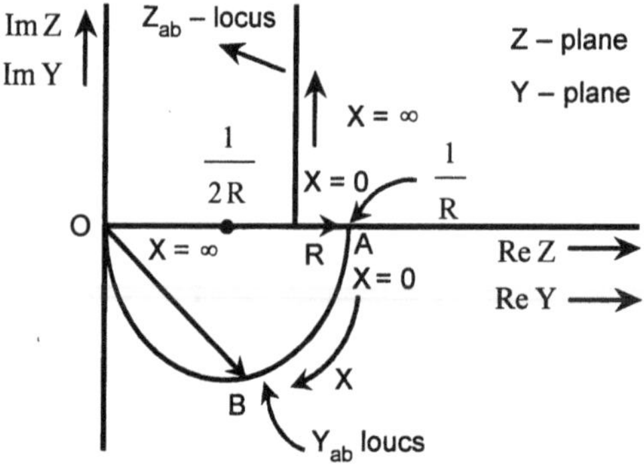

Fig. 3.59(a) Impedance and admittance loci of branch ab

The total admittance of the parallel circuit is given by

$$Y_{cd} = Y_{ab} + Y_1 \quad .....(3.115)$$

Y_1 is a fixed admittance added to the variable admittance Y_{ab} i.e., the semi circle shown in Fig. 3.59(a). Consider O'O representing the admittance $Y_1 = \frac{1}{Z_1} \angle -\theta$ added to $\overrightarrow{OB} = Y_{ab}$ for a particular value of X, as shown in Fig. 3.59(b).

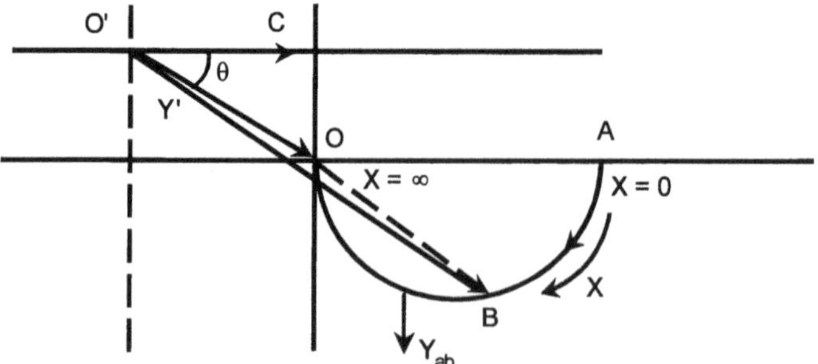

Fig. 3.59(b) Addition of fixed admittance to variable admittance

Now $\vec{O'B} = \vec{O'O} + \vec{OB}$

$\vec{O'B}$ represents the total admittance of the circuit i.e., Y. Thus any point on the Y_{ab} locus represents the total admittance Y_{cd} if the origin 'O' is shifted to O' as shown in Fig. 3.59(b). Thus addition of a fixed branch amounts to shifting the origin 'O' to a new point O' so that $\vec{O'O}$ represents the admittance of the fixed branch with respect to the new origin O'.

The total current locus I in the parallel circuit is also given by the admittance locus w.r.t the new origin O' to a different scale as discussed ealier. O'C in Fig. 3.59(b) represents the voltage vector V. If the impedance Z_{cd} is desired, the circular locus of Y_{ab} can be inverted with respect to the origin O' to result in another circle. The procedure for inversion of a circle is discussed in section 3.8.6.

The case of a fixed branch in series with a parallel branch with one variable element can be handled in a similar manner.

3.8.6 Inversion of circle :

We know that the inverse of a circle results in another circle. How can we find the inverse of a circle ? The procedure is as follows. Consider a circle 'τ' with respect to the origin 'O' as shown in Fig. 3.60.

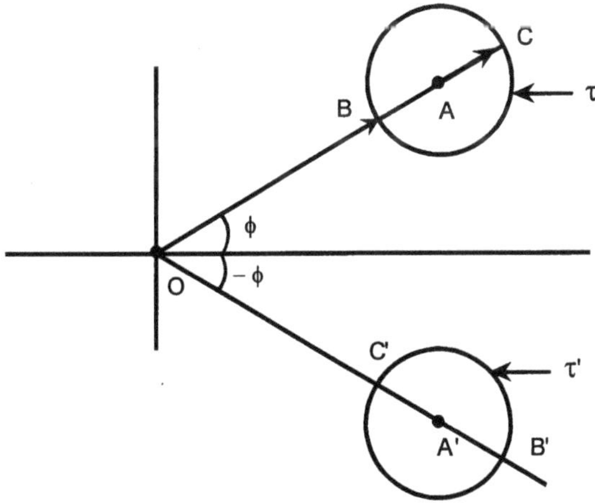

Fig. 3.60 Procedure to find the inverse of a circle

Join the centre 'A' of the circle to the origin O and extend it to cut the circle. Let the points of intersection of the circle with this line be B and C.

Find the inverse of $\vec{OB}$ and $\vec{OC}$.

$$\vec{OB}' = \frac{1}{OB} \angle -\phi$$

and $$\vec{OC}' = \frac{1}{OC} \angle -\phi$$

Locate the points C' and B'. The line C' B' represents the diameter of the inverse circle. Bisect this line to get the centre of this circle. Draw a circle with A' as centre and A'B' as radius. This is the inverse of the circle 'τ'. In this procedure we observe that the point B which is nearer to the origin 'O', goes far away from the origin when it is inverted to get point B'. Similarly the inverse point C' comes closer to the origin since the corresponding point 'C' is far away from the origin. The angle ϕ remains the same except that it has opposite sign. Thus in the process of inversion, the distances between the points are not preserved but the angles are.

Using the procedures discussed in previous section, the impedance, admittance or current locii of more complex circuits can be obtained graphically.

Example 3.14

Obtain the current locus of the circuit shown in Fig. 3.61(a).

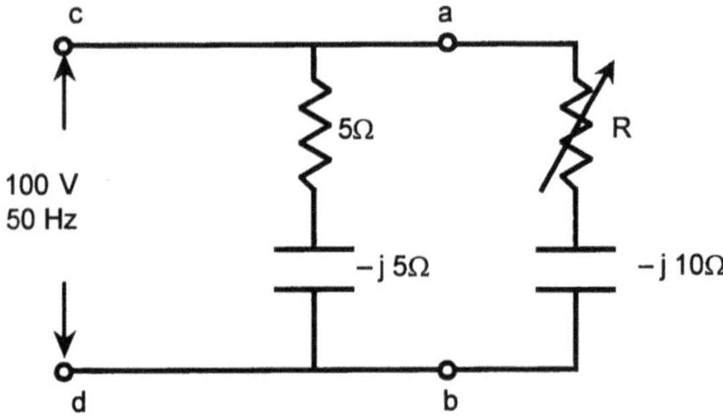

Fig. 3.61(a) Network for example 3.14

Solution :

The impedance locus Z_{ab} is drawn as shown in Fig. 3.61(b). A suitable scale is chosen to draw the locus. Let 1 cm = 2 Ω.

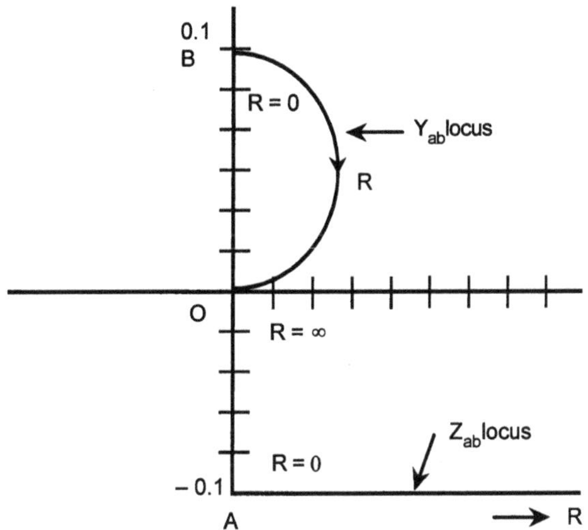

Fig. 3.61(b) Construction of Z_{ab} and Y_{ab} loci

Take a suitable scale for the admittance say, 1 cm = 0.02 S. Mark the point B,

with $$OB = \frac{1}{X_C} = \frac{1}{10} = 0.1$$

on the Y – axis and taking OB as the diameter draw a semi circle to represent the Y_{ab} locus.

Now the fixed admittance $Y_1 = \frac{1}{5 - j5}$ has to be added to this locus to obtain the total admittance locus.

$$Y_1 = \frac{1}{5\sqrt{2}} \angle 45^0 \text{ S} = 0.1414 \angle 45^0 \text{ S}$$

The fixed admittance Y_1 can be added to the variable admittance Y_{ab} by shifting the origin 'O' to O' so that $\overrightarrow{O'O}$ represents Y_1 with respect to the new origin 'O. This is indicated in Fig. 3.61(c).

Now any point such as P represents the total admittance of the circuit when measured with respect to the origin. O'P gives the magnitude and the angle it makes with the O'V axis is the admittance angle. OP reprensts the admittance of the variable branch corresponding to the impedance represented by the point P' on the impedance locus Z_{ab}. Obviously OP' is the impedance of the variable branch a b for a value of R given by R_1 as indicated in Fig. 3.61(c).

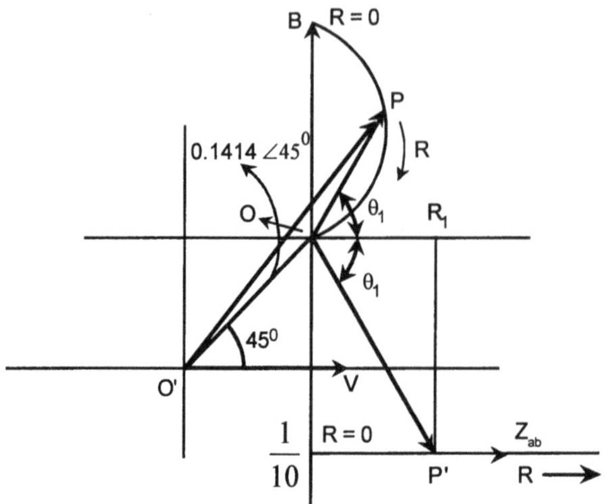

Fig. 3.61(c) Complete admittance and hence the current locus for example 3.14

The total admittance locus also represents the total current locus if the scale chosen for admittance is multiplied by the voltage V = 100 V.

Thus the current scale becomes

$$1 \text{ cm} = 0.02 \times 100$$
$$= 2 \text{ A}$$

The vector $\vec{O'P}$ represents the total current vector w.r.t the voltage vector $\vec{O'V}$.

3.9 MAGNETICALLY COUPLED CIRCUITS :

The mutual inductance was introduced in chapter 1 as a circuit element. When a network consists of these elements in addition to other circuit elements, we know how to write the differential equations governing the behaviour of the network. In this section we will describe methods of solving these networks to obtain their steady state sinusoidal response.

An important physical device based on the principle of mutual inductance is the transformer. An electric power system uses large transformers for increasing and decreasing the voltage amplitudes for economical transmission and distribution of power. Almost all electronic devices and gadgets use smaller transformers. We can say that there is almost no electrical or electronic system which does not use a transformer as one of its components.

In this section we will also study the important characteristics of a linear transformer and an ideal transformer which is an important approximation to a good iron core transformer.

3.9.1 Sinusoidal steady state analysis of magnetically coupled circuits:

The mutual inductance is similar to the self inductance of a coil and hence the behaviour of mutual inductance can be treated in a similar way. The reactance due to the mutual inductance is defined as

$$X_M = \omega M \quad .....(3.116)$$

and the impedance is given by

$$Z_M = j\omega M \quad .....(3.117)$$

Consider the circuit in Fig. 3.62 with the currents and voltage shown.

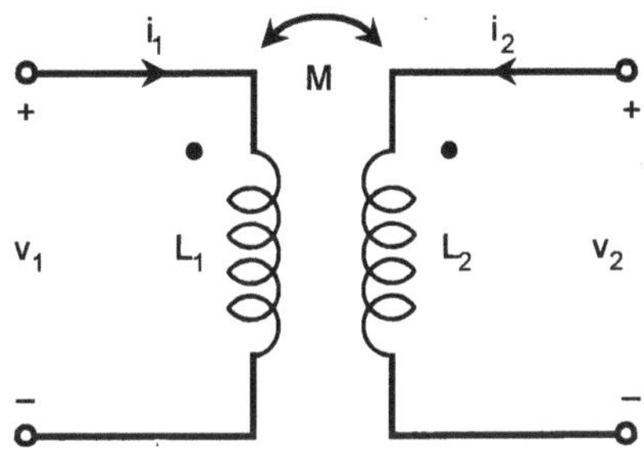

Fig. 3.62 A circuit with mutual inductance

Equations for v_1 and v_2 can be written down easily using the concepts developed earlier in this chapter and the dot convention for mutual inductance introduced in chapter 1. The frequency domain circuit can be obtained as shown in Fig. 3.63.

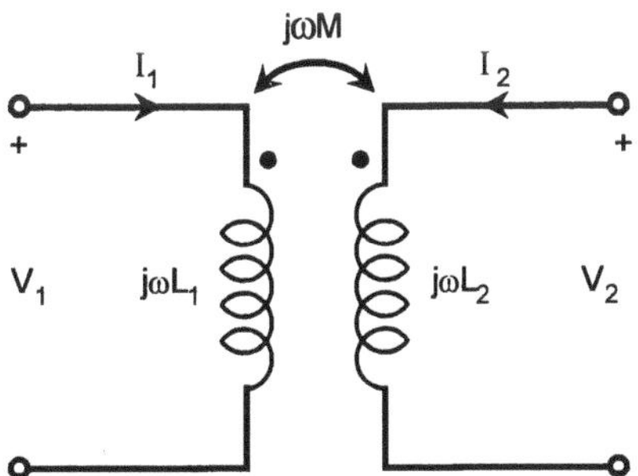

Fig. 3.63 Frequency domain circuit of network in Fig. 3.62

Now the equations for V_1 and V_2 can be written down as

$$V_1 = j\omega L_1 I_1 + j\omega M\, I_2 \quad .....(3.118)$$

$$V_2 = j\omega L_2 I_2 + j\omega M\, I_1 \quad .....(3.119)$$

The voltage due to mutual inductance is taken as positive in eqs. (3.118) and (3.119) because the two currents I_1 and I_2 are entering the dotted terminals of the respective coils.

Let us now see how a more complex network involving coils with mutual inductance can be solved by considering a numerical example.

Example 3.15

Solve for the currents I_1 and I_2 in the circuit of Fig. 3.64. Also find the ratio $\frac{V_2}{V_1}$.

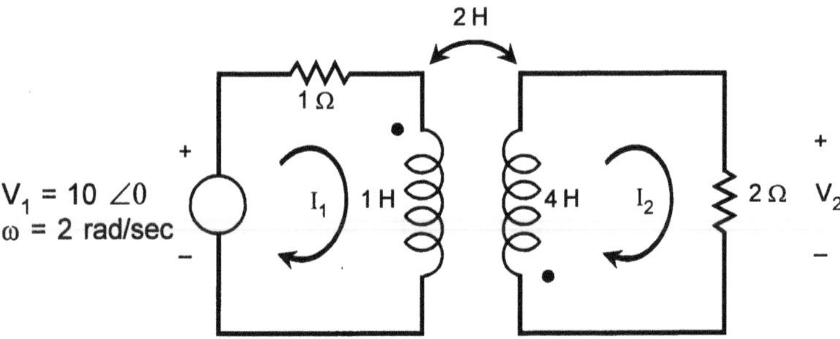

Fig. 3.64 Network for example 3.15

Solution :

Writing the loop equations we get

$$I_1 + j2I_1 + j4I_2 = 10\angle 0^0$$

or $$I_1(1 + j2) + j4I_2 = 10\angle 0^0$$

$$2I_2 + j\,8I_2 + j\,4I_1 = 0$$

or $$I_2(2+j8) + j\,4I_1 = 0$$

Solving the two equations for I_1 and I_2, we have

$$I_1 = \frac{\begin{vmatrix} 10 & j4 \\ 0 & 2+j8 \end{vmatrix}}{\begin{vmatrix} 1+j2 & j4 \\ j4 & 2+j8 \end{vmatrix}}$$

$$I_1 = 6.78\ \angle -4.57^0$$

and $$I_2 = \frac{\begin{vmatrix} 1+j2 & 10 \\ j4 & 0 \end{vmatrix}}{\begin{vmatrix} 1+j2 & j4 \\ j4 & 2+j8 \end{vmatrix}}$$

$$I_2 = 3.29 \angle -170.53^0$$

The ratio $\frac{V_2}{V_1}$ is given by

$$\frac{V_2}{V_1} = \frac{I_2.2}{V_1} = \frac{6.58\angle -170.53^0}{10}$$

$$= 0.658 \angle -170.53^0.$$

Example 3.16

Write a set of mesh equations for the network shown in Fig. 3.65.

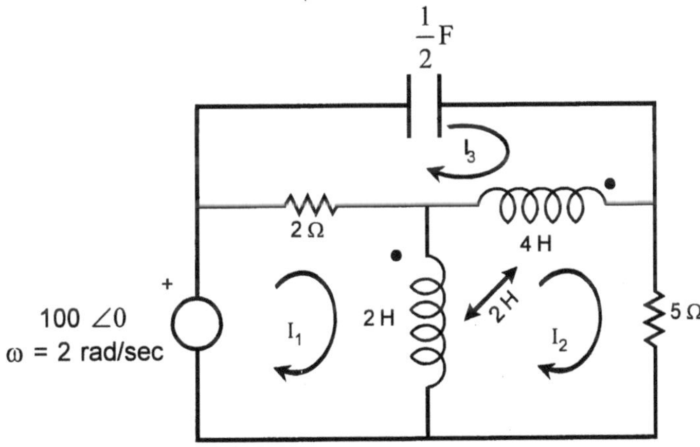

Fig. 3.65 Network for example 3.16

Solution :

The mesh equations can be written down easily as :

$$2\,(I_1 - I_3) + j4\,(I_1 - I_2) + j4\,(I_3 - I_2) = 100$$

$$5I_2 + j4\,(I_2 - I_1) + j8\,(I_2 - I_3) + j4\,(I_2 - I_1) + j4\,(I_2 - I_3) = 0$$

$$-j1\,I_3 + j8\,(I_3 - I_2) + j4\,(I_1 - I_2) + 2\,(I_3 - I_1) = 0$$

Simplify these equations, we have

$$I_1\,(2 + j4) + I_2\,(-j8) + I_3\,(-2 + j4) = 100$$

$$I_1\,(-j8) + I_2\,(5 + j20) + I_3\,(-j12) = 0$$

$$I_1\,(-2 + j4) + I_2\,(-j12) + I_3\,(2 + j7) = 0$$

These three equations can be solved for the three currents I_1, I_2 and I_3.

3.9.2 The linear transformer :

A practical device which may be represented by a network model containing mutual inductance is called the *Linear transformer*. A transformer is a device which contains two coils which are coupled magnetically. A practical linear transformer is used in high frequency circuits such as radio circuits.

Consider the circuit shown in Fig. 3.66. A transformer is shown with two mesh currents. The first mesh contains the source and is called *primary circuit.* The second mesh usually contains the load and is known as *secondary circuit.* The left side coil with self inductance L_1 is called as *primary winding* of the transformer and the right side coil with self inductance L_2 is called as *Secondary winding.* If the flux versus current characteristics of the coil is linear, the transformer is known *linear transformer.* Without a magnetic material, which usually causes a non linear flux Vs current characteristics, the coefficient of coupling between the two coils is very small. The resistors in the circuit are used to represent the resistance of the coils.

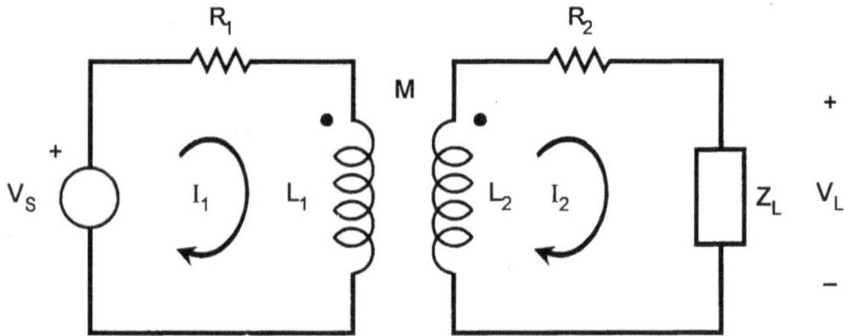

Fig. 3.66 A linear transformer

Writing the mesh equations, we have

$$I_1\,(R_1 + j\omega L_1) - I_2\,j\omega M = V_S \quad .....(3.120)$$

$$-I_1\,j\omega M + I_2(R_2 + j\omega L_2 + Z_L) = 0 \quad .....(3.121)$$

Let $Z_{11} = R_1 + j\omega L_1\,,\; Z_{22} = R_2 + j\omega L_2 + Z_L$

Thus

$$I_1\,Z_{11} - I_2\,j\omega M = V_S \quad .....(3.122)$$

$$-I_1\,j\omega M + I_2\,Z_{22} = 0 \quad .....(3.123)$$

Solving eq. (3.123) for I_2, we have

$$I_2 = I_1 \frac{j\omega M}{Z_{22}} \qquad .....(3.124)$$

Substituting I_2 in eq. (3.122), we get

$$I_1 \left[Z_{11} - \frac{(j\omega M)^2}{Z_{22}} \right] = V_S \qquad .....(3.125)$$

The input impedance defined as

$$Z_{in} = \frac{V_s}{I_1} \qquad .....(3.126)$$

is given by

$$Z_{in} = \frac{V_s}{I_1} = Z_{11} + \frac{\omega^2 M^2}{Z_{22}} \qquad .....(3.127)$$

Several important conclusions can be drawn from eq. (3.127).

1. The expression for Z_{in} is independent of the location of the dots for the two coils. If either of the dots is moved to the other end of the coil, the mutual inductance terms in the equation change their sign, or it is as good as replacing M by (–M) in the above equations. As can be seen from the expression for Z_{in}, this does not change eq. (3.127).
2. The input impedance is simply Z_{11} if the coupling between the coils is reduced to zero. The input impedance is increased by an amount equal to $\frac{\omega^2 M^2}{Z_{22}}$ known as *reflected impedance.*
3. Simplifying eq. (3.127) we have

$$Z_{in} = R_1 + j\omega L_1 + \frac{\omega^2 M^2 R_{22}}{R_{22}^2 + X_{22}^2} - \frac{j\omega^2 M^2 X_{22}}{R_{22}^2 + X_{22}^2} \qquad .....(3.128)$$

where $R_{22} = R_2 + R_L$
$X_{22} = \omega L_2 + X_L$
R_L = Resistive part of Z_L
and X_L = Reactive part of Z_L

which may be positive or negative depending on whether the load is inductive or capacitive.

Since the primary resistance R_1 is increased by a term $\frac{\omega^2 M^2 R_{22}}{R_{22}^2 + X_{22}^2}$ which is positive it follows that the presence of secondary increases the power losses in the primary circuit.

The reflected reactance of the secondary has a sign which is opposite to that of the net reactance of the secondary circuit, X_{22}. The net reactance of secondary is positive if the load is inductive and may have either sign for a capacitive load, depending an the magnitude of the load reactance.

4. If a capacitance C is introduced into the primary network in series with the source and a capacitor of value C is used as a load for a linear transformer with identical primary and secondary coils, we have $R_1 = R_2 = R$ and $L_1 = L_2 = L$ and series resonance occurs in both primary and secondary for a frequency and $\omega_r = \frac{1}{\sqrt{LC}}$. At this frequency the reactances of primary and secondary are zero and there is no reflected reactance. The circuit behaves purely as a resistive circuit. At a slightly higher frequency, the net primary and secondary reactance are both inductive, but the reflected reactance is capacitive. For sufficiently large coefficient of coupling resonance may again occur at this slightly higher frequency. At a slightly lower frequency than ω_r, both the circuits will be capacitive but the reflected reactance will be inductive and cancellation may occur in the primary circuit, making the net impedance purely resistive. Hence at this lower frequency resonance may again occur. Thus unlike a normal series resonance circuit, a resonant circuit using linear transformer may have maximum response over a wider range of frequencies. This properly is made use of in all the communication systems.

5. A linear transformer may be replaced by a T – equivalent circuit or a π equivalent circuit. A T – equivalent circuit is developed here as it is more frequently used than the π - equivalent.

   Consider the linear transformer shown in Fig. 3.67.

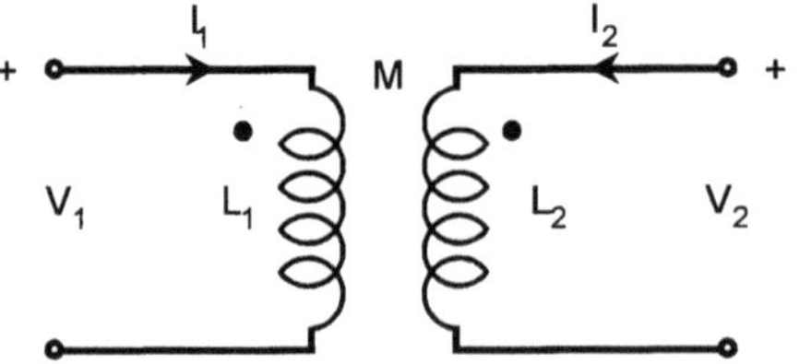

Fig. 3.67 Linear transformer

The loop equations are

$$V_1 = j\omega L_1 I_1 + j\omega M I_2 \quad .....(3.129)$$

$$V_2 = j\omega I_1 + j\omega L_2 I_2 \quad .....(3.130)$$

Consider a T network shown in Fig. 3.68.

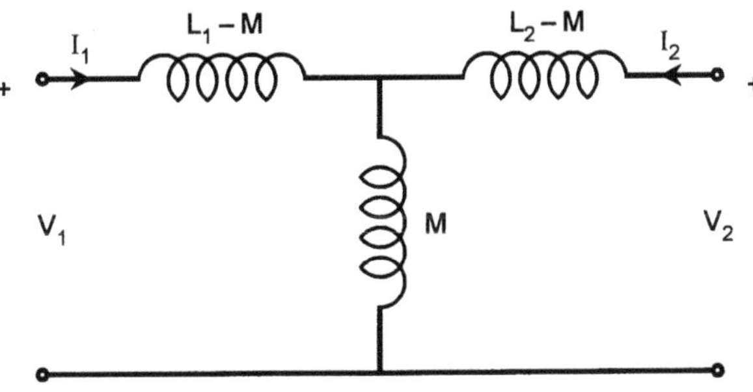

Fig. 3.68 A T equivalent network

Writing the mesh equation for this network we have

$$V_1 = j\omega L_1 I_1 + j\omega M I_2 \qquad .....(3.131)$$

$$V_2 = j\omega M I_1 + j\omega L_2 I_2 \qquad .....(3.132)$$

Which are exactly same as eq. (3.129) and (3.130). If either of the dots on the coils is changed to the other end of the coil the mutual inductance M is changed to (– M), thus making the inductance in the series arms as $L_1 + M$, $L_2 + M$ respectively and the shunt arm inductance to be – M. The three coils shown in the T – equivalent have self inductance only and no mutual inductance.

It is to be noted that the inductance values in the T – equivalent may be negative. Hence it is not possible to construct a practical T equivalent network with negative inductances but as far as mathematical analysis is concerned, the negative inductances are of no consequence.

Example 3.17

Find the T – equivalent of the linear transformer shown in Fig. 3.69(a).

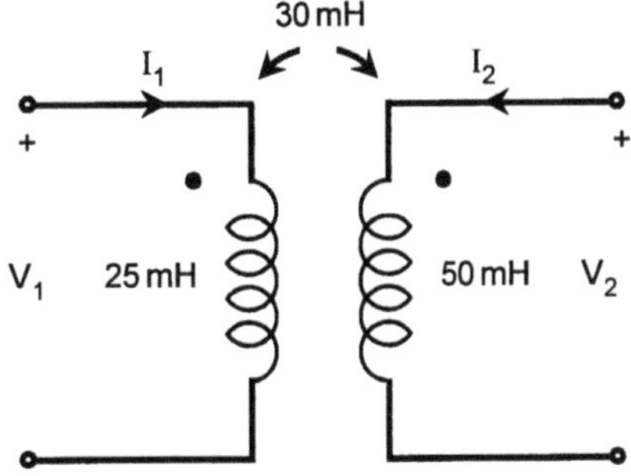

Fig. 3.69(a) Network for example 3.17

Solution :

We have $L_1 = 25$ mH $L_2 = 50$ mH

and $M = 30$ mH

Hence $L_1 - M = -5$ mH

$L_2 - M = 20$ mH

and $M = 30$ mH

The T equivalent network is shown in Fig. 3.69(b).

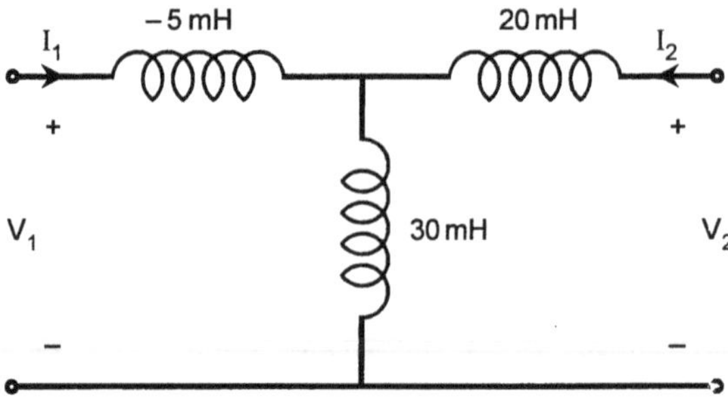

Fig. 3.69(b) T equivalent of network in Fig. 3.69(a)

3.9.3 The ideal transformer :

An ideal transformer is an approximation of a very tightly coupled transformer. The coefficient of such a transformer is almost unity and the primary and secondary reactances are very large in comparison to the other impedances in the circuit in which it is connected. A properly designed iron cored transformer will have these features over a reasonable range of frequencies and load impedances. The analysis of such a transformer can be easily carried out using the model of an ideal transformer.

In the analysis of an ideal transformer, the turns ratio of the primary and secondary coils of the transformer plays an important role. We know that the self inductance of the coil is proportional to the square of the number of turns of the coil. Thus, if we denote the ratio of number of turns in secondary to primary by 'a', we have

$$\frac{L_2}{L_1} = \frac{N_2^2}{N_1^2} = a^2 \qquad .....(3.133)$$

where $$a = \frac{N_2}{N_1} \qquad .....(3.134)$$

Consider an ideal transformer connected to a load as shown in Fig. 3.70.

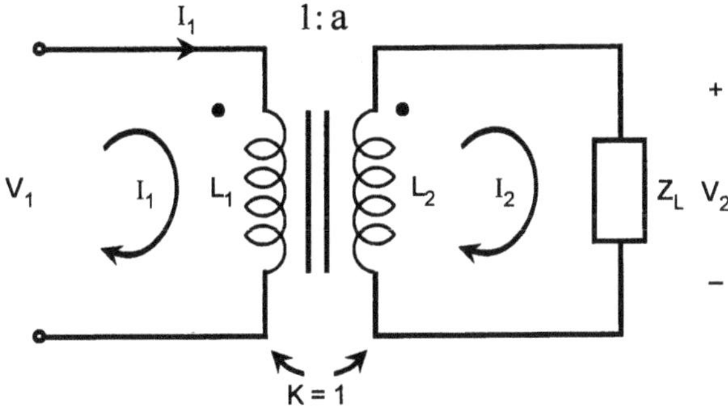

Fig. 3.70 An ideal transformer connected to a load

The symbolic representation of an ideal transformer is : two coils with vertical lines between the coils, to indicate the laminations of iron core, as shown in Fig. 3.70. The turns ratio is shown as 1 : a.

Under steady state sinusoidal conditions, the loop equations can be written as

$$j\omega L_1 I_1 - j\omega M I_2 = V_1 \qquad .....(3.135)$$

$$-j\omega M I_1 + (j\omega L_2 + Z_L) I_2 = 0 \qquad .....(3.136)$$

Let us determine the input impedance of the transformer, which is $Z_{in} = \frac{V_1}{I_1}$.

Solving eqs. (3.135) and (3.136) for I_1, we get

$$I_1 = \frac{V_1}{j\omega L_1 + \dfrac{\omega^2 M^2}{Z_L + j\omega L_2}} \qquad .....(3.137)$$

Since, K = 1 for an ideal transformer, we have $M = \sqrt{L_1 L_2}$. Further, for an ideal transformer the self inductances approach infinite value but their ratio remains finite and is given by eq (3.133). Substituting $M^2 = L_1 L_2$ and $L_2 = a^2 L_1$ in eq. (3.137) we have

$$I_1 = \frac{V_1}{j\omega L_1 + \dfrac{a^2\omega^2 L_1^2}{Z_L + j\omega a^2 L_1}} \qquad .....(3.138)$$

The input impedance is given by

$$Z_{in} = j\omega L_1 + \frac{\omega^2 a^2 L_1^2}{Z_L + j\omega a^2 L_1} \qquad .....(3.139)$$

If now L_1 is made infinity, the expression in eq. (3.139) becomes indeterminate and in order to find the limit as L_1 tends to infinity, we have to combine both the terms in eq. (3.139). Thus,

$$Z_{in} = \frac{j\omega L_1 Z_L - \omega^2 a^2 L_1^2 + \omega^2 a^2 L_1^2}{Z_L + j\omega a^2 L_1} \quad .....(3.140)$$

$$Z_{in} = \frac{j\omega L_1 Z_L}{Z_L + j\omega a^2 L_1} \quad .....(3.141)$$

If L_1 is made to approach infinity, we have

$$Z_{in} = \frac{Z_L}{a^2} \quad .....(3.142)$$

Eq. (3.142) indicates that the input impedance of an ideal transformer is proportional to the load impedance, the proportionality constant being the reciprocal of the square of the turns ratio. This implies that if the load impedance is capacitive, the input impedance also is capacitive. However in a linear transformer, we recall that there was a sign change in the reflected impedance, ie., if the load is capacitive the reflected impedance was inductive. Thus it appears to contradict one of the characteristics of a linear transformer, which it should not, because a linear transformer is a more general model. This can be explained by noting that $\frac{Z_L}{a^2}$ is not the reflected impedance. The actual reflected impedance is infinite so that it cancels the infinite impedance of the primary of an ideal transformer. As can be seen from the eq. (3.140), $\frac{Z_L}{a^2}$ is a small amount by which the exact cancellation of primary reactance and reflected reactance does not occur. The reflected reactance does change sign, so that it cancels the primary reactance.

From eq. (3.142) it is seen that one of the important characteristics of an ideal transformer is its ability to change the magnitude of an impedance, or to change impedance level. An ideal transformer with 1000 primary turns and 10,000 secondary turns will have a turns ratio of $a = 10$. Any impedance connected across the secondary of the transformer will appear at the primary terminals reduced in magnitude by $10^2 = 100$. Thus a 100 Ω resistance looks like a 1 Ω resistance at the primary terminals. If the primary and secondary terminals are interchanged, the turns ratio becomes 0.1 and the load impedance appears magnified by a factor of $\left(\frac{1}{0.1}\right)^2 = 100$.

This property of an ideal transformer is made use of in impedance matching circuits. As will be shown in chapter 6, to achieve maximum power transfer from the source to the load, it is neassary that the resistance of the load be equal to the source resistance. In the circuit of a power amplifier connected to a speaker, the speaker resistance is usually a few ohms where as the power amplifier has an internal resistance of a few thousand ohms. To couple the speaker to the amplifier, we require an ideal transformer so that maximum power is transferred to the speaker. For example if the power amplifier has a resistance of 5000 Ω and the speaker impedance is 8 Ω, for maximum power transfer, we should have,

$$Z_g = \frac{Z_L}{a^2}$$

or $$a^2 = \frac{Z_L}{Z_g} = \frac{8}{4000}$$

and $$a = \frac{1}{22.36}$$

Hence the speaker should be coupled to the power amplifier with an ideal transformer with a turns ratio $\frac{N_1}{N_2} = 22.36$.

Relationship between primary and secondary currents :

The relationship between primary and secondary currents can be obtained by considering eq. (3.136). Thus

$$\frac{\mathbf{I}_2}{\mathbf{I}_1} = \frac{j\omega M}{Z_L + j\omega L_2} \qquad .....(3.143)$$

If we allow L_2 to become infinite and noting that Z_L is small compared to $j\omega L_2$, we have

$$\frac{\mathbf{I}_2}{\mathbf{I}_1} = \frac{j\omega M}{j\omega L_2} = \frac{\sqrt{L_1 L_2}}{L_2} = \sqrt{\frac{L_1}{L_2}}$$

or $$\frac{\mathbf{I}_2}{\mathbf{I}_1} = \frac{1}{a} \qquad .....(3.144)$$

Thus, the ratio of secondary current to primary current is the inverse of the turns ratio. Since

$$a = \frac{N_2}{N_1}$$

$$\frac{\mathbf{I}_2}{\mathbf{I}_1} = \frac{N_1}{N_2}$$

and $$\mathbf{I_1\, N_1 = I_2\, N_2} \qquad .....(3.145)$$

Relationship between primary and secondary voltages :

We know that

$$V_2 = I_2 Z_L$$

and $$V_1 = I_1 Z_{in}$$

$$= I_1 \frac{Z_L}{a^2}$$

Thus the ratio $\frac{V_2}{V_1}$ is given by

$$\frac{V_2}{V_1} = \frac{I_2 Z_L}{I_1 Z_L} \cdot a^2 = \frac{I_2}{I_1} \cdot a^2$$

or $$\frac{V_2}{V_1} = a = \frac{N_2}{N_1} \qquad .....(3.146)$$

Hence the ratio of secondary to primary voltage is equal to the turns ratio.

Now combining the equations (3.144) and (3.146), we have

$$\frac{V_2}{V_1} = \frac{I_1}{I_2}$$

or $$V_1 I_1 = V_2 I_2 \qquad .....(3.147)$$

Eq. (3.147) indicates that the complex volt-amperes in primary and secondary are equal.

Example 3.18

Find the primary current, secondary current and power delivered to the load in the circuit of Fig. 3.71.

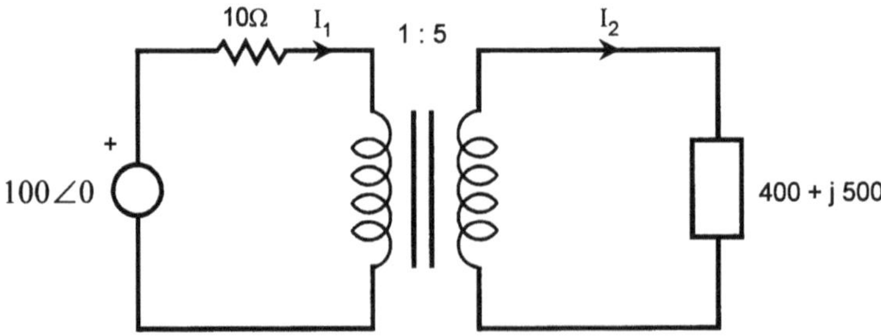

Fig. 3.71 Network for example 3.18

Solution :

The input impedance is given by

$$Z_{in} = 10 + \frac{400 + j500}{5^2}$$

$$= 10 + 16 + j20$$
$$= 26 + j20\ \Omega$$

The primary current I_1 is given by

$$I_1 = \frac{V_I}{Z_{in}}$$

$$= \frac{100\angle 0^0}{26 + j20} = 3.05\ \angle -37.57^0 \text{ amps}$$

Since $\frac{I_1}{I_2} = a = 5$

$$I_2 = \frac{I_1}{5} = \frac{3.05\angle -37.57}{5} = 0.61\ \angle -37.57^0 \text{ amps}$$

The power delivered to the load is given by

$$P_L = I_2^2\ R_L$$
$$= (0.61)^2 \times 400$$
$$= 148.84 \text{ watts}$$

Problems

3.1 Write an expression for a sine wave current having a maximum value of 1.732 amps and a frequency of 1620 kHz. Select t = 0 reference at a point where $\frac{di}{dt}$ is positive and i = + 1.5 amps.

3.2 The time variation of a voltage wave is given by

$$e = 100 \sin 157\, t \text{ volts.}$$

where 't' is time is seconds.

(a) What is the maximum value of the voltage ?

(b) What is the frequency of voltage variation ?

(c) What is the r.m.s value of the voltage ?

3.3 If two voltage v_1 and v_2 are given by

$$v_1 = 100 \sin (\omega t - 30^0)$$

and $v_2 = 10 \sin (\omega t - 60^0)$.

What is the resultant voltage $v = v_1 + v_2$?

3.4 Add the following currents :

$$i_1 = 5 \sin \omega t$$
$$i_2 = 10 \cos (\omega t + 30^0)$$
$$i_3 = -5 \sin (\omega t - 30^0)$$

3.5 Calculate the r.m.s values of the voltage for each of the voltage waveforms shown in Fig. P. 3.1.

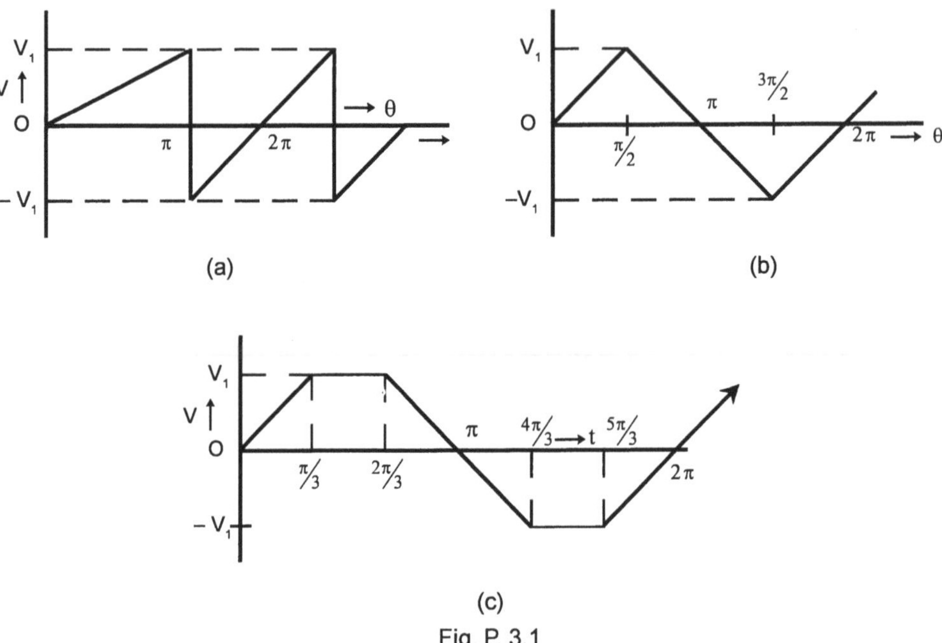

Fig. P. 3.1

3.6 Calculate the half period average value of the voltage waveforms shown in Fig. P. 3.1.

3.7 A current starts abruptly at 10 amps and decrease linearly to zero and then repeats the cycle. Find the rms value of the waveform.

3.8 The voltage of a circuit is

$$v = 200 \sin (\omega t + 30^0)$$

The current is $i = 50 \sin (\omega t + 60^0)$

Calculate the average power, reactive volt amps and apparent power.

3.9 A var meter in a circuit shows 600 VARS and a wattmeter shows 800 watts. Find the apparent power and power factor of the circuit.

3.10 A series circuit has 8 Ω resistance and 20 mH inductance. If 220 volts at 50 Hz, is applied to the combination calculate the current and power.

3.11 Find the form factor of wave forms shown in Fig. P. 3.2.

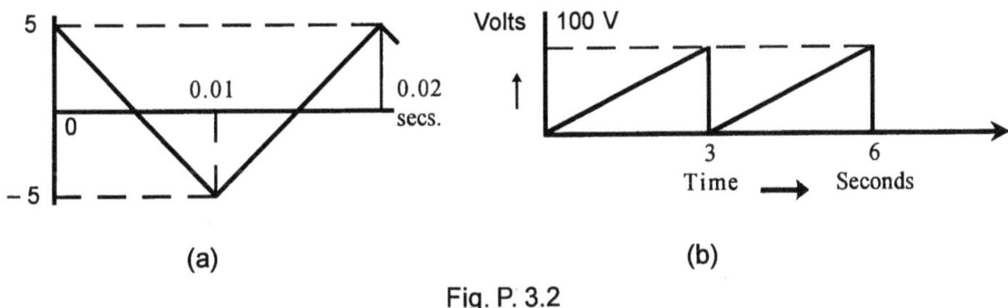

Fig. P. 3.2

3.12 A voltage of 100 sin 100 t is applied to the circuit of Fig. P. 3.3 operating under steady state. Calculate the instantaneous values of the current and voltages across R and C when t = 0.001 sec. Calculate the instantaneous power inputs to R and C at that instant. What are the average power inputs to R and C.

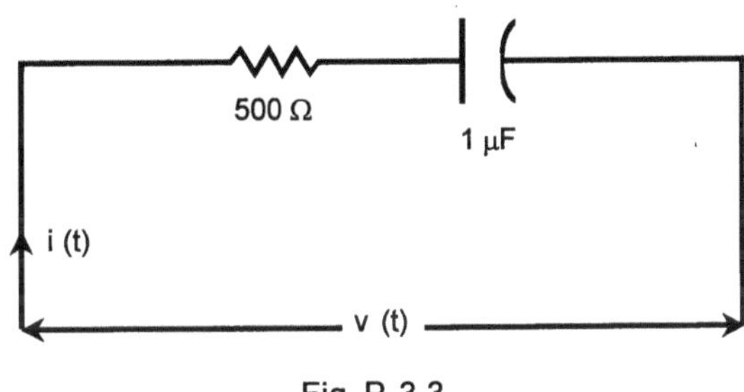

Fig. P. 3.3

3.13 In the series parallel circuit shown in Fig. P. 3.4, the effective value of the voltage across the parallel part of the circuit is 100 volts. Determine the magnitude of the supply voltage V_s and the power supplied by it. Draw the vector diagram.

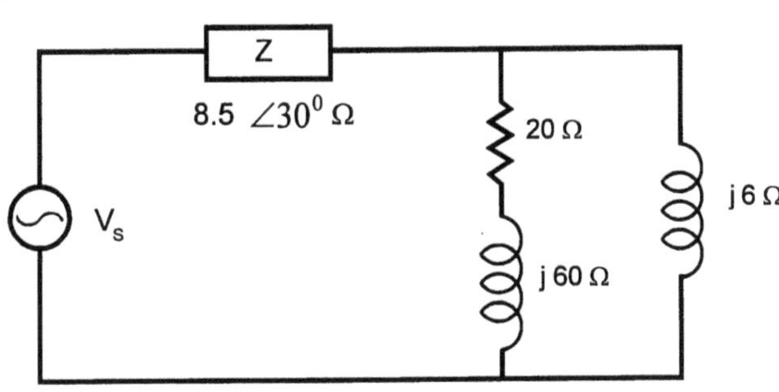

Fig. P. 3.4

3.14 The current in a two ohm resistor has a wave from as shown in Fig. P. 3.5. The maximum value of the current is 5 amps. Calculate the average power dissipated in the resistor for (a) $\theta = 60^0$ and (b) $\theta = 120^0$.

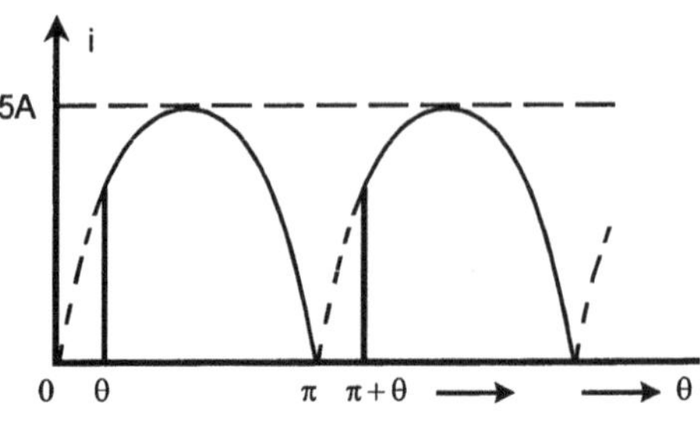

Fig. P. 3.5

3.15 Calculate the equivalent impedance of the circuit shown in Fig. P. 3.6 and the power consumed in each branch.

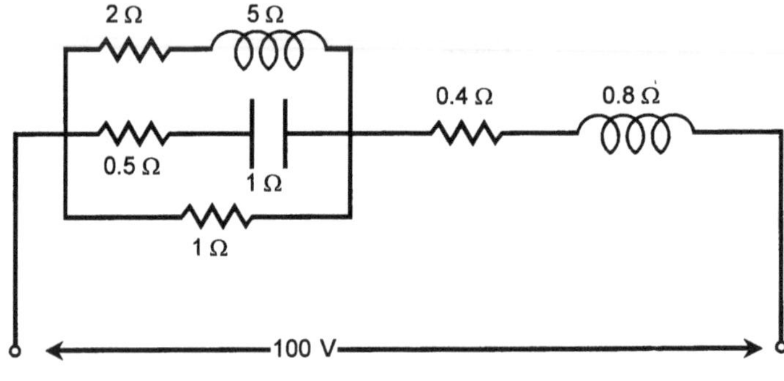

Fig. P. 3.6

3.16 In the of circuit of Fig. P. 3.7, Calculate the supply current, overall power factor and the total power supplied to the circuit.

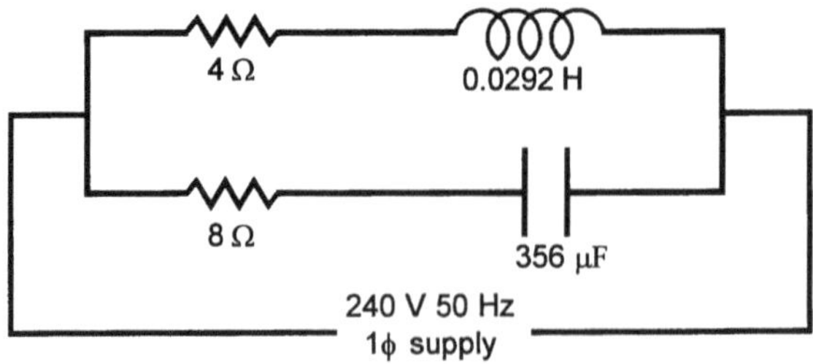

Fig. P. 3.7

3.17 A voltage of 240 V is applied to a pure resistor, a pure capacitor and an inductor in parallel. If the total current and the currents in the resistor, capacitor and inductor are respectively 2.3 A, 1.5 A, 2.0 A and 1.1 A find the overall power factor and the power factor of the inductor.

3.18 A coil has a resistance of 5 Ω and an inductance of 0.05 H. Find a suitable shunt circuit such that the current taken by the combination will be 20 A at 100 V at all frequencies.

3.19 Find the condition that the currents in the two branches of the a.c circuit shown in Fig. P. 3.8 shall remain in quadratic when R_1 and R_2 are varied simultaneously. Determine the frequency at which the total current remains constant. What is the magnitude of the current ?

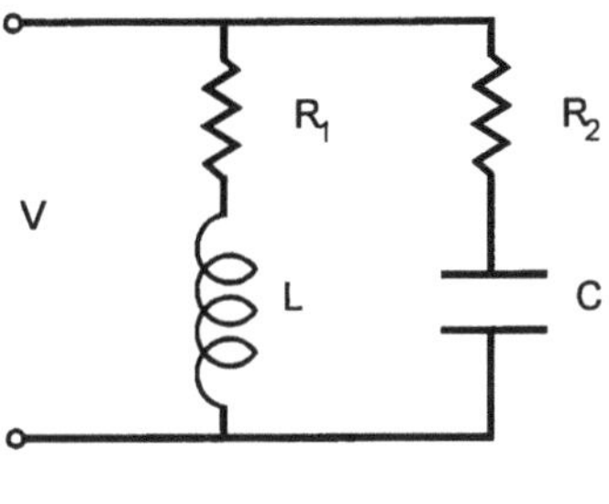

Fig. P. 3.8

3.20 Find the values of R_1 and X_1 if the circuit shown in Fig. P. 3.9 takes a lagging current and dissipates a power of 2 kW.

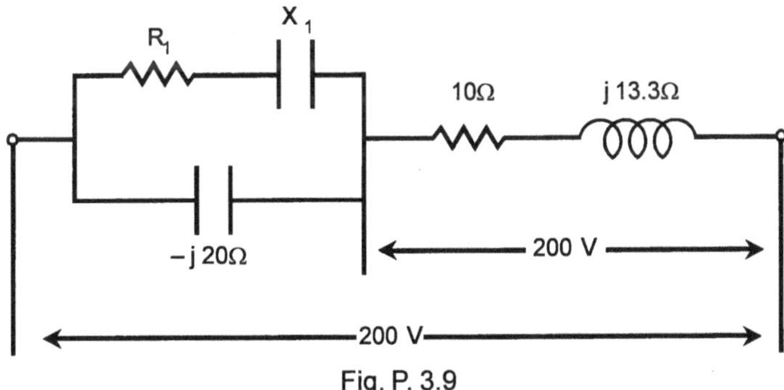

Fig. P. 3.9

3.21 A resistor and a capacitor are in series with a variable inductor. When the circuit is connected to a 200 volt 1 ϕ 50 Hz a.c supply, a maximum current of 0.314 amps was obtained by varying the inductance. The voltage across the capacitance was then 300 volts. Calculate the circuit constants.

3.22 A simple series circuit consisting of a capacitance of reactance 200 ohms and a variable resistance R is connected to a 250 V 50 Hz 1ϕ a.c supply. If R varies between zero and 400 ohms, draw the locus of the extremity of the current phasor. What are the values of current and power factor for maximum power conditions.

3.23 A parallel circuit consists of a 25 Ω inductive reactance in one branch and a resistance of 6.25 Ω in series with a variable capacitance in the other. If the circuit voltage is 250 volts and the capacitor is assumed to vary from 0 to ∞ draw the locus of the current drawn and determine (a) the minimum current and the power factor at which it occurs. (b) the maximum current and the power factor.

3.24 Draw the locus diagram and obtain the value of R_L in the circuit shown in Fig. P. 3.10 which results in resonance for the circuit.

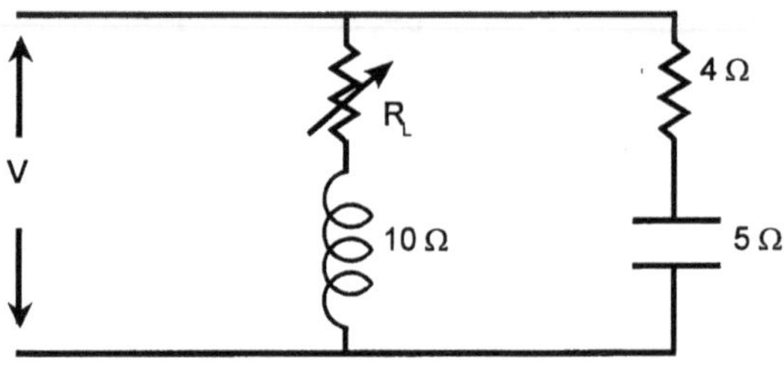

Fig. P. 3.10

3.25 A tuned circuit consists of a coil having an inductance of 200 microhenries and a resistance of 20 Ω in parallel with a series combination of a variable capacitance and a resistance of 80 ohms. It is supplied by a 60 V, 1 MHz source. What is the value of C to give resonance and what is the Q factor of this circuit.

3.26 In a series RLC Circuit R = 1 KΩ, L = 120 mH and C = 12 μμF. If a voltage of 200 V is applied across the combination, determine

(i) resonance frequency
(ii) Q factor
(iii) half power frequencies
(vi) Band width and
(v) the voltage across the inductance and the capacitance

3.27 For the parallel resonant circuit shown in Fig. P. 3.11 obtain the value of capacitance at which maximum impedance occurs at a given frequency. Also obtain the value of inductance at which maximum impedance occurs at a given frequency.

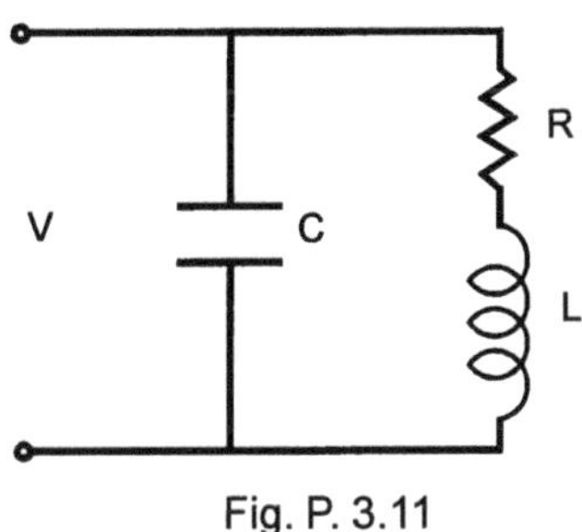

Fig. P. 3.11

3.28 Find the average power delivered to the 5 Ω resistor in Fig. P 3.12 ω = 1000 rad/sec.

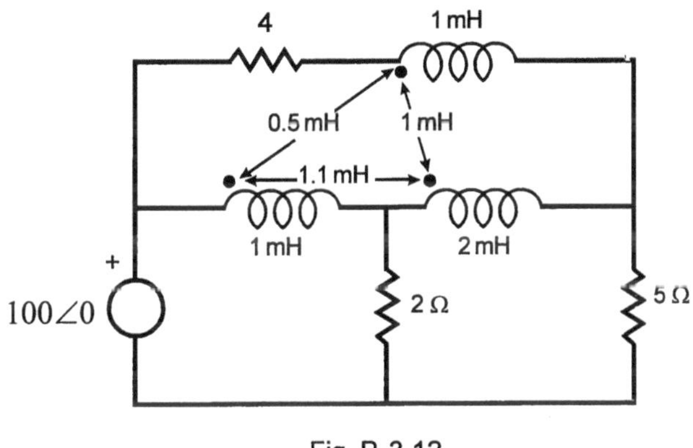

Fig. P. 3.12

3.29 Find the current supplied by the source in Fig. P. 3.13 ω = 100 rad/sec.

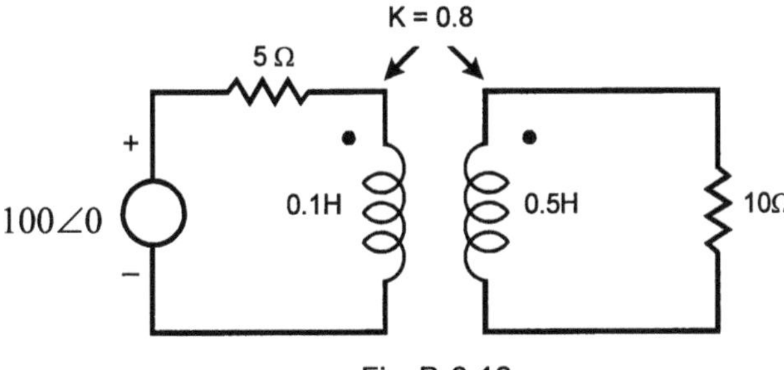

Fig. P. 3.13

3.30 Find the current supplied by the source in Fig. P. 3.14.

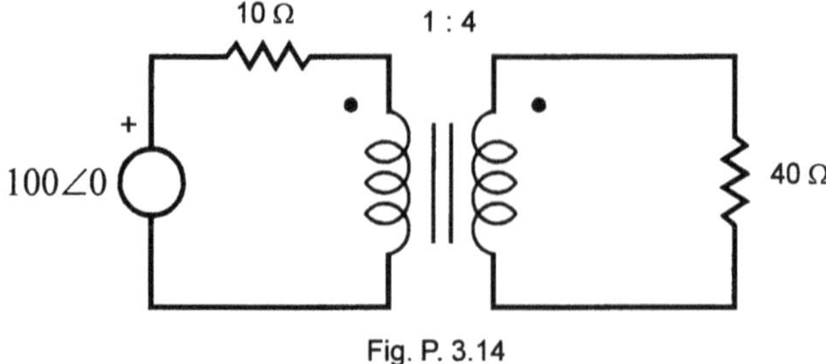

Fig. P. 3.14

3.31 Find the average power delivered to the load Z_L in Fig. P. 3.15.

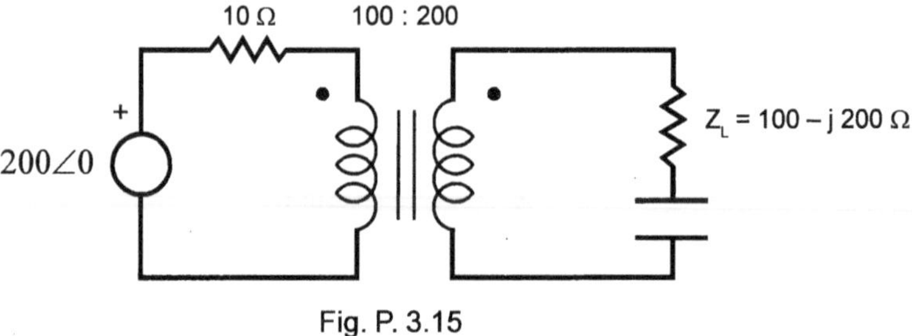

Fig. P. 3.15

3.32 Find $Z_{in}(s) = \dfrac{V_{in}(s)}{I(s)}$ in the network of Fig. P. 3.16.

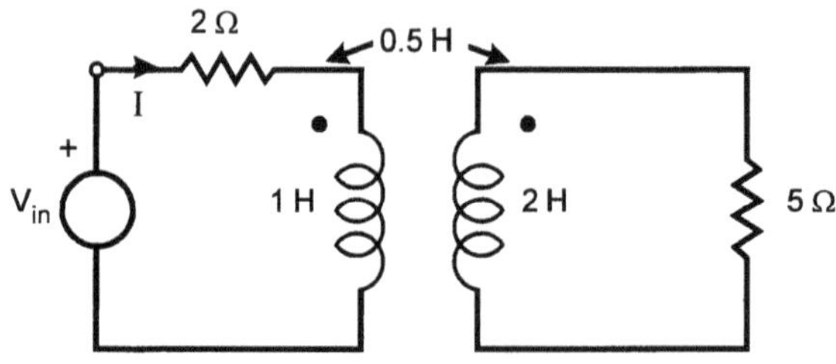

Fig. P. 3.16

4

THREE PHASE CIRCUITS

4.1 INTRODUCTION :

Single phase alternating current systems were considered in the previous chapter because nearly all residential and commercial facilities all over the world are supplied power from 120/240 Volt 60/50 Hz sinusoidal systems. However, the power generated, transmitted and distributed by the power supply companies is invariably a three phase power at 50/60 Hz. While the use of more than three phases is possible, the technical and economic advantages are not significant enough to justify the extra complications. In view of this, except in special cases, three phase systems containing sources which are closely approximated as ideal voltage sources with equal magnitudes of voltages but displaced by 120^0 from each other in phase, are in use.

4.2 ADVANTAGES OF THREE PHASE OVER SINGLE PHASE :

The advantages of three phase working over single phase operation can be summarised as follows :

1. The output from a three phase machine is greater than a single phase machine of the same size.
2. There is considerable economy in the conductor material required for transmission of a given amount of power at a given voltage.
3. Efficiency of three phase generation, transmission is higher compared to single phase while transmitting a given amount of power at a specified voltage.
4. Three phase motors have less vibrations compared to single phase motors.
5. Rectifiers are used for conversion of a.c to d.c. The use of 6 or 12 phase systems as inputs to such rectifiers produces less ripple in the d.c output.

In general, the operating features of three phase appliances are far superior to their single phase counterparts. The complications involved are insignificant compared to the advantages.

4.3 GENERATION OF THREE PHASE VOLTAGES :

While the actual design and construction is more involved, a single phase two pole generator can be represented by a two conductor coil as shown in Fig. 4.1,

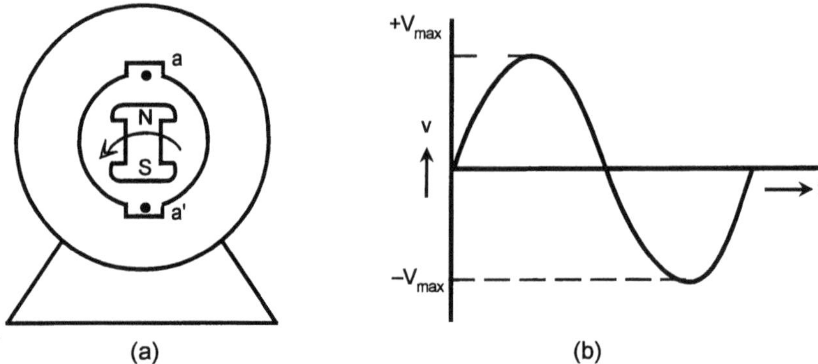

Fig. 4.1 (a) Cross sectional view of a single phase generator
(b) Voltage generated in a single phase generator

a and a' represent the two ends of a coil on a stationary armature. A voltage is generated in this coil as the two pole rotor rotates. This voltage is shown in Fig. 4.1(b). To generate a three phase voltage, two more identical coils b – b' and c – c' displaced by 120^0 and 240^0 degrees from coil a – a' respectively can be installed on the machine as shown in Fig. 4.2(a). As the two pole rotor rotates, the voltages $V_{aa'}$, $V_{bb'}$ and $V_{cc'}$ across the three coils aa', bb' and cc' respectively are shown in Fig. 4.2(b). Here we have used double subscripts to indicate the voltage between the two ends such as a and a', of a coil. In this notation, the end representing the first subscript is assumed to be at a higher potential than the end represented by the second subscript. Thus $V_{aa'}$ represents the voltage rise from end a' to a of the coil a – a'.

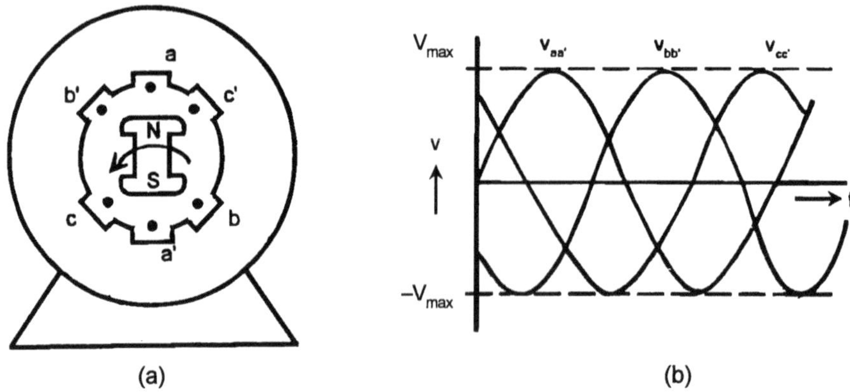

Fig. 4.2 (a) Cross sectional view of a 3 phase generator
(b) Waveform of voltages in the three coils

The three voltages are sinusoidal in nature and are equal in magnitude and displaced from each other by 120^0. These constitute a set of three phase balanced voltages. With $V_{aa'}$ as reference, these voltages can be represented by the following set of expressions.

$$\left.\begin{aligned} v_{aa'} &= V_m \sin \omega t \\ v_{bb'} &= V_m \sin (\omega t - 120^0) \\ v_{cc'} &= V_m \sin (\omega t - 240^0) \end{aligned}\right\} \quad .....(4.1)$$

where $V_m = \sqrt{2}\,V$, is the maximum value of the voltage. The phasor representation of these voltages is given in Fig. 4.3.

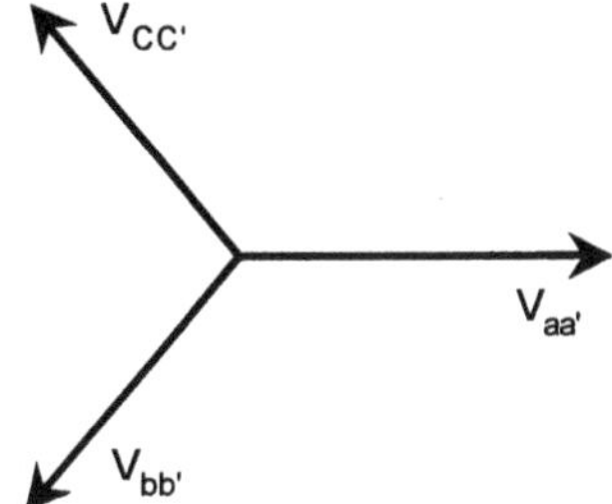

Fig. 4.3 Phasor representation of balanced 3 phase voltages

These voltages in phasor form are given by

$$\left.\begin{aligned} V_{aa'} &- V \angle 0^0 \\ V_{bb'} &= V \angle -120^0 \\ V_{cc'} &= V \angle -240^0 \end{aligned}\right\} \quad .....(4.2)$$

It is easy to verify that the sum of the voltages $v_{aa'}$, $v_{bb'}$ and $v_{cc'}$ is zero. Thus in a three phase balanced system, algebraic sum of the instantaneous voltages is always equal to zero.

Two other forms of representation, called the coil representations are shown in Fig. 4.4. Here all the three coils aa', bb' and cc' are shown side by side as shown in Fig. 4.4(a) or placed at 120^0 from each other as shown in Fig. 4.4(b).

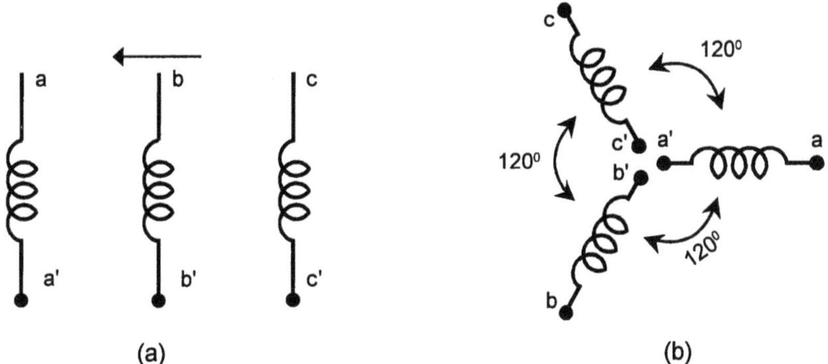

Fig. 4.4 Two forms of coil representation of the three phase set

The arrow indicates the relative motion of the conductors past a point on a pole. Accordingly the order in which the coils move past a pole is abc. This order is called '*phase sequence*'. If however, the poles are rotated in the opposite direction, the order in which the coils go past a pole will be 'acb'. Phase sequence plays an important role in the analysis of 3 phase networks.

4.4 INTERCONNECTION OF THREE PHASE WINDINGS :

While generation of three phase voltages requires 6 terminals, it is always the practice to connect the windings such that a total of only three or 4 wires are needed to supply power. Among the various possible connections, the ones that are very common are the star and the delta connections.

4.4.1 The star connection :

Refer to Fig. 4.4 the identical ends a', b' and c' may be connected to form a common point. This common point is called the '*neutral*'. Four wires can then be brought out to be connected to external circuits as shown in Fig. 4.5. This type of a

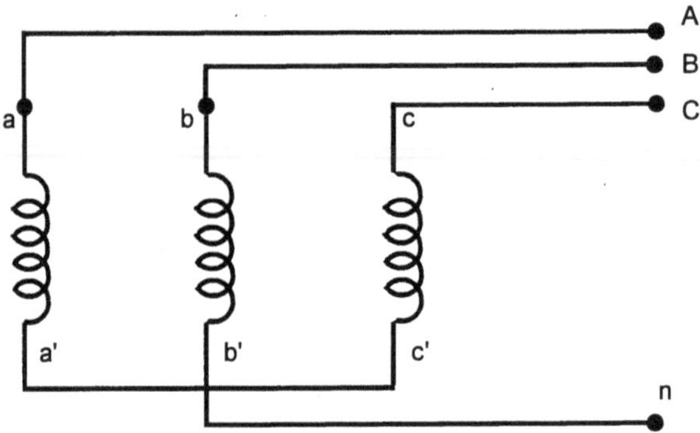

Fig. 4.5 Star connection

connection is called a star or Y connection. The points A, B and C brought out for external connections are called the '*line terminals*' and the point n is called *the neutral*. The voltages V_{AB}, V_{BC} and V_{CA} between the points AB, BC and CA respectively are called '*line voltages*'. The voltage V_{An}, V_{Bn}, and V_{Cn} between A, B or C and n, respectively, are called the '*phase voltages*'. Another way of representing a star connection is to use Fig. 4.4(b) with a', b', and c' joined together.

Voltage Relations in a Star Connected System :

Let us arbitrarily choose $V_{aa'} = V_{An}$ as the reference. Thus

$$V_{An} = V_{aa'} = V_p \angle 0^0 \qquad .....(4.3)$$

where V_p is the rms value of the phase voltage. Then the other two voltages can be written as

$$\left.\begin{aligned} V_{Bn} &= V_p \angle -120^0 \\ V_{Cn} &= V_p \angle -240^0 \end{aligned}\right\} \qquad .....(4.4)$$

These phase voltages can be represented on a phasor diagram as shown in Fig. 4.6. The voltage V_{AB} is calculated from the relations.

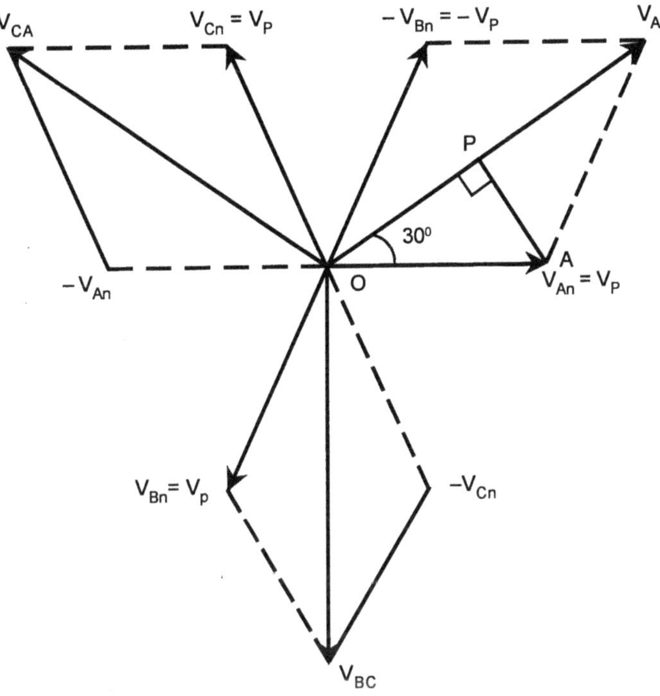

Fig. 4.6 Phasor diagram of a star connection system

$$V_{AB} = V_{An} + V_{nB}$$
$$= V_{An} - V_{Bn} \qquad .....(4.5)$$

In Fig. 4.6,

$$OP = OA\cos 30$$

$$= Vp\frac{\sqrt{3}}{2}$$

and
$$V_{AB} = 2 \, . \, OP$$
$$= \sqrt{3} \; V_P \qquad .....(4.6)$$

The line voltages can thus be represented as

$$\left.\begin{aligned} V_{AB} &= \sqrt{3}\, V_p \angle 30^0 \\ V_{BC} &= \sqrt{3}\, V_p \angle -90^0 \\ V_{CA} &= \sqrt{3}\, V_p \angle -210^0 \end{aligned}\right\} \qquad .....(4.7)$$

Since these voltages are equal in magnitude and displaced by 120^0, the sum of these voltages can be shown to be equal to zero.

Thus $$V_{AB} + V_{BC} + V_{CA} = 0 \quad .....(4.8)$$

If the r.m.s value of each of the line voltages V_{AB}, V_{BC} or V_{CA} is denoted by V_l, then the relation between the r.m.s value of line voltage and the r.m.s value of the phase voltage is given by :

$$V_L = \sqrt{3}\ V_p \quad .....(4.9)$$

It can also be shown that this relationship holds even if the phase sequence is a c b. Thus the line voltage in any balanced star connected system is equal to $\sqrt{3}$ times the phase voltage irrespective of the phase sequence.

Current Relationships in Star Connected System :

The currents flowing in the coils aa', bb' and cc' are known as *phase currents.* Similarly the currents flowing in the lines A, B and C in Fig. 4.5 are termed as line currents. Line currents are the currents which flow in to the network connected to the lines A, B and C. In star connection it is obvious that the line currents are equal to phase currents. Thus, if I_L represents line currents and I_P represents phase currents

$$I_L = I_p \quad .....(4.10)$$

4.4.2 The delta connection :

If the beginning of each coil is connected to the end of another coil, forming a closed circuit as shown in Fig. 4.7 we get a delta connection.

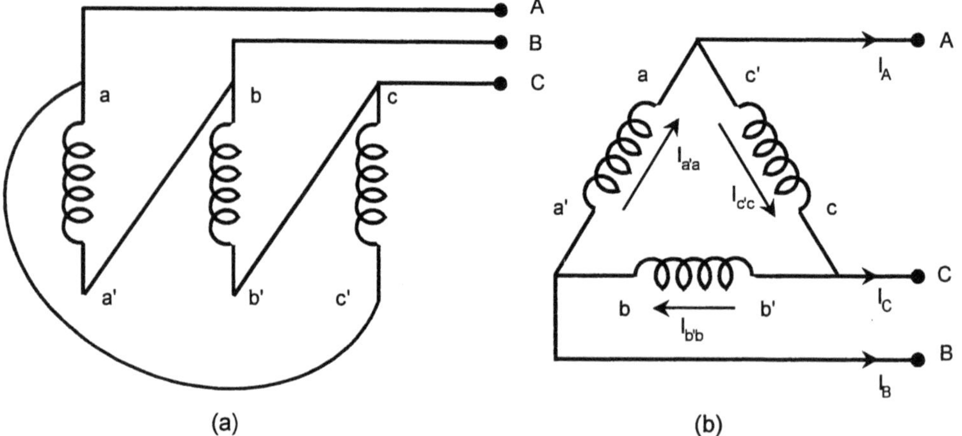

Fig. 4.7 Two alternate forms of Δ connection

Voltage Relationships in a Δ – Connected system :

In Δ – connected system, the phase voltages $V_{aa'}$, $V_{bb'}$ and $V_{cc'}$ are same as the line voltages V_{AB}, V_{BC} and V_{CA}. Thus for a Δ – connected system.

$$V_L = V_p \quad .....(4.11)$$

where V_L is the line voltage and V_p is the phase voltage.

Current Relationships in a Δ – Connected System :

In Fig. 4.7 (b), the currents flowing in coils $I_{a'a}$, $I_{b'b}$ and $I_{c'c}$ are known as *phase currents* and the currents flowing in the lines, I_A, I_B and I_C are known as *line currents*. Applying Kirchhoff's Current Law at each of the three nodes A, B and C we see that

$$I_A = I_{a'a} - I_{c'c}$$
$$I_B = I_{b'b} - I_{a'a} \quad .....(4.12)$$
and $$I_C = I_{c'c} - I_{b'b}$$

If the phase currents $I_{a'a,}$ $I_{b'b}$ and $I_{c'c}$ are equal in magnitude and differ in phase by 120^0 from each other, taking $I_{a'a}$ as reference, they can be represented in phase from as

$$I_{a'a} = I \angle 0^0$$
$$I_{b'b} = I \angle -120^0$$
$$I_{c'c} = I \angle 120^0 \quad .....(4.13)$$

where I is the r.m.s value of the current. The phase currents and line currents are shown in phasor diagram Fig. 4.8.

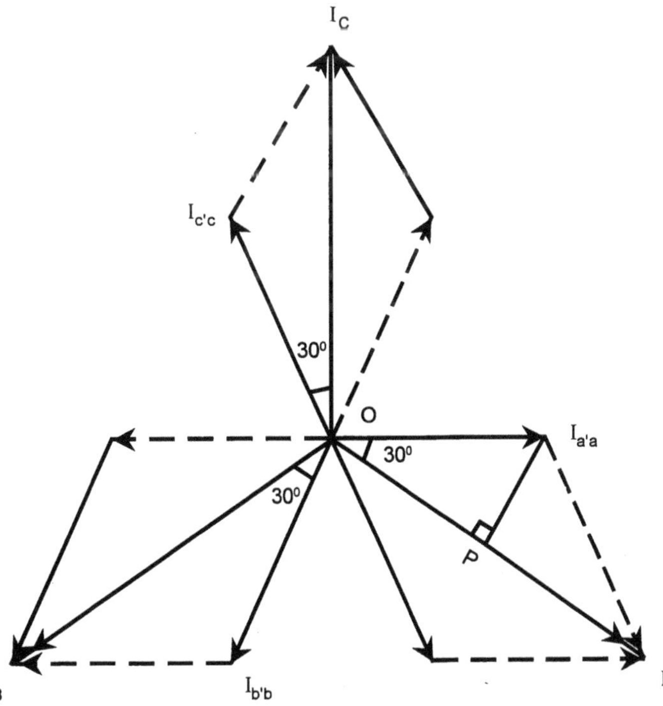

Fig. 4.8 Phasor diagram of a Δ connected system showing phase currents and line currents

From Fig. 4.8

$$OP = I \cos 30 = I \cdot \frac{\sqrt{3}}{2}$$

and $$I_A = 2 \cdot OP = \sqrt{3}\, I$$

It is clear from Fig. 4.8 that the three line currents are equal in magnitude and differ in phase by 120^0 from each other.

The phasor representation of the three line currents are

$$I_A = \sqrt{3}\ I \angle -30^0$$

$$I_B = \sqrt{3}\ I \angle -150^0 \qquad .....(4.14)$$

$$I_C = \sqrt{3}\ I \angle 90^0$$

Thus $$I_L = \sqrt{3}\ I_p \qquad .....(4.15)$$

4.5 ANALYSIS OF 3 PHASE NETWORKS :

A three phase network consisting of 3 impedances connected either in star or Δ, as shown in Fig. 4.9, is called a 3 phase load.

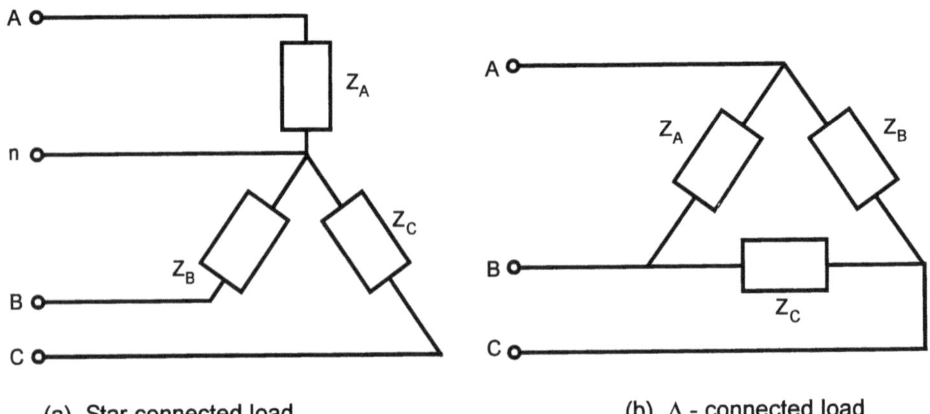

(a) Star connected load (b) Δ - connected load

Fig. 4.9 Three phase networks

These 3 phase loads are supplied from a 3 phase balanced supply. A star connected load can be supplied from a balanced 3 phase 3 wire supply in which the neutral points are not connected, or by a 3 phase 4 wire balanced supply in which the supply neutral and the load neutral are connected together. A delta connected load however can be supplied only from a balanced 3 phase 3 wire supply. We shall now analyze these circuits to determine the currents in the 3 phase loads and power absorbed by them.

4.5.1 Balanced star connected load supplied from balanced 3 phase 4 wire supply :

In Fig. 4.9 if $Z_A = Z_B = Z_C$, then the network is said to be a balanced 3 phase network. This means, if

$$Z_A = R + j\,X$$

then $$Z_B = Z_C = R + j\,X$$

If this condition is not satisfied the load is said to be unbalanced. Let us consider a balanced star connected load shown in Fig. 4.10 supplied from a balanced 3 phase supply of V volts.

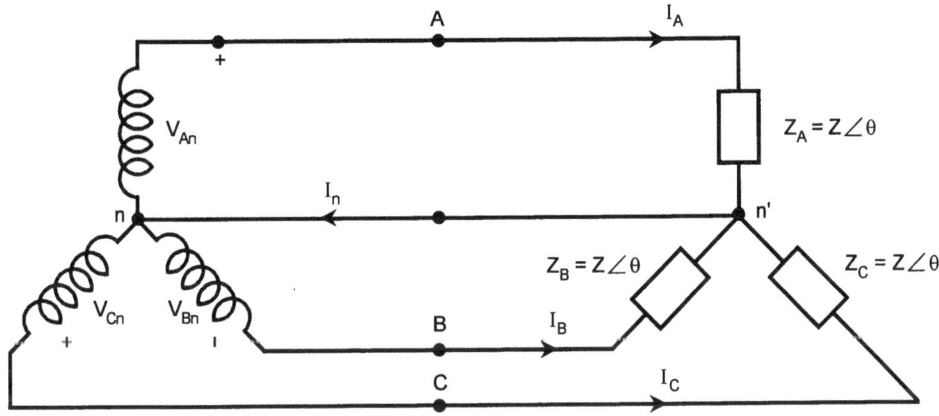

Fig. 4.10 Star connected balanced load supplied from balanced 3 phase4 – wire supply

Since n and n' are connected, the voltages V_{An}, V_{Bn} and V_{Cn} appear directly across the impedances Z_A, Z_B and Z_C respectively and therefore

$$I_A = \frac{V_{An}}{Z_A}$$

$$I_B = \frac{V_{Bn}}{Z_B} \quad .....(4.16)$$

$$I_C = \frac{V_{Cn}}{Z_C}$$

If $$V_{An} = V\,\angle 0^0$$

$$V_{Bn} = V\,\angle -120^0 \quad .....(4.17)$$

$$V_{Cn} = V\,\angle +120^0$$

and $$Z_A = Z_B = Z_C = Z\,\angle\theta^0$$

Then
$$I_A = \frac{V}{Z} \angle -\theta^0$$
$$I_B = \frac{V}{Z} \angle -120-\theta^0$$
$$I_C = \frac{V}{Z} \angle 120-\theta^0 \quad .....(4.18)$$

Thus the three line currents are equal in magnitude and differ in phase by 120^0. Since these are balanced currents, their sum is equal to zero. From Fig. 4.10 we see that the current in the neutral wire is equal to the sum of the line currents and hence $I_{n'n} = 0$. We observe that the line currents differ in phase by θ^0 with the corresponding phase voltages. The phasor diagram is shown in Fig. 4.11.

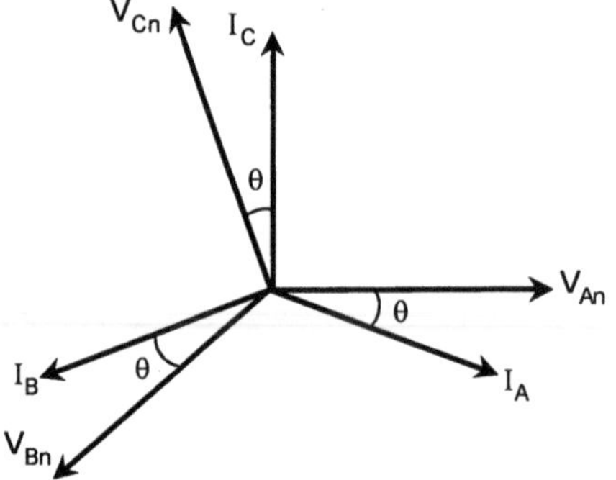

Fig. 4.11 Phasor diagram for a balanced 3 phase load

4.5.2 Balanced star connected load supplied from a balanced 3 phase 3 wire supply :

In Fig. 4.10 if the supply neutral and load neutral are not connected, we have 3 – phase star connected load supplied from a 3 – phase balanced 3 – wire supply as shown in Fig. 4.12.

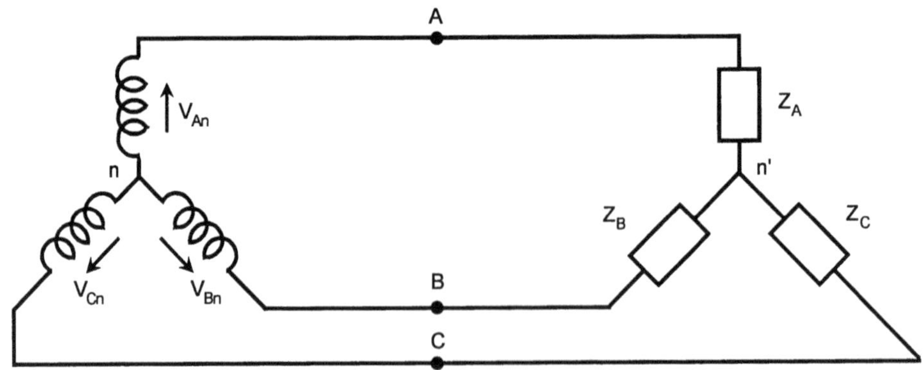

Fig. 4.12 Balanced star connected load supplied from balanced 3 – wire supply

In section 4.5.1 we have observed that the current in the neutral wire was zero if the supply and load are balanced. Hence it is immaterial whether the points n and n' are connected or not. The conditions of the circuit remain the same and the points n and n' are at the same potential even when they are not connected. Thus

$$\begin{aligned} V_{An} &= V_{An'} = V \angle 0^0 \\ V_{Bn} &= V_{Bn'} = V \angle -120^0 \\ V_{Cn} &= V_{Cn'} = V \angle 120^0 \end{aligned} \quad .....(4.19)$$

and

$$I_A = \frac{V}{Z} \angle -\theta$$

$$I_B = \frac{V}{Z} \angle -120^0 - \theta$$

$$I_C = \frac{V}{Z} \angle 120^0 - \theta \quad .....(4.20)$$

The phasor diagram indicating phase voltages V_{An}, V_{Bn}, V_{Cn}; the line voltages V_{AB}, V_{BC}, V_{CA} and the line currents I_A, I_B and I_C, is shown in Fig. 4.13.

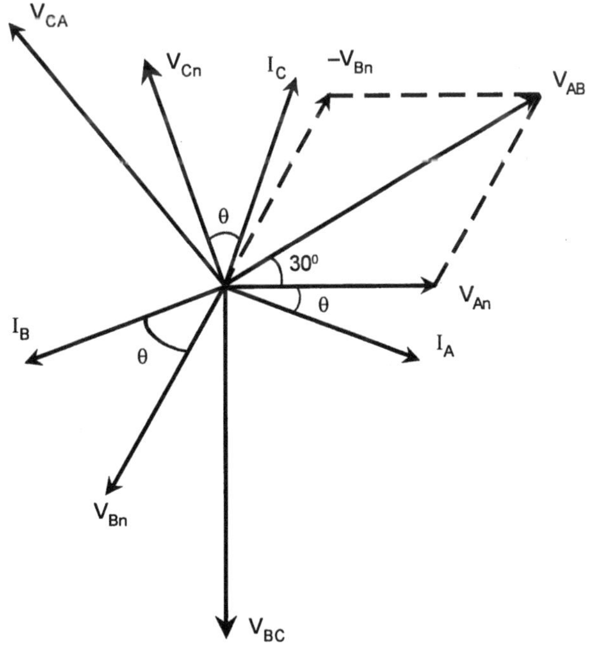

Fig. 4.13 Phasor diagram of a 3 ϕ star connected load supplied from a 3 ϕ, 3 – wire supply

Thus, for a star connected load, the following relationships hold for line and phase quantities.

$$V_L = \sqrt{3}\, V_p$$

$$I_L = I_p \qquad .....(4.21)$$

Example 4.1

The three r.m.s phase voltages of a balanced three phase supply are $V_{An} = 100 \angle 0^0$, $V_{Bn} = 100 \angle -120^0$ and $V_{Cn} = 100 \angle -240^0$.

What are the magnitudes of the line voltages ?

Solution :

Since the system is balanced, the magnitudes of the line voltages V_{AB}, V_{BC} and V_{CA} are equal to V_L.

The voltage V_L is given by

$$V_L = \sqrt{3}\, V_p$$

The r.m.s value of the phase voltage $V_p = 100$ Volts

$$\therefore \qquad V_L = \sqrt{3} \times 100 = 173.2 \text{ Volts}$$

Example 4. 2

If a balanced three phase star connected load of impedance $10 \angle 30^0$ ohms per phase is connected to the supply of example 4.1, what are the line and phase currents ?

Solution :

Taking the phase voltage V_{an} as reference, we have

$$V_{an} = V_p \angle 0^0 = 100 \angle 0^0 \text{ Volts}$$

Impedance Z connected to phase A $= 10 \angle 30^0\ \Omega$

$$I_A = \frac{100\angle 0^0}{10\angle 30^0} = 10 \angle -30^0 \text{ amps}$$

and since the load is balanced, we have

$$I_B = 10 \angle -150^0 \text{ and } I_C = 10 \angle 90^0 \text{ amps}$$

Since it is a star connected load the line currents are same as phase currents.

Example 4. 3

Three impedances each, of 5 + j 12 Ω, connected in star are connected to a 220 V, three phase, 50 Hz supply. Calculate the line currents.

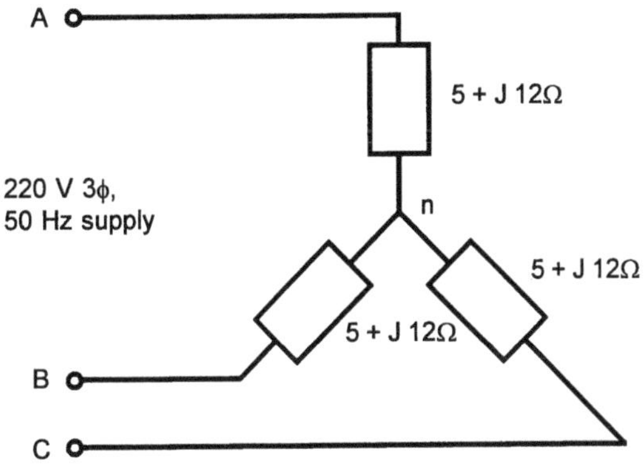

Fig. 4.14 Network for example 4.3

Solution :

Taking V_{AB} as reference

$$V_{AB} = 220 \angle 0^0$$

$$V_{BC} = 220 \angle -120^0$$

$$V_{CA} = 220 \angle 120^0$$

The impedance is $Z = 5 + J\ 12 = 13 \angle 67.38^0$ ohms

The phase voltages are

$$V_{An} = \frac{220}{\sqrt{3}} \angle -30^0$$

$$V_{Bn} = \frac{220}{\sqrt{3}} \angle -150^0$$

$$V_{Cn} = \frac{220}{\sqrt{3}} \angle 90^0$$

and hence the line currents are given by

$$I_A = \frac{V_{An}}{Z} = \frac{220\angle -30^0}{\sqrt{3} \times 13\angle 67.38^0} = 9.77 \angle -97.38$$

Since the load is balanced

$$I_B = 9.77 \angle -217.38^0$$

$$I_C = 9.77 \angle 22.62^0$$

4.5.3 Power in a star connected system :

The power absorbed by a 3 ϕ, star connected load is given by the sum of powers absorbed by the individual impedances in the load. Since the impedances are equal in magnitude, the total power absorbed by the 3 ϕ load is equal to 3 times the power per phase. The angle between the current through the impedance and the voltage across it, i.e., the phase voltage, is same in all the three phases and is equal to the impedance angle 'θ'. Thus the power factor of a balanced load is given by cos θ and the power in each phase is given by

$$\text{Power}/_{\text{phase}} = V_p\, I_p \cos\theta$$

The total power in the three phases is given by

$$P = 3\, V_p\, I_p \cos\theta \text{ watts}$$

Similarly the reactive power and apparent power, respectively, are given by

$$Q = 3\, V_p\, I_p \sin\theta \qquad \text{VARS}$$

and $$S = 3\, V_p\, I_p \qquad \text{VA} \qquad .....(4.22)$$

Since for a star connected system

$$V_p = \frac{V_L}{\sqrt{3}}$$

and $$I_p = I_L$$

we have

$$P = 3 \cdot \frac{V_L}{\sqrt{3}}\, I_L \cos\theta$$

$$= \sqrt{3}\, V_L\, I_L \cos\theta \text{ watts}$$

Similarly $$Q = \sqrt{3}\, V_L\, I_L \sin\theta \text{ VARS} \qquad .....(4.23)$$

and $$S = \sqrt{3}\, V_L\, I_L \qquad \text{VA}$$

Example 4. 4

Calculate the real power, reactive power and apparent power delivered to the load in example 4.3.

Solution :

$$P = \sqrt{3}\, V_L\, I_L \cos\theta$$

$$= \sqrt{3} \times 220 \times 9.77 \times \cos 67.38^0$$

$$= 1431.77 \text{ Watts}$$

$$Q = \sqrt{3}\ V_L\ I_L \sin\theta$$

$$= \sqrt{3} \times 220 \times 9.77 \times \sin 67.38^0 = 3436.39 \text{ Vars}$$

$$S = \sqrt{3}\ V_L\ I_L$$

$$= \sqrt{3} \times 220 \times 9.77 = 3722.8 \text{ VA}$$

4.5.4 Balanced Δ – connected load connected to a 3 – phase balanced supply :

A balanced 3 – phase Δ – connected load is shown in Fig. 4.15.

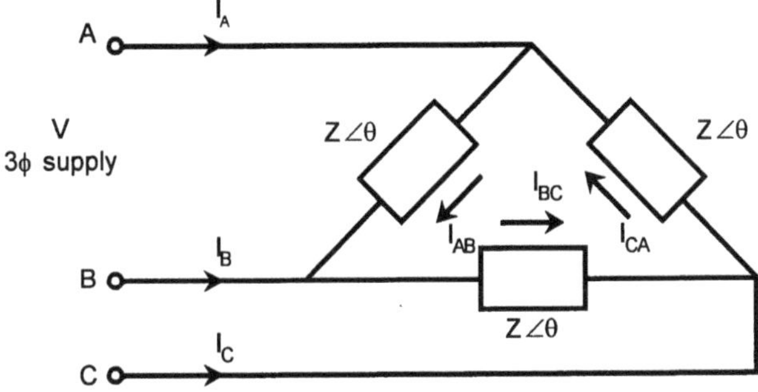

Fig. 4.15 Balanced Δ – connected load

Taking V_{AB} as reference, the three voltages are given by

$$V_{AB} = \angle 0^0$$
$$V_{BC} = \angle -120^0 \qquad \text{.....(4.24)}$$
$$V_{CA} = \angle 120^0$$

Since these voltages directly appear across the impedances in Δ – connected load, we have, for phase currents

$$I_{AB} = \frac{V_{AB}}{Z\angle\theta} = \frac{V\angle 0}{Z\angle\theta}$$

$$= \frac{V}{Z} \angle -\theta \text{ amp}$$

$$I_{BC} = \frac{V_{BC}}{Z\angle\theta} = \frac{V\angle -120}{Z\angle\theta}$$

$$= \frac{V}{Z} \angle(-120-\theta)^0 \text{ amp}$$

and

$$I_{CA} = \frac{V_{CA}}{Z\angle\theta} = \frac{V\angle 120}{Z\angle\theta}$$

$$= \frac{V}{Z} \angle(120-\theta)^0$$

The three phase currents I_{AB}, I_{BC} and I_{CA} are thus equal in magnitude and differ in phase by 120^0. The three currents can be written as

$$I_{AB} = I_p \angle -\theta^0$$
$$I_{BC} = I_p \angle(-120 - \theta)^0 \qquad .....(4.25)$$
$$I_{CA} = I_p \angle(120 - \theta)^0$$

where $I_p = \frac{V}{Z}$ amp

To get the line currents I_A, I_B and I_C, from Fig. 4.15, we have

$$I_A = I_{AB} - I_{CA}$$
$$I_B = I_{BC} - I_{AB} \qquad .....(4.26)$$
$$I_C = I_{CA} - I_{BC}$$

The phasor diagram showing line voltages, phase currents and line currents is given in Fig. 4.16.

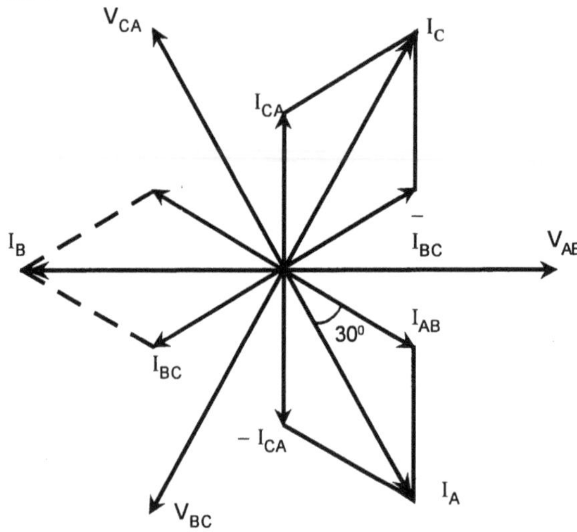

Fig. 4.16 Vector diagram for a balanced Δ connected load

Since I_{AB}, I_{BC} and I_{CA} are equal in magnitude and differ in phase by 120^0, the line currents are

$$I_A = \sqrt{3}\, I_p \angle(-\theta - 30^0)$$
$$I_B = \sqrt{3}\, I_p \angle(-120 - \theta - 30) = \sqrt{3}\, I_p \angle(-150 - \theta)^0 \qquad .....(4.27)$$
$$I_C = \sqrt{3}\, I_p \angle(120 - \theta - 30) = \sqrt{3}\, I_p \angle(90 - \theta)^0$$

Thus the line currents are equal to $\sqrt{3}$ times the phase currents in a Δ – connected system and they lag by 30^0 to the respective phase currents.

4.5.5 Power in a Δ – connected balanced system :

As in a star connected system, the power delivered to the load is same for each phase since the load is balanced and the total power delivered is equal to 3 times the power per phase.

$$P = 3\ V_p\ I_p \cos\theta$$

Since, in a Δ – connected system

$$V_p = V_L$$

and $$I_p = \frac{I_L}{\sqrt{3}}$$

we have $$P = 3 \cdot \frac{V_p I_L}{\sqrt{3}} \cos\theta$$

$$= \sqrt{3}\ V_L\ I_L \cos\theta$$

and similarly

$$Q = \sqrt{3}\ V_L\ I_L \sin\theta$$

and $$S = \sqrt{3}\ V_L\ I_L \qquad .....(4.28)$$

Thus for a 3 phases balanced load, irrespective of whether the load is connected in star or delta, the real power, reactive power and apparent power are respectively,

$$P = \sqrt{3}\ V_L\ I_L \cos\theta$$

$$Q = \sqrt{3}\ V_L\ I_L \sin\theta \qquad .....(4.29)$$

and $$S = \sqrt{3}\ V_L\ I_L$$

where V_L and I_L are line voltage and line current respectively and 'θ' is the impedance angle or the angle between the phase current and the corresponding phase voltage.

Example 4.5

Three impedances each of 3 – j 4 Ω are connected as shown in Fig. 4.17 across a 3 ϕ, 230 V balanced supply. Calculate the line and phase currents in the Δ connected load and the power delivered to the load.

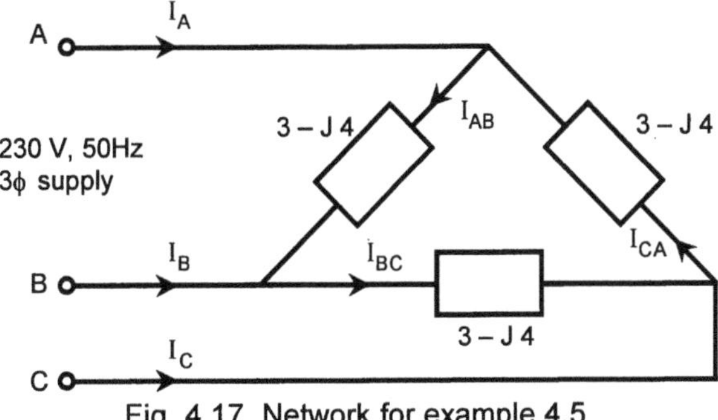

Fig. 4.17 Network for example 4.5

Solution :

Taking V_{AB} as reference

$$V_{AB} = 230 \angle 0^0 \text{ V}$$
$$V_{BC} = 230 \angle -120^0 \text{ V}$$
$$V_{CA} = 230 \angle 120^0 \text{ V}$$

and $Z = 5\angle -53.2^0 \ \Omega$

Therefore the phase currents are given by

$$I_{AB} = \frac{V_{AB}}{Z} = \frac{230\angle 0}{5\angle -53.2} = 46 \angle 53.2^0 \text{ A}$$

and $I_{BC} = 46 \angle -66.8^0$ A

$$I_{CA} = 46\angle -186.8^0 \text{ A}$$

The line currents are given by

$$I_A = \sqrt{3} \times 46\angle(53.2-30)$$
$$= 79.672 \angle 23.2^0 \text{ A}$$
$$I_B = 79.672 \angle -96.8^0 \text{ A}$$

and $I_C = 79.672 \angle -216.8^0$ A

The total power delivered to the load is given by

$$P = \sqrt{3}\ V_L\ I_L \cos\theta$$
$$= \sqrt{3} \times 230 \times 79.672 \cos 53.2^0$$
$$= 19012.9 \text{ Watts}$$

Example 4.6

Each phase of a balanced three phase delta connected load has a 0.2 henry inductor in series with a parallel combination of a 8 μ f capacitor and 100 ohm resistor. If a 3 phase voltage of 200 volts at a frequency of 400 rad/sec is applied to this load, find (a) the phase current (b) line current and (c) total power absorbed by the load.

Solution :

The circuit representing each arm of the delta load is shown in Fig. 4.18.

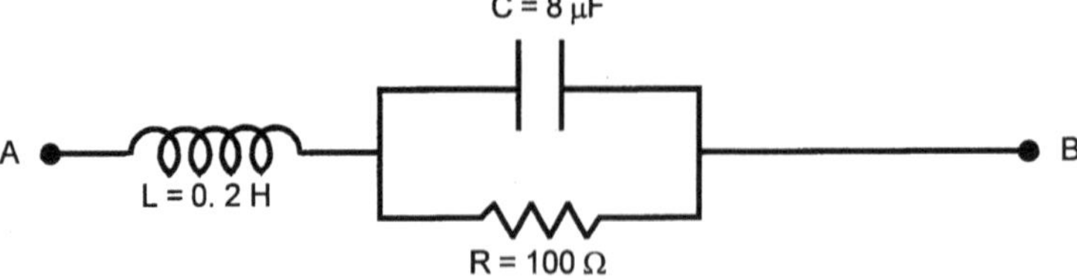

Fig. 4.18 Each arm of a Δ – connected load

$$X_L = 0.2 \times \omega = 0.2 \times 400 = 80 \text{ Ohms.}$$

$$X_C = \frac{1}{\omega C} = \frac{10^6}{400 \times 8} = 312.5 \text{ Ohms.}$$

Resistance R in the parallel branch = 100 ohms.

$$Z_{AB} = j80 + \frac{100 \times (-j\,312.5)}{100 - j312.5}$$

$$= 90.71 - j\,29.03\ \Omega$$

$$= 95.24\ \angle -17.75$$

Line voltage of the supply = 200 volts

∴ Phase voltage across the load = 200 volts

(a) Phase current $I_p = \frac{200}{95.24} = 2.1$ amps

(b) Line current $I_L = 2.1\ \sqrt{3} = 3.637$ amps

(c) Power absorbed by the load $= 3\ I_p^2\ \text{Req}$

$$= 3 \times 2.1^2 \times 90.71$$

$$= 1200 \text{ W}$$

Example 4.7

A balanced 3 phase load draws 100 KW at a lagging power factor of 0.8 from a 400 volt 3ϕ 50 Hz main. Calculate the complex power and the line current.

Solution :

Line current $I_l = \frac{|S|}{\sqrt{3}V_l}$

Real power delivered = 100 KW

Power factor $\cos\theta = 0.8$

∴ Apparent power $S = \frac{100}{0.8} = 125$ KVA

∴ $I_l = \frac{125 \times 1000}{\sqrt{3} \times 400} = 180.43$ amps

Complex power $S = 125 (\cos\theta + j \sin\theta) \times 10^3$

$$= 10^5 + j\,75 \times 10^3 = 125 \times 10^3\ \angle 36.87^0$$

4.6 UNBALANCED THREE PHASE NETWORK :

If the three impedances connected in either star or delta are unequal, the network is said to be unbalanced. Analysis of unbalanced circuits are slightly more involved as the currents and voltages in the 3 phases of the network are unequal. We will analyse these networks in the order of their complexity. Moreover, the phase sequence of the supply plays an important role in the analysis.

4.6.1 Unbalanced star connected load supplied from a 3 – phase balanced 4 – wire supply :

Consider the network shown in Fig. 4.19 let the phase sequence be ABC.

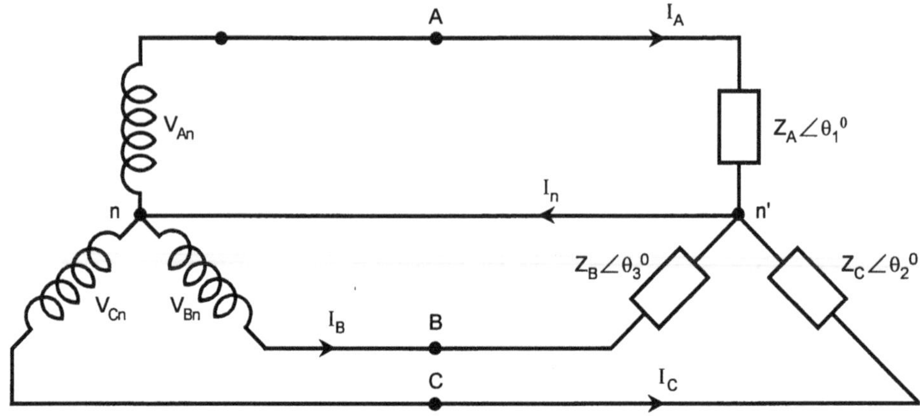

Fig. 4.19 Unbalanced load supplied from a 4 wire system

Since the neutrals are connected, the voltages V_{An}, V_{Bn} and V_{Cn} appear directly across the impedances Z_A, Z_B and Z_C respectively. Hence the currents I_A, I_B and I_C with V_{An} as a reference can be calculated as,

$$I_A = \frac{V_{An}}{Z_A\angle\theta_1} = \frac{V\angle 0}{Z_A\angle\theta_1}$$

$$= \frac{V}{Z_A}\angle -\theta_1^0 \qquad .....(4.30)$$

$$I_B = \frac{V_{Bn}}{Z_B\angle\theta_2} = \frac{V\angle -120}{Z_B\angle\theta_2}$$

$$= \frac{V}{Z_B}\angle(-120-\theta_2)^0 \qquad .....(4.31)$$

$$I_C = \frac{V_{Cn}}{Z_C\angle\theta_3} = \frac{V\angle 120}{Z_C\angle\theta_3}$$

$$= \frac{V}{Z_C} \angle (120 - \theta_3)^0 \qquad .....(4.32)$$

Thus, it is clear that the three line currents are unequal and the phase difference between them is no longer equal. The neutral current is given by,

$$I_n = I_A + I_B + I_C$$

and is a finite value unlike in the case of balanced system where it is equal to zero. The vector diagram for inductive impedances is shown in Fig. 4.20.

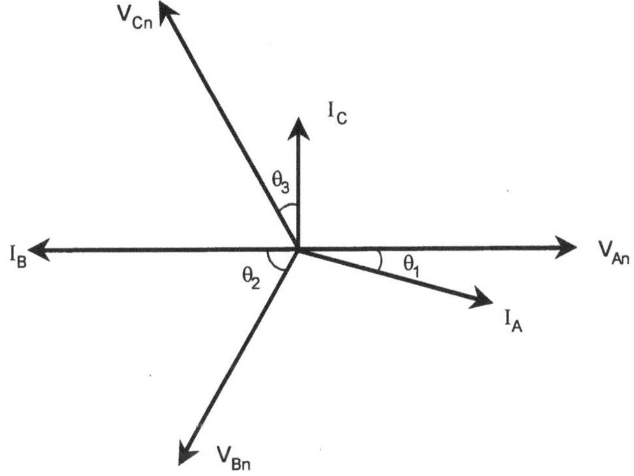

Fig. 4.20 Phasor diagram for unbalanced star connected load

4.6.2 Unbalanced Δ – connected load :

The analysis of unbalanced Δ – connected load is similar to the balanced Δ – connected load. As shown in Fig. 4.21, the voltages appear directly across the impedances and the phase currents in the Δ can be easily calculated.

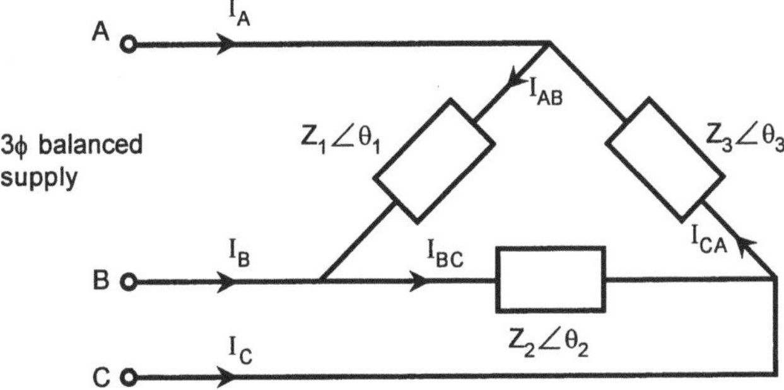

Fig. 4.21 Unbalanced Δ – connected load

If V_{AB} is taken as reference and for a phase sequence ABC, we have

$$V_{AB} = V \angle 0^0$$
$$V_{BC} = V \angle -120^0$$
$$V_{CA} = V \angle 120^0$$

and the phase currents are

$$I_{AB} = \frac{V_{AB}}{Z_1\angle\theta_1} = \frac{V}{Z_1}\angle -\theta_1^{\,0}$$
$$I_{BC} = \frac{V_{BC}}{Z_2\angle\theta_2} = \frac{V}{Z_2}\angle(-120-\theta_2)^0 \qquad .....(4.33)$$
$$I_{CA} = \frac{V_{CA}}{Z_3\angle\theta_3} = \frac{V}{Z_3}\angle(120-\theta_3)^0$$

The three phase currents are unequal and do not differ by 120^0 in phase as in the balanced loads. The line currents can be calculated from the relations

$$I_A = I_{AB} - I_{CA}$$
$$I_B = I_{BC} - I_{AB} \qquad .....(4.34)$$
and $$I_C = I_{CA} - I_{BC}$$

The phasor diagram for this case is given in Fig. 4.22.

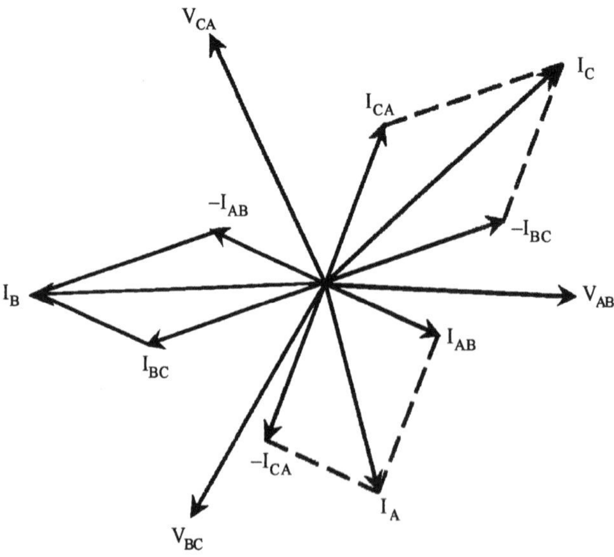

Fig. 4.22 Phasor diagram for unbalanced Δ – connected load

It is clear from the vector diagram that the line currents also are unbalanced.

4.6.3 Unbalanced star connected load supplied from a 3 phase balanced 3 wire supply :

The analysis of unbalanced star connected load supplied from a 3 wire supply can be analysed by the following methods :

(a) Star – delta transformation

(b) Millman's theorem

(c) Loop current method

(d) Node voltage method

(a) *Star – Delta Transformation :*

The star connected load may be transformed into an equivalent Δ – connected load using the relationships between star and delta networks developed earlier for resistive networks. The relationships are repeated here for impedances. Consider the star network of Fig. 4.23(a) and its equivalent Δ – network in Fig. 4.23(b).

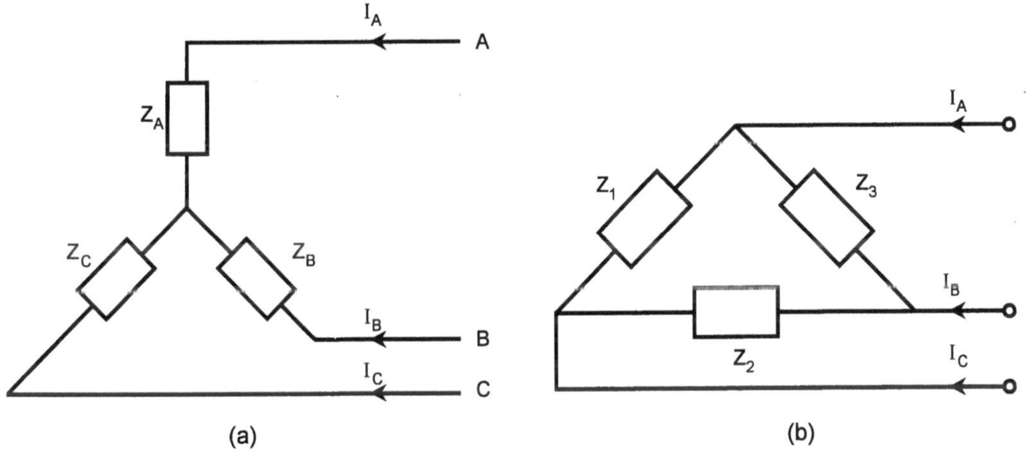

Fig. 4.23 Star Δ – transformation

$$Z_1 = \frac{Z_A Z_C + Z_C Z_B + Z_B Z_A}{Z_B}$$

$$Z_2 = \frac{Z_A Z_C + Z_C Z_B + Z_B Z_A}{Z_A}$$

and $$Z_3 = \frac{Z_A Z_C + Z_C Z_B + Z_B Z_A}{Z_C}$$

When a 3 – phase balanced supply is given to the terminals A, B and C of the network of Fig. 4.23(a), the line currents I_A, I_B and I_C can be easily calculated using the relations developed is section 4.6.2.

(b) *Millman's Theorem :*

The statement of the theorem is given below without proof.

Statement :

Millman's theorem states that if a set of voltage sources V_1, V_2, V_3.....V_n in series with their internal impedances Z_1, Z_2.....Z_n, are connected in parallel across the terminals a, b, the voltage V_{ab} across these terminal is given by

$$V_{ab} = \frac{V_1Y_1 + V_2Y_2 + ... + V_nY_n}{Y_1 + Y_2 + ... + Y_n} \qquad .....(4.35)$$

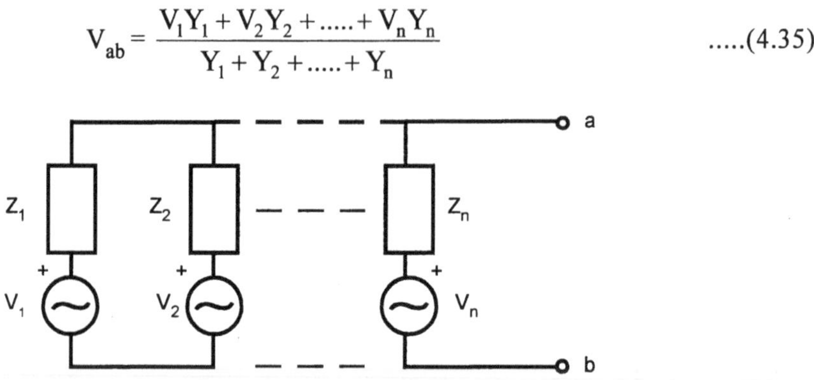

Fig. 4.24 Millman's theorem

where Y'_is are the reciprocals of the impedance Z_i for i = 1, 2, ...n.

Application to Solution of 3 Phase Unbalanced Systems :

Millman's theorem can be used to solve an unbalanced star connected 3 wire system. Referring to Fig. 4.25, it is possible to find the potential of point n' with respect to n using this theorem. Once this potential is obtained, the voltage across each phase of the load can be calculated.

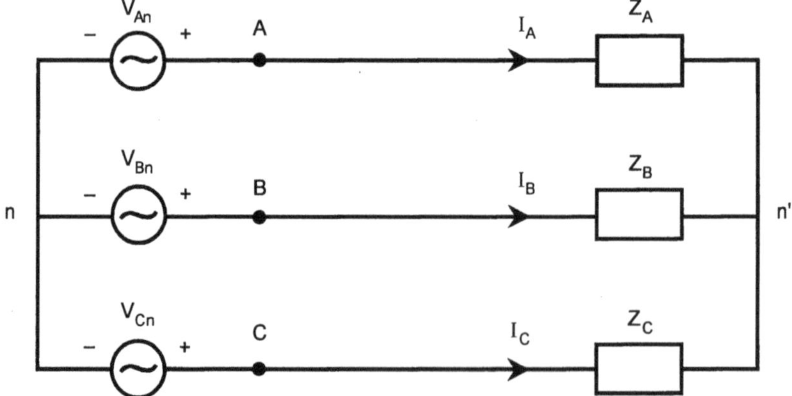

Fig. 4.25 Unbalanced star connected load supplied from a 3 wire system

According to the Millman's theorem the potential difference

$$V_{n'n} = \frac{V_{An}Y_A + V_{Bn}Y_B + V_{Cn}Y_C}{Y_A + Y_B + Y_C} \qquad .....(4.36)$$

where Y_A, Y_B and Y_C are the reciprocals of Z_A, Z_B and Z_C respectively.

Once $V_{n'n}$ is obtained, the voltages across the loads are calculated from the equations

$$\left.\begin{aligned} V_{An'} &= V_{An} - V_{n'n} \\ V_{Bn'} &= V_{Bn} - V_{n'n} \\ V_{Cn'} &= V_{Cn} - V_{n'n} \end{aligned}\right\} \qquad .....(4.37)$$

The currents in Z_a, Z_b and Z_c can now be easily calculated.

Using Loop Current Method :

Consider the circuit of Fig. 4.26

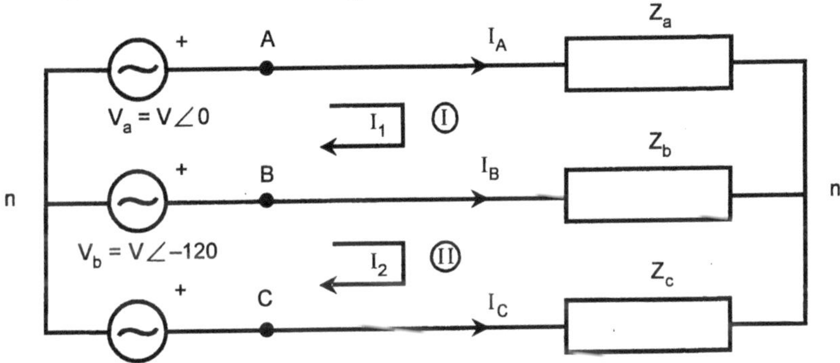

Fig. 4.26 Loop current method

Applying Kirchhoff's Voltage Law to loop I,

$$I_1 Z_a + (I_1 - I_2) Z_b + V \angle -120^0 = V \angle 0^0$$

For loop II

$$(I_2 - I_1) Z_b + I_2 Z_c + V \angle -240^0 = \angle -120^0 \qquad .....(4.38)$$

Solving (4.38), I_1 and I_2 are obtained. Hence the current I_A, I_B and I_C in phases A, B and C would be respectively

$$\left.\begin{aligned} I_A &= I_1 \\ I_B &= I_2 - I_1 \\ \text{and} \quad I_C &= -I_2 \end{aligned}\right\} \qquad .....(4.39)$$

using these currents the phase voltages $V_{an'}$, $V_{bn'}$ and $V_{cn'}$ can be calculated.

Remark :

It is also possible to write equations similar to eq. (4.38) using line voltages instead of phase voltages. The student is advised to write these equations.

Using Node – Voltage method :

Taking the supply neutral as ground, node voltage equation can be written for the node n'. In Fig. 4.26, denoting the voltage of neutral n' with respect to ground as V_n', we can write the following node voltage equation.

$$\frac{V_{n'} - V\angle 0}{Z_a} + \frac{V_{n'} - V\angle -120}{Z_b} + \frac{V_{n'} - V\angle 120}{Z_c} = 0 \quad .....(4.40)$$

Once V_n' is determined, computation of voltages across the loads Z_a, Z_b and Z_c and the currents I_A, I_B and I_C is straight forward.

Example 4.8

A balanced 3 ϕ, 3 wire 50 Hz 100 volt supply is given to a load consisting of three impedances (1 + j1), (1 + j2) and (3 + j4) ohms connected in star as shown in Fig. 4.27. Compute the voltages across and currents in the three phases of the load using (a) Millman's theorem (b) Loop current method. Phase sequence is ABC.

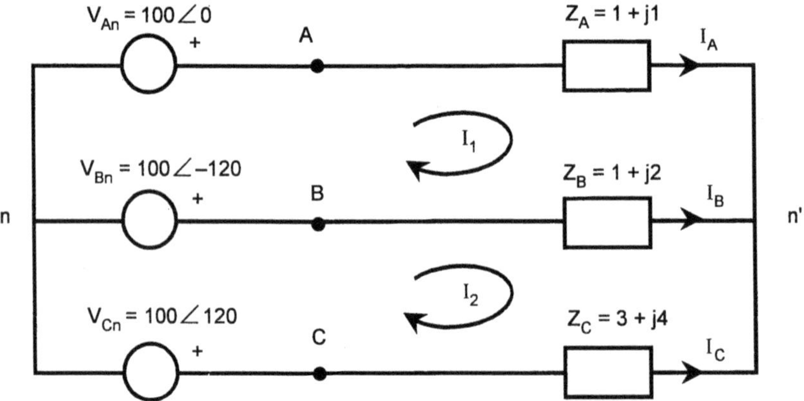

Fig. 4.27 Network for example 4.8

Solution :

(a) *Using Millman's Theorem :*

According to eq. (4.36), the voltage

$$V_{n'n} = \frac{V_{An}Y_A + V_{Bn}Y_B + V_{Cn}Y_C}{Y_A + Y_B + Y_C}$$

Now, $$Y_A = \frac{1}{1+j1} = \frac{1}{\sqrt{2}\angle 45} = 0.707\ \angle -45^0 = 0.5 - j\,0.5\ S$$

$$Y_B = \frac{1}{1+j2} = \frac{1}{\sqrt{5}\angle 63.44^0} = 0.447\ \angle -63.44^0 = 0.2 - j\,0.4\ S$$

$$Y_C = \frac{1}{3+j4} = \frac{1}{5\angle 53.13^0} = 0.2\ \angle -53.13^0 = 0.12 - j\,0.16\ S$$

$$V_{An} = 100\ \angle 0^0$$

$$V_{Bn} = 100\ \angle -120^0 \text{ and } V_{Cn} = 100\ \angle -240^0$$

Substituting in the expression for $V_{n'n}$ we get

$$V_{n'n} = \frac{100\angle 0\left(0.707\angle -45^0\right) + 100\angle -120^0\left(0.447\angle -63.44^0\right) + 100\angle -240^0\left(0.2\angle -53.13^0\right)}{.707\angle -45^0 + 0.447\angle -63.44^0 + 0.2\angle -53.13^0}$$

$$= 23.73\ \angle -13.12^0$$

$$= 23.11 - j\,5.386\ V$$

$$V_{An'} = 100\angle 0\ - 23.11 + j\,5.386 = 76.89 + j\,5.386 = 77.07\angle 4^0\ \ V$$

$$V_{Bn'} = 100\angle -120^0 - 23.11 + j\,5.386\ V = 109.29\angle -132^0\ \ V$$

$$V_{Cn'} = 100\angle 120 - 23.11 + j\,5.386\ V = 117.5\angle 128.48^0\ \ V$$

$$I_A = \frac{V_{An'}}{Z_A} = \frac{77.07\angle 4^0}{\sqrt{2}\angle 45^0} = 54.5\angle -40.9^0$$

$$I_B = \frac{V_{Bn'}}{Z_B} = 48.87\ \angle 164.67$$

$$I_C = \frac{V_{Cn'}}{Z_C} = 23.48\ \angle -75.38^0$$

(b) *Loop current method :*

Let I_1 and I_2 be the loop currents as shown in Fig. 4.27

Applying Kirchhoff's Voltage law to loop I

$$(1+j1)I_1 + (1+j2)(I_1 - I_2) = -V_{Bn} + V_{An} = -100\ \angle -120 + 100\ \angle 0$$

$$(2 + j3)I_1 - (1+j2)\ I_2 = 150 + j\,86.6 \qquad .....(4.41)$$

Similarly for loop II we get

$$(1+j2)(I_2 - I_1) + (3 + j4)\ I_2 = -V_{Cn} + V_{Bn} = -100\ \angle -240^0 + 100\ \angle -120^0$$

$$(4 + j6)I_2 - (1 + j2)\ I_1 = 173.2\ \ \angle -90^0 = -j\,173.2 \qquad .....(4.42)$$

Solving for I_1 and I_2 we get

$$I_1 = 54.5\ \angle -40.9^0\ \text{A}; \quad I_2 = 23.48\ \ \angle -104.63^0\ \text{A}$$

The line currents can be calculated as

$$I_A = I_1 = 54.5 \ \angle -40.9^0 \ \text{A}$$

$$I_B = I_2 - I_1 = 48.87 \ \angle 164.61^0 \ \text{A}$$

$$I_C = -I_2 = 23.48 \ \angle 75.37^0 \ \text{A}$$

Hence the voltages across the loads are

$$V_{An'} = I_A Z_A = 54.5 \ \angle -40.9 \times 1.414 \ \angle 45$$

$$= 77.06 \ \angle 4.1^0 \ \text{V}$$

$$V_{Bn'} = I_B Z_B = 48.87 \ \angle 164.61 \times \sqrt{5} \ \angle 63.44$$

$$= 109.29 \ \angle 228.05^0 \ \text{V or} = 109.29 \angle -132^0 \ \text{V}$$

$$V_{Cn'} = I_C Z_C = 23.48 \ \angle 75.37 \times 5 \ \angle 53.2$$

$$= 117.4 \ \angle 128.57^0 \ \text{V}$$

4.7 MEASUREMENT OF THREE PHASE POWER :

A single phase wattmeter consisting of two coils is used to measure the average power in single phase circuits. One of the coils is called the current coil while the other is called the pressure coil. The current coil is connected in series with the circuit while the pressure coil is connected across the circuit in which power is to be measured. The interaction of the magnetic fields due to the currents in the two coils, produces a deflection of the pointer along a scale calibrated in watts. Measurement of power in three phase circuit is slightly more intricate. Since a three phase system is a combination of three single phase circuits – balanced or unbalanced, it is always possible to use three single phase wattmeters to measure power in a 3 – phase circuit .

However, power in a 3 – phase circuit can be measured using two watt-meters only as described in the next section.

4.7.1 Two wattmeter method :

In this method two single phase wattmeter are used. The current coils are connected in two of the three lines of the three phase circuits. The pressure coils are connected between the two lines in which the current coils are connected and the third line. The connection diagram is shown in Fig. 4.28. It does not matter whether the load is star or delta connected.

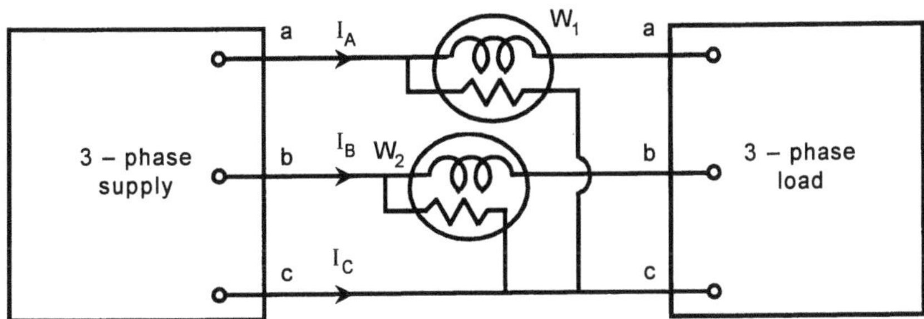

Fig. 4.28 Two wattmeter method

It can be shown that the sum of the two wattmeter readings W_1 and W_2 gives the total power delivered to the load.

Proof :

It can be assumed without loss of generality that the sources and the loads are star connected and the circuit of Fig. 4.28 can be represented as shown in Fig. 4.29.

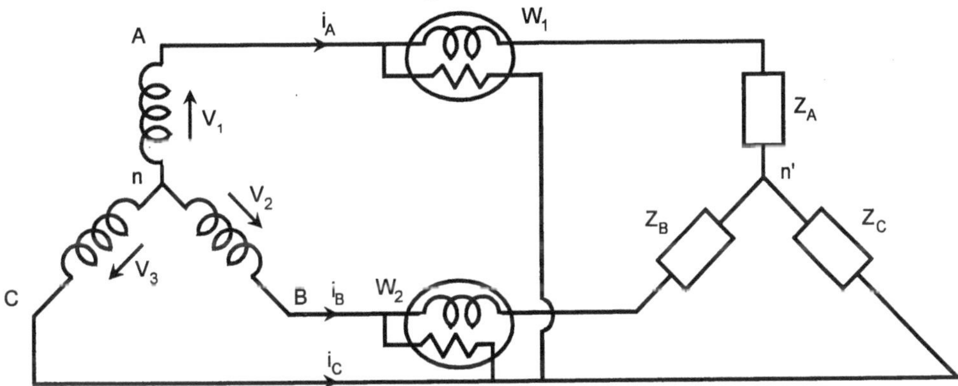

Fig. 4.29 Three wire system – 2 wattmeter method

Let the instantaneous values of currents in phases A, B and C be denoted as i_A, i_B, i_C respectively. Since all the three currents meet at a point,

$$i_A + i_B + i_C = 0 \qquad .....(4.43)$$

The instantaneous total power 'p' delivered to the load is given by

$$p = i_A v_A + i_B v_B + i_C v_C \qquad .....(4.44)$$

where v_A, v_B and v_C are the instantaneous voltages in phases A, B and C.

Since the current i_C is not measured, it is eliminated using

$$i_C = -(i_A + i_B) \qquad .....(4.45)$$

Substituting eqs.(4.46) in (4.45),

$$p = i_A v_A + i_B v_B - v_C (i_A + i_B)$$

$$= i_A (v_A - v_C) + i_B (v_B - v_C) \qquad .....(4.46)$$

Since $v_A - v_c$ is the instantaneous voltage applied to the pressure coil of wattmeter W_1 and the current in the current coil of W_1 is i_A, the first term of eq. (4.46) is the average power measured by the wattmeter W_1. Similarly the second term in eq. (4.46) is the average power measured by the wattmeter W_2. Hence the total average power P is the sum of the powers measured by the two wattmeters.

Thus $\qquad P = W_1 + W_2$

This is the most common method of measuring power in a three phase system whether balanced or not.

4.7.2 Wattmeter readings and power factor in a balanced load :

It is possible to obtain expressions for the wattmeter readings and the value of the power factor of the load in case the system is balanced. Consider the phasor diagram of Fig. 4.30 for a balanced 3 – phase star connected system in which the power is to be measured. In drawing the diagram, it is assumed that the load has lagging power factor, cosθ. From the phasor diagram, we observe that the angle between the current I_A and the voltage V_{AC} is (30 – θ) and the angle between I_B and V_{BC} is (30 + θ). Since the power measured by W_1 is due to current I_A and Voltage V_{AC}

$$W_1 = V_{AC} I_A \cos(30 - \theta)$$

$$= V_L I_L \cos(30 - \theta) \qquad .....(4.47)$$

Similarly

$$W_2 = V_L I_L \cos(30 + \theta) \qquad .....(4.48)$$

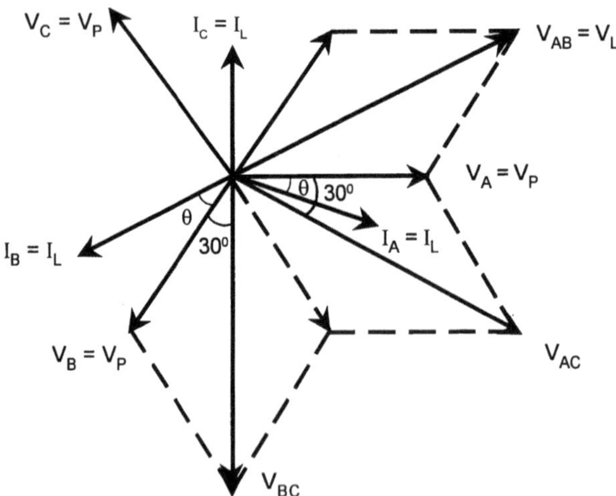

Fig. 4.30 Phasor diagram for 2 wattmeter method

The sum of the wattmeter readings

$$W_1 + W_2 = V_L I_L \; [\cos(30 - \theta) + \cos(30 + \theta)]$$

$$= \sqrt{3}\; V_L I_L \cos\theta \qquad .....(4.49)$$

Thus $W_1 + W_2$ is the total power. Further , the difference of the two readings

$$W_1 - W_2 = V_L I_L [\cos(30 - \theta) - \cos(30 + \theta)]$$

$$= V_L I_L \sin\theta \qquad .....(4.50)$$

$W_1 - W_2$ is a measure of the reactive power. Taking the ratio of the two values $(W_1 - W_2)$ and $(W_1 + W_2)$ we get

$$\frac{W_1 - W_2}{W_1 + W_2} = \frac{\tan\theta}{\sqrt{3}}$$

or $$\tan\theta = \frac{\sqrt{3}(W_1 - W_2)}{W_1 + W_2}$$

or $$\theta = \tan^{-1}\left(\frac{\sqrt{3}(W_1 - W_2)}{W_1 + W_2}\right) \qquad .....(4.51)$$

It is therefore possible to calculate the load power factor, $\cos\theta$. So,

1. If $\theta = 0$, both W_1 and W_2 are equal.
2. If $\theta < 60^0$, W_1 and W_2 are both positive and the total power will be the sum of the readings
3. If $\theta = 60^0$, W_2 will be equal to zero. W_1 will measure the total power.
4. If $\theta > 60^0$, W_2 will be negative. When W_2 shows a negative reading, the pressure coil connections are interchanged to get a positive reading and the power read by the meter is taken as – ve

Some examples are now given to illustrate the concepts developed in this chapter.

Example 4.9

For the network of Fig. 4.31 calculate the line currents and power consumed if (a) the phase sequence is ABC and (b) the phase sequence is ACB.

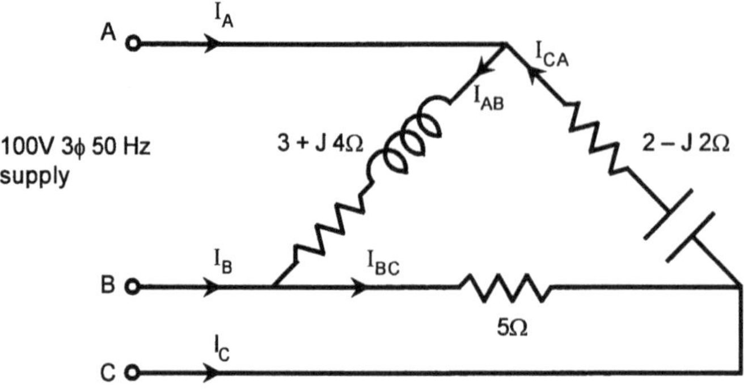

Fig. 4.31 Network for example 4.9

Solution :

(a) When the phase sequence is ABC. The voltages across the 3 phases of the load are

$$V_{AB} = 100\ \angle 0^0\ V$$

$$V_{BC} = 100\ \angle -120^0\ V$$

$$V_{CA} = \angle 120^0\ V$$

The phase currents are

$$I_{AB} = \frac{V_{AB}}{Z_{AB}} = \frac{100\angle 0}{3+j4} = 20\ \angle -53.2^0 = 12 - j\ 16\ A$$

$$I_{BC} = \frac{V_{BC}}{Z_{BC}} = \frac{100\angle -120^0}{5} = 20\ \angle -120^0 = -\ 10 - j\ 17.32\ A$$

$$I_{CA} = \frac{V_{CA}}{Z_{CA}} = \frac{100\angle 120}{2-j2} = 35.35\ \angle 165^0 = -\ 34.145 + j\ 9.149\ A$$

The line currents are

$$I_A = I_{AB} - I_{CA}$$
$$= 12 - j\ 16 + 34.145 - j\ 9.149$$

$$= 46\ .145 - j\ 25.149 = 52.55\ \angle -28.39^0$$

$$I_B = I_{BC} - I_{AB}$$
$$= -\ 10 - j\ 17.32 - 12 + j\ 16$$
$$= -\ 22 - j\ 1.32 = 22.04\ \angle 183.43^0\ A$$

$$I_C = I_{CA} - I_{BC}$$
$$= -34.145 + j\,9.149 + 10 + j\,17.32$$
$$= -24.145 + j\,26.469 = 35.827\ \angle 132.37^0\ \text{A}$$

The power in the three phase are calculated as follows :

$$P_{AB} = V_{AB}\, I_{AB} \cos\theta_{AB}\ ;\quad \text{where } \theta_{AB} \text{ is the angle between } V_{AB} \text{ and } I_{AB}$$
$$= 100 \times 20 \cos 53.2^0 = 1200\ \text{W}$$
$$P_{BC} = V_{BC}\, I_{BC} \cos\theta_{BC} = 100 \times 20 \cos 0^0 = 2000\ \text{W}$$
$$P_{CA} = V_{CA}\, I_{CA} \cos\theta_{CA} = 100 \times 35.35 \cos 45^0 = 2500\ \text{W}$$

The total power in the load

$$P = P_{AB} + P_{BC} + P_{CA} = 1200 + 2000 + 2500 = 5700\ \text{W}$$

(b) When the phase sequence is ACB. The voltages across the 3 phases of the load are

$$V_{AB} = 100\ \angle 0^0\ \text{V}$$
$$V_{BC} = 100\ \angle 120^0\ \text{V}$$
$$V_{CA} = 100\ \angle -120^0\ \text{V}$$

The phase currents are

$$I_{AB} = \frac{V_{AB}}{Z_{AB}} = \frac{100\angle 0}{3 + j4} = 20\angle -53.2^0 = 12 - j\,16\,\text{A}$$

$$I_{BC} = \frac{V_{BC}}{Z_{AB}} = \frac{100\angle 120^0}{5} = 20\angle 120^0 = -10 + j\,17.32\ \text{A}$$

$$I_{CA} = \frac{V_{CA}}{Z_{CA}} = \frac{100\angle -120^0}{2 - j2} = 35.35\ \angle -75 = 9.149 - j34.145\ \text{A}$$

The line currents are given by

$$I_A = I_{AB} - I_{CA}$$
$$= 12 - j\,16 - 9.149 + j\,34.155$$
$$= 2.851 + j\,18.155 = 18.37\ \angle 81.07^0\ \text{A}$$

$$I_B = I_{BC} - I_{AB}$$
$$= -10 + j\,17.32 - 12 + j\,16$$

$$I_B = -22 + j\,33.32 = 39.93\ \angle 123.44^0$$

$$I_C = I_{CA} - I_{BC}$$

$$= 9.149 - j\,34.145 + 10 - j\,17.32$$

$$= 19.149 - j\,51.465 = 54.91\ \angle -69.59^0$$

The power in the 3 phases are

$$P_{AB} = V_{AB}\,I_{AB}\cos\theta_{AB}$$

$$= 100 \times 20 \cos 53.2^0$$

$$= 1200\ \text{W}$$

$$P_{BC} = V_{BC}\,I_{BC}\cos\theta_{BC}$$

$$= 100 \times 20 \cos 0$$

$$= 2000\ \text{W}$$

$$P_{CA} = V_{CA}\,I_{CA}\cos\theta_{CA}$$

$$= 100 \times 35.35 \cos 45^0$$

$$= 2500\ \text{W}$$

and the total power is

$$P = 1200 + 2000 + 2500$$

$$= 5700\ \text{W}$$

Note :

For different phase sequence, we observe that the line currents are different in both magnitudes and phase angles but the power in the individual impedances remains same as the magnitude of the phase currents are same in both cases.

Example 4.10

An unbalanced star connected load is connected across a 3 ϕ, 400 V balanced supply of sequence RYB as shown in Fig. 4.32. Two wattmeters are connected to measure the total power supplied as shown in the figure. Find the readings of the wattmeters.

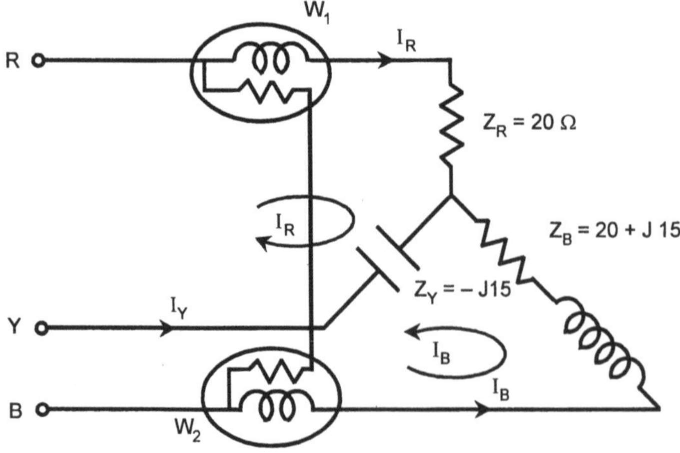

Fig. 4.32 Network for examples 4.10

Solution :

Taking V_{RY} as reference

$$V_{RY} = 400 \angle 0^0 \text{ V}$$

$$V_{YB} = 400 \angle -120^0 \text{ V}$$

$$V_{BR} = 400 \angle 120^0 \text{ V}$$

Writing the loop equations, we get

$$20 I_R - j\,15\,(I_R + I_B) = V_{RY} = 400 \angle 0^0$$

$$(20 - j\,15)\, I_R - j\,15\, I_B = 400 \quad \text{.....(4.52)}$$

$$(20 + j\,15)\, I_B + (I_R + I_B)(-j\,15) = -V_{YB}$$

$$20 I_B - j\,15\, I_R = -400 \angle -120 \quad \text{.....(4.53)}$$

Solving eqs. (4.52) and (4.53) we get

$$I_R = 5.923 \angle 72.57^0 \text{ A}$$

and $$I_B = 19.52 \angle 72.83^0 \text{ A}$$

The vector diagram for the network is shown in Fig. 4.33

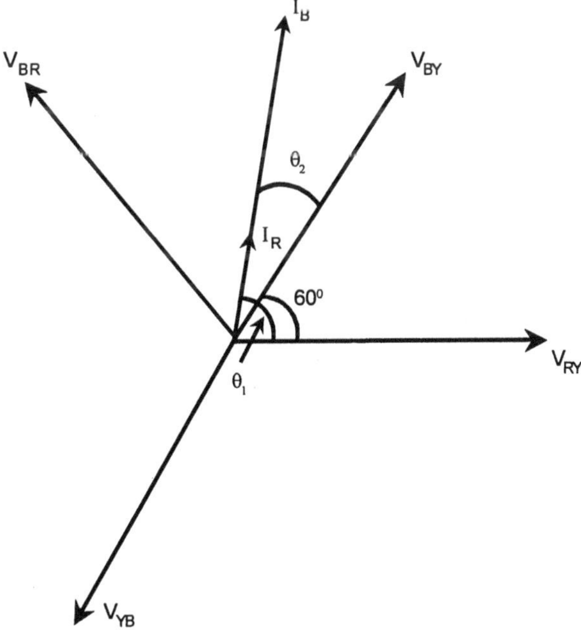

Fig. 4.33 Vector diagram for example 4.10

The wattmeter readings are given by

$$W_1 = V_{RY}\, I_R \cos\theta_1$$
$$= 400 \times 5.923 \cos 72.57^0$$
$$= 709.67 \text{ Watts}$$
$$W_2 = V_{BY}\, I_B \cos\theta_2$$
$$= 400 \times 19.52 \cos (72.83 - 60)^0$$
$$= 7613 \text{ Watts}$$

Total power $= W_1 + W_2 = 8320.73$ Watts

Example 4.11

A 220, 3 ϕ balanced supply is given to a 3 ϕ star connected unbalanced, 4 wire load as shown in Fig. 4.34. Find the line currents, neutral current and the power read by the wattmeter. Phase sequence is RYB

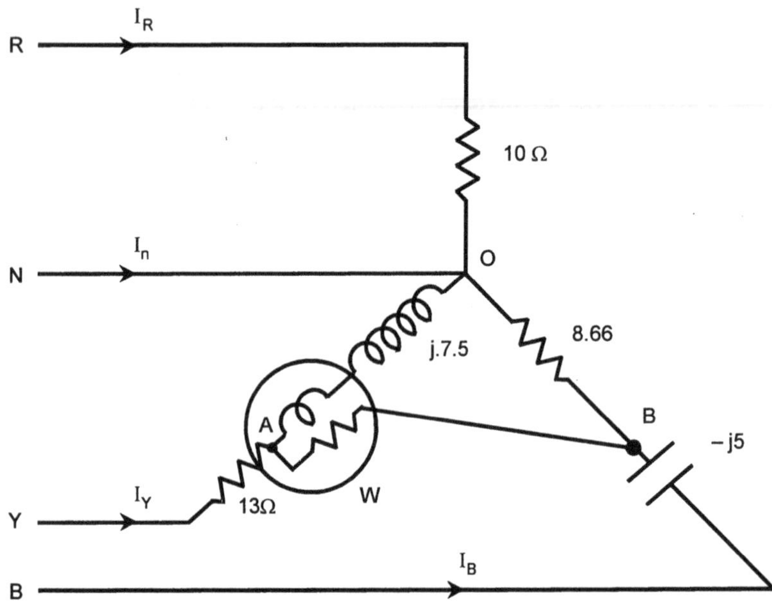

Fig. 4.34 Network for example 4.11

Solution :

Since line voltages are 220 V, the phase voltages of the supply, taking V_{Rn} as reference, are given by

$$V_{Rn} = \frac{220}{\sqrt{3}} \angle 0^0 \text{ V} = 127 \angle 0^0 \text{ V}$$

$$V_{Yn} = 127 \angle -120^0 \text{ V}$$

$$V_{Bn} = 127 \angle 120^0 \text{ V}$$

The line currents can be calculated as

$$I_R = \frac{127\angle 0}{10} = 12.7 \angle 0^0 \text{ A}$$

$$I_Y = \frac{127\angle -120}{13 + j7.5}$$

$$= 8.47 \angle -150^0 \text{ A}$$

$$= -7.335 - j\,4.235 \text{ A}$$

$$I_B = \frac{127\angle 120^0}{8.66 - j5}$$

$$= 12.7 \angle 150^0 \text{ A}$$

$$= -11 + j\,6.35 \text{ A}$$

$$I_n = I_R + I_Y + I_B$$

$$= 12.7 - 7.335 - j\,4.235 - 11 + j\,6.35$$

$$= -5.635 + j\,2.115 = 6.02 \angle 159.45^0 \text{ A}$$

For calculating the power read by the wattmeter W_1 we must know the current flowing through the current coil, the voltage across the pressure coil and the phase angle between these two quantities .

Now the current through the current coil is

$$I_Y = 8.47 \angle -150^0 \text{ A}$$

The voltage across the pressure coil is V_{AB}

and $$V_{AB} = V_{AO} + V_{OB}$$

$$= V_{AO} - V_{BO}$$

$$= I_Y\,(j\,7.5) - I_B\,(8.66)$$

$$= 8.47 \angle -150\;(j\,7.5) - 12.7 \angle 150\;(8.66)$$

$$= 127 - j\,110 \text{ V}$$

$$= 168 \angle -40.9^0 \text{ V}$$

Thus the wattmeter reading is

$$P = 168 \times 8.47 \cos 109.1$$
$$= -465.62 \text{ watts}$$

The vector diagram is given in Fig. 4.35

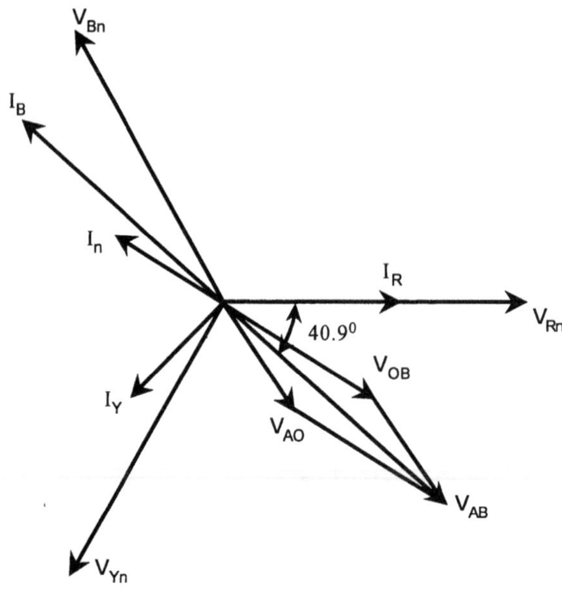

Fig. 4.35 Vector diagram for example 4.11

Problems

4.1 Three inductors each of resistance 2 ohms and an inductive reactance of 8 ohms are connected in star and supplied from a three phase 230 V 50 Hz supply. What are the line and phase currents and voltages ? Also calculate the power input and power factor.

4.2 If the three inductors of problem 4.1 are reconnected in delta, how do the currents and voltages change if the supply is the same. What is the power input and power factor ?

4.3 A balanced Y – connected load with a phase resistance of 12 Ω in series with an inductive reactance of 16 Ω is connected to a balanced 3 – ϕ source with a line voltage of 240 V. Find the line current and complex power absorbed by the load.

4.4 A balanced three phase inductive load is connected to a balanced $3-\phi$ power system. The line voltage is 480 Volts and the line current is 10 A. The angle of the phase impedance of the load is 60^0. Find the complex power S and real power P absorbed by the load.

4.5 A balanced delta connected three phase load absorbs a complex power of 100 kVA with a lagging power factor of 0.8 when the r.m.s. line to line voltage is 2400 volts. Calculate the impedance of each arm of the Δ connected load.

4.6 A 400 V 50 Hz 3 phase supply is given to three identical star connected impedances as shown in Fig. P. 4.1. The watt meter readings are $W_1 = 100$ watts, $W_2 = 150$ watts. Determine the real and imaginary parts of the impedance Z.

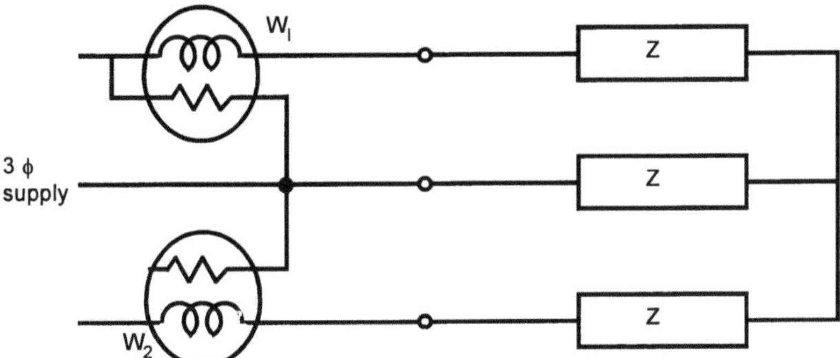

Fig. P. 4.1

4.7 A delta connected load with impedance $Z_{AB} = 10\angle 30^0$ ohms, $Z_{BC} = 25\angle 0^0$ ohms, and $Z_{CA} = 20\angle -30^0$ ohms is connected to a three phase three wire 500 volt system. If the phase sequence is ABC, calculate the line currents and the total power.

4.8 For the network shown in Fig. P. 4.2, calculate the values of elements in the single delta connected equivalent network.

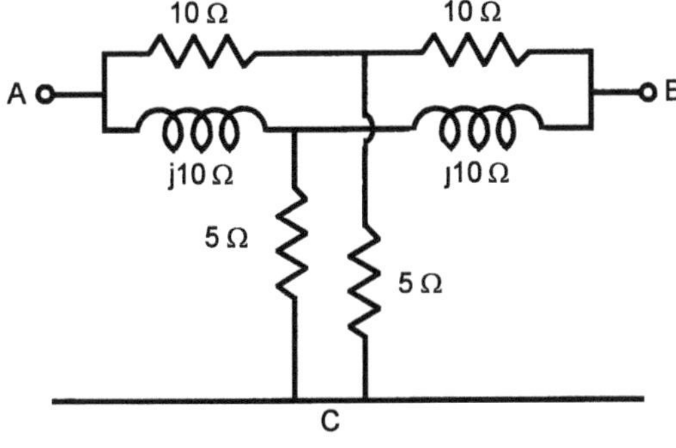

Fig. P. 4.2

4.9 A star connected load with $Z_a = 10\angle 0\ \Omega$, $Z_b = 10\angle 60^0\ \Omega$ and $Z_c = 10\angle -60^0\ \Omega$ is connected to a three phase three wire 200 Volts a b c system. Find the voltages $V_{an'}, V_{bn'}, V_{cn'}$ when n' is load neutral. Use Millman's theorem.

4.10 Three voltmeters having resistances of 10, 10 and 5 kilo ohms are connected in star to lines R, Y and B respectively of a balanced 3 phase three wire supply. The line voltage is 440 volts. Determine the readings of the three voltmeters. The phase sequence is RYB.

4.11 A three phase four wire 400 volts a.c system supplies a star connected load in which $Z_A = 10\angle 0^0\ \Omega$, $Z_B = 15\angle 30^0\ \Omega$ and $Z_C = 10\angle -30^0\ \Omega$. The phase sequence is ABC. A wattmeter W_1 has its current coil in phase A and its pressure coil across A and B. Another wattmeter W_2 has its current coil in phase C and its pressure coil across B and C. Calculate the wattmeter readings and the current through the neutral wire. Also calculate the voltage between supply neutral and load neutral.

4.12 The power delivered to a balanced delta connected load by a 400 volt 3ϕ supply is measured by two wattmeter method. If the readings of the two wattmeters are 2000 and –1500 watts respectively, calculate the magnitude of the impedance in each arm of the delta load and its resistive component.

4.13 A wattmeter is connected as shown in Fig. P. 4.3. Find the reading of the wattmeter. Phase sequence is abc.

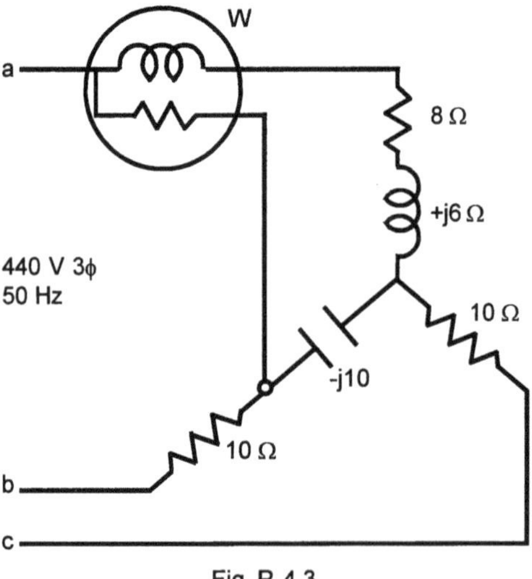

Fig. P. 4.3

4.14 Three capacitors each of 50 μF are connected in star to a 440 volt 3ϕ 50 Hz supply. What will be the value of each of the three phase capacitors to be connected in delta, if the line current is to remain the same.

5

NETWORK TOPOLOGY

5.1 INTRODUCTION :

In the earlier chapters of this book, emphasis was given to the most important concepts and properties of circuits. These were understood using simple circuits containing few elements only. In practice, however, situations occur where it is necessary to handle more complex networks containing several elements. In order to do so, the nature of the elements is set aside, because the Kirchhoff's voltage and current laws do not make any assumptions concerning their nature. Once this is done, the network reduces to a geometrical figure indicating the interconnection of various nodes. The study of the geometrical properties of such figures which are unchanged when the figure is twisted, bent, folded or stretched, squeezed or tied in knots is called *network topology*. The only condition, however, is that no parts of the network should be torn apart or to be joined together. Thus topology deals with the way in which the various elements are interconnected at their terminals without considering the properties and type of the elements connected. We first present a list of definitions and then consider the nature of the interconnection and finally integrate the properties of the elements with the kind of interconnections. This method is considered to be a more systematic approach to the analysis of large electrical networks. Further, a systematic approach in formulating the network equations enables one to use a computer to perform complex analysis and design of very large networks.

5.2 THE CONCEPT OF A NETWORK GRAPH :

When the nature of elements is disregarded, each element is replaced by a line segment between two end points called *the nodes or vertices*. The result is a geometrical

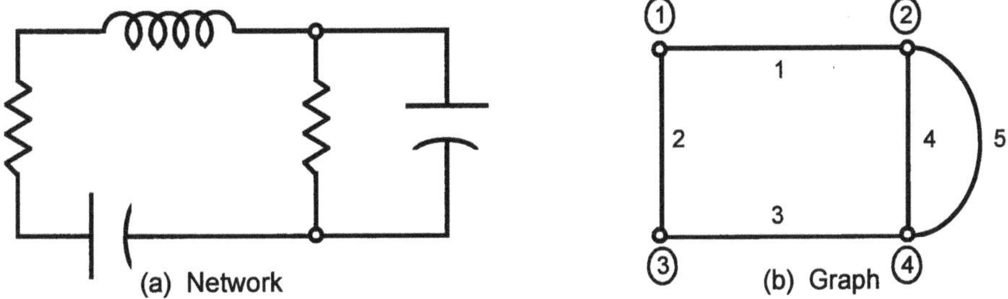

Fig. 5.1 A network and its graph

figure containing these vertices (or nodes) and line segments called the *edges* (or branches). This geometrical figure is called the '*graph*' of the network. As an example, consider the network of Fig. 5.1(a). Each element of the network is replaced by a line segment with two vertices. These vertices are shown as bigger dots. The resulting figure shown in Fig. 5.1(b) is called the *network graph* or simply a *graph*. Usually the nodes are identified by numbers with circles around them. The edges are also likewise indicated by numbers written by the side of the corresponding edges as shown in Fig. 5.1(b).

Let us now consider some definitions useful in the study of graphs.

1. *Graph :* A graph is defined more precisely as a set of nodes together with a set of edges with the condition that each end of an edge terminates at one of the vertices.

2. *Subgraph*: Suppose G is a graph of a network. A graph G_1 is said to be a subgraph if each node of G_1 is in G and each edge of G_1 is in G. According to this definition it is possible to build a graph G_1 by deleting some elements and possibly some vertices as shown in Fig. 5.2(b) and (c).

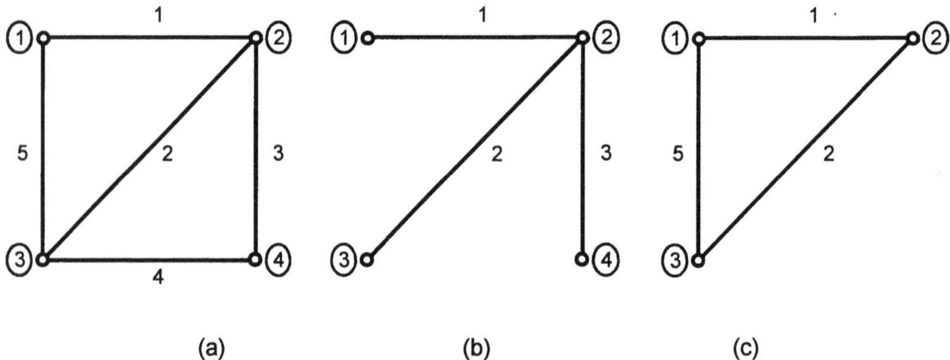

(a) (b) (c)

Fig. 5.2 (a) A graph (b) A subgraph with all nodes and fewer branches (c) Another subgraph with fewer branches and nodes

3. *A path :* A path is a traversal from one node to another node of a graph along the branches such that no node is encountered twice.

4. *Degree of a vertex :* The number of edges terminating at a node is called *the degree of a node*. In Fig. 5.3, the degree of node ② is 4.

where as that of vertex ① is 2.

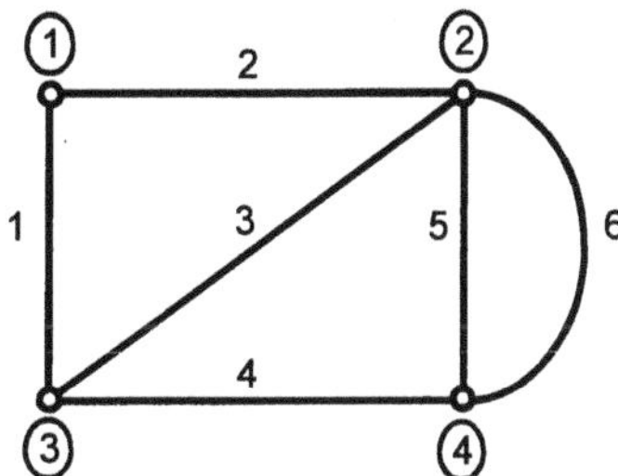

Fig. 5.3 A network graph

5. *Loop* : A loop is a subgraph in which the degree of each node is exactly two. In Fig. 5.3, the subgraph formed by the branches 1, 2 and 3 is a loop. In simple terms a loop is a closed path. A loop can also be defined as a path between a node '*i*' and itself through the branches of the graph.

6. *Connected graph* : A graph is said to be connected if there exists at least one path between any two vertices of the network. By convention, a graph with one vertex is considered to be connected. In any unconnected graph the number of individual connected subgraphs are called *separate parts*. In an unconnected graph there will atleast be two separate parts. The graph of Fig. 5.4(a) is connected while the graph in Fig. 5.4(b) is unconnected with two separate parts.

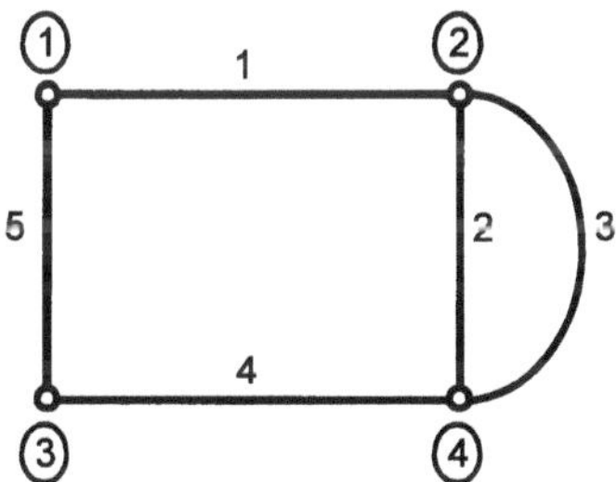

Fig. 5.4(a) Connected graph

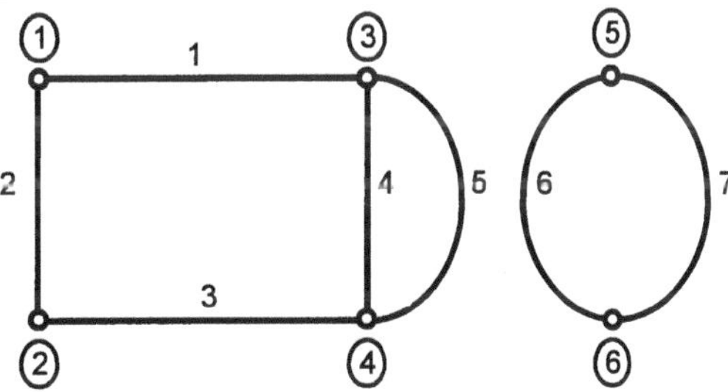

Fig. 5.4(b) Unconnected graph

7. *Oriented graph :* When arbitrary directions are assigned to all the edges of a graph, the graph is said to be oriented. The graph of Fig. 5.5 is said be oriented.

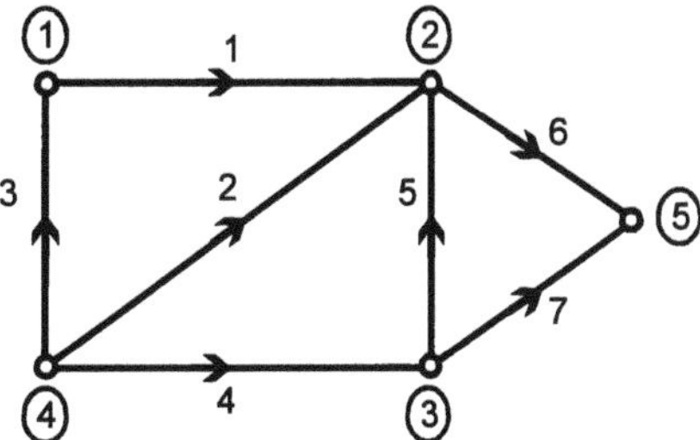

Fig. 5.5 Oriented graph

8. *Planar graph* : A planar graph is one, which can be drawn on the plane in such a way, that no two branches intersect at a point which is not a node. The graph of Fig.5.6(a) is a planar graph where as the graph in Fig.5.6(b) is a nonplanar graph.

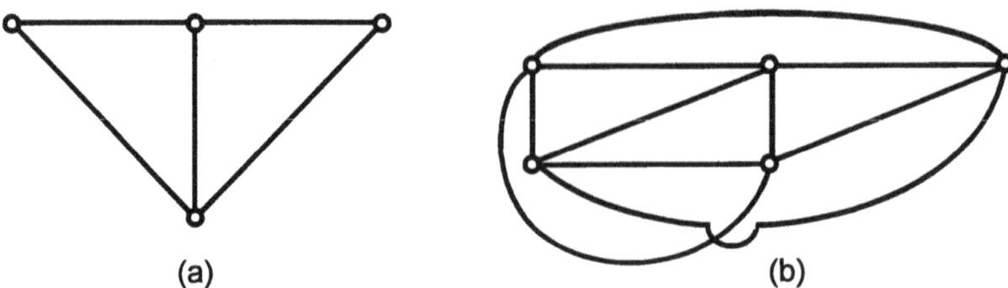

Fig. 5.6(a) Planar graph (b) Nonplanar graph

The graph in Fig. 5.7(a) is also a planar graph though it is drawn such that two of its branches are intersecting. The graph can be redrawn as shown in Fig. 5.7 (b) without any crossing of the branches.

Fig. 5.7 (a) A graph with two branches crossing each other
(b) Same graph redrawn without the crossing of branches

9. *Rank* : The rank of a connected graph is defined as $n - 1$ where 'n' is the number of nodes in the graph. The rank of a graph with 'ℓ' separate parts is the sum of the ranks of the individual parts.

10. *Nullity* : The nullity of a graph is $b - n + \ell$ where b is the number of branches, n is the number of nodes and ℓ is the number of separate parts. If the graph is connected, the nullity is equal to $b - n + 1$.

11. *Cutset* : A cutset is a set of branches of a graph such that when this set of branches are removed from the graph, the rank of the graph is reduced by exactly one and removal of any proper subset of this set does not reduce the rank by one.

As an example consider the graph of Fig. 5.8(a). The number of nodes in the graph is 4 and hence its rank is equal to 3. If the set of edges 1, 2, 3 and 4 are removed, the rank of the resulting subgraph is 2 as shown in Fig. 5.8(b). If any one or more of these edges is replaced as shown in Fig. 5.8(c) the rank of the graph remains to be 3.

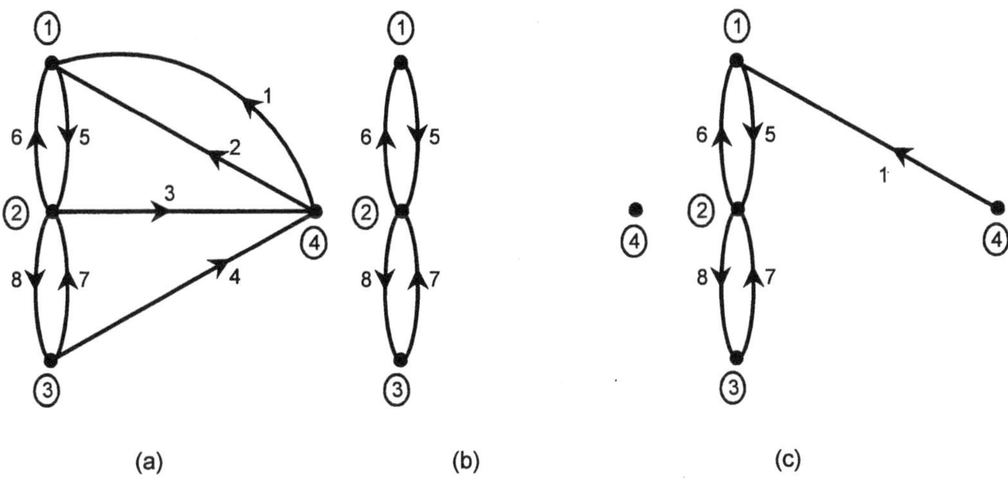

Fig. 5.8(a) A graph (b) The edges 1, 2, 3 and 4 are removed (c) edge 1 is replaced

The set of edges 1, 2, 3 and 4 is known as a cutset. Similarly for the graph in Fig. 5.8(a) different cutsets are

(6, 5, 2 and 1); (8, 7, 3, 2 and 1); (6, 5, 3 and 4)

In simple terms, a cutset cuts the network into two parts. It is customary to indicate a cutset by drawing a line cutting all the edges forming a cutset as shown in Fig. 5.9.

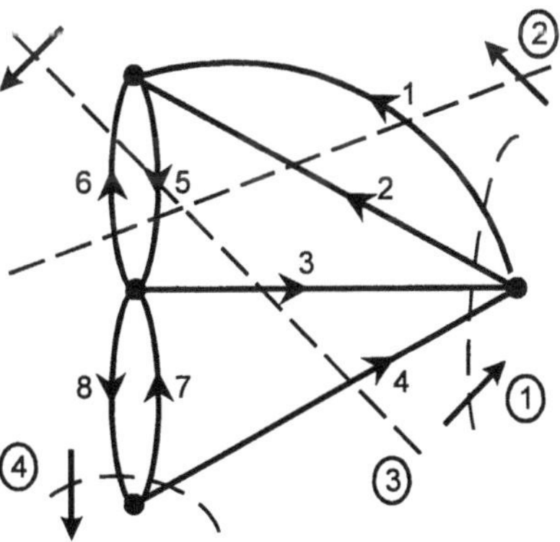

Fig. 5.9 Graph with 4 cutsets indicated

Each cutset has a certain arbitrary direction indicated by an arrow as shown in Fig. 5.9.

12. A *Tree* : The most important concept of graph theory is the tree of a graph. It is defined as a connected sub graph of a given graph which contains all the nodes but no circuits of the original graph. For example a graph and 3 of its trees are shown in Fig. 5.10 (a), (b), (c), and (d):

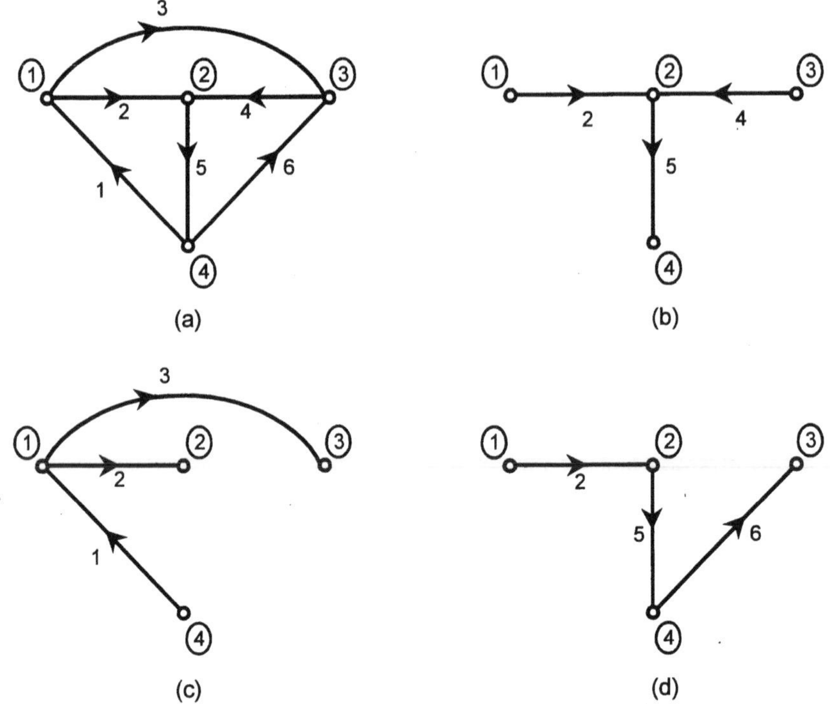

Fig. 5.10 (a) graph (b), (c) and (d) trees of the graph

The structure of a tree resembles a physical tree and hence the name tree is given to the sub graph.

Some important properties of trees :

1. It is a connected sub graph.
2. It contains all the nodes of the original graph.
3. It does not contain any circuits.
4. It has at least one node with a degree 1. If every node has a degree 2 or more, the subgraph will contain a circuit, violating the definition.
5. It has exactly one and only one path between any two nodes. If it has two paths between any two nodes, these two paths from a circuit, violating the definition of a tree.
6. The tree has exactly $n - 1$ branches where 'n' is the numbers of nodes. The branches of a tree are also known as '*twigs*'.

5.3 CHORDS OR LINKS :

The branches of the graph which are removed to form a tree of the graph are called *chords* or *links*. If there are 'b' branches and 'n' nodes in a graph, since the tree contains n – 1 branches, the number of chords is equal to b – (n – 1) or b – n +1. For the graph of Fig. 5.10(a) and the tree given in Fig. 5.10(b), the chords are 1, 3 and 6. Similarly, the chords are 4, 5 and 6; and 1, 3 and 4 for the trees in Fig. 5.10(c) and (d) respectively.

The set of branches which are not in the tree is known as a co – tree or complement of a tree.

If a graph has several separate parts, the concept of tree can be used for each separate part resulting in a tree for each separate part and the combination is known as a *'Forest'*. It is obvious that for a given graph, a number of trees can be drawn. If a graph has n nodes and if there is only one branch connected between any two nodes and further, if every node is connected to every other node, it can be shown that the number of trees that can be drawn for this graph will be n^{n-2}. For a 5 node graph, with every node connected to every other node with exactly one branch, the number of trees will be $5^3 = 125$.

Having defined some important terms used in the graph theory, let us try to formulate the network equations based on Kirchhoff's Current Law (KCL) and Kirchhoff's Voltage Law (KVL) in a systematic way. The graph of a network can be described in an analytical way by means of a set of matrices.

5.3.1. Incidence matrix (A) :

An oriented graph can be described in an analytical way by listing all branches and nodes and indicating how the branches are connected to each of the nodes. This is conveniently done by means of a matrix. Let the oriented graph contain '*b*' branches and '*n*' nodes. Let us arbitrarily assign numbers to the branches and nodes. The *nodes to branch augmented incidence matrix* A_a is a rectangular matrix of *n* rows and *b* columns whose (i, j)th element is defined as

$$a_{ij} = \begin{cases} +1 & \text{If branch '}j\text{' is incident at node '}i\text{' and is directed away from it.} \\ -1 & \text{If branch '}j\text{' is incident at node '}i\text{' and is directed towards it.} \\ 0 & \text{If branch '}j\text{' is not incident at node '}i\text{'.} \end{cases}$$

The augmented incidence matrix A_a of the graph shown in Fig. 5.11 is

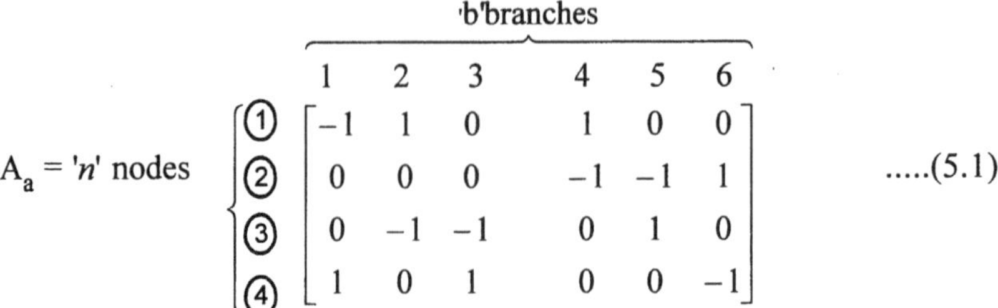

$$A_a = \text{'}n\text{' nodes} \begin{cases} ① \\ ② \\ ③ \\ ④ \end{cases} \overbrace{\begin{bmatrix} -1 & 1 & 0 & 1 & 0 & 0 \\ 0 & 0 & 0 & -1 & -1 & 1 \\ 0 & -1 & -1 & 0 & 1 & 0 \\ 1 & 0 & 1 & 0 & 0 & -1 \end{bmatrix}}^{\text{'b' branches: } 1 \; 2 \; 3 \; 4 \; 5 \; 6} \quad .....(5.1)$$

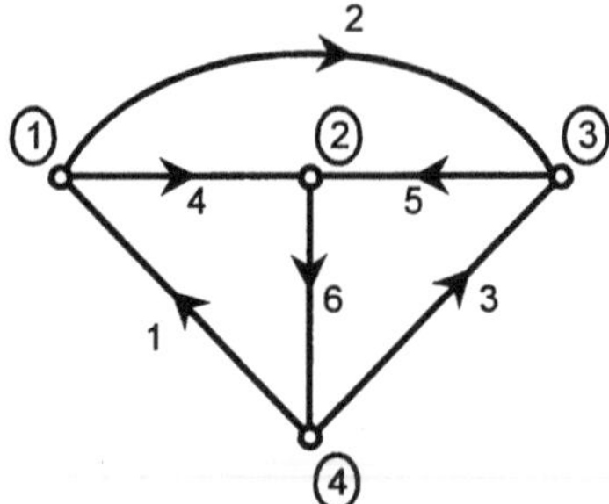

Fig. 5.11 A graph

Since each branch leaves a single node and enters a single node, each column of A_a contains exactly two non – zero elements and one element is +1 and the other is –1. Since adding of all the rows produces a row of zeros, the rows of this matrix are linearly dependent. Hence deleting any one row, still describes the graph completely. The matrix obtained by deleting one row of the augmented incidence matrix A_a is known as the *reduced incidence matrix* or simply the *incidence matrix* and is denoted by A. The node corresponding to the deleted row is known as the *datum node*. Thus the incidence matrix for the graph in Fig. 5.11, with node 4 as reference or *datum node*, is given by

$$\mathbf{A} = \begin{matrix} \\ ① \\ ② \\ ③ \end{matrix} \begin{vmatrix} 1 & 2 & 3 & 4 & 5 & 6 \\ -1 & 1 & 0 & 1 & 0 & 0 \\ 0 & 0 & 0 & -1 & -1 & 1 \\ 0 & -1 & -1 & 0 & 1 & 0 \end{vmatrix} \quad .....(5.2)$$

The node numbers and branch numbers are indicated outside the matrix for easy reference.

If we denote the branch currents by j_k for $k = 1, 2, ...b$, the branch current vector '**j**' is given by

$$\mathbf{j} = \begin{bmatrix} j_1 \\ j_2 \\ j_3 \\ j_4 \\ j_5 \\ j_6 \end{bmatrix}$$

Consider the matrix equation

$$\mathbf{Aj} = \mathbf{0} \qquad \text{(KCL)} \qquad .....(5.3)$$

Expanding eq. (5.3) with A given by eq. (5.2)

$$\begin{aligned} -j_1 + j_2 + j_4 &= 0 \\ -j_4 - j_5 + j_6 &= 0 \\ -j_2 - j_3 + j_5 &= 0 \end{aligned} \qquad .....(5.4)$$

It is easy to observe that eqs. (5.4) are the equations obtained by applying KCL at nodes 1, 2 and 3. Thus $\mathbf{Aj} = \mathbf{0}$ is a statement of KCL for the given network. In other words, eq. (5.3) is the set of node voltage equations for the given network with node ④ as datum node.

Now consider the matrix equation

$$\mathbf{v} = \mathbf{A}^{\mathbf{T}} \mathbf{e} \qquad \text{(KVL)} \qquad .....(5.5)$$

where **v** is the branch voltage vector and
e is the node to datum voltage vector.
$\mathbf{A}^{\mathbf{T}}$ is the transpose of the incidence matrix.

The branch voltages have associated reference directions with respect to the current directions in these branches, indicated by the arrow in the directed graph.

The graph of Fig. 5.11 is redrawn in Fig. 5.12 with the branch voltages and node voltages indicated.

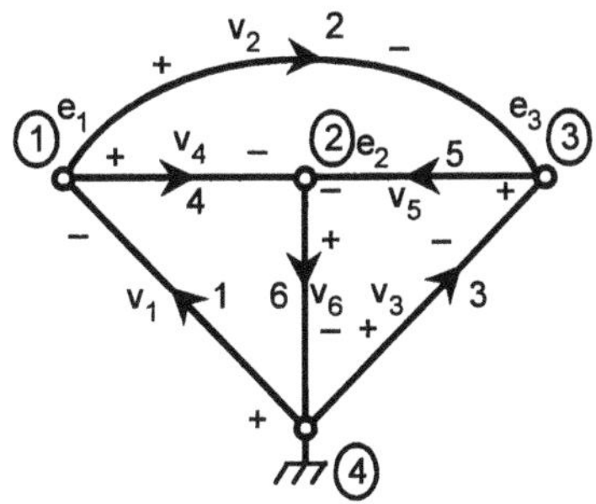

Fig. 5.12 A graph with node and branch voltages indicated

Eq. (5.5) can be written as

$$\begin{bmatrix} v_1 \\ v_2 \\ v_3 \\ v_4 \\ v_5 \\ v_6 \end{bmatrix} = \begin{bmatrix} -1 & 0 & 0 \\ 1 & 0 & -1 \\ 0 & 0 & -1 \\ 1 & -1 & 0 \\ 0 & -1 & 1 \\ 0 & 1 & 0 \end{bmatrix} \begin{bmatrix} e_1 \\ e_2 \\ e_3 \end{bmatrix} \qquad .....(5.6)$$

or

$$\begin{aligned} v_1 &= -e_1 \\ v_2 &= e_1 - e_3 \\ v_3 &= -e_3 \\ v_4 &= e_1 - e_2 \\ v_5 &= -e_2 + e_3 \\ v_6 &= e_2 \end{aligned} \qquad .....(5.7)$$

These six equation can easily be recognised as the expressions for the KVL.

Thus the equations

$$\mathbf{Aj} = \mathbf{0} \qquad \text{(KCL)}$$
$$\mathbf{v} = \mathbf{A^T j} \qquad \text{(KVL)}$$

are the two basic equations for the network analysis using the graph of the network and Kirchhoff's laws. These equations are independent of the nature of the elements in each branch of the network. Unless the branch voltages '**v**' and branch currents '**j**' are related by branch equations for the particular elements of the branches, these equations cannot be solved.

5.3.2 Circuit matrix or tie set matrix B :

Let us now consider another method of specifying the graph analytically. Consider an oriented graph shown in Fig. 5.13(a) and one of its trees shown by thick lines in

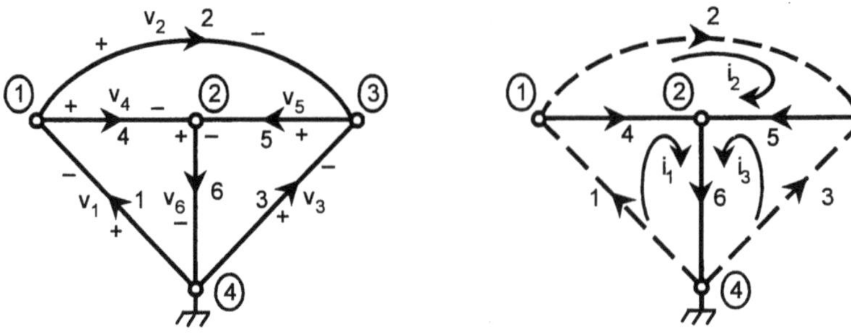

Fig. 5.13 (a) The graph (b) A tree with links shown in dotted lines

Fig. 5.13(b). The links or chords are shown by dotted lines. Let us follow a convention to number the branches. The link branches are numbered from 1 to l and the tree branches are numbered from l + 1 to b. As per the definition of a tree, whenever a chord is replaced, a circuit is formed by that chord and some of the branches of the tree. For example, if chord 1 is replaced as shown in Fig. 5.13(b) a circuit is formed by the branches 1, 4 and 6. This circuit which is formed by exactly one chord and some of the branches of the tree is known as a *fundamental circuit* with respect to the given tree of the graph. The set of branches which form a fundamental circuit is also called a Tieset. Thus there will be as many fundamental circuits as there are links of the graph. These circuits are independent of each other, as each circuit contains atleast one element (corresponding chord) which is not common to any other circuit. The loop obtained by replacing the chords are marked as i_1, i_2 and i_3 in the graph with reference directions agreeing with the direction of the respective chords.

Now we are ready to describe the fundamental circuit matrix B of a given graph with respect to a given tree. The rows of this matrix are the independent loops whose number is equal to the number of chords, $l = (b - n + 1)$. The columns are the branches. But we use a particular order for these branches in defining the columns of the matrix. First the link or chord branches are considered in the order in which the respective loops are taken to define the rows. The tree branches follow after all the link branches are considered as shown in eq. 5.8.

$$\text{B} = \begin{matrix}\text{loop}\\\text{currents}\end{matrix} \begin{matrix} \\ 1 \\ 2 \\ 3 \end{matrix} \overset{\text{branches}}{\overset{\begin{matrix}\overbrace{\quad\text{Links}\quad} & & \overbrace{\text{Tree branches}}\end{matrix}}{\begin{matrix} 1 & 2 & 3 & | & 4 & 5 & 6 \\ \end{matrix}}} \left[\begin{array}{ccc|ccc} 1 & 0 & 0 & 1 & 0 & 1 \\ 0 & 1 & 0 & -1 & 1 & 0 \\ 0 & 0 & 1 & 0 & 1 & 1 \end{array}\right] \quad .....(5.8)$$

Here

$$b_{ij} = \begin{cases} +1 & \text{If } j^{th} \text{ branch is in the } i^{th} \text{ loop and its direction is same as the loop direction.} \\ -1 & \text{If } j^{th} \text{ branch is in the } i^{th} \text{ loop and its direction is opposite to that of the loop direction.} \\ 0 & \text{If } j^{th} \text{ branch is not in th } i^{th} \text{ loop} \end{cases}$$

Since links 1, 2 and 3 are present only in the 1st, 2nd and 3rd loops, the leading 3 × 3 matrix is an identity matrix. Thus the matrix B can be partitioned as

$$\text{B} = [\text{I} \quad \text{B}_1] \quad .....(5.9)$$

where I is the identify matrix and thus the rank of circuit matrix is equal to $b - n + 1$.

Now consider the equation

$$\mathbf{Bv} = \mathbf{0} \qquad \text{.....(5.10)}$$

where v is the branch voltage vector. Substituting for B from eq. (5.8)

$$\begin{bmatrix} 1 & 0 & 0 & 1 & 0 & 1 \\ 0 & 1 & 0 & -1 & 1 & 0 \\ 0 & 0 & 1 & 0 & 1 & 1 \end{bmatrix} \begin{bmatrix} v_1 \\ v_2 \\ v_3 \\ v_4 \\ v_5 \\ v_6 \end{bmatrix} = \begin{bmatrix} 0 \\ 0 \\ 0 \end{bmatrix} \qquad \text{.....(5.11)}$$

Thus

$$\begin{aligned} v_1 + v_4 + v_6 &= 0 \\ v_2 - v_4 + v_5 &= 0 \\ v_3 + v_5 + v_6 &= 0 \end{aligned} \qquad \text{.....(5.12)}$$

These can be recognised as the KVL equations for the three fundamental loops. Thus $\mathbf{Bv} = \mathbf{0}$ defines KVL equations.

Consider the equation

$$\mathbf{j} = \mathbf{B}^T\mathbf{i} \qquad \text{(KCL)} \qquad \text{.....(5.13)}$$

where $\mathbf{j}$ is the branch current vector

$\mathbf{i}$ is the loop current vector

and $\mathbf{B}^T$ is the transpose of the circuit matrix B

or

$$\begin{bmatrix} j_1 \\ j_2 \\ j_3 \\ j_4 \\ j_5 \\ j_6 \end{bmatrix} = \begin{bmatrix} 1 & 0 & 0 \\ 0 & 1 & 0 \\ 0 & 0 & 1 \\ 1 & -1 & 0 \\ 0 & 1 & 1 \\ 1 & 0 & 1 \end{bmatrix} \begin{bmatrix} i_1 \\ i_2 \\ i_3 \end{bmatrix} \qquad \text{.....(5.14)}$$

or

$$\begin{aligned} j_1 &= i_1 \\ j_2 &= i_2 \\ j_3 &= i_3 \\ j_4 &= i_1 - i_2 \\ j_5 &= i_2 + i_3 \\ j_6 &= i_1 + i_3 \end{aligned} \qquad \text{.....(5.15)}$$

These equations can be considered as KCL equations.

In summary we have,

$$\mathbf{Bv = 0} \qquad \text{(KVL)}$$
$$\mathbf{j = B^T i} \qquad \text{(KCL)}$$

5.3.3. Cutset matrix Q :

The third method of specifying the network graph analytically is by using a fundamental cutset matrix. Let us consider a graph and one of its trees as shown in Fig. 5.14.

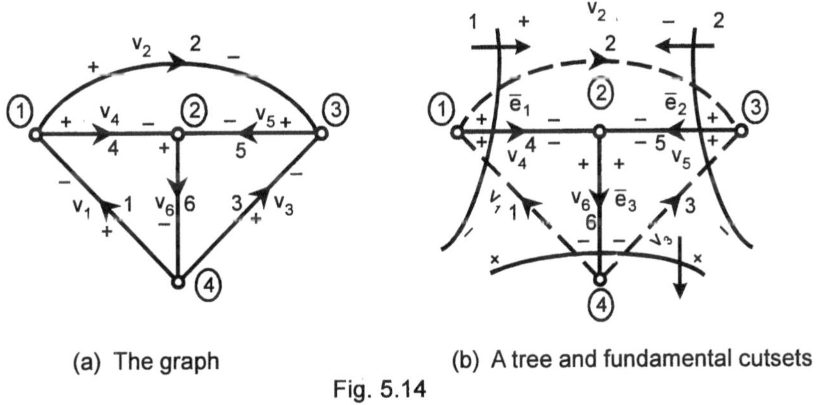

Fig. 5.14

Let us draw lines cutting one of the branches of the tree and some of the chords as shown in Fig. 5.14(b). It is obvious that the branches cut by the lines represent the cutsets for the given graph. Thus, in Fig. 5.14(b), the sets of branches, (1, 2, 4), (2, 3, 5) and (1, 3, 6) represent 3 cutsets. These cutsets are called *fundamental cutsets* because each fundamental cutset contains exactly one branch of the tree and some of the chords with respect to a given tree. These sets are independent of each other because each set contains at least one new element, which is the branch of a tree. Let us also define a direction for each of these cutsets and indicate it by an arrow on each line representing a cutset. The conventional direction for the fundamental cutset is taken to be the same as the direction of the tree branch which defines the particular cutset.

For example the tree branch 4 defines the cutset 1 and the direction of this cutset is taken to agree with the direction of this branch i.e, from left to right as shown in Fig. 5.14(b). Now we are ready to define the fundamental cutset matrix 'Q' for the given graph with respect to a chosen tree. The rows of this matrix are the cutsets and the columns are branches. The columns are numbered from 1 to l to represent the links and $l + 1$ to b to represent the tree branches. The rows are numbered in the order of the cutsets defined by the tree branches $l + 1$ to b. Now the elements of the 'Q' matrix are defined by :

$$q_{ij} = \begin{cases} +1 & \text{If } j^{th} \text{ branch belongs to } i^{th} \text{ cutset and has the same direction.} \\ -1 & \text{If } j^{th} \text{ branch belongs to the } i^{th} \text{ cutset and has opposite direction.} \\ 0 & \text{If } j^{th} \text{ branch does not belong to the } i^{th} \text{ cutset} \end{cases}$$

Using the above definition the Q matrix for the graph given in Fig. 5.14(a) is

$$\mathbf{Q} = \begin{matrix} & 1 & 2 & 3 & 4 & 5 & 6 \\ 1 \\ 2 \\ 3 \end{matrix} \begin{bmatrix} -1 & 1 & 0 & 1 & 0 & 0 \\ 0 & -1 & -1 & 0 & 1 & 0 \\ -1 & 0 & -1 & 0 & 0 & 1 \end{bmatrix} \qquad .....(5.16)$$

The fundamental cutset matrix Q can be partitioned as

$$\mathbf{Q} = \begin{bmatrix} Q_l & I_{n-1} \end{bmatrix}$$

Since the number of branches in a tree is equal to $n - 1$ where n is the number of nodes, the rank of the Q – matrix is equal to $n - 1$.

Now the KCL and KVL can be written in terms of the Q – matrix. Thus

$$\mathbf{Qj = 0} \quad \text{(KCL)} \qquad .....(5.17)$$

and for the graph of Fig. 5.14(a)

$$\begin{bmatrix} -1 & 1 & 0 & 1 & 0 & 0 \\ 0 & -1 & -1 & 0 & 1 & 0 \\ -1 & 0 & -1 & 0 & 0 & 1 \end{bmatrix} \begin{bmatrix} j_1 \\ j_2 \\ j_3 \\ j_4 \\ j_5 \\ j_6 \end{bmatrix} = \begin{bmatrix} 0 \\ 0 \\ 0 \end{bmatrix} \qquad .....(5.18)$$

or

$$-j_1 + j_2 + j_4 = 0$$
$$-j_2 - j_3 + j_5 = 0$$
$$-j_1 - j_3 + j_6 = 0 \qquad .....(5.19)$$

These equations are the KCL equations at nodes 1, 3 and 4.

Now consider,

$$\mathbf{v} = \mathbf{Q}^T \bar{\mathbf{e}} \qquad \text{(KVL)} \qquad ....(5.20)$$

where $\bar{\mathbf{e}}$ is the tree branch voltage vector as shown in Fig. 5.14(b).

Using eq. (5.16) in eq. (5.20), we have

$$\begin{bmatrix} v_1 \\ v_2 \\ v_3 \\ v_4 \\ v_5 \\ v_6 \end{bmatrix} = \begin{bmatrix} -1 & 0 & -1 \\ 1 & -1 & 0 \\ 0 & -1 & -1 \\ 1 & 0 & 0 \\ 0 & 1 & 0 \\ 0 & 0 & 1 \end{bmatrix} \begin{bmatrix} \bar{e}_1 \\ \bar{e}_2 \\ \bar{e}_3 \end{bmatrix} \qquad .....(5.21)$$

or

$$v_1 = -\, e_1 - e_3$$
$$v_2 = \bar{e}_1 - \bar{e}_2$$
$$v_3 = -\, \bar{e}_2 - \bar{e}_3$$
$$v_4 = \bar{e}_1$$
$$v_5 = \bar{e}_2$$
$$v_6 = \bar{e}_3 \qquad .....(5.22)$$

These equations are the KVL equations for the network.

In summary, we have

$$\mathbf{Qj} = \mathbf{0} \qquad \text{(KCL)}$$
$$\mathbf{v} = \mathbf{Q}^T \bar{\mathbf{e}} \qquad \text{(KVL)}$$

5.4 FORMULATION OF NETWORK EQUATIONS :

So far we considered the graph of a network which describes the way in which the various network elements are interconnected. We have not taken cognizance of the types of elements. Unless we link the type of elements to the theory developed so far, it will be of no practical use. Let us precisely do the same in this section.

5.4.1 A typical branch of a network :

Consider the kth branch of a network, which may include an independent voltage source in series with an impedance and an independent current source in parallel with the impedance as shown in Fig. 5.15.

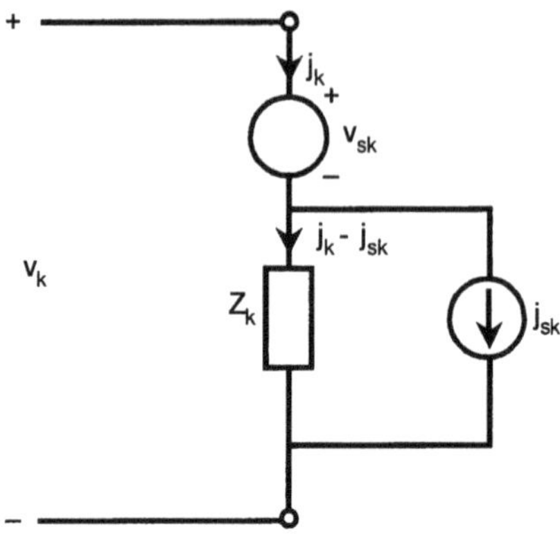

Fig. 5.15 A typical branch of a network

If the branch contains only an impedance, then $v_{sk} = 0$ and $j_{sk} = 0$. If it has only a voltage source in series with Z_k, $j_{sk} = 0$ and finally if it contains only a current source in parallel, then $v_{sk} = 0$. We note that the voltage sources are always associated with a series impedance and current sources are always associated with parallel impedances. If branches contain only independent sources, these branches have to be transformed to satisfy these requirements. These transformations were discussed in chapter 2.

Let us write down the $v - i$ equation for the typical branch shown in Fig. 5.15

$$v_k = v_{sk} + Z_k (j_k - j_{sk}) \quad .....(5.23)$$

The current j_k can also be expressed in terms of v_k as

$$j_k = \frac{v_k - v_{sk}}{Z_k} + j_{sk} \quad .....(5.24)$$

or considering admittance of the branch

$$j_k = j_{sk} + (v_k - v_{sk}) Y_k \quad .....(5.25)$$

These equations can be combined with the KCL and KVL equations derived in the earlier section in terms of A, B and Q matrices to obtain the final equations in the matrix form describing the network. These final equations can then be solved to obtain the desired solution.

5.4.2 Node voltage equations :

Consider a linear, connected network with b branches and n nodes, a typical branch of which is as shown in Fig. 5.15. Let any node be taken as reference and the other nodes numbered as 1, 2, ...$n - 1$. Let the node to datum voltages be $e_1, e_2 \ldots\ldots e_{n-1}$. The branches are also numbered 1, 2, ...b and are assigned arbitrary directions. The KCL equations for this network are

$$\mathbf{Aj} = \mathbf{0}$$

The volt – ampere equation of the kth branch is given by eq. 5.25 :

$$j_k = Y_k v_k + j_{sk} - v_{sk} Y_k \qquad \ldots\ldots(5.26)$$

for $\quad k = 1, 2, \ldots\ldots b$

This can be put in the matrix form as

$$\mathbf{j} = \mathbf{Y_b\, v} + \mathbf{j_s} - \mathbf{Y_b\, v_s} \qquad \ldots\ldots(5.27)$$

where

$$\mathbf{j} = \begin{bmatrix} j_1 \\ j_2 \\ \vdots \\ j_b \end{bmatrix} \quad \mathbf{v} = \begin{bmatrix} v_1 \\ v_2 \\ \vdots \\ v_b \end{bmatrix} \quad \mathbf{Y_b} = \begin{bmatrix} Y_1 & 0 & \ldots & 0 \\ 0 & Y_2 & \ldots & 0 \\ \vdots & & & \\ 0 & \ldots\ldots & & Y_b \end{bmatrix}$$

$$\mathbf{j_s} = \begin{bmatrix} j_{s1} \\ j_{s2} \\ \vdots \\ j_{sb} \end{bmatrix} \qquad \mathbf{v_s} = \begin{bmatrix} v_{s1} \\ v_{s2} \\ \vdots \\ v_{sb} \end{bmatrix}$$

and $\mathbf{j}$ is the branch current vector

$\mathbf{v}$ is the branch voltage vector

$\mathbf{Y_b}$ is the branch admittance matrix, which is a diagonal matrix with branch admittance as the diagonal elements.

$\mathbf{j_s}$ is the branch source current vector.

and $\mathbf{v_s}$ is the branch source voltage vector

Pre – multiplying eq. (5.27) by A and using eq. (5.5), we have

$$\mathbf{Aj} = \mathbf{0} = \mathbf{AY_bA^Te} + \mathbf{Aj_s} - \mathbf{AY_bv_s} \qquad \ldots\ldots(5.28)$$

or

$$\mathbf{AY_bA^Te} = \mathbf{AY_bv_s} - \mathbf{Aj_s} \qquad \ldots\ldots(5.29)$$

In eq. (5.29), $\mathbf{AY_bA^T}$ is a square matrix of order (n – 1) and $\mathbf{AY_bv_s}$ and $\mathbf{A\,j_s}$ are (n –1) dimensional vectors.

Defining $\quad \mathbf{Y_n} \triangleq \mathbf{AY_bA^T} \quad$ and $\quad \mathbf{i_{es}} \triangleq \mathbf{AY_bv_s} - \mathbf{Aj_s}$

we have $\quad \mathbf{Y_ne} = \mathbf{i_{es}} \quad$ ....(5.30)

Eqs. (5.30) are the node voltage equations where **e** is the node voltage vector, $\mathbf{Y_n}$ is called the *node admittance matrix* and $\mathbf{i_{es}}$ is the *equivalent current source vector.* Since $\mathbf{Y_n}$ and $\mathbf{i_{es}}$ are known, the unknown node voltage vector can be obtained as

$$e = \mathbf{Y_n^{-1}}\, \mathbf{i_{es}} \qquad .....(5.31)$$

Let us now illustrate the method with an example.

Example 5.1

Obtain the node voltages for the network in Fig. 5.16(a).

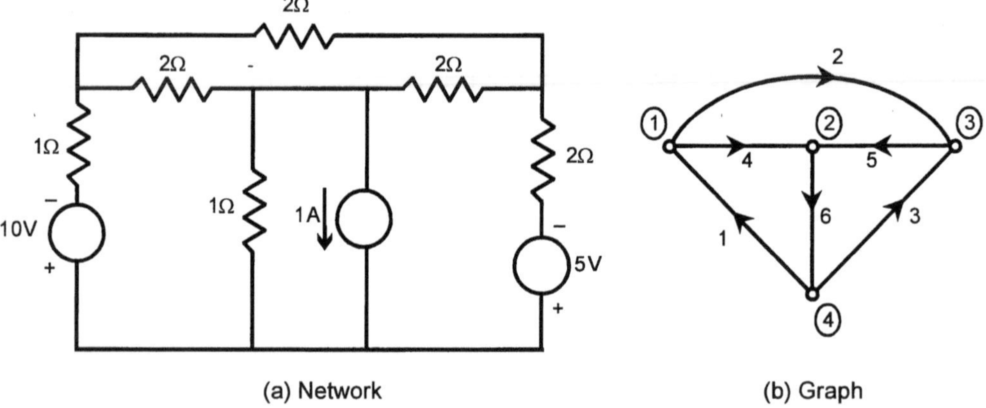

Fig. 5.16(a) Network for example 5.1 (b) its oriented graph

Solution :

Step 1 *:* Draw the graph of the network. Series connected elements and parallel connected elements are taken as single branches. Assign arbitrary directions to the branches excepts for the branches containing sources. For source branches, assign directions to correspond to the directions indicated in Fig. 5.15 for a typical branch.

Step 2 *:* Write down the matrices **A** and $\mathbf{Y_b}$ and vectors $\mathbf{j_s}$ and $\mathbf{v_s}$

$$A = \begin{bmatrix} -1 & 1 & 0 & 1 & 0 & 0 \\ 0 & 0 & 0 & -1 & -1 & 1 \\ 0 & -1 & -1 & 0 & 1 & 0 \end{bmatrix}$$

$$\mathbf{Y_b} = \begin{bmatrix} 1 & 0 & 0 & 0 & 0 & 0 \\ 0 & 0.5 & 0 & 0 & 0 & 0 \\ 0 & 0 & 0.5 & 0 & 0 & 0 \\ 0 & 0 & 0 & 0.5 & 0 & 0 \\ 0 & 0 & 0 & 0 & 0.5 & 0 \\ 0 & 0 & 0 & 0 & 0 & 1 \end{bmatrix}$$

$$\mathbf{v_s} = \begin{bmatrix} 10 \\ 0 \\ 5 \\ 0 \\ 0 \\ 0 \end{bmatrix} \qquad \mathbf{j_s} = \begin{bmatrix} 0 \\ 0 \\ 0 \\ 0 \\ 0 \\ 1 \end{bmatrix}$$

Step 3 : Form $\mathbf{Y_n} = \mathbf{AY_b A^T}$

$$\mathbf{Y_n} - \begin{bmatrix} -1 & 1 & 0 & 1 & 0 & 0 \\ 0 & 0 & 0 & -1 & -1 & 1 \\ 0 & -1 & -1 & 0 & 1 & 0 \end{bmatrix} \begin{bmatrix} 1 & & & & & \\ & 0.5 & & & 0 & \\ & & 0.5 & & & \\ & & & 0.5 & & \\ & 0 & & & 0.5 & \\ & & & & & 1 \end{bmatrix} \begin{bmatrix} -1 & 0 & 0 \\ 1 & 0 & -1 \\ 0 & 0 & -1 \\ 1 & -1 & 0 \\ 0 & -1 & 1 \\ 0 & 1 & 0 \end{bmatrix}$$

$$= \begin{bmatrix} 2 & -0.5 & -0.5 \\ -0.5 & 2 & -0.5 \\ -0.5 & -0.5 & 1.5 \end{bmatrix}$$

The matrix $\mathbf{Y_n}$ can also be written down by inspection. The diagonal elements are the self admittances of the corresponding nodes and the off diagonal elements are the mutual admittances between two corresponding nodes and will always have negative sign. However if the network contains dependent sources, it is difficult to write $\mathbf{Y_n}$ by inspection.

Step 4 : Obtain $\mathbf{i_{es}} = \mathbf{AY_b v_s} - \mathbf{Aj_s}$

$$\mathbf{i_{es}} = \begin{bmatrix} -1 & 1 & 0 & 1 & 0 & 0 \\ 0 & 0 & 0 & -1 & -1 & 1 \\ 0 & -1 & -1 & 0 & 1 & 0 \end{bmatrix} \begin{bmatrix} 1 & & & & & 0 \\ & 0.5 & & & & \\ & & 0.5 & & & \\ & & & 0.5 & & \\ & & & & 0.5 & \\ 0 & & & & & 1 \end{bmatrix} \begin{bmatrix} 10 \\ 0 \\ 5 \\ 0 \\ 0 \\ 0 \end{bmatrix}$$

$$- \begin{bmatrix} -1 & 1 & 0 & 1 & 0 & 0 \\ 0 & 0 & 0 & -1 & -1 & 1 \\ 0 & -1 & -1 & 0 & 1 & 0 \end{bmatrix} \begin{bmatrix} 0 \\ 0 \\ 0 \\ 0 \\ 0 \\ 1 \end{bmatrix} = \begin{bmatrix} -10 \\ -1 \\ -2.5 \end{bmatrix}$$

Step 5 : From the equation

$$\mathbf{Y_n\, e} = \mathbf{i_{es}}$$

$$\begin{bmatrix} 2 & -0.5 & -0.5 \\ -0.5 & 2 & -0.5 \\ -0.5 & -0.5 & 1.5 \end{bmatrix} \begin{bmatrix} e_1 \\ e_2 \\ e_3 \end{bmatrix} = \begin{bmatrix} -10 \\ -1 \\ -2.5 \end{bmatrix}$$

Solving for **e**, we get

$$\begin{bmatrix} e_1 \\ e_2 \\ e_3 \end{bmatrix} = \begin{bmatrix} 2 & -0.5 & -0.5 \\ -0.5 & 2 & -0.5 \\ -0.5 & -0.5 & 1.5 \end{bmatrix}^{-1} \begin{bmatrix} -10 \\ -1 \\ -2.5 \end{bmatrix}$$

$$= \frac{1}{35} \begin{bmatrix} 22 & 8 & 10 \\ 8 & 22 & 10 \\ 10 & 10 & 30 \end{bmatrix} \begin{bmatrix} -10 \\ -1 \\ -2.5 \end{bmatrix} = \frac{1}{35} \begin{bmatrix} -253 \\ -127 \\ -185 \end{bmatrix}$$

The node voltages are $e_1 = \dfrac{-253}{35}$ V, $e_2 = \dfrac{-127}{35}$ V and $e_3 = \dfrac{-185}{35}$ V

If it is desired to find the branch voltages and currents we can use the relations

$$\mathbf{v} = \mathbf{A}^T \mathbf{e} = \begin{bmatrix} \frac{253}{35} \\ -\frac{68}{35} \\ \frac{185}{35} \\ -\frac{126}{35} \\ -\frac{58}{35} \\ -\frac{127}{35} \end{bmatrix}$$

and $$\mathbf{j} = \mathbf{Y}_b \mathbf{v} + \mathbf{j}_s - \mathbf{Y}_b \mathbf{v}_s$$

$$= \begin{bmatrix} -\frac{97}{35} \\ -\frac{34}{35} \\ \frac{1}{7} \\ -\frac{9}{5} \\ -\frac{29}{35} \\ -\frac{92}{35} \end{bmatrix}$$

This completes the analysis of the network. It can be observed that because of the systematic formulation of the problems in matrix form, it is easy to obtain the solution using a digital computer.

5.4.3 Loop current equations :

To obtain the loop current equations, we consider a tree of the given network and the fundamental circuit matrix B of the graph. In order to incorporate the nature of elements and obtain loop current equations in matrix form, we consider the typical branch of any network given by Fig. (5.15). The volt – ampere equation for this branch is given by eq. (5.23).

$$v_k = v_{sk} + Z_k j_k - Z_k j_{sk}$$

By changing k from 1 to b, we get the $v - i$ equations for all the 'b' branches of the network and are given in matrix form by

$$\mathbf{v} = \mathbf{v_s} + \mathbf{Z_b}\,\mathbf{j} - \mathbf{Z_b}\,\mathbf{j_s} \qquad .....(5.32)$$

where $\mathbf{v}$ is the branch voltage vector

$\mathbf{j}$ is the branch current vector

$\mathbf{v_s}$ is the branch source voltage vector

$\mathbf{j_s}$ is the branch source current vector

and $\mathbf{Z_b}$ is the branch impedance matrix, which is a diagonal matrix with the branch impedances on the diagonal

Thus

$$\mathbf{v} = \begin{bmatrix} v_1 \\ v_2 \\ \vdots \\ v_b \end{bmatrix}; \quad \mathbf{j} = \begin{bmatrix} j_1 \\ j_2 \\ \vdots \\ j_b \end{bmatrix}; \quad \mathbf{v_s} = \begin{bmatrix} v_{s1} \\ v_{s2} \\ \vdots \\ v_{sb} \end{bmatrix}; \quad \mathbf{j_s} = \begin{bmatrix} j_{s1} \\ j_{s2} \\ \vdots \\ j_{sb} \end{bmatrix}$$

and

$$\mathbf{Z_b} = \begin{bmatrix} Z_1 & 0 & 0 & \cdots & 0 \\ 0 & Z_2 & 0 & \cdots & 0 \\ \vdots & & & & \\ 0 & 0 & 0 & \cdots & Z_b \end{bmatrix}$$

Pre multiplying eq. (5.32) by the fundamental circuit matrix B, and noting that

$$\mathbf{Bv} = \mathbf{0}$$

We have $\mathbf{Bv} = \mathbf{0} = \mathbf{Bv_s} + \mathbf{BZ_b}\,\mathbf{j} - \mathbf{BZ_b j_s}$ .....(5.33)

Also, we have $\mathbf{j} = \mathbf{B^T i}$, where $\mathbf{i}$ is the loop current vector for the chosen tree. Substituting for '$\mathbf{j}$' in eq. (5.33) and rearranging, we have,

$$\mathbf{BZ_b}\,\mathbf{B^T i} = \mathbf{BZ_b}\,\mathbf{j_s} - \mathbf{Bv_s} \qquad .....(5.34)$$

In eq. (5.34) $\mathbf{BZ_b}\,\mathbf{B^T}$ is a square matrix of dimension $l = b - n + 1$, where l is the number of links and $\mathbf{BZ_b}\,\mathbf{j_s}$ and $\mathbf{Bv_s}$ are $b - n + 1$ dimension vectors.

Defining $\mathbf{BZ_b B^T} \triangleq \mathbf{Z}_l$ and $\mathbf{v_{es}} \triangleq \mathbf{BZ_b j_s} - \mathbf{Bv_s}$

We have $\mathbf{Z}_l \mathbf{i} = \mathbf{v_{es}}$ .....(5.35)

Here **i** is the loop current vector, $\mathbf{Z}_l$ is the loop impedance matrix and $\mathbf{v_{es}}$ is the equivalent source voltage vector. Solving for **i**, we get

$$\mathbf{i} = \mathbf{Z}_l^{-1}\,\mathbf{v_{es}} \qquad ....(5.36)$$

Let us illustrate the method with an example.

Example 5.2 :

Obtain the loop currents in the network of example 5.1 using the branches 1, 2, and 3 as links.

Solution :

The network and its graph are reproduced here for convenience. A tree with branches 1, 2 and 3 as links is also shown.

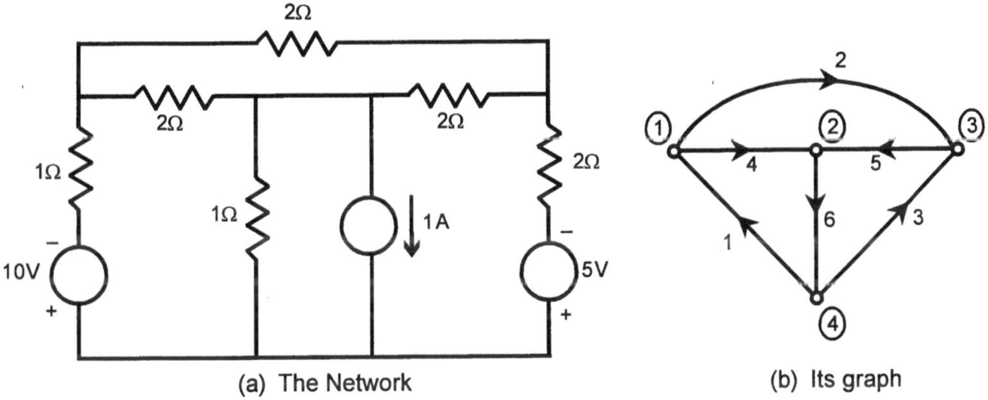

(a) The Network (b) Its graph

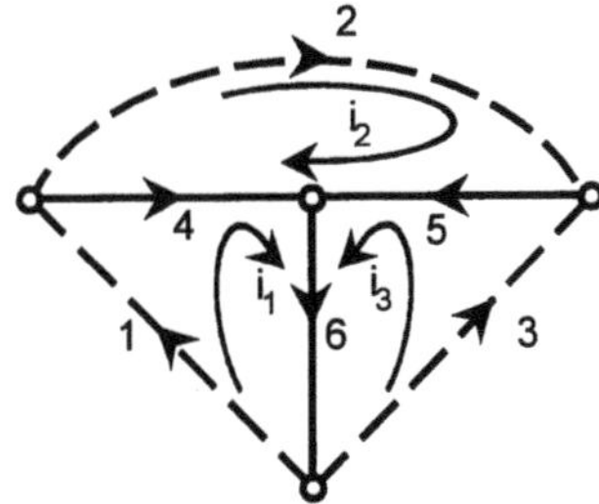

(c) A tree and the fundamental circuits and loop currents

Fig. 5.17 Network, its graph and a tree for example 5.2

Step 1: Draw the graph of the network as in Example 5.1. Draw a tree with links 1, 2, and 3. Draw tree branches with continuous lines and links with dotted lines. Identify the fundamental circuits and mark the loop currents with directions agreeing with the respective link branches in the graph.

Step 2: Write down the matrices **B** and $\mathbf{Z_b}$ and vector $\mathbf{v_s}$ and $\mathbf{j_s}$

$$\mathbf{B} = \begin{bmatrix} 1 & 0 & 0 & 1 & 0 & 1 \\ 0 & 1 & 0 & -1 & 1 & 0 \\ 0 & 0 & 1 & 0 & 1 & 1 \end{bmatrix}$$

$$\mathbf{Z_b} = \begin{bmatrix} 1 & 0 & 0 & 0 & 0 & 0 \\ 0 & 2 & 0 & 0 & 0 & 0 \\ 0 & 0 & 2 & 0 & 0 & 0 \\ 0 & 0 & 0 & 2 & 0 & 0 \\ 0 & 0 & 0 & 0 & 2 & 0 \\ 0 & 0 & 0 & 0 & 0 & 1 \end{bmatrix}$$

$$\mathbf{v_s} = \begin{bmatrix} 10 \\ 0 \\ 5 \\ 0 \\ 0 \\ 0 \end{bmatrix} \quad \text{and} \quad \mathbf{j_s} = \begin{bmatrix} 0 \\ 0 \\ 0 \\ 0 \\ 0 \\ 1 \end{bmatrix}$$

Step 3 : Form $\mathbf{Z}_l = \mathbf{B}\mathbf{Z_b}\,\mathbf{B}^{\mathrm{T}}$

$$\mathbf{Z}_l = \begin{bmatrix} 1 & 0 & 0 & 1 & 0 & 1 \\ 0 & 1 & 0 & -1 & 1 & 0 \\ 0 & 0 & 1 & 0 & 1 & 1 \end{bmatrix} \begin{bmatrix} 1 & 0 & 0 & 0 & 0 & 0 \\ 0 & 2 & 0 & 0 & 0 & 0 \\ 0 & 0 & 2 & 0 & 0 & 0 \\ 0 & 0 & 0 & 2 & 0 & 0 \\ 0 & 0 & 0 & 0 & 2 & 0 \\ 0 & 0 & 0 & 0 & 0 & 1 \end{bmatrix} \begin{bmatrix} 1 & 0 & 0 \\ 0 & 1 & 0 \\ 0 & 0 & 1 \\ 1 & -1 & 0 \\ 0 & 1 & 1 \\ 1 & 0 & 1 \end{bmatrix}$$

$$\mathbf{Z}_l = \begin{bmatrix} 4 & -2 & 1 \\ -2 & 6 & 2 \\ 1 & 2 & 5 \end{bmatrix}$$

As before, $\mathbf{Z}_l$ can be written down by inspection, Z_{ii} for i = 1, 2, 3 are the self impedances of the loops. Z_{ij} is the mutual impedance between loops i and j and is positive if the loop currents are in the same direction; negative if the loop currents are in opposite direction in the common branch.

Step 4 : Obtain $\mathbf{v}_{es} = \mathbf{BZ}_b\,\mathbf{j}_s - \mathbf{Bv}_s$

$$\mathbf{v}_{es} = \begin{bmatrix} 1 & 0 & 0 & 1 & 0 & 1 \\ 0 & 1 & 0 & -1 & 1 & 0 \\ 0 & 0 & 1 & 0 & 1 & 1 \end{bmatrix} \begin{bmatrix} 1 & & & & & \\ & 2 & & & 0 & \\ & & 2 & & & \\ & & & 2 & & \\ & 0 & & & 2 & \\ & & & & & 1 \end{bmatrix} \begin{bmatrix} 0 \\ 0 \\ 0 \\ 0 \\ 0 \\ 1 \end{bmatrix}$$

$$- \begin{bmatrix} 1 & 0 & 0 & 1 & 0 & 1 \\ 0 & 1 & 0 & -1 & 1 & 0 \\ 0 & 0 & 1 & 0 & 1 & 1 \end{bmatrix} \begin{bmatrix} 10 \\ 0 \\ 5 \\ 0 \\ 0 \\ 0 \end{bmatrix}$$

$$= \begin{bmatrix} -9 \\ 0 \\ -4 \end{bmatrix}$$

Step 5 : From the equation

$$\mathbf{Z}_l\mathbf{i} = \mathbf{v}_{es}$$

$$\begin{bmatrix} 4 & -2 & 1 \\ -2 & 6 & 2 \\ 1 & 2 & 5 \end{bmatrix} \begin{bmatrix} i_1 \\ i_2 \\ i_3 \end{bmatrix} = \begin{bmatrix} -9 \\ 0 \\ -4 \end{bmatrix}$$

Solving for **i**, we get

$$= \begin{bmatrix} i_1 \\ i_2 \\ i_3 \end{bmatrix} = \begin{bmatrix} 4 & -2 & 1 \\ -2 & 6 & 2 \\ 1 & 2 & 5 \end{bmatrix}^{-1} \begin{bmatrix} -9 \\ 0 \\ -4 \end{bmatrix}$$

$$= \frac{1}{70} \begin{bmatrix} 26 & 12 & -10 \\ 12 & 19 & -10 \\ -10 & -10 & 20 \end{bmatrix} \begin{bmatrix} -9 \\ 0 \\ -4 \end{bmatrix}$$

$$\mathbf{i} = \begin{bmatrix} \frac{-97}{35} \\ \frac{-34}{35} \\ \frac{1}{7} \end{bmatrix}$$

Thus the loop currents are

$$i_1 = \frac{-97}{35} \text{A}; \quad i_2 = \frac{-34}{35} \text{A} \quad \text{and} \quad i_3 = \frac{1}{7} \text{A}$$

The branches voltages and currents can be found, if desired, using equations

$$\mathbf{j} = \mathbf{B}^T \mathbf{i}$$

and $\mathbf{v} = \mathbf{v_s} + \mathbf{Z_b}\, \mathbf{j} - \mathbf{Z_b}\, \mathbf{j_s}$

Thus

$$\mathbf{j} = \begin{bmatrix} -\frac{97}{35} \\ -\frac{34}{35} \\ \frac{1}{7} \\ -\frac{9}{5} \\ -\frac{29}{35} \\ -\frac{92}{35} \end{bmatrix}$$

and

$$\mathbf{v} = \begin{bmatrix} \frac{253}{35} \\ -\frac{68}{35} \\ \frac{185}{35} \\ -\frac{126}{35} \\ -\frac{58}{35} \\ -\frac{127}{35} \end{bmatrix}$$

This completes the analysis of the network using loop current method.

5.4.4 Cutset equations :

Consider a tree of the network and the fundamental cutset matrix Q of the graph. The v – i equations for all the branches of the network are given by eq. (5.27).

$$\mathbf{j = Y_b v + j_s - Y_b v_s} \qquad .....(5.37)$$

But the KCL equations are given by

$$\mathbf{QJ = 0}$$

Thus, pre multiplying both sides of eq. (5.37) by Q, we have

$$\mathbf{Qj = 0 = QY_b v + Qj_s - QY_b v_s} \qquad .....(5.38)$$

From KVL equation given by

$$\mathbf{v = Q^T \bar{e}} \qquad .....(5.39)$$

where $\bar{\mathbf{e}}$ is the tree branch voltage vector, we have

$$\mathbf{QY_b Q^T \bar{e} = QY_b v_s - Qj_s} \qquad .....(5.40)$$

Defining $\quad \mathbf{QY_b Q^T \triangleq Y_c}$ and $\mathbf{i_{es} \triangleq QY_b v_s - Qj_s}$

We can rewrite equation (5.39) as

$$\mathbf{Y_c \bar{e} = i_{es}} \qquad .....(5.41)$$

or $\quad$ $$\mathbf{\bar{e} = Y_c^{-1} i_{es}} \qquad .....(5.42)$$

Once the tree branch voltage vector $\bar{\mathbf{e}}$ is known, all the branch voltages and currents can be obtained by using eqs. (5.39) and (5.37). This is illustrated by an example.

Example 5.3 :

Obtain the tree branch voltages and hence all the branch voltages and currents for the network shown in Fig. 5.16(a). Consider a tree with branches 4, 5 and 6.

Solution :

Step 1 : Draw the tree of the given graph as shown in Fig. 5.18. Mark the fundamental cutsets with directions agreeing with the directions of the respective tree branches.

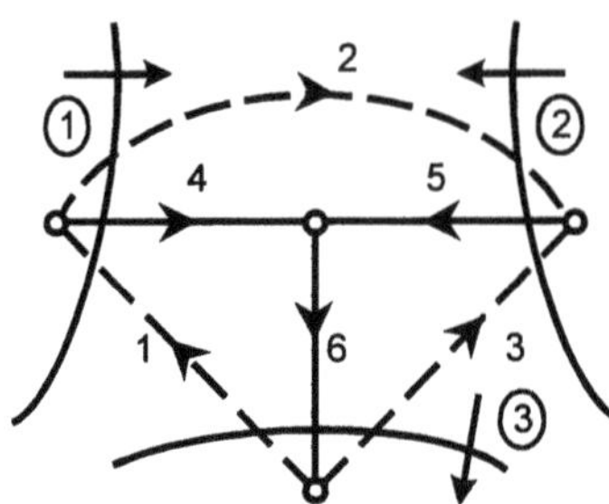

Fig. 5.18 A tree of the given graph and the fundamental cutsets

Step 2 : Write down the matrix $\mathbf{Q}$, $\mathbf{Y_b}$ and vectors $\mathbf{v_s}$ and $\mathbf{j_s}$

$$\mathbf{Q} = \begin{bmatrix} -1 & 1 & 0 & 1 & 0 & 0 \\ 0 & -1 & -1 & 0 & 1 & 0 \\ -1 & 0 & -1 & 0 & 0 & 1 \end{bmatrix}$$

$$\mathbf{Y_b} = \begin{bmatrix} 1 & & & & & \\ & 0.5 & & & 0 & \\ & & 0.5 & & & \\ & & & 0.5 & & \\ & 0 & & & 0.5 & \\ & & & & & 1 \end{bmatrix}$$

$$\mathbf{v_s} = \begin{bmatrix} 10 \\ 0 \\ 5 \\ 0 \\ 0 \\ 0 \end{bmatrix} \qquad \mathbf{j_s} = \begin{bmatrix} 0 \\ 0 \\ 0 \\ 0 \\ 0 \\ 1 \end{bmatrix}$$

Step 3 : From $\mathbf{Y_c} = \mathbf{QY_b\,Q^T}$

Performing the operations using $\mathbf{Q}$ and $\mathbf{Y_b}$ from step 2 we get

$$\mathbf{Y_c} = \begin{bmatrix} 2 & -0.5 & 1 \\ -0.5 & 1.5 & 0.5 \\ 1 & 0.5 & 2.5 \end{bmatrix}$$

The cutset admittance matrix can be written down by inspection also. Y_{ii} is the self admittance of cutset 'i' and is given by the sum of admittances of the branches in that cutset. Y_{ij} is the admittance of the branch common to i^{th} cutset and j^{th} cutset and is negative if the two cutset directions in that branch are opposite.

Step 4 : Obtain $\mathbf{i_{es}} = \mathbf{QY_b\,v_s} - \mathbf{Qj_s}$

Performing the indicated operations, we have

$$\mathbf{i_{es}} = \begin{bmatrix} -10 \\ -2.5 \\ -13.5 \end{bmatrix}$$

Step 5 : From the equation

$$\mathbf{Y_c}\ \bar{\mathbf{e}} = \mathbf{i_{es}}$$

Solving for $\bar{\mathbf{e}}$,

$$\begin{bmatrix} 2 & -0.5 & -0.5 \\ -0.5 & 2 & -0.5 \\ 1 & 0.5 & 2.5 \end{bmatrix} \begin{bmatrix} \bar{e}_1 \\ \bar{e}_2 \\ \bar{e}_3 \end{bmatrix} = \begin{bmatrix} -10 \\ -2.5 \\ -13.5 \end{bmatrix}$$

$$\begin{bmatrix} \bar{e}_1 \\ \bar{e}_2 \\ \bar{e}_3 \end{bmatrix} = \begin{bmatrix} -\dfrac{126}{35} \\ -\dfrac{58}{35} \\ -\dfrac{127}{35} \end{bmatrix}$$

Thus the tree branch voltages are

$$\bar{e}_1 = \frac{-126}{35}\,\text{V};$$

$$\bar{e}_2 = \frac{-58}{35}\,\text{V}$$

and $$\bar{e}_3 = \frac{-127}{35}\,\text{V}$$

The branch voltages and currents can be obtained using eq. (5.39) and (5.37)

$$\mathbf{v} = \mathbf{Q}^T\bar{\mathbf{e}} = \begin{bmatrix} \frac{253}{35} \\ -\frac{38}{35} \\ \frac{185}{35} \\ -\frac{126}{35} \\ -\frac{58}{35} \\ -\frac{127}{35} \end{bmatrix}$$

and $$\mathbf{j} = \mathbf{Y}_b\mathbf{v} + \mathbf{j}_s - \mathbf{Y}_b\mathbf{v}_s$$

$$= \begin{bmatrix} -\frac{97}{35} \\ -\frac{34}{35} \\ \frac{1}{7} \\ -\frac{9}{5} \\ -\frac{29}{35} \\ -\frac{92}{35} \end{bmatrix}$$

This completes the analysis of the network by cutset method. Let us now consider an example with dependent sources present in the network.

Example 5.4 :

Obtain the branch currents in the network shown in Fig. 5.19(a) by loop current method.

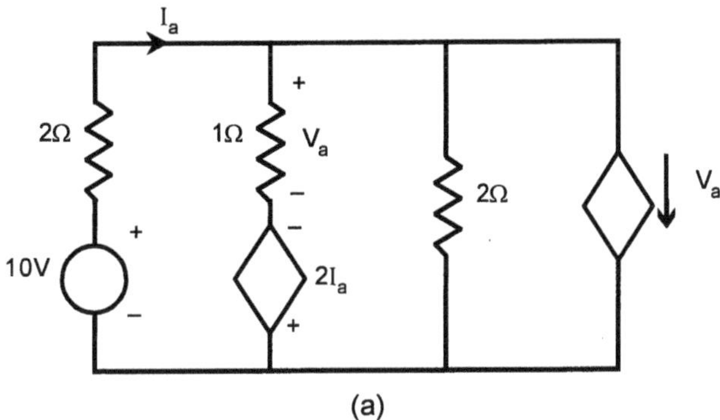

Fig. 5.19(a) Network for example 5.4

Solution :

The graph of the network is shown in Fig. 5.19(b) and a tree is shown in Fig. 5.19(c).

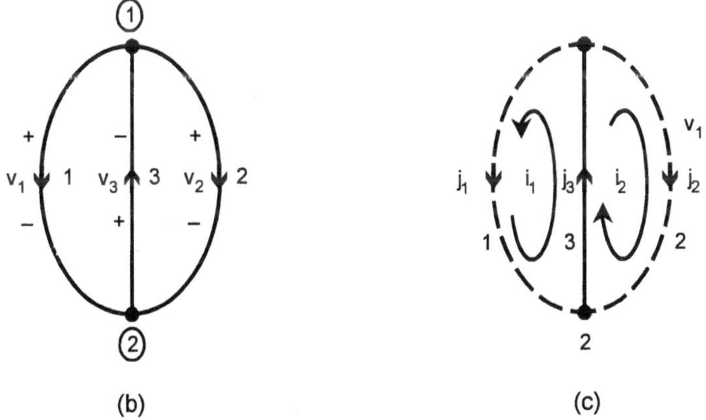

Fig. 5.19(b) Graph of given network (c) A tree and the loop currents

The circuit matrix B is given by

$$B = \begin{bmatrix} 1 & 0 & 1 \\ 0 & 1 & 1 \end{bmatrix}$$

The branch impedance matrix $\mathbf{Z_b}$ cannot be written down by inspection because of the presence of dependent sources. Hence writing down the branch $v - i$ equations we have

$$v_1 = 2j_1 + 10$$

$$v_2 = (j_2 - v_a)2$$

$$= 2j_2 - 2v_a$$

$$v_3 = 2i_a + j_3$$

$$v_a = -j_3$$

and $$i_a = -j_1$$

Eliminating v_a and i_a, we have

$$v_1 = 2j_1 + 10$$

$$v_2 = 2j_2 - 2(-j_3)$$

$$= 2j_2 + 2j_3$$

$$v_3 = 2(-j_1) + j_3$$

$$= -2j_1 + j_3$$

Writing in matrix form

$$\begin{bmatrix} v_1 \\ v_2 \\ v_3 \end{bmatrix} = \begin{bmatrix} 2 & 0 & 0 \\ 0 & 2 & 2 \\ -2 & 0 & 1 \end{bmatrix} \begin{bmatrix} j_1 \\ j_2 \\ j_3 \end{bmatrix} + \begin{bmatrix} 10 \\ 0 \\ 0 \end{bmatrix}$$

From this equation, we observe that

$$\mathbf{Z_b} = \begin{bmatrix} 2 & 0 & 0 \\ 0 & 2 & 2 \\ -2 & 0 & 1 \end{bmatrix} \quad \mathbf{v_s} = \begin{bmatrix} 10 \\ 0 \\ 0 \end{bmatrix} \quad \mathbf{j_s} = \begin{bmatrix} 0 \\ 0 \\ 0 \end{bmatrix}$$

It is to be noted that $\mathbf{Z_b}$ is no longer diagonal because of the presence of dependent sources.

The loop equations are given by

$$\mathbf{B}_l\,\mathbf{i} = \mathbf{v}_{es}$$

where $\mathbf{B}_l = \mathbf{B}\mathbf{Z}_b\,\mathbf{B}^T$

and $\mathbf{v}_{es} = \mathbf{B}\mathbf{Z}_b\,\mathbf{j}_s - \mathbf{B}\mathbf{v}_s$

$$\therefore \quad \mathbf{B}_l = \begin{bmatrix} 1 & 0 & 1 \\ 0 & 1 & 1 \end{bmatrix} \begin{bmatrix} 2 & 0 & 0 \\ 0 & 2 & 2 \\ -2 & 0 & 1 \end{bmatrix} \begin{bmatrix} 1 & 0 \\ 0 & 1 \\ 1 & 1 \end{bmatrix}$$

$$= \begin{bmatrix} 1 & 1 \\ 1 & 5 \end{bmatrix}$$

$$\therefore \quad \mathbf{v}_{es} = -\,\mathbf{B}\mathbf{v}_s$$

$$= -\begin{bmatrix} 1 & 0 & 1 \\ 0 & 1 & 1 \end{bmatrix} \begin{bmatrix} 10 \\ 0 \\ 0 \end{bmatrix}$$

$$= \begin{bmatrix} -10 \\ 0 \end{bmatrix}$$

$$\therefore \quad \begin{bmatrix} 1 & 1 \\ 1 & 5 \end{bmatrix} \mathbf{i} = \begin{bmatrix} -10 \\ 0 \end{bmatrix}$$

$$\mathbf{i} = \begin{bmatrix} 1 & 1 \\ 1 & 5 \end{bmatrix}^{-1} \begin{bmatrix} -10 \\ 0 \end{bmatrix}$$

$$= \frac{1}{4} \begin{bmatrix} 5 & -1 \\ -1 & 1 \end{bmatrix} \begin{bmatrix} -10 \\ 0 \end{bmatrix} = \begin{bmatrix} -12.5 \\ 2.5 \end{bmatrix}$$

Thus $i_1 = -12.5\text{A}$ and $i_2 = 2.5\text{A}$

The branch currents are

$$\mathbf{j} = \mathbf{B}^T \mathbf{i}$$

$$= \begin{bmatrix} 1 & 0 \\ 0 & 1 \\ 1 & 1 \end{bmatrix} \begin{bmatrix} -12.5 \\ 2.5 \end{bmatrix}$$

$$= \begin{bmatrix} -12.5 \\ 2.5 \\ -10 \end{bmatrix}$$

Thus the branch currents are $j_1 = -12.5A$, $j_2 = 2.5A$ and $j_3 = -10A$

Problems

5.1 Draw the graphs of the following networks in Fig. P. 5.1. Are they connected graphs ? Are they planar ?

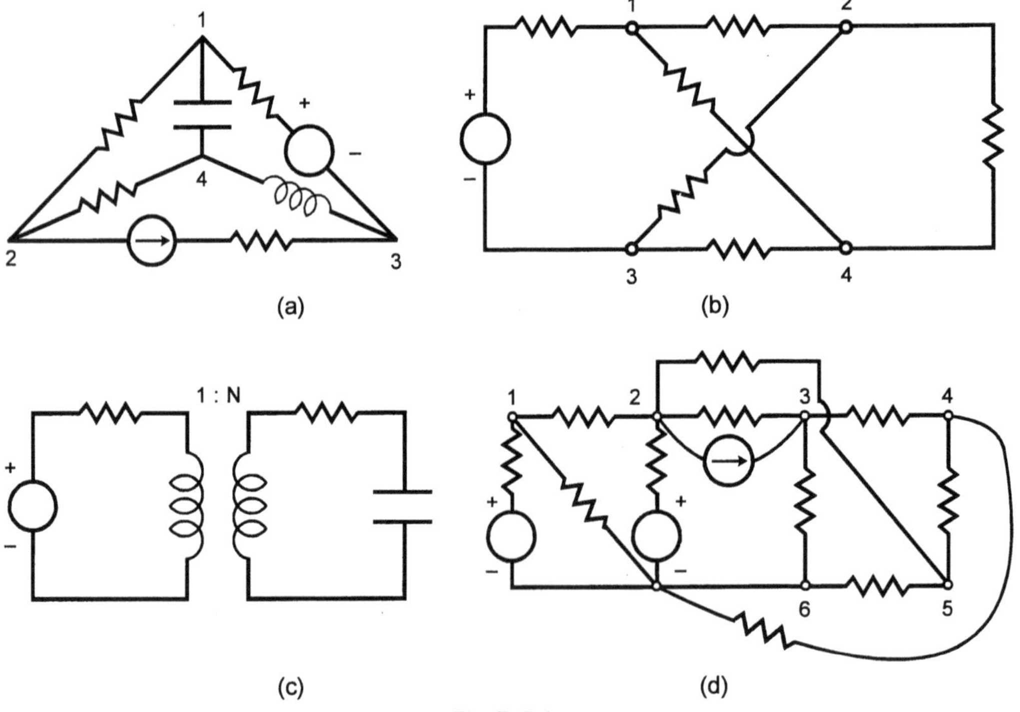

Fig. P. 5.1

5.2 Draw 3 possible trees for the graphs shown in Fig. P. 5.2.

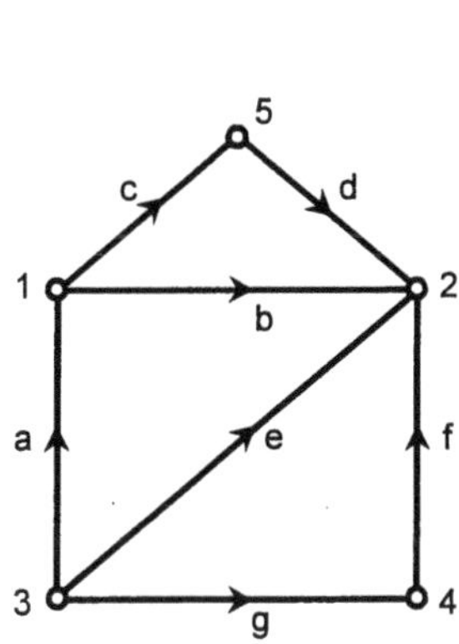

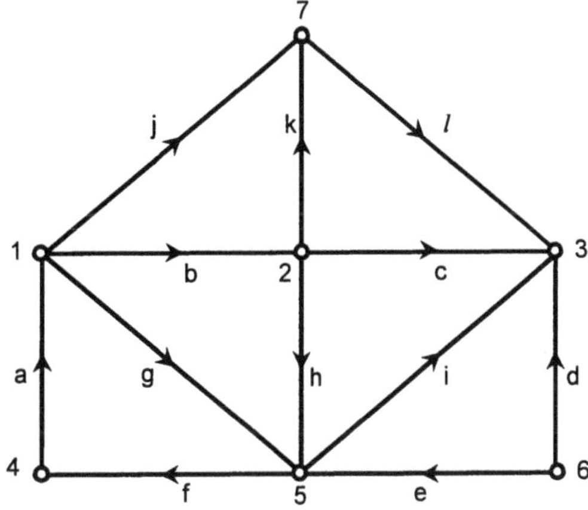

Fig. P. 5.2

5.3 Enumerate the fundamental cutsets for the graphs shown in Fig. P. 5.2 for all the trees chosen in problem 5.2.

5.4 Obtain the incidence matrix for the graph shown in Fig. 5.3. Write the KCL and KVL equations for the graph. Take node 4 as reference node.

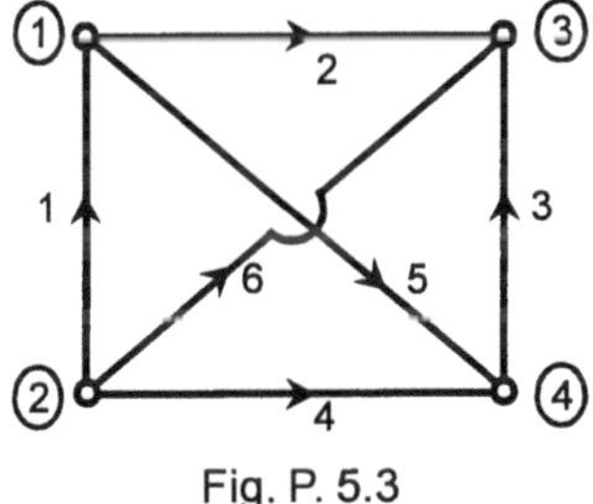

Fig. P. 5.3

5.5 Obtain the fundamental circuit matrix for the graph in Fig. P. 5.4. Choose the tree consisting of branches 6, 7, 8 and 9. Write the KVL and KCL equations for the graph.

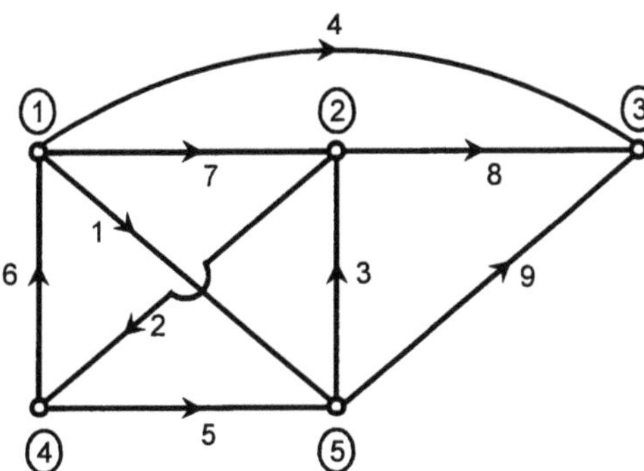

Fig. P. 5.4

5.6 For the tree specified in problem 5.5 and Fig. P. 5.4, obtain fundamental cutset matrix. Write the KVL and KCL equations for the graph.

5.7 For the matrices B and Q obtained in problems P.5.5 and P. 5.6 show that $BQ^T = 0$.

5.8 Write the A, B and Q matrices for the graph in Fig. P. 5.3. For writing the B and Q matrices take the tree consisting of branches 4, 5 and 6. Take the same order of branches in the columns for all the 3 matrices. Show that $AB^T = O$, $BA^T = O$ and $QB^T = O$ where 'O' is a matrix of '0' elements.

5.9 Solve the problem 5.8 by taking any other tree. (*Hint* : Remember to maintain the same order of the branches in the columns in writing all the three matrices).

5.10 Formulate and solve the equations for the networks in Fig. P. 5.5 based on

(a) node voltage method

(b) loop current method and

(c) tree branch voltage method (cutset equations).

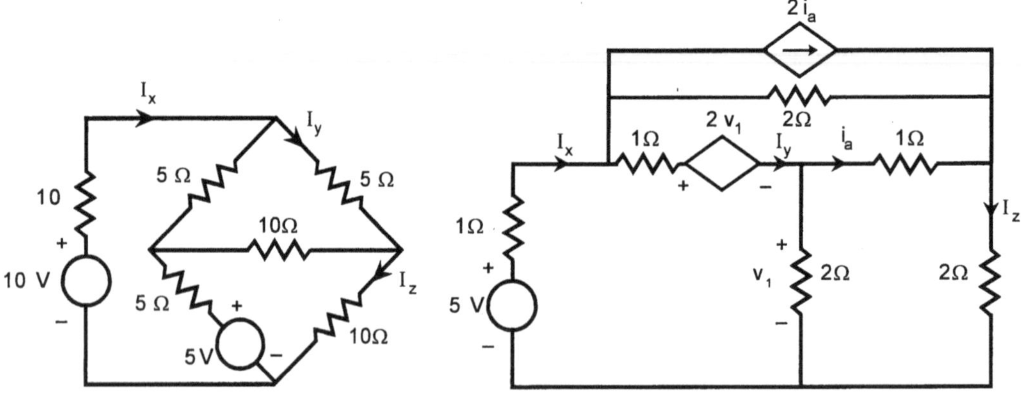

Fig. P. 5.5

6

NETWORK THEOREMS

6.1 INTRODUCTION :

In previous chapters we have described general methods that are useful in solving for the currents and voltages in a given network. In certain networks, we may be interested in the response of a particular part of the network, or we may like to describe the behaviour of a complex network at a set of two terminals only. Further, we may be interested in understanding the nature of the network rather than the actual solution. Certain properties of the network are stated as network theorems and these theorems are used to simplify the analysis of more complex networks.

6.2 GENERAL NETWORK EQUATIONS :

In earlier chapters we have applied Kirchhoff's laws to the networks and formulated the loop and node voltage equations. Let us now formulate these equations in a more general form so that the equations can be written down by inspection. Let us consider a simple 3 loop network to illustrate the method.

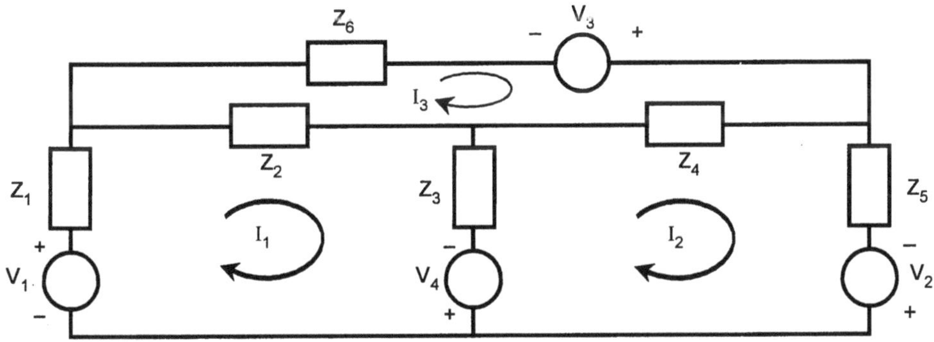

Fig. 6.1 A three loop network

Let us write down the loop equations assuming clockwise loop currents in all the loops.

$$I_1 Z_1 + (I_1 - I_3) Z_2 + (I_1 - I_2) Z_3 = V_1 + V_4$$

$$I_2 Z_5 + (I_2 - I_3) Z_4 + (I_2 - I_1) Z_3 = V_2 - V_4$$

$$I_3 Z_6 + (I_3 - I_2) Z_4 + (I_3 - I_1) Z_2 = V_3 \qquad .....(6.1)$$

The set of eqns. (6.1) can be simplified as :

$$I_1(Z_1 + Z_2 + Z_3) - I_2Z_3 - I_3Z_2 = V_1 + V_4$$
$$-I_1Z_3 + I_2(Z_3 + Z_4 + Z_5) - I_3Z_4 = V_2 - V_4$$
$$-I_1Z_2 - I_2Z_4 + I_3(Z_2 + Z_4 + Z_6) = V_3 \qquad .....(6.2)$$

The set of eqns. of (6.2) can be put in a more general format as,

$$I_1Z_{11} + I_2Z_{12} + I_3Z_{13} = V_{11}$$
$$I_1Z_{21} + I_2Z_{22} + I_3Z_{23} = V_{22}$$
$$I_1Z_{31} + I_2Z_{32} + I_3Z_{33} = V_{33} \qquad .....(6.3)$$

where Z_{ii} is defined as the self impedance of the loop 'i' or the sum of all the impedances in that loop, for i = 1, 2, 3

Z_{ij} is the mutual or common impedance between the ith and j^{th} loops and has a + ve sign if i^{th} and j^{th} loop currents have the same direction in that common impedance and has – ve sign if they are opposite in direction, for i, j = 1, 2, 3

and V_{ii} is the algebraic sum of the source voltage in the ith loop. It is positive if it is a gain in the direction of the ith loop current and is negative if it is a drop, for i = 1, 2, 3

In this formulation it is assumed that all the current sources are converted into voltage sources using source transformation. For a general network with n loops the n equations can be written down as :

$$I_1Z_{11} + I_2Z_{12} + ... + I_nZ_{1n} = V_1$$
$$I_1Z_{21} + I_2Z_{22} + ... + I_nZ_{2n} = V_2$$
$$\vdots$$
$$I_1Z_{n1} + I_2Z_{n2} + ... + I_nZ_{nn} = V_n \qquad .....(6.4)$$

In matrix form eq. (6.4) can be written as :

$$\begin{bmatrix} Z_{11} & Z_{12} & \cdots & Z_{1n} \\ Z_{21} & Z_{22} & \cdots & Z_{2n} \\ \vdots & & & \\ Z_{n1} & Z_{n2} & \cdots & Z_{nn} \end{bmatrix} \begin{bmatrix} I_1 \\ I_2 \\ \vdots \\ I_n \end{bmatrix} = \begin{bmatrix} V_1 \\ V_2 \\ \vdots \\ V_n \end{bmatrix} \qquad .....(6.5)$$

or $\mathbf{ZI} = V \qquad .....(6.6)$

where **Z** is the coefficient matrix in eq.(6.5) and is called *impedance matrix.*

I is the loop current vector

and **V** is the source voltage vector.

The solution for I can be obtained as

$$I = Z^{-1}V \qquad \text{....(6.7)}$$

Or by using *Cramer's rule* to obtain the current I_j,

$$I_j = \frac{\Delta_{1j}}{\Delta_Z}V_1 + \frac{\Delta_{2j}}{\Delta_Z}V_2 + \text{.....} + \frac{\Delta_{nj}}{\Delta_Z}V_n \qquad \text{.....(6.8)}$$

where Δ_{ij} is the cofactor of the (i, j)th element of the Z – matrix and

Δ_Z is the determinant of the impedance matrix Z.

In a similar way the network equations can be formulated in an 'n' node network based on node voltage equations in the form.

$$V_1Y_{11} + V_2Y_{12} + \text{.....} + V_nY_{1n} = I_1$$
$$V_1Y_{21} + V_2Y_{22} + \text{.....} + V_nY_{2n} = I_2$$
$$\vdots$$
$$V_1Y_{n1} + V_2Y_{n2} + \text{.....} + V_nY_{nn} = I_n \qquad \text{.....(6.9)}$$

where Y_{ii} is the self admittance of the node i which is the sum of all admittances connected to that node, for i = 1, 2, ... n

Y_{ij} is the admittance common to the nodes i and j and is negative, for i, j = 1, 2, ...n

I_i is the algebraic sum of the source currents connected to the ith node. It is positive if the source currents are directed towards the node, otherwise it is negative.

Eq. (6.9) can be put in the matrix form as

$$\mathbf{YV} = \mathbf{I} \qquad \text{.....(6.10)}$$

where **Y** is the coefficient matrix in eq. (6.9) and is called the *admittance matrix*.

V is the node voltage vector

and **I** is the source current vector.

Again, the node voltages can be obtained by

$$\mathbf{V} = [\mathbf{Y}]^{-1}\,\mathbf{I}$$

Or by *Cramer's rule*, for jth node voltage

$$V_j = \frac{\Delta_{1j}}{\Delta_y}I_1 + \frac{\Delta_{2j}}{\Delta_y}I_2 + \text{......} + \frac{\Delta_{nj}}{\Delta_y}I_n \ ; \ J = 1, 2, \text{......}n \qquad \text{.....(6.11)}$$

6.3 SUPERPOSITION THEOREM :

This is one of the fundamental theorems applicable to linear networks. Linear networks are those networks which are constructed with linear elements only. The active sources we have considered so far and the passive elements like R, L, C and M are assumed to be operated in their linear ranges. The v – i, relationships for these elements are linear. The Superposition theorem can be stated as follows :

" *In a linear network with several independent sources which include equivalent sources due to initial conditions, and linear dependent sources, the overall response in any part of the network is equal to the sum of the individual responses due to each independent source, considered separately, with all other independent sources reduced to zero* ".

Some comments are in order here :

1. The sources which are considered one at a time making all other sources zero, are the independent sources including sources due to initial conditions only. The dependent sources are retained as they are in the network.
2. When one independent source is considered all other independent sources are reduced to zero.

This means all independent voltage sources are shorted and all independent current sources are open circuited. If the sources contain internal impedances, the sources are replaced by their internal impedances.

Proof :

Consider a linear network in which all the current sources are converted to voltage sources by source transformation. Then, the general equations based on loop current method are given by eq. (6.4) and their solution is given by eq. (6.8). This equation is repeated here for convenience.

$$I_j = \frac{1}{\Delta_Z}(\Delta_{ij}V_1 + \Delta_{2j}V_2 + \ldots\ldots + \Delta_{nj}V_n) \qquad \ldots\ldots(6.12)$$

Case 1 *:*

If the voltage sources V_1, V_2, ... V_n are not in the common branches of any two loops, then each of the term in eq. (6.12) represents the contribution of the source V_i to the current I_j for i = 1, 2, ... n, with all other sources reduced to zero. Hence the total response I_j is the sum of the responses due to individual sources with all other sources reduced to zero.

Case 2 :

If a source V_ℓ occurs in a common branch of loops, say, i and j, then right hand sides of ith and j[th] equation of eq. (6.4) contain this term in V_i and V_j. When this source is considered with all other sources reduced to zero, the expression for I_j in eq. (6.12) will contain two terms as in case 1. This will be true even if more sources occur in common branches of the loops. Thus we can see that the response due to all the sources considered together is the sum of the responses due to each source with all other sources reduced to zero.

The above property of linear networks is called *the additivity property of the linear network.* Further if all the sources are multiplied by a constant, the response also will be multiplied by a constant as is evident from eq. (6.8). If all voltage sources V_1, V_2, ...V_n are multiplied by K_1 then I_j is also multiplied by K.

This property is known as *homogeneity property of linear networks.* These two properties together constitute the Superposition principle. This property is very useful in obtaining the response due to any arbitrary input if it can be considered as a summation of some standard inputs like impulse, step etc.

Another comment about Superposition theorem is quite valid here. It is more a conceptual theorem than a theorem which, when applied to networks with many sources, reduces the complexity of the calculations. This theorem helps proving many other theorems to follow.

Example 6.1

Calculate the current in the 2Ω resistor using Superposition theorem in Fig. 6.2(a).

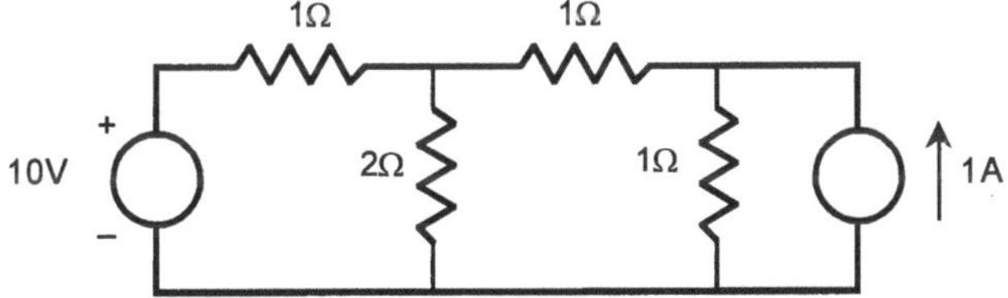

Fig. 6.2(a) Network for example 6.1

Solution :

Considering 10V source to be present and 1A source made equal to zero, i.e., open circuited we have the circuit in Fig. 6.2(b).

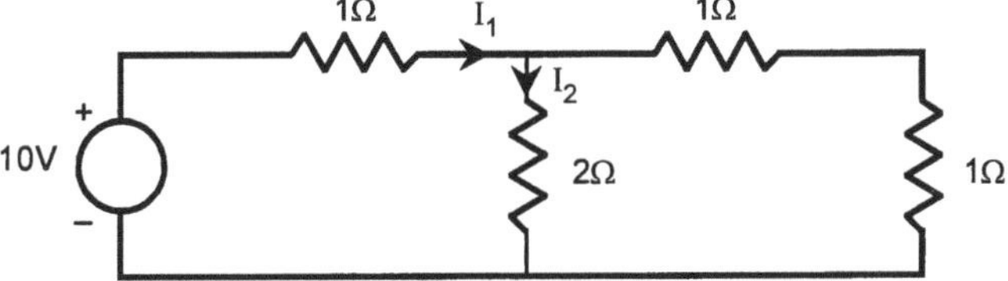

Fig. 6.2(b) Network with only 10V source present

I_1 can be calculated using network reduction

$$I_1 = \frac{10}{1+\frac{2\times 2}{2+2}} = 5\text{ A}$$

I_2 the current in the 2Ω resistor is

$$I_2 = I_1 \times \frac{2}{2+2} = 5 \times \frac{2}{4} = 2.5\text{ A}$$

Now considering the 1A current source only and short circuiting the voltage source, we have the circuit in Fig. 6.2(c).

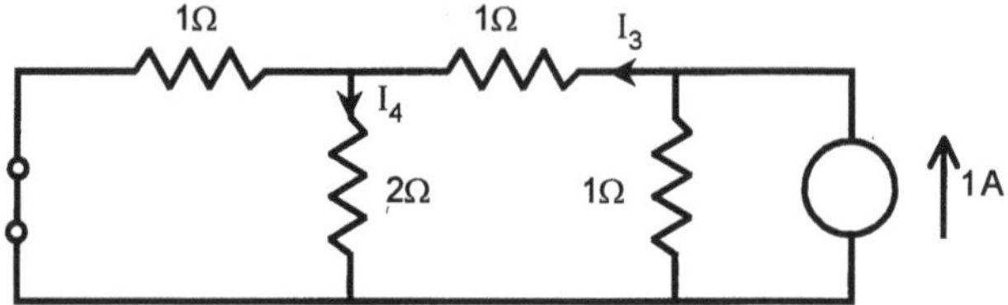

Fig. 6.2(c) Network with 1A current source present

Again using network reduction to find the current I_3, we have

$$I_3 = 1 \times \frac{1}{1+\left(1+\frac{2\times 1}{3}\right)}$$

$$= \frac{1}{1+\frac{5}{3}} = \frac{3}{8}\text{A}$$

and $$I_4 = I_3 \times \frac{1}{3} = \frac{3}{8} \times \frac{1}{3} = \frac{1}{8}\text{A}$$

∴ By Superposition theorem the current in the 2Ω resistor when both the sources are present is the sum of the currents due to individual sources with the other source reduced to zero.

∴ The current in the 2 – Ω resister is

$$= I_2 + I_4$$

$$= \frac{5}{2} + \frac{1}{8}$$

$$= \frac{21}{8}\text{ A}$$

The student is advised to verify this by directly solving the network in Fig. 6.2(a) using loop current method or node voltage method.

Example 6.2

Find the current I in the network shown in Fig. 6.3 using super position theorem.

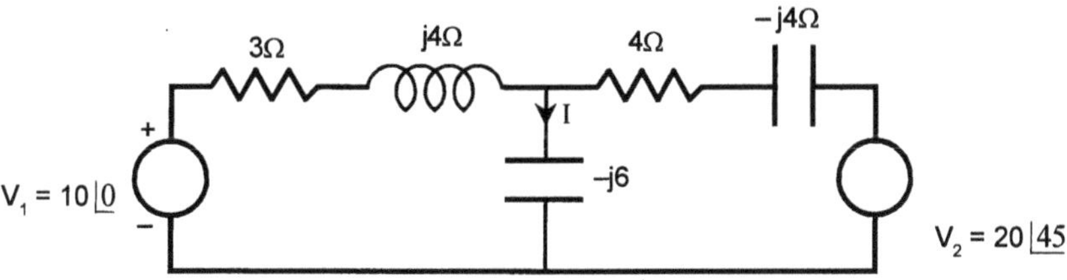

Fig. 6.3 Network for example 6.2

Solution :

To use super position theorem let us consider the voltage source V_1 first with V_2 short circuited. The resulting circuit is shown in Fig. 6.3(a).

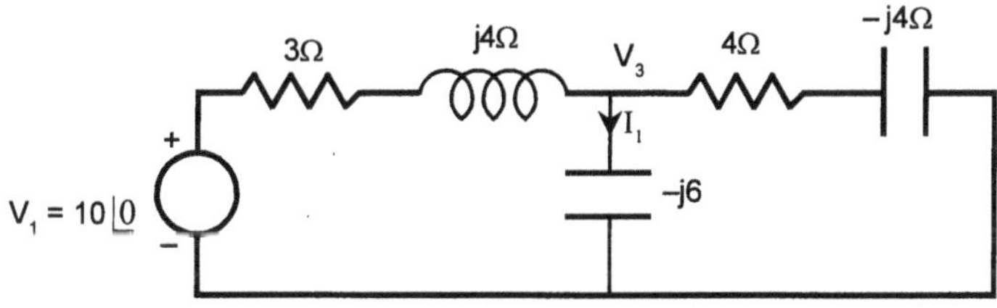

Fig. 6.3(a)

Using node voltage method, we get

$$\frac{V_3 - 10}{3 + j4} + \frac{V_3}{-j6} + \frac{V_3}{4 - j4} = 0$$

$$V_3\left(\frac{1}{3 + j4} - \frac{1}{j6} + \frac{1}{4 - j4}\right) = \frac{10}{3 + j4}$$

Solving for V_3, we get

$$V_3 = 7.19 \angle -81.39 \text{ V}$$

The current I_1 is given by,

$$I_1 = \frac{V_3}{-j6} = \frac{7.19 \angle -81.39}{-j6}$$

$$= 1.198 \angle 8.61 \text{ A}$$

Now using the source V_2 and shorting source V_1, we have,

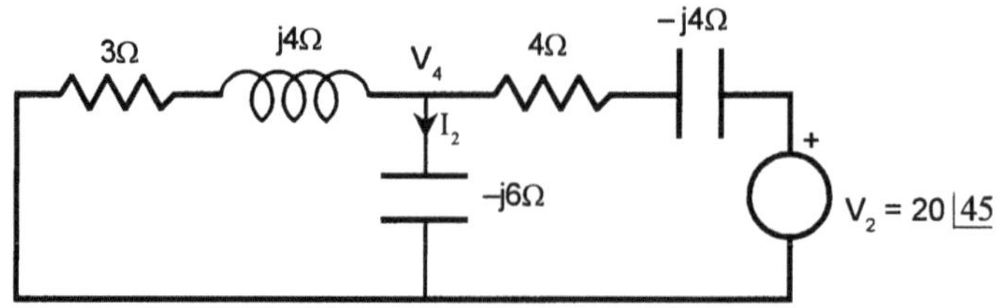

Fig. 6.3(b)

Again using Node voltage method to find V_4, we have,

$$\frac{V_4}{3+j4}+\frac{V_4}{-j6}+\frac{V_4-20\angle 45}{4-j4}=0$$

$$V_4\left(\frac{1}{3+j4}-\frac{1}{j6}+\frac{1}{4-j4}\right)=\frac{20\angle 45}{4-j4}$$

Solving for V_4, we get,

$$V_4 = 12.713\ \angle 61.74\ \text{V}$$

The current I_2 is given by,

$$I_2=\frac{V_4}{-j6}=\frac{12.713\angle 61.74}{-j6}$$

$$= 2.12\ \angle 151.74\ \text{A}$$

By superposition theorem we have,

$$I = I_1 + I_2$$

$$= 1.198\ \angle 8.61 + 2.12\ \angle 151.74$$

$$= 1.363\ \angle 120\ \text{A}$$

Example 6.3

A linear passive network has 3 terminal pairs available as shown in Fig. 6.4. Two voltage sources V_1 and V_2 are applied at terminals 1, 1' and 2, 2' and the current I is measured at the terminals 3, 3'. The values of the current are recorded for different values of V_1 and V_2 as shown in table 6.1. Find the current I when

$$V_1 = 10\text{V and } V_2 = -5\text{V}$$

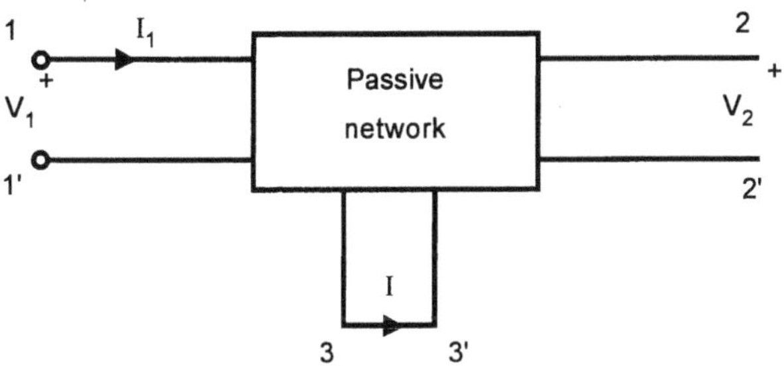

Fig. 6.4 Network for example 6.3

Table 6.1

V_1	V_2	I
2V	0	0.5A
0	5V	– 1A
10V	–5V	?

Since the given network is a passive linear network, it obeys Superposition theorem. The response I depends linearly on the two source voltages. Therefore

$$I = K_1V_1 + K_2V_2$$

where K_1 and K_2 are constants

(i) With $V_1 = 2$, $V_2 = 0$, we have $I = 0.5$ A

Substituting in expression for I

$$0.5 = 2K_1 + 0$$

$$K_1 = 0.25$$

(ii) With $V_1 = 0$ and $V_2 = 5V$, we have $I = -1A$

$$\therefore \quad -1 = 0 + 5K_2$$

$$K_2 = -\frac{1}{5} = -0.2$$

$$\therefore \quad I = 0.25V_1 - 0.2V_2$$

Now when $V_1 = 10V$ and $V_2 = -5V$

$$I = 0.25 \times 10 - 0.2 \times -5$$

$$= 2.5 + 1$$

$$= 3.5 \text{ A}$$

6.4 THEVENIN'S THEOREM :

In many practical situations, one desires to know the response of a particular component of a network rather than the entire network. For example, one is interested in the current in a loudspeaker connected to the terminals of an amplifier. The complex amplifier network can be replaced by a single network with only one source and one impedance connected to the load using two theorems – Thevenin's theorem and Norton's theorem.

Thevenin's theorem was originally formulated for resistive networks by Helmholtz and later enunciated by French Engineer L.C. Thevenin for networks containing impedances. The statement of Thevenin's theorem is :

Any two terminal network consisting of linear impedances and generators may be replaced at the two terminals by a single voltage source acting in series with an impedance. The voltage of the equivalent source is the open circuit voltage measured at the terminals of the network and the impedance, known as Thevenin's equivalent impedance, Z_{Th}, is the impedance measured at the terminals with all the independent sources in the network reduced to zero.

If the independent sources have internal impedances, these sources are replaced by their internal impedances. Reducing a voltage source to zero implies V = 0 and hence it is replaced by a short circuit and reducing a current source to zero implies I = 0 and hence it is replaced by an open circuit. Hence all voltage sources are replaced by short circuits and all current sources are replaced by open circuits. According to this theorem the active network A in Fig. 6.5(a) can be replaced by the simple Thevenin's equivalent network as shown in Fig. 6.5(b).

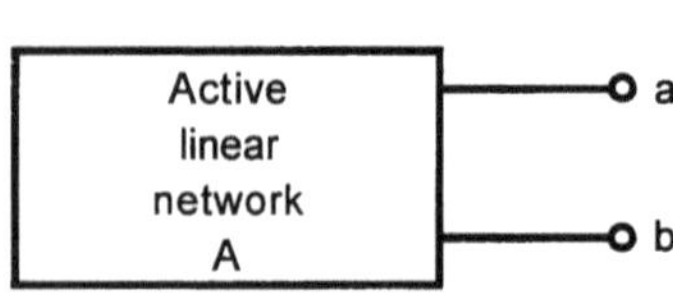

Fig. 6.5(a) General network

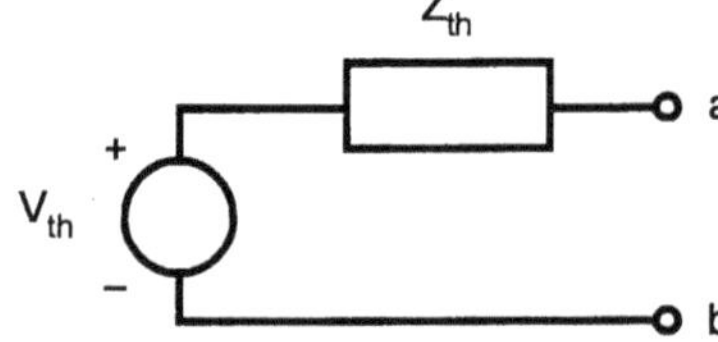

Fig. 6.5(b) Its Thevenin's equivalent

The network in Fig. 6.5(a) or its equivalent in Fig. 6.5(b) will produce exactly the same effect on any impedance or another active network B, connected at the terminals a and b. However the network A must not have any magnetic coupling with network B and must not contain any dependent sources whose controlling elements are present in network B. Thus the network A may contain independent or dependent sources and linear circuit elements which may have initial conditions. On the other hand, the network B may have sources, both dependent and independent and the elements may be linear or nonlinear, time invariant or time varying. However it should not have any magnetic coupling with network A and also if it has controlled sources, the controlling elements should not be in network A.

Proof :

Consider the network of Fig. 6.6(a). Let a load impedance Z_L be connected across the terminals a and b. Let the current flowing into the load impedance due to all the sources in the network A be I. Now, let a voltage source of value V_1 be connected in series with the load as shown in Fig. 6.6(b). Let the current due to V_1 alone, with all other sources in the network A reduced to zero, be equal to I'. Adjust the voltage V_1 such that the net current flowing into the impedance Z_L is zero.

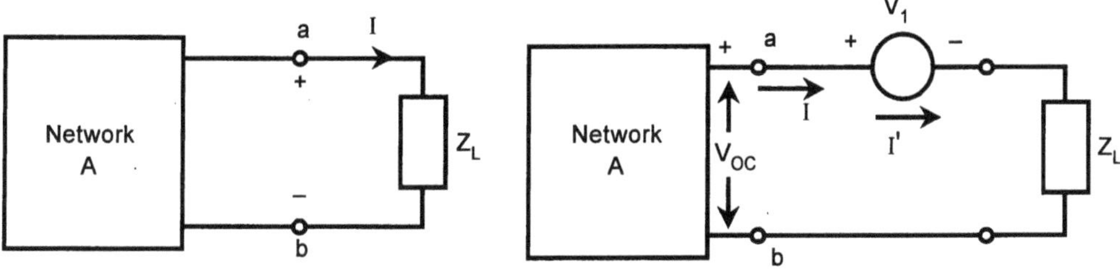

Fig. 6.6(a) Original network with load impedance Z_L

Fig. 6.6(b) Network A in series with a voltage source V_1 and Z_L

By superposition theorem

$$I + I' = 0 \qquad .....(6.13)$$

$$I' = -I \qquad .....(6.14)$$

Further, since the net current from network A is zero, it is as good as open circuited. Also, since the current in Z_L is zero, there is no voltage drop across it. Applying Kirchhoff's voltage law we get

$$V_{OC} - V_1 = 0 \qquad ......(6.15)$$

or

$$V_1 = V_{OC} \qquad ......(6.16)$$

Now reduce all the independent voltages sources in network A to zero. Since there are no sources in the network, the network can be replaced by a single impedance and by definition, it is the Thevenin's equivalent impedance. Hence with the source V_1 only present, the equivalent network is given in Fig. 6.6.

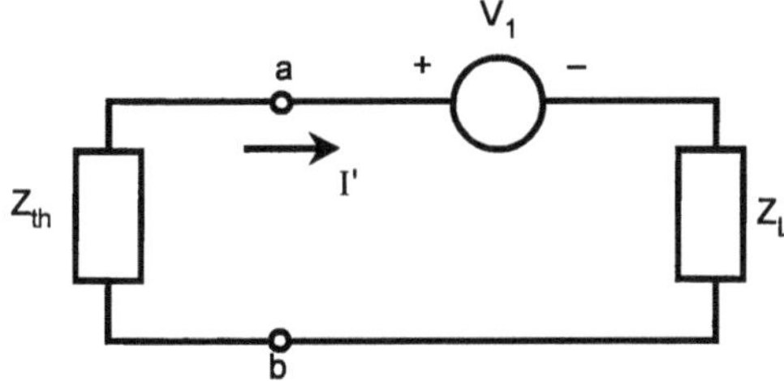

Fig. 6.7 The equivalent network with only V_1 present

The current I' due to the source V_1 alone is given by

$$I' = -\frac{V_1}{Z_{Th} + Z_L} \qquad \text{......(6.17)}$$

But from eqs. (6.14) and (6.16), we have

$$I = -I' = \frac{V_{oc}}{Z_{Th} + Z_L} \qquad \text{.....(6.18)}$$

This equation can be translated as an equivalent network given in Fig. 6.8.

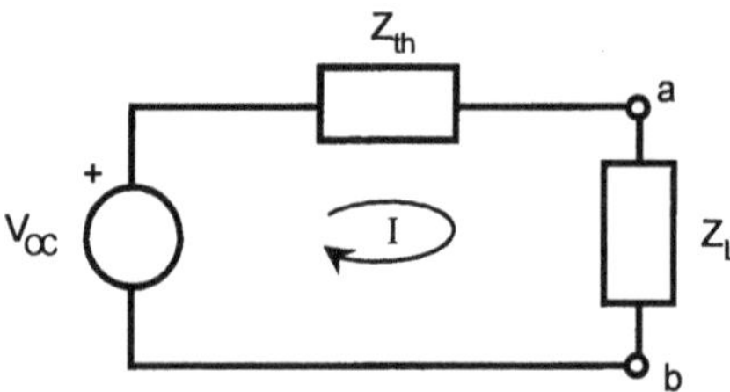

Fig. 6.8 Thevenin's equivalent network

V_{OC} is also known as the Thevenin's equivalent voltage.

6.5 NORTON'S THEOREM :

The theorem due to E.I. Norton is stated as follows :

"*Any two terminal network A consisting of sources and linear impedances can be replaced at its two terminals, by an equivalent network consisting of a single current source in parallel with an impedance. The equivalent current source is the short circuit current measured at the terminals and the equivalent impedance is same as the Thevenin's equivalent impedance*".

The Norton's equivalent is shown in Fig. 6.9. It can be observed that the Norton's equivalent network is the dual of the Thevenin's equivalent network.

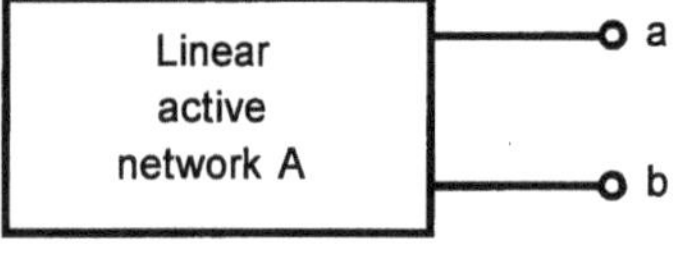

Fig. 6.9 (a) General network

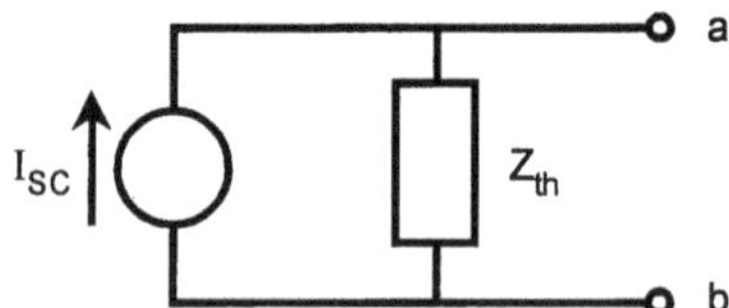

Fig. 6.9 (b) Its Norton's equivalent

The conditions that apply to network A and the network B which can be connected at the terminals a, b are exactly the same as for Thevenin's theorem.

Proof :

Let the voltage across the load Z_L due to all the sources in the network A be V. Now connect a current source across a, b as shown in Fig. 6.10(b) and adjust its current, I_1, such that the voltage across the load is zero.

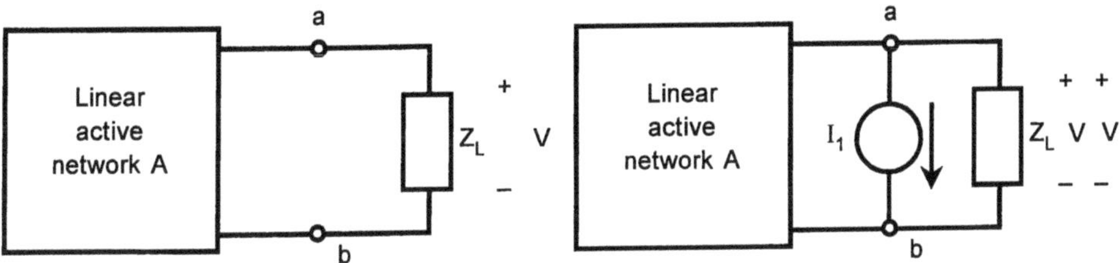

Fig. 6.10 (a) The original network   Fig. 6.10 (b) A current source connected across a, b

By superposition theorem

$$V + V' = 0 \quad \text{......(6.19)}$$

where V' is the voltage across the load Z_L due to the current source I_1 acting alone.

$$\therefore \quad V' = -V \quad \text{.....(6.20)}$$

Since the voltage across a, b is zero, the terminals a, b are as good as shorted and the current I_1 is the short circuit current for the network A.

$$\therefore \quad I_1 = I_{SC} \quad \text{.....(6.21)}$$

Let us now reduce all the independent sources to zero in the network A. The network A, thus can be replaced by a single impedance Z_{Th}, as per the definition. The network reduces to a simple network as given in Fig. 6.11.

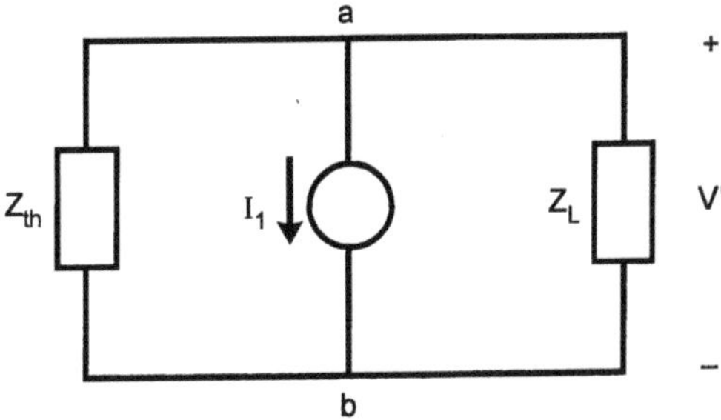

Fig. 6.11 Network with all sources reduced to zero except I_1

$$V' = -I_1 \frac{Z_{Th}}{Z_{Th} + Z_L} \cdot Z_L \quad \text{.....(6.22)}$$

Using eq. (6.20) and (6.21), we have

$$V = -V' = I_{SC} \frac{Z_{Th}}{Z_{Th} + Z_L} \cdot Z_L \qquad .....(6.23)$$

This equation can be translated into a network as shown in Fig. 6.12.

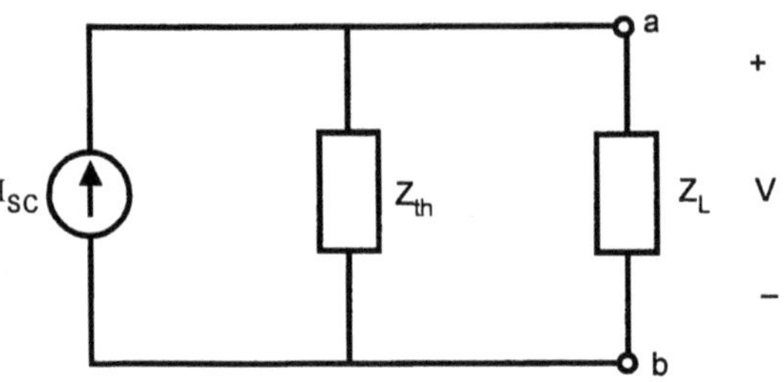

Fig. 6.12 Norton's equivalent of the network

This completes the proof.

***Note*:**

Since a given network can be replaced by a Thevenin's equivalent network or a Norton's equivalent network, relation between the three quantities V_{OC}, I_{SC} and Z_{Th} can be established easily. Consider the Thevenin's equivalent of a network as shown in Fig. 6.13.

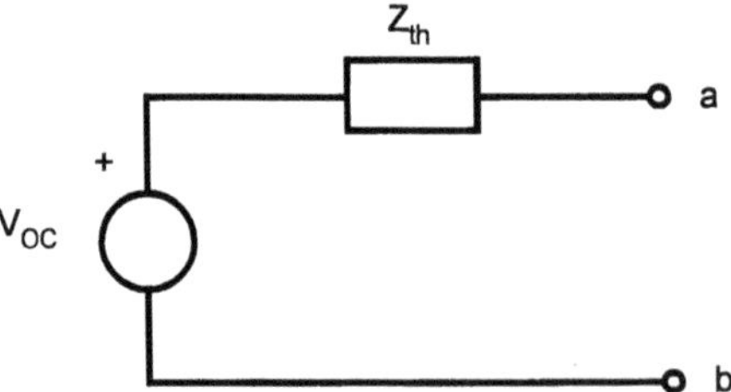

Fig. 6.13 Thevenin's equivalent of a network

We can obtain the Norton's equivalent of the original network by applying the theorem to the equivalent network of Fig. 6.13.

$$I_{sc} = \frac{V_{oc}}{Z_{Th}} \qquad .....(6.24)$$

or $$V_{oc} = I_{sc} Z_{Th} \qquad .....(6.25)$$

It is often convenient to define the Thevenin's equivalent impedance in terms of V_{oc} and I_{sc}.

Thus $$Z_{Th} = \frac{V_{oc}}{I_{sc}} \qquad .....(6.26)$$

This is an important relation which is useful in determining Z_{Th} when the network contains dependent sources, as will be shown in the following examples.

Example 6.4

Calculate the current in the 2Ω resistor of Fig. 6.14(a) using Thevenin's theorem.

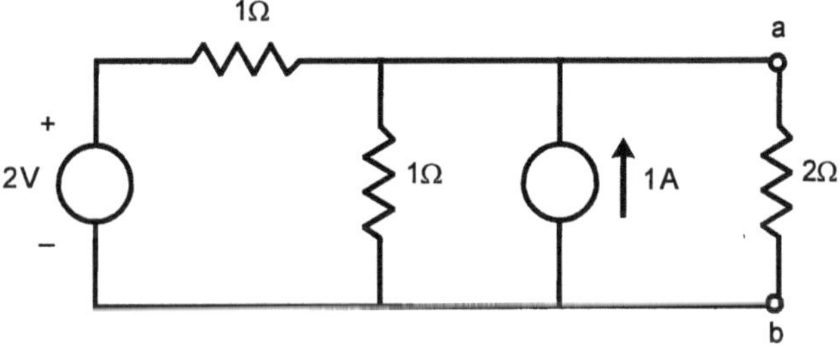

Fig. 6.14(a) Network for example 6.4

Solution :

First step in the solution is to find the open circuit voltage at the terminals a, b. Fig. 6.14(b) shows the network with terminals a, b open circuited

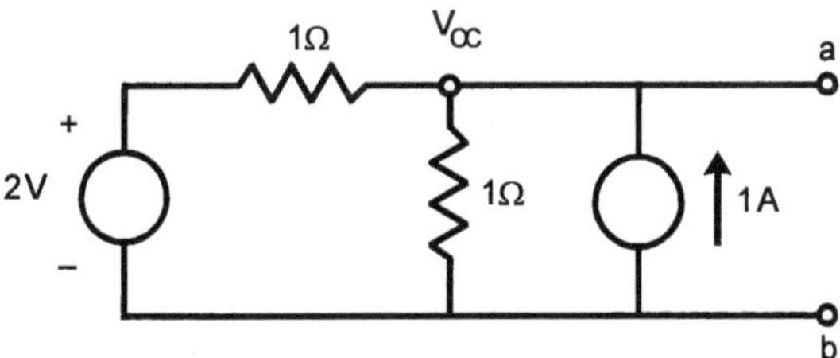

Fig. 6.14(b) Network for finding V_{OC}

Using node voltage method,

$$\frac{V_{oc} - 2}{1} + \frac{V_{oc}}{1} - 1 = 0$$

$$2V_{oc} = 3$$

$$V_{oc} = \frac{3}{2} \text{ Volts}$$

The second step is to find the Thevenin's equivalent impedance. Reducing all sources in the network to zero i.e, short circuiting the voltage source and open circuiting the current source, we have the network in Fig. 6.14(c).

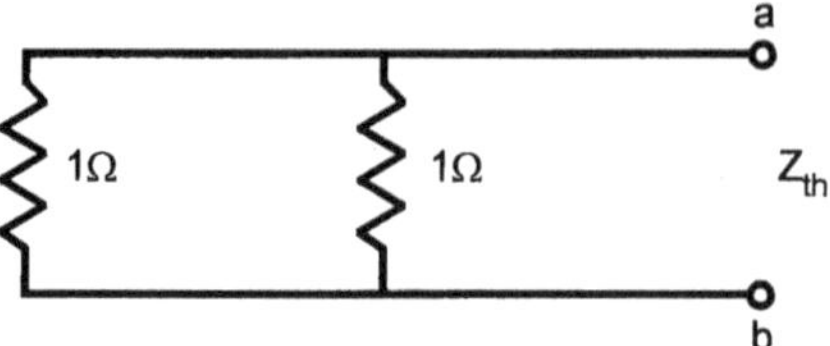

Fig. 6.14(c) Network for finding Z_{Th}

$$Z_{Th} = R_{Th} = \frac{1}{2}\Omega$$

where R_{Th} may be called Thevenin's Equvalent Resistance.

The Thevenin's equivalent network at terminals a, b of the network, with 2Ω resistance connected between them, is given in Fig. 6.14(d).

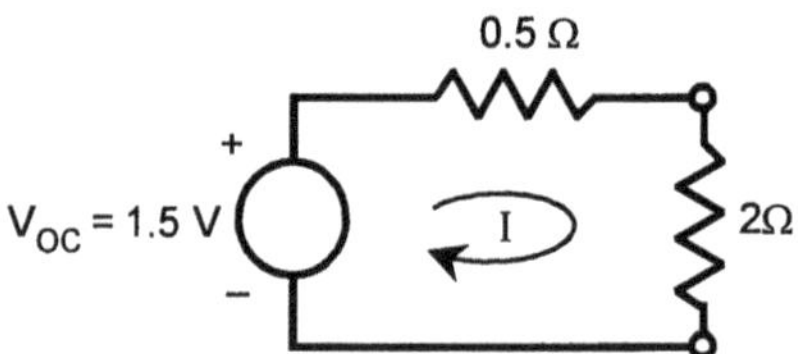

Fig. 6.14(d) Thevenin's equivalent of the network

The current in the 2 Ω resistance is

$$I = \frac{1.5}{2.5} \text{ A}$$

$$= 0.6 \text{ A}$$

Example 6.5

Obtain the Norton's equivalent at the terminals 1, 1' of the network in Fig. 6.15(a).

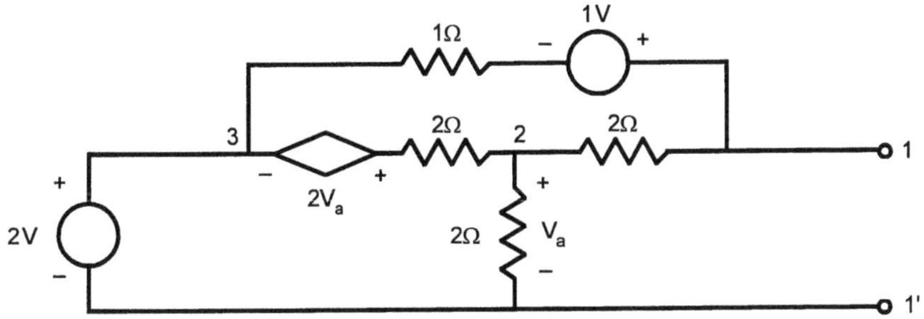

Fig. 6.15(a) Network for example 6.5

Solution :

(i) *Short circuit current at terminals* 1, 1' :

Shorting the terminals 1, 1' as in Fig. 6.15(b),

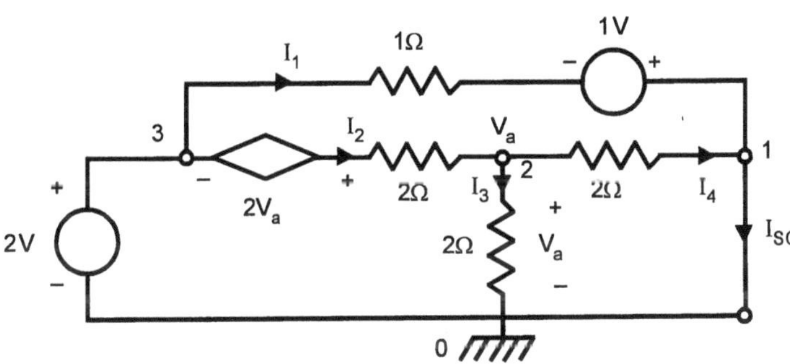

Fig. 6.15(b) Circuit to find I_{SC}

and applying Kirchhoff's voltage law to the loop 0, 3, 2 we have

$$2 + 2V_a - I_2 \cdot 2 = V_a$$

$$V_a - 2I_2 = -2$$

But $$I_3 = \frac{I_2}{2}$$

and $$V_a = 2 \times I_3 = I_2$$

$\therefore \quad I_2 - 2I_2 = -2$

$I_2 = 2A$

and $\quad I_4 = IA$

Applying Kirchhoff's voltage law to loop 1, 3, 0 we have

$$2 - I_1 . 1 + 1 = 0$$

$$I_1 = 3A$$

Applying Kirchhoff's current law at node 1, we have

$$I_{sc} = I_1 + I_4$$

$$= 3 + 1$$

$$= 4A$$

Thus the Norton's equivalent current source is

$$I_{sc} = 4A$$

(ii) *Thevenin's equivalent Resistance* (R_{Th}) :

Open circuiting terminals 1, 1' and reducing the independent voltage sources to zero, we have the circuit shown in Fig. 6.15(c).

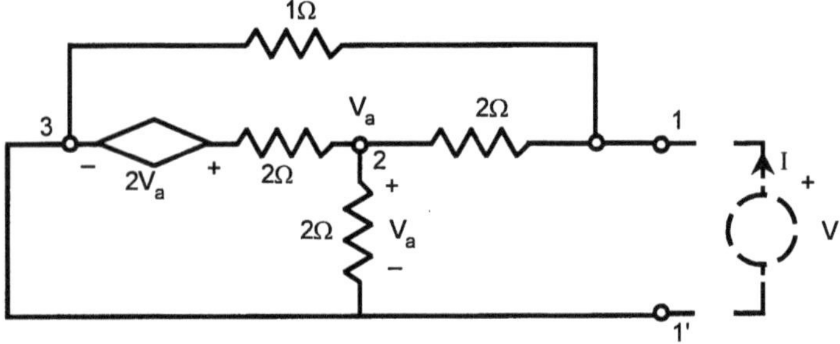

Fig. 6.15(c) Network for calculating Thevenin's equivalent Resistance

Note that the dependent source is retained as it is. In order to find the equivalent resistance of the above network, network reduction can not be used because of the presence of the dependent source. The most common method of obtaining the equivalent resistance is to apply a voltage source, V, at 1, 1' and measuring the current I delivered by it. The Thevenin's resistance then is the ratio of V to I. The second method is to obtain the open circuit voltage V_{oc}, at the terminal 1, 1' of the original network, and then the Thevenin's resistance is given by

$$R_{Th} = \frac{V_{oc}}{I_{sc}}$$

Let us obtain R_{Th} by the first method. Apply a source voltage of V at the terminals 1, 1'.

Applying node voltage method

$$\frac{V_a - 2V_a}{2} + \frac{V_a}{2} + \frac{V_a - V}{2} = 0$$

$$\therefore \quad V_a = V$$

Since $V_a = V$, no current is flowing in the resistance between the terminals 1 and 2. Since node 3 is at ground potential, the current I is the same as the current in the 1Ω resistance between nodes 1 and 3.

$$\therefore \quad I = \frac{V}{1}$$

$$\text{or} \quad \frac{V}{I} = R_{Th} = 1\Omega$$

The Norton's equivalent circuit is given by the network in Fig. 6.16.

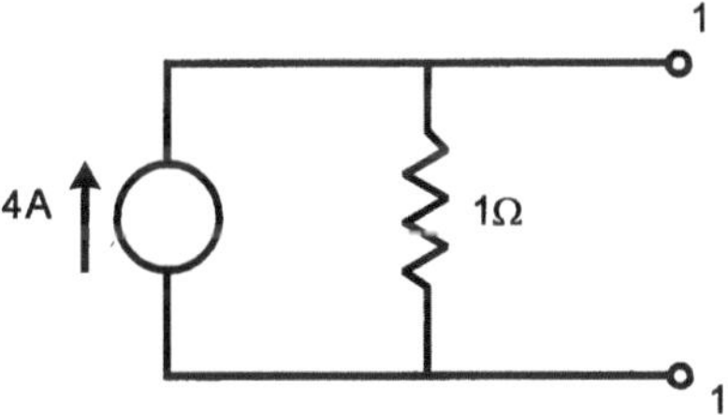

Fig. 6.16 Norton's equivalent of network in Fig. 6.15(a)

Example 6.6

Find the current I in the circuit using Norton's theorem in Fig. 6.17(a).

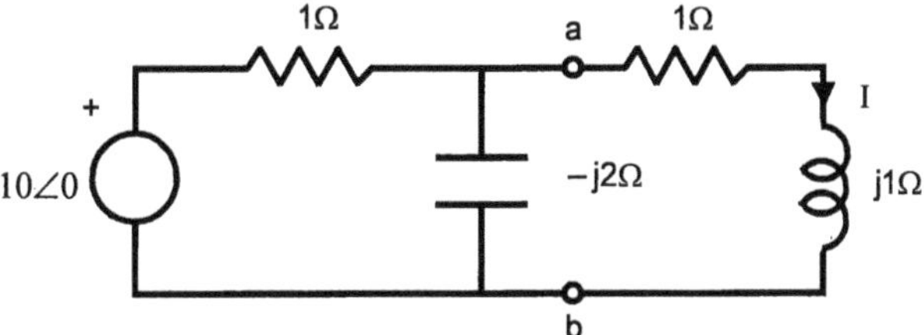

Fig. 6.17(a) Network for example 6.6

Solution :

Finding the Norton's equivalent at the terminals a, b, we get

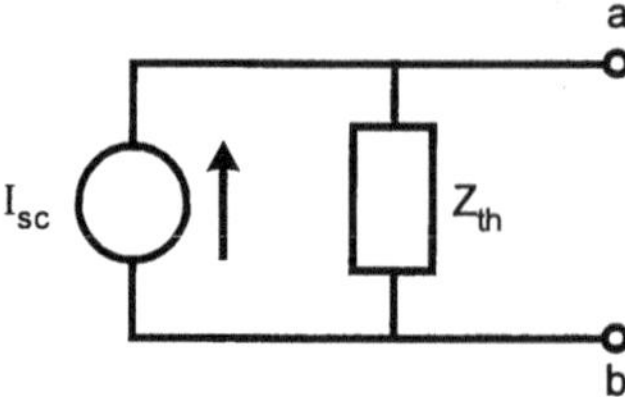

Fig. 6.17(b) Norton's equivalent

$$I_{sc} = \frac{10\angle 0^0}{1} = 10\angle 0^0 \text{ A}$$

$$Z_{Th} = \frac{1\times(-j2)}{1-j2} = 0.8 - j0.4\Omega$$

Reconnecting the impedance 1 + j1 at the terminals a, b of Norton's equivalent circuit as shown in Fig. 6.17(c), we have

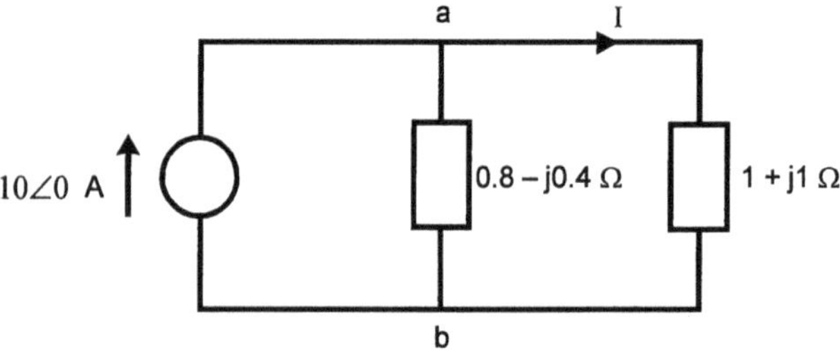

Fig. 6.17(c) Network for calculating current I

$$I = \frac{10\angle 0\times(0.8-j0.4)}{1.8+j0.6}$$

$$= \frac{8.944\angle -26.56}{1.897\angle 18.435}$$

$$= 4.715\angle -45^0 \text{ A}$$

Example 6.7

Find the current I in the circuit using Norton's theorem in Fig. 6.18(a).

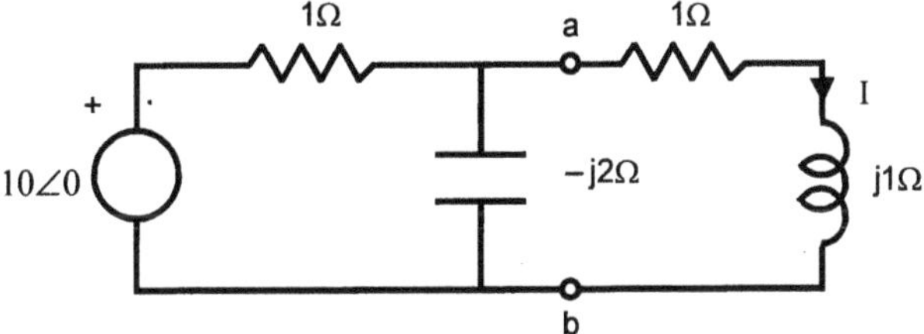

Fig. 6.18 (a) Network for example 6.7

Solution :

Finding the Norton's equivalent at the terminals a, b, we get

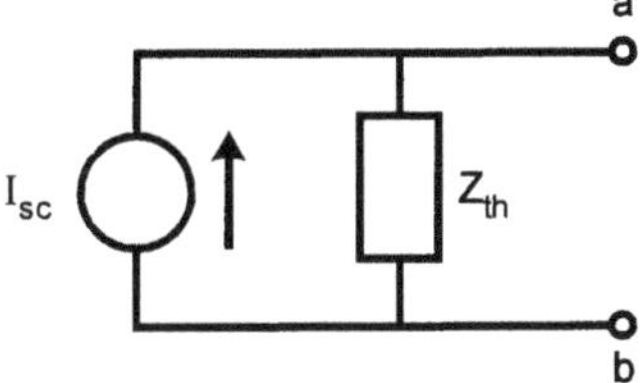

Fig. 6.18 (b) Norton's equivalent

$$I_{sc} = \frac{10\angle 0^0}{1} = 10\angle 0^0$$

$$Z_{Th} = \frac{1\times(-j2)}{1-j2} = 0.8 - j0.4\Omega$$

Reconnecting the impedance 1 + j1 at the terminals a, b of Norton's equivalent circuit as shown in Fig. 6.18(c), we have

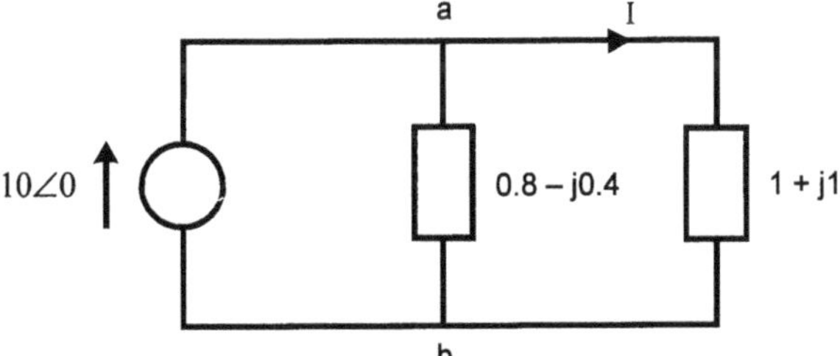

Fig. 6.18 (c) Network for calculating current I

$$I = \frac{10\angle 0\times(0.8-j0.4)}{1.8+j0.6}$$

$$= \frac{8.944\angle -26.56}{1.897\angle 18.435}$$

$$= 4.715\angle -45^0\text{ A}$$

6.6 MILLMAN'S THEOREM :

This theorem gives the voltage at two given terminals between which, a number of voltage sources with their internal impedances, are connected in parallel. The theorem states that :

At two given terminals, if a number of voltage sources V_i, $i = 1, 2, ...n$ with their internal impedance Z_i, $i = 1, 2, ...n$ are connected in parallel, then the voltage at the two terminals is given by

$$V_{ab} = \frac{\sum_{i=1}^{n} V_i Y_i}{\sum_{i=1}^{n} Y_i} \qquad .....(6.27)$$

where $Y_i = \dfrac{1}{Z_i}$ *for* $i = 1, 2, ...n$ *are the admittances.*

Proof :

Consider the network shown in Fig. 6.19.

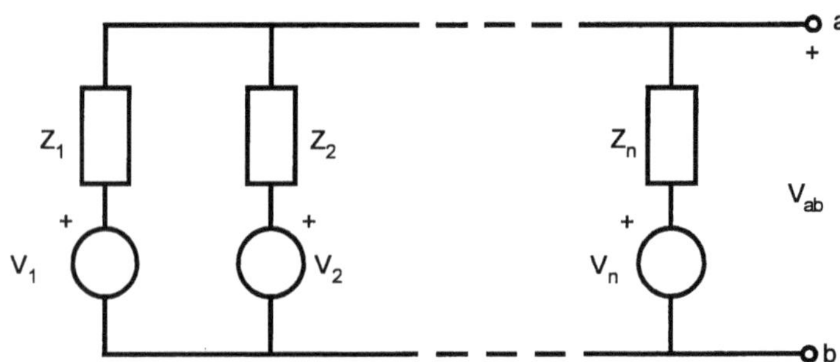

Fig. 6.19 (a) Network to illustrate Millman's theorem

Each branch containing the voltage source and internal impedance can be replaced by its Norton's equivalent as shown in Fig. 6.19(b)

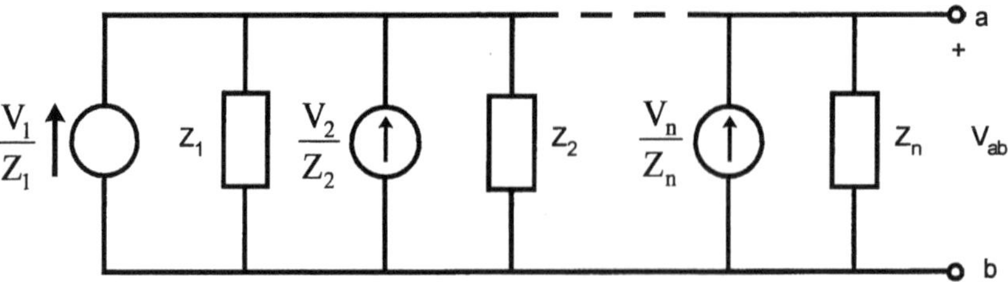

Fig. 6.19(b) Source transformation

Writing $\frac{1}{Z_i} = Y_i$ and using admittances rather than impedances, the current sources in parallel can be combined into a single source and the admittances can be combined into a single admittance as shown in Fig. 6.19(c).

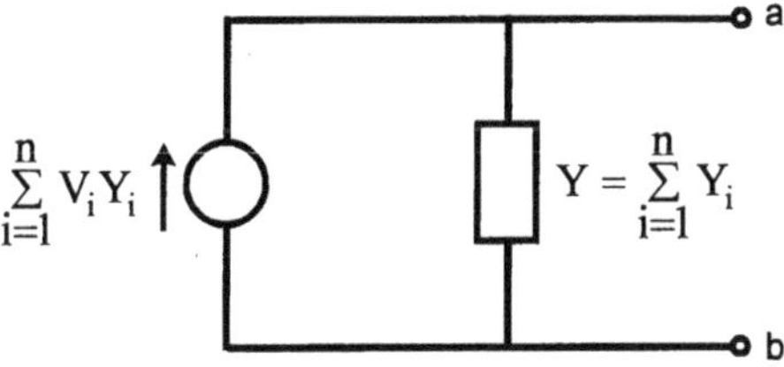

Fig. 6.19(c) Simplified network

Thus
$$V_{ab} = \frac{\sum_{i=1}^{n} V_i Y_i}{\sum_{i=1}^{n} Y_i}$$

Example 6.8

Calculate the current in the 4Ω resistance using Millman's theorem.

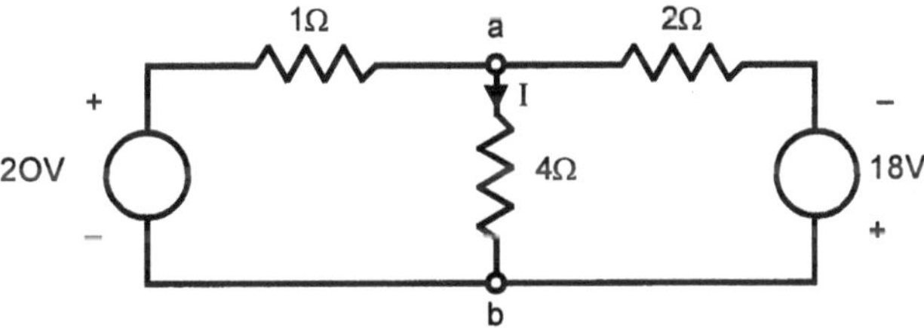

Fig. 6.20 Network for example 6.8

Solution :

The voltage V_{ab} across the terminals a, b, by Millman's theorem, is given by

$$V_{ab} = \frac{20(1) + 0\left(\frac{1}{4}\right) + (-18)\left(\frac{1}{2}\right)}{1 + \frac{1}{4} + \frac{1}{2}}$$

Note that the voltage source in the 4Ω branch is consider to be zero.

$$V_{ab} = \frac{11}{7} \times 4 = \frac{44}{7} \text{ V}$$

The current in the 4Ω resistance is given by

$$I = \frac{V_{ab}}{4} = \frac{44}{7} \times \frac{1}{4} = \frac{11}{7} \text{ A}$$

Example 6.9

Consider a 3 – φ circuit shown in Fig. 6.21.

Find the current in the Y line. Phase sequences is RYB.

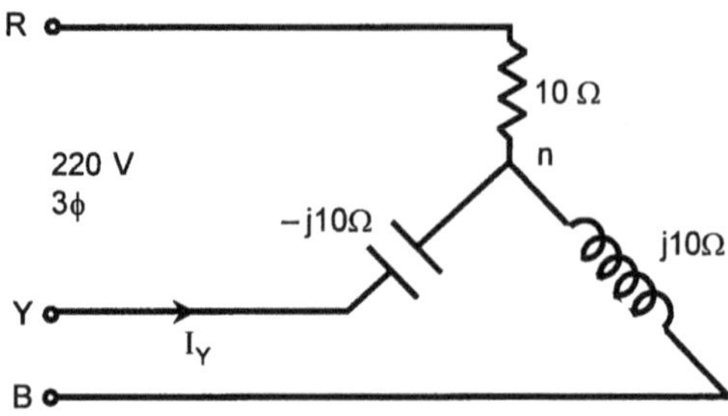

Fig. 6.21 3 phase network for example 6.7

Solution :

Since phase sequence is RYB

$$V_{RY} = 220 \angle 0^0$$

$$V_{YB} = 220\angle -120^0$$

Using Millman's theorem.

$$V_{nY} = \frac{V_{RY}Y_R + V_{BY}Y_B}{Y_R + Y_Y + Y_B}$$

$$= \frac{220\angle 0 \times \frac{1}{10} - 220\angle -120 \times \frac{1}{j10}}{\frac{1}{10} + \frac{1}{-j10} + \frac{1}{j10}}$$

$$= \frac{22 - 22\angle -210}{0.1}$$

$$= 220[1 + 0.886 - j0.5]$$

$$= 220[1.886 - j0.5]$$

$$= 429\angle -15^0 \text{ V}$$

$$I_Y = \frac{V_{Yn}}{Z_Y} = -\frac{V_{nY}}{Z_Y}$$

$$= -\frac{429\angle -15^0}{-j10}$$

$$= 42.9\angle -105^0 \text{ A}$$

6.7 RECIPROCITY THEOREM :

The reciprocity theorem states that the ratio of response to excitation is invariant to an interchange of the positions of the excitation and response in a single source network. However if the excitation is a voltage source, the response should be a current and vice versa. This is a very important property and networks which obey this theorem are called as reciprocal networks.

Proof :

Consider a network in which all the elements are initially relaxed or there are no initial conditions on any energy storing elements. Further there is only one source at terminals 1, 1' of the network and the current is measured in the jth loop at terminals 2, 2' as shown in Fig. 6.22(a).

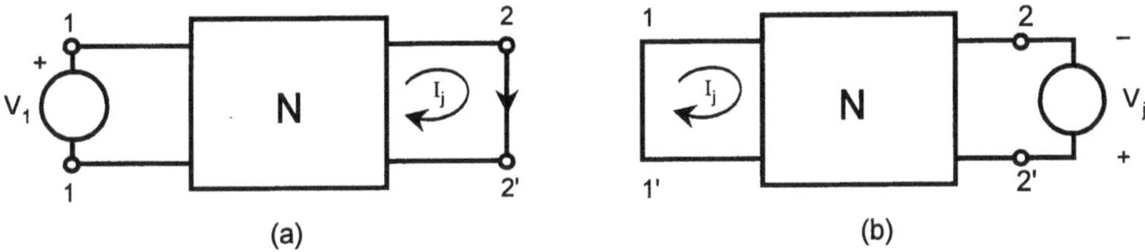

Fig. 6.22 (a) Network N with voltage source V_1 at terminals 1, 1'
(b) Network N with voltage source and current response interchanged

From general network equations based on loop current method developed in section 6.2, the current I_j is given by eq. (6.8) with

$$V_2 = V_3 = \ldots\ldots = V_n = 0,$$

Thus
$$I_j = \frac{\Delta_{1j} V_1}{\Delta_Z} \qquad \ldots\ldots(6.28)$$

When the voltage source and the current response are interchanged, i.e, a voltage source V_j is introduced in the jth loop and current is measured in loop 1, the current I_1 is given by

$$I_1 = \frac{\Delta_{j1}}{\Delta_Z} V_j \qquad \ldots\ldots(6.29)$$

If we have only R L C elements in the network and there are no controlled sources, the impedance matrix Z in the network is symmetric since $Z_{ij} = Z_{ji}$ is the impedance common to the ith and j^{th} loop. Thus in such networks $\Delta_{ij} = \Delta_{ji}$ and therefore

$$\frac{V_j}{I_1} = \frac{V_1}{I_j} = \frac{\Delta_Z}{\Delta_{1j}} \qquad \ldots\ldots(6.30)$$

Therefore the ratio of excitation to response remains same if the position of response and excitation are interchanged. Thus the theorem is proved.

Example 6.10

Using reciprocity theorem and superposition theorem simultaneously, obtain the current I in the network of Fig. 6.23(a) if 5V source acting alone produces a current of 1A in the resistor R.

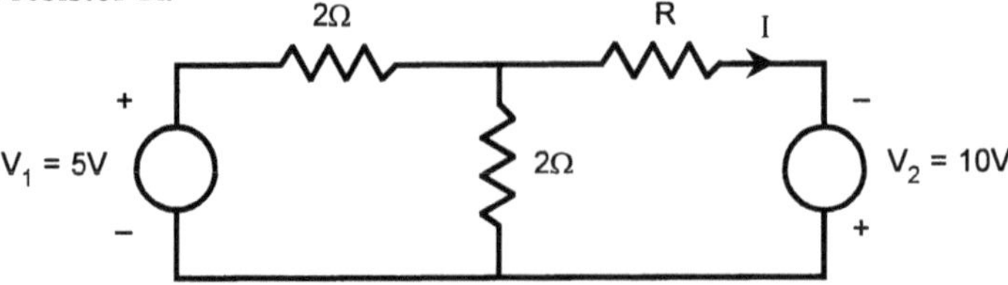

Fig. 6.23(a) Network for example 6.10

Solution :

It is given that the 5V sources alone produces a current I' = 1A in the resistor R as shown in Fig. 6.23(b).

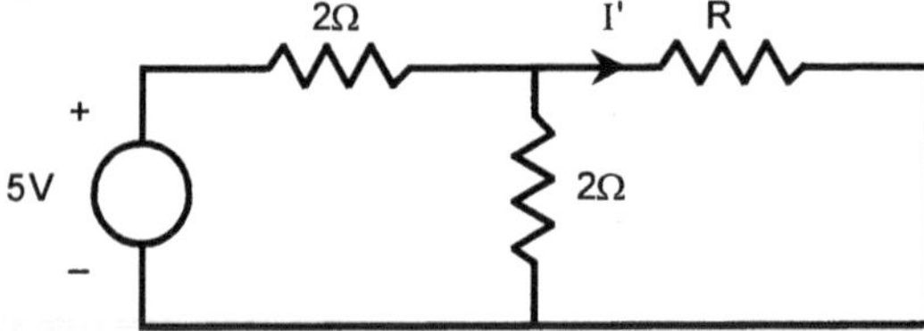

Fig. 6.23(b) Network with 5V source acting alone

Now let the 10V source be acting alone with 5V source reduced to zero as shown in Fig. 6.23(c).

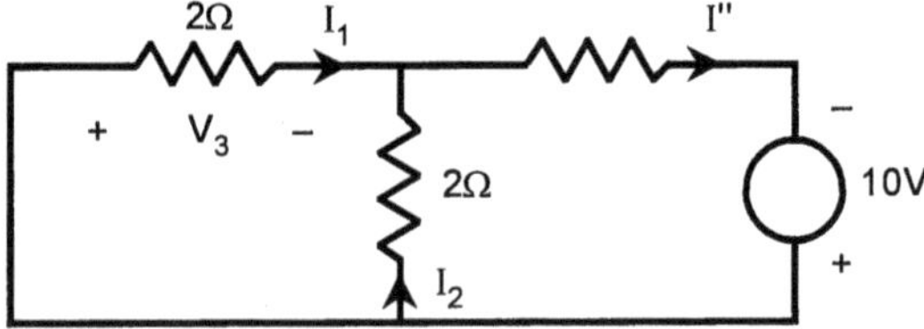

Fig. 6.23 (c) Network in Fig. 6.23 (a) with only 10V source acting alone

Comparing Fig. 6.23(b) and (c) we can observe that Network in Fig. 6.23(c) is obtained by interchanging the source and response in the network of Fig. 6.23(b). Hence using Reciprocity theorem

$$\frac{V_1}{I'} = \frac{V_2}{I_1}$$

$$\therefore \quad I_1 = V_2 \cdot \frac{I'}{V_1} = 10.\frac{1}{5} = 2A$$

$$\therefore \quad V_3 = 2.(2) = 4V$$

$$\text{and} \quad I_2 = \frac{V_3}{2} = \frac{4}{2} = 2A$$

$$\therefore \quad I'' = 4A$$

By superposition theorem the current

$$I = I' + I'' = 1 + 4 = 5A$$

6.8 COMPENSATION THEOREM :

In many practical networks, it is often desired to find the response due to a change in the value of a component. Usually the components are available with a certain tolerance. Hence we would like to find the change in the response if the value of the component is not the exact value. The change in the currents in various branches of the network due to a change in the impedance of a particular branch is given by the compensation theorem. This theorem states that :

" *In a linear network, if the current in a branch is I and the impedance Z in that branch is changed by ΔZ, then the changes in the currents in the branches of the network can be obtained as the currents due to a voltage source of value* IΔZ, *introduced in that branch in a direction opposing the current* I, *with all the sources in the network reduced to zero* ".

Proof :

Consider a network with n loops. The loop equations are given by eq. (6.4)

$$
\begin{aligned}
&I_1Z_{11} + I_2Z_{12} + \ldots\ldots + I_k\, Z_{1k} + \ldots\ldots + I_nZ_{1n} = V_1\\
&\vdots\\
&I_1Z_{k1} + I_2Z_{k2} + \ldots\ldots + I_kZ_{kk} + \ldots\ldots + I_nZ_{kn} = V_k \qquad \ldots\ldots(6.31)\\
&\vdots\\
&I_1Z_{n1} + I_2Z_{n2} + \ldots\ldots + I_kZ_{nk} + \ldots\ldots + I_nZ_{nn} = V_n
\end{aligned}
$$

Let the impedance in a particular branch of kth loop, which is not common with any other loop, be changed by ΔZ_k. This will cause changes in the currents of all the loops. Since all the impedances except Z_{kk} are same, we have the new set of equations.

$$
\begin{aligned}
&I'_1Z_{11} + I'_2\, Z_{12} + \ldots\ldots + I'_k\, Z_{1k} + \ldots\ldots + I'_n\, Z_{1n} = V_1\\
&\vdots\\
&I_1Z_{k1} + I'_2Z_{k2} + \ldots\ldots + I'_k(Z_{kk} + \Delta Z_k) + \ldots\ldots + I'_nZ_{1n} = V_k\\
&\vdots\\
&I'_1Z_{n1} + I'_2Z_{n2} + \ldots\ldots + I'_kZ_{nk} + \ldots\ldots + I'_nZ_{nn} = V_n \qquad \ldots\ldots(6.32)
\end{aligned}
$$

Subtracting the set of eq. (6.31) from corresponding equations of (6.32) after adding $I_k\,\Delta Z_k$ on both sides of the kth loop equation in eq. (6.31), to make the impedance of kth loop same as in k^{th} loop of eq. (6.32) and defining,

$$\Delta I_j = I'_j - I_j \text{ for } j = 1, 2, \ldots\ldots n$$

we have

$$
\begin{aligned}
&\Delta I_1Z_{11} + \Delta I_2Z_{12} + \ldots\ldots + \Delta I_kZ_{1k} + \ldots\ldots + \Delta I_nZ_{1n} = 0\\
&\vdots\\
&\Delta I_1Z_{k1} + \Delta I_2Z_{k2} + \ldots\ldots + \Delta I_k(Z_{kk} + \Delta Z_k) + \ldots\ldots + \Delta I_nZ_{kn} = -\, I_k\Delta Z_k\\
&\vdots\\
&\Delta I_1Z_{n1} + \Delta I_2Z_{n2} + \ldots\ldots + \Delta I_kZ_{nk} + \ldots\ldots + \Delta I_nZ_{nn} = 0 \qquad \ldots\ldots(6.33)
\end{aligned}
$$

Eqs. (6.33) show that the changes, in the loop currents are given by considering the changed network, with all sources reduced to zero and introducing a voltage source of value $I_k \Delta Z_k$ in a direction opposing the current I_k (because of the – ve sign in the kth eq. (6.33)). This proves the compensation theorem.

Now the change in the current in j^{th} loop is given by the Cramer's rule,

$$\Delta I_j = \frac{-\Delta'_{jk} I_k \Delta Z_k}{\Delta'_Z} \quad .....(6.34)$$

where Δ'_Z is the determinant of the altered impedance matrix and Δ'_{jk} is the cofactor of the (j, k)th element of the altered impedance matrix. But it is obvious that

$$\Delta'_{jk} = \Delta_{jk}$$

since the kth row, in which the impedance Z_{kk} is changed, is eliminated in the calculation of Δ'_{jk}.

Thus $$\Delta I_j = \frac{-\Delta_{jk} I_k \Delta Z_k}{\Delta'_Z} \quad ......(6.35)$$

If the change ΔZ_k in the kth loop is small compared to Z_{kk}, then

$$\Delta'_Z \cong \Delta_Z \quad .....(6.36)$$

Thus $$\Delta I_j \cong \frac{\Delta_{jk} I_k \Delta Z_k}{\Delta_Z} \quad .....(6.37)$$

If the cofactor of the original impedance matrix and its determinant are known, the changes in the loop currents can be calculated using equation (6.37).

Example 6.11

Find the change in the current I_3 when the 10 ohm resistor is changed to 9Ω in the network of Fig. 6.24. Use compensation theorem.

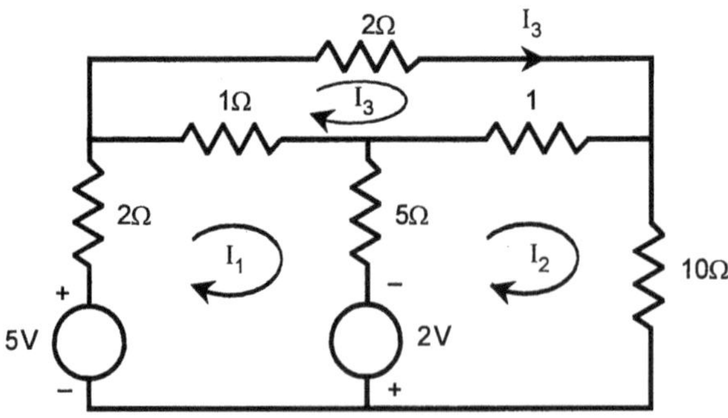

Fig. 6.24 Network for example 6.11

Solution :

Let us first find the current is 10Ω resistance of the original network. The loop equations are

$$8I_1 - 5I_2 - I_3 = 7$$
$$-5I_1 + 16I_2 - I_3 = -2$$
$$-I_1 - I_2 + 4I_3 = 0$$

Solving for I_2 and I_3 we have

$$I_2 = \frac{\begin{vmatrix} 8 & 7 & -1 \\ -5 & -2 & -1 \\ -1 & 0 & 4 \end{vmatrix}}{\begin{vmatrix} 8 & -5 & -1 \\ -5 & 16 & -1 \\ -1 & -1 & 4 \end{vmatrix}} = 7 \cdot \frac{\Delta_{12}}{\Delta_Z} - 2\frac{\Delta_{22}}{\Delta_Z}$$

$$\Delta_{12} = 21 \; ; \qquad \Delta_{22} = 31$$

$$\Delta_Z = 378$$

$$\therefore \qquad I_2 = \frac{1}{378}(7 \times 21 - 2 \times 31) = 0.2249 \text{ A}$$

Similarly $$I_3 = \frac{\begin{vmatrix} 8 & -5 & 7 \\ -5 & 16 & -2 \\ -1 & -1 & 0 \end{vmatrix}}{\Delta_Z} = \frac{1}{\Delta_Z}(7\,\Delta_{13} - 2\Delta_{23})$$

$$\Delta_{13} = 5 + 16 = 21$$
$$\Delta_{23} = -(-8 - 5) = 13$$

$$\therefore \qquad I_3 = \frac{1}{378}(7 \times 21 - 2 \times 13) = 0.32 \text{ A}$$

By Compensation theorem

$$\Delta I_j = \frac{-\Delta_{jk}}{\Delta_z} I_k \,\Delta Z_k$$

Now j = 3, k = 2 and assuming $\Delta_z = \Delta'_z$

$$\Delta I_3 = -\frac{\Delta_{32}}{\Delta_z} I_2 \,\Delta Z_2$$

Since the Z matrix is symmetric

$$\Delta_{32} = \Delta_{23}$$

$$\therefore \qquad \Delta I_3 = \frac{-13}{378} \cdot (0.2249)(9-10) = 7.735 \times 10^{-3}\text{ A}$$

The current I_3 is increased by 7.735 mA due to a decrease in the resistance in 10Ω branch.

If the new impedance matrix is used for better accuracy; we have

$$[Z'] = \begin{bmatrix} 8 & -5 & -1 \\ -5 & 15 & -1 \\ -1 & -1 & 4 \end{bmatrix}$$

and $\qquad \Delta'_Z = 347$

and $\qquad \Delta I_3 = \dfrac{-13}{347}(0.2249)(-1) = 8.43\text{ mA.}$

Example 6.12

In the network of Fig. 6.25, the impedance Z_1 is changed to 5.5 + j5.5. Find the change in the current I_2 and the new current using compensation theorem.

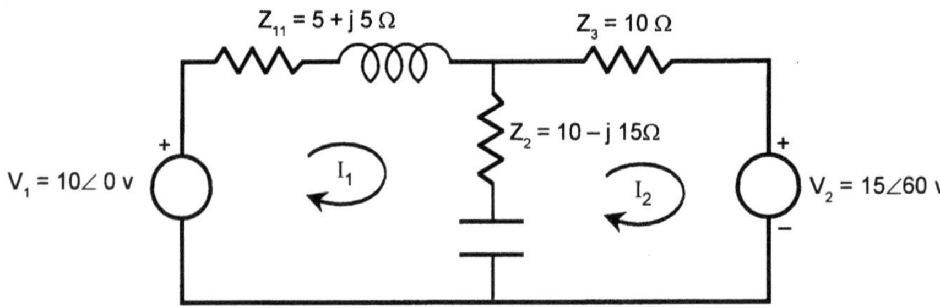

Fig. 6.25 Network for Ex. 6.12

Solution :

First let us calculate the currents I_1 and I_2 in the original network. The loop current equations are,

$$\begin{bmatrix} 15-j10 & -(10-j15) \\ -(10-j15) & 20-j15 \end{bmatrix}\begin{bmatrix} I_1 \\ I_2 \end{bmatrix} = \begin{bmatrix} 10\angle 0 \\ -15\angle 60 \end{bmatrix}$$

Here, $\qquad [z] = \begin{bmatrix} 15-j10 & -(10-j15) \\ -(10-j15) & 20-j15 \end{bmatrix}$

$$\Delta_z = 275 - j\,125$$

$$I_1 = \frac{10\angle 0\Delta_{11} - 15\angle 60\Delta_{21}}{\Delta_z}$$

$$\Delta_{11} = 20 - j15, \qquad \Delta_{21} = (10 - j\,15)$$

$$I_1 = 0.0188 - j\,0.6\text{ A}$$

Similarly, $$I_2 = \frac{10\angle 0\Delta_{12} - 15\angle 60\Delta_{22}}{\Delta_z}$$

and $$\Delta_{12} = 10 - j\,15, \qquad \Delta_{22} = 15 - j\,10$$

$\therefore$ $I_2 = -\,0.0595 - j\,1.008\text{ A}$

By compensation theorem, we have,

$$\Delta I_2 = \frac{\Delta_{21}}{\Delta_2'} I_1\,\Delta\,Z_1$$

Here,

$$\Delta\,Z_1 = 0.5 + j\,0.5\ \Omega$$

$$\Delta_z' \simeq \Delta_z = 275 - j125$$

$$\Delta_{21} = 10 - j\,15$$

$\therefore$ $$\Delta I_2 = -\frac{(10 - j\,15)(0.0188 - j\,0.6)(0.5 + j\,0.5)}{275 - j\,125}$$

$$= -\,0.0065 + j\,0.0245\text{ A}$$

and $$\Delta\,I_2 = I_2' - I_2$$

where $$I_2' = I_2 + D\,I_2$$

$$= -\,0.0595 - j\,1.008 - 0.0065 + j\,0.0245$$

$$= -\,0.066 - j\,0.9835\text{ A}$$

with the changed network, the new current can be calculated using the loop current method. Therefore the actual current is

$$I_2' = -\,0.065 - j\,0.985\text{ A}$$

which agrees closely with the current calculated using compensation theorem.

6.9 TELLEGEN'S THEOREM :

Tellegen's theorem is one of the most general theorems which is applicable to any lumped network with linear or non linear, time varying or time invariant elements. The theorem is based on the two Kirchhoff's laws and is completely independent of the nature of the elements. The theorem is stated as follows :

Consider a network with b branches and n nodes, with each branch having a branch voltage v_k and current i_k for $k = 1, 2, ...b$ with an arbitrarily assigned associated reference directions. Let all the voltages satisfy Kirchhoff's voltage law and all the currents satisfy Kirchhoff's current law, then

$$\sum_{k=1}^{b} v_k i_k = 0 \qquad .....(6.38)$$

Proof :

Let the branch voltages and currents be given an associated reference direction as shown in Fig. 6.26 i.e., the current is taken as positive when it enters the element at the terminal where the voltage is +ve.

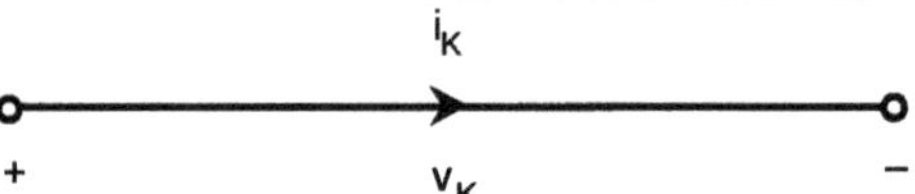

Fig. 6.26 Associated reference directions for voltage and current

Let us also assume that between any two nodes there is only one branch connected. All parallel branches are combined into a single branch. Let the branch voltages and currents be v_k and i_k, for $i = 1, 2, ...b$. Let one of the nodes be taken as reference. Consider 2 nodes p and q between which the kth branch is connected as shown in Fig. 6.27.

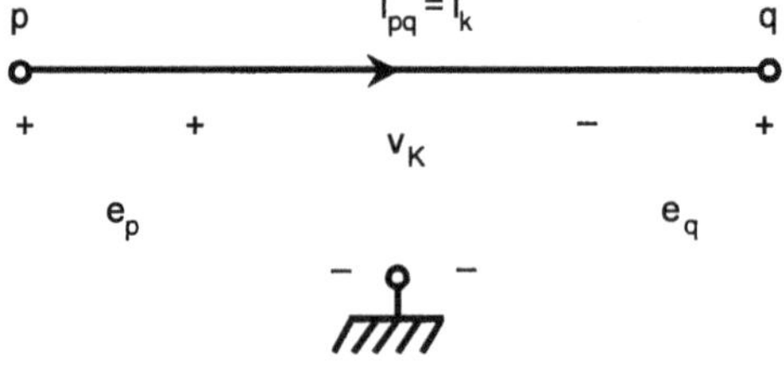

Fig. 6.27 Branch k connected between nodes p and q

Using Kirchhoff's voltage law, we have

$$e_p - v_k - e_q = 0$$

or $$v_k = e_p - e_q \qquad .....(6.39)$$

and $$v_k i_k = (e_p - e_q) i_{pq} \qquad .....(6.40)$$

Eq. (6.40) can also be written as

$$v_k i_k = (e_q - e_p) i_{qp} \qquad .....(6.41)$$

Thus, adding eq. (6.40) and (6.41), we have

$$v_k i_k = \frac{1}{2}\left[(e_p - e_q) i_{pq} + (e_q - e_p) i_{qp}\right] \qquad .....(6.42)$$

If we consider all the 'b' branches of the network, summing $v_k i_k$ for $k = 1, 2, ...b$, and considering n nodes, we have

$$\sum_{k=1}^{b} v_k i_k = \frac{1}{2}\sum_{p=1}^{n}\sum_{q=1}^{n}(e_p - e_q) i_{pq} \qquad .....(6.43)$$

$$= \frac{1}{2}\sum_{p=1}^{n} e_p \sum_{q=1}^{n} i_{pq} - \frac{1}{2}\cdot\sum_{q=1}^{n} e_q \sum_{p=1}^{n} i_{pq} \qquad .....(6.44)$$

But $\sum_{q=1}^{n} i_{pq} = 0$, since all the branch currents satisfy Kirchhoff's current law at all nodes, as the term $\sum_{q=1}^{n} i_{pq}$ represents the sum of all the currents incident at node 'p'.

Similarly $$\sum_{p=1}^{n} i_{pq} = 0$$

$$\therefore \qquad \sum_{k=1}^{b} v_k i_k = 0 \qquad ......(6.45)$$

This proves the Tellegen's theorem.

Alternate Proof :

Tellegen's theorem can also be proved easily, using topological properties of networks discussed in chapter 5.

Let A denote the incidence matrix of the given network

v denote the branch voltage vector

j denote the branch current vector

and e denote the node to datum voltage vector then,

since the voltages and currents of the network satisfy KCL and KVL, from eqs. (5.3) and (5.5), we have

$$AJ = 0$$

$$A^T e = v$$

Now $\qquad \Sigma v_k j_k = v^T j = \left(A^T e\right)^T j$

$$= e^T (AJ) = 0$$

This theorem is more general than it appears. Consider two networks which have similar structure or graph with different elements. If v_{k1} and i_{k1} are the branch voltages and currents of one network and v_{k2} and i_{k2} are the branch voltages and currents of the second network respectively, then

$$\sum_{k=1}^{b} v_{k1} i_{k2} = 0; \qquad \sum_{k=1}^{b} v_{k2} i_{k1} = 0 \qquad .....(6.46)$$

Observe that branch voltages of one network and branch currents of the second network are used in forming the products.

Similarly, in a network if $v_k\ (t_1)$ are the branch voltages at an instant t_1 and $i_k\ (t_2)$ are the currents at another instant of time t_2, then

$$\sum_{k=1}^{b} v_k(t_1) i_k(t_2) = 0 \qquad .....(6.47)$$

Remarks on Tellegen's Theorem :

The product $v_k i_k$ gives the instantaneous power in the kth branch of the network. The summation over all the branches gives the total instantaneous power in a network and according to the Tellegen's theorem the net power in the network is zero indicating that the power is conserved in any network.

Let us consider a network in which there are 'b' branches. Out of these 'b' branches let 'm' branches contain sinusoidal energy sources and b – m branches thus contain passive elements. These 'm' branches can be brought out as shown in Fig. 6.28.

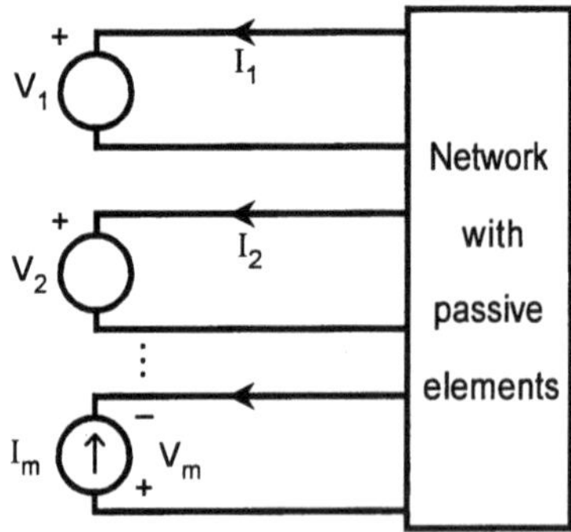

Fig. 6.28 Network to prove conservation of complex power

If the currents I_k for k = 1, 2, ...b satisfy Kirchhoff's current law, their conjugates, I_k^*, also satisfy Kirchhoff's current law. Hence by Tellegen's theorem

$$\sum_{k=1}^{b} V_k I_k^* = 0$$

The term $V_k\ I_k^*$ represents the complex power in the kth branch

$$\therefore \qquad \sum_{k=1}^{m} V_k I_k^* + \sum_{k=m+1}^{b} V_k I_k^* = 0$$

or

$$\sum_{k=m+1}^{b} V_k I_k^* = -\sum_{k=1}^{m} V_k I_k^* \qquad .....(6.48)$$

The quantity $\sum_{k=1}^{m} V_k I_k^*$ represents the complex power delivered to the source branches and $-\sum_{k=1}^{m} V_k I_k^*$, therefore, represents the complex power delivered by the source to the rest of the network containing only passive branches. The quantity on LHS of the eq. (6.48) represents the complex power delivered to the network containing only passive elements. Thus from eq. (6.48) it is clear that the complex power delivered by the source is equal to the complex power received by the passive elements of the network, which implies conservation of complex power.

Example 6.13

The network N is a purely resistive network. When 15V is applied across the terminals 1, 1' with the terminals 2, 2' open circuited, a current of 5A is drawn from the source producing a voltage of 5V at terminals 2, 2'. When the terminals 2, 2' are shorted, the network takes a current of 6A from the same source. Find the Thevenin's equivalent of the network at terminals 2, 2'.

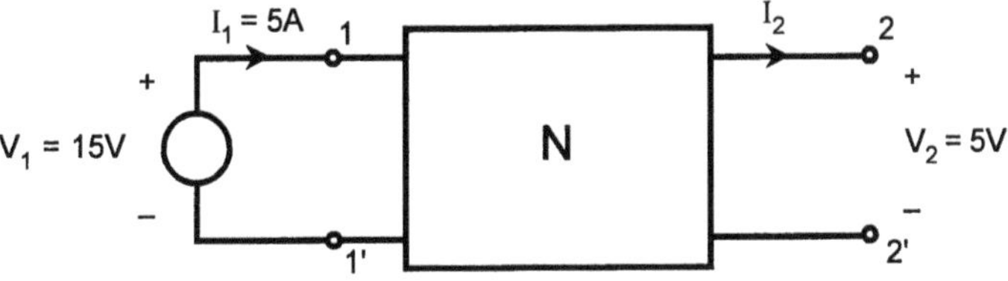

Fig. 6.29(a) Network for example 6.13

Solution :

Let the network contain (n – 2) branches in which the voltages and currents are denoted by V_k, I_k respectively for k = 3, ...n. with associated reference directions. Consider the network N with terminals 2, 2' shorted as shown in Fig. 6.29(b) and let the voltages and currents in this case be V'_k and I'_k for i = 3, ...n respectively.

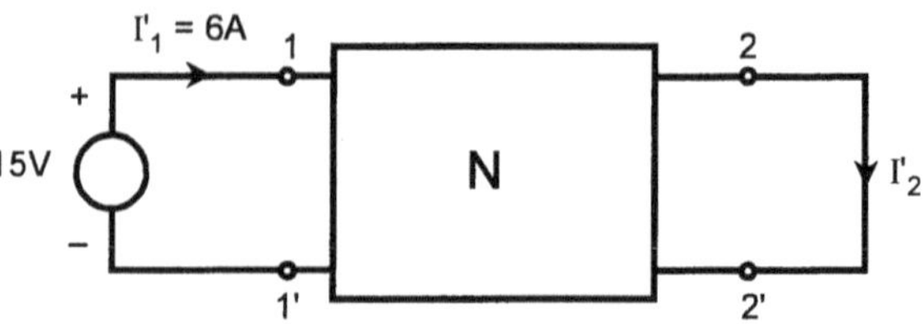

Fig. 6.29(b) Network N with terminals 2, 2' shorted

Using Tellegen's Theorem, taking voltages of network in Fig. 6.29(a) and currents in network of Fig. 6.29(b), since they obey Kirchhoff's laws, we have

$$V_1 \times (-I'_1) + V_2 \times I'_2 + V_3I'_3 + \ldots + V_nI'_n = 0 \qquad \ldots(6.49)$$

$$-90 + 5 \times I'_2 + V_3I'_3 + \ldots + V_nI'_n = 0$$

$$5I'_2 + V_3 I'_3 + \ldots + V_nI'_n = 90 \qquad \ldots(6.50)$$

Similarly using the voltages of network in Fig. 6.29(b) and currents of network in Fig. 6.29(a), we have

$$15(-5) + 0 \times I_2 + V'_3I_3 + \ldots + V'_nI_n = 0 \qquad \ldots(6.51)$$

or $$V'_3I_3 + \ldots + V'_nI_n = 75 \qquad \ldots(6.52)$$

But $$V_3 = I_3R_3, \ldots V_n = I_n \cdot R_n$$

and $$V'_3 = I'_3R_3, \ldots V'_n = I'_n R_n$$

where R_3, R_4, ... R_n are the resistances of the respective branches. Substituting these values in (6.50) and (6.52).

$$5I'_2 + I_3I'_3R_3 + \ldots + I_nI'_nR_n = 90 \qquad \ldots(6.53)$$

and $$I_3I'_3R_3 + \ldots + I_nI'_nR_n = 75 \qquad \ldots(6.54)$$

Subtracting eq. (6.54) from eq. (6.53) we have

$$5I'_2 = 15$$

or $$I'_2 = 3A$$

Thus the short circuit current at terminals 2, 2' is 3A. The open circuit voltage at terminals 2, 2' is given as 5V

$$\therefore \quad R_{Th} = \frac{V_{OC}}{I_{SC}} = \frac{5}{3}\ \Omega$$

∴ The Thevenin's equivalent at terminals 2, 2' of the network N is

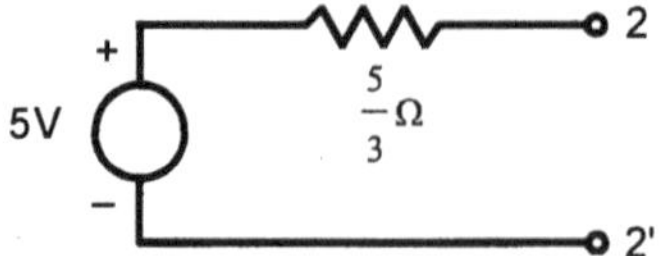

Fig. 6.29(c) Thevenin's equivalent of network in Fig 6.29(a)

6.10 MAXIMUM POWER TRANSFER THEOREM :

This is a very important and useful theorem. When a signal with certain power associated with it is applied to a load, the load should abstract maximum power available in the signal. For example when a loudspeaker is connected to an amplifier, the loudspeaker must take maximum available power from the amplifier to give better output. Thus the load to be connected to the source must be designed to take maximum power from the source. We call this as impedance matching. The optimum load to be connected to a source with some internal impedance is given by the Maximum Power Transfer Theorem.

6.10.1 When source has internal resistance and the load is purely resistive :

Consider a source with internal resistance R_g. Let a variable resistor R_L be connected across the source. Let us find out the value of R_L to obtain maximum power in the load.

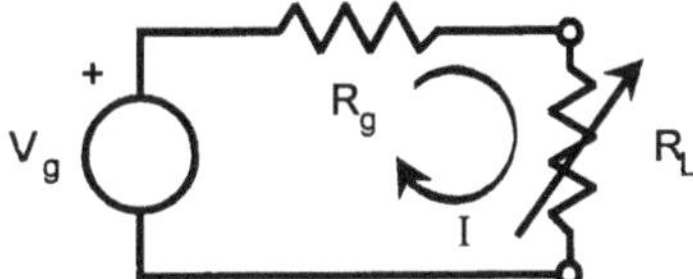

Fig. 6.30 Source with internal resistance R_g and load R_L

The power in the load R_L is given by

$$P = I^2 R_L = \left(\frac{V_g}{R_L + R_g}\right)^2 R_L \qquad .....(6.55)$$

Maximum power is obtained when

$$\frac{dP}{dR_L} = 0 \qquad .....(6.56)$$

Rewriting eq. (6.55) as

$$P = \frac{V_g^2}{\frac{1}{R_L}(R_L + R_g)^2} \qquad .....(6.57)$$

The power P is a maximum when the denominator of eq. (6.57) is a minimum.

Thus if $\quad D = R_L + \frac{R_g^2}{R_L} + 2R_g$

then $\quad \frac{dD}{dR_L} = 0 = 1 - \frac{R_g^2}{R_L^2}$

or $\quad R_L = R_g \qquad .....(6.58)$

Thus the power absorbed by the load is a maximum when the load resistance is equal to the source resistance. If instead of a single source with an internal resistance, we have a complex network with a number of sources and resistances connected to a load at terminals 1, 1', we can replace the network with its Thevenin's equivalent at the terminals 1, 1' and in this case the load resistance must be equal to the Thevenin's equivalent resistance at the terminals 1, 1' for maximum power transfer. The variation of power with load R_L is shown in Fig. 6.31.

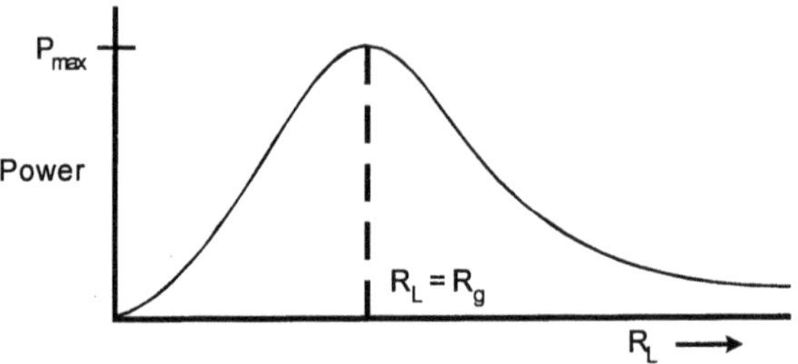

Fig. 6.31 Variation of power with load resistance

The efficiency of power transfer, however, is very poor because

$$\text{efficiency} = \frac{\text{power obserbed by the load}}{\text{power available in the source}} = \frac{I^2(R_L)}{I^2(R_L + R_g)}$$

$$= \frac{I^2.R_L}{2I^2R_L} = 50\%$$

Impedance matching is resorted to in electronic circuits where the signal powers available are very small and the endeavour would be to extract as much power from the sources as possible, rather than how efficiently it is drawn.

6.10.2 When source has an impedance and the load has variable impedance :

This is a more general case. Let the internal impedance of the source be

$$Z_g = R_g + jX_g$$

The reactance X_g may be positive or negative. Consider a load with variable resistance R_L and variable reactance X_L as shown in Fig. 6.32.

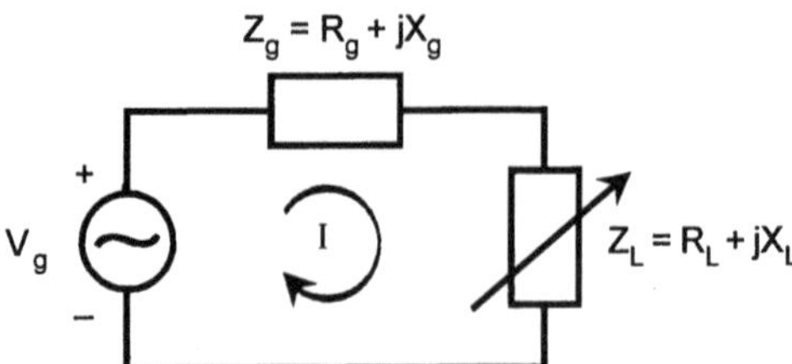

Fig. 6.32 Maximum power transfer theorem for a.c networks with impedances

The power absorbed by the load is given by

$$P = \frac{V_g^{\;2}}{(R_g + R_L)^2 + (X_g + X_L)^2} \times R_L \qquad .....(6.59)$$

Power can be made maximum by first changing the reactance X_L keeping R_L constant and then changing R_L to get further increase in power.

It is obvious that P will be a maximum when

$$X_L = -X_g \qquad .....(6.60)$$

This gives $$P = \frac{V_g^{\;2}}{(R_g + R_L)^2} \cdot R_L \qquad .....(6.61)$$

Now changing R_L to get maximum power, we have from section 6.10.1,

$$R_L = R_g \qquad .....(6.62)$$

Thus $$R_L + jX_L = R_g - jX_g$$

or $$Z_L = Z^*_g \qquad .....(6.63)$$

where (*) indicates complex conjugate.

Thus for a.c circuits the maximum power transfer theorem states that, *The maximum power is abstracted from the source when the load impedance is the complex conjugate of the source impedance.*

6.10.3 When source has impedance and the load is purely resistive :

Consider the circuit in Fig. 6.33.

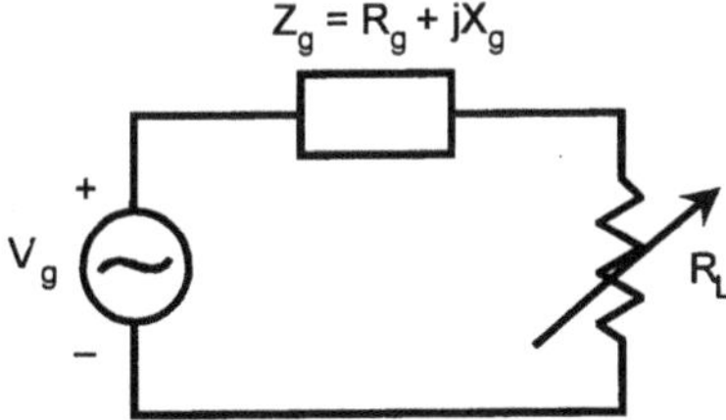

Fig. 6.33 Source having impedance and load purely resistive

The power absorbed by the load R_L is given by

$$P = \frac{V_g^{\;2}}{(R_g + R_L)^2 + X_g^{\;2}} \cdot R_L \qquad .....(6.64)$$

Define $$D = \frac{(R_g + R_L)^2}{R_L} + \frac{X_g^2}{R_L}$$

The power is a maximum when D is minimum

or when $$\frac{dD}{dR_L} = 0$$

$$\frac{-R_g^2}{R_L^2} + 1 - \frac{X_g^2}{R_L^2} = 0 \qquad .....(6.65)$$

$$R_g^2 + X_g^2 = R_L^2$$

or $$R_L = \sqrt{R_g^2 + X_g^2} \qquad .....(6.66)$$

Thus the power in the load is a maximum when the load resistance is equal to the magnitude of the source impedance.

or $$R_L = |Z_g| \qquad .....(6.67)$$

Example 6.14

What load resistance must be connected across the terminals 1, 1' of the network shown in Fig. 6.34(a) to obtain maximum power.

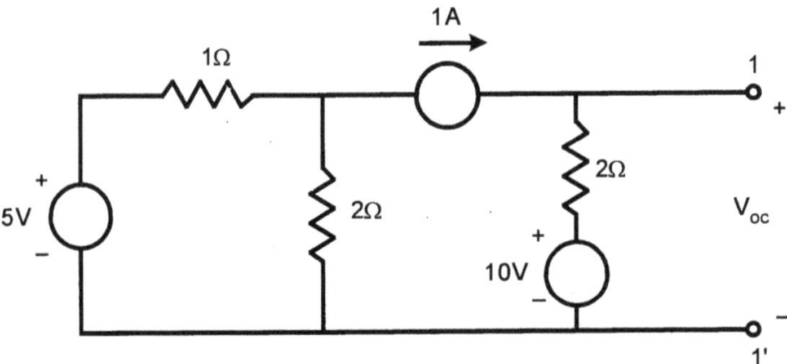

Fig. 6.34(a) Network for example 6.14

Solution :

Let us first find the Thevenin's equivalent of the network at 1, 1'. To obtain V_{OC}, since 1, 1' are open circuited the current in the 2Ω resistance in series with the 10V source is 1A.

$$\therefore \qquad V_{OC} = 10 + 2 \times 1 = 12V$$

To obtain R_{Th}, we open circuit the 1A current source and short circuit the 5V and 10V voltage sources to get the network in Fig. 6.34(b)

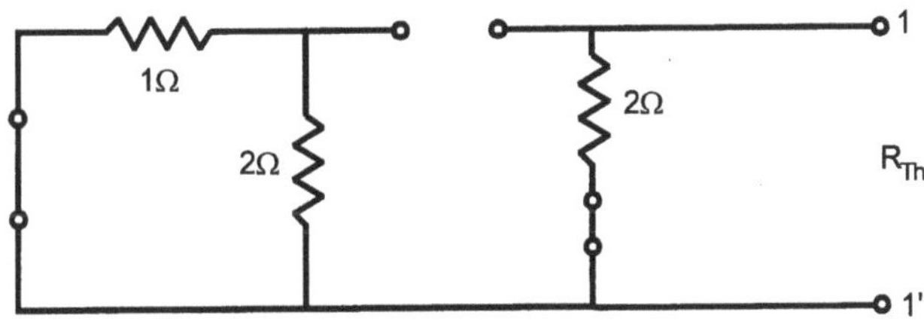

Fig. 6.34(b) Network for finding R_{Th}

It is obvious that $R_{Th} = 2\Omega$

$\therefore$ The Thevenin's equivalent at terminals 1, 1' of the network is

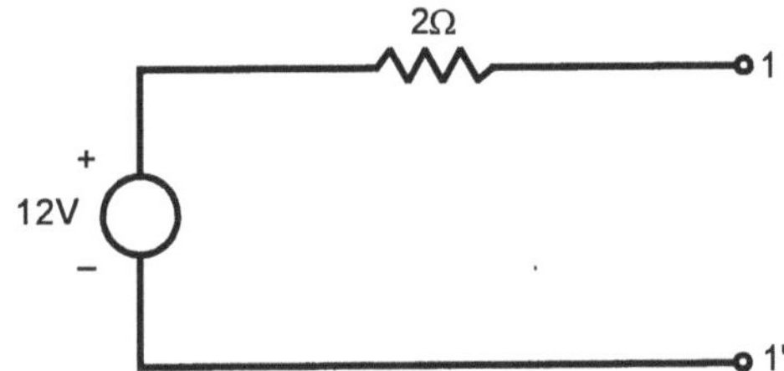

Fig. 6.34(c) Thevenin's equivalent

From maximum power transfer theorem, the load resistance to be connected is equal to $R_{Th} = 2\ \Omega$ and the maximum power absorbed by the load is

$$P_m = \left(\frac{12}{4}\right)^2 \times 2 = 18\ W$$

Example 6.15

Find the load impedance for maximum power transfer in the network of Fig. 6.35. If the load is purely resistive, what will be its value for maximum power transfer? Also find the maximum power taken by the load in both cases.

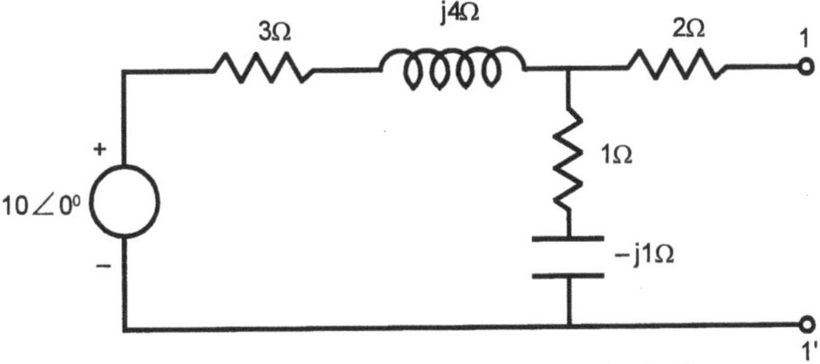

Fig. 6.35 Network for example 6.15

Solution :

Thevenin's equivalent is obtained first.

$$V_{OC} = \frac{10}{4+j3} \times (1-j1) = \frac{10}{5\angle 36.8} \times \sqrt{2}\angle -45^0 = 2.828 \angle -81.8$$

$$Z_{Th} = 2 + \frac{(3+j4)(1-j1)}{4+j3} = 3.31 \angle -11.78\ \Omega$$

(i) When the load is an impedance, the maximum power is transferred to it when

$$Z_L = Z^*_{Th} = 3.31 \angle 11.78\ \Omega = 3.24 + j0.676\ \Omega$$

The power taken by the load is

$$P = \left(\frac{2.828}{3.24+3.24}\right)^2 (3.24) = 0.617 \text{ watts}$$

(ii) When the load is pure resistance, the maximum power is transferred to it when

$$R_L = |Z_{Th}| = 3.31\Omega$$

The power taken by it is

$$P = \left(\frac{2.828}{|6.55 - j0.676|}\right)^2 \times 3.31 = 0.61 \text{ watts}$$

Example 6.16

Obtain the Thevenin's equivalent at the terminals a, b. What will be the maximum power supplied by this circuit to a load resistance ?

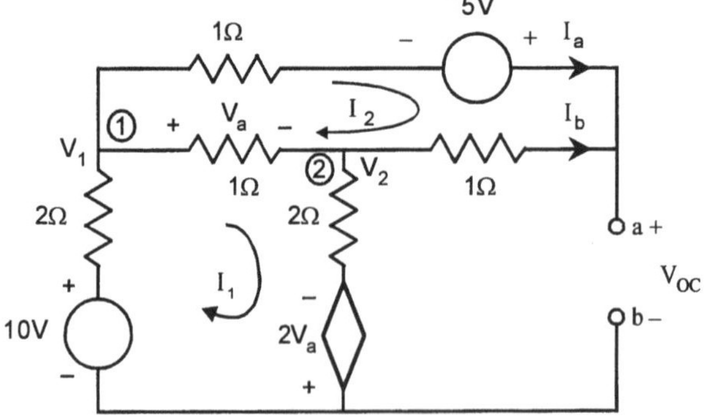

Fig. 6.36(a) Network for example 6.16

Solution :

Open circuit voltage V_{OC}, at the terminals a,b.

Using loop current method, writing loop equations, we have,

Loop I : $2I_1 + 1(I_1 - I_2) + 2I_1 - 2V_a = 10$

But $V_a = (I_1 - I_2) \cdot 1$

$\therefore$ $5I_1 - I_2 - 2(I_1 - I_2) = 10$

$3I_1 + I_2 = 10$

Loop II : $2I_2 + (I_2 - I_1) = 5$

$-I_1 + 3I_2 = 5$

Solving the equations for I_1 and I_2,

$$I_1 = \frac{\begin{bmatrix} 10 & 1 \\ 5 & 3 \end{bmatrix}}{\begin{bmatrix} 3 & 1 \\ -1 & 3 \end{bmatrix}} = \frac{30-5}{9+1} = 2.5 \text{ A}$$

$$I_2 = \frac{\begin{bmatrix} 3 & 10 \\ -1 & 5 \end{bmatrix}}{10} = \frac{15+10}{10} = 2.5 \text{ A}$$

The Thevenin's equivalent voltage

$$V_{OC} = -2V_a + 2I_1 + I_2 \cdot 1$$
$$= -2(I_1 - I_2) + 2I_1 + I_2$$
$$= 3I_2$$
$$= 3 \times 2.5 = 7.5 \text{ V}$$

Short circuit current, I_{SC}.

Shorting the terminals a,b and writing node voltage equations, we have

Node 1 : $$\frac{V_1 - 10}{2} + \frac{V_1 - V_2}{1} + \frac{V_1 + 5}{1} = 0$$

$$V_1[0.5 + 1 + 1] - V_2 = 0$$
$$5V_1 - 2V_2 = 0$$

Node 2 : $$\frac{V_2 - V_1}{1} + \frac{V_2 + 2V_a}{2} + \frac{V_2}{1} = 0$$

But $$V_a = V_1 - V_2$$

$$-V_1 + V_2 (1 + 0.5 + 1) + V_1 - V_2 = 0$$

$$1.5\ V_2 = 0$$

$$V_2 = 0$$

∴ $$V_1 = 0$$

Since $$V_2 = 0,\ I_b = 0$$

Since $$V_1 = 0,\ I_a = \frac{5}{1} = 5\text{ A}$$

∴ $$I_{SC} = I_a + I_b = 5\text{ A}$$

$$R_{Th} = \frac{V_{OC}}{I_{SC}} = \frac{7.5}{5} = 1.5\ \Omega$$

The Thevenin's equivalent circuit is shown in Fig. 6.36(b).

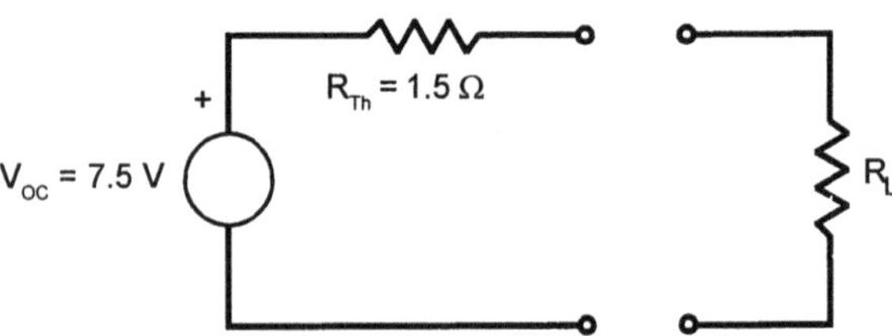

Fig. 6.36(b) Thevenin's equivalent circuit for the network of example 6.16

Maximum Power is transferred when $R_L = R_{Th} = 1.5\ \Omega$

Maximum Power supplied to the load $= \left(\frac{V_{OC}}{3}\right)^2 \times R_L$

$$= \left(\frac{7.5}{3}\right)^2 \times 1.5$$

$$= 9.375\text{ W}$$

Problems

6.1 Using Superposition theorem, calculate the current i_s in the network shown in Fig. P.6.1.

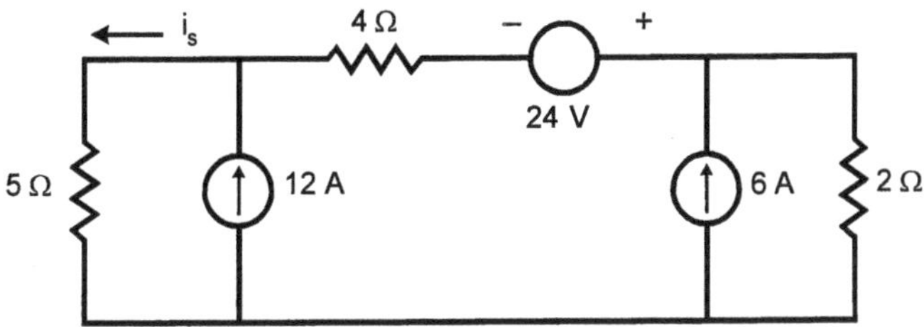

Fig. P. 6.1

6.2 Calculate the current i_y in the network of Fig. P.6.2 using Superposition theorem.

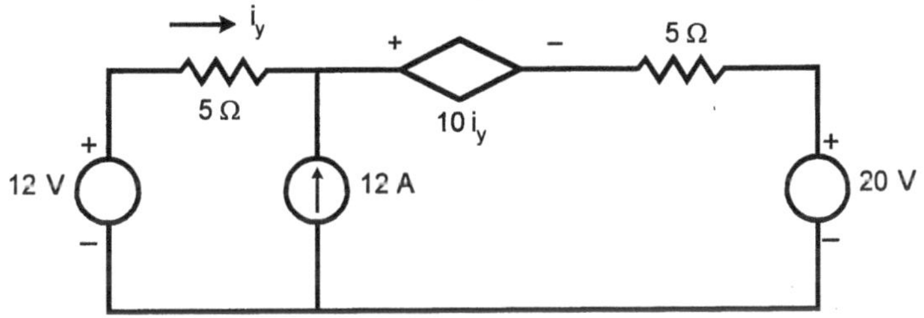

Fig. P. 6.2

6.3 Use Superposition theorem to calculate the currents i_a, i_b and i_c in the network of Fig. P.6.3.

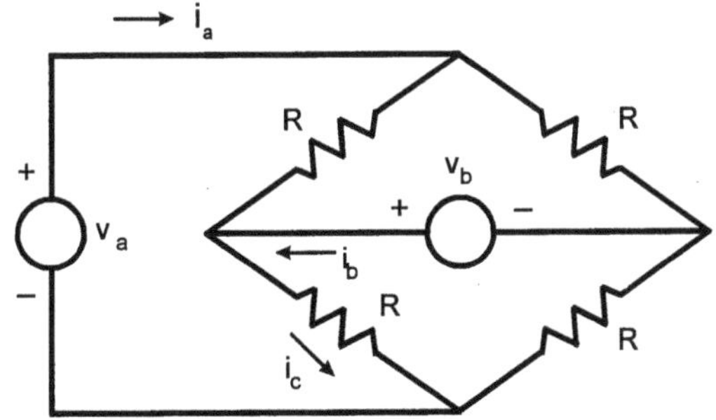

Fig. P. 6.3

6.4 Calculate the current in Z_l in the network of Fig. P.6.4 using Superposition theorem.

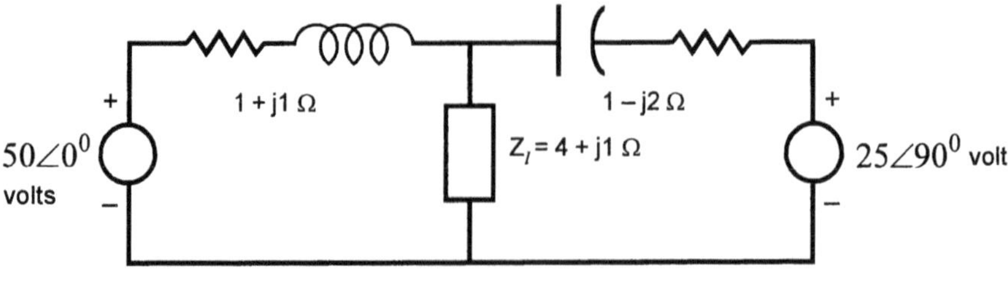

Fig. P. 6.4

6.5 Find the voltages across the two current sources in the network of Fig. P.6.5. using Superposition theorem.

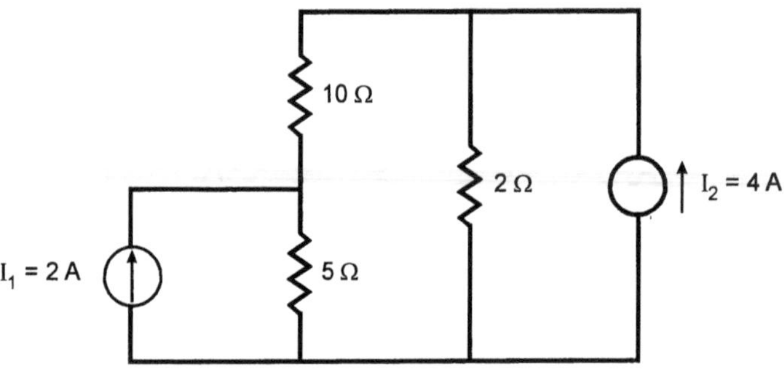

Fig. P. 6.5

6.6 Determine the Thevenin's equivalent circuit with respect to terminals a and b for each of the networks of Fig. P. 6.6.

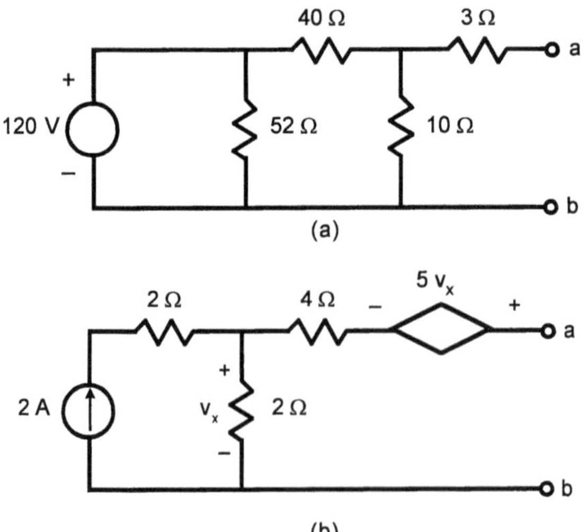

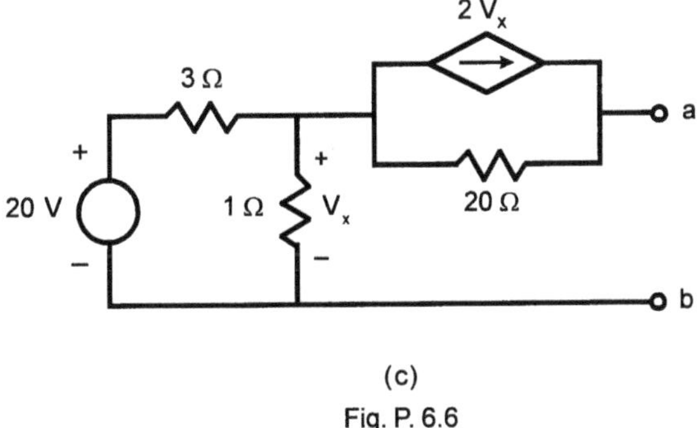

(c)

Fig. P. 6.6

6.7 Construct the Thevenin's equivalent circuit with respect to terminals a and b for the network of Fig. P.6.7.

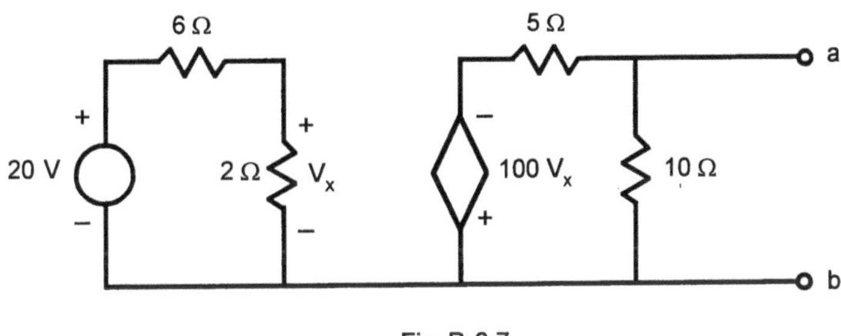

Fig. P. 6.7

6.8 Obtain the Thevenin's equivalent network at the terminals a and b of the network given in Fig. P.6.8. Hence determine the values of R_L and X_c which result in maximum power transfer to the load connected to terminals a and b.

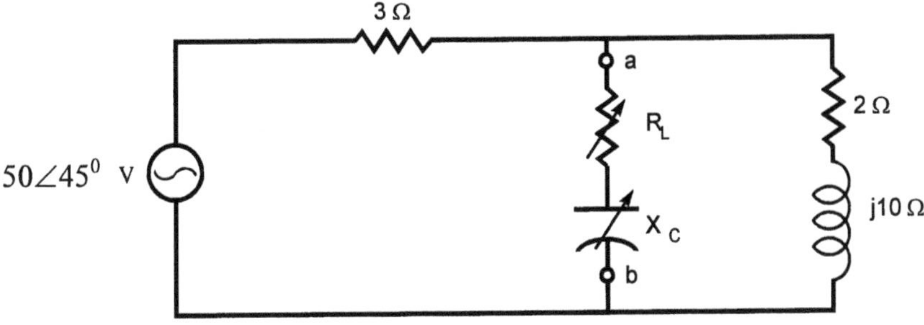

Fig. P. 6.8

6.9 Determine the change in the current in the impedance Z = (3 + j 4) Ω in the circuit of Fig. P.6.9 when the resistance of 5 Ω is changed to 6 Ω, using compensation theorem.

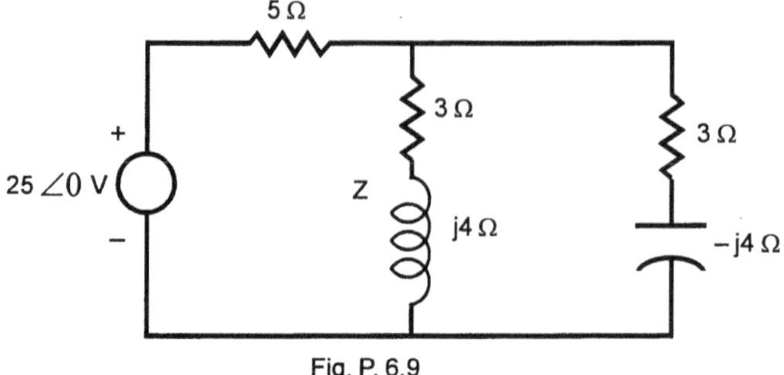

Fig. P. 6.9

6.10 In the network of Fig. P.6.4, calculate the voltage across the Z_l using Millman's theorem.

6.11 In the network shown in Fig. P.6.10, the load consists of fixed capacitive reactance of 15 Ω and a variable resistance R_1. Determine

(a) The value of R_1 for which the power transferred is a maximum

(b) The value of maximum power.

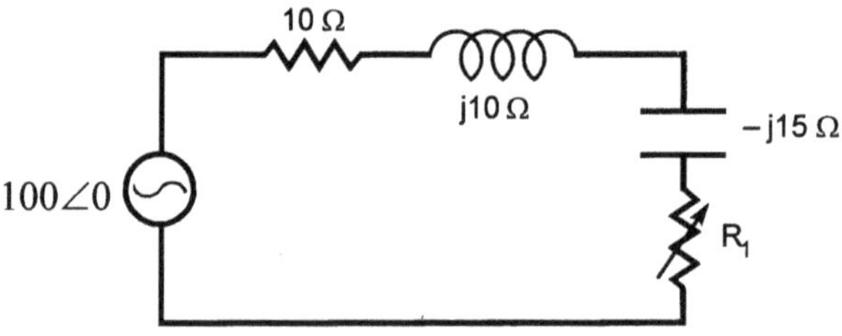

Fig. P. 6.10

6.12 Using Thevenin's theorem, find the current through the galvanometer if AB = 3 Ω, BC = 7 Ω,CD = 3 Ω, and DA = 12 Ω, in a Wheatstone bridge, if the galvanometer possess a resistance of 50 Ω and is placed across BD while the supply of 100 Volts DC is connected across A and C. The source resistance is 4 Ω.

6.13 Construct a Norton equivalent for the network to the left to the terminals a and b in Fig. P.6.11.

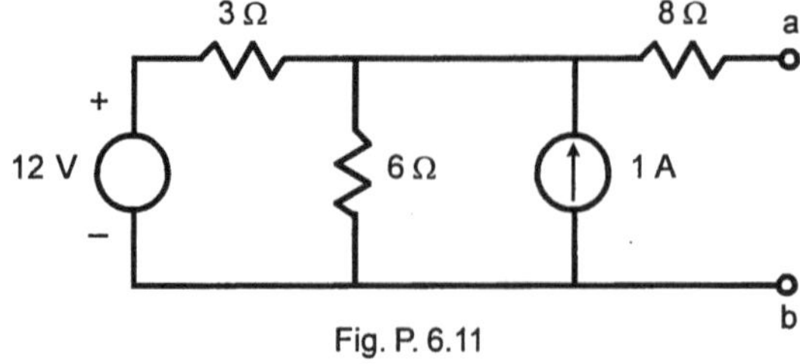

Fig. P. 6.11

6.14 Show that reciprocity holds for the voltage V and current I in the networks shown in Fig. P.6.12(a) and P.6.12(b).

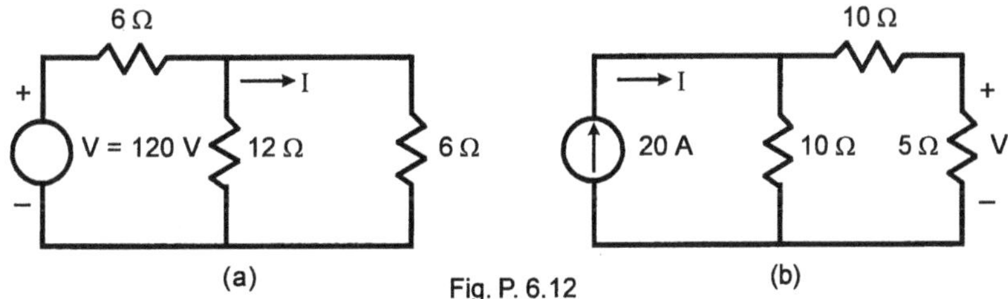

Fig. P. 6.12

6.15 A two terminal RLC network consisting of sinusoidal sources of the same frequency, is connected to a variable load. The load consumes a maximum power of 80 watts when its impedance is 20 – j 15 Ω. Set up the Thevenin's and Norton's equivalent for the two terminal network.

6.16 Use Thevenin's theorem to find the power in the 10 Ω resistor connected across the terminals a, b shown in Fig. P.6.13.

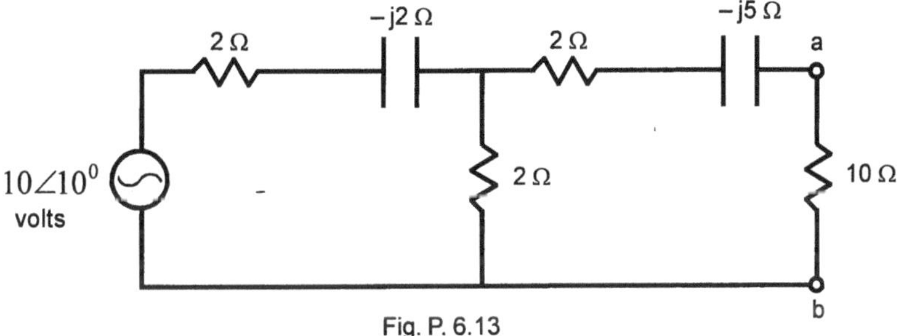

Fig. P. 6.13

6.17 Determine the current in the 1 Ω resistor across A, B of the network shown in Fig. P.6.14, using

(a) Superposition Theorem

(b) Thevenin's Theorem

(c) Norton's Theorem

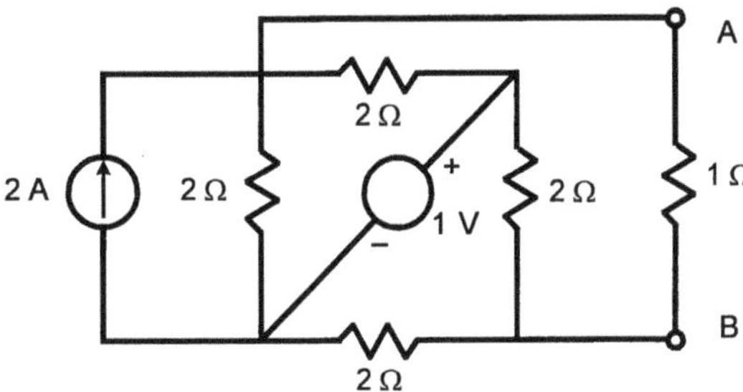

Fig. P. 6.14

6.18 Verify Reciprocity theorem for the circuit of Fig. P.6.15.

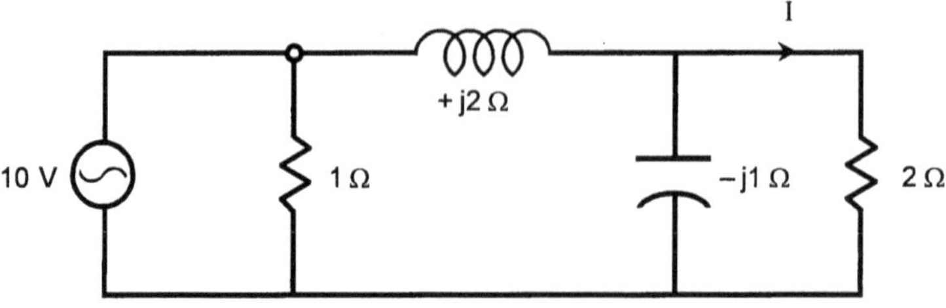

Fig. P. 6.15

6.19 Apply Millman's theorem and determine the currents in the three phase circuit of Fig. P.6.16.

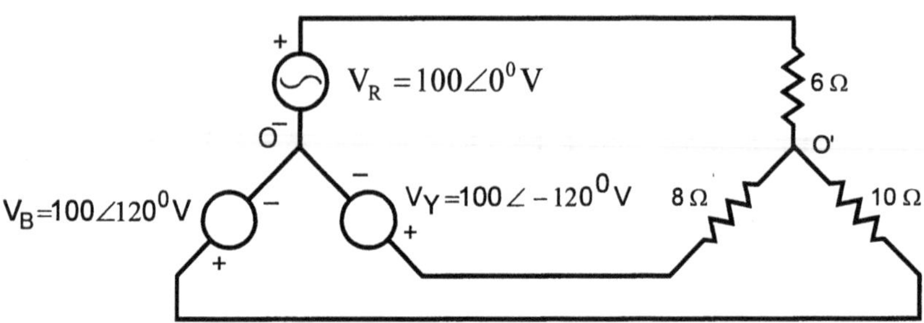

Fig. P. 6.16

6.20 Using Reciprocity theorem and Superposition theorem simultaneously, obtain the voltage across the resistance R of the network shown in Fig. P. 6.17 if the 10 A current source acting alone produces 2V across the resistance R.

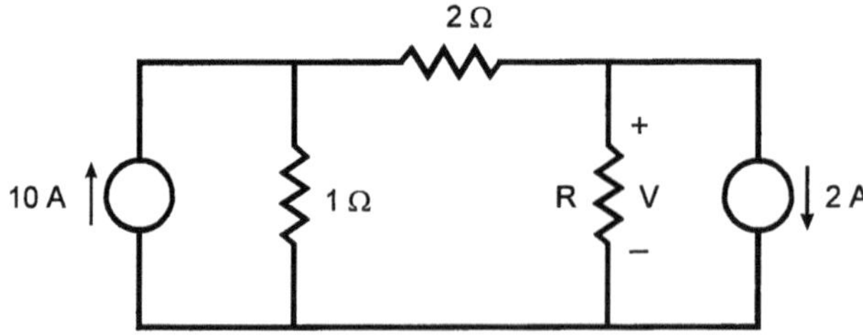

Fig. P. 6.17

6.21 Two networks are given in Fig. P. 6.18 (a) and (b). Verify Tellegen's theorem for the networks in (a) and (b). Take the voltages of branches in network (a) and corresponding branch currents in network (b).

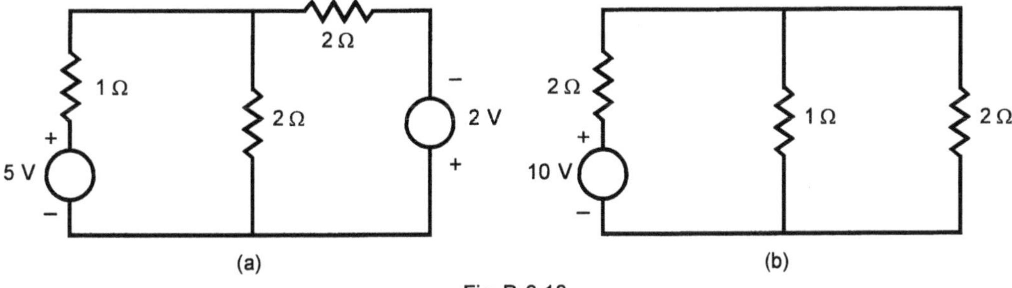

Fig. P. 6.18

Show that the sum of the products of branch voltages of network (a) and corresponding branch currents of network (b) is also equal to zero. Similarly show that the sum of the products of branch currents in network (a) and corresponding branch voltages in network (b) is also equal to zero.

6.22 The network N in Fig. P. 6.19 has a DC voltage source. The current I_x with different combinations of the sources is shown in Table P. 6.1

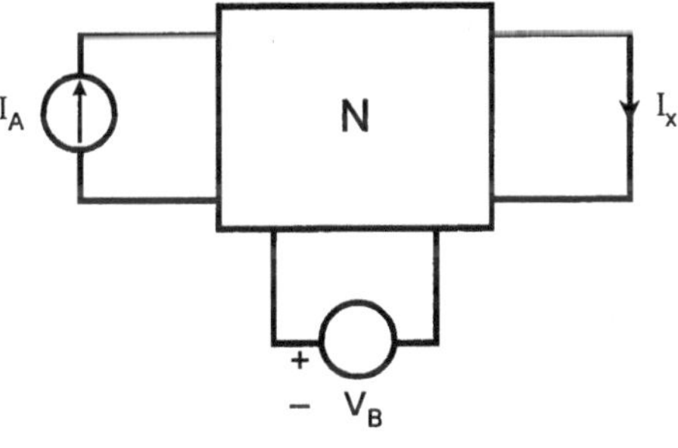

Fig. P. 6.19

Table P. 6.1

I_A	V_B	I_X
0 A	2V	1 A
5 A	0V	−2 A
5 A	2V	−4 A

If $I_A = -2$ A and $V_B = 10$ V what is the value of I_x.

7

DIFFERENTIAL EQUATIONS AND INITIAL CONDITIONS IN RLC NETWORKS

7.1 INTRODUCTION :

For purely resistive networks, application of the basic laws of Kirchhoff yield algebraic equations, where as, for RLC networks they yield differential equations, since the volt ampere equations for the energy storage elements contain derivatives and integrals of current or voltage. In order to get the response of an RLC network, one has to solve these differential equations. The solution of differential equations, unlike the solution of resistive networks, is a function of time and not a constant. More over, the solutions involve arbitrary constants which are to be evaluated by knowing the initial energy storage in the elements L and C of the network. The solution of a multiloop or multinode network yields simultaneous differential equations and the solution of such equations is much more involved. In this chapter, we will first apply Kirchhoff's laws to obtain the differential equations of networks containing energy storage elements. Then the initial conditions in the network will be evaluated so that the constants in the solution of these equations can be obtained.

7.2 FORMULATION OF NETWORK EQUATIONS :

The network equations can be formulated for networks containing RLC elements, in the same way as for resistive networks, using Kirchhoff's laws. Let us recall the volt – ampere relations of the elements R, L and C, namely,

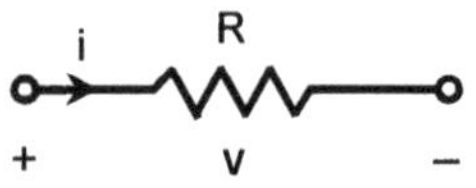

$$v = i\,R \qquad i = \frac{v}{R} \qquad .....(7.1a)$$

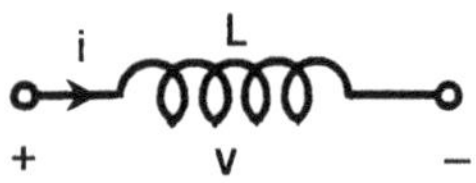

$$v = L\frac{di}{dt} \qquad i = \frac{1}{L}\int_{-\infty}^{t} v\,dt \qquad .....(7.1b)$$

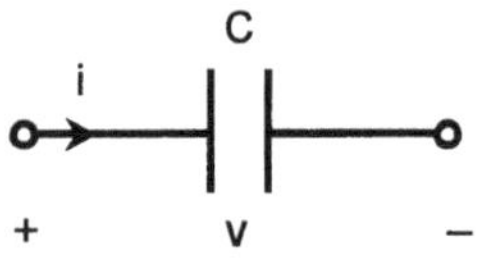

$$v = \frac{1}{C}\int_{-\infty}^{t} i\,dt \qquad i = C\frac{dv}{dt} \qquad .....(7.1c)$$

Using the equations (7.1), the equations for any network can be formulated based on mesh current method or node voltage method which were discussed in chapter 2. Some examples illustrate the two methods of obtaining the differential equations for the given network.

Example 7.1

Write the loop equation for the network of Fig. 7.1.

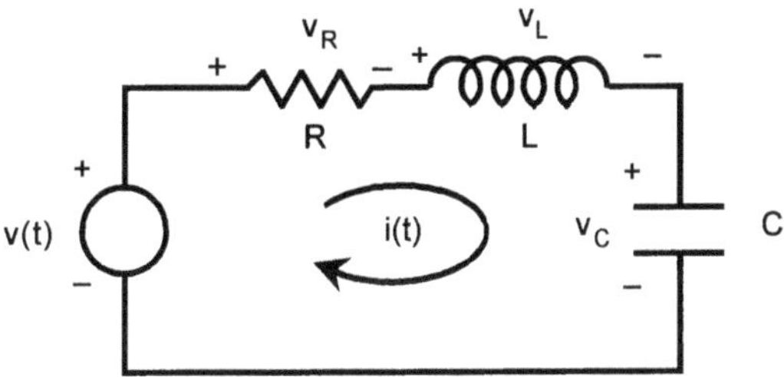

Fig. 7.1 Network for example 7.1

Solution :

Applying Kirchhoff's voltage law around the loop, with loop current i (t), We get

$$v_R(t) + v_L(t) + v_C(t) = v(t)$$

$$i(t)R + L\frac{di(t)}{dt} + \frac{1}{C}\int_{-\infty}^{t} i(t)dt = v(t) \qquad .....(7.2)$$

This is an integro – differential equation which contains both differential and integral of the variable i (t). It is often more convenient to convert this equation into a differential equation , by differentiating eq. (7.2).

Thus
$$R\frac{di}{dt} + L\frac{d^2i}{dt^2} + \frac{1}{C}i(t) = \frac{dv(t)}{dt} \qquad .....(7.3)$$

This is a second order differential equation which can be solved to get the response i (t). The solution of this equation is discussed later in chapter 9.

Example 7.2

Write the loop equations for the network of Fig. 7.2.

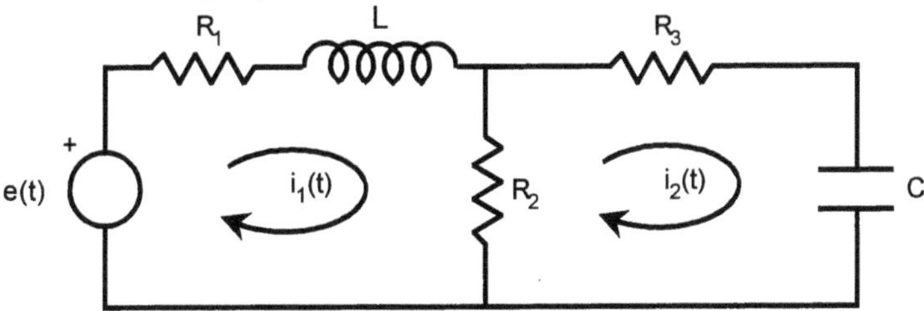

Fig. 7.2 Network for example 7.2

Solution :

The two mesh equation can be written directly as

$$R_1\, i_1\,(t) + L\frac{di_1(t)}{dt} + R_2\big(i_1(t) - i_2(t)\big) = e(t) \qquad .....(7.4)$$

$$R_3\, i_2\,(t) + \frac{1}{C}\int_{-\infty}^{t} i_2(t)dt + R_2\big(i_2(t) - i_1(t)\big) = 0 \qquad .....(7.5)$$

These equations can be converted into differential equations in the same way as in example 7.1. Note that the equations are simultaneous differential equations which can be solved to obtain the responses $i_1(t)$ or $i_2(t)$.

Example 7.3

Obtain the node voltage equations for the network in Fig. 7.3.

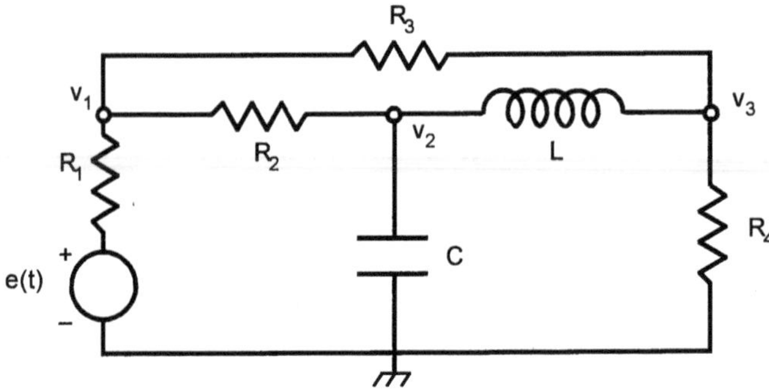

Fig. 7.3 Network for example 7.3

Solution :

Identifying the 3 node voltages with respect to the datum node and writing the node voltage equations, we get

$$\frac{v_1(t) - e(t)}{R_1} + \frac{v_1(t) - v_2(t)}{R_2} + \frac{v_1(t) - v_3(t)}{R_3} = 0 \qquad .....(7.6)$$

$$\frac{v_2(t) - v_1(t)}{R_2} + C\frac{dv_2}{dt} + \frac{1}{L}\int_{-\infty}^{t}[v_2(t) - v_3(t)]dt = 0 \qquad .....(7.7)$$

$$\frac{1}{L}\int_{-\infty}^{t}[v_3(t) - v_2(t)]dt + \frac{v_3(t)}{R_4} + \frac{v_3(t) - v_1(t)}{R_3} = 0 \qquad .....(7.8)$$

In the next example we will write the equations for network involving coupled elements.

Example 7.4

Write the loop equations for the network in Fig. 7.4.

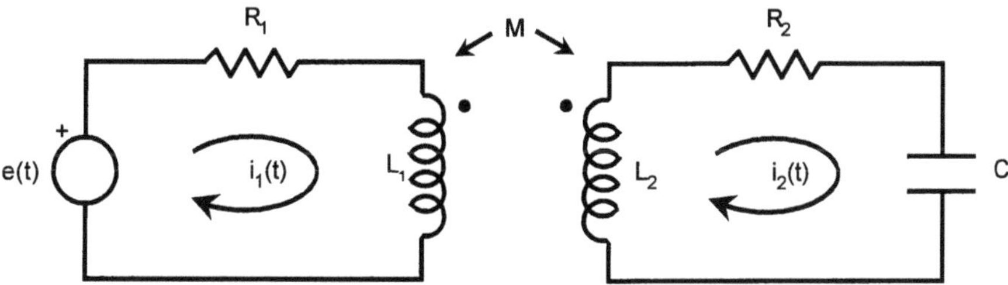

Fig. 7.4 Coupled network for example 7.4

Solution :

Recalling the dot convention introduced in chapter 1, the equation for loop 1 in Fig. 7.4 is

$$R_1 i_1(t) + L_1 \frac{di_1(t)}{dt} - M \frac{di_2}{dt} = e(t) \qquad .....(7.9)$$

For loop 2

$$R_2 i_2(t) + \frac{1}{C}\int_{-\infty}^{t} i_2(t)dt + L_2 \frac{di_2(t)}{dt} - M\frac{di_1}{dt} = 0 \qquad .....(7.10)$$

Example 7.5

A more complex network is considered in this example where 3 coils are mutually coupled. Write the loop equations for network in Fig. 7.5.

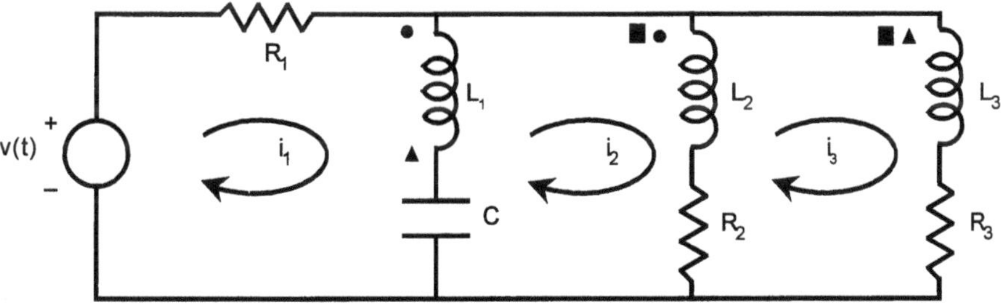

Fig. 7.5 A coupled circuit for example 7.5

Solution :

The mutual inductances between the three coils 1, 2 and 3 are designated as M_{12}, M_{13}, and M_{23} suppressing the time variable 't' in i_1 (t), i_2 (t) or i_3 (t) for convenience, we get :

$$R_1 i_1 + L_1 \frac{d(i_1 - i_2)}{dt} + M_{12}\frac{d(i_2 - i_3)}{dt} - M_{13}\frac{di_3}{dt} + \frac{1}{C}\int_{-\infty}^{t}(i_1 - i_2)dt = v \qquad .....(7.11)$$

$$L_1\frac{d(i_2-i_1)}{dt}-M_{12}\frac{d(i_2-i_3)}{dt}+M_{13}\frac{di_3}{dt}+L_2\frac{d(i_2-i_3)}{dt}$$

$$-M_{12}\frac{d(i_2-i_1)}{dt}+M_{23}\frac{di_3}{dt}+R_2(i_2-i_3)+\frac{1}{C}\int_{-\infty}^{t}(i_2-i_1)dt=0 \qquad .....(7.12)$$

$$R_2(i_3-i_2)+L_2\frac{d(i_3-i_2)}{dt}-M_{12}\frac{d(i_1-i_2)}{dt}-M_{13}\frac{di_3}{dt}$$

$$+L_3\frac{di_3}{dt}-M_{13}\frac{d(i_1-i_2)}{dt}-M_{23}\frac{d(i_3-i_2)}{dt}+R_3i_3=0 \qquad .....(7.13)$$

All the above examples illustrate the method of formulating the equations in networks involving R, L, C and M elements. Let us now consider an interesting property, called *duality*, in networks.

7.3 DUALITY :

The general mesh equations and node equations have certain similarities. This similarity is based on the property called *duality*. If mesh equations of one network are numerically equal to the node equations of second network, then the two networks are said to be duals of each other.

Consider the two networks is Fig. 7.6 (a) and (b). They are entirely different

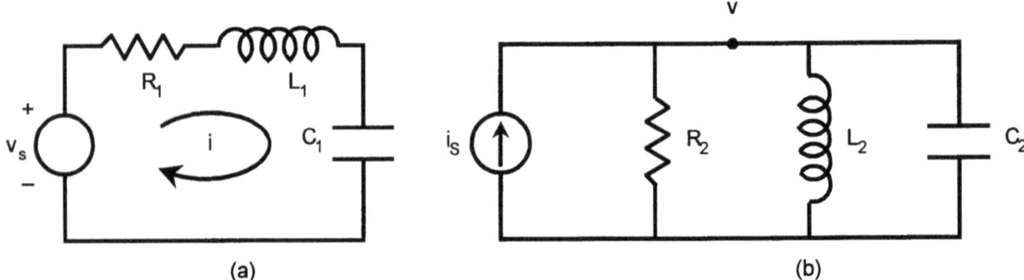

Fig. 7.6 Network to illustrate duality

networks in physical appearance. Writing down mesh equation for the network in Fig. 7.6(a), We get

$$R_1i+L_1\frac{di}{dt}+\frac{1}{C_1}\int idt=v_s \qquad .....(7.14)$$

Similarly writing down the node voltage equation for network in Fig. 7.6 (b) we get

$$\frac{v}{R_2} + C_2\frac{dv}{dt} + \frac{1}{L_2}\int v(t)dt = i_s(t) \qquad .....(7.15)$$

Consider the form of these equations. They look exactly similar in their mathematical operations. Only the symbols used for variables and coefficients in the equations are different. If in the eq. (7.14) we make

$$\begin{aligned} i &= v \\ R_1 &= \frac{1}{R_2} = G_2 \\ L_1 &= C_2 \\ C_1 &= L_2 \\ v_s &= i_s \end{aligned} \qquad .....(7.16)$$

we get equation (7.15). Thus if the solution of eq. (7.14) is known, the solution for eq. (7.15) can be easily obtained by making the changes as indicated in eq. (7.16).

The following quantities are called *dual quantities* in the eqs. (7.14) and (7.15).

$$Ri \longleftrightarrow \frac{V}{R} \text{ or } Gv$$

$$L\frac{di}{dt} \longleftrightarrow C\frac{dv}{dt}$$

$$\frac{1}{C}\int idt \longleftrightarrow \frac{1}{L}\int vdt$$

Table 7.1 gives the dual pairs. Duality is a unique one to one transformation and hence the dual of a dual network gives the original network. The property of duality is applicable to planar networks only. Non planar networks have three or more currents passing through a single element and the dual would require a single dual element between three or more nodes, which is not possible. Hence nonplanar networks have no duals.

Table 7.1

Original Network	Dual Network
i	v
v	i
R	G
L	C
C	L
series branches	parallel branches
parallel branches	series branches
short circuits	open circuits
mesh currents	node voltages
node voltages	mesh currents

For a given network, the dual can be obtained in the following manner :

1. Write down the mesh equations for the given network assuming all clockwise mesh currents.
2. Replace the quantities in these equations by the corresponding dual quantities as given in table 7.1.
3. The resulting node equations are interpreted as the equations of the dual network.

Let us consider an example to illustrate this method.

Example 7.6

Obtain the dual of the network of Fig. 7.7(a).

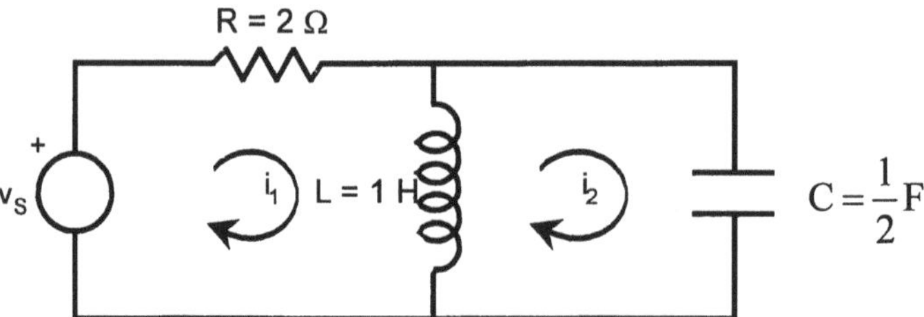

Fig. 7.7(a) Network for example 7.6

Solution :

Writing down the mesh equations,

$$2i_1 + \frac{d(i_1 - i_2)}{dt} = v_s(t) \qquad .....(7.17)$$

$$2\int i_2 dt + \frac{d(i_2 - i_1)}{dt} = 0 \qquad .....(7.18)$$

Transforming eqs. (7.17) and (7.18) by replacing the quantities by their duals

$$2v_1 + \frac{d(v_1 - v_2)}{dt} = i_s(t) \qquad .....(7.19)$$

$$2\int v_2 dt + \frac{d(v_2 - v_1)}{dt} = 0 \qquad ....(7.20)$$

Identifying two nodes, 1 and 2, and a datum node, 0, in the dual network, these equations can be interpreted as

a conductance of 2 S connected between node 1 or 0

a capacitance of 1 F connected between node 1 and 2

a current source i_S (t) numerically equal to v_s (t) connected between nodes 1 and 0

and an inductance of $\frac{1}{2}$ H connected between nodes 2 and 0.

The network can be drawn as shown in Fig. 7.7(b). Writing the node voltage equations for the network will result in eqs. (7.19) and (7.20).

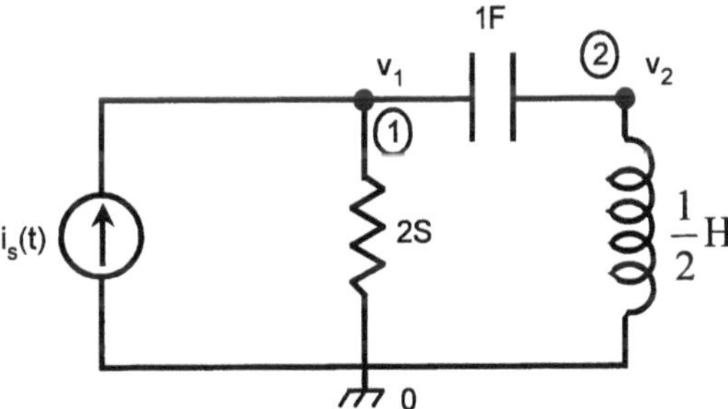

Fig. 7.7(b) Dual of network in Fig. 7.7(a)

The procedure discussed above is a round about way of obtaining a dual network. A geometrical method described below accomplishes the same result without the need to write equations. Each mesh current in the original network transforms to a node voltage in the dual . Hence a node can be located inside each of the meshes by placing a dot in the centre of the mesh. The elements common to two meshes appears in dual form between the corresponding nodes in the dual circuit. These elements are located geometrically by drawing lines between the two nodes in such a way that they cut each element that exists between the two corresponding meshes. Each line is then replaced by the dual of the element it cuts. This takes care of only those elements which are common to two meshes. The remaining elements are on the periphery of the original network. In the dual network they represent elements which are connected between the corresponding node and the ground.

Hence a ground node can be placed outside the network and each node in the dual is joined to this ground node by lines passing through these outside elements. These lines are then replaced by the dual elements they pass through.

To complete the process of constructing a dual, we have to determine the polarities of the sources in the dual network. The conventional direction for mesh currents is clockwise and conventional polarity of a node voltage is positive with respect to ground. We state the following rules for determining the polarity of voltage and current sources in dual networks.

Rule 1 : If a voltage source in the original network produces a clockwise current in the mesh, the corresponding dual element is a current source, whose direction is towards the node representing the corresponding mesh.

Rule 2 : If a current source in the original network produces a current in the clockwise direction in the mesh, the voltage source representing this current source in the dual network will have a polarity such that the node representing this mesh is positive.

The above geometrical procedure is now illustrated with an example.

Example 7.7

Obtain the dual of the network in Fig. 7.8 using geometrical method.

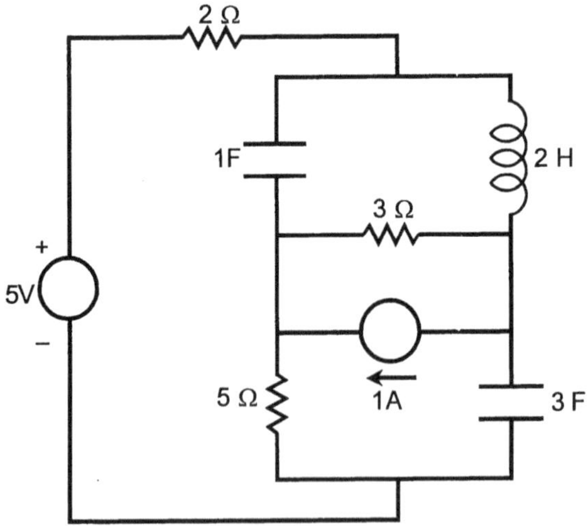

Fig. 7.8 Network for example 7.7

Solution :

The network given is redrawn in Fig. 7.9 (a) and dots are placed in each of the meshes to represent the nodes in the dual network. A ground node is placed outside the network. Each of these nodes is joined by lines drawn to cut the elements in going from one node to the other. Each element is cut only by one line. Also, each line cuts only one element. These line are then replaced by the duals of the element they cut. Lastly the polarities of sources are determined by using the rules given earlier. The dual network thus obtained is redrawn separately in Fig. 7.9(b).

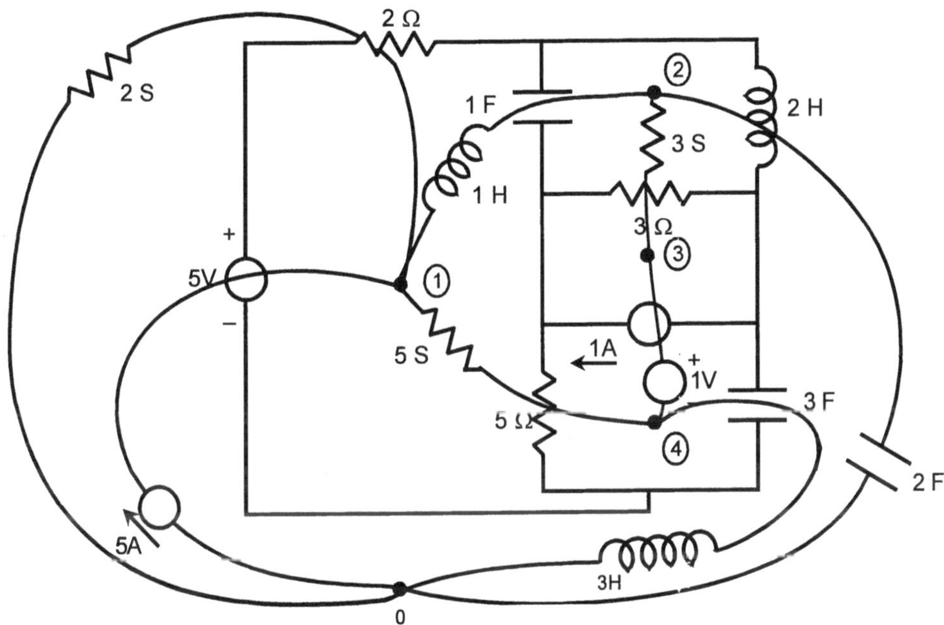

Fig. 7.9(a) Geometrical method of obtaining a dual

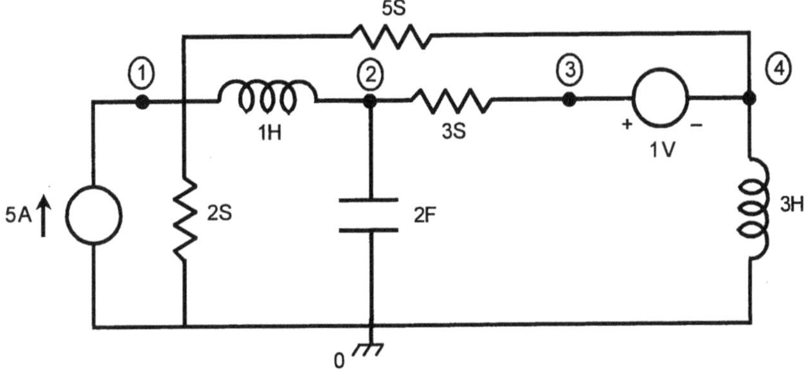

Fig. 7.9(b) The dual of network in Fig. 7.8

The nodes are marked in both the figures by encircled numbers and the order in which these nodes are taken is not important. The polarity of current source is taken to point towards node 1, as the voltage source in the original network produces a clockwise mesh current in mesh 1. The polarity of the voltage source in the dual network is chosen to make the node 3 positive since the 1 A current source in the original network produces a clockwise mesh current in mesh 3.

7.4 INITIAL CONDITIONS IN NETWORKS :

The solution of a differential equation involves arbitrary constants which have to be evaluated using the initial state or initial energy in the network. The condition or state of the network is given by the amounts of energy stored in various elements of the network. When we change the state of the network from one energy state to another, the change does not take place instantaneously. The changes occur slowly and the network passes through a transient state in which all the voltages and currents are affected. The change of state is usually effected by either closing a switch or opening a switch in a circuit. The conditions obtained in the network immediately after the closing or opening of a switch, are known as initial conditions in the network. These initial conditions are dependent on the conditions existing just before closing or opening of the switch. The instant at which the condition of the network is changed is usually taken as $t = 0$. To distinguish the instants of time just before and just after the change of the state, these instants are taken as $t = 0^-$ and $t = 0^+$ respectively.

The behaviour of the network during the transition from one state to another, is termed as natural behaviour of the network. This is often referred to as transient behaviour also. Independent of the way the network is excited, the network responds in its characteristic or natural way. A simple way of observing this natural behaviour is to give some initial energy to a capacitor or inductor and leave the network to itself. The system will decay from this energy state to zero energy state. This behaviour can be best understood by considering the example of a simple pendulum. We can give some initial energy to the pendulum by displacing it slightly and leaving it. The pendulum will oscillate and finally come to rest at its equilibrium position.

The v – i relationships of energy storage elements L and C contain integrals or derivatives. Integration represents 'memory' in a circuit. The integral is the summation of contributions received by the element. The element remembers these contributions and hence these elements are known as 'memory' elements. The capacitance, for example, remember the charges which have been stored in it, and the inductance, remembers the flux linkages stored in it. The response of the elements depends on these stored values at $t = 0$.

The derivative, on the other hand, represents prediction of the future. If we know the derivative of the current in an inductance, we can calculate the current at a future instant. Similarly if we know the derivative of voltage in a capacitance we can know the voltage at a future instant. But, this requires the values of current or voltage at the initial instant. Thus the initial conditions in the network play a very important role in determining the response of a network when energy storage elements are present in the network.

7.4.1 Behaviour of R, L and C at t = O :

The initial conditions in the network are the conditions obtained at $t = O^+$. i.e.., immediately after the switching operation at $t = O$. The condition immediately prior to switching operation is the condition at $t = O^-$. Let us now obtain the behaviour of the elements R, L and C at $t = O^+$.

Resistance :

The volt ampere relation for resistance is

$$v = Ri \qquad .....(7.21)$$

The voltage across the resistance is dependent only on the instantaneous value of the current and hence there is no change in the behaviour of resistance at $t = O^-$ and $t = O^+$.

Inductance :

The current in the inductance is given by

$$i(t) = \frac{1}{L}\int_{-\infty}^{t} v\,dt \qquad .....(7.22)$$

Eq. (7.22) can be written as

$$i(t) = \frac{1}{L}\int_{-\infty}^{O^-} v\,dt + \frac{1}{L}\int_{O^-}^{t} v\,dt \qquad .....(7.23)$$

The first term on right hand side of eq. (7.23) represents the current at $t = O^-$. Let this current be I_0.

At $t = O^+$ eq. (7.23) can be written as

$$i(O^+) = I_0 + \frac{1}{L}\int_{O^-}^{O^+} v\,dt \qquad .....(7.24)$$

If the voltage 'v' is finite, since the integral is from O^- to O^+, the second term in eq. (7.24) is equal to zero. Hence eq. (7.24) becomes,

$$i(O^+) = I_0 + O \qquad .....(7.25)$$

If $I_0 = O$, eq. (7.25) implies that the inductance behaves like an open circuit since the current is zero.

If $I_0 \neq O$, $i(O^+)$ is given as a sum of two currents, I_O and O, eq. (7.25) implies that the inductance can be represented by a current source of value I_O in parallel with an open circuit as shown in Fig. 7.10.

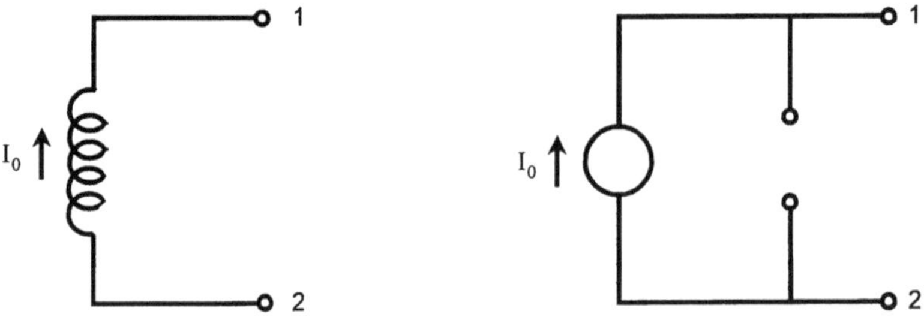

(a) Inductor with initial current I_0 (b) Its equivalent at $t = O^+$

Fig. 7.10. Behaviour of inductance at $t = O^+$

If the initial current $I_0 = O$, then the inductance behaves like an open circuit at $t = O^+$.

(a) Inductor with zero initial current I_0 (b) Its equivalent at $t = O^+$

Fig 7.11. Inductance with no initial current

Capacitance :

The voltage across a capacitance is given by

$$v(t) = \frac{1}{C}\int_{-\infty}^{t} i\,dt \qquad .....(7.26)$$

This equation can be written as

$$v(t) = \frac{1}{C}\int_{-\infty}^{O^-} i\,dt + \frac{1}{C}\int_{O^-}^{t} i\,dt \qquad .....(7.27)$$

If the voltage across the capacitor is V_0 at $t = O^-$, the first term on right side of eq. (7.27) is equal to V_0 and the voltage across the capacitance at $t = O^+$ is given by

$$v(O^+) = V_0 + \frac{1}{C}\int_{O^-}^{O^+} i\,dt \qquad .....(7.28)$$

If the current through the capacitance is finite, the second term in eq. (7.28) is zero. Thus

$$v(O^+) = V_0 + O \qquad .....(7.29)$$

Since $v(O^+)$, is given as a sum of two voltages, V_0 and O, eq. (7.29) can be represented as a series circuit consisting of a voltage source of value V_0 and a short circuit as shown in Fig. 7.12.

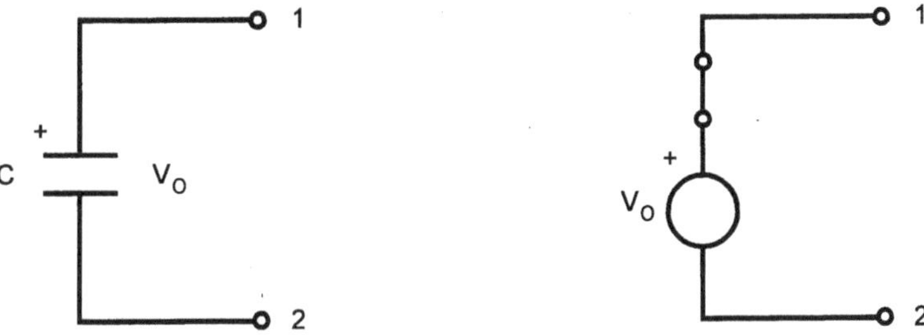

(a) A charged capacitor (b) Its equivalent at $t = O^+$

Fig. 7.12 A charged capacitor and its equivalent at $t = O^+$

If $V_0 = O$ then the behaviour of the capacitance at $t = O^+$ can be represented by a short circuit as shown in Fig. 7.13.

(a) Capacitor with zero initial voltage (b) Its equivalent at $t = O^+$.

Fig. 7.13 Uncharged capacitor and its equivalent at $t = O^+$

7.4.2 Behaviour of R, L and C at $t = \infty$:

If a circuit remains in a certain energy state for a long time (usually this time may be only a few seconds or for some networks only milliseconds) the currents and voltages in the network become constants for networks in which the excitations are either constant or decay to zero with time. Since the currents and voltages are constant, their derivatives are zero. Hence, the voltage across an inductance and current through a capacitance at t = ∞ . are given by

$$v_L (t) = L\frac{di}{dt} = O$$

and
$$i_C (t) = C\frac{dv}{dt} = O.$$

It is clear that the inductance behaves like a short circuit since the voltage across it is zero and the capacitor behaves like an open circuit since the current through it is zero at t = ∞ . The behaviour of the three elements at t = O^+ and t = ∞ is summarised in Fig. 7.14.

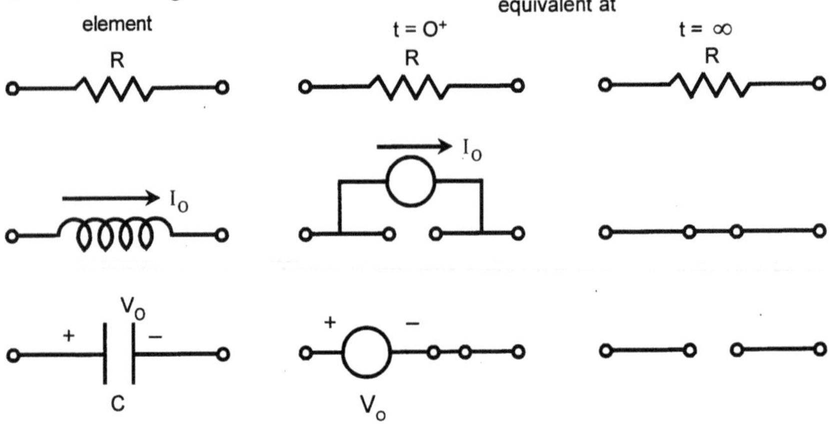

Fig. 7.14 Summary of behaviour of R, L and C at t = O^+ and t = ∞

The t = ∞ condition is very useful in determining the initial conditions in a network when a network is in a particular energy state for a long time and the state is changed by a switch, at t = O. In order to calculate the values of currents and voltages at t = O^+, we require the currents and voltages at t = O^- for the altered network. t = O^- for altered network is same as t = ∞ for the original network before the operation of the switch.

It is clear form the summary presented in Fig 7.14 that the currents and voltages in the energy storing elements L and C are important in determining the initial conditions in a network. The inductance opposes any change in the current and therefore it tries to maintain the current at t = O^+, the same as at t = O^-. Similarly, the capacitor opposes any sudden change in the voltage across it and hence tries to maintain the same voltage at t = O^+, as the voltage at t = O^-. In order to calculate the currents and voltages at t = O^+ and t = ∞ , the elements R, L and C are replaced by their equivalents at t = O^+ and t = ∞ respectively. The network then contains only d.c. sources and resistances. The methods discussed in chapter 2 can now be applied to determine the initial conditions or final conditions. This procedure is illustrated by a few examples.

Example 7.8

Compute the currents and voltages in the inductors and capacitors at $t = O^+$ and $t = \infty$ when the switch is closed at $t = O$ in Fig. 7.15(a).

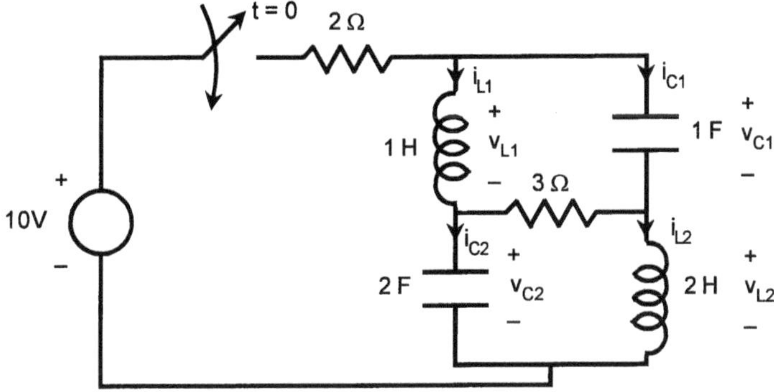

Fig. 7.15(a) Network for example 7.8

Solution :

At $t = O^+$ the equivalent circuit is given in Fig. 7.15(b), assuming no initial energies in inductors and capacitors.

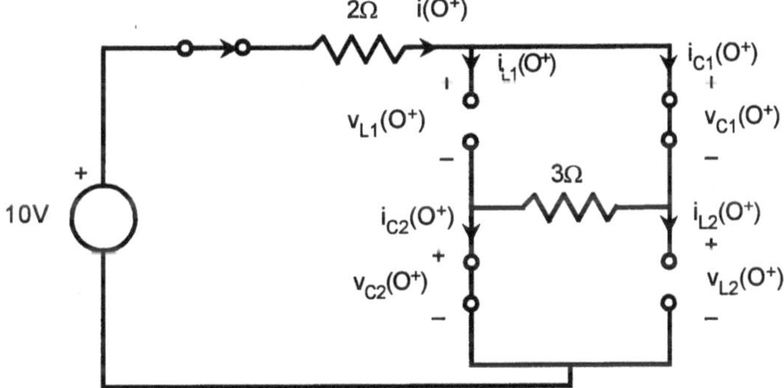

Fig. 7.15(b) Network at $t = O^+$

From Fig. 7.15(b)

$$i\,(O^+) = \frac{10}{5} = 2\text{ A}$$

$$i_{C1}\,(O^+) = i_{C2}\,(O^+) = i\,(O^+) = 2\text{A}$$

$$i_{L1}\,(O^+) = i_{L2}\,(O^+) = 0$$

$$v_{C1}\,(O^+) = v_{C2}\,(O^+) = 0$$

The voltages across the inductors are same as the voltages across the 3 Ω resistor.

$$v_{L1}(O^+) = v_{L2}(O^+) = 3i(O^+) = 6\text{V}$$

Similarly at $t = \infty$, the equivalent circuit of the network in Fig. 7.15(a) is given in Fig. 7.15(c).

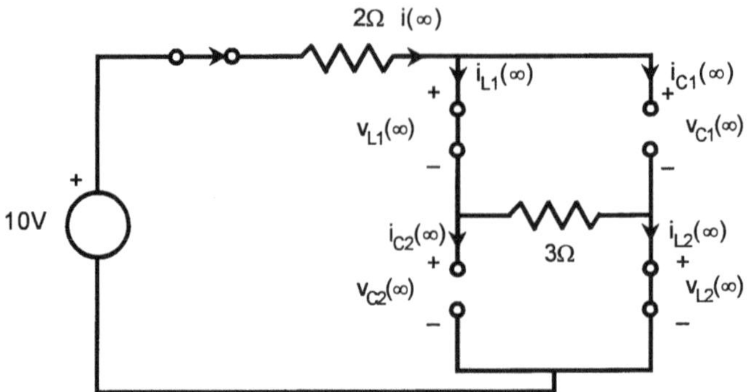

Fig. 7.15(c) Network at t = ∞

From Fig. 7.15 (c), it is easy to calculate the various currents and voltages.

$$i(\infty) = i_{L1}(\infty) = i_{L2}(\infty) = \frac{10}{5} = 2A$$

$$v_{L1}(\infty) = v_{L2}(\infty) = O$$

$$v_{C1}(\infty) = v_{C2}(\infty) = 3i(\infty) = 6V$$

Before we take up more examples, it is pertinent to mention that we can calculate not only the initial values of the variables, namely, currents and voltages, but also their derivatives. In solving higher order differential equations, we need the initial values of the variables and also their derivatives. If it is an nth order differential equation we need 'n' initial conditions, namely, the variable and its n – 1 derivatives at $t = O^+$.

In the following examples, we illustrate the procedure to obtain the initial conditions in a given network.

Example 7.9

Find $i(O^+)$, $\frac{di}{dt}(O^+)$ and $\frac{d^2i(O^+)}{dt^2}$ in the network shown in Fig. 7.16 (a).

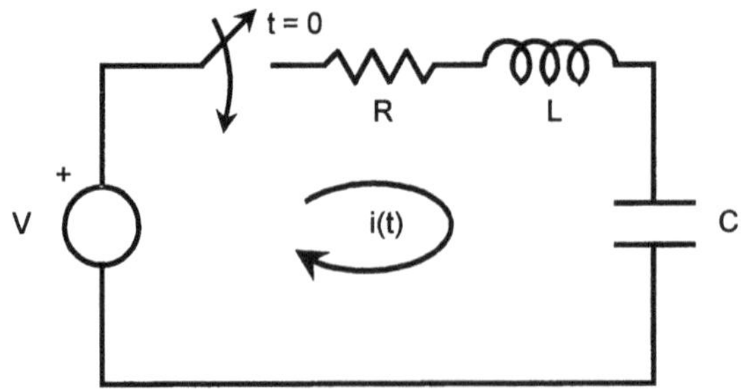

Fig. 7.16(a) Network for example 7.9

Solution :

To find i (O^+) we simply write the equivalent of the network at $t = O^+$.

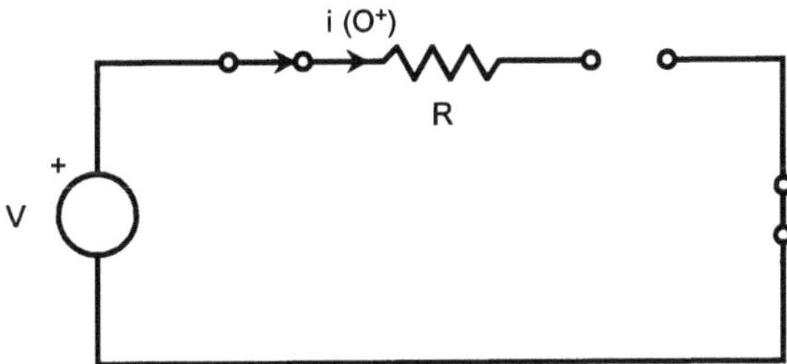

Fig. 7.16(b) Network at $t = O^+$

From Fig. 7.16 (b) it is clear that

$$i(O^+) = O$$

To obtain the derivatives at $t = O^+$ we write the loop equation for network in Fig. 7.16 (a) which is valid for any time 't' after closing the switch at t = O.

$$Ri + L\frac{di}{dt} + \frac{1}{C}\int_{-\infty}^{t} idt = V \quad .....(7.30)$$

In particular, this equation is also valid at $t = O^+$. Thus

$$Ri(O^+) + L\frac{di}{dt}(O^+) + \frac{1}{C}\int_{-\infty}^{O^+} idt = V \quad .....(7.31)$$

In eq. (7.31), $i(O^+) = O$ and $\frac{1}{C}\int_{-\infty}^{O^+} idt$ is the voltage across the capacitor at $t = O^+$.

Since the capacitor is a short circuit at $t = O^+$ the voltage across it is zero. Thus

$$L\frac{di}{dt}\left(O^+\right) = V$$

$$\frac{di}{dt}\left(O^+\right) = \frac{V}{L}$$

Differentiating eq. (7.30) (Eq. (7.30) is a general equation valid for any t, and eq. (7.31) is valid at only $t = O^+$. Hence differentiation can be performed only on eq. (7.30) and not on eq. (7.31).

$$R\frac{di}{dt}+L\frac{d^2i}{dt^2}+\frac{i}{C}=O \qquad .....(7.32)$$

This is also valid at t = O⁺. Hence

$$R\frac{di}{dt}(O^+)+L\frac{d^2i(O^+)}{dt^2}+\frac{i(O^+)}{C}=O \qquad .....(7.33)$$

Solving for $\frac{d^2i}{dt^2}(O^+)$, we get

$$\frac{d^2i}{dt^2}(O^+)=-\frac{1}{L}\left[\frac{i(O^+)}{C}+R\frac{di}{dt}(O^+)\right]$$

$$=-\frac{R}{L}\cdot\frac{V}{L}=\frac{-VR}{L^2}$$

Thus, we have

$$i\,(O+)=O,\ \frac{di}{dt}\left(O^+\right)=\frac{V}{L} \quad \text{and} \quad \frac{d^2i}{dt^2}(O^+)=\frac{-VR}{L^2}$$

Example 7.10

The switch in Fig. 7.17(a) is changed from position 1 to 2 at t = O.

Solve for $i, \frac{di}{dt}, \frac{d^2i}{dt^2}$ at $t = O^+$

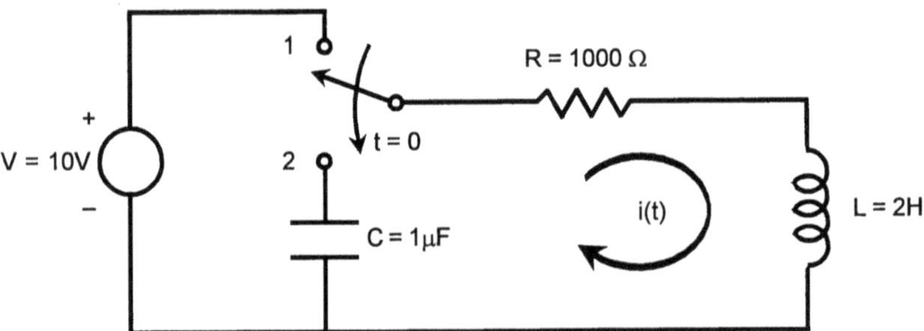

Fig. 7.17(a) Network for example 7.10

Solution :

The switch in Fig. 7.17(a) is assumed to be in position 1 for a long time before it is changed to position 2 at t = O. To calculate the initial conditions in the network, we have to first find the current in inductance and voltage across the capacitor just before changing the switch from position 1 to 2. i.e., at

$t = O^-$. This condition is same as the condition $t = \infty$ for the network with switch at position 1, as this condition is assumed to exist for a long time.

Hence at $t = O^-$ the equivalent circuit is obtained by replacing the elements by their equivalents at $t = \infty$ with switch at position 1 as shown in Fig. 7.17(b).

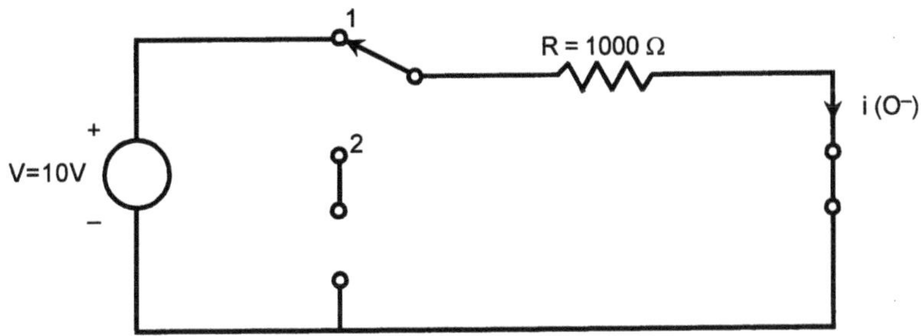

Fig. 7.17(b) Equivalent at t = O⁻

$$i(O^-) = \frac{V}{R} = \frac{10}{1000} = 10 \text{ mA}$$

This current flows through the inductor and hence the current through the inductor at $t = O^-$ is equal to 10 mA. The voltage across the capacitor is zero at $t = O^-$ as can be verified from Fig. 7.17(b).

To find the initial conditions, we write the equivalent at $t = O^+$, as shown in Fig. 7.17(c).

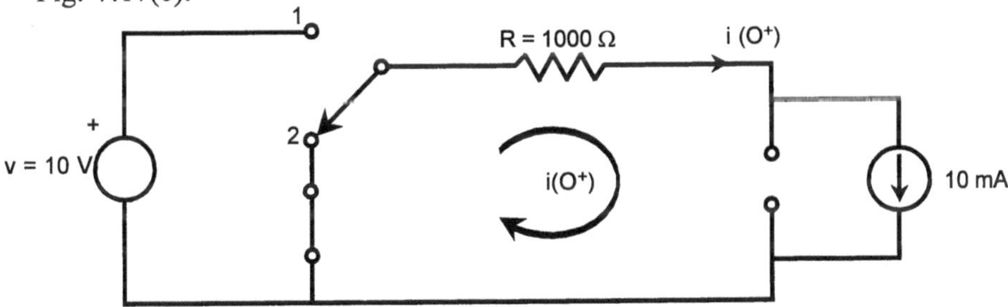

Fig. 7.17 (c) Network at t = O⁺

From Fig. 7.17(c)

$$i(O^+) = 10 \text{ mA}$$

To find the derivatives, we need to write the differential equations for the network in Fig. 7.17 (a) with switch in position 2.

Thus, writing the loop equation

$$Ri + L\frac{di}{dt} + \frac{1}{C}\int_{-\infty}^{t} i\,dt = O \qquad \text{.....(7.34)}$$

At $t = O^+$, equation (7.34) becomes,

$$Ri(O^+) + L\frac{di}{dt}(O^+) + \frac{1}{C}\int_{-\infty}^{O+} idt = O$$

At $t = O^+$, $i(O^+) = 10$ mA

and $\frac{1}{C}\int_{-\infty}^{O+} idt = O$ ($\because$ voltage across the capacitor at $t = O^+$ is zero)

$$10 \times 10^{-3} \times 1000 + 2\,\frac{di}{dt}\left(O^+\right) = O$$

$$\frac{di}{dt}(O^+) = \frac{-10}{2} = -5 \text{ A/sec.}$$

To get the second derivative of i (t) at $t = O^+$ we differentiate eq. (7.34) to yield

$$R\frac{di}{dt} + L\frac{d^2i}{dt^2} + \frac{i}{C} = O \qquad .....(7.35)$$

At $t = O^+$

$$R\frac{di}{dt}(O^+) + L\frac{d^2i}{dt^2}(O^+) + \frac{i(O^+)}{C} = O$$

$$\frac{d^2i(O^+)}{dt^2} = -\frac{i(O^+)}{LC} - \frac{R}{L}\frac{di}{dt}(O^+)$$

Using the values of i (O+) and $\frac{di}{dt}(O^+)$

$$\frac{d^2i(O^+)}{dt^2} = \frac{-10 \times 10^{-3}}{2 \times 1 \times 10^{-6}} - \frac{10^3}{2}(-5)$$

$$= -5000 + 2500$$

$$= -2500 \text{ A/sec}^2.$$

Example 7.11

In the network of Fig. 7.18 determine $v_K(O^+), \frac{dv_K}{dt}(O^+)$ and $\frac{d^2v_K}{dt^2}(O^+)$. The network was in steady state when the switch K is opened at t = O.

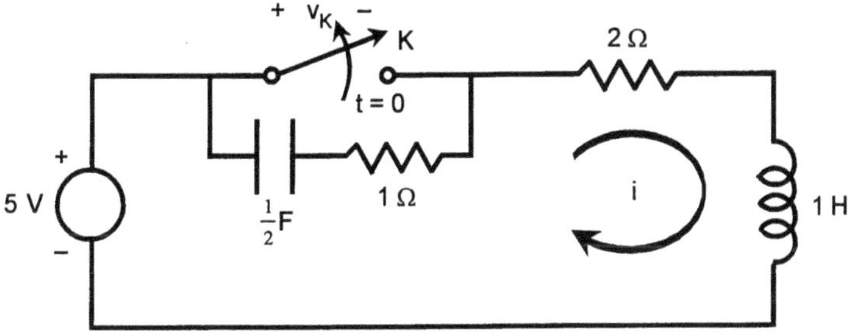

Fig. 7.18 (a) Network for example 7.11

Solution :

Let us calculate the current through the inductor at $t = O^-$. The capacitor – resistor branch is shorted by the switch K and hence it need not be considered at $t = O^-$.

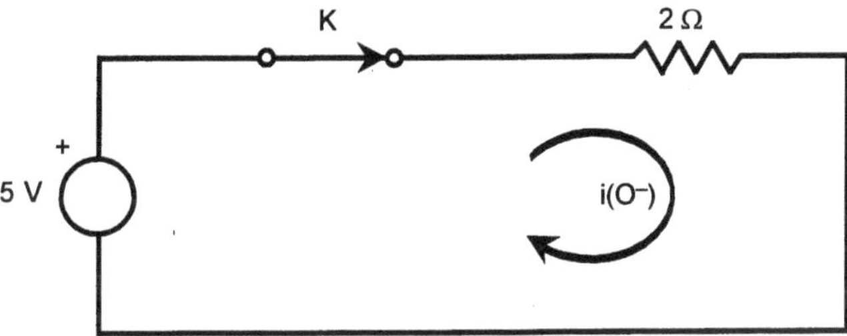

Fig. 7.18 (b) The network at $t = O^-$

From Fig. 7.18(b), $i(O^-) = \frac{5}{2} = 2.5$ A

Now let us consider the network at $t = O^+$. Since a current of 2.5 A was passing through the inductor at $t = O^-$, it is replaced by a current source of 2.5 A in parallel with an open circuit. The capacitor has no voltage across it at $t = O^-$ the and hence it is replaced with a short circuit at $t = O^+$. The network at $t = O^+$ is :

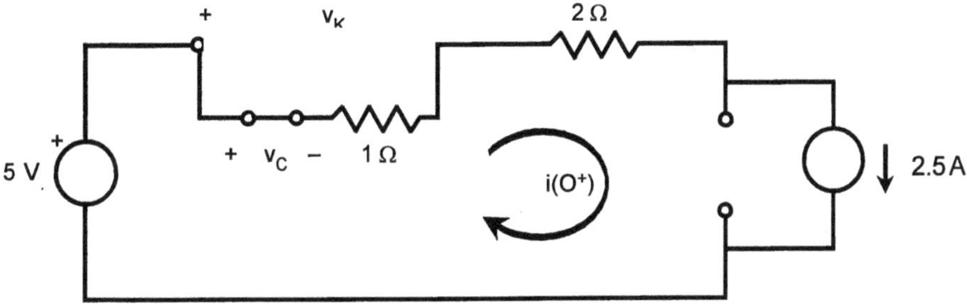

Fig. 7.18(c) The network at $t = O^+$

From Fig. 7.18 (c)

$$i(O^+) = 2.5 \text{ A} \quad \text{and} \quad v_C(O^+) = O$$

$$v_K(O^+) = (2.5)(1) = 2.5 \text{ V}$$

To determine the derivatives of $v_K(t)$ at $t = O^+$. We have to write the integro – differential equation for the network in Fig. 7.18(d) which is valid for any time $t > O$.

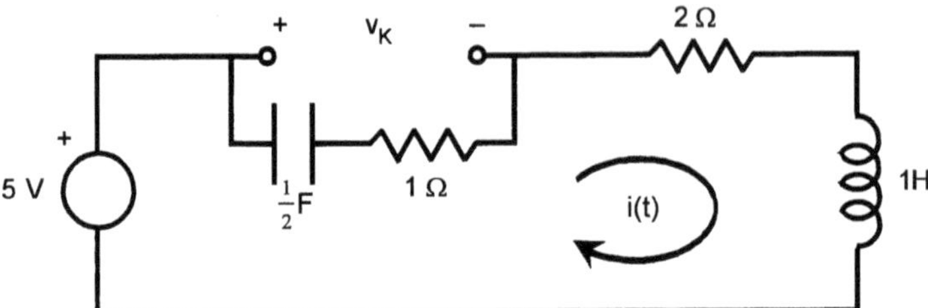

Fig. 7.18 (d) Network to determine the derivatives of V_K at $t = O^+$

$$v_K = 2\int_{-\infty}^{t} i\,dt + i \qquad .....(7.36)$$

Differentiating eq. (7.36), we get

$$\frac{dv_K}{dt} = 2i + \frac{di}{dt} \qquad .....(7.37)$$

At $t = O^+$

$$\frac{dv_K}{dt}(O^+) = 2i(O^+) + \frac{di}{dt}(O^+) \qquad .....(7.38)$$

To determine $\frac{dv_K}{dt}(O^+)$, we need to know $\frac{di}{dt}(O^+)$ in eq. (7.38). Writing the loop equation for the network in Fig. 7.18 (d).

$$3i + \frac{di}{dt} + 2\int_{-\infty}^{t} i\,dt = 5 \qquad .....(7.39)$$

At $t = O^+$

$$3i(O^+) + \frac{di}{dt}O^+ + 2\int_{-\infty}^{O^+} i\,dt = 5 \qquad .....(7.40)$$

Since the third term on the left hand side of eq. (7.40) represents the voltage across the capacitor at $t = O^+$, its value is zero. Substituting the value $i(O^+)$ in eq. (7.40), we get

$$3(2.5) + \frac{di}{dt}(O^+) = 5$$

$$\frac{di}{dt}(O^+)\ 5 - 7.5 = -2.5 \text{ A/sec.}$$

Using this value in eq. (7.38),

$$\frac{dv_K(O^+)}{dt} = 2\,(2.5) - 2.5$$

$$= 2.5 \text{ V/sec.}$$

Differentiating eq. (7.37)

$$\frac{d^2v_K}{dt^2} = 2\frac{di}{dt} + \frac{d^2i}{dt^2} \qquad .....(7.41)$$

At $t = O^+$

$$\frac{d^2v_K}{dt^2}(O^+) = 2\frac{di}{dt}(O^+) + \frac{d^2i}{dt^2}(O^+) \qquad .....(7.42)$$

Since $\frac{d^2i}{dt^2}(O^+)$ is not known in eq. (7.42), it has to be evaluated by using eq. (7.39).

Differentiating eq. (7.39), we get

$$3\frac{di}{dt} + \frac{d^2i}{dt^2} + 2i = 0 \qquad .....(7.43)$$

At $t = O^+$

$$3\frac{di}{dt}(O^+) + \frac{d^2i}{dt^2}(O^+) + 2i\,(O^+) = 0 \qquad .....(7.44)$$

Substituting the known values of $\frac{di}{dt}(O^+)$ and $i\,(O^+)$ in eq. (7.44), we get

$$\frac{d^2i}{dt^2}(O^+) = -2(2.5) - 3\,(-2.5)$$

$$= -5 + 7.5$$

$$= 2.5 \text{ A/sec}^2$$

From eq. (7.42)

$$\frac{d^2v_K}{dt^2}(O^+) = 2\,(-2.5) + 2.5$$

$$= -5 + 2.5$$

$$= -2.5 \text{ V/sec}^2$$

7.4.3 Special cases :

Two special cases occur in networks for which rules developed in section 7.4.1 cannot be applied. The first one is illustrated in Fig. 7.19.(a).

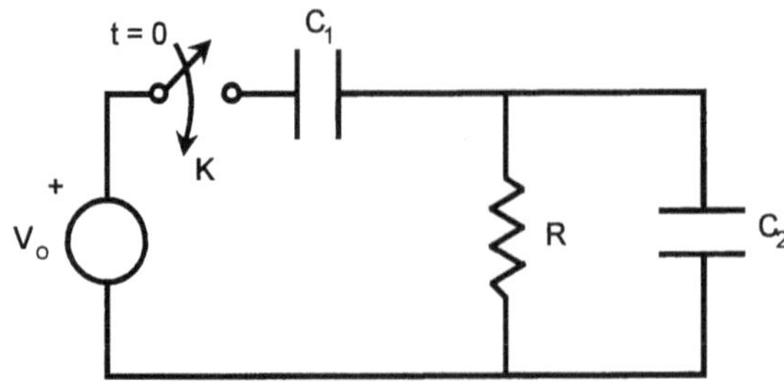

Fig. 7.19 (a) Network to illustrate the special case of capacitor loop

Let us assume that the capacitors in Fig. 7.19 (a) are uncharged at t = O when the switch K is closed. At $t = O^+$ the capacitors behave like short circuits and hence the network at $t = O^+$ is given in Fig. 7.19 (b).

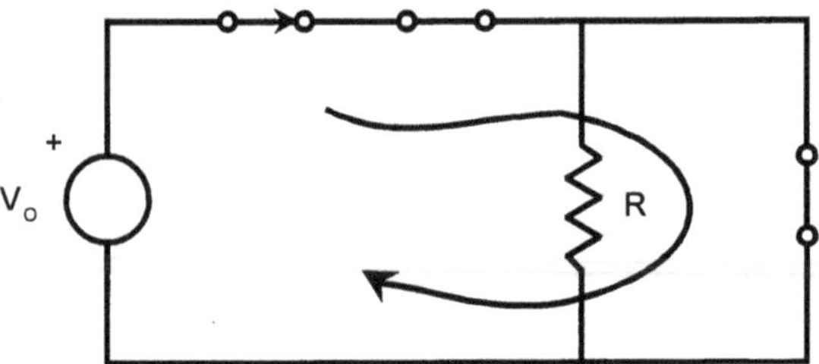

Fig. 7.19 (b) The equivalent network at $t = O^+$

We have the unusual situation of a voltage source being short circuited at $t = O^+$ which results in an infinite current in the loop indicated by the arrow in Fig. 7.19 (b). Thus an impulse of current flows between the instants $t = O^-$ and $t = O^+$ depositing a finite charge q on the two capacitors during this period. Note that the current flowing through the two capacitors is same (No current flows in the resistor during $t = O^-$ to $t = O^+$ as it is short circuited) and hence the charge on the capacitors is the same. In this case, the voltages across the capacitors will change instantaneously since there is an impulse of current flowing through them and these voltages together with the voltage source V must obey Kirchhoff voltage law in the loop indicated in Fig. 7.19(b). To describe the behaviour of the circuit at $t = O^+$ mathematically, we can write the loop equation :

$$\frac{1}{C_1}\int_{-\infty}^{O^+} i\,dt + \frac{1}{C_2}\int_{-\infty}^{O^+} i\,dt = V_O \qquad .....(7.45)$$

Since the voltages across the capacitors are zero at $t = O^-$, eq. 7.45 can be written as

$$\frac{1}{C_1}\int_{O^-}^{O^+} i\,dt + \frac{1}{C_2}\int_{O^-}^{O^+} i\,dt = V_O \qquad .....(7.46)$$

In eq. (7.46) the term $\int_{0^-}^{0^+} idt$ would be zero if the current i is finite as discussed in section 7.4.1. The voltages across the capacitors would be zero and they behave like short circuits. But in the special case where i is infinite during the interval O^- to O^+, this term $\int_{0^-}^{0^+} idt$ represents the charge deposited on the capacitors due to the flow of infinite current. Thus

$$q(O^+) = \int_{0^-}^{0^+} idt \qquad .....(7.47)$$

Thus eq. (7.46) becomes

$$\frac{q(O^+)}{C_1} + \frac{q(O^+)}{C_2} = V_O \qquad .....(7.48)$$

or

$$q(O^+)\left[\frac{C_1 + C_2}{C_1 C_2}\right] = V_O \qquad ....(7.49)$$

$$q(O^+) = V_O \frac{C_1 C_2}{C_1 + C_2} \qquad .....(7.50)$$

Knowing q (O^+), the voltages across the two capacitors $V_{C1}(O^+)$ and $V_{C2}(O^+)$ can be calculated as :

$$V_{C_1}(O^+) = \frac{q(O^+)}{C_1} = V_O \frac{C_1 C_2}{C_1 + C_2} \frac{1}{C_1} = V_O \frac{C_2}{C_1 + C_2} \qquad .....(7.51)$$

$$V_{C_2}(O^+) = \frac{q(O^+)}{C_2} = \frac{V_O C_1}{C_1 + C_2} \qquad .....(7.52)$$

To summarise this special case :

Normally the capacitors behave like short circuits at $t = O^+$, if they are uncharged at $t = O^-$. If they are charged initially at $t = O^-$, they behave like short circuits in series with a voltage source of value equal to the voltage at $t = O^-$. Thus we say that the voltage across a capacitor cannot be changed instantaneously. This is true only when the current through the capacitor is finite during the interval $t = O^-$ and $t = O^+$. But in the special case where a loop can be traced through the network containing only capacitors and voltage sources, an impulse of current flows through this loop and hence there is a finite amount of charge transferred to these capacitors. Hence the voltages across the capacitors change instantaneously to satisfy the Kirchhoff's voltage law around this loop. Note that if a resistance is introduced in this loop, this situation will not occur. A finite current will flow during the interval $t = O^-$ to $t = O^+$ and the voltage across the capacitors do not change at $t = O^+$.

If the capacitors in the network in Fig. 7.19(a) have initial voltages across them at $t = O^-$, say $V_{C_1}(O^-)$ and $V_{C_2}(O^-)$, the voltages across them at $t = O^+$ can be calculated using eq. (7.45). Rewriting eq. (7.45) as

$$\frac{1}{C_1}\int_{-\infty}^{O^-} i dt + \frac{1}{C_1}\int_{O^-}^{O^+} i dt + \frac{1}{C_2}\int_{-\infty}^{O^-} i dt + \frac{1}{C_2}\int_{O^-}^{O^+} i dt = V_0 \quad .....(7.53)$$

The first and third term in eq. (7.53) are the initial voltages across the capacitors at $t = O^-$ and hence eq. (7.53) can be written as

$$v_{C_1}(O^-) + \frac{q(O^+)}{C_1} + v_{C_2}(O^+) + \frac{q(O^+)}{C_2} = V_O \quad .....(7.54)$$

$$q(O^+)\left[\frac{C_1 + C_2}{C_1C_2}\right] = V_O - v_{C_1}(O^-) - v_{C_2}(O^-) \quad .....(7.55)$$

$$q(O^+) = \frac{C_1C_2}{C_1 + C_2}[V_O - v_{C_1}(O^-) - v_{C_2}(O^-)] \quad .....(7.56)$$

Using eq. (7.54) and (7.56) the voltages across the capacitors at $t = O^+$ can be calculated as

$$v_{C_1}(O^+) = v_{C_1}(O^-) + \frac{q(O^+)}{C_1}$$

$$= v_{C_1}(O^-) + \frac{C_2}{C_1 + C_2}(V_O - v_{C_1}(O^-) - v_{C_2}(O^-)) \quad .....(7.57)$$

and similarly

$$v_{C_2}(O^+) = v_{C_2}(O^-) + \frac{q(O^+)}{C_2}$$

$$= v_{C_2}(O^-) + \frac{C_1}{C_1 + C_2}(V_O - v_{C_1}(O^-) - v_{C_2}(O^-)) \quad .....(7.58)$$

The second special case is a dual of the first case indicated in Fig. 7.20(a). This network consists of a node at which only inductors and current sources are connected.

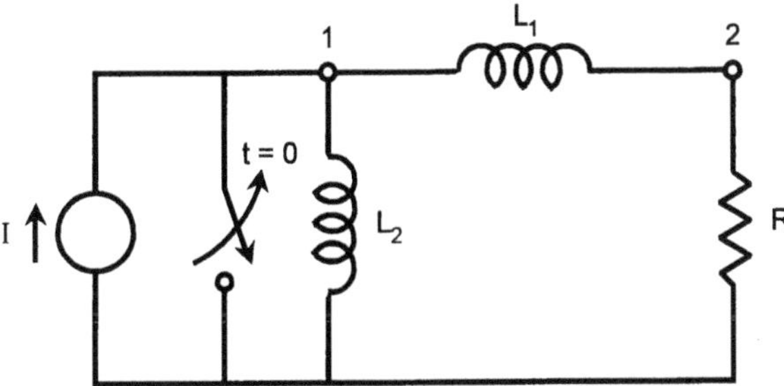

Fig. 7.20 (a) Network to illustrate the second special case of inductors at a node

Assuming no initial currents in the inductors at $t = 0^+$, the equivalent at $t = 0^+$ can be written as in Fig. 7.20(b).

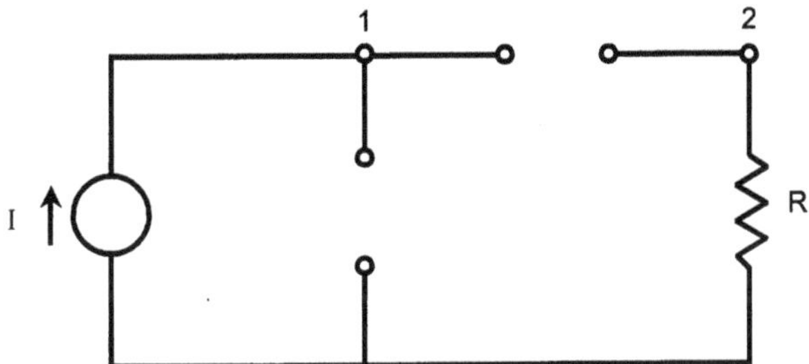

Fig. 7.20(b) The equivalent network at t = O+

This gives rise to the special situation where the current source is open circuited and hence an infinite voltage results across the terminals of the two inductors. When infinite voltage occurs across the terminals of an inductor, the current through it has to change instantaneously as this infinite voltage causes a finite flux linkages in the inductor. Hence the circuit in Fig. 7.20(b) will not be valid and currents in inductors will flow satisfying the Kirchhoff's current law at the node. Since the voltages across the inductors are infinite, the voltage across the resistor and hence the node voltage at node 2 is negligible. Writing the Kirchhoff's current law at node 1, we get, at $t = 0^+$

$$\frac{1}{L_1}\int_{-\infty}^{0^+} v\,dt + \frac{1}{L_2}\int_{-\infty}^{0^+} v\,dt = I \qquad .....(7.59)$$

Assuming no initial currents in inductors, eq. (7.59) can be written as

$$\frac{1}{L_1}\int_{0^-}^{0^+} v\,dt + \frac{1}{L_2}\int_{0^-}^{0^+} v\,dt = I \qquad .....(7.60)$$

$$\frac{\psi(0^+)}{L_1} + \frac{\psi(0^+)}{L_2} = I \qquad .....(7.61)$$

where $\Psi(O^+)$ is the flux linkage.

Hence $\Psi(O^+) = I \dfrac{L_1 L_2}{L_1 + L_2}$ ....(7.62)

and $$i_{L1}(O^+) = \frac{\psi(O^+)}{L_1} = \frac{I\, L_2}{L_1 + L_2}$$ .....(7.63)

or $$i_{L2}(O^+) = \frac{\psi(O^+)}{L_2} = \frac{I\, L_1}{L_1 + L_2}$$ .....(7.64)

The case where the inductors have initial currents can be handled similar to the case of capacitors with initial charges.

To summarise the second special case :

Normally, the inductors behave like open circuits at $t = O^+$, thus resisting an instantaneous change in the current. But in the special case where, at a node, only inductors and current sources are connected, the current in the inductors will change instantaneously, as infinite voltages are produced across them between $t = O^-$ and $t = O^+$. The currents at $t = O^+$ in the inductors can be calculated by writing Kirchhoff's current law equation at the node as in eq. (7.59) and calculating the flux linkages in the inductors using eq. (7.62).

Some examples illustrate the concepts presented in section 7.4.3.

Example 7.12

Find the voltages across the three capacitors at $t = O^+$. The initial voltages across the capacitors are indicated in Fig. 7.21.

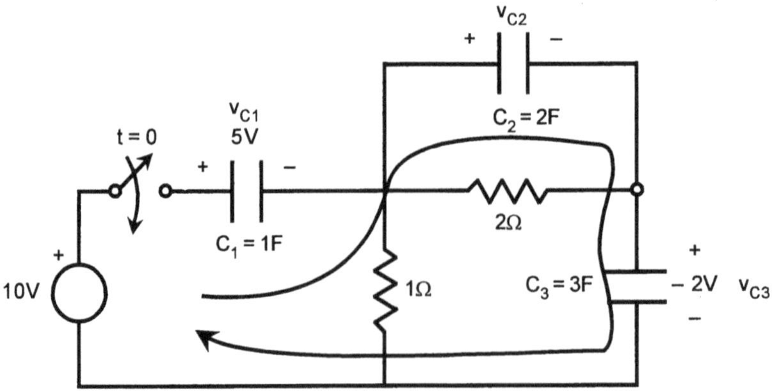

Fig. 7.21 Network for example 7.12

Solution :

As a loop is formed by the three capacitors and the voltage source, this problem comes under the first special case.

Denoting the voltages across the three capacitors as v_{C_1}, v_{C_2} and v_{C_3} and writing Kirchhoff's law equation around the loop.

$$v = v_{C_1} + v_{C_2} + v_{C_3}$$

At $t = O^+$

$$\frac{1}{C_1}\int_{-\infty}^{O^+} idt + \frac{1}{C_2}\int_{-\infty}^{O^+} idt + \frac{1}{C_3}\int_{-\infty}^{O^+} idt = V$$

$$\frac{1}{1}\int_{-\infty}^{O^-} idt + \frac{1}{1}\int_{O^-}^{O^+} idt + \frac{1}{2}\int_{-\infty}^{O^-} idt + \frac{1}{2}\int_{O^-}^{O^+} idt + \frac{1}{3}\int_{-\infty}^{O^-} idt + \frac{1}{3}\int_{O^-}^{O^+} idt = 10$$

$$5 + \frac{q(O^+)}{1} + O + \frac{q(O^+)}{2} + (-2) + \frac{q(O^+)}{3} = 10$$

$$q(O^+)\frac{11}{6} = 7$$

$$q(O^+) = \frac{42}{11} C$$

$$v_{C_1}(O^+) = 5 + \frac{42}{11} = \frac{97}{11} \text{ V}$$

$$v_{C_2}(O^+) = O + \frac{42}{11} \times \frac{1}{2} = \frac{21}{11} \text{ V}$$

$$v_{C_3}(O^+) = -2 + \frac{42}{11} \times \frac{1}{3} = -\frac{8}{11} \text{ V}$$

Caution : Note the polarities of v_{C_1}, v_{C_2} and v_{C_3} at $t = O^-$.

Example 7.13

The network in Fig. 7.22 (a) was in steady state before the switch K is opened at $t = O$. Calculate the current through the inductor at $t = O^+$.

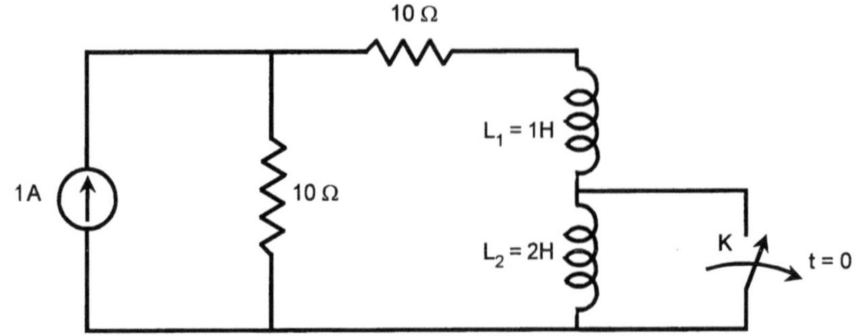

Fig. 7.22 (a) Network for example 7.13

Solution :

The network was in steady state with switch K closed. Hence the inductor L_1 behaves like a short circuit (behaviour at t = ∞). L_2 is shorted and hence no current passes through it. The network at $t = O^-$ is shown in Fig. 7.22(b).

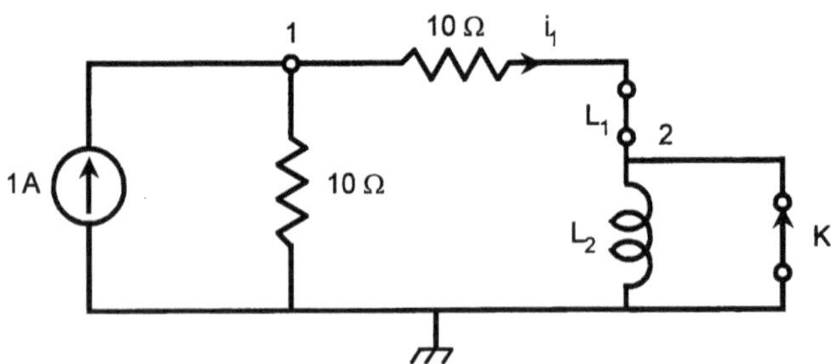

Fig. 7.22 (b) Network at $t = O^-$

The current i_1 through L_1 is given by

$$i_1(O^-) = 1 \times \frac{10}{20} = 0.5A$$

Thus the current through L_1 at $t = O^-$ is 0.5 A. At $t = O^+$, since only inductors are connected at node 2 with the initial current through L_1 acting as current source, this comes under the second special case. Writing the Kirchhoff's current law equation at node 2, at $t = O^+$, and ignoring the other node voltages in comparison with the impulse voltage present at node 2,

$$\frac{1}{L_1}\int_{-\infty}^{O^+} v_2 dt + \frac{1}{L_2}\int_{-\infty}^{O^+} v_2 dt = 0$$

where v_2 is the voltage of node 2 w.r.t ground.

$$\frac{1}{1}\int_{-\infty}^{O^-} v_2 dt + \frac{1}{1}\int_{O^-}^{O^+} v_2 dt + \frac{1}{2}\int_{-\infty}^{O^-} v_2 dt + \frac{1}{2}\int_{O^-}^{O^+} v_2 dt = O \qquad .....(7.65)$$

Since the first term in eq. (7.65) represents the current at $t = O^-$ in the inductor L_1, going away form the node 2, it is equal to – 0.5 A. The third term in eq. (7.65) is the current in the inductor L_2 at $t = O^-$ and is equal to zero. Thus eq. (7.65) can be written as,

$$-0.5 + \frac{\psi(O^+)}{1} + 0 + \frac{\psi(O^+)}{2} = 0$$

$$\frac{3}{2}\psi\left(O^+\right) = 0.5$$

$$\psi\ (O^+) = \frac{1}{3} \text{ weber} - \text{turns}$$

Hence the current through L_1 is given by

$$i_{L_1}(O^+) = -\left[i_{L_1}(O^-) + \frac{\psi\left(O^+\right)}{1}\right] = -\left[-0.5 + \frac{1}{3}\right]$$

$$= \frac{1}{2} - \frac{1}{3} = \frac{1}{6} \text{ A}$$

The current through L_2 is given by

$$i_{L_2}(O^+) = \frac{\psi(O^+)}{2} = \frac{1}{6} \text{ A}$$

which is same as $i_{L_1}(O^+)$ as the two inductors are in series at $t = O^+$.

Problems

7.1 Write the integro differential equation for the voltage v in Fig. P. 7.1 and convert it in to a differential equation.

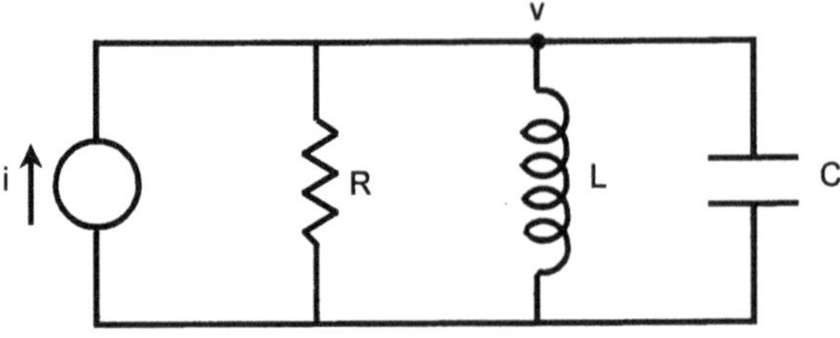

Fig. P. 7.1

7.2 Write the integro differential equation for the currents $i_1(t)$, $i_2(t)$ and $i_3(t)$ and convert the same into differential equations.

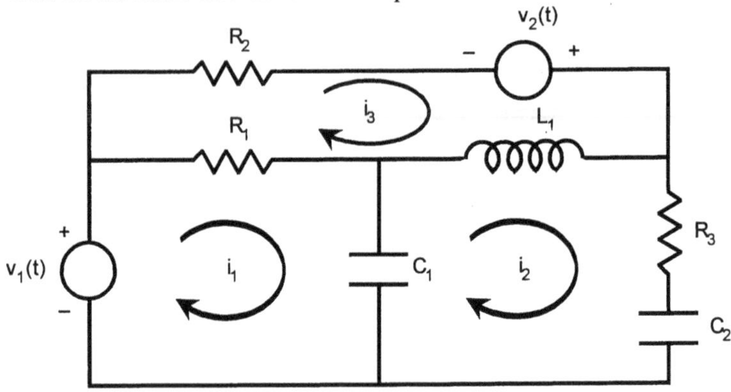

Fig. P. 7.2

7.3 Write the differential equations describing the network in Fig. P. 7.3.

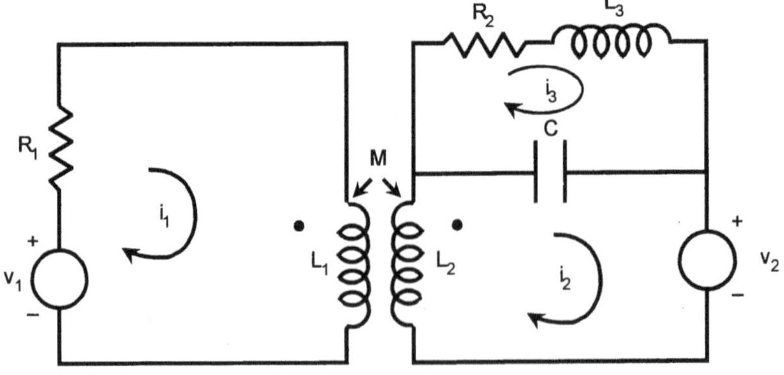

Fig. P. 7.3

7.4 Write the differential equations for the loop currents in Fig. P. 7.4. The mutual inductances between L_1 and L_2, L_2 and L_3, L_1 and L_3 are respectively M_{12}, M_{23} and M_{13}.

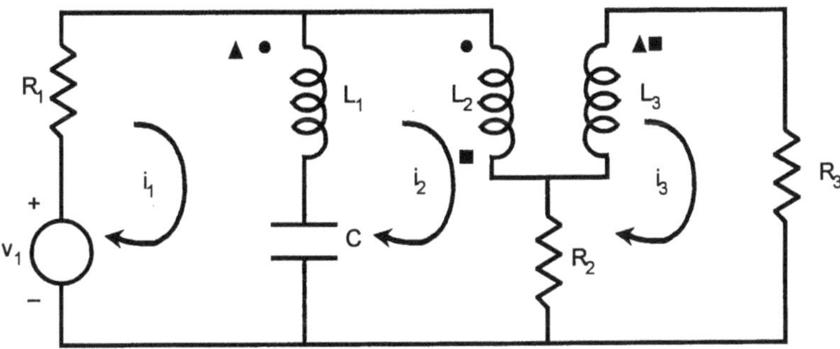

Fig. P. 7.4

7.5 Write down the duals of the networks by first writing the mesh equations, and then converting them to dual equations for the networks in Fig. P. 7.5. The dual of mutual inductance in Fig. P. 7.5(c) can be taken as a capacitance whose value may be +ve or – ve depending on the polarity dots.

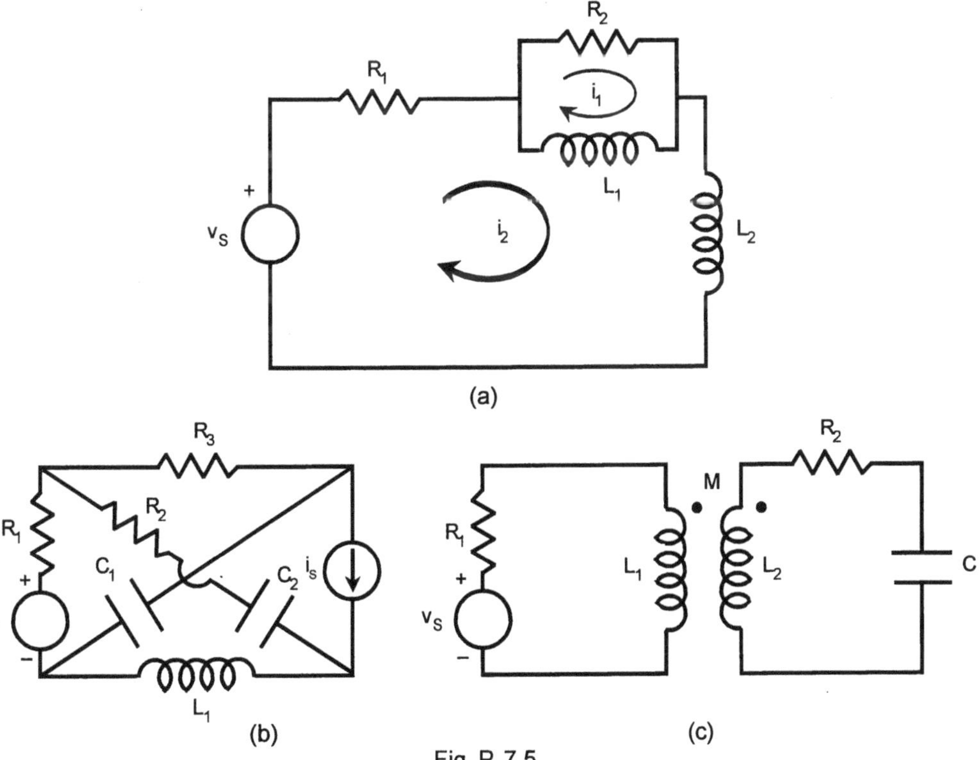

Fig. P. 7.5

7.6 Write down the duals of the networks in Fig. P. 7.6 using the geometrical procedure described in section 7.3.

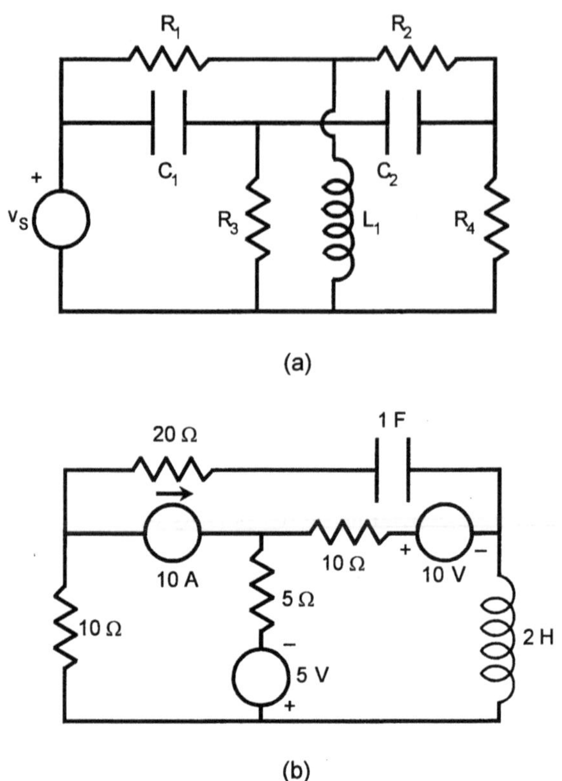

Fig. P. 7.6

7.7 For the circuit shown in Fig. P. 7.7 find $i(0^+), \frac{di}{dt}(0^+), \frac{d^2i}{dt^2}(0^+), i(\infty)$.

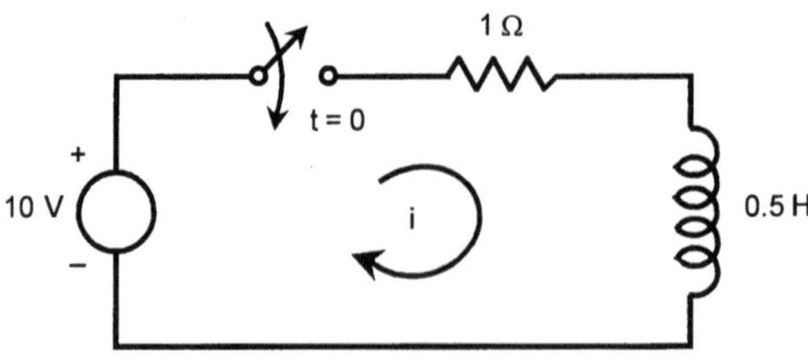

Fig. P. 7.7

7.8 For the circuit shown in Fig. P. 7.8 find $v(0+), \frac{dv}{dt}(0+), \frac{d^2v}{dt^2}(0^+)$ and $v(\infty)$.

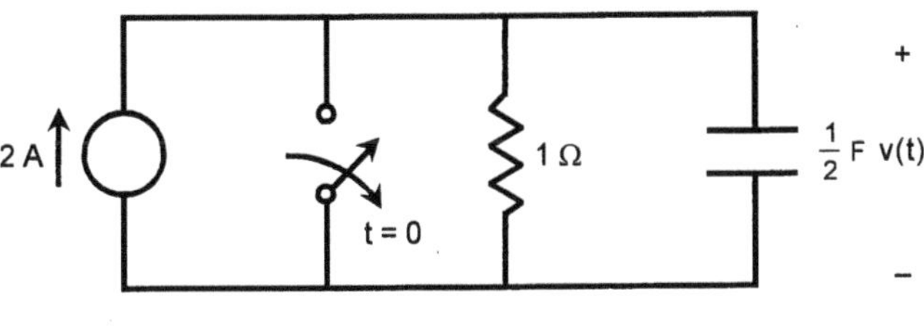

Fig. P. 7.8

7.9 The switch in Fig. P. 7.9 is in closed position for a long time and at t = 0 it is opened. Find v_k, $\frac{dv_k}{dt}$ at $t = 0^+$.

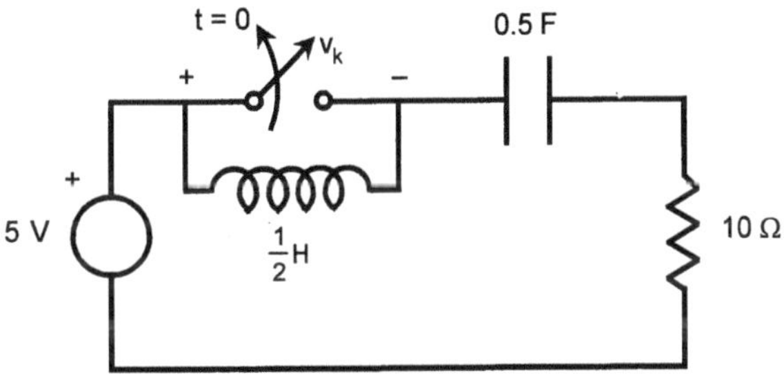

Fig. P. 7.9

7.10 In the network shown in Fig. P. 7.10 the switch K is opened at t = 0. Find $v(0^+)$, $\frac{dv}{dt}(0^+)$ and $v(\infty)$.

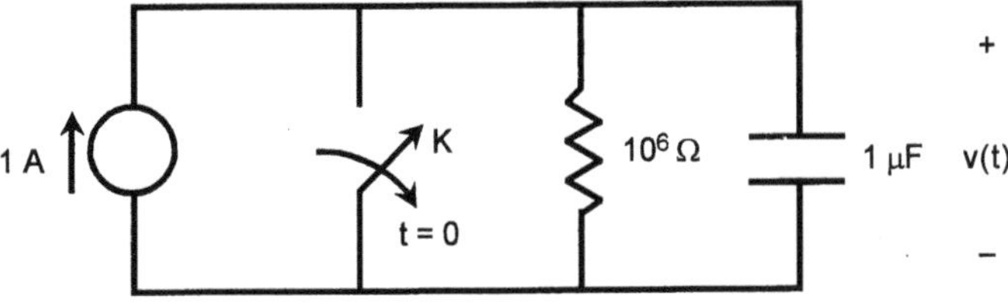

Fig. P. 7.10

7.11 For the network shown in Fig. P. 7.11. Find $\frac{d^2 i_1}{dt^2}$ at $t = 0^+$.

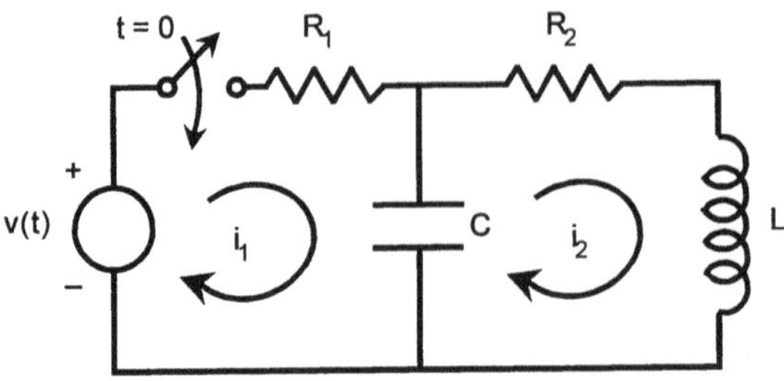

Fig. P. 7.11

7.12 Find v_a at $t = 0^+$ in Fig. P. 7.12. Also find the voltages across the capacitors at $t = 0^+$ and $t = \infty$. If 5 Ω resistor is made zero, what will be these values.

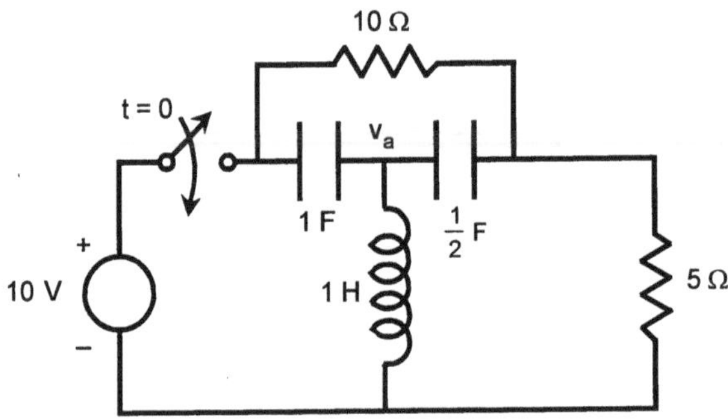

Fig. P. 7.12

7.13 Find the currents i_1, i_2 and i_3 at $t = 0^+$ in Fig. P. 7.13. The switch K was in position 1 for a long time and at t = 0 it is thrown to position 2.

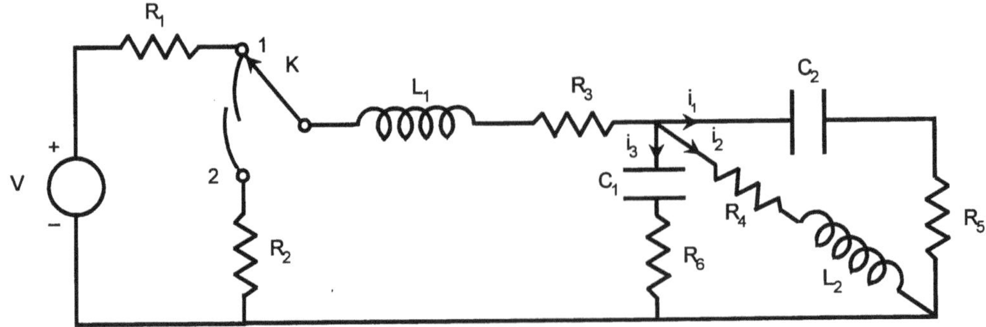

Fig. P. 7.13

7.14 In the network shown in Fig. P. 7.14 a steady state is reached with the switch K closed. At t = 0 the switch is opened.

(a) What is the voltage across C at $t = 0^-$?

(b) Find $i_1(0^+)$, $i_2(0^+)$, $\frac{di_1}{dt}(0^+)$, $\frac{di_2}{dt}(0^+)$.

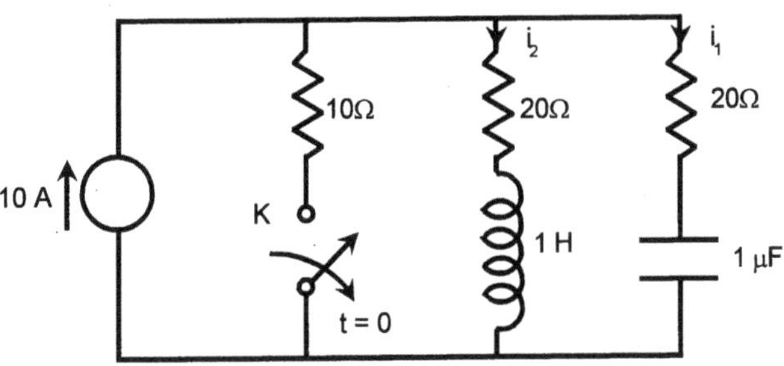

Fig. P. 7.14

7.15 A capacitor C_1 = 1F is charged to 10 V and at t = 0 it is connected across another uncharged capacitor C_2 – 2F. What is the voltage across the parallel combination at $t = 0^+$? What is the total energy in the two capacitors before and after they are connected in parallel. Give the reason for any difference between the two values. What are the charges in the capacitors before and after they are connected in parallel ?

7.16 A charged capacitor C_1 is connected to a series combination of R_1 and C_2 at t = 0 as shown in Fig. P. 7.15. What are the voltages v_{C_1} and v_{C_2} at $t = 0^+$ if

(a) $R_1 = 5\ \Omega$ (b) $R_1 = 1\ \Omega$ and (c) $R_1 = 0\ \Omega$

Fig. P. 7.15

7.17 An inductor L_1 with an initial current of 10A is connected to a parallel combination of R_1 and L_2 at t = 0 as shown in Fig. P. 7.16. Find i_{L1} and i_{L2} at $t = 0^+$ if

(a) $R_1 = 10\ \Omega$ (b) $R_1 = 1\ \Omega$ and (c) $R_1 = \infty$.

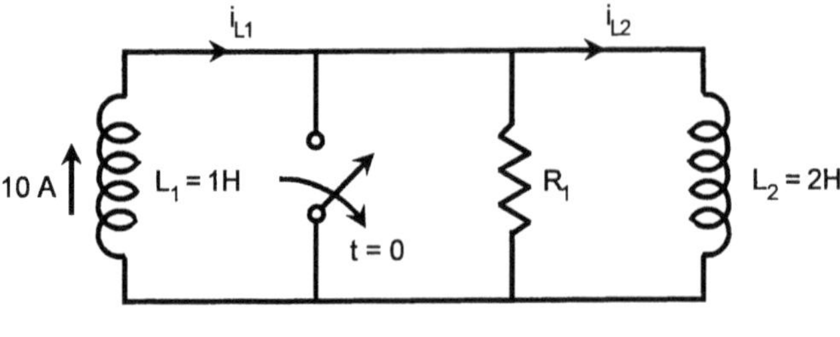

Fig. P. 7.16

7.18 In the network shown in Fig. P. 7.17. Find $v_1(0^+)$, $\frac{dv_1}{dt}(0^+)$, and $\frac{d^2v_1}{dt^2}(0^+)$.

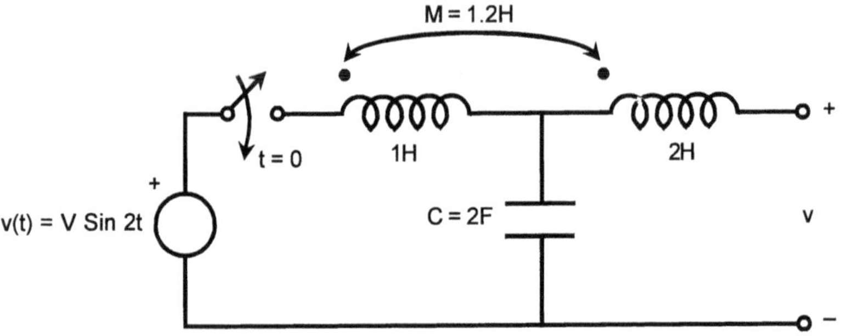

Fig. P. 7.17

7.19 In the network of Fig. P. 7.18. Find $v_c(0^+)$.

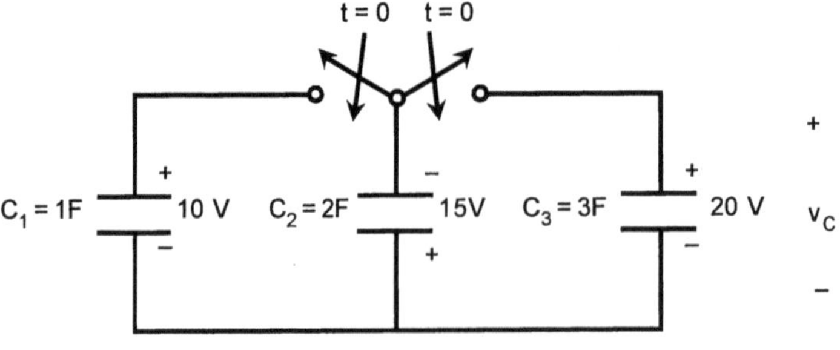

Fig. P. 7.18

7.20 In the Fig. P. 7.19. Find $i(0^+)$.

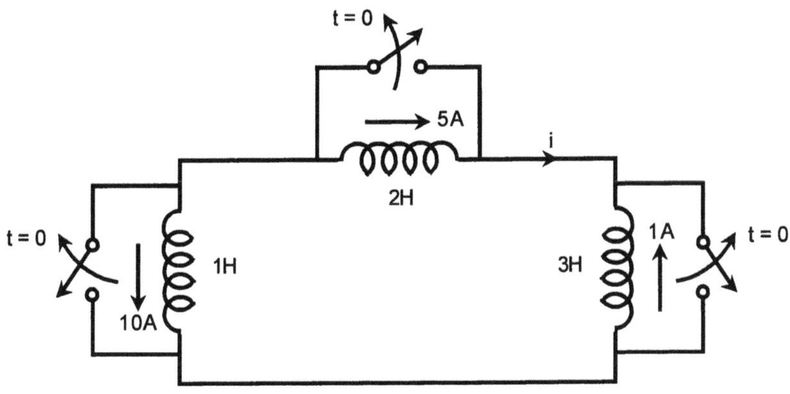

Fig. P. 7.19

8

TIME RESPONSE OF R L C NETWORKS

8.0 RESPONSE OF R L C NETWORKS :

In section 8.2, the network equations were formulated based on Kirchhoff's voltage law and current law. Application of these laws results in a set of differential equations with constant coefficients. These equations have to be solved to get the response of the network when the network is subjected to different types of excitations. The initial conditions are evaluated using the concepts developed in section 8.4. From the general solution of the network, the particular solution applicable to the network is obtained by using these initial conditions. In this chapter we will develop methods of determining the complete response of simple RL, RC, R L C networks.

8.1 TYPES OF ENERGY SOURCES :

We will consider various types of energy sources as excitations for the simple networks and evaluate their response. The most common type of energy sources used to excite the network are :

1. *Internal energy excitation :* The initial energy present in the energy storing elements L and C is used as the excitation. No other external source is present in the network.

2. *Unit impulse excitation :* The external source present in the network is a unit impulse source which was introduced in chapter 1. The symbol for this source is $\delta(t)$. The impulse function can be defined mathematically by considering a pulse of width 'h' and height $\frac{1}{h}$ and making $h \rightarrow 0$ as shown in Fig. 8.1

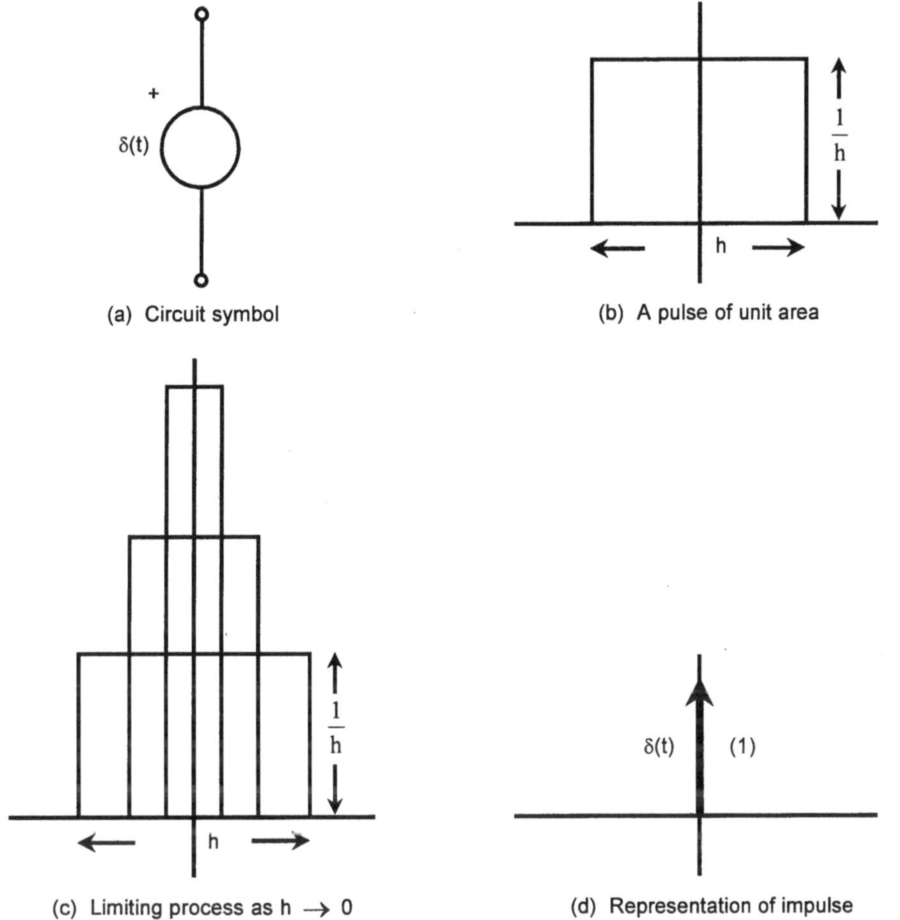

(a) Circuit symbol

(b) A pulse of unit area

(c) Limiting process as h → 0

(d) Representation of impulse

Fig. 8.1 The symbol and limiting process in the definition of an impulse function

In Fig. 8.1(b) the area enclosed by the pulse is unity. If 'h' is halved, the height is doubled but the area remains unity as shown in Fig. 8.1(c). As h → 0, the pulse height goes to infinity but the area remains unity. In the limit, the pulse approaches the impulse function as shown in Fig. 8.1(d). The unit impulse function goes to infinity at $t = O^-$ and returns to zero at $t = O^+$ enclosing a finite area of unity. This is an ideal source and hence can not be realised in practice. If the height of the pulse is taken to be $\frac{A}{h}$ instead of $\frac{1}{h}$, the area under the pulse will be A units and hence under the limit the impulse encloses an area of A units. This area is known as the *strength of the impulse* and the *impulse function of strength* A units is denoted by or $A\delta(t)$.

3. *Unit step excitation :* This is the most common type of excitation. At t = 0 the excitation goes to unity and remains constant at this value. For t < 0 it is zero. Thus a unit step function is defined as

$$u(t) = 0 \quad \text{for} \quad t < 0$$
$$= 1 \quad \text{for} \quad t \geq 0 \qquad .....(8.1)$$

A d.c source with a switch closed at t = 0 is a practical example of a unit step source as shown in Fig. 8.2.

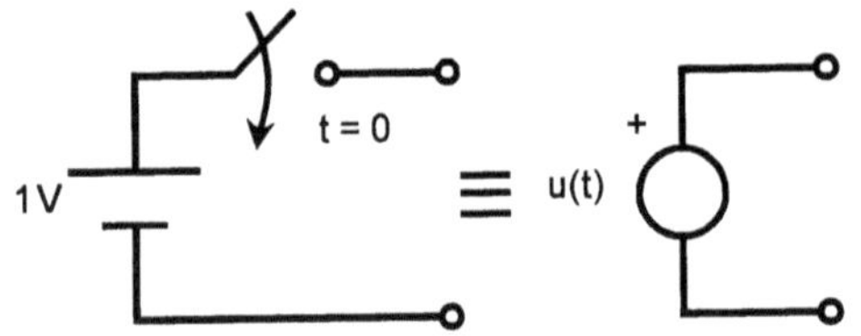

(a) Practical unit step voltage source

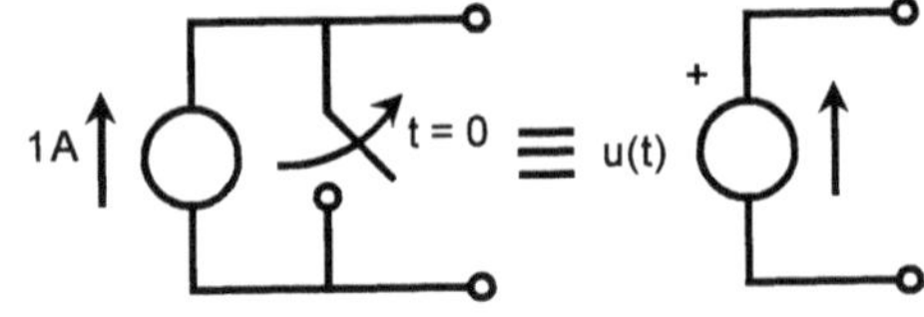

(b) Practical unit step current source

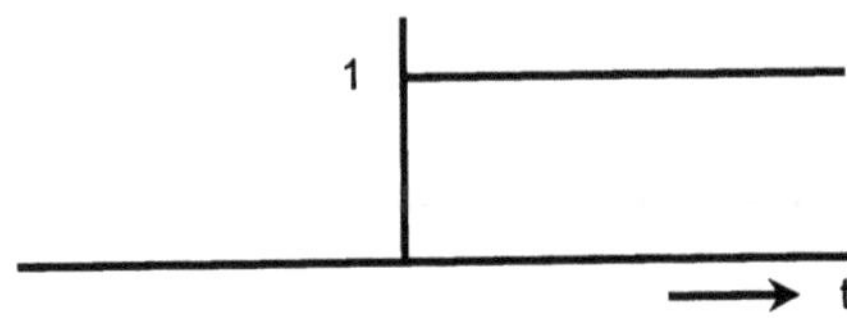

(c) Graphical representation of a step source

Fig. 8.2 Unit step source representations

4. *Sinusoidal excitation :* The sinusoidal excitation is another frequently used source in networks and is defined by

$$f(t) = 0 \quad \text{for} \quad t < 0$$
$$= F\sin(\omega t + \theta) \quad \text{for} \quad t \geq 0 \qquad .....(8.2)$$

where F is the amplitude
ω is the angular frequency
and θ is the phase.

This is graphically represented as shown in Fig. (8.3) for $\theta = 0$.

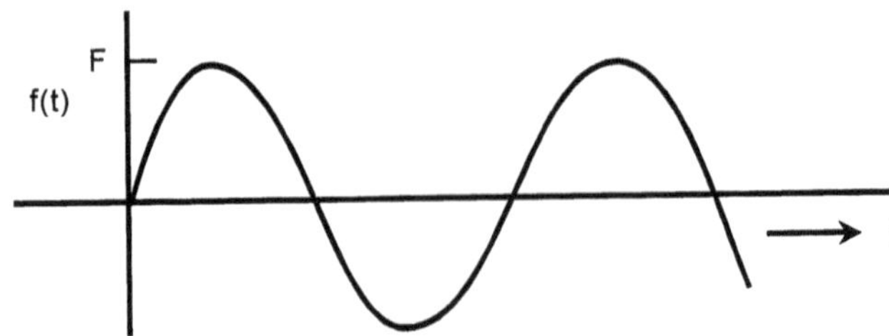

Fig. 8.3 Graphical representation of sinusoidal source

5. *Exponential excitation :* This is a useful excitation defined by

$$f(t) = 0 \quad \text{for} \quad t < 0 \qquad .....(8.3)$$
$$= Fe^{\alpha t} \quad \text{for} \quad t \geq 0$$

This is graphically represented as shown in Fig. 8.4, for (a) $\alpha > 0$ and (b) $\alpha < 0$

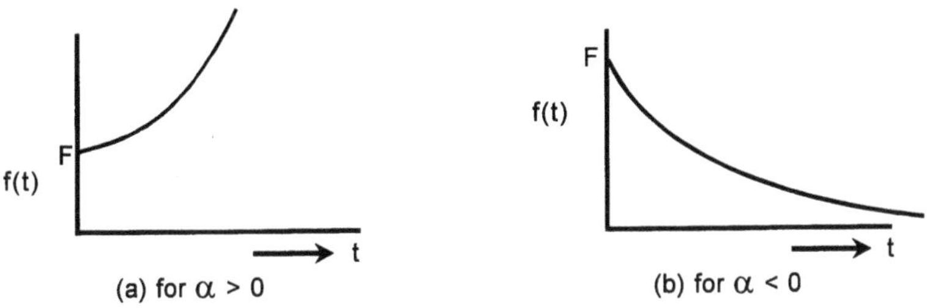

Fig. 8.4 Graphical representation of exponential excitation

6. *Polynomial source :* This is a general type of excitation defined by

$$f(t) = a_0 + a_1 t + a_2 t^2 + ... + a_n t^n \qquad .....(8.4)$$

8.2 RESPONSE OF RL NETWORKS :

The response of RL networks for internal energy, impulse, step and sinusoidal excitation will be obtained in this section. The polynomial and exponential excitations are considered later in the chapter.

8.2.1 Internal energy excitation :

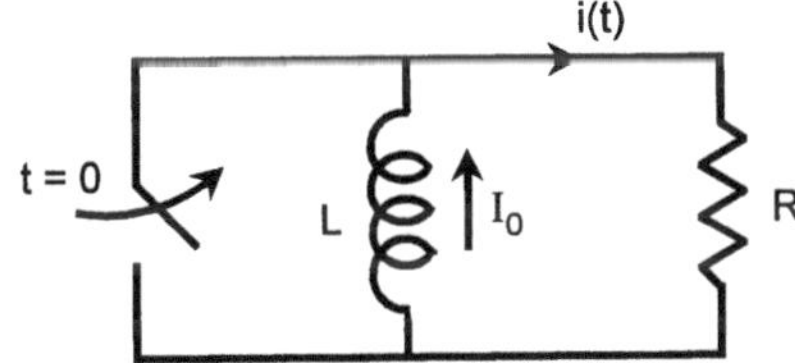

Fig. 8.5 An RL circuit excited by internal energy

In Fig. 8.5 an RL circuit is shown with an initial current of I_0 at $t = O^-$. At $t = 0$, the switch is opened and the objective is to find the current in the network for $t > 0$.

The differential equation for the network can be written as

$$Ri + L\frac{di}{dt} = 0 \qquad .....(8.5)$$

and the initial condition is $i(O^-) = I_0$. Since the current in the inductor cannot change instantaneously, $i(O^+) = I_0$. Eq. (8.5) is a first order homogeneous differential equation with constant coefficients. This can be solved by separating the variables.

By rearranging the eq. (8.5), we get

$$\frac{di}{i} = \frac{R}{L}dt \qquad .....(8.6)$$

Integrating eq. (8.6) to give

$$ln\, i = -\frac{R}{L}t + k \qquad .....(8.7)$$

which simplifies to

$$i = e^{\left(-\frac{R}{L}t+k\right)} \qquad .....(8.8)$$

Denoting e^k as another constant K, eq. (8.8) can be written as

$$i(t) = Ke^{-\frac{R}{L}t} \qquad .....(8.9)$$

This is the response of the network and is known as the *general solution.* If the constant K is evaluated, the solution becomes a particular solution. This constant can be evaluated using the initial condition, i.e., at $t = O^+$ $i(t) = I_o$. Substituting this value in eq. (8.9), we get

$$I_0 = Ke^0$$

or $K = I_0$

Thus the particular solution for the network is

$$i(t) = I_0\, e^{-\frac{R}{L}t} \qquad .....(8.10)$$

The response is plotted in Fig. 8.6

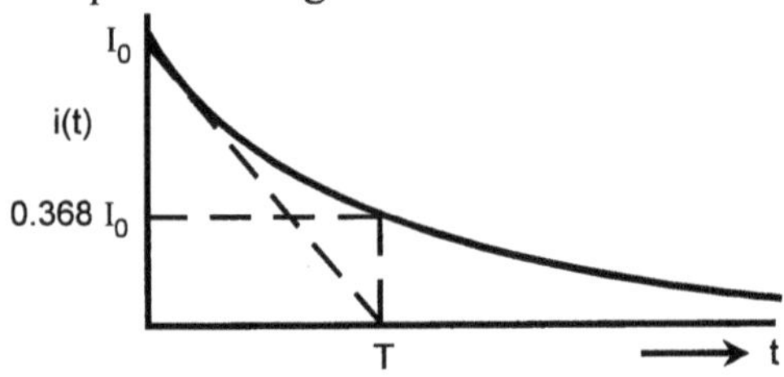

Fig. 8.6 Response of RL circuit

Time constant : An important concept in defining the response of a network is the time constant. In eq. (8.10) for $t = T = \frac{L}{R}$ Sec., the current is

$$i(T) = 0.368\, I_0 \qquad .....(8.11)$$

ie., the current decays to 36.8 per cent of the initial value I_0 in $T = \frac{L}{R}$ Sec. This time is known as the *time constant of the network.* The time constant indicates how fast the response is decaying to zero. When t = 4T, the response decays to approximately 2 per cent of the initial value.

A second interpretation can also be given to the time constant. The rate of change of response at t = 0 is given by

$$\frac{di}{dt}(0) = -\frac{R}{L}I_0$$

$$= -\frac{I_0}{L/R}$$

$$= -\frac{I_0}{T} \qquad .....(8.12)$$

Eq. (8.12) implies that, if the rate of change of current is maintained constant at a value given by eq. (8.12), the response would reach zero value in one time constant 'T'. This is illustrated in Fig. 8.6. Hence the time constant can also be defined as the time required for the response to reach zero value if the initial rate of change of current is maintained constant. In Fig. 8.7, the responses of networks

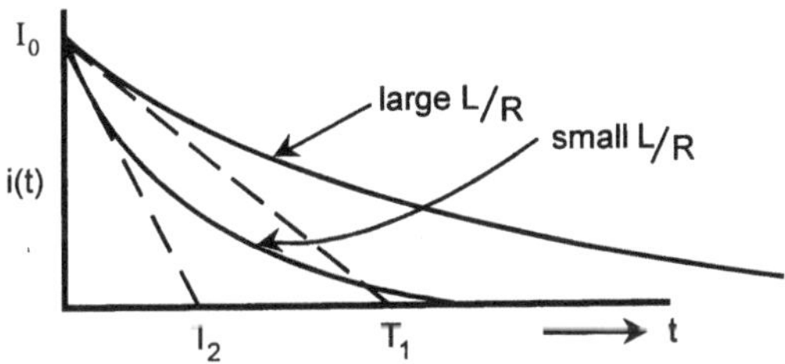

Fig. 8.7 Response of networks with different time constants

with different time constants are shown. The time constants of networks can be changed by changing the ratio $\frac{L}{R}$ of the network.

8.2.2 Unit impulse excitation :

The RL network with a unit impulse voltage excitation is shown in Fig. 8.8.

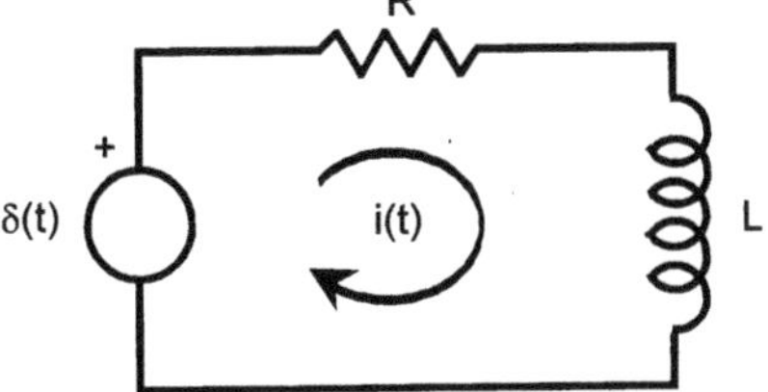

Fig. 8.8 RL network with impulse excitation

The differential equation for the network is given by

$$Ri + L\frac{di}{dt} = \delta(t). \qquad .....(8.13)$$

for t > 0, eq. (8.13) reduces to

$$Ri + L\frac{di}{dt} = 0 \qquad .....(8.14)$$

This equation is identical to eq. (8.5) and hence the general solution is

$$i(t) = Ke^{\frac{-R}{L}t} \qquad .....(8.15)$$

However, the initial condition due to the presence of impulse source has to be evaluated. Since at t = 0, the inductance behaves like an open circuit, the entire impulse voltage appears across the inductance and hence the current in it will change instantaneously. The current at $t = 0^+$ is given by

$$i(0^+) = \frac{1}{L}\int_{0^-}^{0^+} \delta(t)dt \qquad .....(8.16)$$

$$= \frac{1}{L} \qquad \because \left(\int_{0^-}^{0^+} \delta(t)dt = 1\right)$$

Using this value, the constant K in eq. (8.15) can be evaluated as

$$K = \frac{1}{L}$$

Thus the response of the network is given by

$$i(t) = \frac{1}{L} e^{-\frac{R}{L}t} \qquad .....(8.17)$$

This, obviously, is same as the response of the RL network with an initial current of $\frac{1}{L}$. Thus the effect of an impulse source is only to inject an initial current of $\frac{1}{L}$ Amps into the inductance. If an impulse of strength V is used instead of unit impulse, the response would be

$$i(t) = \frac{V}{L} e^{-\frac{R}{L}t} \qquad .....(8.18)$$

8.2.3 Unit step excitation :

An R L network excited by a unit step excitation is shown in Fig. 8.9.

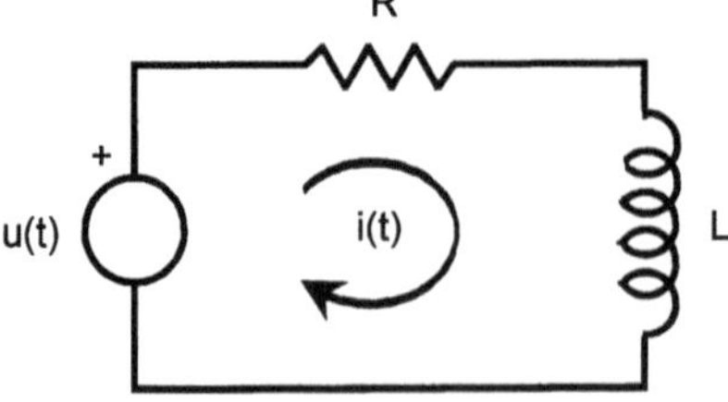

Fig. 8.9 RL network excited by unit step excitation

The differential equation for the network of Fig. 8.9 is

$$Ri + L\frac{di}{dt} = u(t) \qquad .....(8.19)$$

for $t \geq 0$, $u(t) = 1$

Hence eq. (8.19) reduces to

$$Ri + L\frac{di}{dt} = 1 \qquad .....(8.20)$$

The solution for this constant coefficient, linear, first order differential equation can be written as

$$i(t) = i_h(t) + i_p(t) \qquad .....(8.21)$$

Where $i_h(t)$ is called the *homogeneous solution,* which is obtained by solving the homogeneous differential equation

$$Ri + L\frac{di}{dt} = 0 \qquad .....(8.22)$$

and $i_p(t)$ is the particular integral.

The solution of eq. (8.22) is

$$i_h(t) = Ke^{-\frac{R}{L}t} \qquad .....(8.23)$$

where K is a constant to be determined using initial conditions in the network.

The particular integral can be obtained by using any of the standard mathematical techniques available. We will use a simple method to evaluate the particular integral which uses the fact that the particular integral for a given v(t) on the right hand side of the differential equation is of the same form except for an undetermined coefficient. Some trial functions are listed in table 8.1.

Table 8.1

	v(t)	**Choice for particular integral**
1.	V(a constant)	A
2.	$ae^{\alpha t}$	$Be^{\alpha t}$
3.	at^n	$b_0t^n + b_1t^{n-1} + ... + b_{n-1}t + b_n$
4.	$V \operatorname{Cos}\omega t$ or $V \operatorname{Sin}\omega t$	$A\operatorname{Cos}\omega t + B\operatorname{Sin}\omega t$

The trial solution is then substituted in the differential equation, and a set of algebraic equations is obtained by equating coefficients of like terms on both sides of the equality sign. The undetermined coefficients are then determined by solving this set of equations.

Using the above procedure, let

$$i_p(t) = A \qquad .....(8.24)$$

a constant, since the right hand side of eq. (8.20) is a constant. Since any solution must satisfy the differential equation, substituting eq. (8.24) in eq. (8.20) we get

$$RA + L . O = 1 \qquad .....(8.25)$$

$$A = +\frac{1}{R} \qquad .....(8.26)$$

Thus, the solution of eq. (8.20) is

$$i(t) = K\,e^{-\frac{R}{L}t} + \frac{1}{R} \qquad .....(8.27)$$

Assuming that no current was flowing in the network at $t = O^-$, the current at $t = O^+$ is

$$i(O^+) = 0 \qquad .....(8.28)$$

Using this initial condition in eq. (8.27),

$$i(O^+) = O = K + \frac{1}{R}$$

$$K = -\frac{1}{R} \qquad .....(8.29)$$

Hence $$i\,(t) = -\frac{1}{R}\,e^{-\frac{R}{L}t} + \frac{1}{R}$$

$$= \frac{1}{R}\left(1 - e^{-\frac{R}{L}t}\right) \qquad .....(8.30)$$

This is the response of the network of Fig. 8.9.

The variation of current i(t) is shown in the graph of Fig. (8.10). The response is seen to rise exponentially and approach the value $\frac{1}{R}$ at $t = \infty$.

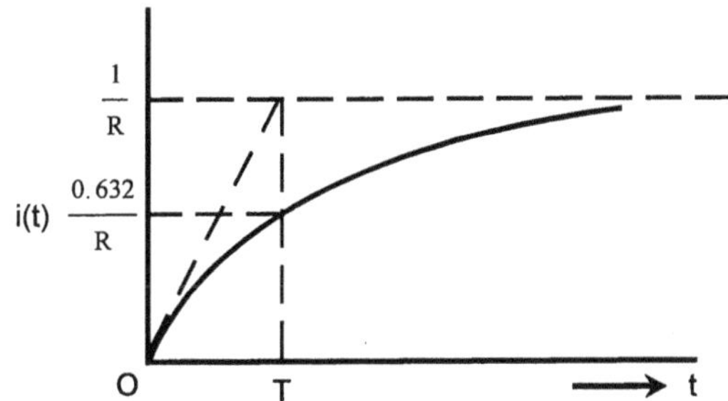

Fig. 8.10 Response of R-L network for a step input

Recall the definition of a time constant in an R L circuit in which the current was exponentially decaying. In a similar manner, we can define the time constant for this network also, in which the current is exponentially increasing. Thus

at $$t = T = \frac{L}{R}$$

eq. (8.30) becomes

$$i(T) = \frac{L}{R}\,(1 - e^{-1})$$

$$= \frac{0.632}{R} \qquad \text{.....(8.31)}$$

Now consider the response of the network at $t = \infty$. From eq. (8.30)

$$i(\infty) = \frac{1}{R}$$

Using this value in eq. (8.31)

$$i(t) = 0.632\ i\,(\infty).$$

Thus the time constant for a network, in which the response is increasing exponentially, can be defined as the time required for the response to reach 63.2% of the final value or its value at $t = \infty$. For an R-L network.

$$T = \frac{L}{R}$$

A different interpretation can be given for the time constant in terms of the rate of rise of the response at $t = 0$.

Thus, from eq. (8.30)

$$\frac{di}{dt} = -\frac{1}{R}\left(-\frac{R}{L}\right)e^{-\frac{R}{L}t}$$

at t = 0

$$\frac{di}{dt}(0) = \frac{1}{R} \cdot \frac{R}{L}$$

$$= \frac{\frac{1}{R}}{\frac{L}{R}}$$

$$= \frac{\frac{1}{R}}{T} \qquad .....(8.32)$$

Thus, if the rate of rise of current is maintained constant at its value at t = 0, given by eq. (8.32), the response reaches the value $i(\infty) = \frac{1}{R}$ in one time constant.

Hence, the time constant for an exponentially increasing response can also be defined as the time required for the response to reach the final value if the rate of rise of the response is maintained constant at its value at t = 0. This is shown clearly in Fig. 8.10.

The response of an R-L network excited by a unit step function is shown to contain two terms in its solution.

a) The homogeneous solution, $i_h(t)$
b) The particular integral, $i_p(t)$

The homogeneous solution $i_h(t)$ is also known as a *force tree solution* since it does not depend on the forcing function u(t). It is obtained by making the right hand side of eq. (8.20) to be zero. This part of the solution depends entirely on the parameters of the network and it decays to zero as $t \rightarrow \infty$. Thus it exists only for a short time and therefore it is also known as the *transient solution.*

On the other hand, the particular integral depends on the nature of the forcing function and is of the same general form as the forcing function. Hence this solution is also known as the *forced solution.* Unlike the force free solution, this does not approach zero as $t \rightarrow \infty$ unless the forcing function itself approaches zero as $t \rightarrow \infty$. Hence this part of the solution is also known as the *steady state solution.* Thus the total solution can also be written as

$$i(t) = i_{tr}(t) + i_{ss}(t)$$

where $i_{tr}(t)$ is the transient response

and $i_{ss}(t)$ is the steady state response.

8.2.4 Sinusoidal excitation :

Consider the network in Fig. 8.11 which shows an R-L network excited by a sinusoidal source.

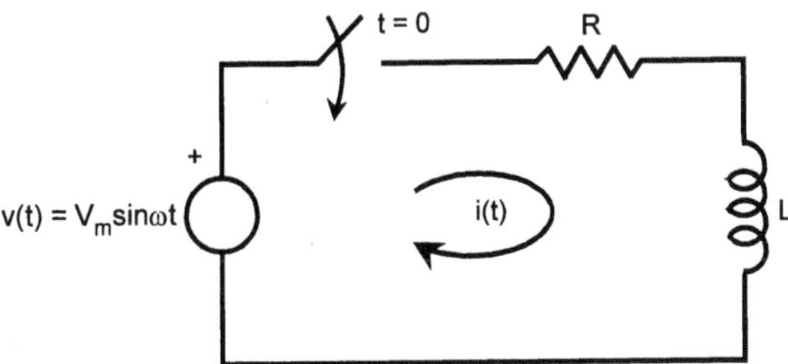

Fig. 8.11 R-L network excited by a sinusoidal source

The differential equation for this network is

$$Ri + L\frac{di}{dt} = V_m \text{Sin}\omega t \qquad .....(8.33)$$

The homegeneous solution can be obtained in the usual way as

$$i_h(t) = Ke^{-\frac{R}{L}t} \qquad .....(8.34)$$

The particular integral can be obtained by assuming a solution as given in table 8.1

$$i_p(t) = A\text{Cos}\omega t + B\text{Sin}\omega t. \qquad .....(8.35)$$

Substituting this trial solution in eq. (8.33)

$$RA\text{Cos}\omega t + RB\text{Sin}\omega t + L(-\omega A\text{Sin}\omega t + \omega B\text{Cos}\omega t)$$
$$= V_m\text{Sin}\omega t \qquad .....(8.36)$$

Collecting the terms of Cosωt and Sinωt

$$(RA + \omega LB)\text{Cos}\omega t + (RB - \omega LA)\text{Sin}\omega t$$
$$= V_m\text{Sin}\omega t \qquad .....(8.37)$$

Equating the coefficients of Sinωt and Cosωt on both sides of eq. (8.37), we get

$$RA + \omega LB = 0 \qquad .....(8.38)$$

$$RB - \omega LA = V_m \qquad .....(8.39)$$

Solving eq. (8.38) and (8.39) for A and B

we get

$$A = -\frac{\omega L V_m}{R^2 + \omega^2 L^2} \qquad .....(8.40)$$

and $$B = \frac{V_m R}{R^2 + \omega^2 L^2} \qquad .....(8.41)$$

Therefore, the particular integral is given by

$$i_p(t) = -\frac{\omega L V_m}{R^2+\omega^2L^2}\text{Cos}\omega t + \frac{V_m R}{R^2+\omega^2 L^2}\text{Sin}\omega t$$

$$= \frac{V_m}{R^2+\omega^2L^2}[R\text{Sin}\,\omega t - \omega L\text{Cos}\,\omega t]$$

$$= \frac{V_m}{\sqrt{R^2+\omega^2L^2}}\text{Sin}(\omega t - \phi) \qquad .....(8.42)$$

Where $\quad \text{Tan}\,\phi = \dfrac{\omega L}{R}$

The complete solution is given by

$$i(t) = Ke^{-\frac{R}{L}t} + \frac{V_m}{\sqrt{R^2+\omega^2L^2}}\text{Sin}(\omega t - \phi) \qquad .....(8.43)$$

The constant K can be determined using the initial conditions in the network. Assuming that there was no current in the inductance at $t = O^-$, the current at $t = O^+$ will also be equal to zero. Using this value in eq. (8.43) at t = 0, we get

$$O = K + \frac{V_m}{\sqrt{R^2+\omega^2L^2}}\text{Sin}(-\phi)$$

$$K = \frac{V_m}{\sqrt{R^2+\omega^2L^2}} \cdot \frac{\omega L}{\sqrt{R^2+\omega^2L^2}} \qquad \left(\because \tan\phi = \frac{\omega L}{R}, \sin\phi = \frac{\omega L}{\sqrt{R^2+\omega^2L^2}}\right)$$

$$= \frac{\omega L V_m}{R^2+\omega^2L^2}$$

$$\therefore \quad i(t) = \frac{\omega L V_m}{R^2+\omega^2L^2}e^{-\frac{R}{L}t} + \frac{V_m}{\sqrt{R^2+\omega^2L^2}}\text{Sin}(\omega t - \phi) \qquad .....(8.44)$$

Thus the response for a sinusoidal excitation also consists of a transient part which decays with time and a steadystate term which is of the same form as the excitation.

8.3 RESPONSE OF RC NETWORKS :

Now, let us consider an RC network excited by internal energy, impulse, step and sinusoidal excitations. Since the differential equation governing this network is essentially of the same type, as for the RL network considered in previous section, the detailed method of solution is not given.

8.3.1 Internal energy excitation :

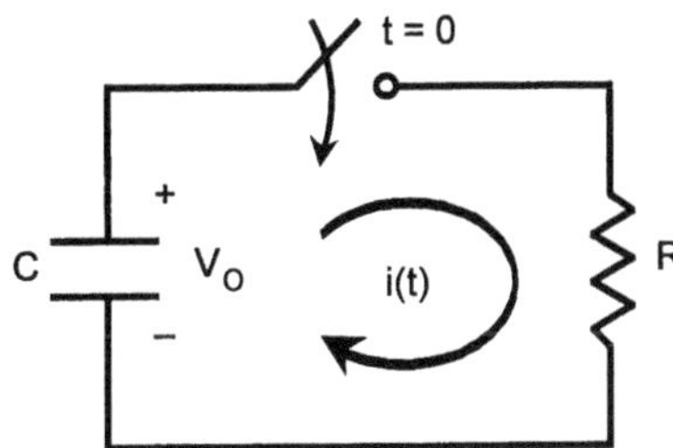

Fig. 8.12 RC circuit excited by internal energy

Writing the loop equation, we get

$$Ri + \frac{1}{C}\int_{-\infty}^{t} i\,dt = 0 \qquad .....(8.45)$$

Differentiating eq. (8.45) to get a differential equation,

$$R\frac{di}{dt} + \frac{i}{C} = 0 \qquad .....(8.46)$$

The solution for this homogeneous equation is easily obtained as

$$i(t) = Ke^{-\frac{1}{RC}t} \qquad .....(8.47)$$

Where K is a constant to be determined using initial state of the network.

At $t = O^+$, the network of Fig. 8.12 reduces to :

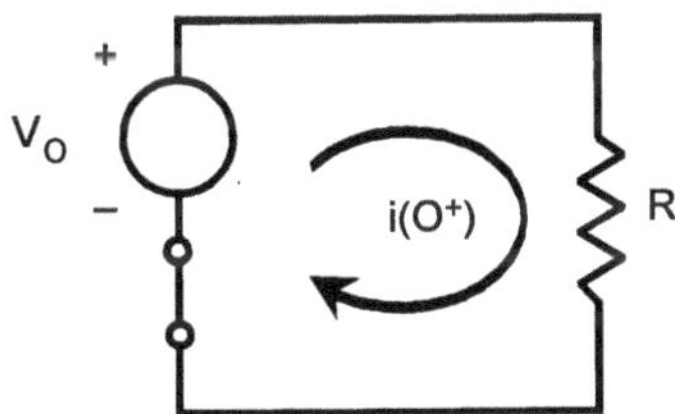

Fig. 8.13 Network of Fig. 8.12 at $t = O^+$

From Fig. 8.13

$$i(O^+) = \frac{V_o}{R} \qquad .....(8.48)$$

Using this initial condition in eq. (8.47) we get

$$i(O^+) = \frac{V_o}{R} = K$$

$$\therefore \qquad i(t) = \frac{V_o}{R}e^{-\frac{1}{RC}t} \qquad .....(8.49)$$

This is plotted in Fig. 8.14

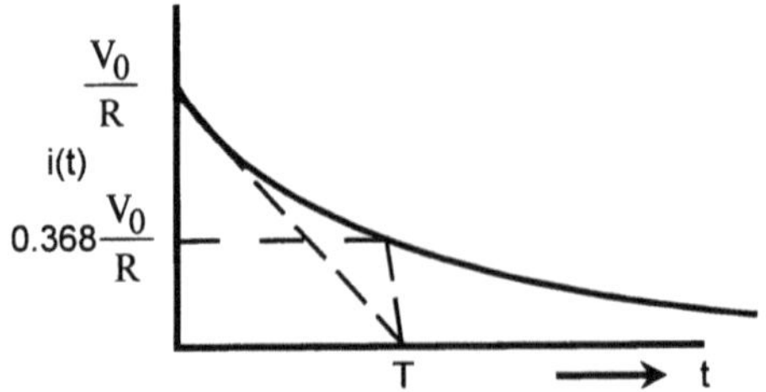

Fig. 8.14 Response of RC circuit to internal excitation

The time constant for RC circuit is defined in the same way as for an RL circuit. Time constant is the time required for the response to reach 0.368 of the initial value. It can also be defined, as in the case of RL network, to be the time required for the response to go to zero if the initial rate of decay of the response is maintained constant. This is shown in Fig. 8.14. Obviously.

$$T = RC \qquad .....(8.50)$$

8.3.2 Impulse excitation :

Consider the RC network excited by an impulse current source as shown in Fig. 8.15.

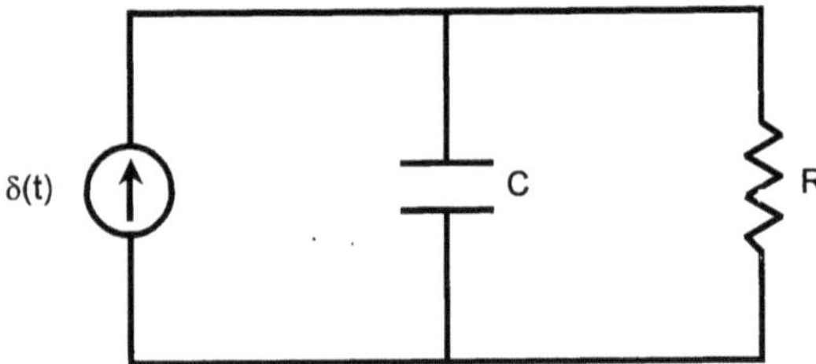

Fig. 8.15 An RC circuit excited by an impulse current source

Since impulse current source is effective only during $t = O^-$ to $t = O^+$ and C tends to act like a short circuit during this time, the entire impulse current passes through the capacitor. Due to this impulse current, the voltage across the capacitor changes instantaneously and is given by

$$V_C(O^+) = \frac{1}{C}\int_{O^-}^{O^+} \delta(t)dt$$

$$= \frac{1}{C} \qquad .....(8.51)$$

Hence the effect of the impulse source is only to transfer some energy to the capacitor in the internal O^- to O^+ and then disappear ! Hence the network for t > 0 is given in Fig. 8.16.

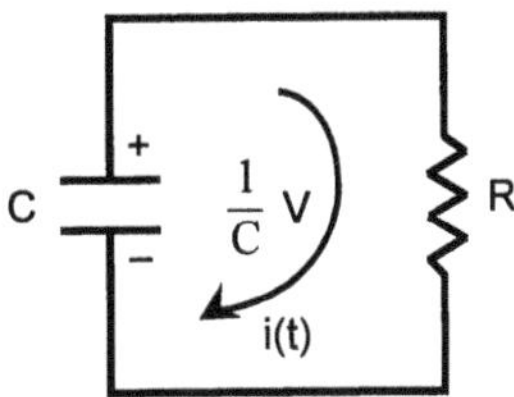

Fig. 8.16 The network of Fig. 8.15 for t > 0

This network is same as network in Fig. 8.12 except for the initial voltage across the capacitor. The initial voltage across the capacitor is $\frac{1}{C}$ volts instead of V_0. Hence the response of the network is

$$i(t) = \frac{1}{CR} e^{-\frac{1}{RC}t} \qquad .....(8.52)$$

8.3.3 Unit step excitation :

Consider the network shown in Fig. 8.17

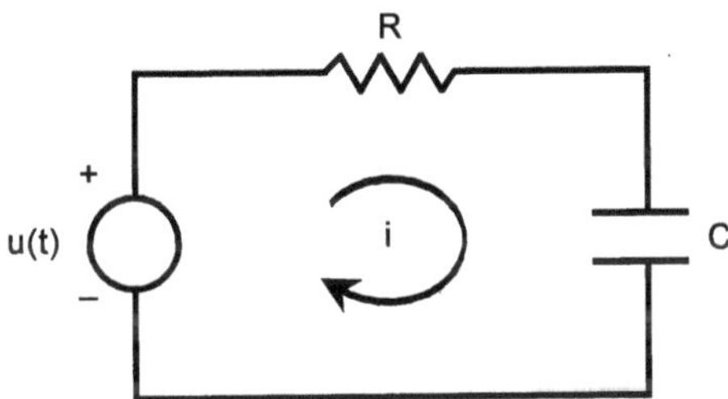

Fig. 8.17 RC network excited by a unit step source

The differential equation for this network is given by

$$Ri + \frac{1}{C} \int_{-\infty}^{t} i\,dt = u(t) = 1 \qquad .....(8.53)$$

To get a differential equation, we differentiate eq. (8.53) thus,

$$R\frac{di}{dt} + \frac{i}{C} = 0 \qquad .....(8.54)$$

This equation is same as eq. (8.46) and the solution is

$$i(t) = Ke^{-\frac{1}{RC}t} \qquad .....(8.55)$$

Now, to determine the constant K, we have to evaluate the initial condition.

At t = 0^+, assuming an uncharged capacitance, the network of Fig. 8.17 is represented as in Fig. 8.18.

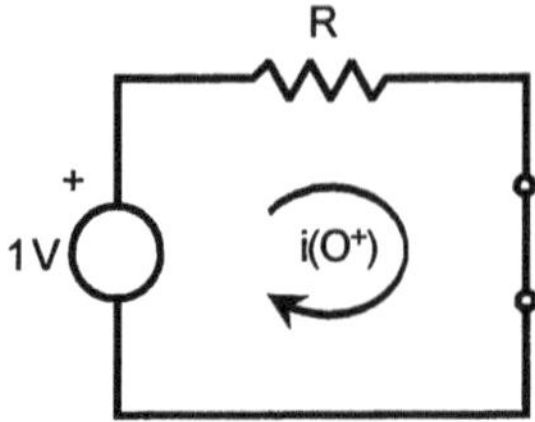

Fig. 8.18 The RC network of Fig. 8.17 at t = 0^+

Thus $$i(0^+) = \frac{1}{R}$$

Using this initial condition in eq. (8.55) we get

$$i(t) = \frac{1}{R} e^{-\frac{1}{RC}t} \qquad .....(8.56)$$

This response is also a decaying exponential with a time constant T = RC.

Instead of taking the current in the network as the response, if we consider the voltage across the capacitor as the response.

$$v_C(t) = \frac{1}{C} \int_{-\infty}^{t} i\,dt \qquad .....(8.57)$$

Since the voltage across the capacitor is 0 at t = 0^+,

$$v_C(t) = \frac{1}{C} \int_{0^+}^{t} i\,dt = \frac{1}{C} \int_{0}^{t} \frac{1}{R} e^{-\frac{1}{RC}t} dt = \frac{1}{RC} \left. \frac{e^{-\frac{1}{RC}t}}{-\frac{1}{RC}} \right|_0^t$$

$$= -\left[e^{-\frac{1}{RC}t} - 1 \right] = 1 - e^{-\frac{1}{RC}t} \qquad .....(8.58)$$

The voltage across the capacitor is plotted in Fig. 8.19. This is a rising exponential and at t = ∞, the voltage reaches the value of IV , which is the source voltage. Thus the capacitor gets charged to the voltage of the source at t = ∞ .

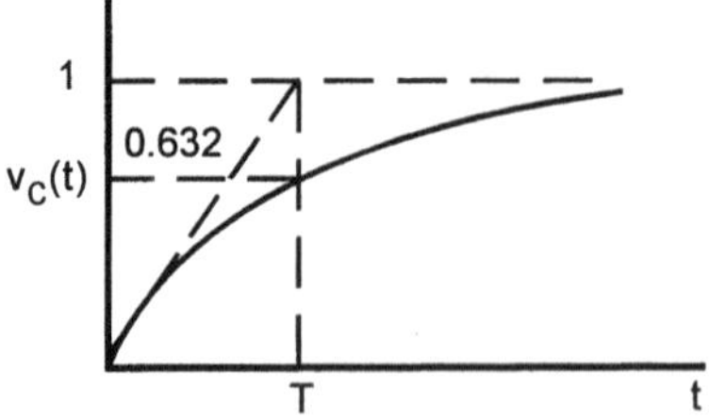

Fig. 8.19 The voltage across the capacitor

The time constant for this exponentially increasing response is defined in a similar manner to that of the RL network. The time constant is defined as the time required for the response to reach 63.2% of the final value and is equal to RC.

We also note that the response consists of a transient part given by

$$v_{C\,tr} = -\,e^{-\frac{1}{RC}t}$$

and a steady state part given by

$$v_{C\,ss} = 1$$

and the total response is given by

$$v_C(t) = v_{C\,tr} + v_{Css}$$

8.3.4 Sinusoidal excitation :

Consider the network in Fig. 8.20

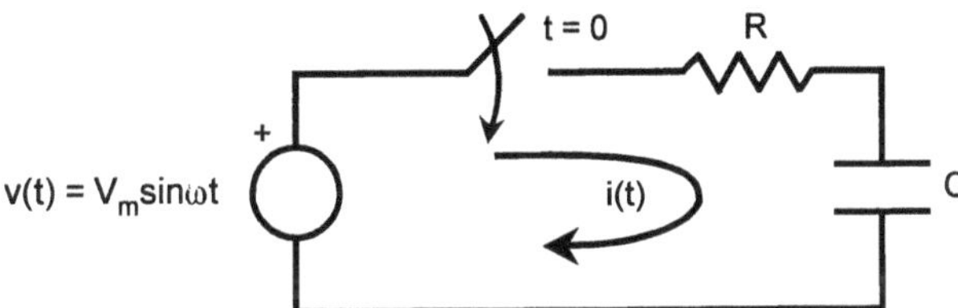

Fig. 8.20 RC network excited by a sinusoidal source

The loop equation for this network is given by

$$Ri + \frac{1}{C}\int_{-\infty}^{t} i\,dt = V_m \,\mathrm{Sin}\,\omega t \qquad .....(8.59)$$

Differentiating, we get

$$R\frac{di}{dt} + \frac{i}{C} = V_m\,\omega\,\mathrm{Cos}\,\omega t \qquad .....(8.60)$$

The homogeneous solution is given by

$$i_h(t) = K\,e^{-\frac{1}{RC}t} \qquad .....(8.61)$$

The particular integral is obtained as in section 8.2.4 and is given by

$$i_p\,(t) = \frac{V_m}{\sqrt{R^2 + \frac{1}{\omega^2 C^2}}}\,\mathrm{Sin}\,(\omega t + \phi) \qquad .....(8.62)$$

where $\tan \phi = \dfrac{1}{\omega CR}$

The complete solution, then, is given by

$$i(t) = Ke^{-\frac{1}{RC}t} + \frac{V_m}{\sqrt{R^2 + \frac{1}{\omega^2 C^2}}} \cdot \text{Sin}\,(\omega t + \phi) \qquad .....(8.63)$$

The constant K can be evaluated using the initial condition in the network. Assuming that there was no charge on the capacitor at $t = 0^-$, the current at $t = 0^+$ can be easily evaluated as

$$i(0^+) = 0 \qquad .....(8.64)$$

Using this initial value in eq. (8.63) we get,

$$K = -\frac{V_m}{\sqrt{R^2 + \frac{1}{\omega^2 C^2}}} \cdot \text{Sin}\,\phi$$

$$= -\frac{V_m}{\sqrt{R^2 + \frac{1}{\omega^2 C^2}}} \cdot \frac{\frac{1}{\omega C}}{\sqrt{R^2 + \frac{1}{\omega^2 C^2}}}$$

$$= -\frac{V_m/\omega C}{R^2 + 1/\omega^2 C^2} \qquad .....(8.65)$$

The complete solution is given by

$$i(t) = -\frac{V_m/\omega C}{R^2 + \frac{1}{\omega^2 C^2}} e^{-\frac{1}{RC}t} + \frac{V_m}{\sqrt{R^2 + \frac{1}{\omega^2 C^2}}} \text{Sin}\,(\omega t + \phi) \qquad .....(8.66)$$

It is seen from eq. (8.66) that the response of an RC circuit also consists of a transient part which dies down to zero and a steady state part which is of similar form as that of the excitation.

In section 8.2 and 8.3 we have considered the RL and RC networks excited with internal energy, impulse, step and sinusoidal excitation. The responses due to two other types of excitation discussed in section 8.1 namely, polynomial and exponential type of excitations are illustrated in the following examples.

Example 8.1

Find i(t) in the network of Fig. 8.21.The inductor has no current at $t = 0^-$

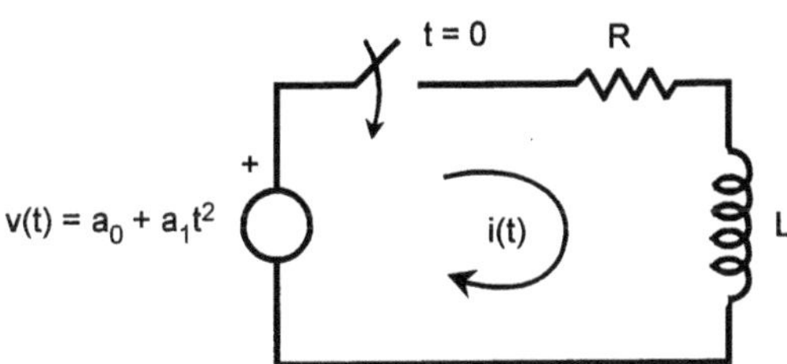

Fig. 8.21 Network for example 8.1

Solution :

The loop equation for the network is

$$Ri + L\frac{di}{dt} = a_0 + a_1t^2 \qquad .....(8.67)$$

The homogeneous solution is

$$i_h(t) = K e^{-\frac{R}{L}t}$$

The particular integral is obtained by assuming a trial solution as given in Table 8.1.

$$i_p(t) = b_0 + b_1t + b_2t^2$$

Substituting the trial solution in eq.

$$R(b_o + b_1t + b_2t^2) + L\,[b_1 + 2b_2t] = a_0 + a_1t^2 \qquad .....(8.68)$$

Equating the coefficients of like terms on both sides of eq. (8.68), we get

$$Rb_0 + Lb_1 = a_0$$

$$Rb_1 + 2Lb_2 = 0 \qquad .....(8.69)$$

$$Rb_2 = a_1$$

Solving the set of equations (8.69) for b_0, b_1 and b_2 we get

$$b_0 = \frac{1}{R}\left(a_0 + \frac{2L^2a_1}{R^2}\right)$$

$$b_1 = -\frac{2L}{R^2}a_1$$

and $$b_2 = \frac{a_1}{R}$$

The complete response of the network is given by

$$i(t) = Ke^{-\frac{R}{L}t} + \frac{1}{R}\left(a_0 + \frac{2L^2a_1}{R^2}\right) - \frac{2L}{R^2}a_1t + \frac{a_1}{R}t^2 \qquad .....(8.70)$$

The constant K can be evaluated noting that i $(O^+) = 0$

$$O = K + \frac{1}{R}\left(a_o + \frac{2L^2a_1}{R^2}\right)$$

$$\text{or } K = -\frac{1}{R}\left(a_0 + \frac{2L^2a_1}{R^2}\right)$$

Thus

$$i(t) = -\frac{1}{R}\left(a_0 + \frac{2L^2a_1}{R^2}\right)e^{-\frac{R}{L}t} + \frac{1}{R}\left(a_0 + \frac{2L^2a_1}{R^2}\right) - \frac{2L}{R^2}a_1t + \frac{a_1}{R}t^2 \qquad .....(8.71)$$

Example 8.2

Find the current i(t) in the network of Fig. 8.22. The capacitor is uncharged at $t = O^-$

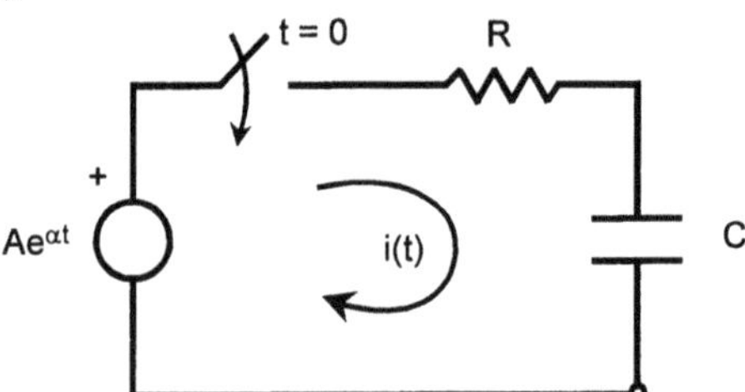

Fig. 8.22 RC circuit excited by an exponential source

The loop equation can be written as

$$Ri + \frac{1}{C}\int_{-\infty}^{t} i\,dt = Ae^{\alpha t} \qquad .....(8.72)$$

Differentiating,

$$R\frac{di}{dt} + \frac{i}{C} = A\,\alpha\,e^{\alpha t} \qquad .....(8.73)$$

The homogeneous solution can be obtained as usual.

$$i_h(t) = Ke^{-\frac{1}{RC}t} \qquad .....(8.74)$$

The particular integral can be obtained by assuming a trial solution as indicated in table 8.1. However, caution is to be exercised here since the homogeneous solution is also of the same form as that of the excitation function. The trial solution can be assumed as

$$i_p(t) = B\,e^{\alpha t} \qquad .....(8.75)$$

provided $\alpha \neq \frac{1}{RC}$

This is because, if $a = -\frac{1}{RC}$ the homogenous solution and the particular integral will be exactly the same! This condition is also known as condition of resonance.

Now, assuming $\alpha \neq -\frac{1}{RC}$, substituting eq. (8.75) in eq. (8.73), we get

$$RB\alpha e^{\alpha t} + \frac{B}{C}e^{\alpha t} = A\alpha e^{\alpha t}$$

or $$B\left(R\alpha + \frac{1}{C}\right) = A\alpha$$

$$B = \frac{A\alpha}{R\alpha + \frac{1}{C}} \qquad .....(8.76)$$

The complete solution is given by

$$i(t) = Ke^{-\frac{1}{RC}t} + \frac{A\alpha}{R\alpha + \frac{1}{C}}e^{\alpha t} \qquad .....(8.77)$$

The constant K is to be evaluated using the initial condition in the network.

At $t = O^+$, the network in Fig. 8.22 reduces to :

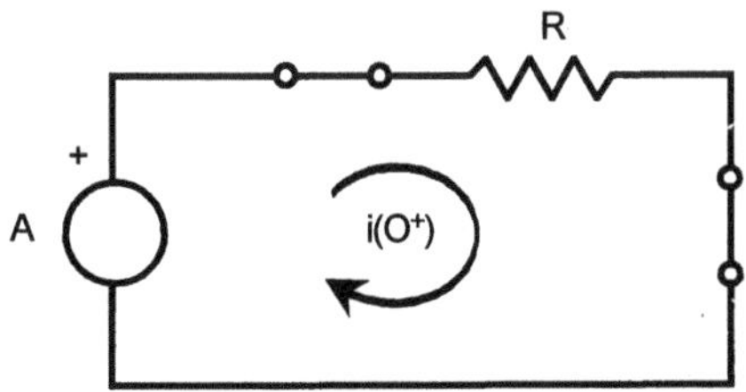

Fig. 8.23 Network of Fig. 8.22 at $t = O^+$

From Fig. 8.23,

$$i(O^+) = \frac{A}{R} \qquad .....(8.78)$$

Substituting eq. (8.78) in eq. (8.77), we get

$$\frac{A}{R} = K + \frac{A\alpha}{R\alpha + \frac{1}{C}}$$

$$K = \frac{A}{R}\left(1 - \frac{\alpha}{\alpha + \frac{1}{RC}}\right) \qquad .....(8.79)$$

Thus the complete solution is given by

$$i\,(t) = \frac{A}{R}\left(1 - \frac{\alpha}{\alpha + \frac{1}{RC}}\right)e^{-\frac{1}{RC}t} + \frac{A\alpha}{R\alpha + \frac{1}{C}}e^{\alpha t}$$

Example 8.3

Find i (t) in the network of Fig. 8.24. A current of 2A was flowing at $t = 0^-$ in the inductor.

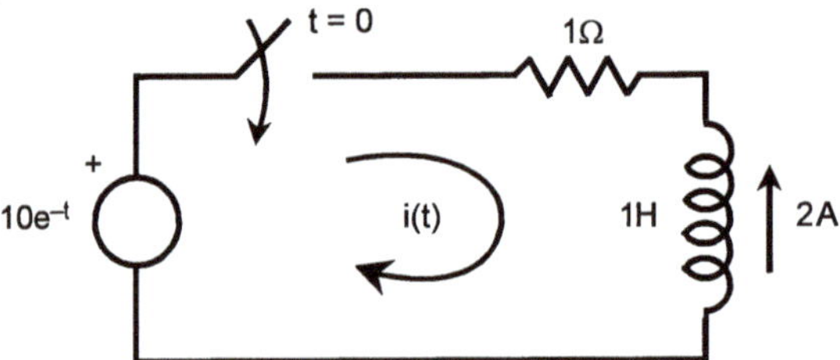

Fig. 8.24 Network for example 8.3

Solution :

The loop equation for current i (t) is

$$i + \frac{di}{dt} = 10e^{-t} \qquad .....(8.80)$$

The complementary solution is

$$i_h(t) = Ke^{-t}$$

Since the homogenous solution is of the same form as the excitation, the trial solution indicated in Table 8.1 can not be used. The technique to be used for obtaining the particular integral is, to assume the complete solution to be of the same form as i_h (t) but the arbitrary constant is replaced by a function of time. Thus, the solution is assumed as

$$i\,(t) = K(t)e^{-t} \qquad .....(8.81)$$

This is substituted in eq. (8.80) to yield,

$$K(t)e^{-t} + \mathring{K}(t)e^{-t} - K(t)e^{-t} = 10e^{-t}$$

which gives

$$\mathring{K}(t) = 10 \quad .....(8.82)$$

Integrating eq. (8.82), we get

$$K(t) = 10t + c \quad .....(8.83)$$

where c is a constant to be determined using the initial condition.

At $t = 0^+$, we have

$$i(0^+) = -2\text{ A}$$

Using this value in eq. (8.81)

$$-2 = K(0^+)$$

Using this value of $K(0^+)$ in eq. (8.83), we get

$$-2 = 0 + c$$

or $c = -2$

Thus the complete solution is given by

$$i(t) = (10t - 2)e^{-t}$$

Example 8.4

Find $v_C(t)$ in the network of Fig. 8.25

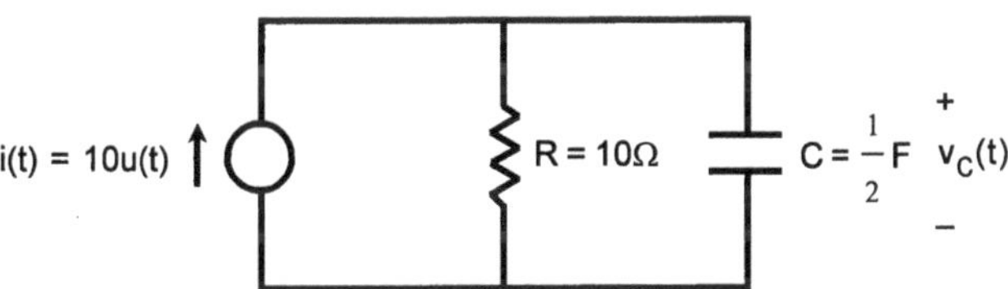

Fig. 8.25 Network for example 8.4

Solution :

Method 1 : Consider the dual of this network given in Fig. 8.26

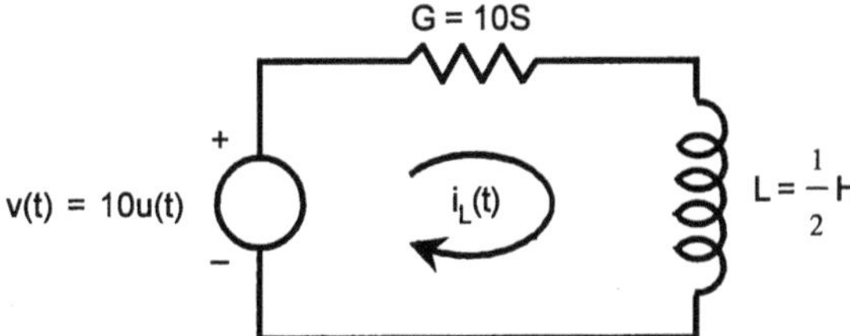

Fig. 8.26 Dual of the network in Fig. 8.25

This is an RL circuit excited by a step voltage source of 10 V, the solution for which is already obtained in section 8.2.3. The solution for a unit step input is given by eq. (8.30). Since the input in the present network is 10 units, the response is given by

$$i_L(t) = \frac{10}{\frac{1}{10}}\left(1 - e^{-\frac{t}{10\times\frac{1}{2}}}\right) = 100\left(1 - e^{-\frac{t}{5}}\right)$$

This is the response of the dual network. By changing the current $i_L(t)$ to its dual quantity, $v_C(t)$, we get

$$v_C(t) = 100\left(1 - e^{-\frac{t}{5}}\right)$$

which is the required solution.

Method 2 : Writing the node voltage equation for the network in Fig. 8.25

$$\frac{1}{2}\frac{dv_C}{dt} + \frac{v_C}{10} = 10u(t)$$

$$\frac{dv_C}{dt} + \frac{v_C}{5} = 20 \qquad \text{.....(8.84)}$$

The homogeneous and particular integrals of eq. (8.84) can be obtained in the usual manner as

$$v_{ch}(t) = Ke^{-\frac{1}{5}t}$$

$$V_{CSS}(t) = 100$$

The complete solution is

$$V_C(t) = v_{ch}(t) + V_{CSS}(t) = Ke^{-\frac{1}{5}t} + 100 \qquad \text{.....(8.85)}$$

At $t = 0^+$, $v_C(0^+) = 0$

Substituting in eq. (8.85) and solving for K, we get

$$K = -100$$

∴ The complete solution is

$$v_c(t) = -100e^{-\frac{1}{5}t} + 100$$

or

$$v_C(t) = 100\left(1 - e^{-\frac{1}{5}t}\right)$$

8.4 RESPONSE OF RLC NETWORKS :

In section 8.2 and 8.3, the response of RL, RC networks was obtained for different common excitations. The equations for all these networks were first order constant coefficient differential equations. The solution for these equations contained one arbitrary constant, which was determined by using the initial condition in the network. This is usually the current through the inductance or voltage across the capacitance at $t = O^+$. For RLC networks, the equations will be second order differential equations with constant coefficients. The solution will contain two arbitrary constants which require two initial conditions to be known. These initial conditions are the current through the inductance and voltage across the capacitance at $t = O^+$. These initial conditions can be transformed to either $i(O^+)$ and $\frac{di}{dt}(O^+)$ or $v(O^+)$ and $\frac{dv}{dt}(O^+)$ so that the two arbitrary constants can be evaluated. If in these RLC circuits, R is made equal to zero, we get the special case of LC networks.

8.4.1 Internal energy excitation :

Consider the network of Fig. 8.27 in which a series RLC network is shown with the current in inductance to be I_0 and voltage across the capacitance to be V_0 at $t = O^-$.

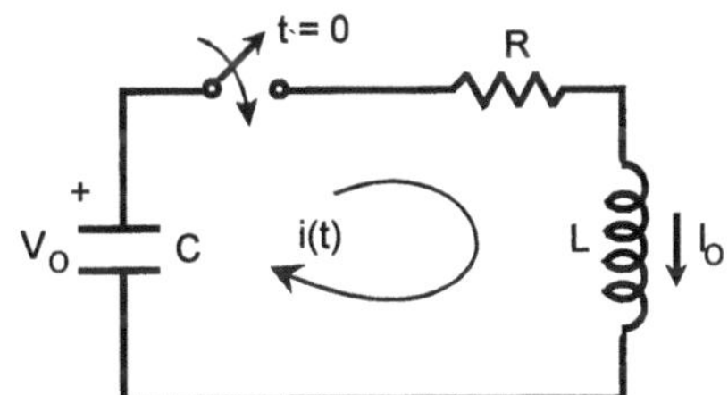

Fig. 8.27 RLC Network with internal energy excitation

Since the current in the inductance and voltage across the capacitance do not change instantaneously, the current in inductance and voltage across the capacitance will remain I_0 and V_0 respectively at $t = O^+$. Writing the loop equation for the network in Fig. 8.27, we get

$$L\frac{di}{dt} + Ri + \frac{1}{C}\int_{-\infty}^{t} i\,dt = 0 \qquad .....(8.86)$$

Differentiating eq. (8.86) to eliminate the integral, we get

$$L\frac{d^2i}{dt} + R\frac{di}{dt} + \frac{i}{C} = 0 \qquad .....(8.87)$$

This is a constant coefficient, homogenous, linear differential equation which has only a complementary function as solution. The auxiliary equation is given by

$$Lm^2 + Rm + \frac{1}{C} = 0 \qquad .....(8.88)$$

Eq. (8.88) has two roots for m, say m_1, m_2

$$m_1, m_2 = \frac{-R \pm \sqrt{R^2 - 4\frac{L}{C}}}{2L} = -\frac{R}{2L} \pm \sqrt{\left(\frac{R}{2L}\right)^2 - \frac{1}{LC}} \qquad .....(8.89)$$

Three different cases arise, depending on the values of R, L and C.

Case 1 : $\frac{R}{2L} > \frac{1}{\sqrt{LC}}$

In this case, the two roots m_1, m_2 are real, distinct and negative. The solution is then given by

$$i(t) = K_1 e^{m_1 t} + K_2 e^{m_2 t} \qquad .....(8.90)$$

The constants K_1 and K_2 are evaluated using the initial conditions.

In eq. (8.86), at $t = O^+$ we get

$$L\frac{di}{dt}(O^+) + Ri(O^+) + \frac{1}{C}\int_{-\infty}^{O^+} idt = 0 \qquad .....(8.91)$$

We note that $i(O^+) = I_0$ and $\frac{1}{C}\int_{-\infty}^{O^+} idt$ is the voltage across the capacitor at $t = O^+$ and is equal to $-V_0$ for the polarity shown.

Using these values, we get

$$L\frac{di}{dt}(O^+) + R I_0 - V_0 = 0$$

$$\therefore \quad \frac{di}{dt}(O^+) = \frac{V_0 - RI_0}{L} \qquad .....(8.92)$$

Using the values of $i\,(O^+)$ and $\frac{di}{dt}\,(O^+)$ in eq. (8.90) the constants K_1 and K_2 can be evaluated. For simplicity let $I_0 = 0$. Then the response given by eq. (8.90) can be plotted as shown in Fig. 8.28.

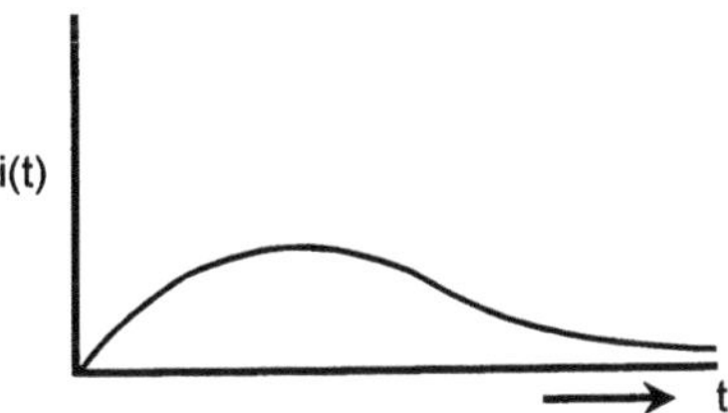

Fig. 8.28 Response of RLC Network for m_1, m_2 real and distinct

Since both m_1 and m_2 are negative the response decays to zero at $t = \infty$. At $t = 0$ and $t = \infty$, the current is zero, the current rises to a maximum in an exponential way and then decays to zero exponentially. This type of response is termed as over damped response. The reason and meaning for this will be explained later.

Case 2 : $$\frac{R}{2L} = \frac{1}{\sqrt{LC}}$$

Then $$m_1 = m_2 = -\frac{R}{2L}$$

The solution is given by

$$i\,(t) = e^{\frac{-R}{2L}t}\,(k_1 t + k_2) \qquad .....(8.93)$$

The constants k_1 and k_2 can be evaluated by using the initial conditions as before.

Again, assuming $I_0 = 0$, the response can be plotted as shown in Fig. 8.29. This type of response is termed as critically damped response.

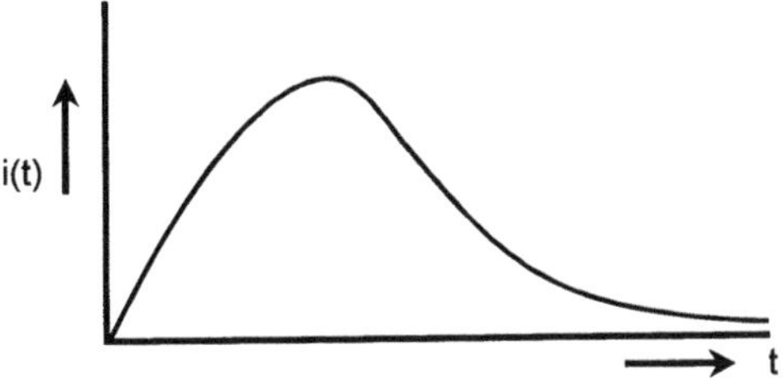

Fig. 8.29 Response of RLC Network for $m_1 = m_2$

Case 3 : $\frac{R}{2L} < \frac{1}{\sqrt{LC}}$

In this case, m_1 and m_2 are complex and conjugate

$$m_1, m_2 = \frac{R}{2L} \pm \sqrt{\frac{1}{LC} - \frac{R^2}{4L^2}}$$

$$= \alpha \pm j\beta$$

The solution is given by

$$i(t) = K_1 e^{(-\alpha+j\beta)t} + K_2 e^{(-\alpha-j\beta t)} \quad \text{.....(8.94)}$$

$$= e^{-\alpha t}\left(K_1 e^{j\beta t} + K_2 e^{-j\beta t}\right) \quad \text{.....(8.95)}$$

Noting that $e^{j\beta t} = \cos\beta t + j\sin\beta t$

and $e^{-j\beta t} = \cos\beta t - j\sin\beta t$

Eq. (8.95) becomes

$$i(t) = e^{-\alpha t}\left[(K_1 + K_2)\cos\beta t + j(K_1 - K_2)\sin\beta t\right] \quad \text{.....(8.96)}$$

Replacing the constants $K_1 + K_2$ and $j(K_1 - K_2)$ by new constants A and B respectively, eq. (8.96) can be written as :

$$i(t) = e^{-\alpha t}(A\cos\beta t + B\sin\beta t) \quad \text{.....(8.97)}$$

Eq. (8.97) has two more alternate forms which are useful

$$i(t) = Ce^{-\alpha t}\sin(\beta t + \phi) \quad \text{.....(8.98)}$$

and

$$i(t) = Ce^{-\alpha t}\cos(\beta t - \theta) \quad \text{.....(8.99)}$$

where $C = \sqrt{A^2 + B^2}$

$$\phi = \tan^{-1}\frac{A}{B} \quad \text{and} \quad \theta = \tan^{-1}\frac{B}{A}$$

The constants in eqs. (8.97), (8.98) or (8.99) can be evaluated by using the initial conditions as discussed earlier.

If $I_0 = 0$, the response can be plotted as shown in Fig. 8.30. This response is termed as under damped response.

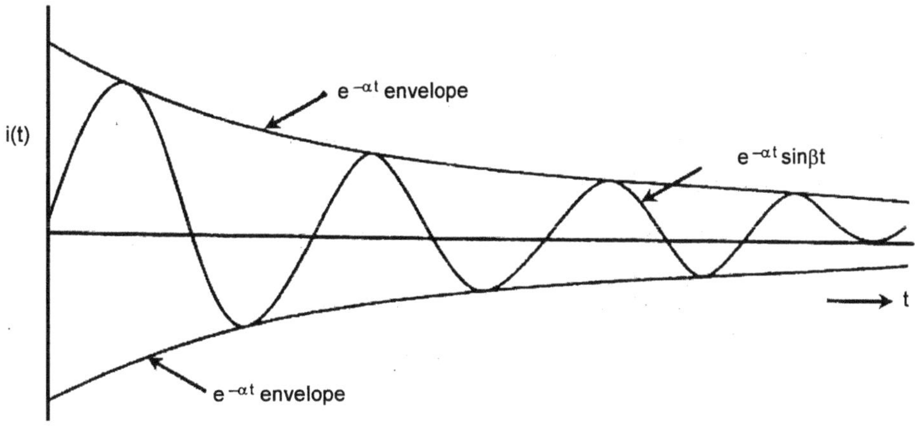

Fig. 8.30 Response of RLC network for complex roots m_1 and m_2

Example 8.5

Find the current i(t) in the network shown in Fig. 8.31.

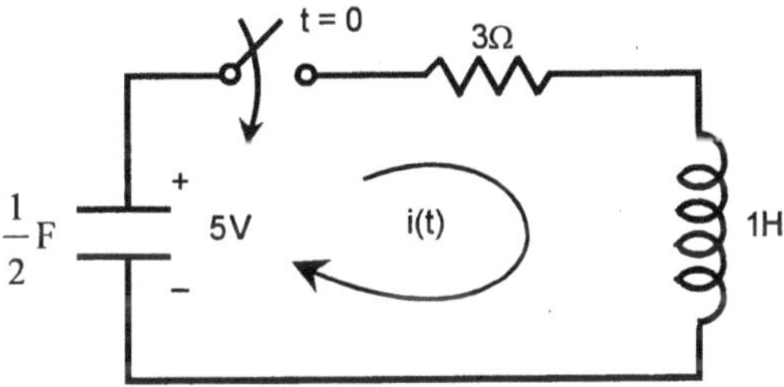

Fig. 8.31 Network for example 8.5

Solution :

The differential equation for the current i (t) is given by

$$\frac{di}{dt} + 3\,i + 2\int_{-\infty}^{t} i\,dt = 0 \qquad .....(8.100)$$

Differentiating,

$$\frac{d^2i}{dt^2} + 3\,\frac{di}{dt} + 2\,i = 0 \qquad .....(8.101)$$

The auxiliary equation or the characteristic equation is given by

$$m^2 + 3m + 2 = 0$$

which gives $m_1 = -2$ and $m_2 = -1$

Hence the solution of eq. (8.101) is

$$i\,(t) = K_1\, e^{-2t} + K_2\, e^{-t} \qquad .....(8.102)$$

The constants K_1 and K_2 are evaluated using the initial conditions. Since there was no current in the inductance at $t = O^-$, $i\,(O^+) = 0$. At $t = O^+$ eq. (8.100) yields.

$$\frac{di}{dt}\,(O^+) + 3\,i\,(O^+) + 2\int_{-\infty}^{O^+} i dt = 0$$

$$\frac{di}{dt}\,(O^+) + (-5) = 0$$

$$\frac{di}{dt}\,(O^+) = 5 \text{ A/sec}$$

Using the values of $i\,(O^+) = 0$ and $\frac{di}{dt}\,(O^+) = 5$ in eq. (8.102),

$$O = K_1 + K_2$$

$$5 = -2\,K_1 - K_2$$

Solving for K_1 and K_2

$$K_1 = -5 \text{ and } K_2 = 5$$

Thus the solution is

$$i\,(t) = -5\, e^{-2t} + 5\, e^{-t} \qquad .....(8.103)$$

The individual terms of eq. (8.103) and their sum are plotted in Fig. 8.32.

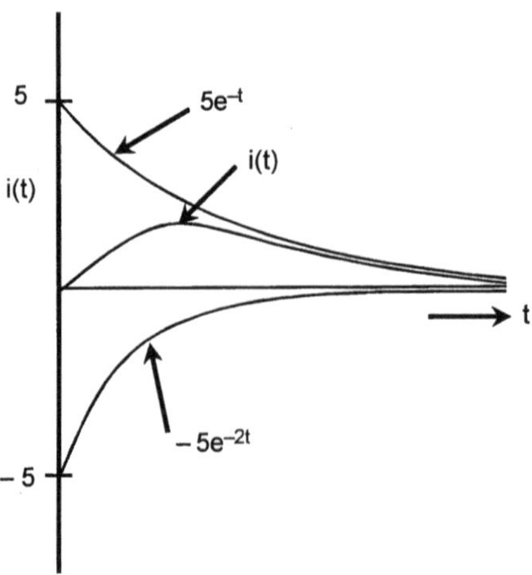

Fig. 8.32 The plot of i (t) and its components for example 8.5

Example 8.6

Solve for the current i (t) in the RLC network of Fig. 8.33.

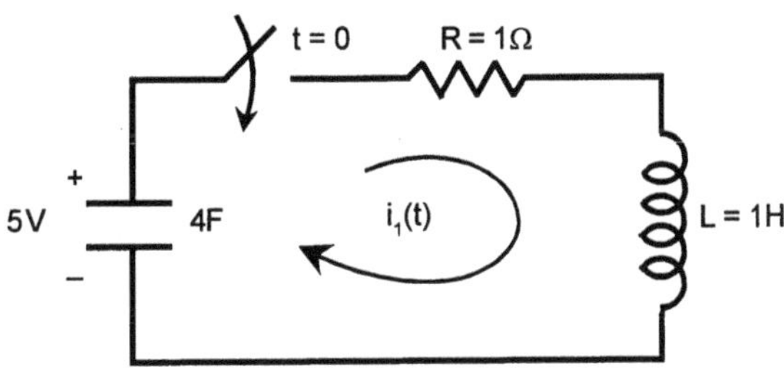

Fig. 8.33 Network for example 8.6

Solution :

In this network $\frac{R}{2L} = \frac{1}{2}$

$$\frac{1}{\sqrt{LC}} = \frac{1}{\sqrt{1 \times 4}} = \frac{1}{2}$$

Thus $\frac{R}{2L} = \frac{1}{\sqrt{LC}}$

Hence the current is given by eq. (8.93)

$$i(t) = e^{-\frac{1}{2}t}(K_1 t + K_2)$$

The initial conditions can be evaluated as

$$i(O^+) = 0$$

$$\frac{di}{dt}(O^+) = +5 \text{ A/sec}$$

Using these values, K_1 and K_2 can be evaluated as

$$K_2 = 0 \text{ and } K_1 = 5$$

$$\therefore \qquad i(t) = 5\,t\;e^{-\frac{1}{2}t}$$

Example 8.7

Find i (t) in the network of Fig. 8.34.

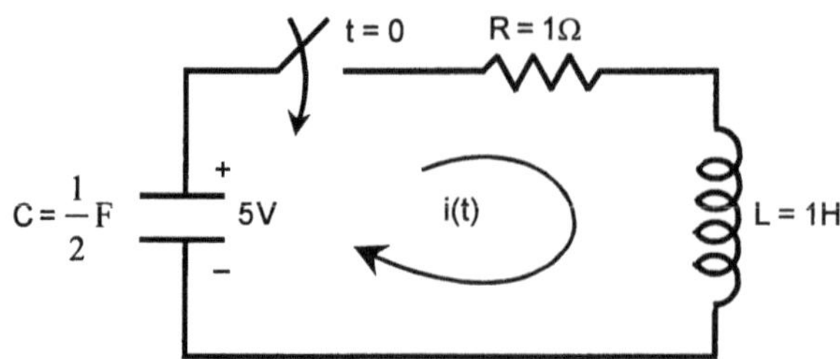

Fig. 8.34 Network for example 8.7

Solution :

In this network

$$\frac{R}{2L} = \frac{1}{2}$$

and $$\frac{1}{\sqrt{LC}} = \frac{1}{\sqrt{1\left(\frac{1}{2}\right)}} = \sqrt{2}$$

Thus $$\frac{R}{2L} < \frac{1}{\sqrt{LC}}$$

∴ The solution is, from eq. (8.97)

$$i\,(t) = e^{-\frac{1}{2}t}\left(A\cos\frac{\sqrt{7}}{2}t + B\sin\frac{\sqrt{7}}{2}t\right)$$

Evaluating the constants using the initial conditions

$$i\,(O^+) = 0$$

$$\frac{di}{dt}\,(O^+) = +\,5\text{A/sec}$$

we get $$A = O,\ \ B = \frac{10}{\sqrt{7}}$$

Thus $$i\,(t) = \frac{10}{\sqrt{7}}\,e^{-\frac{1}{2}t}\,\sin\,\frac{\sqrt{7}}{2}\,t$$

The above three examples illustrate the three different cases discussed in section 8.4.

The special case where R = 0 is discussed in example 8.8.

Example 8.8

Solve for the current i (t) in the LC network shown in Fig. 8.35.

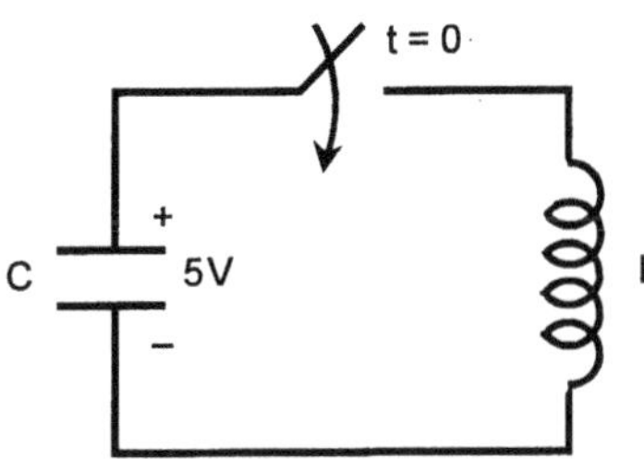

Fig. 8.35 The LC Network for example 8.8

Solution :

Since R = 0, the roots of the characteristic equation are, from eq. (8.89)

$$m_1 = 0 + \frac{j}{\sqrt{LC}} \qquad m_2 = -\frac{j}{\sqrt{LC}}$$

Since $\alpha = 0$, the solution from eq. (8.97), is

$$i\,(t) = A \cos \frac{1}{\sqrt{LC}}\, t + B \sin \frac{1}{\sqrt{LC}}\, t$$

Using the initial condition

$$i\,(O^+) = 0$$

$$\frac{di}{dt}\,(O^+) = \frac{5}{L} \text{ A/sec}$$

we get $\quad A = 0$

$$B = 5\sqrt{\frac{C}{L}}$$

and $\quad i\,(t) = 5\sqrt{\dfrac{C}{L}}\, \sin \dfrac{1}{\sqrt{LC}}\, t$

Thus the response is a sinusoid of fixed amplitude. This type of response is known as *undamped oscillatory response*.

We recall that a simple pendulum in perfect vacuum, has exactly the same response when it is given an initial displacement from its equilibrium position. The potential energy at t = 0 is slowly converted to kinetic energy. Kinetic energy is a maximum when the pendulum passes through the equilibrium point at which the potential energy is zero. The pendulum swings to the opposite side converting the kinetic energy to potential energy again. When the potential energy reaches the initial energy given to the pendulum at t = 0, the kinetic energy is zero and the process repeats. As there is no air damping these are sustained oscillations. In a similar manner, the electrostatic energy in the capacitor at t = 0 is converted to electromagnetic energy in the inductor when i (t) is a maximum and this electromagnetic energy is converted back to electrostatic energy when i (t) = 0. Thus the energy swaps between the capacitor and inductor and thereby produces sustained oscillations. If R is not equal to zero, these oscillations will be damped and we get damped oscillations. If the damping is large enough, the oscillations will disappear and we get over damped response. The critical value of R at which the damping is just enough to stop oscillations, is the critically damped case. Thus the three cases discussed in section 8.4.1 are respectively, overdamped, critically damped and under damped.

8.4.2 Impulse excitation :

Find the current i (t) in the network of Fig. 8.36.

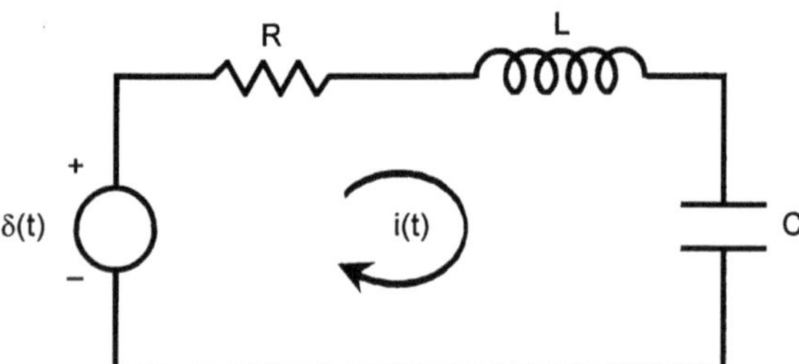

Fig. 8.36 RLC Network with impulse excitation

Since the inductance tries to behave like an open circuit during t = O^- to O^+, the entire impulse voltage appears across the inductance and the current instantaneously changes from O to I_0, given by

$$I_0 = \frac{1}{L}\int_{O^-}^{O^+} \delta(t)$$

$$= \frac{1}{L} A$$

Thus for t > 0, the network in Fig. 8.36 behaves like an RLC circuit with internal energy in the inductance, as shown in Fig. 8.37.

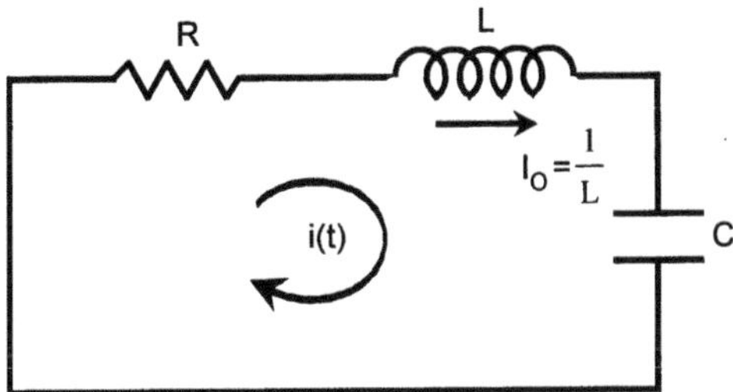

Fig. 8.37 The Network of Fig. 8.36 for t > 0

This network is similar to the network considered is section 8.4.1 with $V_0 = 0$ and $I_0 = \frac{1}{L}$ and hence the solution can be obtained in a similar way.

8.4.3 Unit step excitation :

Consider the RLC series network excited by a step excitation as shown in Fig. 8.38. The current in L and voltage across C are assumed to be zero at $t = 0^-$.

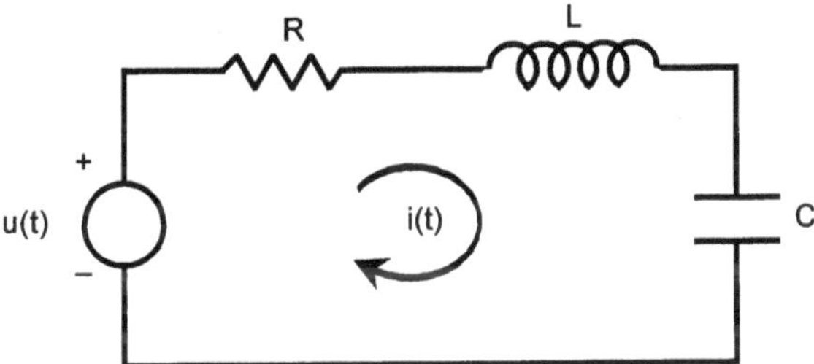

Fig. 8.38 RLC Network excited by a unit step

The differential equation for the current i (t) is

$$Ri + L\frac{di}{dt} + \frac{1}{C}\int_{-\alpha}^{t} idt = u(t) \qquad .....(8.104)$$

Differentiating,

$$L\frac{d^2i}{dt^2} + R\frac{di}{dt} + \frac{i}{C} = 0 \qquad .....(8.105)$$

This is exactly same as eq. (8.87) and hence the solution is also the same. The only difference is in the initial conditions of the network due to the presence of the step source. The current at $t = O^+$ is zero because of the presence of the inductance.

$$\therefore \qquad i(O^+) = 0 \qquad .....(8.106)$$

Applying the conditions at t = O^+ to eq. (8.104)

$$\therefore \quad R\,i(O^+) + L\frac{di}{dt}(O^+) + \frac{1}{C}\int_{-\infty}^{O^+} idt = 1$$

$$\therefore \quad L\frac{di}{dt}(O^+) = 1 \left(\text{Since } i(O^+) = 0 \text{ and } v_C(O^+) = \frac{1}{C}\int_{-\infty}^{O^+} idt = 0 \right)$$

$$\therefore \quad \frac{di}{dt}(O^+) = \frac{1}{L} \qquad .....(8.107)$$

Depending on the values of the R, L and C elements, we have three different cases as explained in section 8.4.1 and the constants have to be evaluated using eqs. (8.106) and (8.107).

We observe that the current in the circuit, although excited by a constant voltage source, has only transient response which dies down to zero as $t \to \infty$ and no steady state response. This can be explained by noting that the capacitor behaves like an open circuit at t = ∞.

Example 8.9

Find the current i (t) and the voltage across capacitor v_C (t) in the network of Fig. 8.39. The network is unenergised at t = O^-.

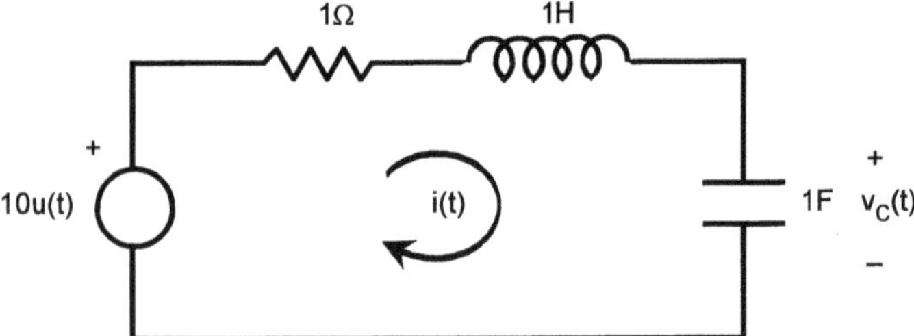

Fig 8.39 Network for example 8.9

Solution :

The differential equation for current i (t) is

$$\frac{di}{dt} + i + \int_{-\alpha}^{t} idt = 10$$

Differentiating,

$$\frac{d^2i}{dt^2} + \frac{di}{dt} + i = 0$$

The auxiliary equation is

$$m^2 + m + 1 = 0$$

The roots are $m_{1,2} = \frac{-1}{2} \pm j\sqrt{\frac{3}{2}}$

The solution is

$$i(t) = e^{-\frac{1}{2}t}\left(A\sin\frac{\sqrt{3}}{2}t + B\cos\frac{\sqrt{3}}{2}t\right)$$

The initial conditions are

$$i(O^+) = 0$$

$$\frac{di}{dt}(O^+) = 10 \text{ A/sec}$$

Using these initial conditions, the constants can be evaluated as

$$B = 0, \quad A = \frac{20}{\sqrt{3}}$$

Hence the current i (t) is

$$i(t) = \frac{20}{\sqrt{3}} e^{-\frac{1}{2}t} \sin \frac{\sqrt{3}}{2} t$$

The voltage across the capacitor v_C (t) is obtained from

$$v_C(t) = \frac{1}{C}\int_{-\infty}^{t} i\,dt = \int_{-\infty}^{O} i\,dt + \int_{O}^{t} i\,dt$$

The term $\int_{-\infty}^{O} i\,dt$ is the voltage across the capacitor at t = 0 and in the problem it is equal to zero.

$$v_C(t) = \int_{O}^{t} \frac{20}{\sqrt{3}} e^{-\frac{1}{2}t} \sin \frac{\sqrt{3}}{2} t \; dt$$

This integral can be evaluated as

$$v_C(t) = \frac{10}{\sqrt{3}}\left[\sqrt{3} - 2e^{-\frac{1}{2}t}\sin\left(\frac{\sqrt{3}}{2}t + 60\right)\right]$$

$$= 10 - \frac{20}{\sqrt{3}} e^{-\frac{1}{2}t} \sin\left(\frac{\sqrt{3}}{2}t + 60\right)$$

The response v_C (t) has both transient and steady state components. The second term on the right hand side of v_C (t) is the transient term which decays to zero as $t \to \infty$. The first term is the steady state component. The capacitance gets charged to the source voltage of 10V at $t = \infty$. As capacitance behaves like an open circuit at $t = \infty$ the entire source voltage appears across the capacitor.

8.4.4 Sinusoidal excitation :

Consider the RLC network in Fig. 8.40 with a sinusoidal excitation. The network is initially unenergised.

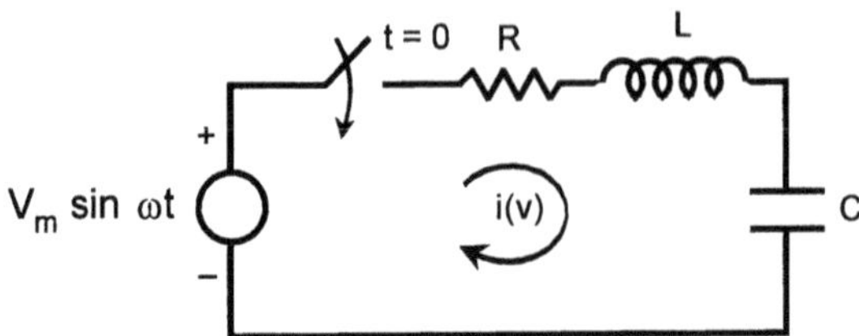

Fig. 8.40 RLC Network with sinusoidal excitation

Writing down the differential equation for the current i (t), we get

$$L\frac{di}{dt} + Ri + \frac{1}{C}\int_{-\alpha}^{t} i\,dt = V_m \sin \omega t \qquad .....(8.108)$$

Differentiating,

$$L\frac{d^2i}{dt^2} + R\frac{di}{dt} + \frac{i}{C} = V_m\,\omega \cos \omega t \qquad .....(8.109)$$

The homogeneous or force-free solution, i_h (t) is obtained by making the right hand side of eq. (8.109) equal to zero. The characteristic equation is given by

$$Lm^2 + Rm + \frac{1}{C} = 0$$

and the roots are

$$m_{1,2} = -\frac{R}{2L} \pm \sqrt{\frac{R^2}{4L^2} - \frac{1}{LC}}$$

The homogeneous solution, again, depends on the values of R, L and C and will be obtained as under damped, over damped or critically damped responses as discussed in section 8.1.4.

The particular integral or forced response is obtained by assuming a solution as given in table 8.1, namely,

$$i_{ss}(t) = A \sin \omega t + B \cos \omega t \qquad .....(8.110)$$

Proceeding exactly in the same way as discussed in section 8.3.4, constants A and B can be evaluated and the solution is given by,

$$i_{ss}(t) = \frac{V_m}{\sqrt{R^2 + \left(\omega L - \frac{1}{\omega C}\right)^2}} \sin(\omega t - \phi) \qquad .....(8.111)$$

where $\phi = \tan^{-1} \dfrac{\omega L - \dfrac{1}{\omega C}}{R}$

The total solution is given by

$$i(t) = i_h(t) + i_{ss}(t)$$

The arbitrary constants in $i_h(t)$ have to be evaluated using the initial conditions. This is illustrated in example 8.10.

Example 8.10

Find the current i (t) in the network Fig. 8.41. for t > 0. At t = 0^- the network was unenergized.

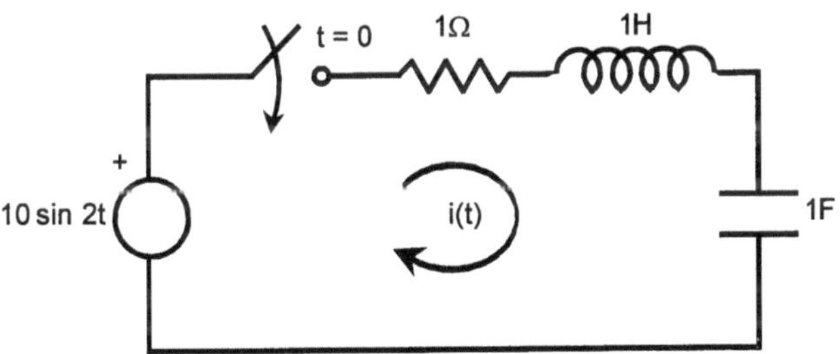

Fig. 8.41. Network for example 8.6

Solution :

The differential equation for i (t) is

$$\frac{di}{dt} + i + \int_{-\infty}^{t} i\,dt = 20 \sin 2t$$

Differentiating,

$$\frac{d^2i}{dt^2} + \frac{di}{dt} + i = 40 \cos 2t$$

The roots of the characteristic equation are

$$m_{1,2} = \frac{-1 \pm \sqrt{1-4}}{2} = -\frac{1}{2} \pm j\frac{\sqrt{3}}{2}$$

The homogeneous solution is

$$i_h(t) = e^{-\frac{1}{2}t}\left(K_1 \sin\frac{\sqrt{3}}{2}t + K_2 \cos\frac{\sqrt{3}}{2}t\right)$$

The particular solution is obtained by substituting the values of R, L, C, ω and V_m in eq. (8.111).

$$i_{ss}(t) = \frac{10}{\sqrt{1+\left(2-\frac{1}{2}\right)^2}} \sin\left(2t - \tan^{-1}\frac{3}{2}\right)$$

$$= \frac{20}{\sqrt{13}} \sin\left(2t - \tan^{-1}\frac{3}{2}\right)$$

The total solution is

$$i(t) = i_h(t) + i_{ss}(t)$$

$$= e^{-\frac{1}{2}t}\left(K_1 \sin\frac{\sqrt{3}}{2}t + K_2 \cos\frac{\sqrt{3}}{2}t\right) + \frac{20}{\sqrt{13}} \sin\left(2t - \tan^{-1}\frac{3}{2}\right) \qquad .....(8.112)$$

The initial conditions are

$$i(O^+) = 0$$

and from eq. (8.112), at $t = O^+$

$$\frac{di}{dt}(O^+) + i(O^+) + 0 = 0$$

$$\frac{di}{dt}(O^+) = 0$$

Substituting the values of $i(O^+)$ and $\frac{di}{dt}(O^+)$ in eq. (8.112), we get

$$K_1 = \frac{-100}{13\sqrt{3}} = -4.441$$

and $$K_2 = \frac{60}{13} = 4.615$$

Thus, the total solution is

$$i\,(t) = e^{-\frac{1}{2}t}\left(-4.441\sin\frac{\sqrt{3}}{2}t + 4.615\cos\frac{\sqrt{3}}{2}t\right)$$

$$+\ 5.547\ \sin\left(2t - \tan^{-1}\frac{3}{2}\right)$$

Example 8.11

Find the current through the capacitor i_C (t) in the circuit of Fig. 8.42. Assume the circuit to be unenergised at $t = O^-$.

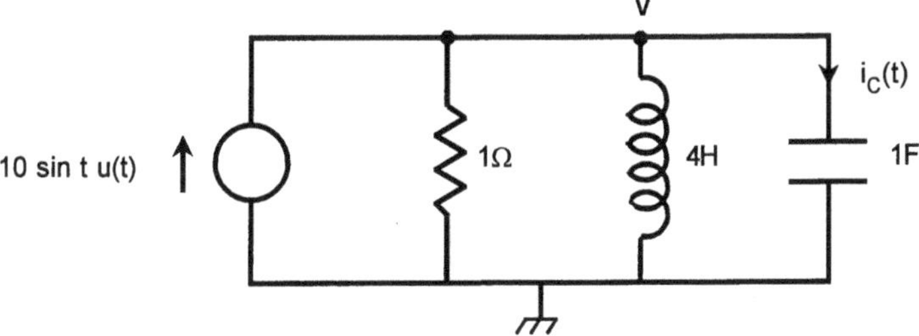

Fig. 8.42 Parallel R L C Network for example 8.11

Solution :

Let the voltage across the parallel combination be v (t), writing the node voltage equation

$$\frac{v}{1} + \frac{1}{4}\int_{-\infty}^{t} v\,dt + \frac{dv}{dt} = 10\sin t$$

Differentiating,

$$\frac{d^2v}{dt^2} + \frac{dv}{dt} + \frac{v}{4} = 10\cos t \qquad .....(8.113)$$

The characteristic equation is

$$m^2 + m + \frac{1}{4} = 0$$

The roots are $m_1 = m_2 = \dfrac{-1}{2}$

The homogeneous solution is

$$v_h\,(t) = e^{-\frac{1}{2}t}\,(K_1\,t + K_2)$$

where K_1 and K_2 are constants to be determined using initial conditions.

The particular integral can be obtained in the usual way by assuming a solution of the from

$$v_{ss}(t) = A \sin t + B \cos t$$

Substituting the solution in eq. (8.113) and evaluating the constants A and B, we get the solution as

$$v_{ss}(t) = \frac{8}{5}(4 \sin t - 3 \cos t) = 8 \sin (t - 36.8^0)$$

The total solution is

$$v(t) = e^{-\frac{1}{2}t}(K_1 t + K_2) + 8 \sin (t - 36.8^0)$$

The initial conditions can be evaluated as

$$v(O^+) = 0 \quad \text{and} \quad \frac{dv}{dt}(O^+) = 0$$

Using these initial conditions, K_1 and K_2 can be evaluated as

$$K_1 = -4, \quad K_2 = 4.8$$

Thus $$v(t) = e^{-\frac{1}{2}t}(-4t + 4.8) + 8 \sin (t - 36.8^0)$$

The current through the capacitance is

$$i_c(t) = C\frac{dv}{dt}$$

$$= \frac{-1}{2}e^{-\frac{1}{2}t}(-4t + 4.8) + e^{-\frac{1}{2}t}(-4) + 8 \cos (t - 36.8^0)$$

$$= 2e^{-\frac{1}{2}t}(t - 3.2) + 8 \cos (t - 36.8^0)$$

Problems

8.1 Find i(t) in Fig. P. 8.1. The inductor has a current of 2 A at t = 0.

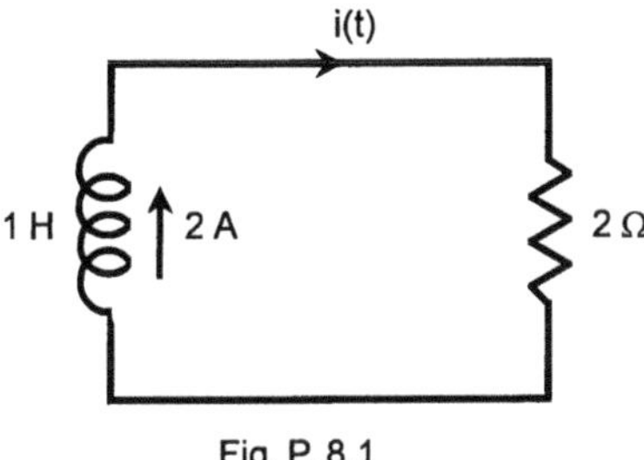

Fig. P. 8.1

8.2 Find v(t) in Fig. P. 8.2. The capacitor has an initial voltage of 10V at t = 0. What is the initial energy in the capacitor ? How much energy is consumed by the resistor from t = 0 to ∞ ?

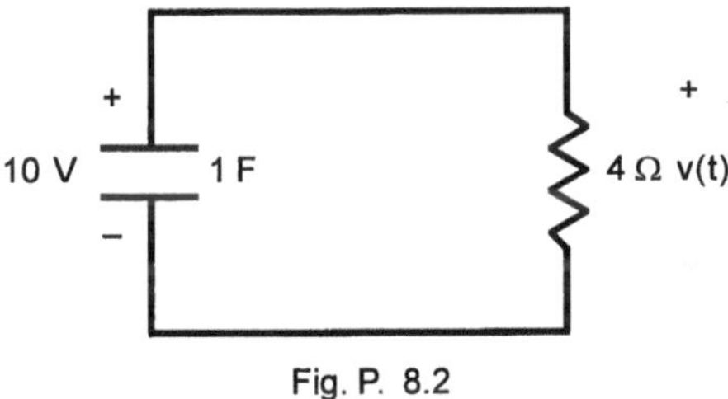

Fig. P. 8.2

8.3 Find i(t) in Fig. P. 8.3. for t > 0.

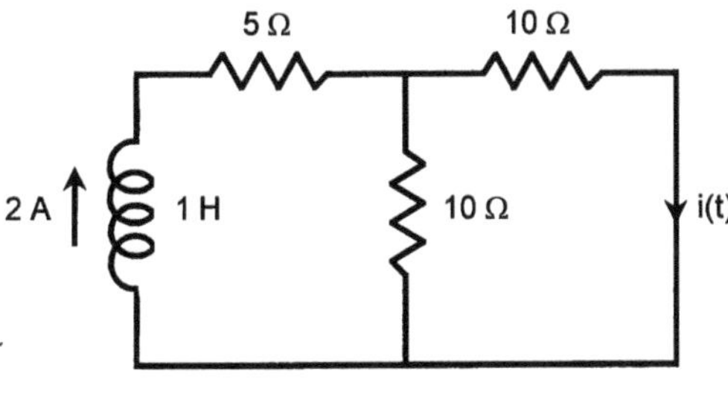

Fig. P. 8.3

8.4 Find i(t) in Fig. P. 8.4. for t > 0.

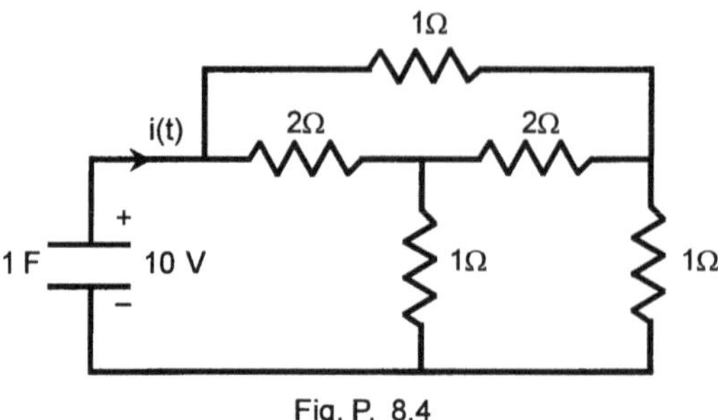

Fig. P. 8.4

8.5 Find i(t) in Fig. P. 8.5 for t > 0. The switch was open for a long time so that steadystate conditions were present before it is closed at t = 0.

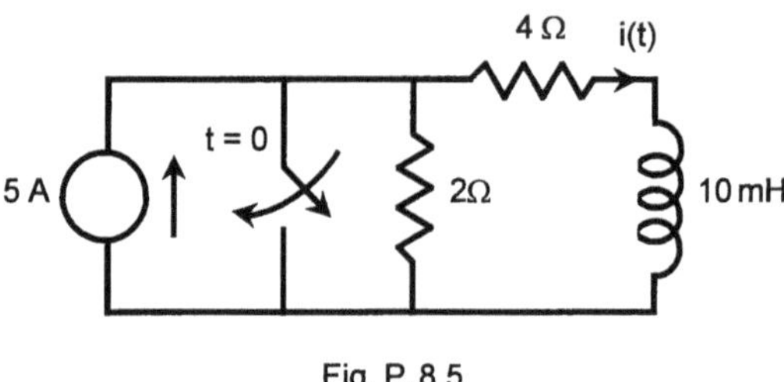

Fig. P. 8.5

8.6 Find $v_1(t)$ and $v_2(t)$ for t > 0 in the network of Fig. P. 8.6. The switch was in position 1 for a long time before it is thrown to position 2.

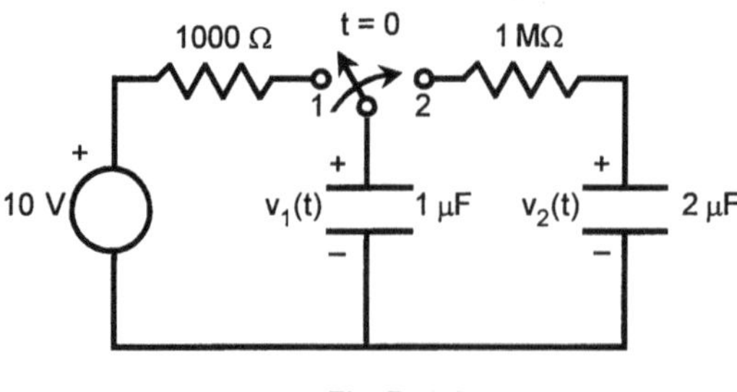

Fig. P. 8.6

8.7 Find v(t) when switch in Fig. P. 8.7 is opened at t = 0.

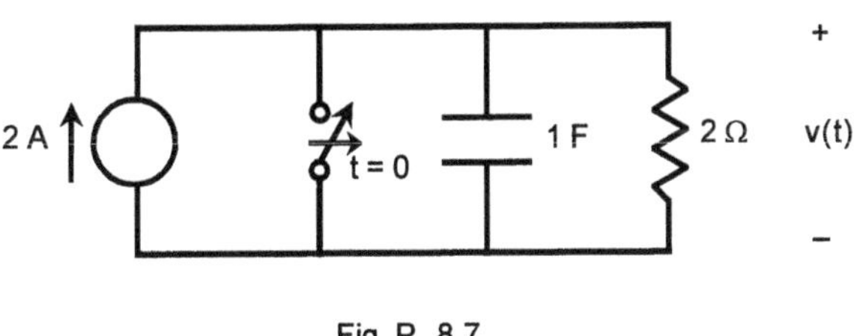

Fig. P. 8.7

8.8 Find i(t) when switch in Fig. P. 8.8 is opened at t = 0.

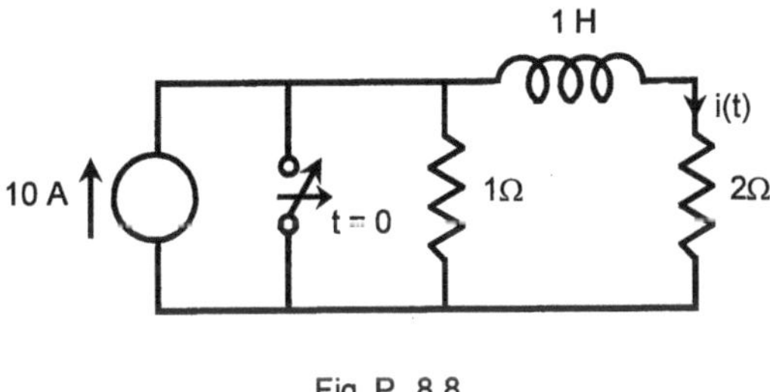

Fig. P. 8.8

8.9 Find $v_c(t)$ if the capacitor has an initial voltage of 10 V in Fig. P. 8.9.

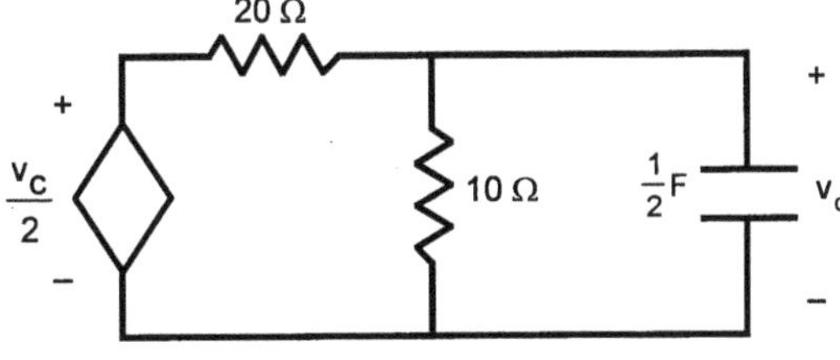

Fig. P. 8.9

8.10 In the circuit of Fig. P. 8.10 the current in the inductance was 5 A just before the occurrence of the impulse. Find i(t) for $t > 0$.

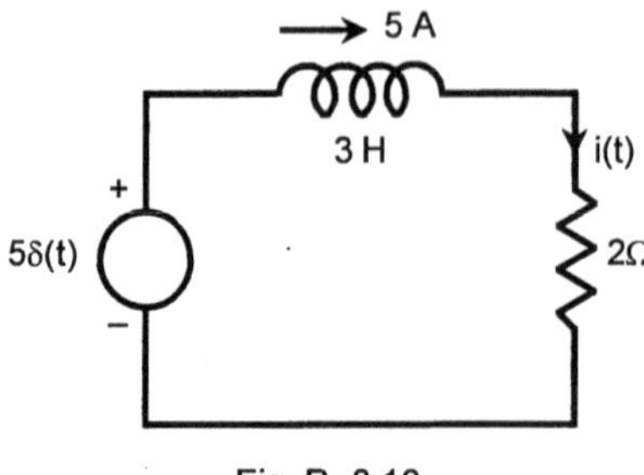

Fig. P. 8.10

8.11 The current in a series R - L circuit is $i(t) = 100e^{-\frac{t}{10}}u(t)$. The values of R and L are 2 Ω and 20 H respectively. If an impulse voltage occurs at t = 10 sec. and reduces the current instantaneously to zero, what is the strength of this impulse.

8.12 In the circuit of Fig. P. 8.11 the switch is closed at $t = 0^+$, just after the impulse is over. What is the voltage v(t) and the current i(t) in the switch for $t > 0$?

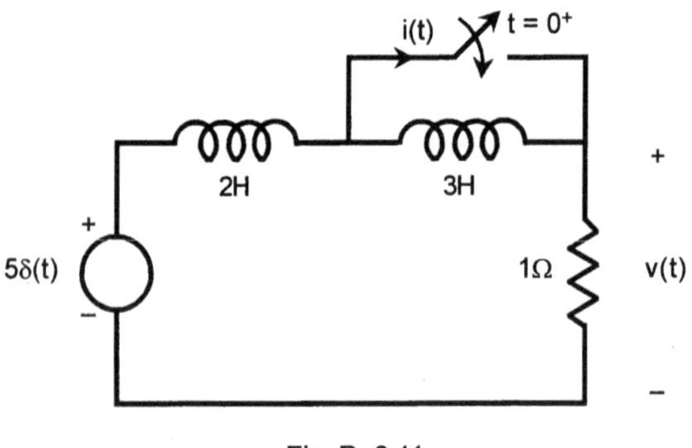

Fig. P. 8.11

8.13 Find i(t) in the circuit of Fig. P. 8.12.

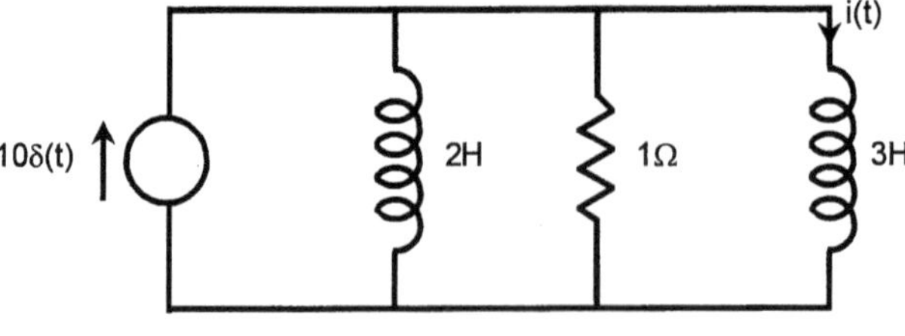

Fig. P. 8.12

8.14 Find the current i(t) for $t \geq 0$ in the circuit of Fig. P. 8.13.

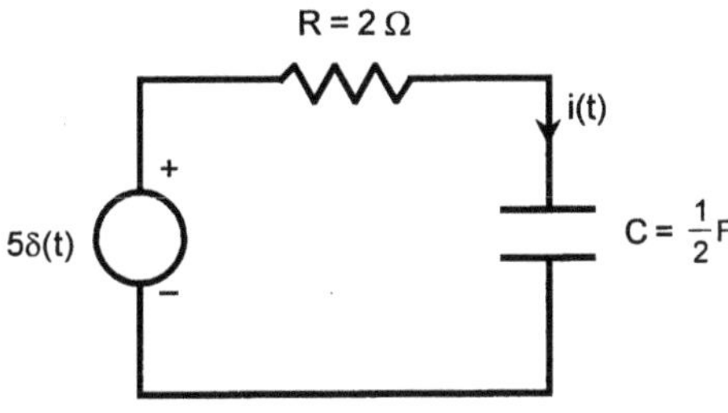

Fig. P. 8.13

8.15 Find v(t) for $t > 0$ in Fig. P. 8.14.

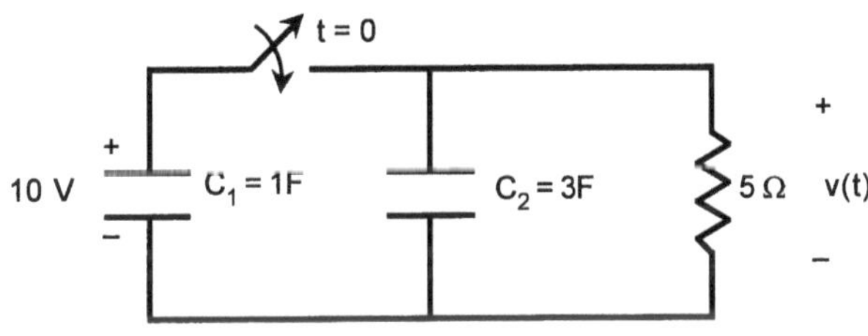

Fig. P. 8.14

8.16 Find i(t) for $t > 0$ in Fig. P. 8.15. The capacitor has an initial voltage of 10V just before the impulse occurs.

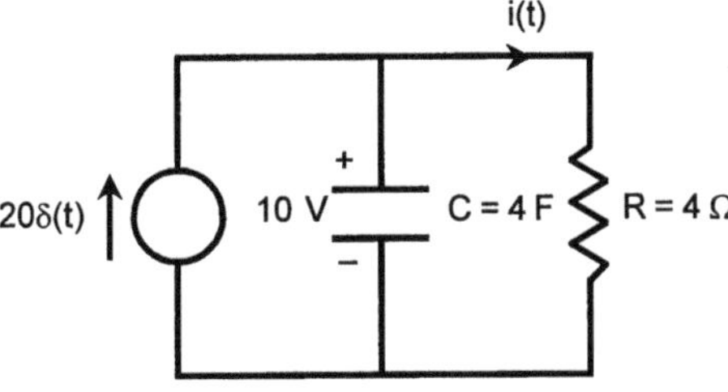

Fig. P. 8.15

8.17 The switch in the circuit of Fig. P. 8.16 is opened at t = 0. Find i(t) and $v_k(t)$.

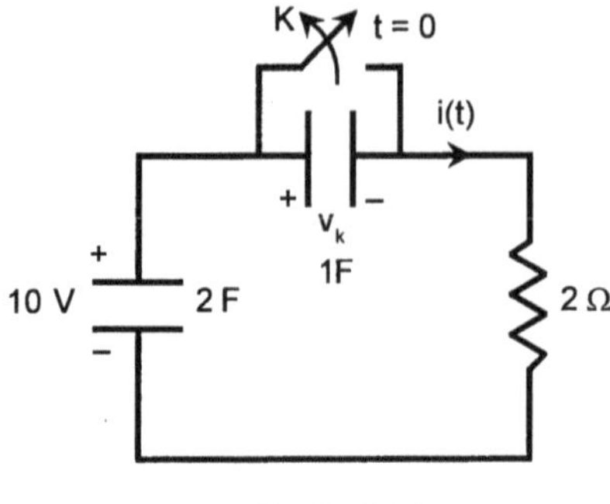

Fig. P. 8.16

8.18 Find the value of $v_c(t)$ at t = 0.3 sec and t = 1 sec in the circuit of Fig. P. 8.17.

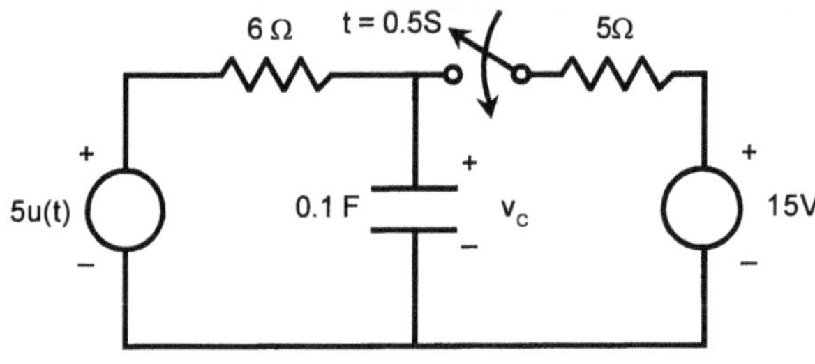

Fig. P. 8.17

8.19 Find i(t) in the circuit of Fig. P. 8.18.

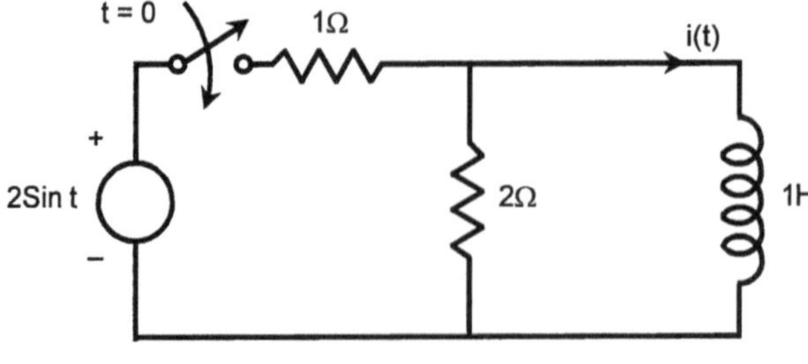

Fig. P. 8.18

8.20 Find the voltage across the capacitor for t > 0 in Fig. P. 8.19.

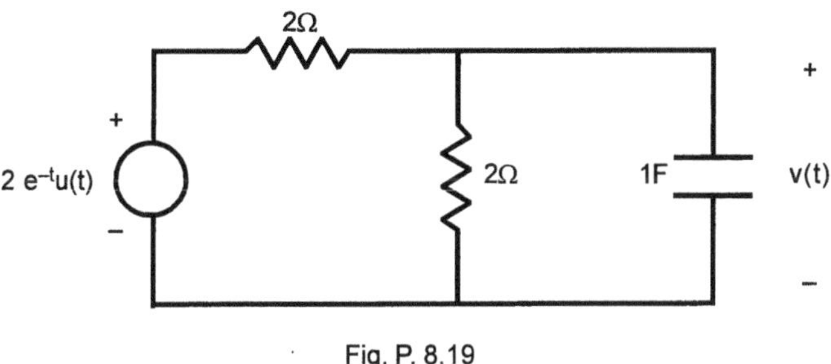

Fig. P. 8.19

8.21 Find the current through the inductor for t > 0 in Fig. P. 8.20.

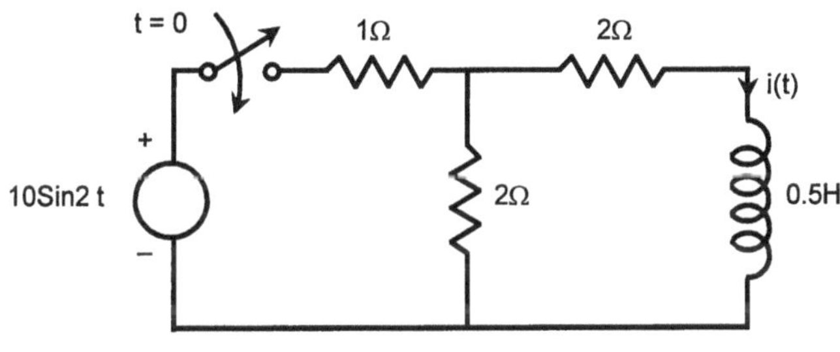

Fig. P. 8.20

8.22 If the initial value of the current in the circuit of Fig. P.8.21 is 10 A, find i(t). The capacitance is uncharged at t = 0.

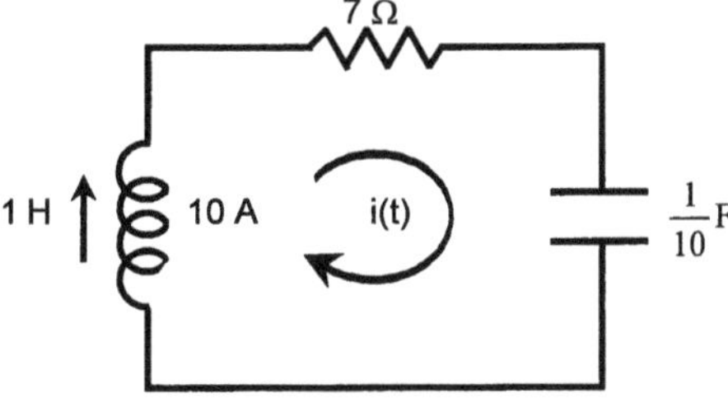

Fig. P.8.21

8.23 The switches in the circuit of Fig. P.8.22 are simultaneously operated at t = 0. Find v(t).

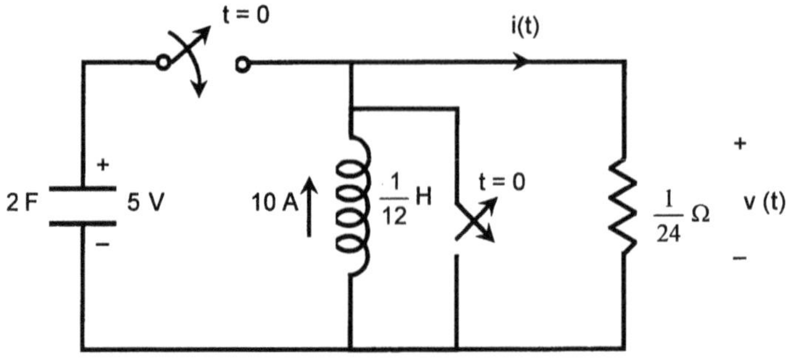

Fig. P.8.22

8.24 Find the current i(t) in the circuit of Fig. P.8.23 if the initial voltage across the capacitor is 75 V and initial current in the inductor is 5 A.

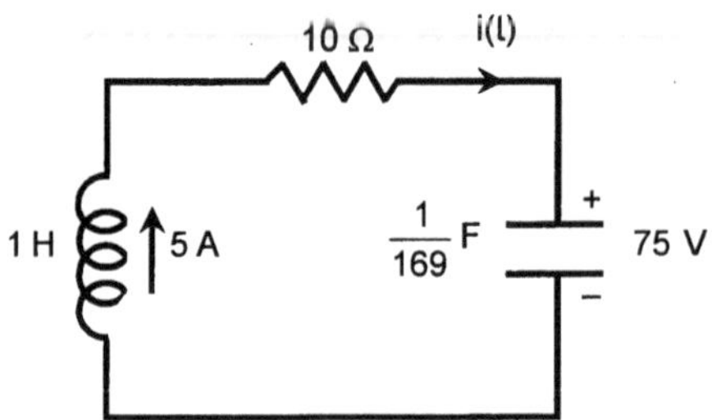

Fig. P.8.23

8.25 The switch was in position 1 for a long time so that steady state conditions were obtained in the circuit of Fig. P.8.24. At t = 0 the switch is thrown to position 2. Find i(t) for t > 0.

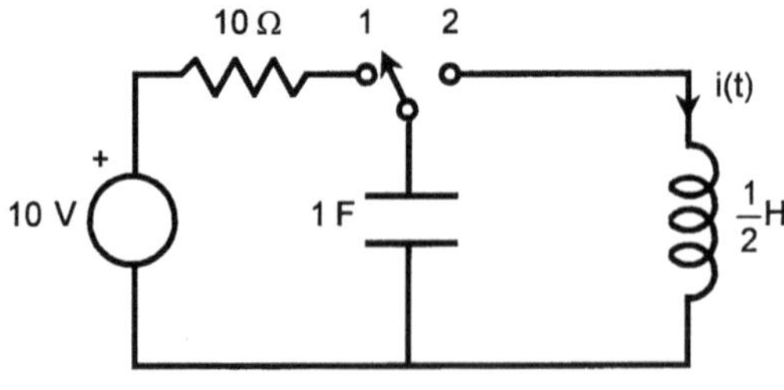

Fig. P.8.24

8.26 The switch in the circuit of Fig. P.8.25 closed for a long time and at t = 0 it opens. Find the voltage across the capacitor, v_C (t) for t > 0.

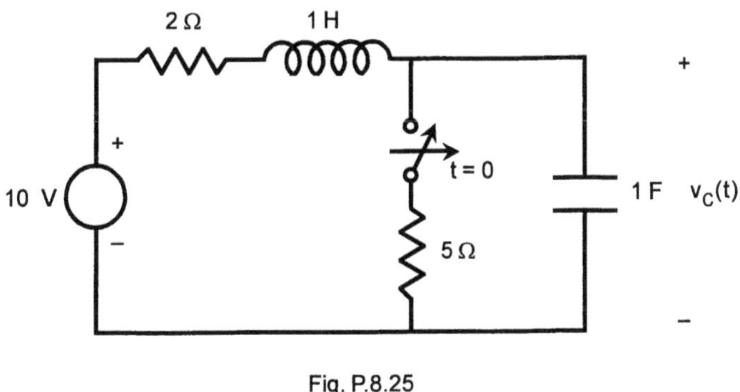

Fig. P.8.25

8.27 Find v(t) in Fig. P.8.26.

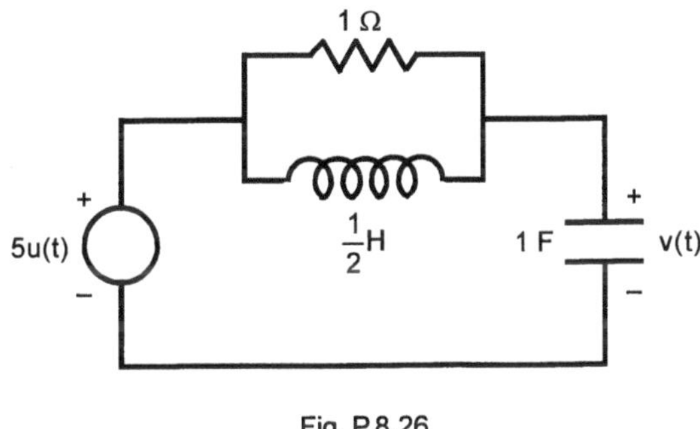

Fig. P.8.26

8.28 Find i(t) in the circuit of Fig.P. 8.27.

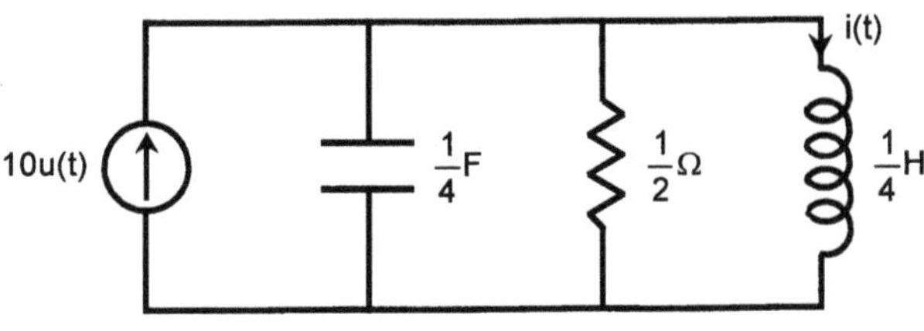

Fig. P. 8.27

8.29 Find i(t) in Fig. P.8.28.

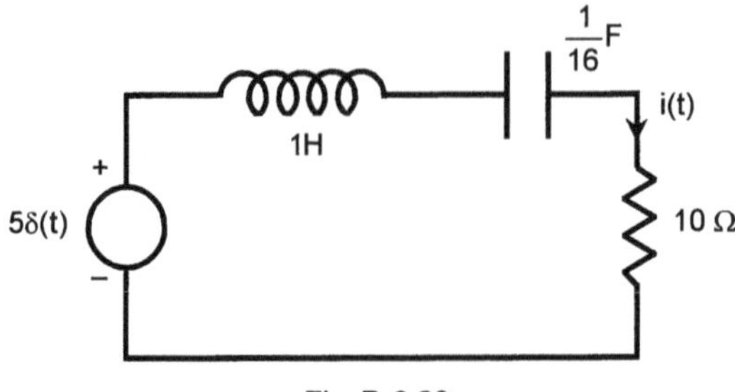

Fig. P. 8.28

8.30 Find v(t) in Fig. P.8.29.

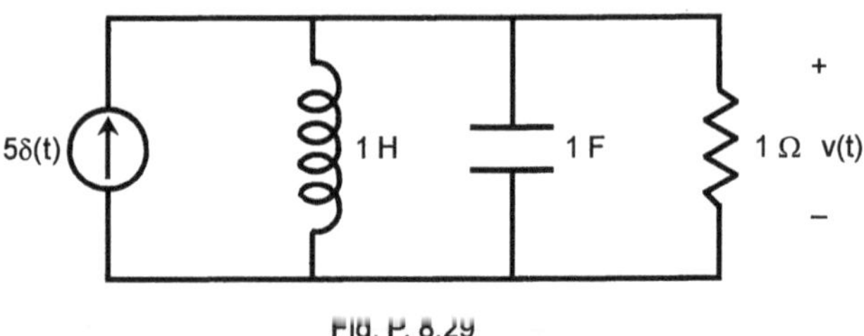

Fig. P. 8.29

8.31 Find the voltage across the capacitance for t > 0 in the circuit of Fig. 8.30.

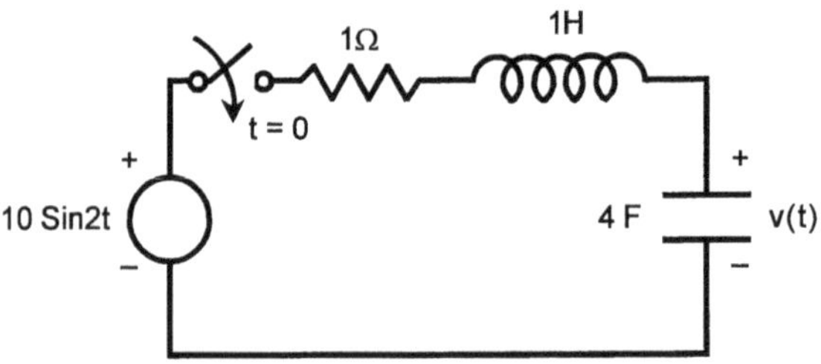

Fig. P. 8.30

8.32 Find the current i(t) in the circuit of Fig. P. 8.31.

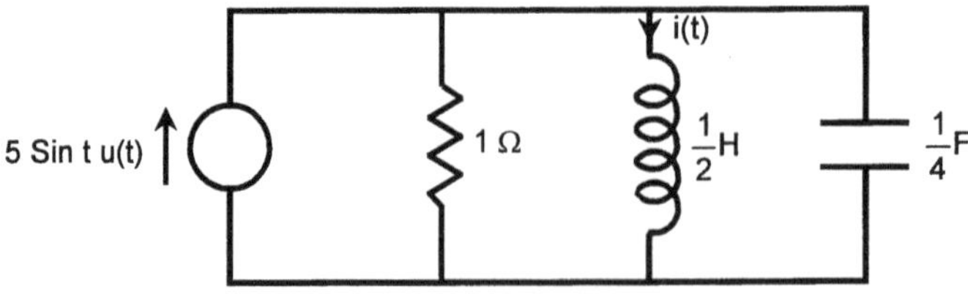

Fig. P. 8.31

8.33 Find the current i(t) in he circuit of Fig. 8.32.

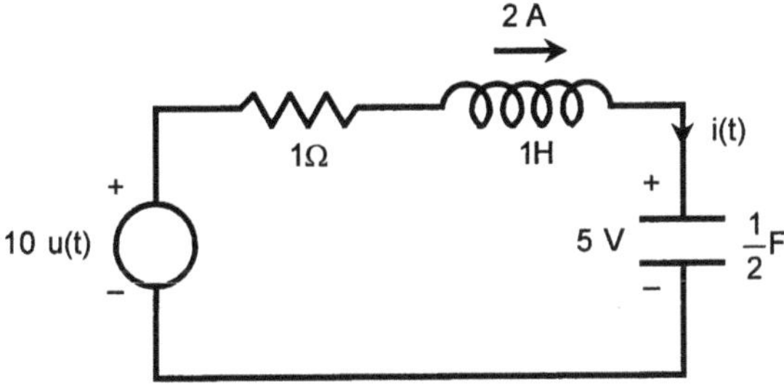

Fig. P. 8.32

8.34 The switch was closed at t = 0 when steady state conditions were reached in the circuit to the left side of the switch. Find the zero state and zero input response v(t) for t > 0 in Fig. P. 8.33.

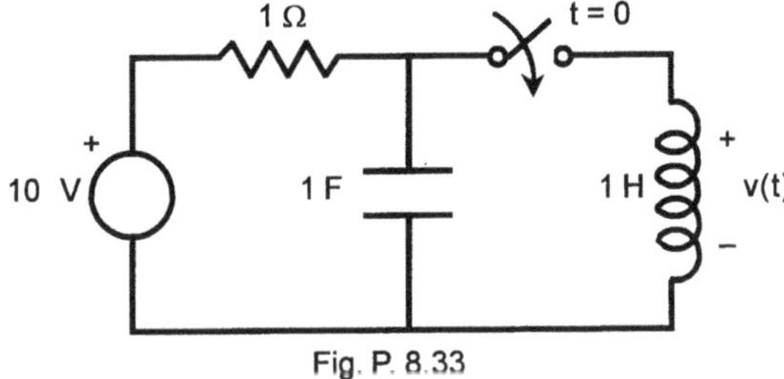

Fig. P. 8.33

8.35 In the circuit of Fig. P. 8.34 find

(a) i(t) for t > 0

(b) The total energy lost in the resistance

(c) The energy stored in C_1 and C_2 and check for energy balance

and (d) What happens to these quantities when R is made to approach zero value.

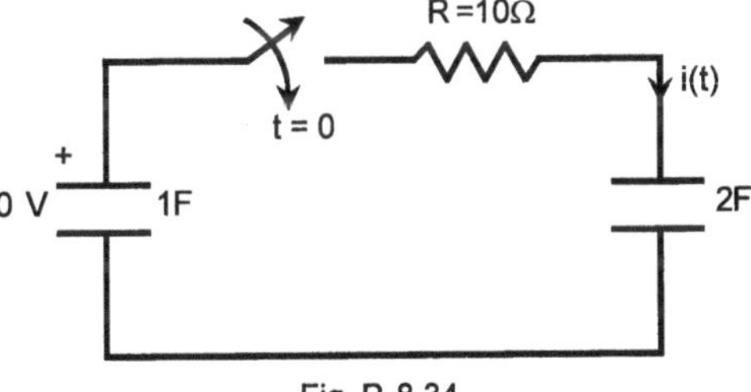

Fig. P. 8.34

8.36 The switch in the circuit of Fig. P. 8.35 is closed at t = 0. Find the energy supplied by the source, the energy lost in the resistance and the final energy stored in the capacitance .

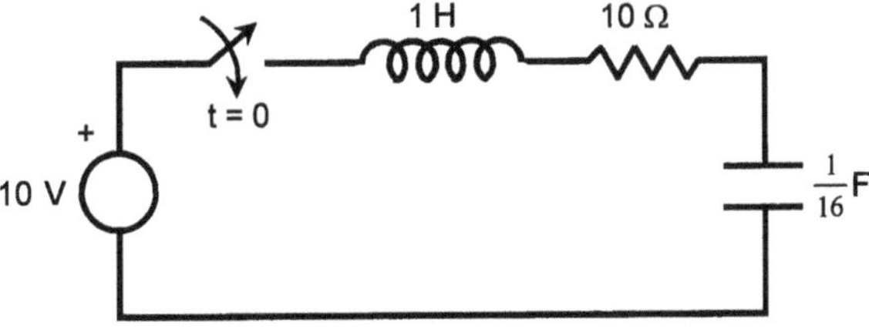

Fig. P. 8.35

8.37 Find v(t) in Fig. P. 8.36 if the switch K is closed at t = 0.

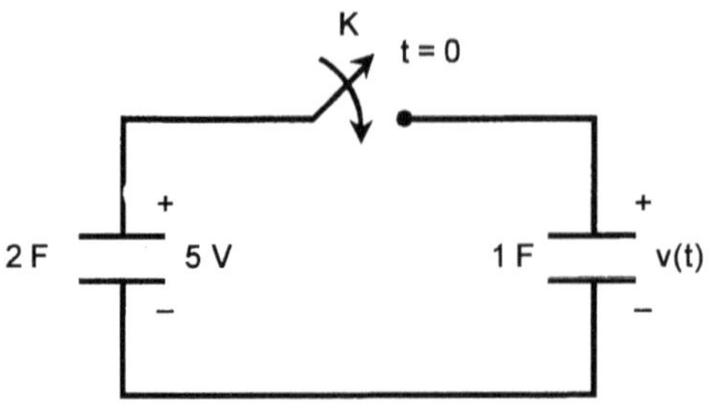

Fig. P. 8.36

8.38 Find i(t) in Fig. P. 8.37 if the switch K is opened at t = 0.

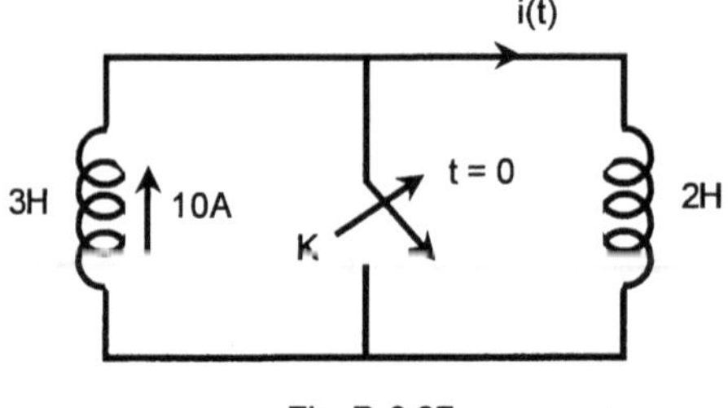

Fig. P. 8.37

9

LAPLACE TRANSFORM METHOD OF ANALYSIS OF NETWORKS

9.1 INTRODUCTION :

The roots of Laplace transform technique of solving the differential equations lie in the operational calculus invented by Oliver Heaviside. He was an engineer and was looking for practical methods of solution of electrical networks rather than the mathematical rigour. He was able to choose a correct solution from a number of alternatives using heuristic approach. The mathematical rigour was supplied by Bromvich Giorgi and others based on the earlier work of Laplace. The transformation which was used to solve the differential equations was called the *Laplace transform.*

The Laplace transformation method offers several advantages over the classical methods of solution of differential equations in general and electrical networks in particular. The following advantages make the application of Laplace transformation to networks almost compulsive.

1. The solution of differential equations become systematic, simple and routine. The differential equations are converted into algebraic equations, the solutions of which are easy and simple.
2. The method gives the total solution – the homogeneous solution and particular integral – at one stroke.
3. Initial conditions are taken care of in the beginning only, instead of at the end as in classical methods.
4. The special cases, occurring whenever capacitance loops occur or junctions with only inductances are present, do not trouble us anymore as in classical method. One need not worry about conditions at $t = O^+$ being different from conditions at $t = O^-$. The initial conditions are considered to be conditions at $t = O^-$ rather than at $t = O^+$.
5. Solution of multiloop or multinode networks becomes very difficult by classical methods. They pose no such problem if Laplace transform method is used.

An example of a similar transformation is logarithm, which converts multiplications and divisions into additions and subtractions. The anti logarithm gives back the result. Similarly, the Laplace transform converts the differential equations into algebraic equations, on which algebraic manipulations can be performed to get the solution in transform domain. The time domain solution is then obtained by taking inverse Laplace transform. The disadvantage of this method, as in logarithms, is that the physical insight is lost because of the transformation.

9.2 THE LAPLACE TRANSFORMATION :

The Laplace transform, F(s), of a time function f(t) is defined as

$$L\,[f(t)] = F(s) = \int_{0^-}^{\infty} f(t)e^{-st}dt0 \qquad .....(9.1)$$

where $s = \sigma + j\omega$ is a complex number, usually called as *Complex frequency*. The symbol L is used to indicate that it is the Laplace transform of f(t). The real part σ is known as *the neper frequency* and is measured in nepers/sec. and the imaginary part ω is known as *the radian* or *real frequency* and is measured in radians/sec.

The lower limit of the integral in eq. (9.1) is taken as O^- rather than O, to distinguish between two instants just before or just after the condition of the network is changed at t = O. If the function f(t) is continuous at t = O, i.e., $f(O^-) = f(O^+)$, it does not really matter whether we take O^- or O^+ as the limit. By taking O^- we can include functions like impulses which occur frequently in electrical networks.

Now the question arises, whether we can transform any function f(t). For a function to have Laplace transform, it is sufficient that

$$\int_{0^-}^{\infty} |f(t)|e^{-\sigma t}dt \text{ is finite} \qquad .....(9.2)$$

for a real, positive σ. The evaluation of the integral in eq. (9.1) looks formidable, but in practice, it is easy. Laplace transforms of various functions are evaluated and tabulated so that, they need not be evaluated time and again. The function f(t) and its transform F(s) are called a *Laplace transform pair*. Using the uniqueness property of this transformation, inverse transformation can be easily obtained.

The limitation imposed by eq. (9.2) is satisfied by most of the functions encountered in engineering practice. Hence it is not a serious restriction for the use of Laplace transform in the solution of networks.

Now, the next important question is, how to obtain the inverse transform or the function f(t) from F(s). The inverse relationship is given by

$$f(t) = L^{-1}F(s) = \frac{1}{2\pi j}\int_{\sigma-j\infty}^{\sigma+j\infty} F(s)e^{st}ds \qquad .....(9.3)$$

The symbol L^{-1} is used to indicate the inverse Laplace transform of F(s). This is a complex integration i.e., contour integration along the vertical line $s = \sigma_1 - j\infty$

to $\sigma_1 + j\infty$ as shown in Fig. 9.1.

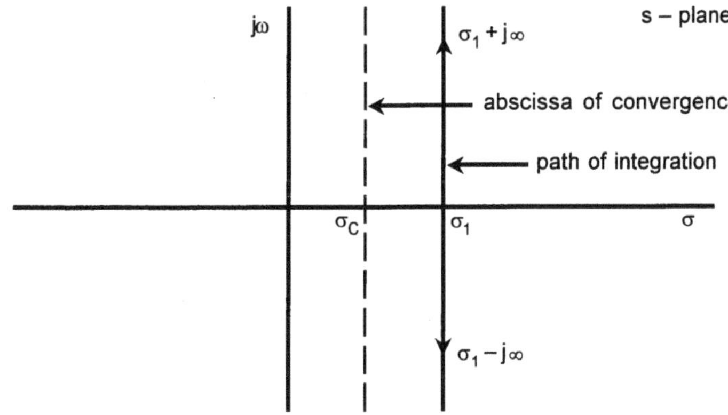

Fig. 9.1 Complex s – plane showing the path of integration and abscissa of convergence

For the evaluation of the integral in eq. (9.3) it is required that $\sigma_1 > \sigma_c$ as shown in Fig. 9.1. σ_c is known as *the abscissa of absolute convergence.* One need not worry unnecessarily about this inversion integral as we rarely use it to find the inverse. Since the Laplace transform of a function f(t) is unique, we may use the table of transform pairs, just as we use tables of logarithms and antilogarithm, for obtaining the time function f(t).

9.3 LAPLACE TRANSFORM OF SOME IMPORTANT TIME FUNCTIONS :

Let us now obtain the Laplace transform of some important time functions.

(i) *Unit step function :*

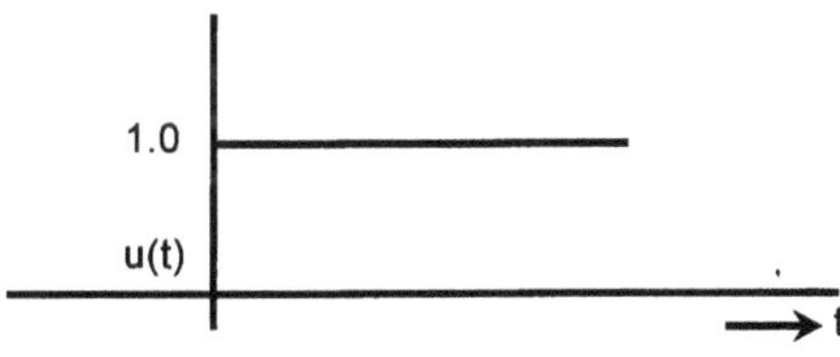

Fig. 9.2 Unit step function

$$u(t) = 1 \qquad t \geq 0$$
$$= 0 \qquad t < 0$$

$$L[u(t)] = \int_{0^-}^{\infty} u(t)e^{-st}dt = \int_0^{\infty} e^{-st}dt$$

$$= \left.\frac{e^{-st}}{-s}\right|_0^{\infty}$$

$$= \frac{1}{s} \qquad .....(9.4)$$

Similarly the Laplace transform of Ku(t) is $\frac{K}{s}$.

Here in the evaluation of e^{-st} at $t = \infty$, we have assumed that we can choose a $\sigma_1 > 0$ so that $e^{-(\sigma_1 + j\omega)\infty} \rightarrow 0$. For this function the abscissa of absolute convergence is $\sigma_c = 0$.

(ii) *Unit impulse*

$$f(t) = \delta(t)$$

$$F(s) = \int_{0^-}^{\infty} \delta(t)e^{-st}dt$$

$$= \int_{0^-}^{0^+} \delta(t)e^{-st}dt$$

The function $u_0(t)$ exists only at $t = 0$ and has an area of unity enclosed between instants O^- and O^+. The integral of any function multiplied by an impulse is given by

$$\int_{0^-}^{0^+} \delta(t)g(t)dt = g(0)$$

Using this expression, we have

$$F(s) = e^{-st}\Big|_{t=0} = 1 \qquad .....(9.5)$$

(iii) $f(t) = e^{at}$; a is a constant

$$L[f(t)] = F(s) = \int_{0^-}^{\infty} e^{at}e^{-st}dt$$

$$= \int_{0^-}^{\infty} e^{at}e^{-st}dt$$

$$= \frac{e^{-(s-a)t}}{-(s-a)}\Bigg|_{0^-}^{\infty}$$

If σ_1 is chosen greater than a, real part of $(s - a)$ is positive and $e^{-(s-a)} \rightarrow 0$ as $t \rightarrow \infty$.

$$\therefore \qquad F(s) = \frac{1}{s-a} \qquad .....(9.6)$$

(iv) $f(t) = t^n$

$$F(s) = \int_{0^-}^{\infty} t^n e^{-st}dt$$

Integrating by parts,

$$F(s) = t^n \frac{e^{-st}}{-s}\Bigg|_{0^-}^{\infty} - \int_{0^-}^{\infty} nt^{n-1}\frac{e^{-st}}{-s}dt$$

The first term on the right hand side vanishes at both the limits, for $\sigma_1 > 0$ and hence

$$F(s) = \frac{n}{s}\int_{0^-}^{\infty} t^{n-1} e^{-st} dt$$

$$= \frac{n}{s} L\left[t^{n-1}\right]$$

Repeating the procedure for $L[t^{n-1}]$; we get

$$F(s) = \frac{n(n-1)}{s^2} L\left[t^{n-2}\right]$$

$$= \frac{n(n-1)(n-2).....1}{s^n} L\left[t^0\right]$$

$$= \frac{n!}{s^{n+1}} \qquad ....(9.7)$$

9.4 SOME THEOREMS OF LAPLACE TRANSFORMATION :

Let us now consider some important and useful theorems of Laplace transformation which will help us in evaluating transforms of some complicated functions.

1. Linearity :

If $f_1(t)$ and $f_2(t)$ are Laplace transformable and a and b are two constants then

$$L[af_1(t) + bf_2(t)] = aL[f_1(t)] + bL[f_2(t)]$$

$$= aF_1(s) + bF_2(s) \qquad .....(9.8)$$

Proof :

$$L[af_1(t) + bf_2(t)] = \int_{0^-}^{\infty} [af_1(t) + bf_2(t)]e^{-st} dt$$

$$= a\int_{0^-}^{\infty} f_1(t)e^{-st} dt + b\int_{0^-}^{\infty} f_2(t)e^{-st} dt$$

$$= aF_1(s) + bF_2(s)$$

Let us use this theorem to evaluate the transform of $\sin\omega_0 t$

$$f(t) = \sin\omega_0 t = \frac{e^{j\omega_0 t} - e^{-j\omega_0 t}}{2j}$$

$$L[f(t)] = F(s) = \int_{0^-}^{\infty} \frac{e^{j\omega_0 t} - e^{-j\omega_0 t}}{2j} e^{-st} dt$$

Using eq. (9.7)

$$= \frac{1}{2j}\left[L[e^{j\omega_0 t}] - L[e^{-j\omega_0 t}]\right]$$

$$= \frac{1}{2j}\left[\frac{1}{s - j\omega_0} - \frac{1}{s + j\omega_0}\right]$$

$$= \frac{\omega_0}{s^2 + \omega_0^2} \quad .....(9.9)$$

Similarly $L[\cos\omega_0 t] = \dfrac{s}{s^2 + \omega_0^2}$ .....(9.10)

2. Complex translation or shifting theorem :

If $L[f(t)] = F(s)$, then $L[e^{-at}f(t)] = F(s + a)$

Proof :

$$L[e^{-at}f(t)] = \int_{0^-}^{\infty} e^{-at} f(t) e^{-st} dt$$

$$= \int_{0^-}^{\infty} f(t) e^{-(s+a)t} dt$$

In this integral s is replaced by (s + a) and hence

$$L[e^{-at}f(t)] = F(s + a) \quad .....(9.11)$$

F(s + a) is a shifted function in Laplace transform domain and hence multiplication by e^{-at} of a time function results in shifting of F(s) in transform domain.

Using this theorem, we can evaluate the transform of $f(t) = te^{-at}$

$$L[te^{-at}] = L[t]\Big|_{s \to s+a}$$

From eq. (9.6) with n = 1

$$L[te^{-at}] = \frac{1}{s^2}\Big|_{s \to s+a}$$

$$= \frac{1}{(s + a)^2} \quad .....(9.12)$$

Similarly $L[e^{-at}\cos\omega_0 t] = \dfrac{s + a}{(s + a)^2 + \omega_0^2}$ .....(9.13)

and $L[e^{-at}\sin\omega_0 t] = \dfrac{\omega_0}{(s + a)^2 + \omega_0^2}$ .....(9.14)

3. Real shifting theorem :

If $L[f(t)] = F(s)$, then $L[f(t-T)u(t-T)] = e^{-sT}F(s)$ .....(9.15)

Proof :

$$L[f(t-T)u(t-T)] = \int_{0^-}^{\infty} f(t-T)u(t-T)e^{-st}dt$$

Let $t - T = \tau$

$$= \int_{-T}^{\infty} f(\tau)u(\tau)e^{-s(\tau+T)}d\tau$$

Since $u(\tau) = 1$ for $\tau \geq 0$
$= 0$ for $\tau < 0$

We have

$$L[f(t-T)u(t-T)] = e^{-sT}\int_{0}^{\infty} f(\tau)e^{-s\tau}d\tau$$

$$= e^{-sT}F(s)$$

Some explanation regarding the function f(t – T) u(t – T) is necessary here. If f(t) is as shown in Fig. 9.3(a), different function f(t) u(t), f(t) u(t – T), f(t – T) u(t) and f(t – T) u(t – T) are shown in Fig. 9.3(b), (c), (d) and (e) respectively. The theorem is obviously applicable to function in Fig. 9.3(e) only. This is called *shifted time function.*

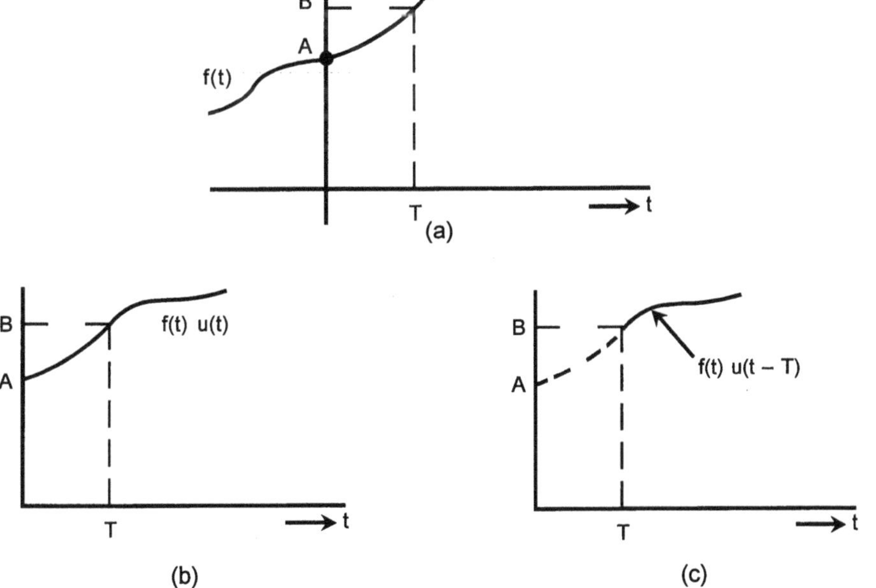

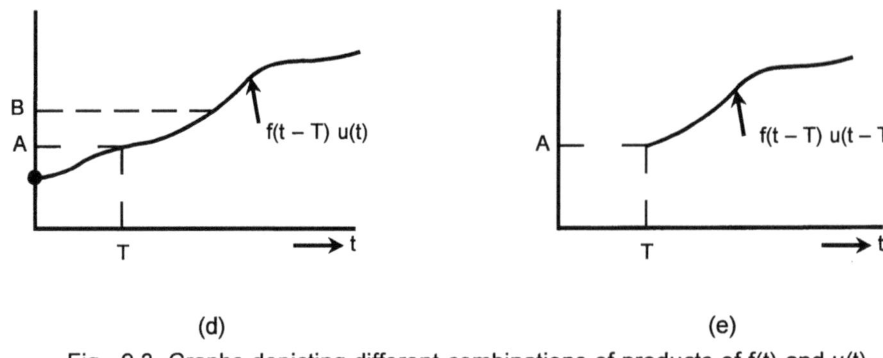

Fig. 9.3 Graphs depicting different combinations of products of f(t) and u(t)

(b)	f(t)u(t)	The function f(t) is considered only for $t \geq 0$
(c)	f(t) u(t – T)	The function f(t) is considered only for $t \geq T$
(d)	f(t – T)u(t)	The function f(t) is shifted by an amount T to right and the resulting function is considered only for $t \geq 0$
(e)	f(t – T) u(f – τ)	The function f(t) is shifted by an amount T and the resulting function is considered only for $t \geq T$.

4. Differentiation theorem :

$$\text{If } L[f(t)] = F(s) \text{ then } L\left[\frac{df(t)}{dt}\right] = sF(s) - f(O^-)$$

Proof :

$$L\left[\frac{df(t)}{dt}\right] = \int_{O^-}^{\infty} \frac{df(t)}{dt} e^{-st} dt$$

Integrating by parts, we have

$$L\left[\frac{df(t)}{dt}\right] = f(t)e^{-st}\Big|_{O^-}^{\infty} + \int_{O^-}^{\infty} sf(t)e^{-st} dt$$

$$= -f(O^-) + sF(s) \quad .....(9.16)$$

provided that $\underset{t \to \infty}{Lt} f(t)e^{-st} = 0$. This is true if f(t) and its successive derivatives are finite at $t = \infty$ and $\sigma > 0$.

Using this theorem we can obtain the transforms of higher derivatives also.

$$L\left[\frac{d^2f(t)}{dt^2}\right] = L\left[\frac{d}{dt}\left(\frac{d}{dt}f(t)\right)\right]$$

$$= sL\left[\frac{d}{dt}f(t)\right] - \frac{df}{dt}(O^-)$$

$$= s^2F(s) - sf(O^-) - \frac{df}{dt}(O^-)$$

The general expression for an nth derivative is

$$L\left[\frac{d^nf(t)}{dt^n}\right] = s^nF(s) - s^{n-1}f(O^-) - s^{n-2}\frac{df}{dt}(O^-) ..... - \frac{d^{n-1}f}{dt^{n-1}}(O^-) \quad .....(9.17)$$

5. Integration theorem :

If $L[f(t)] = F(s)$ then $L\left[\int_{O^-}^{t} f(t)dt\right] = \frac{F(s)}{s}$ .....(9.18)

Proof :

$$L\left[\int_{O^-}^{t} f(t)dt\right] = \int_{O^-}^{\infty}\left[\int_{O^-}^{t} f(t)dt\right]e^{-st}dt$$

Integrating by parts, we have

$$= \left[\int_{O^-}^{t} f(t)dt \cdot \frac{e^{-st}}{-s}\right]\Bigg|_{O^-}^{\infty} + \int_{O^-}^{\infty}\frac{f(t)e^{-st}}{s}dt$$

The term e^{-st} approaches zero as $t \to \infty$ and $\int_{O^-}^{t} f(t)dt \to 0$ as $t \to O^-$. Hence

$$L\left[\int_{O^-}^{t} f(t)dt\right] = \frac{F(s)}{s}$$

In network theory, we encounter integrals of type $\int_{-\infty}^{t} f(t)dt$. This integral can be written as

$$\int_{-\infty}^{t} f(t)dt = \int_{-\infty}^{O^-} f(t)dt + \int_{O^-}^{t} f(t)dt$$

The first term on the right hand side of the above equation is a constant. Hence

$$L\left[\int_{-\infty}^{t} f(t)dt\right] = \frac{\int_{-\infty}^{0^-} f(t)dt}{s} + \frac{F(s)}{s} \qquad .....(9.19)$$

The quantity $\int_{-\infty}^{0^-} f(t)dt$ is the charge at $t = 0^-$, $q(0^-)$, if f(t) is a current and it is flux linkages at $t = 0^-$, $\psi(0^-)$ if f(t) is a voltage.

6. Multiplication by t :

$$\text{If } L[f(t)] = F(s) \text{ then } L[tf(t)] = -\frac{dF(s)}{ds} \qquad .....(9.20)$$

Proof :

$$L[tf(t)] = \int_{0^-}^{\infty} tf(t)e^{-st}dt$$

$$= -\int_{0^-}^{\infty} f(t)\frac{d}{ds}e^{-st}dt$$

Since the differentiation is w.r.t 's' and integration is w.r.t 't', we have

$$L[tf(t)] = -\frac{d}{ds}F(s)$$

Using this theorem we can evaluate transforms of some complex functions. For example,

$$L[te^{-at}] = -\frac{d}{ds}L[e^{-at}]$$

$$= -\frac{d}{ds}\cdot\frac{1}{s+a}$$

$$= \frac{1}{(s+a)^2}$$

7. Division by t :

$$\text{If } L[f(t) = F(s), \text{ then } L\left[\frac{f(t)}{t}\right] = \int_{s}^{\infty} F(s)ds$$

Proof :

$$L\left[\frac{f(t)}{t}\right] = \int_{0^-}^{\infty} \frac{f(t)}{t} e^{-st} dt$$

$$= \int_{0^-}^{\infty} f(t)\left[\int_{s}^{\infty} e^{-st} ds\right] dt$$

$$\left[\because \int_{s}^{\infty} e^{-st} ds = \frac{e^{-st}}{-t}\bigg|_{s}^{\infty} = \frac{e^{-st}}{t}\right]$$

Interchanging the two integrals, we have

$$= \int_{s}^{\infty}\left[\int_{0^-}^{\infty} f(t) e^{-st} dt\right] ds$$

$$= \int_{s}^{\infty} F(s) ds \qquad .....(9.21)$$

8. Initial value theorem :

If f(t) and $\frac{df(t)}{dt}$ are Laplace transformable then

$$f(0^+) = \underset{t \to 0^+}{Lt} f(t) = \underset{s \to \infty}{Lt} sF(s) \qquad .....(9.22)$$

Proof :

$$\underset{s \to \infty}{Lt} \int_{0^-}^{\infty} \frac{df(t)}{dt} e^{-st} dt = \underset{s \to \infty}{Lt} [sF(s) - f(0^-)]$$

$$\underset{s \to \infty}{Lt} \int_{0^-}^{0^+} \frac{df(t)}{dt} e^{0} dt + \underset{s \to \infty}{Lt} \int_{0^+}^{\infty} \frac{df(t)}{dt} e^{-st} dt$$

$$= \underset{s \to \infty}{Lt} [sF(s) - f(0^-)] \qquad .....(9.23)$$

Since s and t are independent, the integral and limiting process can be interchanged in eq. (9.23), resulting in

$$f(0^+) - f(0^-) + \int_{0^+}^{\infty} \frac{df(t)}{dt} \underset{s \to \infty}{Lt} e^{-st} dt = \underset{s \to \infty}{Lt} sF(s) - f(0^-)$$

But $\underset{s \to \infty}{Lt} e^{-st} = 0$

and hence $\underset{s \to \infty}{Lt} sF(s) - f(0^-) = f(0^+) - f(0^-)$

or $f(0^+) = \underset{s \to \infty}{Lt} sF(s)$

9. Final value theorem :

If f(t) and $\frac{df(t)}{dt}$ are Laplace transformable then

$$\underset{t \to \infty}{\text{Lt}} f(t) = \underset{s \to 0}{\text{Lt}} sF(s) \qquad .....(9.24)$$

Proof :

Consider $\underset{s \to 0}{\text{Lt}} \int_{0^-}^{\infty} \frac{df(t)}{dt} e^{-st} dt = \underset{s \to 0}{\text{Lt}} sF(s) - f(0^-)$

Interchanging the integral and the limit

$$\int_{0^-}^{\infty} \frac{df(t)}{dt} \underset{s \to 0}{\text{Lt}} e^{-st}dt = \underset{s \to 0}{\text{Lt}} sF(s) - f(0^-)$$

$$f(t)\Big|_{0^-}^{\infty} = \underset{s \to 0}{\text{Lt}} sF(s) - f(0^-)$$

$\therefore$ $$f(\infty) - f(0^-) = \underset{s \to 0}{\text{Lt}} sF(s) - f(0^-)$$

or $$f(\infty) = \underset{s \to 0}{\text{Lt}} sF(s)$$

It is to be noted that, the result is valid only if $\underset{t \to \infty}{\text{Lt}} f(t)$ exists.

10. Convolution theorem :

If $f_1(t)$ and $f_2(t)$ are Laplace transformable and $L[f_1(t)] = F_1(s)$ and $L[f_2(t)] = F_2(s)$, then

$$L\left[\int_{0^-}^{t} f_1(\tau)f_2(t-\tau)d\tau\right] = F_1(s) \, . \, F_2(s) \qquad ....(9.25)$$

Proof :

$$L\left[\int_{0^-}^{t} f_1(\tau)f_2(t-\tau)d\tau\right] = \int_{0^-}^{\infty} \left\{\int_{0^-}^{t} f_1(\tau)f_2(t-\tau)d\tau \; e^{-st}dt \qquad .....(9.26)\right.$$

Introduce a function $u(t-\tau)$ defined by

$$u(t-\tau) = 1 \quad \tau \le t$$
$$= 0 \quad \tau > t$$

in the second integral on RHS of eq. (9.26).

Since τ changes from 0 to t, $u(t-\tau)$ is unity during this interval and for $\tau > t$, $u(t-\tau)$ is zero, the upper limit in second integral can be changed to ∞ without affecting the value of the integral. Thus

$$L\left[\int_{0^-}^{t} f_1(\tau)f_2(t-\tau)d\tau\right] = \int_{0^-}^{\infty}\left[\int_{0^-}^{\infty} f_1(\tau)f_2(t-\tau)u(t-\tau)d\tau\right]e^{-st}dt$$

Interchanging the order of integrations

$$= \int_{0^-}^{\infty} f_1(\tau)\left[\int_{0^-}^{\infty} f_2(t-\tau)u(t-\tau)e^{-st}dt\right]d\tau$$

Using the shifting theorem, we have

$$L\left[\int_{0^-}^{t} f_1(\tau)f_2(t-\tau)d\tau\right] = \int_{0^-}^{\infty} f_1(\tau)\ e^{-s\tau}F_2(s)d\tau = F_1(s)\ F_2(s)$$

This theorem is a very important theorem with many applications. This enables us to evaluate the response of a network for any arbitrary input in terms of the impulse response of the network. Therefore let us consider this convolution integral in more detail.

Convolution of two function $f_1(t)$ and $f_2(t)$ is indicated by $f_1(t) * f_2(t)$. Thus

$$f(t) = f_1(t) * f_2(t) = \int_{0^-}^{t} f_1(\tau)\,f_2(t-\tau)d\tau = \int_{0^-}^{t} f_1(t-\tau)\,f_2(\tau)d\tau \quad .....(9.27)$$

Let us first try to understand the operation involved in convolving two functions $f_1(t)$ and $f_2(t)$ to obtain f(t). This process involves 4 steps : 1. folding, 2. translating, 3. multiplying and 4. integrating. These four steps are described by the word *'Faltung'* in German language. Thus the convolution integral is also known as *Faltung integral*. Let us take a few examples to illustrate these four steps.

Example 9.1

Convolve $f_1(t) = e^{-at}u(t)$ and $f_2(t) = u(t)$ where u(t) is a unit step function.

Solution :

Let us illustrate the process graphically

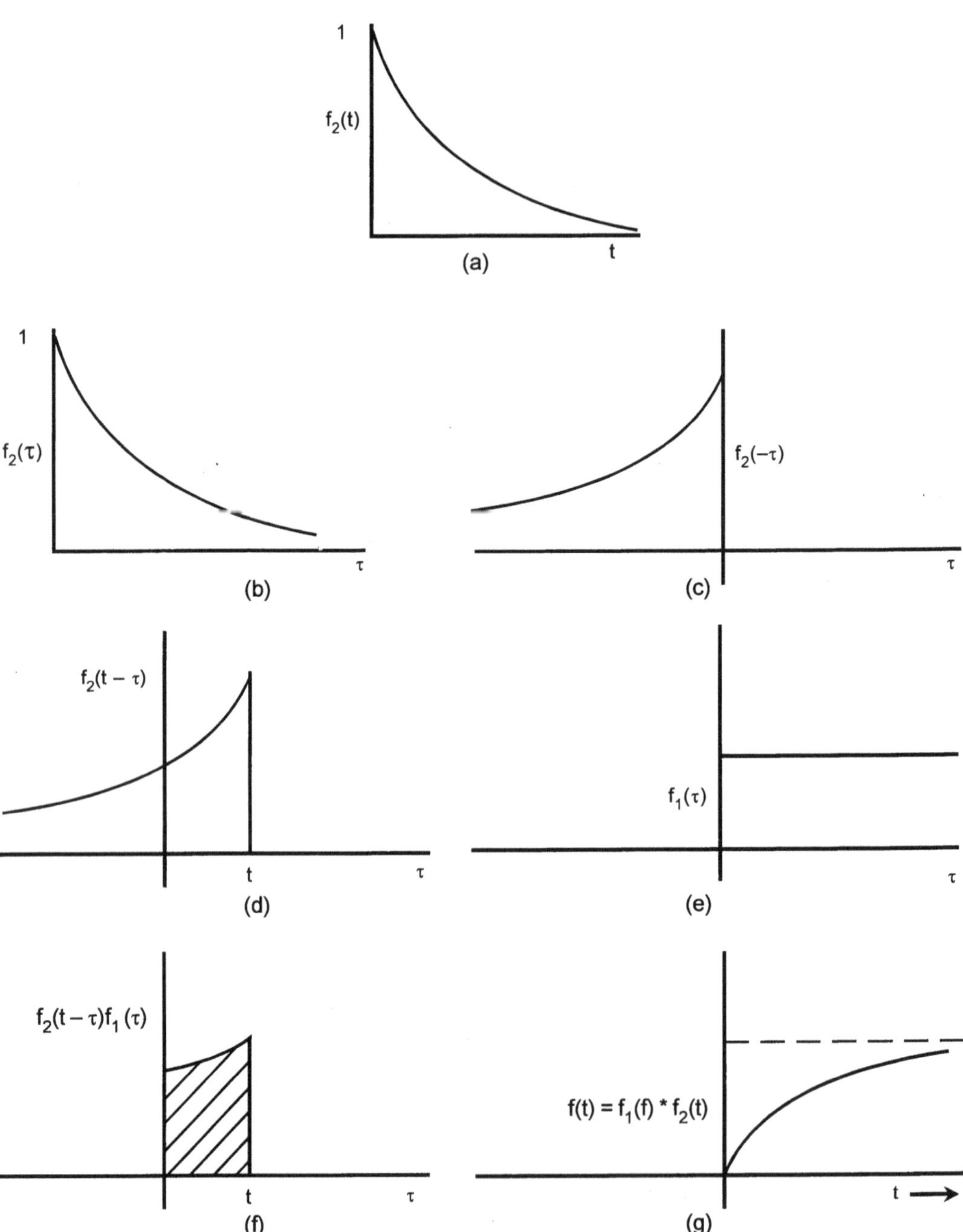

Fig. 9.4 Steps involved in convolution of two functions

(a) In step (a) $f_2(t)$ is shown

(b) In step (b) the variable t is changed to a dummy variable τ.

(c) In step (c) the function $f_2(\tau)$ is ***folded*** to obtain $f_2(-\tau)$.

(d) In step (d) the function $f_2(-\tau)$ is ***translated*** or shifted to right by an amount t to obtain $f_2(t-\tau)$.

(e) In step e, $f_1(t)$ is shown as a function of dummy variable τ.

(f) In this step, the two function in Fig. 9.4(a) and (e) are ***multiplied*** and the resulting function is shown in Fig 9.4(f).

(g) The area under the curve in Fig. 9.4(f) gives the value of the ***integral***, $\int_{0^-}^{t} f_1(\tau)\, f_2(t-\tau)d\tau$ for any given t. The value of t can be changed from 0 to any desired value to obtain the curve f(t) shown in Fig. 9.4 (g).

The same steps can be repeated for the function $f_1(t)$ to get $\int_{0^-}^{t} f_1(t-\tau)f_2(\tau)d\tau$. This is indicated in Fig. 9.5. The result is exactly same as in the previous case.

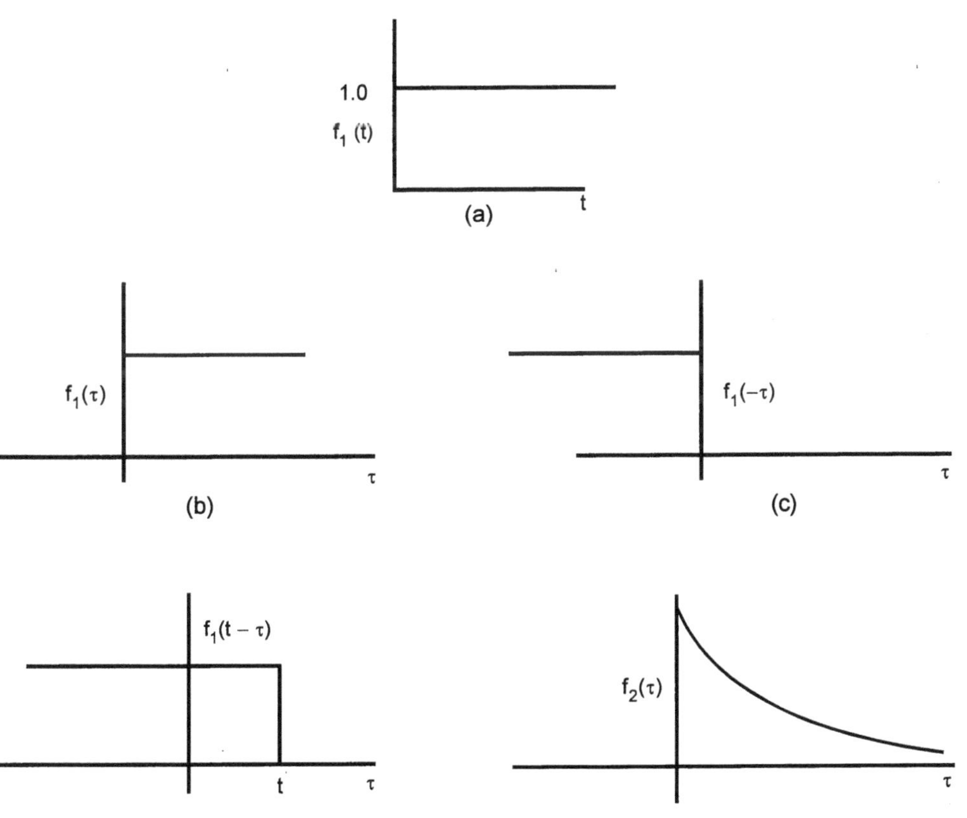

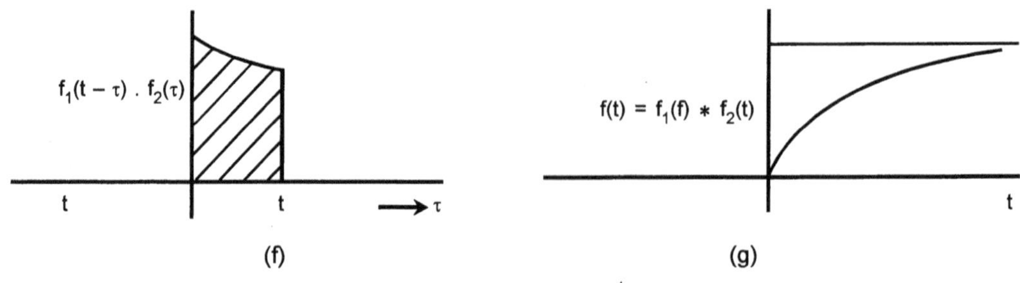

Fig. 9.5 Illustration of evaluation of $\int_{0^-}^{t} f_1(t-\tau) f_2(\tau) d\tau$

Example 9.2

Convolve two functions $f_1(t)$ and $f_2(t)$ given in Fig. 9.6(a) and (b).

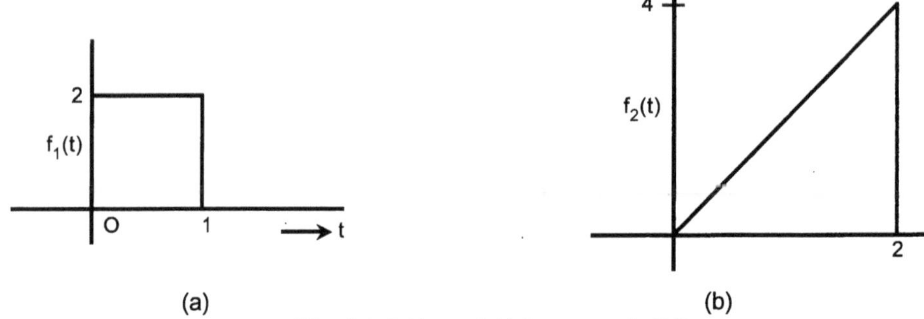

Fig. 9.6 $f_1(t)$ are $f_2(t)$ for example 9.2

Solution :

The various steps involved in the evaluation of convolution are shown in Fig. 9.6(c) to (i).

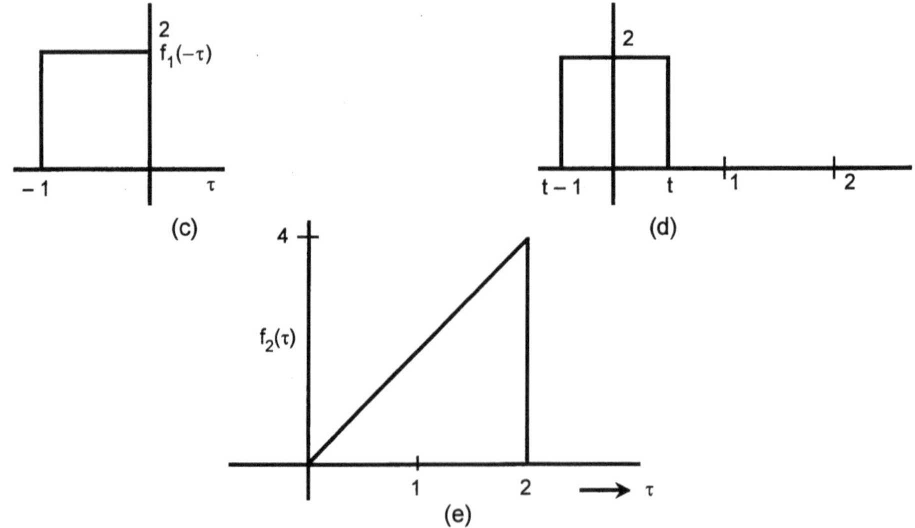

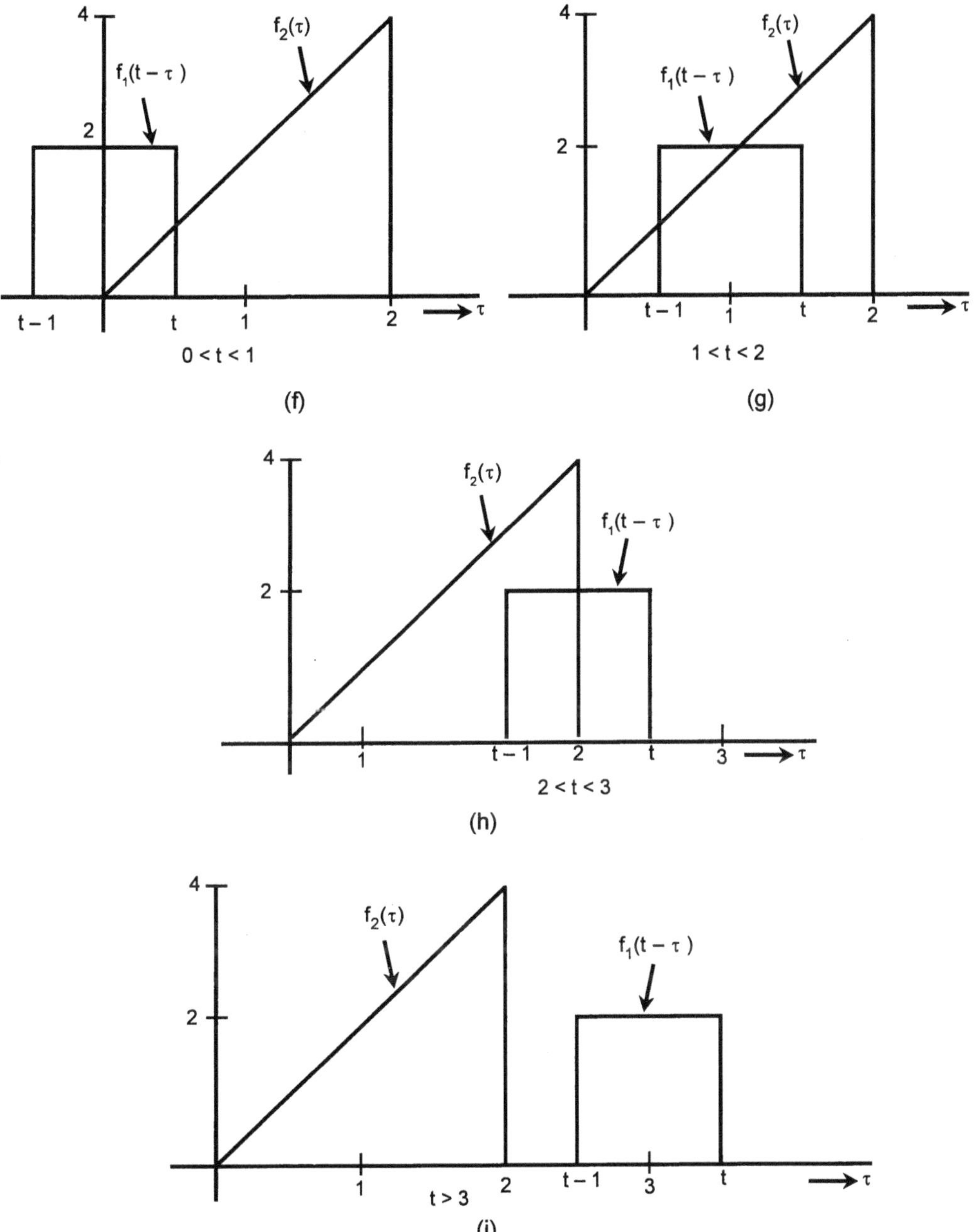

Fig. 9.6 Various steps involved in the evaluation of convolution

In evaluating the convolution integral, as t change from 0 to ∞, the limits of integration change depending on the range of values of 't'. Hence the integration has to be performed over different limits for different ranges of 't'.

(a) $0 \leq t \leq 1$

From Fig. 9.6(f)

$$f_1(t-\tau) = 2$$

$$f_2(\tau) = 2\tau$$

$$f(t) = f_1(t) * f_2(t) = \int_0^{\infty} f_1(t-\tau) f_2(\tau) d\tau$$

$$= \int_0^t 2.2\tau \, d\tau$$

$$= \left. \frac{4\tau^2}{2} \right|_0^t$$

$$= 2t^2$$

This expression is valid only for $0 \leq t \leq 1$ and the curve is a parabola.

(b) $1 \leq t \leq 2$,

From Fig. 9.6(g)

$$f(t) = \int_{t-1}^{t} 2.2\tau \, d\tau$$

$$= 4 \left. \frac{\tau^2}{2} \right|_{t-1}^{t}$$

$$= 2[t^2 - (t-1)^2] = 4t - 2; \quad 1 \leq t \leq 2$$

This is a straight line.

(c) $2 \leq t \leq 3$,

From Fig. 9.6(h)

$$f(t) = \int_{t-1}^{2} 22\tau d\tau$$

$$= 4 \left. \frac{\tau^2}{2} \right|_{t-1}^{2}$$

$$= 2[4 - (t-1)^2]$$

$$= 6 - 2t^2 + 4t; \ 2 \leq t \leq 3$$

This again is a parabola.

(d) $t \geq 3$,

From Fig. 9.6(i),

There is no common area for the product $f_1(t - \tau)\, f_2(\tau)$ and hence $f(t) = 0$.

The resulting function f(t) is plotted in Fig. 9.6(j).

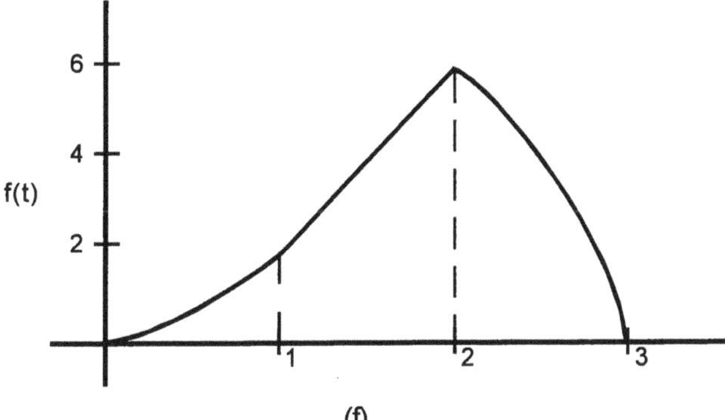

(f)
Fig. 9.6 The result $f(t) = f_1(t) * f_2(t)$

Example 9.3

Convolve the two functions $f_1(t)$ and $f_2(t)$ given in Fig. 9.7(a) and (b).

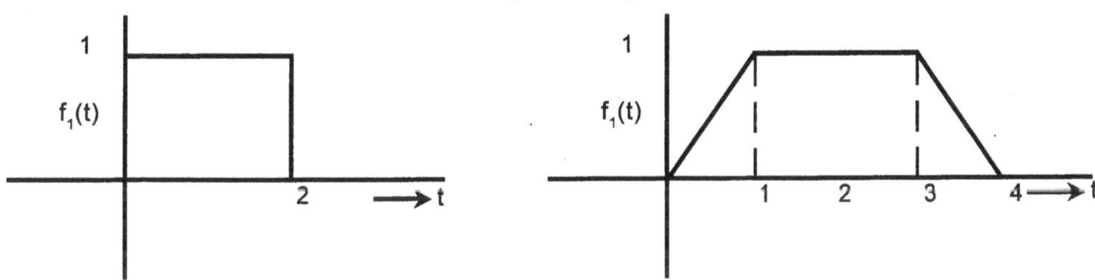

(a) (b)
Fig. 9.7 $f_1(t)$ and $f_2(t)$ for example 9.3

Solution :

This involves more complex functions. In this case not only the limits of integration change but also the functions themselves change. The steps are illustrated in Fig. 9.7 (c) to (i).

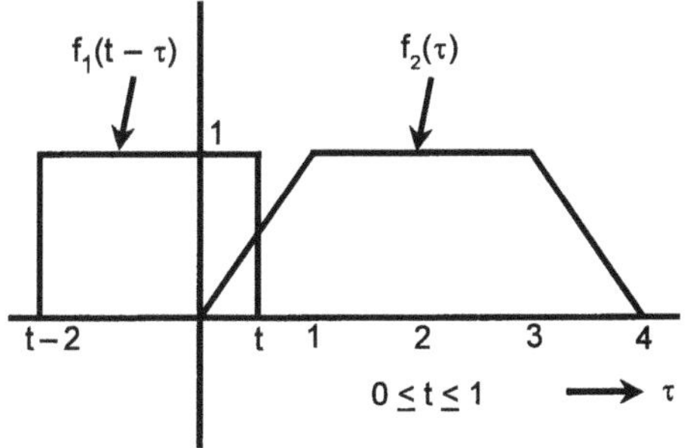

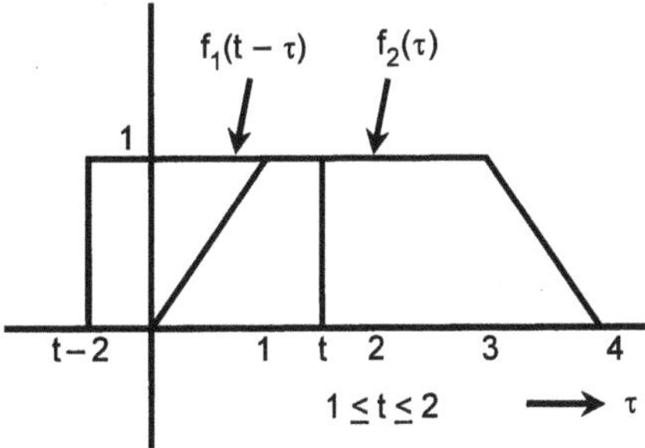

(c) (d)

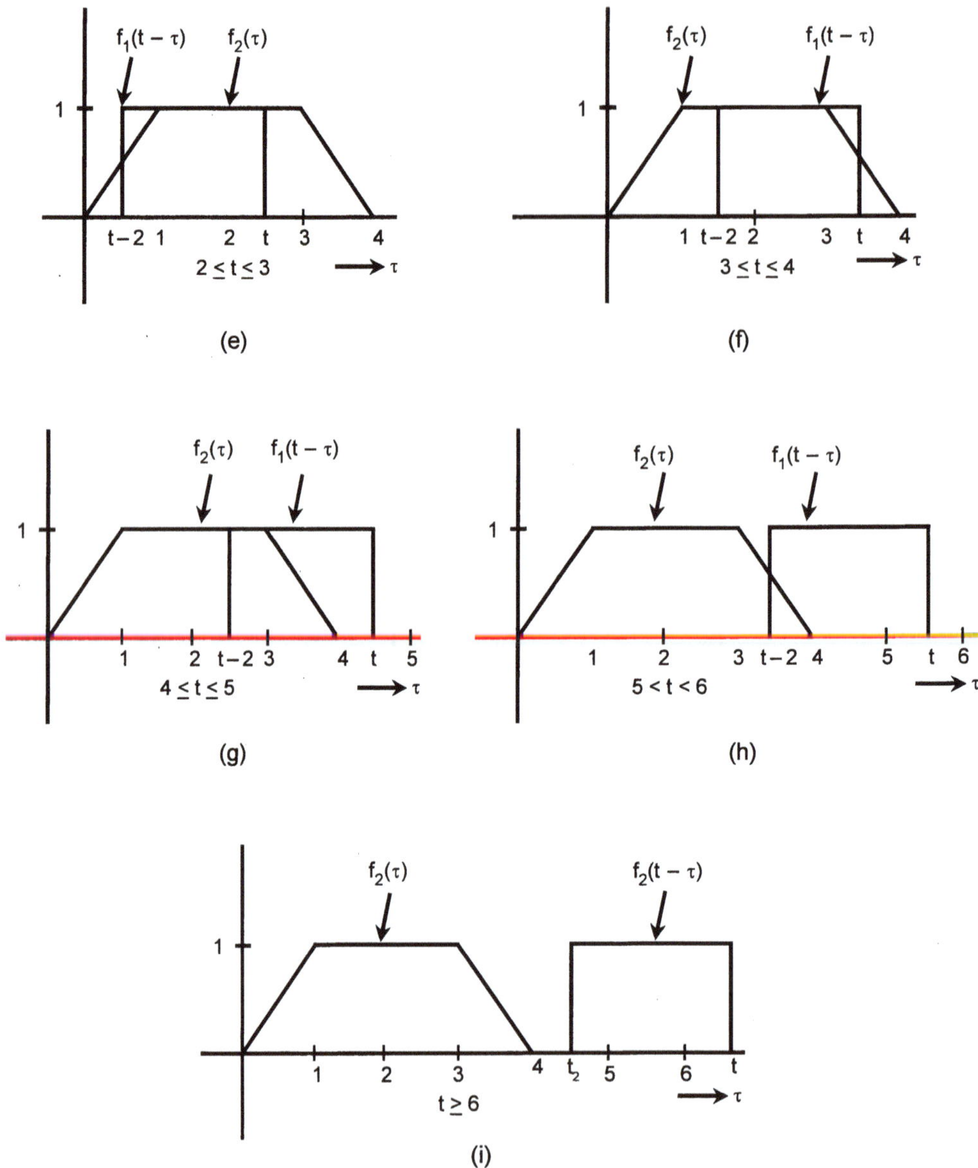

Fig. 9.7 Steps in the convolution of $f_1(t)$ and $f_2(t)$ for example 9.3

(a) For $0 \le t \le 1$

Refer Fig. (c).

The integration is performed between limits 0 to t

$$f_1(t) = 1$$

$$f_2(t) = t$$

$$f(t) = \int_0^t 1.\tau\, d\tau$$

$$= \frac{\tau^2}{2}\Big|_0^t$$

$$= \frac{t^2}{2}\Big|_0^t \text{(parabola)}$$

(b) For $1 \le t \le 2$

Refer to Fig. (d)

The integration is performed between limits 0 to t. But the variation of $f_2(t)$ between 0 and 1 and 1 and t is different.

$$f_1(\tau) = 1 \quad 0 < \tau < t$$
$$f_2(\tau) = \tau \quad 0 \le \tau \le 1$$
$$= 1 \quad 1 \le \tau \le t$$

Hence we have

$$f(t) = \int_0^1 1.\tau\, d\tau + \int_1^t 1.1 d\tau$$

$$= \frac{\tau^2}{2}\Big|_0^1 + \tau\Big|_1^t$$

$$= \frac{1}{2} + t - 1$$

$$= t - \frac{1}{2} \quad \text{(straight line)}$$

(c) For $2 \le t \le 3$

Refer Fig. (e)

The integration is performed from $t - 2$ to t.

$$f_1(\tau) = 1 \quad t-2 \le \tau \le t$$
$$f_2(\tau) = \tau \quad t-2 \le \tau \le 1$$
$$= 1 \quad 1 \le \tau \le t$$

Hence we have

$$f(t) = \int_{t-2}^1 1.\tau d\tau + \int_1^t 1.1 d\tau$$

$$= \frac{\tau^2}{2}\Big|_{t-2}^1 + \tau\Big|_1^t$$

$$= \frac{1}{2} - \frac{(t-2)^2}{2} + t - 1$$

$$= -\frac{t^2}{2} + 3t - 2.5 \quad \text{(parabola)}$$

(d) For $3 \leq t \leq 4$

Refer to Fig. (f)

The limits of integration are t – 2 to t

$$\begin{aligned} f_1(\tau) &= 1 && t-2 \leq \tau \leq t \\ f_2(\tau) &= 1 && t-2 \leq \tau \leq 3 \\ &= -\tau + 4 && 3 \leq \tau \leq t \end{aligned}$$

Hence

$$f(t) = \int_{t-2}^{3} 1.1\, d\tau + \int_{3}^{t} 1(-\tau + 4) d\tau$$

$$= \tau \Big|_{t-2}^{3} + \left(\frac{-\tau^2}{2} + 4\tau\right)\Big|_{3}^{t}$$

$$= 3 - t + 2 - \frac{t^2}{2} + 4t + \frac{9}{2} - 12$$

$$= -\frac{t^2}{2} + 3t - 2.5 \quad \text{(parabola)}$$

Note that f(t) is same as in step (c)

(e) For $4 \leq t \leq 5$

Refer to Fig (g)

Limits of integration are from t – 2 to 4

$$\begin{aligned} f_1(\tau) &= 1 && t-2 \leq \tau \leq 4 \\ f_2(\tau) &= 1 && t-2 \leq \tau \leq 3 \\ &= -\tau + 4 && 3 \leq \tau \leq 4 \\ &= 0 && \tau \geq 4 \end{aligned}$$

Hence, we have

$$f(t) = \int_{t-2}^{3} 1.1\, d\tau + \int_{3}^{4} 1(-\tau + 4) d\tau$$

$$= \tau \Big|_{t-2}^{3} + \left(\frac{-\tau^2}{2} + 4\tau\right)\Big|_{3}^{4}$$

$$= 3 - (t-2) - 8 + 16 + 4.5 - 12$$

$$= -t + 5.5 \quad \text{(straight line)}$$

(f) For $5 \leq t > 6$

Refer to Fig. (h)

Limits of integration are from $t - 2$ to 4

$$f_1(\tau) = 1 \qquad t - 2 \leq \tau \leq 4$$
$$f_2(\tau) = -\tau + 4 \qquad t - 2 \leq \tau \leq 4$$
$$= 0 \qquad \tau \geq 4$$

Hence we have

$$f(t) = \int_{t-2}^{4} 1 \cdot (-\tau + 4)d\tau$$

$$= \left(-\frac{\tau^2}{2} + 4\tau\right)\Bigg|_{t-2}^{4}$$

$$= -8 + 16 + \frac{(t-2)^2}{2} - 4(t-2)$$

$$= \frac{t^2}{2} - 6t + 18 \quad \text{(parabola)}$$

(g) For $t > 6$

Refer to Fig. (i)

$$f_2(\tau) = 0 \qquad \text{for} \quad \tau > 4$$

Hence there is no common area in the product of $f_1(t - \tau)$ and $f_2(\tau)$.

$\therefore \qquad f(t) = 0$

The result of convolution of $f_1(t)$ and $f_2(t)$ is plotted in Fig. 9.7(j).

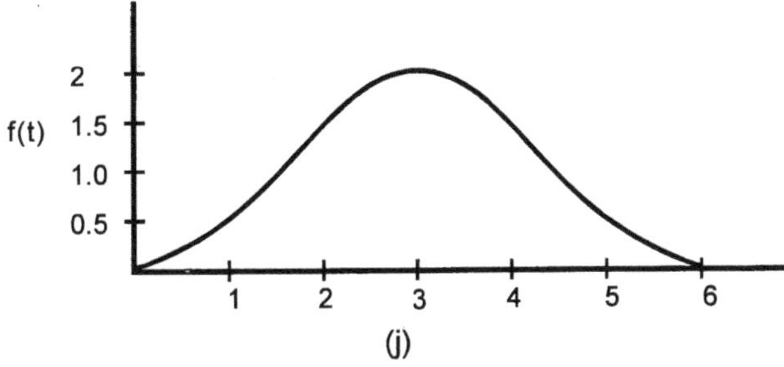

(j)

Fig. 9.7 The result of convolving $f_1(t)$ and $f_2(t)$ for example 9.3

Let us now go back to the theorems.

11. Time scaling :

If $L[f(t)] = F(s)$ then $L[f(at)] = \frac{1}{a}F\left(\frac{s}{a}\right)$ .....(9.28)

Proof :

By definition

$$L[f(at)] = \int_{0^-}^{\infty} f(at)e^{-st}dt$$

Let $at = \tau$

$a\,dt = d\tau$

$$\therefore \quad L[f(at)] = \int_{0^-}^{\infty} f(\tau)e^{-s\frac{\tau}{a}}\frac{d\tau}{a}$$

$$= \frac{1}{a}\int_{0^-}^{\infty} f(\tau)\, e^{-\left(\frac{s}{a}\right)\tau}\, d\tau$$

$$= \frac{1}{a}F\left(\frac{s}{a}\right)$$

9.5 WAVEFORM SYNTHESIS :

Laplace transforms of transient functions, which exist only over a given interval, can be evaluated by using the concept of waveform synthesis to define the functions in terms of unit step, unit ramp and other functions. If f(t) is any given function, $f(t - \tau)$ is the same function shifted to the right by an amount equal to τ. We will synthesise any given function in terms of the shifted step and ramp functions.

1. *A pulse :*

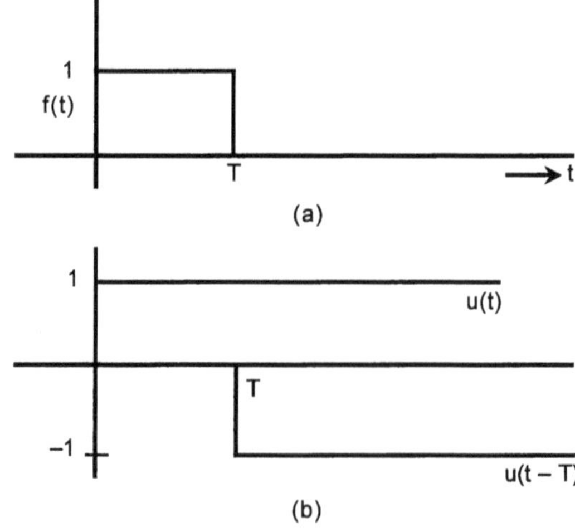

Fig. 9.8 A pulse synthesised in terms of step functions

The pulse in Fig. 9.8(a) is shown to be a sum of two step functions in Fig. 9.8(b). Thus, the pulse can be written as

$$f(t) = u(t) - u(t - T) \qquad .....(9.29)$$

If we wish to find the Laplace transform of the pulse in Fig. 9.8, it can be easily found by using the eqs. (9.4) and (9.15) in eq. (9.29). Hence

$$F(s) = L[u(t)] - L[u(t - T)]$$

$$= \frac{1}{s} - \frac{1}{s}e^{-Ts}$$

$$= \frac{1}{s}(1 - e^{-Ts}) \qquad .....(9.30)$$

2. *A triangular pulse :*

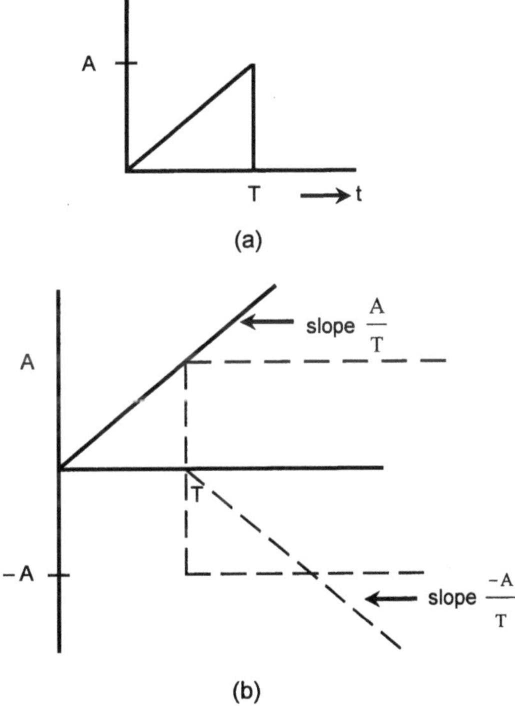

Fig. 9.9 (a) Triangular pulse and (b) its synthesis using ramp and step functions

$$f(t) = \frac{A}{T}t - \frac{A}{T}(t - T)u(t - T) - Au(t - T)$$

The Laplace transform of f(t) is

$$F(s) = \frac{A}{Ts^2} - \frac{A}{Ts^2}e^{-Ts} - \frac{A}{s}e^{-Ts}$$

3. *A trapezoidal pulse :*

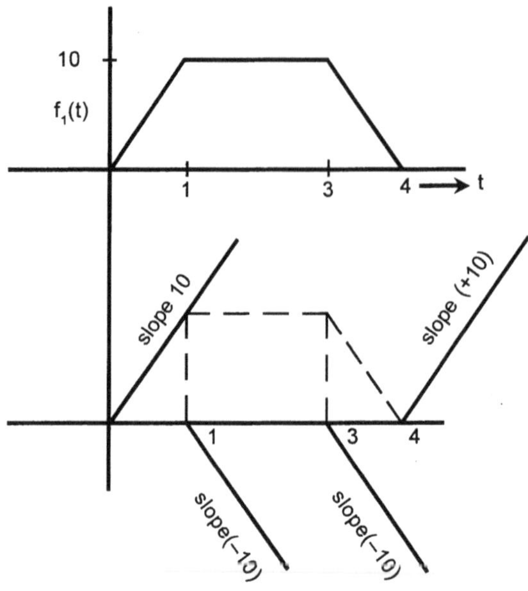

Fig. 9.10 A trapezoidal function and its synthesis

$$f(t) = 10t - 10(t-1)u(t-1) - 10(t-3)u(t-3) + 10(t-4)u(t-4)$$

and $$F(s) = \frac{10}{s^2}[1 - e^{-s} - e^{-3s} + e^{-4s}]$$

4. *A sinusoidal pulse :*

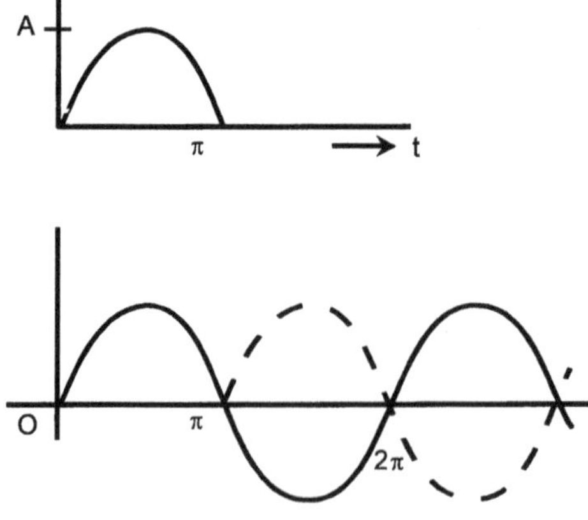

Fig. 9.11 A sinusoidal pulse and its synthesis using shifted sinusoid

$$f(t) = A\,\sin t + A\,\sin(t-\pi)u(t-\pi)$$

and $$F(s) = \frac{A}{s^2+1} + \frac{A}{s^2+1}\cdot e^{-\pi s} = \frac{A}{s^2+1}(1 + e^{-\pi s})$$

Let us now work out an example to understand the procedure.

Example 9.4

Find the Laplace transform of the function shown in Fig. 9.12(a) using waveform synthesis.

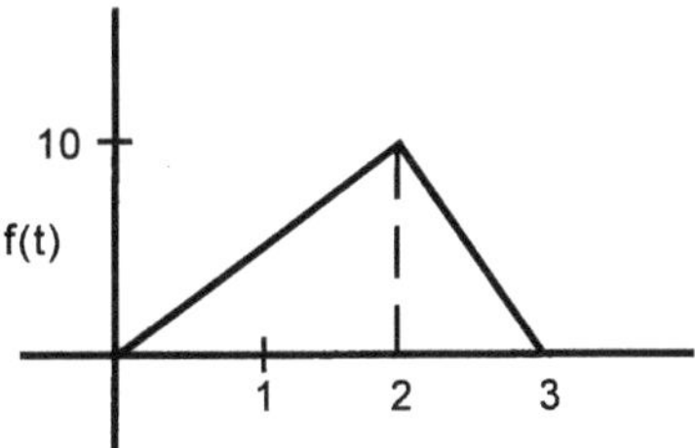

Fig. 9.12 (a) f(t) for example 9.4

Solution :

The synthesis of f(t) in Fig. 9.12 is shown in Fig. 9.12(b).

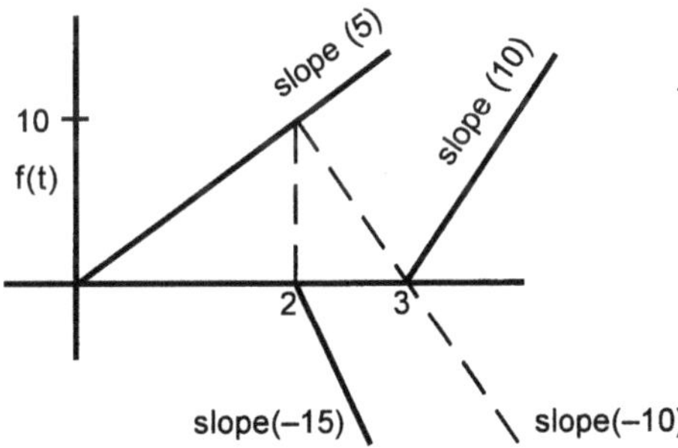

Fig. 9.12 (b) Synthesis of f(t) of Fig. 9.12(a)

From Fig. 9.12(b), we have

$$f(t) = 5t - 15(t-2)u(t-2) + 10(t-3)u(t-3)$$

and $$F(s) = \frac{5}{s^2}(1 - 3e^{-2s} + 2e^{-3s})$$

9.6 GATE FUNCTION :

Laplace transform of transient functions can also be obtained easily, using the concept of Gate function. The Gate function is shown in Fig. 9.13.

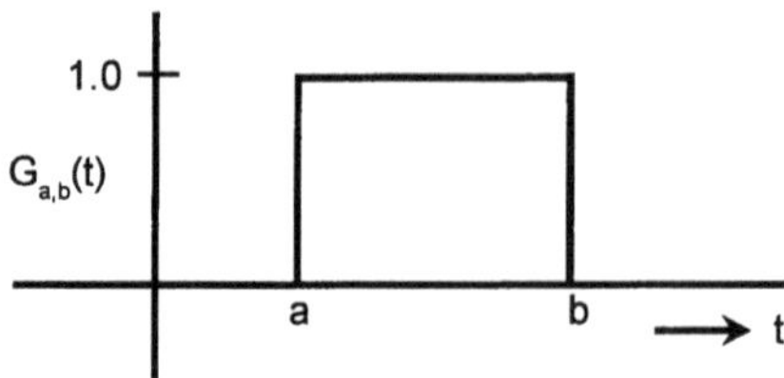

Fig. 9.13 A Gate function

$$G_{a,\,b}(t) = u(t - a) - u(t - b)$$

When a function f(t) is multiplied by a Gate function, that part of the function f(t), which lies in the interval (a, b), only remains and the remaining part becomes zero. It means that when a function passes through a gate only that part remains which goes through the gate. For example, when a gate function is applied to the function of Fig. 9.14(a) with a gate of Fig. 9.14(b), the function shown in Fig. 9.11(c) results.

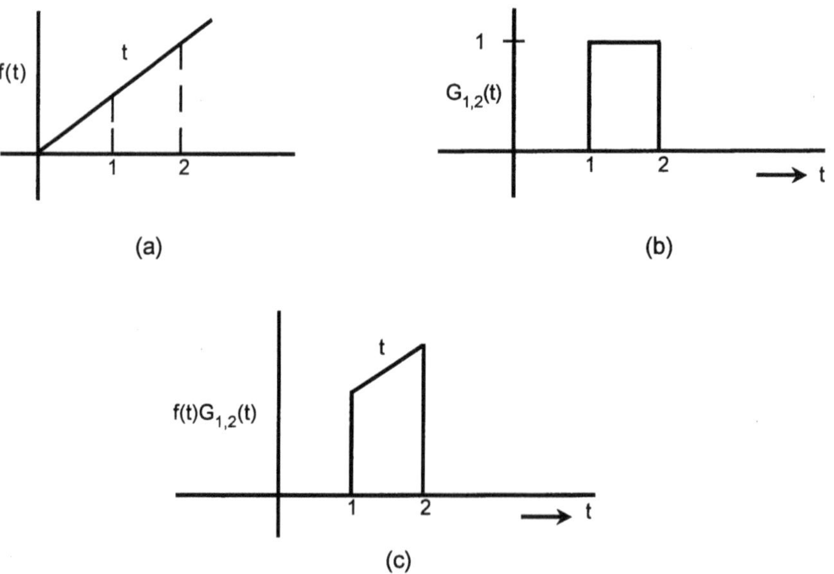

Fig. 9.14 (a) Given f(t) (b) gate $G_{1,\,2}(t)$ and (c) product of $G_{1,2}(t)$ and f(t)

The use of gate function is illustrated using an example.

Example 9.5

Find the Laplace transform of a triangular pulse shown in Fig. 9.15(a).

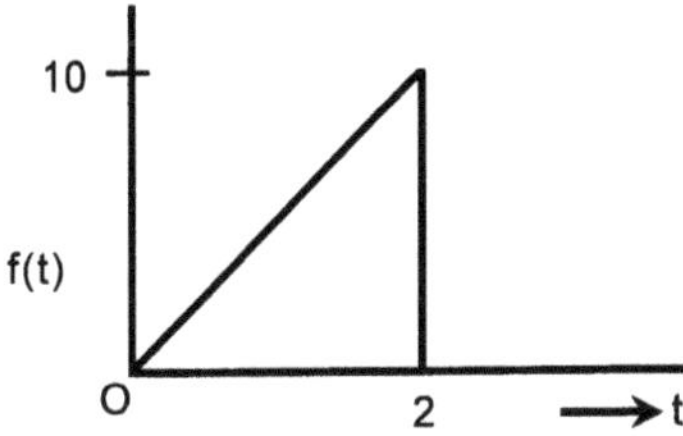

Fig. 9.15 (a) Triangular pulse

Solution :

Let us consider a function as shown in Fig. 9.15(b) and a gate as shown in Fig. 9.15(c).

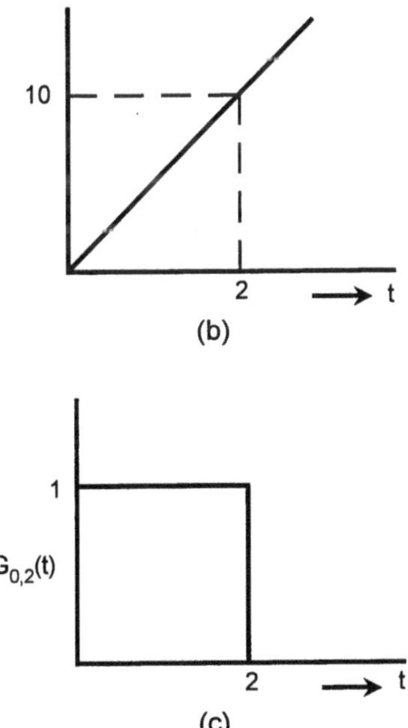

Fig. 9.15 (b) A ramp extending to infinity (c) A gate of width (0, 2)

When the ramp extending to infinity is multiplied by the gate of Fig. 9.15(c), we get the triangular pulse of Fig. 9.15(a).

Thus we have,

$$f(t) = 5t\,[u(t) - u(t-2)]$$
$$= 5t\,u(t) - 5t\,u(t-2) \qquad .....(9.31)$$

The Laplace transform of first term of the right hand side of eq. (9.31) can be easily obtained but the second term is not in the proper form to apply shifting theorem. A little mathematical manipulation will enable us to obtain its transform. Eq. (9.31) can be written as

$$f(t) = 5t\,u(t) - 5(t-2+2)u(t-2)$$
$$= 5t\,u(t) - 5(t-2)u(t-2) - 10u(t-2)$$

The Laplace transform of f(t) can now be obtained as

$$F(s) = \frac{5}{s^2} - \frac{5}{s^2}e^{-2s} - \frac{10}{s}e^{-2s}$$

Example 9.6

Find the Laplace transform of the half period of the sine function shown in Fig. 9.16(a).

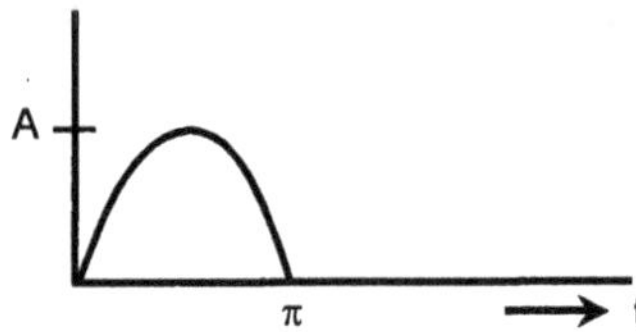

Fig. 9.16 (a) Half period of a sine function

Solution :

Consider an infinite sinusoidal function and a gate function as shown in Fig 9.16(b) and (c) respectively.

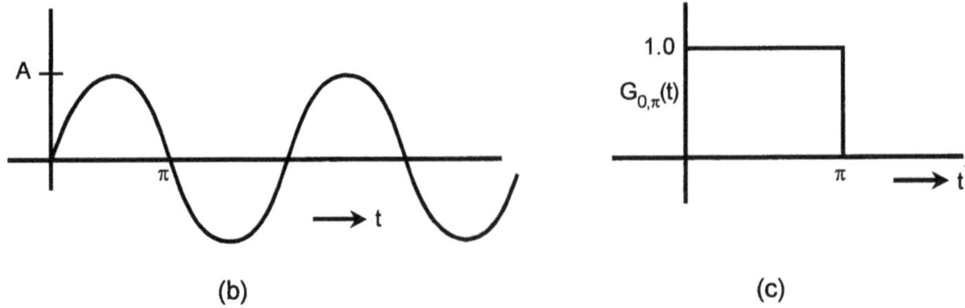

Fig. 9.16 (b) Infinite sinusoidal and (c) gate function

Thus

$$\begin{aligned} f(t) &= A \sin t\,[u(t) - u(t - \pi)] \\ &= A \sin t\, u(t) - A \sin(t - \pi + \pi)u(t - \pi) \\ &= A \sin t\, u(t) - A[\sin(t - \pi) \cos \pi + \cos(t - \pi) \sin \pi]u(t - \pi) \\ &= A \sin t\, u(t) + A \sin(t - \pi)u(t - \pi) \end{aligned}$$

$$\therefore \qquad F(s) = \frac{A}{s^2 + 1} + \frac{A}{s^2 + 1}e^{-\pi s}$$

$$= \frac{A}{s^2 + 1}(1 + e^{-\pi s})$$

The advantage of gate function over waveform synthesis is that one can blindly apply a gate function of required interval to get the transient function, but function so obtained has to be manipulated before we can take the Laplace transform of the resulting function. But if one is a little imaginative, waveform synthesis gives the function which is directly amenable to transformation.

9.7 PERIODIC FUNCTION :

It is easier to find the Laplace transform of a transient function since the limits of integration will be finite. If a function extends upto infinity we assume that

$$\int f(t)\, e^{-st}dt \rightarrow 0 \text{ as } t \rightarrow \infty$$

and the Laplace transform of f(t) can be evaluated. But if we consider a periodic function, the condition stated above is not satisfied and hence it is difficult to evaluate the Laplace transform of such functions. A method based on waveform synthesis is presented here to evaluate the Laplace transform of a periodic function in terms of the Laplace transform of one cycle of the periodic function.

Consider the periodic function f(t) where first cycle is designated as $f_1(t)$. Let the Laplace transform of $f_1(t)$ be $F_1(s)$.

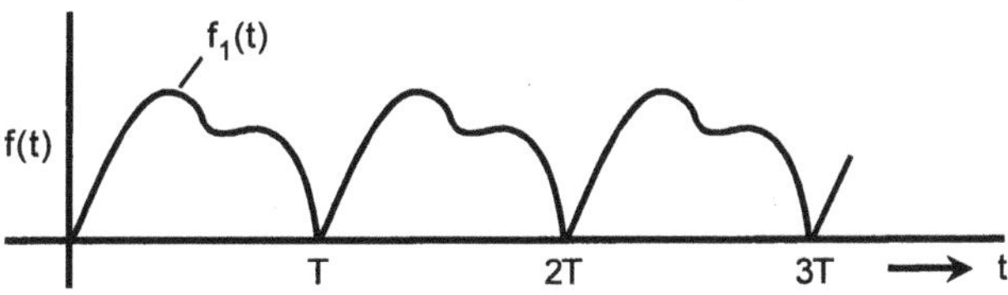

Fig. 9.17 A periodic function

The second cycle of the periodic function f(t) can be thought of as the first cycle $f_1(t)$ delayed or shifted by an amount T. Similarly the third cycle is $f_1(t)$ shifted by 2T and so on. Hence f(t) can be written as

$$f(t) = f_1(t)u(t) + f_1(t - T)u(t - T) + f_1(t - 2T)u(t - 2T) + ... \quad .....(9.32)$$

Taking the Laplace transform on both sides, we have

$$F(s) = F_1(s) + F_1(s)e^{-Ts} + F_1(s)e^{-2Ts} + ...$$
$$= F_1(s)[1 + e^{-Ts} + e^{-2Ts} + ...] \quad .....(9.33)$$

The term in brackets of eq. (9.33) can be identified as $\frac{1}{1-e^{-Ts}}$, using Binomial theorem. Thus

$$F(s) = \frac{F_1(s)}{1-e^{-Ts}} \quad .....(9.34)$$

Eq. (9.34) relates the Laplace transform of a periodic function to the Laplace transform of its first cycle.

Example 9.7

Find the Laplace transform of a periodic square wave shown in Fig. 9.18.

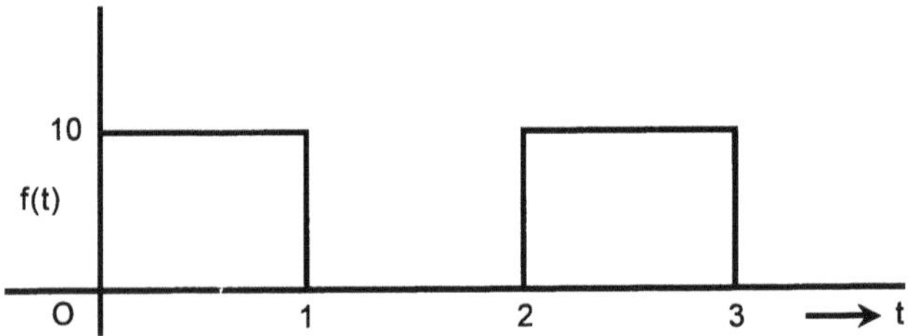

Fig. 9.18 Periodic square wave

Solution :

The Laplace transform of one period of the square wave can be easily obtained as follows :

$$f_1(t) = 10[u(t) - u(t - 1)]$$

$$F_1(s) = \frac{10}{s}\left(1-e^{-s}\right)$$

$\therefore$ The Laplace transform of a periodic waveform with period 2 is given by

$$F(s) = \frac{10}{s}\frac{1-e^{-s}}{1-e^{-2s}}$$

A table of transforms of some important functions and the theorems is given in table 9.1.

Table 9.1 Table of Laplace Transforms

	f(t)	F(s)
1.	$u(t)$	$\frac{1}{s}$
2.	$\delta(t)$	1
3.	t	$\frac{1}{s^2}$
4.	t^n	$\frac{n!}{s^{n+1}}$
5.	e^{at}	$\frac{1}{s-a}$
6.	$\sin\omega_0 t$	$\frac{\omega_0}{s^2+\omega_0^2}$
7.	$\cos\omega_0 t$	$\frac{s}{s^2+\omega_0^2}$
8.	$af_1(t) + bf_2(t)$	$aF_1(s) + bF_2(s)$
9.	$\frac{df(t)}{dt}$	$sF(s) - f(O^-)$
10.	$\frac{d^n f(t)}{dt^n}$	$s^nF(s) - s^{n-1}f(O^-) - s^{n-2}\frac{df}{dt}(O^-)$ ..... $- s\frac{d^{n-2}f}{dt^{n-2}}(O^-) - \frac{d^{n-1}f}{dt}(O^-)$
11.	$\int_0^t f(t)dt$	$\frac{F(s)}{s}$
12.	$\int_{-\infty}^t f(t)dt$	$\frac{F(s)}{s} + \frac{\int_{-\infty}^{0^-} f(t)dt}{s}$
13.	$e^{-at}f(t)$	$F(s+a)$
14.	$f(t-T)u(t-T)$	$F(s)e^{-ST}$
15.	$tf(t)$	$-\frac{d}{ds}F(s)$
16.	$\frac{f(t)}{t}$	$\int_s^\infty F(s)ds$
17.	Periodic function with $f_1(t)$ as first cycle and period T	$\frac{F_1(s)}{1-e^{-Ts}}$
18.	$f_1(t) * f_2(t)$	$F_1(s) . F_2(s)$
19.	$f(at)$	$\frac{1}{a}F\left(\frac{s}{a}\right)$

9.8 APPLICATION TO SOLUTION OF DIFFERENTIAL EQUATIONS :

Let us consider an nth order differential equation :

$$\frac{d^n y}{dt^n} + a_1 \frac{d^{n-1} y}{dt^{n-1}} + a_2 \frac{d^{n-2} y}{dt^{n-2}} + \ldots\ldots + a_{n-1} \frac{dy}{dt} + a_n y = v(t) \qquad \ldots\ldots(9.35)$$

Taking the Laplace transform of this equation using table 9.1.

$$s^n Y(s) - s^{n-1} y(O) - s^{n-2} y^1(O) \ldots\ldots y^{(n-1)}(O) + a_1[s^{n-1} Y(s) - s^{n-2} y(O) - s^{n-3} y^1(O) - \ldots\ldots - y^{(n-2)}(O)] + \ldots\ldots + a_n Y(S) = V(s)$$

$$\therefore \quad Y(s)[s^n + a_1 s^{n-1} + a_2 s^{n-2} + \ldots\ldots + a_n] = V(s) + s^{n-1} y(O) + s^{n-2} y^1(O) + \ldots\ldots + y^{(n-2)}(O) + \ldots\ldots$$

$$\therefore \quad Y(s) = \frac{V(s) + s^{n-1} y(O) + s^{n-2} y(O) \ldots\ldots + y^{(n-1)}(O) + \ldots\ldots}{s^n + a_1 s^{n-1} + a_2 s^{n-2} + \ldots\ldots a_n} \qquad \ldots\ldots(9.36)$$

In this equation $y(O)$, $y^{(1)}(O)$, $y^{(2)}(O)$ ..... $y^{(n-1)}(O)$ are the initial conditions for the given different equation. Thus we see that the advantages listed in section 9.1, viz, the differential equation is converted into algebraic equation, initial condition are used in the beginning of the problem, the total solution is obtained at one stroke, are well substantiated.

But the problem is not still completely solved. The time function y(t) is required to be found by inverse transformation. It is not readily obtained by inspection or referring to the table of transforms and using the uniqueness property. The Laplace transform function has to be expressed in a simpler form so that the table of transforms can be used and the inverse can be written down by inspection.

The function Y(s), as seen in eq. (9.36) is usually of the form

$$Y(s) = \frac{N(s)}{D(s)}$$

where N(s) is the numerator polynomial in s

D(s) is the denominator polynomial in s

As a first step in simplifying Y(s), we check to see if the order of the polynomial N(s) is less than that of D(s). If it is not, the numerator is divided by the denominator to get

$$\frac{N(s)}{D(s)} = k_0 + k_1 s + \ldots\ldots + k_{m-n} s^{m-n} + \frac{N_1(s)}{D(s)} \qquad \ldots\ldots(9.37)$$

where 'm' is the order of the numerator polynomial and n is the order of the denominator polynomial. Now the order of $N_1(s)$ is one less than the order of D(s).

Now D(s) can be factored and $\frac{N_1(s)}{D(s)}$ can be written as

$$\frac{N_1(s)}{D(s)} = \frac{N_1(s)}{(s-s_1)(s-s_2)\ .....(s-s_n)}$$

where $s_1, s_2\ .....\ s_n$ are the roots of D(s) = 0. D(s) = 0 is known as *the characteristic equation* and $s_1, s_2\ .....\ s_n$ can be real or complex and distinct or repeated.

Let us now consider Heaviside's partial fraction expansion method to simplify $\frac{N_1(s)}{D(s)}$.

9.9 HEAVISIDE'S PARTIAL FRACTION EXPANSION TECHNIQUE :

(i) Roots of D(s) = 0 are distinct. Then $\frac{N_1(s)}{D(s)}$ can be written as :

$$\frac{N_1(s)}{D(s)} = \frac{K_1}{s-s_1} + \frac{K_2}{s-s_2} + ... + \frac{K_n}{s-s_n} \qquad .....(9.38)$$

To find the constant K_1, we multiply eq. (9.38) by $(s - s_1)$ and let $s \rightarrow s_1$ on both sides

$$\underset{s \rightarrow s_1}{\text{Lt}} \frac{N_1(s)(s-s_1)}{D(s)} = \left[\frac{K_1(s-s_1)}{(s-s_1)} + \frac{K_2(s-s_1)}{s-s_2} + ... + \frac{K_n(s-s_1)}{s-s_n}\right]_{s \rightarrow s_1}$$

or $$K_1 = \underset{s \rightarrow s_1}{\text{Lt}} \frac{N_1(s)(s-s_1)}{D(s)}$$

In general $$K_j = \underset{s \rightarrow s_j}{\text{Lt}} \frac{N_1(s)(s-s_j)}{D(s)} \qquad .....(9.39)$$

Example 9.8

Find $L^{-1}\left[\frac{10s}{(s+1)(s+2)}\right]$

Solution :

$$F(s) = \frac{10s}{(s+1)(s+2)}$$

$$= \frac{K_1}{s+1} + \frac{K_2}{s+2}$$

$$K_1 = F(s) \cdot (s+1)\Big|_{s=-1} = \frac{10s}{s+2}\Big|_{s=-1}$$

$$= \frac{-10}{1} = -10$$

$$K_2 = F(s)\,(s+2)\Big|_{s=-2}$$

$$= \frac{10s}{s+1}\Big|_{s=-2}$$

$$= \frac{-20}{-2+1} = 20$$

$$\therefore \qquad F(s) = \frac{-10}{s+1} + \frac{20}{s+2}$$

Using table 9.1, we have

$$f(t) = -\,10e^{-t} + 20e^{-2t}$$

(ii) *Repeated roots :*

Let the jth root s_j be repeated r – times.

Let $$\frac{N_1(s)}{D(s)} = \frac{R(s)}{(s-s_j)^r}$$

where $$R(s) = \frac{N_1(s)}{D(s)}\,(s-s_j)^r$$

We can write

$$\frac{R(s)}{(s-s_j)^r} = \frac{K_{j1}}{s-s_j} + \frac{K_{j2}}{(s-s_j)^2} + \ldots\ldots + \frac{K_{jp}}{(s-s_j)^p} \;\ldots\ldots\; \frac{K_{jr-1}}{(s-s_j)^{r-1}} + \frac{K_{jr}}{(s-s_j)^r} \qquad \ldots\ldots(9.40)$$

Multiplying eq. (9.40) by $(s-s_j)^r$, we have

$$R(s) = K_{j1}(s-s_j)^{r-1} + K_{j2}(s-s_j)^{r-2} \ldots\ldots + K_{jp}(s-s_j)^{r-p} + \ldots\ldots + K_{jr} \qquad \ldots\ldots(9.41)$$

If we now let $s \rightarrow s_j$ in eq. (9.41) we get

$$K_{jr} = R(s)\Big|_{s=s_j} = \frac{N_1(s)}{D(s)}\,(s-s_j)^r\Big|_{s=s_j}$$

To evaluate $K_{j,\,r-1}$, we differentiate eq. (9.41) once and make $s = s_j$. Then all terms except $K_{j\,r-1}$ will vanish and

$$K_{j\,r-1} = \frac{d}{ds}R(s)\Big|_{s=s_j} = \frac{d}{ds}\frac{N_1(s)}{D(s)}\,(s-s_j)^r \qquad \ldots\ldots(9.42)$$

This procedure can be generalised to get the general term in eq. (9.40)

$$K_{jp} = \frac{1}{(r-p)!}\frac{d^{r-p}}{ds^{r-p}}R(s)\bigg|_{s=s_j}$$

$$= \frac{1}{(r-p)!}\frac{d^{r-p}}{ds^{r-p}}\left[\frac{N_1(s)}{D(s)}(s-s_j)^r\right]_{s=s_j} \qquad \text{.....(9.43)}$$

Example 9.9

Find $f(t) = L^{-1}F(s)$ where

$$F(s) = \frac{s+1}{s(s+2)^3}$$

Solution :

$$F(s) = \frac{K_1}{s} + \frac{K_{21}}{s+2} + \frac{K_{22}}{(s+2)^2} + \frac{K_{23}}{(s+2)^3}$$

$$K_1 = F(s)\cdot s\Big|_{s=0}$$

$$= \frac{s+1}{(s+2)^3}\bigg|_{s=0} = \frac{1}{8}$$

$$K_{23} = F(s)\cdot(s+2)^3\Big|_{s=-2}$$

$$= \frac{s+1}{s}\cdot\bigg|_{s=-2}$$

$$= \frac{1}{2}$$

$$K_{22} = \frac{1}{(3-2)!}\frac{d}{ds}\left[\frac{s+1}{s}\right]\bigg|_{s=-2}$$

$$= \frac{1}{1!}\left[\frac{-1}{s^2}\right]\bigg|_{s=-2}$$

$$= -\frac{1}{4}$$

$$K_{21} = \frac{1}{(3-1)!}\frac{d^2}{ds^2}\left(\frac{s+1}{s}\right)\bigg|_{s=-2}$$

$$= \frac{1}{2}\left[\frac{2}{s^3}\right]_{s=-2}$$

$$= -\frac{1}{8}$$

$$\therefore \qquad F(s) = \frac{1}{8s} - \frac{1}{8(s+2)} - \frac{1}{4(s+2)^2} + \frac{1}{2(s+2)^3}$$

Using the table of transforms

$$f(t) = \frac{1}{8}u(t) - \frac{1}{8}e^{-2t} - \frac{1}{4}te^{-2t} + \frac{1}{4}\,t^2e^{-2t}$$

Example 9.10

Find $L^{-1}\dfrac{1}{s^2+2s+2}$

Solution :

$$\text{Let } F(s) = \frac{1}{s^2+2s+2}$$

The denominator of F(s) has complex roots

$$s = -1 \pm j1$$

When F(s) has complex roots the procedure given for distinct roots can be used. But a simpler method is given below. F(s) can be written as

$$F(s) = \frac{1}{(s+1)^2+1}$$

Using shifting theorem given in item 13 and item 6 of table 9.1, we have

$$f(t) = e^{-t}\sin t$$

Whenever complex roots occur, they occur in conjugate pairs if the coefficients of D(s) are real. The partial fraction expansion can be modified to include quadratic terms. The procedure is illustrated in example 9.11.

Example 9.11

Find $L^{-1}\dfrac{2s}{(s+1)(s^2+2s+2)}$

Solution :

$$F(s) = \frac{2s}{(s+1)(s^2+2s+2)}$$

$$= \frac{K_1}{s+1} + \frac{K_2 s + K_3}{(s^2+2s+2)}$$

$$K_1 = F(s) \cdot (s+1)\Big|_{s=-1}$$

$$= \frac{2s}{s^2+2s+2}\Bigg|_{s=-1}$$

$$= \frac{-2}{1} = -2$$

To find K_2 and K_3, we can write

$$F(s) = \frac{K_1\left(s^2+2s+2\right)+\left(K_2 s+K_3\right)(s+1)}{(s+1)\left(s^2+2s+1\right)}$$

Comparing the coefficients of like powers of s on both sides of above equation

$$K_1 + K_2 = 0 \text{ (coefficients of } s^2)$$

or $$K_2 = -K_1 = 2$$

$$2K_1 + K_2 + K_3 = 2 \text{ (coefficients of s)}$$

$$K_3 = 2 - 2 + 4$$
$$= 4$$

$$\therefore \quad F(s) = \frac{-2}{s+1} + \frac{2s+4}{(s+1)^2+1}$$

To use shifting theorem for second term in the above equation, we have to modify F(s) as

$$F(s) = \frac{-2}{s+1} + \frac{2s+2+2}{(s+1)^2+1}$$

$$= \frac{-2}{s+1} + \frac{2(s+1)}{(s+1)^2+1} + \frac{2}{(s+1)^2+1}$$

Now using table 9.1, we have

$$f(t) = -2e^{-t} + 2e^{-t}\cos t + 2e^{-t}\sin t$$

With this background, we are now in a position to apply the Laplace transform technique to solve the networks.

9.10 APPLICATION TO NETWORKS :

Now equipped with the definition of Laplace transform and its inverse, we are in a position to solve the network problem.

Example 9.12

Find the current i(t) in the network shown in Fig. 9.19.

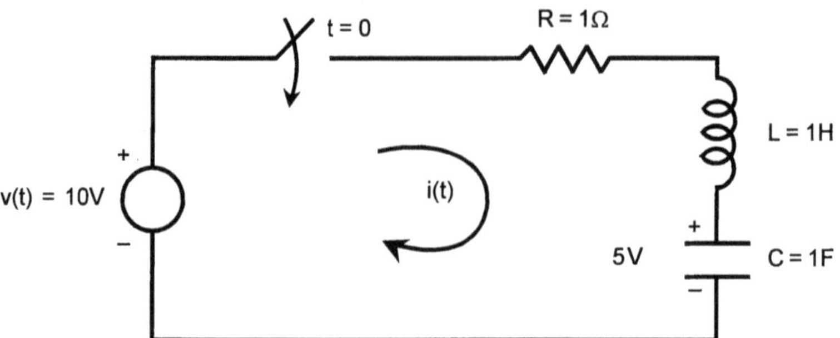

Fig. 9.19 Network for example 9.12

Solution :

Let us write down the integro differential equation for the network using Kirchhoff's voltage law.

$$Ri + L\frac{di}{dt} + \frac{1}{C}\int_{-\infty}^{t} idt = v(t) \qquad .....(9.44)$$

If L[i(t)] = I(s) and L[v(t)] = V(s) we can transform each term of eq. (9.44) using table 9.1. Thus we have

$$RI(s) + L[sI(s) - i(O^{-})] + \frac{I(s)}{Cs} + \frac{\int_{-\infty}^{O^{-}} idt}{Cs} = V(s) \qquad .....(9.45)$$

In this equation $\int_{-\infty}^{O^{-}} idt$ is the charge accumulated in the capacitor at t = O and therefore $\frac{q(O^{-})}{C}$ is the voltage across the capacitor at $t = O^{-}$. Since there was no current in the inductance at $t = O^{-}$, $i(O^{-}) = 0$ and hence eq. (9.45) becomes

$$RI(s) + LsI(s) + \frac{I(s)}{Cs} + \frac{5}{s} = V(s) = \frac{10}{s} \qquad .....(9.46)$$

$$\therefore \quad I(s)\left[1 + s + \frac{1}{s}\right] = \frac{10}{s} - \frac{5}{s} = \frac{5}{s} \qquad .....(9.47)$$

$$I(s) = \frac{5}{s^2 + s + 1} \qquad .....(9.48)$$

We have obtained the expression for the Laplace transform of i(t). The next step would be to find its inverse to obtain the required solution.

Using the technique presented in the previous section

$$I(s) = \frac{5}{\left(s+\frac{1}{2}\right)^2 + \frac{3}{4}} = \frac{5\cdot\frac{\sqrt{3}}{2}\cdot\frac{2}{\sqrt{3}}}{\left(s+\frac{1}{2}\right)^2 + \left(\frac{\sqrt{3}}{2}\right)^2}$$

From table 9.1,

$$i(t) = \frac{10}{\sqrt{3}} e^{-\frac{1}{2}t} \sin\frac{\sqrt{3}}{2}t \qquad .....(9.49)$$

This is the desired solution .

The following observations are in order to emphasize the efficacy of the method.

1. The differential eq. (9.44) is transformed into an algebraic equation as eq. (9.45).
2. The initial condition are introduced in the beginning only as in eq. (9.45).
3. The conditions at $t = O^-$ are used rather than at $t = O^+$ as in classical solution, thus avoiding evaluation of initial conditions at $t = O^+$.
4. Algebraic manipulations are done to get the required solution in the transform domain as in eqs. (9.46) to (9.48).
5. Inverse transformation gives the final result and the total solution is obtained in time domain. In this example, however, there is no steady state solution.
6. If multiloop or multinode networks are to be solved, application of this method results in simultaneous algebraic equations and they can be easily solved. This is not the case with classical methods where solution of simultaneous differential equations is involved.
7. Since conditions at $t = O^-$ are only considered capacitance loops and inductance nodes will not bother us as will be demonstrated later.

The solution of networks will be much more simplified if we can avoid writing down the differential equations and then transforming them. The given network in time domain can be transformed into a network in transform domain directly so that the network appears like a resistive network, and the methods used for solving resistive networks can be used. This is illustrated in the next section.

9.11 TRANSFORM NETWORKS :

Let us now consider the transformation of the network itself into transform domain.

9.11.1 Resistance :

The v – i relationship for a resistance is

$$v(t) = i(t)R \qquad .....(9.50)$$

Taking the Laplace transform of eq. (9.50)

$$V(s) = I(s)R$$

or

$$R = \frac{V(s)}{I(s)} \qquad .....(9.51)$$

The quotient of V(s) and I(s) is defined as the transform impedance, and is denoted by the symbol Z(s). Thus for a resistance

$$\frac{V(s)}{I(s)} = Z(s) = R$$

Transform admittance can also be defined as the ratio of $\frac{I(s)}{V(s)}$ and is denoted by Y(s)

$$\frac{I(s)}{V(s)} = Y(s) = \frac{1}{Z(s)} = \frac{1}{R} \qquad .....(9.52)$$

The time domain and transform domain representation of a resistor are indicated in Fig. 9.20.

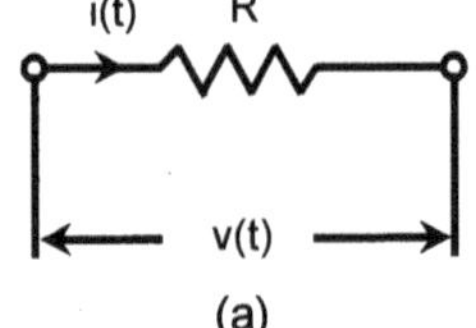

(a)

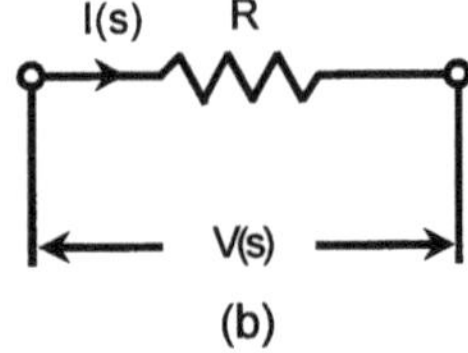

(b)

Fig. 9.20 Representation of a resistor in time and transform domains

9.11.2 Inductance :

The v – i relationship is given by

$$v(t) = L\frac{di(t)}{dt} \qquad .....(9.53)$$

Taking the transform eq. (9.53), we have

$$V(s) = L[sI(s) - i(O^-)] \quad .....(9.54)$$

$$LsI(s) = V(s) + Li(O^-) \quad .....(9.55)$$

In eq. (9.55), V(s) is the transform of the applied voltage and $Li(O^-)$ is the transform of the voltage due to initial current $i(O^-)$. Designating the resultant voltage by $V_1(s)$, we have

$$\frac{V_1(s)}{I(s)} = Ls = Z(s) \quad .....(9.56)$$

Thus the transform impedance of an inductance is Ls. The time domain and transform domain network representations of inductance are given in Fig. 9.21(a) and (b).

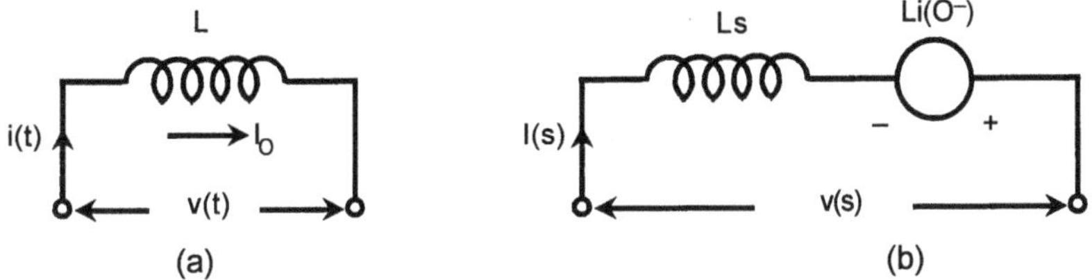

Fig. 9.21 Time domain and transform domain representation of the inductance

An alternate representation can also be obtained by solving for I(s) in eq. (9.55)

$$I(s) = \frac{V(s)}{Ls} + \frac{Li(O^-)}{Ls}$$

$$= V(s)\,Y(s) + \frac{i(O^-)}{s}$$

where $Y(s) = \dfrac{1}{Ls}$, is defined as the transform admittance of an inductor.

The alternate representation of inductor is given in Fig. 9.22.

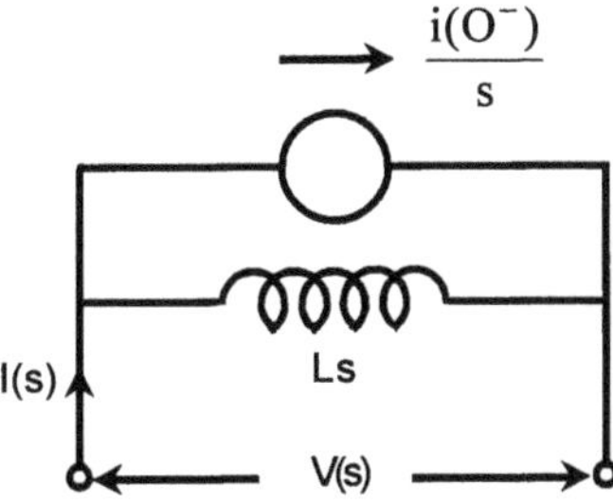

Fig. 9.22 Alternate representation of inductance in transform domain

In Fig. 9.21(b), $Li(O^-)$ is a constant and hence it can be visualised as the transform of an impulse voltage source. Similarly in Fig. 9.22, the source representing the initial condition is a step current source.

9.11.3 Capacitance :

The v – i relationships for a capacitor are given by,

$$i = C\frac{dv}{dt} \qquad .....(9.57)$$

or

$$v = \frac{1}{C}\int_{-\infty}^{t} idt \qquad .....(9.58)$$

The transform of these equations result in

$$I(s) = C\left(sV(s) - v(O^-)\right)$$

$$= Cs\ V(s) - Cv(O^-) \qquad .....(9.59)$$

and

$$V(s) = \frac{I(s)}{Cs} + \frac{v\left(O^-\right)}{s} \qquad .....(9.60)$$

The transform impedance, and admittance of a capacitor are given by

$$Z(s) = \frac{1}{Cs} \qquad .....(9.61)$$

and

$$Y(s) = Cs \qquad .....(9.62)$$

The initial condition is represented by a step voltage source in eq. (9.60) and an impulse current source in eq. (9.59). The equivalent network representations are given in Fig. 9.23 (a) and (b).

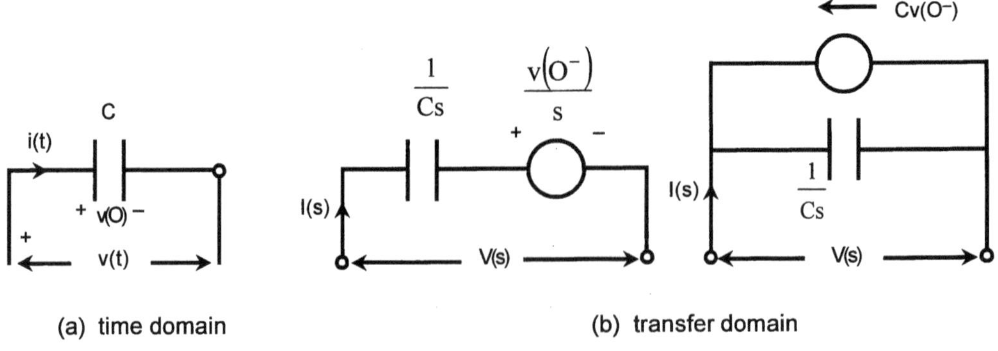

(a) time domain (b) transfer domain

Fig. 9.23 Representation of a capacitor

9.11.4 Solution of transform networks :

Using the transform domain representation of the elements and the transforms of the sources present in the network, a transform domain network can be obtained in which the variables are the transforms of currents and voltages. The network appears like a resistive network and the solution can be obtained using any of the techniques developed for solving resistive networks. However the solution obtained is in terms of the transforms of the currents and voltages, and it can be inverted to get the solution in time domain. The method of solution is illustrated by some examples.

Example 9.13

Find the voltage across the capacitor in the network of Fig. 9.24(a).

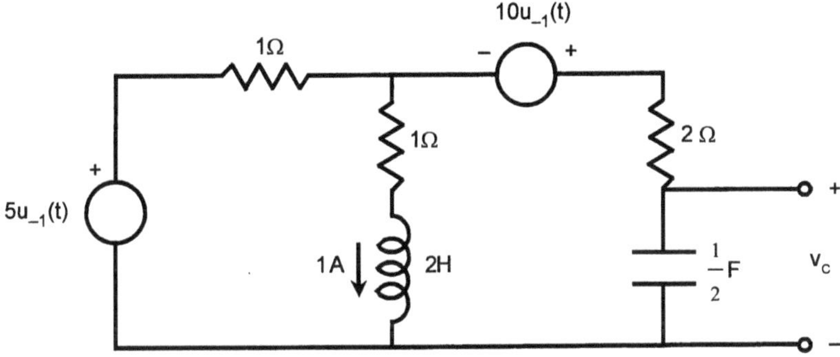

Fig. 9.24 (a) Network for example 9.13

Solution :

The transform network is shown in Fig. 9.24(b)

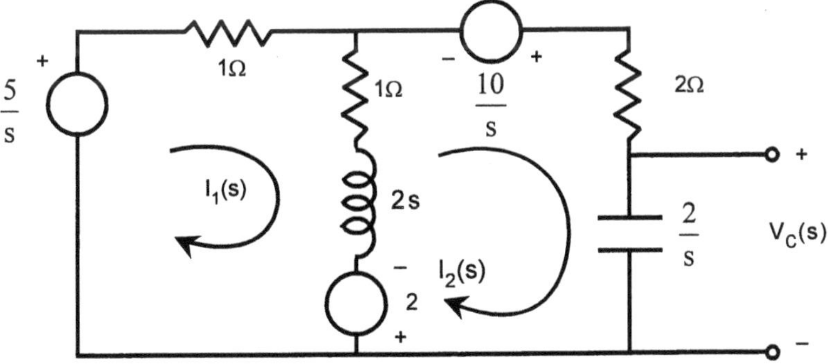

Fig. 9.24 (b) Transform network

Writing the loop equations, we get

$$(2 + 2s)\, I_1(s) - (1 + 2s)\, I_2(s) = \frac{5}{s} + 2$$

$$\left(3+2s+\frac{2}{s}\right) I_2(s) - I_1(s)(1 + 2s) = \frac{10}{s} - 2$$

Solving for $I_2(s)$

$$I_2(s) = \frac{\begin{vmatrix} 2+2s & \dfrac{5+2s}{s} \\ -(1+2s) & \dfrac{10-2s}{s} \end{vmatrix}}{\begin{vmatrix} 2+2s & -(1+2s) \\ -(1+2s) & \dfrac{2s^2+3s+2}{s} \end{vmatrix}}$$

$$I_2(s) = \frac{\dfrac{(2+2s)(10-2s)}{s} + \dfrac{(5+2s)(1+2s)}{s}}{\dfrac{(2+2s)(2s^2+3s+2)}{s} - (1+2s)^2}$$

$$= \frac{-4s^2+16s+20+4s^2+12s+5}{4s^3+10s^2+10s+4-s-4s^3-4s^2}$$

$$= \frac{28s+25}{6s^2+9s+4}$$

$$= \frac{14}{3}\left[\frac{s+\dfrac{25}{28}}{s^2+1.5s+\dfrac{2}{3}}\right]$$

$$= \frac{14}{3} \cdot \frac{s+\dfrac{25}{28}}{\left(s+\dfrac{3}{4}\right)^2+\left(\sqrt{\dfrac{5}{48}}\right)^2}$$

To bring it into standard form,

$$= \frac{14}{3}\left[\frac{s+\dfrac{3}{4}-\dfrac{3}{4}+\dfrac{25}{28}}{\left(s+\dfrac{3}{4}\right)^2+\left(\sqrt{\dfrac{5}{48}}\right)^2}\right]$$

$$= \frac{14}{3}\left[\frac{s+3/4}{\left(s+\frac{3}{4}\right)^2+\left(\sqrt{\frac{5}{48}}\right)^2}+\frac{\frac{1}{7}}{\left(s+\frac{3}{4}\right)^2+\left(\sqrt{\frac{5}{48}}\right)^2}\right]$$

Taking the inverse transform, we have

$$i_2(t) = \frac{14}{3}\cdot\left[e^{\frac{-3}{4}t}\cos\sqrt{\frac{5}{48}}t+\frac{1}{7}\sqrt{\frac{48}{5}}.e^{-\frac{3}{4}t}\sin\sqrt{\frac{5}{48}}t\right]$$

$$= 4.67e^{-0.75t}[\cos 0.323t + 0.443\sin 0.323t]$$

$$= 5.108e^{-0.75t}\sin(0.323t + 66.1^0)$$

Since we are interested in the voltage across the capacitor,

$$V_C(s) = I_2(s)\cdot\frac{2}{s}$$

$$= \frac{28}{3}\cdot\frac{s+\frac{25}{28}}{s\left(s^2+\frac{3}{2}s+\frac{2}{3}\right)}$$

Applying Heaviside's partial fraction expansion technique, we have

$$V_C(s) = \frac{28}{3}\left[\frac{K_1}{s}+\frac{K_2s+K_3}{s^2+\frac{3}{2}s+\frac{2}{3}}\right]$$

Evaluating K_1, K_2 and K_3 we have

$$V(s) = \frac{28}{3}\left[\frac{75}{56s}+\frac{-\frac{75}{56}s-\frac{113}{112}}{s^2+\frac{3}{2}s+\frac{2}{3}}\right]$$

Inverse transforming we obtain

$$v_e(t) = \frac{25}{2}u_{-1}(t) - \frac{25}{2}e^{-0.75t}[\cos 0.323t + 0.01\sin 0.323t]$$

$$\simeq \frac{25}{2}[1 - e^{-0.75t}\cos 0.323t]$$

It can be observed that the homogeneous part of the solution for $v_c(t)$ is

$$v_{ch}(t) = -\frac{25}{2}e^{-0.75t}\cos 0.323t$$

and the steady state part of the solution is

$$v_{css}(t) = \frac{25}{2} \text{ volts}$$

The steady state solution can be verified by drawing the equivalent circuit at $t = \infty$ as shown in Fig. 9.24(c).

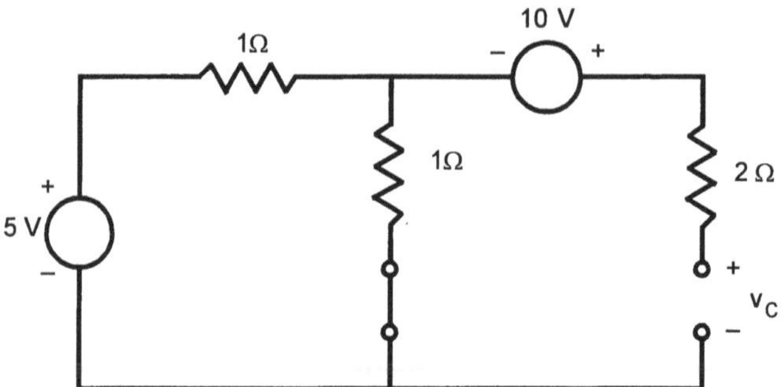

Fig. 9.24 (c) The equivalent circuit at $t = \infty$

From Fig. 9.24(c) it is clear that

$$v_{css}(t) = 12.5 \text{ v}$$

This example shows that the solution of multiloop networks also becomes routine by the use of Laplace transform method. Further, the transient and steady state solutions are simultaneously obtained. There is no necessity of evaluating the constants in the homogenous solution as the initial conditions are incorporated in the beginning itself.

Example 9.14

Obtain the voltage across the capacitors in Fig. 9.25(a) when the switches are closed simultaneously at $t = 0$.

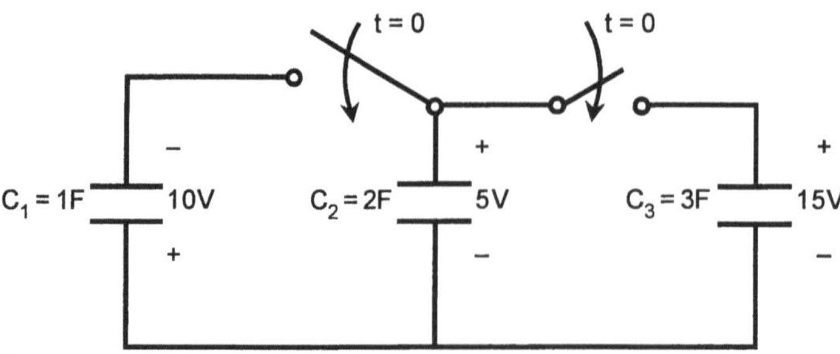

Fig. 9.25 (a) Network for example 9.14

Solution :

This is a problem to illustrate the method when capacitance loops occur. The classical method requires evaluation of initial conditions at $t = O^+$. The conditions at $t = O^+$ are different from conditions at $t = O^-$ since, as soon as the switches are closed impulse current flows in the capacitors. But if we use Laplace transform method, we need not worry about these impulses. Draw the Laplace transform domain circuit as shown in Fig. 9.25(b) and solve as usual.

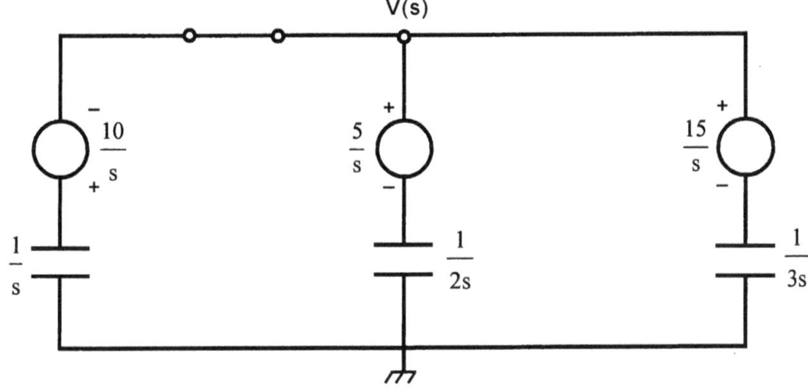

Fig. 9.25 (b) Laplace transform domain circuit

Writing the node voltage equation, we have

$$\frac{V(s)+\frac{10}{s}}{\frac{1}{s}}+\frac{V(s)-\frac{5}{s}}{\frac{1}{2s}}+\frac{V(s)-\frac{15}{s}}{\frac{1}{3s}}=0$$

$$V(s)[s + 2s + 3s] = -10 + 10 + 45 = 45$$

$$V(s) = \frac{45}{6s}$$

Taking the inverse transform, we have

$$v(t) = \frac{15}{2}u(t)$$

All the capacitors will have a common voltage of 7.5 V for $t > 0$.

The current through C_3 is given by

$$I_3(s) = \frac{V(s)-\frac{15}{s}}{\frac{1}{3s}}$$

$$= 3s \cdot \frac{15}{2s} - 45$$

$$= -\frac{45}{2}$$

Taking the inverse transform, we have

$$i_3(t) = -\frac{45}{2}\delta(t) \text{ ; where } \delta(t) \text{ is a unit impulse function}$$

Similarly $\quad i_2(t) = 5\ \delta(t)$

and $\quad i_1(t) = 17.5\ \delta(t)$

Try solving this problem in time domain !

We have seen in chapter 4 that whenever exponential inputs are applied to an RC or RL network, a condition of resonance occurs if the homogeneous response is of the same form as the input. The forced response was evaluated by using the method of variation of parameters. But we will show that if we use Laplace transform method no such problems arise and the method is straight forward.

Example 9.15

Solve for the current in the network of Fig. 9.26(a)

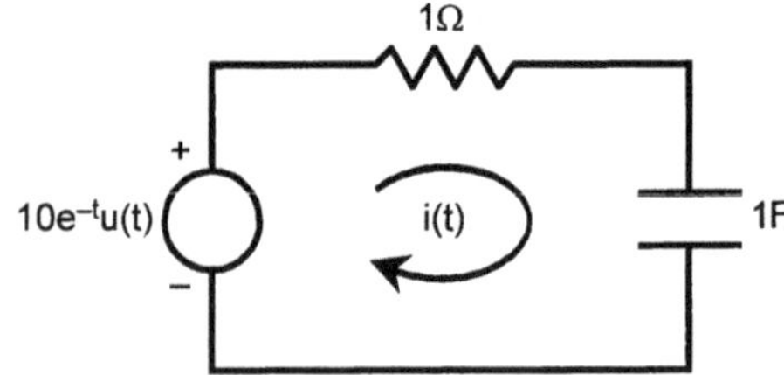

Fig. 9.26 (a) RC network with exponential excitation

Solution :

Draw the Laplace transform domain circuit as shown in Fig. 9.26(b).

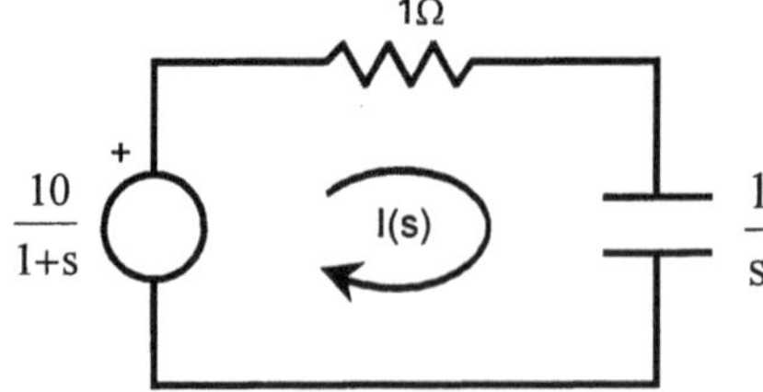

Fig. 9.26 (b) Laplace transform domain circuit

Applying Kirchhoff's law we have

$$I(s)\left[1+\frac{1}{s}\right] = \frac{10}{1+s}$$

$$I(s) = \frac{10s}{(s+1)^2} = \frac{10(s+1-1)}{(s+1)^2}$$

$$I(s) = \frac{10}{s+1} - \frac{10}{(s+1)^2}$$

Taking inverse Laplace transform, we have

$$i(t) = 10e^{-t} - 10te^{-t}$$

Try solving this problem in time domain !

9.12 TRANSFER FUNCTION :

Consider a network excited by a single source. If we are interested in the current or voltage in a particular branch of the network only, we can call this quantity as the output. Now let us introduce a useful term called Transfer function of the network. Transfer function for a network is defined as the ratio of the Laplace transform of the output to the Laplace transform of the input. This ratio is valid only if there is only one input to the system and hence all initial conditions which can also be represented by equivalent sources are taken to be zero. Thus a transfer function for a circuit is defined with all initial conditions in the network taken to be zero.

The output can be a current or voltage and the input source can be a voltage source or current source. The transfer function is usually denoted by T(s). Let the input be a voltage source and be denoted by V(s) and let the output be a current denoted by I(s). Then

$$T(s) = \frac{I(s)}{V(s)} \qquad .....(9.63)$$

$$\therefore \quad I(s) = T(s) \cdot V(s) \qquad .....(9.64)$$

If we know the *transfer function of the network*, the response current can be evaluated for any given input V(s) using eq. (9.64). We need not solve the network for every input. Let the input be a unit impulse for which

$$V(s) = 1$$

$$\therefore \quad I(s) = T(s)$$

Taking the inverse Laplace transform and denoting $L^{-1}[T(s)]$ by h(t), we have

$$i(t) = h(t) \qquad .....(9.65)$$

This is a very important relationship, which shows that the impulse response is nothing but the Laplace inverse of the transfer function of the system. Since the response to any input can be obtained provided we know the transfer function, it should be possible to find the response of the network if we know the impulse response of the network ! Taking the Laplace inverse of eq. (9.64) using convolution theorem, we have

$$i(t) = \int_0^t h(\tau)v(t-T)dT \qquad .....(9.66)$$

where $L^{-1}[T(s)] = h(t)$

Thus using eq. (9.66), the response to any input can be obtained provide we know the response to the impulse. This is the importance of impulse response of a system. All we should know about the network is its impulse response for finding the response to any other input.

Example 9.16

Find the transfer function $T(s) = \dfrac{V_C(s)}{V(s)}$ for the network shown in Fig. 9.27. Also find the response for a unit step input.

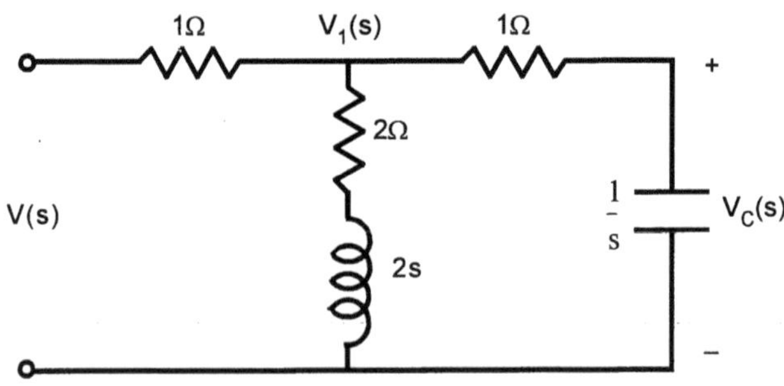

Fig. 9.27 Network for example 9.16

Solution :

Writing the node voltage equation we have

$$\frac{V_1(s)-V(s)}{1} + \frac{V_1(s)}{2+2s} + \frac{V_1(s)}{1+\frac{1}{s}} = 0$$

$$V_1(s)\left[1+\frac{1}{2(1+s)}+\frac{s}{s+1}\right] = V(s)$$

$$\frac{V_1(s)}{V(s)} = \frac{2(s+1)}{4s+3}$$

But $$V_C(s) = \frac{V_1(s)}{1+\frac{1}{s}} \times \frac{1}{s}$$

$$= \frac{2(s+1)}{4s+3}$$

$$\therefore \quad \frac{V_C(s)}{V(s)} = \frac{2(s+1)}{4s+3} \cdot \frac{1}{s+1}$$

∴ The transfer function is given by

$$T(s) = \frac{2}{4s+3}$$

The response for a unit step input can be obtaining by multiplying T(s) by the Laplace transform of the input i.e., $\frac{1}{s}$ and taking the inverse Laplace transform.

$$V_C(s) = T(s) \cdot V(s)$$

$$= \frac{2}{4s+3} \cdot \frac{1}{s}$$

$$= \frac{1}{2}\left[\frac{k_1}{s+\frac{3}{4}} + \frac{k_2}{s}\right]$$

$$k_1 = -\frac{4}{3}$$

$$k_2 = \frac{4}{3}$$

$$\therefore \quad V_C(s) = \frac{1}{2} \cdot \frac{4}{3}\left[\frac{1}{s} - \frac{1}{s+\frac{3}{4}}\right]$$

$$\therefore \quad V_C(t) = \frac{2}{3}\left[1 - e^{-\frac{3}{4}t}\right] \qquad .....(9.67)$$

The impulse response of the system is $L^{-1}[T(s)]$

$$\therefore \quad h(t) = \frac{1}{2}e^{-\frac{3}{4}t}$$

The response also can be obtained in time domain by using convolution of h(t) with $u_{-1}(t)$.

$$v_C(t) = \int_0^t \frac{1}{2} e^{-\frac{3}{4}\tau} u_{-1}(t-\tau) d\tau$$

$$= \frac{1}{2} \int_0^t e^{-\frac{3}{4}\tau} d\tau$$

$$= \frac{1}{2} \cdot \left. \frac{e^{-\frac{3}{4}\tau}}{-3/4} \right|_0^t$$

$$= \frac{2}{3}\left[1 - e^{-\frac{3}{4}t}\right]$$

This is same as eq. (9.67).

9.12.1 Poles and zeroes of transfer function :

The general form of transfer function of a network is a ratio of polynomials in s given by

$$T(s) = \frac{N(s)}{D(s)}$$

$$= \frac{a_0 s^m + a_1 s^{m-1} + \ldots + a_{m-1} s + a_m}{b_0 s^n + b_1 s^{n-1} + \ldots + b_{n-1} s + b_n} \quad \ldots\ldots(9.68)$$

For a physical network, the coefficients a_i and b_i are real and positive. The polynomial N(s) and D(s) can be factored and T(s) can be written as

$$T(s) = H \frac{(s-z_1)(s-z_2)\ldots\ldots(s-z_m)}{(s-p_1)(s-p_2)\ldots\ldots(s-p_n)} \quad \ldots\ldots(9.69)$$

where $H = \frac{a_0}{b_0}$ is known as a *scale factor* and $z_1, z_2 \ldots\ldots z_m, p_1, p_2 \ldots\ldots p_n$ are complex frequencies. It is observed that when the variable s takes the values $z_1, z_2 \ldots\ldots z_m$, T(s) becomes zero. These values of s, which make the function T(s) = 0, are known as *'zeros'* of the transfer function. Similarly if s takes on values equal to $p_1, p_2 \ldots\ldots p_n$, the function T(s) becomes infinity. Thus these values of s for which the function blows up to infinity are known as *'poles'* of the transfer function. If poles and zeros together with the scale factor H are specified, the complete transfer function is known. The poles and zeros can be distinct, repeated, real or complex in nature. In addition to the poles and zeros at finite values of s, we may have poles or zeros at $s = \infty$. If m > n.

$$\underset{s \to \infty}{\text{Lt}} \ T(s) = H \cdot s^{m-n} = \infty$$

Thus there are (m – n) poles at $s = \infty$

If $m < n$

$$\underset{s \to \infty}{\text{Lt}} \ T(s) = \frac{H}{s^{n-m}} = 0 \text{ since } n - m > 0$$

Thus there are (n – m) zeros at infinity. If we consider the poles and zeros at infinity in addition to the n poles and m zeros, the total number of poles is always equal to the total number of zeros. The poles and zeros can be plotted in the s – plane. For example if

$$T(s) = \frac{(s+1)^2 (s+2)(s^2+s+1)}{s^3 (s+3)(s^2+2s+4)}$$

The zeros are : double zeros at – 1, zero at – 2 and complex conjugate zeros at $-\frac{1}{2} \pm j\frac{\sqrt{3}}{2}$

The poles are : 3rd order pole at origin, simple pole at – 3 and complex conjugate poles at $-1 \pm j\sqrt{3}$

These poles and zeros are plotted in Fig. 9.28.

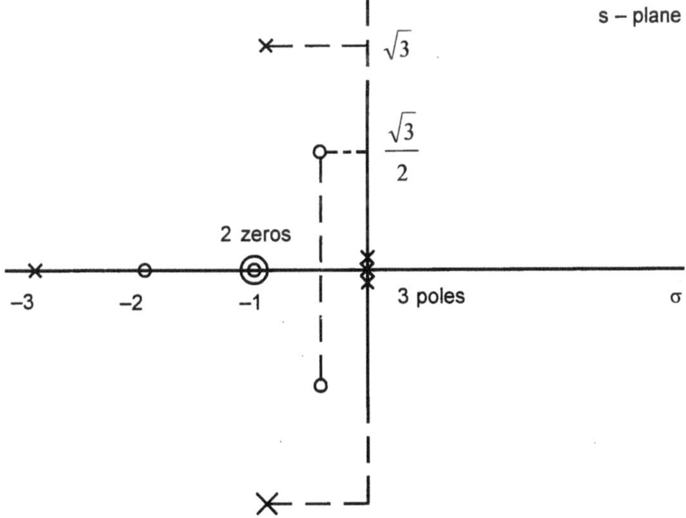

Fig. 9.28 Pole zero plot in s plane

A small circle is used to indicate a zero where as 'X' is used to indicate pole on the diagram. Except for a scale factor, the pole zero plot completely specifies a transfer function.

9.12.2 Significance of poles and zeros :

Let the transfer function be given by

$$T(s) = \frac{I(s)}{V(s)} = \frac{K \prod_{i=1}^{m} (s - z_i)}{\prod_{j=1}^{n} (s - p_j)} \quad .....(9.70)$$

where the symbol $\prod_{i=1}^{m}$ is used for multiplication.

Thus $\quad I(s) = T(s) \,.\, V(s)$

If V(s) has 'q' poles given by p_{vi}, I(s) can be expanded in partial fractions as

$$I(s) = \sum_{i=1}^{q} \frac{K_i}{s - p_{vi}} + \sum_{j=1}^{n} \frac{K_j}{s - p_j} \quad .....(9.71)$$

Assuming that the poles are distinct, the response i(t) is given by

$$i(t) = \sum_{i=1}^{q} K_i e^{p_{vi} t} + \sum_{j=1}^{n} K_j e^{p_j t} \quad .....(9.72)$$

The first set of terms on the right hand side of eq. (9.72) represent the response due to the input or the forced response of the system and the second set of terms are the terms corresponding to the homogeneous response. Thus the complex frequencies $s_i = p_{vi}$ are the complex frequencies corresponding to forced oscillations and $s_j = p_j$ are the complex frequencies corresponding to the free or natural oscillations of the system. The poles of the transfer function, therefore, determine the nature of the response where as the zeros, together with the poles, determine the magnitude of the response since they determine the values of K_i and K_j in eq. (9.71).

The time domain behaviour of the network can be determined from the pole zero plot of the transfer function. Let the response I(s) be given by

$$I(s) = T(s)V(s) = \frac{N(s)}{D(s)}$$

$$= H \frac{(s - z_1)(s - z_2).....(s - z_m)}{(s - p_1)(s - p_2).....(s - p_n)} \quad .....(9.73)$$

The poles and zeros of the above equation are either real, imaginary or complex. The complex poles occur in complex conjugate pairs so that the product of these two factors can be represented by a quadratic. Let the quadratic be given by

$$Q(s) = s^2 + as + b \qquad \text{.....(9.74)}$$

$$s_{1,2} = \frac{-a \pm \sqrt{a^2 - 4b}}{2}$$

$$= -\frac{a}{2} \pm \frac{1}{2}\sqrt{a^2 - 4b} = \frac{-a}{2} \pm j\sqrt{b}\sqrt{1 - \frac{a^2}{4b}}$$

Let us define two quantities ξ and ω_n, the damping factor and natural frequency respectively and express the roots in terms of these quantities.

Let $\qquad s_{1,2} = -\xi\,\omega_n \pm j\omega_n\sqrt{1-\xi^2}$

Thus $\qquad \xi\,\omega_n = \frac{a}{2}$

$\qquad \sqrt{b} = \omega_n \quad$ or $\quad \omega_n^2 = b$

and $\qquad a = 2\xi\,\omega_n$

In terms of these definitions, eq. (9.74) can be written as

$$Q(s) = s^2 + 2\xi\omega_n s + \omega_n^2$$

and general form of response due to there complex conjugate poles is given by

$$e^{-\xi\omega_n t}\sin\left(\omega_n\sqrt{1-\xi^2}\,t + \theta\right) \qquad \text{.....(9.75)}$$

If $\xi < 1$, the roots are complex and we have damped oscillations. If $\xi = 1$ we have

$$s_{1,2} = -\omega_n, -\omega_n$$

and therefore we have a critically damped response. If $\xi > 1$, the roots are real and distinct and hence we have overdamped response. The response is not oscillatory. ω_n is known as the natural frequency of the response and $\omega_d = \omega_n\sqrt{1-\xi^2}$ is known as the damped natural frequency. This is the actual frequency of the oscillations as given in eq. (9.75). If $\xi = 0$, we have

$$s_{1,2} = \pm j\omega_n$$

which results in purely oscillatory response with the frequency given by the undamped natural frequency, ω_n, of the network.

With this background, let the poles and zeros of eq. (9.73) be of the form

$$s_{1,2} = -\xi\omega_n \pm j\omega_n\sqrt{1-\xi^2}, \quad \xi < 1 \quad .....(9.76)$$

$$= \xi\omega_n \pm \omega_n\sqrt{\xi^2 - 1}, \quad \xi > 1 \quad .....(9.77)$$

$$= -\omega_n \quad \xi = 1 \quad .....(9.78)$$

$$= \pm j\omega_n \quad \xi = 0 \quad .....(9.79)$$

For different locations of the poles, we get different types of responses. These responses are shown in Fig. 9.29.

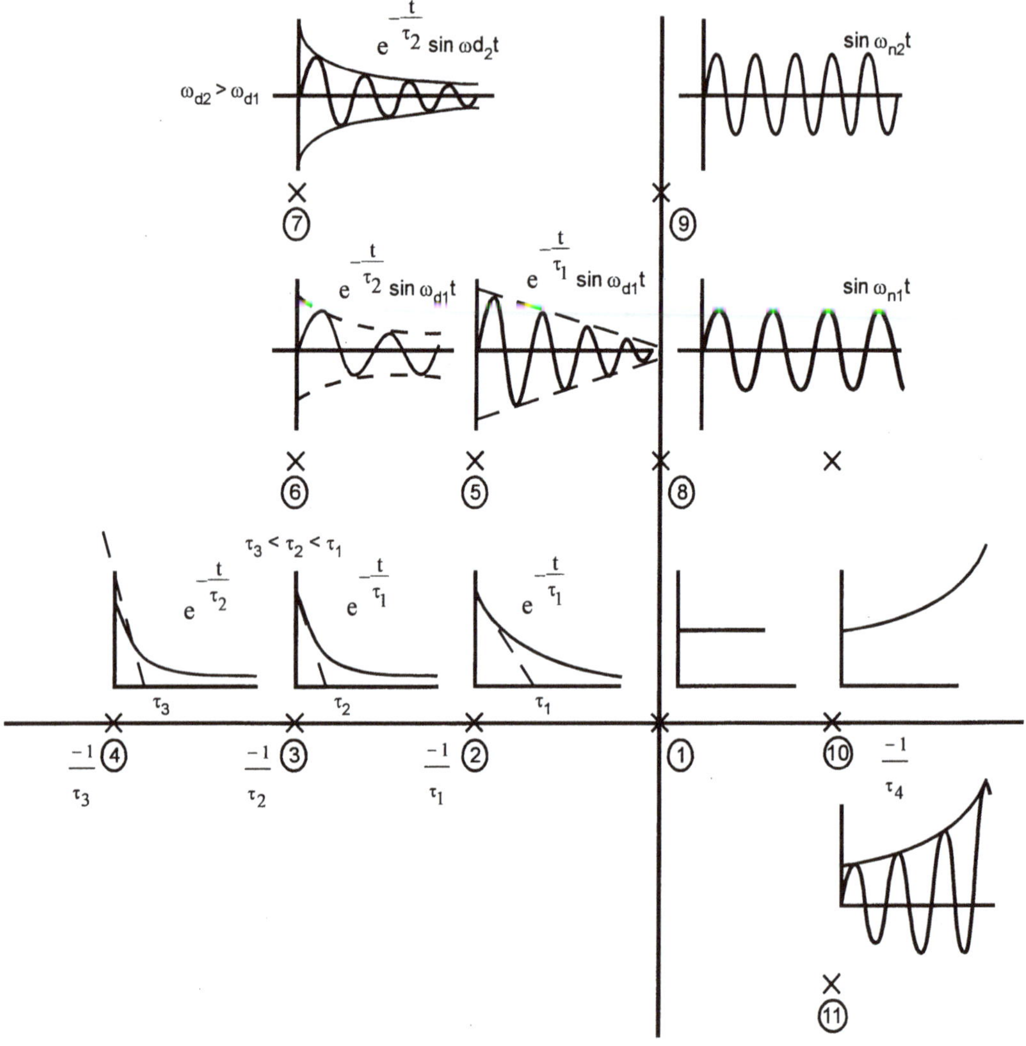

Fig. 9.29 Pole locations and corresponding responses. For conjugate poles response is due to the combination

Notes :

1 : Pole at the origin. Response is a constant.

2, 3 and 4 : The poles are real and negative. Response is exponentially decaying with time constants τ_1, τ_2 and τ_3 with $\tau_3 < \tau_2 < \tau_1$. It is observed that a farther the location of the pole from the origin, faster the response dies down. Hence poles nearer the origin are more effective. If the pole is at the origin as at 1, the response is a constant. It can be treated as an exponential function with infinite time constant.

5 and 6 : Exponentially decaying sinusoidal oscillations or damped oscillations. The frequency of oscillation is $\omega_{d1} = \omega_n\sqrt{1-\zeta^2}$. In 6 the oscillations have same frequency ω_{d1} compared to the one at 5 but they die down faster.

7 : Damped oscillations. The damping is same as at 6 but the frequency of oscillation is increased. $\omega_{d2} > \omega_{d1}$.

8 and 9 : Undamped oscillation with natural frequency ω_{n1} and ω_{n2} ; $\omega_{n2} > \omega_{n1}$.

10 : Real and positive roots – exponentially increasing response.

11 : Exponentially increasing sinusoidal oscillations.

9.13 NETWORK FUNCTIONS :

The passive network with one pair of terminals available for connecting the source is defined as one port network. The one port network shown in Fig. 9.30 has only two quantities V(s) and I(s) which can be measured.

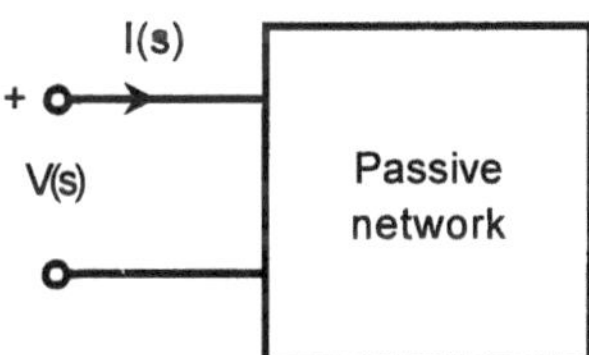

Fig. 9.30 One port network

The box contains only passive elements with no initial conditions but may contain controlled sources. Thus we can describe the network in terms of its driving point impedance Z(s) as

$$Z(s) = \frac{V(s)}{I(s)} \qquad .....(9.80)$$

Similarly the admittance Y(s) is given by

$$Y(s) = \frac{1}{Z(s)} = \frac{I(s)}{V(s)} \qquad .....(9.81)$$

A common term, immittance, is used to describe impedance and admittance.

A passive network with one pair of terminals to connect the driving source and another pair of terminals to connect a load is described as a two port network. A passive 2 – port network is represented in Fig. 9.31.

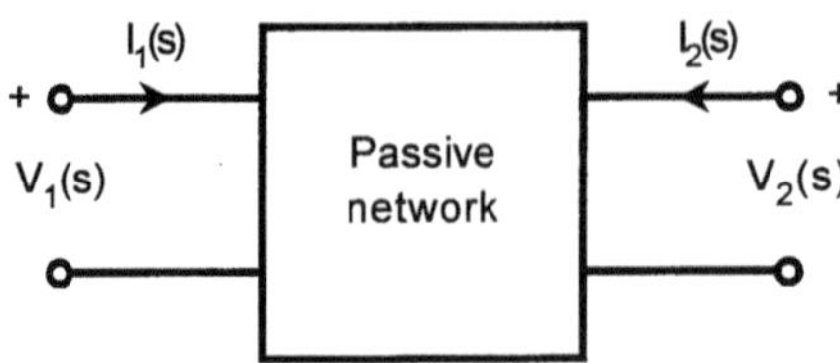

Fig. 9.31 A two port network

Transfer functions are used to describe the behaviour of two port networks. Since there are two quantities at the input port and two quantities at the output port, we have four different types of transfer functions :

$$Z_{12}(s) = \frac{V_2(s)}{I_1(s)} \quad : \text{ Transfer impedance} \qquad .....(9.82)$$

$$Y_{12}(s) = \frac{I_2(s)}{V_1(s)} \quad : \text{ Transfer admittance} \qquad .....(9.83)$$

$$G_{12}(s) = \frac{V_2(s)}{V_1(s)} \quad : \text{ Voltage transfer ratio} \qquad .....(9.84)$$

$$\alpha_{12}(s) = \frac{I_2(s)}{I_1(s)} \quad : \text{ Current transfer ratio} \qquad .....(9.85)$$

All these transfer functions, for a given network, have a form given by

$$T(s) = \frac{N(s)}{D(s)} = \frac{a_0 s^m + a_1 s^{m-1} + ... + a_{m-1} s + a_m}{b_0 s^n + b_1 s^{n-1} + ... + b_{n-1} s + b_n} \qquad .....(9.86)$$

Since these are function of a network containing elements which have real values only, the coefficients a_i and b_j are all real.

9.13.1 Restrictions on poles and zeros of driving point functions :

Consider the driving point function T(s) given by eq. (9.86). It may be a driving point impedance Z(s) or driving point admittance Y(s). Since the coefficients a_i and b_j are real the function T(s) is real when 's' is real. Such functions are known as real functions. Since the coefficients are real in N(s) and D(s), if a complex root $\alpha + j\beta$ occurs in either of them, its conjugate also must be a root. For, the polynomial

$$(s - \alpha + j\beta)(s - \alpha - j\beta) = (s - \alpha)^2 + \beta^2$$

will have all real coefficients. If the conjugate root doesnot occur, some of the coefficients will be imaginary. Thus if a complex root occurs in either N(s) or D(s), its conjugate also must necessarily occur.

Further, we also observe that networks with only passive elements are stable i.e., if an initial condition is given to one of the elements of the network, the voltages and currents are bounded. By bounded we mean that the responses never become infinite even if we wait for a long time. We have seen that the nature of response depends on the location of the poles of T(s) or roots of D(s). If the response is to be bounded the following terms only must be present.

$$e^{-at},\ e^{-at}\sin\omega_n t,\ \sin\omega_n t,\ u_{-1}(t), t^n e^{-at},\ t^n e^{-at}\sin\omega_n t.....\text{etc}$$

We observe that the terms $t^n e^{-at}$, $t^n e^{-at}\sin\omega_n t$ tend to zero as $t \to \infty$, if a is > 0

The roots of D(s) corresponding to these responses are

$$(s + a);\ (s + a)^2 + \omega_n^2,\ (s^2 + \omega_n^2),\ s,\ (s + a)^n,\ [(s + a)^2 + \omega_n^2]^n.....\text{etc}$$

If the polynomial D(s) is made up of products of terms given above where 'a' is positive and ω_n^2 is positive, D(s) will have all coefficients positive. D(s) cannot contain roots which have positive values or positive real parts if they are complex, since these terms will give rise to exponentially increasing functions. Thus the poles must lies in the left half of s – plane. The LHS roots can be simple or multiple. No roots shall occur on the RHS of the s – plane. Then what about the boundary between the LHS and RHS of the s – plane. i.e., imaginary axis?

Roots can occur on the $j\omega$ axis but they must be simple. This is clear from the following :

Nature of pole			Type of response
simple pole on $j\omega$ axis	:	$s^2 + \omega_n^2$	$\sin\omega_n t$
multiple pole on $j\omega$ axis	:	$(s^2 + \omega_n^2)^n$	$t^n \sin\omega_n t$
simple pole at origin	:	s	$u_{-1}(t)$
multiple pole at origin	:	s^n	t^{n-1}

From the above, we see that multiple poles on $j\omega$ axis result in responses which tend to infinity as $t \to \infty$. Hence they cannot be factors of D(s).

If T(s) represents Z(s), the denominator D(s) must satisfy the conditions discussed above. Since $Y(s) = \frac{1}{Z(s)}$. The denominator of Y(s) i.e., N(s) must also satisfy these requirements. Hence both D(s) and N(s) must have real and positive coefficients. Since all the factors which comprise D(s) or N(s) have positive coefficients their products cannot result in any cancellation and hence no power of 's' can be missing in between highest and lowest powers of s. But if D(s) or N(s) contains only $s^2 + \omega^2$ type of terms, all odd powers of s can be missing and if it has a root at the origin in addition to poles on $j\omega$ axis only, all even powers of s can be missing.

One more interesting property of these polynomials can be observed by considering the behaviour of the network at $s \to \infty$ and $s \to 0$ i.e., at very high frequencies and at very low frequencies. Since the value of reactance $X_L = sL$ and $X_C = \frac{1}{Cs}$ change with frequency, as s tends to infinity, the impedance due to inductance dominates all other elements. If neither inductance is present for impedance computation or capacitance is present for admittance computation, the network may be dominated by a resistance. Thus at high frequencies the network may be represented by one of the following single element equivalent networks.

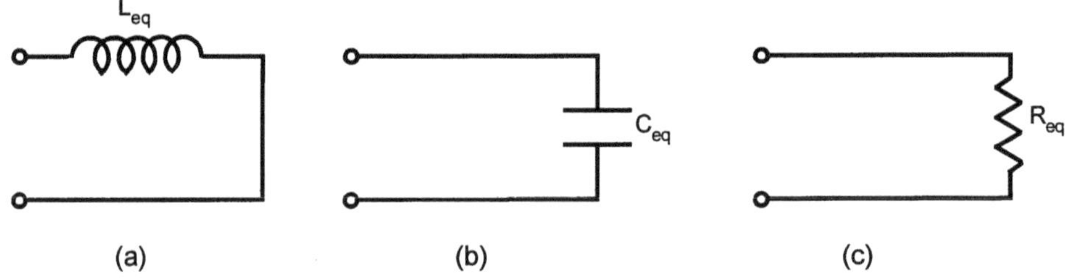

Fig. 9.32 Equivalent network as $s \to \infty$

Similarly at very low frequencies also the network can be represented by single elements whose immittances are given by

$$Y_L = \frac{1}{sL_{eq}}, \; Z_C = \frac{1}{C_{eq}s} \quad \text{or} \quad Z = R_{eq}$$

If these conditions must apply to immittance functions, say W(s)

$$\underset{s \to \infty}{\text{LT}} W(s) = \underset{s \to \infty}{\text{LT}} \frac{a_0 s^m}{b_0 s^n} = \frac{a_0}{b_0} s^{m-n} \Big|_{s \to \infty}$$

If this should represent a single element as explained before, this term must reduce to either

$$\frac{a_0}{b_0}, \; \frac{a_0}{b_0 s} \text{ or } \frac{a_0 s}{b_0}$$

which means

$$|m - n| \leq 1$$

This condition can be stated as follows. For immittance driving point functions, the highest powers of s in numerator and denominator may at most differ by 1.

Similarly at low frequencies

$$\underset{s \to \infty}{\text{Lt}} W(s) = \frac{a_{m-1}s + a_m}{b_{n-1}s + b_n}$$

In order that this may represent one of the single element equivalents shown in Fig. 9.32, the above term must reduce to

$$\frac{a_{m-1}s}{b_n} \text{ or } \frac{a_m}{b_{n-1}s} \text{ or } \frac{a_m}{b_n}$$

This means that the lowest powers of s in N(s) and D(s) may at most differ by 1.

To summarise, the necessary conditions to be satisfied by N(s) and D(s) of driving point immittance functions are tabulated in table 9.2.

Table 9.2

Necessary conditions to be satisfied by driving point immittance functions

1.	The coefficients in the polynomials N(s) and D(s) must be positive and real.
2.	Poles and zeros must occur in conjugate pairs.
3.	The poles and zeros must occurs only on the LHS of s – plane i.e., the poles and zeros must be –ve if real or must have –ve real parts if complex.
4.	If the poles and zeros occur on jω axis, they must be simple.
5.	The polynomials N(s) and D(s) cannot have any powers of s missing between highest and lowest powers unless all odd or even powers of s are missing.
6.	The highest powers of s in N(s) and D(s) can at most differ by 1
7.	The lowest powers of s in N(s) and D(s) can at most differ by 1.

9.13.2 Restrictions on locations of poles and zeros of transfer functions :

Similar considerations of stability and behaviour of network at $s = \infty$ and $s = 0$ result in some restrictions on the locations of poles and zeros of transfer functions $G_{12}(s)$, $\alpha_{12}(s)$, $Z_{12}(s)$ or $Y_{12}(s)$. But since $Z_{12}(s) \neq \frac{1}{Y_{12(s)}}$, the conditions applicable for denominators of these functions do not apply for the numerators. The necessary conditions to be satisfied by transfer functions are summarised in Table 9.3. The proofs are omitted.

Table 9.3

Necessary conditions to be satisfied by transfer functions

1.	The coefficients in the polynomial N(s) and D(s) must be real. The coefficients of D(s) must also be positive.
2.	Poles and zeros must occur in conjugate pairs if complex.
3.	The real parts of poles must be negative or zero. If zero, the poles must be simple.
4.	The polynomial N(s) can have negative coefficients or some powers of s may be missing between highest and lowest powers of s.
5.	The polynomial D(s) must not have any terms missing between highest and lowest powers of 's'. All odd powers of 's' or all even powers of s may be missing.
6.	The degree of N(s) may be as small as zero and is independent of the degree of D(s).
7.	For $G_{12}(s)$ and $\alpha_{12}(s)$, the maximum degree of N(s) is the degree of D(s). For $Z_{12}(s)$ and $Y_{12}(s)$, the maximum degree of N(s) is one more than the degree of D(s).

Computation of driving point functions and transfer functions are illustrated by some examples.

Example 9.17

Find the driving point impedance of the network in Fig. 9.33.

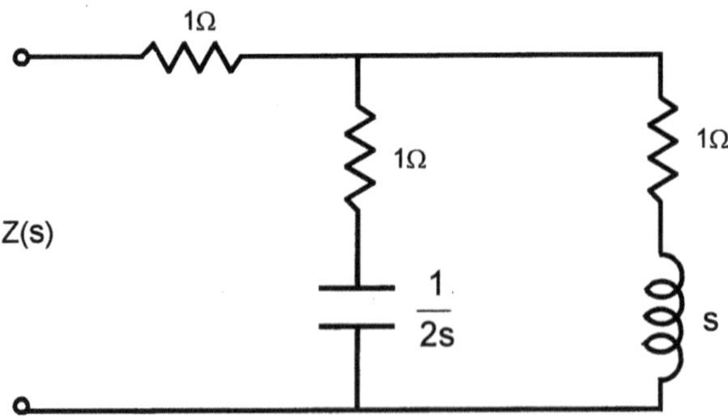

Fig. 9.33 Network for example 9.17

Solution :

By network reduction technique, we have

$$Z(s) = 1 + \frac{\left(1+\frac{1}{2s}\right)(1+s)}{1+\frac{1}{2s}+1+s}$$

$$= 1 + \frac{(2s+1)(s+1)}{2s^2+4s+1}$$

$$= \frac{2s^2+4s+1+2s^2+3s+1}{2s^2+4s+1}$$

$$Z(s) = \frac{4s^2+7s+2}{2s^2+4s+1}$$

Example 9.18

Find $G_{12}(s)$ and $Z_{12}(s)$ for the network of Fig. 9.34.

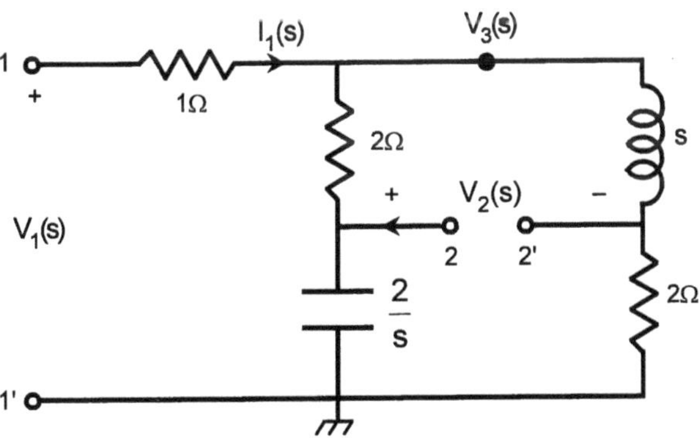

Fig. 9.34 A two part network for example 9.18

Solution :

By definition $\quad G_{12}(s) = \frac{V_2(s)}{V_1(s)}$

Writing node voltage equation

$$\frac{V_3(s)-V_1(s)}{1}+\frac{V_3(s)}{2+\frac{2}{s}}+\frac{V_3(s)}{s+2}=0$$

$$V_3(s)\left[1+\frac{s}{2s+2}+\frac{1}{s+2}\right]=V_1(s)$$

$$V_3(s)=V_1(s)\cdot\frac{2(s+1)(s+2)}{2s^2+6s+4+s^2+2s+2s+2}$$

$$=V_1(s)\frac{2(s+1)(s+2)}{3s^2+10s+6}$$

But $$V_2(s)=s\cdot\frac{V_3(s)}{s+2}-\frac{V_3(s)}{2+\frac{2}{s}}\cdot 2$$

$$=V_3(s)\left[\frac{s}{s+2}-\frac{2s}{2(s+1)}\right]$$

$$=V_3(s)\left[\frac{2s^2+2s-2s^2-4s}{2(s+1)(s+2)}\right]$$

$$=V_3(s)\cdot\frac{-2s}{2(s+1)(s+2)}$$

Substituting for $V_3(s)$, we have

$$V_2(s)=\frac{2(s+1)(s+2)}{3s^2+10s+6}\times V_1(s)\cdot\frac{-2s}{2(s+1)(s+2)}$$

$$= -\frac{2sV_1(s)}{3s^2 + 10s + 6}$$

$$G_{12}(s) = \frac{V_2(s)}{V_1(s)} = \frac{-2s}{3s^2 + 10s + 6}$$

Calculation of $Z_{12}(s)$

We have $$Z_{12}(s) = \frac{V_2(s)}{I_1(s)}$$

$$I_1(s) = \frac{V_3(s) - V_1(s)}{1}$$

$$= \frac{2(s+1)(s+2)}{3s^2 + 10s + 6} \cdot V_1(s) - V_1(s)$$

$$= V_1(s)\left[\frac{2s^2 + 6s + 4 - 3s^2 - 10s - 6}{3s^2 + 10s + 6}\right]$$

$$= -V_1(s)\,\frac{s^2 + 4s + 2}{3s^2 + 10s + 6}$$

But $$\frac{V_2(s)}{V_1(s)} = \frac{-2s}{3s^2 + 10s + 6}$$

$\therefore$ $$I_1(s) = \frac{V_2(s)(3s^2 + 10s + 6)}{2s} \cdot \frac{s^2 + 4s + 2}{3s^2 + 10s + 6}$$

$$= V_2(s) \cdot \frac{s^2 + 4s + 2}{2s}$$

$\therefore$ $$Z_{12}(s) = \frac{V_2(s)}{I_1(s)} = \frac{2s}{s^2 + 4s + 2}$$

Problems

9.1 Find the Laplace transform of the following time functions :

(i) $t\,e^{-at}\,\text{Sin}\omega t$ (ii) $t\,e^{-at}\,\text{Cos}\omega t$ (iii) $\text{Sin}(\omega t + \theta)$

(iv) $\text{Cos}(\omega t + \theta)$ (v) $\text{Sinh}\,\omega t$ (vi) $\text{Cosh}\,\omega t$

(vii) $\text{Cos}^2\,\omega t$

9.2 Find the inverse Laplace transform of the following functions :

(i) $\dfrac{(s+1)(s+3)}{s(s+2)(s+4)}$ (ii) $\dfrac{s^2+1}{s(s^2+2)(s^2+4)}$

(iii) $\dfrac{5}{s^2(s+1)(s+2)^2}$ (iv) $\dfrac{1}{(s^3+2s^2+2s+1)}$

(v) $\dfrac{s}{(s+1)(s^2+2s+2)}$ (vi) $\dfrac{s^2}{(s^2+1)^2}$

9.3 Obtain the Laplace transform of the functions shown in Fig. P. 9.1

(a) Using waveform synthesis (b) Using gate function

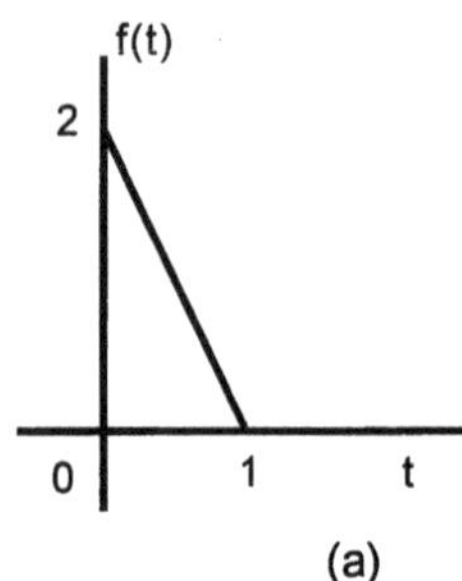

(a)

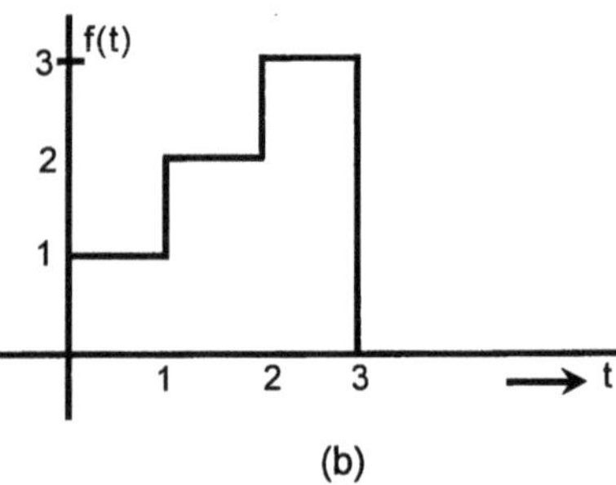

(b)

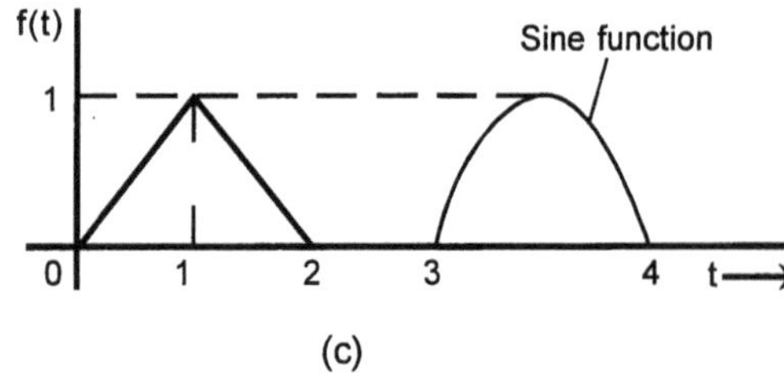

(c)

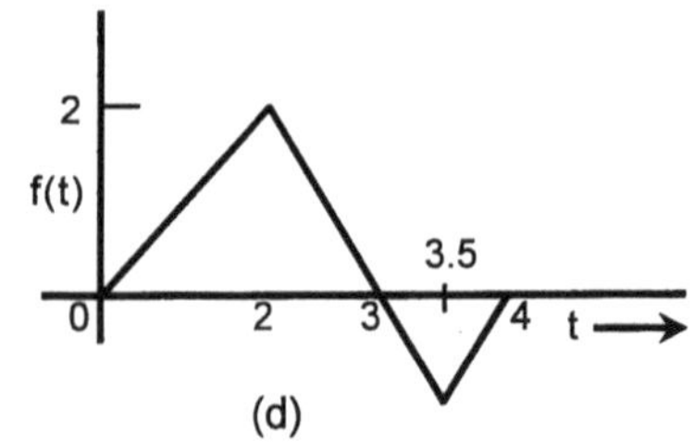

(d)

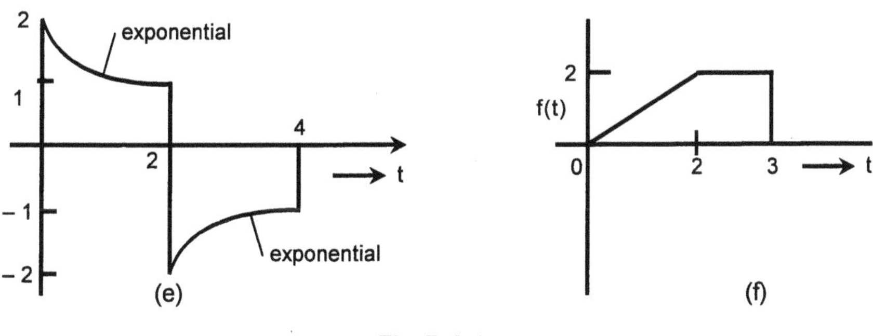

Fig. P. 9.1

9.4 Find the Laplace transform of the waveform shown in Fig. P. 9.2.

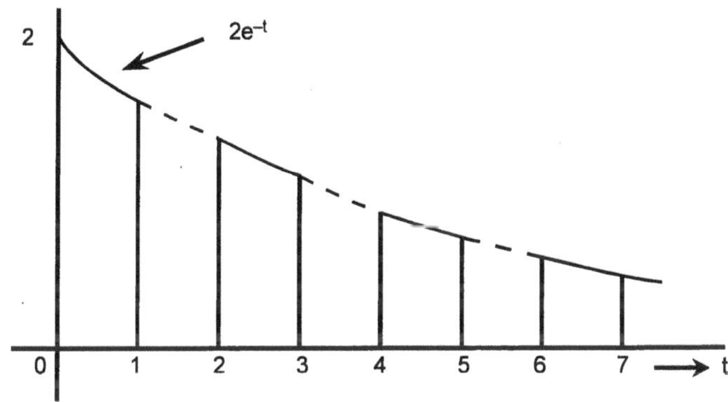

Fig. P. 9.2

9.5 Find the Laplace transforms of the periodic waveform shown in Fig. P. 9.3.

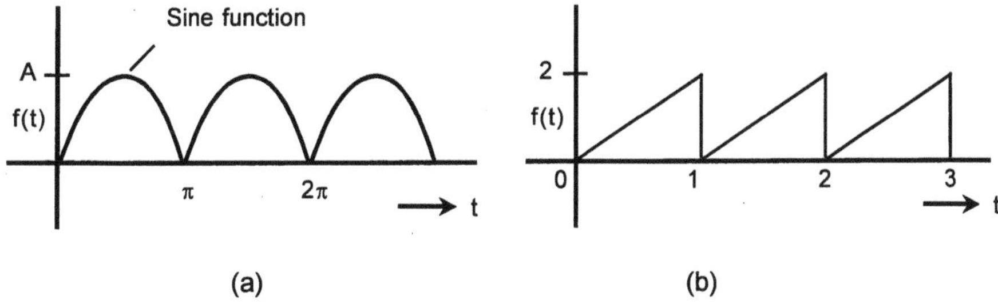

Fig. P. 9.3

9.6 Find $f(t) = L^{-1}[F_1(s) \cdot F_2(s)]$ by using convolution theorem for the following functions

(i) $F_1(s) = \dfrac{1}{s}$; $F_2(s) = \dfrac{1}{s+2}$

(ii) $F_1(s) = \dfrac{1}{s+3}$; $F_2(s) = \dfrac{1}{s+4}$

(iii) $F_1(s) = \dfrac{1}{s(s+1)}$; $F_2(s) = \dfrac{1}{s^2+1}$

(iv) $F_1(s) = \dfrac{1}{s^2}$; $F_2(s) = \dfrac{s}{s^2+4}$

9.7 For each of the waveform shown in Fig. P. 9.4 (a). Find the function obtained by convolving it with each of the functions shown in Fig. P. 9.4(b).

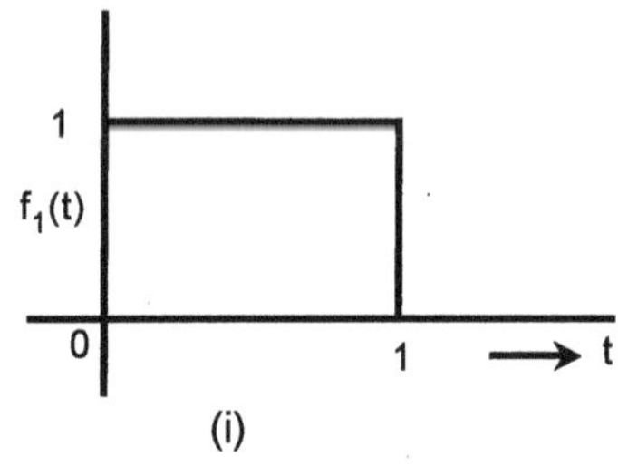

(i)

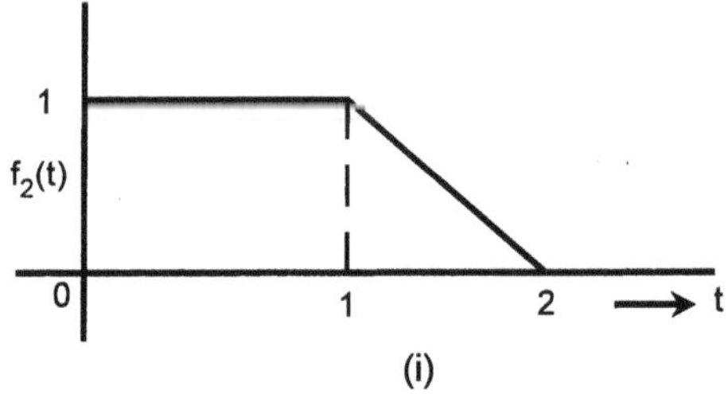

(i)

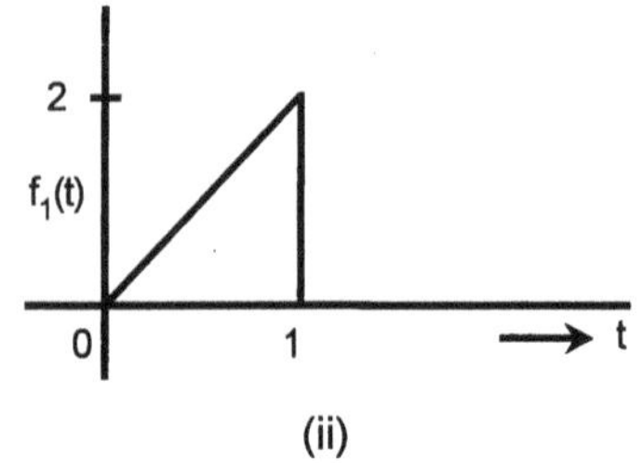

(ii)

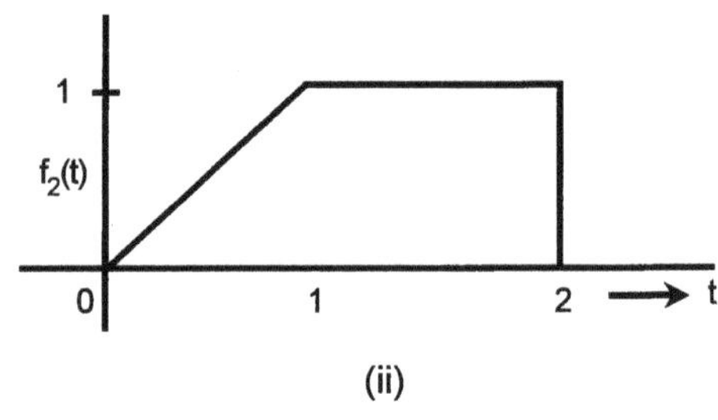

(ii)

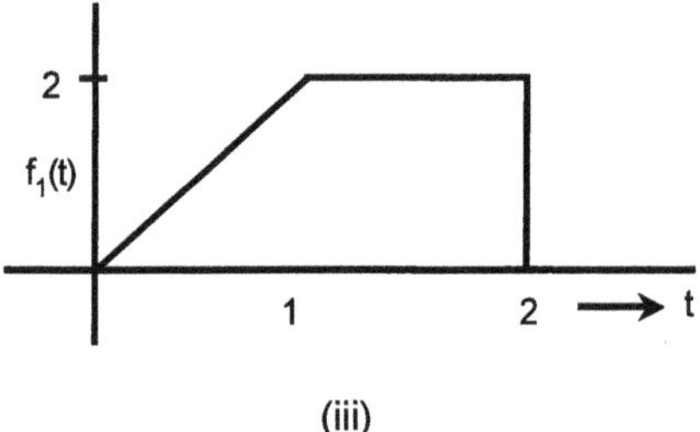

(iii)

(a)

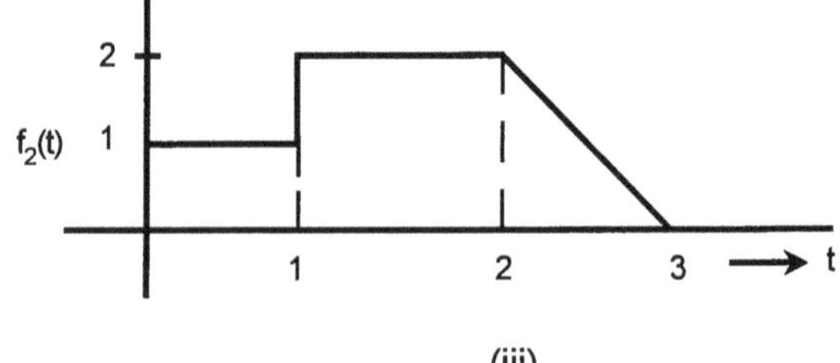

(iii)

(b)

Fig. P. 9.4

9.8 Find f(0), $\frac{df(0)}{dt}$ and $f(\infty)$ for the following functions without actually finding the time functions

(i) $F(s) = \frac{s+2}{s(s+1)(s+3)}$ (ii) $F(s) = \frac{s+3}{(s+1)(s+5)}$

(iii) $F(s) = \frac{s^2+6}{s^2+8}$ (iv) $F(s) = \frac{s^2+5}{2s(s^2+3s+6)}$

9.9 Solve for the current i(t) when the switch K is closed at t = 0 in the network of Fig. P. 9.5.

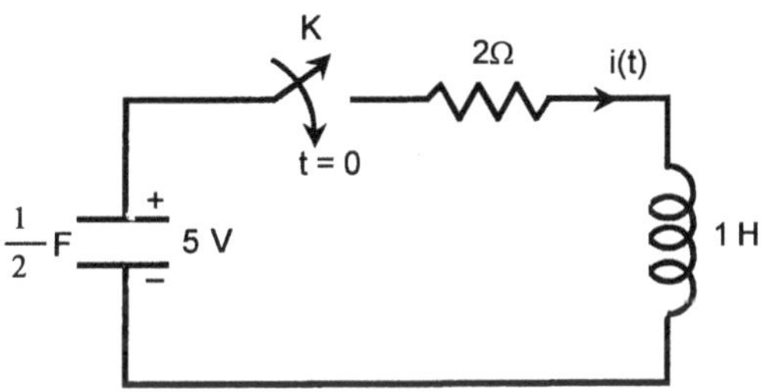

Fig. P. 9.5

9.10 The switch K in Fig.P. 9.6 is in position 1 for a long time and then at t = 0 it is moved to position 2. Find i(t).

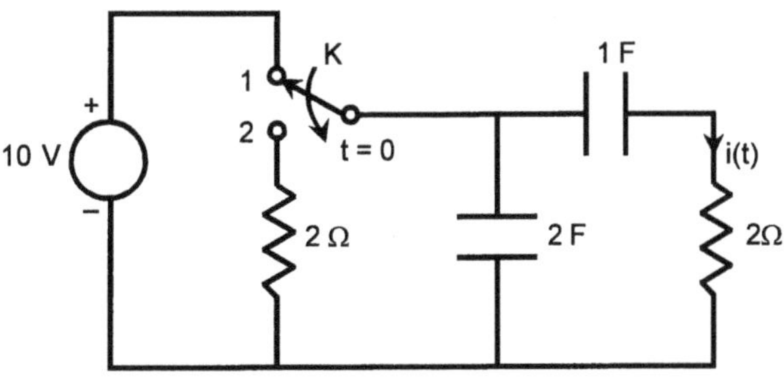

Fig. P. 9.6

9.11 Find the current $i_1(t)$ in Fig. P. 9.7 for $t > 0$ if the switch K is closed at $t = 0$.

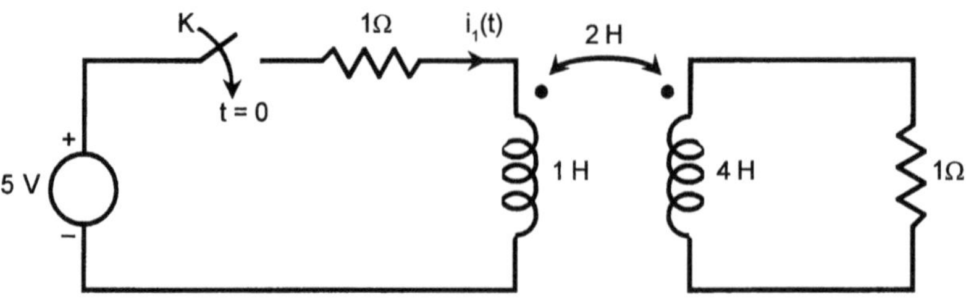

Fig. P. 9.7

9.12 Find v(t) in the circuit of Fig. P. 9.8.

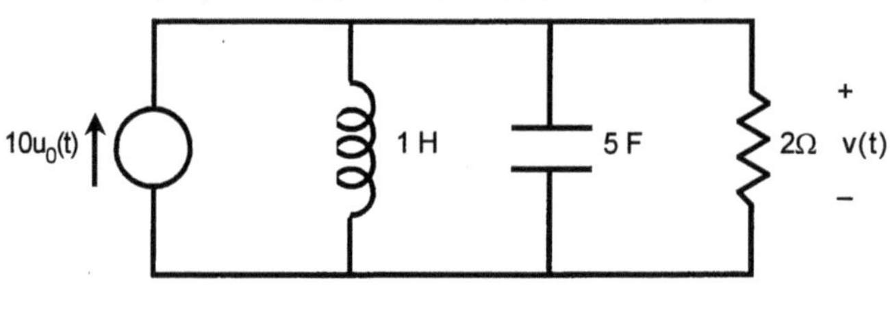

Fig. P. 9.8

9.13 Find v(t) in Fig. P. 9.9 if the switch K is closed at $t = 0$.

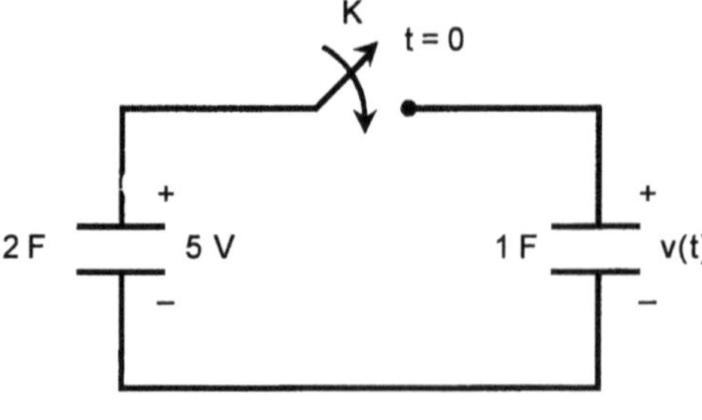

Fig. P. 9.9

9.14 Find i(t) in Fig. P. 9.10 if the switch K is opened at t = 0.

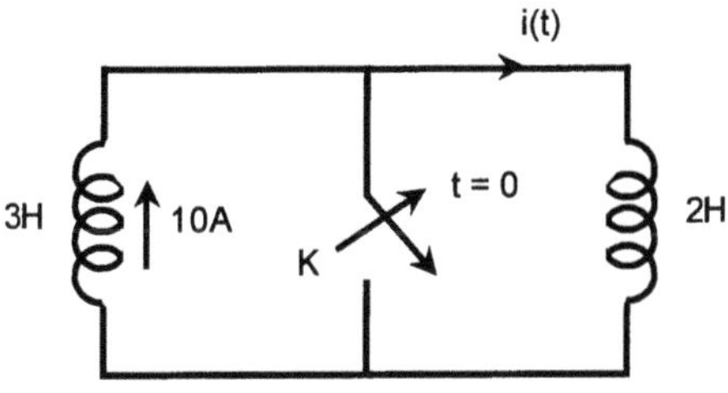

Fig. P. 9.10

9.15 Find i(t) in Fig.P.9.11 when the switches across the inductors are opened simultaneously at t = 0

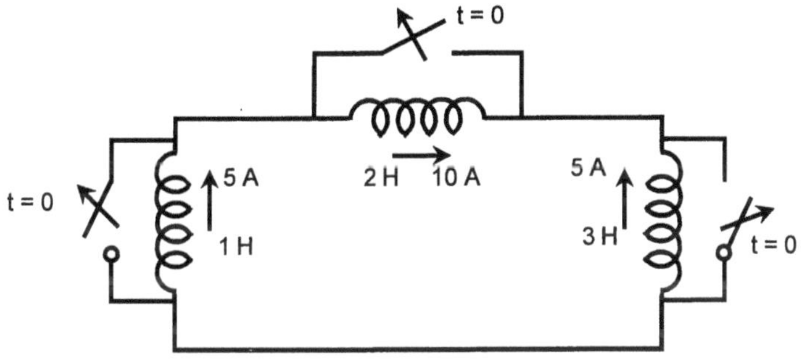

Fig. P. 9.11

9.16 Find the currents in the inductors for t > 0 in the circuit of Fig.P. 9.12. Inductors were having currents as shown in the Fig.P. 9.12 at $t = 0^-$.

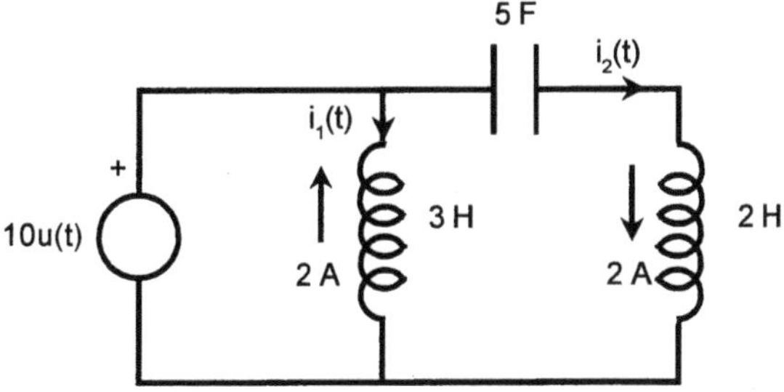

Fig. P. 9.12

9.17 Find v(t) in Fig. P. 9.13 for t > 0

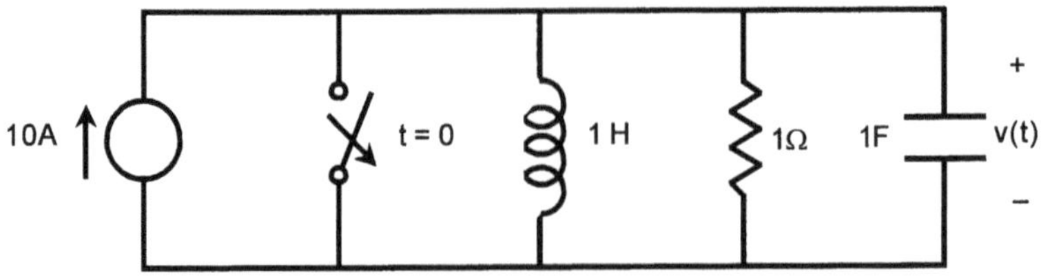

Fig. P. 9.13

9.18 Find i(t) for t > 0 in Fig. P. 9.14. The initial conditions and inductor and capacitor are as shown.

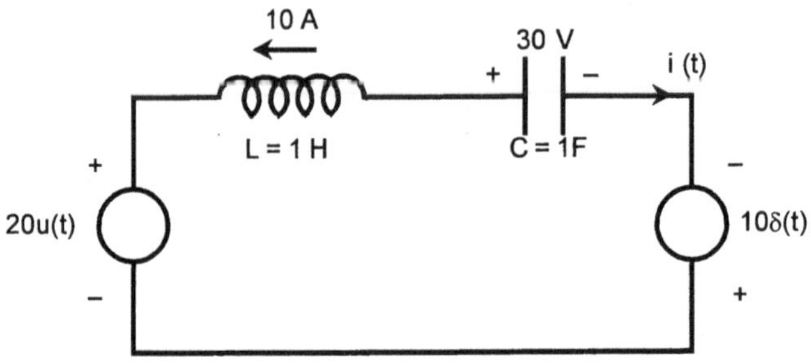

Fig. P. 9.14

9.19 Find i(t) and plot it for t > 0 for the network shown in Fig. P. 9.15 (b) when a voltage of waveform shown in Fig. P. 9.15 (a) is applied to it.

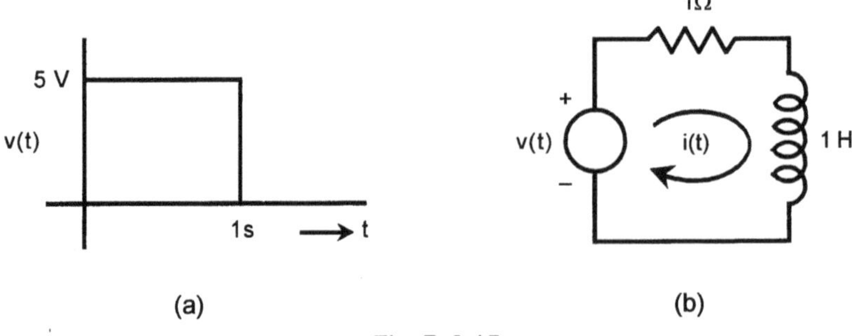

Fig. P. 9.15

9.20 Find v(t) for the circuit in Fig. P. 9.16.

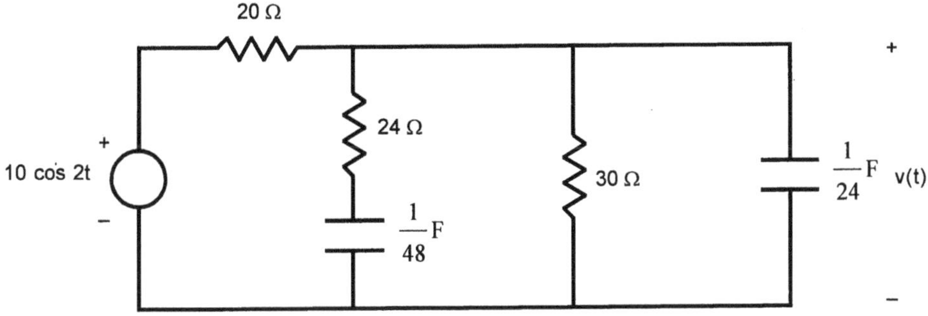

Fig. P. 9.16

9.21 For the circuit in Fig.P. 9.17 find the transfer function $H(s) = \frac{I_2(s)}{V(s)}$. Also find the impulse response h(t). If v(t) = 5u(t) find $i_2(t)$.

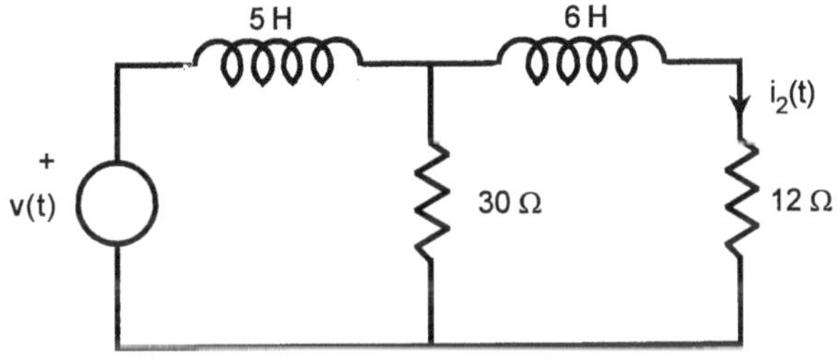

Fig. P. 9.17

9.22 For the network shown in Fig. P. 9.18 find the voltage ratio transfer function $G_{12}(s)$.

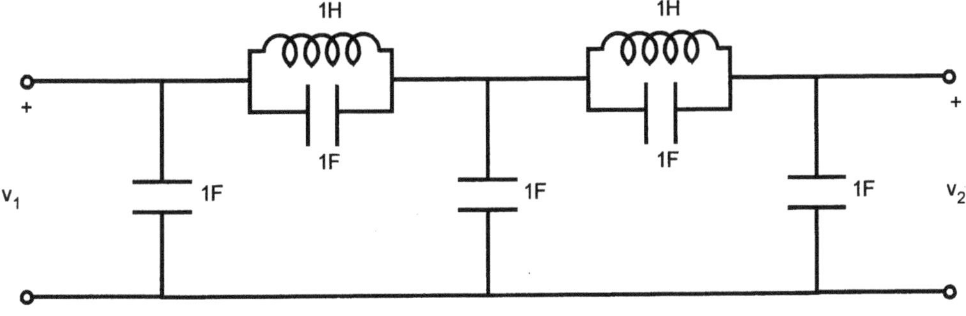

Fig. P. 9.18

9.23 For the network shown in Fig. P. 9.19 find Z_{11} and locate the poles and zeroes in s – plane.

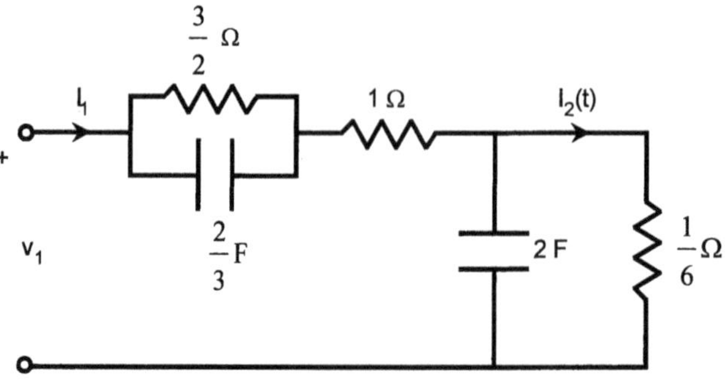

Fig. P. 9.19

9.24 Find $Y_{12}(s) = \dfrac{I_2(s)}{V_1(s)}$ for the circuit in Fig. P. 9.19. Locate its poles and zeros in s – plane.

10

TWO PORT NETWORKS

10.1 INTRODUCTION :

In this chapter we will consider the representation of two port networks which are made up of passive elements and controlled sources. Many times we may not be interested in the internal structure of this network but are satisfied with its behaviour at the given set of terminals. Such networks are represented usually by a rectangular box with the terminals brought out. If a network has only one pair of terminals available for connection to a source and no other terminals are available for external use, we call that network as a two terminal or one port network. The two terminals at which the source is connected is known as a port of entry into the network. Similarly if a network has two pairs of terminals, one pair for conection to a source and one pair for connection to a load, the network is known as a 4 terminal or 2 port network. We may have multiport networks also. These networks are shown in Fig. 10.1(a), (b) and (c).

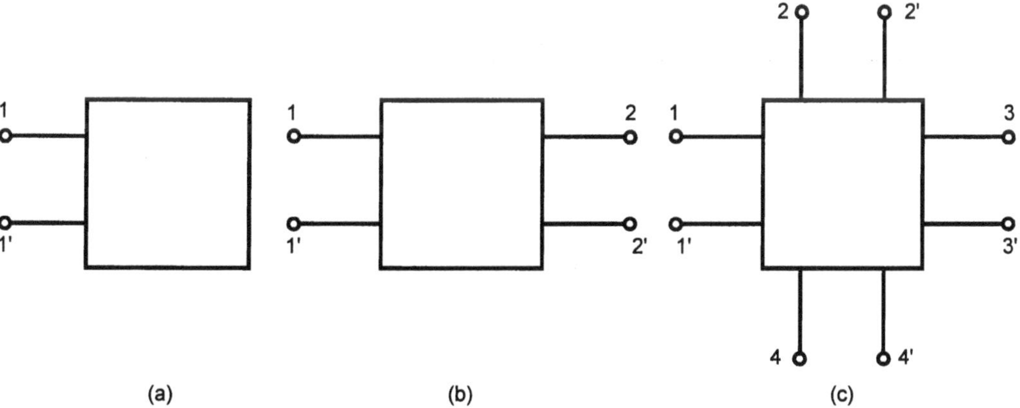

Fig. 10.1 (a) One port network (b) two port network (c) multiport network

In this chapter we will concentrate on the properties and representation of two port networks.

10.2 TWO PORT NETWORKS :

Consider a passive two port network shown in Fig. 10.2.

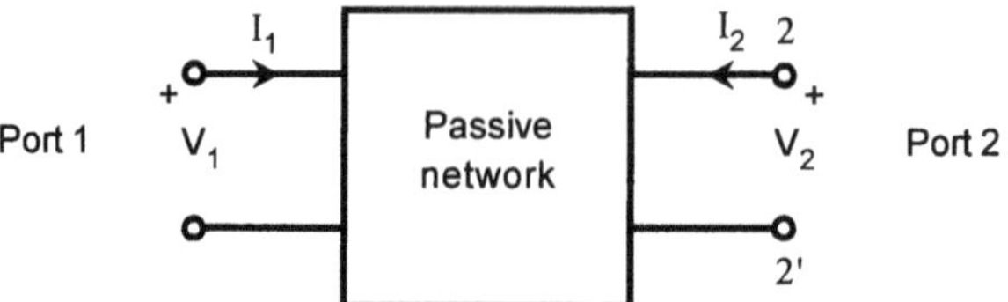

Fig. 10.2 A two port network

In the network shown, two currents and two voltage are identified and eventhough other currents and voltages are present inside the network, they are not available for measurement. Now, only two variables out of the 4 variables are independent and if they are specified others can be calculated. Thus there are six different ways of choosing the independent variables. These are listed in table 10.1.

Table 10.1 Two port parameters

Name of the parameters	**independent variables**	**dependent variables**	**Defining equations**
Open circuit impedance	I_1, I_2	V_1, V_2	$V_1 = I_1 z_{11} + I_2 z_{12}$ $V_2 = I_1 z_{21} + I_2 z_{22}$
Short circuit admittance	V_1, V_2	I_1, I_2	$I_1 = V_1 y_{11} + V_2 y_{12}$ $I_2 = V_1 y_{21} + V_2 y_{22}$
Transmission	V_2, I_2	V_1, I_1	$V_1 = AV_2 - BI_2$ $I_1 = CV_2 - DI_2$
Inverse transmission	V_1, I_1	V_2, I_2	$V_2 = A'V_1 - B'I_1$ $I_2 = C'V_1 - D'I_1$
Hybrid	I_1, V_2	V_1, I_2	$V_1 = h_{11} I_1 + h_{12} V_2$ $I_2 = h_{21} I_1 + h_{22} V_2$
Inverse hybrid	V_1, I_2	I_1, V_2	$I_1 = g_{11} V_1 + g_{12} I_2$ $V_2 = g_{21} V_1 + g_{22} I_2$

These different descriptions of the two port networks are useful in designing networks like filters, attenuators, transmission lines which are inserted between a source and a load.

10.3 OPEN CIRCUIT IMPEDANCE PARAMETERS :

Consider the passive network shown in Fig. 10.2. Using the node voltages V_1 and V_2 in addition to other node voltages V_3, V_4,.....V_n inside the network, we can write

$$\begin{aligned}
V_1Y_{11} + V_2Y_{12} + ... + V_nY_{1n} &= I_1 \\
V_1Y_{21} + V_2Y_{22} + ... + V_nY_{2n} &= I_2 \\
V_1Y_{31} + V_2Y_{32} + ... + V_nY_{3n} &= 0 \\
\vdots \quad \vdots \quad \vdots \quad \vdots \qquad \vdots \quad &\quad \vdots \\
V_1Y_{n1} + V_2Y_{n2} + ... + V_nY_{nn} &= 0
\end{aligned} \qquad .....(10.1)$$

Solving for V_1 and V_2 we have

$$V_1 = \frac{\Delta_{11}}{\Delta} I_1 + \frac{\Delta_{21}}{\Delta} I_2$$

$$V_2 = \frac{\Delta_{12}}{\Delta} I_1 + \frac{\Delta_{22}}{\Delta} I_2 \qquad .....(10.2)$$

where Δ_{ij} are the cofactors and Δ is the determinant of the coefficient matrix in eq. (10.1). In eq. (10.2) the coefficients $\frac{\Delta_{ij}}{\Delta}$ must have the dimensions of impedance and denoting

$$\frac{\Delta_{ij}}{\Delta} = z_{ji} \quad \text{for} \quad i, j = 1, 2$$

We write

$$V_1 = z_{11}I_1 + z_{12}I_2$$

$$V_2 = z_{21}I_1 + z_{22}I_2 \qquad .....(10.3)$$

which can be put in the matrix form

$$\begin{bmatrix} V_1 \\ V_2 \end{bmatrix} = \begin{bmatrix} z_{11} & z_{12} \\ z_{21} & z_{22} \end{bmatrix} \begin{bmatrix} I_1 \\ I_2 \end{bmatrix} \qquad .....(10.4)$$

Now if either I_1, or I_2 is made zero the four parameters can be obtained as

$$z_{11} = \left.\frac{V_1}{I_1}\right|_{I_2=0} \quad \text{open circuit driving point impedance of port 1}$$

$$z_{21} = \left.\frac{V_2}{I_1}\right|_{I_2=0} \quad \text{open circuit forward transfer impedance}$$

$$z_{12} = \left.\frac{V_1}{I_2}\right|_{I_1=0} \quad \text{open circuit reverse transfer impedance}$$

$$z_{22} = \left.\frac{V_2}{I_2}\right|_{I_1=0} \quad \text{open circuit driving point impedance of port 2}$$

Since these z – parameters are obtained by open circuiting the ports, we call them open circuit impedance parameters. In terms of these parameters the network inside the box can be represented by a simple network. Two such equivalent representations are shown in Fig. 10.3(a) and (b). They are known as two source equivalent and single source equivalent circuits respectively.

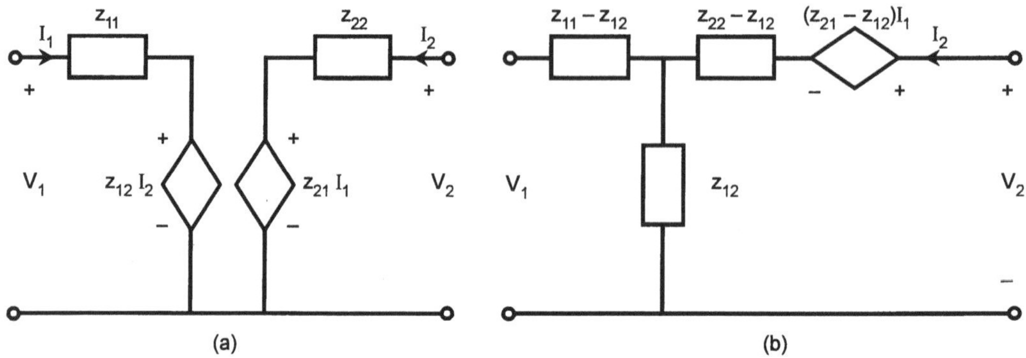

Fig. 10.3 (a) Two source equivalent (b) Single source equivalent

For obtaining the single source equivalent eqs. (10.3) are modified as

$$V_1 = (z_{11} - z_{12})I_1 + z_{12}(I_1 + I_2)$$

$$V_2 = (z_{21} - z_{12})I_1 + (z_{22} - z_{12})I_2 + z_{12}(I_1 + I_2)$$

It is easy to verify that these equations represent the network in Fig. 10.3(b). If the network is reciprocal, as per Reciprocity theorem discussed in chapter 6,

$$\left.\frac{V_2}{I_1}\right|_{I_2=0} = \left.\frac{V_1}{I_2}\right|_{I_1=0}$$

or $$z_{21} = z_{12} \qquad .....(10.5)$$

For reciprocal networks, the single source equivalent shown in Fig. 10.3(b) reduces to a passive network with no dependent source. A symmetrical network is defined as a network which can be divided into two halves, with each half as a mirror image of the other. In other words, the network looks exactly the same whether viewed from port 1 or port 2. For such networks

$$z_{22} = z_{11} \qquad .....(10.6)$$

10.4 SHORT CIRCUIT ADMITTANCE PARAMETERS :

The short circuit parameters are defined by the equations

$$I_1 = V_1 y_{11} + V_2 y_{12}$$

$$I_2 = V_1 y_{21} + V_2 y_{22} \qquad .....(10.7)$$

or in matrix form

$$\begin{bmatrix} I_1 \\ I_2 \end{bmatrix} = \begin{bmatrix} y_{11} & y_{12} \\ y_{21} & y_{22} \end{bmatrix} \begin{bmatrix} V_1 \\ V_2 \end{bmatrix} \qquad .....(10.8)$$

The Y – parameters can be obtained by short circuiting one of the ports and supplying the other port with a voltage source and making measurements as shown below :

$y_{11} = \left.\dfrac{I_1}{V_1}\right|_{V_2=0}$ Short circuit driving point admittance of port 1

$y_{21} = \left.\dfrac{I_2}{V_1}\right|_{V_2=0}$ Short circuit forward transfer admittance

$y_{12} = \left.\dfrac{I_2}{V_2}\right|_{V_1=0}$ Short circuit reverse transfer admittance

$y_{22} = \left.\dfrac{I_2}{V_2}\right|_{V_1=0}$ Short circuit driving point admittance of port 2

As before the two source and single source equivalent circuits are shown in Fig. 10.4 (a) and (b).

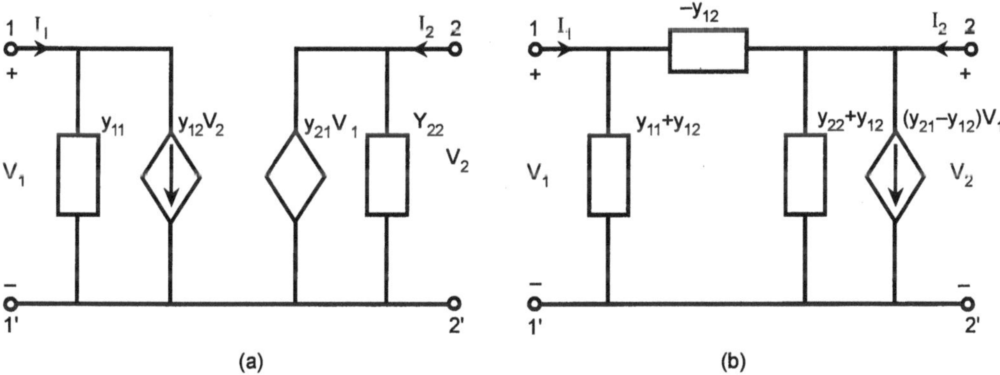

Fig. 10.4 (a) Two source equivalent (b) Single source equivalent

To obtain the single source equivalent, eq. (10.7) is modified as

$$I_1 = (y_{11} + y_{12})V_1 - y_{12}(V_1 - V_2)$$

$$I_2 = (y_{21} - y_{12})V_1 + (y_{22} + y_{12})V_2 - y_{12}(V_2 - V_1)$$

If the network is a reciprocal network, then

$$y_{12} = y_{21} \qquad .....(10.9)$$

and in addition, if it is symmetrical

$$y_{11} = y_{22} \qquad .....(10.10)$$

Example 10.1

Obtain the z – parameters of the network in Fig. 10.5.

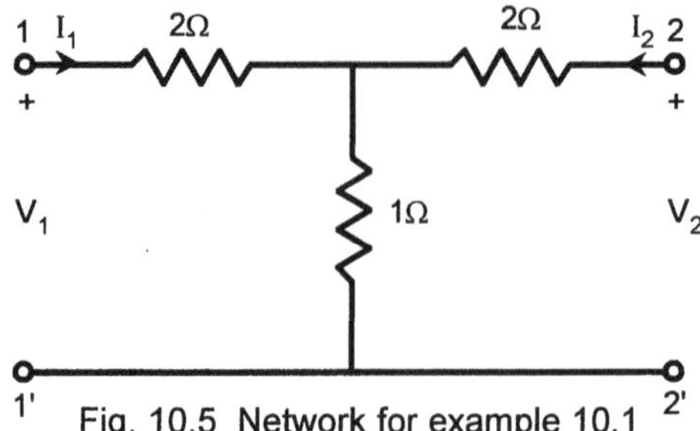

Fig. 10.5 Network for example 10.1

Solution :

Open circuiting port 2, we have

$$z_{11} = \left.\frac{V_1}{I_1}\right|_{I_2=0} = 3\ \Omega$$

$$z_{21} = \left.\frac{V_2}{I_1}\right|_{I_2=0} = 1\ \Omega$$

Similarly open circuiting port 1, and applying a voltage V_2 at port 2, we have

$$z_{22} = \left.\frac{V_2}{I_2}\right|_{I_1=0} = 3\ \Omega$$

$$z_{12} = \left.\frac{V_1}{I_2}\right|_{I_1=0} = 1\ \Omega$$

∴ The impedance parameter matrix is

$$[z] = \begin{bmatrix} 3 & 1 \\ 1 & 3 \end{bmatrix}$$

Example 10.2

Obtain the z – parameters of the network in Fig. 10.6(a).

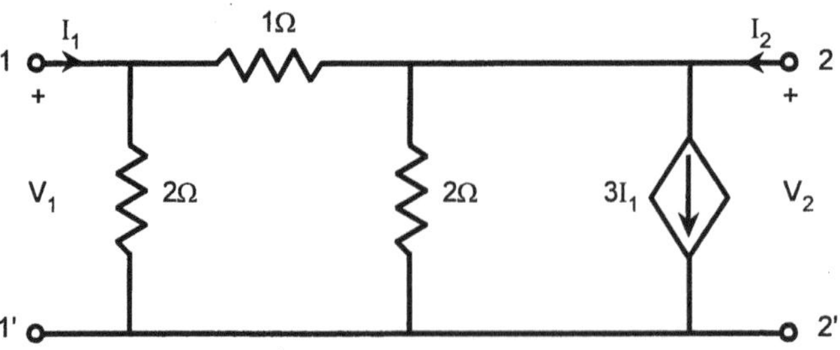

Fig. 10.6 (a) Network for example 10.2

Solution :

Open circuiting port 2, and writing node voltage equation at node 2, we have

$$\frac{V_2}{2}+\frac{V_2-V_1}{1}+3I_1=0$$

and at node 1,

$$\frac{V_1}{2}+\frac{V_1-V_2}{1}=I_1$$

Eliminating V_2 from the above two equations, we get

$$z_{11}=\frac{V_1}{I_1}=-\frac{6}{5}\ \Omega$$

and
$$z_{21}=\frac{V_2}{I_1}=-\frac{14}{5}\ \Omega$$

Next, open circuiting port 1, we have $I_1 = 0$ and therefore the dependent source current is zero.

$$\therefore \qquad z_{22}=\frac{V_2}{I_2}=\frac{3(2)}{3+2}=\frac{6}{5}\Omega$$

and
$$V_1=I_2\left(\frac{2}{5}\right)\times 2$$

$$=\frac{4I_2}{5}$$

or
$$z_{12}=\frac{V_1}{I_2}=\frac{4}{5}\ \Omega$$

Thus the impedance parameters are given by the matrix

$$[z]=\begin{bmatrix}-\frac{6}{5} & \frac{4}{5}\\ -\frac{14}{5} & \frac{6}{5}\end{bmatrix}$$

Note that some of the z – parameters are – ve because of the presence of dependent source. If it is a passive network with no dependent sources, all the z – parameters will be positive as in example 10.1.

Example 10.3

Find the y – parameters of the network in Fig. 10.6(b).

Solution :

Short circuit port 2 as shown in Fig. 10.6 (b). As a result the 2 Ω resistor and the current source are shorted and we have

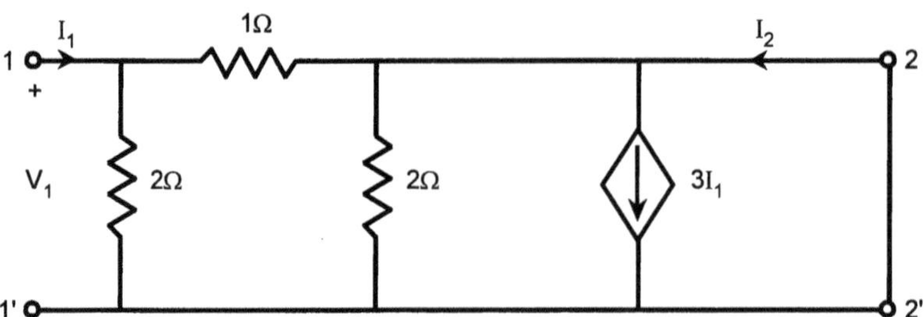

Fig. 10.6 (b) Network of Fig. 10.6 with port 2 shorted

$$I_1 = \frac{V_1}{2} + \frac{V_1}{1}$$

or $$\frac{I_1}{V_1} = y_{11} = 1.5 \text{ S}$$

and $$I_2 = 3I_1 - \frac{V_1}{1}$$

But $$I_1 = 1.5\ V_1$$

$\therefore$ $$I_2 = 3(1.5\ V_1) - V_1$$
$$= 3.5\ V_1$$

$$y_{21} = \frac{I_2}{V_1} = 3.5 \text{ S}$$

Similarly shorting port 1 as shown in Fig. 10.6(c), we have

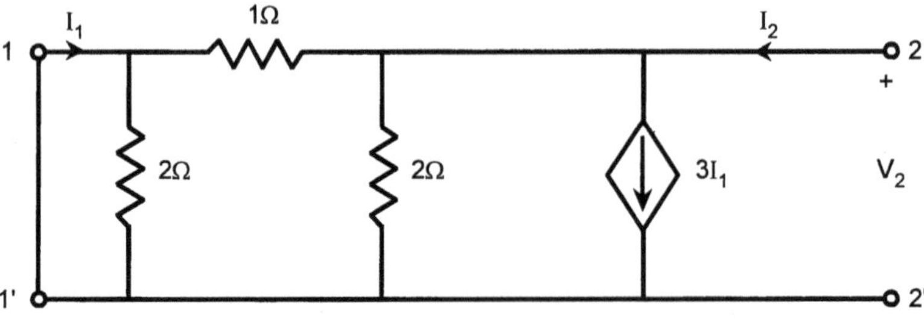

Fig. 10.6 (c) Network of Fig. 10.6 with port 1 shorted

$$I_2 = \frac{V_2}{2} + \frac{V_2}{1} + 3I_1$$

But $$I_1 = -\frac{V_2}{1}$$

$\therefore$ $$y_{12} = \frac{I_1}{V_2} = -1\,S$$

and $$I_2 = \frac{V_2}{2} + \frac{V_2}{1} - 3\,V_2$$
$$= -1.5\,V_2$$

$\therefore$ $$y_{22} = \frac{I_2}{V_2} = -1.5\,S$$

$\therefore$ The admittance parameters are given by the matrix

$$[y] = \begin{bmatrix} 1.5 & -1 \\ 3.5 & -1.5 \end{bmatrix}$$

10.5 TRANSMISSION PARAMETERS :

The defining equations for transmission parameters are

$$V_1 = AV_2 - BI_2$$
$$I_1 = CV_2 - DI_2 \qquad .....(10.11)$$

The negative sign in eq. (10.11) is associated with the current I_2. These parameters were first used to define transmission line parameters in which the current in port 2 is always taken to be going away from the network. Since the reference directions assumed for defining all the parameters are as indicated in Fig. 10.2, the current I_2 is taken as negative in this case.

To define the parameters, we take $I_2 = 0$, or open circuit the port 2.

$$A = \left.\frac{V_1}{V_2}\right|_{I_2=0} \text{ open circuit reverse voltage gain}$$

$$C = \left.\frac{I_1}{V_2}\right|_{I_2=0} \text{ open circuit transfer admittance.}$$

Similarly, short circuiting port 2 we have $V_2 = 0$ and

$$B = \left.\frac{V_1}{-I_2}\right|_{V_2=0} \text{ short circuit transfer impedance}$$

and $$D = \left.\frac{I_1}{-I_2}\right|_{V_2=0} \text{ short circuit reverse current gain}$$

These parameters are also known as ABCD parameters. The inverse ABCD parameters relate the output port quantities in terms of input port quantities.

$$V_2 = A'V_1 - B'I_1$$
$$I_2 = C'V_1 - D'I_1 \qquad .....(10.12)$$

These inverse parameters are useful for transmission in opposite direction and have similar definitions and properties as for ABCD parameters.

10.6 HYBRID PARAMETERS OR h - PARAMETERS :

Hybrid parameters are widely used to represent electronic circuits. The defining equations are

$$V_1 = h_{11}I_1 + h_{12}V_2$$
$$I_2 = h_{21}I_1 + h_{22}V_2 \qquad .....(10.13)$$

The h – parameters can be obtained by making $V_2 = 0$ or short circuiting port 2. Thus

$$h_{11} = \left.\frac{V_1}{I_1}\right|_{V_2=0} \quad \text{short circuit input impedance}$$

$$h_{21} = \left.\frac{I_2}{I_1}\right|_{V_2=0} \quad \text{short circuit current gain}$$

Open circuiting port 1, we get

$$h_{12} = \left.\frac{V_1}{V_2}\right|_{I_1=0} \quad \text{open circuit reverse voltage gain}$$

$$h_{22} = \left.\frac{I_2}{V_2}\right|_{I_1=0} \quad \text{open circuit output admittance.}$$

The parameters have mixed dimensions and hence are termed as hybrid parameters.

If we relate I_1 and V_2 in terms of V_1 and I_2 we have g – parameters. Thus we have

$$I_1 = g_{11}V_1 + g_{12}I_2$$
$$V_2 = g_{21}V_1 + g_{22}I_2 \qquad .....(10.14)$$

$$g_{11} = \left.\frac{I_1}{V_1}\right|_{I_2=0} \quad \text{open circuit input admittance}$$

$$g_{21} = \left.\frac{V_2}{V_1}\right|_{I_2=0} \quad \text{open circuit voltage gain}$$

$$g_{12} = \left.\frac{I_1}{I_2}\right|_{V_1=0}$$ short circuit current gain

$$g_{22} = \left.\frac{V_2}{I_2}\right|_{V_1=0}$$ short circuit input impedance of port 2.

Example 10.4

Obtain the h – parameters of the network in Fig. 10.7.

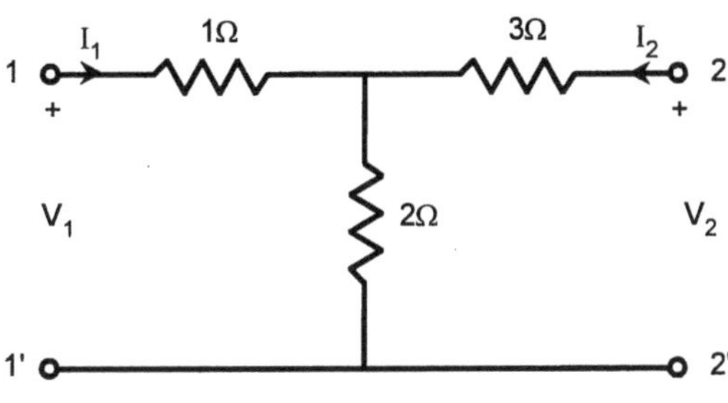

Fig. 10.7 Network for example 10.4

Solution :

Short circuit port 2

$$h_{11} = \left.\frac{V_1}{I_1}\right|_{V_2=0} = 1 + \frac{3 \times 2}{3+2} = \frac{11}{5}\Omega$$

$$I_2 = -I_1 \cdot \frac{2}{5}$$

$$h_{21} = \left.\frac{I_2}{I_1}\right|_{V_2=0} = -\frac{2}{5}$$

Open circuit port 1 and apply a voltage V_2 at port 2.

$$h_{22} = \left.\frac{I_2}{V_2}\right|_{I_1=0} = \frac{1}{3+2} = \frac{1}{5}\text{ S}$$

$$V_1 = \frac{V_2}{5} \cdot 2$$

$$\therefore \quad \frac{V_1}{V_2} = h_{12} = \frac{2}{5}$$

Thus the h – parameters are

$$[h] = \begin{bmatrix} 2.2 & 0.4 \\ -0.4 & 0.2 \end{bmatrix}$$

Example 10.5

Find the ABCD parameters of the network shown in Fig. 10.8.

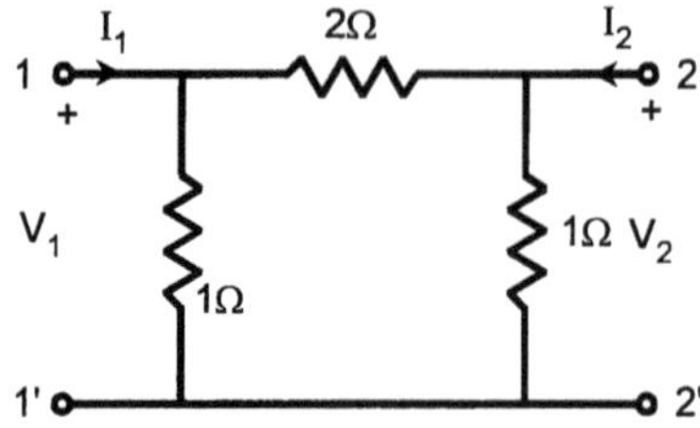

Fig. 10.8 Network for example 10.5

Solution :

Open circuit port 2. Thus $I_2 = 0$

$$I_1 = \frac{V_1}{1} + \frac{V_1}{3} = V_1 \cdot \frac{4}{3}$$

$\therefore$ $$V_2 = I_1 \cdot \frac{1}{4} \times 1 = \frac{1}{4} I_1$$

$\therefore$ $$C = \left.\frac{I_1}{V_2}\right|_{I_2=0} = 4S$$

and $$V_2 = \frac{1}{4} \cdot \frac{4V_1}{3} = \frac{V_1}{3}$$

$\therefore$ $$A = \left.\frac{V_1}{V_2}\right|_{I_2=0} = 3$$

Now short circuit port 2. i.e., $V_2 = 0$

$$I_1 = \frac{V_1}{1} + \frac{V_1}{2}$$

$$I_1 = \frac{3}{2} V_1$$

and $$I_2 = -\frac{V_1}{2}$$

$$B = \frac{V_1}{-I_2} = 2\Omega$$

and $$I_1 = \frac{3}{2} \cdot (-2I_2) = -3I_2$$

$\therefore$ $$D = \frac{I_1}{-I_2} = 3$$

$\therefore$ The transmission parameters are

$$[T] = \begin{bmatrix} 3 & 2 \\ 4 & 3 \end{bmatrix}$$

10.7 RELATIONSHIPS BETWEEN DIFFERENT SETS OF PARAMETERS :

The six parameter sets defined in earlier sections may be used to describe the same network. Hence they are all inter related. For a given network it may be easier to obtain a particular set of parameters but the actual parameters required may be different. Hence it is desirable to be able to convert one set of parameters to another set.

10.7.1 Relation between y and z – parameters :

We know that

$$\begin{bmatrix} I_1 \\ I_2 \end{bmatrix} = \begin{bmatrix} y_{11} & y_{12} \\ y_{21} & y_{22} \end{bmatrix} \begin{bmatrix} V_1 \\ V_2 \end{bmatrix}$$

Solving for V_1 and V_2, we have

$$\begin{bmatrix} V_1 \\ V_2 \end{bmatrix} = \begin{bmatrix} y_{11} & y_{12} \\ y_{21} & y_{22} \end{bmatrix}^{-1} \begin{bmatrix} I_1 \\ I_2 \end{bmatrix}$$

and hence

$$\begin{bmatrix} z_{11} & z_{12} \\ z_{21} & z_{22} \end{bmatrix} = \begin{bmatrix} y_{11} & y_{12} \\ y_{21} & y_{22} \end{bmatrix}^{-1}$$

$$\therefore \quad z_{11} = \frac{y_{22}}{\Delta y}\ ;\ z_{12} = -\frac{y_{12}}{\Delta y}$$

$$z_{22} = \frac{y_{11}}{\Delta y}\ ;\ z_{21} = -\frac{y_{21}}{\Delta y} \qquad .....(10.15)$$

where $\Delta y = y_{11}y_{22} - y_{12}y_{21}$

Similarly $$\begin{bmatrix} y_{11} & y_{12} \\ y_{21} & y_{22} \end{bmatrix} = \begin{bmatrix} z_{11} & z_{12} \\ z_{21} & z_{22} \end{bmatrix}^{-1}$$

and $$y_{11} = \frac{z_{22}}{\Delta z}\ ;\ y_{21} = \frac{-z_{12}}{\Delta z}$$

$$y_{22} = \frac{z_{11}}{\Delta z};\ y_{21} = \frac{-z_{21}}{\Delta z}$$

10.7.2 Transmission parameters to z – parameters :

We have $V_1 = AV_2 - BI_2$

$$I_1 = CV_2 - DI_2$$

Now from the second equation:

$$V_2 = \frac{I_1}{C} + \frac{D}{C}I_2 \quad .....(10.16)$$

From first equation

$$V_1 = A\left[\frac{I_1}{C} + \frac{D}{C}I_2\right] - BI_2 = \frac{A}{C}I_1 + \left(\frac{AD}{C} - B\right)I_2$$

$$= \frac{A}{C}I_1 + \left(\frac{AD - BC}{C}\right)I_2 \quad .....(10.17)$$

Equations (10.16) and (10.17), when put in the matrix form, give

$$\begin{bmatrix} V_1 \\ V_2 \end{bmatrix} = \begin{bmatrix} \frac{A}{C} & \frac{AD - BC}{C} \\ \frac{1}{C} & \frac{D}{C} \end{bmatrix} \begin{bmatrix} I_1 \\ I_2 \end{bmatrix} \quad .....(10.18)$$

The coefficient matrix is identified as the matrix of impedance parameters. Therefore

$$z_{11} = \frac{A}{C} \; ; z_{22} = \frac{D}{C} \; ; z_{12} = \frac{AD - BC}{C} \; ; z_{21} = \frac{1}{C}$$

10.7.3 h – parameters to z – parameters :

The current voltage relations in terms of h parameters are

$$V_1 = h_{11}I_1 + h_{12}V_2$$

$$I_2 = h_{21}I_1 + h_{22}V_2 \quad .....(10.19)$$

From the second equation

$$V_2 = \frac{1}{h_{22}}(I_2 - h_{21}I_1) \quad .....(10.20)$$

Substituting for V_2 in the first equation of (10.19) we get

$$V_1 = h_{11}I_1 + \frac{h_{12}}{h_{22}}(I_2 - h_{21}I_1)$$

$$= \left(h_{11} - \frac{h_{12}h_{21}}{h_{22}}\right)I_1 + \frac{h_{12}}{h_{22}}I_2 \quad .....(10.21)$$

Eqs. (10.20) and (10.21) can be put in to the following matrix equation :

$$\begin{bmatrix} V_1 \\ V_2 \end{bmatrix} = \begin{bmatrix} h_{11} - \dfrac{h_{12}h_{21}}{h_{22}} & \dfrac{h_{12}}{h_{22}} \\ -\dfrac{h_{21}}{h_{22}} & \dfrac{1}{h_{22}} \end{bmatrix} \begin{bmatrix} I_1 \\ I_2 \end{bmatrix} \qquad .....(10.22)$$

The coefficient matrix is again the matrix of z – parameters. So

$$z_{11} = \frac{h_{11}h_{22} - h_{12}h_{21}}{h_{22}}$$

$$z_{12} = \frac{h_{12}}{h_{22}}$$

$$z_{21} = -\frac{h_{12}}{h_{22}}$$

and $$z_{22} = \frac{1}{h_{22}}$$

10.7.4 Inverse Hybrid parameters to z – parameters :

The relations between currents and voltages using inverse hybrid parameters given by equations (10.14) will be used to obtain the relationships between 'g' and 'z' parameters.

From the first equation of (10.14), we have

$$V_1 = \frac{1}{g_{11}} I_1 + \left(-\frac{g_{12}}{g_{11}} \right) I_2 \qquad .....(10.23)$$

Substituting in the second equation of (10.14), we have

$$V_2 = \frac{g_{12}}{g_{11}} I_1 + \left(g_{22} - \frac{g_{12}g_{21}}{g_{11}} \right) I_2 \qquad .....(10.24)$$

Accordingly

$$z_{11} = \frac{1}{g_{11}}; \qquad z_{12} = -\frac{g_{12}}{g_{11}}$$

and $$z_{22} = \frac{g_{11}g_{22} - g_{12}g_{21}}{g_{11}}; \qquad z_{21} = \frac{g_{21}}{g_{11}}$$

Table 10.2 shows all similar relationships between various sets of parameters. In this table, Δ is the determinant of the corresponding matrix of the parameters. For example $\Delta_Z = z_{11}z_{22} - z_{12}z_{21}$.

10.8 RECIPROCITY AND SYMMETRY :

What are the conditions for reciprocity and symmetry for various sets of parameters ?

For z – parameters, we have

$$z_{12} = z_{21} \quad \text{(Reciprocity)}$$

$$z_{11} = z_{22} \quad \text{(Symmetry)}$$

For y – parameters, we have

$$y_{12} = y_{21} \quad \text{(Reciprocity)}$$

$$y_{11} = y_{22} \quad \text{(Symmetry)}$$

Using these relations and interrelations of various parameters as given in table 10.2 we can deduce the relationships in other parameter sets.

For example from the first row of table 10.2, we have

$$z_{12} = \frac{\Delta_T}{C}$$

and $$z_{21} = \frac{1}{C}$$

For reciprocal networks,

$$z_{12} = z_{21}$$

$$\therefore \quad \frac{\Delta_T}{C} = \frac{1}{C}$$

or $$AD - BC = 1$$

Similarly, for symmetry .....(10.25)

$$z_{11} = z_{22}$$

$$\therefore \quad \frac{A}{C} = \frac{D}{C}$$

or $$A = D \quad .....(10.26)$$

For h – parameters, we have

$$z_{12} = \frac{h_{12}}{h_{22}} \quad z_{11} = \frac{\Delta_h}{h_{22}}$$

and $$z_{21} = -\frac{h_{21}}{h_{22}} \quad z_{22} = \frac{1}{h_{22}}$$

Table : 10.2 Relationships among the network parameters

Matrices →	[Z]	[Y]	[T]	[T']	[H]	[G]
[Z]	$\begin{matrix} z_{11} & z_{12} \\ z_{21} & z_{22} \end{matrix}$	$\begin{matrix} \frac{y_{22}}{Y} & -\frac{y_{12}}{Y} \\ -\frac{y_{21}}{Y} & \frac{y_{11}}{Y} \end{matrix}$	$\begin{matrix} \frac{A}{C} & \frac{T}{C} \\ \frac{1}{C} & \frac{D}{C} \end{matrix}$	$\begin{matrix} \frac{D'}{C'} & \frac{1}{C'} \\ \frac{T'}{C'} & \frac{A'}{C'} \end{matrix}$	$\begin{matrix} \frac{H}{h_{22}} & \frac{h_{12}}{h_{22}} \\ -\frac{h_{21}}{h_{22}} & \frac{1}{h_{22}} \end{matrix}$	$\begin{matrix} \frac{1}{g_{11}} & -\frac{g_{12}}{g_{11}} \\ \frac{g_{21}}{g_{11}} & \frac{G}{g_{11}} \end{matrix}$
[Y]	$\begin{matrix} \frac{z_{22}}{Z} & -\frac{z_{12}}{Z} \\ -\frac{z_{21}}{Z} & \frac{z_{11}}{Z} \end{matrix}$	$\begin{matrix} y_{11} & y_{12} \\ y_{21} & y_{22} \end{matrix}$	$\begin{matrix} \frac{D}{B} & -\frac{T}{B} \\ -\frac{1}{B} & \frac{A}{B} \end{matrix}$	$\begin{matrix} \frac{A'}{B'} & -\frac{1}{B'} \\ \frac{T'}{B'} & \frac{D'}{B'} \end{matrix}$	$\begin{matrix} \frac{1}{h_{11}} & -\frac{h_{12}}{h_{11}} \\ \frac{h_{21}}{h_{11}} & \frac{H}{h_{11}} \end{matrix}$	$\begin{matrix} \frac{G}{g_{22}} & \frac{g_{12}}{g_{22}} \\ -\frac{g_{21}}{g_{22}} & \frac{1}{g_{22}} \end{matrix}$
[T]	$\begin{matrix} \frac{z_{11}}{z_{21}} & \frac{Z}{z_{21}} \\ \frac{1}{z_{21}} & \frac{z_{22}}{z_{21}} \end{matrix}$	$\begin{matrix} -\frac{y_{22}}{y_{21}} & -\frac{1}{y_{21}} \\ -\frac{Y}{y_{21}} & -\frac{y_{11}}{y_{21}} \end{matrix}$	$\begin{matrix} A & B \\ C & D \end{matrix}$	$\begin{matrix} \frac{D'}{T'} & \frac{B'}{T'} \\ \frac{C'}{T'} & \frac{A'}{T'} \end{matrix}$	$\begin{matrix} -\frac{H}{h_{21}} & -\frac{h_{11}}{h_{21}} \\ -\frac{h_{22}}{h_{21}} & -\frac{1}{h_{21}} \end{matrix}$	$\begin{matrix} \frac{1}{g_{21}} & \frac{g_{22}}{g_{21}} \\ \frac{g_{11}}{g_{21}} & \frac{G}{g_{21}} \end{matrix}$
[T']	$\begin{matrix} \frac{z_{22}}{z_{12}} & \frac{Z}{z_{12}} \\ \frac{1}{z_{12}} & \frac{z_{11}}{z_{12}} \end{matrix}$	$\begin{matrix} -\frac{y_{11}}{y_{12}} & -\frac{1}{y_{12}} \\ -\frac{Y}{y_{12}} & -\frac{y_{11}}{y_{12}} \end{matrix}$	$\begin{matrix} \frac{D}{T} & \frac{B}{T} \\ \frac{C}{T} & \frac{A}{T} \end{matrix}$	$\begin{matrix} A' & B' \\ C' & D' \end{matrix}$	$\begin{matrix} \frac{1}{h_{12}} & \frac{h_{11}}{h_{12}} \\ \frac{h_{22}}{h_{12}} & \frac{H}{h_{12}} \end{matrix}$	$\begin{matrix} -\frac{G}{g_{12}} & -\frac{g_{22}}{g_{12}} \\ -\frac{g_{11}}{g_{12}} & \frac{1}{g_{12}} \end{matrix}$
[H]	$\begin{matrix} \frac{Z}{z_{22}} & \frac{z_{12}}{z_{22}} \\ -\frac{z_{21}}{z_{22}} & \frac{1}{z_{22}} \end{matrix}$	$\begin{matrix} \frac{1}{y_{11}} & -\frac{y_{12}}{y_{11}} \\ \frac{y_{21}}{y_{11}} & \frac{Y}{y_{11}} \end{matrix}$	$\begin{matrix} \frac{B}{D} & \frac{T}{D} \\ -\frac{1}{D} & \frac{C}{D} \end{matrix}$	$\begin{matrix} \frac{B'}{A'} & \frac{1}{A'} \\ -\frac{T'}{A'} & \frac{C'}{A'} \end{matrix}$	$\begin{matrix} h_{11} & h_{12} \\ h_{21} & h_{22} \end{matrix}$	$\begin{matrix} \frac{g_{22}}{G} & \frac{g_{12}}{G} \\ -\frac{g_{21}}{G} & \frac{g_{11}}{G} \end{matrix}$
[G]	$\begin{matrix} \frac{1}{z_{11}} & -\frac{z_{12}}{z_{11}} \\ \frac{z_{21}}{z_{11}} & \frac{Z}{z_{11}} \end{matrix}$	$\begin{matrix} \frac{Y}{y_{22}} & \frac{y_{12}}{y_{22}} \\ -\frac{y_{21}}{y_{22}} & \frac{1}{y_{22}} \end{matrix}$	$\begin{matrix} \frac{C}{A} & -\frac{T}{A} \\ \frac{1}{A} & \frac{B}{A} \end{matrix}$	$\begin{matrix} \frac{C'}{D'} & -\frac{1}{D'} \\ \frac{T'}{D'} & \frac{B'}{D'} \end{matrix}$	$\begin{matrix} \frac{h_{22}}{H} & -\frac{h_{12}}{H} \\ -\frac{h_{21}}{H} & \frac{h_{11}}{H} \end{matrix}$	$\begin{matrix} g_{11} & g_{12} \\ g_{21} & g_{22} \end{matrix}$

Note : Upper case italic implies "determinant" of the concerned matrix

For reciprocity we get

$$\frac{h_{12}}{h_{22}} = -\frac{h_{21}}{h_{22}}$$

or $$h_{12} = -h_{21} \qquad ....(10.27)$$

and for symmetry

$$\frac{\Delta_h}{h_{22}} = \frac{1}{h_{22}}$$

or $$h_{11}h_{22} - h_{12}h_{21} = 1 \qquad ....(10.28)$$

For g and A' B' C' D' parameters we can easily deduce the following relationships.

For passivity or reciprocity

$$g_{12} = -g_{21} \qquad .....(10.29)$$

$$A'D' - B'C' = 1 \qquad .....(10.30)$$

and for symmetry

$$g_{11}g_{22} - g_{12}g_{21} = 1 \qquad .....(10.31)$$

and $$A' = D' \qquad .....(10.32)$$

Example 10.6

Find the z and y parameters of the network in Fig. 10.9

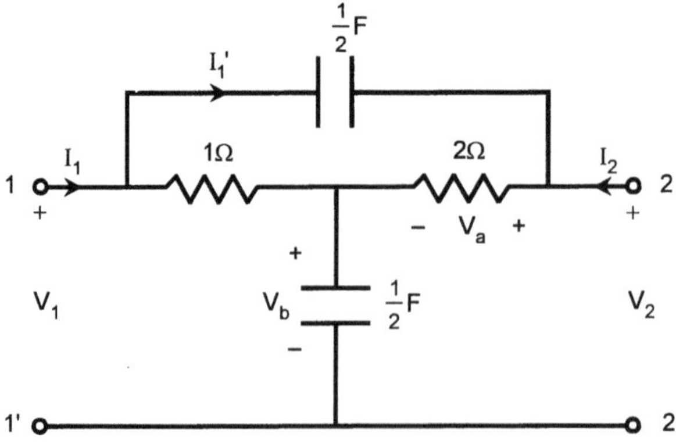

Fig. 10.9 Network for example 10.6

Solution :

Let us first calculate the z – parameters. By definition

$$Z_{11} = \left.\frac{V_1}{I_1}\right|_{I_2 = 0}$$

Since $I_2 = 0$, 2Ω resistor and capacitor between the terminals 1 and 2 are in series and the combination is in parallel with 1 Ω resistor. Z_{11} is the equivalent impedance at terminals 1, 1' and is given by

$$Z_{11} = \frac{2}{s} + \frac{1 \times \left(2 + \frac{2}{s}\right)}{1 + 2 + \frac{2}{s}} = \frac{2}{s} + \frac{2s + 2}{3s + 2} = \frac{2s^2 + 8s + 4}{s(3s + 2)}$$

Similarly $Z_{22} = \left.\frac{V_2}{I_2}\right|_{I_1 = 0}$ is the equivalent impedance at the terminals 2, 2' with the terminals 1, 1' open circuted. Hence,

$$Z_{22} = \frac{2}{s} + \frac{2\left(1 + \frac{2}{s}\right)}{3 + \frac{2}{s}} = \frac{2}{s} + \frac{2(s + 2)}{3s + 2} = \frac{2s^2 + 10s + 4}{s(3s + 2)}$$

Since the network has no dependent source, it is a reciprocal network. Hence

$$Z_{12} = Z_{21} = \left.\frac{V_2}{I_1}\right|_{I_2 = 0}$$

But $\qquad V_2 = V_b + V_a = I_1 \times \frac{2}{s} + I_1' 2$

and $\qquad I_1' = \frac{I_1 \times 1}{1 + 2 + \frac{2}{s}} = \frac{sI_1}{3s + 2}$

$\therefore \qquad V_2 = I_1\left[\frac{2}{s} + \frac{2s}{3s + 2}\right]$

or $\qquad Z_{21} = \frac{V_2}{I_1} = \frac{2s^2 + 6s + 4}{s(3s + 2)}$

The impedance parameter matrix [Z] is given by

$$[Z] = \begin{bmatrix} \dfrac{2s^2+s+4}{s(3s+2)} & \dfrac{2s^2+6s+4}{s(3s+2)} \\ \dfrac{2s^2+6s+4}{s(3s+2)} & \dfrac{s^2+10s+4}{s(3s+2)} \end{bmatrix}$$

The Y-parameter matrix is obtained by inverting the Z-parameter matrix [Z]. Thus

$$[Y] = [Z]^{-1} = \begin{bmatrix} \dfrac{s^2+5s+2}{2(s+3)} & -\dfrac{s^2+3s+2}{2(s+3)} \\ -\dfrac{s^2+3s+2}{2(s+3)} & \dfrac{s^2+4s+2}{2(s+3)} \end{bmatrix}$$

Example 10.7

Obtain the Z – parameters of the network shown in Fig. 10.10. Determine whether the network is (a) Reciprocal (b) Symmetric.

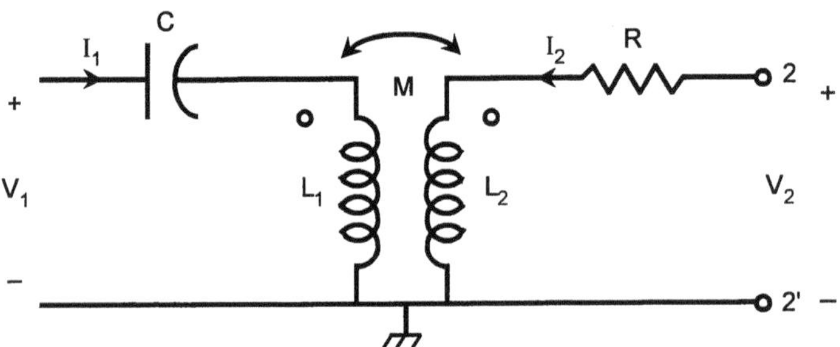

Fig. 10.10 Circuit for example 10.7

Solution :

Applying Kirchhoff's Voltage law, the following equations are obtained

$$V_1(s) = \frac{1}{Cs}I_1(s) + I_1(s)sL_1 + sMI_2(s)$$

$$= \left(\frac{1}{Cs} + sL_1\right)I_1(s) + sMI_1(s)$$

$$V_2(s) = RI_2(s) + sL_2I_2(s) + sMI_1(s)$$
$$= sMI_1(s) + (R + sL_2)I_2(s)$$

Putting the equations in the matrix form

$$\begin{bmatrix} V_1(s) \\ V_2(s) \end{bmatrix} = \begin{bmatrix} \left(\dfrac{1}{Cs} + sL_1\right) & sM \\ sM & R + sL_2 \end{bmatrix} \begin{bmatrix} I_1(s) \\ I_2(s) \end{bmatrix}$$

The z – parameter matrix is the coefficient matrix in the above equation. Since $z_{12} = z_{21} = sM$, the network is reciprocal. However since $z_{11} \neq z_{22}$, the network is not *symmetric.*

Example 10.8

Determine the y – parameters of the network shown. Hence determine the h – parameters.

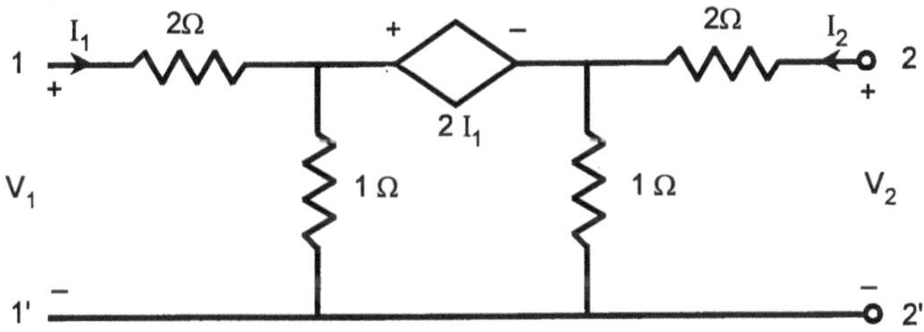

Fig. 10.11 Network for example 10.8

Solution :

y – parameters :

Short circuit the port 2, 2'. For the resulting network shown in Fig. 10.11(a); the loop equations are :

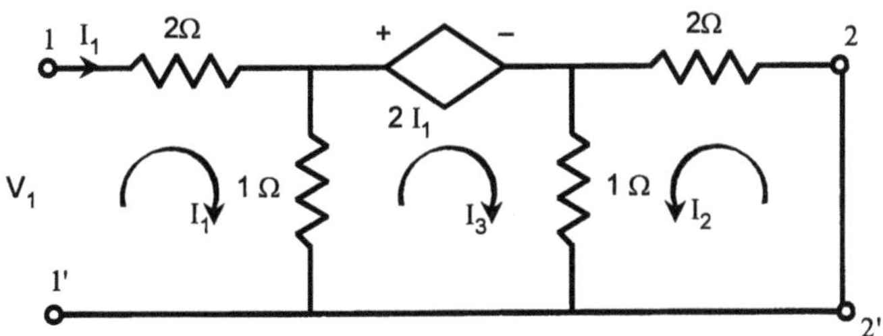

Fig. 10.11(a) Port (2, 2') is short circuited

$$3I_1 - I_3 = V_1 \quad .....(10.33)$$

$$2I_3 - I_1 + I_2 = -2I_1 \quad .....(10.34)$$

or $$I_1 + I_2 + 2I_3 = 0 \quad .....(10.35)$$

and $$3I_2 + I_3 = 0 \quad .....(10.36)$$

Solving eqs. (10.33), (10.34) and (10.36) for I_1 and I_2

$$I_1 = \frac{5V_1}{18}$$

and $$I_2 = \frac{V_1}{18}$$

Thus $$y_{11} = \frac{I_1}{V_1} = \frac{5}{18} \text{ S}$$

and $$y_{21} = \frac{I_2}{V_1} = \frac{1}{18} \text{ S}$$

Now shorting the port (1, 1')

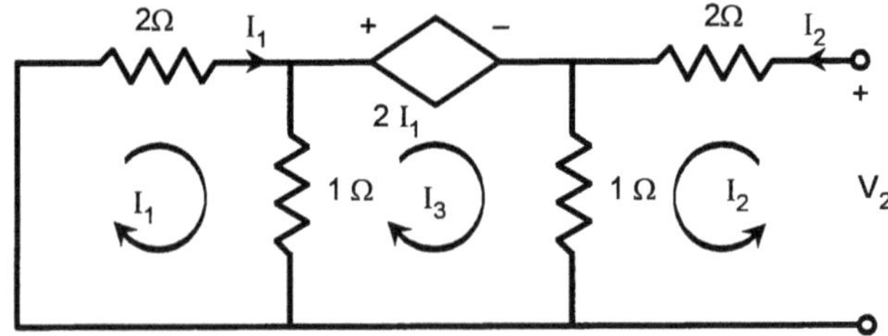

Fig. 10.11(b) Port (1, 1') shorted

Writing loop equations,

$$3I_2 + I_3 = V_2 \quad .....(10.37)$$

$$2I_3 + I_2 - I_1 = -2I_1 \quad .....(10.38)$$

or $$I_1 + I_2 + 2I_3 = 0 \quad .....(10.39)$$

$$3I_1 - I_3 = 0 \quad .....(10.40)$$

Solving eqs. (10.37), (10.38) and (10.40) for I_1 and I_2, we have

$$I_1 = \frac{-V_2}{18} \qquad I_2 = \frac{7}{18}V_2$$

Thus $$y_{12} = \frac{I_1}{V_2} = \frac{-1}{18} S$$

and $$y_{22} = \frac{I_2}{V_2} = \frac{7}{18} S$$

The y parameter matrix for the network is

$$\begin{bmatrix} \frac{5}{18} & \frac{-1}{18} \\ \frac{1}{18} & \frac{7}{18} \end{bmatrix}$$

The h – parameters can be obtained using the Table 10.2

$$h_{11} = \frac{1}{y_{11}} = \frac{18}{5}$$

$$h_{12} = \frac{-y_{12}}{y_{11}} = \frac{\frac{1}{18}}{\frac{5}{18}} = \frac{1}{5}$$

$$h_{21} = \frac{y_{21}}{y_{11}} = \frac{1}{5}$$

$$h_{22} = \frac{\Delta_y}{y_{11}} = \frac{y_{11}y_{22} - y_{12}y_{21}}{Y_{11}} = \frac{\frac{5}{18}\cdot\frac{7}{18} + \frac{1}{18}\cdot\frac{1}{18}}{\frac{5}{18}}$$

$$= \frac{36}{18} \times \frac{1}{5} = \frac{2}{5}$$

∴ The h – parameter matrix is

$$[h] = \begin{bmatrix} \frac{18}{5} & \frac{1}{5} \\ \frac{1}{5} & \frac{2}{5} \end{bmatrix}$$

10.9 INTERCONNECTION OF TWO PORT NETWORKS :

In the design of networks wherever possible, it is preferable to design and construct smaller blocks, establish their properties and then interconnect them together. These interconnections may be cascade, series, parallel or series parallel in nature. The parameters resulting after the combinations may be calculated where the parameters of each smaller block are known. The way in which these parameters can be calculated will now be discussed.

10.9.1 Cascade connection of networks :

Two, two – port networks A and B are said to be connected in cascade if the out put port of A becomes the input port of the network B. This connection may also be viewed as tandem connection. The connection is shown in Fig. 10.12. Our interest is to determine the parameters of the entire network if the parameters of A and B are given.

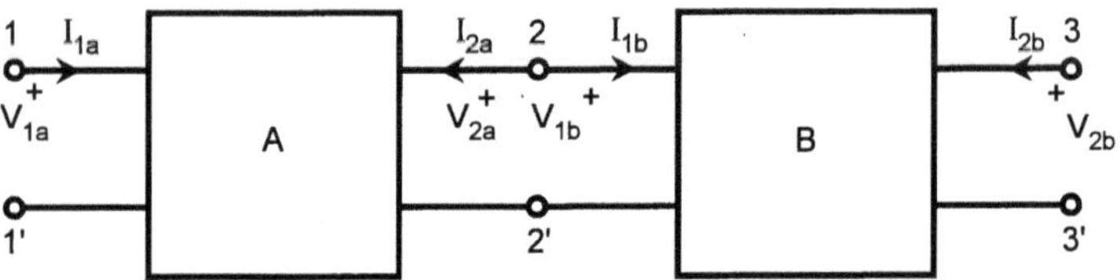

Fig. 10.12 Two port networks in cascade

(i) *If the neworks are connected in cascade, it will be conveninent to analyse the networks in terms of ABCD parameters.*

Let the network A be governed by the equations,

$$\begin{bmatrix} V_{1a} \\ I_{1a} \end{bmatrix} = \begin{bmatrix} A_a & B_a \\ C_a & D_a \end{bmatrix} \begin{bmatrix} V_{2a} \\ -I_{2a} \end{bmatrix} \quad .....(10.41)$$

Similarly the relationships for network B are

$$\begin{bmatrix} V_{1b} \\ I_{1b} \end{bmatrix} = \begin{bmatrix} A_b & B_b \\ C_b & D_b \end{bmatrix} \begin{bmatrix} V_{2b} \\ -I_{2b} \end{bmatrix} \quad .....(10.42)$$

where 1, 1' is the input port and 2, 2' is the output port of A. Similarly 2, 2' and 3, 3' are respectively the input and output ports of B. The two networks are connected at 2, 2'. This cascade connection requires the relation between the sets (V_{1a}, I_{1a}) and (V_{2b}, I_{2b}). Since the networks are connected at port 2, 2',

$$V_{2a} = V_{1b} \ ; -I_{2a} = I_{1b}$$

Hence vector

$$\begin{bmatrix} V_{2a} \\ -I_{2a} \end{bmatrix} = \begin{bmatrix} V_{1b} \\ I_{1b} \end{bmatrix} \quad .....(10.43)$$

Using this relation, equations (10.41) and (10.42) can be combined in the form

$$\begin{bmatrix} V_{1a} \\ I_{1a} \end{bmatrix} = \underbrace{\begin{bmatrix} A_a & B_a \\ C_a & D_a \end{bmatrix} \begin{bmatrix} A_b & B_b \\ C_b & D_b \end{bmatrix}} \begin{bmatrix} V_{2b} \\ -I_{2b} \end{bmatrix} \quad .....(10.44)$$

If A, B, C and D are the parameters of the entire network we have

$$\begin{bmatrix} V_{1a} \\ I_{1a} \end{bmatrix} = \begin{bmatrix} A & B \\ C & D \end{bmatrix} \begin{bmatrix} V_{2b} \\ -I_{2b} \end{bmatrix} \quad .....(10.45)$$

Comparing (10.44) and (10.45), the following relationship is obtained.

$$\begin{bmatrix} A & B \\ C & D \end{bmatrix} = \begin{bmatrix} A_a & B_a \\ C_a & D_a \end{bmatrix} \begin{bmatrix} A_b & B_b \\ C_b & D_b \end{bmatrix} \quad .....(10.46)$$

To generalize, when a number of networks are connected in cascade, the matrix of ABCD parameters of entire group is obtained by multiplying the matrices of the respective ABCD parameters in the order they are connected. Since the matrices are multiplied in a chain, these parameters are also known as chain parameters.

Example 10.6

Two networks A and B shown in Fig. 10.10 are connected in cascade by connecting the ports 2 and 3. What are the transmission parameters of the entire network ?

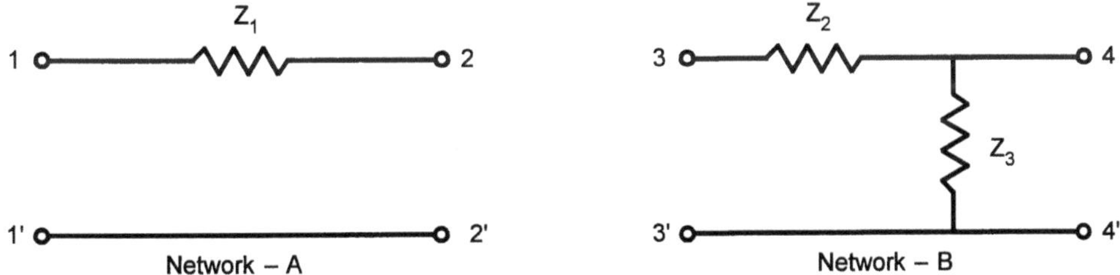

Fig. 10.13 Networks for example 10.6

Solution :

Transmission parameters of A :

The governing equations are

$$\begin{bmatrix} V_1 \\ I_1 \end{bmatrix} = \begin{bmatrix} A_a & B_a \\ C_a & D_a \end{bmatrix} \begin{bmatrix} V_2 \\ -I_2 \end{bmatrix}$$

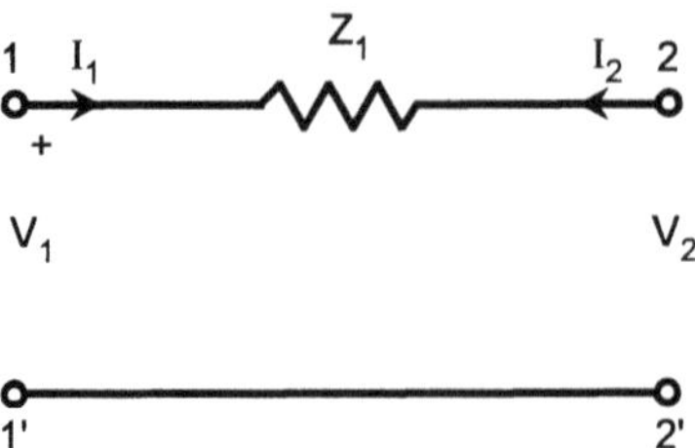

Fig. 10.13 (a) Network A

From Fig. 10.13(a),

$$I_1 = -I_2$$

$$V_1 = V_2 - I_2 Z_1$$

Combining the two equations,

$$\begin{bmatrix} V_1 \\ I_1 \end{bmatrix} = \begin{bmatrix} 1 & Z_1 \\ 0 & 1 \end{bmatrix} \begin{bmatrix} V_2 \\ -I_2 \end{bmatrix} \quad \text{.....(10.47)}$$

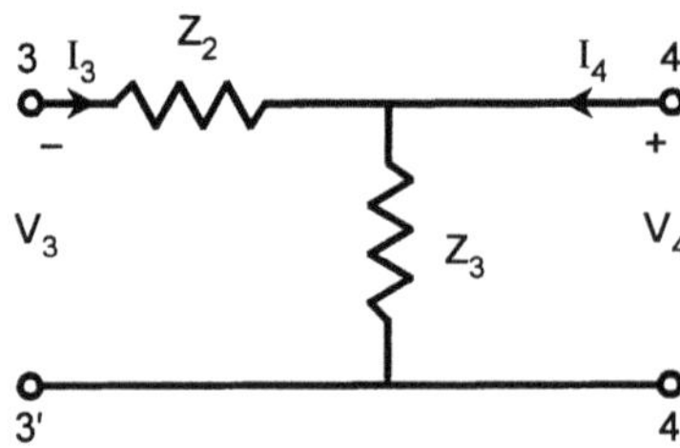

Fig. 10.13 (b) Network B

From Fig. 10.13(b),

$$V_3 = I_3Z_2 + V_4 \quad \text{.....(10.48)}$$

Also $$I_3 = \frac{V_4}{Z_3} - I_4 \quad \text{.....(10.49)}$$

Substituting for I_3 in eq. (10.48) we have

$$V_3 = \left(\frac{V_4}{Z_3} - I_4\right) Z_2 + V_4$$

$$= \left(1 + \frac{Z_2}{Z_3}\right) V_4 - Z_2 I_4 \quad \text{.....(10.50)}$$

Combining (10.49) and (10.50) we have

$$\begin{bmatrix} V_3 \\ I_3 \end{bmatrix} = \begin{bmatrix} \dfrac{Z_3 + Z_2}{Z_3} & Z_2 \\ \dfrac{1}{Z_3} & 1 \end{bmatrix} \begin{bmatrix} V_4 \\ -I_4 \end{bmatrix} \quad .....(10.51)$$

For the interconnected network, using the relation in eq. (10.46) we get the ABCD parameters of the overall network as,

$$\begin{bmatrix} A & B \\ C & D \end{bmatrix} = \begin{bmatrix} 1 & Z_1 \\ 0 & 1 \end{bmatrix} \begin{bmatrix} \dfrac{Z_2 + Z_3}{Z_3} & Z_2 \\ \dfrac{1}{Z_3} & 1 \end{bmatrix}$$

$$= \begin{bmatrix} \dfrac{Z_1 + Z_2 + Z_3}{Z_3} & Z_1 + Z_2 \\ \dfrac{1}{Z_3} & 1 \end{bmatrix} \quad .....(10.52)$$

Example 10.7

Two identical sections of the network shown in Fig. 10.14 are connected in cascade. Obtain the chain parameters of the resulting circuit.

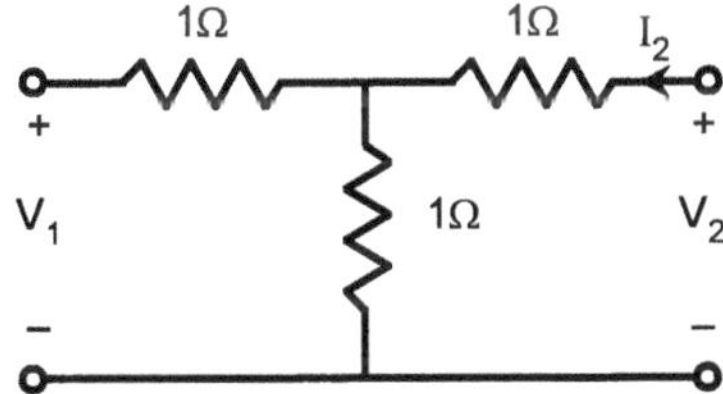

Fig. 10.14 Network of example 10.7

Solution :

The ABCD parameters of each section are calculated as follows :

Section 1.

$$A_1 = \left.\frac{V_1}{V_2}\right|_{I_2=0} = \frac{2}{1} = 2$$

$$C_1 = \left.\frac{I_1}{V_2}\right|_{I_2=0} = 1$$

Section 2.

$$D_1 = \left.\frac{I_1}{-I_2}\right|_{V_2=0} = 2$$

$$B_1 = \left.\frac{V_1}{-I_2}\right|_{V_2=0} = 3$$

Writing in the matrix form :

$$\begin{bmatrix} V_1 \\ I_1 \end{bmatrix} = \begin{bmatrix} 2 & 3 \\ 1 & 2 \end{bmatrix} \begin{bmatrix} V_2 \\ I_2 \end{bmatrix}$$

The second section being identical to the first, it also has the same parameters and hence the same coefficient matrix.

The ABCD parameter matrix for the entire network after interconnection is given by

$$\begin{bmatrix} A & B \\ C & D \end{bmatrix} = \begin{bmatrix} 2 & 3 \\ 1 & 2 \end{bmatrix} \begin{bmatrix} 2 & 3 \\ 1 & 2 \end{bmatrix} = \begin{bmatrix} 7 & 12 \\ 4 & 7 \end{bmatrix}$$

10.9.2 Series connection networks :

Let two networks A and B be identified by their input ports 1, 1' and 3, 3' respectively and output ports 2, 2' and 4, 4' respectively as shown in Fig. 10.15.

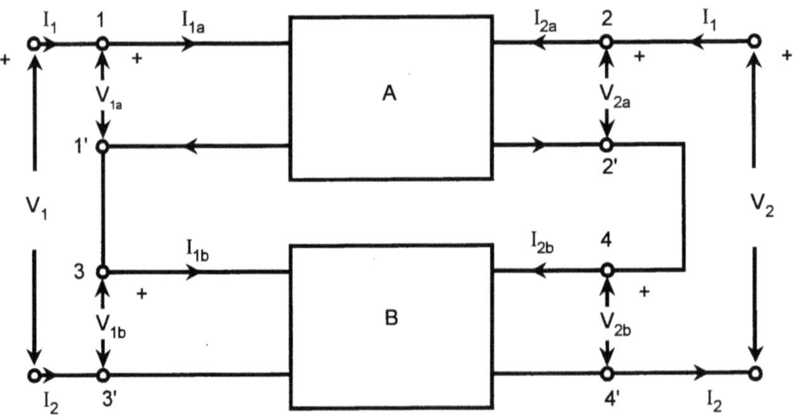

Fig. 10.15 Series connected networks

If the terminal 1' of input port 1, 1' is connected to the terminal 3 of B and the terminal 2' of output port of A is connected to terminal 4 of B resulting in the input port

1, 3' and an output port 2, 4' for the combination, the resulting network is said to be series connected. To determine the parameters of the entire network, the use of z – parameters will be convenient. The z – parameters for the network A are given by

$$\begin{bmatrix} V_{1a} \\ V_{2a} \end{bmatrix} = \begin{bmatrix} z_{11a} & z_{12a} \\ z_{21a} & z_{22a} \end{bmatrix} \begin{bmatrix} I_{1a} \\ I_{2a} \end{bmatrix} \qquad .....(10.53)$$

Similarly, for the network B

$$\begin{bmatrix} V_{1b} \\ V_{2b} \end{bmatrix} = \begin{bmatrix} z_{11b} & z_{12b} \\ z_{21b} & z_{22b} \end{bmatrix} \begin{bmatrix} I_{1b} \\ I_{2b} \end{bmatrix} \qquad .....(10.54)$$

Since $I_{1a} = I_{1b} = I_1$

$I_{2a} = I_{2b} = I_2$

we have

$$\begin{bmatrix} V_{1a} \\ V_{2a} \end{bmatrix} = \begin{bmatrix} z_{11a} & z_{12a} \\ z_{21a} & z_{22a} \end{bmatrix} \begin{bmatrix} I_1 \\ I_2 \end{bmatrix} \qquad .....(10.55)$$

$$\begin{bmatrix} V_{1b} \\ V_{2b} \end{bmatrix} = \begin{bmatrix} z_{11b} & z_{12b} \\ z_{21b} & z_{22b} \end{bmatrix} \begin{bmatrix} I_1 \\ I_2 \end{bmatrix} \qquad .....(10.56)$$

Adding eqs. (10.55) and (10.56) we have

$$\begin{bmatrix} V_{1a} + V_{1b} \\ V_{2a} + V_{2b} \end{bmatrix} = \begin{bmatrix} z_{11a} + z_{11b} & z_{12a} + z_{12b} \\ z_{21a} + z_{21b} & z_{22a} + z_{22b} \end{bmatrix} \begin{bmatrix} I_1 \\ I_2 \end{bmatrix} \qquad .....(10.57)$$

But $V_{1a} + V_{1b} = V_1$

and $V_{2a} + V_{2b} = V_2$

Hence

$$\begin{bmatrix} V_1 \\ V_2 \end{bmatrix} = \begin{bmatrix} z_{11a} + z_{11b} & z_{12a} + z_{12b} \\ z_{21a} + z_{21b} & z_{22a} + z_{22b} \end{bmatrix} \begin{bmatrix} I_1 \\ I_2 \end{bmatrix} \qquad .....(10.58)$$

Eq. (10.58) shows that the corresponding z – parameters for the series connected networks can be added together to get the overall z – parameters.

Thus for the overall network

$$\begin{bmatrix} z_{11} & z_{12} \\ z_{21} & z_{22} \end{bmatrix} = \begin{bmatrix} z_{11a} + z_{11b} & z_{12a} + z_{12b} \\ z_{21a} + z_{21b} & z_{22a} + z_{22b} \end{bmatrix} \begin{bmatrix} I_1 \\ I_2 \end{bmatrix} \qquad .....(10.59)$$

Note :

The series connection, however, assumes that the performance of any individual network does not get affected or altered due to the inter connection. This is confirmed in the following way.

First, the terminals 1' and 3 are connected together on the input side. The voltage between 2' and 4 is measured. If this voltage is zero, the terminals 2' and 4 are connected. This completes the series connection. If the voltage between 2' and 4 is not zero, these networks cannot be connected in series. However, the networks can still be connected in series as shown in Fig. 10.16 using isolation transformers.

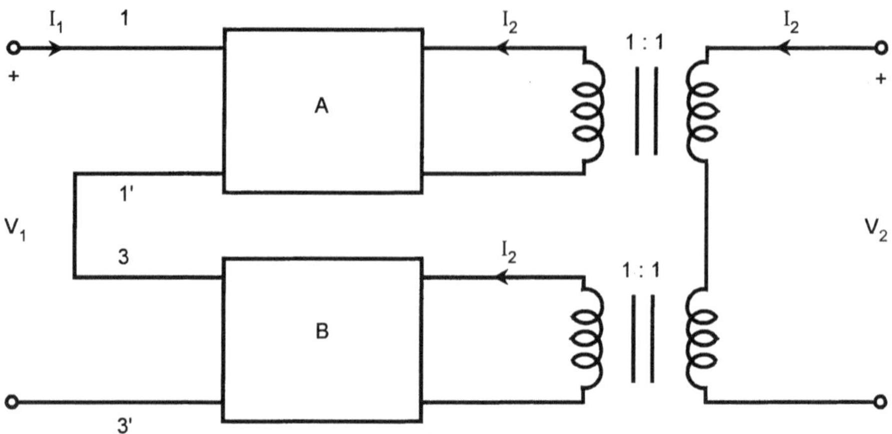

Fig. 10.16 Series connection using 1 : 1 isolation transformers

10.9.3 Parallel connection of networks :

Two networks A and B are said to be connected in parallel if the terminals of input port of A are connected to their corresponding input terminals of B and output port terminals of A are connected to the corresponding output terminals of B. This type of connection is shown in Fig. 10.17.

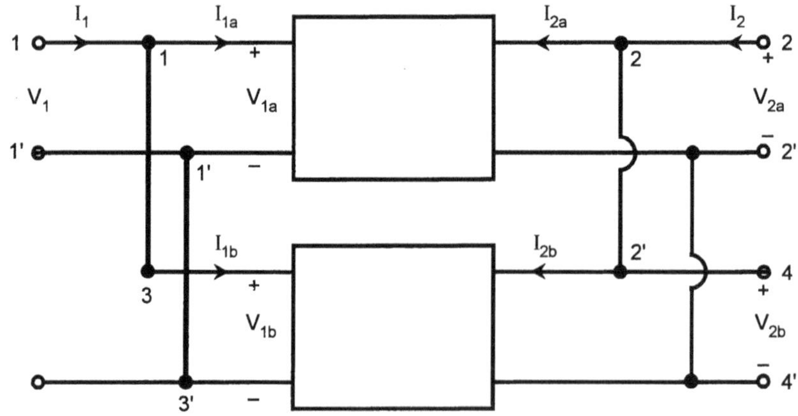

Fig. 10.17 Parallel connection of networks

For the two networks A and B we have

$$\begin{bmatrix} I_{1a} \\ I_{2a} \end{bmatrix} = \begin{bmatrix} y_{11a} & y_{12a} \\ y_{21a} & y_{22a} \end{bmatrix} \begin{bmatrix} V_{1a} \\ V_{2a} \end{bmatrix} \quad .....(10.60)$$

and

$$\begin{bmatrix} I_{1b} \\ I_{2b} \end{bmatrix} = \begin{bmatrix} y_{11b} & y_{12b} \\ y_{21b} & y_{22b} \end{bmatrix} \begin{bmatrix} V_{1b} \\ V_{2b} \end{bmatrix} \quad .....(10.61)$$

But $V_1 = V_{1a} = V_{1b}$

$V_2 = V_{2a} = V_{2b}$

and $I_1 = I_{1a} + I_{1b}$

$I_2 = I_{2a} + I_{2b}$

Using these equations and adding eqs. (10.60) and (10.61) we have

$$\begin{bmatrix} I_1 \\ I_2 \end{bmatrix} = \begin{bmatrix} y_{11a} + y_{11b} & y_{12a} + y_{12b} \\ y_{21a} + y_{21b} & y_{22a} + y_{22b} \end{bmatrix} \begin{bmatrix} V_1 \\ V_2 \end{bmatrix} \quad .....(10.62)$$

Thus for parallel connected networks the overall parameters are given by the addition of the individual y – parameters as shown in eq. (10.62).

As in the case of series connection, it is assumed that the performance of the individual networks is not affected as a consequence of the parallel connection. To assure that this condition is achieved, a test similar to the one conducted in the case of series connection is performed. The input ports are connected in parallel. The output ports are short circuited and the voltage between the output ports is measured as shown in Fig. 10.18. If the voltage V_r is zero, then the corresponding terminals at the output port can be connected in parallel. If V_r is not zero, then isolation transformers have to be used as shown in Fig. 10.19.

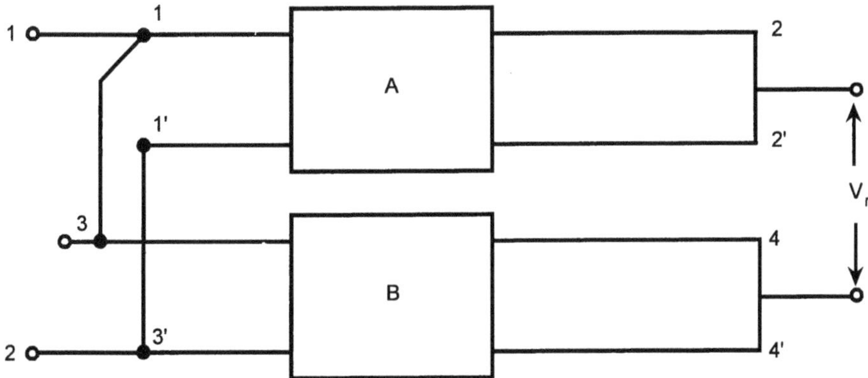

Fig. 10.18 Test for parallel connection

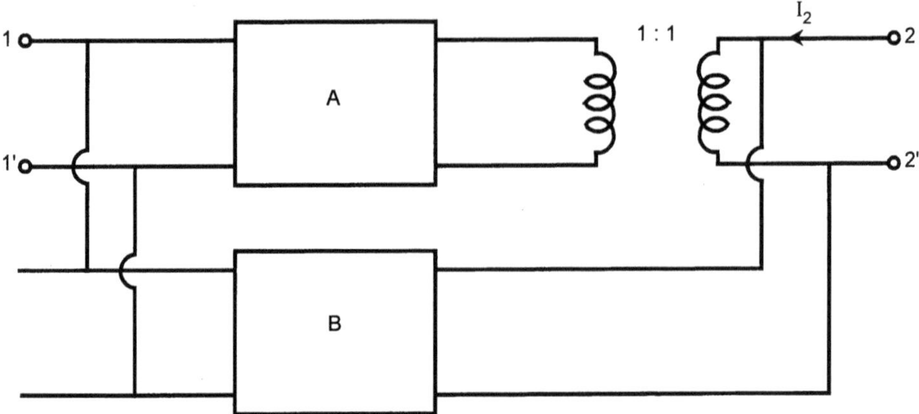

Fig. 10.19 Parallel connection using isolation transformers

Example 10.8

Two identical sections of the network shown in Fig. 10.20 are connected in parallel. Determine the y – parameters of the resulting network. Check your result by an alternate method.

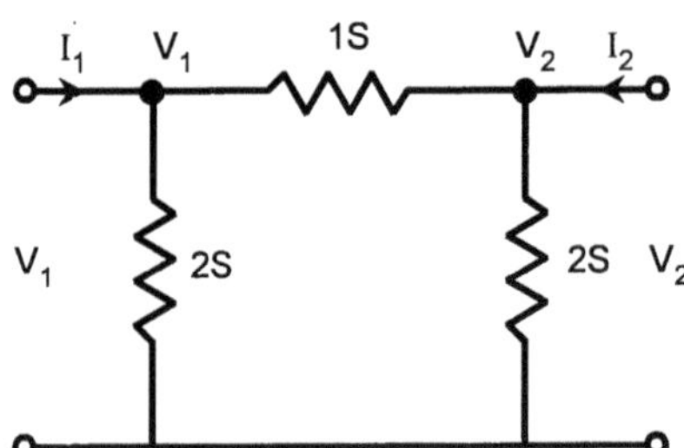

Fig.10.20 Network for example 10.8

Solution :

The y – parameters of each network :

$$I_1 = (V_1 - V_2)1 + 2V_1$$
$$= 3V_1 - V_2$$

Similarly $\quad I_2 = -V_1 + 3V_2$

From the above equations, the matrix of admittance parameters

$$[y] = \begin{bmatrix} 3 & -1 \\ -1 & 3 \end{bmatrix}$$

$\therefore$ The matrix of y – parameters of the combined network is given by

$$[y] = \begin{bmatrix} 3 & -1 \\ -1 & 3 \end{bmatrix} + \begin{bmatrix} 3 & -1 \\ -1 & 3 \end{bmatrix} = \begin{bmatrix} 6 & -2 \\ -2 & 6 \end{bmatrix}$$

Check :

Since the series arms of each of the π networks are in parallel, the equivalant admittance = 2 S. Also each of the vertical arms of the two π networks are in parallel,

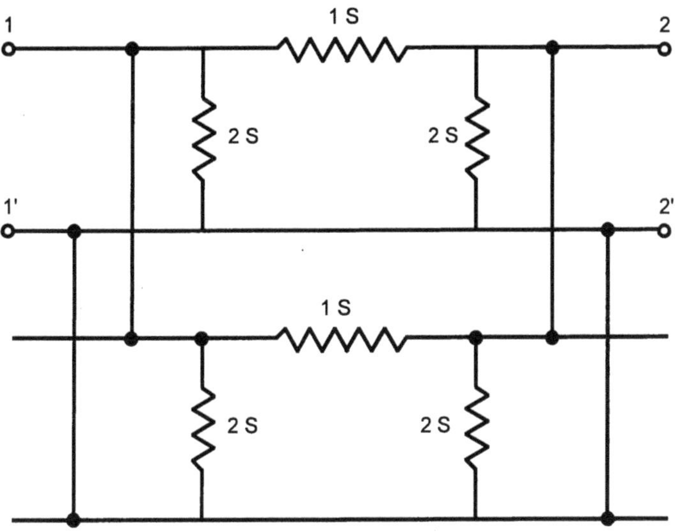

Fig. 10.21 Combined network

the admittance of each vertical arm in the equivalant π is = 4 S. The resulting network is shown in Fig. 10.22.

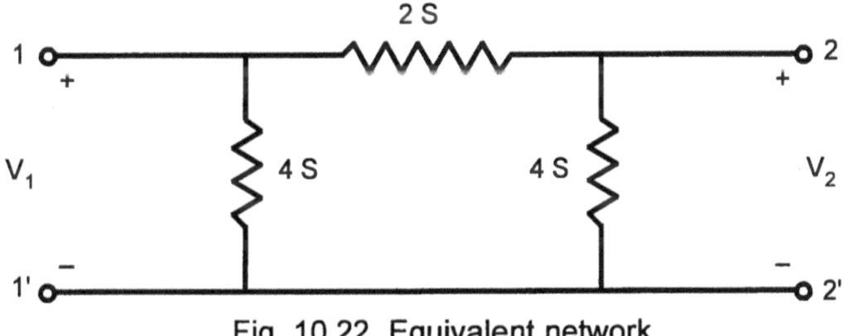

Fig. 10.22 Equivalent network

The admittance parameter matrix is

$$= \begin{bmatrix} 6 & -2 \\ -2 & 6 \end{bmatrix}$$

which is the same as the one obtained earlier.

10.9.4 Series parallel connections :

The series parallel connection is obtained when two networks A and B are connected together in series at the input port and in parallel at the output port as shown in Fig. 10.23.

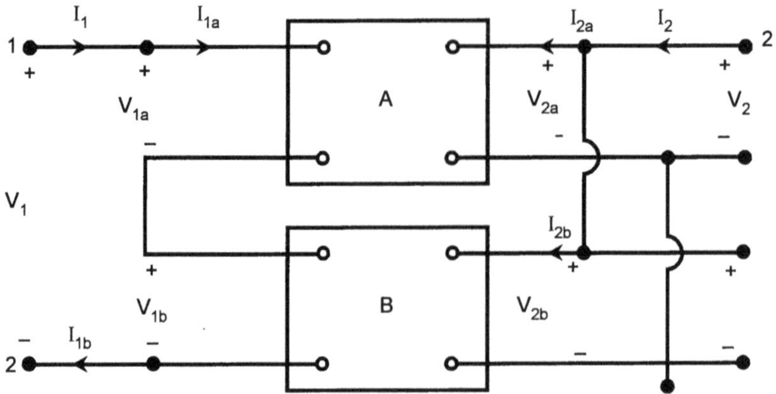

Fig. 10.23 Series parallel connection of A and B

To obtain the parameters of the resulting network the h - parameters of individual networks are used.

For the network A, the equations are

$$\left.\begin{aligned} V_{1a} &= h_{11a}I_{1a} + h_{12a}V_{2a} \\ I_{2a} &= h_{21a}I_{1a} + h_{22a}V_{2a} \end{aligned}\right\} \qquad \text{.....(10.63)}$$

For the network B, the equations are

$$\left.\begin{aligned} V_{1b} &= h_{11b}I_{1b} + h_{12b}V_{2b} \\ I_{2b} &= h_{21b}I_{1b} + h_{22b}V_{2b} \end{aligned}\right\} \qquad \text{.....(10.64)}$$

For the series connection at the input side, the relations to be satisfied are

$$\left.\begin{aligned} V_1 &= V_{1a} + V_{1b} \\ \text{and} \quad I_1 &= I_{1a} = I_{1b} \end{aligned}\right\} \qquad \text{.....(10.65)}$$

and the relations for a parallel connection at the output end are

$$\left.\begin{aligned} V_2 &= V_{2a} = V_{2b} \\ \text{and} \quad I_2 &= I_{2a} + I_{2b} \end{aligned}\right\} \qquad \text{.....(10.66)}$$

Using eqs. (10.65) and (10.66), the following relations for the overall network are obtained.

$$\begin{bmatrix} V_1 \\ I_2 \end{bmatrix} = \begin{bmatrix} h_{11} & h_{12} \\ h_{21} & h_{22} \end{bmatrix} \begin{bmatrix} I_1 \\ V_2 \end{bmatrix}$$

$$= \begin{bmatrix} h_{11a} + h_{11b} & h_{12a} + h_{12b} \\ h_{21a} + h_{21b} & h_{22a} + h_{22b} \end{bmatrix} \begin{bmatrix} I_1 \\ V_2 \end{bmatrix}$$

Thus for series parallel connection, the overall parameters are obtained by adding the corresponding h – parameters

10.9.5 Parallel – series connection :

This connection is the opposite of the series – parallel connection. The input side connections are in parallel while the connections at the output end are in series as shown in Fig. 10.24. It is natural, therefore, to use the inverse hybrid or g – parameters. The overall parameters can be shown to be the addition of corresponding g – parameters of the individual networks.

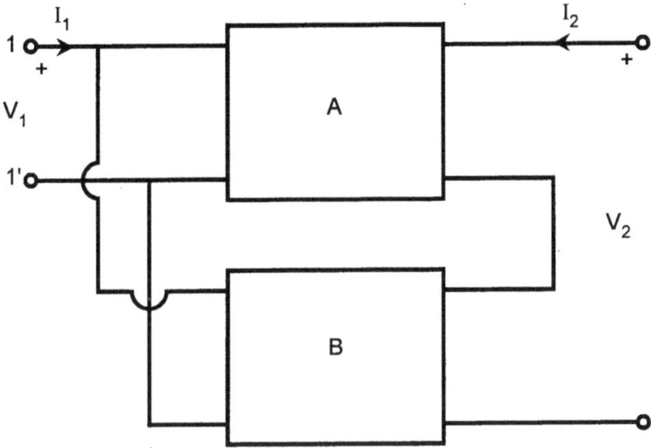

Fig. 10.24 Parallel – Series connection

Example 10.9

Find the Y and Z parameters of the network in Fig. 10.25(a)

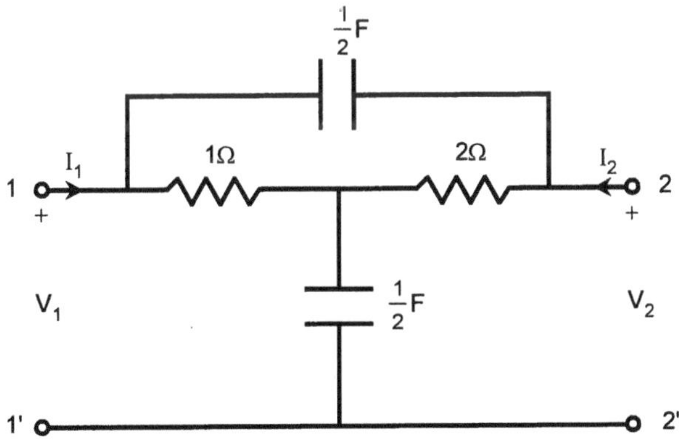

Fig. 10.25 (a) Network for example 10.9

Solution :

To find the y – parameters, we treat the network as two subnetworks connected in parallel as shown in Fig. 10.25(b) and (c).

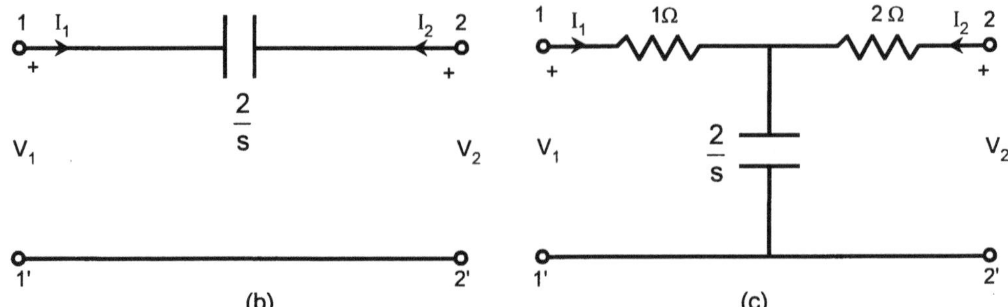

Fig. 10.25 (b) and (c) Two subnetworks when connected in parallel result in the network of Fig 10.25(a)

Since capacitors and resistors are involved, we will use Laplace transforms and evaluate the y - parameters of the two sub-networks.

Network (Fig. 10.25(b)) : Since the network is symmetrical and reciprocal we have

$$y_{11a}(s) = y_{22a}(s) = \frac{s}{2} \qquad y_{21a}(s) = y_{12a}(s) = -\frac{s}{2}$$

Network (Fig. 10.25(c)) : Short circuiting terminals 2, 2', we have

$$\left.\frac{I_1}{V_1}\right|_{V_2=0} = y_{11b} = \frac{1}{1+\dfrac{2.\dfrac{2}{s}}{2+\dfrac{2}{s}}} = \frac{1}{1+\dfrac{4}{2s+2}} = \frac{2(s+1)}{2s+6} = \frac{s+1}{s+3}$$

$$y_{21b} = \left.\frac{I_2}{V_1}\right|_{V_2=0}$$

$$I_2 = -\frac{I_1 \times \dfrac{2}{s}}{2+\dfrac{2}{s}} = -\frac{1}{s+1} I_1$$

$$= -\frac{1}{s+1}\left(\frac{s+1}{s+3}\right) V_1$$

$$y_{21b} = \frac{I_2}{V_1} = -\frac{1}{s+3}$$

$$\therefore \qquad y_{21b} = y_{12b} = -\frac{1}{s+3}$$

Short circuiting port 1, 1', we have

$$\left.\frac{I_2}{V_2}\right|_{V_1=0} = y_{22b} = \frac{1}{2+\cfrac{1.\frac{2}{s}}{1+\frac{2}{s}}} = \frac{1}{2+\cfrac{2}{s+2}}$$

$$= \frac{s+2}{2s+6}$$

$$= \frac{s+2}{2(s+3)}$$

Thus the y – parameters of the original network are

$$y_{11} = y_{11a} + y_{11b} = \frac{s}{2} + \frac{s+1}{s+3}$$

$$= \frac{s^2+5s+2}{2(s+3)}$$

$$y_{12} = y_{21} = y_{21a} + y_{21b} = -\frac{s}{2} - \frac{1}{s+3} = -\frac{s^2+3s+2}{2(s+3)}$$

$$y_{22} = y_{22a} + y_{22b} = \frac{s}{2} + \frac{s+2}{2(s+3)}$$

$$= \frac{s^2+4s+2}{2(s+3)}$$

Thus the matrix of y – parameters is,

$$\begin{bmatrix} \dfrac{s^2+5s+2}{2(s+3)} & -\dfrac{s^2+3s+2}{2(s+3)} \\ -\dfrac{s^2+3s+2}{2(s+3)} & \dfrac{s^2+4s+2}{2(s+3)} \end{bmatrix}$$

z – parameters :

The network can be treated as series connection of the two networks shown in Fig. 10.26(a) and (b).

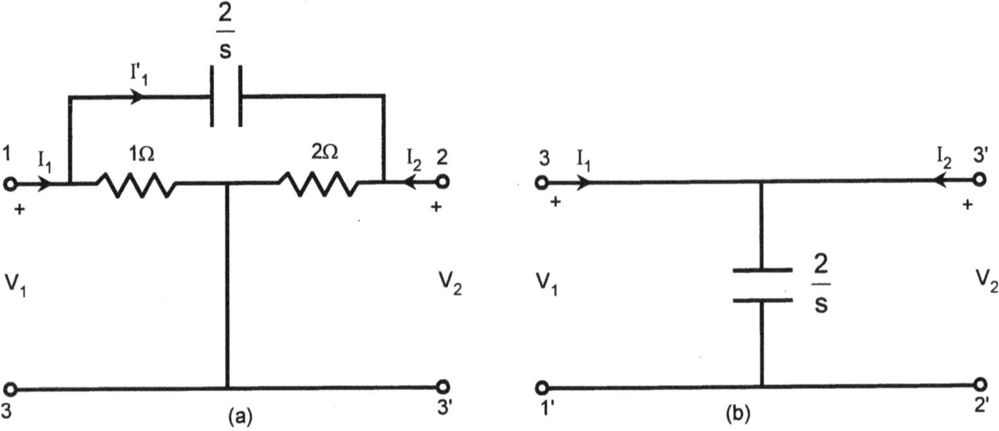

Fig. 10.26 (a) and (b) Subnetworks, when connected in series results in the network of Fig. 10.25(a)

It is obvious that, when the two networks are connected in series (the terminals marked 3 in the two networks are connected together on input side and terminals marked 3' and are connected together on output side), the original network of Fig. 10.26(b) results. Calculating the z – parameters of the individual networks and adding the corresponding parameters we can obtain the overall z – parameters.

Network (a) :

Open circuiting the port 2, 3', we have

$$z_{11d}(s) = \frac{1\left(2+\frac{2}{s}\right)}{1+2+\frac{2}{s}} = \frac{2(s+1)}{3s+2}$$

In Fig. 10.26(a),

$$I'_1 = \frac{I_1 \times 1}{1+2+\frac{2}{s}} = \frac{I_1 \cdot s}{3s+2}$$

and $$V_2 = 2\,.\,I'_1 = \frac{2I_1 s}{3s+2}$$

$$\therefore \quad z_{21d}(s) = \frac{V_2}{I_1} = \frac{2s}{3s+2} = z_{12d}$$

Similarly open circuiting port (1, 3)

$$z_{22d}(s) = \frac{2\left(1+\frac{2}{s}\right)}{3+\frac{2}{s}} = \frac{2(s+2)}{3s+2}$$

Network (b) :

Open circuit port (3',2')

$$z_{11e}(s) = \frac{2}{s}$$

$$z_{21e}(s) = \frac{2}{s}$$

Since the network is symmetrical and reciprocal

$$z_{22e}(s) = z_{11e}(s) = \frac{2}{s}$$

$$z_{21e}(s) = z_{12e}(s) = \frac{2}{s}$$

The overall parameters are

$$z_{11} = z_{11d} + z_{11e} = \frac{2(s+1)}{3s+2} + \frac{2}{s}$$

$$= \frac{2s^2+2s+6s+4}{s(3s+2)} = \frac{2s^2+8s+4}{s(3s+2)}$$

$$z_{21} = z_{12} = z_{21d} + z_{21e} = \frac{2s}{3s+2} + \frac{2}{s}$$

$$= \frac{2s^2+6s+4}{s(3s+2)}$$

$$z_{22} = z_{22_d} + z_{22_e} = \frac{2(s+2)}{3s+2} + \frac{2}{s}$$

$$= \frac{2s^2+4s+6s+4}{s(3s+2)} = \frac{2s^2+10s+4}{s(3s+2)}$$

Thus the z – parameter matrix is

$$\begin{bmatrix} \frac{2s^2+8s+4}{s(3s+2)} & \frac{2s^2+6s+4}{s(3s+2)} \\ \frac{2s^2+6s+4}{s(3s+2)} & \frac{2s^2+10s+4}{s(3s+2)} \end{bmatrix}$$

Example 10.10

Obtain the z – parameters of the network shown in Fig. 10.27. Determine whether the network is (a) Reciprocal (b) Symmetric.

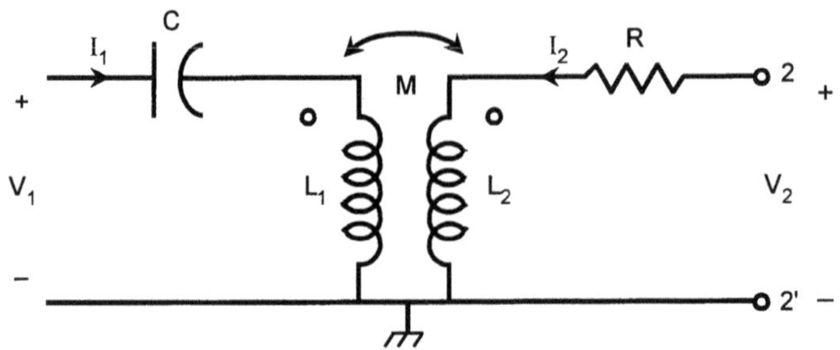

Fig. 10.27 Circuit for example 10.10

Solution :

Applying Kirchhoff's Voltage law, the following equations are obtained

$$V_1(s) = \frac{1}{Cs} I_1(s) + I_1(s)sL_1 + sMI_2(s)$$

$$= \left(\frac{1}{Cs} + sL_1\right) I_1(s) + sMI_1(s)$$

$$V_2(s) = RI_2(s) + sL_2I_2(s) + sMI_1(s)$$

$$= sMI_1(s) + (R + sL_2)I_2(s)$$

Putting the equations in the matrix form

$$\begin{bmatrix} V_1(s) \\ V_2(s) \end{bmatrix} = \begin{bmatrix} \left(\frac{1}{Cs} + sL_1\right) & sM \\ sM & R + sL_2 \end{bmatrix} \begin{bmatrix} I_1(s) \\ I_2(s) \end{bmatrix}$$

The z – parameter matrix is the coefficient matrix in the above equation. Since $z_{12} = z_{21} = sM$, the network is reciprocal. However since $z_{11} \neq z_{22}$, the network is not *symmetric.*

Example 10.11

Determine the y – parameters of the network shown. Hence determine the h – parameters.

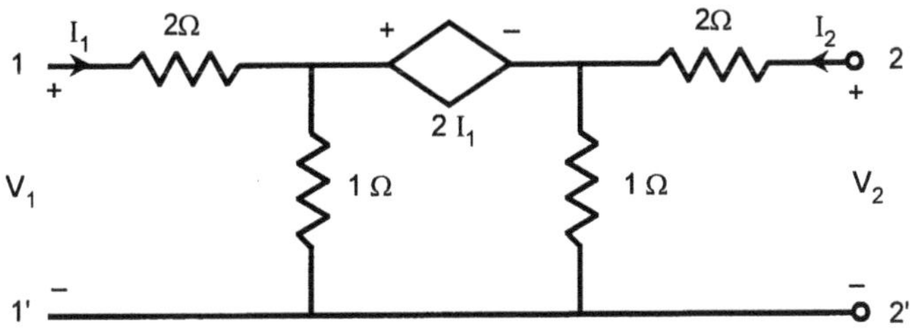

Fig. 10.28 Network for example 10.11

Solution :

y – parameters :

Short circuit the port 2, 2'. For the resulting network shown in Fig. 10.28(a); the loop equations are :

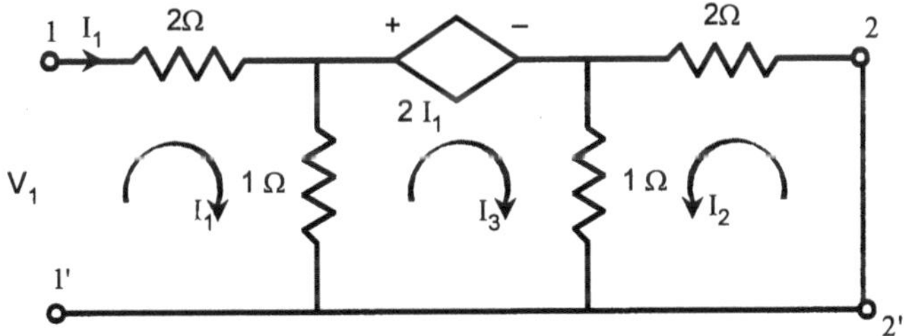

Fig. 10.28 (a) Port (2, 2') is short circuited

$$3I_1 - I_3 = V_1 \quad .....(10.67)$$

$$2I_3 - I_1 + I_2 = -2I_1 \quad .....(10.68)$$

Simplifying

$$I_1 + I_2 + 2I_3 = 0 \quad .....(10.69)$$

$$3I_2 + I_3 = 0 \quad .....(10.70)$$

Solving eqs. (10.67), (10.69) and (10.70) for I_1 and I_2

$$I_1 = \frac{5V_1}{18}$$

and $$I_2 = \frac{V_1}{18}$$

Thus $$y_{11} = \frac{I_1}{V_1} = \frac{5}{18}\ S$$

and $$y_{21} = \frac{I_2}{V_1} = \frac{1}{18}\ S$$

Now shorting the port (1, 1')

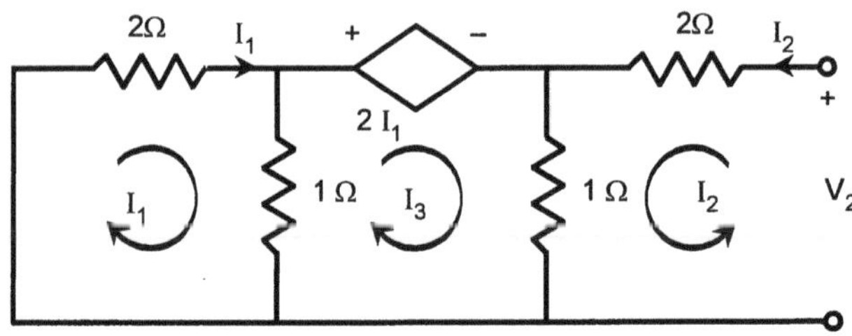

Fig. 10.28 (b) Port (1, 1') shorted

Writing loop equations,

$$3I_2 + I_3 = V_2 \qquad \text{.....(10.71)}$$

$$2I_3 + I_2 - I_1 = -2I_1 \qquad \text{.....(10.72)}$$

Simplfying

$$I_1 + I_2 + 2I_3 = 0 \qquad \text{.....(10.73)}$$

$$3I_1 - I_3 = 0 \qquad \text{.....(10.74)}$$

Solving eqs. (10.71), (10.73) and (10.74) for I_1 and I_2, we have

$$I_1 = \frac{-V_2}{18} \qquad I_2 = \frac{7}{18}V_2$$

Thus $$y_{12} = \frac{I_1}{V_2} = \frac{-1}{18}S$$

and $$y_{22} = \frac{I_2}{V_2} = \frac{7}{18}\ S$$

The y parameter matrix for the network is

$$\begin{bmatrix} \frac{5}{18} & \frac{-1}{18} \\ \frac{1}{18} & \frac{7}{18} \end{bmatrix}$$

The h – parameters can be obtained using the Table 10.2

$$h_{11} = \frac{1}{y_{11}} = \frac{18}{5}$$

$$h_{12} = \frac{-y_{12}}{y_{11}} = \frac{\frac{1}{18}}{\frac{5}{18}} = \frac{1}{5}$$

$$h_{21} = \frac{y_{21}}{y_{11}} = \frac{1}{5}$$

$$h_{22} = \frac{\Delta_y}{y_{11}} = \frac{y_{11}y_{22} - y_{12}y_{21}}{Y_{11}}$$

$$= \frac{\frac{5}{18} \cdot \frac{7}{18} + \frac{1}{18} \cdot \frac{1}{18}}{\frac{5}{18}}$$

$$= \frac{36}{18} \times \frac{1}{5}$$

$$= \frac{2}{5}$$

∴ The h – parameter matrix is

$$[h] = \begin{bmatrix} \frac{18}{5} & \frac{1}{5} \\ \frac{1}{5} & \frac{2}{5} \end{bmatrix}$$

Problems

10.1 Find the z and y parameters if they exist for the network shown in Fig. P. 10.1.

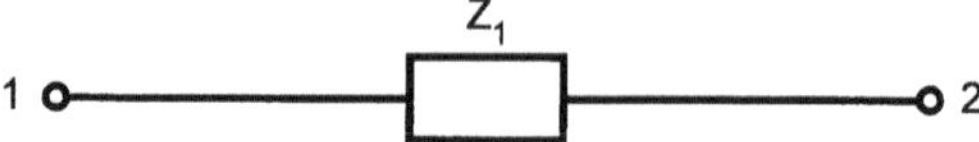

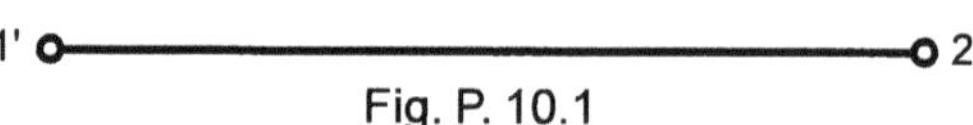

Fig. P. 10.1

10.2 Obtain the z and y parameters of the 2 port network shown in Fig. P.10.2.

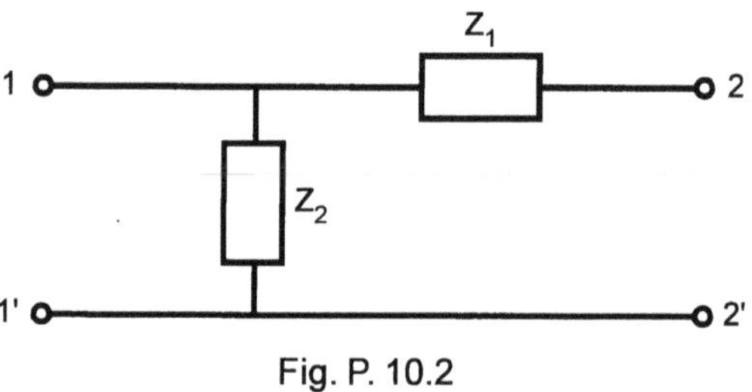

Fig. P. 10.2

10.3 For the resistive network of Fig. P. 10.3, obtain the y and z parameters.

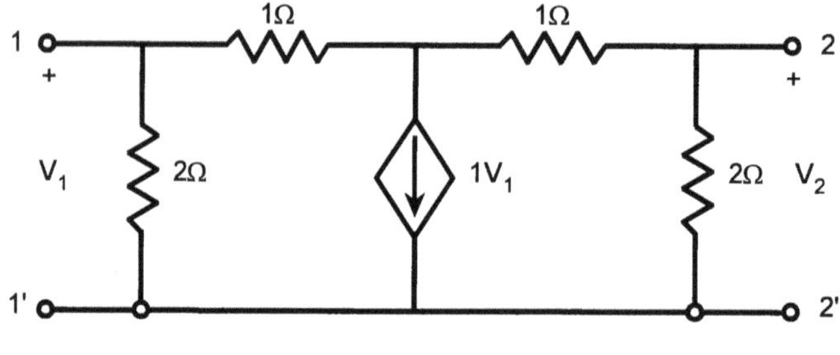

Fig. P. 10.3

10.4 A given two port network has the following open circuits impedance matrix.

$$\begin{bmatrix} z_{11} & z_{12} \\ z_{21} & z_{22} \end{bmatrix} = \begin{bmatrix} -j120 & -j160 \\ -j160 & -j80 \end{bmatrix}$$

What are the short circuit admittance parameters ?

10.5 A network has two input terminals a, b and two output terminals c, d. The input impedance with c, d open circuited is 400 ohms and with c, d short circuited is 250 ohms. The impedance across c, d with a, b open is 200 ohms. Determine its equivalent T – network parameters.

10.6 Obtain the ABCD parameters for the network of Fig. P. 10.4.

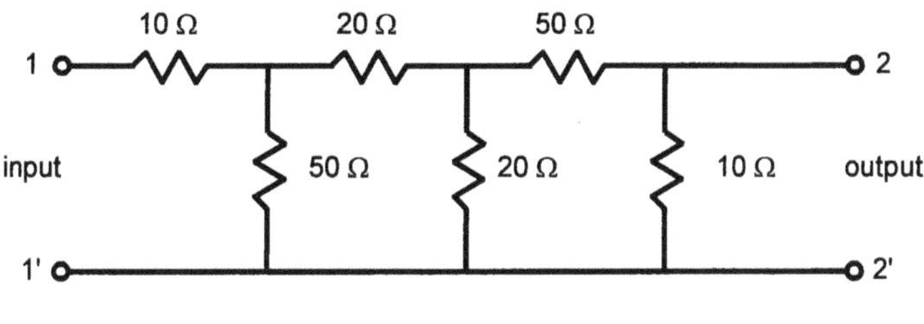

Fig. P. 10.4

10.7 Obtain the short circuit admittance parameters of the network of Fig. P. 10.5 and thereby obtain the A, B, C, D parameters.

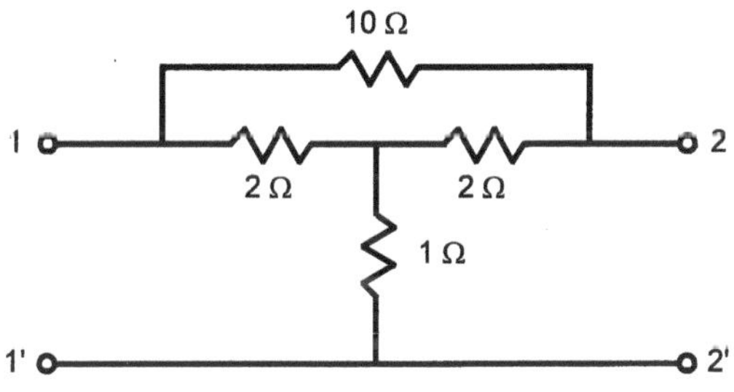

Fig. P. 10.5

10.8 Calculate the values of the A, B, C, D parameters for the network of Fig. P. 10.6. Hence determine the short circuit admittance parameters?

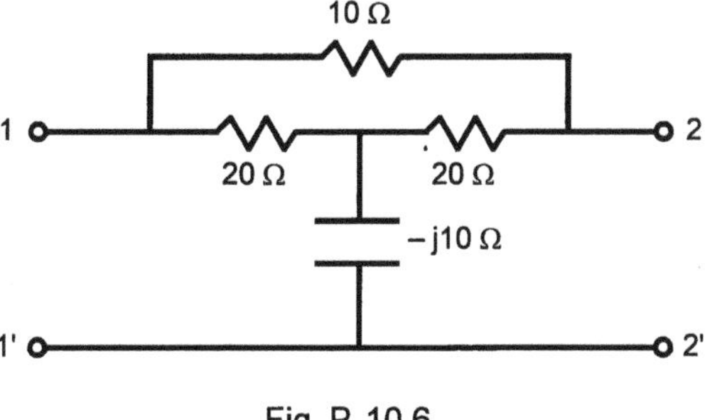

Fig. P. 10.6

10.9 Obtain the values of y_a, y_b and y_c in the passive network of Fig. P.10.7 for which the measurements made were as follows :

(i) With pair 2 – 2' shorted, a voltage of 10 $\angle 0^0$ volts applied at the terminals 1 – 1', resulted in the current $I_1 = 2.5 \angle 0^0$ amps and $I_2 = -0.5 \angle 0^0$ amps.

(ii) With the terminal pair 1 – 1' shorted and the same voltage applied at 2 – 2' resulted in $I_2 = 1.5 \angle 0^0$.

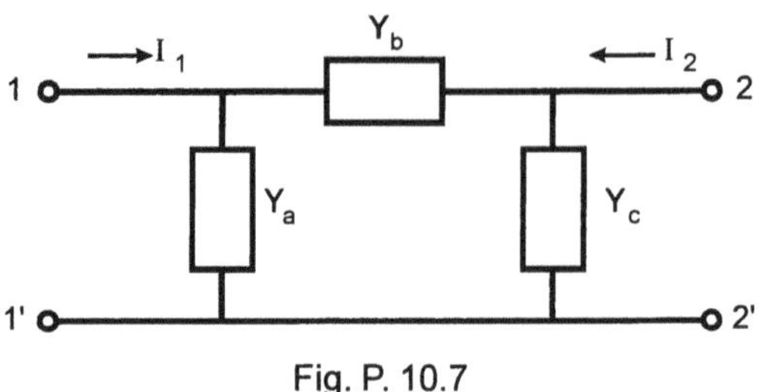

Fig. P. 10.7

10.10 The network of Fig. P. 10.8, contains a voltage controlled source and a current controlled source as shown. With the elements values as specified, calculate the z and y parameters.

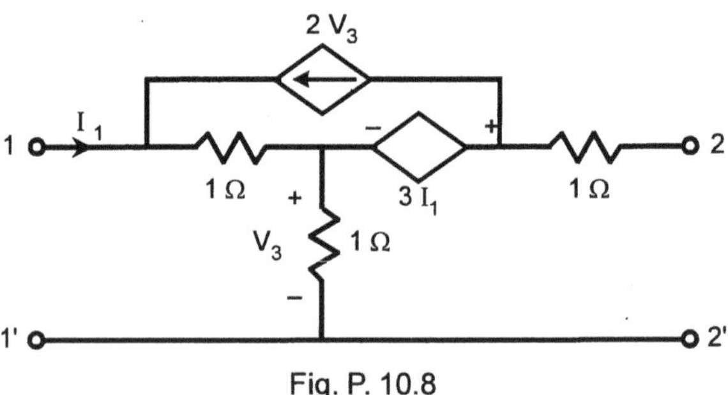

Fig. P. 10.8

10.11 Give a possible z, y and h parameter equivalent circuits for the two port networks shown in Fig. P. 10.9.

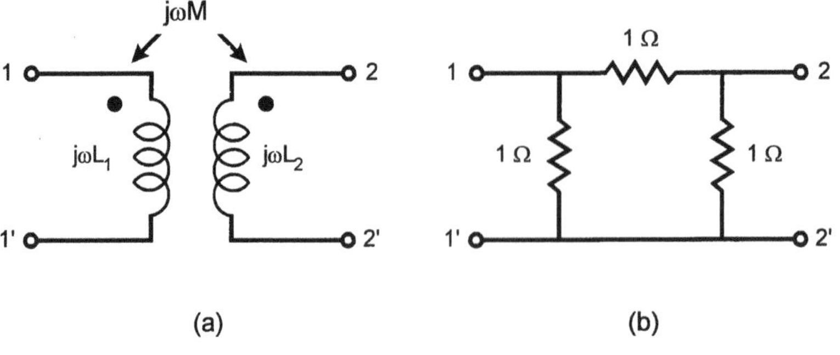

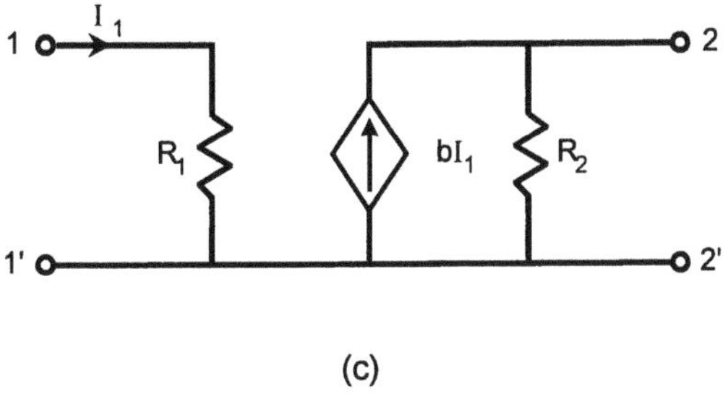

(c)

Fig. P. 10.9

10.12 The network of Fig. P. 10.10 represents a transistor operating over a given range of frequencies. Determine the h and g parameters.

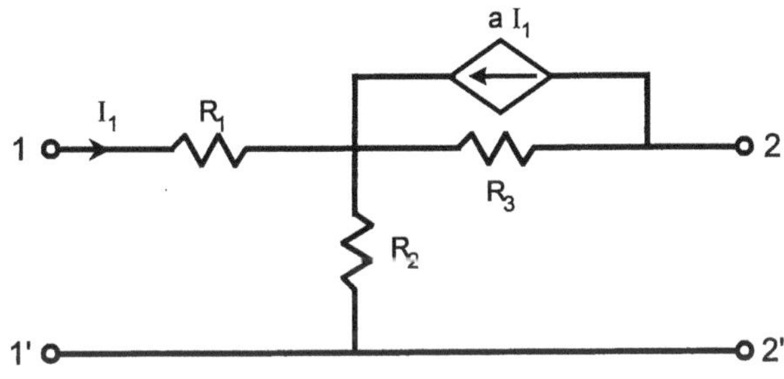

Fig. P. 10.10

10.13 In the circuit of Fig. P.10.11, it is given that $V_s = 2$ V, $R_1 = 2\ \Omega$; $y_{11} = y_{22} = 2$ S and $y_{12} = y_{21} = -1$ S. Find the value of R_L for maximum power to be transferred to R_L.

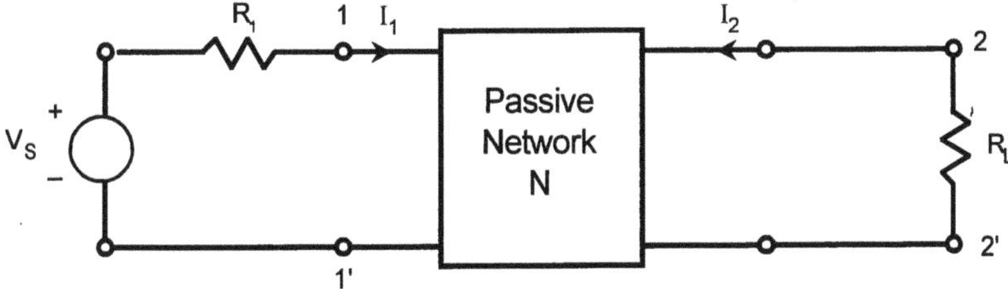

Fig. P. 10.11

10.14 For the network of Fig. P. 10.12, obtain the z – parameters. Therefrom, determine the h – parameters.

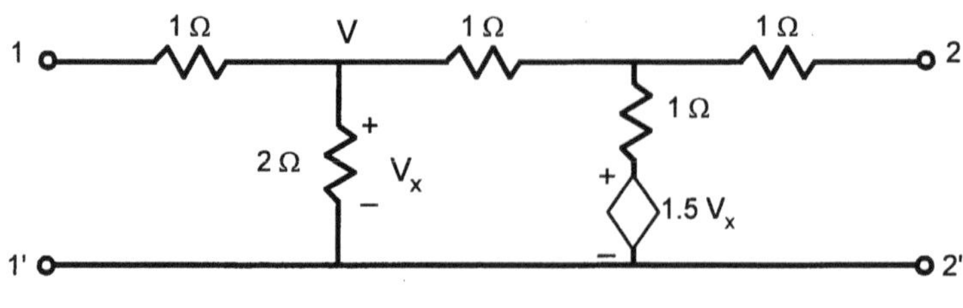

Fig. P. 10.12

10.15 Obtain the h – parameters of the network of Fig. P.10.13.

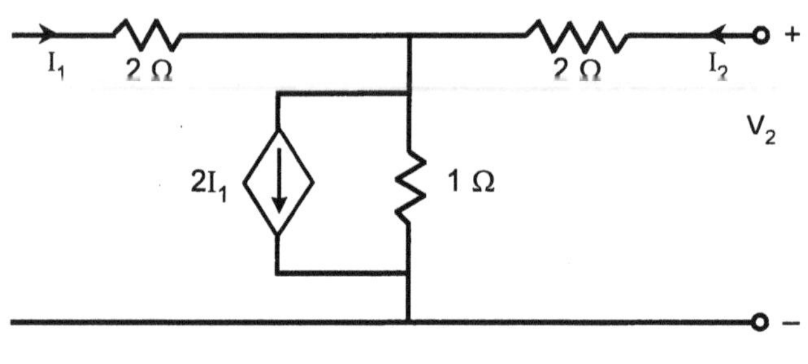

Fig. P. 10.13

10.16 Find the impedance parameters of the network of Fig. P. 10.14.

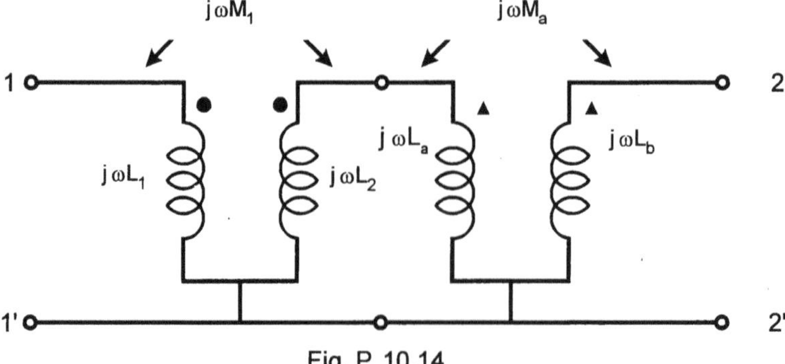

Fig. P. 10.14

10.17 The two currents of a 2 port network are

$$I_1 = 2V_1 - V_2$$
$$I_2 = -V_1 + 4V_2$$

What is the equivalent π – network ?

10.18 For the two port network of Fig. P. 10.15, find the h and g parameters.

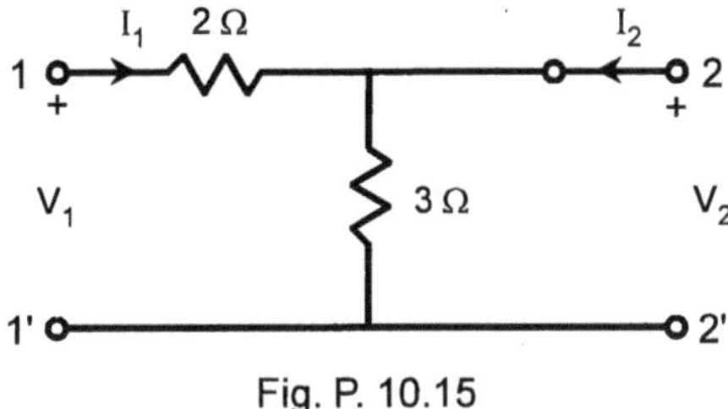

Fig. P. 10.15

10.19 Determine the impedance parameters of the symmetrical lattice shown in Fig. P.10.16.

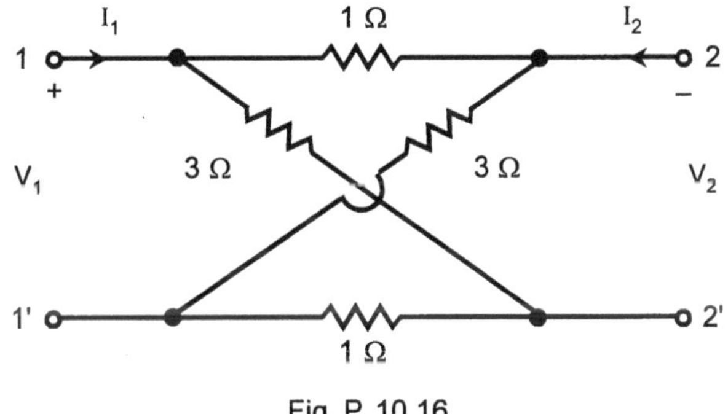

Fig. P. 10.16

10.20 Determine the y – parameters for the network shown in Fig. P.10.17 in which a resistive T and a resistive π network are connected in parallel.

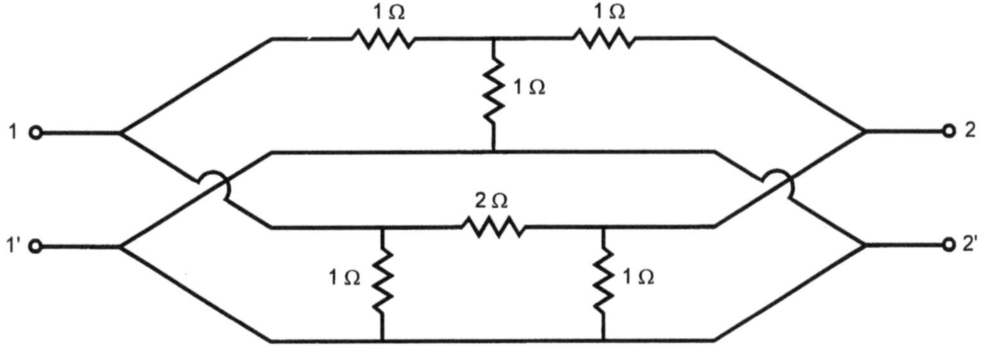

Fig. P. 10.17

10.21 The network N in Fig. P. 10.18 is terminated at port 2 with an impedance $Z_L = \frac{1}{Y_L}$. Show that the voltage ratio transfer function $G_{12} = \frac{V_2}{V_1}$ is given by

$$G_{12} = \frac{-y_{21}}{y_{22} + Y_L}.$$

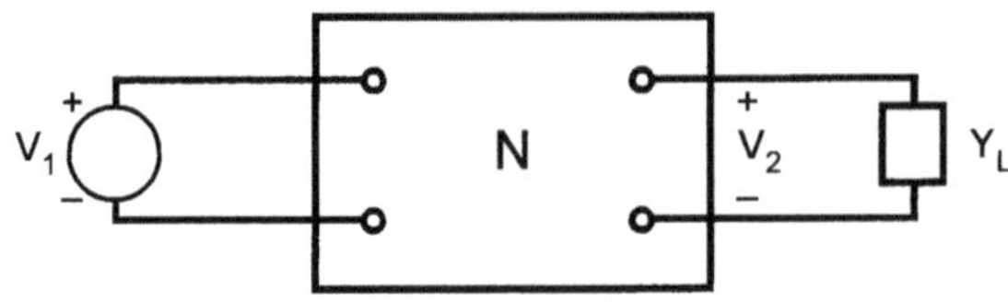

Fig. P. 10.18

10.22 For the two port network shown in Fig. P.10.19, obtain the y – parameters.

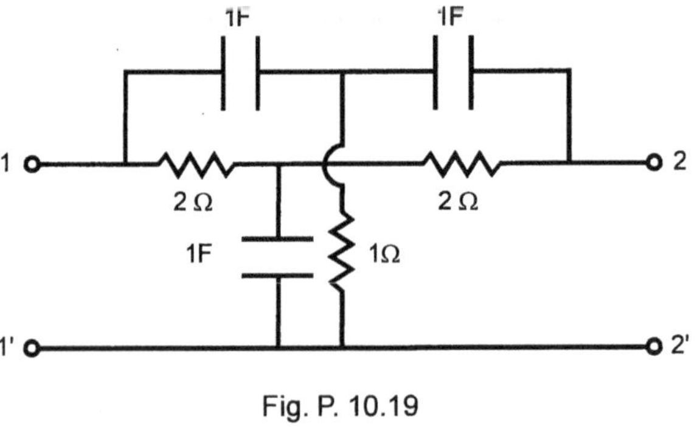

Fig. P. 10.19

10.23 Obtain y – parameters of the network in Fig. P. 10.20, treating the given network as two parallel connected two port nerworks.

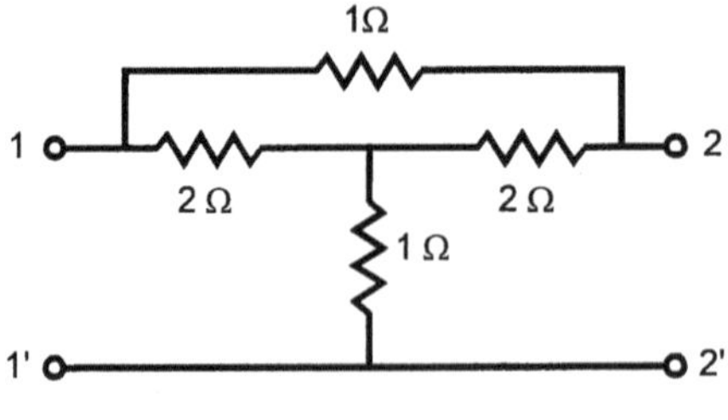

Fig. P. 10.20

10.24 Obtain the z – parameters of the network in Fig. P. 10.21, treating the given network as two series connected networks.

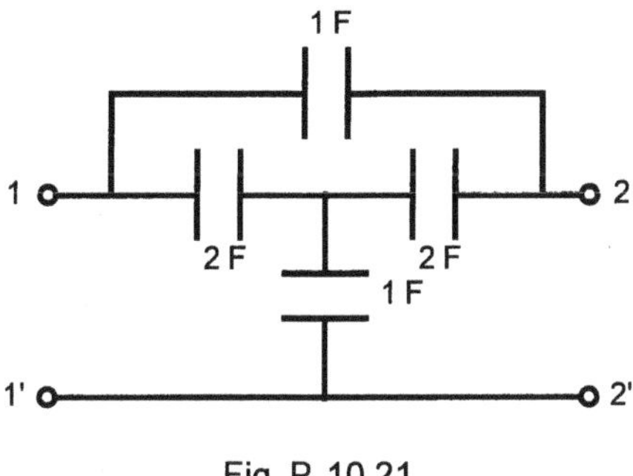

Fig. P. 10.21

10.25 Two Two – port networks have z – parameters

$$[z]_a = \begin{bmatrix} z_{11a} & z_{12a} \\ z_{21a} & z_{22a} \end{bmatrix} \text{ and } [z]_b = \begin{bmatrix} z_{11b} & z_{12b} \\ z_{21b} & z_{22b} \end{bmatrix}$$

Show that the open circuit transfer impedance of the network obtained by cascading the two networks is given by $z_{12} = \dfrac{z_{12a} + z_{12b}}{z_{11b} + z_{22a}}$

11

NETWORK FILTERS AND ATTENUATORS

11.1 INTRODUCTION :

In chapter 3 we have seen that a series or parallel RLC network can be made frequency selective. They are called resonant circuits which select or reject a narrow band of frequencies. Purely reactive networks can be designed to freely pass a designed band of frequencies. Such networks are known as '*Filters*'. These reactive networks were first studied by G. A. Campbell and O. J. Zobel of the Bell Telephone laboratories.

Filter networks are very commonly used in power supplies, radio receivers, telephone lines etc. Ideal filters would pass a band of frequencies without any reduction in magnitude and suppress entirely all other frequencies. Filter networks can be designed to separate different frequency bands available in an alternating current signal. The set of frequencies, for which there is no change in magnitude of the signal or there is zero attenuation of the signal, is called pass band and the set of frequencies with infinite attenuation is called stop band or attenuation band. The frequencies which separate these two bands are defined as cut off frequencies.

11.2 SYMMETRICAL NETWORKS :

Filters are usually constructed by using T or π networks which are symmetrical. T and π sections can be used to construct a ladder network. Symmetrical T and π sections can themselves be considered as made up of unsymmetrical L sections connected in one way for T section and opposite way for a π section as shown in Fig. 11.1. These L sections are known as half sections.

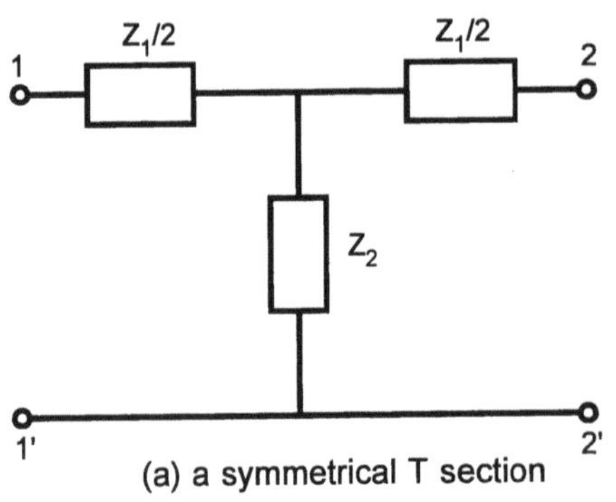

(a) a symmetrical T section

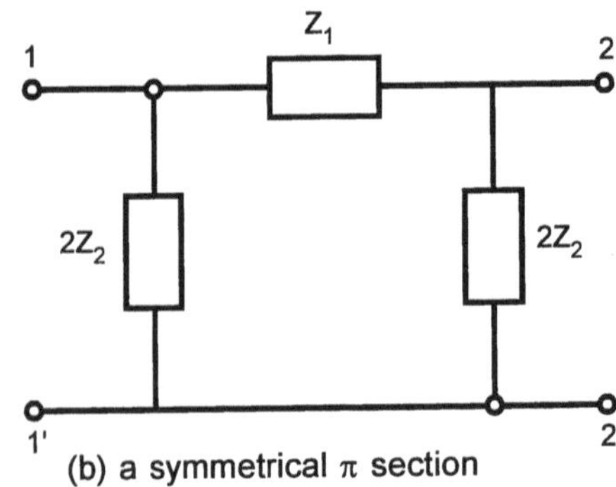

(b) a symmetrical π section

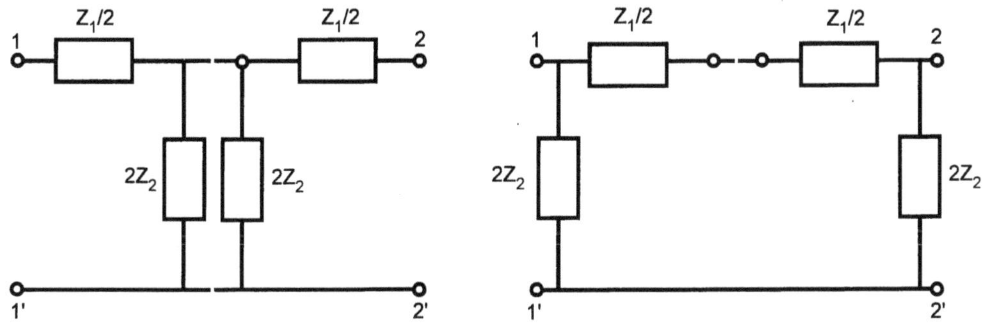

(c) T– section constructed from 2 L – sections (d) π – section constructed from 2 L – sections

Fig. 11.1 Symmetrical T and π sections and their decomposition into L – sections

These T and π networks can be combined to form a ladder network. Except for the terminating L half sections, the remainder of the ladder network can be constructed using either T networks or π networks as shown in Fig. 11.2.

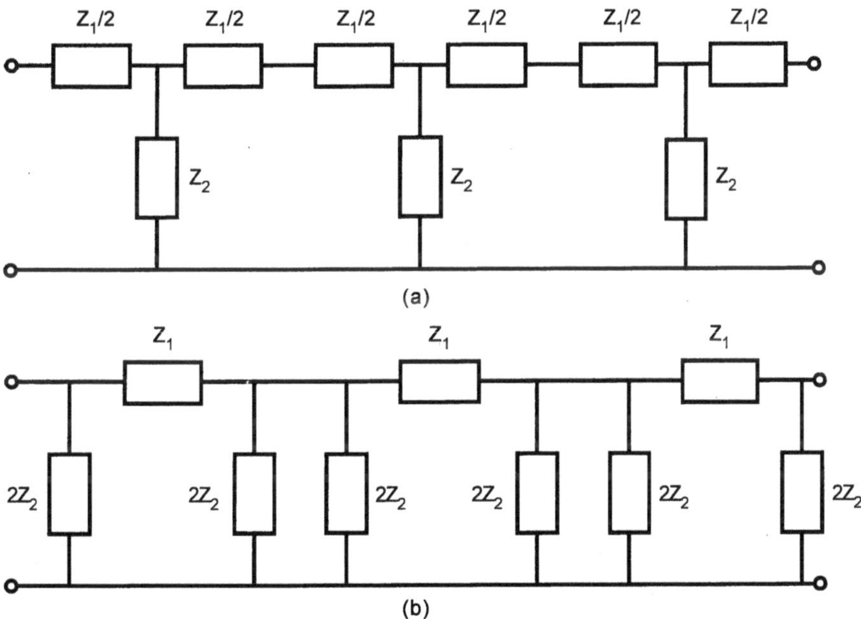

Fig. 11.2 Ladder network built from (a) T – sections (b) π – sections

11.2.1 Image impedances of networks :

The filter networks are most commonly interposed between a generator with an internal impedance Z_{1i} and a load impedance Z_{2i} as shown in Fig. 11.3. The load impedance has to be matched to the generator impedance for maximum power transfer. Thus it is required that the impedance Z_{1i} looking to the right of terminals 1,1' be equal to the generator impedance and the impedance looking into to the left of

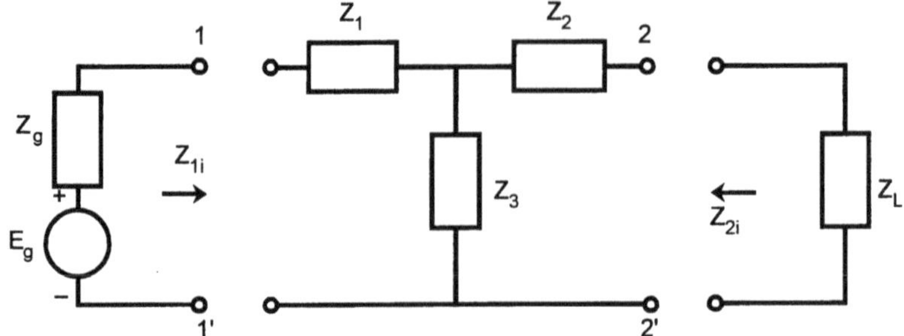

Fig. 11.3 Filter network inserted between generator and load

terminals 2, 2' be equal to the load impedance Z_L. Thus the impedance into the terminals 1,1' in either direction are equal. The same is true at the other terminals 2, 2'. These impedances Z_{1i} and Z_{2i} are known as *image impedances.*

Let us now derive expressions for the image impedances Z_{1i} and Z_{2i} of the T sections shown in Fig. 11.3.

$$Z_{1i} = Z_1 + \frac{Z_3(Z_2 + Z_{2i})}{Z_3 + Z_2 + Z_{2i}} \qquad .....(11.1)$$

and

$$Z_{2i} = Z_2 + \frac{Z_3(Z_1 + Z_{1i})}{Z_1 + Z_3 + Z_{1i}} \qquad .....(11.2)$$

Eqs. (11.1) and (11.2) can be solved for Z_{1i} and Z_{2i} to yield,

$$Z_{1i} = \frac{\sqrt{(Z_1 + Z_3)(Z_1Z_2 + Z_2Z_3 + Z_3Z_1)}}{Z_2 + Z_3} \qquad .....(11.3)$$

and

$$Z_{2i} = \frac{\sqrt{(Z_2 + Z_3)(Z_1Z_2 + Z_2Z_3 + Z_3Z_1)}}{Z_1 + Z_3} \qquad .....(11.4)$$

The expressions for Z_{1i} and Z_{2i} can be expressed in simple form if we consider the open circuit impedances of the T – network at the terminals 1, 1' and 2, 2'. With terminals 2, 2' open, the impedance measured at 1, 1' is given by

$$Z_{1OC} = Z_1 + Z_3 \qquad .....(11.5)$$

Similarly $$Z_{2OC} = Z_2 + Z_3 \qquad .....(11.6)$$

With terminals 2, 2' short circuited, the impedance measured at terminals 1, 1' is

$$Z_{1SC} = Z_1 + \frac{Z_2 Z_3}{Z_2 + Z_3}$$

$$Z_{1SC} = \frac{Z_1 Z_2 + Z_2 Z_3 + Z_1 Z_3}{Z_2 + Z_3} \quad .....(11.7)$$

Similarly $$Z_{2SC} = \frac{Z_1 Z_2 + Z_2 Z_3 + Z_1 Z_3}{Z_1 + Z_3} \quad .....(11.8)$$

Using eqs. (11.5) and (11.8) in eqs. (11.3) and (11.4) we get

$$Z_{1i} = \sqrt{Z_{10C} Z_{1SC}} \quad .....(11.9)$$

and $$Z_{2i} = \sqrt{Z_{20C} Z_{2SC}} \quad .....(11.10)$$

Thus the image impedances of T – network can easily by obtained by measuring the open circuit impedances Z_{10C}, Z_{20C} and short circuit impedances Z_{1SC}, Z_{2SC}.

11.2.2 Characteristic impedance of symmetrical networks :

If the T – section of Fig. 11.3 is symmetrical, i.e., $Z_1 = Z_2$, it is clear that

$$Z_{1i} = Z_{2i}$$

The image impedance is called the characteristic impedance or the iterative impedance and is denoted by Z_0. For a symmetrical T – network terminated in an impedance Z_0, the input impedance will be equal to Z_0. If several such symmetrical T networks each with characteristic impedance Z_0 are connected in tandem or cascade, the input impedance will still be Z_0. Thus, we can say that Z_0 is iterated at the input of each T – network and is appropriately termed also as iterative impedance.

For a symmetrical T network terminated in Z_0, as in Fig. 11.4(a) we have,

$$Z_{1i} = \frac{Z_1}{2} + \frac{Z_2\left(\frac{Z_1}{2} + Z_0\right)}{\frac{Z_1}{2} + Z_2 + Z_0} \quad .....(11.11)$$

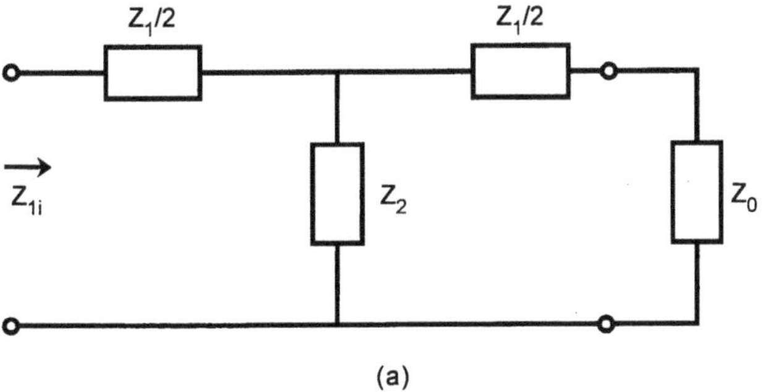

(a)

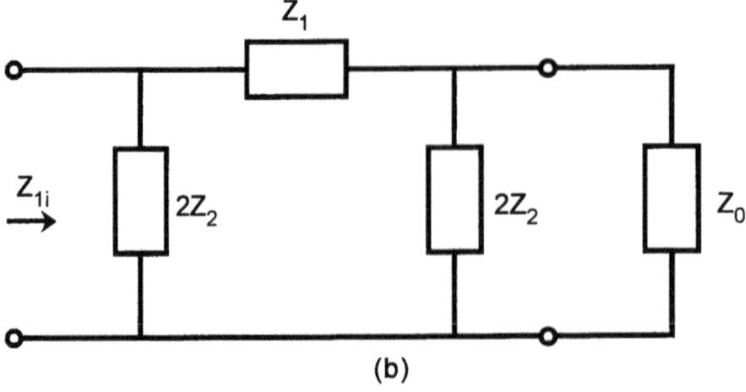

(b)

Fig. 11.4 Symmetrical T and π sections terminated in Z_0

By a suitable choice of the impedances Z_1 and Z_2, the input impedance Z_{1i} can be made equal to Z_0. Thus from eq. (11.11) making $Z_{1i} = Z_0$, we have

$$Z_0 = \frac{\frac{Z_1^2}{4} + Z_1 Z_2 + Z_0 Z_2 + \frac{Z_1}{2} Z_0}{\frac{Z_1}{2} + Z_2 + Z_0}$$

Solving for Z_0, we get

$$Z_0^2 = \frac{Z_1^2}{4} + Z_1 Z_2$$

Denoting the characteristic impedance of a T – network by Z_{OT}, we have

$$Z_{OT} = \sqrt{\frac{Z_1^2}{4} + Z_1 Z_2}$$

$$= \sqrt{Z_1 Z_2 \left(1 + \frac{Z_1}{4Z_2}\right)} \qquad .....(11.12)$$

Similarly for the π – section of Fig. 11.4(b) we get

$$Z_{1i} = \frac{\left[Z_1 + \frac{2Z_2 Z_0}{2Z_2 + Z_0}\right] 2Z_2}{Z_1 + \frac{2Z_2 Z_0}{2Z_2 + Z_0} + 2Z_2}$$

and setting $Z_{1i} = Z_0$ and solving for Z_0 we get

$$Z_{0\pi} = \sqrt{\frac{Z_1 Z_2}{1 + \frac{Z_1}{4Z_2}}} \qquad .....(11.13)$$

where $Z_{0\pi}$ is the characteristic impedance of a symmetrical π network. As discussed in section 11.2.1, we can show that the characteristic impedance for a T or π section can also be given as

$$Z_O = \sqrt{Z_{oc} Z_{sc}} \quad .....(11.14)$$

where Z_{OC} is the open circuit impedance and Z_{SC} is the short circuit impedance either at terminals 1, 1' or 2, 2'.

11.3 MEASURE OF ATTENUATION – NEPER AND DECIBEL :

In filter circuits, the performance of the circuit is measured in terms of the ratio of the magnitudes of input current to output current.

If the two image impedances Z_{1i} and Z_{2i} are equal, then the ratio of magnitudes of currents is same as ratio of magnitudes of input and output voltages as shown in the following :

$$|Z_{1i}| = \left|\frac{V_1}{I_1}\right|$$

$$|Z_{2i}| = \left|\frac{V_2}{I_2}\right|$$

If $\quad |Z_{1i}| = |Z_{2i}|$

Then $\quad \left|\frac{V_1}{I_1}\right| = \left|\frac{V_2}{I_2}\right|$

or $\quad \left|\frac{I_1}{I_2}\right| = \left|\frac{V_1}{V_2}\right| \quad .....(11.15)$

Filter networks are usually connected in cascade as shown in Fig. 11.5 and the overall performance is measured as

$$\left|\frac{V_1}{V_2}\right| \times \left|\frac{V_2}{V_3}\right| \times ..... \times \left|\frac{V_{n-1}}{V_n}\right| = \left|\frac{V_1}{V_n}\right| \quad ....(11.16)$$

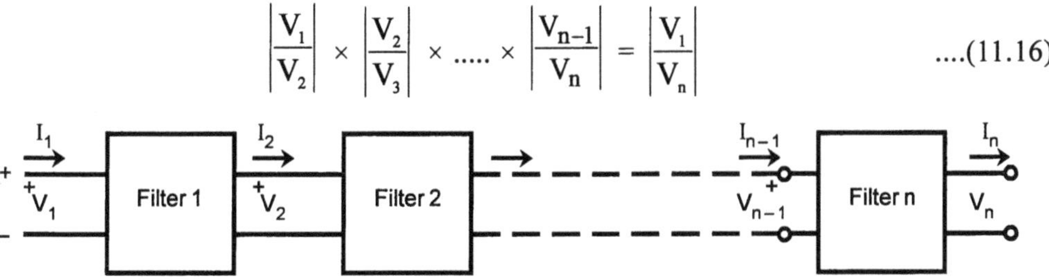

Fig. 11.5 Cascade of filters

If the ratios of voltages are defined by

$$\frac{V_1}{V_2} = e^a, \quad \frac{V_2}{V_3} = e^b, \; ..... \; \frac{V_{n-1}}{V_n} = e^n,$$

then $$\left|\frac{V_1}{V_n}\right| = e^a \cdot e^b \; ..... \; e^n$$

$$= e^{\,a+b+.....+n} \qquad .....(11.17)$$

If we take natural logarithm of eq. (11.17) we have

$$ln\left|\frac{V_1}{V_n}\right| = a + b + ...+ n$$

Thus the process of multiplication is converted into addition, if we define, under the condition of equal image impedances for each of the networks,

$$\left|\frac{V_1}{V_n}\right| = \left|\frac{I_1}{I_2}\right| = e^N \qquad .....(11.18)$$

The unit of N has been given the name 'neper' and is defined as

$$\text{N (nepers)} = ln\left|\frac{V_1}{V_2}\right| = ln\left|\frac{I_1}{I_2}\right| \qquad .....(11.19)$$

Hence if $V_1 = eV_2$ or $I_1 = eI_2$ where e is the base of natural logarithm, the two voltages or currents differ by 1 neper. N is a measure of the attenuation of voltages in a circuit and the overall attenuations of a series of networks in cascade can be obtained by adding, rather than multiplying, the individual attenuations measured in nepers. Often we are interested in power gain or loss in a network and under the condition of equal image impedances the ratio of input and output powers is given by

$$\frac{P_1}{P_2} = \left|\frac{I_1^2 Z_{1i}}{I_2^2 Z_{2i}}\right| = \left|\frac{I_1^2}{I_2^2}\right| \qquad .....(11.20)$$

Hence $$\frac{P_1}{P_2} = e^{2N} \qquad .....(11.21)$$

A similar unit is used widely by telephone industry, which uses logarithm to base 10 rather than natural logarithm and one tenth of this unit is defined as decibel in honour of Alexander Graham Bell. The decibel, which is commonly abbreviated as db, is given by

$$\text{db} = 10 \log \frac{P_1}{P_2} \qquad .....(11.22)$$

If equal image impedances are considered we can write

$$\text{db} = 10 \log \frac{P_1}{P_2} = 10 \log \left|\frac{I_1}{I_2}\right|^2 = 10 \log \left|\frac{V_1}{V_2}\right|^2$$

$$= 20 \log \left|\frac{I_1}{I_2}\right| = 20 \log \left|\frac{V_1}{V_2}\right| \quad .....(11.23)$$

A relationship between the two units neper and decibel can be established by taking the logarithm of eq. (11.18), multiplying by 20 and comparing it with eq. (11.23).

Thus $\qquad \text{db} = 20 \log e^N$

$$= 20\ N \log e$$

$$= 8.686\ N$$

or $\qquad 1 \text{ neper} = 8.686 \text{ db} \quad \text{or} \quad 1 \text{ db} = \dfrac{1}{8.686} = 0.115 \text{ nepers}$

11.4 PROPAGATION CONSTANT :

In order to obtain easily, the change in the voltage or current at the output relative to the input, of a cascaded network in terms of individual networks, we had expressed the ratio of magnitudes of voltages or currents as exponentials. The ratio of only magnitudes does not give a complete picture of the performance of the network. The phase angles of the currents and voltages must also be considered to evaluate the performance of the network. Assuming that the network is terminated in its characteristic impedance Z_0 we will define the ratio of currents expressed in complex or vector form

$$\frac{\mathbf{I}_1}{\mathbf{I}_2} = e^{\gamma} \quad .....(11.24)$$

where $\qquad \gamma = \alpha + j\beta \qquad .....(11.25)$

is a complex number

Hence

$$\frac{\mathbf{I}_1}{\mathbf{I}_2} = e^{\alpha + j\beta} \quad .....(11.26)$$

If we express $\dfrac{\mathbf{I}_1}{\mathbf{I}_2} = A\ \angle\theta$

then from eq. (11.26)

$$A = e^{\alpha} \text{ and } \beta = \theta$$

the symbol γ in eq. (11.24) is defined as the propagation constant. The real part of γ, i.e.., α is known as the attenuation constant and is measured in nepers. Since α is the ratio of I_1 and I_2 or V_1 and V_2 it gives the attenuation produced in the signal as it passes from the input to the output through the network. The imaginary part of γ, i.e.,

β is defined as the phase constant and is measured in radians. β gives the change in phase angle produced by the network between the input and output quantities.

The advantage of expressing the ratios of currents and voltages as exponentials is that when several networks are connected in cascade, the overall effect can be obtained by summation rather than the multiplication.

It is obvious that if 'n' networks with propagation constants γ_1, γ_2, ...γ_n are connected in cascade, the overall propagation constant γ is given by

$$\gamma = \gamma_1 + \gamma_2 + \ldots\ldots + \gamma_n \qquad \ldots\ldots(11.27)$$

11.5 PROPAGATION CONSTANT OF SYMMETRICAL NETWORKS :

Consider the symmetrical T – network terminated in its characteristic impedance Z_0 as shown in Fig. 11.6.

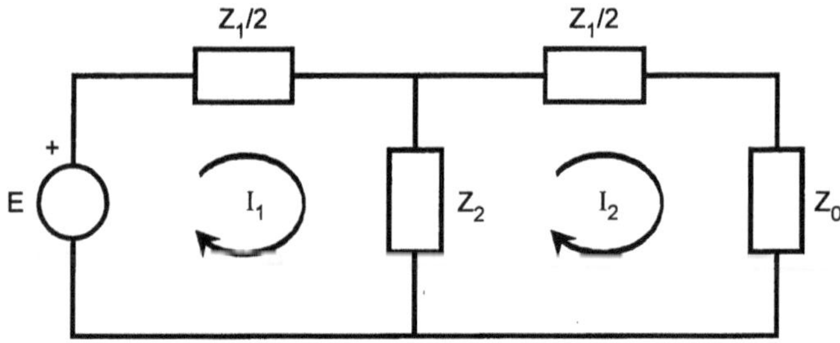

Fig. 11.6 A symmetrical T – network terminated in Z_0

Writing down the loop equations, we have

$$I_1\left(\frac{Z_1}{2} + Z_2\right) - I_2 Z_2 = E \qquad \ldots\ldots(11.28)$$

$$-I_1 Z_2 + I_2\left(\frac{Z_1}{2} + Z_2 + Z_0\right) = 0 \qquad \ldots\ldots(11.29)$$

From eq. (11.29), we get

$$\frac{I_1}{I_2} = \frac{\frac{Z_1}{2} + Z_2 + Z_0}{Z_2} \qquad \ldots\ldots(11.30)$$

But $\frac{I_1}{I_2}$ is defined as e^{γ}

Therefore from eq. (11.30),

$$\frac{Z_1}{2} + Z_2 + Z_0 = e^{\gamma} Z_2 \qquad \ldots\ldots(11.31)$$

But from eq. (11.12), we have

$$Z_0^2 = \frac{Z_1^2}{4} + Z_1 Z_2 \qquad .....(11.32)$$

Eliminating Z_0 in eqs. (11.31) and (11.32), we have

$$\left[Z_2\left(e^{\gamma}-1\right)-\frac{Z_1}{2}\right]^2 = \frac{Z_1^2}{4} + Z_1 Z_2 \qquad .....(11.33)$$

Simplification of eq. (11.33) leads to

$$Z_2(e^{\gamma} - 1)^2 - Z_1 e^{\gamma} = 0$$

$$e^{2\gamma} + 1 - 2e^{\gamma} = \frac{Z_1}{Z_2} e^{\gamma}$$

Dividing by e^{γ}, we get

$$e^{\gamma} + e^{-\gamma} = \frac{Z_1}{Z_2} + 2$$

$$\frac{e^{\gamma}+e^{-\gamma}}{2} = 1 + \frac{Z_1}{2Z_2}$$

$$\cos h\, \gamma = 1 + \frac{Z_1}{2Z_2} \qquad .....(11.34)$$

This is an important relation for the propagation constant γ in terms of the symmetrical T network impedances Z_1 and Z_2. Further, it can be shown that eq. (11.34) holds good for a symmetrical π network also. Noting that,

$$\cos h\, \gamma = 1 + 2 \sin h^2 \frac{\gamma}{2}$$

We can write,

$$\sin h \frac{\gamma}{2} = \sqrt{\frac{Z_1}{4Z_2}} \qquad .....(11.35)$$

We can also get an explicit expression for γ by substituting the value of Z_{OT} in eq. (11.31).

$$e^{\gamma} = \frac{\frac{Z_1}{2} + Z_2 + \sqrt{\frac{Z_1^2}{4} + Z_1 Z_2}}{Z_2} \qquad .....(11.36)$$

$$= 1 + \frac{Z_1}{2Z_2} + \sqrt{\left(\frac{Z_1}{2Z_2}\right)^2 + \frac{Z_1}{Z_2}} \qquad .....(11.37)$$

Taking natural logarithm on both sides of eq (11.37),

$$\gamma = ln\left[1+\frac{Z_1}{2Z_2}+\sqrt{\left(\frac{Z_1}{2Z_2}\right)^2+\frac{Z_1}{Z_2}}\right] \qquad .....(11.38)$$

11.6 FILTER CHARACTERISTICS :

Ideal characteristics of filter require that it passes a desired band of frequencies without any decrease in magnitude and stop or attenuate completely other frequencies. The propagation constant $\gamma = \alpha + j\beta$ is a function of frequency as given by eq. (11.31) and therefore can throw light on the performance of the filter.

$$\text{Since} \quad \frac{I_1}{I_2} = e^{\gamma} = e^{\alpha + j\beta} \quad .....(11.39)$$

$$\left|\frac{I_1}{I_2}\right| = e^{\alpha} \quad .....(11.40)$$

If α in eq. (11.39) is zero, $|I_1| = |I_2|$ and there is no reduction in the current magnitude as the signal passes through the filter. Thus in pass band the attenuation is zero and the propagation constant is purely imaginary. When α is positive I_2 is smaller in magnitude then I_1 and hence there is attenuation of the signal. Therefore in attenuation or stop band of frequencies, α is positive. The nature of variation of γ can be best studied from the eq. (11.35).

$$\sin h\,\frac{\gamma}{2} = \sqrt{\frac{Z_1}{4Z_2}} \quad .....(11.41)$$

Substituting $\gamma = \alpha + j\beta$ and expanding we have,

$$\sin h\left(\frac{\alpha + j\beta}{2}\right) = \sin h\,\frac{\alpha}{2}\cos\frac{\beta}{2} + j\cos h\,\frac{\alpha}{2}\sin\frac{\beta}{2} \quad .....(11.42)$$

If we consider, filter networks consisting of purely reactance elements, we can study the nature of γ as the ratio $\frac{Z_1}{4Z_2}$ varies from $-\infty$ to $+\infty$. We observe that if Z_1 and Z_2 are reactances, the ratio $\frac{Z_1}{4Z_2}$ is +ve and real if Z_1 and Z_2 are of same type and it is –ve and real if they are of opposite type. In either case $\frac{Z_1}{4Z_2}$ is real.

Case 1: $\frac{Z_1}{4Z_2} > 0$

In this case the impedances Z_1 and Z_2 are of the same type. Since $\frac{Z_1}{4Z_2}$ is real the imaginary part of eq. (11.42) must be zero. Hence we have

$$\cos h\,\frac{\alpha}{2}\sin\frac{\beta}{2} = 0 \quad ......(11.43)$$

and $$\sinh\frac{\alpha}{2}\cos\frac{\beta}{2} = \sqrt{\frac{Z_1}{4Z_2}} \quad \text{......(11.44)}$$

Eq. (11.43) is satisfied only for

$$\sin\frac{\beta}{2} = 0\,; \qquad \text{or } \frac{\beta}{2} = n\pi \quad \text{for } n = 0, 1, 2, ...$$

$$\text{or} \quad \beta = 2n\pi \qquad \text{for } n = 0, 1, 2, ...$$

For these values of β, $\cos\frac{\beta}{2} = \pm 1$ and from eq. (11.44), we have

$$\sinh\frac{\alpha}{2} = \sqrt{\frac{Z_1}{4Z_2}}$$

or $$\alpha = 2\sinh^{-1}\sqrt{\frac{Z_1}{4Z_2}} \quad \text{.....(11.45)}$$

Hence $\frac{Z_1}{4Z_2} > 0$ implies that Z_1 and Z_2 are of the same type and it always corresponds to stop band or attenuation band.

Case 2 : $\frac{Z_1}{4Z_2} < 0$

This implies that Z_1 and Z_2 are of opposite type. This condition makes $\sinh\frac{\gamma}{2}$ purely imaginary and hence from eq. (11.42) we have

$$\sinh\frac{\alpha}{2}\cos\frac{\beta}{2} = 0 \quad \text{.....(11.46)}$$

and $$\cosh\frac{\alpha}{2}\sin\frac{\beta}{2} = \sqrt{\frac{Z_1}{4Z_2}} \quad \text{.....(11.47)}$$

Eq. (11.46) is satisfied for either

$$\sinh\frac{\alpha}{2} = 0\,; \qquad \alpha = 0$$

or $$\cos\frac{\beta}{2} = 0\,; \qquad \beta = (2n-1)\pi; \text{ for } n = 0, 1, 2, ...$$

Case a : $\sinh\frac{\alpha}{2} = 0\,; \qquad \alpha = 0$

This corresponds to pass band since $\alpha = 0$. Substituting $\alpha = 0$ in eq. (11.47), we have,

$$\sin\frac{\beta}{2} = \sqrt{\frac{Z_1}{4Z_2}}$$

or,

$$\beta = 2 \sin^{-1} \sqrt{\frac{Z_1}{4Z_2}} \qquad .....(11.48)$$

Since the limiting value of $\sin \frac{\beta}{2}$ is 1, and $\frac{Z_1}{4Z_2}$ is negative, $\frac{Z_1}{4Z_2}$ must be in the limits

$$-1 < \frac{Z_1}{4Z_2} < 0 \qquad .....(11.49)$$

Case b : $\cos \frac{\beta}{2} = 0;\ \beta = (2n - 1)\pi$ for $n = 0, 1, 2, ...$

Using this value of β in eq. (11.47), we have

$$\cos h \frac{\alpha}{2} = \sqrt{\frac{Z_1}{4Z_2}} \qquad .....(11.50)$$

Hence, attenuation in the stop band is given by

$$\alpha = 2 \cos h^{-1} \sqrt{\frac{Z_1}{4Z_2}} \qquad .....(11.51)$$

Since $\cos h \frac{\alpha}{2}$ cannot be less than unity and $\frac{Z_1}{4Z_2}$ is negative, eq. (11.50) implies that

$$\frac{Z_1}{4Z_2} < -1$$

To summarise :

(i) When $\frac{Z_1}{4Z_2} > 0$ we have stop band

(ii) When $-1 < \frac{Z_1}{4Z_2} < 0$ we have pass band and

(iii) When $-\infty < \frac{Z_1}{4Z_2} < -1$ we have stop band

The two frequencies which separate the pass band from the two stop bands are known as cut off frequencies and are obtained when :

$$\frac{Z_1}{4Z_2} = 0 \quad \text{or} \quad Z_1 = 0$$

$$\frac{Z_1}{4Z_2} = -1 \quad \text{or} \quad Z_1 = -4\,Z_2 \qquad .....(11.52)$$

11.7 NATURE OF CHARACTERISTIC IMPEDANCE IN PASS AND STOP BANDS :

The characteristic impedance for a T – network from eq. (11.12), is

$$Z_{OT} = \sqrt{Z_1 Z_2\left(1+\frac{Z_1}{4Z_2}\right)}$$

If Z_1 and Z_2 are pure reactances, $Z_1 = j\,X_1$ and $Z_2 = j\,X_2$. If they are inductive reactances, X_1 and X_2 are positive and if they are capacitive reactances X_1 and X_2 are negative. Substituting these values in the expression for Z_{OT}, we get

$$Z_{OT} = \sqrt{-X_1 X_2\left(1+\frac{X_1}{4X_2}\right)} \qquad .....(11.53)$$

We have seen in section 11.6 that if X_1 and X_2 are of same type, it results in a stop band. In this case, since X_1 and X_2 are of same sign, Z_{OT} will be purely imaginary quantity.

If X_1 and X_2 are of opposite sign and $-1 < \frac{X_1}{4X_2} < 0$, a pass band results.

Thus in pass band, from eq. (11.53), Z_{OT} is a real quantity. Since Z_{OT} is real, it will be able to absorb power from the source. Hence if the network is terminated in a pure resistance of value equal to Z_0, the filter can transmit power from the source to the load without any attenuation. However, if X_1 and X_2 are of opposite sign, but, $\frac{X_1}{4X_2} < -1$, we have again a stop band and Z_{OT} in eq. (11.53) is purely imaginary. We conclude that Z_{OT} is resistive in the pass band and reactive in the stop band. We note that eq. (11.53) equally holds good for T and π networks. Further, form eqs. (11.12) and (11.13) we have

$$Z_{OT}\,Z_{0\pi} = Z_1\,Z_2 \qquad .....(11.54)$$

If Z_1 and Z_2 are pure reactances, the product $Z_1\,Z_2$ is a real constant. Hence all the conclusions regarding pass and stop bands and characteristic impedance hold good for π networks also.

11.8 CONSTANT K - FILTERS :

If the impedances Z_1 and Z_2 in the T or π networks are pure reactances and of opposite sign, we have

$$Z_1 Z_2 = K^2 \quad .....(11.55)$$

where K is a constant and is independent of frequency. Filters which satisfy this relation are called as constant – K filters. These filters, either of T or π configuration, are also known as prototype because more complex filters can be derived from them.

11.8.1 Constant - K low pass filter :

A constant – K low pass filter is shown in Fig. 11.7.

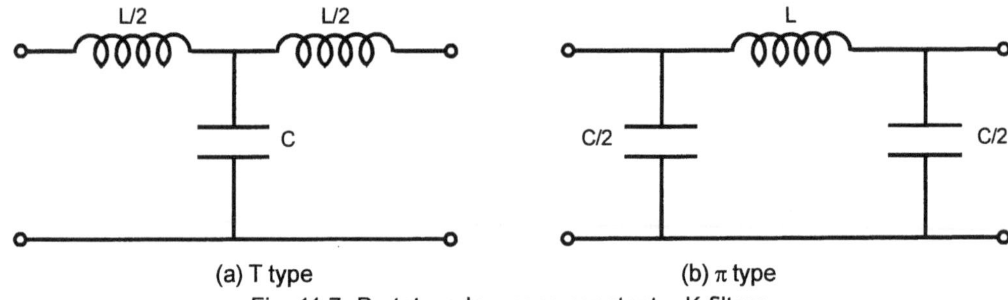

Fig. 11.7 Prototype low pass constant – K filters

The behaviours of this filter can be studied by observing that

$$Z_1 = j\,\omega L$$

and $$Z_2 = \frac{-j}{\omega C}$$

Hence $$Z_1\, Z_2 = j\omega L\left(\frac{-j}{\omega C}\right) = \frac{L}{C}$$

$$= K^2$$

Since the product $Z_1 Z_2$ is independent of frequency, it is a constant – K filter. If this real constant is denoted as R_K,

$$Z_1 Z_2 = R_K{}^2$$

R_K is called as the design impedance or nominal impedance of the filter.

Thus $$R_K{}^2 = \frac{L}{C}$$

or $$R_K = \sqrt{\frac{L}{C}} \quad .....(11.56)$$

The characteristic impedances Z_{OT} is given by

$$Z_{OT} = \sqrt{\frac{Z_1^2}{4} + Z_1 Z_2}$$

Substituting the values of Z_1 and Z_2 we get

$$Z_{OT} = \sqrt{\frac{(j\omega L)^2}{4} + j\omega L \cdot \frac{1}{j\omega C}}$$

$$= \sqrt{-\frac{\omega^2 L^2}{4} + \frac{L}{C}}$$

$$= \sqrt{\frac{L}{C}}\sqrt{1 - \frac{\omega^2 LC}{4}} \qquad .....(11.57)$$

If $\frac{\omega^2 LC}{4} < 1$, Z_{0T} is real and it is imaginary if $\frac{\omega^2 LC}{4} > 1$. The frequency at which Z_{0T} changes from real to imaginary is called as the cut – off frequency and is denoted by w_C

Thus $$\frac{\omega_C^2 LC}{4} = 1$$

$$\omega_C = \frac{2}{\sqrt{LC}}$$

or $$f_C = \frac{1}{\pi\sqrt{LC}} \qquad .....(11.58)$$

From eq. (11.57), it is seen that Z_{0T} is real for frequencies ranging from $f_C = 0$ to $f_C = \frac{1}{\pi\sqrt{LC}}$ and hence this filter passes frequencies upto f_C without attenuation. All frequencies above f_C are in the attenuation band and hence this filter is known as low pass filter.

If the cut off frequency f_C and design impedance R_K for a filter are given we can design the filter from eqs. (11.56) and (11.58), we have

$$L = \frac{R_K}{\pi f_C} \qquad .....(11.59)$$

and $$C = \frac{1}{\pi R_K f_C} \qquad .....(11.60)$$

11.8.2 Phase shift of a constant – K LP filter :

From eq. (11.48), in pass band, the phase shift is given by

$$\beta = 2\ \sin^{-1}\sqrt{\frac{Z_1}{4Z_2}}$$

$$= 2\ \sin^{-1}\sqrt{\frac{\omega L}{4.\dfrac{1}{\omega C}}}$$

$$= 2\ \sin^{-1}\sqrt{\frac{\omega^2 LC}{4}}$$

$$= 2\ \sin^{-1}\frac{f}{f_C} \qquad .....(11.61)$$

The variation β with frequency is shown in Fig. 11.8(a)

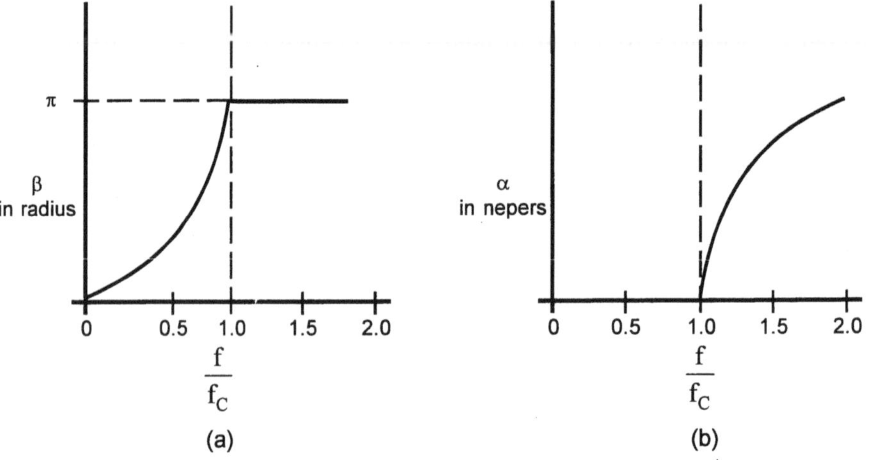

Fig. 11.8 Variation of phase angle and attenuation of constant K low pass filter with frequency

11.8.3 Attenuation of a constant – K LP filter :

In the stop band, from eq. (11.51), we have

$$\alpha = 2\ \cos h^{-1}\sqrt{\frac{Z_1}{4Z_2}}$$

$$= 2\ \cos h^{-1}\frac{f}{f_C} \qquad .....(11.62)$$

The variation of α with frequency is shown in Fig. 11.8(b)

11.8.4 Characteristic impedance of a constant – K LP filter :

From eq. (11.57) we have

$$Z_{0T} = R_K \sqrt{1-\left(\frac{\omega}{\omega_C}\right)^2} \qquad .....(11.63)$$

At $\omega = 0$, $Z_{0T} = R_K$ and at $\omega = \omega_C$, $Z_{0T} = 0$. If the load impedance is chosen as R_K, it will match only at $\omega = 0$ and varies in the entire pass band. Hence power transfer is higher at lower frequencies and drops to zero at cut off frequency f_C. The variation of Z_{0T} is shown in Fig. 11.9.

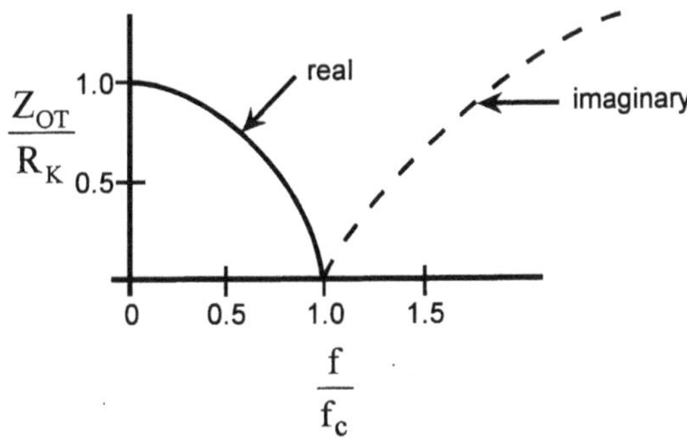

Fig. 11.9 Variation of $\frac{Z_{OT}}{R_K}$ with frequency

11.8.5 Constant – K high pass filter :

If the positions of inductance and capacitance in low pass filter are interchanged, we get a high pass constant K – filter, as shown in Fig. 11.10.

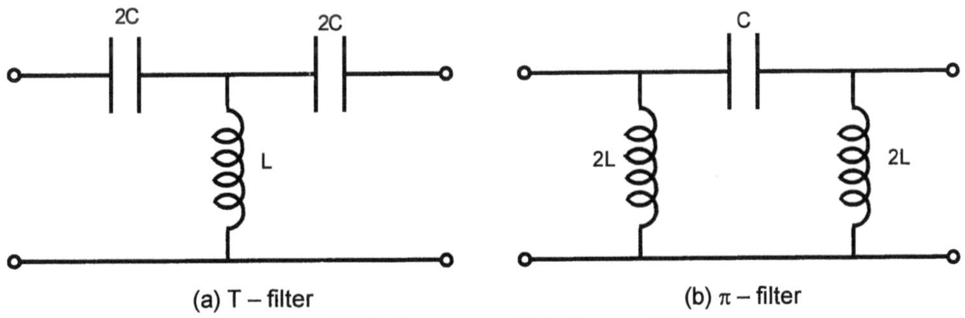

(a) T – filter (b) π – filter

Fig. 11.10 Constant – K high pass filter

Here also

$$Z_1 Z_2 = \frac{1}{j\omega C} \cdot j\omega L$$

$$Z_1 Z_2 = \frac{L}{C} = R_K^2$$

$$R_K = \sqrt{\frac{L}{C}} \qquad .....(11.64)$$

Z_{0T} is given by

$$Z_{0T} = \sqrt{\frac{Z_1^2}{4} + Z_1 Z_2}$$

$$= \sqrt{-\frac{1}{4\omega^2 C^2} + \frac{L}{C}}$$

$$= \sqrt{\frac{L}{C}} \sqrt{1 - \frac{1}{4\omega^2 LC}}$$

Z_{0T} will be real when $\frac{1}{4\omega^2 LC} < 1$ and is imaginary when $\frac{1}{4\omega^2 LC} > 1$

Hence the cut off frequency is given by

$$\frac{1}{4\omega^2 LC} = 1$$

or $$\omega_C^2 = \frac{1}{4LC}$$

Thus $$f_C = \frac{1}{4\pi\sqrt{LC}} \qquad .....(11.65)$$

If R_K and f_C are given, the element values can be obtained as

$$L = \frac{R_K}{4\pi f_C} \qquad .....(11.66)$$

$$C = \frac{1}{4\pi R_K f_C} \qquad .....(11.67)$$

The phase shift β in pass band is given by

$$\beta = 2 \sin^{-1} \sqrt{\frac{\frac{1}{\omega C}}{4.\omega L}} = 2 \sin^{-1} \sqrt{\frac{1}{4.\omega^2 LC}}$$

$$\beta = 2 \sin^{-1} \sqrt{\frac{f_C^{\ 2}}{f^2}}$$

$$= 2 \sin^{-1} \frac{f_C}{f} \qquad .....(11.68)$$

The attenuation in stop band is given by

$$\alpha = 2 \cos h^{-1} \sqrt{\frac{Z_1}{4Z_2}}$$

$$= 2 \cos h^{-1} \frac{f_C}{f} \qquad .....(11.69)$$

Likewise Z_{0T} is obtained as

$$Z_{0T} = R_K \sqrt{1 - \frac{\omega_C^{\ 2}}{\omega^2}} \qquad .....(11.70)$$

The variations of β, α and Z_{0T} are shown in Fig. 11.11.

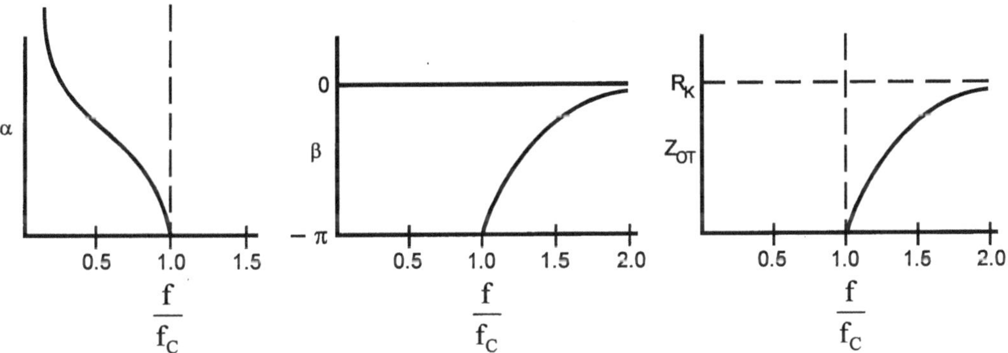

Fig. 11.11 Variation of α, β and Z_{0T} for constant – K HP filter

11.9 m-DERIVED FILTERS :

There are two disadvantages of the constant – K filter.

1. The attenuation does not increase sharply beyond cut off frequency. Hence frequencies just beyond cut off frequency are not sufficiently attenuated compared to frequencies inside the passband.

2. The characteristic impedances varies widely over the passband frequencies, so that impedance matching over the entire passband becomes impossible.

Hence a high value of attenuation just above cut off frequency and a constant impedance in the passband are desirable for a good filter. Using the prototype T or π filter, it is possible to design a filter with these desirable characteristics.

11.9.1 m–Derived T–filter :

Let us consider a modified T–filter with series arm impedance replaced by $\frac{mZ_1}{2}$ as shown in Fig. 11.12.

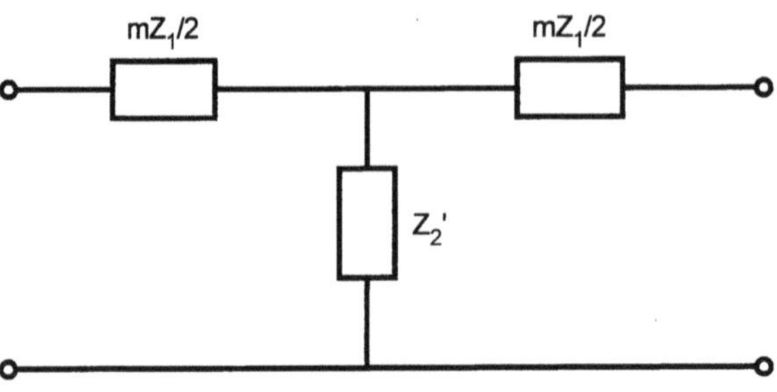

Fig. 11.12 Modified T–filter

If this network has to have the same characteristic impedance Z_{0T}, the shunt arm impedance also has to be modified. Let it be equal to Z_2'. Under the equivalence of Z_0 for the prototype T–network and modified T–network, we have

$$Z_{0T} = \sqrt{\frac{Z_1^2}{4} + Z_1 Z_2} = \sqrt{\frac{m^2 Z_1^2}{4} + m Z_1 Z_2'}$$

$$\frac{Z_1^2}{4} + Z_1 Z_2 = \frac{m^2 Z_1^2}{4} + m Z_1 Z_2'$$

Solving for Z_2', we have

$$Z_2' = \frac{Z_2}{m} + \frac{1-m^2}{4m} . Z_1 \qquad .....(11.71)$$

Thus, if the shunt arm is made up of two impedances, $\frac{Z_2}{m}$ and $\frac{1-m^2}{4m} Z_1$, in series, then the characteristic impedance of the two network will be the same. It is clear that the term $\frac{1-m^2}{4m}$ must be positive for realising Z_2' and hence $0 < m < 1$.

The m–derived T–filter is shown in Fig. 11.13.

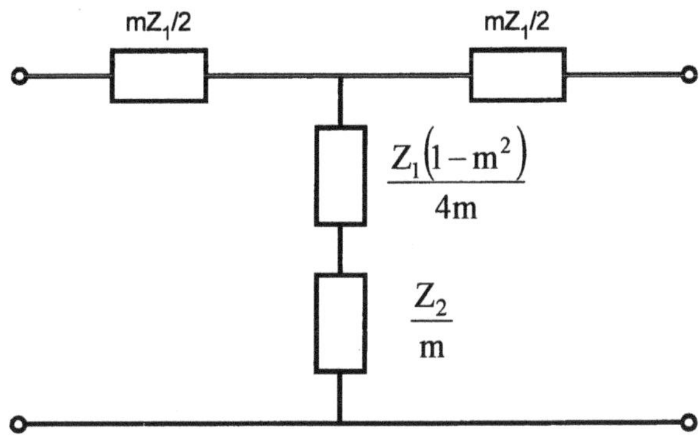

Fig. 11.13 m–derived T–filter

11.9.2 m–Derived π–filter :

Now consider a π–section in which the shunt arms are divided by m.

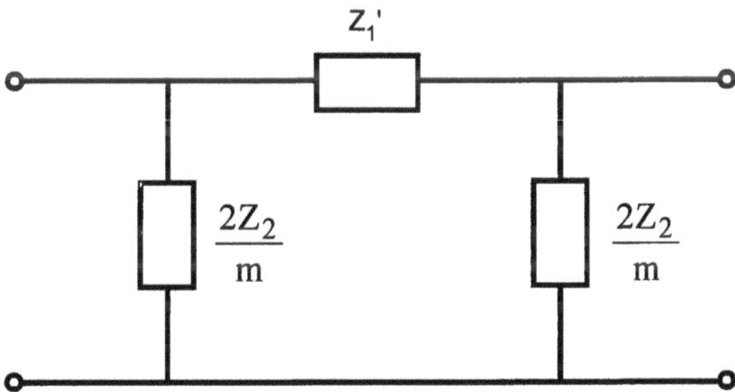

Fig. 11.14 m – derived π – section

For Z_0 – equivalence of the prototype and modified π filter, we have

$$Z_{0\pi} = \sqrt{\frac{Z_1 Z_2}{1+\frac{Z_1}{4Z_2}}} = \sqrt{\frac{Z_1'\frac{Z_2}{m}}{1+\frac{mZ_1'}{4Z_2}}}$$

or,
$$\frac{Z_1Z_2}{1+\frac{Z_1}{4Z_2}} = \frac{Z_1'\frac{Z_2}{m}}{1+\frac{mZ_1'}{4Z_2}}$$

Solving for Z_1', we get

$$Z_1' = \frac{Z_1Z_2}{\frac{Z_2}{m}+\frac{Z_1(1-m^2)}{4m}}$$

$$= \frac{1}{\frac{1}{Z_1m}+\frac{1}{\frac{4m}{1-m^2}Z_2}} \qquad .....(11.72)$$

It is clear from eq. (11.72) that the series arm Z_1' is made up of two impedances in parallel, one being mZ_1 and the other $\frac{4m}{1-m^2}Z_2$. In this case also for physical realisability $0 < m < 1$.

11.9.3 m-Derived low pass filter :

The m – derived low pass filter, in both T and π forms, is shown in Fig. 11.15.

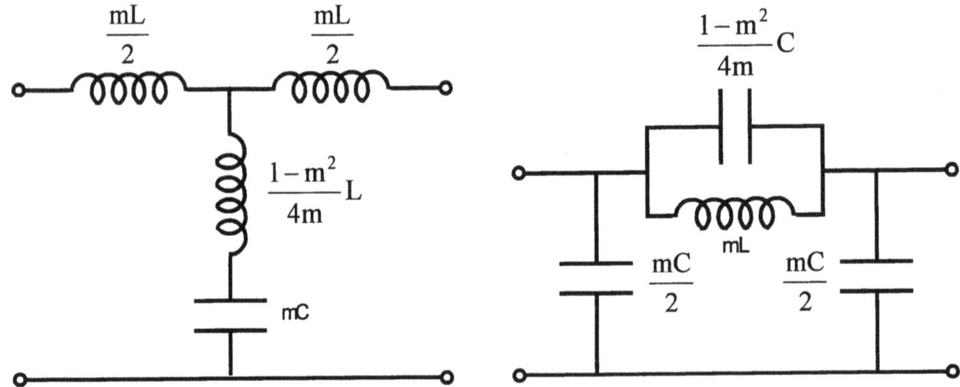

Fig. 11.15 m–derived T and π filters

The shunt arm in T–filter contains an LC resonant circuit and its impedance is zero at the resonance frequency. At this frequency it acts like a short circuit and therefore output will be zero, thus producing infinite attenuation. This condition is satisfied when

$$\frac{1}{\omega mC} = \frac{1-m^2}{4m}\,\omega L$$

or
$$\omega^2 = \frac{4}{(1-m^2)LC}$$

If we denote the frequency at which infinite attenuation is produced, by f_∞,

$$f_\infty = \frac{1}{\pi\sqrt{LC(1-m^2)}}$$

But for a prototype low pass filter,

$$f_C = \frac{1}{\pi\sqrt{LC}}$$

$\therefore$
$$f_\infty = \frac{f_C}{\sqrt{1-m^2}}$$

Solving for m, we get

$$m = \sqrt{1-\left(\frac{f_C}{f_\infty}\right)^2} \qquad .....(11.73)$$

This expression for m holds good even for m derived π – section. For a π – section the parallel combination of L, C elements results in parallel resonance and at resonant frequency it offers infinite impedance. It behaves like an open circuit and the signal is zero at the output and thus we have infinite attenuation at this frequency.

The variation of attenuation over the stop band for a low pass m – derived filter is dependent on the sign of the impedances. The impedance Z_2' behaves like a capacitance for $f < f_\infty$. Thus the series and parallel arm impedances are of opposite type for $f_C < f < f_\infty$ and are of same type for $f > f_\infty$. The attenuation in these two cases is given by eqs. (11.51) and (11.45) respectively. The attenuation α is given by,

$$\alpha = 2\cosh^{-1}\sqrt{\left(\frac{Z_1}{4Z_2}\right)} \quad \text{for} \quad f_C < f < f_\infty \qquad .....(11.74)$$

$$\alpha = 2\sinh^{-1}\sqrt{\left(\frac{Z_1}{4Z_2}\right)} \quad \text{for} \quad f > f_\infty \qquad .....(11.75)$$

For α m–derived T–filter, we have

$$\left|\frac{Z_1}{4Z_2}\right| = \frac{m\omega L}{4\left[\frac{1}{m\omega C} - \frac{\omega L(1-m^2)}{4m}\right]}$$

$$\left|\frac{Z_1}{4Z_2}\right| = \frac{m^2\omega^2 LC}{4\left[1 - \frac{\omega^2 LC}{4}(1-m^2)\right]}$$

Substituting $\omega = 2\pi f$

$$\left|\frac{Z_1}{4Z_2}\right| = \frac{m^2\pi^2 f^2 LC}{\left[1 - \pi^2 f^2 LC(1-m^2)\right]} \quad .....(11.76)$$

But $$f_C = \frac{1}{\pi\sqrt{LC}}$$

and $$f_\infty = \frac{f_C}{\sqrt{1-m^2}}$$

Substituting the value of f_C in eq. (11.76) and simplifying, we get

$$\left|\frac{Z_1}{4Z_2}\right| = \frac{\frac{m^2 f^2}{f_C^2}}{1-\left(\frac{f}{f_\infty}\right)^2} \quad .....(11.77)$$

Substituting eq. (11.77) in eqs. (11.74) and (11.75) we get

$$\alpha = 2\cosh^{-1}\frac{\frac{mf}{f_C}}{\sqrt{1-\left(\frac{f}{f_\infty}\right)^2}} \; ; \quad \text{for} \quad f_C < f < f_\infty \quad .....(11.78)$$

and $$\alpha = 2\sinh^{-1}\frac{\frac{mf}{f_C}}{\sqrt{\left(\frac{f}{f_\infty}\right)^2 - 1}} \; ; \quad \text{for} \quad f > f_\infty \quad .....(11.79)$$

The phase shift β in the pass band can similarly be obtained from eq. (11.48) as

$$\beta = 2 \sin^{-1} \sqrt{\left|\frac{Z_1}{4Z_2}\right|}$$

$$= 2 \sin^{-1} \frac{\frac{mf}{f_C}}{\sqrt{1 - \frac{f^2(1-m^2)}{f_C^2}}} \quad \text{.....(11.80)}$$

In the attenuation band, β has the value π upto $f = f_\infty$, but for $f > f_\infty$, the value of β drops down to zero. This is because the shunt arm becomes inductive above resonance frequency and the impedances in series and shunt arms will be of same type. The variation of α and β are shown in Fig. 11.16(a) and (b).

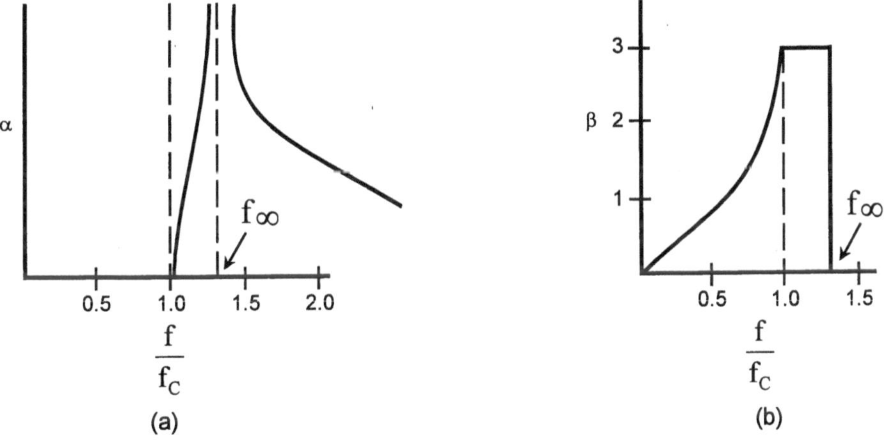

Fig. 11.16 Variation of α and β with Frequency for m – Derived Low Pass Filter

The characteristic impedance of a low pass T – filter was shown to be

$$Z_{0T} = R_K \sqrt{1 - \left(\frac{f}{f_C}\right)^2}$$

Similarly the characteristic impedances of low pass π – filter can be shown to be

$$Z_{0\pi} = \frac{R_K}{\sqrt{1 - \left(\frac{f}{f_C}\right)^2}}$$

The variation of Z_{0T} and $Z_{0\pi}$ with frequency are shown in Fig. 11.17.

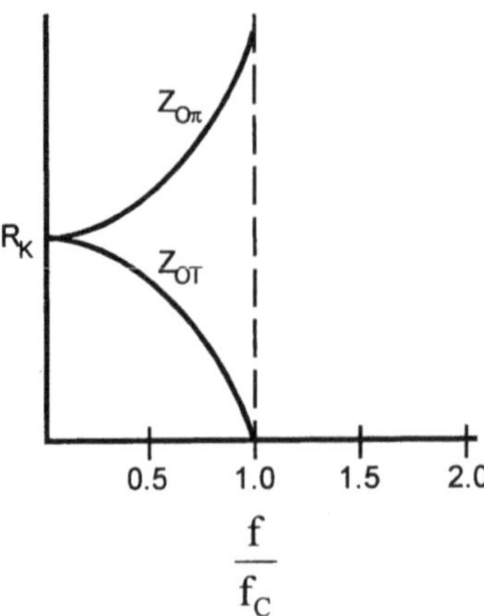

Fig. 11.17 Variation of Z_{0T} and $Z_{0\pi}$ with frequency for low pass T and π filters

11.9.4 m – Derived high pass filter :

The T and π m – derived filters are shown in Fig. 11.18.

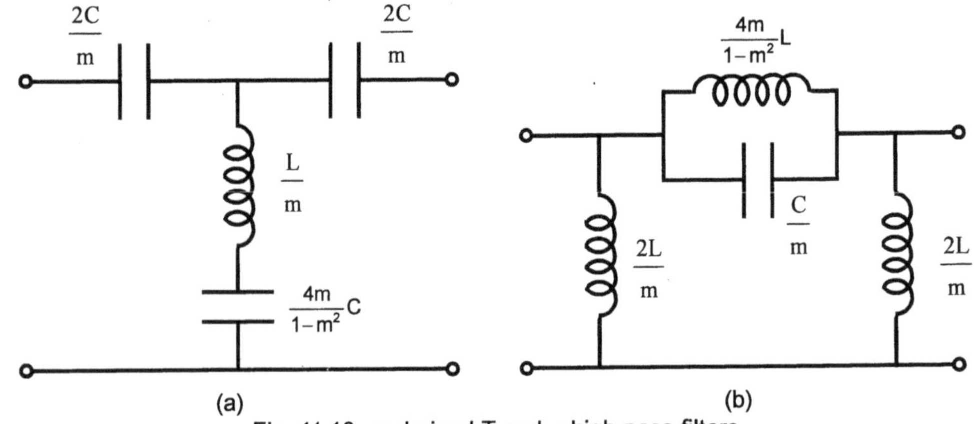

Fig. 11.18 m derived T and π high pass filters

The series resonance in the shunt arm of T section and parallel resonance in the series arm of π section, produces infinite attenuation at the resonance frequency, which is same for both the networks. The resonant frequency ω_∞ is given by

$$\omega_\infty^2 = \frac{1}{\frac{L}{m} \times \frac{4m}{1-m^2} C}$$

$$= \frac{1-m^2}{4LC}$$

or $$f_\infty = \frac{\sqrt{1-m^2}}{4\pi\sqrt{LC}}$$

But $$f_C = \frac{1}{4\pi\sqrt{LC}}$$

$\therefore$ $$f_\infty = f_C\sqrt{1-m^2}$$

and $$m = \sqrt{1-\left(\frac{f_\infty}{f_C}\right)^2} \quad .....(11.81)$$

The variation of α in the stop band and β in pass band can be obtained as in the case of a low pass filter. These characteristics are drawn in Fig. 11.19.

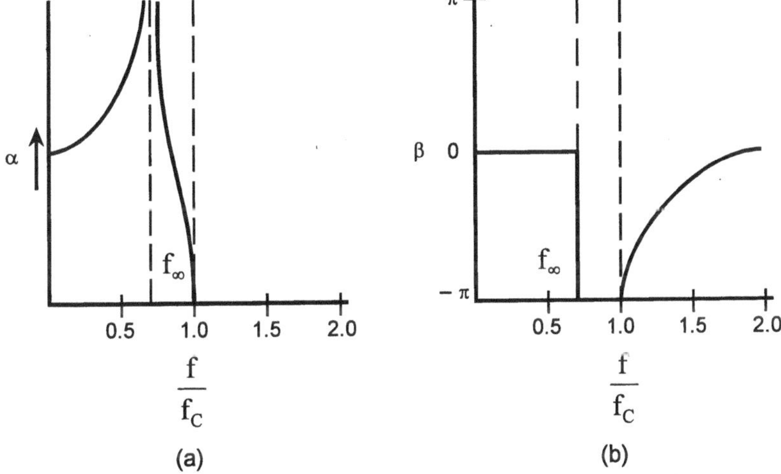

Fig. 11.19 Variation of α and β with frequency for m – derived high pass filters

The characteristic impedance for a high pass T network can be obtained as

$$Z_{0T} = R_K\sqrt{1-\left(\frac{f_C}{f}\right)^2} \quad .....(11.82)$$

Similarly for a π network

$$Z_{0\pi} = \frac{R_K}{\sqrt{1-\left(\frac{f_C}{f}\right)^2}} \quad .....(11.83)$$

The variation of Z_{0T} and $Z_{0\pi}$ are shown in Fig. 11.20 for a high pass filter.

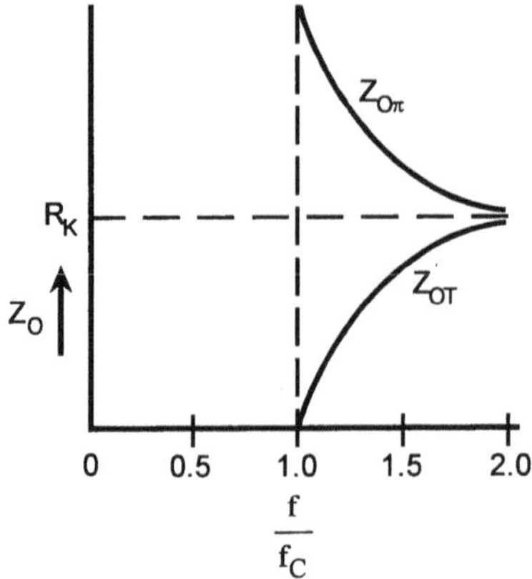

Fig. 11.20 Variation of Z_{0T} and $Z_{0\pi}$ with frequency for a high pass filter

It is observed from Fig. 11.16(a) and 11.19(a) that the attenuation can be increased in the stop band near to the cut off frequency, but the attenuation beyond the peak attenuation point is much less. For a low pass filter with m = 0.6, the comparison of prototype and m – derived filters are shown in Fig. 11.21.

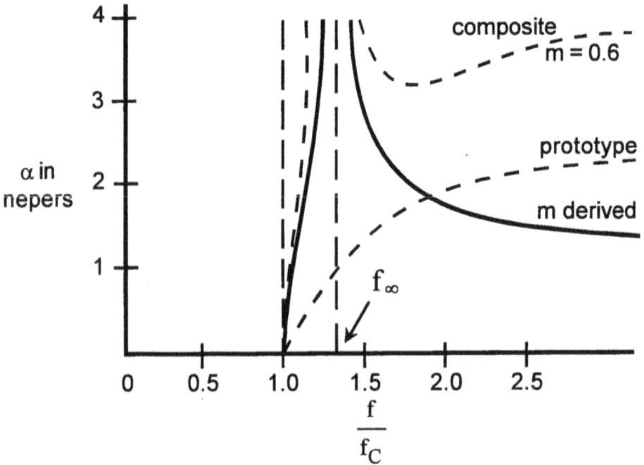

Fig. 11.21 Comparison of variation of attenuation m – derived LP, prototype LP filter and a combination of the two called composite filter

To improve the attenuation characteristic in the stop band, a combination of the prototype and m – derived sections can be used. Any desired attenuation characteristic can be obtained by a suitable choice of prototype and a number of m – derived sections connected in cascade. Such a combination of filters is known as a composite filter. However the characteristic impedances, still, is not a constant over the entire pass band and hence terminating half sections are used as impedance transformation networks. The half section, some times called as L – section, should be capable of changing its characteristics in such a way that the filter is approximately matched to the load at all frequencies in the pass band. An m – derived half section was shown to possess such properties by Zobel.

11.9.5 Terminating half or L – section :

Consider the π – section shown in Fig. 11.22(b) derived by rearranging the elements of a m – derived T– section shown in Fig. 11.22(a).

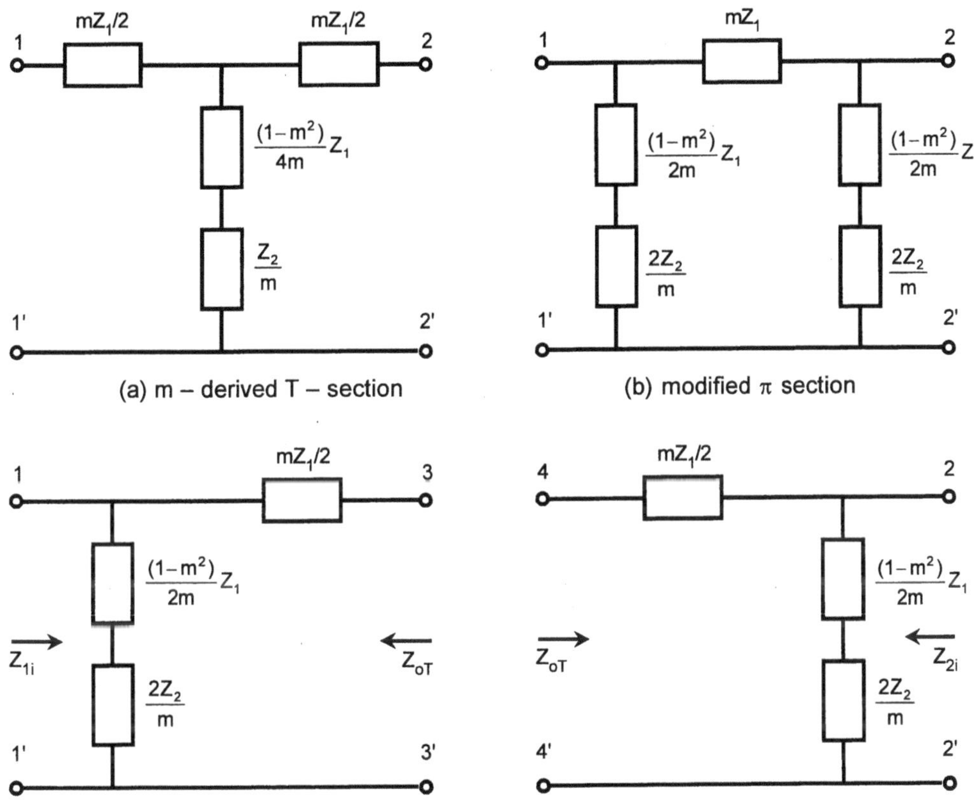

Fig. 11.22 m – derived T – section and modified π section

The modified π section shown in Fig. 11.22(b) is split into 2 section as shown in Fig. 11.22(c). The image impedance of the left half section at the terminals 1,1' is given by eq. (11.9).

$$Z_{1i} = \sqrt{Z_{1OC} Z_{1SC}} \qquad .....(11.84)$$

From Fig. 11.22(c)

$$Z_{1OC} = \frac{1-m^2}{2m} Z_1 + \frac{2Z_2}{m}$$

and $$Z_{1SC} = \frac{\left(\frac{mZ_1}{2}\right)\left(\frac{1-m^2}{2m}Z_1 + \frac{2Z_2}{m}\right)}{\frac{mZ_1}{2} + \frac{1-m^2}{2m}Z_1 + \frac{2Z_2}{m}}$$

$$= \frac{\left(\frac{mZ_1}{2}\right)\left(\frac{1-m^2}{2m}Z_1 + \frac{2Z_2}{m}\right)}{\frac{Z_1}{2m} + \frac{2Z_2}{m}}$$

Substituting the value of Z_{1OC} and Z_{1SC} in eq. (11.84)

$$Z_{1i} = \sqrt{\frac{\left(\frac{1-m^2}{2m}Z_1 + \frac{2Z_2}{m}\right)^2 \frac{mZ_1}{2}}{\frac{Z_1}{2m} + \frac{2Z_2}{m}}}$$

$$= \sqrt{\frac{\frac{4Z_2^{\ 2}}{m^2}\left(1+\left(1-m^2\right)\frac{Z_1}{4Z_2}\right)^2 \frac{mZ_1}{2}}{\frac{2Z_2}{m}\left(1+\frac{Z_1}{4Z_2}\right)}}$$

$$= \left[1+\left(1-m^2\right)\frac{Z_1}{4Z_2}\right]\sqrt{\frac{Z_1Z_2}{1+\frac{Z_1}{4Z_2}}}$$

$$= \left[1+\left(1-m^2\right)\frac{Z_1}{4Z_2}\right] Z_{0\pi} \qquad \text{.....(11.85)}$$

Eq. (11.85) shows that the image impedance at terminals 1, 1' of the half section of the modified π – section is equal to $Z_{0\pi}$ multiplied by a factor $1 + (1- m^2)\ \frac{Z_1}{4Z_2}$. Thus it is possible to change this impedance by changing the value of m.

The variation of Z_{1i} for various values of m is plotted in Fig. 11.23 over the pass band frequencies of a low pass filter. For a low pass filter eq. (11.85) may be written as

$$Z_{1i} = \frac{R_K\left[1-\left(1-m^2\right)\left(\frac{f}{f_C}\right)^2\right]}{\sqrt{1-\left(\frac{f}{f_C}\right)^2}} \qquad .....(11.86)$$

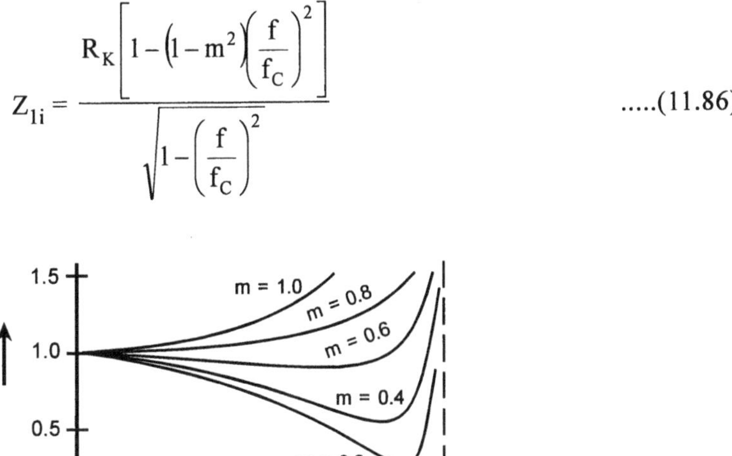

Fig. 11.23 Variation of $\frac{Z_{1i}}{R_K}$ of the half section over the pass band frequencies for various values of m

It may be observed from Fig. 11.23 that if m = 0.6, the image impedance Z_{1i} is fairly constant over 85% of the pass band and is equal to the design impedance R_K. Thus a source with internal impedance R_K can be matched satisfactorily over most of the passband. This can also be shown to be true for a high pass filter.

Let us find the image impedance, Z_{2i}, at the terminals 3, 3' of the half section in Fig. 11.22(c).

$$Z_{2i} = \sqrt{Z_{2OC}Z_{2SC}}$$

$$= \sqrt{\left(\frac{mZ_1}{2}+\frac{1-m^2}{2m}Z_1+\frac{2Z_2}{m}\right)\left(\frac{mZ_1}{2}\right)}$$

$$= \sqrt{\left(\frac{Z_1}{2m}+\frac{2Z_2}{m}\right)\left(\frac{mZ_1}{2}\right)}$$

$$= \sqrt{\frac{2Z_2}{m}\left(1+\frac{Z_1}{4Z_2}\right)\frac{mZ_1}{2}}$$

$$= \sqrt{Z_1 Z_2 \left(1 + \frac{Z_1}{4Z_2}\right)} \qquad .....(11.87)$$

This is same as Z_{0T}. Hence

$$Z_{2i} = Z_{0T} \qquad .....(11.88)$$

Similarly the image impedance at terminals 4, 4' is same as Z_{0T} and the image impedance at terminals 2, 2' is same as given in eq. (11.86). Thus a generator of impedance R_K can be connected to the terminals 1, 1' of the half section shown in Fig. 11.22(c) and a load of value R_K can be connected to the terminals 2, 2' of the half section shown in Fig. 11.22(c) so that the load and generator are perfectly matched. Between the terminals 3, 3' and 4, 4' we may insert prototype and m derived sections of the filter with design impedance R_K. The two m derived half sections are added at either end of any filter to provide the matching over the entire pass band. These half sections are known as terminating half sections.

m – derived half π – section can be obtained in a similar manner. The m – derived π – section is rearranged as a T – section and split into 2 – half sections and used at the two ends to get good matching characteristics. In this case also m = 0.6 results in reasonably good matching over the pass band. The m derived half π – sections are shown in Fig. 11.24.

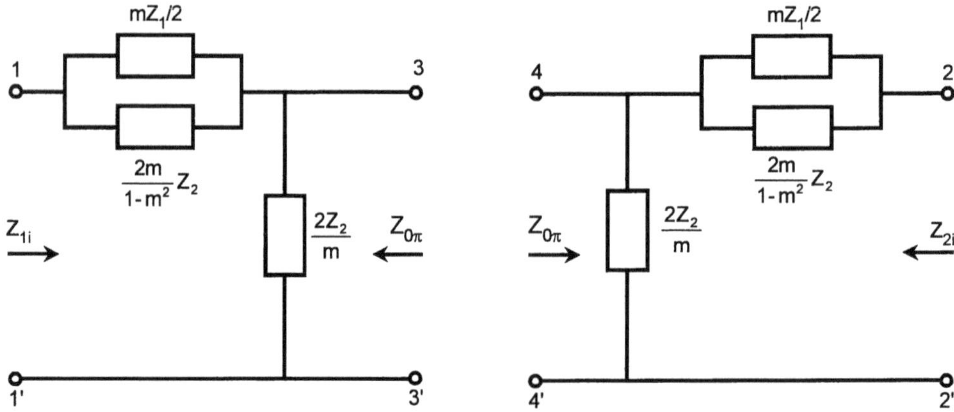

Fig. 11.24 Terminating half sections to be used with π section filters

A filter with one or more prototype constant K – filters, one or more m – derived sections and terminating half sections at the ends is known as a composite filter.

11.10 BANDPASS FILTERS :

In some applications, a certain range of frequencies are to be passed without attenuation, and on either side, the frequencies have to be attenuated. This type of filter is known as band pass filter. This can be achieved by cascading a low pass filter with cut off frequency f_{C1}, with a high pass filter with a cut off frequency f_{C2}, where $f_{C1} > f_{C2}$. But it is more economical to combine low pass and high pass sections into a single filter.

11.10.1 Bandpass constant K – filter :

This can be either a T – section or a π – section as shown in Fig. 11.25.

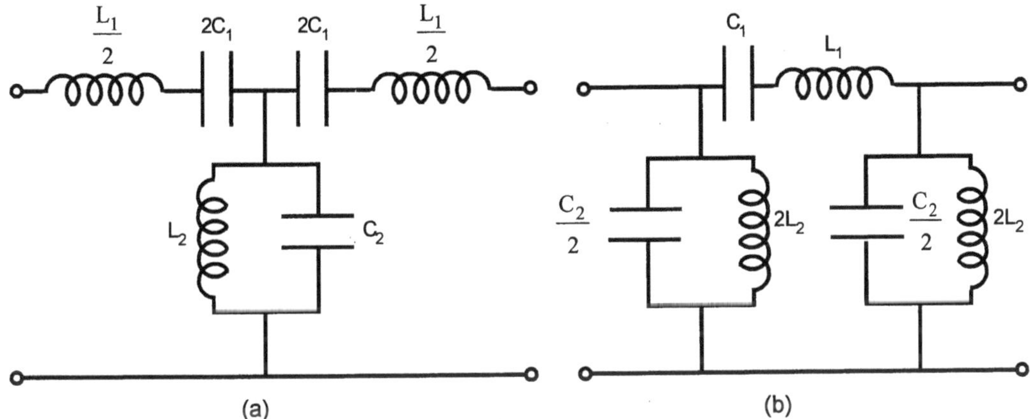

Fig. 11.25 T and π – section band pass constant K filters

The band pass filter is a combination of a series resonant circuit in the series arm and a parallel resonant circuit in the shunt arm. If the two resonance frequencies are equal, we have

$$\omega_0^2 = \frac{1}{L_1C_1} = \frac{1}{L_2C_2}$$

or $$L_1 C_1 = L_2 C_2 \qquad .....(11.89)$$

The impedances Z_1 and Z_2 of the series and shunt arms are respectively,

$$Z_1 = j\left(\omega L_1 - \frac{1}{\omega C_1}\right)$$

$$Z_1 = j\left(\frac{\omega^2 L_1 C_1 - 1}{\omega C_1}\right) \qquad .....(11.90)$$

and $$Z_2 = \frac{j\omega L_2\left(\frac{-j}{\omega C_2}\right)}{j\left(\omega L_2 - \frac{1}{\omega C_2}\right)}$$

$$= \frac{j\omega L_2}{1-\omega^2 L_2 C_2} \quad .....(11.91)$$

We will show that $Z_1 Z_2 = R_K{}^2$, so that this is also a constant K – filter. Multiplying Z_1 and Z_2, we get

$$Z_1 Z_2 = \frac{j(\omega^2 L_1 C_1 - 1)}{\omega C_1} \times \frac{j\omega L_2}{1-\omega^2 L_2 C_2}$$

Since $L_1 C_1 = L_2 C_2$, we have

$$Z_1 Z_2 = \frac{L_2}{C_1} = \frac{L_1}{C_2} = R_K{}^2 \quad .....(11.92)$$

Since $Z_1 Z_2 = R_K{}^2$, the theory developed in section 11.8 still holds good. At the cut off frequencies

$$Z_1 = -4 Z_2$$

or $$Z_1{}^2 = -4 Z_1 Z_2 = -4 R_K{}^2$$

and $$Z_1 = \pm 2j R_K \quad .....(11.93)$$

Thus the impedance Z_1 at lower cut off frequency f_1 is negative of Z_1 at upper cut off frequency f_2.

Therefore reactances of the series arm at ω_1 and ω_2 are

$$\frac{1}{\omega_1 C_1} - \omega_1 L_1 = \omega_2 L_1 - \frac{1}{\omega_2 C_1}$$

$$1 - \omega_1{}^2 L_1 C_1 = \frac{\omega_1}{\omega_2}(\omega_2{}^2 L_1 C_1 - 1) \quad .....(11.94)$$

But $$L_1 C_1 = \frac{1}{\omega_0^2}$$

Substituting this value in eq. (11.94), we get

$$1 - \frac{\omega_1^2}{\omega_0^2} = \frac{\omega_1}{\omega_2}\left(\frac{\omega_2^2}{\omega_0^2} - 1\right) = \frac{\omega_1\omega_2}{\omega_0^2} - \frac{\omega_1}{\omega_2}$$

$$\omega_2\,\omega_0^2 - \omega_1^2\,\omega_2 = \omega_1\,\omega_2^2 - \omega_1\,\omega_0^2$$

$$\omega_0^2\,(\omega_2 + \omega_1) = \omega_1\,\omega_2\,(\omega_1 + \omega_2)$$

or $$\omega_0^2 = \omega_1\,\omega_2$$

Thus the resonant frequency is the geometric mean of the two cut off frequencies. If the filter is terminated in a given design resistance R_K and the cut off frequencies f_1 and f_2 are given, the filter components can be designed.

From eq. (11.93), at lower cut off frequency f_1

$$Z_1 = -2j\,R_K$$

$$\left(\frac{-j}{\omega_1 C_1} + j\omega_1 L_1\right) = -2j\,R_K$$

$$1 - \omega_1^2\,L_1\,C_1 = 2R_K\,\omega_1\,C_1$$

$$1 - \frac{\omega_1^2}{\omega_0^2} = 2R_K\,\omega_1\,C_1$$

or $$1 - \frac{f_1^2}{f_0^2} = 4\pi\,R_K\,C_1\,f_1$$

But $$f_0^2 = f_1\,f_2$$

$\therefore$ $$\frac{f_2 - f_1}{f_2} = 4\pi\,R_K\,C_1\,f_1$$

$\therefore$ $$C_1 = \frac{f_2 - f_1}{4\pi R_K f_1 f_2} \qquad \text{......(11.95)}$$

But $$L_1\,C_1 = \frac{1}{\omega_0^2} = \frac{1}{4\pi^2 f_1 f_2}$$

From eq. (11.95), we get

$$L_1 = \frac{1}{4\pi^2 f_1 f_2} \cdot \frac{4\pi R_K f_1 f_2}{f_2 - f_1} = \frac{R_K}{\pi(f_2 - f_1)} \qquad \text{.....(11.96)}$$

The shunt arm element values can be easily calculated from eq. (11.92).

$$L_2 = C_1\,R_K^2 = R_K \cdot \frac{(f_2 - f_1)}{4\pi f_1 f_2} \qquad \text{.....(11.97)}$$

$$C_2 = \frac{L_1}{R_K^2} = \frac{1}{\pi R_K (f_2 - f_1)} \qquad \text{.....(11.98)}$$

Thus the design of a band pass filter is accomplished.

The variation of attenuation in the stop band and phase shift in pass band are shown in Fig. 11.26(a) and (b). The variation of Z_{0T} and $Z_{0\pi}$ are shown in Fig. 11.26(c).

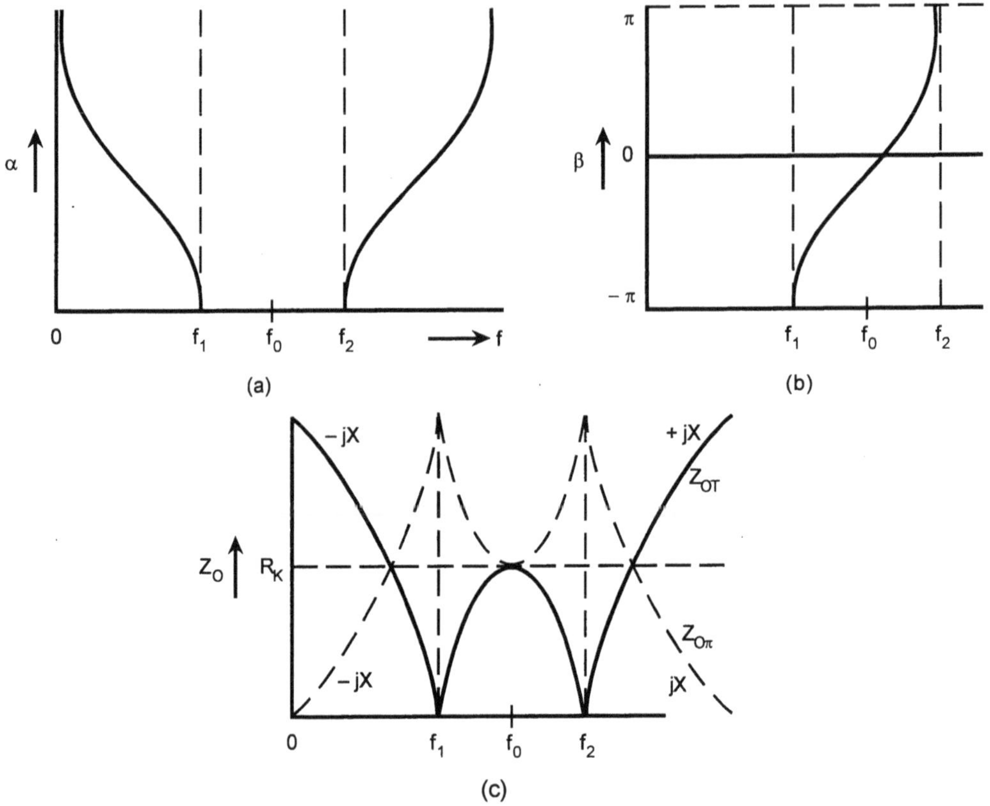

Fig. 11.26 (a) Variation of attenuation α (b) Variation of phase shift β (c) Variation of Z_{0T} and $Z_{0\pi}$

11.10.2 Bandpass m – derived filter :

We can obtain an m – derived bandpass filler from a prototype bandpass filter exactly in the same manner as discussed in section 11.9. The result is shown in Fig. 11.27 in which the shunt arm consists of a series resonant and an anti – resonant circuit in series. The anti – resonance frequency of this shunt arm is f_0 but the total shunt arm becomes resonant at two frequencies, one below f_0 and the other above f_0. This makes the shunt arm a short circuit and therefore infinite attenuation is produced at two frequencies on either side of f_0. These frequencies are termed as, $f_{\infty 1}$ and $f_{\infty 2}$, the frequencies at which infinite attenuation is produced. Thus the m – derived section could be used to produce high attenuation on both sides of the cut off frequency.

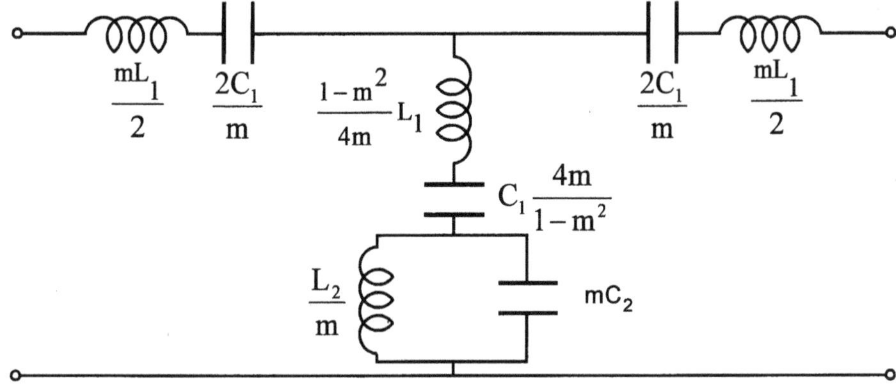

Fig. 11.27 Band pass m-derived filter

If we denote the reactance of series resonant circuit in the shunt arm by X_r and the reactance of anti resonant circuit in the shunt arm by X_{ar}, at one of the infinite frequencies $\omega_{\infty 1}$ the reactances are equal and opposite in sign.

Hence, we have

$$j\omega_\infty\left(\frac{1-m^2}{4m}\right)L_1 - \frac{j}{\omega_\infty\left[\frac{4m}{1-m^2}\right]C_1} = \frac{\left(j\omega_\infty\frac{L_2}{m}\right)\left(-\frac{j}{\omega_\infty mC_2}\right)}{j\left(\frac{1}{\omega_\infty mC_2}-\frac{\omega_\infty L_2}{m}\right)}$$

$$\frac{1-m^2}{4}\left(\omega_\infty{}^2L_1C_1-1\right) = \frac{\omega_\infty^2L_2C_1}{\omega_\infty^2L_2C_2-1} \qquad .....(11.99)$$

Since $\qquad L_1\,C_1 = L_2\,C_2 = \frac{1}{\omega_0^2}$

Eq. (11.99) can be written as

$$\frac{1-m^2}{4}\left(\frac{f_\infty^2}{f_0^2}-1\right)^2 = 4\pi^2\,f_\infty{}^2\,L_2\,C_1 \qquad .....(11.100)$$

From eqs. (11.95) and (11.97), we get

$$L_2C_1 = \frac{(f_2-f_1)^2}{16\pi^2f_1^2f_2^2} = \frac{(f_2-f_1)^2}{16\pi^2f_0^4}$$

Substituting the value of $L_2\,C_1$, in eq. (11.100), we get

$$\frac{1-m^2}{4}\left(\frac{f_\infty^2-f_0^2}{f_0^2}\right)^2 = 4\pi^2f_\infty^2\frac{(f_2-f_1)^2}{16\pi^2f_0{}^4}$$

$$(1 - m^2)\left(f_\infty^{\,2} - f_0^{\,2}\right)^2 = f_\infty^{\,2}\ (f_2 - f_1)^2$$

or $$(1 - m^2)\left(f_\infty^{\,2} - f_1 f_2\right)^2 = f_\infty^{\,2}\ (f_2 - f_1)^2 \qquad .....(11.101)$$

Taking square root on both sides and rearranging, we have

$$f_\infty^{\,2} \pm f_\infty \frac{(f_2 - f_1)}{\sqrt{1 - m^2}} - f_1 f_2 = 0$$

Solving for f_∞, we get

$$f_\infty = \pm \frac{f_2 - f_1}{2\sqrt{1 - m^2}} \pm \sqrt{\frac{(f_2 - f_1)^2}{4(1 - m^2)} + f_1 f_2} \qquad .....(11.102)$$

Since the term with the radical sign is larger than $\frac{f_2 - f_1}{2\sqrt{1 - m^2}}$, out of the four combination of eq. (11.102) only two are positive and hence are valid. Therefore

$$f_{\infty 1} = \sqrt{\frac{(f_2 - f_1)^2}{4(1 - m^2)} + f_1 f_2} - \frac{f_2 - f_1}{2\sqrt{1 - m^2}} \qquad .....(11.103)$$

$$f_{\infty 2} = \sqrt{\frac{(f_2 - f_1)^2}{4(1 - m^2)} + f_1 f_2} + \frac{f_2 - f_1}{2\sqrt{1 - m^2}} \qquad .....(11.104)$$

$\therefore$ $$f_{\infty 2} - f_{\infty 1} = \frac{f_2 - f_1}{\sqrt{1 - m^2}}$$

Solving for m, we get

$$m = \sqrt{1 - \frac{(f_2 - f_1)^2}{(f_{\infty 2} - f_{\infty 1})^2}} \qquad .....(11.105)$$

Further, taking the product of $f_{\infty 2}$ and $f_{\infty 1}$ we get

$$f_{\infty 1}\ f_{\infty 2} = \frac{(f_2 - f_1)^2}{4(1 - m^2)} + f_1\ f_2 - \frac{(f_2 - f_1)^2}{4(1 - m^2)}$$

$$= \frac{f_2^2 + f_1^2 - 2f_1 f_2 + 4f_1 f_2 - 4m^2 f_1 f_2 - f_2^2 - f_1^2 + 2f_1 f_2}{4(1 - m^2)}$$

$$= \frac{4f_1 f_2 (1 - m^2)}{4(1 - m^2)} = f_1\ f_2 = f_0^{\,2} \qquad .....(11.106)$$

Thus we see that $f_{\infty 1}\, f_{\infty 2} = f_0^2$.

If $f_{\infty 1}$ is chosen at a desired point $f_{\infty 2}$ is automatically fixed. From the values of f_1, f_2, $f_{\infty 1}$ and $f_{\infty 2}$ we can calculate the value of m. This completes the design of a m – derived band pass filter. To match the impedances over the pass band, m – derived half sections can be designed and used as terminating half sections.

11.11 BANDSTOP OR BAND ELIMINATION FILTER :

A bandstop filter attenuates a given range of frequencies and passes all other frequencies. Hence this filter characteristic may be obtained by cascading a low pass filter with a high pass filter in which the cut off frequency of LP filter is lower than the cut off frequency of HP filter.

11.11.1 Constant K Bandstop filter :

If the series arm and shunt arm of the bandpass filter are interchanged we get a bandstop filter. The T and π configurations of a prototype constant K bandstop filter are shown in Fig. 11.28.

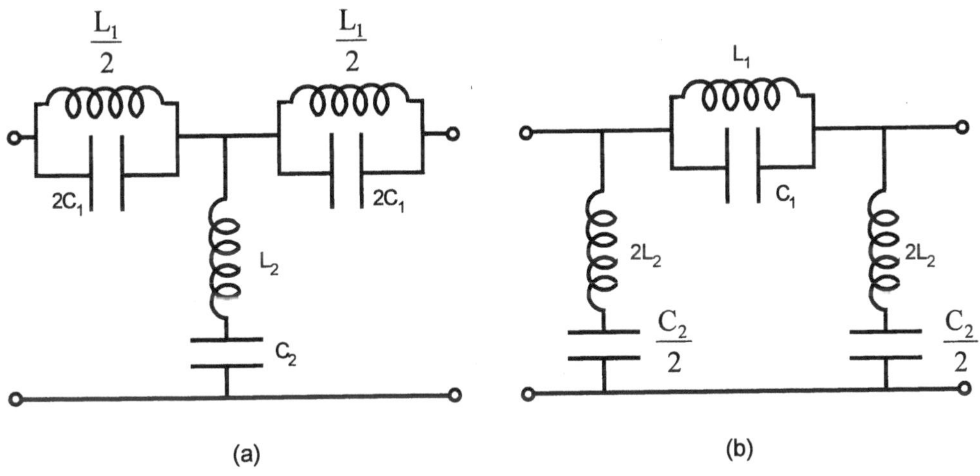

Fig. 11.28 Constant K bandstop (a) T filter (b) π filter

The series and shunt arms are made anti resonant and resonant respectively at the same frequency f_0. Thus

$$\frac{L_1}{C_2} = \frac{L_2}{C_1} = R_K^2 \quad .....(11.107)$$

and $$f_0^2 = f_1\, f_2 \quad .....(11.108)$$

At the cut off frequencies

$$Z_1 = -4\, Z_2$$

$$\therefore \quad Z_1\, Z_2 = -4\, Z_2^2 = R_K^2$$

or $$Z_2 = \pm \frac{jR_K}{2}$$

If the filter is terminated in a load resistance R_K then at the lower cut off frequency

$$Z_2 = j\left(\frac{1}{\omega_1 C_2} - \omega_1 L_2\right) = \frac{jR_K}{2}$$

$$1 - \omega_1^2 L_2 C_2 = \frac{R_K \omega_1 C_2}{2}$$

But $$L_2 C_2 = \frac{1}{\omega_0^2}$$

$\therefore$ $$1 - \frac{\omega_1^2}{\omega_0^2} = \frac{R_K \omega_1 C_2}{2}$$

or $$1 - \frac{f_1^2}{f_0^2} = \pi R_K C_2 f_1$$

$\therefore$ $$C_2 = \frac{1}{\pi R_K f_1}\left(\frac{f_0^2 - f_1^2}{f_0^2}\right)$$

Substituting $f_0^2 = f_1 f_2$ and simplifying, we get,

$$C_2 = \frac{1}{\pi R_K}\left(\frac{f_2 - f_1}{f_1 f_2}\right) \quad .....(11.109)$$

Also $$L_2 C_2 = \frac{1}{4\pi^2 f_0^2} = \frac{1}{4\pi^2 f_1 f_2}$$

$\therefore$ $$L_2 = \frac{1}{4\pi^2 f_1 f_2} \cdot \frac{f_2 f_1}{f_2 - f_1} \cdot \pi R_K = \frac{R_K}{4\pi(f_2 - f_1)} \quad .....(11.110)$$

From eq. (11.107), we get

$$C_1 = \frac{1}{4\pi R_K (f_2 - f_1)} \quad .....(11.111)$$

$$L_1 = \frac{R_K}{\pi} \frac{(f_2 - f_1)}{f_1 f_2} \quad .....(11.112)$$

Thus we have all the desired relations to design a constant K – band stop filter.

11.11.2 m - Derived bandstop filter :

The m – derived bandstop filter is shown in Fig. 11.29.

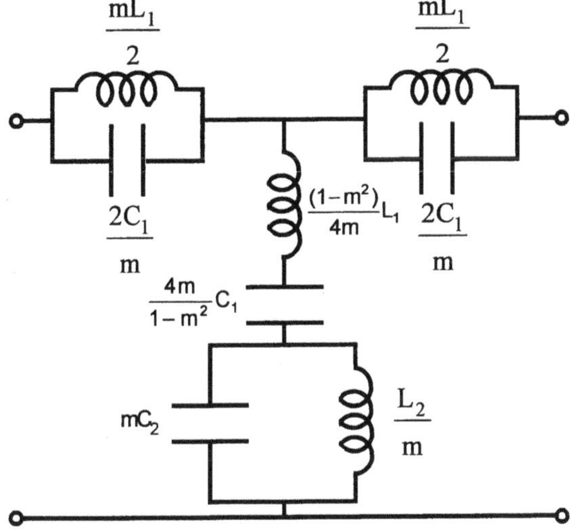

Fig. 11.29 m – derived bandstop filter

Proceeding on similar lines as in section 11.10.2 we can show that

$$m = \sqrt{1-\left(\frac{f_{\infty 2}-f_{\infty 1}}{f_2-f_1}\right)^2} \quad .....(11.113)$$

and $$f_{\infty 1} f_{\infty 2} = f_0^{\,2} \quad ......(11.114)$$

This value of m, together with the design equations for prototype band stop filter enables us to design the m – derived band stop filter.

Example 11.1

A low pass π section filter consists of an inductance of 25 mH in the series arm and two capacitors of 0.2 μF in the shunt arms. Calculate the cut off frequency, design impedance, attenuation at 5 KHZ and phase shift at 2 KHZ. Also find the characteristic impedance at 2 KHZ.

Solution :

The π section is shown in Fig. 11.30.

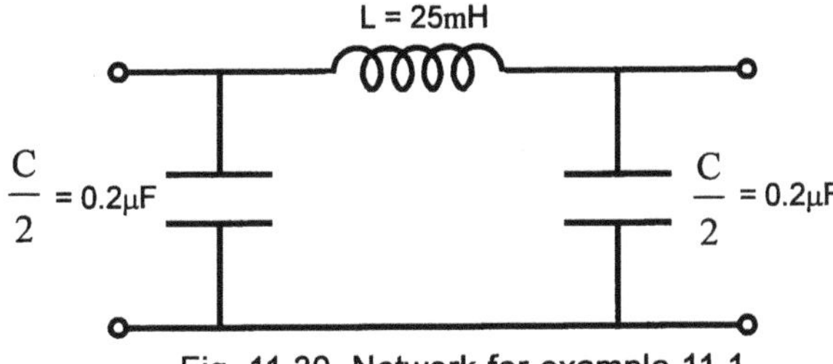

Fig. 11.30 Network for example 11.1

(a) The cut off frequency is given by eq. (11.88).

$$f_C = \frac{1}{\pi\sqrt{LC}}$$

In the given network L = 25mH and C = 2 × 0.2 = 0.4 μF

Hence $$f_C = \frac{1}{\pi\sqrt{25\times10^{-3}\times0.4\times10^{-6}}}$$

$$= 3.183 \text{ KHZ}$$

(b) Design impedance is given by

$$R_K = \sqrt{\frac{L}{C}} = \sqrt{\frac{25\times10^{-3}}{0.4\times10^{-6}}}$$

$$= 250 \text{ Ohms.}$$

(c) Attenuation at 5 KHz

The frequency 5 KHz is in the attenuation band and α is given by eq. (11.62)

$$\alpha = 2 \cos h^{-1} \frac{f}{f_C}$$

$$= 2 \cos h^{-1} \left(\frac{5}{3.183}\right)$$

$$= 2.0465 \text{ nepers}$$

$$= 17.78 \text{ db}$$

(d) Phase shift at 2 KHz

The frequency 2 KHz falls in passband and β is given by eq. (11.61).

$$\beta = 2 \sin^{-1} \frac{f}{f_C}$$

$$= 2 \sin^{-1} \frac{2}{3.183}$$

$$= 77.85^0$$

(e) Characteristic impedance $Z_{0\pi}$ at 2 KHz

$$Z_{0\pi} = \frac{R_K}{\sqrt{1-\left(\frac{f}{f_C}\right)^2}}$$

$$= \frac{250}{\sqrt{1-\left(\frac{2}{3.183}\right)^2}}$$

$$= 321.36 \text{ Ohms}$$

Example 11.2

A T section low pass filter has an inductance of 30 mH in each of the series arms and a shunt arm capacitance of 0.25 μF. Calculate the cut off frequency, characteristic impedance, ratio of input and output voltages, and phase shift at (a) 1 KHz and (b) 5 KHz.

Solution :

$$\frac{L}{2} = 30 \text{ mH}$$

$$L = 60 \text{ mH}$$

$$C = 0.25 \text{ μF}$$

(i) The cut off frequency

$$f_C = \frac{1}{\pi\sqrt{LC}}$$

$$= \frac{1}{\pi\sqrt{60\times10^{-3}\times0.25\times10^{-6}}}$$

$$= 2.6 \text{ KHZ}$$

(a) at 1 KHZ

$$Z_{0T} = R_K\sqrt{1-\left(\frac{\omega}{\omega_C}\right)^2}$$

Since $R_K = \sqrt{\frac{L}{C}} = \sqrt{\frac{60\times10^{-3}}{0.25\times10^{-6}}}$

$$= 490\ \Omega$$

$\therefore$ $Z_{0T} = 490\sqrt{1-\left(\frac{1}{2.6}\right)^2}$

$$= 452.3\ \Omega$$

Since 1 KHZ is in the passband $\alpha = 1$

and $$\frac{V_{in}}{V_{out}} = 1$$

The phase shift β is given by

$$\beta = 2 \sin^{-1}\left(\frac{\omega}{\omega_C}\right)$$

$$= 2 \sin^{-1}\left(\frac{1}{2.6}\right) = 45.24^0$$

(b) at 5 KHZ

Since this frequency is in stop band, Z_{0T} is imaginary.

The attenuation α is given by

$$\alpha = 2 \cos h^{-1}\left(\frac{\omega}{\omega_C}\right)$$

$$= 2 \cos h^{-1}\left(\frac{5}{2.6}\right) = 2.543 \text{ nepers}$$

The voltage ratio is

$$\frac{V_{in}}{V_{out}} = e^{\alpha} = e^{2.543} = 10.71$$

The phase shift in the attenuation band is equal to π rad.

Example 11.3

A prototype high pass filter has a cut off frequency of 8 KHz and nominal impedance of 600 ohms. Calculate the values of inductance and capacitance used in the filter.

Solution :

$$f_C = 8 \times 10^3 \text{ HZ}$$

$$R_K = 600\ \Omega$$

$$L = \frac{R_K}{4\pi f_C}$$

$$= \frac{600}{4\pi \times 8 \times 10^3}$$

$$= 5.97 \text{ mH}$$

$$C = \frac{1}{4\pi R_K f_C}$$

$$= \frac{1}{4\pi \times 600 \times 8 \times 10^3}$$

$$= 0.0165 \mu F$$

The high pass filter is given in Fig. 11.31

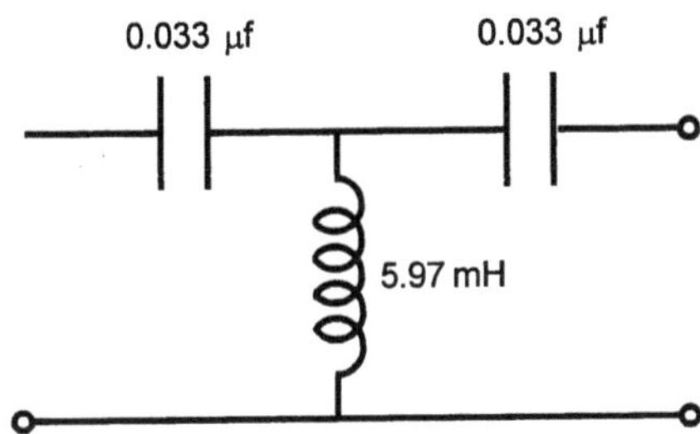

Fig. 11.31 HP filter for problem 11.3

Example 11.4

Design an m – derived T – section low pass filter having a cut off frequency f_C = 2000 HZ, design impedance R_K = 600 Ω and frequency of infinite attenuation f_∞ = 2050 HZ.

Solution :

$$R_K = 600\ \Omega$$
$$f_C = 2000\ HZ$$
$$f_\infty = 2050\ HZ$$

LP prototype filter :

$$L = \frac{R_K}{\pi f_C}$$

$$= \frac{600}{\pi \times 2000}$$

$$= 95.49\ mH$$

$$C = \frac{1}{\pi R_K f_C}$$

$$= \frac{1}{\pi \times 600 \times 2 \times 10^3}$$

$$= 0.265\ \mu F$$

m – derived LP filter :

$$m = \sqrt{1-\left(\frac{f_C}{f_\infty}\right)^2}$$

$$= \sqrt{1-\left(\frac{2000}{2050}\right)^2}$$

$$= 0.2195$$

$$\frac{mL}{2} = \frac{0.2195 \times 95.49}{2} = 10.48 \text{ mH}$$

$$mC = 0.2195 \times 0.265 = 0.0582 \ \mu F$$

$$\frac{1-m^2}{4m} L = 103.51 \text{ mH}$$

The m – derived T – section low pass filter is shown in Fig. 11.32.

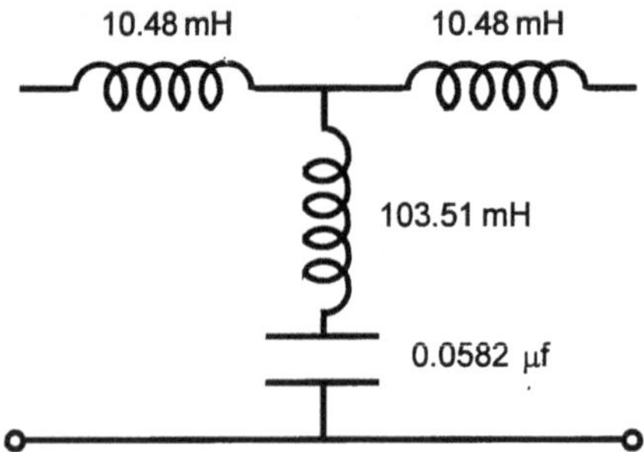

Fig. 11.32 m – derived T section H.P filter for example 11.4

Example 11.5

Design an m – derived π section high pass filter with a cut off frequency $f_C = 10$ KHz, $R_K = 600\ \Omega$ and $f_\infty = 8$ KHz

Solution :

$f_C = 10$ KHz
$R_K = 600\ \Omega$
$f_\infty = 8$ KHz

$$m = \sqrt{1-\left(\frac{f_\infty}{f_C}\right)^2}$$

$$= \sqrt{1-\left(\frac{8}{10}\right)^2} = 0.6$$

The prototype π section high pass filter :

$$L = \frac{R_K}{4\pi f_C}$$

$$= \frac{600}{4\pi \times 10 \times 10^3} = 4.775 \text{ mH}$$

$$C = \frac{1}{4\pi R_K f_C}$$

$$= \frac{1}{4\pi \times 600 \times 10 \times 10^3}$$

$$= 0.0133 \ \mu\text{F}$$

The π – section H P filter elements are

$$\frac{2L}{m} = \frac{2 \times 4.775}{0.6} = 15.92 \text{ mH}$$

$$\frac{C}{m} = \frac{0.0133}{0.6} = 0.0222 \ \mu\text{F}$$

$$\frac{4m}{1-m^2} L = \frac{4 \times 0.6}{1-0.36} \times 4.775 = 17.92 \text{ mH}$$

The filter circuit is given in Fig. 11.33

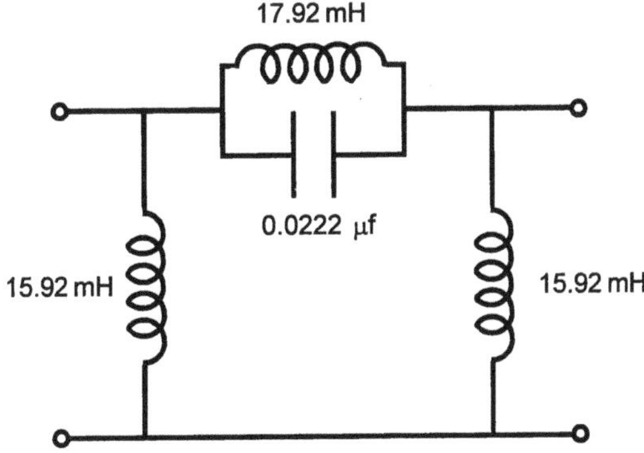

Fig. 11.33 m – derived HP π section filter for example 11.5

Example 11.6

Design a composite H.P T section filter to operate into a load impedance of 600 Ω, cut off frequency 1.5 KHz and f_∞ = 1.4 KHz.

Solution :

$$f_C = 1.5 \text{ KHz}$$

$$f_\infty = 1.4 \text{ KHz}$$

$$R_K = 600\Omega$$

(a) Prototype HP filter

$$L = \frac{R_K}{4\pi f_C} = \frac{600}{4\pi \times 1.5}$$

$$= 31.831 \text{ mH}$$

$$C = \frac{1}{4\pi R_K f_C}$$

$$= 0.0884 \ \mu\text{F}$$

(b) m – derived section

$$m = \sqrt{1 - \left(\frac{1.4}{1.5}\right)^2}$$

$$= 0.359$$

The elements of m – derived T section are

$$\frac{2C}{m} = \frac{2 \times 0.0884}{0.359}$$

$$= 0.4925 \ \mu\text{F}$$

$$\frac{L}{m} = \frac{31.831}{0.359}$$

$$= 88.67 \text{ mH}$$

$$\frac{4m}{1-m^2} C = \frac{4 \times 0.359}{1-(0.359)^2} \times 0.0884$$

$$= 0.1457 \mu\text{F}$$

(c) Terminating half section

$$m = 0.6$$

$$\frac{2C}{m} = \frac{2 \times 0.0884}{0.6} = 0.295 \ \mu\text{F}$$

$$\frac{2L}{m} = \frac{2\times 31.831}{0.6}$$
$$= 106.1 \text{ mH}$$

$$\frac{2m}{1-m^2}\ C = \frac{2\times 0.6}{1-0.36} \times 0.0884\ \mu F$$
$$= 0.1658\ \mu F$$

The composite filter is shown in Fig. 11.34 (a).

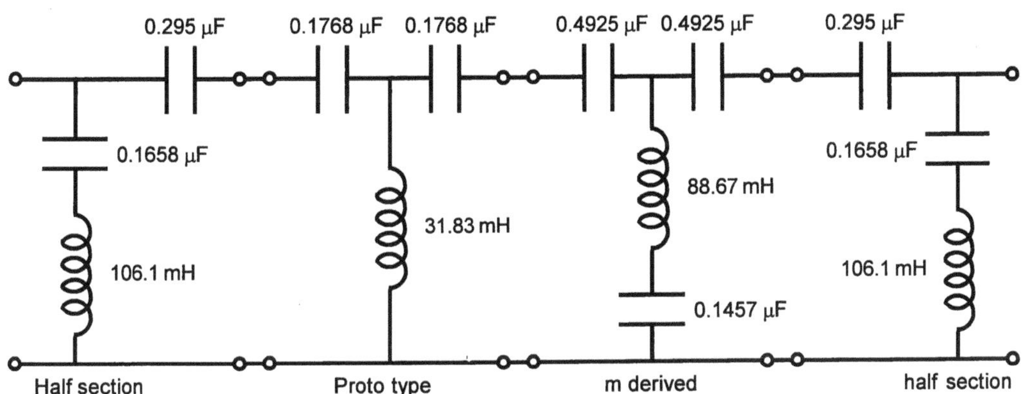

Fig. 11.34 (a) Composite filter for example 11.6

Combining the elements we get the final filter as shown in Fig. 11.34 (b)

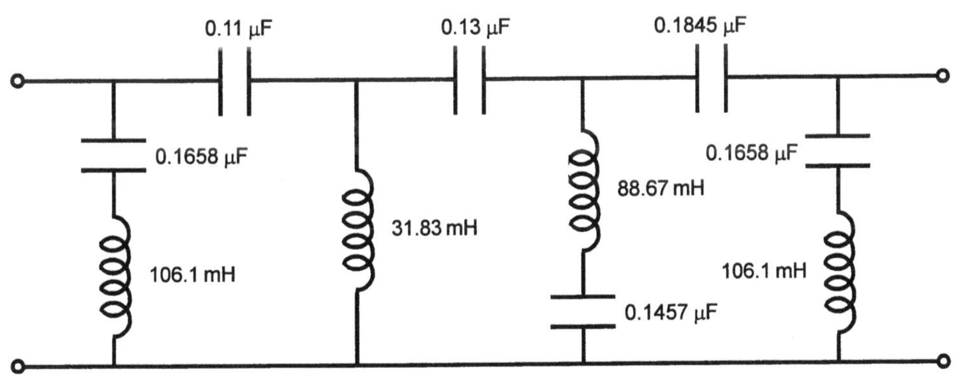

Fig. 11.34 (b) Simplified H.P filter

Example 11.7

Design a composite low pass filter to meet the following specifications.

$R_K = 600\ \Omega$

$f_C = 2000$ HZ

$$f_{\infty 1} = 2100 \text{ HZ}$$
$$f_{\infty 2} = 2400 \text{ HZ}$$

It is to have infinite attenuation at infinite frequency.

Solution :

$$R_K = 600 \text{ Ohms}$$
$$f_C = 2000 \text{ HZ}$$

Since it has infinite attenuation at 2 frequencies and at infinity, the filter will have 1 prototype, 2 m derived sections and half sections.

(a) LP prototype – T section

$$L = \frac{R_K}{\pi f_C}$$

$$= \frac{600}{\pi \times 2000}$$

$$= 95.5 \text{ mH}$$

$$C = \frac{1}{\pi R_K f_C}$$

$$= \frac{1}{\pi \times 600 \times 2000}$$

$$= 0.265 \ \mu\text{F}$$

(b) m – derived T section with

$$f_\infty = 2100 \text{ HZ}$$

$$m_1 = \sqrt{1 - \left(\frac{f_C}{f_{\infty 1}}\right)^2}$$

$$= \sqrt{1 - \left(\frac{2000}{2100}\right)^2}$$

$$= 0.305$$

$$\frac{m_1 L}{2} = \frac{0.305 \times 95.5}{2}$$

$$= 14.56 \text{ mH}$$

$$mC = 0.305 \times 0.265$$

$$= 0.081 \ \mu\text{F}$$

$$\frac{1-m^2}{4m}\,L = \frac{1-(0.305)^2}{4\times 0.305}\times 95.5$$

$$= 71 \text{ mH}$$

(c) m – derived section for $f_{\infty 2}$ = 2400 Hz

$$m_2 = \sqrt{1-\left(\frac{f_C}{f_\infty}\right)^2}$$

$$= \sqrt{1-\left(\frac{2000}{2400}\right)^2}$$

$$= 0.553$$

$$\frac{m_1 L}{2} = \frac{0.553\times 95.5}{2}$$

$$= 26.4 \text{ mH}$$

$$mC = 0.553 \times 0.265$$

$$= 0.1465\ \mu F$$

$$\frac{1-m^2}{4m}\,L = \frac{1-(0.553)^2}{4\times 0.553}\times 95.5$$

$$= 29\,.\,97 \text{ mH}$$

(d) Terminating half section

$$m = 0.6$$

$$\frac{mL}{2} = \frac{0.6\times 95.5}{2}$$

$$= 28.65\text{mH}$$

$$\frac{mC}{2} = \frac{0.6\times 0.265}{2}$$

$$= 0.0795\ \mu F$$

$$\frac{1-m^2}{2m}\,L = \frac{1-(0.6)^2}{2\times 0.6}\times 95.5$$

$$= 50.93 \text{ mH}$$

The composite filter is given in Fig. 11.35

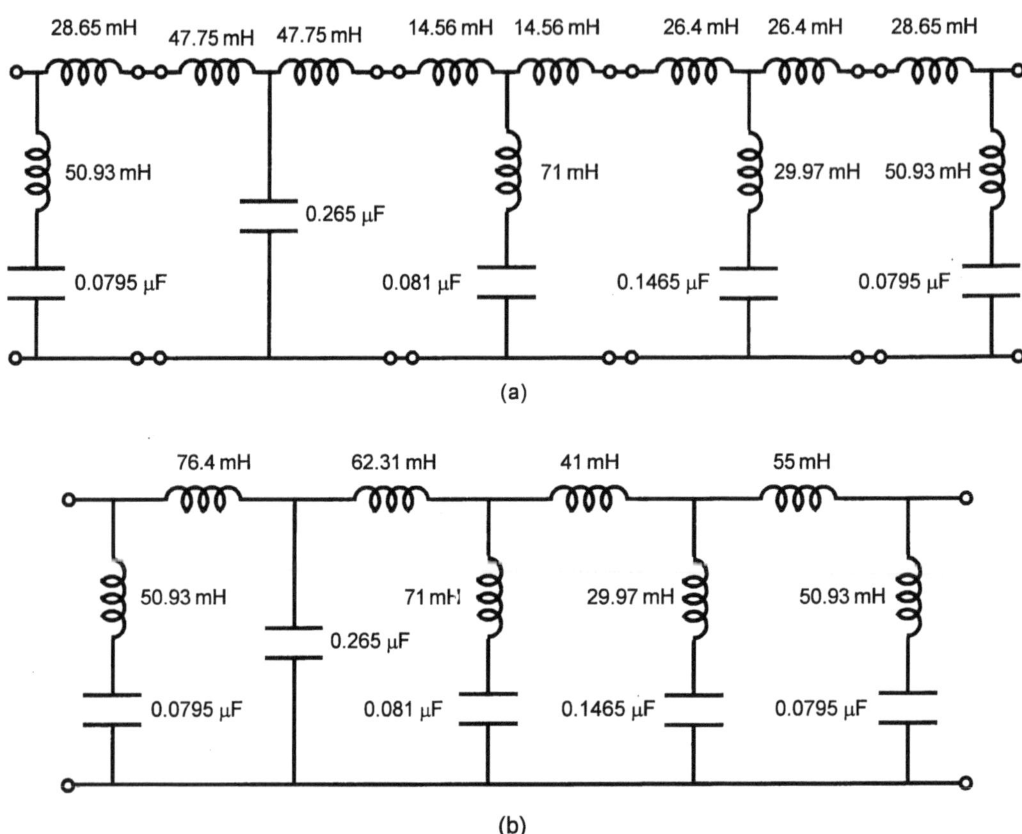

Fig. 11.35 (a) Composite filter (b) Simplified network for example 11.7.

Example 11.8

Design a prototype band pass filter with cut off frequencies 1.5 KHZ and 5 KHZ and a design impedances of 500 Ω.

Solution :

$$f_1 = 1.5 \text{ KHZ}$$

$$f_2 = 5 \text{ KHZ}$$

$$R_K = 500 \ \Omega$$

$$C_1 = \frac{f_2 - f_1}{4\pi R_K f_1 f_2}$$

$$= \frac{3.5 \times 10^3}{4\pi \times 500 \times 1.5 \times 5 \times 10^6}$$

$$= 0.0743\ \mu F$$

$$L_1 = \frac{R_K}{\pi(f_2 - f_1)}$$

$$= \frac{500}{\pi(3.5) \times 10^3}$$

$$= 45.47\ mH$$

$$C_2 = \frac{1}{\pi R_K (f_2 - f_1)}$$

$$C_2 = \frac{1}{\pi \times 500 \times 3.5 \times 10^3}$$

$$= 0.1819\ \mu F$$

$$L_2 = \frac{R_K(f_2 - f_1)}{4\pi f_1 f_2}$$

$$= \frac{500 \times 3.5 \times 10^3}{4\pi \times 1.5 \times 5 \times 10^6}$$

$$= 18.57\ mH$$

The filter network is given in Fig. 11.36

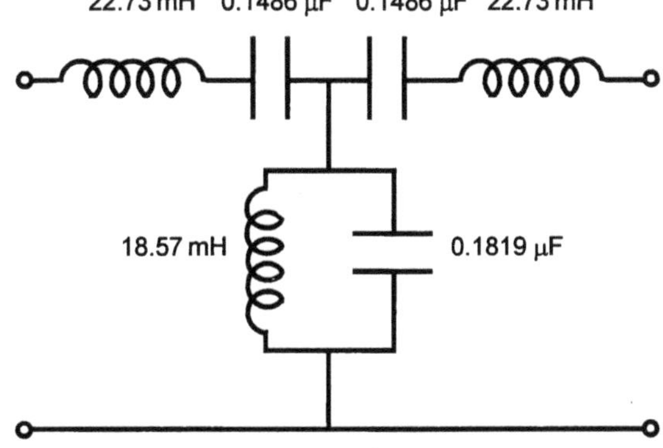

Fig. 11.36 Constant K bandpass filter for example 11.8

11.12 ATTENUATORS :

An attenuator is a two port network used to reduce the signal level by a given amount. These networks are usually resistive networks designed for the purpose of introducing a given loss when inserted between two networks. Since it is a resistive network, the propagation constant γ is real and is equal to the attenuation constant α. Attenuators are used in place of transformers for matching circuits of different resistances, as volume controls in radio circuits etc. A fixed attenuator is also known as a 'pad'.

11.12.1 Symmetrical attenuators :

Symmetrical attenuators are designed so that both the pairs of terminals are matched to the same resistance. This resistance is known as the characteristic resistance, R_0, of the attenuator. The attenuation is measured in nepers or decibels. Recall that the decibel was defined as

$$DB = 20 \log_{10} \frac{I_1}{I_2} \qquad .....(11.115)$$

Defining $$N = \frac{I_1}{I_2} \qquad .....(11.116)$$

$$DB = 20 \log_{10} N$$

and $$N = \text{antilog} \frac{DB}{20} \qquad .....(11.117)$$

If the attenuation to be produced by the attenuator network is given in DB, the factor N can be calculated. There are different types of attenuators among which the important ones are : T, π, bridged T and lattice networks. We shall discuss the design of these networks in the following sections.

11.12.2 Symmetrical T – attenuator :

Consider the symmetrical T – network shown in Fig. 11.37. The network is terminated in its image impedance R_0.

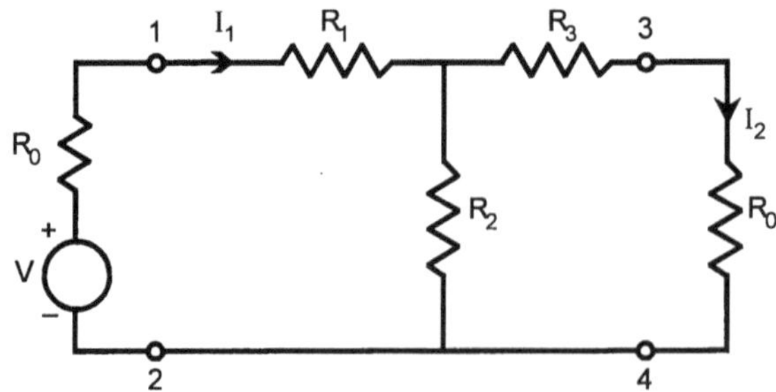

Fig. 11.37 Symmetrical T – attenuator

Writing the loop equation for loop 2

$$I_2 (R_1 + R_2 + R_0) - I_1R_2 = 0$$

$$\therefore \quad \frac{I_1}{I_2} = \frac{R_1 + R_2 + R_0}{R_2}$$

But $$\frac{I_1}{I_2} = N$$

$$\therefore \quad N = \frac{R_1 + R_2 + R_0}{R_2} \quad .....(11.118)$$

We know that the impedance looking into terminals 1, 2 when the network is terminated in image impedance R_0, is equal to R_0 and therefore

$$R_0 = R_1 + \frac{R_2(R_0 + R_1)}{R_1 + R_2 + R_0}$$

Using eq. (11.118),

$$R_0 = R_1 + \frac{R_0 + R_1}{N}$$

Solving for R_1, we get

$$R_1 = \frac{R_0(N-1)}{N+1} \quad .....(11.119)$$

Substituting, the value of R_1 from eq. (11.119) in eq. (11.118), we obtain

$$R_2 = \frac{2R_0N}{N^2 - 1} \quad .....(11.120)$$

The attenuation introduced by this network can be calculated from the formula

$$\alpha = \cos h^{-1}\left(1 + \frac{Z_1}{2Z_2}\right) = \cos h^{-1}\left(1 + \frac{R_1}{R_2}\right) \text{ nepers} \quad .....(11.121)$$

11.12.3 Symmetrical π - attenuator :

The symmetrical π – attenuator is shown in Fig. 11.38.

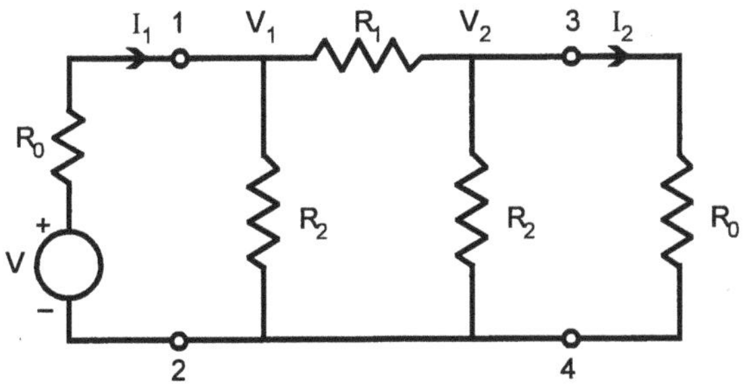

Fig. 11.38 Symmetrical π – attenuator

Writing the node voltage equation at node 3, we have

$$V_2\left(\frac{1}{R_1}+\frac{1}{R_2}+\frac{1}{R_0}\right)-\frac{V_1}{R_1}=0$$

Solving for $\frac{V_1}{V_2}$

$$\frac{V_1}{V_2}=R_1\left[\frac{1}{R_1}+\frac{1}{R_2}+\frac{1}{R_0}\right]$$

But $$\frac{V_1}{V_2}=\frac{I_1}{I_2}=N$$

$\therefore$ $$N=\frac{R_0R_2+R_1R_0+R_1R_2}{R_2R_0} \qquad .....(11.122)$$

But $$R_0=\frac{R_2\left[R_1+\dfrac{R_0R_2}{R_0+R_2}\right]}{R_2+R_1+\dfrac{R_0R_2}{R_0+R_2}}$$

$$=\frac{R_2[R_1R_0+R_1R_2+R_0R_2]}{R_2(R_0+R_2)+(R_1R_0+R_1R_2+R_0R_2)} \qquad .....(11.123)$$

Using eq. (11.122) in eq. (11.123), we have

$$R_0=\frac{R_2^2R_0N}{R_2(R_0+R_2)+NR_2R_0}$$

Solving for R_2, we get

$$R_2=R_0\left(\frac{N+1}{N-1}\right) \qquad .....(11.124)$$

Substituting the value of R_2 in eq. (11.122) and solving for R_1, we get

$$R_1=\frac{R_0\left(N^2-1\right)}{2N} \qquad .....(11.125)$$

These are the design equations for a symmetrical π attenuator.

11.12.4 Symmetrical bridged T – attenuator :

The third symmetrical attenuator which is commonly used is the Bridged T – attenuator shown in Fig. 11.39.

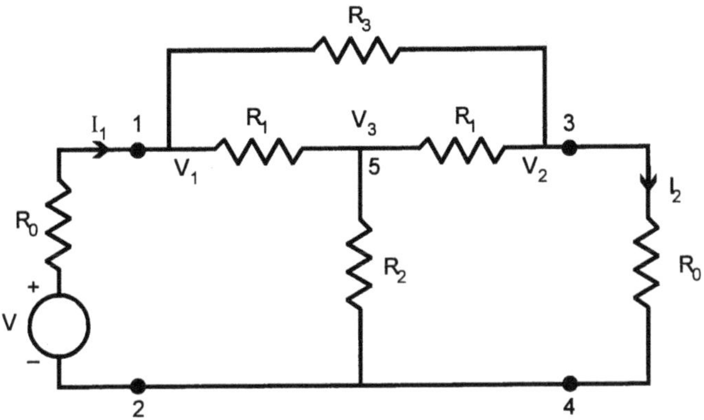

Fig. 11.39 Bridged T attenuator

Let us first analyse the network to express the characteristic resistance R_0 in term of R_1, R_2 and R_3.

Since the bridge T – attenuator is a symmetrical network, the characteristic resistance is given by

$$R_0 = \sqrt{R_{SC}R_{OC}}$$

R_{SC} is the short circuit resistance and R_{OC} is the open circuit impedance at terminals 1, 2 of the network in Fig. 11.39. Thus

$$R_{SC} = \frac{R_3\left(R_1 + \dfrac{R_1R_2}{R_1 + R_2}\right)}{R_3 + R_1 + \dfrac{R_1R_2}{R_1 + R_2}}$$

$$= \frac{R_3\left(2R_1R_2 + R_1^2\right)}{R_3R_1 + R_3R_2 + R_1^2 + 2R_1R_2}$$

$$R_{OC} = R_2 + \frac{R_1\left(R_1 + R_3\right)}{2R_1 + R_3}$$

$$= \frac{2R_1R_2 + R_2R_3 + R_1^2 + R_1R_3}{2R_1 + R_3}$$

and $$R_0 = \sqrt{\frac{R_3\left(2R_1R_2 + R_1^2\right)}{2R_1 + R_3}} \qquad .....(11.126)$$

Usually $R_1 = R_0$ and only R_2 and R_3 are made variable. Substituting the value of R_1 eq. (11.106), we get

$$R_0 = \sqrt{\frac{R_3\left(2R_0R_2 + R_0^2\right)}{2R_0 + R_3}}$$

$$R_0^2 = \frac{2R_0R_2R_3 + R_0^2R_3^2}{2R_0 + R_3}$$

Solving for R_0, we get

$$R_0 = \sqrt{R_2R_3} \qquad .....(11.127)$$

Referring to Fig. 11.39, writing node voltage equations at nodes 3 and 5, and with $R_1 = R_0$

$$V_2\left(\frac{1}{R_0} + \frac{1}{R_0} + \frac{1}{R_3}\right) - \frac{V_1}{R_3} - \frac{V_3}{R_0} = 0 \qquad .....(11.128)$$

and
$$V_3\left(\frac{2}{R_0} + \frac{1}{R_2}\right) - \frac{V_2}{R_0} - \frac{V_1}{R_0} = 0 \qquad .....(11.129)$$

Eliminating V_3 from eq. (11.128) and (11.129), we get

$$\frac{V_1}{V_2} = \frac{\dfrac{2R_3 + R_0}{R_0R_3} - \dfrac{R_2}{R_0\left(2R_2 + R_0\right)}}{\dfrac{1}{R_3} + \dfrac{R_2}{R_0\left(2R_2 + R_0\right)}}$$

Noting that $\dfrac{V_1}{V_2} = \dfrac{I_1}{I_2} = N$, we have

$$N = \frac{\left(2R_2 + R_0\right)\left(2R_3 + R_0\right) - R_2R_3}{2R_0R_2 + R_0^2 + R_2R_3}$$

Substituting $R_0^2 = R_2\,R_3$ and simplifying we get

$$N = 1 + \frac{R_3}{R_0} \qquad .....(11.130)$$

Since $R_0^2 = R_2\,R_3$, eq. (11.130) can also be written as

$$N = 1 + \frac{R_0}{R_2} \qquad .....(11.131)$$

From eqs. (11.130) and (11.131) we have

$$R_3 = R_0\,(N - 1) \qquad .....(11.132)$$

and
$$R_2 = \frac{R_0}{N - 1} \qquad .....(11.133)$$

The bridged T – attenuator is more economical and convenient to use as a variable attenuator because, only two resistances have to be changed to change the attenuation without affecting R_0, whereas in T and π attenuators 3 resistances have to be changed.

11.12.5 Symmetrical lattice attenuator :

The symmetrical lattice attenuator is shown in Fig. 11.40(a). It is redrawn as a bridge network in Fig. 11.40(b).

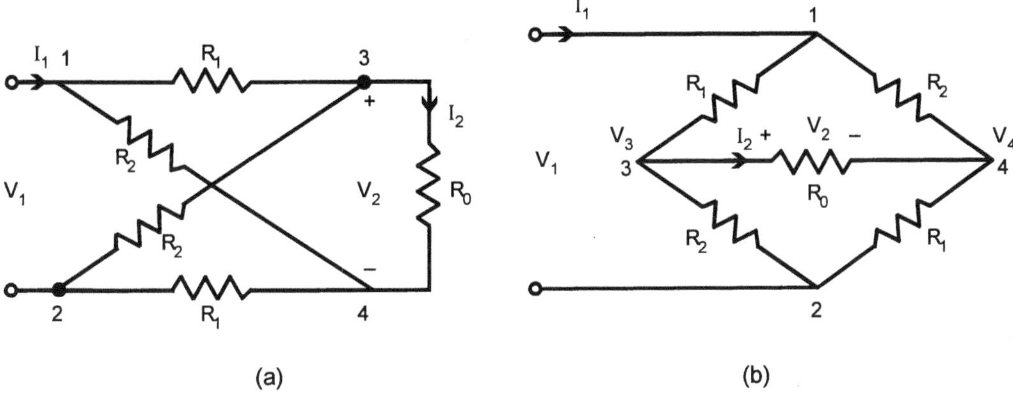

Fig. 11.40 (a) Symmetrical lattice attenuator (b) Redrawn as a bridge circuit

From Fig. 11.40(b) we have

$$Z_{SC} = \frac{R_1R_2}{R_1+R_2} + \frac{R_1R_2}{R_1+R_2}$$

$$= \frac{2R_1R_2}{R_1+R_2} \quad .....(11.134)$$

$$Z_{OC} = \frac{(R_1+R_2)}{2} \quad .....(11.135)$$

$$\therefore \quad R_0 = \sqrt{Z_{SC}.Z_{OC}}$$

$$= \sqrt{\frac{2R_1R_2}{R_1+R_2}.\frac{(R_1+R_2)}{2}} = \sqrt{R_1R_2} \quad .....(11.136)$$

Referring to Fig. 11.40(a), and writing node voltage equations for nodes 3 and 4, taking node 2 as reference, we have

$$\frac{V_3-V_4}{R_0} + \frac{V_3-V_1}{R_1} + \frac{V_3}{R_2} = 0 \quad .....(11.137)$$

and

$$\frac{V_4-V_3}{R_0} + \frac{V_4}{R_1} + \frac{V_4-V_1}{R_2} = 0 \quad .....(11.138)$$

Noting that $V_3 - V_4 = V_2$, eq. (11.137) and (11.138) become

$$\frac{V_2}{R_0} + V_3\left(\frac{1}{R_1} + \frac{1}{R_2}\right) = \frac{V_1}{R_1} \qquad .....(11.139)$$

$$-\frac{V_2}{R_0} + V_4\left(\frac{1}{R_1} + \frac{1}{R_2}\right) = \frac{V_1}{R_2} \qquad .....(11.140)$$

Subtracting eq. (11.139) from eq. (11.140) and again replacing $(V_3 - V_4)$ by V_2,

$$\frac{2V_2}{R_0} + V_2\left(\frac{1}{R_1} + \frac{1}{R_2}\right) = V_1\left(\frac{1}{R_1} - \frac{1}{R_2}\right)$$

or $$V_2\left[\frac{2R_1R_2 + R_0R_1 + R_2R_0}{R_0R_1R_2}\right] = V_1\left[\frac{R_2 - R_1}{R_1R_2}\right]$$

Thus $$\frac{V_1}{V_2} = N = \frac{1}{R_0} \cdot \frac{(2R_1R_2 + R_0R_1 + R_2R_0)}{R_2 - R_1}$$

From eq. (11.136), $R_0^2 = R_1 R_2$ and

$$N = \frac{2\sqrt{R_1R_2} + R_1 + R_2}{R_2 - R_1} = \frac{\left(\sqrt{R_1} + \sqrt{R_2}\right)^2}{\left(\sqrt{R_2} - \sqrt{R_1}\right)\left(\sqrt{R_2} + \sqrt{R_1}\right)}$$

$$N = \frac{1 + \sqrt{\dfrac{R_1}{R_2}}}{1 - \sqrt{\dfrac{R_1}{R_2}}}$$

or, in terms of R_0 and R_1

$$N = \frac{1 + \dfrac{R_1}{R_0}}{1 - \dfrac{R_1}{R_0}} \qquad .....(11.141)$$

from which we get

$$R_1 = R_0\left(\frac{N-1}{N+1}\right) \qquad .....(11.142)$$

and $$R_2 = R_0 \frac{(N+1)}{N-1} \qquad .....(11.143)$$

These are the required design equations for a symmetrical lattice attenuator.

Example 11.9

Design a T – attenuator pad to give an attenuation of 20 db. The characteristic resistance is 500 Ω.

Solution :

It is given that

$$DB = 20$$

$$R_0 = 500\ \Omega$$

$$\therefore \quad N = \text{Antilog}\ \frac{D}{20}$$

$$= \text{Antilog}\ \frac{20}{20}$$

$$= 10$$

From eq. (11.119) and (11.120)

$$R_1 = \frac{R_0(N-1)}{(N+1)} = \frac{500(10-1)}{10+1} = 409.1\ \Omega$$

and $$R_2 = \frac{R_0\ 2N}{N^2-1} = \frac{500\times2\times10}{100-1} = 101.01\ \Omega$$

The T – attenuator is given in Fig. 11.41.

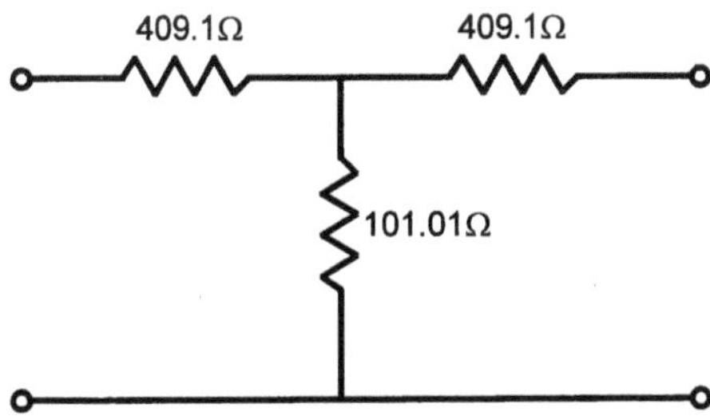

Fig 11.41 T – attenuator for example 11.9

Problems

11.1 Obtain the image impedance of the network shown in Fig. P. 11.1.

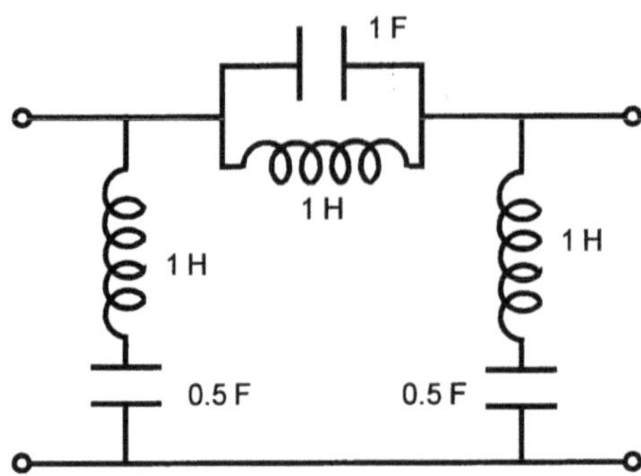

Fig. P. 11.1

11.2 Determine the image impedance of the network shown in Fig. P. 11.2.

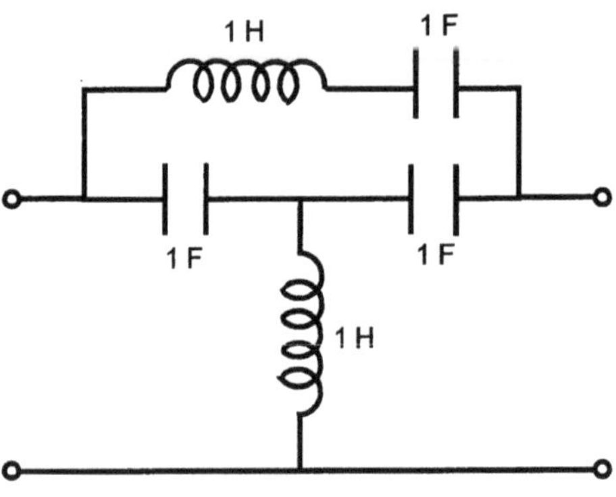

Fig. P. 11.2

11.3 Obtain the propagation constant, attenuation, passband and critical frequencies of the filter shown in Fig. P. 11.3.

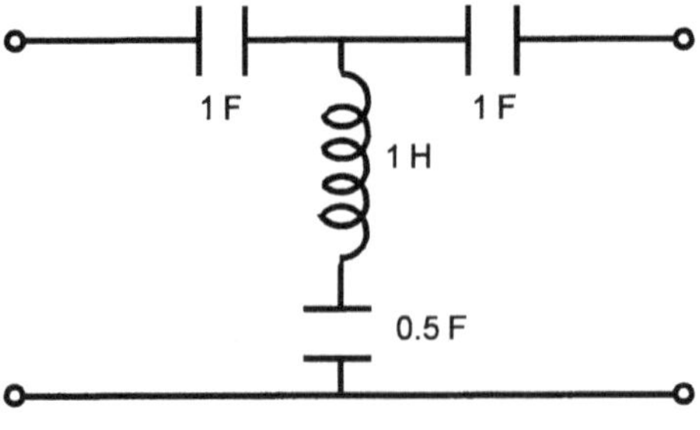

Fig. P. 11.3

11.4 Design a constant K prototype low pass (a) T – Section (b) π – Section filter to operate with a terminating load of 600 Ω and a cut off frequency of 5 KHz. Find the phase shift, attenuation and the characteristic impedance at (i) 2.5 KHz and (ii) 10 KHz in each case.

11.5 Design a constant K prototype high pass filter with a cut off frequency of 5 KHz working into an impedance of 600 Ω. Calculate the phase shift, attenuation and the characteristic impedance at (a) 2.5 KHz and (b) 8 KHz.

11.6 Design an m-derived T-section low pass filter having a cut off frequency of 2 KHz. Take R_K = 600 Ω and f_∞ = 2.1 KHz. Find the attenuation at f = 2.05 KHz and 2.5 KHz.

11.7 Design an m-derived π-section high pass filter having a cut off frequency of 10 KHz, R_k = 600 Ω and f_∞ = 9.5 KHz. Find the attenuation at f = 9.8 KHz and 5 KHz.

11.8 Design a composite low pass T-section filter with the following specifications f_C = 2 KHz, f_∞ = 2.05 KHz and R_K = 600 Ω.

11.9 Design a composite high pass π – section filter with the specifications given below:
f_C = 2 KHz, f_∞ = 1.9 KHz and R_K = 600 Ω.

11.10 Design a high pass composite filter to operate into a load of 60 Ω and with a cut off frequency of 1.5 MHz. The filter should consist of a prototype, and m – derived section with m = 0.4 and terminating half sections. What is the frequency of infinite attenuation.

11.11 Design a bandpass T–section filter with cut off frequencies of 5 KHz and 10 KHz operating into a load of 500 Ω.

11.12 Design a bandstop T–section filter with cut off frequencies 10 KHz and 20 KHz. The design impedance R_K = 600 Ω.

11.13 Design a T– attenuator to give an attenuation of 10 db. The characteristic resistance is 600 Ω.

11.14 Design a symmetrical lattice attenuator to have a characteristic impedance of 100 Ω and an attenuation of 12 db.

11.15 Design a bridged T - attenuator which produces an attenuation of 10 db and working between two 100 Ω impedances.

11.16 A symmetrical π – attenuator has a series arm of value 371 Ω and two shunt arms of value 91.67 Ω. Find the attenuation produced by the network when it operates into a load of 75 Ω.

11.17 Design a symmetrical lattice attenuator with a characteristic resistance of 75 Ω and producing an attenuation of 12 db.

12

FOURIER SERIES AND FOURIER TRANSFORMS

12.1 INTRODUCTION :

The analysis of networks with sinusoidal inputs is made easy by the consideration of frequency domain circuits. The need for analysis of networks with nonsinusoidal periodic signals arose in communication networks. Fourier has shown that any nonsinusoidal periodic signal could be resolved into sinusoidal components of different frequencies. The networks were required to pass certain of these frequencies to get desired output.

In this chapter we will consider the basic theorem due to Fourier and apply it to periodic functions. He has shown that any periodic waveform can be considered as an infinite summation of harmonic sinusoides. This sum of infinite terms is called the Fourier series. Certain symmetry conditions of the waveforms simplify the Fourier analysis. Later, the Fourier theorem is extended to nonperiodic waveforms to result in Fourier transform of a function. The convergence of Fourier series and Fourier transform is also discussed. Application of these concepts to simple networks excited by periodic waveforms and aperiodic waveforms is also given.

12.2 FOURIER SERIES :

In the year 1822 Fourier showed that any arbitrary function could be expanded in terms of infinite series of sinusoidal components. The exact conditions under which this could be done were given by Norbert Wiener in 1933, almost a hundred years later. Let us first consider the case of periodic functions.

All periodic functions, satisfying certain conditions, with a period T, can be expanded into an infinite series, called Fourier series of the form

$$\begin{aligned} f(t) = {} & a_0 + a_1 \cos\omega_0 t + a_2 \cos 2\omega_0 t + \ldots + a_n \cos n\omega_0 t \\ & + \ldots + b_1 \sin\omega_0 t + b_2 \sin 2\omega_0 t + \ldots + b_n \sin n\omega_0 t \\ & + \ldots \end{aligned} \qquad \ldots(12.1)$$

where $\omega_0 = \dfrac{2\pi}{T}$ is the frequency of the given waveform and is known as the fundamental frequency. The a^s are b^s are constants to be found so that eq. (12.1) is satisfied. The process of determining the coefficients is known as Fourier analysis. Dealing with infinite number of sinusoidal components may seem impractical but it is observed that most of the series converge rather rapidly and it is adequate to represent the given waveform by a finite number of terms.

Fourier analysis is carried out by the use of orthogonality properties of sinusoidal functions. The set of periodic functions $f_n(t)$ for n = 1, 2,..... , with period T, are said to be orthogonal if they satisfy the following conditions.

$$\int_0^T f_i(t) f_j(t)\, dt = 0 \quad \text{if } i \neq j$$

$$= \text{a constant if } i = j \qquad .....(12.2)$$

There are many functions which satisfy these conditions and it is therefore possible to represent any periodic function in terms of this set of infinite functions. Sinusoidal functions also satisfy these conditions as can be verified easily by simple integration. Thus the set of functions sin $n\omega_0 t$, cos $n\omega_0 t$ for n = 1, 2, ... satisfy the conditions:

$$\int_0^T \sin n\omega_0 t \cos m\omega_0 t\, dt = 0 \quad \text{for all n, m}$$

$$\int_0^T \sin n\omega_0 t \sin m\omega_0 t\, dt = \int_0^T \cos n\omega_0 t \cos m\omega_0 t\, dt$$

$$= 0 \qquad \text{for all } n \neq m$$

$$= \frac{T}{2} \qquad \text{for } n = m \qquad .....(12.3)$$

Now, integrating, eq. (12.1) on both sides over a period T, we get

$$\int_0^T f(t)\, dt = \int_0^T a_0\, dt + \int_0^T a_1 \cos\omega_0 t\, dt + \int_0^T a_2 \cos 2\omega_0 t\, dt$$

$$+ \,.....\, + \int_0^T b_1 \sin\omega_0 t\, dt + \int_0^T b_2 \sin 2\omega_0 t\, dt \; + ...$$

All the integrals on the right side of the equation except the first are zero and

$$\int_0^T a_0\, dt = a_0 T$$

Thus $$a_0 = \frac{1}{T} \int_0^T f(t)\, dt \qquad .....(12.4)$$

we observe that a_0 is nothing but the average value of the given function f(t) over a period. a_0 is also called as the DC term.

To find a_n, we multiply eq. (12.1) by cos $n\omega_0 t$ and integrate over a period and use orthogonality property of the sine functions.

$$\int_0^T f(t)\cos n\omega_0 t\ dt = \int_0^T a_0 \cos n\omega_0 t\ dt + \int_0^T a_1 \cos \omega_0 t \cos n\omega_0 t\ dt$$

$$+ \ldots\ldots + \int_0^T a_n \cos n\omega_0 t . \cos n\omega_0 t\ dt + \ldots\ldots$$

$$+ \int_0^T b_1 \sin \omega_0 t \cos n\omega_0 t\ dt + \ldots\ldots +$$

$$+ \int_0^T b_n \sin \omega_0 t \cos n\omega_0 t\ dt + \ldots\ldots \qquad \ldots\ldots(12.5)$$

From the orthogonality property, we observe that all the terms on the right side of eq. (12.5) except the term

$$\int_0^T a_n \cos^2 n\omega_0 t\ dt$$

are zero and from eq. (12.2),

$$\int_0^T a_n \cos^2 n_0\omega t\ dt = a_n \frac{T}{2} \qquad \ldots\ldots(12.6)$$

Thus $$a_n = \frac{2}{T}\int_0^T f(t)\cos n\omega_0 t\ dt \qquad \text{for } n = 1, 2, \ldots\ldots \qquad \ldots\ldots(12.7)$$

Similarly b_n could be found by multiplying eq. (12.1) by sin $n\omega_0 t$ and integrating over a period. The expression for b_n is,

$$b_n = \frac{2}{T}\int_0^T f(t)\sin n\omega_0 t\ dt \qquad \text{for } n = 1, 2, \ldots\ldots \qquad \ldots\ldots(12.8)$$

Let us take an example to illustrate the computation of Fourier coefficients.

Example 12.1

Consider the periodic square function shown in Fig. 12.1. Find the Fourier series.

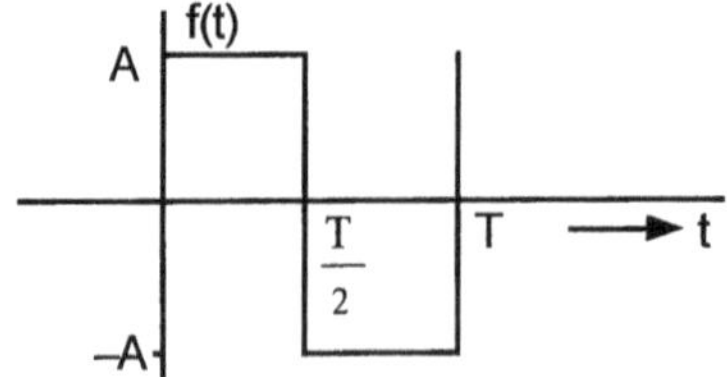

Fig. 12.1 Square waveform

Solution :

The function f(t) is defined by

$$f(t) = A \qquad 0 < t < \frac{T}{2}$$

$$= -A \qquad \frac{T}{2} < t < T$$

$$a_0 = \frac{1}{T}\int_0^T f(t)\,dt$$

$$= \frac{1}{T}\int_0^{\frac{T}{2}} A\,dt + \frac{1}{T}\int_{\frac{T}{2}}^{T}(-A)\,dt$$

$$= \frac{A}{T}\left[\frac{T}{2}-0\right] - \frac{A}{T}\left[T-\frac{T}{2}\right]$$

$$= 0$$

As can be easily observed, the average value of the given function over a period is zero.

$$a_n = \frac{2}{T}\int_0^T f(t)\cos n\omega_0 t\,dt$$

$$= \frac{2}{T}\int_0^{\frac{T}{2}} A\cos n\omega_0 t\,dt + \frac{1}{T}\int_{\frac{T}{2}}^{T} -A\cos n\omega_0 t\,dt$$

$$= \frac{2A}{T}\left[\frac{\sin n\omega_0 t}{n\omega_0}\Bigg|_0^{\frac{T}{2}} - \frac{\sin n\omega_0 t}{n\omega_0}\Bigg|_{\frac{T}{2}}^{T}\right]$$

$$= \frac{2A}{Tn\omega_0}\left[\sin n\omega_0\frac{T}{2} + \sin n\omega_0\frac{T}{2} - \sin n\omega_0 T\right]$$

But $\quad \omega_0 = \frac{2\pi}{T} \quad$ or $\quad \omega_0 T = 2\pi$

Thus $\quad a_n = \frac{2A}{2\pi n}\left[2\sin n\pi - \sin 2n\pi\right]$

$= 0 \qquad$ for all n = 1, 2, ...

Similarly

$$b_n = \frac{2}{T}\int_0^T f(t)\sin n\omega_0 t\, dt$$

$$= \frac{2}{T}\int_0^{\frac{T}{2}} A\sin n\omega_0 t\, dt + \frac{2}{T}\int_{\frac{T}{2}}^{T} -A\sin n\omega_0 t\, dt$$

$$= \frac{2A}{T}\left[-\frac{\cos n\omega_0 t}{n\omega_0}\Bigg|_0^{\frac{T}{2}} + \frac{\cos n\omega_0 t}{n\omega_0}\Bigg|_{\frac{T}{2}}^{T}\right]$$

$$= \frac{2A}{n\omega_0 T}\left[1 - \cos n\omega_0\frac{T}{2} - \cos n\omega_0\frac{T}{2} + \cos n\omega_0 T\right]$$

$$= \frac{A}{\pi n}[2 - 2\cos n\pi]$$

$$= \frac{2A}{\pi n}[1 - \cos n\pi]$$

$$= \frac{4A}{\pi n} \qquad \text{for n = 1, 3, 5.....}$$

$$= 0 \qquad \text{for n = 2, 4, 6.....}$$

Hence the Fourier series for the wavefrom shown is Fig. 12.1 is

$$f(t) = \frac{4A}{\pi}\left[\sin\omega_0 t + \frac{1}{3}\sin 3\omega_0 t + \frac{1}{5}\sin 5\omega_0 t + ...\right]$$

The individual sinusoidal components of f(t) for n = 1, and 3 and their sum are shown in Fig. 12.2. With partial sum of only two terms, the square waveform is seen to be approached quite closely.

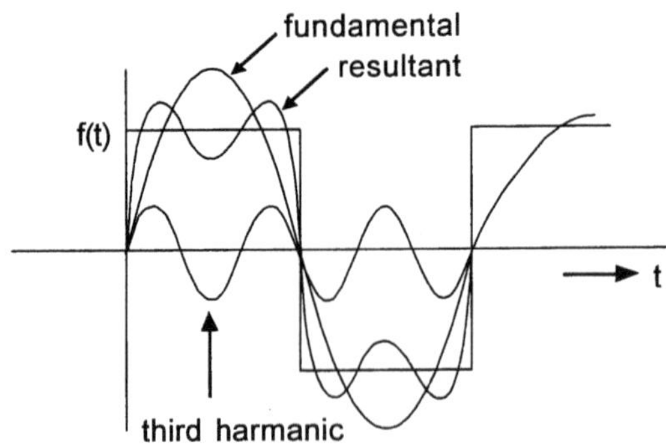

Fig. 12.2 Partial sum of two terms in Fourier series for a square waveform

The addition of more and more terms will result in much closer approximation to the given waveform, except at the points of discontinuity. The square waveform is not a physically realisable function and the Fourier series does not converge to the actual form near the discontinuities.

12.3 CONVERGENCE OF FOURIER SERIES :

Can we find the Fourier series for any given periodic function ? Can we approximate the given function with a smaller number of terms of the series. These questions arise naturally and are answered by studying the convergence properties of the Fourier series.

We can find the Fourier series provided we can evaluate the coefficients a_n and b_n. From the definition of these constants, and noting that sine and cosine functions vary between +1 and – 1 only, the integrals can be evaluated if the following condition is satisfied.

$$\int_0^T |f(t)|\, dt < \infty \qquad .....(12.9)$$

A more useful condition would be

$$\int_0^T [f(t)]^2\, dt < \infty \qquad .....(12.10)$$

which can be interpreted as the energy in the waveform for one cycle. This condition implics that thc cncrgy for onc cyclc of thc waveform must be finite. This condition is known as the weak Dirichlet condition. All physical waveforms satisfy this condition. According to this condition the functions are not restricted to be finite. Impulse functions could exist with finite areas under them. Hence it will be possible to find the Fourier coefficients when impulses are present in the waveform. But, wherever the function is infinite, the Fourier series doesnot converge uniformly everywhere and additional conditions are to be satisfied. These conditions are known as strong Dirichlet conditions. For a uniformly convergent Fourier series, the following additional conditions must be satisfied.

(i) The function f(t) must remain finite

and (ii) the function f(t) must have finite number of maxima and minima in a period. It can have a finite number of finite discontinuities in a period.

The reasons for these conditions are obvious. If the function is infinite at some point, the corresponding series would also be infinite and therefore will not converge. Similarly if there are infinite number of maxima and minima in a given region, a Fourier term of infinite frequency is required to approximate the function. Thus the series would require infinite number of terms and it would not converge at infinity.

All physically realizable waveforms satisfy the strong Dirichlet conditions. Hence their Fourier series could be found and the series can be summed. Some idealieasd waveforms, which we use, satisfy only weak Dirichlet condition and hence even though we can find the Fourier series, it is not possible to sum them. The next important question is : is it possible to obtain a good approximation for the given function by considering only a few terms of the Fourier series.

If a function has finite number of discontinuities in a period, the series converges as $\frac{1}{n}$. If there are finite discontinuities in a period of the derivative of a function, the series converges as $\frac{1}{n^2}$ and so on. If more derivatives exist for a function, more rapidly the series will converge. Thus, smoother the function, fewer will be the terms required in the Fourier series. A sine function has infinite derivatives and hence it requires only one term in its Fourier series!

A discontinuity in a function requires infinite number of terms in its Fourier series. At the discontinuity, the Fourier series produces a high frequency oscillation as shown in Fig. 12.3. At the discontinuity there is an overshoot of about 18 percent. This overshoot takes place right at the discontinuity, if infinite number of terms are taken and it will be spread over a considerable portion of the waveform, if only finite number of terms are considered. This phenomenon of nonuniform convergence was first studied by Sir Willard Gibbs and the characteristic overshoot is named after him as Gibbs phenomenon.

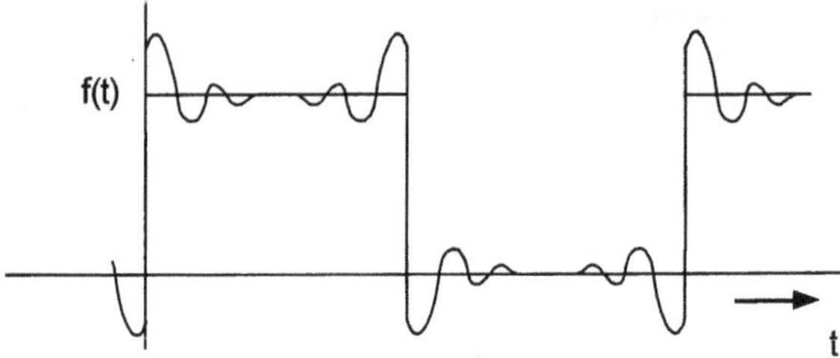

Fig. 12.3 Gibb's Phenomena

12.4 SYMMETRY CONDITIONS IN FOURIER SERIES :

In the example 12.1, we observe that,

$$a_0 = 0 \; ; \; a_n = 0 \qquad \text{for all n}$$
$$b_n = 0 \qquad \text{for n even.}$$

Can we predict which coefficients are going to be zero apriori ? If we can do this it would save a lot of effort in obtaining the Fourier series of a given function. These coefficients are zero because of some sysmmetry present in the waveform. Usually the symmetry is obvious by inspection of the waveform. Simple mathematical tests are also available to determine the nature of symmetry. There are four types of symmetry, possessed by the waveforms :

1. Even function symmetry
2. Odd function symmetry
3. Rotation or Half wave symmetry
4. Quarter wave symmetry

12.4.1 Even function symmetry :

If a waveform satisfies the condition

$$f(t) = f(-t)$$

the waveform is said to be possessing even function symmetry. An even function is symmetric about the y-axis. If the function for positive time is folded about the y-axis, it falls exactly on the function for negative time. A sketch of an even function is given in Fig. 12.4. A cosine function is an example of an even function. A series consisting of all cosine terms is also an even function. But a sine function is an odd function and hence even if one sine term is present, it spoils the symmetry. Thus we can conclude that for an even function

$$b_n = 0 \qquad \text{for all n}$$

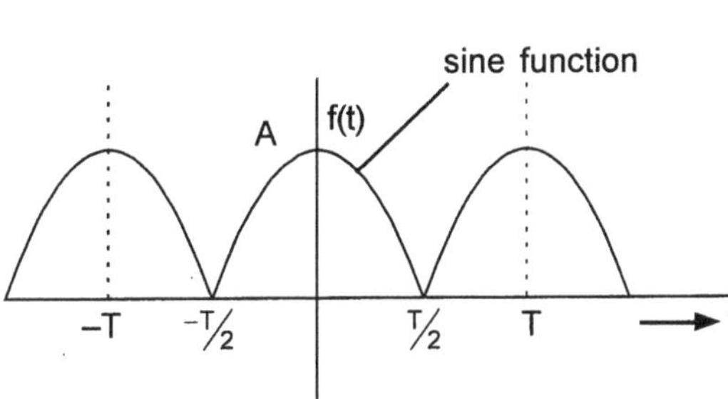

Fig. 12.4 Example of an even function

The definition of a_0, a_n and b_n given by eqs. (12.4), (12.7) and (12.8) can also be written as:

$$a_0 = \frac{1}{T}\int_{-T/2}^{T/2} f(t)\,dt\,; \qquad a_n = \frac{2}{T}\int_{-T/2}^{T/2} f(t)\cos n\omega t\,dt$$

$$b_n = \frac{2}{T}\int_{-T/2}^{T/2} f(t)\sin n\omega_0 t\,dt \qquad .....(12.11)$$

Recall the following mathematical identities.

$$\int_{-a}^{+a}\phi(x)dx = 2\int_{0}^{a}\phi(x)dx \qquad \text{if } \phi(x) \text{ is an even function}$$

$$= 0 \qquad \text{if } \phi(x) \text{ is an odd function} \qquad .....(12.12)$$

Applying the conditions to eq. (12.11), we get

$$a_0 = \frac{2}{T}\int_{0}^{T/2} f(t)\,dt \qquad .....(12.13)$$

Since f(t) and cos $n\omega_0 t$ are both even, their product is also even and hence

$$a_n = \frac{4}{T}\int_0^{T/2} f(t)\cos n\omega_0 t \, dt \qquad .....(12.14)$$

Since f(t) is even and sin $\omega_0 t$ is odd, their product in eq. (12.11) is odd. Hence

$$b_n = 0 \qquad .....(12.15)$$

Eqs. (12.13) and (12.14) imply that we integrate the waveform over half the period only and double the value to obtain a_0 and a_n.

Example 12.2

Find the Fourier series of the function shown in Fig. 12.4 with T = 2π. The waveform is reproduced in Fig. 12.5 for convenience.

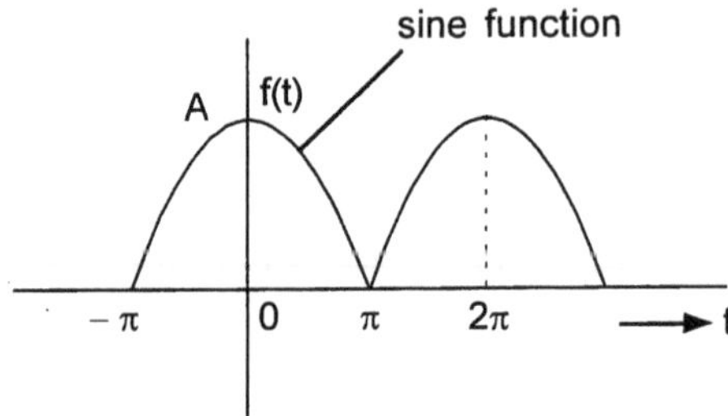

Fig. 12.5 Waveform for example 12.2

Solution :

Since the function is having even symmetry, we can immediately write the following:

$$b_n = 0$$

$$a_0 = \frac{2}{T}\int_0^{T/2} f(t)\, dt$$

$$a_n = \frac{4}{T}\int_0^{T/2} f(t)\cos n\omega_0 t \, dt$$

Since $T = 2\pi, \quad \omega_0 = \frac{2\pi}{T} = 1$

and $f(t) = A\cos\frac{t}{2} \qquad -\pi \le t \le \pi$

Thus $a_0 = \frac{2}{2\pi}\int_0^{\pi} A\cos\frac{t}{2}\, dt$

$$= \frac{A}{\pi} 2\left(\sin\frac{t}{2}\right)\Bigg|_0^{\pi}$$

$$= \frac{2A}{\pi}$$

$$a_n = \frac{4}{2\pi} \int_0^{\pi} A\cos\frac{t}{2}\cos nt\, dt$$

$$= \frac{2A}{\pi}\left[\int_0^{\pi} \frac{1}{2}\left\{\cos\left(n+\frac{1}{2}\right)t + \cos\left(n-\frac{1}{2}\right)t\right\} dt\right]$$

$$= \frac{A}{\pi}\left[\frac{\sin\frac{2n+1}{2}t}{\frac{2n+1}{2}} + \frac{\sin\frac{2n-1}{2}t}{\frac{2n-1}{2}}\right]_0^{\pi}$$

$$= \frac{2A}{\pi}\left[\frac{\sin\frac{(2n+1)\pi}{2}}{2n+1} + \frac{\sin\frac{(2n-1)\pi}{2}}{2n-1}\right]$$

We observe that

$$\sin(2n+1)\frac{\pi}{2} = -1 \qquad \text{for n } 1, 3, 5.....$$

$$= +1 \qquad \text{for } n = 2, 4, 6.....$$

Similarly $\quad \sin(2n-1)\frac{\pi}{2} = 1 \qquad \text{for } n = 1, 3, 5$

$$= -1 \qquad \text{for } n = 2, 4, 6$$

Hence $\quad a_n = \frac{2A}{\pi}.(-1)^n\left[\frac{1}{2n+1} - \frac{1}{2n-1}\right]$

$$= \frac{2A}{\pi}(-1)^n\left[\frac{-2}{4n^2-1}\right]$$

$$= \frac{4A(-1)^{n+1}}{\pi(4n^2-1)} \qquad \text{for } n = 1, 2, ...$$

Thus the Fourier series of the given waveform is

$$f(t) = \frac{2A}{\pi}\left[1+\frac{2}{3}\cos t-\frac{2}{15}\cos 2t+.....\right]$$

12.4.2 Odd function symmetry

If the waveform of f(t) satisfies the condition

$$f(t) = -f(-t)$$

the function is said to possess odd function symmetry. Sinωt is an example of an odd function. Again, even if one cosine term is present in the Fourier series of such a function, the series is no longer an odd function and it can not represent the given waveform. Since for every positive value of the function there is a corresponding negative value of the function in a given period, its average value must be zero. Thus $a_0 = 0$. Using eq. (12.12) and the fact that f(t) cos $n\omega_0 t$ is an odd function, we get

$$a_n = 0 \qquad \text{for all n} \qquad .....(12.16)$$

Also, since f(t) sin $n\omega_0 t$ is an even function, using eq. (12.12), we have

$$b_n = \frac{4}{T}\int_0^{\frac{T}{2}} f(t)\sin n\omega_0 t\ dt \qquad .....(12.17)$$

A sketch of an odd function is given in Fig. 12.6.

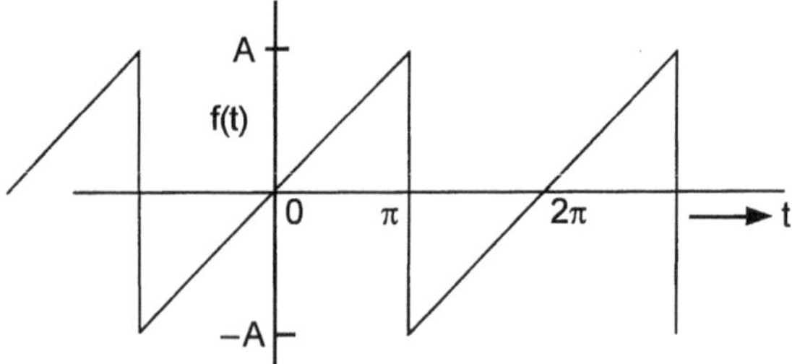

Fig. 12.6 Example of an odd function

Example 12.3

Obtain the Fourier series of the waveform shown is Fig. 12.6.

Solution :

Since the wavefrom has odd function symmetry,

$$a_0 = a_n = 0$$

$$b_n = \frac{4}{T}\int_0^{\frac{T}{2}} f(t)\sin n\omega_0 t\ dt$$

Here $\omega_0 = \frac{2\pi}{T} = 1$

$$f(t) = \frac{A}{\pi}t \quad \text{for} \quad 0 \le t < \pi$$

Therefore

$$b_n = \frac{4}{2\pi}\int_0^{\pi}\frac{A}{\pi}t \sin nt\, dt$$

$$= \frac{2A}{\pi^2}\left[-t\frac{\cos nt}{n}\Big|_0^{\pi} + \frac{\sin nt}{n^2}\Big|_0^{\pi}\right]$$

$$= \frac{2A}{\pi n}\left[-\cos n\pi\right] = (-1)^{n+1}\frac{2A}{n\pi}$$

Thus the Fourier series of the given waveform is

$$f(t) = \frac{2A}{\pi}\left[\sin t - \frac{1}{2}\sin 2t + \frac{1}{3}\sin 3t.....\right]$$

12.4.3 Halfwave or rotation symmetry

Whenever positive half cycle of the waveform is exactly same as the negative half cycle, we have half wave symmetry. Mathematically, this can be stated as

$$f(t) = -f\left(t \pm \frac{T}{2}\right) \qquad .....(12.18)$$

An example of a function with half wave symmetry is given in Fig. 12.7.

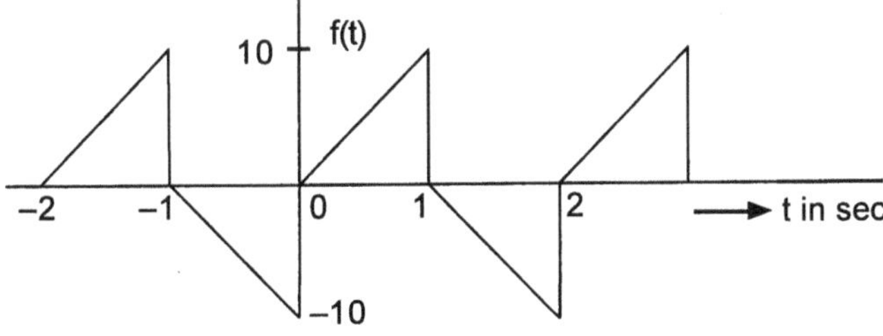

Fig. 12.7 An example of a function with half wave symmetry

It is obvious that the average value of this waveform is zero. Thus

$$a_0 = 0 \qquad .....(12.19)$$

The a_n coefficients are given by,

$$a_n = \frac{2}{T}\int_0^T f(t)\cos n\omega_0 t\, dt$$

$$= \frac{2}{T}\int_0^{\frac{T}{2}} f(t)\cos n\omega_0 t\, dt + \frac{2}{T}\int_{\frac{T}{2}}^{T} f(t)\cos n\omega_0 t\, dt$$

If t is replaced by $t + \frac{T}{2}$ in the second integral, the limits can be changed to 0 to $\frac{T}{2}$ without changing the value of the integral. Thus we have

$$a_n = \frac{2}{T}\int_0^{T/2} f(t)\cos n\omega_0 t\, dt + \frac{2}{T}\int_0^{T/2} f\left(t+\frac{T}{2}\right)\cos n\omega_0\left(t+\frac{T}{2}\right) dt$$

But $$f\left(t+\frac{T}{2}\right) = -f(t)$$

$\therefore$ $$a_n = \frac{2}{T}\int_0^{T/2} f(t)\cos n\omega_0 t\, dt - \frac{2}{T}\int_0^{T/2} f(t)\left[\cos n\omega_0 t\cos n\omega_0\frac{T}{2} - \sin n\omega_0 t\sin n\omega_0\frac{T}{2}\right] dt$$

$$= \frac{2}{T}\int_0^{T/2} f(t)\cos n\omega_0 t\, dt - \frac{2}{T}\int_0^{T/2} f(t)(-1)^n\cos n\omega_0 t\, dt$$

$(\because \omega_0 T = 2\pi)$

$$= \frac{4}{T}\int_0^{T/2} f(t)\cos n\omega_0 t\, dt \qquad \text{for n odd}$$

$$= 0 \qquad \text{for n even}$$

Similarly we can show that

$$b_n = \frac{4}{T}\int_0^{T/2} f(t)\sin n\omega_0 t\, dt \qquad \text{for n odd}$$

$$= 0 \qquad \text{for n even}$$

To summarise, for half wave symmetry

$$a_0 = 0$$

$$a_n = \frac{4}{T}\int_0^{T/2} f(t)\cos n\omega_0 t\, dt \qquad n = \text{odd} \qquad .....(12.20)$$

$$= 0 \qquad n = \text{even}$$

$$b_n = \frac{4}{T}\int_0^{T/2} f(t)\sin n\omega_0 t \, dt \qquad n = \text{odd}$$

$$= 0 \qquad n = \text{even} \qquad .....(12.21)$$

The waveform produced by the rotating electrical machines have this property and hence this symmetry is also known as rotation symmetry.

Example 12.4

Obtain the Fourier series of the waveform shown in Fig. 12.7.

Solution :

Since the waveform has rotation symmetry,

$$a_0 = 0$$

$$f(t) = 10\,t \quad \text{for } 0 < t < 1$$

$$T = 2 \text{ sec}$$

$$\omega_0 = \frac{2\pi}{T} = \pi \text{ rad/sec}$$

$$a_n = \frac{4}{2}\int_0^1 10t \cos n\pi t \, dt$$

$$= 20\left[t\frac{\sin n\pi t}{n\pi}\bigg|_0^1 + \frac{\cos n\pi t}{n^2\pi^2}\bigg|_0^1 \right]$$

$$= \frac{20}{n^2\pi^2}[\cos n\pi - 1] \quad \text{for n odd}$$

$$= -\frac{40}{n^2\pi^2} \qquad \text{for n odd}$$

From eq. (12.20)

$$a_n = 0 \qquad \text{for n even}$$

$$b_n = \frac{4}{2}\int_0^1 10t \sin n\pi t \, dt$$

$$= 20\left[-t\frac{\cos n\pi t}{n\pi}\bigg|_0^1 + \frac{\sin n\pi t}{n^2\pi^2}\bigg|_0^1 \right]$$

$$= \frac{-20}{n\pi}\cos n\pi \quad \text{for n odd}$$

$$b_n = \frac{20}{n\pi} \qquad \text{for n odd}$$

from eq. (12.21)

$$b_n = 0 \qquad \text{for n even}$$

The Fourier series of the given function is

$$f(t) = \frac{-40}{\pi^2}\left[\cos \pi t + \frac{1}{9}\cos 3\pi t + \frac{1}{25}\cos 5\pi t + ...\right]$$

$$+ \frac{20}{\pi}\left[\sin \pi t + \frac{1}{3}\sin 3\pi t + \frac{1}{5}\sin 5\pi t + ...\right]$$

12.4.4 Quarterwave symmetry :

If a function possesses rotation symmetry in addition to either odd symmetry or even symmetry, the function is said to have quarter wave symmetry.

(a) *Even function and rotation symmetry :*

The expression for the Fourier coefficients can be easily obtained as

$$a_0 = 0$$

$$a_n = \frac{8}{T}\int_0^{T/4} f(t)\cos n\omega_0 t \, dt \qquad \text{for n odd}$$

$$= 0 \qquad \text{for n even} \qquad .....(12.22)$$

$$b_n = 0 \qquad \text{for all n}$$

(b) *Odd function and rotation symmetry :*

$$a_0 = 0$$

$$a_n = 0 \qquad \text{for all n}$$

$$b_n = \frac{8}{T}\int_0^{T/4} f(t)\sin n\omega_0 t \, dt \qquad \text{for n odd} \qquad .....(12.23)$$

$$= 0 \qquad \text{for n even}$$

Example 12.5

Obtain the Fourier series of the waveform given in Fig. 12.8.

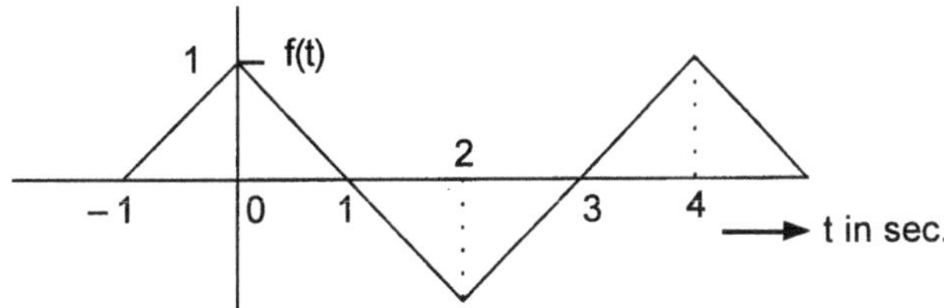

Fig. 12.8 Waveform for example 12.5

Solution :

The given waveform possesses even function symmetry and also rotation symmetry. Hence it has quarter wave symmetry.

Thus $a_0 = 0$

$$a_n = \frac{8}{T}\int_0^{T/4} f(t)\cos n\omega_0 t\, dt \qquad \text{for n odd}$$

$$= 0 \qquad \text{for n even}$$

$$b_n = 0 \qquad \text{for all n}$$

we have $f(t) = -t + 1$ for $0 \le t \le 1$

and $T = 4$ sec

$$\therefore \qquad \omega_0 = \frac{2\pi}{4} = \frac{\pi}{2} \text{ rad/sec.}$$

Evaluating a_n, we have

$$a_n = \frac{8}{4}\int_0^1 (-t+1)\cos n\frac{\pi}{2}t\, dt$$

$$= 2\left[-(-t+1)\frac{\sin\frac{n\pi}{2}t}{\frac{n\pi}{2}}\Bigg|_0^1 - \frac{\cos\frac{n\pi t}{2}}{\left(\frac{n\pi}{2}\right)^2}\Bigg|_0^1 \right]$$

$$= \frac{8}{n^2\pi^2}\left[1-\cos\frac{n\pi}{2}\right] \qquad \text{for n odd}$$

$$= \frac{8}{n^2\pi^2} \qquad \text{for n odd}$$

$$= 0 \qquad \text{for n even}$$

The Fourier series is given by

$$f(t) = \frac{8}{\pi^2}\left[\cos\frac{\pi}{2}t + \frac{1}{9}\cos\frac{3\pi}{2}t + \frac{1}{25}\cos\frac{5\pi}{2}t + ...\right]$$

12.4.5 Concealed symmetry :

Sometimes rotation symmetry is concealed or hidden when the waveform has a d.c value. If the d.c value is subtracted from the original waveform, the symmetry can be discovered. An example of a waveform with no apparent rotation symmetry is given in Fig. 12.9.

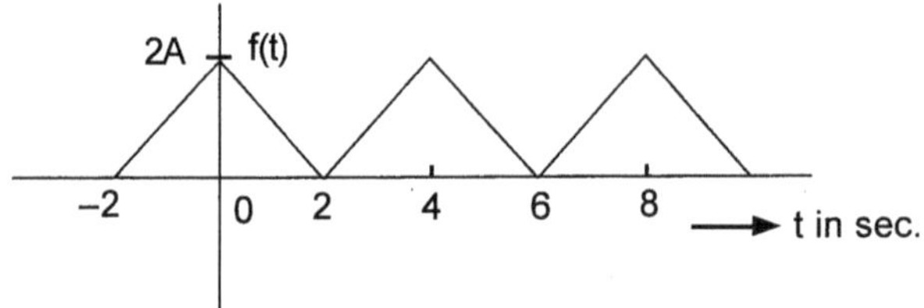

Fig. 12.9 Waveform with hidden rotation symmetry

The average value of this waveform is A. If this average value is subtraced from the original waveform, which amounts to shifting the 't' axis up by A, the rotation symmetry is revealed. The resulting waveform $f_1(t)$ is shown in Fig. 12.10.

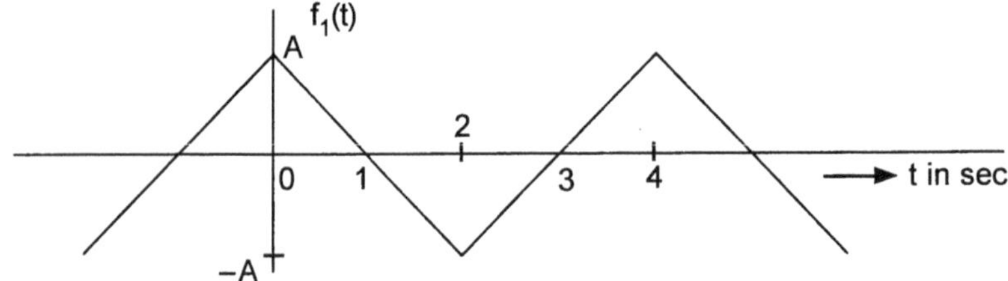

Fig. 12.10 Waveform of Fig. 12.9 with 't' axis shifted up by 'A'

The Fourier series of the waveform shown in Fig. 12.10 can now be obtained by using the symmetry conditions. From example 12.5, we have the fourier series of $f_1(t)$ as

$$f_1(t) = \frac{8A}{\pi^2}\left[\cos\frac{\pi}{2}t + \frac{1}{9}\cos\frac{3\pi}{2}t + \frac{1}{25}\cos\frac{5\pi}{2}t + ...\right]$$

To get the Fourier series of the given waveform we add the average value. Thus

$$f(t) = A + \frac{8A}{\pi^2}\left[\cos\frac{\pi}{2}t + \frac{1}{9}\cos\frac{3\pi}{2}t + \frac{1}{25}\cos\frac{5\pi}{2}t + ...\right]$$

Example 12.6

Obtain the Fourier series of the function shown in Fig. 12.11.

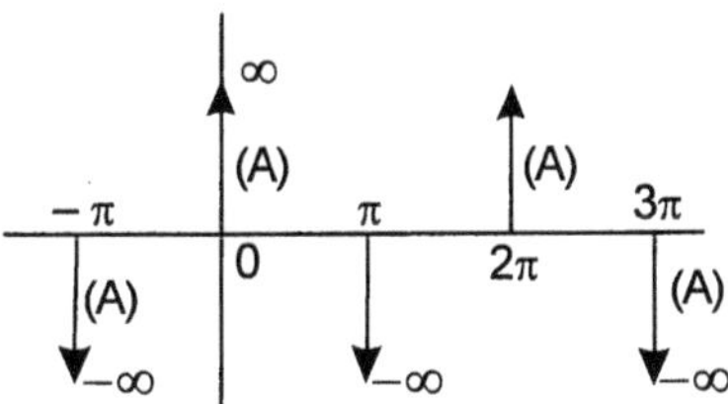

Fig. 12.11 An inmpulse train waveform for example 12.6

Solution :

It has even function and halfwave symmetry. Hence

$$a_0 = 0$$

$$b_n = 0 \qquad \text{for all n}$$

Since we have impulses let us integrate over the full period

$$a_n = \frac{2}{2\pi}\int_{0^-}^{2\pi^-} f(t)\cos nt\, dt \quad \text{(where } 2\pi^- \text{ has the same meaning as } O^-\text{, that is, just before the instant } t = 2\pi)$$

$$= \frac{1}{\pi}\left[\int_{0^-}^{\pi^-} Au_0(t)\cos nt\, dt + \int_{\pi-}^{2\pi^-} -Au_0(t-\pi)\cos nt\, dt\right]$$

$$= \frac{A}{\pi}[1-\cos n\pi]$$

$$= \frac{2A}{\pi} \qquad \text{for n odd}$$

$$= 0 \qquad \text{for n even}$$

Thus the Fourier series is

$$f(t) = \frac{2A}{\pi}[\cos t + \cos 3t + \cos 5t + ...]$$

If a waveform has a period equal to 2π, ω_0 is equal to 1. If the waveform has a period other than 2π, it is computationally convenient to take the period as 2π and evaluate the Fourier coefficients with $\omega_0 = 1$. In the final expression for f(t) the appropriate frequency has to be introduced in the cos $n\omega_0 t$, sin $n\omega_0 t$ terms. For example, if the time period of the waveform of Fig. 12.6 is 2×10^{-3} sec., instead of 2π, the Fourier series of this function is given by

$$f(t) = \frac{2A}{\pi}\left[\sin \pi \times 10^3 t - \frac{1}{2}\sin 2\pi \times 10^3 t + \frac{1}{3}\sin 3\pi \times 10^3 + ...\right]$$

Important note : The Fourier series coefficients are independent of the time period or the frequency of the given waveform.

12.5 COMPLEX FOURIER SERIES :

The Fourier series in eq. (12.1) can be written in a more compact form as

$$f(t) = a_0 + \sum_{n=1}^{\infty}\left(a_n \cos n\omega_0 t + b_n \sin n\omega_0 t\right) \qquad .....(12.24)$$

$$= a_0 + \sum_{n=1}^{\infty}\overline{C}_n \sin\left(n\omega_0 t + \theta_n\right) \qquad .....(12.25)$$

where $\overline{C}_n = \sqrt{a_n^2 + b_n^2}$ .....(12.26)

and $\theta_n = \tan^{-1}\dfrac{a_n}{b_n}$ .....(12.27)

This form of Fourier series is known as sine series.

Similarly eq. (12.24) can also be written as

$$f(t) = a_0 + \sum_{n=1}^{\infty}\overline{C}_n \cos\left(n\omega_0 t + \theta_n\right) \qquad .....(12.28)$$

where $\overline{C}_n$ is again given by eq. (12.26) and

$$\theta_n = -\tan^{-1}\frac{b_n}{a_n} \qquad .....(12.29)$$

This form of Fourier series is known as cosine series.

From eqs. (12.25) and (12.28) we observe that a non sinusoidal periodic function contains sinusoidal signals of frequencies 0, ω_0, $2\omega_0$, $3\omega_0$, ... $n\omega_0$..... where ω_0 is the frequency of the given waveform and is known as the fundamental frequency. The other frequencies are integral multiples of the fundamental frequency and are known as harmonic frequencies. Thus a non-sinusoidal periodic function consists of, in general, a d.c term, a fundamental frequency term and terms of harmonic frequencies. The plots of nth harmonic term $\overline{C}_n$ Vs ω and θ_n Vs ω are known as amplitude spectrum and phase spectrum. For example the amplitude and phase spectrum of the waveform of Fig. 12.7 are given in Fig. 12.12.

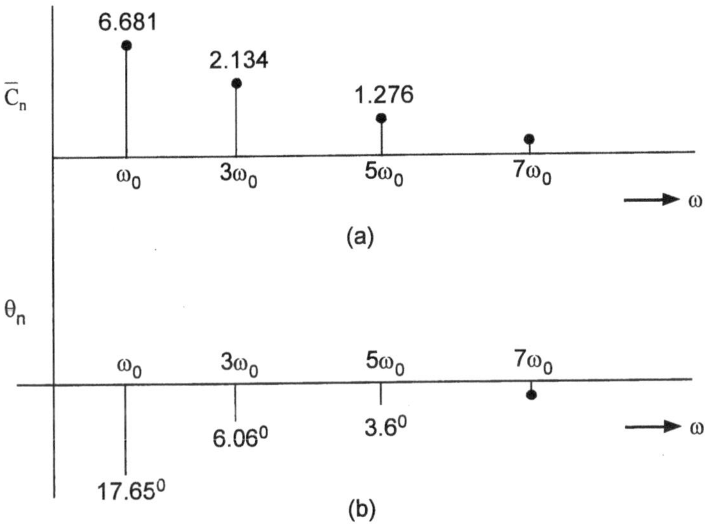

Fig. 12.12 (a) Amplitude spectrum (b) phase spectrum of waveform of Fig. 12.7

Another form of the Fourier series is the complex Fourier series, which can be obtained by expressing $\sin n\omega_0 t$ and $\cos n\omega_0 t$ in eq. (12.25) in terms of exponential functions. Noting that

$$\sin n\omega_0 t = \frac{e^{jn\omega_0 t} - e^{-jn\omega_0 t}}{2j}$$

and

$$\cos n\omega_0 t = \frac{e^{jn\omega_0 t} + e^{-jn\omega_0 t}}{2}$$

eq. (12.25) can be written as

$$f(t) = a_0 + \sum_{n=1}^{\infty}\left[a_n \frac{e^{jn\omega_0 t} + e^{-jn\omega_0 t}}{2} + b_n \frac{e^{jn\omega_0 t} - e^{-jn\omega_0 t}}{2j}\right]$$

$$= a_0 + \sum_{n=1}^{\infty}\left[e^{jn\omega_0 t}\left(\frac{a_n - jb_n}{2}\right) + e^{-jn\omega_0 t}\left(\frac{a_n + jb_n}{2}\right)\right]$$

$$= a_0 + \sum_{n=1}^{\infty}\left(c_n e^{jn\omega_0 t} + c_n^* e^{-jn\omega_0 t}\right) \qquad \text{.....(12.30)}$$

Where c_n^* is the complex conjugate of c_n and

$$c_n = \frac{a_n - jb_n}{2} = \frac{2}{2T}\int_0^T f(t)\left(\frac{e^{jn\omega_0 t} + e^{-jn\omega_0 t}}{2} - j\frac{e^{jn\omega_0 t} - e^{-jn\omega_0 t}}{2j}\right)dt$$

$$\therefore \qquad c_n = \frac{1}{T}\int_0^T f(t)e^{-jn\omega_0 t}\,dt \qquad \text{.....(12.31)}$$

Noting that the value of c_n given by eq. (12.31) for n = 0 is

$$c_0 = \frac{1}{T}\int_0^T f(t)\,dt = a_0$$

and using –n instead of n in eq. (12.31) yields

$$c_{-n} = c_n^* = \frac{1}{T}\int_0^T f(t)e^{jn\omega_0 t}\,dt$$

We can write eq. (12.30) as

$$f(t) = c_0 + \sum_{n=1}^{\infty}\left(c_n e^{-jn\omega_0 t} + c_{-n} e^{-jn\omega_0 t}\right) \qquad \text{.....(12.32)}$$

Taking the summation from $n = -\infty$ to $n = \infty$ we can write eq. (12.32) in a more compact form as

$$f(t) = \sum_{n=-\infty}^{\infty} c_n e^{jn\omega_0 t} \qquad \text{.....(12.33)}$$

In this equation n = 0 gives the term c_0, n = 1 to ∞ yields the first term and $n = -1$ to $-\infty$ yields the second term of the summation in eq. (12.32). Eq. (12.33) is known as the exponential form of Fourier series. The Fourier coefficients c_n are given by eq. (12.31).

12.5.1 Some comments about complex fourier series :

The complex Fourier series in eq. (12.33) expresses a non sinusoidal periodic function of time as a series of terms comprising positive and negative complex exponentials. The actual sinusoidal harmonics in the series are obtained by combining pairs of positive and negative terms for each frequency. It is to be observed that the amplitude of the harmonic is twice the amplitude of either of the corresponding exponential terms alone. The frequency associated with the term $e^{jn\omega_0 t}$, $n\,\omega_0$ is referred to as positive frequency and that of $e^{-jn\omega_0 t}$, $-n\omega_0$ is referred to as negative frequency. If we plot the frequency spectrum of complex Fourier series, we have frequency components for 0, $\pm\,\omega_0$, $\pm\,2\,\omega_0$, $\pm\,3\omega_0$,..... $\pm\,n\omega_0$,..... with spectrum ranging over positive and negative frequencies. The idea of negative frequency appears to be strange as the frequency in real life is always positive. We loosely call $n\,\omega_0$ as the frequency but in actual practice the pair of exponential terms, $e^{jn\omega_0 t}$ and $e^{-jn\omega_0 t}$ together constitute a $\cos n\omega_0 t$ or a $\sin n\omega_0 t$ term.

Example 12.7

Find the exponential Fourier series for the waveform shown is Fig. 12.13. Draw the Fourier spectrum.

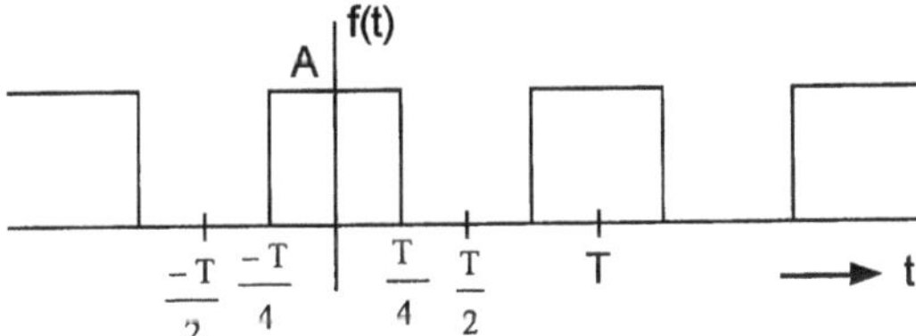

Fig. 12.13 A square wave for example 12.7

Solution :

$$f(t) = \sum_{n=-\infty}^{\infty} c_n e^{jn\omega_0 t}$$

$$\omega_0 = \frac{2\pi}{T}$$

$$c_n = \frac{1}{T}\int_{-\frac{T}{2}}^{\frac{T}{2}} f(t)e^{-jn\omega_0 t}\,dt$$

(Note : The integration can be performed over a complete cycle starting from any arbitrary time. The value of the integral does not change.)

$$c_n = \frac{1}{T}\int_{\frac{-T}{4}}^{\frac{T}{4}} Ae^{-jn\omega_0 t}dt$$

$$= \frac{A}{T}\left[\frac{e^{-jn\omega_0 t}}{-jn\omega_0}\right]_{\frac{-T}{4}}^{\frac{T}{4}}$$

$$= \frac{A}{n\pi}\left[\frac{-e^{-jn\omega_0\frac{T}{4}} + e^{+jn\omega_0\frac{T}{4}}}{2j}\right]$$

$$= \frac{A}{n\pi}\sin\frac{n\pi}{2}$$

$$= \frac{A}{2}\left(\frac{\sin\frac{n\pi}{2}}{\frac{n\pi}{2}}\right)$$

Noting that

$$\underset{x\to 0}{\text{Lt}}\ \frac{\sin x}{x} = 1, \text{ we have}$$

$$c_0 = \underset{n\to 0}{\text{Lt}}\ \frac{A}{2}\frac{\sin\frac{n\pi}{2}}{\frac{n\pi}{2}} = \frac{A}{2}$$

Thus the complex Fourier series is

$$f(t) = \sum_{n=-\infty}^{\infty}\frac{A}{2}\left(\frac{\sin\frac{n\pi}{2}}{\frac{n\pi}{2}}\right)e^{jn\omega_0 t}$$

The spectrum is obtained by plotting c_n for n ranging from $-\infty$ to ∞. The amplitude spectrum is shown in Fig. 12.14.

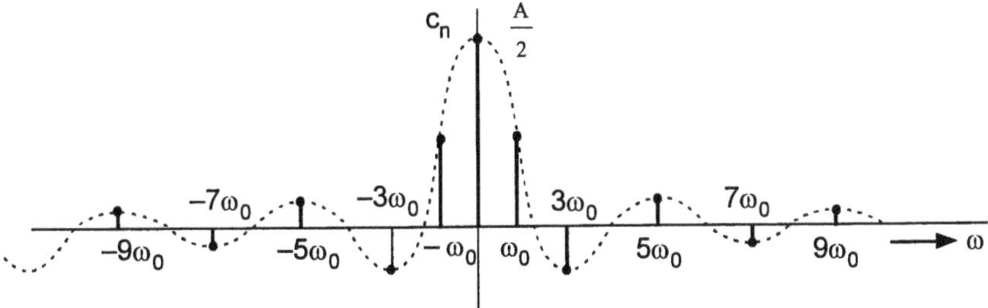

Fig. 12.14 Amplitude spectrum of waveform of Fig. 12.13

The trigonometric form of Fourier series can be easily obtained as

$$f(t) = \frac{A}{2} + \sum_{n=1}^{\infty} A \frac{\sin\frac{n\pi}{2}}{\frac{n\pi}{2}} \cos n\omega_0 t$$

The Amplitude spectrum of this series is given in Fig. 12.15.

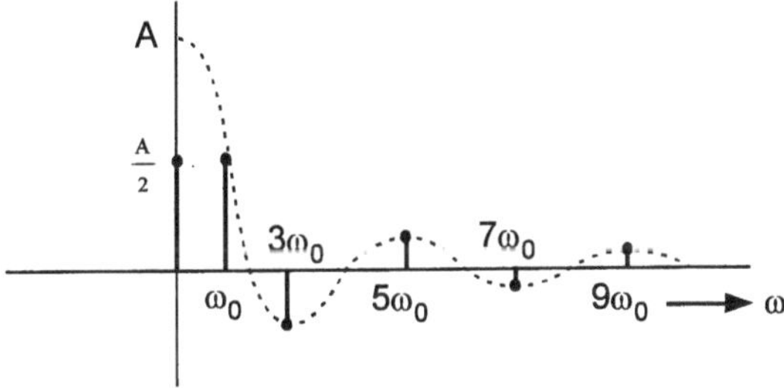

Fig. 12.15 Amplitude spectrum of waveform in Fig. 12.13 for trigonometric form of Fourier series

Comparing the two spectra given in Fig. 12.14 and 12.15, we observe that the spectrum of Fig. 12.15 is obtained by folding the spectrum in Fig. 12.14 about the vertical axis and adding the magnitudes at $n\omega_0$ and $-n\omega_0$ for each value of n. Though the conventional frequency ω is associated with the sine (cosine) series, it is more convenient to use the exponential spectra in practice. Since the spectrum is defined only at discrete values of 'n', this spectrum is known as discrete spectrum.

Example 12.8

Obtain the exponential Fourier series for the waveform shown in Fig. 14.16

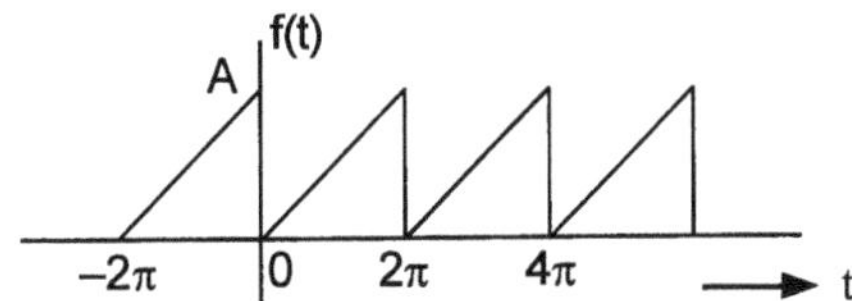

Fig. 12.16 Waveform for example 12.8

Solution :

$$c_0 = \frac{A}{2}$$

$$c_n = \frac{1}{2\pi}\int_0^{2\pi} \frac{A}{2\pi} te^{-jnt}\,dt$$

$$= \frac{A}{4\pi^2}\left[t\frac{e^{-jnt}}{-jn}\Bigg|_0^{2\pi} - \frac{e^{-jnt}}{(-jn)^2}\Bigg|_0^{2\pi} \right]$$

$$= \frac{-A}{2\pi jn}e^{-jn2\pi} + \frac{A}{4\pi^2 n^2}\left[e^{-jn2\pi} - 1\right]$$

$$= \frac{jA}{2\pi n}$$

$$f(t) = \frac{A}{2} + \sum_{\substack{n=-\infty \\ n\neq 0}}^{\infty} \frac{jA}{2\pi n} e^{jnt}$$

The frequency spectrum is given in Fig. 12.17.

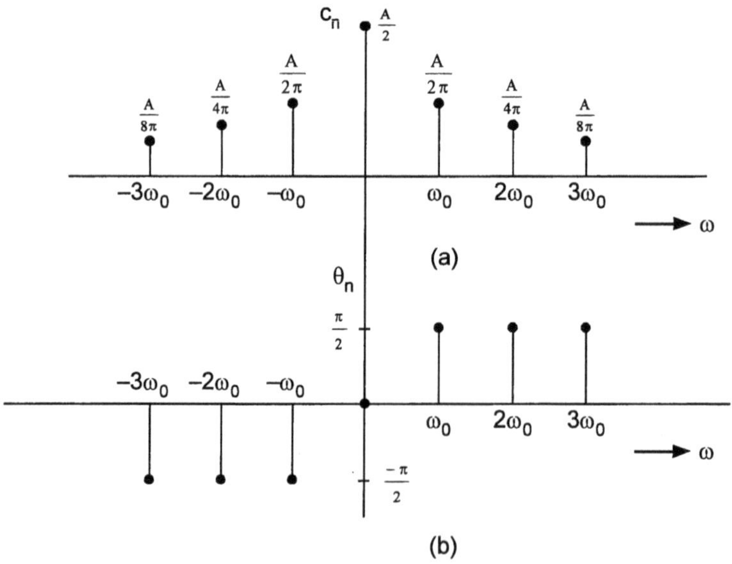

Fig. 12.17 (a) Amplitude spectrum (b) Phase spectrum for waveform in Fig. 12.16

Example 12.9

Evaluate the exponential Fourier series of the wave form shown in Fig. 12.18 (a) and thereby obtain the trigonometric Fourier series of the waveform shown in Fig. 12.18 (b).

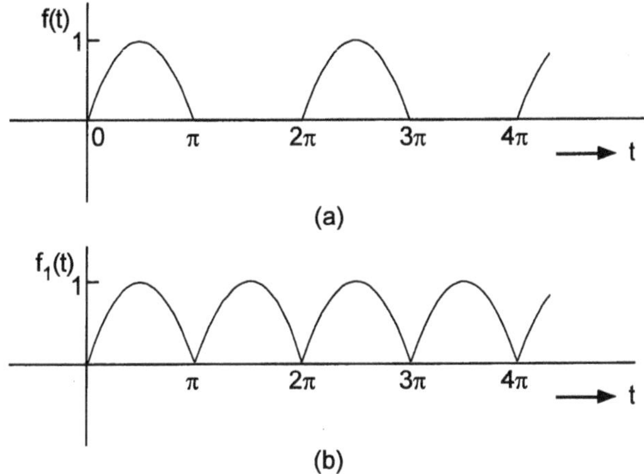

Fig. 12.18 (a) Half rectified sine wave (b) Full rectified sine wave

Solution :

For the given waveform, T = 2π and ω_0 = 1 rad/sec and c_0 can be easily evaluated as :

$$c_0 = \frac{1}{\pi}$$

$$c_n = \frac{1}{T}\int_0^T f(t)e^{-jnt}dt$$

$$= \frac{1}{2\pi}\int_0^\pi \sin t\, e^{-jnt}\, dt$$

$$= \frac{-1}{\pi(n^2-1)} \qquad \text{for } n = 2, 4, 6.....$$

$$= 0 \qquad \text{for } n = 3, 5, 7.....$$

$$= \text{indeterminate for } n = 1$$

Evaluating c_1 separately, we have

$$c_1 = \frac{1}{2\pi}\int_0^\pi \sin t\, e^{-jt}dt = \frac{-j}{4}$$

and $$c_{-1} = c_1^* = \frac{j}{4}$$

Thus the exponential Fourier series is :

$$f(t) = \frac{1}{\pi} + \frac{j}{4}e^{-jt} - \frac{j}{4}e^{jt} - \sum_{\substack{n=-\infty \\ (\text{even})}}^{\infty} \frac{1}{\pi(n^2-1)}e^{jnt}$$

The trigonometric Fourier series can be obtained easily by combining pairs of exponential terms of same frequency and is given by

$$f(t) = \frac{1}{\pi} + \frac{\sin t}{2} - \sum_{n=\text{even}}^{\infty} \frac{2}{\pi(n^2-1)}\cos nt \qquad \text{.....(12.34)}$$

Now, the waveform given in Fig. 12.18(b) can be considered as a summation of Fig. 12.18(a) and the waveform, $f_2(t)$, obtained by shifting the waveform of Fig. 12.18(a) by π, as shown in Fig. 12.18(c).

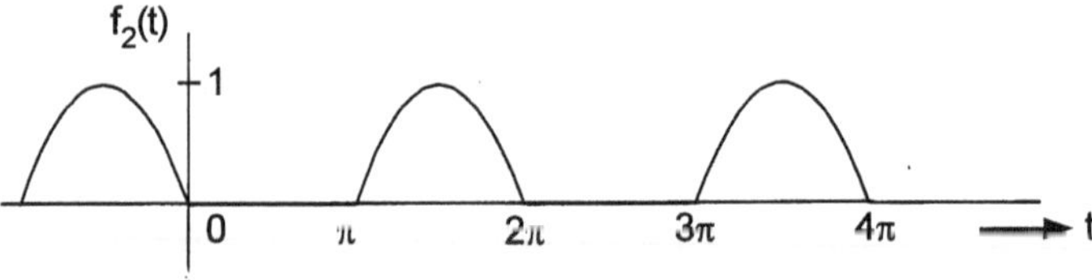

Fig. 12.18 (c) Waveform obtained by shifting the function in Fig.12.18(b) by π

Fourier series of $f_2(t)$ can be obtained by substituting $t = t - \pi$ in eq. (12.34).

$$f_2(t) = f(t-\pi) = \frac{1}{\pi} + \frac{\sin(t-\pi)}{2} - \sum_{n=\text{even}}^{\infty} \frac{2}{\pi(n^2-1)}\cos n(t-\pi)$$

Now the Fourier series of $f_1(t)$ can be obtained by adding the Fourier series of $f(t)$ and $f_2(t)$.

Thus, we have

$$f_1(t) = \frac{1}{\pi} + \frac{\sin t}{2} - \sum_{n=\text{even}}^{\infty} \frac{2}{\pi(n^2-1)}\cos nt$$

$$+ \frac{1}{\pi} - \frac{1}{2}\sin t - \sum_{n=\text{even}}^{\infty} \frac{2}{\pi(n^2-1)}\cos nt$$

$$= \frac{2}{\pi} - \sum_{n=\text{even}}^{\infty} \frac{4}{\pi(n^2-1)}\cos nt$$

It can be observed that $f_1(t)$ is an even function and has a period π and $\omega_0 = 2$. Therefore the Fourier series of this function contains only cosine terms and has all the harmonics present.

12.6 FOURIER TRANSFORMS :

So far, we have considered periodic signals and obtained their representation by either sinusoidal signals or exponential signals. Can this signal analysis be extended to nonperiodic signals. It will be shown in this section, by a simple limiting process, that it is indeed possible to represent a nonperiodic signal as a continuous sum of exponential signals.

Consider the non periodic signal f(t) shown in Fig. 12.19 (a). Let us construct a periodic signal $f_p(t)$ consisting of f(t) repeating itself after a period T as shown in Fig. 12.19 (b). The period T is chosen long enough so that there is no overlap between the repeating pulses.

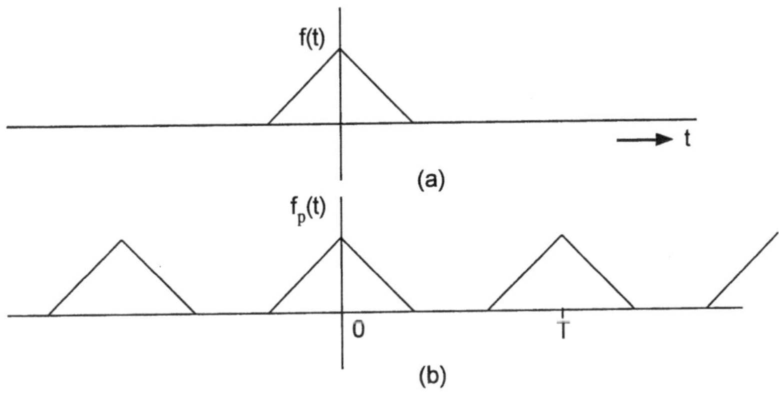

Fig. 12.19 (a) Non periodic signal f(t) (b) Periodic signal constructed using f(t) with period T

The signal $f_p(t)$ is a periodic signal and hence it can be represented by an exponential Fourier series. If we let the period T tend to infinity, the periodic function $f_p(t)$ tends to the non periodic function f(t). In other words, any nonperiodic signal can be considered as a periodic signal which repeats itself after a period of infinity and therefore can be represented by a Fourier series. Thus

$$\underset{T \to \infty}{\text{Lt}} \; f_p\,(t) = f(t).$$

Hence the Fourier series representing $f_p(t)$ will also represent f(t) in the limit. The exponential Fourier series for $f_p(t)$ is :

$$f_p(t) = \sum_{n=-\infty}^{\infty} F_n\, e^{jn\omega_0 t}, \quad \omega_0 = \frac{2\pi}{T} \qquad .....(12.35)$$

and

$$F_n = \frac{1}{T}\int_{\frac{-T}{2}}^{\frac{T}{2}} f_p(t)\, e^{-jn\omega_0 t}\, dt \qquad .....(12.36)$$

As $T \to \infty$, the fundamental frequency ω_0 tends to zero and the spectrum will consist of infinite number of frequency components which are infinitesimally small. The spectrum exists for every value of ω instead of at discrete values only and therefore it is a continuous function of ω.

Let us call the new distribution function as $F(n\omega_0)$, defined by

$$F(n\omega_0) = F_n T = \int_{-\frac{T}{2}}^{\frac{T}{2}} f_p(t) e^{-jn\omega_0 t}\, dt \qquad .....(12.37)$$

and

$$f(t) = \sum_{n=-\infty}^{\infty} \frac{F(n\omega_0)}{T} e^{jn\omega_0 t} \qquad .....(12.38)$$

If we now apply the limiting process, we have :

As $\qquad T \to \infty$

$$\omega_0 = \frac{2\pi}{T} \to \Delta\omega \to d\omega$$

and $\qquad n\omega_0 \to n\Delta\omega \to \omega$.

Eqs. (12.37) and (12.38) become

$$F(\omega) = \int_{-\infty}^{\infty} f(t) e^{-j\omega t}\, dt \qquad .....(12.39)$$

$$f(t) = \sum_{n=-\infty}^{\infty} \left[\frac{F(n\Delta\omega)}{2\pi} \omega_0 \right] e^{jn\omega_0 t}$$

$$= \sum_{n=\infty}^{\infty} \frac{F(n\Delta\omega)}{2\pi} \Delta\omega\, e^{jn\Delta\omega t} \qquad .....(12.40)$$

Here, f(t) is represented as a sum of exponentials of frequencies 0, $\pm\Delta\omega$, $\pm 2\Delta\omega$, $\pm 3\Delta\omega$, ... The amplitude of the component of frequency n $\Delta\omega$ is $\frac{F(n\Delta\omega)}{2\pi}\Delta\omega$ and is infinitesimally small and approaches zero as $T \to \infty$, but the relative strength of the component is given by $F(n\Delta\omega)$. Hence in the limit the summation is replaced by an integral and f(t) is given by

$$f(t) = \frac{1}{2\pi} \int_{-\infty}^{\infty} F(\omega) e^{j\omega t}\, d\omega . \qquad .....(12.41)$$

Thus we have shown that a non-periodic function f(t) can be represented by eternal exponential functions given by

$$f(t) = \frac{1}{2\pi}\int_{-\infty}^{\infty} F(\omega)e^{j\omega t}\,d\omega \qquad .....(12.42)$$

where $$F(\omega) = \int_{-\infty}^{\infty} f(t)e^{-j\omega t}\,dt \qquad .....(12.43)$$

Eqs. (12.42) and (12.43) are known as Fourier transform pairs. Eq. (12.43) is the direct Fourier transform of f(t) and eq. (12.42) is the inverse Fourier transform of F(ω). Symbolically these transform are written as

$$F(\omega) = \mathfrak{F}[f(t)] \text{ and } f(t) = \mathfrak{F}^{-1}[F(\omega)]$$

The transform pair is also represented as

$$f(t) \leftrightarrow F(\omega)$$

which means F(ω) is the Fourier transform of f(t) and f(t) is the inverse Fourier transform of F(ω).

12.6.1 Existence of the fourier transform :

From eq. (12.43)

$$F(\omega) = \int_{-\infty}^{\infty} f(t)e^{-j\omega t}\,dt$$

or $$F(\omega) \le \int_{-\infty}^{\infty} |f(t)|\,dt, \qquad \text{since } \left|e^{-j\omega t}\right| = 1$$

Hence if the right hand side of this equation is finite, F(ω) is finite and existence of F(ω) is guaranteed. Thus if the function f(t) is absolutely integrable, its Fourier transform exists. It means

$$\int_{-\infty}^{\infty} |f(t)|\,dt < \infty \qquad .....(12.44)$$

Further, it can also be shown that, if

$$\int_{-\infty}^{\infty} |f(t)|^2\,dt < \infty \qquad .....(12.45)$$

the Fourier transform exists. If f(t) is a voltage and is applied across a 1 ohm resistance,

$$\int_{-\infty}^{\infty} f(t)^2\,dt$$

represents the energy dissipated in the resistor. Thus this quantity is also defined as the energy of a signal f(t). If a function has finite energy, it is said to be Fourier transformable.

Example 12.10

Find the Fourier transform of the function shown in Fig. 12.20.

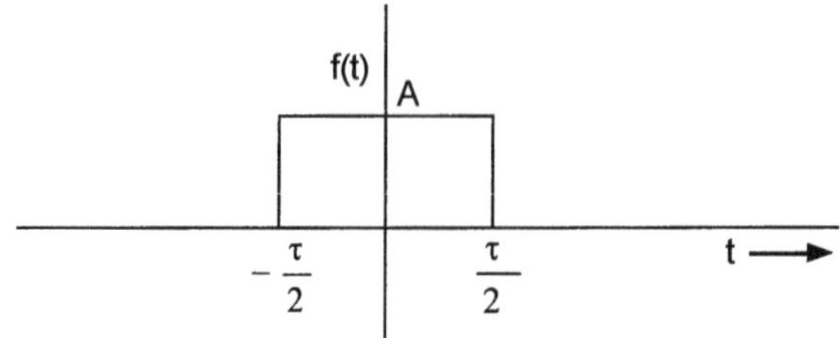

Fig. 12.20 A square pulse for example 12.10

Solution :

$$F(\omega) = \int_{-\infty}^{\infty} f(t) e^{-j\omega t}\, dt$$

$$= \int_{\frac{-\tau}{2}}^{\frac{\tau}{2}} A e^{-j\omega t}\, dt$$

$$= A \left. \frac{e^{-j\omega t}}{-j\omega} \right|_{\frac{-\tau}{2}}^{\frac{\tau}{2}}$$

$$= \frac{A}{j\omega}\left[e^{j\frac{\omega\tau}{2}} - e^{-j\frac{\omega\tau}{2}} \right]$$

$$= \frac{2A}{\omega} \sin\frac{\omega\tau}{2}$$

$$= A\tau \left(\frac{\sin\frac{\omega\tau}{2}}{\frac{\omega\tau}{2}} \right) \qquad .....(12.46)$$

The frequency spectrum F(ω) is plotted in Fig. 12.21.

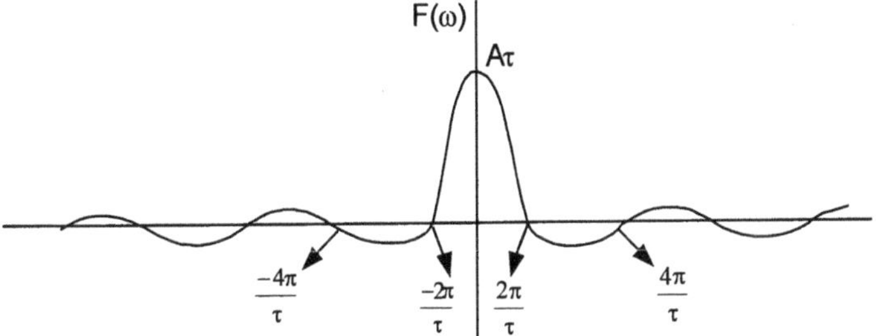

Fig. 12.21 Continuous frequency spectrum of a square pulse

The function $\frac{\sin x}{x}$ occurs frequently in communication theory and hence a special notation is used to represent this function. Let us define

$$\text{Sinc }(x) = \frac{\sin \pi x}{\pi x} \qquad .....(12.47)$$

Eq. (12.46) can be written interms of eq. (12.47) as

$$F(\omega) = A\tau \text{ sinc } \frac{\omega\tau}{2\pi}$$

The sinc(x) is plotted in Fig. 12.22.

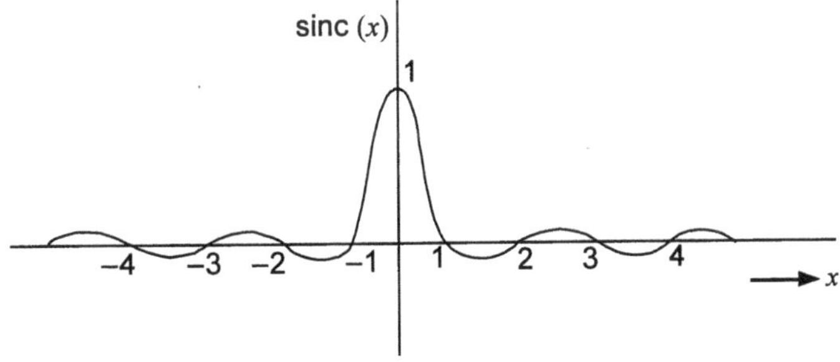

Fig. 12.22 The sinc(x) function

Note that $\frac{\sin x}{x}$ tends to 1 as $x \to 0$. Also sinc(x) = 0 for all integral values of x.

Let us find the frequency spectrum of certain important functions in the following examples.

Example 12.11

Find the Fourier transform of the function in Fig. 12.23.

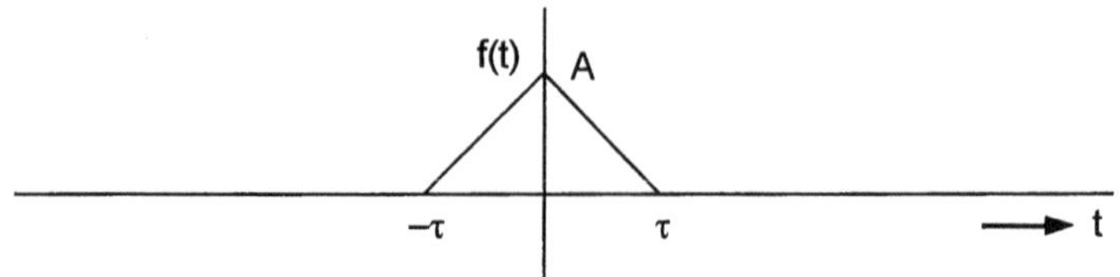

Fig. 12.23 A triangular pulse

Solution :

$$f(t) = A\left(1-\frac{|t|}{\tau}\right) \qquad \text{for } |t| \le \tau$$

$$= 0 \qquad \text{for } |t| > \tau$$

The Fourier transform F(ω) of this function is given by

$$F(\omega) = \int_{-\tau}^{0} A\left(1+\frac{t}{\tau}\right)e^{-j\omega t}\,dt + \int_{0}^{\tau} A\left(1-\frac{t}{\tau}\right)e^{-j\omega t}dt$$

$$= A\left[\left(1+\frac{t}{\tau}\right)\frac{e^{-j\omega t}}{-j\omega}+\frac{1}{j\omega\tau}\frac{e^{-j\omega t}}{-j\omega}\right]_{-\tau}^{0}$$

$$+A\left[\left(1-\frac{t}{\tau}\right)\frac{e^{-j\omega t}}{-j\omega}-\frac{1}{j\omega\tau}\frac{e^{-j\omega t}}{-j\omega}\right]_{0}^{\tau}$$

$$= A\left[-\frac{e^{j\omega\tau}}{\omega^2\tau}-\frac{1}{j\omega}+\frac{1}{\omega^2\tau}-\frac{e^{-j\omega\tau}}{\omega^2\tau}+\frac{1}{j\omega}+\frac{1}{\omega^2\tau}\right]$$

$$= \frac{A}{\omega^2\tau}\left[-\left(e^{j\omega\tau}+e^{-j\omega\tau}\right)+2\right]$$

$$= \frac{2A}{\omega^2\tau}\left[1-\cos\omega\tau\right]$$

$$= \frac{2A}{\omega^2\tau}.2\sin^2\frac{\omega\tau}{2}$$

$$= A\tau \left(\frac{\sin \frac{\omega\tau}{2}}{\frac{\omega\tau}{2}} \right)^2$$

$$= A\tau \ \text{sinc}^2 \left(\frac{\omega\tau}{2\pi} \right)$$

The continuous spectrum is plotted in Fig. 12.24.

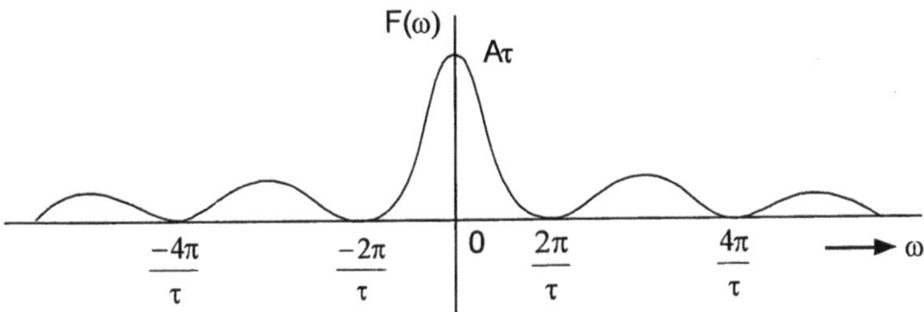

Fig. 12.24 Continuous spectrum of a triangular pulse shown in Fig. 12.23

Example 12.12

Find the Fourier spectrum of the exponential function shown in Fig. 12.25.

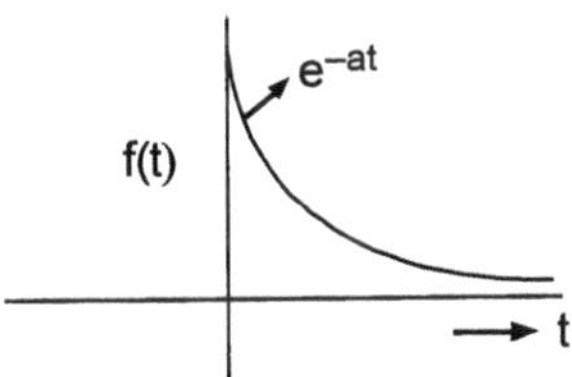

Fig. 12.25 Exponential function

Solution :

$$f(t) = e^{-at}u_{-1}(t) \qquad \text{where } u_{-1}(t) \text{ is a unit step function.}$$

$$F(\omega) = \int_{-\infty}^{\infty} e^{-at}u_{-1}(t)e^{-j\omega t}\,dt$$

$$= \int_{0}^{\infty} e^{-(a+j\omega)t}\,dt$$

$$= \left. \frac{e^{-(a+j\omega)t}}{-(a+j\omega)} \right|_0^{\infty}$$

$$= \frac{1}{a+j\omega} \qquad \text{for } a > 0$$

For a < 0 the Fourier transform doesnot exist. We note that F(ω) is a complex function and

$$|F(\omega)| = \frac{1}{\sqrt{a^2 + \omega^2}}$$

and
$$\theta(\omega) = -\tan^{-1}\frac{\omega}{a}$$

|F(ω)| and θ(ω) which are known as amplitude and phase spectrum respectively, are plotted in Fig. 12.26.

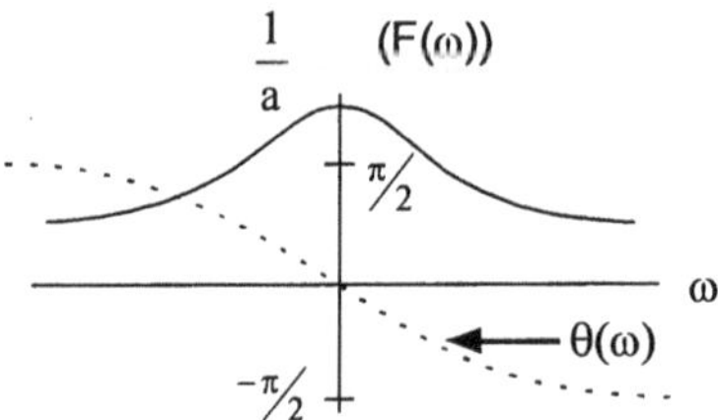

Fig. 12.26 Amplitude and phase spectrum of exponential function

Example 12.13

Find the Fourier transform of the signum function Sgn (t) shown in Fig. 12.27.

$$\text{Sgn}(t) = 1 \qquad t > 0$$
$$= -1 \qquad t > 0$$

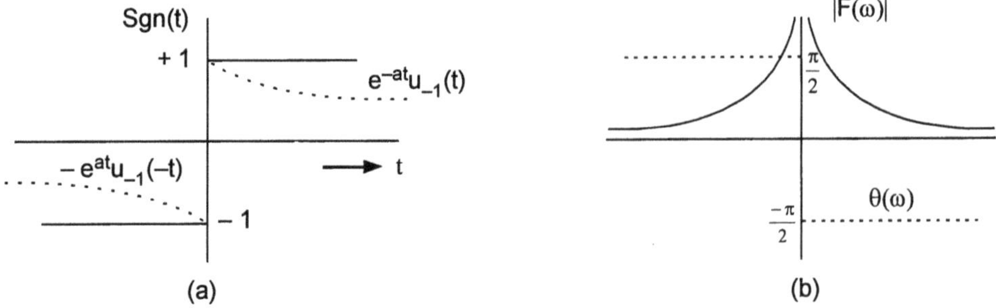

Fig. 12.27 (a) Signum function and (b) its spectrum

Solution :

The signum function can be approximated by

$$f(t) = \text{Sgn}(t) = \underset{a \to 0}{\text{Lt}} \left[e^{-at}u_{-1}(t) - e^{at}\, u_{-1}(-t)\right]$$

$$F(\omega) = \underset{a \to 0}{\text{Lt}} \left[-\int_{-\infty}^{0} e^{at} e^{-j\omega t}\, dt + \int_{0}^{\infty} e^{-at}\, e^{-j\omega t}\, dt\right]$$

$$= \underset{a \to 0}{\text{Lt}} \left[-\frac{1}{a - j\omega} + \frac{1}{a + j\omega}\right]$$

$$= \frac{2}{j\omega}$$

$$|F(\omega)| = \frac{2}{\omega}$$

and
$$\theta(\omega) = \frac{-\pi}{2} \qquad \text{for } \omega > 0$$

$$= \frac{\pi}{2} \qquad \text{for } \omega < 0$$

The magnitude and phase spectrum are shown in Fig. 12.27 (b).

Example 12.14

Find the Fourier transform of the impulse function $u_0(t)$ shown in Fig. 12.28(a).

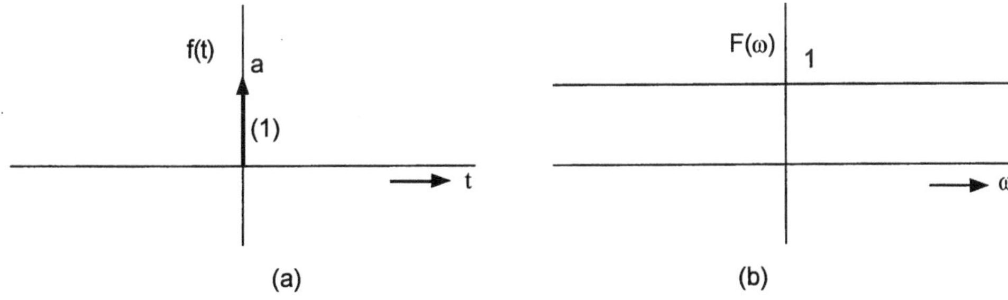

Fig. 12.28 (a) Unit inpulse function and (b) its spectrum

Solution :

$$f(t) = u_0(t)$$

$$F(\omega) = \int_{0^-}^{0^+} u_0(t) e^{-j\omega t}\, dt = 1$$

Thus we see that the Fourier transform of a unit impulse function is a constant and hence it has all frequencies with constant amplitude of unity. Since the Fourier transform is real, $\theta(\omega) = 0$.

Example 12.15

Find the Fourier transform of a constant f(t) = 1, shown in Fig. 12.29(a).

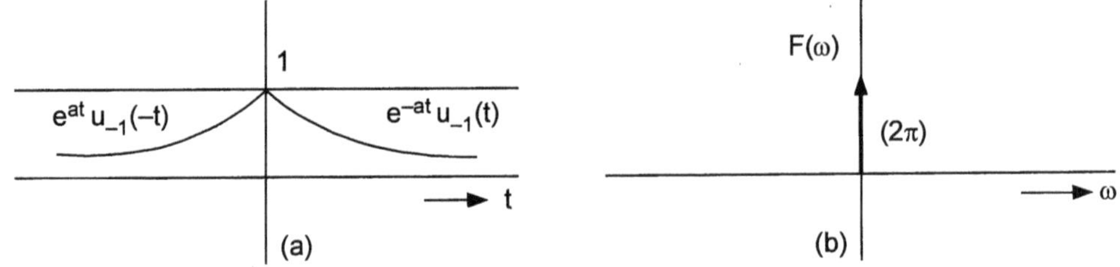

Fig. 12.29 (a) A constant function and (b) its spectrum

Solution :

$$f(t) = \underset{a \to 0}{\text{Lt}} \left[e^{at}u_{-1}(-t) + e^{-at}u_{-1}(t)\right]$$

$$F(\omega) = \underset{a \to 0}{\text{Lt}} \left[\int_{-\infty}^{0} e^{at}e^{-j\omega t}\,dt + \int_{0}^{\infty} e^{-at}e^{-j\omega t}\,dt\right]$$

$$= \underset{a \to 0}{\text{Lt}} \left[\frac{1}{a - j\omega} + \frac{1}{a + j\omega}\right] \qquad \text{for } a > 0$$

$$= \underset{a \to 0}{\text{Lt}} \frac{2a}{a^2 + \omega^2}$$

If $\omega \neq 0$, the limit $a \to 0$ gives $F(\omega) = 0$.

If $\omega = 0$, the limit $a \to 0$ gives rise to $F(\omega) = \infty$. Thus $F(\omega)$ is an impulse function. The strength of the impulse can be found by finding the area under the curve of $F(\omega)$ from $-\infty$ to $+\infty$.

The area under the curve is $= \int_{-\infty}^{\infty} F(\omega)d\omega$

$$= \int_{-\infty}^{\infty} \frac{2a}{a^2 + \omega^2}\,d\omega$$

Let $\omega = a \tan\theta$

$d\omega = a \sec^2\theta \, d\theta$

$$= \int_{\frac{-\pi}{2}}^{\frac{\pi}{2}} \frac{2a}{a^2(1+\tan^2\theta)}(a\sec^2\theta)\,d\theta$$

$$= \int_{\frac{-\pi}{2}}^{\frac{\pi}{2}} 2d\theta$$

$$= 2\,\theta \Big|_{\frac{-\pi}{2}}^{\frac{\pi}{2}} = 2\pi$$

Thus the Fourier transform of a constant of value 1 is

$$F(\omega) = 2\pi\, u_0(\omega)$$

12.7 FOURIER TRANSFORM PROPERTIES :

We shall now consider some important properties of Fourier transforms.

(a) *Linearity* :

The Fourier transform of a function is a linear operation. This implies that if

$$f_1(t) \leftrightarrow F_1(\omega) \text{ and } f_2(t) \leftrightarrow F_2(\omega)$$

then $\quad k_1 f_1(t) + k_2\, f_2(t) \leftrightarrow k_1 F_1(\omega) + k_2 F_2(\omega)$

The proof of this theorem is straight forward.

(b) *Fourier transform of a Real function f(t)* :

If f(t) is a real function, the magnitude function $|F(\omega)|$ is an even function and phase spectrum $\theta(\omega)$ is an odd function of ω.

Proof : If f(t) is a real function

$$F(\omega) = \int_{-\infty}^{\infty} f(t)e^{-j\omega t}\, dt$$

$$F(-\omega) = \int_{-\infty}^{\infty} f(t)e^{j\omega t}\, dt$$

$$= F^*(\omega)$$ where * means complex conjugate

Hence if

$$F(\omega) = |F(\omega)|e^{j\theta(\omega)}$$

$$F(-\omega) = |F(\omega)|e^{-j\theta(\omega)}$$

This means

$$|F(\omega)| = |F(-\omega)| \qquad .....(12.48)$$

and $$\theta(\omega) = -\theta(-\omega) \qquad .....(12.49)$$

Thus $|F(\omega)|$ is an even functional $\theta(\omega)$ is an odd function of ω when f(t) is a real function.

(c) *Symmetry :*

If $$f(t) \leftrightarrow F(\omega)$$

then $$F(t) \leftrightarrow 2\pi\, f(-\omega) \qquad .....(12.50)$$

Proof : From the definition of Fourier transform

$$F(\omega) = \int_{-\infty}^{\infty} f(t)e^{-j\omega t}\, dt$$

Changing ω to x, which does not alter the value of integral, we have

$$F(x) = \int_{-\infty}^{\infty} f(t)e^{-jxt}\, dt$$

Now let $t = -\omega$

$$dt = -\, d\omega$$

$$F(x) = -\int_{\infty}^{-\infty} f(-\omega)e^{j\omega x}\, d\omega = \int_{-\infty}^{\infty} f(-\omega)e^{j\omega x}\, d\omega$$

Now change x to t which does not alter the value of the integral. Hence

$$F(t) = \int_{-\infty}^{\infty} f(-\omega)e^{j\omega t}\, d\omega \qquad .....(12.51)$$

By the definition of inverse Fourier transform, eq. (12.51) can be written as

$$F(t) \leftrightarrow 2\pi\, f(-\omega)$$

Note that if the given function f(t) is an even function,

i.e., $$f(t) = f(-t)$$

then $$F(t) \leftrightarrow 2\pi\, f(\omega)$$

Consider the pulse and its Fourier transform shown in Fig. 12.30 (a) and (b) respectively.

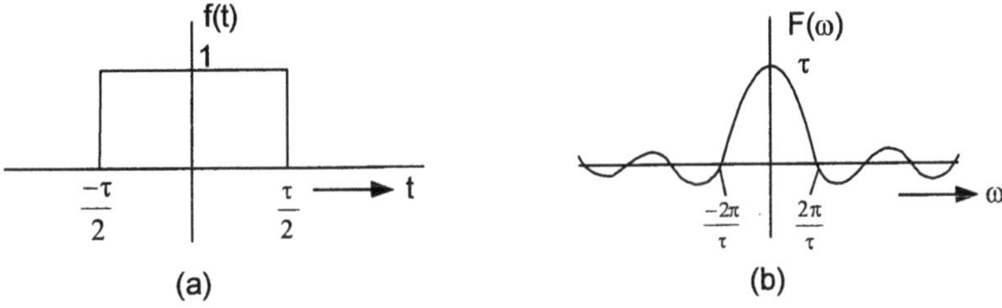

Fig. 12.30 (a) A Pulse and (b) its Fourier transform

Consider the waveform of Fig. 12.30(b) as a time function, as shown in Fig. 12.30(c). From the sysmmetry property, its Fourier transform is given by Fig. 12.30(d)

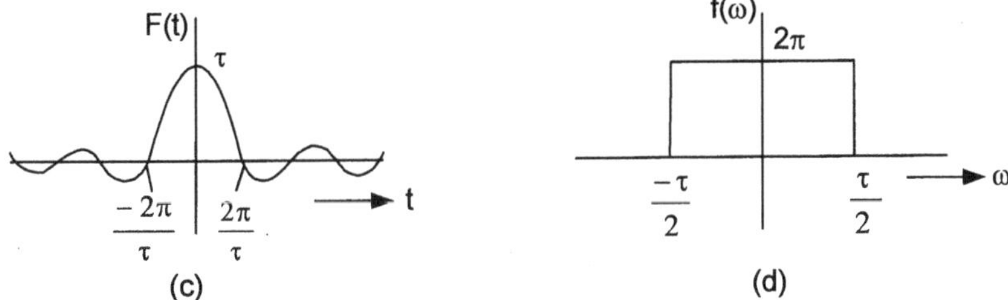

Fig. 12.30 (c) The function F(t) and (d) its transform

Example 12.16

If $\mathcal{F}\,[e^{-at}\,u_{-1}(t)] = \dfrac{1}{a + j\omega}$ find the $\mathcal{F}\left(\dfrac{1}{a + jt}\right)$.

From eq. (12.49)

$$\mathcal{F}\left(\frac{1}{a + jt}\right) = 2\pi[e^{-at}\,u_{-1}(t)]_{t\,=\,-\omega}$$

$$= 2\pi\, e^{a\omega}\, u_{-1}(-\omega)$$

Example 12.17

Find the Fourier transform of a unit step function $u_{-1}(t)$.

Solution :

The unit step function can be considered as a sum of a constant function and a signum function.

$$u_{-1}(t) = \frac{1}{2}\,(1 + \text{sgn}\,(t))$$

From the linearity property,

$$\mathcal{F}\,[u_{-1}(t)] = \frac{1}{2}\left[2\pi u_0(\omega)+\frac{2}{j\omega}\right] = \pi u_0(\omega)+\frac{1}{j\omega}$$

The spectrum of $u_{-1}(t)$ is shown in Fig. 12.31.

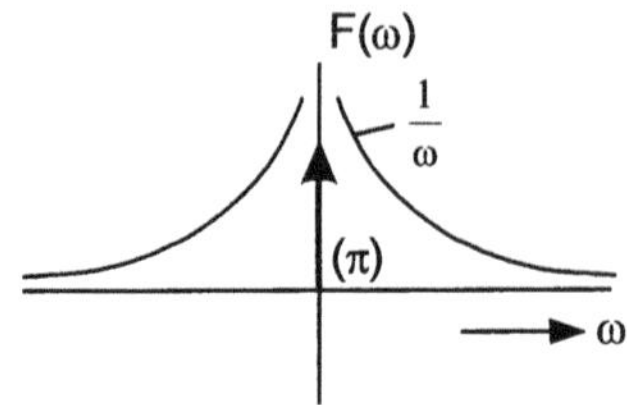

Fig. 12.31 Spectrum of a unit step function

(d) *Scaling :*

If $f(t) \leftrightarrow F(\omega)$, then for a real constant 'a'

$$f(at) \leftrightarrow \frac{1}{|a|}\, F\left(\frac{\omega}{a}\right) \qquad .....(12.52)$$

Proof :

$$\mathcal{F}\,[f(at)] = \int_{-\infty}^{\infty} f(at)e^{-j\omega t}\,dt$$

Let $a > 0$ and let $at = \tau$

$$\mathcal{F}\,[f(at)] = \int_{-\infty}^{\infty} f(\tau)e^{-j\omega\frac{\tau}{a}}\,\frac{d\tau}{a}$$

$$= \frac{1}{a}\int_{-\infty}^{\infty} f(\tau)e^{-j\left(\frac{\omega}{a}\right)\tau}\,d\tau$$

$$= \frac{1}{a}F\left(\frac{\omega}{a}\right) \qquad .....(12.53)$$

Let $a < 0$ and let $at = \tau$

$$\mathcal{F}\,[f(at)] = \int_{\infty}^{-\infty} f(\tau)e^{-j\frac{\omega}{a}\tau}\,\frac{d\tau}{a}$$

$$= -\frac{1}{a}\int_{-\infty}^{\infty} f(\tau)e^{-j\left(\frac{\omega}{a}\right)\tau}\,d\tau$$

$$= \frac{1}{-a}\,F\left(\frac{\omega}{a}\right) \qquad .....(12.54)$$

From eqs. (12.53) and (12.54), we have

$$\mathcal{F}\,[f(at)] = \frac{1}{|a|}F\left(\frac{\omega}{a}\right)$$

The function f(at) represents the function f(t) compressed (expanded) by a factor 'a' and its frequency spectrum $F\left(\frac{\omega}{a}\right)$ will be expanded (compressed) by the same factor 'a'. This is expected because, it the function is compressed, it is varying fast and frequencies of its components will also be increased. Consider the unit pulse and its Fourier spectrum shown in Fig. 12.32(a).

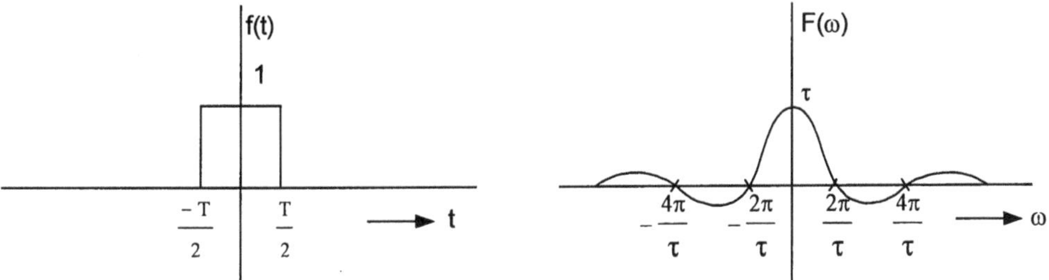

Fig. 12.32 (a) Unit pulse and its frequency spectrum

If the function f(t) is expanded by a factor 2. The resulting function $f\left(\frac{t}{2}\right)$ and its Fourier transform F(2ω) are shown in Fig. 12.32 (b).

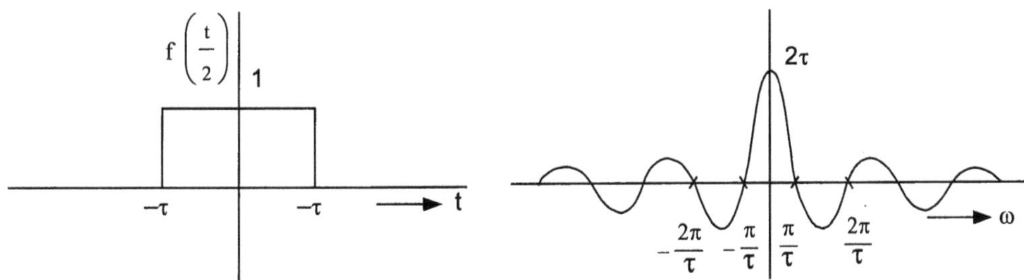

Fig. 12.32 (b) Unit pulse and its frequency spectrum

In the limit if T is made infinity, f(t) approaches a constant function of value unity and its Fourier transform will have an impulse at ω = 0.

(e) *Shifting in time domain :*

If $f(t) \leftrightarrow F(\omega)$ then

$$f(t-\tau) \leftrightarrow F(\omega)\, e^{-j\omega\tau} \qquad .....(12.55)$$

Proof :

$$\mathcal{F}[f(t-\tau)] = \int_{-\infty}^{\infty} f(t-\tau)e^{-j\omega t}\, dt$$

Let $t - \tau = T$

$$\mathcal{F}[f(t-\tau)] = \int_{-\infty}^{\infty} f(T)e^{-j\omega(T+\tau)}\, dT$$

$$= e^{-j\omega\tau} F(\omega)$$

Example 12.18

Find the Fourier transform of the pulse shown in Fig 12.33.

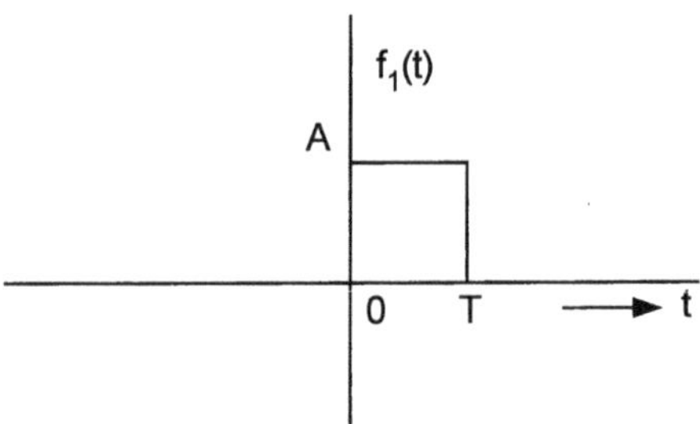

Fig. 12.33 Pulse for example 12.18

Solution :

The pulse shown in Fig. 12.33 is the result of shifting the pulse f(t) of Fig. 12.20 by $\frac{\tau}{2}$. Therefore

$$f_1(t) = f\left(t - \frac{\tau}{2}\right)$$

Hence by time shifting property

$$F_1(\omega) = e^{-j\omega\frac{\tau}{2}} F(\omega)$$

$$= A\tau\left(\frac{\sin\frac{\omega\tau}{2}}{\frac{\omega\tau}{2}}\right)e^{-j\omega\frac{\tau}{2}}$$

Hence we see that the magnitude spectrum does not change when the time shift takes place but the phase spectrum changes.

$$|\mathfrak{F}_1(\omega)| = A\tau\left(\frac{\sin\frac{\omega\tau}{2}}{\frac{\omega\tau}{2}}\right)$$

and $$\theta_1(\omega) = \frac{-\omega\tau}{2}$$

(f) ***Shifting in frequency domain :***

If $f(t) \leftrightarrow F(\omega)$

then $f(t)\ e^{j\omega_0 t} \leftrightarrow F(\omega - \omega_0)$ .....(12.56)

Proof :

$$\mathfrak{F}\left[f(t)e^{j\omega_0 t}\right] = \int_{-\infty}^{\infty} f(t)e^{j\omega_0 t}e^{-j\omega t}dt$$

$$= \int_{-\infty}^{\infty} f(t)e^{-j(\omega-\omega_0)t}\, dt$$

$$= F(\omega - \omega_0)$$

Thus $f(t)\ e^{j\omega_0 t} \leftrightarrow F(\omega - \omega_0)$.

Significance of frequency shifting :

Frequency shifting is of great significance in communication systems. Multiplications of the time function by $e^{j\omega_0 t}$ amounts to multiplying by a sinusoidal signal of frequency ω_0, a process known as modulation. This translates or shifts the entire frequency spectrum of f(t) by the frequency of the sinusoid.

Example 12.19

Find the Fourier transform of $f(t) = e^{j\omega_0 t}$.

Solution :

Consider the function

$$f_1(t) = 1$$

Its Fourier transform was obtained earlier and is given by

$$F_1(\omega) = 2\pi\ u_0(\omega)$$

Now the given function can be written as

$$f(t) = f_1(t)\, e^{j\omega_0 t}$$

From the frequency shifting theorem

$$F(\omega) = F_1(\omega - \omega_0)$$

$$= 2\pi\ u_0(\omega - \omega_0)$$

Example 12.20

Find the Fourier transform of $f(t) = \cos \omega_0 t$.

Solution :

$$f(t) = \frac{e^{j\omega_0 t} + e^{-j\omega_0 t}}{2}$$

From example 12.19, we have

$$F(\omega) = \frac{2\pi}{2}\left[u_0(\omega - \omega_0) + u_0(\omega + \omega_0)\right]$$

$$= \pi[u_0(\omega - \omega_0) + u_0\,(\omega + \omega_0)]$$

Similarly

(i) $$\mathcal{F}\,(\sin \omega_0\, t) = \frac{2\pi}{2j}\left[u_0(\omega - \omega_0) - u_0(\omega + \omega_0)\right]$$

$$= j\pi\ [u_0(\omega + \omega_0) - u_0(\omega - \omega_0)]$$

(ii) $\mathfrak{F}[\sin(\omega_0 t + \phi)] = \mathfrak{F}\left[\dfrac{e^{j(\omega_0 t+\phi)} - e^{-j(\omega_0 t+\phi)}}{2j}\right]$

$$= 2\pi\left[\frac{u_0(\omega-\omega_0)e^{j\phi} - u_0(\omega+\omega_0)e^{-j\phi}}{2j}\right]$$

$$= j\pi\left[u_0(\omega+\omega_0)e^{-j\phi} - u_0(\omega-\omega_0)e^{j\phi}\right]$$

(iii) $\mathfrak{F}[\cos\omega_0 t\, u_{-1}(t)] = \dfrac{1}{2}\left[e^{j\omega_0 t}u_{-1}(t) + e^{-j\omega_0 t}u_{-1}(t)\right]$

Since $\mathfrak{F}\{u_{-1}(t)] = \pi u_0(\omega) + \dfrac{1}{j\omega}$

We have

$$\mathfrak{F}[\cos\omega_0 t\, u_{-1}(t)] = \frac{1}{2}\left[\pi u_0(\omega-\omega_0) + \frac{1}{j(\omega-\omega_0)} + \pi u_0(\omega+\omega_0) + \frac{1}{j(\omega+\omega_0)}\right]$$

$$= \frac{\pi}{2}\left[u_0(\omega-\omega_0) + u_0(\omega+\omega_0)\right] - \frac{j\omega}{\omega^2-\omega_0^2}$$

(iv) $\mathfrak{F}[\sin\omega_0 t\, u_{-1}(t)] = \dfrac{\pi}{2j}\left[u_0(\omega-\omega_0) - u_0(\omega+\omega_0)\right] - \dfrac{\omega_0}{\omega^2-\omega_0^2}$

The functions $\sin\omega_0 t$, $\cos\omega_0 t$ and their frequency spectra are shown in Fig. 12.34

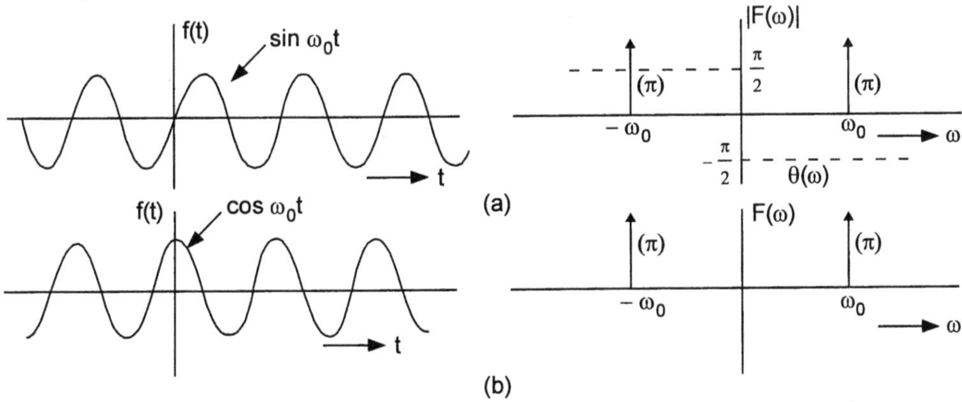

Fig. 12.34 (a) sin ω_0t and its spectrum (b) cos ω_0t and its spectrum

Note that the Fourier transforms of sin $\omega_0 t$ and cos $\omega_0 t$ consist of only two impulses at $\omega = \omega_0$ and $\omega = -\omega_0$. Their magnitude spectra are same but phase spectra are different. This is obvious because the sine function can be considered as a cosine function shifted by $\frac{\pi}{2}$.

(g) ***Time differentiation :***

If $f(t) \leftrightarrow F(\omega)$ then

$$\frac{df}{dt} \leftrightarrow j\omega F(\omega) \text{ if } \mathcal{F}\left(\frac{df}{dt}\right) \text{ exists.} \quad .....(12.57)$$

Proof :

We have, by definition

$$\mathcal{F}\left(\frac{df}{dt}\right) = \int_{-\infty}^{\infty} \frac{df}{dt} e^{-j\omega t}\, dt$$

Integrating by parts, we have,

$$\mathcal{F}\left[\frac{df}{dt}\right] = f(t)e^{-j\omega t}\Big|_{-\infty}^{\infty} - \int_{-\infty}^{\infty} f(t)(-j\omega)e^{-j\omega t}\, dt$$

Since f(t) is Fourier transformable $\underset{t \to \pm\infty}{\text{Lt}} f(t) = 0$

and hence

$$\mathcal{F}\left[\frac{df}{dt}\right] = j\omega \int_{-\infty}^{\infty} f(t)e^{-j\omega t}\, dt$$

$$= j\omega F(\omega)$$

Further, we can show that

$$\mathcal{F}\left[\frac{d^n f}{df^n}\right] = (j\omega)^n F(\omega) \quad .....(12.58)$$

(h) ***Time integration :***

If $f(t) \leftrightarrow F(\omega)$ and $F(0) = 0$ then

$$\int_{-\infty}^{t} f(\tau)d\tau \leftrightarrow \frac{F(\omega)}{j\omega} \quad .....(12.59)$$

Proof :

We have

$$\mathfrak{F}\left[\int_{-\infty}^{t} f(t)dt\right] = \int_{-\infty}^{\infty}\left(\int_{-\infty}^{t} f(\tau)d\tau\right)e^{-j\omega t}\,dt$$

Integration by parts yields,

$$\mathfrak{F}\left[\int_{-\infty}^{t} f(t)dt\right] = \int_{-\infty}^{t} f(\tau)d\tau \left.\frac{e^{-j\omega t}}{-j\omega}\right|_{-\infty}^{\infty} - \int_{-\infty}^{\infty} \frac{f(t)e^{-j\omega t}}{-j\omega}dt \qquad .....(12.60)$$

The first term on the right hand side is,

$$\int_{-\infty}^{\infty} f(\tau)d\tau\left(\frac{e^{-j\omega t}}{-j\omega}\right)_{t=\infty}$$

But $\underset{\omega\to 0}{\text{Lt}} \int_{-\infty}^{\infty} f(\tau)e^{-j\omega\tau}d\tau = \int_{-\infty}^{\infty} f(\tau)d\tau = F(0)$

If $F(0) = 0$ then eq. (12.60) yields

$$\mathfrak{F}\left[\int_{\infty}^{t} f(\tau)d\tau\right] = \frac{1}{j\omega}F(\omega)$$

Example 12.21

Find the Fourier transform of the triangular pulse shown in Fig. 12.35(a).

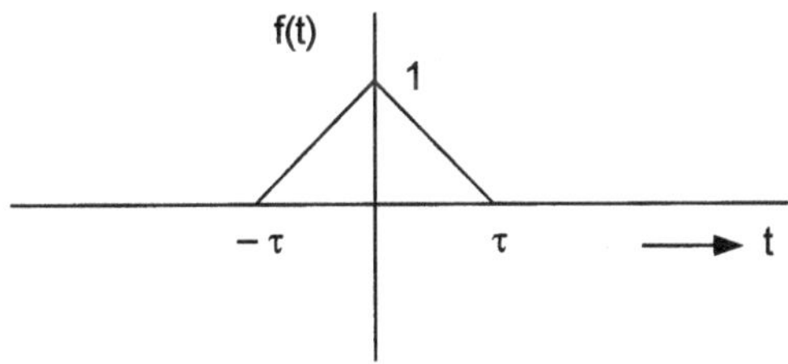

Fig. 12.35 (a) Triangular pulse

Solution :

We have already obtained the Fourier Transform of this pulse in example 12.11. Let us apply differentiation theorem and obtain its Fourier transform in a simple way.

The pulse is differentiated until impulses are obtained. In this case we differentiate twice. The results are shown in Fig. 12.35(b) and (c).

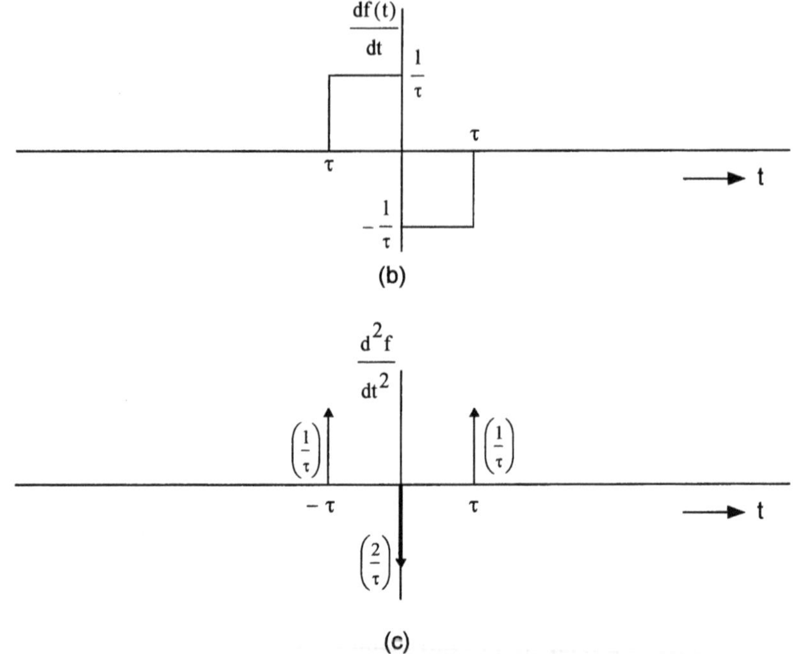

Fig. 12.35 (b) Waveform of $\frac{df(t)}{dt}$ (c) waveform of $\frac{d^2f(t)}{dt}$

Fourier transform of the function in Fig. 12.35(c) is given by

$$\mathcal{F}\left[\frac{d^2f}{dt^2}\right] = (j\omega)^2 F(\omega) = \frac{1}{\tau}\left[e^{j\omega\tau} - 2 + e^{-j\omega\tau}\right]$$

$$F(\omega) = \frac{1}{\tau\omega^2}\left[2 - 2\cos\omega\tau\right]$$

$$= \frac{4}{\omega^2\tau}\sin^2\frac{\omega\tau}{2}$$

$$= \tau\left(\frac{\sin\frac{\omega\tau}{2}}{\frac{\omega\tau}{2}}\right)^2$$

$$= \tau\,\mathrm{sinc}^2\left(\frac{\omega\tau}{2\pi}\right)$$

which is the desired result.

(i) *Frequency differentiation :*

This is a dual of the time differentiation.

If $f(t) \leftrightarrow F(\omega)$ then

$$-jt\, f(t) \leftrightarrow \frac{d}{d\omega} F(\omega) \qquad .....(12.61)$$

Proof :

$$F(\omega) = \int_{-\infty}^{\infty} f(t) e^{-j\omega t}\, dt$$

$$\frac{d}{d\omega} F(\omega) = \frac{d}{d\omega}\left[\int_{-\infty}^{\infty} f(t) e^{-j\omega t}\, dt\right]$$

$$= \int_{-\infty}^{\infty} f(t) \frac{d}{d\omega} e^{-j\omega t}\, dt$$

$$= \int_{-\infty}^{\infty} [-jtf(t)] e^{-j\omega t}\, dt$$

$$= \mathfrak{F}\,[-jtf(t)]$$

(j) *Convolution :*

If $f_1(t) \leftrightarrow F_1(\omega)$ are $f_2(t) \leftrightarrow F_2(\omega)$ then the Fourier transform pair of the convolution of $f_1(t)$ and $f_2(t)$ is

$$f_1(t) * f_2(t) \leftrightarrow F_1(\omega)\,.\,F_2(\omega) \qquad .....(12.62)$$

Proof :

$$\mathfrak{F}\,[f_1(t) * f_2(t)] = \int_{-\infty}^{\infty}\left[\int_{-\infty}^{\infty} f_1(\tau) f_2(t-\tau)\, d\tau\right] e^{-j\omega t}\, dt$$

Interchanging the integrals

$$\mathfrak{F}\,[f_1(t) * f_2(t)] = \int_{-\infty}^{\infty} f_1(\tau)\left[\int_{-\infty}^{\infty} f_2(t-\tau) e^{-j\omega t}\, dt\right] d\tau$$

Using Time shifting property given by eq. (12.55), we have

$$\mathcal{F}\,[f_1(t) * f_2(t)] = \int_{-\infty}^{\infty} f_1(\tau) F_2(\omega) e^{-j\omega\tau}\, d\tau$$

$$= F_2(\omega) \int_{-\infty}^{\infty} f_1(\tau) e^{-j\omega\tau}\, d\tau$$

$$= F_2(\omega) \, . \, F_1(\omega).$$

Similarly we can show that the convolution in frequency domain amounts to multiplication in time domain

i.e., $F_1(\omega) * F_2(\omega) \leftrightarrow 2\pi\, f_1(t) \, . \, f_2(t)$ .....(12.63)

Using the concept of convolution, the response of any linear system to an arbitrary input can be found. Consider the linear system shown in Fig. 12.36. Let the impulse response of the system be h(t). If the input is $v_1(t)$ and the output is $v_2(t)$, we have

$$v_2(t) = v_1(t) * h(t) \qquad .....(12.64)$$

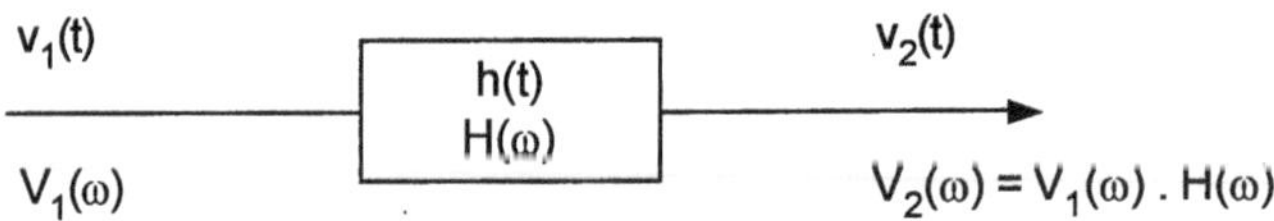

Fig. 12.36 A linear system

Taking the Fourier transform of eq. (12.64) we have

$$V_2(\omega) = V_1(\omega) \, . \, H(\omega) \qquad .....(12.65)$$

and $v_2(t)$ can be obtained as

$v_2(t) = \mathcal{F}^{-1}\, V_2(\omega)$ where $\mathcal{F}^{-1}$ stands for inverse Fourier transform.

H(ω) is called as the system function.

(k) *Parseval's theorem :*

If f(t) ↔ F(ω) then the energy in the signal, defined by

$$\int_{-\infty}^{\infty} f^2(t) dt = \frac{1}{2\pi} \int_{-\infty}^{\infty} | f(\omega)|^2\, d\omega \qquad .....(12.66)$$

Proof :

$$\int_{-\infty}^{\infty} f^2(t) dt = \int_{-\infty}^{\infty} f(t) . \left[\frac{1}{2\pi} \int_{-\infty}^{\infty} F(\omega) e^{j\omega t}\, d\omega\right] dt \qquad .....(12.67)$$

$$\left(\because \;\; f(t) = \frac{1}{2\pi} \int_{-\infty}^{\infty} F(\omega) e^{j\omega t}\, d\omega\right)$$

Interchanging the order of integration in eq. (12.67) we get

$$\int_{-\infty}^{\infty} f^2(t)dt = \frac{1}{2\pi}\int_{-\infty}^{\infty} F(\omega)\left[\int_{-\infty}^{\infty} f(t)e^{j\omega t}\, dt\right]d\omega$$

But $\int_{-\infty}^{\infty} f(t)e^{j\omega t}\, dt = F(-\omega)$

$$\therefore \quad \int_{-\infty}^{\infty} f^2(t)dt = \frac{1}{2\pi}\int_{-\infty}^{\infty} F(\omega)F(-\omega)d\omega$$

We observe that, if f(t) is a real function, $F(-\omega) = F^*(\omega)$ and therefore we have

$$\int_{-\infty}^{\infty} f^2(t)dt = \frac{1}{2\pi}\int_{-\infty}^{\infty} |F(\omega)|^2\, d\omega$$

If f(t) is either a voltage or current in a 1 Ω resistance, $\int_{-\infty}^{\infty} f^2(t)dt$ is the energy lost in the resistor and therefore $\int_{-\infty}^{\infty} f^2(t)dt$ is referred to as the energy in the signal.

(I) *Fourier transform of a general periodic function :*

Let f(t) be a periodic function with period T and $\omega_0 = \frac{2\pi}{T}$. The Fourier series of this function can be written down as

$$f(t) = \sum_{n=-\infty}^{\infty} c_n e^{jn\omega_0 t}$$

Using the linearity property of Fourier transforms we have

$$\mathfrak{F}[f(t)] = \mathfrak{F}\left[\sum_{n=-\infty}^{\infty} c_n e^{jn\omega_0 t}\right]$$

$$= \sum_{n=-\infty}^{\infty} c_n\, \mathfrak{F}\left(e^{jn\omega_0 t}\right)$$

$$= 2\pi\sum_{n=-\infty}^{\infty} c_n u_0(\omega - n\omega_0) \qquad .....(12.68)$$

This shows that f(t) has a discrete spectrum consisting of impulses located at $\omega = n\omega_0$ for n = ... –2, –1, 0, 1, 2..... The strength of each impulse is 2π times the value of the corresponding Fourier coefficient appearing in the complex Fourier series expansion of the given periodic function.

12.8 APPLICATION TO NETWORKS :

12.8.1 Periodic input functions :

If a periodic input is given to a simple network, the output of the network can be found easily by considering the Fourier series of the given periodic function, finding the response due to individual terms of this series and adding these responses. We will illustrate this by an example.

Example 12.22

Find the steady state response of the network in Fig. 12.37(a) to a periodic input shown in Fig. 12.37(b).

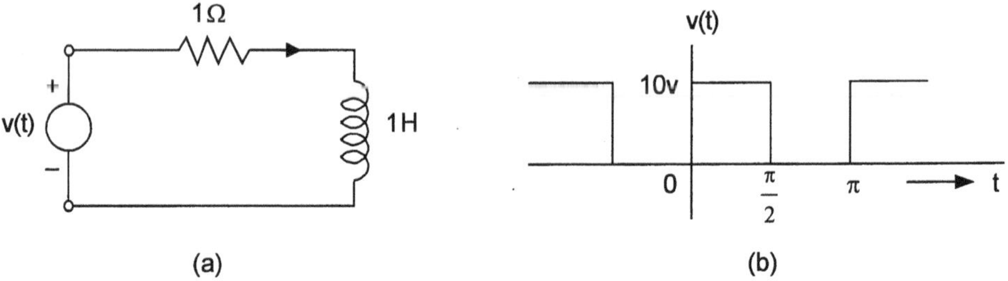

Fig. 12.37 (a) The network and (b) the input

Solution :

The Fourier series of v(t) is given by

$$v(t) = 5 + \frac{20}{\pi} \sum_{n=1,3,5}^{\infty} \frac{\sin 2nt}{n}$$

The nth term of this series is

$$v_n(t) = \frac{20}{\pi n} \sin 2nt$$

In vector notation

$$V_n = \frac{20}{\pi n} \cdot \frac{1}{\sqrt{2}} \angle 0$$

we have $\quad Z_n = 1 + j\omega_0 n$

$\because \quad \omega_0 = 2$ we have

$Z_n = 1 + jn$

$$I_n = \frac{V_n}{Z_n} = \frac{20}{\sqrt{2}\pi n} \angle 0 \; \frac{1}{1+j2n}$$

$$= \frac{20}{\sqrt{2}\pi n\sqrt{1+4n^2}} \angle -\tan^{-1} 2n$$

$$\therefore \quad i_n(t) = \frac{20}{\pi n\sqrt{1+4n^2}} \sin\left(2nt - \tan^{-1} 2n\right)$$

The response due to d.c term is

$i_0 = 5A$

The complete steady state response is given by

$$i(t) = 5 + \frac{20}{\pi} \sum_{n=1,3,5}^{\infty} \frac{1}{n\sqrt{1+4n^2}} \sin\left(2nt - \tan^{-1} 2n\right)$$

If the voltage is applied at t = 0 and the initial condition in the network is given, natural response also could be found out in the usual way.

12.8.2 Pulse input functions :

If the input to a network is a pulse, the response could be found out using Fourier transforms. The is illustrated using an example.

Example 12.23

Find the response of the circuit in Fig. 12.38.

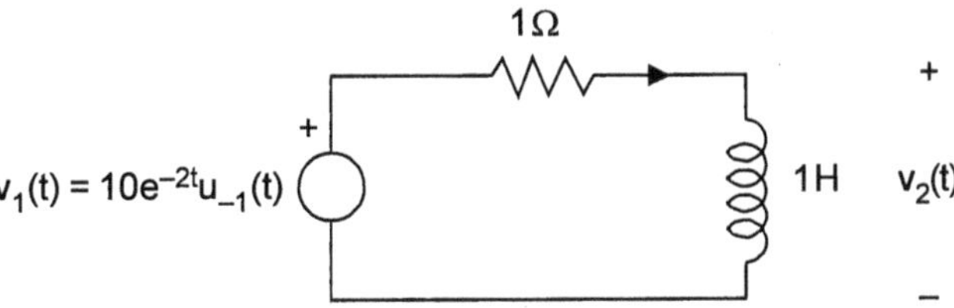

Fig. 12.38 Network for example 12.23

Solution :

The system function for the network is given by

$$H(\omega) = \frac{j\omega}{1 + j\omega}$$

The Fourier transform of the input function is

$$\mathcal{F}[v_1(t)] = V_1(\omega) = \frac{10}{2 + j\omega}$$

Thus from eq. (12.62), we have

$$V_2(\omega) = V_1(\omega) \,.\, H(\omega)$$

$$= \frac{10}{2 + j\omega} \cdot \frac{j\omega}{1 + j\omega}$$

$$= \frac{k_1}{2 + j\omega} + \frac{k_2}{1 + j\omega}$$

Evaluating k_1 and k_2, we have

$$V_2(\omega) = \frac{20}{j\omega + 2} - \frac{10}{j\omega + 1}$$

Inverse transform of $V_2(\omega)$ is

$$v_2(t) = 10[2e^{-2t} - e^{-t}]u_{-1}(t)$$

Problems

12.1 Find the Fourier series of the waveform shown in Fig. P. 12.1. What is the Fourier series of this waveform if the period is changed to 1 micro second.

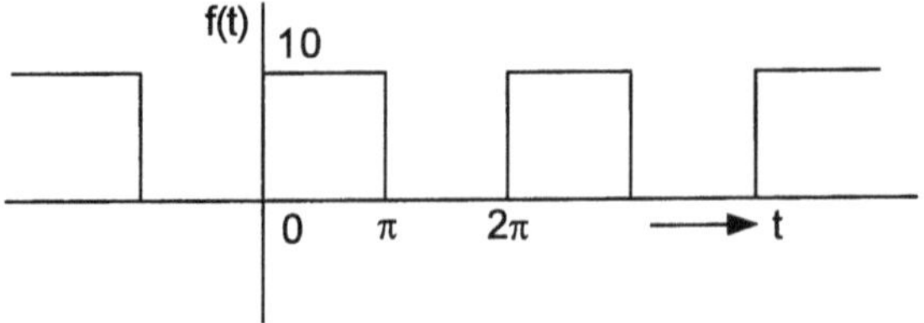

Fig. P. 12.1

12.2 Find the Fourier series of the impulse train shown in Fig. P. 12.2.

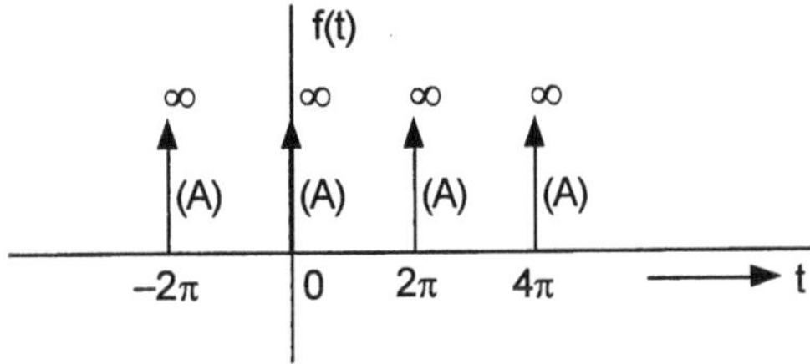

Fig. P. 12.2

12.3 Find the Fourier series of the waveform shown in Fig. P. 12.3 (a). Plot the frequency

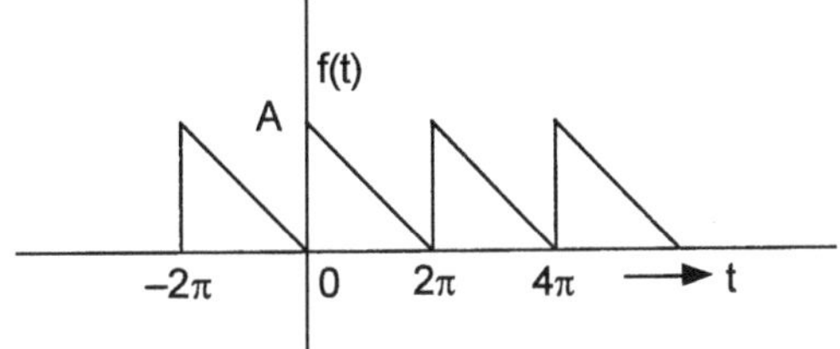

Fig. P. 12.3 (a)

spectrum. From the above result, find the Fourier series of its mirror image shown in Fig. P. 12.3 (b). Plot the frequency spectrum.

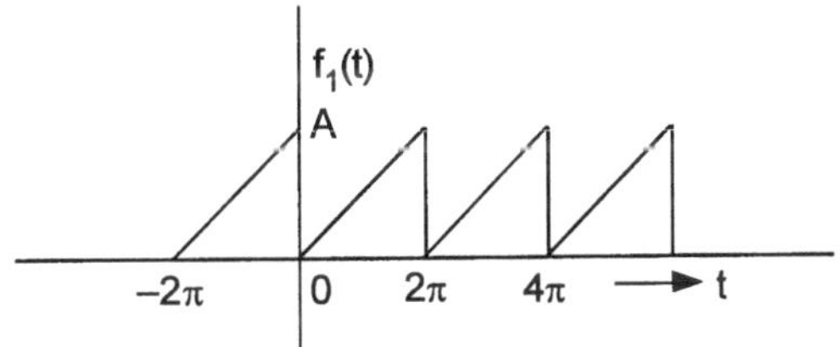

Fig. P. 12.3 (b)

12.4 Find the Fourier series of the impulse train shown in Fig. P. 12.4.

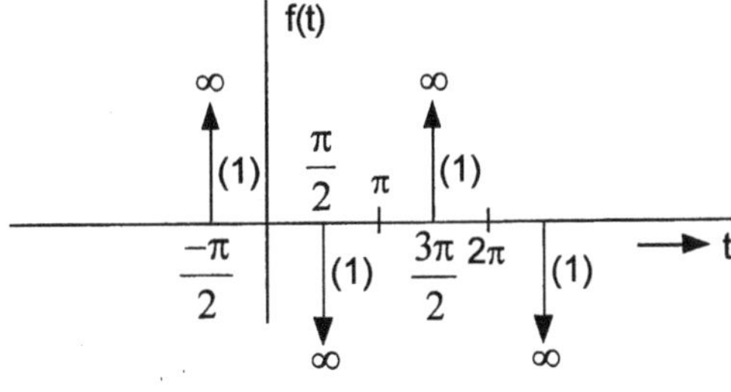

Fig. P. 12.4

12.5 Find the Fourier series of the full wave rectified sine wave shown in Fig. P. 12.5.

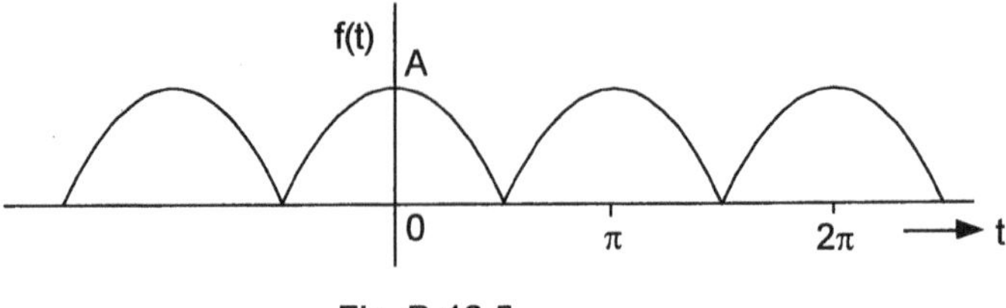

Fig. P. 12.5

If the period of the above waveform is changed to 2π instead of π, what is its Fourier series.

12.6 Find the Fourier series of the waveform shown in Fig. P. 12.6(a). Hence find the Fourier series of the waveform shown in Fig. P. 12.6(b).

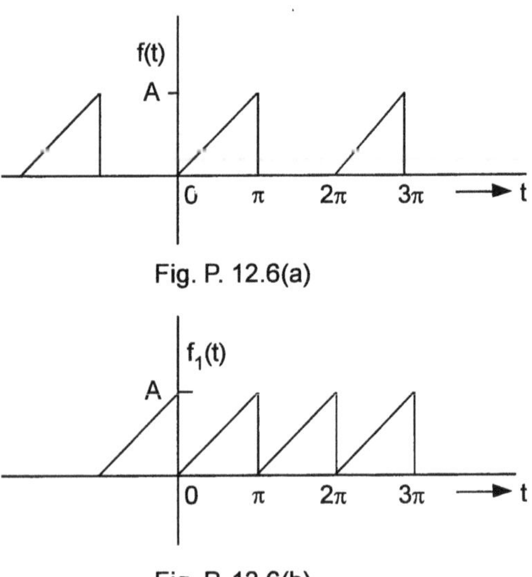

Fig. P. 12.6(a)

Fig. P. 12.6(b)

12.7 Find the exponential Fourier series of the waveform shown in Fig. P. 12.7. Convert it to trignomatric form. Plot the spectrum.

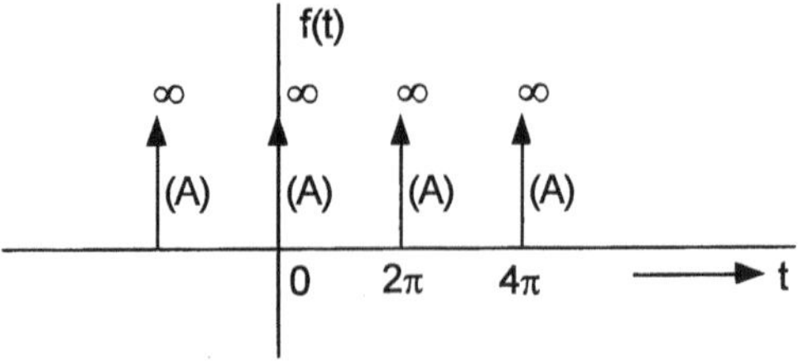

Fig. P. 12.7

12.8 Find the complex Fourier series of the waveform shown in Fig. P. 12.8. Plot the spectrum.

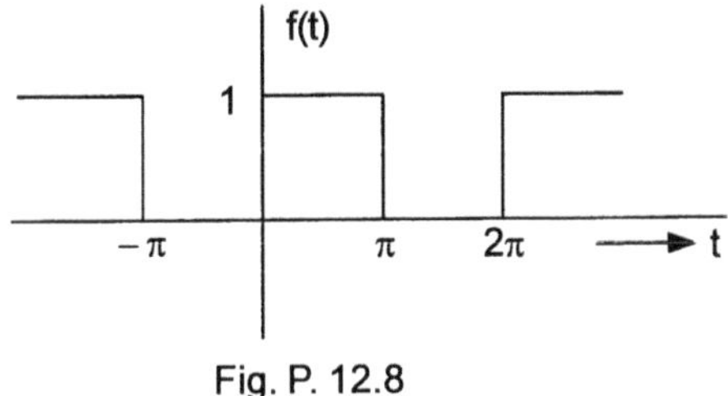

Fig. P. 12.8

12.9 Find the coefficients of the complex Fourier series for the waveform shown in Fig. P. 12.9.

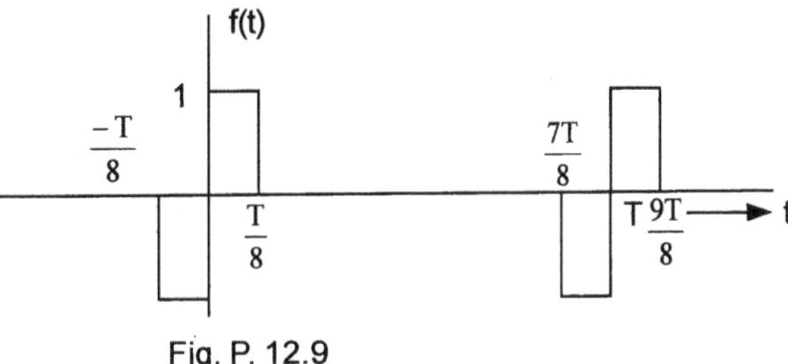

Fig. P. 12.9

12.10 Find the coefficients of the complex Fourier series for the waveform shown in Fig. P. 12.10.

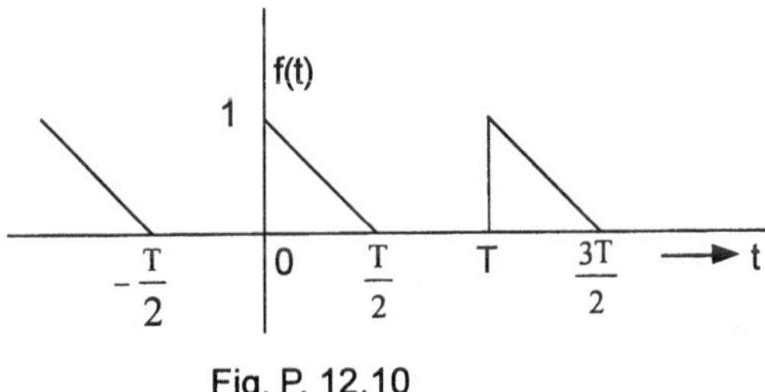

Fig. P. 12.10

12.11 Find the Fourier transform of the pulse shown in Fig. P. 12.11.

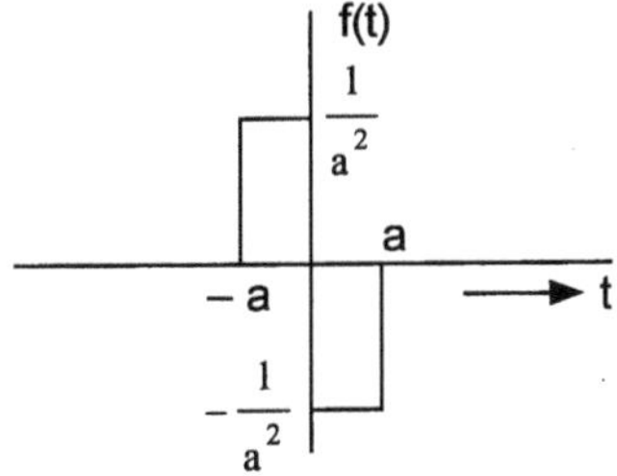

Fig. P. 12.11

12.12 Find the Fourier transform of the pulse shown in Fig. P. 12.12.

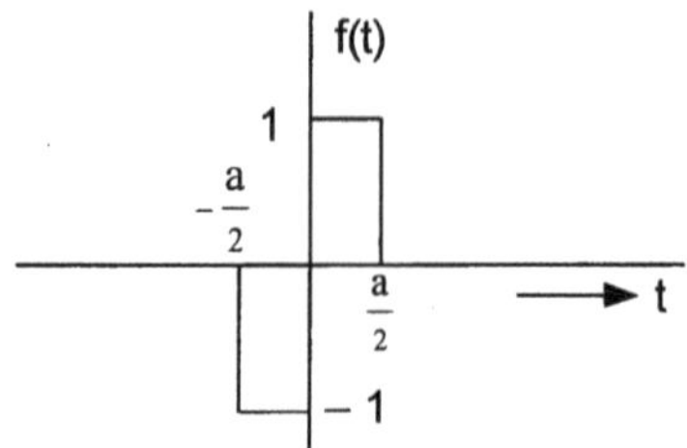

Fig. P. 12.12

12.13 Find the Fourier transforms of the following functions.

(a) $u_0(t - t_0)$ (b) $\sin^2 3t$ (c) $\cos(2t + 45)$

(d) $te^{-4t}u(t)$ (e) $e^{-at}\sin \omega_0 t\ u(t)$

12.14 Find the Fourier transform of the function shown in Fig. P. 12.13.

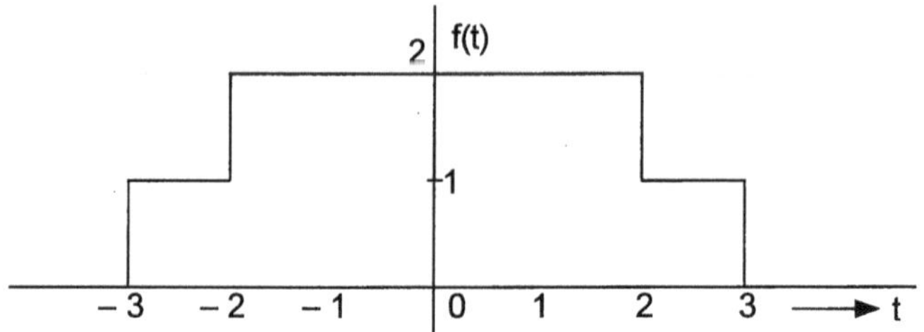

Fig. P. 12.13

12.15 Find the Fourier transform of

(a) sgn (t) . $u_0(t - 1)$ (b) sgn (t – 1) . $u_0(t)$

12.16 If the Fourier trasform of f(t) shown in Fig. P. 12.14 (a) is F(ω). Find the Fourier transform of $f_1(t)$ shown in Fig. P. 12.14(b) in terms of F(ω).

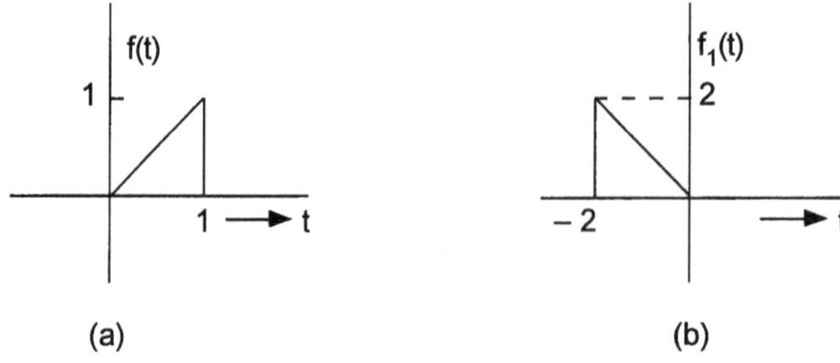

Fig. P. 12.14

12.17 Find the Fourier transform of $f(t) = e^{-a|t-t_0|}$. Plot the continuous spectrum of amplitude and phase.

12.18 If $\mathfrak{F}\,[f(t)] = F(\omega)$ find the Fourier transform of

(a) $f(t - t_0) + f(t + t_0)$

(b) $\cos \omega_0 t\, f(t)$

(c) $\sin \omega_0 t\, f(t)$

12.19 Find the Fourier transform of the waveform shown in Fig. P. 12.15 using differentiation theorem.

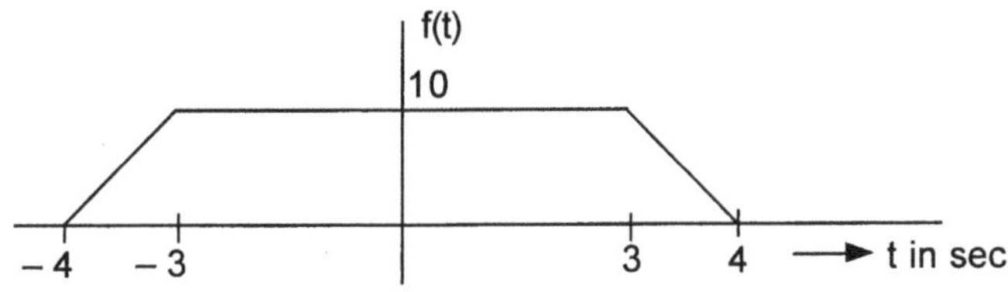

Fig. P. 12.15

12.20 If $\mathfrak{F}\,[f_1(t)] - F_1(\omega)$ and $\mathfrak{F}\,[f_2(t)] = F_2(\omega)$ show that

$$\mathfrak{F}\,[f_1(t)\,.\,f_2(t)] = \frac{1}{2\pi} F_1(\omega)\,.\,F_2(\omega).$$

12.21 If $\mathfrak{F}\,[f(t)] = F(\omega)$ using the result of problem 12.20, show that

$$\int_{-\infty}^{\infty} f^2(t)\, dt = \frac{1}{2\pi} \int_{-\infty}^{\infty} |F(\omega)|^2\, d\omega$$

12.22 Find the Fourier transform of the periodic function shown in Fig. P. 12.16.

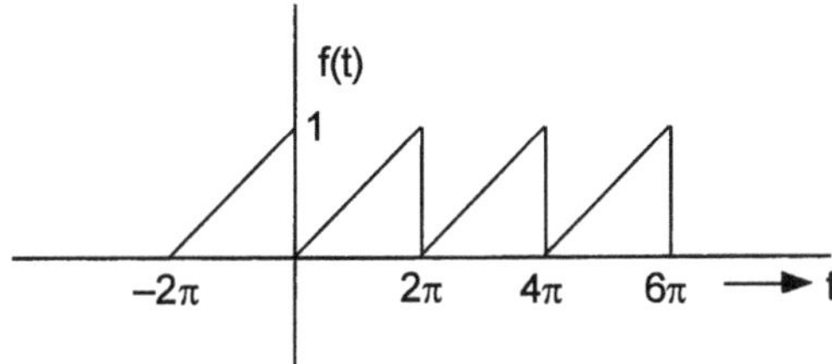

Fig. P. 12.16

12.23 The circuit shown in Fig. P. 12. 17 is a filter for the half wave rectified sine wave. Find the d.c. component and the amplitude of the fundamental component.

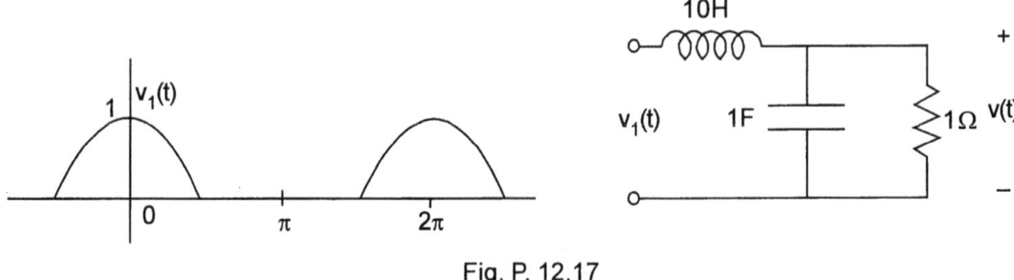

Fig. P. 12.17

12.24 Find the current i(t) in the circuit of Fig. P. 12. 18.

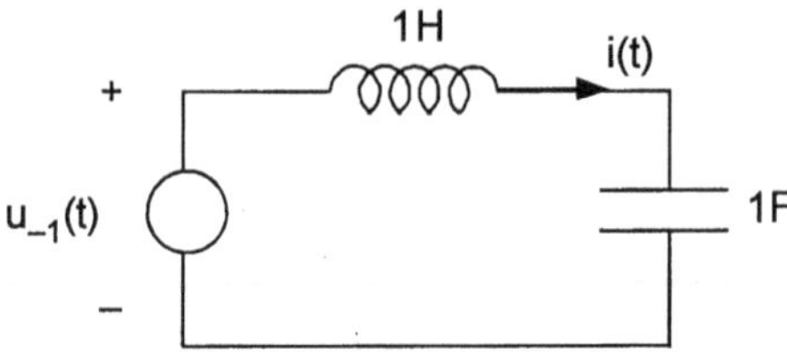

Fig. P. 12.18

12.25 Find the unit impulse response of the circuit shown in Fig. P. 12.19.

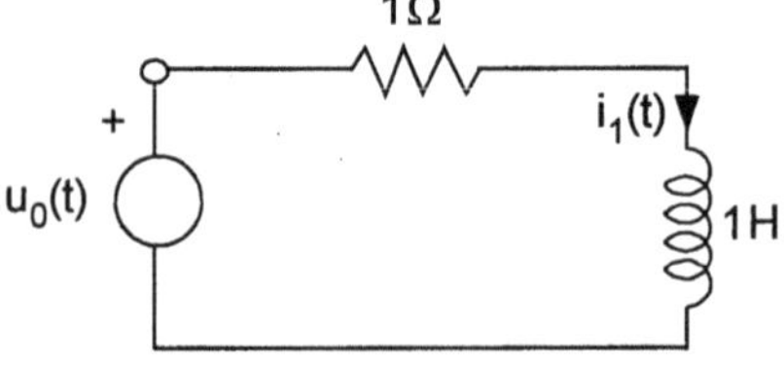

Fig. P. 12.19

Answers to Problems

Chapter 1

1.1 (a) 10A (b) –10A (c) – 10A, 10A

1.2 (a) 30V (b) – 30V (c) – 30V (d) 30V

1.3 (a) 75W (b) – 75W (c) 150W

1.4 b and c

1.6 Resistance of value 24 Ω.

1.7 (a) $30e^{-t}$V, $300e^{-2t}$ W

(b) $-20e^{-t}$V, $-200e^{-2t}$ W

(c) $100(1 - e^{-t})$V, $1000\,(e^{-t} - e^{-2t})$W

1.9 $1000\,\pi \cos 100\pi t$ A, $10^3 \sin^2 100\pi t$ J

1.10 $7 - 5 \cos 1000\,t$ A

1.11 5 mH

1.13

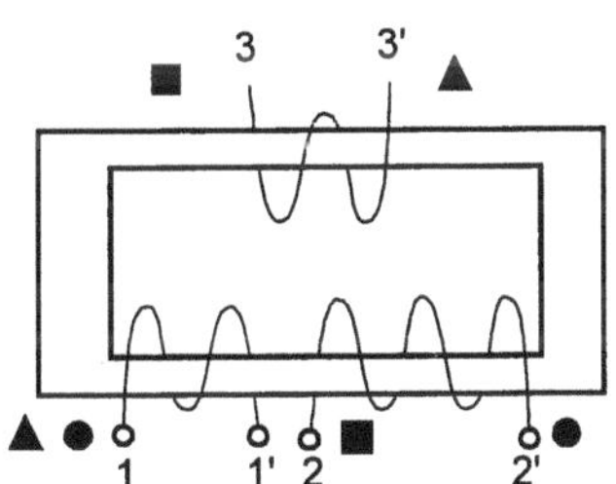

1.14 (a)

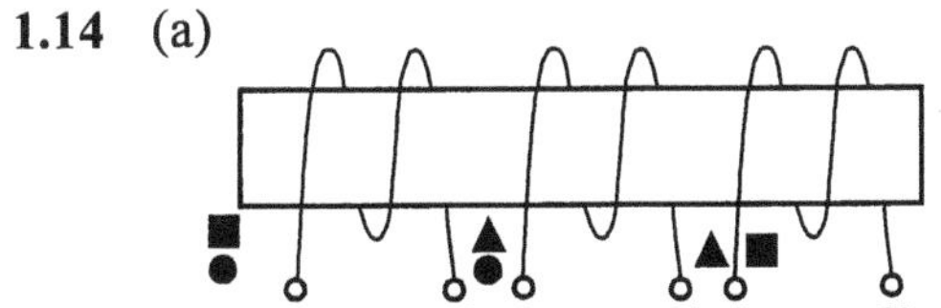

(b)

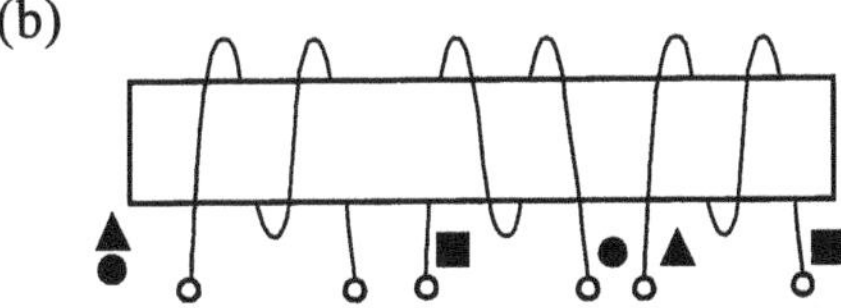

1.15 K = 0.75 e = 0.45(cos t + cos 2t) mv

1.16 (a) 0.27 A/sec (b) 0.4325 AT (c) 0.135 AT/sec
(d) 0.135 V (e) 0.187 J

1.17 (a) 0.932 C (b) 0.271 V/sec (c) 1.864 V (d) 0.8686 J

1.18

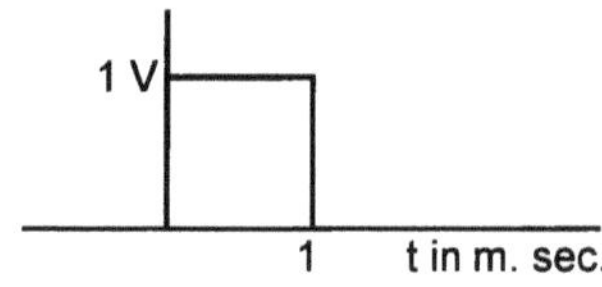

1.19

$$i(t) = 20 \cos 2t \quad 0 \le v \le 5V$$
$$= 10 \cos 2t \quad 5 \le v \le 15$$

1.20

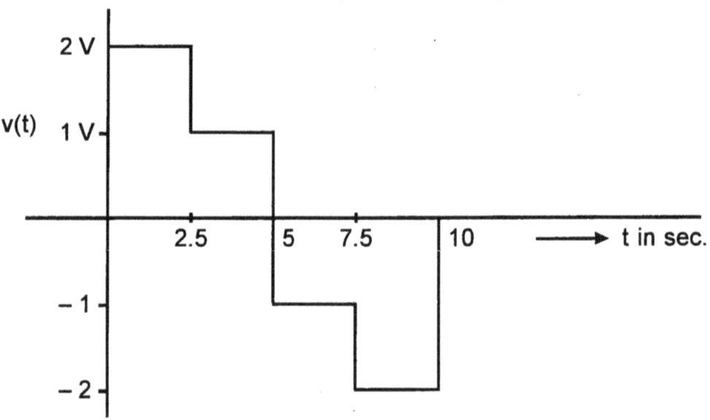

Chapter 2

2.1 $\frac{11}{6}\Omega$

2.2 $\frac{13}{4}\Omega$

2.3 (a) v = 2i + 5 (b) 2i + 10 (c) 2i + 2 (d) 2i

2.4 (a)

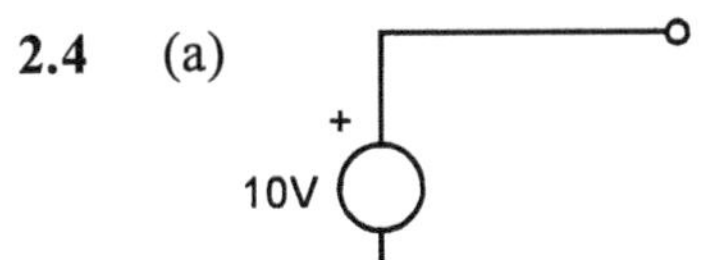

(b)

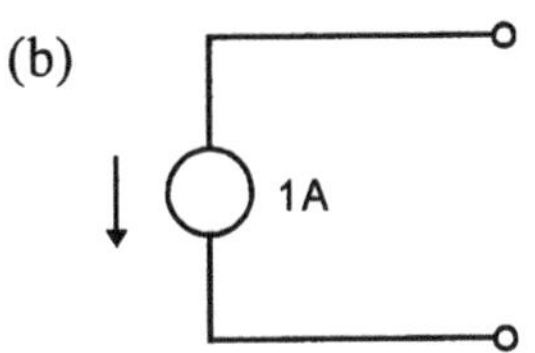

(c)

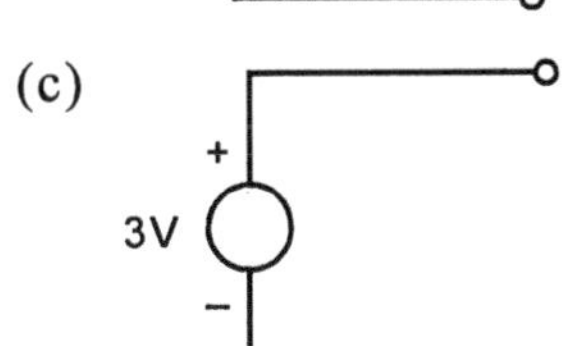

(d)

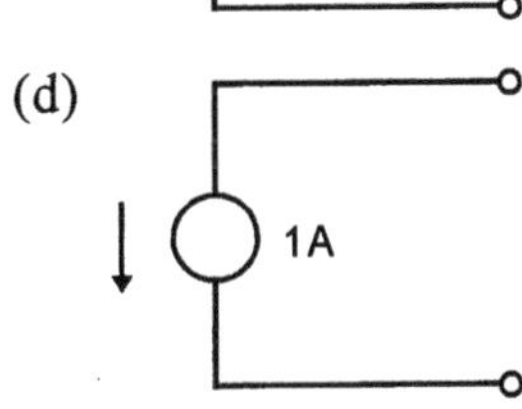

2.5 (a)

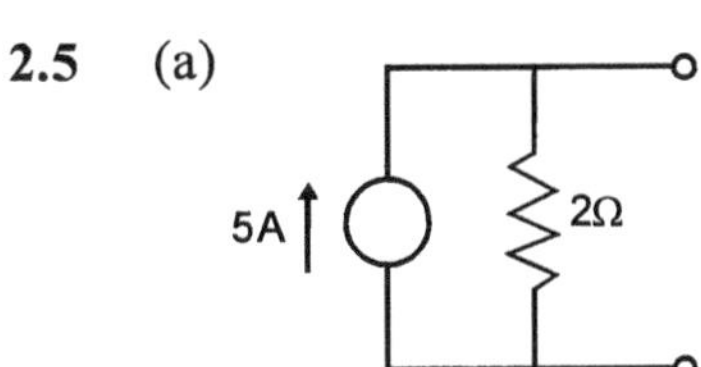

(b)

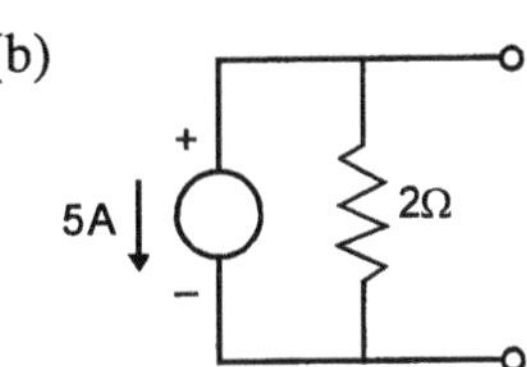

(c)

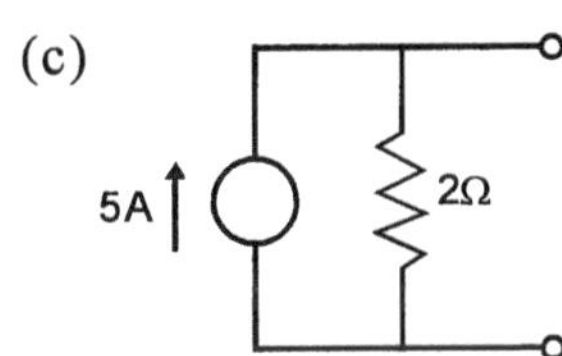

(d) not possible

2.6 (a)

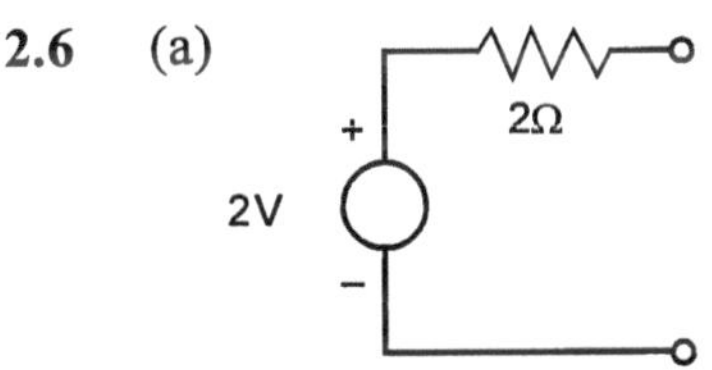

(b)

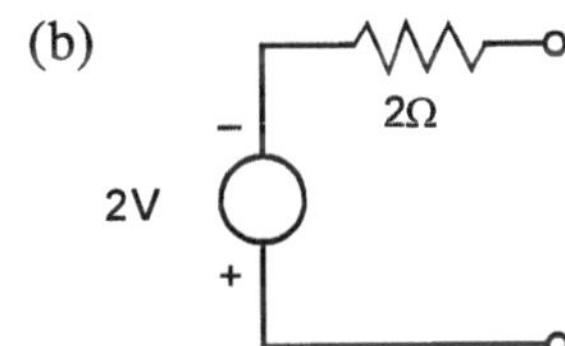

(c) not possible

2.7 (a)

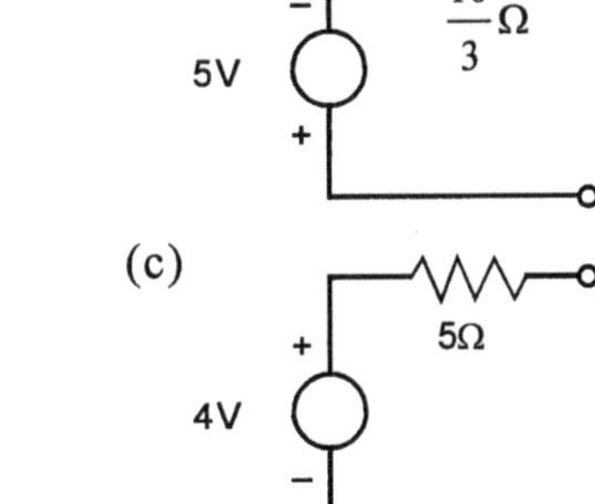

(b)

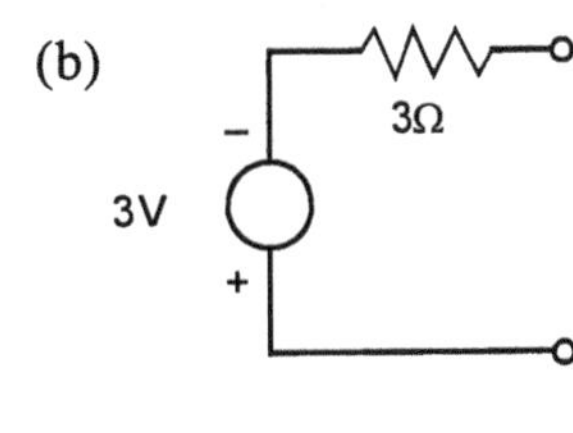

(c)

2.8 2V

2.9 $\frac{55}{8}$V

2.10 12 Ω

2.11 6.5 Ω

2.12 $\frac{22}{37}$ Ω

2.13 $\frac{5}{6}$ Ω

2.14 $\frac{3}{4}$ Ω

2.15 1.269 Ω

2.16 –40V

2.17 $\frac{17}{11}$V, $\frac{13}{11}$V

2.18 $\frac{8}{9}$A, $\frac{5}{9}$W

2.19 2.5W

2.20 33W

2.21 $\frac{-15}{8}$A

2.22 $\frac{95}{8}$A

2.26 0.247 mwb

2.27 0.1257 T

2.28 12.2 A

2.29 6.52 A

2.30 1.945 A

2.31 (a) 1080 AT and 3600 AT (b) 884, 442 AT/Wb

2.32 3.75 A

Chapter 3

3.1 $i(t) = 1.732 \sin(10178.76\, t + 60^0)$

3.2 100V, 24.987 HZ, 70.7 V

3.3 $v(t) = 108.3 \sin(\omega t - 32.77^0)$

3.4 $i(t) = 11.97 \sin(\omega t + 111.2^0)$

3.5 (a) $\frac{V_1}{\sqrt{3}}$ (b) $\frac{V_1}{\sqrt{3}}$ (c) 0.745 V_1

3.6 (a) $\frac{V_1}{2}$ (b) $\frac{V_1}{2}$ (c) $\frac{2V_1}{3}$

3.7 5.77 amps

3.8 4330 W, 2500 VAR, 5000 VA

3.9 1000 VA, 0.8

3.10 21.63 amps, 3740.25 W

3.11 (a) 1.154 (b) 1.154

3.12 9.975×10^{-3} amp ; $v_R = 4.987$ V ; $v_C = 4.983$ V ;
$p_R = 0.049$ watts; $P_R = 0.025$ W
$p_C = 0.0497$ watts ; $P_C = 0$

3.13 $V_s = 223.71$ V; P = 2488.6 W

3.14 (a) 20.11 W (b) 4.89 W

3.15 Z = 0.9751 + j 0.5543 Ω, P(2Ω) = 214.36 W ; P (0.5Ω) = 1243.3W ;
P(1Ω) = 3108W; P(0.4Ω) = 3179.8 W

3.16 $24\angle -17.15^0$ amps; 0.956 lag; 5506 W

3.17 0.88, 0.5 lag

3.18 R = 5 Ω ; C = 2000 μf

3.19 $R_1R_2 = \frac{L}{C}$; $\omega = \frac{1}{\sqrt{LC}}$; $I = V\sqrt{\frac{C}{L}}$

3.20 $R_1 = 5.768$ Ω ; $V_i = 3.72$ Ω

3.21 R = 636.94 Ω ; L – 3.04 H ; C = 3.3 μf

3.25 Q = 62.8 ; C = 1266.67 μf or 0.0311 μf

3.26 (i) 132.6 kHz ; (ii) 100 ; (iii) 131.26 kHz
(iv) 1.32 kHz ; (v) 20 kV

3.27 $C = \frac{L}{R^2 + \omega^2 L^2}$; $L = \frac{1}{2\omega^2 C}\left(1 + \sqrt{1 + 4\omega^2 R^2 C^2}\right)$

3.28 212.23 W

3.29 $0.0855 \angle 75.32^0$

3.30 8 amps

3.31 268.39 watts

3.32 $\frac{(1.75\,s^2 + 9s + 10)}{(2s + 5)}$

Chapter 4

4.1 $V_L = 230$ V ; $V_{ph} = 132.8$V ; $I_l = I_{ph} = 16.1$ amp ;
P = 1555 watts, p.f = 0.2425 lag

4.2 P = 4664 watts ; 0.2425 lag

4.3 6.93 amps; 1728.12 + j 2304.28

4.4 4157 + j 7200 ; 4157 w

4.5 $172.78\angle 36.86\ \Omega$

4.6 571.6 + j 198 Ω

4.7 64.95 – j 37.5 A, – 53.3 + j 7.68 A; – 11.65 + j 29.82 ; 42.475 kW.

4.8 40 j ; 8 + 4 j ; 8 + 4 j ohms

4.9 $173.2\angle 0,\ 100\angle -90,\ 100\ \angle 90^0$

4.10 290.99 V, 290.99, 190.5 Volts

4.11 8000 W, 4619 W ; 10.94 amps ; zero

4.12 Z = 78.9 Ω ; R = 6.487 Ω

4.13 6663.23 W

4.14 16.67 μf

Chapter 5

5.4 $$A = \begin{bmatrix} -1 & 1 & 0 & 0 & 1 & 0 \\ 1 & 0 & 0 & 1 & 0 & 1 \\ 0 & -1 & -1 & 0 & 0 & -1 \end{bmatrix}$$

5.5 $$B = \left[\begin{array}{ccccc|cccc} 1 & 0 & 0 & 0 & 0 & 0 & -1 & -1 & 1 \\ 0 & 1 & 0 & 0 & 0 & 1 & 1 & 0 & 0 \\ 0 & 0 & 1 & 0 & 0 & 0 & 0 & 1 & -1 \\ 0 & 0 & 0 & 1 & 0 & 0 & -1 & -1 & 0 \\ 0 & 0 & 0 & 0 & 1 & -1 & -1 & -1 & 1 \end{array}\right]$$

$B\underline{v} = \underline{0}$; $\underline{j} = B^T \underline{i}$

5.6 $Q = \left[\begin{array}{ccccc|cccc} 0 & -1 & 0 & 0 & 1 & 1 & 0 & 0 & 0 \\ 1 & -1 & 0 & 1 & 1 & 0 & 1 & 0 & 0 \\ 1 & 0 & -1 & 1 & 1 & 0 & 0 & 1 & 0 \\ -1 & 0 & 1 & 0 & -1 & 0 & 0 & 0 & 1 \end{array}\right]$

$Q\underline{j} = 0$; $\underline{v} = Q^T\underline{e}$

5.10 (a) $I_x = 0.4313$ A
$I_y = 0.314$ A
$I_z = 0.4118$ A

(b) $I_x = 1.2264$ A
$I_y = 0.3774$ A
$I_z = 0.66$ A
$i_a = -0.189$ A

Chapter 6

6.1 5.4545 amp

6.2 – 3.4 amp

6.3 $\frac{V_a}{R}$; $\frac{V_b}{R}$, $\frac{V_a + V_b}{2R}$

6.4 $6.39\angle -57.5^0$ amps

6.5 9.41 V, 8.2364 V

6.6 $V_{Th} = 24$V, $R_{Th} = 11\Omega$; $V_{Th} = 24$ V, R_{Th} 16Ω ; $V_{Th} = 205$ V, $R_{Th} = 50.75\Omega$

6.7 333 V, 3.33 Ω

6.8 $45.6\ \angle 60.26$ V, 2.64 + j 0.72 ; 2.64 Ω, – j 0.72 Ω

6.9 – 0.1343 + j 0.179 A

6.10 $26.35\angle -43.46^0$ V

6.11 (a) 11.18 Ω, (b) 236 watts

6.12 0.544 amp

6.13 1 A, 10 Ω

6.15 80V, 20 + j15 ; $3.2\ \angle -36.87$A, 20 + j 15 Ω

6.16 0.984 W

6.17 0.667 amp

6.19 $14.39\angle 3.67^0$, $12.89\angle 231.77^0$, $11.21\angle 124.73^0$

6.20 0.8V

6.22 –5A

Chapter 7

7.1 $$\frac{v}{R}+\frac{1}{L}\int_{-\infty}^{t} v\,dt + C\frac{dv}{dt} = i$$

7.2 $$R_1(i_1-i_2)+\frac{1}{C_1}\int_{-\infty}^{t}(i_1-i_2)dt = v_1$$

$$\frac{1}{C_1}\int_{-\infty}^{t}(i_2-i_1)dt + L_1\frac{d(i_2-i_1)}{dt}+R_3 i_2+\frac{1}{C_2}\int_{-\infty}^{t} i_2\,dt = 0$$

$$R_2 i_3 + L_1\frac{d(i_3-i_2)}{dt}+R_1(i_3-i_1) = v_2$$

7.3 $$R_1 i_1 + L_1\frac{di_1}{dt} - M\frac{di_2}{dt} = v_1$$

$$L_2\frac{di_2}{dt}+\frac{1}{C}\int_{-\infty}^{t}(i_2-i_3)dt - M\frac{di_2}{dt} = -v_2$$

$$R_2 i_3 + L_3\frac{di_3}{dt}+\frac{1}{C}\int_{-\infty}^{t}(i_3-i_2)dt = 0$$

7.4 $$R_1 i_1 + L_1\frac{d(i_1-i_3)}{dt}+\frac{1}{C}\int_{-\infty}^{t}(i_1-i_2)dt + M_{12}\frac{di_2}{dt} - M_{13}\frac{di_3}{dt} = v_1$$

$$L_1\frac{d(i_2-i_1)}{dt}+\frac{1}{C}\int_{-\infty}^{t}(i_2-i_1)dt + L_2\frac{di_2}{dt}+R_2(i_2-i_3)$$

$$+M_{23}\frac{di_3}{dt}+M_{13}\frac{di_3}{dt}-M_{12}\frac{di_2}{dt}-M_{12}\frac{d(i_2-i_1)}{dt} = 0$$

$$i_3 R_3 + R_2(i_3-i_2)+L_3\frac{di_3}{dt}-M_{13}\frac{d(i_1-i_2)}{dt}+M_{23}\frac{di_2}{dt} = 0$$

7.5 (a)

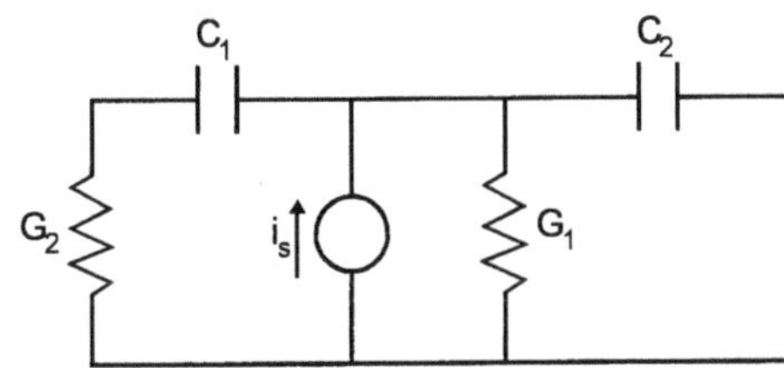
C1
C2
G2
is
G1

(b)

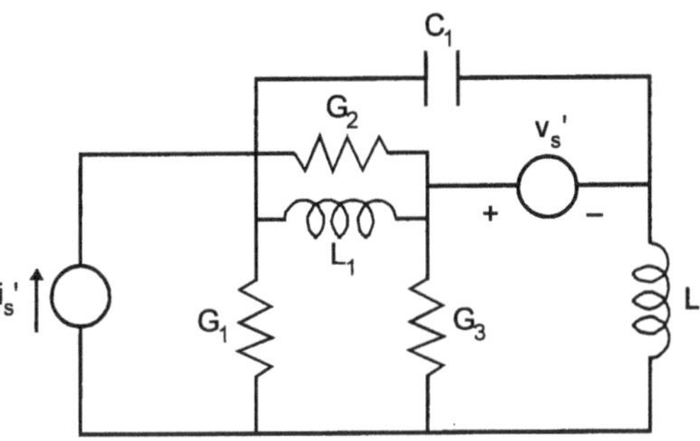
C1
G2
vs'
+
–
L1
is'
G1
G3
L

(c)

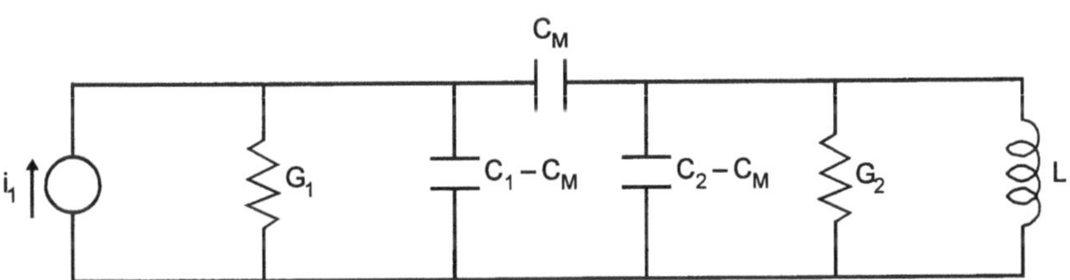
CM
i1
G1
C1 – CM
C2 – CM
G2
L

7.6 (a)

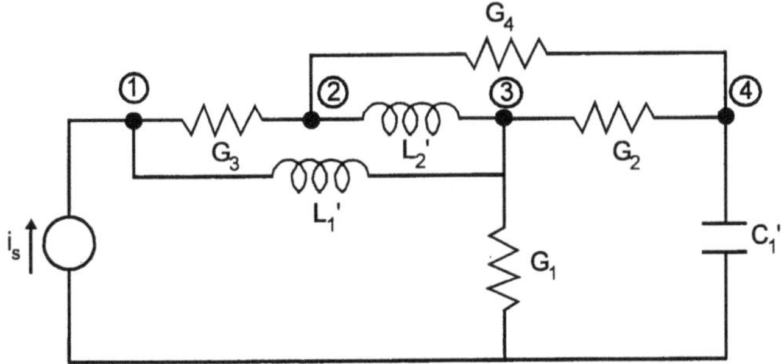
G4
1
2
3
4
G3
L2'
G2
L1'
is
G1
C1'

(b)

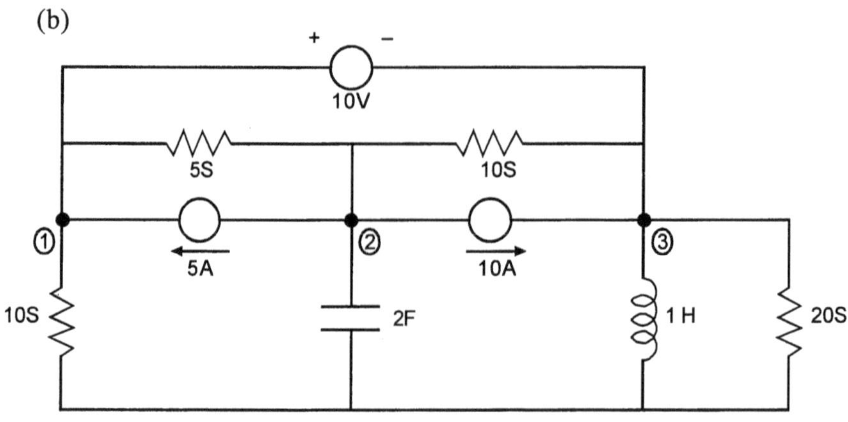

7.7 0, 20 A/sec, – 40 A/sec^2, 10A

7.8 0, 4 V/sec, – 8 V/sec^2, 2 V

7.9 0, 0

7.10 0, 10^6 V/s, 10^6 V

7.11 $\frac{1}{R_1} \quad \frac{d^2v(0^+)}{dt^2} - \frac{1}{R_1^2C}\left[\frac{dv}{dt}(0^+) - \frac{v(0^+)}{R_1C}\right]$

7.12 (a) $v_a(0^+) = 10V,\ v_{C_1}(0^+) = v_{C_2}(0^+) = 0$

$v_{C_1}(\infty) = 10V \qquad v_{C_2}(\infty) = -\frac{10}{3}V$

(b) $v_a(0^+) = \frac{20}{3}V,\ v_{C_1}(0^+) = \frac{10}{3}V \quad v_{C_2}(0^+) = \frac{20}{3}V$

$v_{C_1}(\infty) = 10V \qquad v_{C_2}(\infty) = 0$

7.13 $i_1(0^+) = i_3(0^+) = 0,\ i_2(0^+) = \frac{V}{R_1 + R_3 + R_4}$

7.14 (a) $v_C(0^-) = \frac{200}{3}V$ (b) $\frac{20}{3}A,\ \frac{10}{3}A,\ -\frac{400}{3}A/s,\ \frac{400}{3}A/s$

7.15 $\frac{10}{3}V$, 50J, $\frac{50}{3}J$, 10C, 0C, $\frac{10}{3}C$, $\frac{20}{3}C$

7.16 (a) 10V, 0V (b) 10V, 0V (c) 2V, 2V

7.17 (a) 10A, 0 (b) 10A, 0 (c) $\frac{10}{3}A,\ \frac{10}{3}A$

7.18 0, – 2.4V volts/sec, 0

7.19 $\frac{20}{3}V$

7.20 $-0.5A$

Chapter 8

8.1 $2\,e^{-2t}$

8.2 $10e^{-\frac{t}{4}}$, 50J, 50J

8.3 e^{-10t}

8.4 $8.421\,e^{-0.8421t}$

8.5 $1.667\,e^{-400t}$

8.6 $3.33(1 + 2e^{-1.5t})$, $3.33\,(1 - e^{-1.5t})$

8.7 $4\,(1 - e^{-0.5t})$

8.8 $3.33\,(1 - e^{-3t})$

8.9 $10e^{-0.25t}$

8.10 $6.667\,e^{-0.667t}$

8.11 735.8

8.12 $v(t) = e^{-0.5t}$, $i(t) = e^{-\frac{1}{2}t} - 1$

8.13 $3.33\,e^{-0.833t}$

8.14 $2.5\,u_0(t) - 2.5e^{-t}$

8.15 $2.5\,e^{-0.05t}$

8.16 $3.75\,e^{-\frac{1}{16}t}$

8.17 $5e^{-0.75t}$, $6.667\,(1 - e^{-0.75t})$

8.18 1.967 V, 9.2346 V

8.19 $0.923\,e^{-0.667t} + 1.1094\sin(t - 56.31^0)$

8.20 $t\,e^{-t}$

8.21 $0.822\ e^{-5.33t} + 2.341\sin(2t - 20.56^0)$

8.22 $-6.667\ e^{-2t} + 16.667\ e^{-5t}$

8.23 $0.216\ e^{-0.523t} + 4.784\ e^{-11.477t}$

8.24 $e^{-5t}(5\cos 12t - 8.33\sin 12t)$

8.25 $14.14\sin 1.414t$

8.26 $10 - 2.86\ e^{-t} - 1.43\ te^{-t}$

8.27 $5 + e^{-0.5t}(1.89\sin 1.323\,t - 5\cos 1.323t)$

8.28 $10(1 - e^{-4t} - 4t\ e^{-4t})$

8.29 $6.667\ e^{-8t} - 1.667\ e^{-2t}$

8.30 $5\ e^{-0.5t}\cos 0.866\,t - 2.89\ e^{-0.5t}\sin 0.866t$

8.31 $0.2768\ e^{-0.5t} + 1.1766t\ e^{-0.5t} - 0.2768\cos 2t - 0.519\sin 2t$

8.32 $0.615\left[7\sin t - 4\cos t + e^{-2t}(4\cos t + 0.5\sin 2t)\right]$

8.33 $e^{-0.5t}(2\cos 1.323\,t + 3.024\sin 1.323\,t)$

8.34 $e^{-0.5t}(1\cos 0.866\,t + 5.78\sin 0.866\,t)$

8.35 $i(t) = 10e^{-0.15t}$, $W_R = 3333.33$ J $W_{C_1}(0) = 5000$ J

$W_{C_2}(0) = 0$, $W_C(0) = 5000$ J, $W_C(\infty) = 1666.67$ J

$W_C(0) - W_C(\infty) = W_R$ J, when $R = 0$, $W_C(0) = 5000$ J

$W_C(\infty) = 1666.67$ J, Energy lost = 3333.33 J

8.36 6.25 J, 3.125 J, 3.125 J

8.37 $\frac{10}{3}V$

8.38 $6u_{-1}(t)$

Chapter 9

9.1 (i) $\dfrac{2\omega(s+a)}{\left[(s+a)^2+\omega^2\right]^2}$ (ii) $\dfrac{(s+a)^2-\omega^2}{\left[(s+a)^2+\omega^2\right]^2}$ (iii) $\dfrac{\omega\cos\theta+s\sin\theta}{s^2+\omega^2}$

(iv) $\dfrac{s\cos\theta-\omega\sin\theta}{s^2+\omega^2}$ (v) $\dfrac{\omega}{s^2-\omega^2}$ (vi) $\dfrac{s}{s^2-\omega^2}$

(vii) $\dfrac{s^2+2\omega^2}{s(s^2+4\omega^2)}$

9.2 (i) $\dfrac{1}{8}(1+2e^{-2t}+3e^{-4t})$

(ii) $\dfrac{1}{8}(1+2\cos\sqrt{2}\,t-3\cos 2t)$

(iii) $-2.5\,(1-0.5t-2e^{-t}+e^{-2t}+0.5t\,e^{-2t})$

(iv) $e^{-t}-e^{-0.5t}(\cos 0.866\,t-0.577\sin 0.866\,t)$

(v) $e^{-t}(-1+\cos t+\sin t)$

(vi) $0.5\,(\sin t+t\cos t)$

9.3 (a) $\dfrac{-2}{s^2}+\dfrac{2}{s}+\dfrac{2e^{-s}}{s^2}$

(b) $\dfrac{1}{s}(1+e^{-s}+e^{-2s}-3e^{-3s})$

(c) $\dfrac{1}{s^2}(1-2e^{-s}+e^{-2s})+\dfrac{e^{-3s}}{s^2+\pi^2}+\dfrac{e^{-4s}}{s^2+\pi^2}$

(d) $\dfrac{1}{s^2}(1-3e^{-2s}+4e^{-3.5s}-2e^{-4s})$

(e) $\dfrac{2}{s+\alpha}[1-(1+e^{-2\alpha})\,e^{-2s}+e^{-2(\alpha+2s)}]$

(f) $\dfrac{1}{s^2}(1-e^{-2s}-2s\,e^{-3s})$

9.4 $\dfrac{2}{s+1}\left[\dfrac{1}{1+e^{-(s+1)}}\right]$

9.5 (a) $\dfrac{A}{(s^2+1)}\dfrac{\left(1+e^{-\pi s}\right)}{(1-e^{-\pi s})}$

(b) $\left[\dfrac{2}{s^2}-\dfrac{2}{s^2}e^{-s}-\dfrac{2}{s}e^{-s}\right]\dfrac{1}{1-e^{-s}}$

9.6 (i) $\dfrac{1}{2}\left(1-e^{-2t}\right)$

(ii) $e^{-3t} - e^{-4t}$

(iii) $1 - \dfrac{1}{2}(e^{-t} + \cos t + \sin t)$

(iv) $\dfrac{1}{4}(1-\cos 2t)$

9.8 (i) $0, 1, \dfrac{2}{3}$

(ii) 1, –3, 0

(iii) ∞, does not exist

(vi) 0.5, – 1.5, 0.417

9.9 $5\, e^{-t} \sin t$

9.10 $\dfrac{10}{3}\left(e^{-t} - e^{-0.25t}\right)$

9.11 $5 - e^{-0.2\, t}$

9.12 $2e^{-0.05t}(\cos 0.444\, t - 0.1125 \sin 0.444\, t)$

9.13 $\dfrac{10}{3}$V

9.14 $6u(t)$

9.15 $\frac{5}{3}u(t)$

9.16 $i_1(t) = \frac{10}{3}t - 2,\ i_2(t) = 2 \cos 0.316\ t + 15.81 \sin 0.316\ t$

9.17 $11.547\ e^{-0.5t} \sin 0.866\ t$

9.18 $-10 \sin t$

9.19 $5(1 - e^{-t})\ u(t) - 5u(t - 1)\ [1 - e^{-(t-1)}]$

9.20 $-0.8\ e^{-t} - 1.6\ e^{-4t} + 2.4\ (\cos 2\ t + \sin 2\ t)$

9.21 $\frac{1}{s^2 + 13s + 12}$; $0.091\ (e^{-t} - e^{-12t})$; $0.417\ u_{-1}(t) - 0.4545e^{-t} + 0.0379\ e^{-12t}$

9.22 $\frac{(s^2 + 1)^2}{5s^4 + 5s^2 + 1}$

9.23 $\frac{(s+4)(s+2)}{(s+1)(s+3)}$, poles : $s = -1, -3$ zeros : $s = -4, -2$

9.24 $\frac{3(s+1)}{(s+2)(s+4)}$, poles : $s = -2, -4$ zeros : $s = -1$

Chapter 10

10.1 $\begin{bmatrix} \frac{1}{z_1} & -\frac{1}{z_1} \\ -\frac{1}{z_1} & \frac{1}{z_1} \end{bmatrix}$

10.2 $[z] = \begin{bmatrix} z_2 & z_2 \\ z_2 & z_1 + z_2 \end{bmatrix}$; $[y] = \begin{bmatrix} \frac{1}{z_1} + \frac{1}{z_2} & -\frac{1}{z_1} \\ -\frac{1}{z_1} & \frac{1}{z_1} \end{bmatrix}$

10.3 $[z] = \begin{bmatrix} \frac{2}{3} & \frac{1}{3} \\ 0 & 1 \end{bmatrix}$; $[y] = \begin{bmatrix} \frac{3}{2} & -\frac{1}{2} \\ 0 & 1 \end{bmatrix}$

10.4 $\begin{bmatrix} -\frac{j}{200} & \frac{j}{100} \\ \frac{j}{100} & -\frac{j3}{400} \end{bmatrix}$

10.5 226.8 Ω, 173.2 Ω, 26.8 Ω

10.6 A = 20.8, B = 178.73, C = 0.68 ; D = 5.9

10.7 $[y] = \begin{bmatrix} \frac{19}{40} & -\frac{9}{40} \\ -\frac{9}{40} & \frac{19}{40} \end{bmatrix}$; $\begin{bmatrix} A & B \\ C & D \end{bmatrix} = \begin{bmatrix} \frac{19}{9} & \frac{40}{9} \\ \frac{280}{9} & \frac{19}{9} \end{bmatrix}$

10.8 A = 1.195 + j 0.244, B = 8.78 + j 0.976, C= 0.0488 + j 0.061, D = 1.195 + j 0.244

$[y] = \begin{bmatrix} 0.25 & 0.25 \\ 0.25 & -0.299 - j0.061 \end{bmatrix}$

10.9 $y_a = 0.2S$, $y_b = 0.05S$, $y_c = 0.1S$

10.10 $[z] = \begin{bmatrix} 4 & 3 \\ 4 & 2 \end{bmatrix}$ $[y] = \begin{bmatrix} -\frac{1}{2} & \frac{3}{4} \\ 1 & -1 \end{bmatrix}$

10.11 (a) $[y] = \begin{bmatrix} j\omega L_1 & j\omega M \\ j\omega M & j\omega L_2 \end{bmatrix}$, $[y] = \begin{bmatrix} \dfrac{j\omega L_2}{|z|} & -\dfrac{j\omega M}{|z|} \\ -\dfrac{j\omega M}{|z|} & \dfrac{j\omega L_1}{|z|} \end{bmatrix}$ where $|z| = \omega^2 (M^2 - L_1L_2)$

$$[h] = \begin{bmatrix} \dfrac{|z|}{j\omega L_2} & \dfrac{M}{L_2} \\ -\dfrac{M}{L_2} & \dfrac{1}{j\omega L_2} \end{bmatrix}$$

(b) $[y] = \begin{bmatrix} 2 & -1 \\ -1 & 2 \end{bmatrix}$, $[z] = \begin{bmatrix} \frac{2}{3} & \frac{1}{3} \\ \frac{1}{3} & \frac{2}{3} \end{bmatrix}$, $[h] = \begin{bmatrix} \frac{1}{2} & \frac{1}{2} \\ -\frac{1}{2} & \frac{3}{2} \end{bmatrix}$

(c) $[z] = \begin{bmatrix} R_1 & 0 \\ bR_2 & R_2 \end{bmatrix}$, $[y] = \begin{bmatrix} \dfrac{1}{R_1} & 0 \\ \dfrac{-b}{R_1} & \dfrac{1}{R_2} \end{bmatrix}$, $[h] = \begin{bmatrix} R_1 & 0 \\ -b & \dfrac{1}{R_2} \end{bmatrix}$

10.12 $h_{11} = \dfrac{R_1R_2 + R_1R_3 + R_2R_3(1+a)}{R_2 + R_3}$, $h_{21} = \dfrac{aR_3 - R_2}{R_2 + R_3}$

$h_{12} = \dfrac{R_2}{R_2 + R_3}$, $h_{22} = \dfrac{1}{R_2 + R_3}$

$g_{11} = \dfrac{1}{R_1 + R_2}$; $g_{21} = \dfrac{R_2 - aR_3}{R_1 + R_2}$; $g_{12} = -\dfrac{R_2}{R_1 + R_2}$

$g_{22} = \dfrac{R_1R_2 + R_1R_3 + R_2R_3(1+a)}{R_1 + R_2}$

10.13 $\dfrac{5}{8}\ \Omega$

10.14 $[z] = \begin{bmatrix} 5 & 2 \\ 5 & 4 \end{bmatrix}$; $[h] = \begin{bmatrix} 2.5 & 0.5 \\ -1.25 & 0.25 \end{bmatrix}$

10.15 $[h] = \begin{bmatrix} \frac{4}{3} & \frac{1}{3} \\ \frac{1}{3} & \frac{1}{3} \end{bmatrix}$

10.16 $[z] = j\omega \begin{bmatrix} \dfrac{L_1L_a + L_1L_2 - M_1^2}{L_2 + L_a} & \dfrac{M_1M_a}{L_2 + L_a} \\ \dfrac{M_1M_a}{L_2 + L_a} & \dfrac{L_bL_2 + L_aL_b - M_a^2}{L_2 + L_a} \end{bmatrix}$

10.17

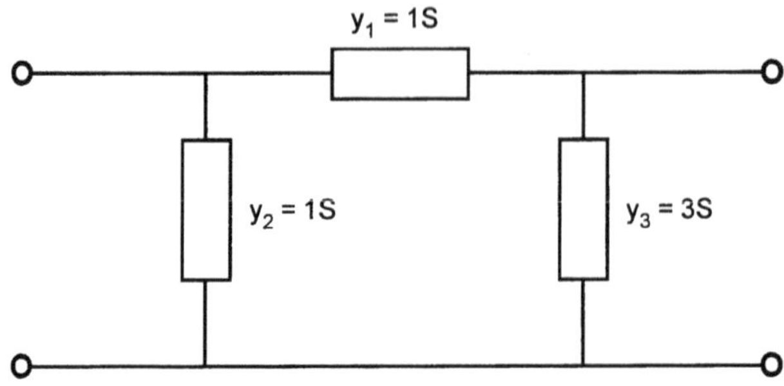

10.18 $[h] = \begin{bmatrix} 2 & 1 \\ -1 & \frac{1}{3} \end{bmatrix}$ $g = \begin{bmatrix} \frac{1}{5} & -\frac{3}{5} \\ \frac{3}{5} & \frac{6}{5} \end{bmatrix}$

10.19 $[z] = \begin{bmatrix} 2 & 1 \\ 1 & 2 \end{bmatrix}$

10.20 $[y] = \begin{bmatrix} \frac{13}{6} & \frac{-5}{6} \\ \frac{-5}{6} & \frac{13}{6} \end{bmatrix}$

10.22 $\begin{bmatrix} \frac{4s^3+12s^2+8s+1}{4(s+1)(2s+1)} & -\frac{4s^3+4s^2+2s+1}{4(s+1)(2s+1)} \\ -\frac{4s^3+4s^2+2s+1}{4(s+1)(2s+1)} & \frac{4s^3+12s^2+8s+1}{4(s+1)(2s+1)} \end{bmatrix}$

10.23 $[y] = \begin{bmatrix} \frac{11}{8} & -\frac{9}{8} \\ -\frac{9}{8} & \frac{11}{8} \end{bmatrix}$

10.24 $[z] = \begin{bmatrix} \frac{11}{8s} & \frac{9}{8s} \\ \frac{9}{8s} & \frac{11}{8s} \end{bmatrix}$

Chapter 11

11.1 $\dfrac{2-\omega^2}{\sqrt{2\omega^4-7\omega^2+4}}$

11.2 $\sqrt{\dfrac{(2\omega^2-1)(\omega^2-1)}{\omega^2(\omega^2-3)}}$

11.3 (i) $\omega<1.414$ stop band ; $\alpha=2\sinh^{-1}\sqrt{\dfrac{1}{2(2-\omega^2)}}$

(ii) $1.414<\omega<1.58$ stop band ; $\alpha=2\cosh^{-1}\sqrt{\dfrac{1}{2(2-\omega^2)}}$

(iii) $\omega>1.58$ pass band $\alpha=0$

11.4

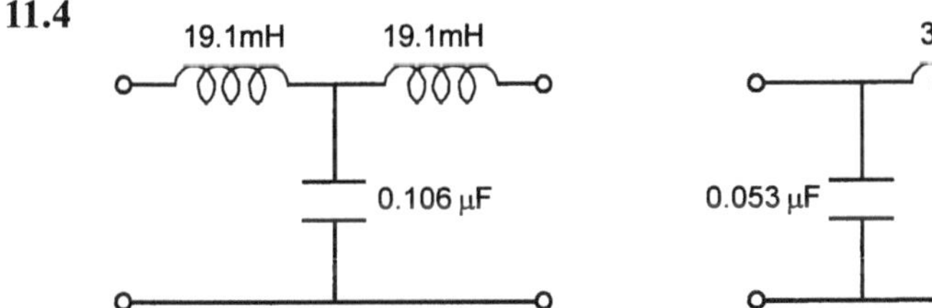

(i) at f = 2.5 kHz $\alpha=0,\ \beta=60^0,$ $z_{0T}=519.62\ \Omega$ $z_0\pi=692.8\ \Omega$

(ii) at f = 10 kHz $\alpha=22.84$ db , $\beta=\pi,$ z_0 is imaginary

11.5

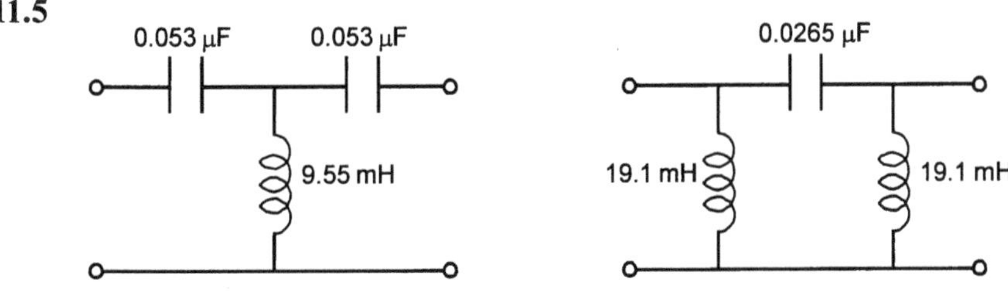

(i) at f = 2.5 kHz $\alpha=22.84$ db, $\beta=\pi$, z_{0T} and $z_{0\pi}$ are imaginary

(ii) at f = 8 kHz $\alpha=0,\ \beta=77.36^0$ $z_{0T}=468.37\ \Omega,\ z_{0\pi}=768.62\ \Omega$

11.6

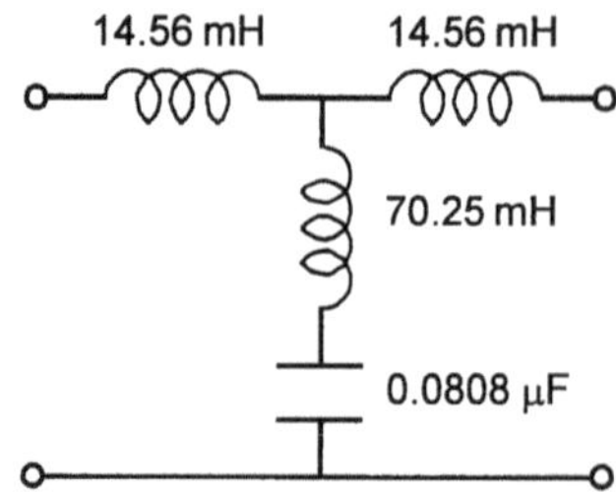

(i) at f = 2.05 kHz α = 15.773 db

(ii) at f = 2.5 kHz α = 9.736 db

11.7

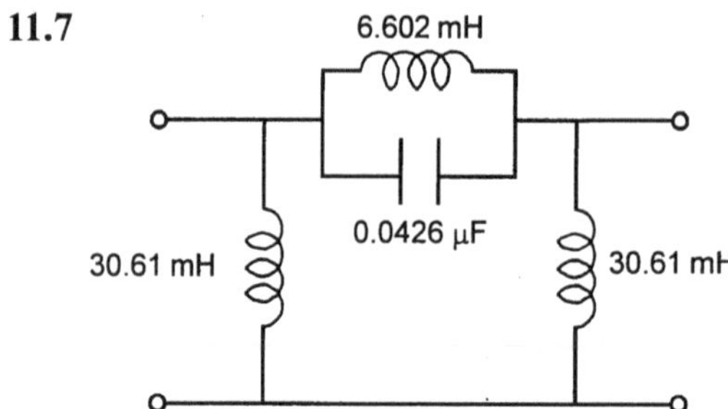

(i) at f = 9.8 kHz α = 13.07 db

(ii) at f = 5 kHz α = 6.55 db

11.8

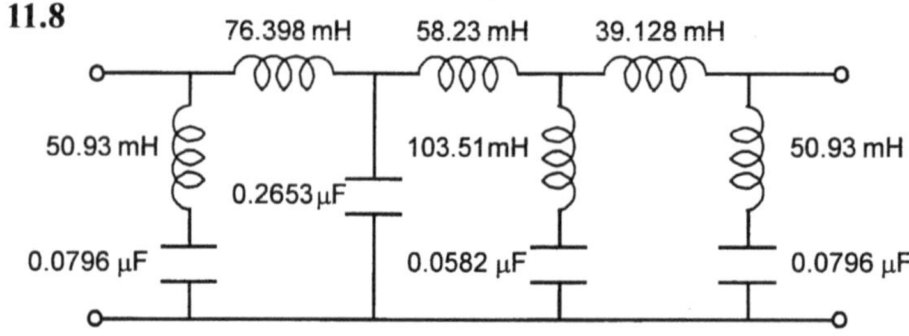

11.9

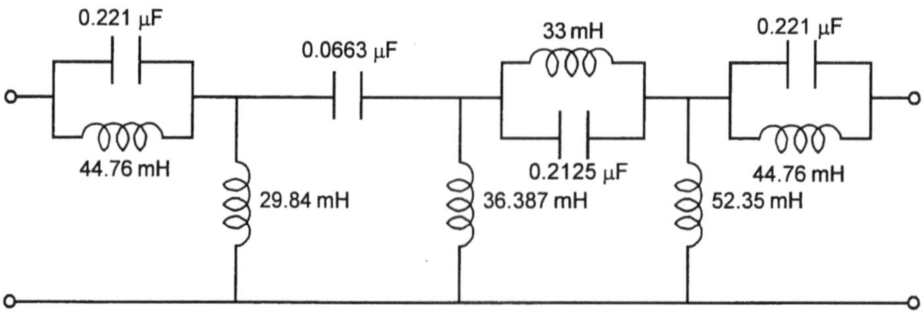

11.10

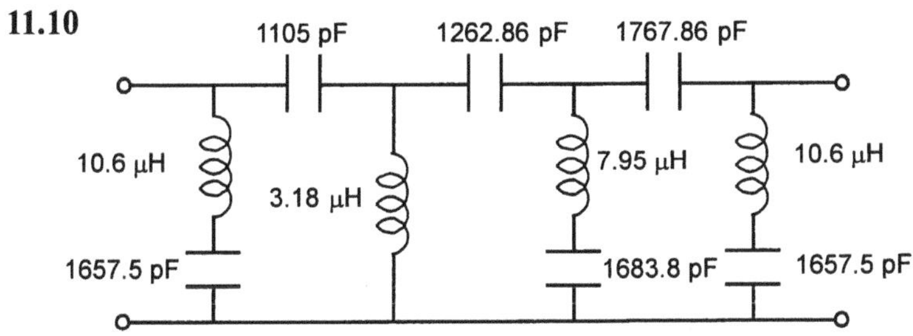

11.11

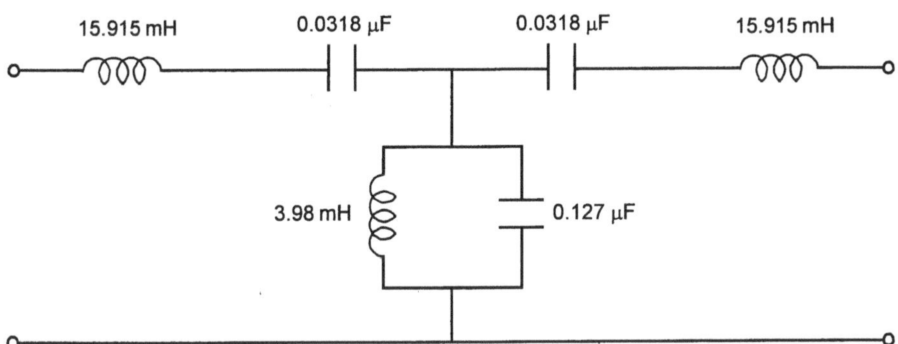

11.12

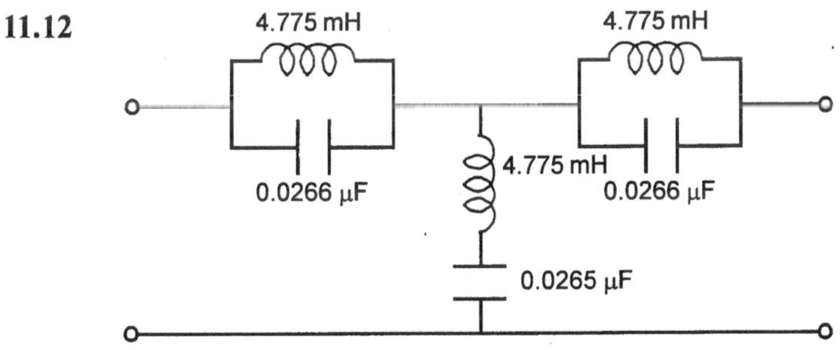

11.13

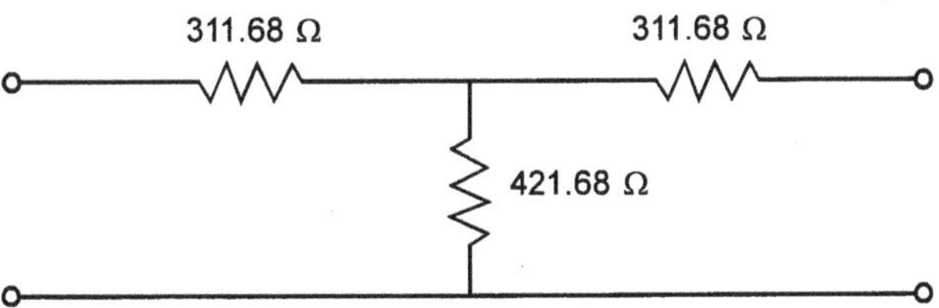

11.14

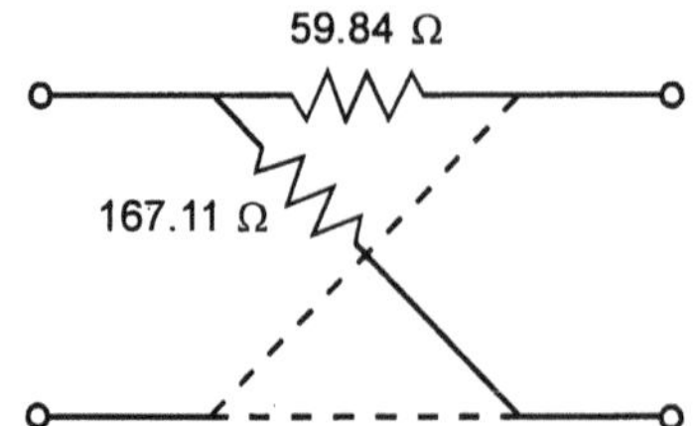

11.15

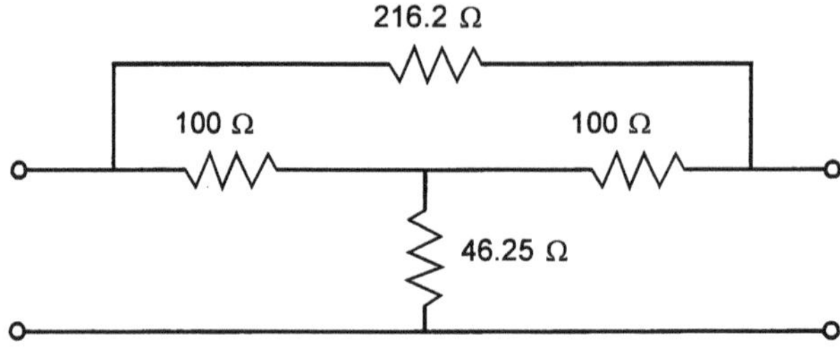

11.16 20db

11.17

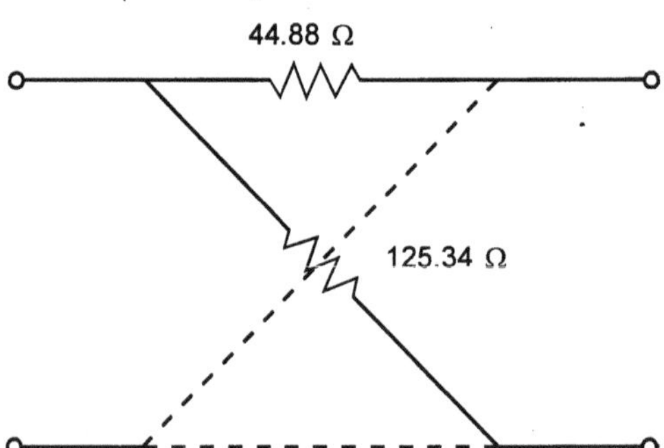

Chapter 12

12.1 **(i)** $f(t) = 5 + \frac{20}{\pi}\left(\sin t + \frac{\sin 3t}{3} + \frac{\sin 5t}{5} + \ldots\right)$

(ii) $f(t) = 5 + \frac{20}{\pi}\left(\sin 2\pi \times 10^6 t + \frac{\sin 6\pi \times 10^6 t}{3} + \frac{\sin 10\pi \times 10^6}{5} t \ldots\right)$

12.2 $f(t) = \frac{A}{2\pi} + \frac{A}{\pi}(\cos t + \cos 2t + \cos 3t + \ldots)$

12.3 **(i)** $f(t) = \frac{A}{2} + \frac{A}{\pi}\left(\sin t + \frac{\sin 2t}{2} + \frac{\sin 3t}{3} + \ldots\right)$

(ii) $f(t) = \frac{A}{2} - \frac{A}{\pi}\left(\sin t + \frac{\sin 2t}{2} + \frac{\sin 3t}{3} + \ldots\right)$

12.4 $f_1(t) = \frac{-2}{\pi}(\sin t - \sin 3t + \sin 5t.....)$

12.5 $f(t) = \frac{2A}{\pi}\left(1 + \frac{2}{3}\cos 2t - \frac{2}{15}\cos 4t \ldots\right)$

12.6 **(i)** $f(t) = \frac{A}{4} - \frac{2A}{\pi^2}\left(\cos t + \frac{1}{9}\cos 3t + \frac{1}{25}\cos 5t + \ldots\right)$

(ii) $f(t) = \frac{A}{2} - \frac{A}{\pi}\left(\sin 2t + \frac{\sin 4t}{2} + \frac{\sin 6t}{3} + \ldots\right)$

12.7 **(i)** $f(t) = \frac{A}{2\pi}\left(1 + e^{jt} + e^{-jt} + e^{2jt} + e^{-2jt} + \ldots\right)$

(ii) $f(t) = \frac{A}{2\pi} + \frac{A}{\pi}(\cos t + \cos 2t + \ldots)$

12.8 $f(t) = -\frac{1}{3\pi j}e^{-j3t} - \frac{1}{\pi j}e^{-jt} + \frac{1}{2} + \frac{1}{\pi j}e^{jt} + \frac{1}{3\pi j}e^{3jt} + \ldots$

12.9 $C_n = \frac{1}{j\pi n}\left(1 - \cos\frac{n\pi}{4}\right)$

12.10 $C_n = \frac{1}{2n^2\pi^2}(1 - \cos n\pi) - j\frac{1}{2n\pi}$

12.11 $F(\omega) = \frac{1}{j\omega a^2}(2\cos \omega a - 2)$

12.12 $F(\omega) = \dfrac{2}{j\omega}\left(1 - \cos\dfrac{\omega a}{2}\right)$

12.13 (a) $F(\omega) = e^{-j\omega t_0}$

(b) $\dfrac{\pi}{2}\left[2u_0(\omega) - u_0(\omega+6) - u_0(\omega-6)\right]$

(c) $\dfrac{\pi}{\sqrt{2}}\left[(1+j)u_0(\omega-2) + (1-j)u_0(\omega+2)\right]$

(d) $\dfrac{1}{(j\omega+4)^2}$

(e) $\dfrac{\omega_0}{(a+j\omega)^2 + \omega^2}$

12.14 $F(\omega) = \dfrac{2}{\omega}(\sin 3\omega + \sin 2\omega)$

12.15 (a) $e^{-j\omega}$ (b) -1

12.16 $4F(-2\omega)$

12.17 $\dfrac{2a}{a^2+\omega^2}e^{-j\omega t_0}$

12.18 (a) $2F(\omega)\cos\omega t_0$ (b) $\dfrac{1}{2}\left[F(\omega+\omega_0) + F(\omega-\omega_0)\right]$

(c) $\dfrac{j}{2}\left[F(\omega+\omega_0) - F(\omega-\omega_0)\right]$

12.19 $F(\omega) = \dfrac{20}{\omega^2}(\cos 3\omega - \cos 4\omega)$

12.22 $F(\omega) = \pi A u_0(\omega) + jA \sum_{\substack{n=-\infty \\ n\neq 0}}^{\infty} u_0(\omega - n\omega_0)$

12.23 DC term $= \dfrac{1}{\pi}$ Fundamental $= \dfrac{1}{27}$

12.24 $i(t) = \sin t$

12.25 $i(t) = e^{-t}u_{-1}(t)$

Multiple Choice Questions (MCQs) for Competitive Examinations

Chapter 1 - Network Variables and Elements

1.1 When a periodic triangular voltage of peak amplitude 1V and frequency 0.5Hz is applied to a parallel combination of 1Ω resistance and 1F capacitance, the current through the voltage source has wave-form.

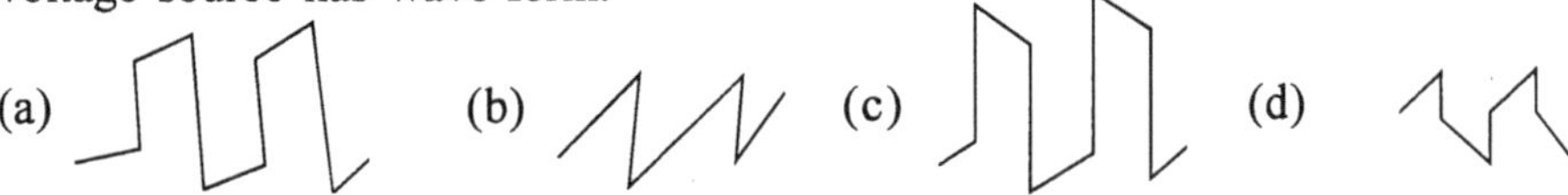

1.2 In the circuit shown in figure it is desired to have a constant direct current i (t) through the ideal inductor L. The nature of the voltage source v (t) must be

(a) constant voltage
(b) Linearly increasing voltage
(c) an ideal impulse
(d) exponentially increasing voltage

i(t)
v(t)
L

1.3 A practical current source is usually represented by

(a) a resistance in series with an ideal current source
(b) a resistance in parallel with an ideal current source
(c) a resistance in parallel with an ideal voltage source
(d) None of the above

1.4 The V-I relation for the network shown in the given box is : V = 4I – 9. If now a resistor R = 2Ω is connected across it, then the value of I will be

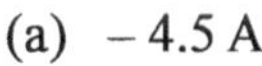

(a) – 4.5 A   (b) – 1.5 A
(c) 1.5 A   (d) 4.5 A

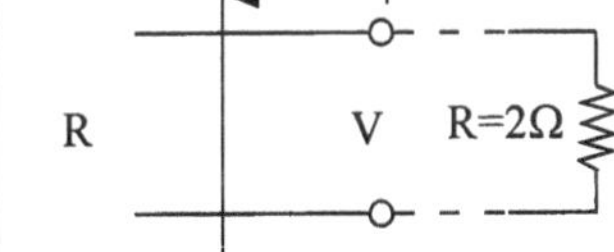

1.5 A current of the waveform shown in the given figure passes through a pure inductance of 3 mH. The instantaneous power, in Watts, during 0 < t < 2 ms is :

(a) 25000 t   (b) 50000 t
(c) 75000 t   (d) 100000 t

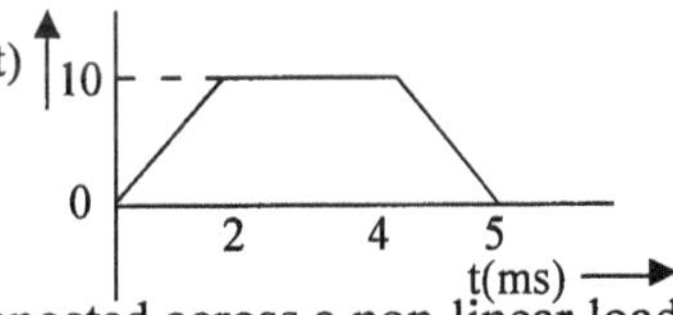

1.6 A 10 V battery with an internal resistance of 1 Ω is connected across a non-linear load whose v-i characteristic is given by $7i = v^2 + 2v$.
The current delivered by the battery is

(a) 2.5 A   (b) 5 A   (c) 6 A   (d) 7 A

1.7 The hot resistance of the filament of a bulb is higher than the cold resistance because the temperature coefficient of the filament if

(a) negative   (b) infinite   (c) zero   (d) positive

1.8 If a capacitor is energised by a symmetrical square wave current source, then the steady-state voltage across the capacitor will be a

(a) square wave   (b) triangular wave   (c) step function   (d) impulse function

1.9 A simple equivalent circuit of the 2-terminal network shown in figure, is

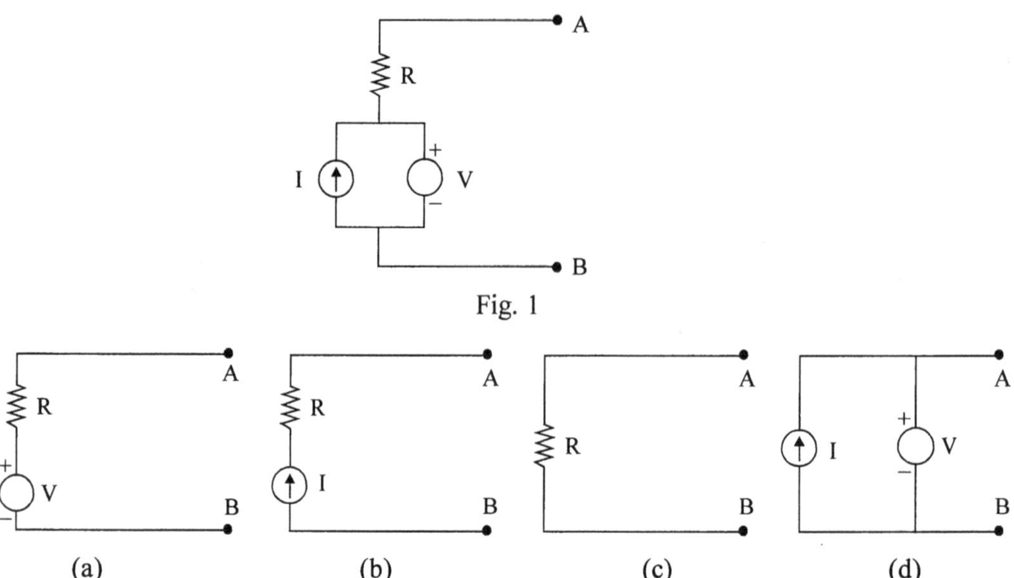

1.10 If a pulse voltage v(t) of 4V magnitude and 2 secs duration is applied to a pure inductor of 1H, with zero initial current, the current (in amps) drawn at t = 3 secs will be

(a) zero

(b) 2

(c) 4

(d) 8

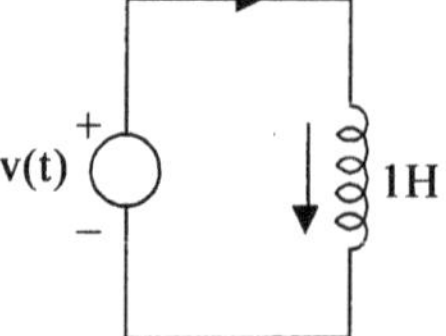

1.11 When a unit impulse voltage is applied to an inductor of 1 H, the energy supplied by the source is

(a) ∞ (b) 1 J (c) $\frac{1}{2}$ J (d) 0

1.12 The voltage applied across a capacitance is triangular in waveform. The waveform of the current is

(a) triangular (b) trapezoidal (c) sinusoidal (d) rectangular

1.13 A two-terminal black box contains a single element which can be R, L, C or M (mutual inductance). As soon as the box is connected to a dc voltage source, a finite non-zero current is observed to flow through the element. The element is a/an

(a) resistance (b) inductance (c) capacitance (d) mutual inductance

1.14 When a resistor R is connected to a current source, it consumes a power of 18 W. When the same R is connected to a voltage source having the same magnitude as the current source, the power absorbed by R is 4.5 W. The magnitude of the current source and the value of R are

(a) $\sqrt{18}$ A and 1Ω (b) 3A and 2Ω (c) 1A and 18Ω (d) 8A and 0.5 Ω

Chapter 2 – Kirchhoff's Laws and Resistive Networks

2.1 When all the resistances in the circuit are of one ohm each, the equivalent resistance across the points A and B will be

(a) 1 Ω

(b) 0.5 Ω

(c) 2 Ω

(d) 1.5 Ω

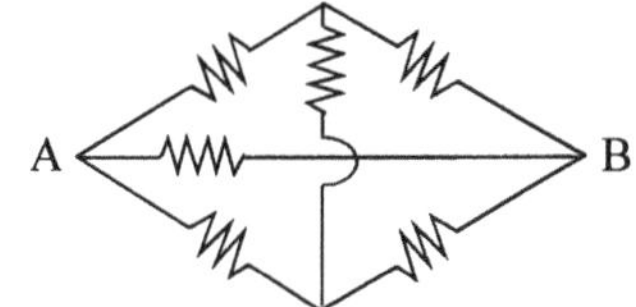

2.2 For a given voltage, four heating coils will produce maximum heat when connected

(a) all in parallel

(b) all in series

(c) with two parallel pairs in series

(d) not identifiable on the basis of the given data

2.3 Two incadescent light bulbs of 40 W and 60 W rating are connected in series across the mains. Then

(a) the bulbs together consume 100 W

(b) the bulbs together consume 50 W

(c) the 60W bulb glows brighter

(d) the 40W bulb glows brighter

2.4 Consider the star network shown in figure. The resistance between terminals A and B with C open is 6Ω, between terminals B and C with A open is 11Ω, and between terminals C and A with B open is 9Ω. Then

(a) $R_A = 4\Omega$, $R_B = 2\Omega$, $R_C = 5\Omega$

(b) $R_A = 2\Omega$, $R_B = 4\Omega$, $R_C = 7\Omega$

(c) $R_A = 3\Omega$, $R_B = 3\Omega$, $R_C = 4\Omega$

(d) $R_A = 5\Omega$, $R_B = 1\Omega$, $R_C = 10\Omega$

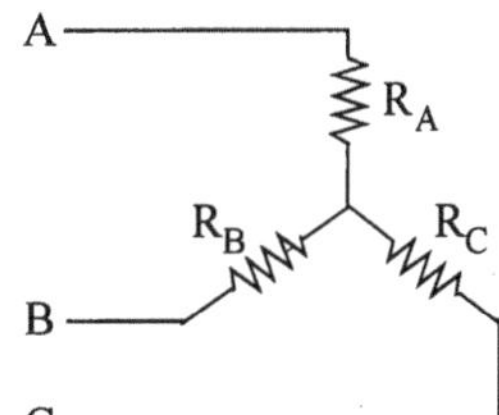

2.5 The driving point impedance of the infinite ladder network shown in the figure is : (Given : $R_1 = 2\Omega$ and $R_2 = 1.5\ \Omega$)

(a) $3\ \Omega$

(b) $3.5\ \Omega$

(c) $\frac{3}{3.5}\ \Omega$

(d) $ln\left(1+\frac{3}{3.5}\right)\ \Omega$

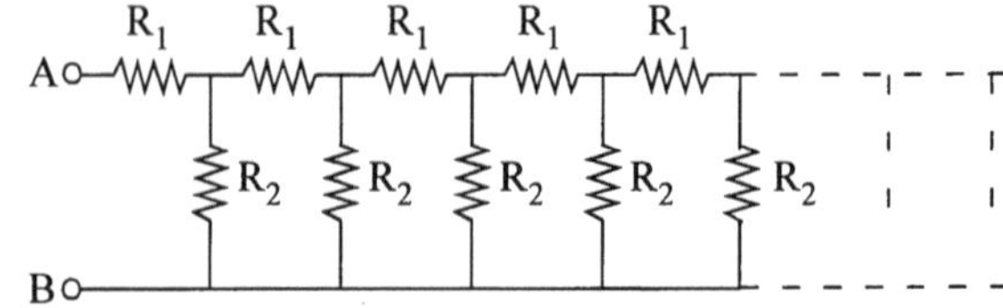

2.6 In the circuit shown in the given figure, if $I_1 = 1.5$ A, then I_2 will be

(a) 0.5 A

(b) 1.0 A

(c) 1.5 A

(d) 3.0 A

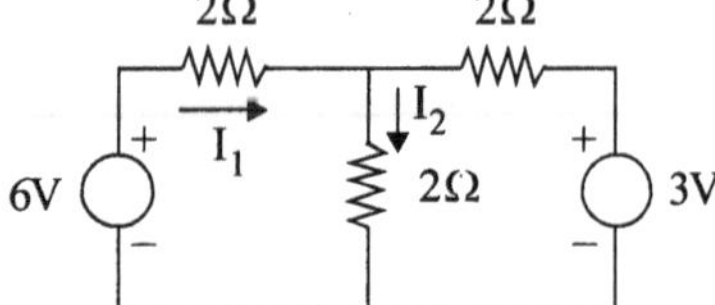

2.7 In the network shown in figure, the effective resistance faced by the voltage source is

(a) $4\ \Omega$

(b) $3\ \Omega$

(c) $2\ \Omega$

(d) $1\ \Omega$

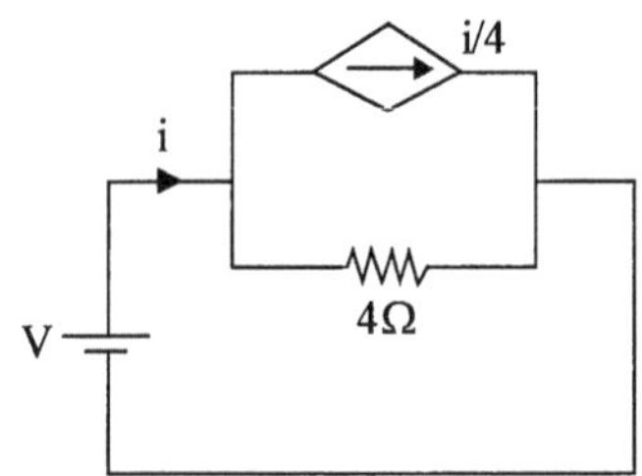

2.8 Five cells are connected in series in a row and then four such rows are connected in parallel to feed the current to a resistive load of 1.25 Ω. Each cell has emf of 1.5 V with internal resistance of 0.2 Ω. The current through the load will be

(a) 3.33 A (b) 23.33 A (c) 5 A (d) 1 A

2.9 Four resistances 80 Ω, 50 Ω, 25 Ω and R are connected in parallel. Current through 25 Ω resistance is 4 A. Total current of the supply is 10 A. The value of R will be

(a) 66.66 Ω (b) 40.25 Ω (c) 36.36 Ω (d) 76.56 Ω

2.10 In the network shown in the given figure, the value of v_x would be

(a) $-\frac{8}{9}$ V

(b) $\frac{8}{9}$ V

(c) $\frac{16}{9}$ V

(d) $-\frac{16}{9}$ V

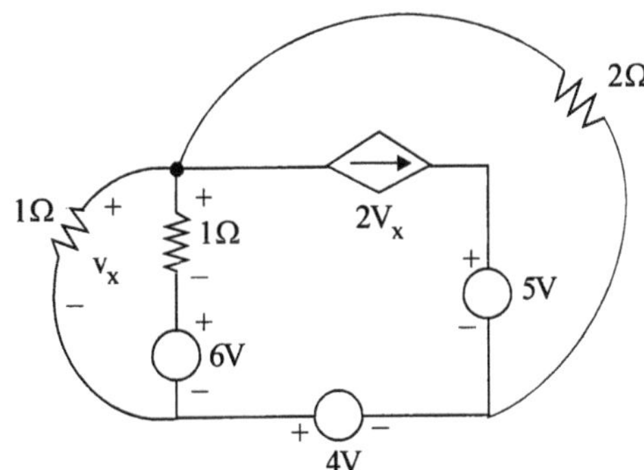

2.11 The voltage across the 1 kΩ resistor between the nodes A and B of the network shown in the given figure is :

(a) 2 V

(b) 3 V

(c) 4 V

(d) 8 V

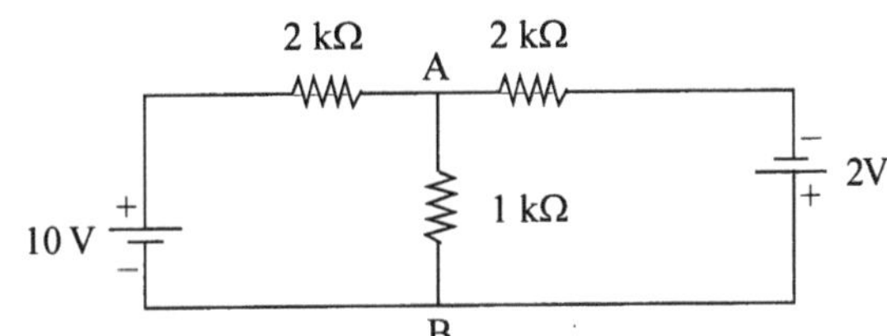

2.12 The current in the given circuit with a dependant voltage source is

(a) 10 A

(b) 12 A

(c) 14 A

(d) 16 A

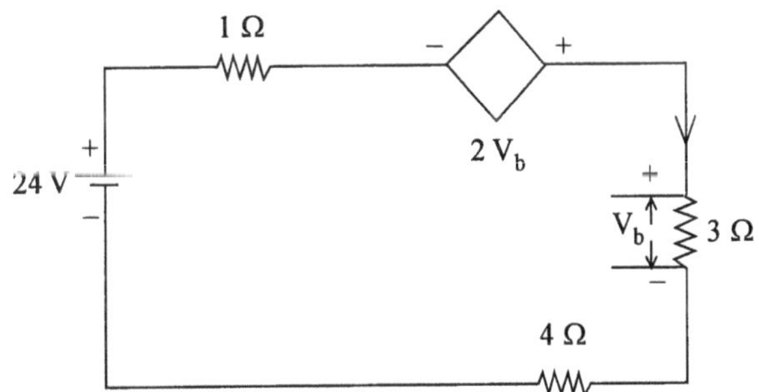

2.13 The value of resistance 'R' shown in the given figure is

(a) 3.5 Ω

(b) 2.5 Ω

(c) 1 Ω

(d) 4.5 Ω

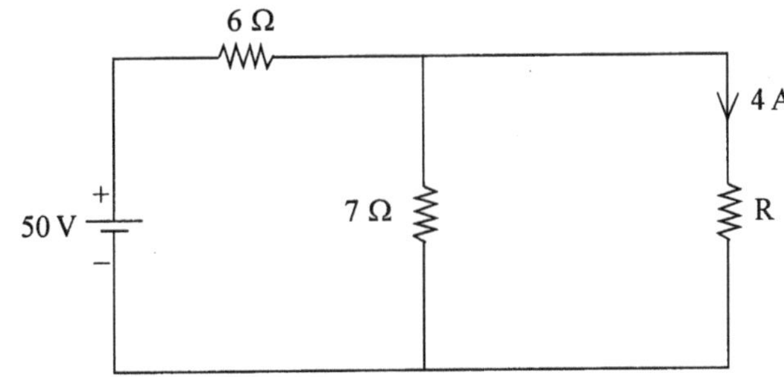

2.14 For the circuit shown in the given figure the current I is given by

(a) 3 A

(b) 2 A

(c) 1 A

(d) zero

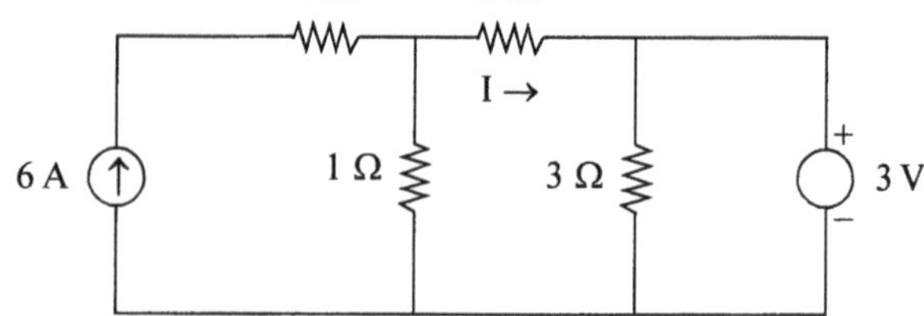

2.15 The value of V in the circuit shown in the given figure is

(a) 1 V

(b) 2 V

(c) 3 V

(d) 4 V

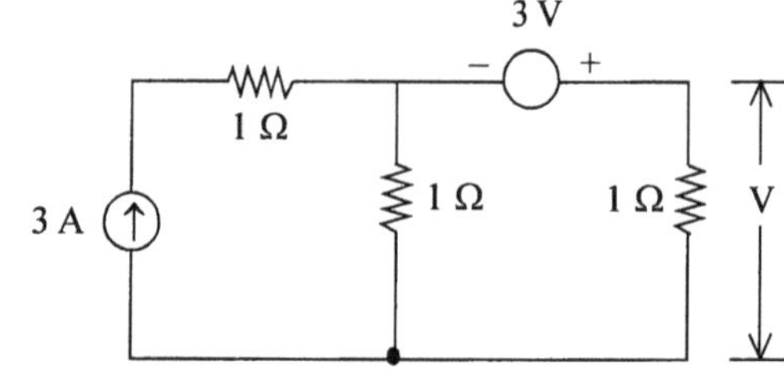

2.16 The current through 120 ohm resistor in the circuit shown in the figure is

(a) 1 A

(b) 2 A

(c) 3 A

(d) 4 A

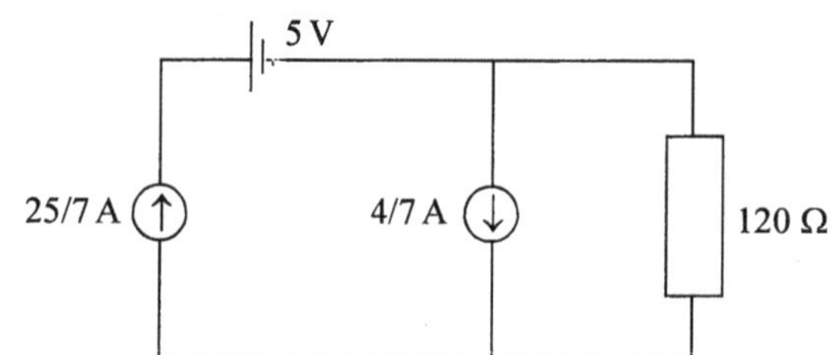

2.17 For the circuit given in figure above the power delivered by the 2 volt source is given by

(a) 4 W

(b) 2 W

(c) – 2 W

(d) – 4 W

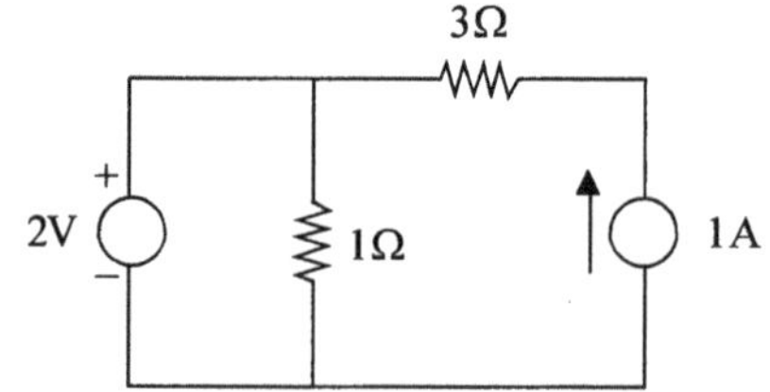

2.18 In the figure shown, if we connect a source of 2V, with internal resistance of 1Ω at A'A, with positive terminal at A', then the current through R is

(a) 2 A

(b) 1.66 A

(c) 1. A

(d) 0.625 A

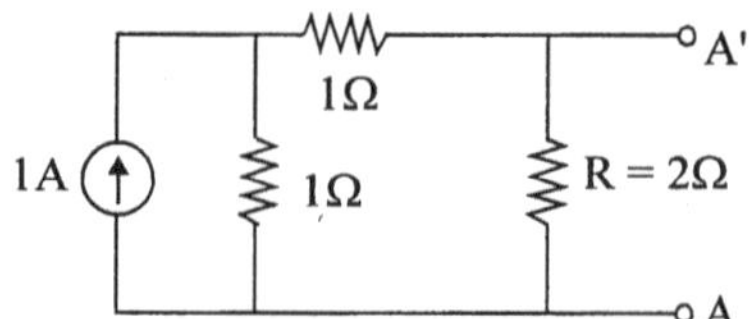

2.19 In the circuit shown the value of I is

(a) 1 A

(b) 2 A

(c) 4 A

(d) 8 A

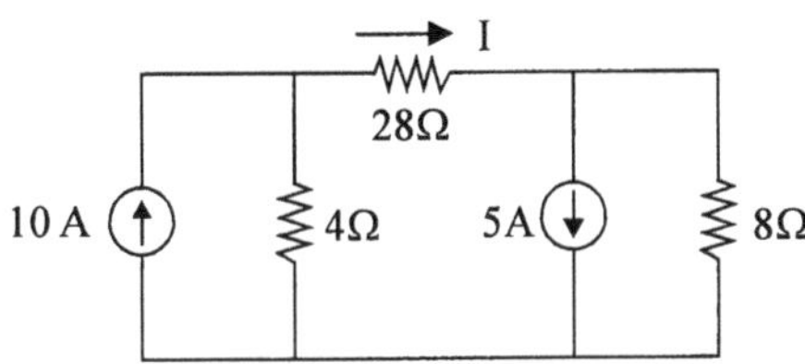

2.20 A 35 V source is connected to a series circuit of 600 ohm and R as shown. If a voltmeter of internal resistance 1.2 kilo ohms is connected across 600 ohm resistor, it reads 5 V. The value of R is

(a) 1.2 k Ω

(b) 2.4 k Ω

(c) 3.6 k Ω

(d) 7.2 k Ω

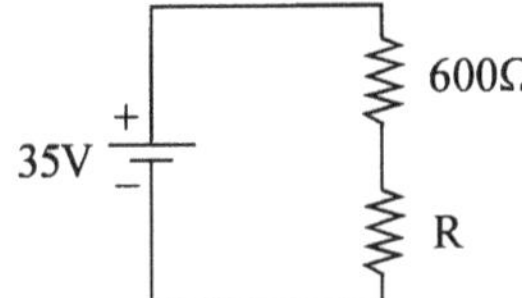

2.21 The current in resistor R shown in figure will be :

(a) 0.2 A

(b) 0.4 A

(c) 0.6 A

(d) 0.8 A

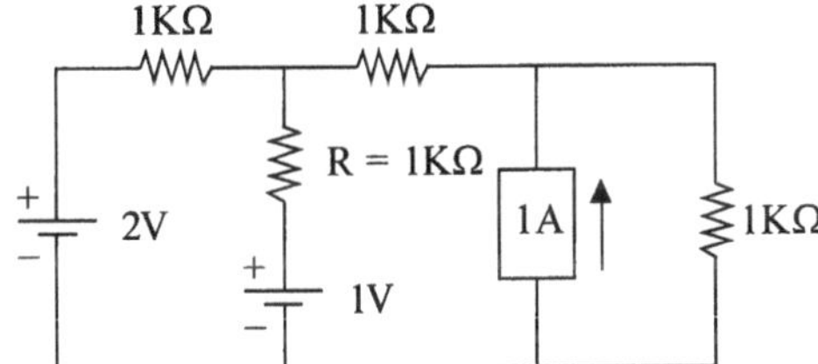

2.22 The circuit shown is a linear time invariant one and the sources are ideal. Choose from the answers given below, the values of voltage across and current through 1Ω resistor

(a) –5V, –5A

(b) 1 V, 6 A

(c) 1 V, 1 A

(d) 5 V, 5 A

2.23 The figures show two different sets of input and output variables for the same two-port resistive network N. I_x is

(a) 12 A

(b) 8 A

(c) 4 A

(d) $\frac{3}{2}$ A

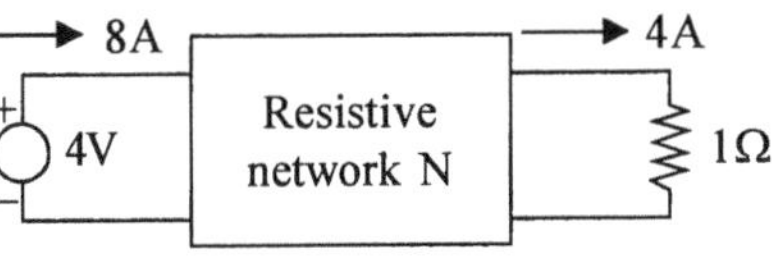

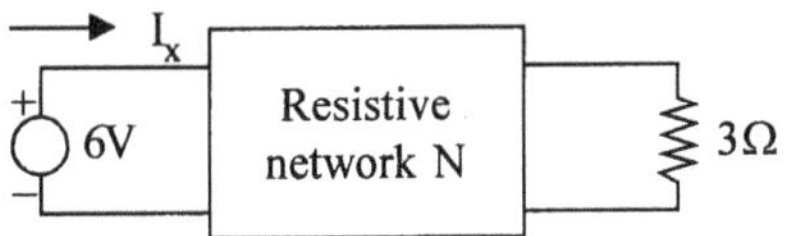

2.24 For the circuit shown in the figure, the current X is 3A. The power delivered by the dependant current source D is

(a) 50 watts

(b) 2250 watts

(c) 2300 watts

(d) 1500 watts

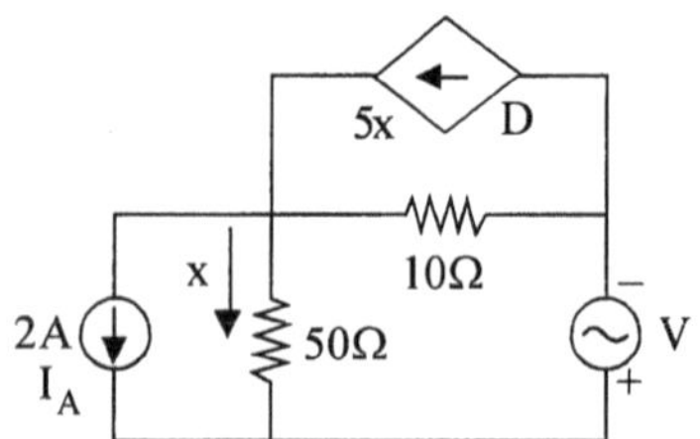

2.25 The voltage transfer ratio for the network shown in the figure is

(a) $\frac{1}{13}$ (b) $\frac{2}{13}$

(c) $\frac{3}{13}$ (d) $\frac{4}{13}$

2.26 In the circuit shown the power delivered by the 10 A source and 5A source are respectively

(a) 120W, 320W

(b) 320W, 120W

(c) 400W, 200W

(d) 200W, 400W

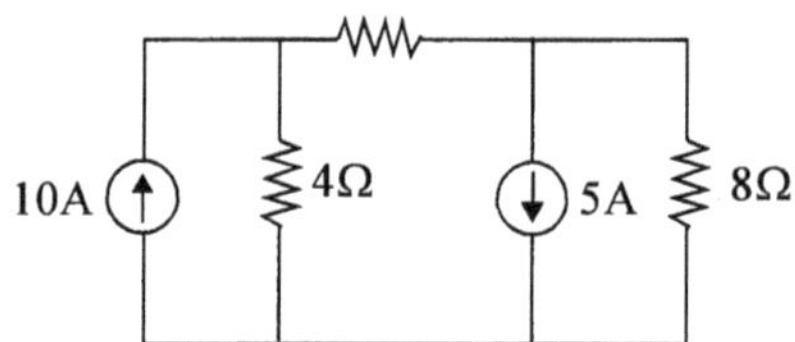

2.27 For the circuit shown in the given figure, the current through R, when $V_A = 0$ and $V_B = 15$ V is 1 amp. Now, if both V_A and V_B are increased by 15 volts, then the current thought R will be

(a) 1 amp

(b) 1/2 amp

(c) 3 amp

(d) 1/3 amp

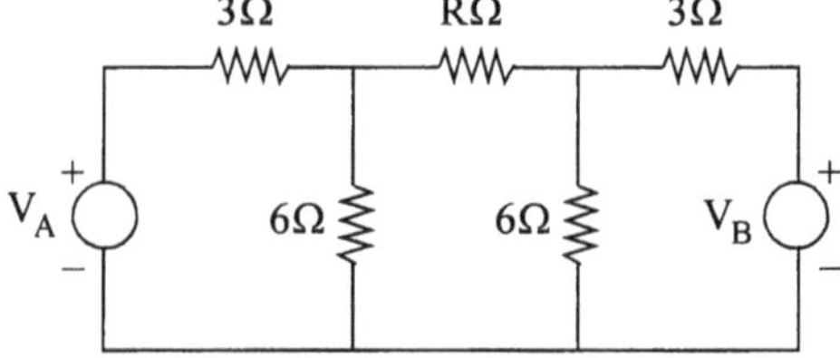

2.28 In the circuit shown in the given figure, current I is

(a) $-\frac{2}{5}$ A (b) $\frac{24}{5}$ A

(c) $\frac{18}{5}$ A (d) $\frac{2}{5}$ A

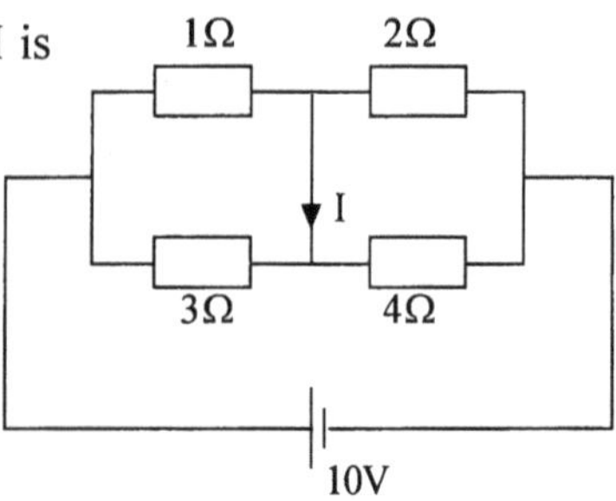

2.29 A resistor R is connected to a voltage source V_s having an internal resistance R_S. A voltmeter of resistance R_m is connected across the terminals of the resistor R. The voltmeter will read a voltage of

(a) $\dfrac{V_s R_s R_m}{R_s R_m + R_s R + R R_m}$ (b) $\dfrac{V_s R}{R + R_s}$

(c) $\dfrac{V_s R_m}{R R_s + R_m}$ (d) $\dfrac{V_s R R_m}{R_s R + R_s R_m + R R_m}$

2.30 Which one of the following is the ratio $\dfrac{V_{24}}{V_{13}}$ of the network shown in the given figure ?

(a) $\dfrac{1}{3}$ (b) $\dfrac{2}{3}$ (c) $\dfrac{3}{4}$

(d) $\dfrac{4}{3}$

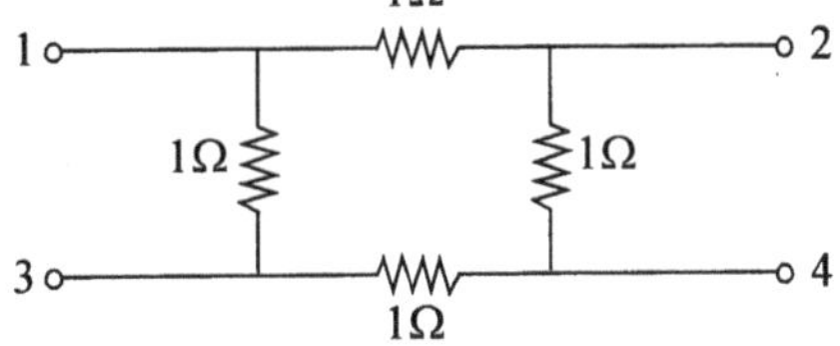

2.31 In the circuit shown in the given figure, the current I in the 2 ohm resistor is

(a) zero

(b) –2A

(c) 2 A

(d) 1 A

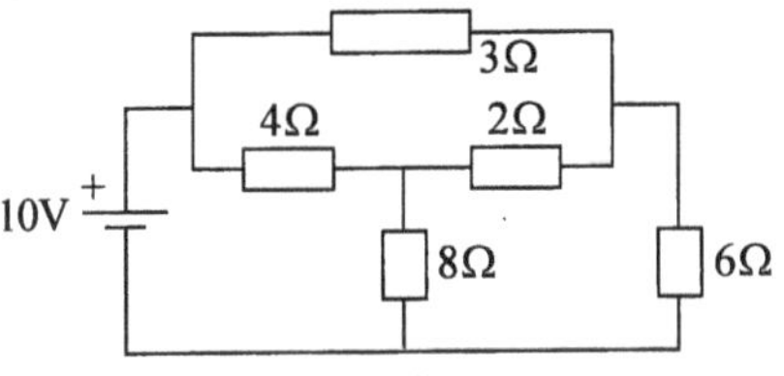

2.32 The total power consumed in the circuit shown in the figure is

(a) 10 W

(b) 12 W

(c) 16 W

(d) 20 W

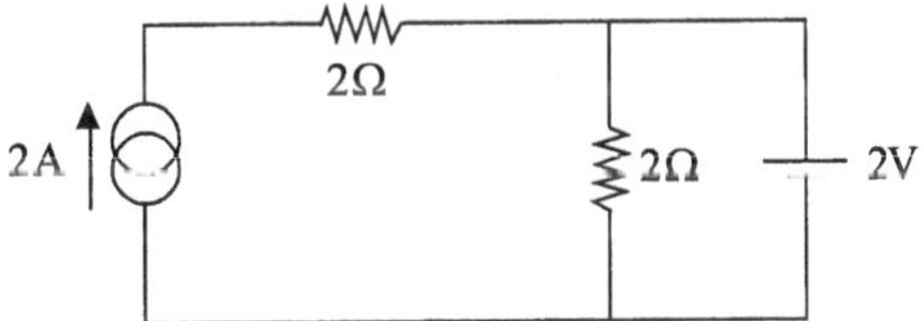

2.33 In the circuit shown in the figure, if I =2, then the value of the battery voltage V will be

(a) 5 V

(b) 3 V

(c) 2 V

(d) 1 V

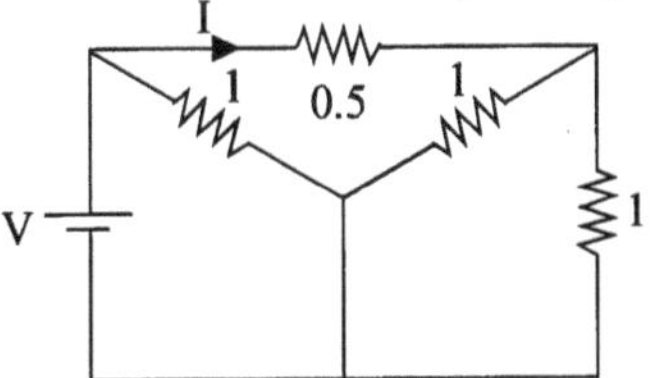

2.34 The effective resistance between the terminals A and B in the circuit shown in the figure is

(a) R

(b) R – 1

(c) R/2

(d) $\dfrac{6}{11}$ R

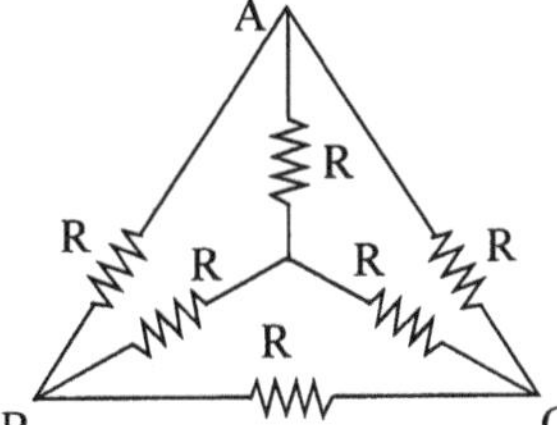

2.35 A network is shown in the given figure. Which one of the following equation would represent the equation for loop 3 ?

(a) $-I_1 + 4I_2 + 11\ I_3 = 0$

(b) $I_1 + 4I_2 + 11\ I_3 = 0$

(c) $-I_1 - 4I_2 + 11\ I_3 = 0$

(d) $I_1 + 4I_2 + 6\ I_3 = 0$

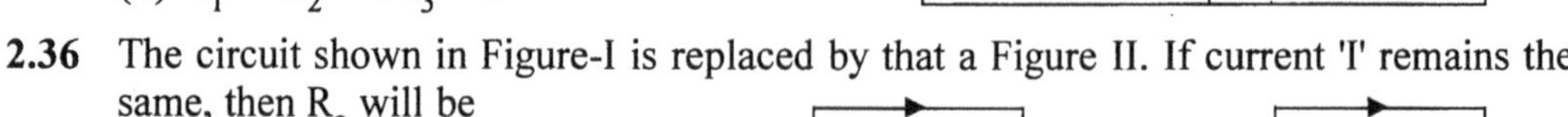

2.36 The circuit shown in Figure-I is replaced by that a Figure II. If current 'I' remains the same, then R_o will be

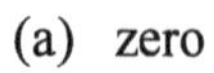

(a) zero

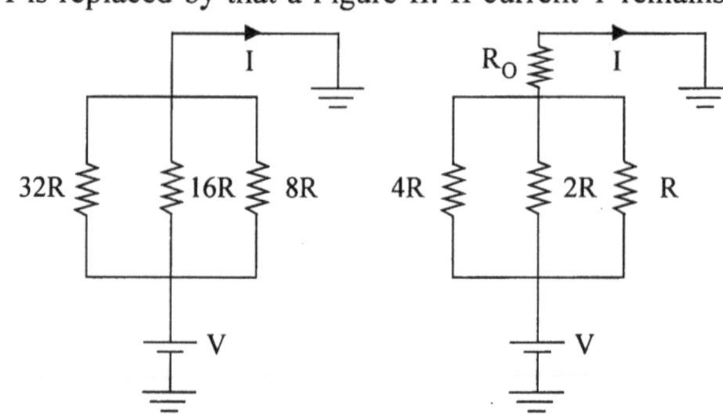

(b) R

(c) 2R

(d) 4R

2.37 In the circuit shown in the figure, for R = 20 Ω the current 'I' is 2 A. When R is 10 Ω, the current 'I' would be

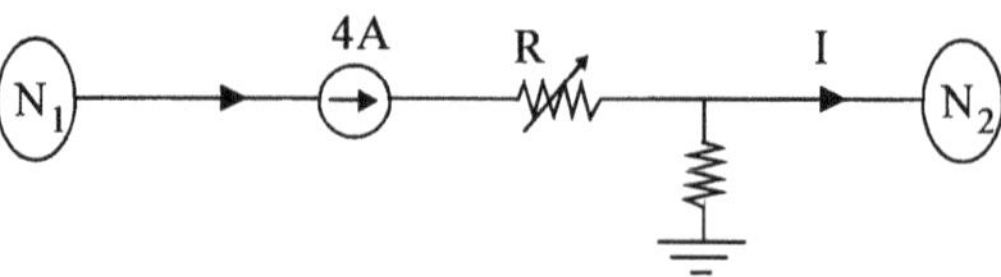

(a) 1 A (b) 2 A (c) 2.5 A (d) 3 A

2.38

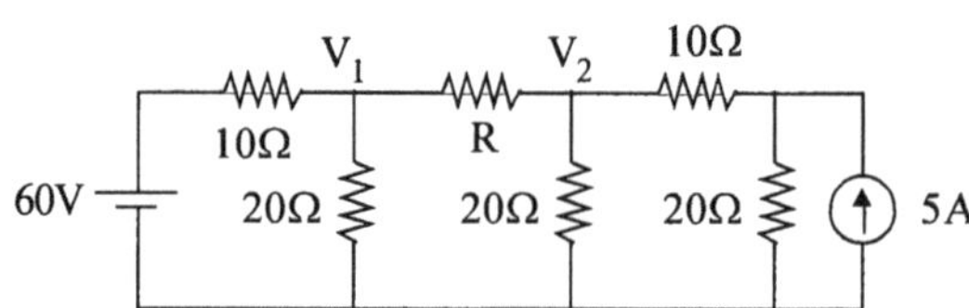

In the above circuit, $V_1 = 40$ V when R is 10 Ω. When R is zero, the value of V_2 will be

(a) 40 V

(b) 30 V

(c) 20 V

(d) 10 V

2.39 For the circuit shown in the figure, the current 'I' is

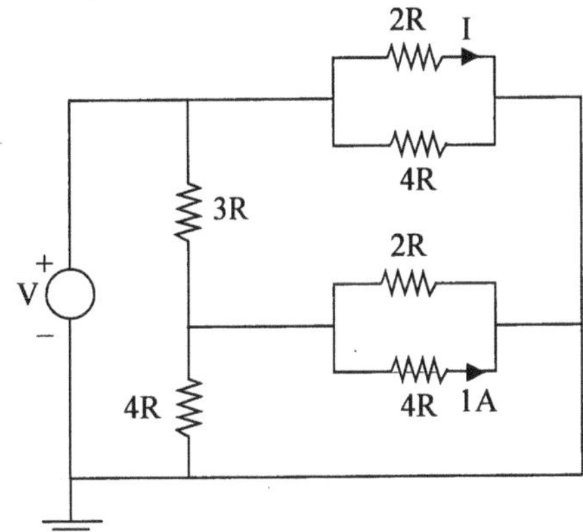

(a) indeterminable due to inadequate data

(b) zero

(c) 4A

(d) 8 A

2.40 A network contains only independent current sources and resistors. If the values of all resistors are doubled, the values of the node voltages

(a) will become half
(b) will remain unchanged
(c) will become double
(d) cannot be determined unless the circuit configuration and the values of the resistors are known

2.41 For the given circuit, the current I is

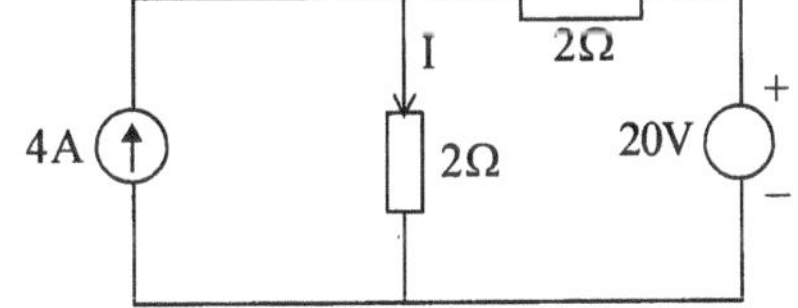

(a) 2 A (b) 5 A
(c) 7 A (d) 9 A

2.42 In the circuit shown in the given figure, power dissipated in the 5 Ω resistor is

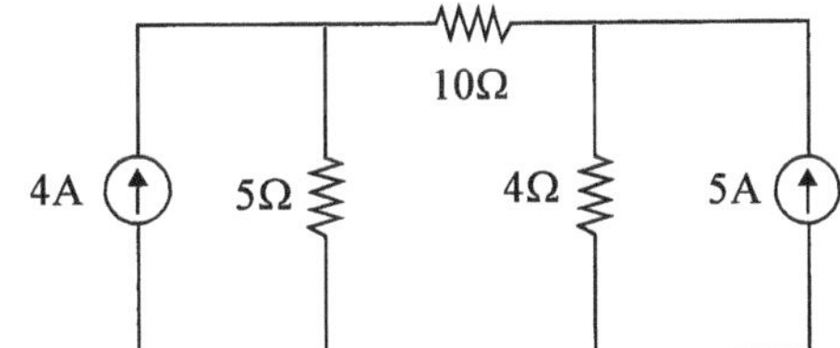

(a) zero

(b) 80 W

(c) 125 W

(d) 405 W

2.43 For the equivalent ≡ Δ circuit shown in the given figure, the values of R_{AB} and R_{BC} are respectively

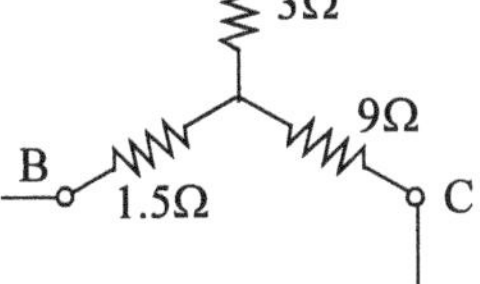

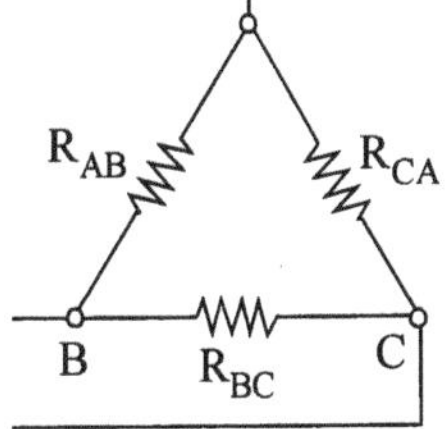

(a) 5 Ω and 15 Ω

(b) 15 Ω and 30 Ω

(c) 30 Ω and 5 Ω

(d) 20 Ω and 35 Ω

2.44 Viewed from the terminals A, B the following circuit shown in figure can be reduced to an equivalent circuit of a single voltage source in series with a single resistor with the following parameters :

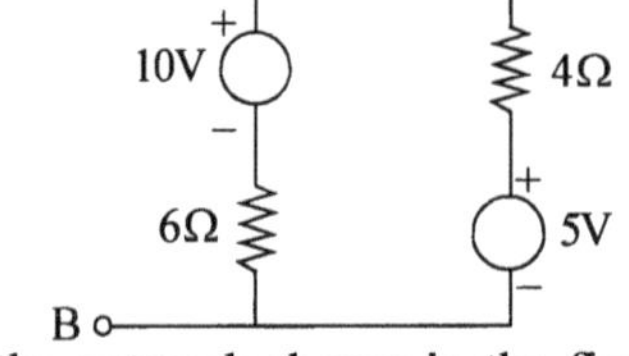

(a) 5 volt source in series with 10 Ω resistor

(b) 7 volt source in series with 2.4 Ω resistor

(c) 15 volt source in series with 2.4 Ω resistor

(d) 1 volt source in series with 10 Ω resistor

2.45 Obtain potential of node B with respect to node G in the network shown in the figure.

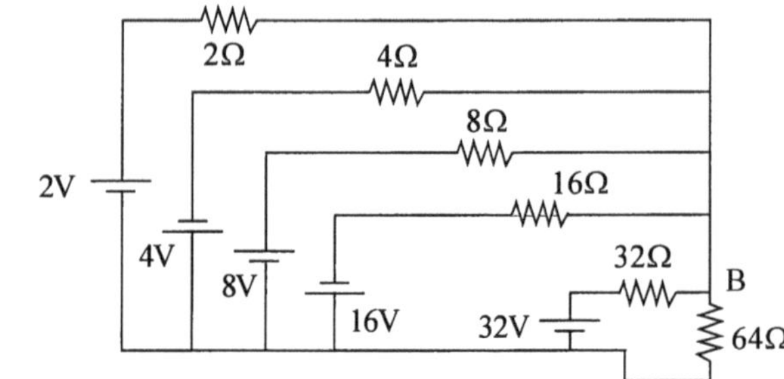

(a) $\frac{64}{63}$ V

(b) 1 V

(c) $\frac{63}{64}$ V

(d) $\frac{32}{63}$ V

2.46 What is the value of R so that i = 2A

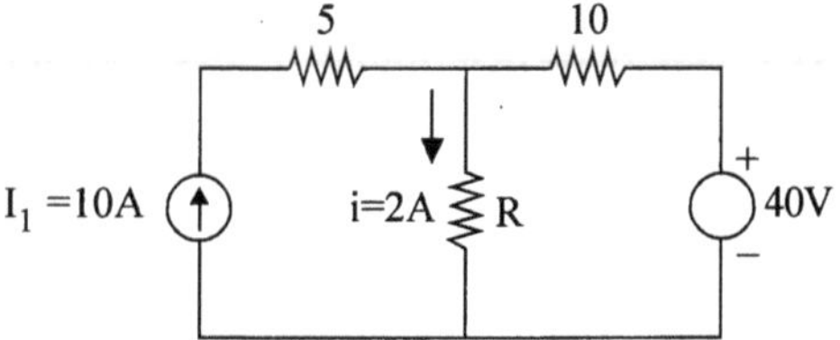

(a) 5 Ω

(b) 10 Ω

(c) 40 Ω

(d) 60 Ω

2.47 What will be the power consumed by the voltage source, current source and resistance respectively ?

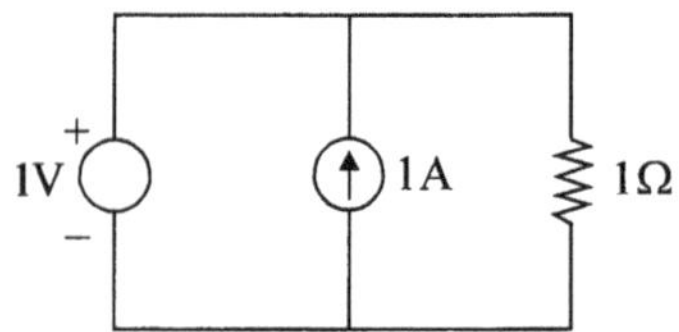

(a) 1W, 1W, 2W

(b) 0W, – 1W, 1W

(c) 1W, 0W, 1W

(d) 0W, 0W, 0W

2.48 In the circuit shown × is an element which absorbs power. During a particular operation, it sets up a current of 1A in the direction shown and absorbs a power Px. It is possible that × can absorb the same power Px for another current I. The value of this current is

(a) $3-\sqrt{14}$ A

(b) $3+\sqrt{14}$ A

(c) 5 A

(d) None of these.

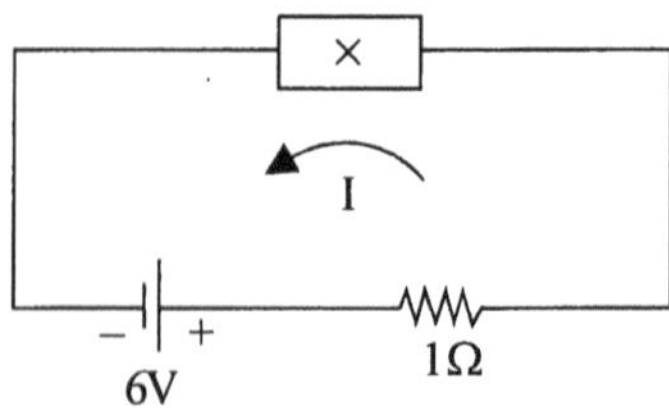

CHAPTER 3 – Single Phase Circuits

3.1 In the circuit shown in the given figure, the voltmeter indicates 30 V. The reading of the ammeter will be

(a) 20 A

(b) $10\sqrt{2}$ A

(c) 10 A

(d) zero

3.2 If a voltmeter placed across the 3 ohm resistor in the circuit given in the figure reads 45 volts, then the reading of ammeter A will be

(a) 10.1 A

(b) 13.3 A

(c) 16.1 A

(d) 19.4 A

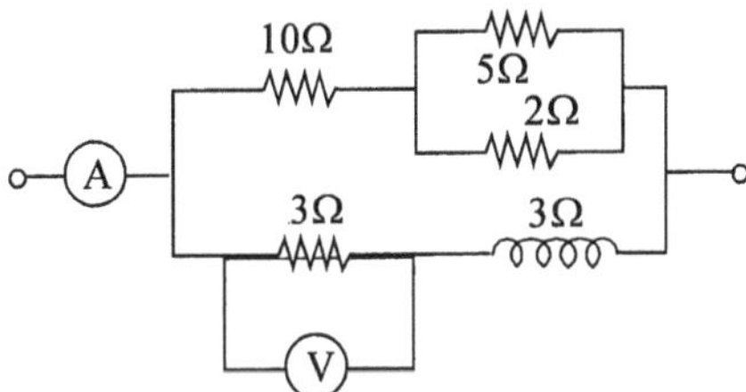

3.3 The RMS value of a half-wave rectified symmetrical square wave current of 2A is

(a) $\sqrt{2}$ A (b) 1 A (c) $1/\sqrt{2}$ A (d) $\sqrt{3}$ A

3.4 The current in the circuit shown in figure is

(a) 5 A

(b) 10 A

(c) 15 A

(d) 25 A

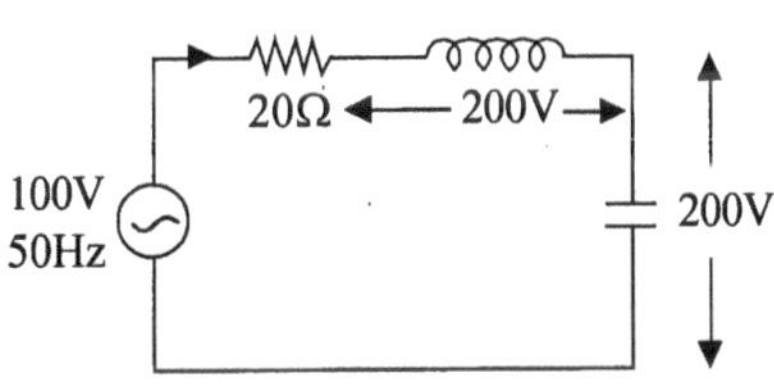

3.5 Currents i_1, i_2 and i_3 meet at a junction (node) in a circuit. All currents are marked as entering the node. If $i_1 = -6 \sin(\omega t)$ mA and $i_2 = 8 \cos(\omega t)$ mA, then i_3 will be

(a) $10 \cos(\omega t + 36.87)$ mA (b) $14 \cos(\omega t + 36.87)$ mA

(c) $-14 \sin(\omega t + 36.87)$ mA (d) $-10 \cos(\omega t + 36.87)$ mA

3.6 A fixed capacitor of reactance $-j\,0.02\ \Omega$ is connected in parallel across a series combination of a fixed inductor of reactance J 0.01 Ω and a variable resistance R. As R is varied from zero to infinity, the locus diagram of the admittance of this L-C-R circuit will be

(a) a semi-circle of diameter j 100 and center at zero

(b) a semi-circle of diameter j 50 and center at zero

(c) a straight line inclined at an angle

(d) a straight line parallel to the x-axix

3.7 The voltage phasor of a circuit is $10\angle 15^0$ V and the current phasor is $2\angle -45^0$ A. The active and the reactive power in the circuit are

(a) 10 W and 17.32 V Ar (b) 5 W and 8.66 V Ar

(c) 20 W and 60 V Ar (d) $20\sqrt{2}$ W and $10\sqrt{2}$ V Ar.

3.8 A sinusoidal source of voltage V and frequency f, is connected to a series circuit of variable resistance R, and a fixed reactance X. The locus of the tip of the current phasor I, as R is varied from 0 to ∞ is

(a) a semi-circle with a diameter of V/X

(b) a straight line with a slope of R/X

(c) an ellipse with V/R as major axis

(d) a circle of radius R/X and origin at $(0, V/2)$

3.9 A water boiler at home is switched on to the a.c. mains supplying power at 230 V/50 Hz. The frequency of instantaneous power consumed by the boiler is

(a) 0 Hz (b) 50 Hz (c) 100 Hz (d) 150 Hz

3.10 In the circuit shown in fig, what value of C will cause a unity power factor at the ac source ?

(a) 68.1 μ F

(b) 165 μ F

(c) 0.681 μ F

(d) 6.81 μ F

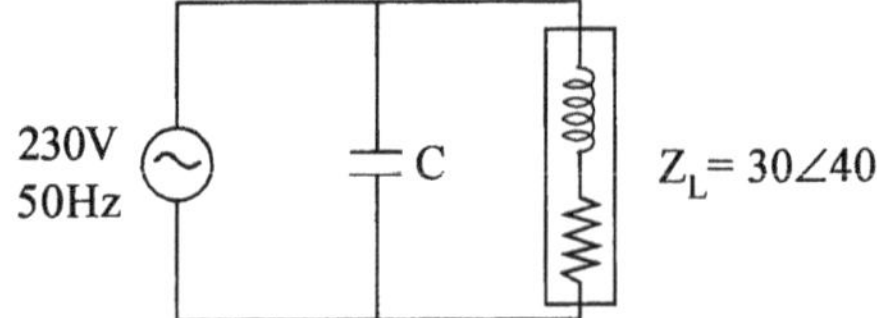

3.11 In a series RLC circuit at resonance, the magnitude of the voltage developed across the capacitor.

(a) is always zero

(b) can never be greater than the input voltage

(c) can be greater than the input voltage, however, is 90^0 out of phase with the input voltage

(d) can be greater than the input voltage, and is in phase with the input voltage.

3.12 A 240 V single-phase ac source is connected to a load with an impedance of $10\angle 60^0\ \Omega$. A capacitor is connected in parallel with the load. If the capacitor supplies 1250 VAR, the real power supplied by the source is

(a) 3600 W (b) 2880 W

(c) 2400 W (d) 1200 W

3.13 The given figure indicates the locus for the total current I taken by a two-branch-parallel circuit fed from a constant voltage ac source $\overline{V}$, when one element of the circuit is varied. Each branch contains two elements (R, L or C) in series. Consider the follwoing possible combinations.

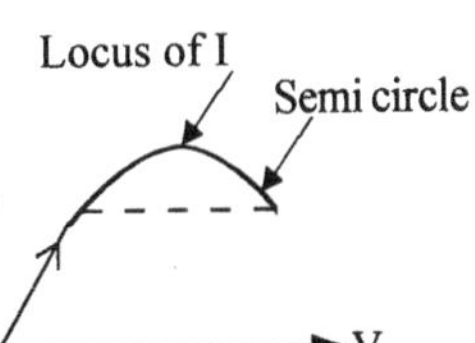

1. Fixed R and C.
2. Fixed R and variable L.
3. Variable R and fixed L
4. Fixed R and variable C.

The arrangements of the two branches are respectively

(a) 1 and 2 (b) 2 and 3 (c) 1 and 4 (d) 2 and 4

3.14 The current read by the ammeter A in the ac circuit shown in the given figure is

(a) 9 A

(b) 5 A

(c) 3 A

(d) 1 A

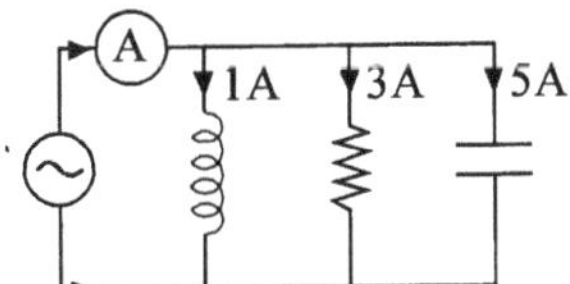

3.15 A series RLC circuit, consisting of R = 10 ohms, X_L = 20 ohms and X_c = 20 ohms is connected across an ac supply of 100 V (rms). The magnitude and phase angle (with reference to supply voltage) of the voltage across the inductive coil are respectively.

(a) 100 V; 90° (b) 100 V; – 90°

(c) 200 V; – 90° (d) 200 V; 90°

3.16 The sinusoidal steady-state voltage gain of the network shown in the given figure will have magnitude equal to 0.707 at an angular frequency of

(a) zero

(b) RC rad/s

(c) 1/RC rad/s

(d) 1 rad/s

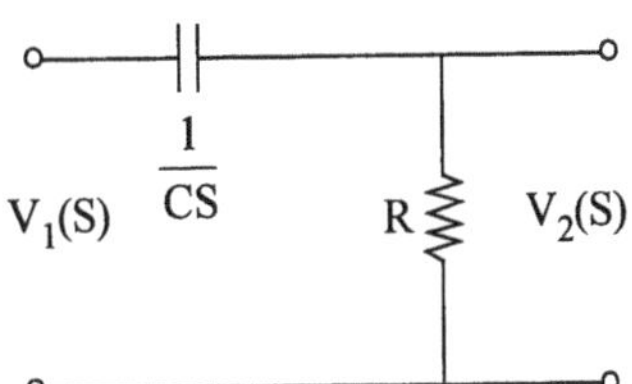

3.17 The phase angle of the current 'I' with respect to the voltage V_1 in the circuit shown in the figure is :

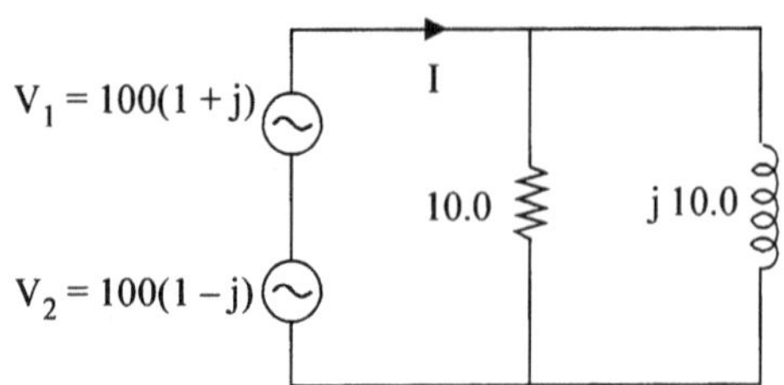

(a) 0° (b) + 45° (c) –45° (d) –90°

3.18 A voltage $\overline{V}$ is applied to an ac circuit resulting in the delivery of a current $\overline{I}$. Which of the following expression would yield the true power delivered by the source ?

1. Real part of $\overline{V}\ \overline{I}^*$. 2. Real part of VI

3. I^2 times the real part of $\dfrac{\overline{V}}{\overline{I}}$

Select the correct answer using the codes given below :

Codes :

(a) 1 alone (b) 1 and 3 (c) 2 and 3 (d) 3 alone.

3.19 The current through the current coil of a wattmeter is given by, i = (1 + 2 sin ω t) V. and the voltage across the pressure coil is, v = (2 + 3 sin 2 ω t) V.

The wattmeter will read

(a) 8.00 W (b) 5.05 W (c) 2.0 W (d) 1.0 W

3.20 In the circuit shown in the figure, v = 2cos 2t, Z_2 = 1 + j : C_1 is so chosen that i = cos 2t. The value of C_1 is

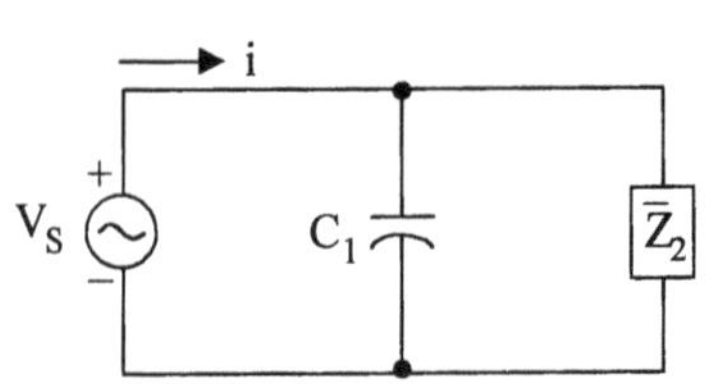

(a) 2 F

(b) 1 F

(c) 0.5 F

(d) 0.25 F

3.21 A function f(t) = sin 1.1t + sin 3.3t has the time period of

(a) $\dfrac{\pi}{1.1}$ (b) $\dfrac{2\pi}{1.1}$ (c) $\dfrac{2\pi}{3.3}$ (d) $\dfrac{2\pi}{2.2}$

3.22 A 10 μF capacitor is fed from an ac voltage source containing a fundamental and a third harmonic of value one-third of fundamental. The third harmonic current flowing through the capacitor expressed as percentage of the fundamental under steady-state condition will be

(a) zero (b) 100 (c) 30 (d) 90

3.23 In a two-element series circuit, the applied voltage and the resulting current are respectively, $v(t) = 50 + 50 \sin (5 \times 10^3 t)$ V and $i(t) = 11.2 \sin (5 \times 10^3 t + 63.4^o)$A.

The nature of the elements would be

(a) R – L (b) R – C (c) L – C (d) neither R, nor L, nor C

3.24 An ac source of 200 V rms supplies active power of 600 W and reactive power of 800 V AR. The rms current drawn from the source is

(a) 10 A (b) 5 A (c) 3.75 A (d) 2.5 A

3.25 The reactive power drawn from the source in the network shown in the given figure is

(a) 300 VAR

(b) 200 VAR

(c) 100 VAR

(d) zero

3Ω –j10Ω +j10Ω

100∠10°V

3.26 In the case of the RLC circuit shown in the given figure, the voltage across the R, L and C would be respectively

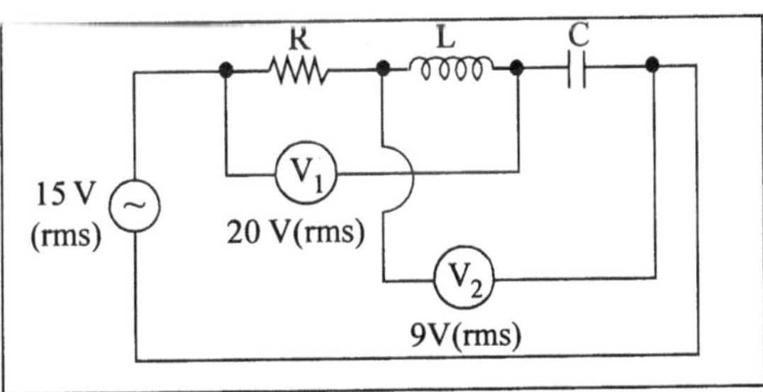

(a) 12 V, 16 V and 7 V or 25 V (b) 16 V, 12 V and 7 V or 25 V

(c) 7 V, 16 V and 12 V (d) 16 V, 12 V and 25 V

3.27 A resistance of 'R' Ω and inductance of 'L' H are connected across 240 V, 50 Hz supply. Power dissipated in the circuit is 300 W and the voltage across R is 100 V. In order to improve the power factor to unity, the capacitor that is to be connected in series should have a value of

(a) 43.7 μF (b) 0.437 μF (c) 437 μF (d) 4.37 μF

3.28 For the network shown in the given figure, the voltage V_B will be

(a) (j 5.33) V

(b) (5.33) V

(c) (–j 5.33) V

(d) (j 3.33) V

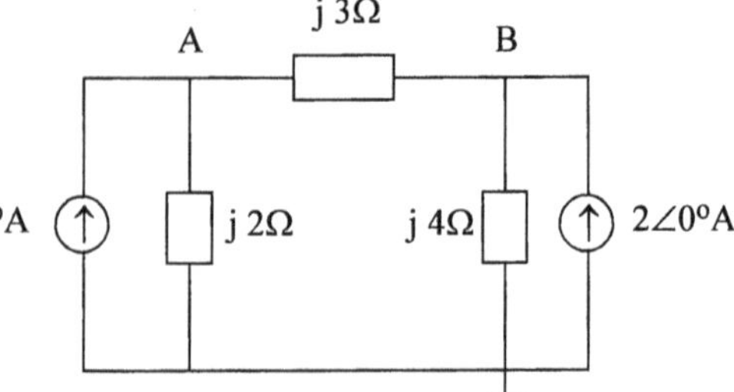

3.29 The system function $H(s) = \frac{1}{s+1}$. For an input signal cos t, the steady state response is

(a) $\frac{1}{\sqrt{2}} \cos\left(t - \frac{\pi}{4}\right)$

(b) cos t

(c) $\cos\left(t - \frac{\pi}{4}\right)$

(d) $\frac{1}{\sqrt{2}} \cos t$

3.30 In the given RC circuit, the current i(t) = 2 cos 5000 t A. The applied voltage v (t) is

(a) 28.28 cos (5000 t – 45°) V

(b) 28.28 cos (5000 t + 45°) V

(c) 28.28 sin (5000 t – 45°) V

(d) 28.28 sin (5000 t + 45°) V

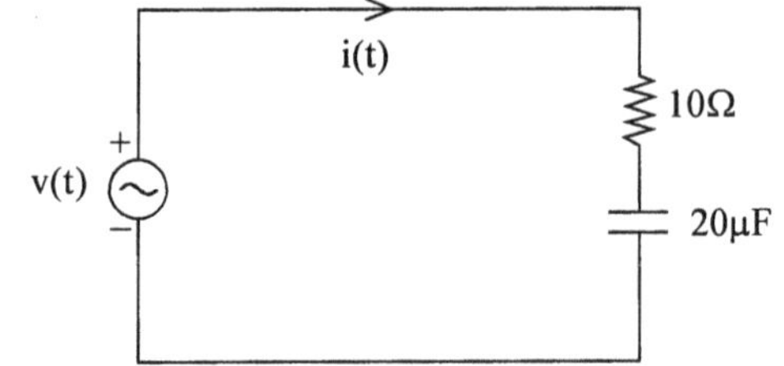

3.31 In the circuit shown in the figure, the current through the inductor L is

(a) 0 A

(b) 3 A

(c) 4 A

(d) 8 A

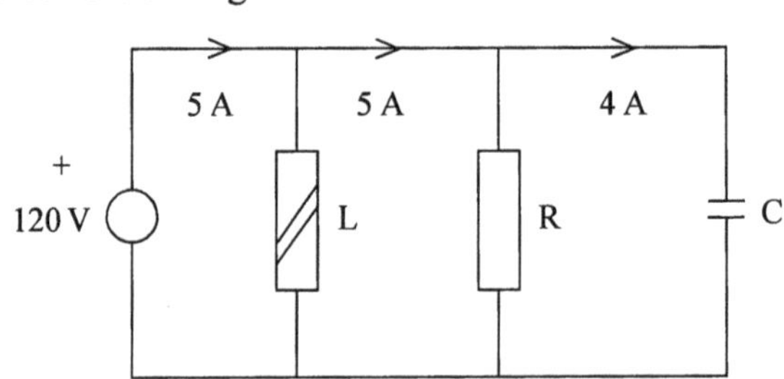

3.32 Three currents i_1, i_2 and i_3 are approaching a node. If i_1 = 10 sin (400t + 60°) A, and i_2 = 10 sin (400 t – 60°)A, the i_3 the

(a) 0

(b) 10 (sin 400 t) A

(c) – 10 (sin 400 t) A

(d) $-5\sqrt{3}$ (3 sin 400 t) A

3.33 The source in the circuit shown is a sinusoidal source. The voltage across various elements are marked in the figure. The input voltage is

(a) 10 V

(b) 5 V

(c) 27 V

(d) 24 V

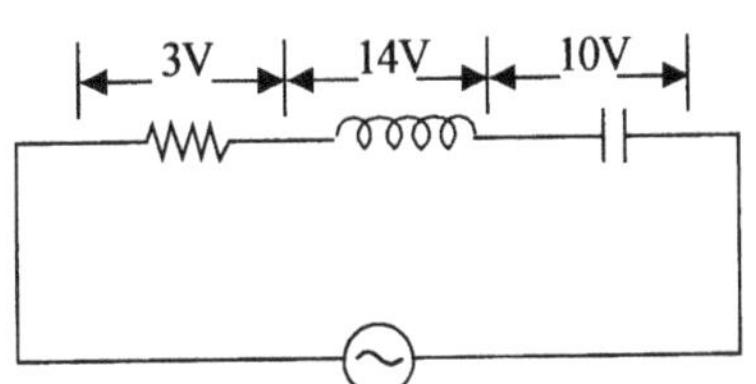

3.34 An alternating current source having voltage E = 110 sin ($\omega t + \pi/3$) is connected in an a.c circuit. If the current drawn from the circuit varies as i (t) = 5 sin ($\omega t + \pi/3$), the impedance of the circuit will be ...

(a) 22 ohm (b) 16 ohms (c) 30.8 ohm (d) none of the above.

3.35 In an LCR circuit, supplied from an ac source, the reactive power is proportional to the

(a) average energy stored in the electric field

(b) average energy stored in the magnetic field

(c) sum of the average energy stored in the electric field and that stored in the magnetic field

(d) difference between the average energy stored in the electric field and that stored in the magnetic field

3.36 The voltage ratio transfer function for the network shown in the given figure under sinusoidal steady state conditions is G (jω)

Match List I with List II and select the correct answer using the codes given below the lists :

List I (frequency)	List II [G (jω)]
1. 0	A. $1 \angle 0^\circ$
2. $\frac{1}{RC}$	B. $0 \angle -90^\circ$
3. ∞	C. $0.707 \angle -45^\circ$

R

INPUT OUTPUT

C

Codes :

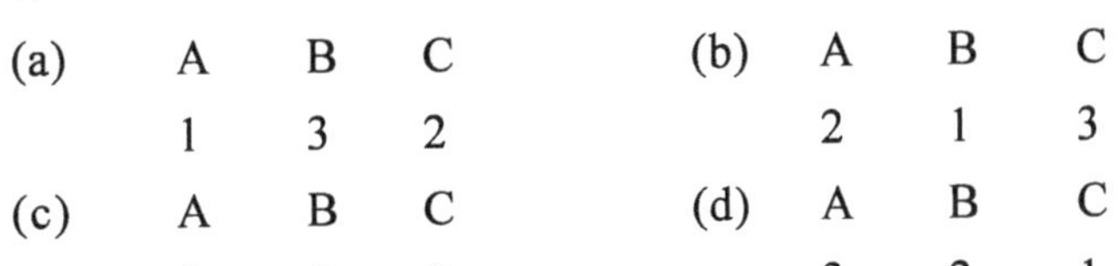

	A	B	C		A	B	C
(a)	1	3	2	(b)	2	1	3
(c)	1	2	3	(d)	3	2	1

3.37 The admittance locus of the circuit shown in figure I, is

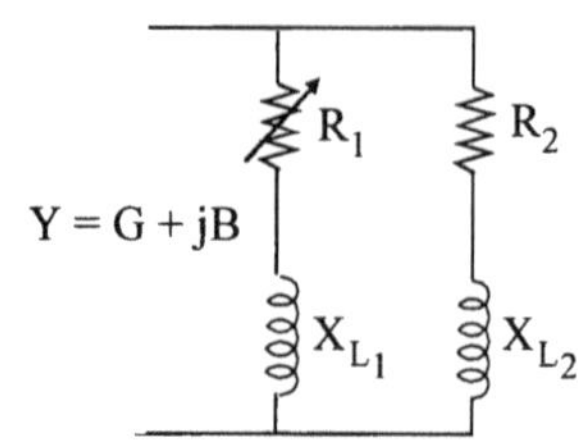

(a)

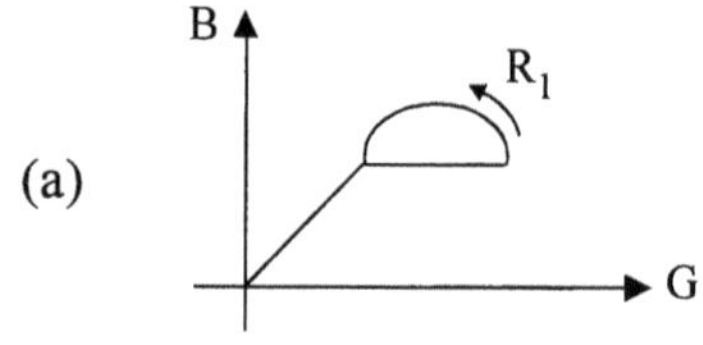

(b)

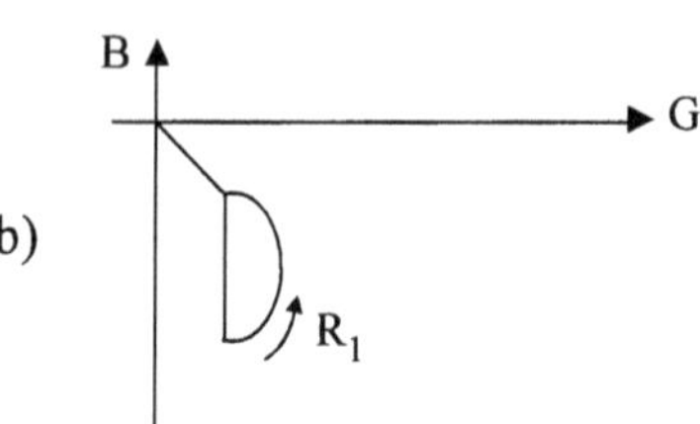

(c)

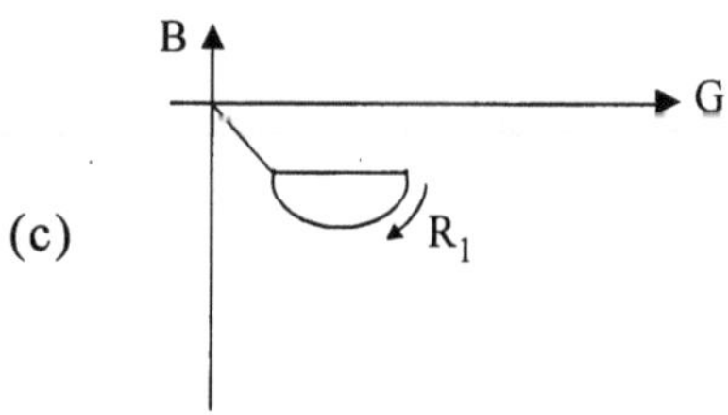

(d)

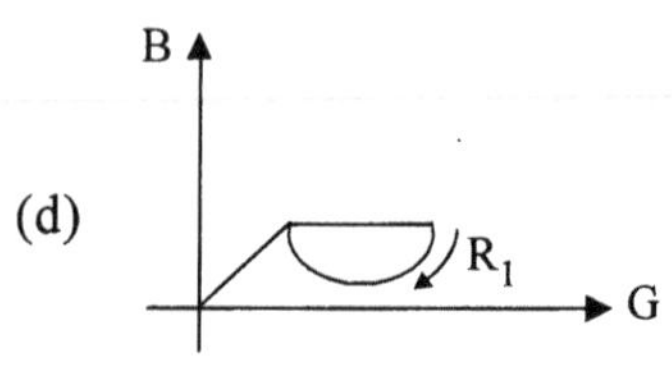

3.38 In the circuit shown, $V_s = 250 \sin 400\,t$; $V_{2m} = 200V$. If $R = 100\ \Omega$, value of L is

(a) $\frac{1}{3}$ (b) $\frac{1}{4}$

(c) $\frac{2}{3}$ (d) $\frac{3}{4}$

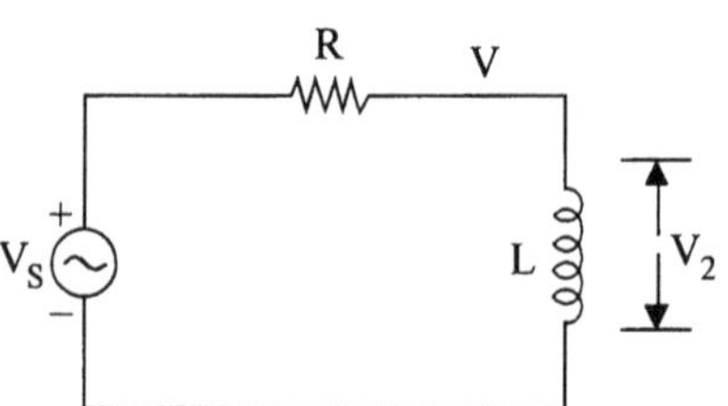

3.39 What will be the value of X_c so that the circuit is noninductive.

(a) 2.5

(b) 5.0

(c) 10.0

(d) 20.0

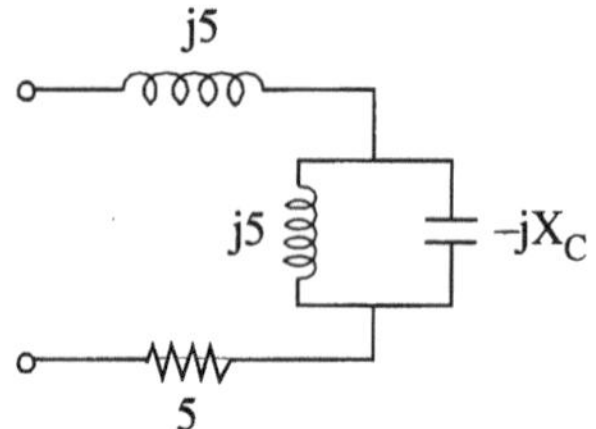

3.40 In a series R-L-C circuit excited by a voltage, e = E sin ωt, where LC < $\left(\frac{1}{\omega^2}\right)$

(a) current lags the applied voltage
(b) current leads the applied voltage
(c) current is in phase with the applied voltage
(d) voltages across L and C are equal

3.41 An alternating current source having a voltage e (t) = 100 sin $\left(\omega t + \frac{\pi}{6}\right)$ is connected in an ac circuit. If the current drawn from the circuit is i (t) = 5 sin $\left(\omega t - \frac{\pi}{6}\right)$, the complex power in the circuit is

(a) 125 + j 216.5 (b) 250 + j 433 (c) 500 + j 0 (d) 250 + j 0

3.42 Which one of the following networks is the Y equivalent of the Δ circuit shown in Figure 1 ?

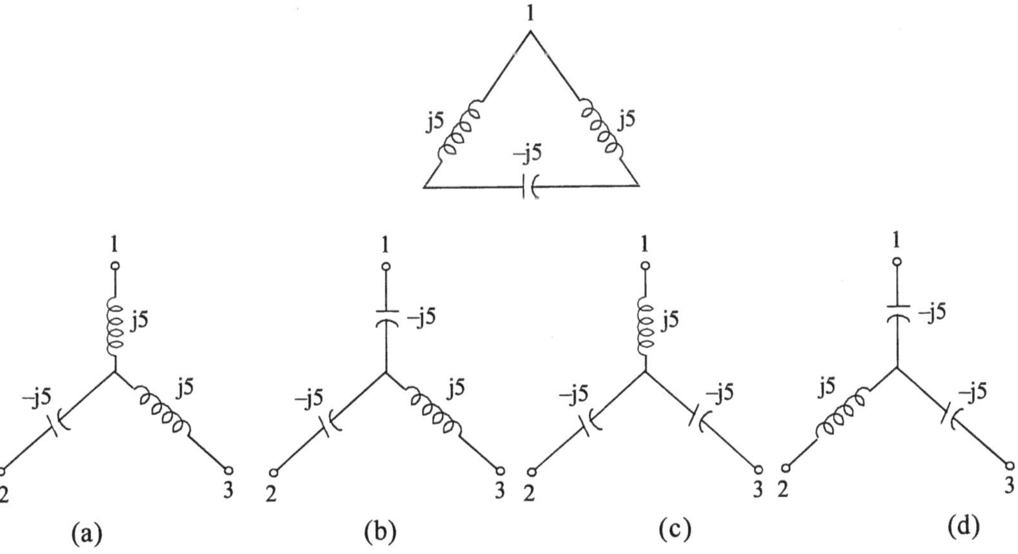

3.43 A two-terminal black box contains one of the R, L, C, elements. The black box is connected to a 220 volts ac supply. The current through the source is I. When a capacitance of 0.1 F is inserted in series between the source and the box, the current through the source is 2I. The element is

(a) a resistance
(b) an inductance
(c) a capacitance of 0.5 F
(d) not readily identifiable from the given data

3.44 Two impedance are connected in series. The 3 voltmeters, one connected across each impedance and one across the combination, read equal value. The phase angle between the voltages across the two impedances is

(a) 30° (b) 60° (c) 90° (d) 120°

3.45 In the given figure, the effective value of the waveform is

(a) 5.0

(b) 2.5

(c) $\sqrt{2.5}$

(d) $\sqrt{5.0}$

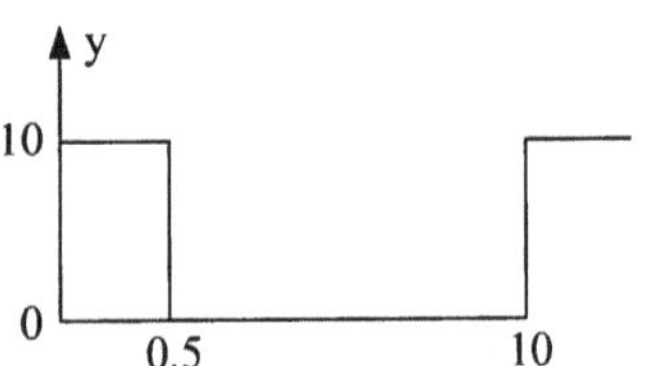

3.46 RMS value of f (t) = 10 (1 + sin ωt) is

(a) 10 (b) $\frac{10}{\sqrt{2}}$ (c) $\sqrt{150}$ (d) $\frac{10}{\sqrt{2}}$

3.47 A particular current is made up of two components : a 10 A dc a sinusoidal current of peak value of 14.14 A. The average value of the resultant current is

(a) zero (b) 24. 14 A (c) 10 A (d) 14.14 A

3.48 In the circuit shown in the figure, if R_o is adjusted such that $|V_{AB}| = |V_{BC}|$, then

(a) $\theta = 2\tan^{-1}\left(\frac{2|V_{BD}|}{|V|}\right)$

(b) $|V_{BC}| = |V_{DC}|$

(c) $|V_{AB}| = |V_{AD}|$

(d) $\theta = \tan^{-1}\left(\frac{|V_{BD}|}{|V|}\right)$

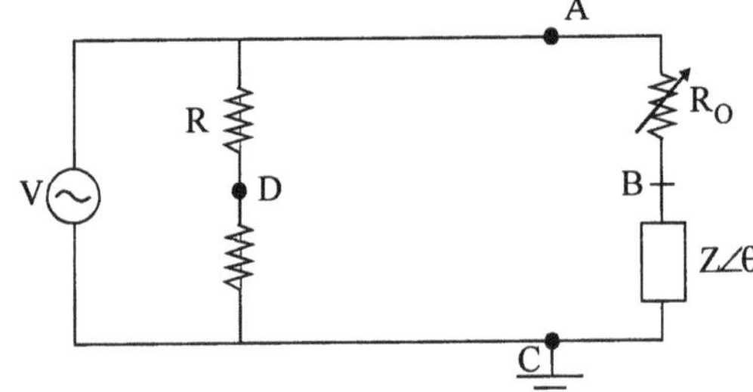

3.49 In the circuit shown in the figure, if the current in resistance 'R' is Nil, then

(a) $\frac{\omega L_1}{R_1} = \frac{1}{\omega C_4 R_4}$

(b) $\frac{\omega L_1}{R_1} = \omega C_4 R_4$

(c) $\tan^{-1}\frac{\omega L_1}{R_1} + \tan^{-1}\omega C_4 R_4 = 0$

(d) $\tan^{-1}\frac{\omega L_1}{R_1} + \tan^{-1}\frac{1}{\omega C_4 R_4} = 0$

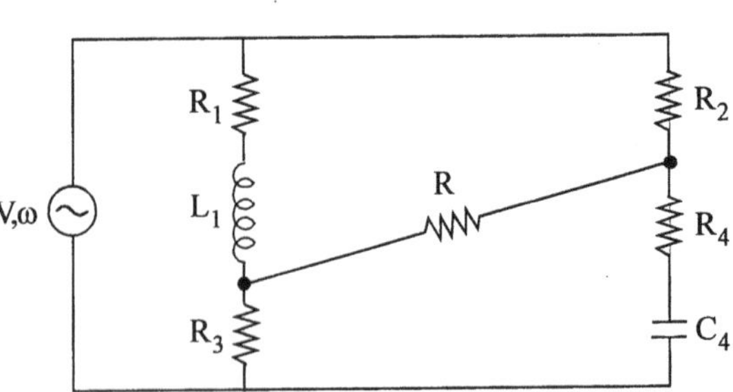

3.50 The average value of the periodic function v(t) of the given figure is

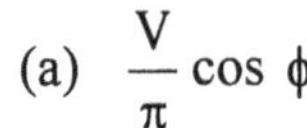
(a) $\frac{V}{\pi}\cos\phi$

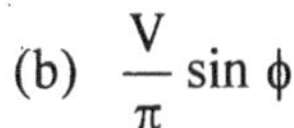
(b) $\frac{V}{\pi}\sin\phi$

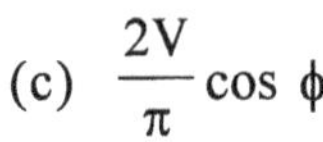
(c) $\frac{2V}{\pi}\cos\phi$

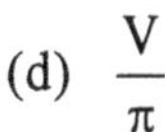
(d) $\frac{V}{\pi}$

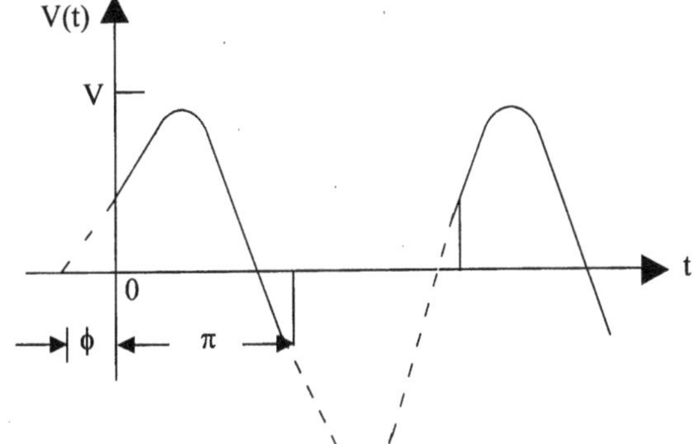

3.51 A capacitor used for power factor correction in single-phase circuit decreases

(a) the power factor

(b) the line current

(c) both the line current and the power factor

(d) the line current and increases power factor

3.52 Consider the following statements :

If a network has an impedance $(1 - j)$ at a specific frequency, the circuit would consist of series

1. R and C
2. R and L
3. R, L and C

Which of these statements are correct ?

(a) 1 and 2 (b) 1 and 3 (c) 1, 2 and 3 (d) 2 and 3

3.53 A series R-L-C circuit when excited by a 10 V sinusoidal voltage source of variable frequency, exhibits resonance at 100 Hz and has a 3 dB bandwidth of 5 Hz. The voltage across the inductor L at resonance is

(a) 10 V (b) $10\sqrt{2}$ V (c) $10/\sqrt{2}$ V (d) 200 V

3.54 A circuit with a resistor, inductor and a capacitor in series is resonant at f_0 Hz. If all the component values are now doubled, the new resonant frequency is

(a) $2 f_0$ (b) Still f_0 (c) $f_0/4$ (d) $f_0/2$

3.55 A coil (which can be modelled as a series RL circuit) has been designed for high-Q performance at a rated voltage and a specified frequency. If the frequency of operation is doubled, and the coil is operated at the same rated voltage, then the Q-factor and the active power P consumed by the coil will be affected as follows :

(a) P is doubled, Q is halved (b) P is halved, Q is doubled

(c) P remains constant, Q is doubled (d) P decreases 4 times, Q is doubled

3.56 A series R-L-C circuit has R = 50 Ω; L = 100 μ H and C – 1 μ F. The lower half power frequency of the circuit is

(a) 30.55 kHz (b) 3.055 kHz (c) 51.92 kHz (d) 1.92 kHz

3.57 Consider the following statements with respect to a series R–L–C circuit under resonance condition :

1. All the applied voltage appears across R.
2. There is no voltage across either L or C.
3. The voltage across L and C equal and equal to their maximum values.

Of these statements

(a) 1 alone is correct (b) 2 alone is correct

(c) 1 and 3 are correct (d) 1 and 2 are correct

3.58 The resonant frequency of the series circuit shown in the given figure is :

(a) $\frac{1}{4\pi}$ Hz (b) $\frac{1}{4\pi\sqrt{2}}$ Hz

(c) $\frac{1}{2\pi\sqrt{2}}$ Hz (d) $\frac{1}{4\pi\sqrt{3}}$ Hz

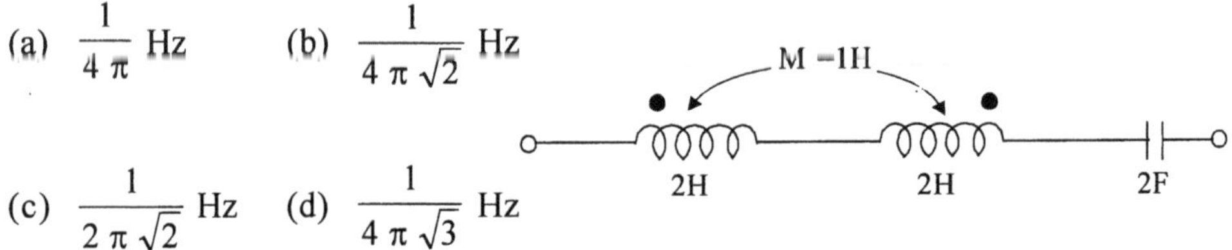

3.59 An RLC resonant circuit has a resonance frequency of 1.5 MHz and a bandwidth of 10 kHz. If C = 150 pF, then the effective resistance of the circuit will be.

(a) 29.5 Ω (b) 14.75 Ω (c) 9.4 Ω (d) 4.7 Ω

3.60 For a series RLC circuit, the power factor at the lower half power frequency is

(a) 0.5 lagging (b) 0.5 leading (c) unity (d) 0.707 leading

3.61 If the resonant frequency of the circuit shown in Figure-I is 1 kHz, the resonant frequency of the circuit shown in Figure-II will be

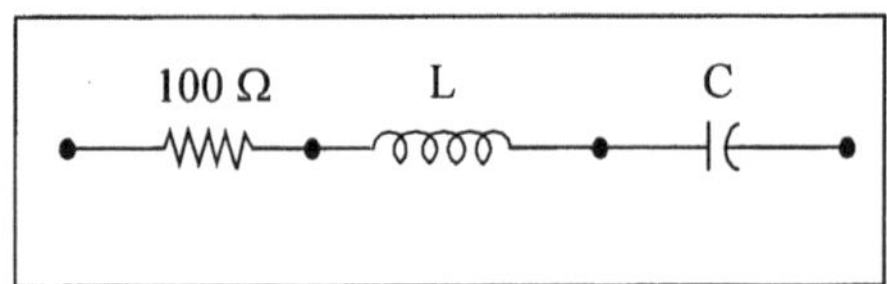

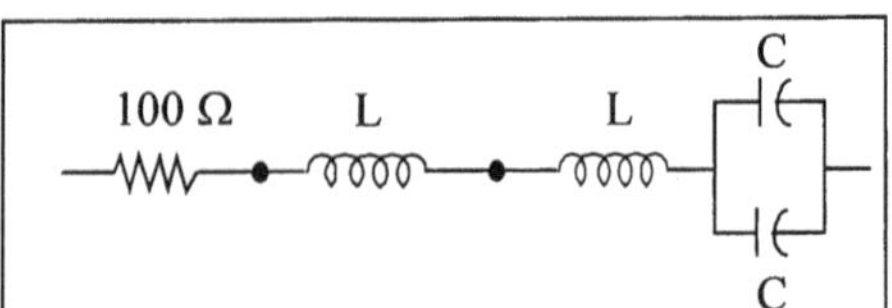

(a) 4 kHz (b) 2 kHz (c) 0.5 kHz (d) 0.25 kHz

3.62 The value of the capacitance 'C' in the given ac circuit to make it a constant resistance circuit OR for the supply current to be independent of its frequency is

(a) $\frac{1}{16}$ F (b) $\frac{1}{12}$ F

(c) $\frac{1}{8}$ F (d) $\frac{1}{4}$ F

3.63 The resonant frequency of the given series circuit is

(a) $\frac{1}{2\pi\sqrt{3}}$ Hz (b) $\frac{1}{4\pi\sqrt{3}}$ Hz

(c) $\frac{1}{4\pi\sqrt{2}}$ Hz (d) $\frac{1}{\pi\sqrt{2}}$ Hz

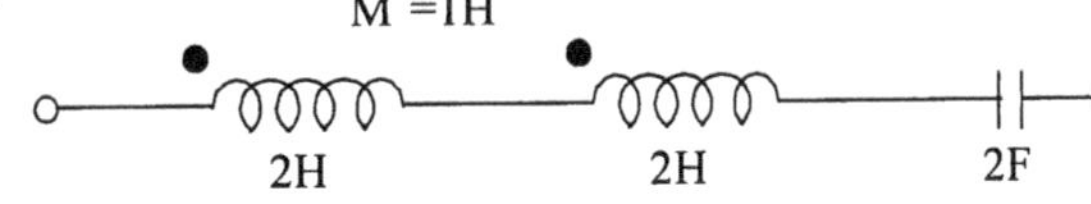

3.64 A parallel circuit consists of two branches. One branch has R_L and L connected in series and the other branch has R_C and C connected in series. Consider the following statements:

1. The two branch currents will be in quadrature if $R_L R_C = L/C$.
2. The impedance of the whole circuit is independent of frequency, if $R_L = R_C$ and $\omega = \frac{1}{\sqrt{LC}}$
3. The circuit is in resonance for all the frequencies if $R_L = R_C$.
4. The two branch currents will be in phase at $\omega = \frac{1}{\sqrt{LC}}$.

Which of the above statements are correct ?

(a) 1 and 2 (b) 2 and 3 (c) 1 and 3 (d) 3 and 4

3.65 At resonant frequency a R-L-C series circuit draws maximum current due to the reason that

(a) the difference between capacitive reactance and inductive reactance is zero
(b) the impedance is more than the resistance
(c) the voltage across the capacitor equals the applied voltage
(d) the power factor is less than unity

3.66 In the given circuit, at resonance, I_R in Amperes is equal to

(a) 0

(b) 10

(c) 5

(d) 0.5

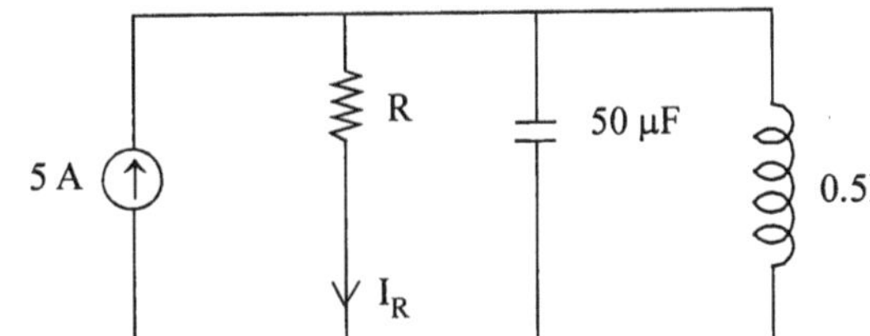

3.67 Consider the following statements about the quality factor of a R-L-C circuit :

1. For the critically damped circuit, the quality factor Q = 1/2
2. Higher the value of quality factor higher will be the bandwidth of the circuit
3. Higher the value of quality factor lower will be the bandwidth of the circuit
4. For under damped circuits the value of Q is great than 1/2

Which of these statements are correct ?

(a) 1 and 2 (b) 1 and 3 (c) 2 and 4 (d) 1, 3 and 4

3.68 Consider the following statements regarding the frequency response curve of a series RLC circuit :

1. At half-power frequencies, the current in the circuit is one-half of the current at resonant frequency
2. At half-power frequencies, the power factor angle of the circuit is 45°.
3. At resonant frequency, the power factor angle of the circuit is 90°.
4. Maximum power occurs at resonant frequency

Of these statements :

(a) 1, 2 and 4 are correct (b) 1, 2 and 3 are correct

(c) 2 and 4 are correct (d) 1 and 4 are corrects

3.69 A coil having a resistance of 5 ohms and inductance of 0.1H connected in series with a condenser of capacitance 50 μF. A constant alternating voltage of 200 volts is applied to the circuit. The voltage across the coil at resonance is

(a) 200 volts (b) 1788 volts (c) 1800 volts (d) 2000 volts

3.70 An RLC series circuit has f_1 and f_2 as the half power frequencies and f_o as the resonance frequency. The Q-factor of the circuit is given by

(a) $\dfrac{f_1 + f_2}{2f_0}$ (b) $\dfrac{f_1 - f_0}{f_2 - f_0}$ (c) $\dfrac{f_0}{f_1 - f_2}$ (d) $\dfrac{f_2 + f_1}{f_0}$

3.71 The circulating current in a parallel LC circuit at any resonant frequency is

(a) directly proportional to frequency (b) inversely proportional to frequency

(c) independent of frequency (d) none of the above

3.72 In the following parallel circuit, resonance will never occur, if

(a) $R_1{}^2 = R_2{}^2 = \dfrac{L}{C}$ (b) $R_1{}^2 < \dfrac{L}{C}$

(c) $R_1{}^2 < \dfrac{L}{C}$ and $R_2{}^2 < \dfrac{L}{C}$ (d) $R_1{}^2 < \dfrac{L}{C}$ and $R^2{}_2 > \dfrac{L}{C}$

3.73 In a series resonance circuit consisting of R, L and C if L_O is the value of L at resonance, L_A is the value of L at the lower half frequency f_A, L_B is the value of L at the upper half frequency f_B, Q_O is the quality factor at resonance, then the value of $\frac{L_O}{L_B - L_A}$ is

(a) $\frac{Q_O}{2}$ (b) $2Q_o$ (c) $\frac{2}{Q_O}$ (d) $\frac{Q_O}{4}$

3.74 A high Q coil has

(a) large bandwidth (b) high losses (c) low losses (d) flat response

3.75 A choke coil having a resistance of R ohms and an inductance of L Heneries is shunted by a capacitor of C farads. The dynamic impedance of the circuit at resonance will be

(a) R/LC (b) C/RL (c) L/RC (d) 1/RLC

3.76 An electric circuit contains R, L and C in series with a voltage source. The current through the circuit is I_0. The frequencies at which the current would reduce to 0.707 I_0 is given by f_{01} and f_{02}. The resonant frequency of the circuit is the

(a) geometric mean of f_{01} and f_{02} (b) arithmetic mean of f_{01} and f_{02}

(c) difference of f_{01} and f_{02} (d) harmonic mean of f_{01} and f_{02}

3.77 Consider the following statements regarding the situation at resonant frequency :

1. For a series RLC circuit, current is minimum.
2. For a series RLC circuit, voltage across C is minimum.
3. For a series RLC circuit, current is maximum.
4. For a parallel RLC circuit, total impedance is maximum.

Of the statements

(a) 1 and 2 are correct (b) 2 and 3 are correct

(c) 3 and 4 are correct (c) 1 and 4 are correct

3.78 In a parallel RLC circuit, if L = 4 H, C = 0.25 F and R = 4 Ω, then the value of Q at resonance will be

(a) 1 (b) 10 (c) 20 (d) 40

3.79 A series RLC circuit is excited by an ac voltage v (t) = 1 sin t. If L - 10 H and C = 0.1F, then the peak value of the voltage across R will be

(a) 0.707 (b) 1

(c) 1.414 (d) indeterminate as the value of R is not given

3.80 The locus of the tip of the voltage phasor (V_R) across the resistance (R) in a series RLC resonant circuit is given by

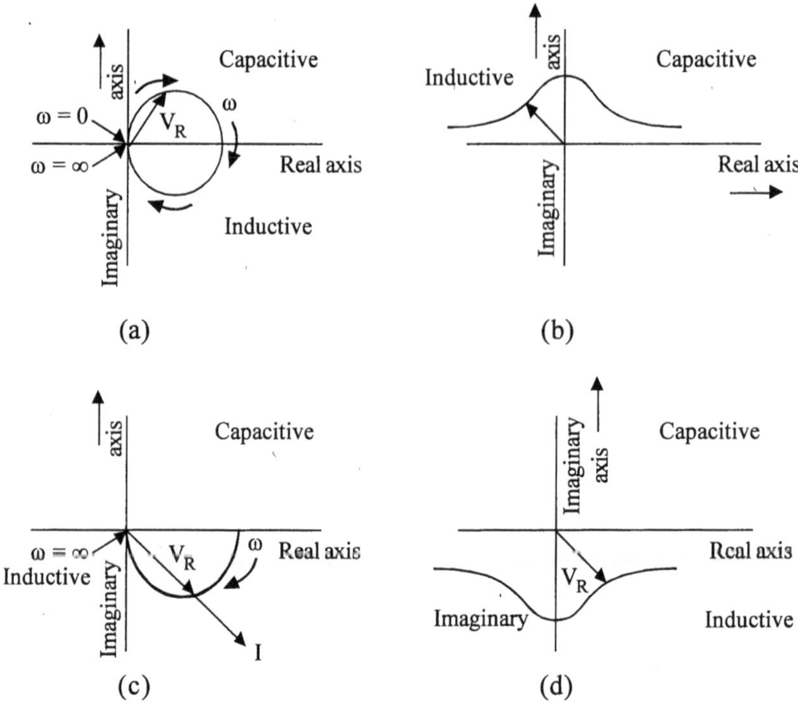

3.81 The response of a series RLC circuit fed from a fixed rms voltage and variable frequency source is represented graphically in the given figure. Match List I with List II and select the correct answer using the codes given below the lists :

List I (curve)	**List II (quantity)**
A. AA	1. Current
B. BB	2. Impedance
C. CC	3. Capacitive reactance
D. DD	4. Net reactance
	5. Inductive reactance

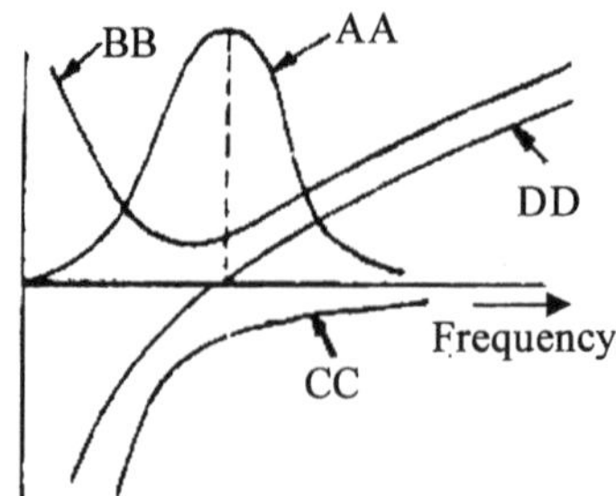

Codes :

(a)	A	B	C	D	(b)	A	B	C	D
	2	1	3	5		1	2	3	5
(c)	A	B	C	D	(d)	A	B	C	D
	1	2	3	4		1	2	4	3

3.82 The effective inductance of the circuit across the terminals A, B in the fig shown below is

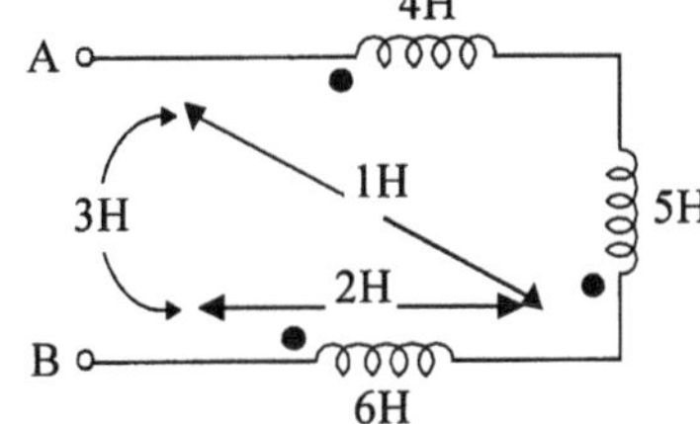

(a) 9 H

(b) 21 H

(c) 11 H

(d) 6 H

3.83 In the circuit shown in figure. It is found that the input ac voltage (v_i) and current i are in phase. The coupling coefficient is $K = \frac{M}{\sqrt{L_1 L_2}}$, where M is the mutual inductance between the two coils. The value of K and dot polarity of the coil P - Q are

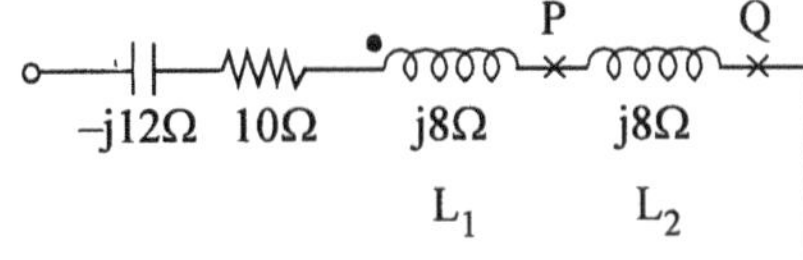

(a) K = 0.25 and dot at P

(b) K = 0.5 and dot at P

(c) K = 0.25 and dot at Q

(d) K = 0.5 and dot at Q

3.84 Given two coupled inductors L_1 and L_2 their mutual inductance M satisfies

(a) $M = \sqrt{L_1^2 + L_2^2}$ (b) $M > \frac{(L_1 + L_2)}{2}$ (c) $M > \sqrt{L_1 L_2}$ (d) $M \le \sqrt{L_1 L_2}$

3.85 The impedance seen by the source in the circuit in figure is given by

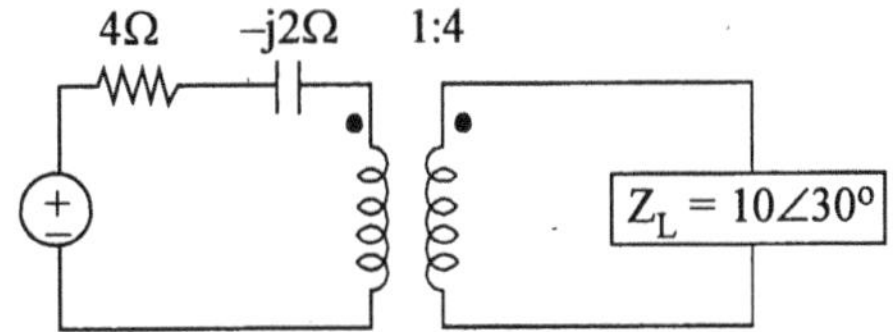

(a) (0.54 + j0.313) ohms

(b) (4 – j2) ohms

(c) (4.54 – j1.69) ohms

(d) (4 + j2) ohms

3.86 For the ideal transformer shown in the given figure

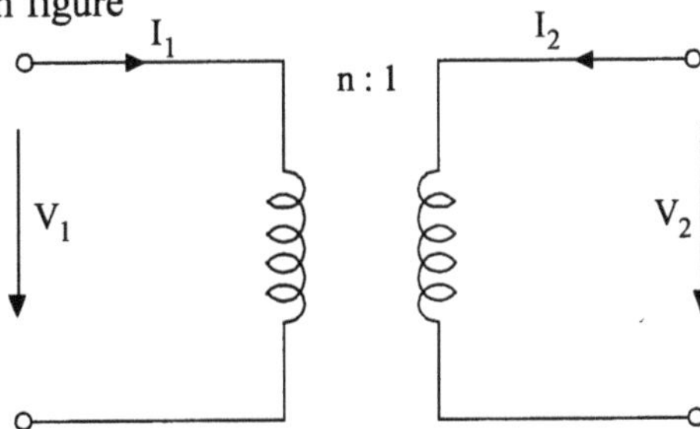

(a) $v_1 = nv_2$, $i_2 = -ni_1$

(b) $v_2 = nv_1$, $i_2 = ni_1$

(c) $v_1 = nv_2$, $i_1 = \frac{1}{n} i_2$

(d) $v_1 = nv_2$, $i_2 = \frac{-1}{n} i_1$

3.87 Consider the circuit shown in the given figure. For maximum power transfer to the load, the primary to secondary turns ratio must be

(a) 9 : 1

(b) 3 : 1

(c) 1 : 3

(d) 1 : 9

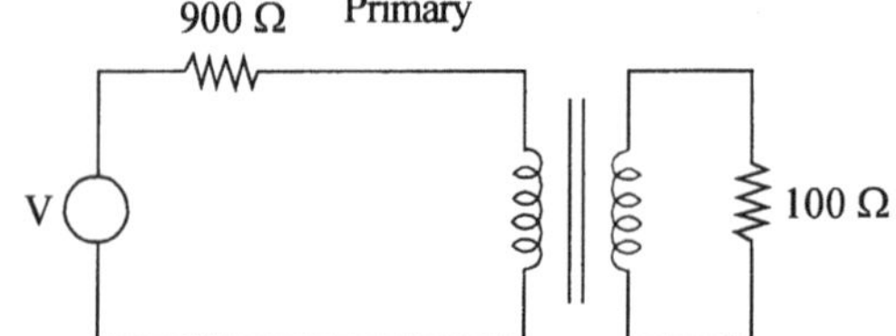

3.88 Two inductive coils with self-inductances L_1 and L_2 are magnetically coupled in series opposing and in parallel aiding respectively. The mutual inductance between the coils is M. The equivalent inductances in the two cases are respectively.

(a) $L_1 + L_2 + 2\,M, \dfrac{L_1L_2 - M^2}{L_1 + L_2 - 2M}$ (b) $L_1 + L_2 - 2\,M, \dfrac{L_1L_2 - M^2}{L_1 + L_2 + 2M}$

(c) $L_1 + L_2 - 2\,M, \dfrac{L_1L_2 - M^2}{L_1 + L_2 - 2M}$ (d) $L_1 + L_2 + 2\,M; \dfrac{L_1L_2 - M^2}{L_1 + L_2 + 2M}$

3.89 The coupling between two magnetically coupled coils is said to be ideal if the coefficient of coupling is

(a) zero (b) 0.5 (c) 1 (d) 2

3.90 Two identical coils of negligible resistance, when connected in series across a 50 Hz fixed voltage source, draw a current of 10 A. When the terminals of one of the coils are reversed, the current drawn is 8 A. The coefficient of coupling between the two coils is

(a) 1/100 (b) 1/9 (c) 4/10 (d) 8/10

3.91 The mutual inductance between two coupled coils is 10 mH. If the turns in one coil are doubled and that in the other arc halved, then the mutual inductance will be

(a) 5 mH (b) 10 mH (c) 14 mH (d) 20 mH

3.92 If an ideal centre-tapped 1:4 transformer is loaded as shown in the figure, the impedance measured across the terminals AA would be

(a) $\dfrac{3Z}{16}$

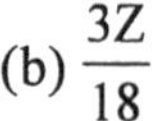

(b) $\dfrac{3Z}{18}$

(c) $\dfrac{2Z}{3}$

(d) $\dfrac{Z}{6}$

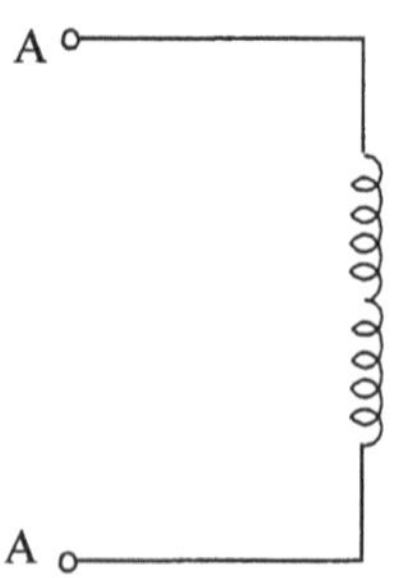

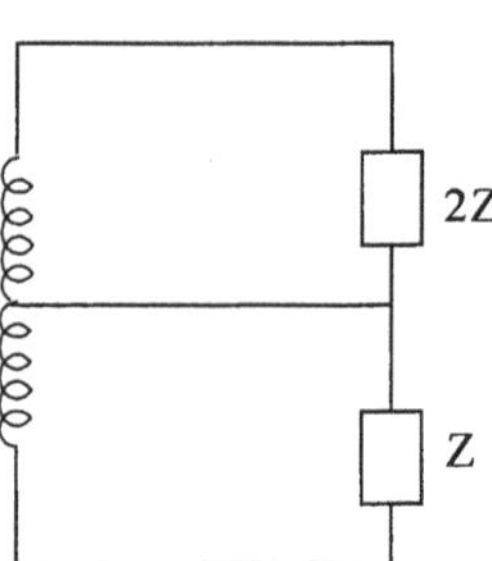

3.93 The equivalent inductance of two coils A and B connected as in the given figure is given by

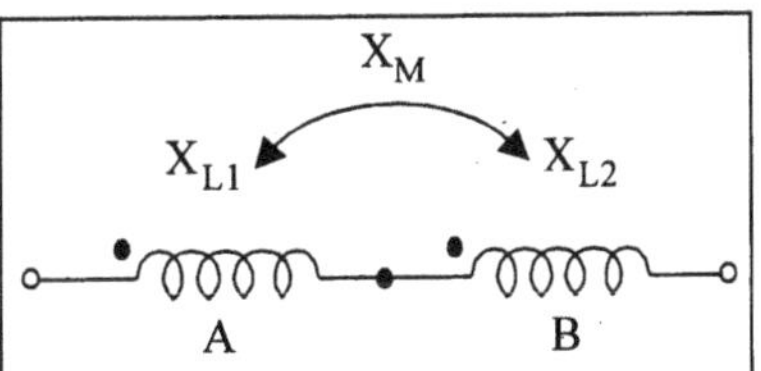

(a) 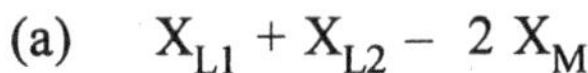 $X_{L1} + X_{L2} - 2\,X_M$

(b) $X_{L1} + X_{L2} + X_M$

(c) $X_{L1} + X_{L2} - X_M$

(d) $X_{L1} + X_{L2} + 2\,X_M$

3.94 A linear transformer and its T-equivalent circuit are shown in Figure-I and Figure-II respectively. The values of inductance L_a, L_b and L_c respectively.

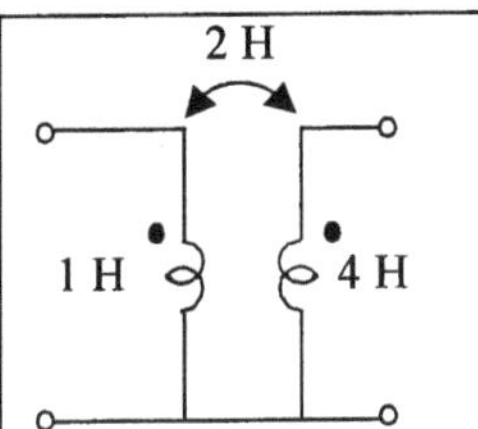

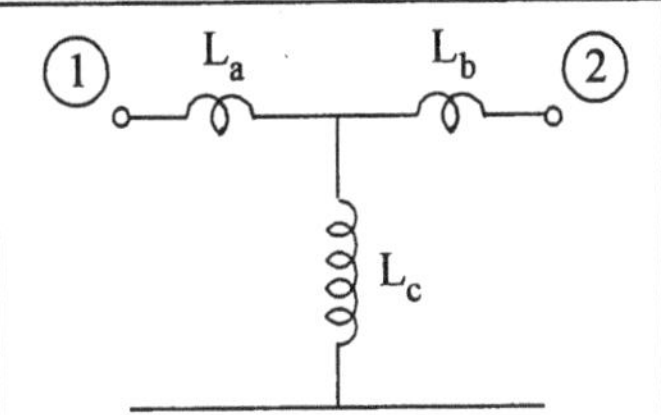

(a) 1H, –2H and 2H

(b) –1H, 2H and 2H

(c) 3H, 6H and – 2H

(d) 3H, 6H and 2H

3.95 The inductance matrix of a system of two mutually coupled inductors shown in figure

is given by $L = \begin{bmatrix} 5 & -4 \\ -4 & 7 \end{bmatrix}$

When the inductors are connected as shown in figure, the equivalent inductance of the system is given by

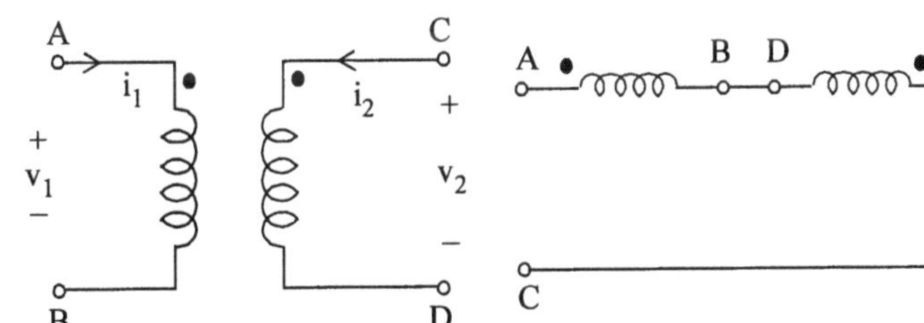

(a) 20 H

(b) 4 H

(c) 16 H

(d) 8H

3.96 Two coupled coils with $L_1 = L_2 = 0.6$ H have a coupling coefficient of K = 0.8. The turns ratio $\dfrac{N_1}{N_2}$ is

(a) 4 (b) 2 (c) 1 (d) 0.5

3.97 In the circuit shown in the given figure, the switch S is closed at t = 0. The induced voltage V_2 will have a maximum value of

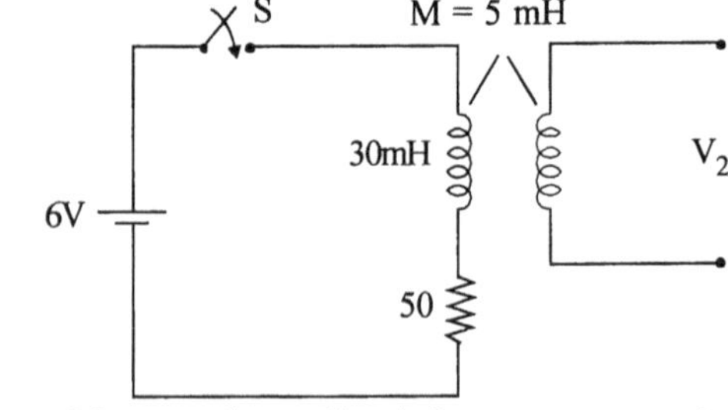

(a) 0.6 V

(b) 1 V

(c) 3.78 V

(d) 6 V

3.98 The equivalent inductance of two inductances with mutual coupling in between connected in parallel having the following configuration

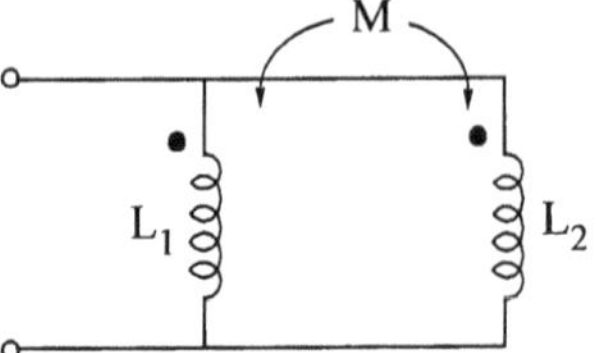

(a) $\dfrac{L_1L_2 + M^2}{L_1 + L_2 - 2M}$ (b) $\dfrac{L_1L_2 + M^2}{L_1 + L_2 + 2M}$

(c) $\dfrac{L_1L_2 - M^2}{L_1 + L_2 + 2M}$ (d) $\dfrac{L_1L_2 - M^2}{L_1 + L_2 - 2M}$

3.99 The input impedance of the circuit shown in the figure, if the respective coil impedance $Z_1 = (5 + j8)\ \Omega$ and $Z_2 = (3 + j8)\ \Omega$ is

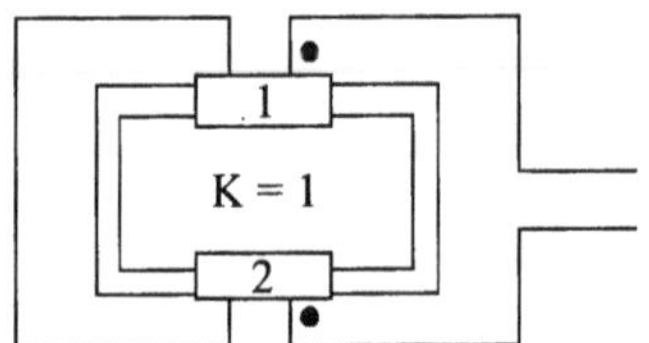

(a) $(8 + 16\,j)\ \Omega$

(b) $(2 + j0)\ \Omega$

(c) $(15 + 64\,j)\ \Omega$ (d) $(8 + 0\,j)\ \Omega$

3.100 When two coupled coils of equal self-inductance are connected in series in one way, the net inductance is 12 mH, and when they are connected in the other way, the net inductance is 4 mH. The maximum value of net inductance when they are connected in parallel in a suitable way is

(a) 2 mH (b) 3 mH (c) 4 mH (d) 6 mH

3.101 Two perfectly coupled coils each of one Henry self inductance are connected in parallel so as to oppose each other. The overall inductance in Henry is

(a) 2 (b) 1 (c) $\dfrac{1}{2}$ (d) zero

3.102 The voltmeter in the circuit shown in the given figure is ideal. The transformer has two identical windings with perfect coupling. The reading on the voltmeter will be

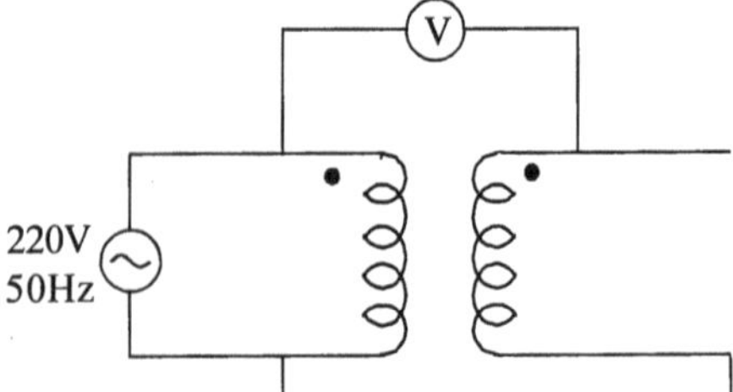

(a) 440 V

(b) 220 V

(c) 110 V

(d) zero

3.103 In the transformer shown in the figure, the inductance measured across the terminal 1 and 2 was 4 H with open terminals 3 and 4. It was 3 H when the terminal 3 and 4 were short circuited. The coefficient of coupling would be

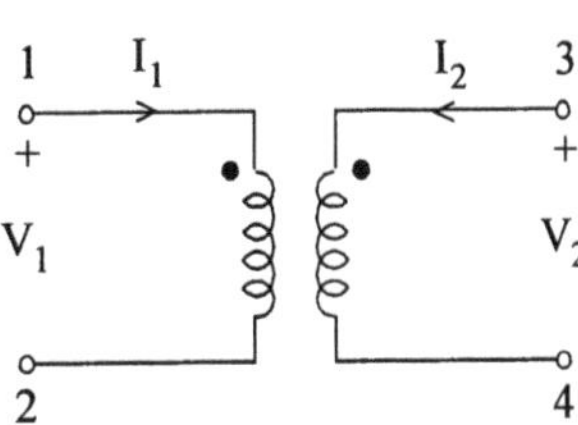

(a) 1

(b) 0.707

(c) 0.5

(d) inderminate due to unsufficient data

3.104 Consider the following statements :

In the circuit shown in the figure, if the equivalent impedance between x-x is Z_{eq}, then

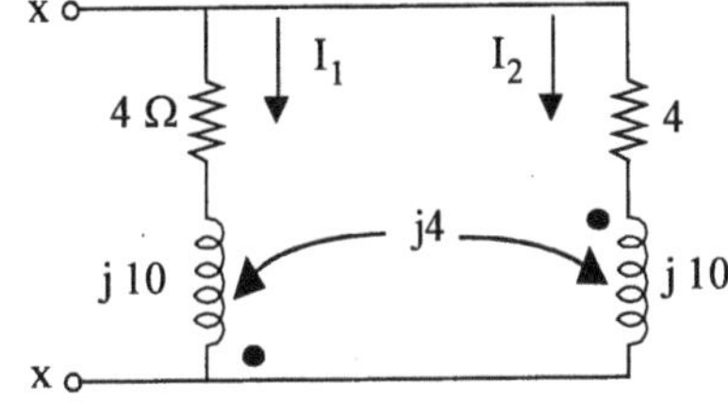

1. $Z_{eq} = 2 + j5$
2. $Z_{eq} = 2 + j3$
3. $I_1 = -I_2$
4. $I_1 = I_2$

Of these statements

(a) 1 alone is true (b) 2 and 4 are correct

(c) 2 and 3 are correct (d) 1 and 4 are correct

3.105 Consider the following statements regarding the circuit shown in the figure. If the power consumed by 5 Ω resistor is 10 W, then

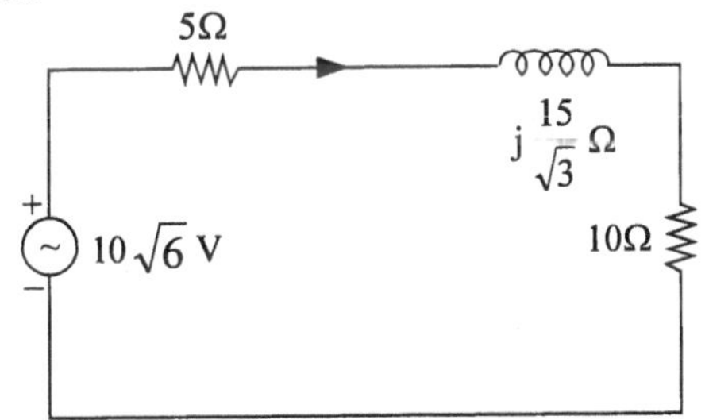

1. $|I| = \sqrt{2}$ A.
2. the total impedance of the circuit is 5 Ω
3. $\cos\phi = 0.866$

Which of these statements are correct ?

(a) 1 and 3

(b) 2 and 3

(c) 1 and 2

(d) 1, 2 and 3

3.106 The equivalent inductance at the terminals a, b in the figure is

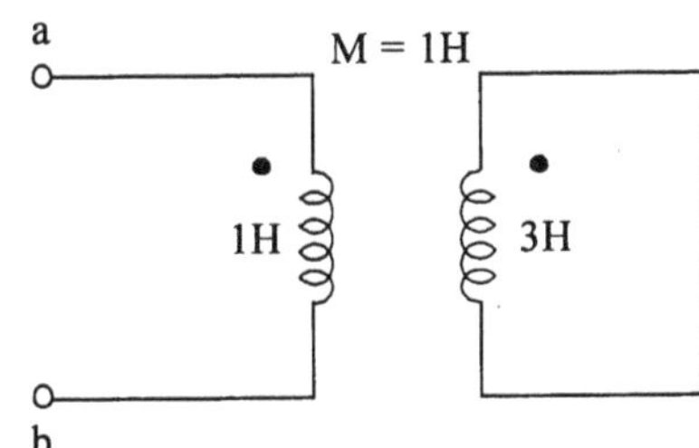

(a) $\frac{3}{2}$ H

(b) $\frac{2}{3}$ H

(c) 2 H

(d) $\frac{1}{3}$ H

3.107 Two current waveforms as shown in the Figure-I and Figure-II, are passed through identical resistors of 1 Ω. The ratio of heat produced in these resistors in a given time by current of Figure-I to Figure-II is

(a) 2 : 1

(b) 1 : 2

(c) 1 : 1

(d) 1 : $\sqrt{2}$

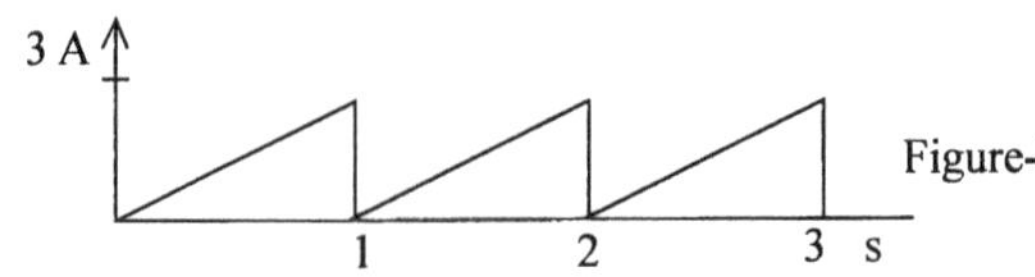

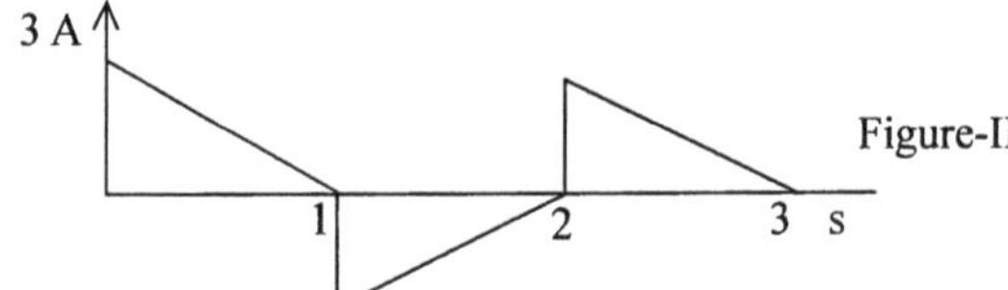

3.108 The current waveform in a pure resistor of 10 Ω is shown in the given figure. Power dissipated in the resistor is

(a) 7.29 W

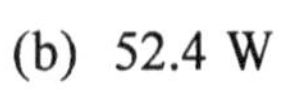
(b) 52.4 W

(c) 135 W

(d) 270 W

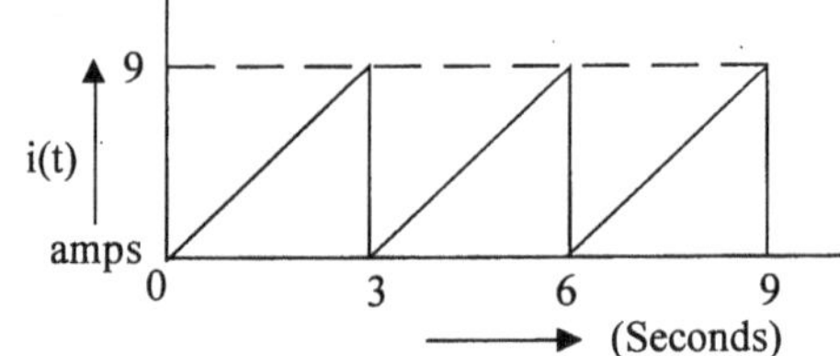

3.109 Current having waveform shown is flowing in a resistance of 10 ohms. The average power is

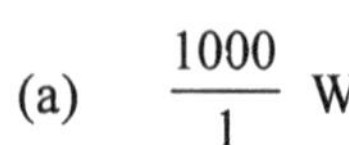
(a) $\frac{1000}{1}$ W (b) $\frac{1000}{2}$ W

(c) $\frac{1000}{3}$ W (d) $\frac{1000}{4}$ W

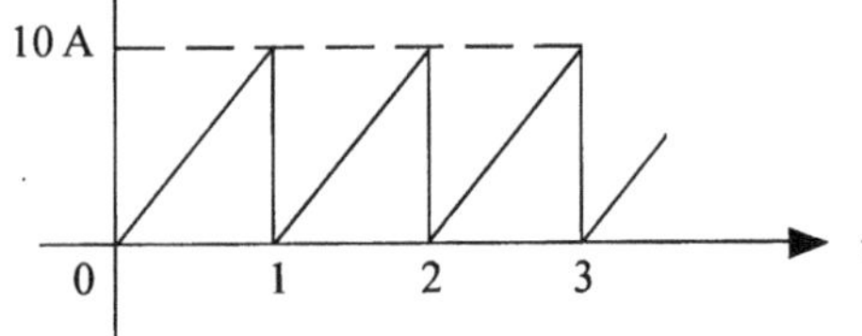

CHAPTER 4 – Three Phase Circuits

4.1 The line-to-line input voltage to the 3 phase, 50 Hz, ac circuit shown in figure is 100 V rms. Assuming that the phase sequence is RYB, the wattmeters would read

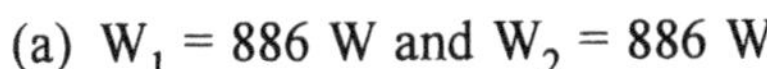
(a) W_1 = 886 W and W_2 = 886 W

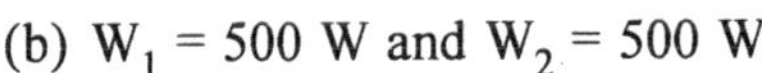
(b) W_1 = 500 W and W_2 = 500 W

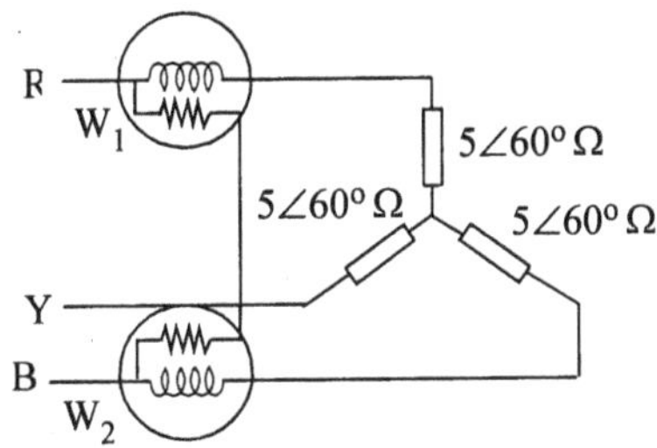

(c) W_1 = 0 W and W_2 = 1000 W

(d) W_1 = 250 W and W_2 = 750 W

4.2 The minimum number of wattmeter(s) required to measure 3-phase, 3-wire balanced or unbalanced power is

(a) 1 (b) 2 (c) 3 (d) 4

4.3 The power delivered to a three-phase load can be measured by the use of 2 wattmeters only when the

(a) load is balanced (b) 3-phase load is connected to the source through 3 wires

(c) load is unbalanced (d) 3-phase load is connected to the source through 4 wires

4.4 A 3-phase star-connected symmetrical load consumes P watts of power from a balanced supply. If the same load is connected in delta to the same supply, the power consumption will be

(a) P (b) $\sqrt{3}P$ (c) 3P (d) not determinable from the given data

4.5 In two-wattmeter method of power measurement, one of the wattmeters will show negative reading when the load pf angle is strictly

(a) less than 30° (b) less than 60°

(c) greater than 30° (d) greater than 60°

4.6 Which of the following are the necessary conditions for an entire three-phase system to be balanced ?

1. The line voltages are equal in magnitude
2. The phase differences between successive line voltages are equal
3. The impedance in each of the phases are identical

Select the correct answer using the codes given below :

Codes :

(a) 1, 2 and 3 (b) 1 and 3 (c) 1 and 2 (d) 2 and 3

4.7 An alternator is delivering power to a balanced load at unity power factor. The phase angle between the line voltage and the line current is

(a) 90° (b) 60° (c) 30° (d) 0°

4.8 E_{oa}, E_{ob} and E_{oc} are three phase voltage while E_{ab}, E_{bc} and E_{ca} are the line voltages of a balanced three-phase system having a-b-c phase sequence. In relation to E_{oc}, E_{bc} would

(a) lag by 30°

(b) lead by 30°

(c) have the same phase

(d) have no definite phase relationship

4.9 A 3-phase, 3-wire supply feeds a load consisting of three equal resistors connected in star. If one of the resistors is open circuited, then the percentage reduction in the load will be

(a) 75 (b) 66.66 (c) 50 (d) 3.33

4.10 Match List-I (Readings obtained while measuring 3-phase power by two-wattmeter method) with List-II (Power factors for the load) and select correct answer using the codes given below the lists :

List-I	List-II
A. Both the wattmeters read equal values of power but of opposite sign	1. Unity
B. Both the wattmeters read equal values of power and both are of positive sign	2. zero
C. One wattmeter reads zero and the other reads the complete power	3. 0.5
	4. 0.866

Codes :

	A	B	C		A	B	C
(a)	2	3	4	(b)	3	1	2
(c)	1	4	3	(d)	2	1	3

4.11 In the measurement of power on balanced load by two-Wattmeter method in a 3-phase circuit, the readings of the Wattmeters are 3 kW and 1 kW respectively, the latter being obtained after reversing the connections of the current coil. The power factor of the load is

(a) 0.277 (b) 0.554 (c) 0.625 (d) 0.866

4.12 When two-Wattmeter method of measurement of power is used to measure power in a balanced three phase circuit, if the Wattmeter reading is zero, then

(a) power consumed in the circuit is zero

(b) power factor of the circuit is zero

(c) power factor is unity

(d) power factor is 0.5

4.13 Which of the following statement is true about two wattmeter method for power measurement in three phase current.

(a) Power can be measured using two wattmeter method only for star connected three phase circuits

(b) When two meters show identical readings, the power factor is 0.5

(c) When power factor is unity, one of the wattmeter reads zero

(d) When the readings of the two wattmeters are equal but of opposite sign, the power factor is zero.

4.14 Consider the following statements :

A 3-phase balanced supply system is connected to a 3-phase unbalanced load. Power supplied to this load can be measured using

1. two wattmeters.
2. one wattmeter.
3. three wattmeters.

Which of these statements is/are correct ?

(a) 1 and 2 (b) 1 and 3 (c) 2 and 3 (d) 3 alone

Chapter 5 – Network Topology

5.1 The graph of an electrical network has N nodes and B branches. The number of links, L, with respect to the choice of a tree, is given by

(a) $B - N + 1$ (b) $B + N$ (c) $N - B + 1$ (d) $N - 2B - 1$

5.2 A connected network of $N > 2$ nodes has at most one branch directly connecting any pair of nodes. The graph of the network.

(a) must have at least N branches for one or more closed paths to exist

(b) can have an unlimited number of branches

(c) can only have at most N branches

(d) can have a minimum number of branches not decided by N

5.3 Match List-I (Loop concept) with List-II (Junction concept) and select the correct answer using the codes given below the List :

	List-I		List-II
A.	Mesh	1.	Number of nodes
B.	Outside mesh	2.	node voltage
C.	Mesh current	3.	Reference node
D.	Number of meshes	4.	node

Codes :

	A	B	C	D
(a)	3	4	1	2
(b)	3	4	2	1
(c)	4	3	2	1
(d)	4	3	1	2

5.4 If the number of branches in a network is 'B', the number of nodes is 'N' and the number of independant loops is 'L', then the number of independent node equations will be

(a) $N + L - 1$ (b) $B - 1$ (c) $B - N$ (d) $N - 1$

5.5 An electric circuit with 10 branches and 7 nodes will have

(a) 3 loop equations (b) 4 loop equations

(c) 7 loop equations (d) 10 loop equations

5.6 A connected planar network has 4 nodes and 5 elements. The number of meshes in its dual network is

(a) 4 (b) 3 (c) 2 (d) 1

5.7 The number of branches in a tree for a network of n nodes and b branches is

(a) $b - n + 1$ (b) $n - 1$

(c) $b + n - 1$ (c) n

5.8 A network has 10 nodes and 17 branches in all. The number of independent node pair voltages would be

(a) 7 (b) 9 (c) 10 (d) 45

5.9 Which of the following oriented graphs have the same fundamental loop matrix ? Select the correct answer using the codes given below :

1. 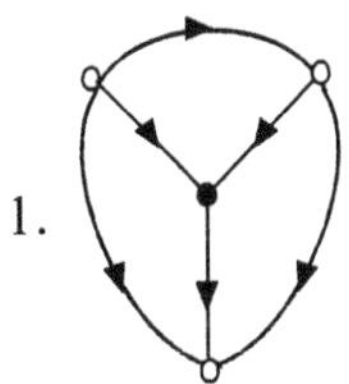

2.

3. 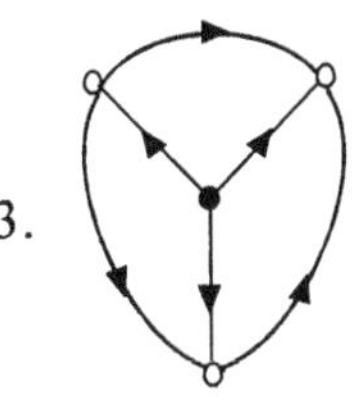

4. 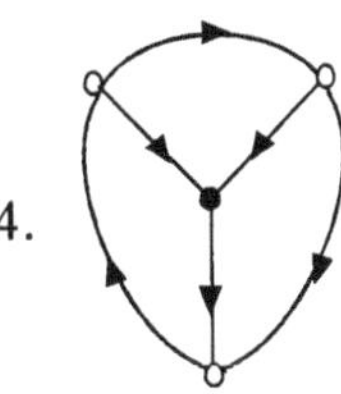

Codes :

(a) 1 and 2 (b) 2 and 3 (c) 1, 3 and 4 (d) 2, 3 and 4

5.10 Consider the following statements :

1. One and only one path exists between any pair of vertices of a tree.
2. The number of fundamental cut sets are the same as the rank of the graph.
3. The cut set is a minimal set of edges removal of which from the graph reduces the rank of the graph by one.
4. The rank of a graph is equal to the number of vertices of the graph.

Of these statements

(a) 2 and 4 are correct (b) 1, 2 and 3 are correct

(c) 2 and 3 are correct (d) 1 and 4 are correct.

5.11 Which one of the following represents the total number of trees in the graph given in figure ?

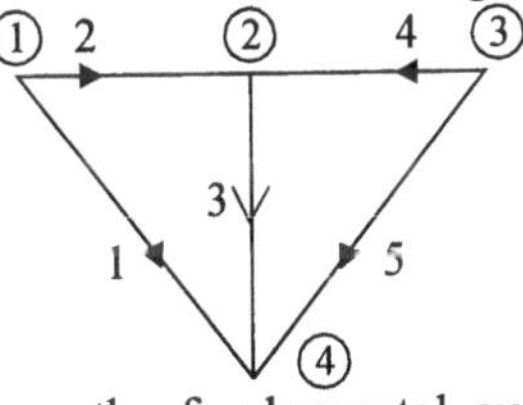

(a) 4 (b) 6

(c) 5 (d) 8

5.12 In the graph and the tree shown in the given figure, the fundamental cut set for the branch 2 is

(a) 2, 1, 5 (b) 2, 6, 8, 7

(c) 2, 1, 3, 4, 5 (d) 2, 3, 4

5.13 Match List I with List II with reference of the graph shown in the given figure and its particular tree of a circuit and select the correct answer using the codes given below the lists :

	List I (Branches)		**List II**
A.	1, 2, 3, 4	1.	Twigs
B.	4, 5, 6, 7	2.	Links
C.	1, 2, 3, 8	3.	Fundamental cutset
D.	1, 4, 5, 6, 7	4.	Fundament loop

Codes :

	A	B	C	D
(a)	3	1	2	4
(b)	2	3	1	4
(c)	3	2	4	1
(d)	1	4	3	2

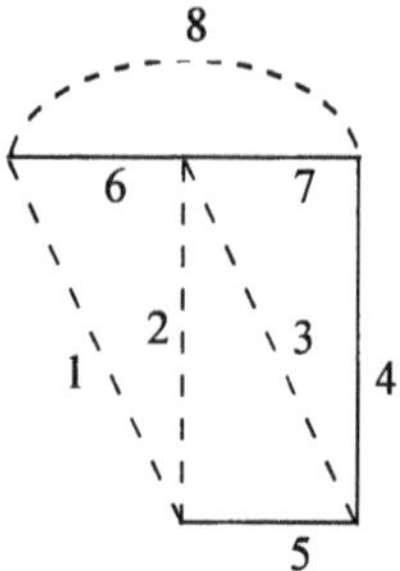

5.14 In graph shown in the figure, for the tree with branches b, d and f, the fundamental loops wound include

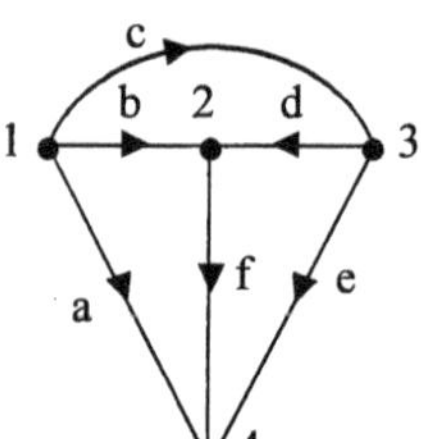

(a) abc, def, bdea

(b) cea, bdea, abc

(c) cdb, def, bfa

(d) abde, def, cdb

5.15 For the graph shown in the given figure, the incidence matrix A is given by

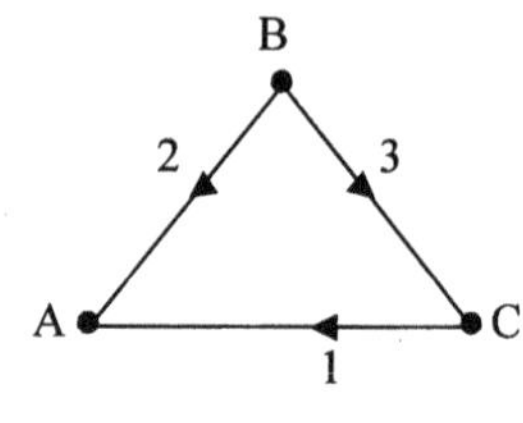

(a) $\begin{bmatrix} -1 & -1 & 0 \\ 0 & 1 & 1 \\ -1 & 0 & -1 \end{bmatrix}$ (b) $\begin{bmatrix} 1 & 0 & -1 \\ 1 & 1 & 0 \\ 0 & -1 & 1 \end{bmatrix}$

(c) $\begin{bmatrix} -1 & -1 & 0 \\ 0 & 1 & 1 \\ 1 & 0 & -1 \end{bmatrix}$ (d) $\begin{bmatrix} 1 & 0 & 1 \\ -1 & 1 & 0 \\ 0 & -1 & 1 \end{bmatrix}$

5.16 The graph of a network has six branches with three tree branches. The MINIMUM number of equations required for the solution of the network is

(a) 2 (b) 3 (c) 4 (d) 5

5.17 In the graph shown in the figure, one possible tree is formed by the branches 4, 5, 6, 7. Then one possible fundamental cut set is

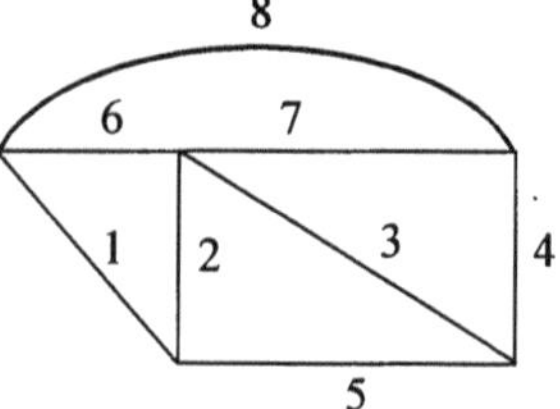

(a) 1, 2, 3, 8

(b) 1, 2, 5, 6

(c) 1, 5, 6, 8

(d) 1, 2, 3, 7, 8

5.18 For a connected planner graph of 'v' vertices and 'e' edges, the number of meshes is

(a) e – v (b) v – 1 (c) e – v – 1 (d) e – v +1

5.19 For the graph shown in the given figure one set of fundamental cut-set would be

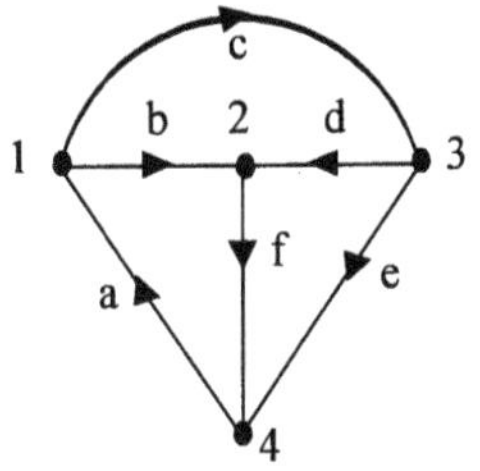

(a) abc, cde, afe

(b) afdc, cde, abde

(c) cdfe, afe, bdf

(d) cbd, abde, cde

5.20 Consider the graph and tree (dotted) of the given figure. The fundamental loops include the set of lines

(a) (1, 5, 3), (5, 4, 2) and (3, 4, 6)

(b) (1, 2, 4, 3), (1, 2, 6), (3, 4, 6) and (1, 5, 4, 6)

(c) (1, 5, 3), (5, 4, 2), (3, 4, 6) and (2, 4, 3, 1)

(d) (1, 2, 4, 3) and (3, 4, 6)

5.21

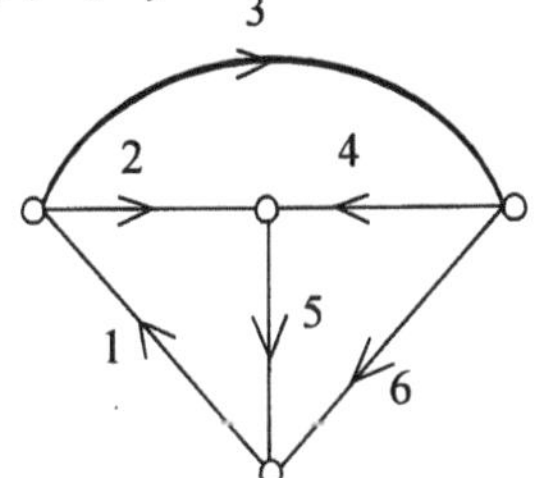

Which one of the following is a cut set of the graph shown in the above figure ?

(a) 1, 2, 3 and 4 (b) 2, 3, 4 and 6

(c) 1, 4, 5 and 6 (d) 1, 3, 4 and 5

5.22 Match List X with List Y for the tree branches 1, 2, 3 and 8 of the graph shown in the given figure and select the correct answer using the codes given below :

List X	List Y
A. Fundamental circuit	1. 3, 5, 6, 7, 9
B. Fundamental cut set	2.1, 3, 8, 7
C. Rank of circuit matrix	3. 3, 2, 4, 6, 7
D. Rank of incidence matrix	4. 5
	5. 4
	6. 3

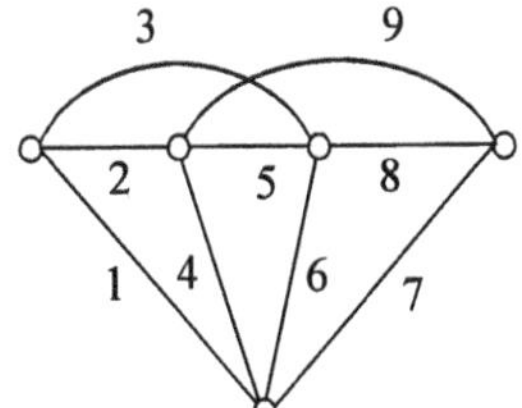

Codes :

	A	B	C	D
(a)	2	1	5	6
(b)	2	1	4	5
(c)	3	2	4	5
(d)	3	1	4	6

Chapter 6 – Network Theorems

6.1 The circuit shown in figure is equivalent to a load of

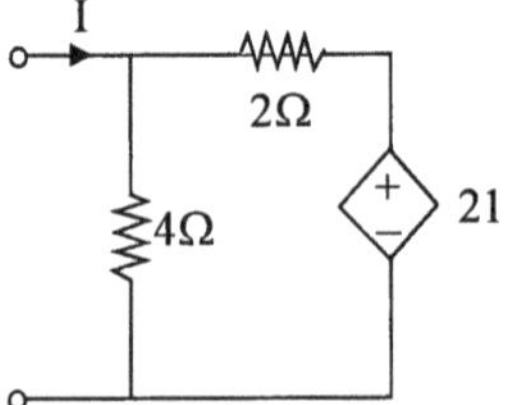

(a) $\frac{4}{3}$ ohms

(b) $\frac{8}{3}$ ohms

(c) 4 ohms

(d) 2 ohms

6.2 If R_s in the circuit shown in the given figure is variable between 20 Ω and 80 Ω then the maximum power transferred to load R_L will be

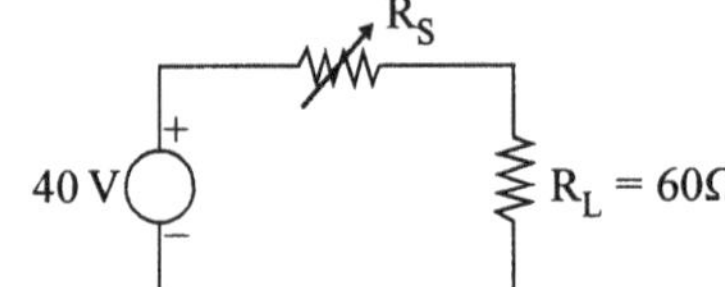

(a) 15 W

(b) 13.33 W

(c) 6.67 W

(d) 2.4 W

6.3 A network is composed of two subnetworks N_1 and N_2 as shown in the given figure.
If the subnetwork N_1 contains only linear, bilateral, time invariant elements, then it can be replaced by its Thevenin equivalent even if the subnetwork N_2 contains

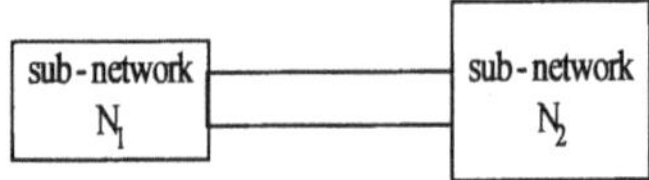

(a) a two-terminal element which is non-linear

(b) a non-linear inductance mutually coupled to an element in N_1

(c) an element which is linear, but mutually coupled to some element in N_1

(d) a dependent source, the value of which depends upon the voltage or current in some element in N_1

6.4 Under conditions of maximum power transfer from as ac source to a variable load

(a) the load impedance must also be inductive, if the generator impedance is inductive

(b) the sum of the source and load impedances is zero

(c) the sum of the source reactance and load reactance is zero

(d) the load impedance has the same phase angle as the generator impedance

6.5 In the delta equivalent of the given star-connected circuit $\overline{Z}_{QR}$ is equal to

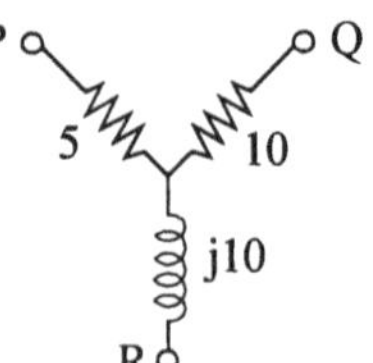

(a)  40 Ω (b) (20 + j 10) Ω

(c) $5 + j\left(\frac{10}{3}\right)$ Ω (d) (10 + j 30) Ω

6.6 Consider the two circuits I and II shown in the following figures :

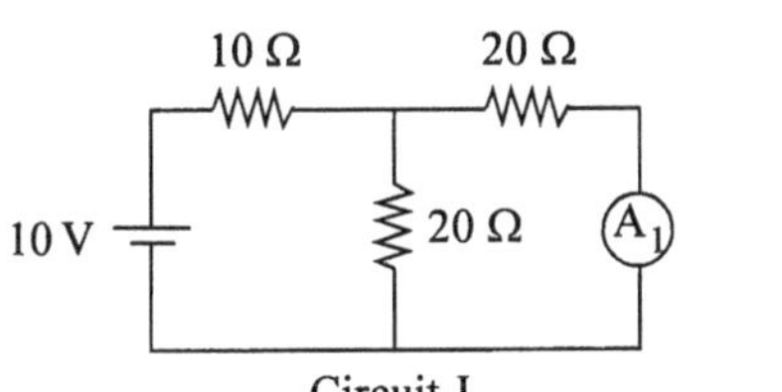

Circuit-I

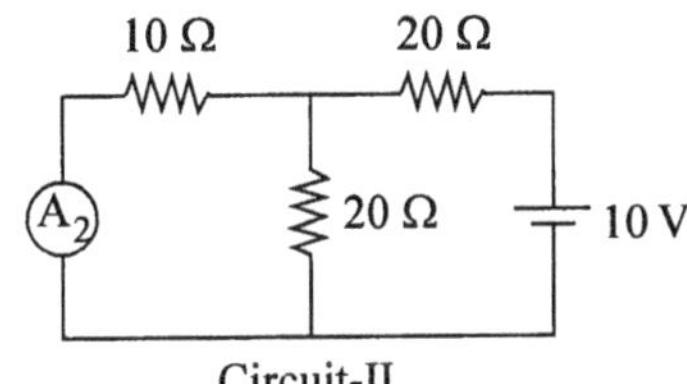

Circuit-II

Which one of the following statements regarding the current flowing through the ammeters A_1 and A_2 is correct ?

(a) The currents in A_1 and A_2 are of the same value and equal 0.25 A
(b) The currents in A_1 and A_2 are respectively 0.25 A and 2.5 A
(c) The current in both the ammeters is of the same value and equals 2.5 A
(d) The currents in A_1 and A_2 are respectively 2.5 A and 0.25 A

6.7 For the circuit shown in the given figure, the Thevenin equivalent impedance across terminals CD is given by

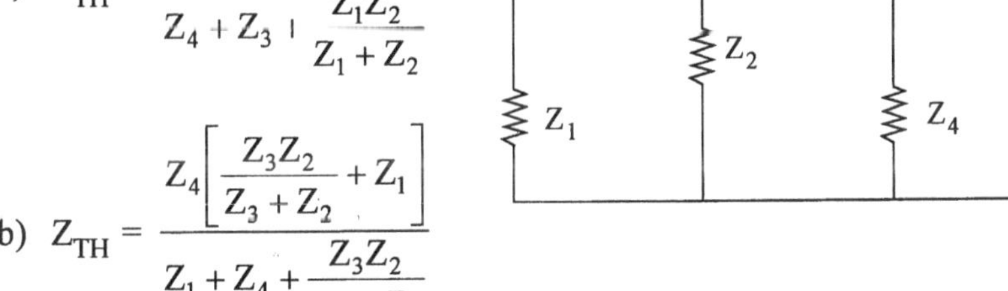

(a) $Z_{TH} = \dfrac{Z_4\left[\dfrac{Z_1Z_2}{Z_1+Z_2}+Z_3\right]}{Z_4+Z_3+\dfrac{Z_1Z_2}{Z_1+Z_2}}$

(b) $Z_{TH} = \dfrac{Z_4\left[\dfrac{Z_3Z_2}{Z_3+Z_2}+Z_1\right]}{Z_1+Z_4+\dfrac{Z_3Z_2}{Z_3+Z_2}}$

(c) $Z_{TH} = \dfrac{Z_1+Z_2+Z_3Z_2}{Z_1+Z_2+Z_4}$ (d) $Z_{TH} = \dfrac{Z_3+(Z_1+Z_2Z_4)}{Z_1+Z_2+Z_3+Z_4}$

6.8 A sinusoidal voltage source $50\angle 0^\circ$ with an internal impedance $10+j20$ is connected to a load (z_1) which is variable in both resistance and reactance. Match List I with List II for maximum power transfer and select the correct answer using the codes given below the lists

	List I		**List II**
A.	Load impedance	1.	62.5
B.	Total impedance	2.	$10+j20$
C.	Current	3.	2.5
D.	Maximum power	4.	20
		5.	$10-j20$
		6.	$2.5-j2.5$

Codes :

(a)	A	B	C	D	(b)	A	B	C	D
	5	2	3	4		5	4	3	1
(c)	A	B	C	D	(d)	A	B	C	D
	2	4	6	1		2	5	6	4

6.9 An impedance match is desired at the 1–1' port of the two-port network shown in the given figure. The match will be obtained when Z_g equals

(a) $Z_1 + Z_3$ (b) $Z_1 + \frac{Z_3(Z_2 + Z_L)}{Z_2 + Z_3 + Z_L}$

(c) $\frac{Z_1(Z_2 + Z_3)}{Z_1 + Z_2 + Z_3}$ (d) Z_1

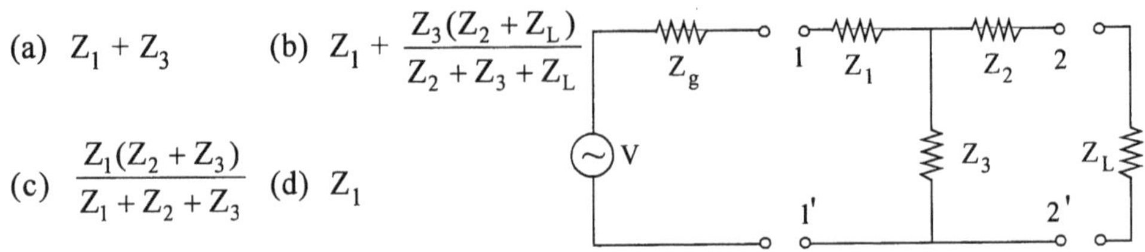

6.10 From a sinusoidal voltage source V_s of impedance $Z_s = R_s + jX_s$, power is drawn by a load $Z_L = R_L + jX_L$. The condition for maximum power in Z_L is given in List-II for the constraints shown in List-I. Match List-I with List II and select the correct answer using the codes given below the Lists :

List - I	**List - II**
A. X_s = zero	1. $Z^2_L = R^2_s + X^2_s$
B. X_L = zero	2. $Z_L = R_L - j X_s$
C. R_L fixed	3. $Z_L = R_s - jX_s$
D. X_L fixed	4. $Z_L = R_s$
	5. $Z_L = [R_s^2 + (Xs + X_L)^2]^{1/2} + jX_L$

Codes :

(a)	A	B	C	D	(b)	A	B	C	D
	4	1	2	3		2	3	4	5
(c)	A	B	C	D	(d)	A	B	C	D
	4	1	2	5		1	3	4	5

6.11 For the network shown in the figure, if $V_s = V_1$ and V = 0, then I = –5 A and if V_s = 0, and V = 1, then I = 1/2 A. The values of I_{sc} and R_i of the Norton's equivalent across AB would be respectivley

(a) – 5A and 2 Ω

(b) 10 A and 0.5 Ω

(c) 5 A and 2 Ω

(d) 2.5 A and 5 Ω

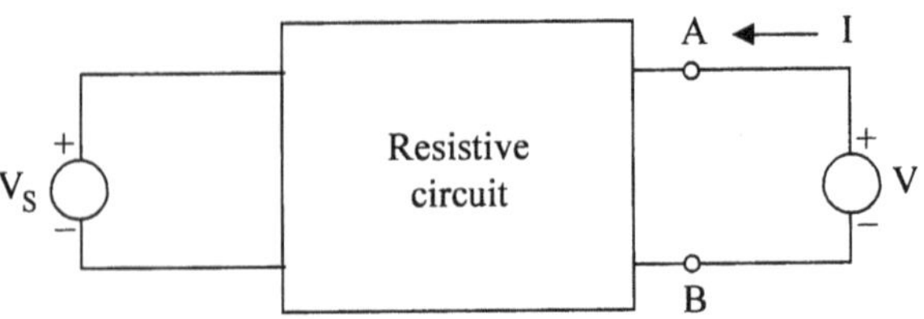

6.12 The Thevenin equivalent of a network is as shown in the given figure. For maximum power transfer to the variable and purely resistive load R_L, its resistance should be

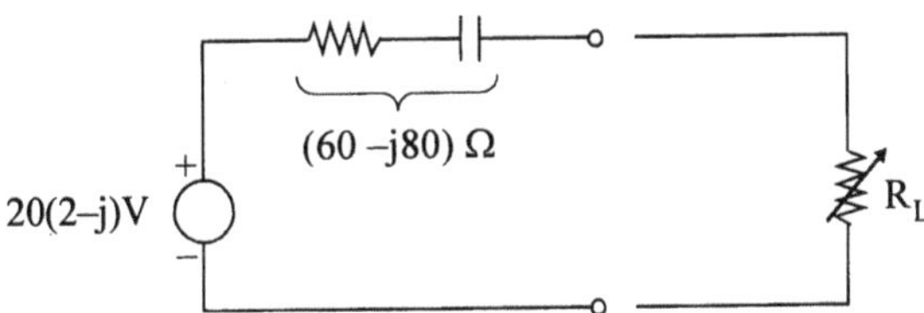

(a) 60 Ω
(b) 80 Ω
(c) 100 Ω
(d) infinity

6.13 In a balanced Wheatstone bridge, if the positions of detector and source are interchanged, the bridge will still remain balanced. This inference can be drawn from

(a) reciprocity theorem (b) duality principle
(c) compensation theorem (d) equivalence theorem

6.14 In the network shown in the given figure, the Thevenin source and the impedance across terminals A-B will be respectively

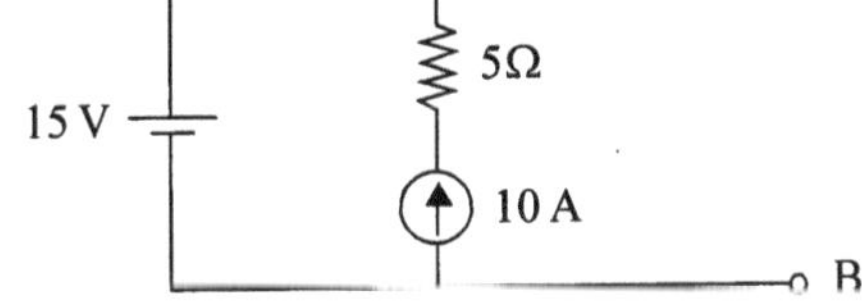

(a) 15 V and 13.33 Ω
(b) 50 V and 15 Ω
(c) 115 V and 20 Ω
(d) 100 V and 25 Ω

6.15 If a network has all linear elements except for a few non-linear ones, then superposition theorem

(a) cannot hold at all
(b) always holds
(c) may hold on careful selection of element values, source waveform and response
(d) holds in case of direct current excitations

6.16 The circuit shown in Figure-I is replaced by its Norton's equivalent in Figure-II. The value of I' will be

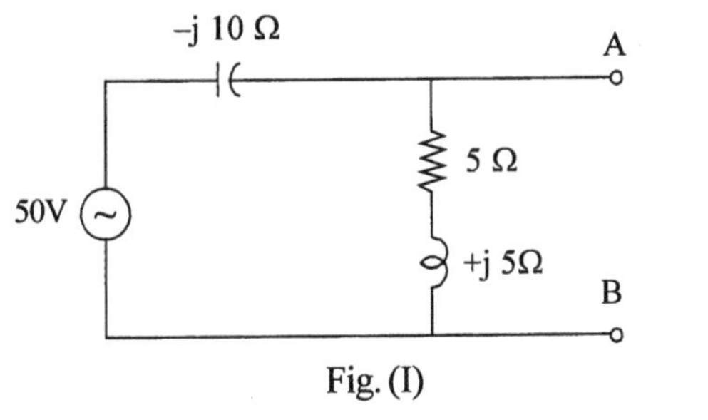

Fig. (I)

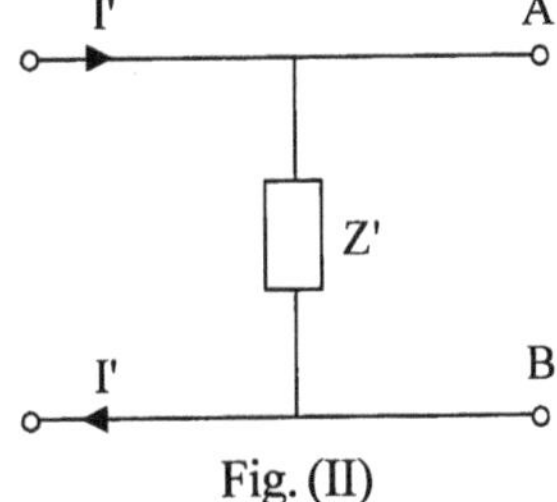

Fig. (II)

(a) 2.5 ∠45° A (b) 5 ∠90° A (c) 10 ∠– 90° A (d) 15 ∠–45° A

6.17 A certain network N feeds a load resistance R as shown in Figure-I. It consumes a power of 'P' W. If an identical network is added as shown in Figure-II, the power consumed by R will be

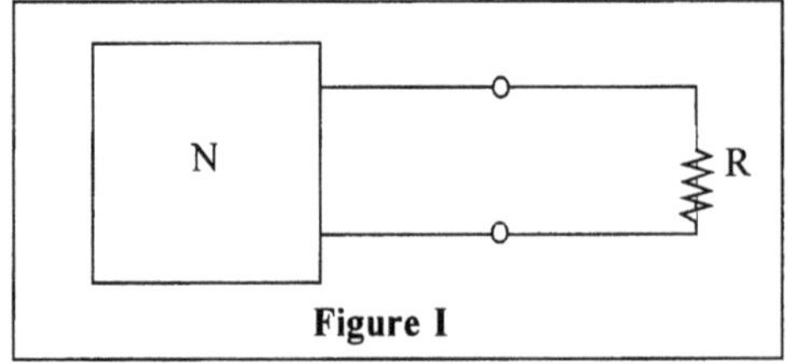

Figure I

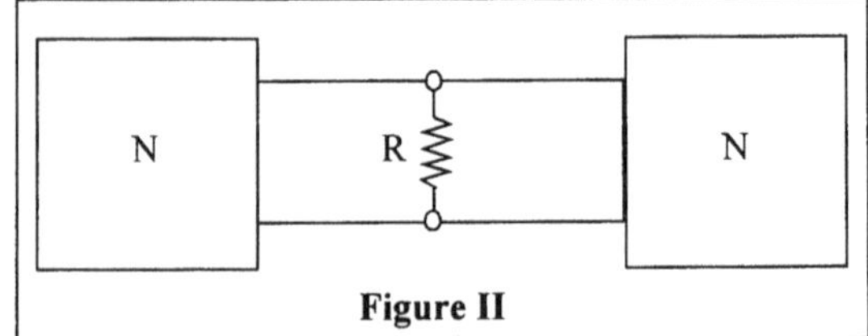

Figure II

(a) less than P (b) equal to P (c) between P and 4P (d) more than 4P

6.18 For the circuit shown in the given figure, when the voltage E is 10 V, the current i is 1 A. If the applied voltage across terminal C-D is 100 V, the short circuit current flowing through the terminal A-B will be

(a) 0.1 A

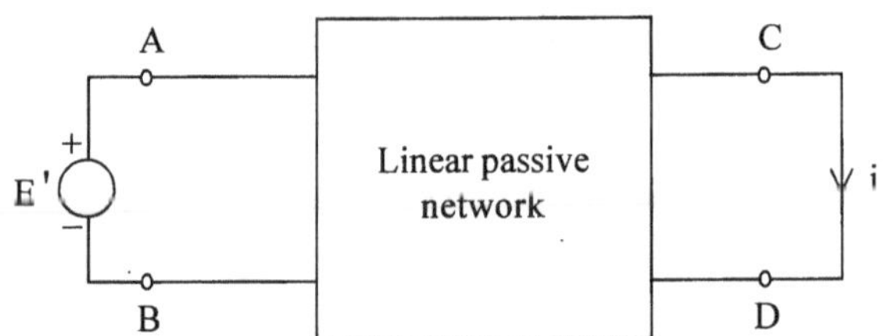

(b) 1 A

(c) 10 A

(d) 100 A

6.19 The Thevenin's equivalent resistance R_{th} for the given network is

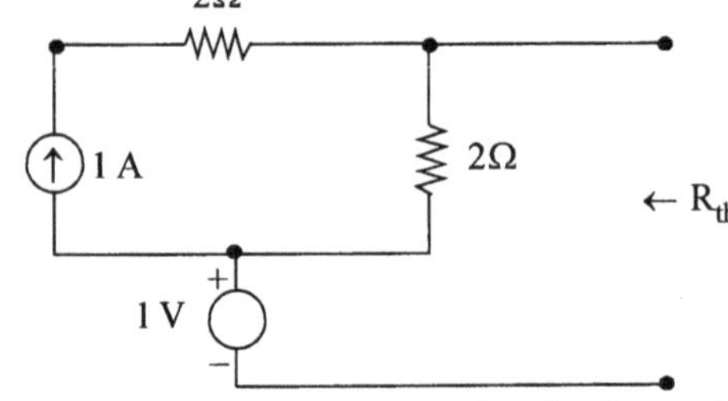

(a) 1 Ω

(b) 2 Ω

(c) 4 Ω

(d) infinity

6.20 The Norton's equivalent of circuit shown in Figure I is drawn in the circuit shown in Figure II. The values of I_{SC} and R_{eq} in Figure II are respectivley.

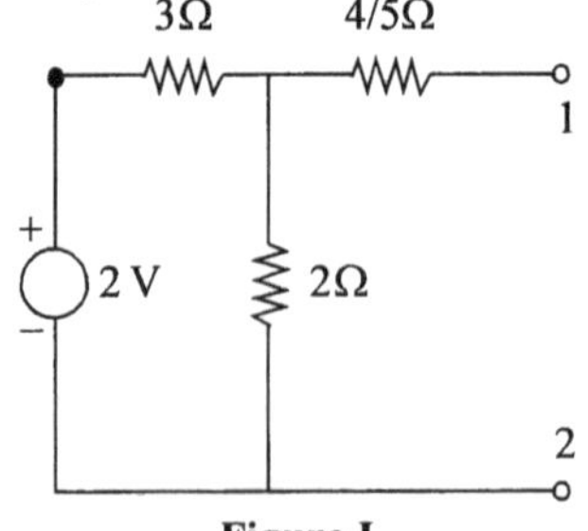

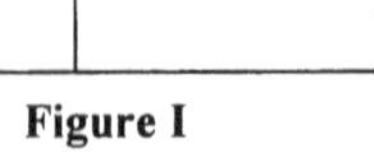
Figure I

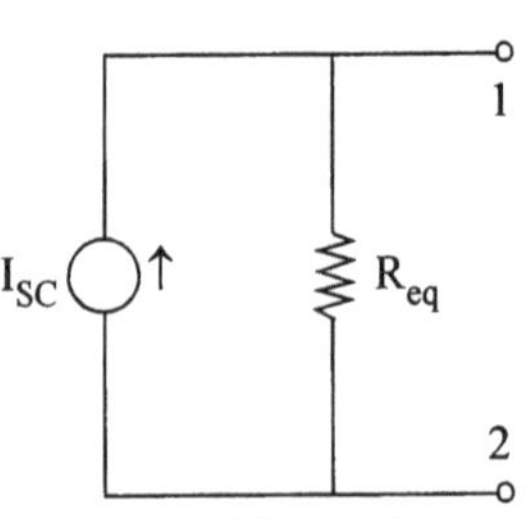

Figure II

(a) $\frac{5}{2}$ A and 2 Ω

(b) $\frac{2}{5}$ A and 1 Ω

(c) $\frac{4}{5}$ A and $\frac{12}{5}$ Ω

(d) $\frac{2}{5}$ A and 2 Ω

6.21 The resistance seen from the terminals A and B of the device whose characteristic is shown in the figure is

(a) $-5\ \Omega$

(b) $-\frac{1}{5}\ \Omega$

(c) $\frac{1}{5}\ \Omega$

(d) $5\ \Omega$

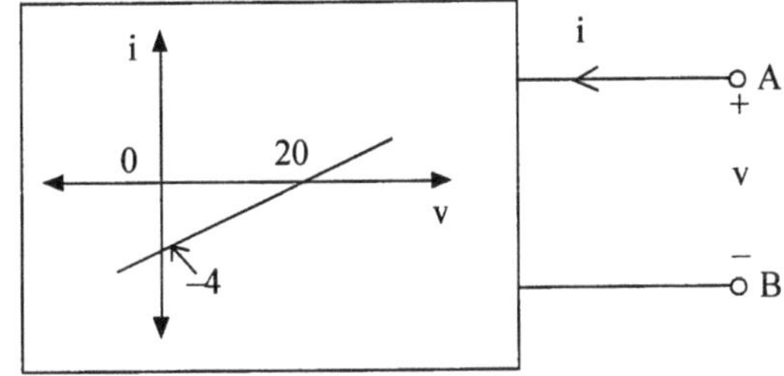

6.22 If the combined generator and line impedance is (5 + j10) ohm, then for the maximum power transfer to a load impedance from a generator of constant generated voltage, the load impedance is given by which one of the following

(a) (5 + j10) Ω (b) (5 – j10) Ω (c) (5 + j5) Ω (d) 5 Ω

6.23 Superposition theorem is not applicable for

(a) voltage calculations (b) bilateral elements

(c) power calculations (d) passive elements

6.24 In the lattice network, find the value of R for the maximum power transfer to the load.

(a) 5 Ω

(b) 6.5 Ω

(c) 8 Ω

(d) 9 Ω

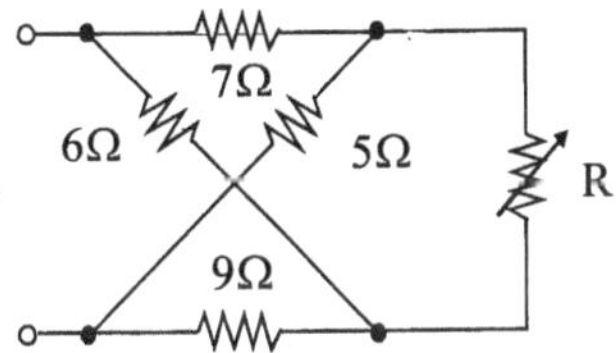

6.25 The value of R_x so that power dissipated in is maximum

(a) 33.4 K

(b) 17.6 K

(c) 10 K

(d) 5 K

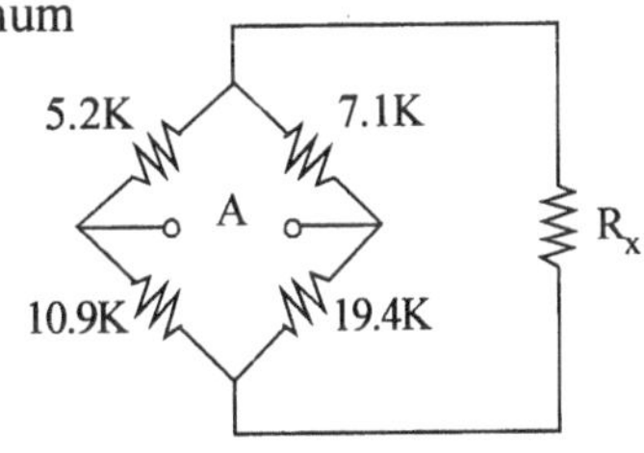

6.26 An ac source of voltage E_s and an internal impedance of $Z_s = (R_s + jX_s)$ is connected to a load of impedance $Z_L = (R_L + jX_L)$. Consider the following conditions in this regard :

1. $X_L = X_s$, if only X_L is varied.
2. $X_L = -X_s$, if only X_L is varied.
3. $R_L = \sqrt{R_s^{\,2} + (X_s + X_L)^2}$, if only R_L is varied
4. $|Z_L| = |Z_s|$, if the magnitude alone of Z_L is varied, keeping the phase angle fixed.

Among these conditions, those which are to be satisfied for maximum power transfer from the source to the load would include

(a) 2 and 3 (b) 1 and 3 (c) 1, 2 and 4 (d) 2, 3 and 4

6.27 The equivalent circuit of the following circuit is :

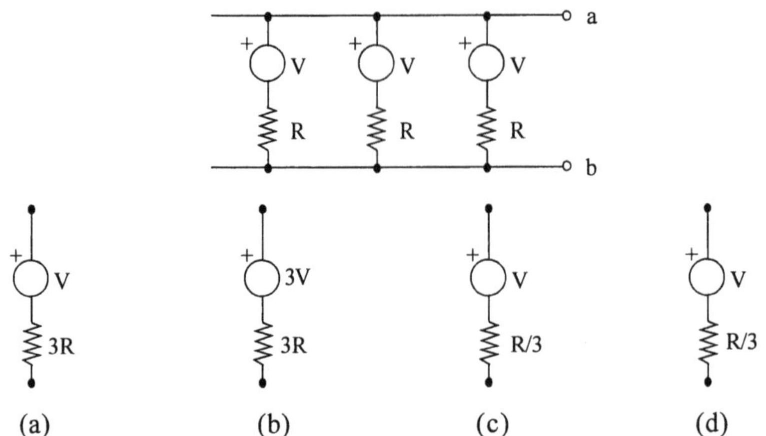

6.28 In a linear network, the ratio of voltage excitation to current response is unaltered when the position of excitation and response are interchanged. This assertion stems from the

(a) principle of duality (b) reciprocity theorem
(c) principle of superposition (d) equivalence theorem

6.29 If all the elements in a particular network are linear, then the superposition theorem would hold, when the excitation is

(a) dc only (b) ac only (c) either ac or dc (d) an impulse

6.30 The circuit shown in the figure, the voltage across the 2 ohm resistor is

(a) 6 V

(b) 4 V

(c) 2 V

(d) zero

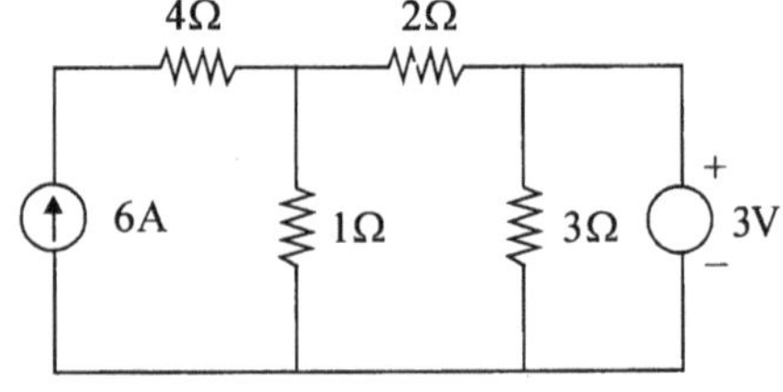

6.31 Four networks are shown below in figures (1), (2), (3) and (4) below :

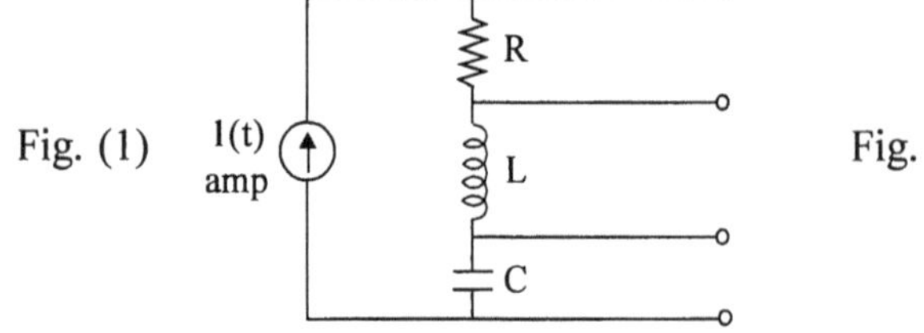

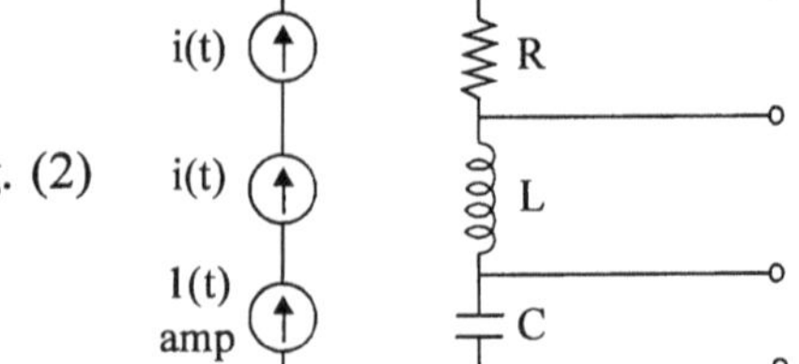

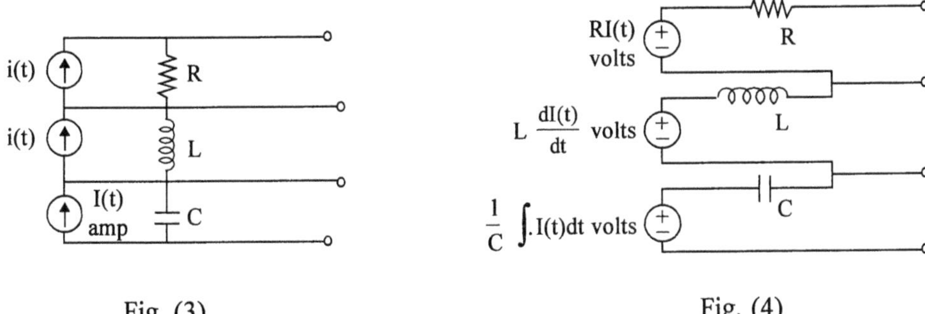

Fig. (3) Fig. (4)

Of these networks,

(a) all the four networks are equivalent
(b) no two networks are equivalent
(c) networks shown in Figures (2), (3) and (4) are equivalent.
(d) networks shown in Figures (3) and (4) are equivalent.

6.32 A voltage source with an internal resistance R_S, supplies power to a load R_L. The power delivered to the load varies with R_L as

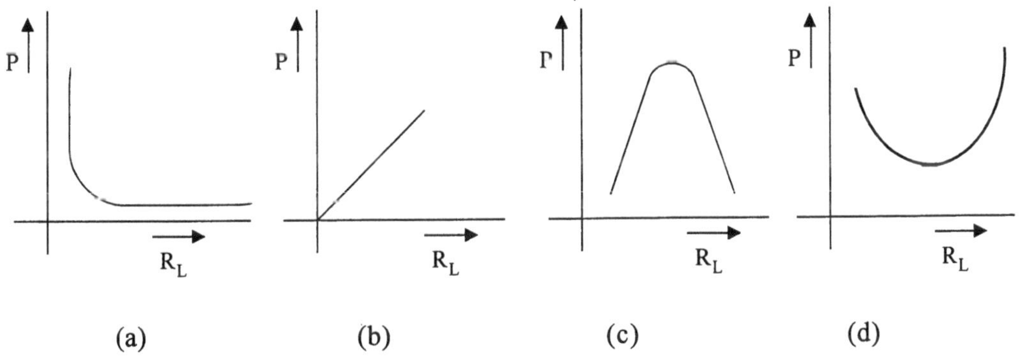

(a) (b) (c) (d)

6.33 Consider the following statements :
The transfer impedances and admittances of a network remain constant when the position of excitation and response are intercharged if the network

1. is linear
2. consists of bilateral elements
3. has high impedance or admittance as the case may be.
4. is resonant.

Of these statements :

(a) 1 and 2 are correct (b) 1, 3 and 4 are correct
(c) 2 and 4 are correct (d) 1, 2, 3 and 4 are correct

6.34 The Thevenin impedance across the terminals AB of the given network is

(a) $\frac{10}{3}\ \Omega$

(b) $\frac{20}{9}\ \Omega$

(c) $\frac{13}{4}\ \Omega$

(d) $\frac{11}{5}\ \Omega$

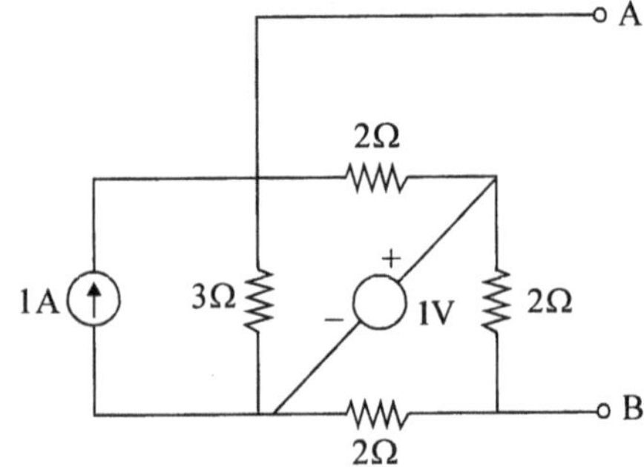

6.35 The value of R which will enable the circuit to deliver maximum power to the terminal a and b in the following circuit diagram is

(a) $\frac{5}{9}$ ohms

(b) $\frac{5}{8}$ ohms

(c) $\frac{5}{6}$ ohms

(d) $\frac{5}{3}$ ohms

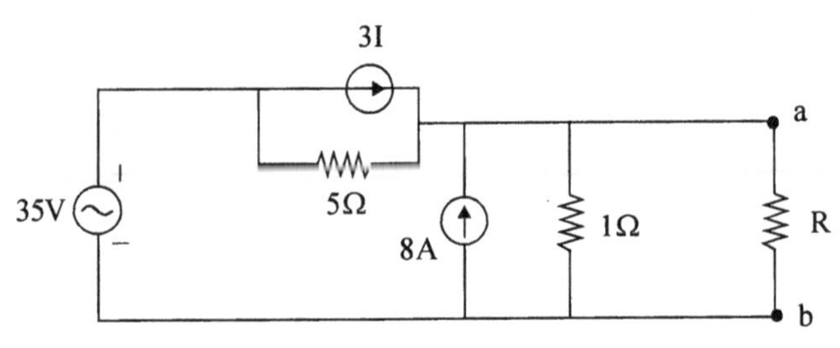

6.36 The Norton's equivalent of the circuit shown is :

(a) $5\sqrt{2}\ \angle 45,\ 1+j1\Omega$

(b) $5\sqrt{2}\ \angle{-45},\ 1-j1\Omega$

(c) $\frac{10}{\sqrt{2}}\ \angle 45,\ 1\ \Omega$

(d) $\frac{10}{\sqrt{2}}\ \angle{-45},\ 1\ \Omega$

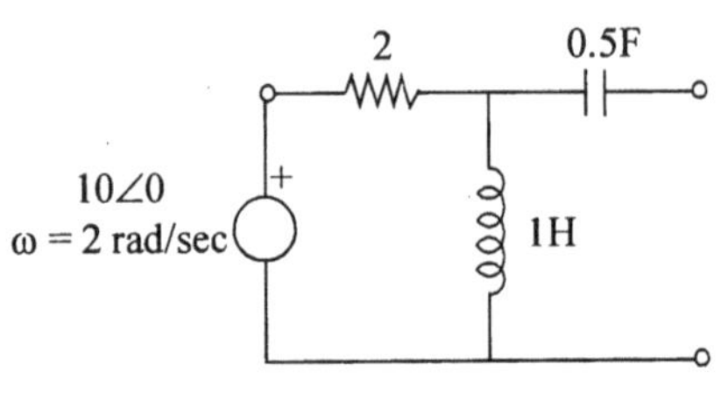

6.37 In the figure shown, if we connect a source of 2 V, with internal resistance of 1 Ω at A'A, with positive terminal at A', then the current through R will be

(a) 2 A

(b) 1.66 A

(c) 1 A

(d) 0.625 A

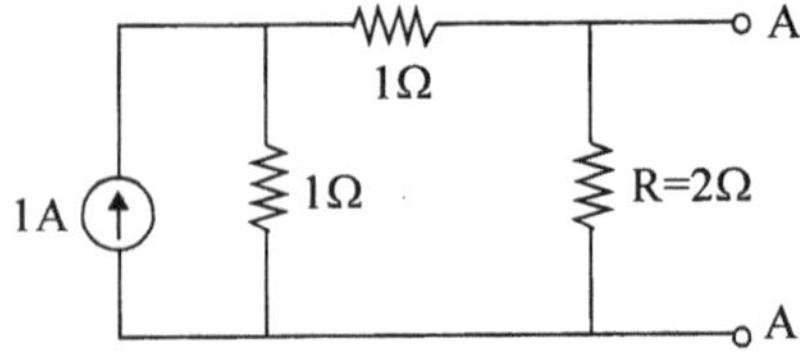

6.38 A 24 V battery of internal resistance r = 4 ohm is connected to a variable resistance R. The rate of heat dissipation in the resistor is maximum when the current drawn from the battery is I. Current drawn from the battery will be $\frac{I}{2}$, when R is equal to

(a) 8 ohms (b) 12 ohms (c) 16 ohms (d) 20 ohms

6.39 If the networks shown in figures I and II are equivalent at terminals A-B, then the values of V (in volts) and Z (in ohms), will be

	V	Z
(a)	100	12
(b)	60	12
(c)	100	30
(d)	60	30

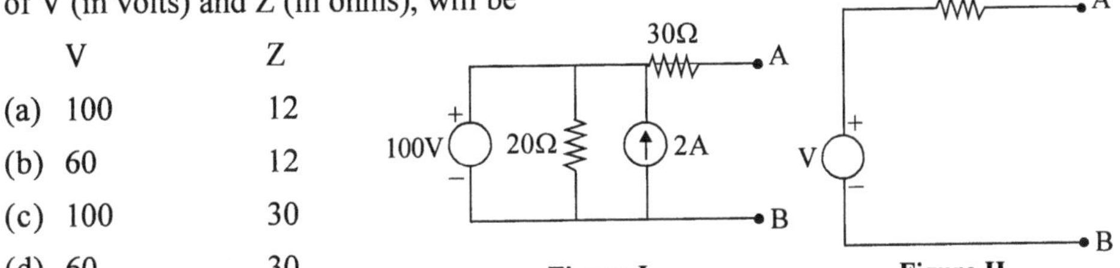

Figure I **Figure II**

6.40 Match list I with List II and select the correct answer using the codes given below the lists :

List I	List II
(Network Theorems)	(Most distinguished property of networks)
A. Reciprocity	1. Impedance matching
B. Tellegen's	2. Passive
C. Superposition	3. $\sum_{k=0}^{b} V_{jk}(t_1)\, i_{jk}(t_2) = 0$
D. Maximum power transfer	4. Linear
	5. Non linear

Codes :

	A	B	C	D
(a)	1	2	3	4
(b)	1	2	3	5
(c)	2	3	4	1
(d)	2	3	5	1

6.41 For the circuit shown in the given figure, the voltage V_{AB} is

(a) 6 V

(b) 10 V

(c) 25 V

(d) 40 V

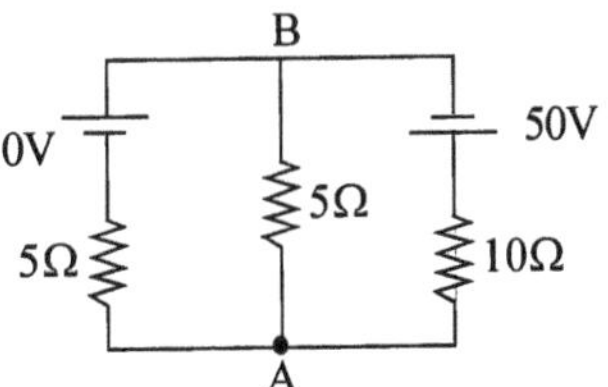

6.42 Which one of the following circuits is the delta equivalent of the star circuit given in the figure I

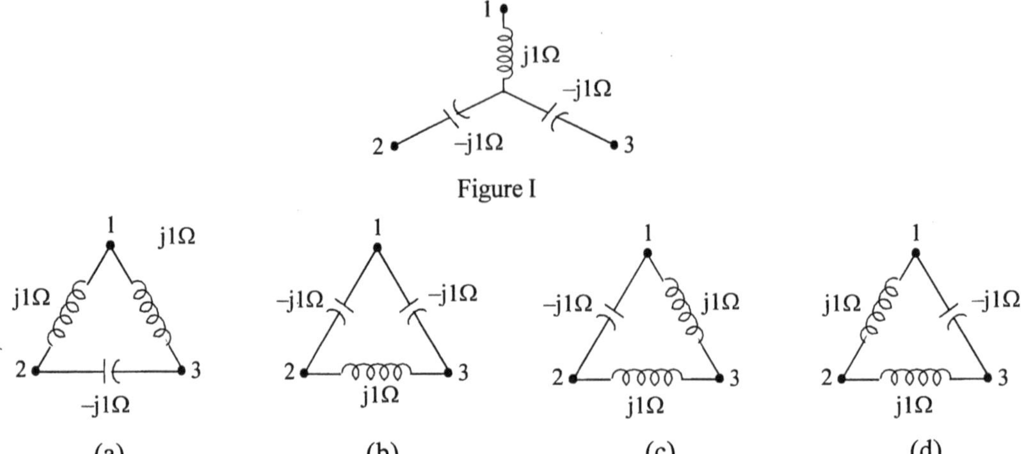

6.43 Which one of the following impedance values of load will cause maximum power to be transferred to the load for the network shown in the given figure ?

(a) $(2 + j2)\ \Omega$

(b) $(2 - j2)\ \Omega$

(c) $-j2\ \Omega$

(d) $2\ \Omega$

2Ω j2Ω j2Ω
V_S –j2 Z_L

6.44 A delta connection contains three equal impedances of 60 ohms. The impedances of the equivalent star connection will be

(a) 15 Ω each (b) 20 Ω each (c) 30 Ω each (d) 40 Ω each

6.45 The network shown in the given figure current i = 0 when E = 4 V, I = 2 A and i = 1 A when E = 8 V, I = 2 A. The Thevenin voltage and the resistance looking into the terminals AB are

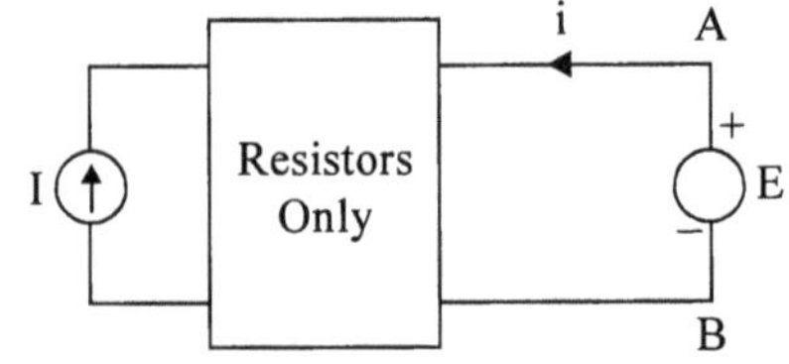

(a) 4 V, 2 Ω

(b) 4 V, 4 Ω

(c) 8 V, 2 Ω

(d) 8 V, 4 Ω

6.46 The Thevenins equivalent resistance of the circuit at the terminals A, B is

(a) $\frac{5}{3}\ \Omega$

(b) $\frac{2}{3}\ \Omega$

(c) 2 Ω

(d) 3 Ω

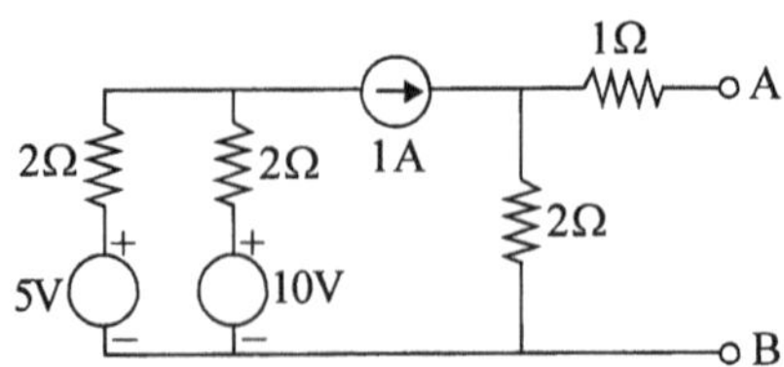

6.47 The Thevenin equivalent of the network shown in figure I is 10 V is series with a resistance of 2Ω. If now, a resistance of 3 Ω is connected across AB as shown in figure II, the Thevenin equivalent of the modified network across AB will be :

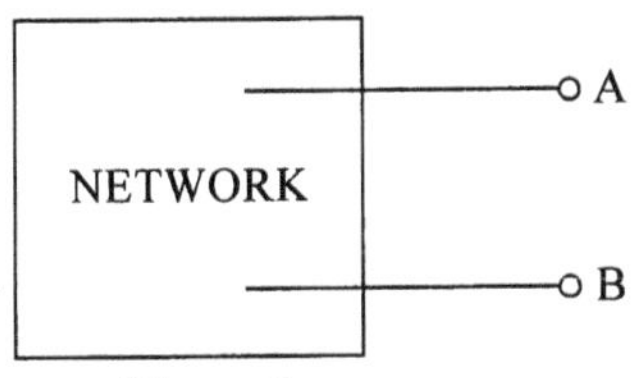

Figure I

Figure II

(a) 10 V in series with 1.2 Ω resistance (b) 6 V series with 1.2 Ω resistance

(c) 10 V in series with 5 Ω resistance (c) 6 V series with 5 Ω resistance

6.48 A certain network consists of two ideal voltage sources and a large number of ideal resistors. The power consumed in one of the resistors is 4 W when either of the two sources is active and the other is replaced by a short-circuit. The power consumed by the same resistor when both the sources are simultaneously active would be

(a) zero or 16 W (b) 4 W or 8 W

(c) zero or 8 W (d) 8 W or 16 W

6.49 The circuit shown in the figure, the effective resistance faced by the voltage source is

(a) 1 Ω

(b) 2 Ω

(c) 3 Ω

(d) 3.3 Ω

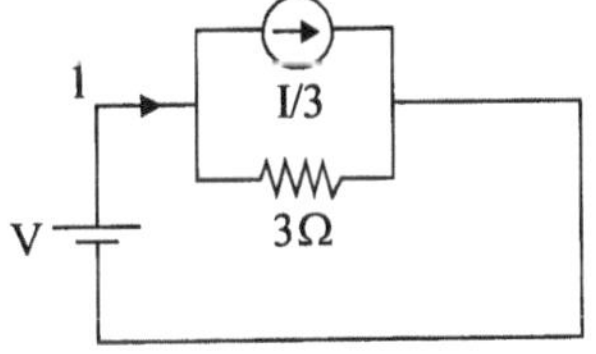

6.50 A dc current source is connected as shown in figure below.

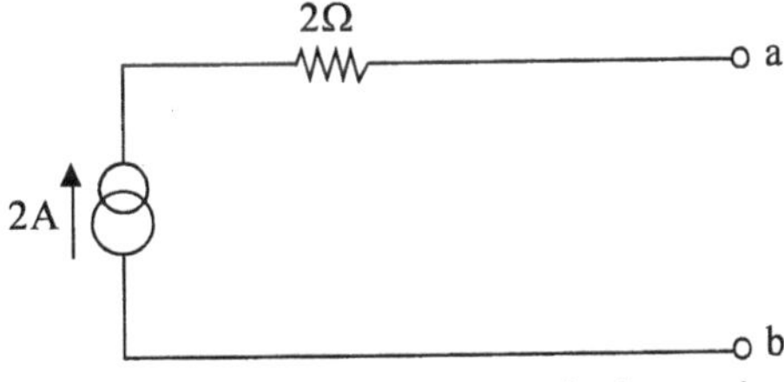

The Thevenin's equivalent of the network at terminals a – b

(a) will be

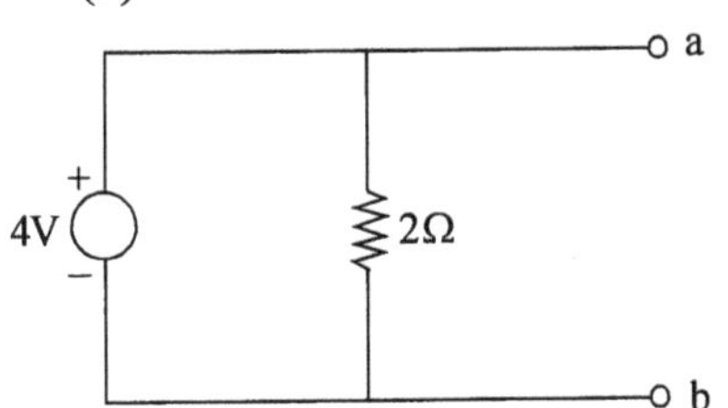

(b) will be

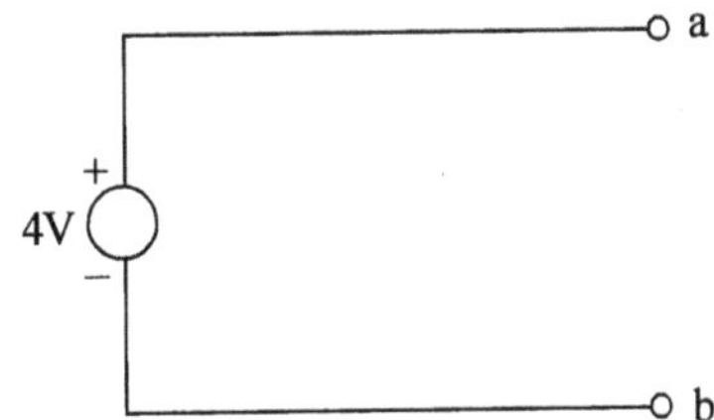

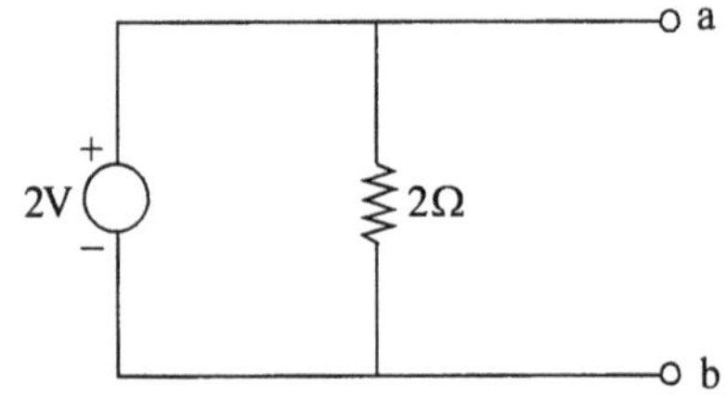

(c) will be (d) is NOT feasible

6.51 A battery charger can drive a current of 5 A into a 1 ohm resistance connected at its output terminals. If it is able to charge an ideal 2 V battery at 7 A rate, then its Thevenin's equivalent will be

(a) 7.5 V in series with 0.5 ohm
(b) 12.5 V in series with 1.5 ohms
(c) 7.5 V in parallel with 0.5 ohm
(d) 1.25 V in parallel with 1.5 ohms

6.52 If a resistance 'R' of 1Ω is connected across the terminals AB as shown in the given figure, then the current flowing through R will be

(a) 1 A
(b) 0.5 A
(c) 0.25 A
(d) 0.125 A

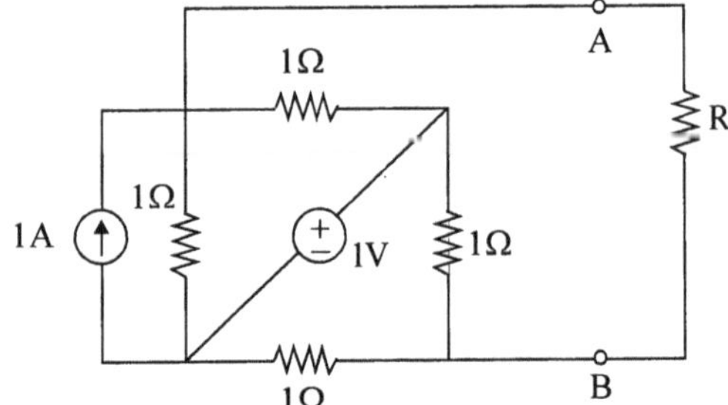

6.53 If the π-network of Figure-I and T-network of Figure-II are equivalent, then the values of R_1, R_2 and R_3 will be respectivley

(a) 9 Ω, 6 Ω and 6 Ω
(b) 6 Ω, 6 Ω and 9 Ω
(c) 9 Ω, 6 Ω and 9 Ω
(d) 6 Ω, 9 Ω and 6 Ω

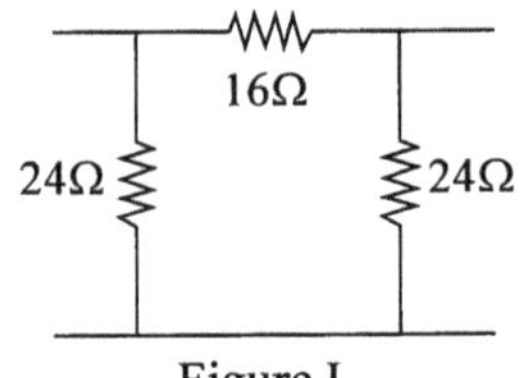

Figure I

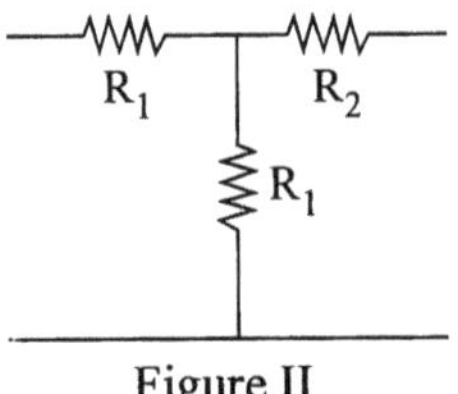

Figure II

6.54 The circuit shown in the figure, the power dissipated in 30 Ω resistor will be maximum if the value of R is

(a) 30 Ω
(b) 16 Ω
(c) 9 Ω
(d) zero

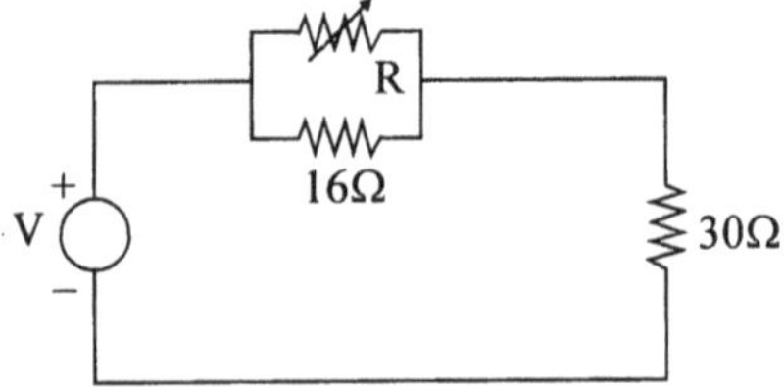

6.55 In a two-terminal network, the open-circuit voltage measured at the given terminals by an electronic voltmeter is 100 V. A short-circuit current measured at the same terminals by an ammeter of negligible resistance is 5 A. If a load resistor of 80 Ω is connected at the same terminals, then the current in the load resistor will be

(a) 1 A (b) 1.25 A (c) 6 A (d) 6.25 A

6.56 Consider the following statements :

1. Tellegen's theorem is applicable to any lumped network.
2. The reciprocity theorem is applicable to linear bilateral networks.
3. Thevenin's theorem is applicable to two-terminal linear active networks.
4. Norton's theorem is applicable to two-terminal linear active networks.

Which of these statements are correct ?

(a) 1, 2 and 3 (b) 1, 2, 3 and 4 (c) 1, 2 and 4 (d) 3 and 4

6.57

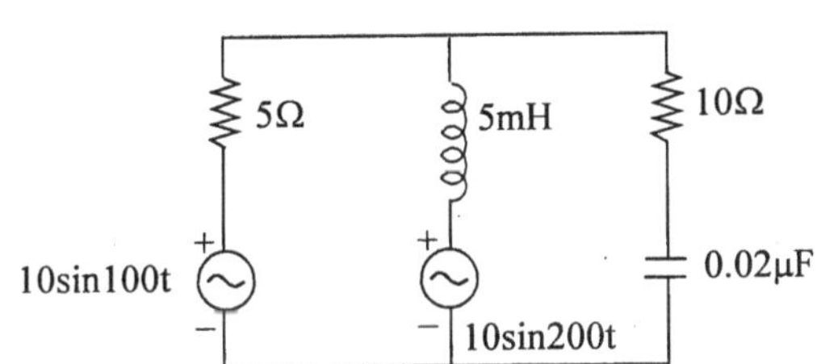

Which one of the following theorems can be conveniently used to calculate the power consumed by the 10 Ω resistor in the network shown in the above figure ?

(a) Thevenin's theorem (b) Maximum power transfer theorem
(c) Millman's theorem (d) Superposition theorem

6.58 The circuit shown in the given figure. R_L will absorb maximum power when its value is

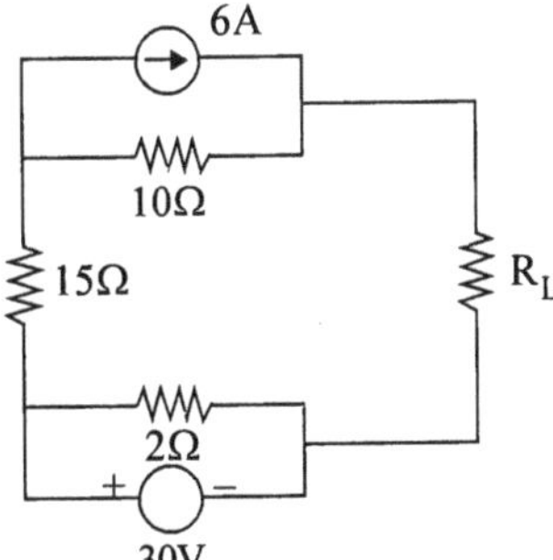

(a) 2.75 Ω

(b) 7.5 Ω

(c) 25 Ω

(d) 27 Ω

6.59 In a linear circuit, the superposition principle can be applied to calculate the

(a) voltage and power (b) voltage and current
(c) current and power (d) voltage, current and power

CHAPTER 7 – Differential Equations and Initial Conditions in RLC Networks

7.1 In the series RC circuit shown in Figure the voltage across C starts increasing when the d.c. source is switched on. The rate of increase of voltage across C at the instant just after the switch is closed (i.e., at $t = 0^+$), will be

(a) zero

(b) Infinity

(c) RC

(d) 1/RC

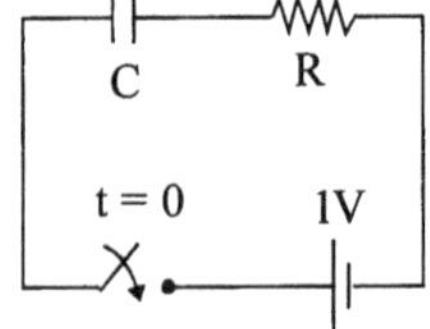

7.2 The network shown in the given figure, there is no initial current through L_2 and no initial voltage across the C. The switch 'S' is closed at t = 0. The current i_{Li} in inductor L_1 and the voltage V_c across C at = 0^+ and t = ∞ will be

	$i_{L1}(0^+)$	$i_{Li}(\infty)$	$V_c(0^+)$	$V_c(\infty)$
(a)	1/3 A	1/3 A	2/3 V	2/3 V
(b)	0	1/3 A	0	1 V
(c)	1/3 A	0	2/3	0
(d)	0	1/3 A	0	2/3 V

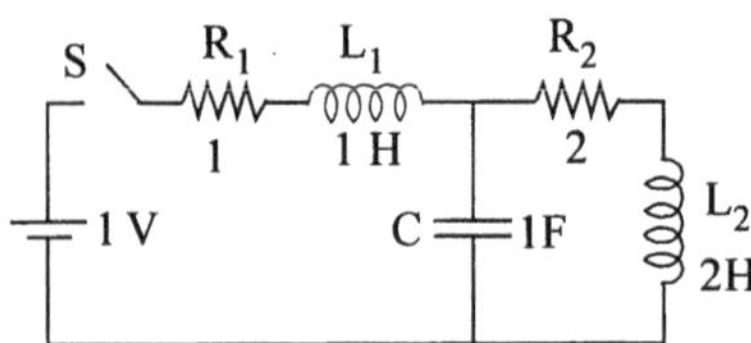

7.3 For the circuit given in the figure V_o = 2 V and the inductor is initially relaxed. The switch S is closed at t = 0. The value of V at $t = 0^+$ is

(a) 3 V

(b) 2 V

(c) 0.5 V

(d) 0.25 V

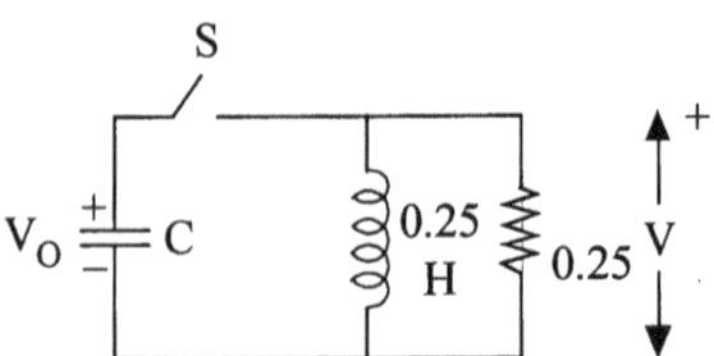

7.4 The circuit shown in the given figure, S is open for a long time and steady state is reached. S is closed at t = 0. The current I at t = 0^+ is

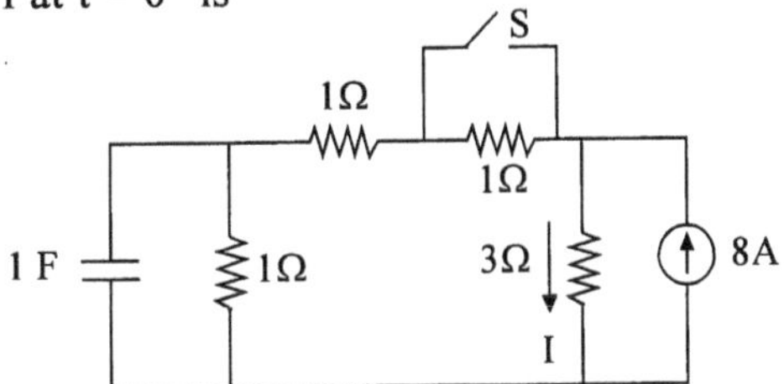

(a) 4 A

(b) 3 A

(c) 2 A

(d) 1 A

7.5 In the circuit shown, switch is opened at t = 0. Prior to that switch was closed. i(t) at t = 0^+ is

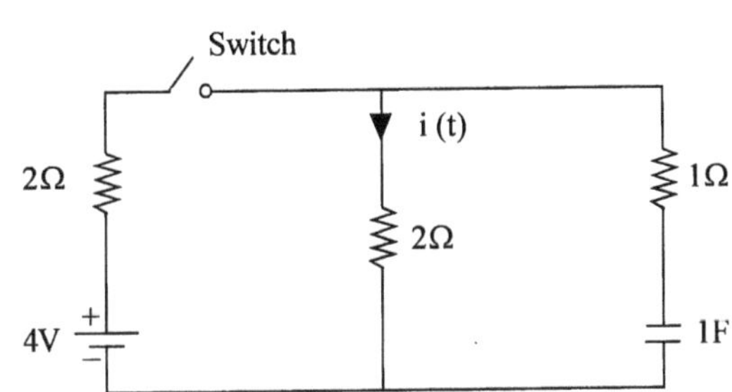

(a) $\frac{2}{3}$ A

(b) $\frac{3}{2}$ A

(c) $\frac{1}{3}$

(d) 1 A

7.6 Initially, the circuit shown in the given figure was relaxed. If the switch is closed at t = 0, the values of i(0^+), $\frac{di}{dt}$ (0^+) and $\frac{d^2i}{dt^2}$ (0^+) will respectivley be

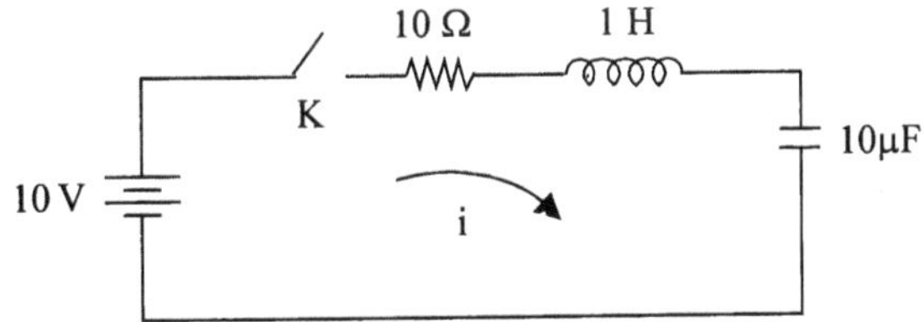

(a) 0, 10 and – 100

(b) 0, 10 and 100

(c) 10, 100 and 0

(d) 100, 0 and 10

7.7 The steady state in the circuit, shown in the given figure is reached with S open. S is closed at t = 0. The current i at t = 0^+ is

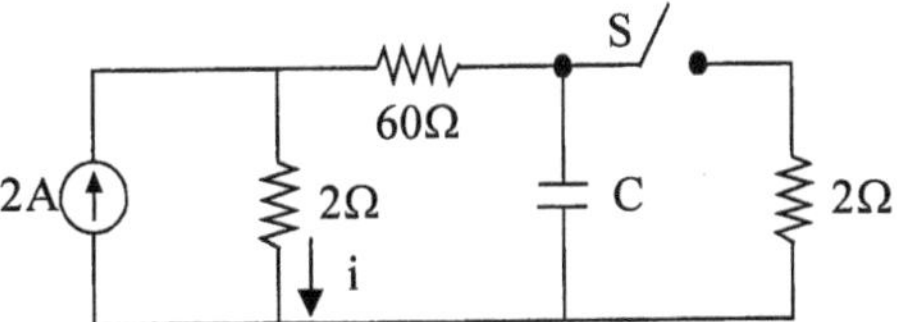

(a) 1 A

(b) 2 A

(c) 3 A

(d) 4 A

7.8 For the circuit shown in the given fig., if $C = 20\ \mu F$, $v(0^-) = -50$ V and $\frac{dv(0^+)}{dt} = 500$ V/S, then R is

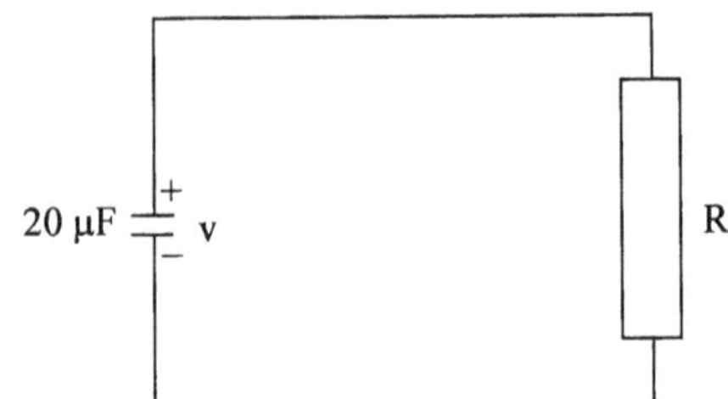

(a) 2 K

(b) 3 K

(c) 5 K

(d) 10 K

7.9 The circuit shown in the given figure, the switch is moved from position A to B at time t = 0. The current i through the inductor satisfies the following conditions :

$i(0) = -8A, \frac{di}{dt}(t = 0) = 3A/s, i(\infty) = 4A.$

The value of R is

(a) 0.5 ohm

(b) 2 ohm

(c) 4 ohm

(d) 12 ohm

7.10 The network shown in the figure, the circuit was initially in the steady-state condition with the switch K closed. At the instant when the switch is opened, the rate of decay of current through the inductance will be

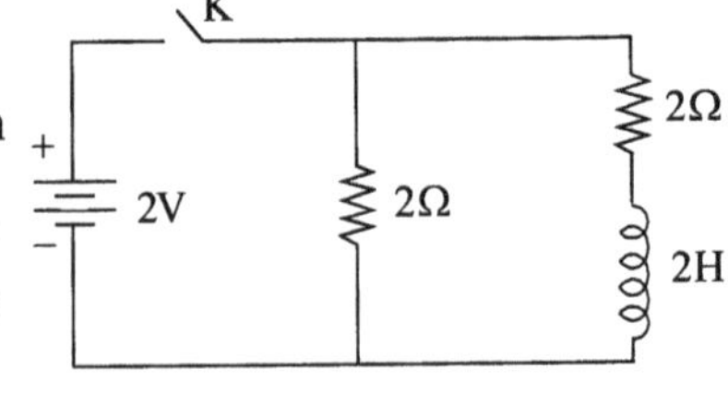

(a) zero (b) 0.5 A/s

(c) 1 A/s (d) 2 A/s

7.11 In the circuit shown, the switch closes at t = 0. The current at t = 0 is (current through inductance for t < 0 is 0)

(a) 1 A

(b) 2 A

(c) $\frac{1}{2}$ A

(d) 0 A

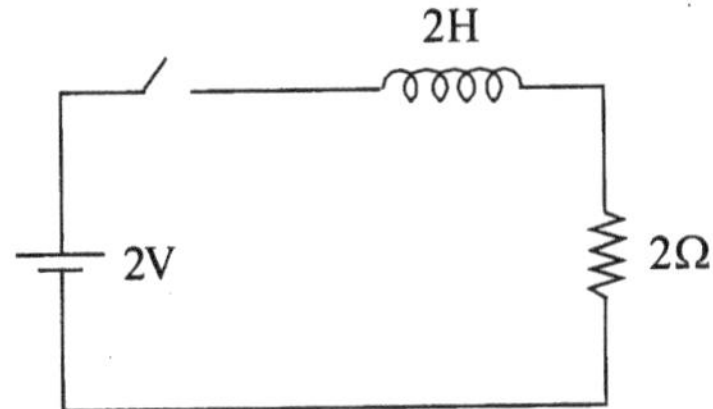

7.12 In the circuit shown, the switch closes at t = 0. The voltage across 4 μF capacitor in ideal condition changes to

(a) 0

(b) 16 V

(c) 15 V

(d) 24 V

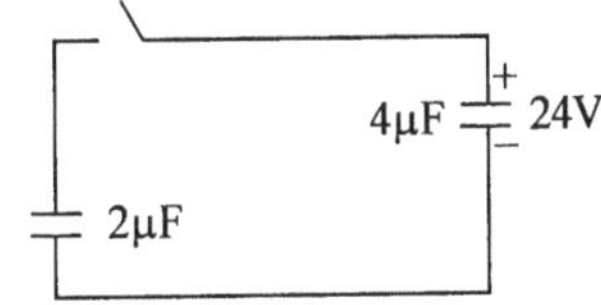

7.13 The steady state current through the 1 H inductance in the circuit, shown in the given figure is

(a) zero

(b) 3 A

(c) 5 A

(d) 6 A

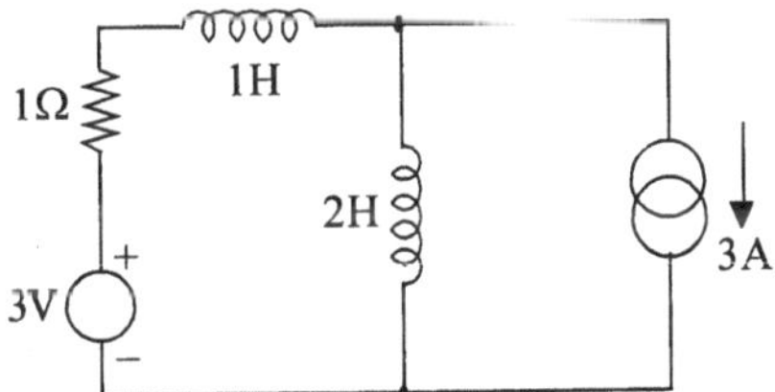

7.14 The network shown in the given figure, the switch K is closed at t = 0 with the capacitor uncharged. The value for $\frac{di(t)}{dt}$ at $t = 0^+$ will be

(a) 100 amp/sec

(b) – 100 amp/sec

(c) 1000 amp/sec

(d) – 1000 amp/sec

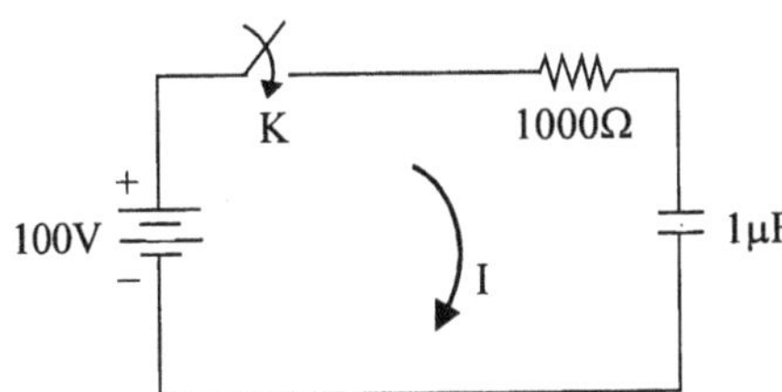

7.15 In the initially relaxed network, the switch is closed at $t = 0$. $i(0^+)$, $i(\infty)$, $v(0^+)$, $v(\infty)$ are respectively

	$i(0^+)$	$i(\infty)$	$v(0^+)$	$v(\infty)$
(a)	0	2A	0	5 V
(b)	0	2.5 A	5 V	10 V
(c)	5 A	2.5 A	5 V	10 V
(d)	0	0	0	10 V

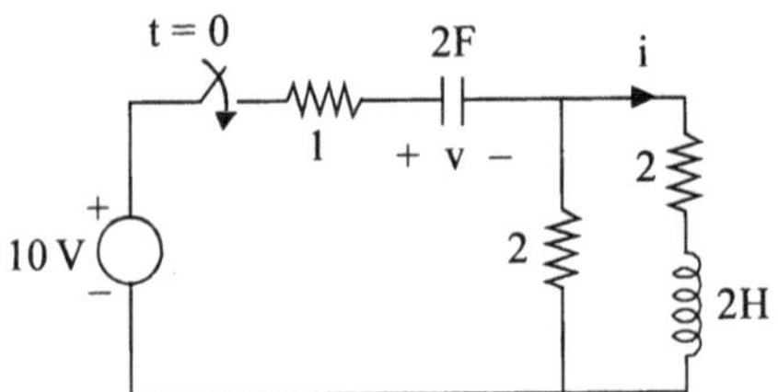

7.16 A circuit consisting of a 1 Ω resistor and a 2 F capacitor in series is excited from a voltage source with the voltage expressed as $3\,e^{-t}$, as shown in the given figure. If the $i(0^-)$ and $v_c(0^-)$ both are zero, then the values of $i(0^+)$ and $i(\infty)$ will be respectively.

(a) 3 A and 1.5 A

(b) 1.5 A and zero

(c) 3 A and zero

(d) 1.5 A and 3 A

7.17 After keeping it open for a long time, the switch 'S' in the circuit, shown in the given figure is closed at $t = 0$. The capacitor voltage $v_C(0^+)$ and inductor current $i_L(0^+)$ will be

(a) 60 V and – 0.3 A

(b) 150 V and zero

(c) zero and 0.3 A

(d) 90 V and – 0.3 A

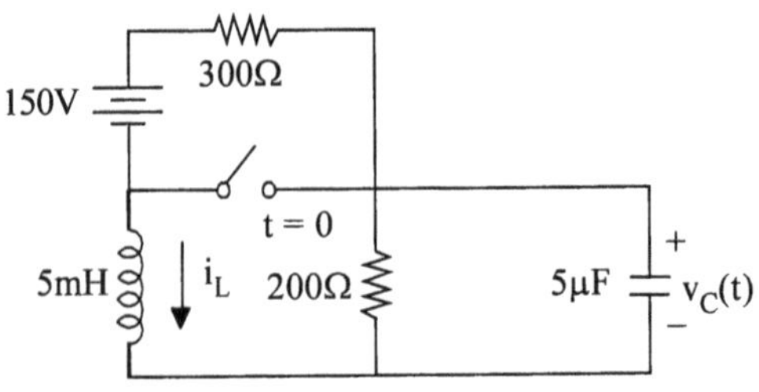

7.18 In the circuit shown, i (t) is a unit step current. The steady-state value of v (t) is

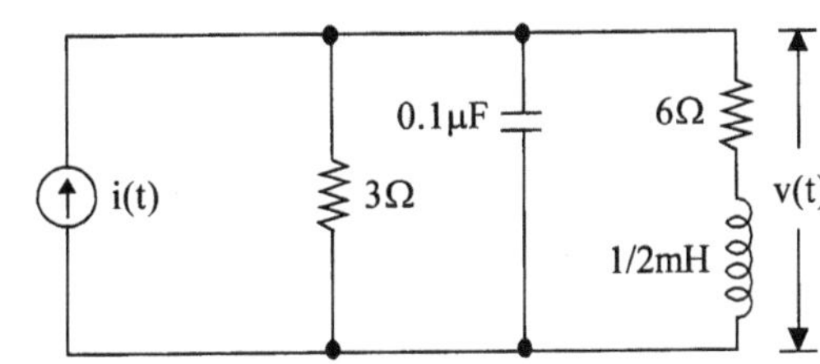

(a) 2 V

(b) 3 V

(c) 6 V

(d) 9 V

7.19 On closing switch 'S', the circuit in the given figure is in steady-state. The current in the inductor after opening switch 'S' will

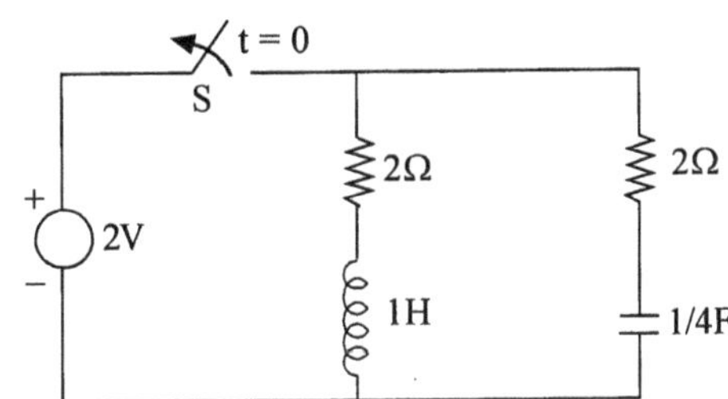

(a) decay exponentially with a time constant of 2 s

(b) decay exponentially with a time constant of 0.5 s

(c) consist of two decaying exponents each with a time constant of 0.5 s

(d) be oscillatory

7.20 In the circuit shown in the figure, i(t) is a unit step current. The steady-state value of v(t) is

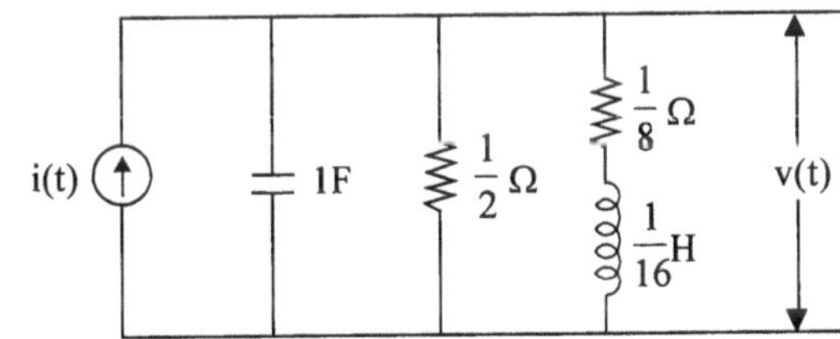

(a) 2.5 V

(b) 1 V

(c) 0.1 V

(d) zero

7.21

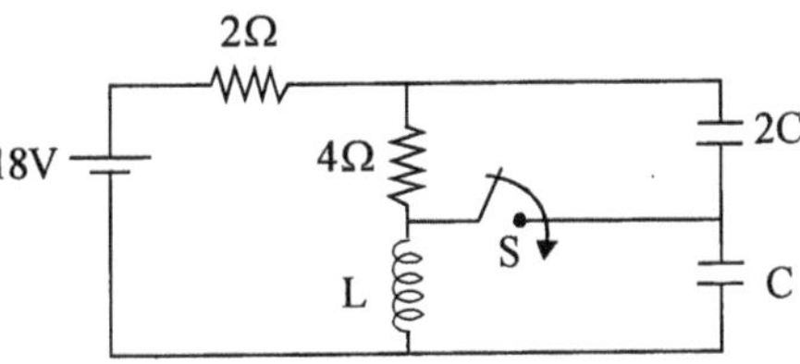

The circuit shown in the above figure, steady-state was reached when the switch S was open. The switch was closed at t = 0. The initial value of the current through the capacitor 2 C is

(a) zero (b) 1 A (c) 2 A (d) 3 A

7.22 The circuit shown in the given figure, when the switch is at position A, the current $i(t) = I \sin(\omega t + 30^o)$ A. When switch is moved to position B at time t = 0, the power dissipated at the switching instant in the resistor R remains unchanged

The value of I and the element x would respectively,be

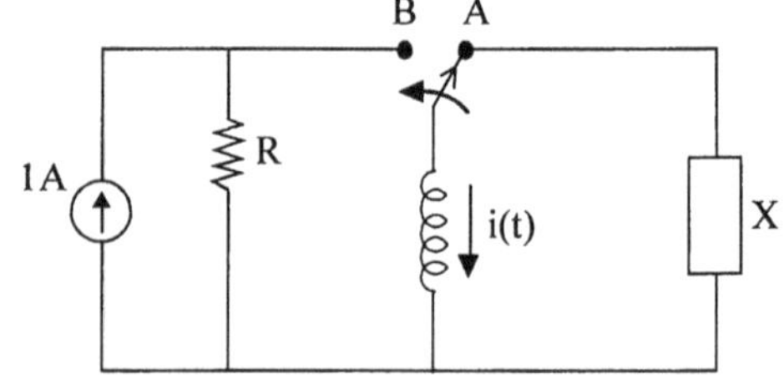

(a) 1 A and resistor

(b) 2 A and capacitor

(c) 3 A and resistor

(d) 4 A and capacitor

7.23 The circuit shown in the given figure, the values of $i(0^+)$ and $i(\infty)$, will be, respectively

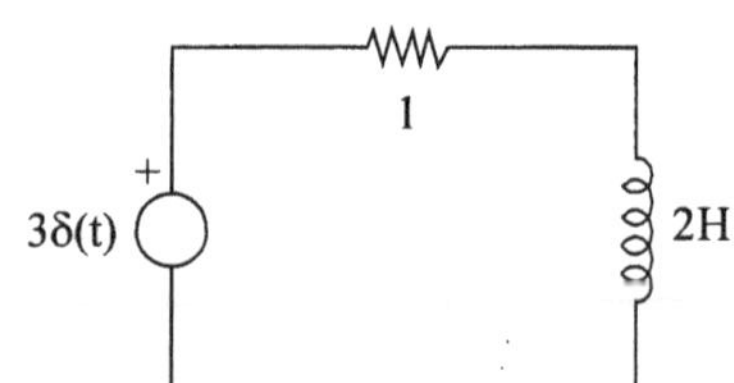

(a) 0, $\frac{3}{2}$ A

(b) 3 A, 0 A

(c) 1.5 A, 0 A

(d) 1.5 A, 3 A

7.24 The circuit shown in the given figure is in steady-state with switch 'S' open. The switch is closed at t = 0. The values of $v_c(0^+)$ and $v_c(\infty)$ will be respectively.

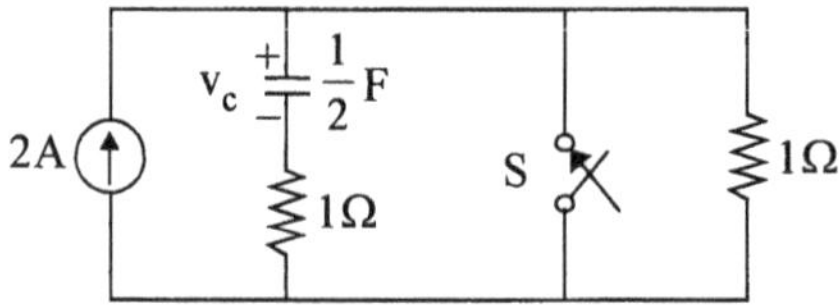

(a) 2 V, 0 V (b) 0 V, 2 V (c) 2 V, 2 V (d) 0 V, 0 V

7.25 The circuit shown in the given figure, if the switch is opened as t = 0, then the voltage $v(0^+)$ and its derivative $\left.\frac{dv}{dt}\right|_{t=0^+}$ will be respectivley

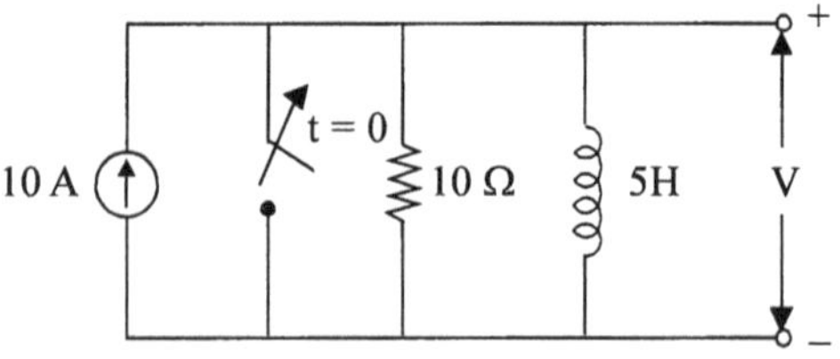

(a) 10 V and 50 V/s

(b) 10 V and – 50 V/s

(c) 100 V and 200 V/s

(d) 100 V and – 200 V/s

Chapter 8 – Time Reponse of RLC Networks

8.1 The circuit shown in the given figure, $C_1 = C_2 = 2F$ and the capacitor C_1 has a voltage of 20 V when S is open.

If the switch S is closed at t = 0, the voltage v_{c2} will be a

(a) fixed voltage of 20 V

(b) fixed voltage of 10 V

(c) fixed voltage of –10 V

(d) sinusoidal voltage

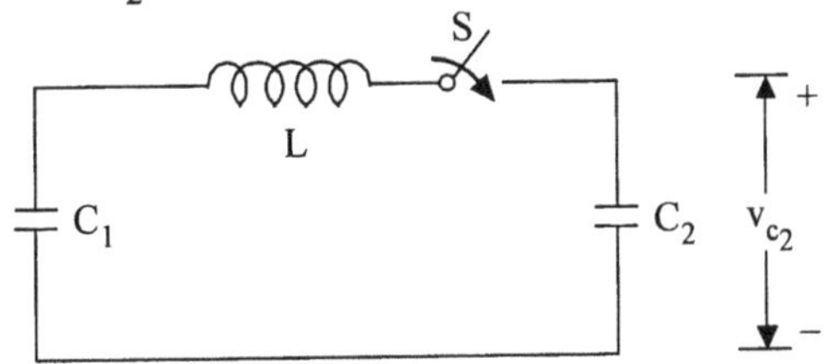

8.2 A rectangular voltage pulse of magnitude V and duration T is applied to a series combination of resistance R and capacitance C. The maximum voltage developed across the capacitor is

(a) V[1 – exp (–T/RC)]   (b) VT/RC   (c) V   (d) V exp (–T/RC)

8.3 An ideal voltage source will charge an ideal capacitor

(a) In infinite time   (b) exponentially   (c) Instantaneously   (d) none of the above

8.4 The impulse response of an R-L circuit is a

(a) rising exponential function   (b) decaying exponential function

(c) step function   (d) parabolic function

8.5 The v-i characteristics as seen from the terminal-pair (A, B) of the network of figure (a) is shown in figure (b). If an inductance of value 6 mH is connected across the terminal pair (A, B), the time constant of the system will be

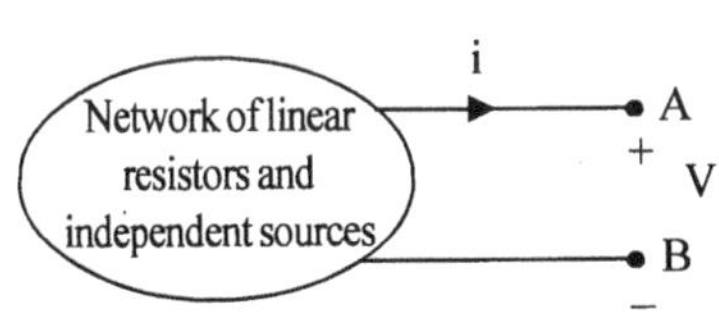

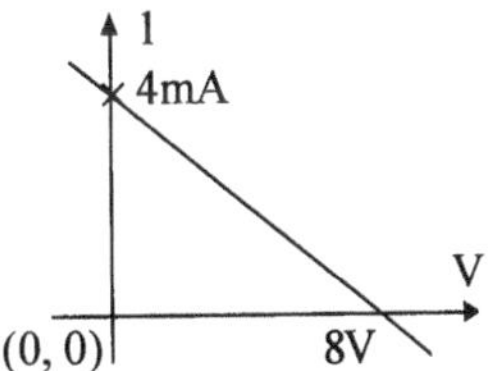

(a) 3 μsec   (b) 12 sec

(c) 32 sec   (d) unknown, unless the actual network is specified

8.6 A current impulse, 5δ (t), is forced through a capacitor C. The voltage, $v_C(t)$ across the capacitor is given by :

(a) 5t   (b) 5u (t) – C   (c) $\frac{5}{C}$ t   (d) $\frac{5u(t)}{C}$

8.7 Consider the circuit shown in figure. If the frequency of the source is 50 Hz, then a value of t_o which results in transient free response is

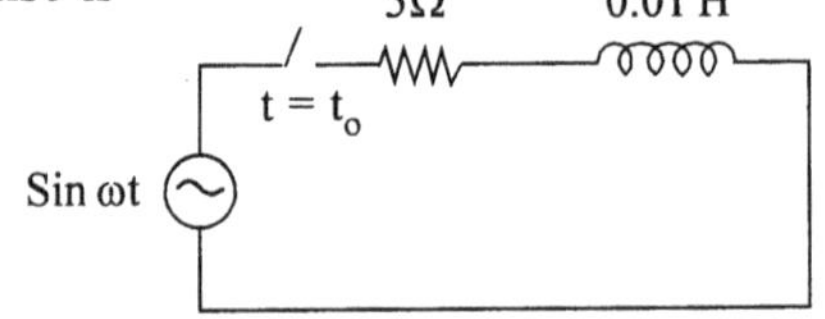

(a) 0 ms

(b) 1.78 ms

(c) 2.71 ms

(d) 2.91 ms

8.8 A unit step voltage is applied at t = 0 to a series RL circuit with zero initial conditions.

(a) It is possible for the current to be oscillatory.

(b) The voltage across the resistor at $t = 0^+$ is zero.

(c) The energy stored in the inductor in the steady state is zero.

(d) The resistor current eventually falls to zero.

8.9 A voltage waveform $V(t) = 12t^2$ is applied across a 1H inductor for $t \geq 0$, with initial current through it being zero. The current through the inductor for $t \geq 0$ is given by

(a) 12t (b) 24t (c) $12t^3$ (d) $4t^3$

8.10 A first order linear system is initially relaxed. For a unit step signal u(t), the response is $v_1(t) = (1 - e^{-3t})$ for t > 0. If a signal 3 u(t) + δ(t) is applied to the same initially relaxed system, the response will be

(a) $(3 - 6\ e^{-3t})\ u(t)$ (b) $(3 - 3\ e^{-3t})\ u(t)$

(c) 3 u(t) (d) $(3 + 3\ e^{-3t})\ u(t)$.

8.11 In the given circuit, the value of R that will give critical damping is

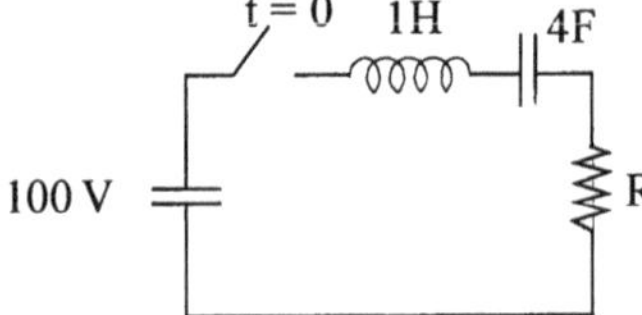

(a) 1 Ω

(b) 2 Ω

(c) 4 Ω

(d) 10 Ω

8.12 A step voltage is applied to an underdamped series RLC circuit with variable R. Which of the following statements correctly describes the behaviour of the circuit ?

1. If R is increased, the steady-state voltage across C will be reduced.
2. If R is increased, the frequency of transient oscillations across C will be reduced
3. If R is reduced, the transient oscillations will die down faster.
4. If R is reduced to zero, the peak amplitude of the voltage across C will be double the input step voltage.

Select the correct answer using the codes given below :

Codes :

(a) 1 and 2 (b) 2 and 3 (c) 2 and 4 (d) 1,3 and 4

8.13 For the network shown in figures, A and B to be duals, it is necessary that R, L and C are respectively equal to

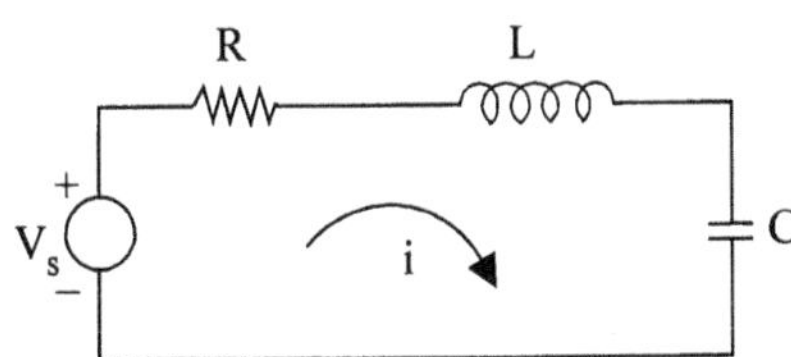

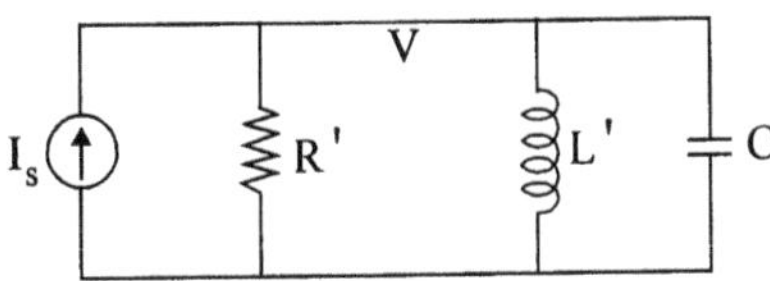

(a) 1/R, C and L (b) 1/R, I/L and I/C
(c) 1/R, 1/L and C (d) R, L and C

8.14 Which of the following statement (s) is/are true of the circuit shown in the given figure.

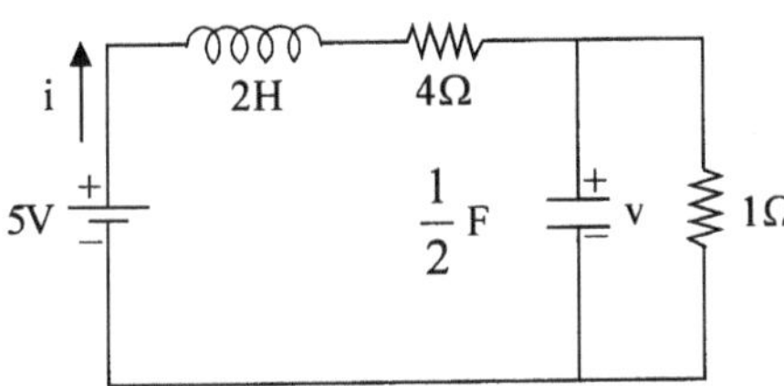

1. It is first order circuit with steady-state values of $v = \frac{5}{3}$ V, $i = \frac{5}{3}$ A
2. It is second order circuit with steady-state values of v = 1 ; i = 1A.
3. The network function V(s)/I(s) has one pole.
4. The network function V (s)/I(s) has two poles.

Select the correct answer using the codes given below :

Codes :

(a) 2 and 3 (b) 2 and 4 (c) 2 alone (d) 1 alone.

8.15 After closing the switch 'S' at t = 0, the current i (t) at any instant 't' in the network shown in the given figure will be

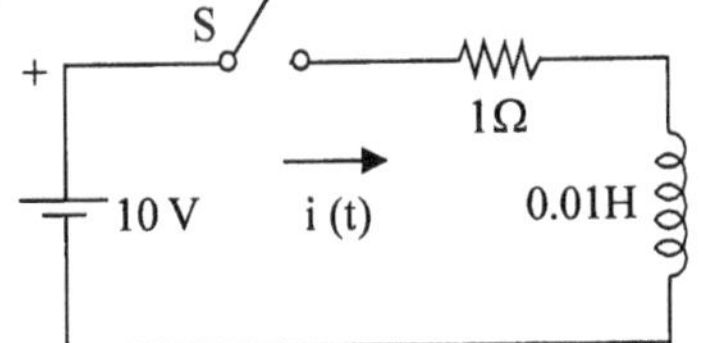

(a) $10 + 10\ e^{100t}$ (b) $10 - 10\ e^{100t}$
(c) $10 + 10\ e^{-100t}$ (d) $10 - 10\ e^{-100t}$

8.16 A system is represented by $\frac{dy}{dt} + 2y = 4t\ u(t)$

The ramp component in the forced response

(a) t u (t) (b) 2t u(t) (c) 3t u(t) (d) 4t u (t)

8.17 If the unit step response of a network is $(1 - e^{-\alpha t})$, then its unit impulse response will be

(a) $\alpha\ e^{-\alpha t}$ (b) $\frac{1}{\alpha}\ e^{-t/\alpha}$ (c) $\frac{1}{\alpha}\ e^{-t/\infty}$ (d) $(1 - \alpha)\ e^{-\alpha t}$

8.18 The network shown in the given figure, if the voltage v at the time considered is 20 V, then dv/dt at that time will be

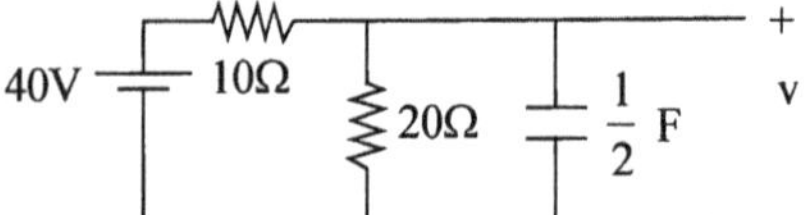

(a) 1 V/s (b) 2 V/s
(c) – 2 V/s (d) zero

8.19 A pulse of unit amplitude and width 'a' is applied to a series RL circuit as shown in the figure. The current i (t) as 't' tends to infinity will be

(a) zero
(b) 1 A
(c) a value between zero and one depending upon the width of the pulse
(d) infinite

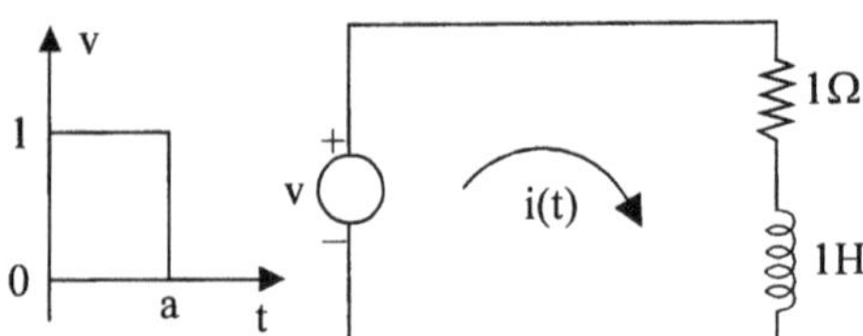

8.20 A second order system is given by $\frac{d^2y}{dt^2} + 12\frac{dy}{dt}\ 100y = 0$

The damped natural frequency in rad/sec is

(a) 100 (b) 10 (c) $\sqrt{44}$ (d) 8

8.21 For loop (1) of the network shown in the given figure, the correct loop equation is :

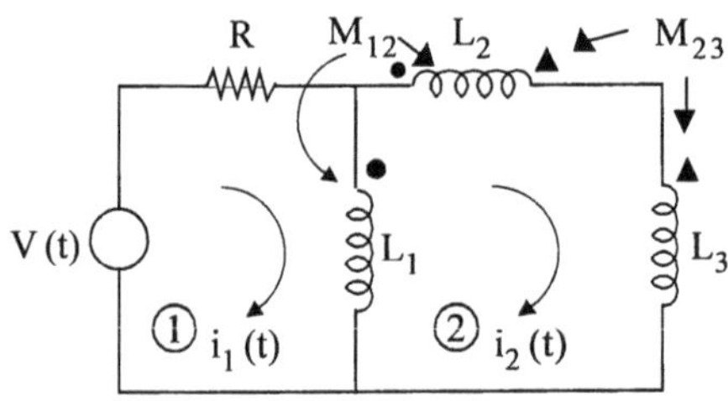

(a) $v(t) = Ri_1(t) + L_1\frac{d}{dt}[i_1(t) - i_2(t)] + M_{12}\frac{di_2}{dt}(t)$

(b) $v(t) = Ri_1(t) + L_1\frac{d}{dt}[i_1(t) - i_2(t)] + M_{12}\frac{di_2}{dt}(t) - M_{23}\frac{di_2}{dt}(t)$

(c) $L_1\frac{d}{dt}[i_1(t) - i_1(t)] + (L_2 + L_3) + \frac{di_2}{dt}(t) = 2(M_{12} + M_{23})\frac{di_2}{dt} - M_{12}\frac{di_1}{dt}$

(d) $L_1\frac{d}{dt}[i_2(t) - i_1(t)] + (L_2 + L_3)\frac{di_2}{dt} = M_{12}\frac{di_1}{dt} - 2(M_{12}+M_{23})\frac{di_2}{dt}$

8.22 If $i(t) = \frac{1}{4}(1 - e^{-2t})\,u(t)$, where u(t) is a unit step voltage, then the complex frequencies associated with i(t) would include

(a) s = 0 and s = j2 (b) s = j2 and s = – j2
(c) s = – j2 and s = – 2 (d) s = 0 and s = – 2

8.23 The time constant of the network shown in the figure is

(a) CR

(b) 2CR

(c) $\frac{CR}{4}$

(d) $\frac{CR}{2}$

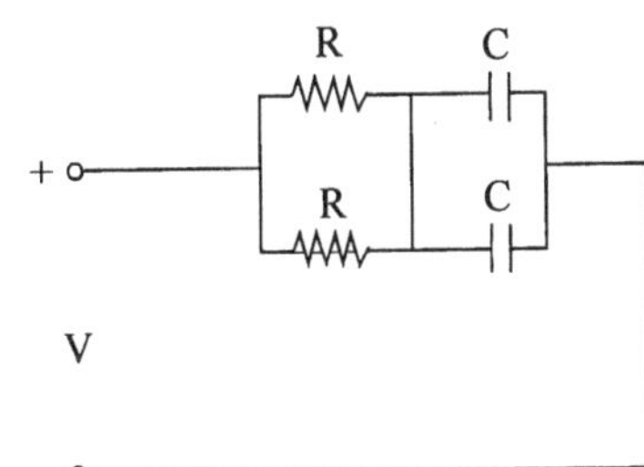

8.24 An initially relaxed RC-series network with R = 2 MΩ and C = 1 μF is switched on to a 10 V step input. The voltage across the capacitor after 2 seconds will be

(a) zero (b) 3.68 V (c) 6.32 V (d) 10 V

8.25 An initially relaxed 100 mH inductor is switched 'ON' at t = 1 sec to an ideal 2 A dc current source. The voltage across the inductor would be

(a) zero (b) $0.2\delta(t)$ V (c) $0.2\,\delta(t-1)$ V (d) $0.2\,t\,u(t-1)$ V

8.26 Consider the following statements ;

The impulse response of a linear network can be used to determine the

1. step response
2. response of the sinusoidal input
3. elements of the network uniquely
4. interconnection of network elements

Which of these statements are correct ?

(a) 1 and 2 (b) 2 and 3

(c) 3 and 4 (d) 1 and 4

8.27 The circuit shown in the given figure, the response current i(t) is

(a) $\frac{V}{R}\exp\left(-\frac{t}{RC}\right)$

(b) $\frac{V}{R}\,\delta(t)$

(c) $\frac{V}{R}\left[\delta(t)-\frac{1}{RC}\exp\left(-\frac{t}{RC}\right)\right]$

(d) $\frac{V}{R}\left[\delta(t)-\exp\left(-\frac{t}{RC}\right)\right]$

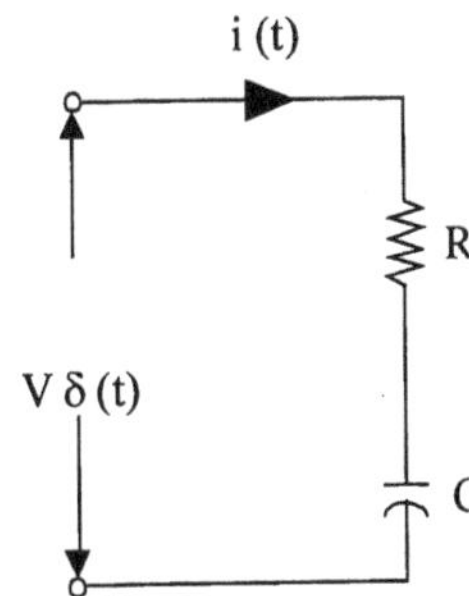

8.28 A voltage $v(t) = 6\ e^{-2t}$ is applied at $t = 0$ to a series R – L circuit with L = 1 H. If $i(t) = 6\ [\exp(-2t) - \exp(-3t)$ then R will have a value of

(a) $\frac{2}{3}\ \Omega$ (b) $1\ \Omega$ (c) $3\ \Omega$ (d) $\frac{1}{3}\ \Omega$

8.29 For a second order system, if both the roots of the characteristic equation are real, then the value of damping ratio will be

(a) less than unity (b) equal to unity (c) equal to zero (d) great than unity

8.30 For the circuit shown in the given figure, the steady state current through L and the voltage across C_2 are respectivley

(a) zero and RI

(b) I and zero

(c) zero and zero

(d) I and RI

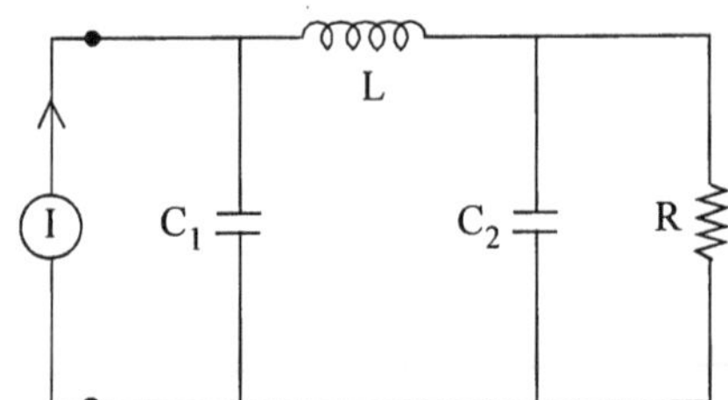

8.31 The response i of a network is expressed by the differential equation $\frac{d^2i}{dt^2} + i = v$. If $v = A\ e^{2t}$, the dominant solution of i for $t > 0$ of the nature

(a) $K_1\ e^t$ (b) $K_1\ e^{-t}$ (c) $K_1\ e^{2t}$ (d) $K_2 \cos t + K_3 \sin t$

8.32 A unit step $u(t - 5)$ is applied to the RL network

The current i is given by

R = 1Ω

(a)  $1 - e^{-t}$

(b) $[1 - e^{-(t-5)}]\ u(t - 5)$

v = u(t – 5) + – i L = 1H

(c) $(1 - e^{-t})\ u\ (t - 5)$

(d) $1 - e^{-(t-5)}$

8.33 The response of an initially relaxed system to a unit ramp excitation is $(t + e^{-t})$. Its step response will be

(a) $\frac{1}{2}\ t^2 - e^{-t}$ (b) $1 - e^{-t}$ (c) $- e^{-t}$ (d) t

8.34 The response of a network is $i(t) = K\,t\,e^{-\alpha t}$ for $t \geq 0$ where α is real positive. The value of 't' at which the i(t) will become maximum, is

(a) α (b) $2\,\alpha$ (c) $\frac{1}{\alpha}$ (d) α^2

8.35 The impulse response of an LTI system is given by 5u (t). If the input to the system is given by ε^{-t} then the output of the system is given by

(a) $5\,(1 - \varepsilon^{-t})$ (b) $(1 - 5\varepsilon^{-t})\,u(t)$ (c) $5 - \varepsilon^{-t}\,u(t)$ (d) $5\,u(t)\,\varepsilon^{-t}$

8.36 The circuit shown in the given figure, the switch is closed at t = 0.

The current through the capacitor will decrease exponentially with a time constant :

(a) 0.5 s

(b) 1 s

(c) 2 s

(d) 10 s

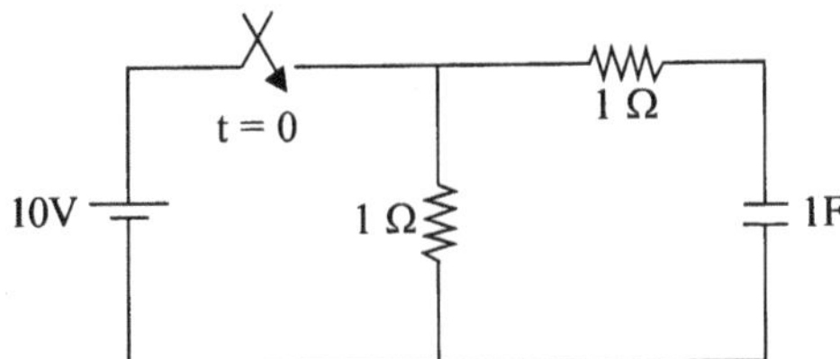

8.37 The number of turns of a coil having a time constant T are doubled. Then the new time constant will be

(a) T (b) 2T (c) 4T (d) T/2

8.38 Consider the following data :

1. Input applied for $t < t_o$
2. Input applied for $t \geq t_o$
3. State of the network at $t = t_o$
4. State of the network at $t < t_o$

Among these, those needed for determining the response of a linear network for $t > t_o$ would include

(a) 1, 3 and 4 (b) 2, 3 and 4 (c) 2 and 3 (d) 2 and 4

8.39 The transient response of the initially relaxed network shown in the figure is :

(a) $i = \frac{V}{R}\,e^{-t/RC}$ (b) $i = \frac{V}{R}\,e^{t/RC}$

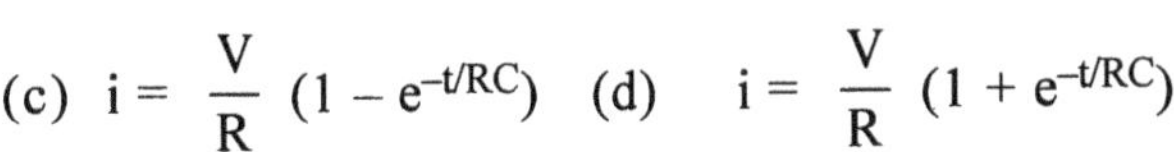

(c) $i = \frac{V}{R}\,(1 - e^{-t/RC})$ (d) $i = \frac{V}{R}\,(1 + e^{-t/RC})$

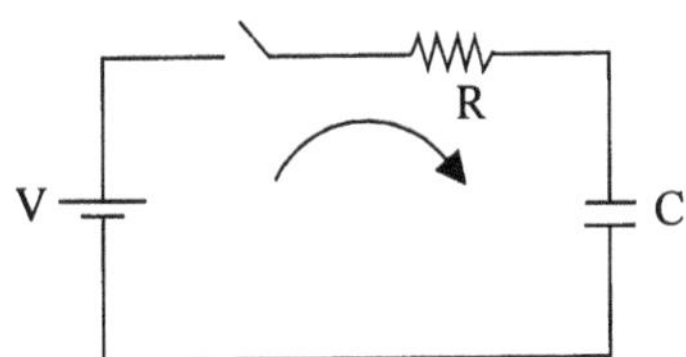

8.40 Two coils having equal resistance but different inductances are connected in series. The time constant of the series combination is the

(a) sum of the time constants of the individual coils

(b) average of the time constants of individual coils

(c) geometric mean of the time constants of the individual coils

(d) product of the time constants of the individual coils

8.41 The circuit shown in the figure, the switch 'S' has been open for a long time. It is closed at t = 0. For t > 0 the current flowing through the inductor will be given by

(a) $i_L(t) = 1.2 + 0.8\,e^{-2t}$

(b) $i_L(t) = 0.8 + 1.2\,e^{-2t}$

(c) $i_L(t) = 1.2 - 0.8\,e^{-2t}$

(d) $i_L(t) = 0.8 - 1.2\,e^{-2t}$

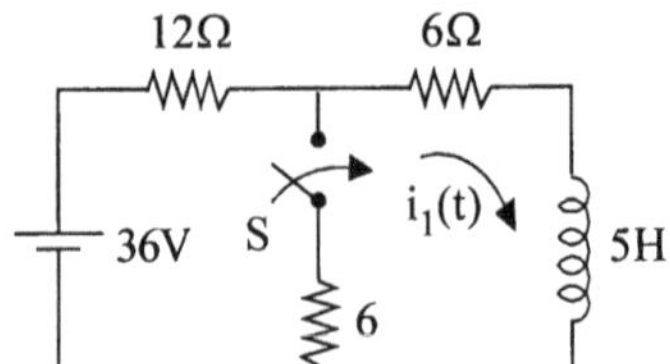

8.42 A unit impulse input to a linear network has a response r (t) and a unit step input to the same network has response s (t). The response r (t).

(a) equals $\dfrac{ds(t)}{dt}$

(b) equals the integral of s (t)

(c) is the reciprocal of s (t)

(d) has no relation with s (t)

8.43 Consider the following statements :
A unit impulse δ (t) is mathematically defined as

1. $\delta(t) = 0,\ t \neq 0$ 2. $\int_{0+}^{\infty} \delta(t)\,dt = 1$ 3. $\int_{-\infty}^{+\infty} \delta(t)\,dt = 1$

Of these statements

(a) 2 and 3 are correct

(b) 1 and 2 are correct

(c) 1, 2 and 3 are correct

(d) 1 and 3 are correct

8.44 At a certain current, the energy stored in an iron-cored coil is 1000 J and its copper loss is 2000 W. The time constant (in seconds) of the coil is

(a) 0.25 (b) 0.5 (c) 1.0 (d) 2.0

8.45 If an R-L circuit having impedance angle ϕ is switched on when the applied sinusoidal voltage wave is passing through an angle θ, there will be no switching transient of

(a) $\theta - \phi = 0$ (b) $\theta + \phi = 0$ (c) $\theta - \phi = 90^\circ$ (d) $\theta + \phi = 90^\circ$

8.46 In the given circuit, the switch S_1, is initially in a closed position and switch S_2 is in the open position. If S_1 is suddenly opened with simultaneous closure of S_2, then the expression for instantaneous current would be

(a) $2\frac{V}{R}\left\{1-e^{-\frac{Rt}{L}}\right\}$ (b) $2\frac{V}{R}\left\{1-e^{-\frac{Rt}{2L}}\right\}$

(c) $\frac{V}{R}\left\{2-e^{-\frac{Rt}{L}}\right\}$ (d) $2\frac{V}{R}\left\{2-e^{-\frac{Rt}{2L}}\right\}$

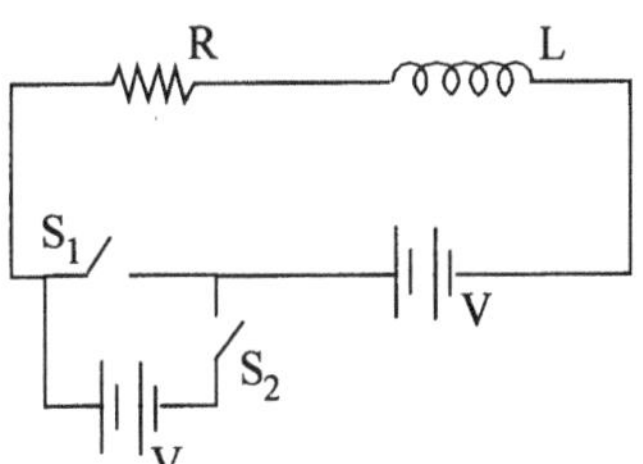

8.47 The time constant of the circuit in seconds is

(a) 0.5

(b) 1.0

(c) 0.25

(d) 0.1

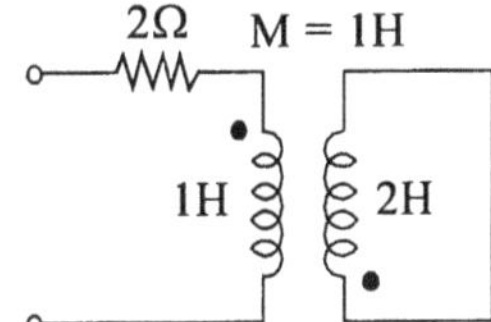

8.48 The voltage across R after t = 0 and t = 1 sec, will be

(a) 100 V, 632 V

(b) 0 V, 63.2 V

(c) 100 V, 36.8 V

(d) 0 V, 26.8

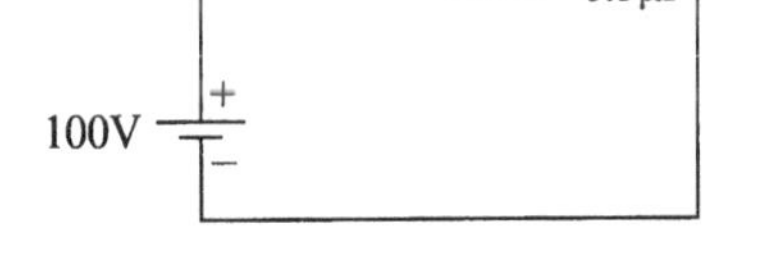

8.49 For a high pass RC circuit, when subjected to a unit step input voltage, the voltage across the capacitor will be

(a) $1 - e^{-t/RC}$ (b) $e^{-t/RC}$ (c) $e^{+t/RC}$ (d) 1

8.50 The time constant of the network shown in the given figure is given by

(a) $\dfrac{L}{R_3 + \dfrac{R_1 R_2}{R_1 + R_2}}$ (b) $\dfrac{L}{R_1 + R_2 + R_3}$

(c) $\dfrac{L}{\dfrac{1}{R_1} + \dfrac{1}{R_2} + \dfrac{1}{R_3}}$ (d) $\dfrac{L}{\dfrac{R_1 R_2}{R_1 + R_2}}$

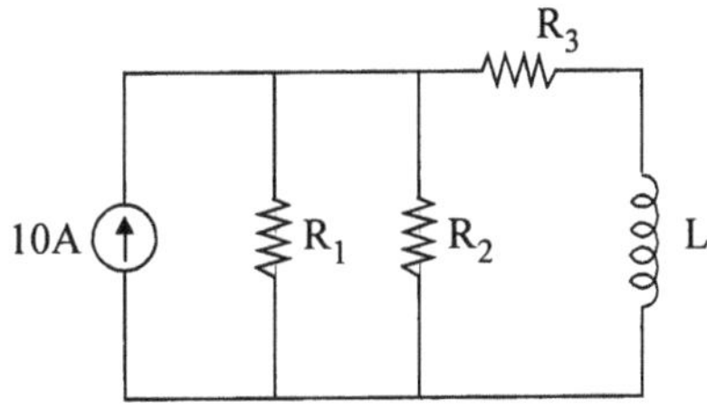

8.51 A series RLC circuit is overdamped when

(a) $\frac{R^2}{4L^2} > \frac{1}{LC}$ (b) $\frac{R^2}{4L^2} = \frac{1}{LC}$ (c) $\frac{R^2}{4L^2} < \frac{1}{LC}$ (d) R = infinity

8.52 The circuit shown in the given figure has been in the steady-state when the switch S is opened. The current i after the switch is opened is given by

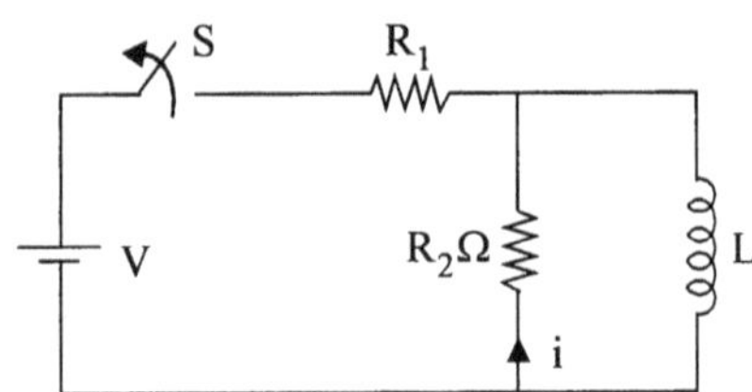

(a) $\frac{V}{R_2} e^{-R_1 t/L}$

(b) $\frac{V}{R_1} e^{-\frac{1}{L/R_2} t}$

(c) $\frac{VR_2}{R_1 + R_2} e^{-L(R_1 + R_2).t/R_1 R_2}$

(d) $\frac{V}{R_1 + R_2} e^{-Lt/R_2}$

8.53 The network shown in the given figure, the capacitor C_1 is initially charged to a voltage V_0 before the switch S in the circuit is closed. In the steady-state

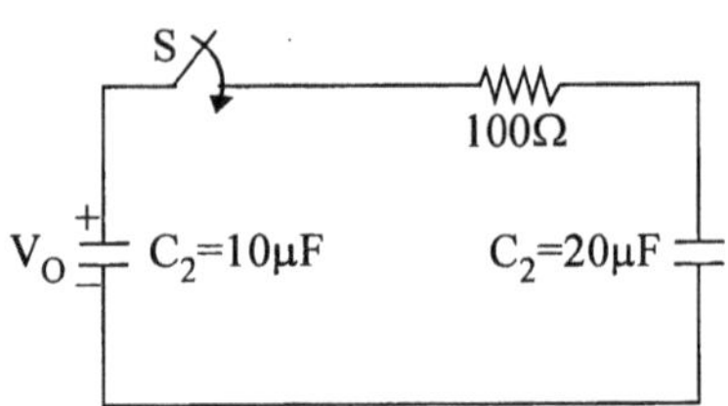

(a) C_1 and C_2 are charged to equal voltages

(b) C_1 and C_2 are charged with equal coulombs

(c) C_1 and C_2 are discharged fully

(d) C_2 alone is charged to voltage V_0

8.54 The dual of the network shown in the figure I, is

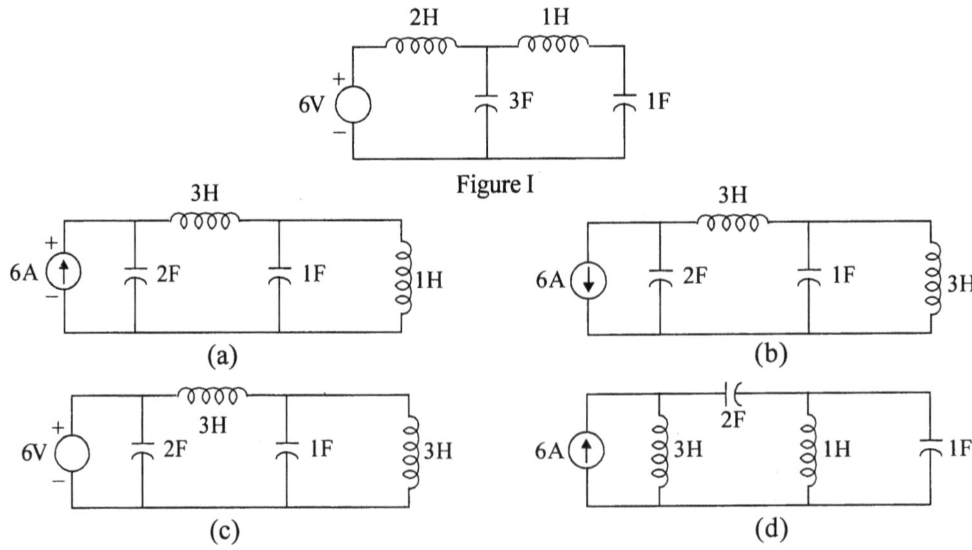

8.55 The impulse response of a first order system is Ke^{-2t}. If the input signal is sin 2t, then the steady state response will be given by

(a) $\dfrac{K}{2\sqrt{2}} \sin\left(2t + \dfrac{\pi}{4}\right)$ (b) $\dfrac{K}{4} \sin 2t$

(c) $\dfrac{K}{2\sqrt{2}} \sin\left(2t - \dfrac{\pi}{4}\right)$ (d) $\dfrac{1}{2\sqrt{2}} \sin\left(2t - \dfrac{\pi}{4}\right) + Ke^{-2t}$

8.56 Match List I with List II and select the correct answer using the codes given below the lists :

List I	**List II**
(Networks switched on to a dc supply)	(Shapes of i(t) curves)

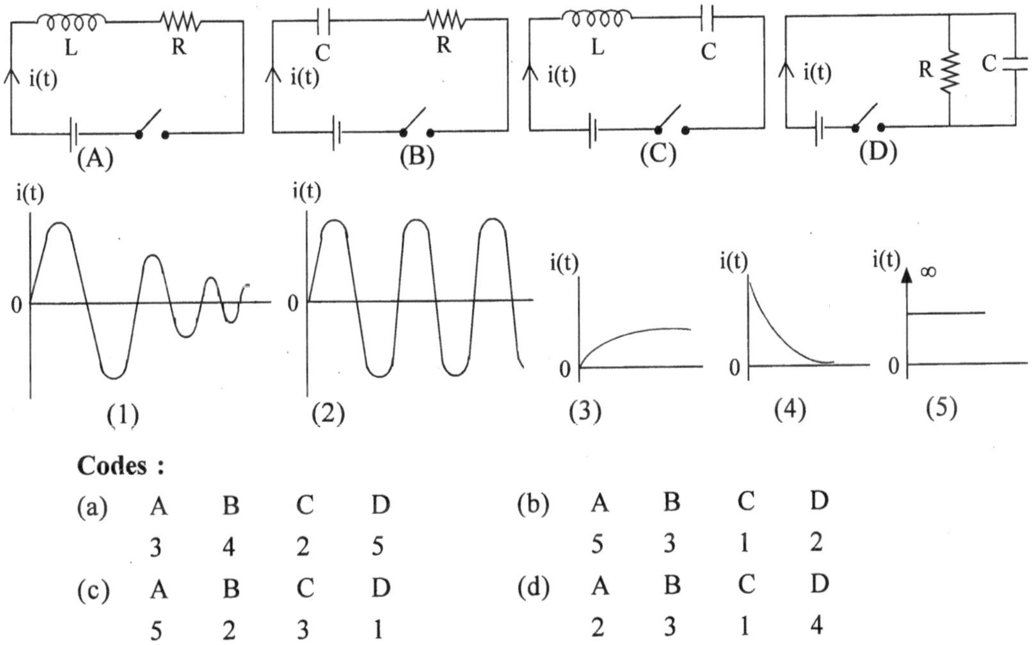

Codes :

	A	B	C	D			A	B	C	D
(a)	3	4	2	5		(b)	5	3	1	2
(c)	5	2	3	1		(d)	2	3	1	4

8.57 An RLC series circuit has R = 1 ohm, L = 1 H and C = 1 F. The damping ratio of the circuit will be

(a) more than unity (b) unity (c) 0.5 (d) zero

8.58 A step function voltage is applied to an RLC series circuit having R = 2 Ω, L = 1 H and C = 1 F. The transient current response of the circuit would be

(a) over damped (b) critically damped

(c) underdamped

(d) over, under or critically damped depending upon the magnitude of the step voltage

8.59 When a current source of value 1 is suddenly connected across a two-terminal relaxed RC network at time t = 0, the observed nature of the voltage across the current source is shown in the given figure.

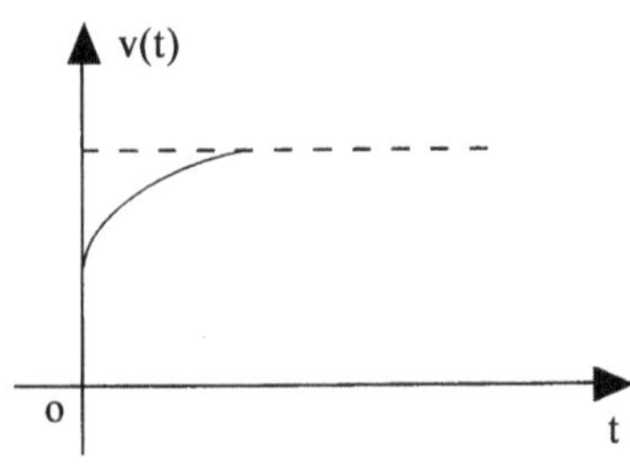

The RC network is

(a) a series combination of R and C

(b) a parallel combination of R and C

(c) a series combination of R and parallel combination of R and C

(d) a pure capacitor

8.60 Double integration of a unit step function would lead to

(a) an impulse (b) a parabola (c) a ramp (d) a doublet

8.61 Consider the following units :

1. sec^{-1} 2. rad^2/sec^2 3. sec 4. ohm

The units of $\frac{R}{L}, \frac{1}{LC}$, CR and $\sqrt{\frac{L}{C}}$ are respectively.

(a) 1, 2, 4 and 3 (b) 3, 2, 1 and 4

(c) 2, 4, 1 and 3 (d) 1, 2, 3 and 4

8.62 The circuit shown in the given figure, a capacitor is used to develop voltage pulses of short duration and large duty cycle across R_2 when the switch 'S' is closed. Which one of the following combinations be best suited for this purpose ?

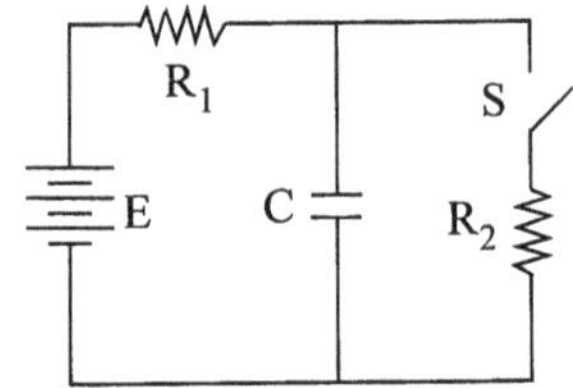

(a) R_1 and R_2 large but C of small value

(b) R_1 and R_2 small but C of large value

(c) R_1 and C large but R_2 small

(d) R_2 and C large but R_1 small

8.63 The circuit shown in the given figure, switch 'S' is closed at time t = 0. After some time when the current in the inductor was 6 A, the rate of change of current through it was 4 A/s. The value of the inductor is

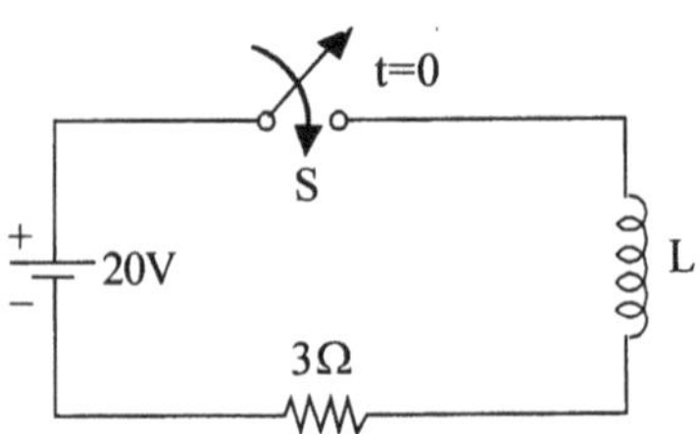

(a) indeterminate

(b) 1.5 H

(c) 1.0 H

(d) 0.5 H

8.64 Match List I with List II and select the correct answer using the codes given below the lists :

A. 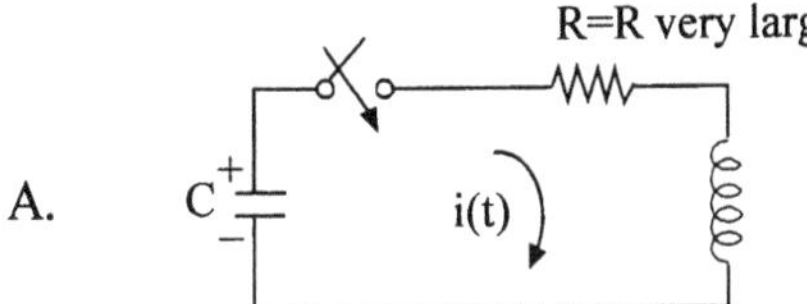

1. 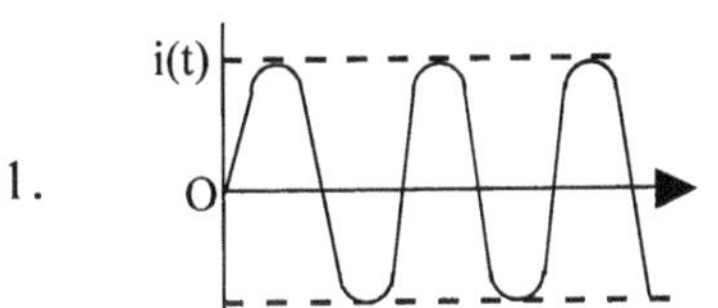

B. 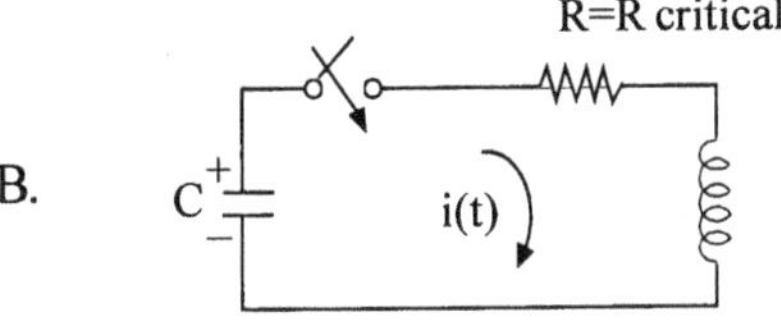

2. 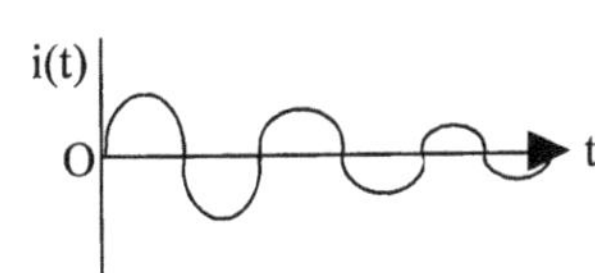

C. 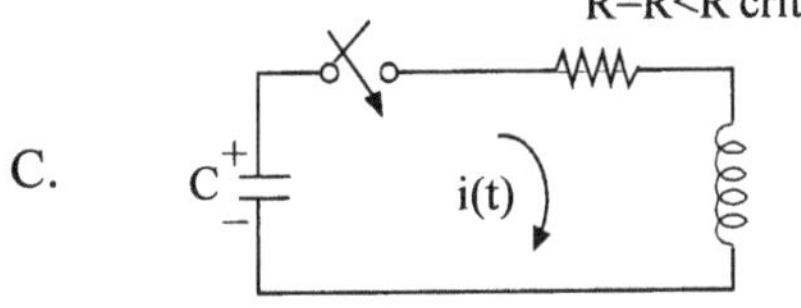

3. 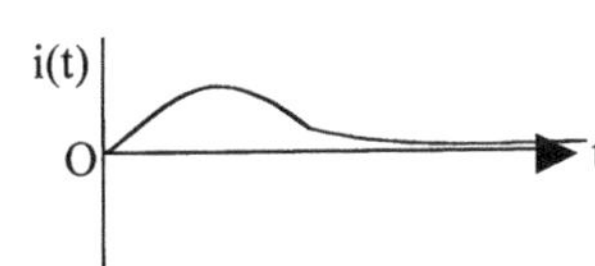

D. 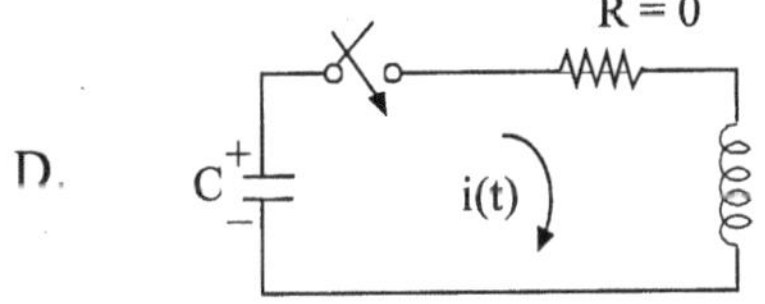

4. 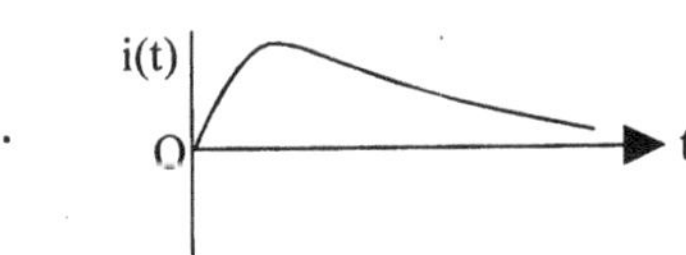

List I (C = Charged Capacitor) **List II**

Codes :

	A	B	C	D
(a)	4	3	1	2
(b)	3	4	1	2
(c)	3	4	2	1
(d)	4	3	2	1

8.65 The time constant associated with the capacitor charging in the circuit shown in the given figure is

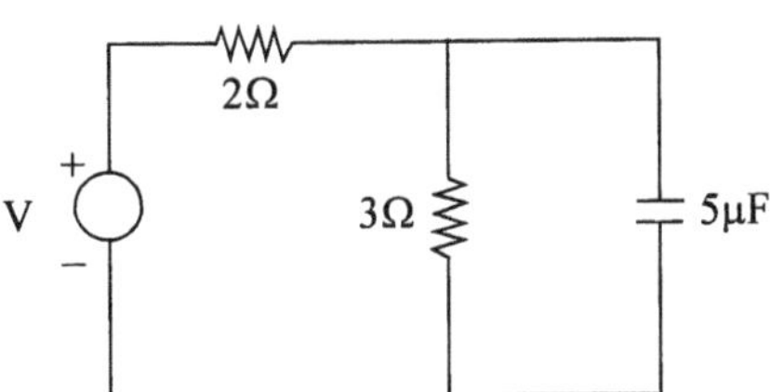

(a) 6 μs

(b) 10 μs

(c) 15 μs

(d) 25 μs

8.66 A series RL circuit is initially relaxed. A step voltage is applied to the circuit. If τ is the time constant of the circuit, the voltage across R and L will be the same at time t equal to

(a) $\tau \log_e 2$ (b) $\tau \log_e \frac{1}{2}$ (c) $\frac{1}{\tau} \log_e 2$ (d) $\frac{1}{\tau} \log_e \frac{1}{2}$

8.67 If the step response of a causal, linear time-invariant system is a(t), then the response of the system to a general input x(t) would be

(a) $\int_{0^+}^{t} \frac{d\,a(\tau)}{d\tau}\, x(t-\tau)\, d\tau$ (b) $a(0)\,x(t) + \int_{0^+}^{t} \frac{d\,a(\tau)}{d\tau}\, x(t-\tau)\, d\tau$

(c) $x(0)\,a(t) + \int_{0^+}^{t} x(\tau)\, a(t-\tau)\, d\tau$ (d) $x(0)\,a(t) + \int_{0^+}^{t} \frac{d\,a(\tau)}{d\tau}\, x(t-\tau)\, d\tau$

8.68 A resistor R of 1 Ω and two inductors L_1 and L_2 of inductances 1 H and 2 H, respectively, are connected in parallel. At some time, the currents through L_1 and L_2 are 1 A and 2 A, respectively. The current through R at time $t = \infty$ will be

(a) zero (b) 1 A (c) 2 A (d) 3 A

8.69 The dual of a parallel R-C circuit is a

(a) series R-C circuit (b) series R-L circuit

(c) parallel R-C circuit (d) parallel R-L circuit

8.70 A rectangular current pulse of duration T and magnitude I has the Laplace transform

(a) I/s (b) (I/s) exp (–Ts) (c) (I/s) exp (Ts) (d) (I/s) [1– exp (–Ts)]

8.71 Let s(t) be the step response of a linear system with zero initial conditions ; then the response of this system to an input u (t) is

(a) $\int_{0}^{t} s(t-\tau)\, u(\tau)\, d\tau$ (b) $\frac{d}{dt}\left[\int_{0}^{t} s(t-\tau)\, u(\tau)\, d\tau\right]$

(c) $\int_{0}^{t} s(t-\tau)\left[\int_{0}^{t} u(\tau_1)\, d\tau_1\right] d\tau$ (d) $\int_{0}^{t} s(t-\tau)^2\, u(\tau)\, d\tau$

8.72 A first order, low pass filter is given with R = 50 Ω and C = 5 μ F. What is the frequency at which the gain of the voltage transfer function of the filter is 0.25 ?

(a) 4.92 kHz (b) 2.46 kHz (c) 0.49 kHz (d) 24.6 kHz

8.73 Consider the following statements :

If the energy sources are connected to the unenergised system as shown in figures A and B, at t = 0, then

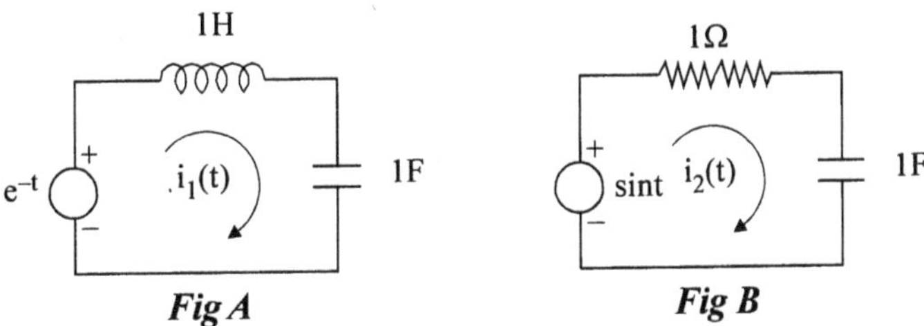

Fig A *Fig B*

1. $i_1(t)$ and $i_2(t)$ are equal
2. the circuit in Fig. A has one free frequency and two forced frequencies.
3. the circuit in Fig. B has one free frequency and two forced frequencies.
4. the circuit in Fig. A has two free frequencies and one forced frequency

On these statements

(a) 1, 2 and 3 are correct (b) 1, 3 and 4 are correct

(c) 1 and 3 are correct (d) 4 alone is correct

8.74 The network shown in the figure is described by the equation :

$$\frac{di(t)}{dt} + 3i(t) = i_s \text{ and } i(0) = 0$$

Match List I with List II and select the correct answer using the codes given below the Lists :

List-I [i_s (t)]	**List-II [i (t)]**
A. u (t)	1. $\frac{1}{\sqrt{10}}[\cos(t-\phi) - e^{-3t}\cos\phi]$
B. e^{2t} u (t)	2. $\frac{1}{j+3}[e^{jt} - e^{-3t}]$
C. e^{jt} u (t)	3. $\frac{1}{5}(e^{2t} - e^{-3t})$
D. cos t u (t)	4. $\frac{1}{3}(1 - e^{-3t})$

Codes :

	A	B	C	D		A	B	C	D
(a)	4	3	1	2	(b)	3	4	2	1
(c)	3	4	1	2	(d)	4	3	2	1

8.75 The network shown in the figure, the switch 'S' is closed and a steady state is attained. If the switch is opened at t = 0, then the current i (t) through the inductor will be

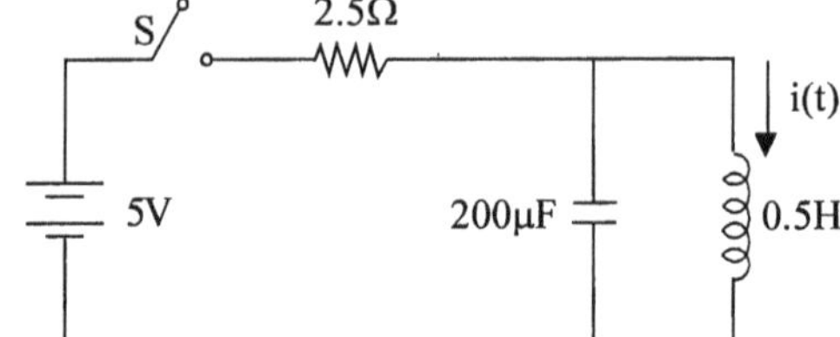

(a) cos 50 t A

(b) 2 A

(c) 2 cos 100 t A

(d) 2 sin 50 t A

CHAPTER 9 – Laplace Transform Method of Analysis of Networks

9.1 The network shown in Fig, 1 if the 1 F capacitor had an initial voltage of 2 V, then which of the following would represent the s-domain equivalent circuits ?
Select the correct answer using the codes given below :

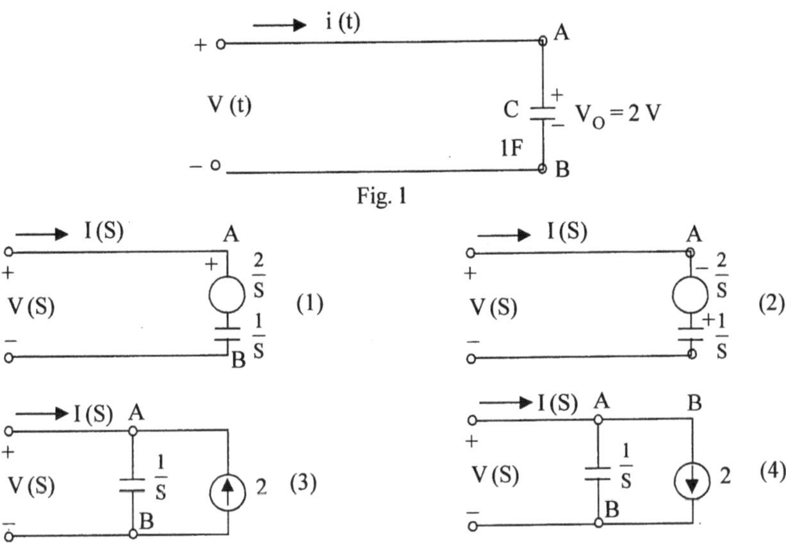

Fig. 1

Codes :

(a) 1 and 3 (b) 1 and 4 (c) 2 and 3 (d) 2 and 4

9.2 The response shown in the given figure is the Laplace transform of the function

(a) $\dfrac{\omega}{(s+\alpha)^2+\omega^2}$

(b) $\dfrac{\alpha}{(s+\alpha)^2+\omega^2}$

(c) $\dfrac{s+\alpha}{(s+\alpha)^2+\omega^2}$

(d) $\dfrac{s}{(s+\alpha)^2+\omega^2}$

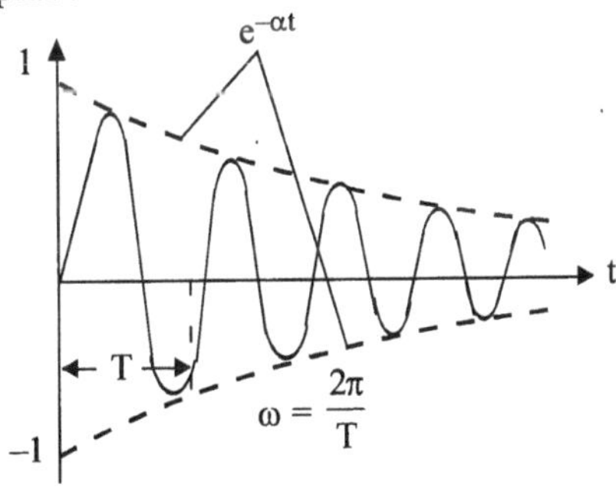

9.3 In the circuit shown in Figure-I, the switch 'K' was initially at position '1' and a current 'I' was flowing through the inductor 'L' and a voltage 'V_0' existed across the capacitor 'C'.

If at t = 0, the switch 'K' is put on the position '2' in the circuit shown in Figure-I, which one of the following transformed circuits will give i(t) for t > 0 ?

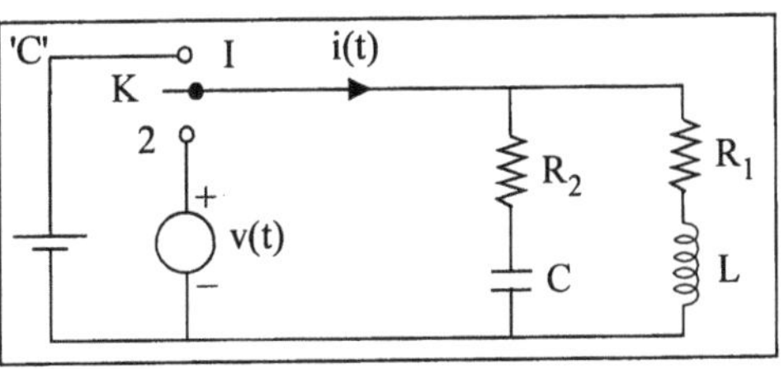

Figure-I

(a)

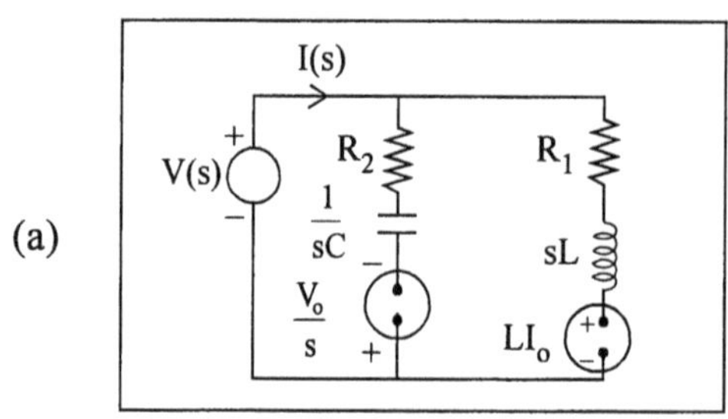

(b)

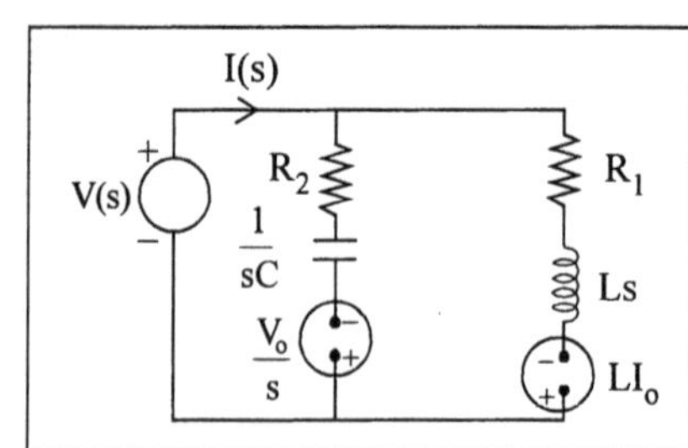

(c)

(d)

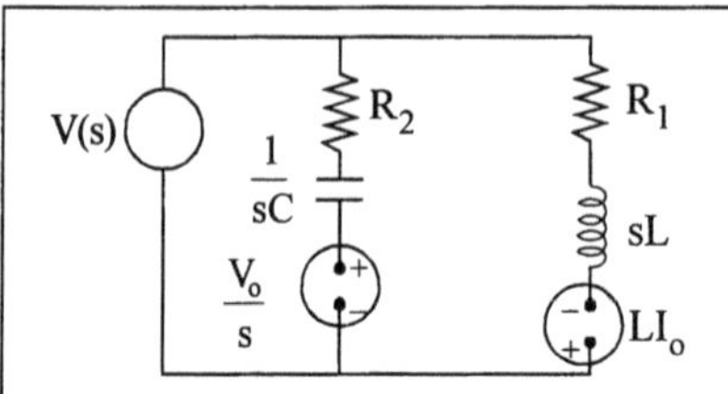

9.4 Match List-I [Input voltage v(t)] with List-II [I(s), the Laplace transform of i(t)] for the given circuit and select the correct answer using codes given below the lists :

List-I	List-II
A. Unit step	1. $\frac{s}{s+1}$
B. Unit ramp	2. $\frac{1}{s+1}$
C. Unit impulse	3. $\frac{s}{(s^2+1)(s+1)}$
D. sin t	4. $\frac{1}{s(s+1)}$

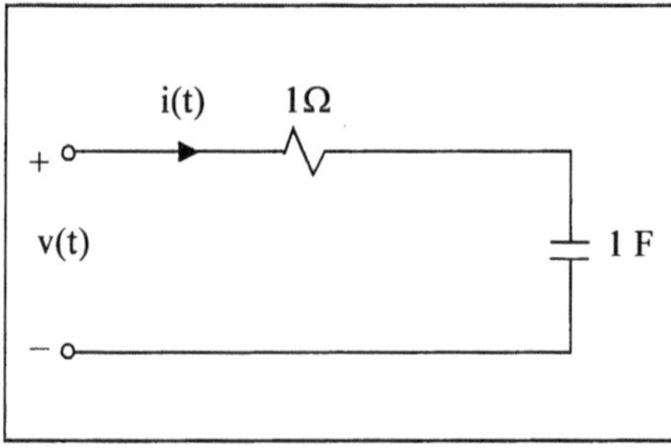

Codes :

	A	B	C	D
(a)	2	4	1	3
(b)	2	1	4	3
(c)	3	1	4	2
(d)	3	4	1	2

9.5 The driving point impedance $Z(s) = \frac{s+2}{s+3}$. The system is initially at rest. For a voltage signal of unit step, the current i(t) through the impedance Z is given by

(a) $2 - e^{-t}$ (b) $\frac{3}{2} - \frac{1}{2} e^{-3t}$ (c) $\frac{3}{2} - \frac{1}{2} e^{-2t}$ (d) $3 - 2e^{-2t}$

9.6 The Laplace transform of $(t^2 - 2t)\, u(t-1)$ is

(a) $\frac{2}{s^3} e^{-s} - \frac{2}{s^2} e^{-s}$ (b) $\frac{2}{s^3} e^{-2s} - \frac{2}{s^2} e^{-s}$

(c) $\frac{2}{s^3} e^{-s} - \frac{1}{s} e^{-s}$ (d) None of the above

9.7 Consider the voltage waveform shown in the given figure. The equation for v(t) is

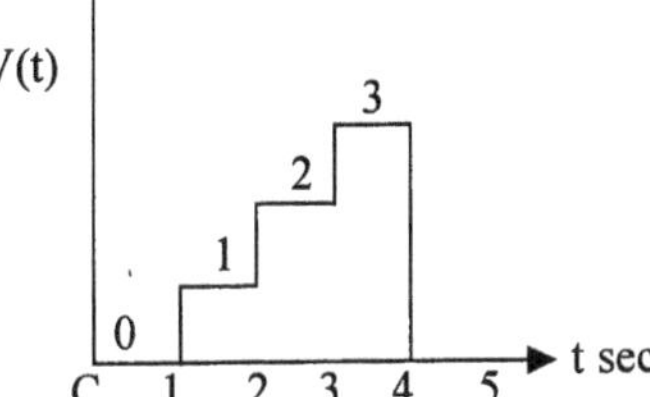

(a) $u(t-1) + u(t-2) + u(t-3)$
(b) $u(t-1) + 2u(t-2) + 3u(t-3)$
(c) $u(t) + u(t-1) + u(t-2) + u(t-4)$
(d) $u(t-1) + u(t-2) + u(t-3) - 3u(t-4)$

9.8 Match List I with List II and select the correct answer using the codes given below the lists :

	List I (Time function)		List II (Laplace transforms)
A.	$[af_1(t) + bf_2(t)]$	1.	$aF_1(s) + bF_2(s)$
B.	$[e^{-at} f(t)]$	2.	$sF(s) + f(0-)$
C.	$\left[\frac{df(t)}{dt}\right]$	3.	$\frac{1}{s} F(s)$
D.	$\left[\int_0^t f(x)dx\right]$	4.	$sF(s) - f(0-)$
		5.	$aF_1(s) + a_2 F_2(s)$
		6.	$F(s+a)$

Codes :

	A	B	C	D
(a)	6	2	3	4
(b)	1	6	4	3
(c)	2	5	3	4
(d)	1	6	3	4

9.9 A transient current in network is : $i(t) = 2e^{-t} - e^{-5t}, t \geq 0$.

The pole-zero configuration of I (s) is

	Poles	Zeros
(a)	1, 5	9
(b)	–1, –5	–9
(c)	2, –1	–1, –5
(d)	2, –1	1, 5

9.10 Match List I with List II and select the correct answer using the codes given below the lists :

List I (Location of poles on s-plane) **List II (Type of response)**

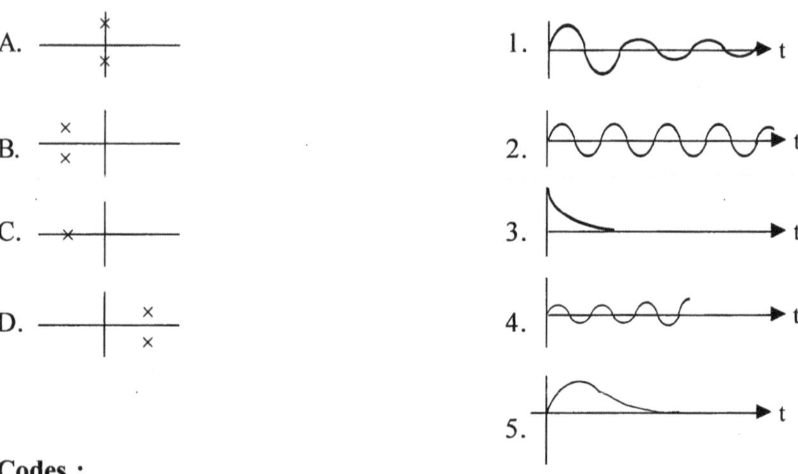

Codes :

	A	B	C	D
(a)	4	1	2	3
(b)	5	1	4	3
(c)	2	3	5	4
(d)	2	1	3	4

9.11 Consider the following statements :

1. Transfer impedance is the reciprocal of transfer admittance.
2. One can derive transfer impedance of a network if its driving-point impedance and admittance are known.
3. Driving-point impedance is the ratio of the Laplace transform of voltage and current time functions at the input

Of these statements

(a) 1,2 and 3 are correct (b) 1 and 2 are correct
(c) 2 and 3 are correct (d) 3 alone is correct

9.12 The pole-zero configuration of a network transfer function is shown in the given figure. The magnitude of the transfer function will

jw
S-Plane
−2 −1 1 2 σ
X-Pole
O-Zero

(a) decrease with frequency

(b) increase with frequency

(c) initially increase and then decrease with frequency

(d) be independent of frequency

9.13 The impedance Z(s) of the one-port network shown in the figure is given by

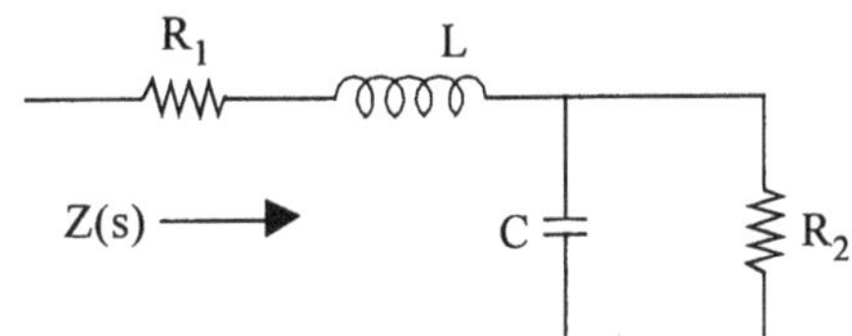

(a) $\dfrac{s^2 + s\dfrac{L}{C} + \dfrac{R_1}{LC}}{s + \dfrac{1}{LC}}$ (b) $L\dfrac{s^2 + s\dfrac{L + R_1R_2C}{R_2LC} + \dfrac{R_1 + R_2}{R_2LC}}{s + \dfrac{1}{R_2C}}$

(c) $L\dfrac{s^2 + s\dfrac{L + C}{R_1R_2} + \dfrac{R_1}{LC}}{s + \dfrac{1}{R_1C}}$ (d) $\dfrac{s^2 + s\dfrac{L + R_2C}{R_1LC} + \dfrac{R_1 + R_2}{R_1LC}}{s + \dfrac{1}{R_1C}}$

9.14 Match List-I with List-II and select the correct answer using the codes given below the List :

List-I (*Function*)	List-II (*Laplace transform*)
A. Unit ramp	1. s
B. Unit step	2. 1
C. Unit impulse	3. 1/s
D. Unit double	4. $1/s^2$

Codes :

	A	B	C	D
(a)	4	3	2	1
(b)	3	4	1	2
(c)	4	3	1	2
(d)	3	4	2	1

9.15 The Laplace transform of the function i(t) is : $I(s) = \dfrac{10s+4}{s(s+1)(s^2+4s+5)}$

Its final value will be

(a) 4/5 (b) 5/4 (c) 4 (d) 5

9.16 Which one of the following pairs of poles and responses is correctly matched ?

	Poles	Responses
(a)	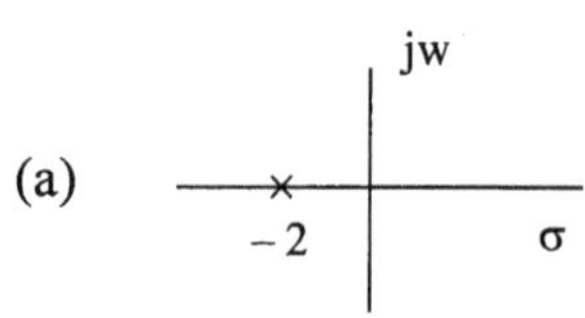	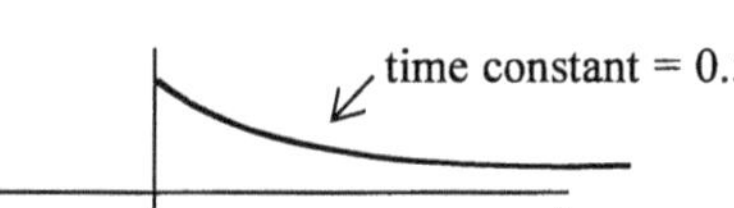
(b)	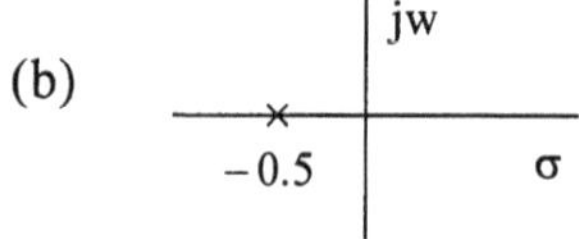	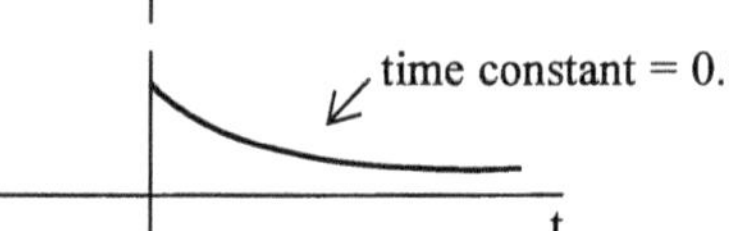
(c)	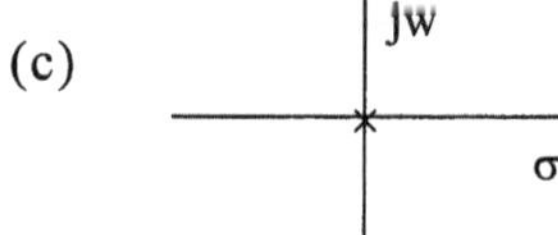	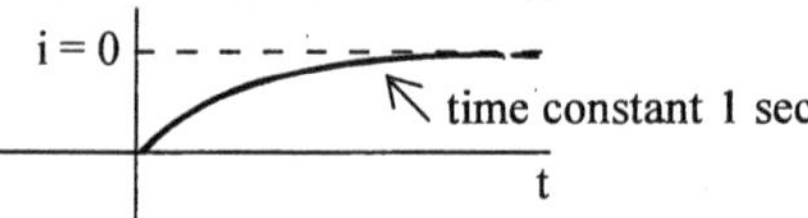
(d)	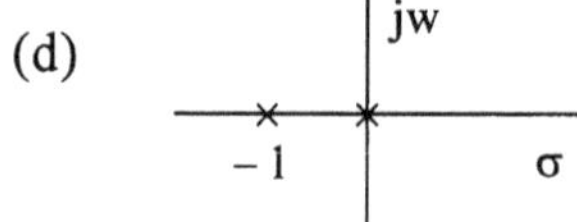	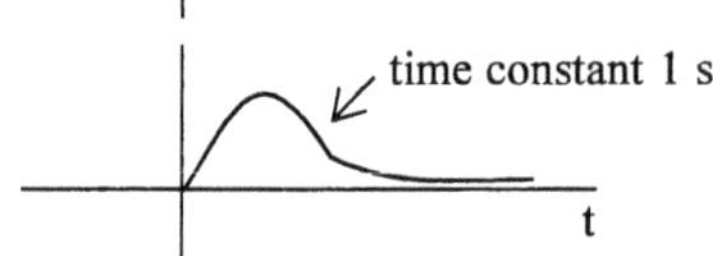

9.17 For the circuit shown in the given figure, if the input impedance Z_1 at port 1 is given by

$$Z_1 = \frac{K_1(s+2)}{(s+5)}$$

then the input impedance Z_2 at port 2 will be

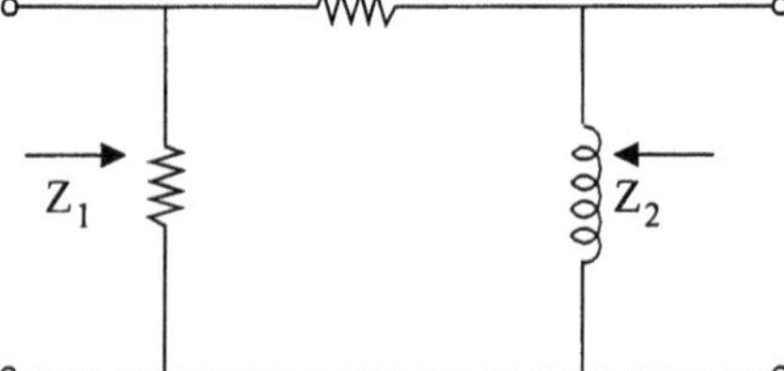

(a) 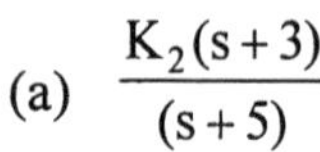 $\dfrac{K_2(s+3)}{(s+5)}$ (b) $\dfrac{K_2(s+2)}{(s+3)}$

(c) $\dfrac{K_2 s}{(s+5)}$ (d) $\dfrac{K_2 s}{(s+2)}$

9.18 For $V(s) = \dfrac{s+2}{s(s+1)}$, the initial and final value of v (t) will be respectively

(a) 1 and 1 (b) 2 and 2 (c) 2 and 1 (d) 1 and 2

9.19 For the function $L[f(t)] = \dfrac{3s+1}{s(s^2+4s+5)}$, $\left.\dfrac{df}{dt}\right|_{t=0^+}$ will be

(a) 3 (b) $\frac{1}{3}$ (c) zero (d) $\frac{2}{3}$

9.20 The Laplace transform of the function f(t) is F(s). u(t) represents the unit step function. The inverse Laplace transform of e^{-s} F(s) is

(a) f (t) u (t – 1) (b) f (t – 1) u (t) (c) f (t – 1) u (t – 1) (d) $\dfrac{f(t)}{(t-1)}$

9.21 If the unilateral Laplace transform X(s) of a signal x(t) is $\dfrac{7s+10}{s(s+2)}$, then the initial and final values of the signal would be respectivley

(a) 3.5 and 5 (b) zero and 7 (c) 5 and zero (d) 7 and 5

9.22 Consider the following functions for the rectangular voltage pulse shown in the given figure :

1. v (t) = u (t – a) – u (t – b)
2. v (t) = u (b – t) – u (a – t)
3. v (t) = u (b – t) . u (t – a)
4. v (t) = u (a – t) . u (t – b)

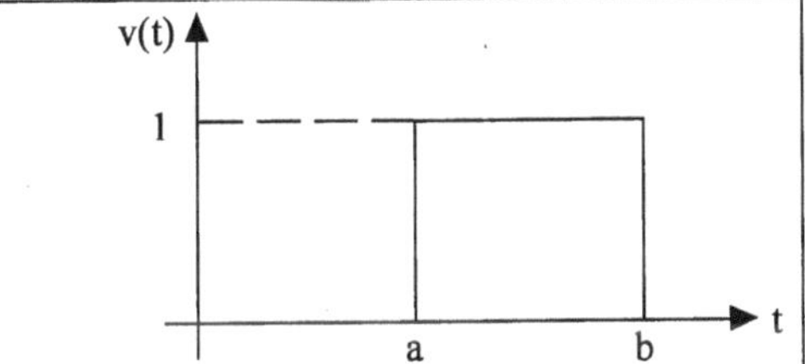

Which of these functions describe the pulse shown in the given figure ?

(a) 1, 2 and 3 (b) 1, 2 and 4 (c) 2, 3 and 4 (d) 1, 3 and 4

9.23 A pole of driving point admittance function implies

(a) zero current for a finite value of driving voltage

(b) zero voltage for a finite value of driving current

(c) an open circuit conditions

(d) None of (a), (b) and (c) mentioned in the question

9.24 Figure I shows a network. Its pole-zero diagram for the network function $\dfrac{V_2(s)}{I_1(s)} = Z(s)$ is given in Figure II. If $\lim_{s \to 0} Z(s) = 1\ \Omega$, then the values of R, L and C will be

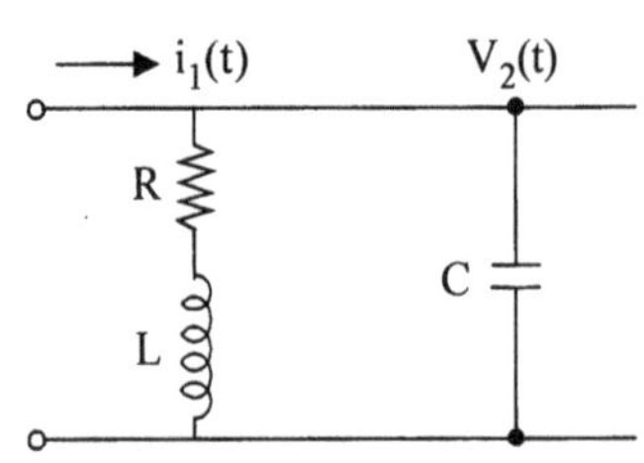

Fig. I

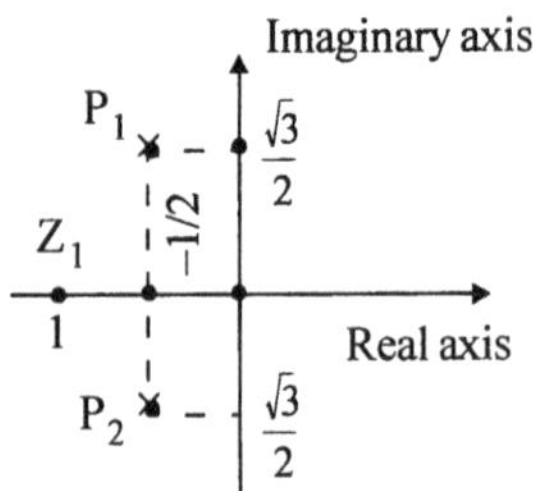

Fig. II

	R (in ohms)	L (in H)	C (in F)
(a)	1	1	1
(b)	1	0.5	0.5
(c)	0.5	1	1
(d)	1	1	0.5

9.25 Consider the following statements regarding the T-network shown in the figure : The zeros of the short-circuit input impedance Z_{is} across 11' are at

1. ω = zero
2. ω = infinity
3. $\omega = \dfrac{2}{\sqrt{LC}}$

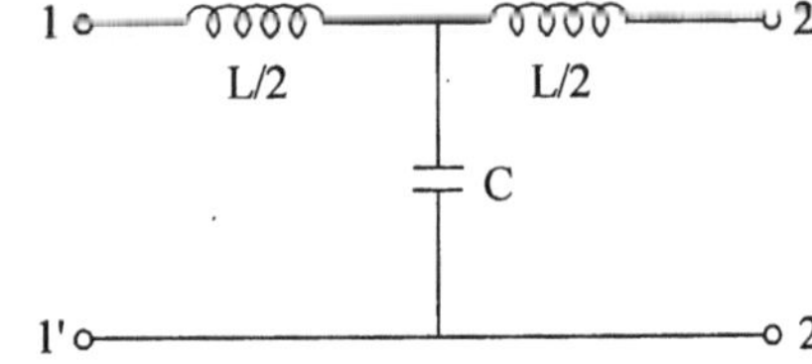

Of these statements

(a) 1 alone is correct (b) 1 and 2 are correct
(c) 2 and 3 are correct (d) 1 and 3 are correct

9.26 Which of the following pairs are correctly matched ?

	Function		**Laplace Transform**
1.	te^{at}	:	$\dfrac{1}{(s-a)^2}$
2.	$1 - e^{at}$	:	$\dfrac{a}{s(s-a)}$
3.	$e^{-at}\sin\omega t$	:	$\dfrac{\omega}{(s+a)^2+\omega^2}$
4.	$1 - \cos\omega t$	:	$\dfrac{\omega^2}{s(s^2+\omega^2)}$

Select the correct answer using the codes given below :

(a) 1, 2 and 3 (b) 1, 3 and 4 (c) 1, 2 and 4 (d) 2, 3 and 4

9.27 With symbols having the usual meanings, the Laplace transform of u(t – a) is

(a) $\frac{1}{s}$ (b) $\frac{1}{s-a}$ (c) $e^{-as}\left(\frac{1}{s}\right)$ (d) $e^{as}\left(\frac{1}{s}\right)$

9.28 If $F(s) = \frac{1(s+1)}{s(s+k)}$ and f(t) as $t \to \infty$ is $\frac{1}{2}$, then the value of k is

(a) $\frac{1}{2}$ (b) 1 (c) 2 (d) ∞

9.29 The Laplace transform of v (t) shown in the figure is

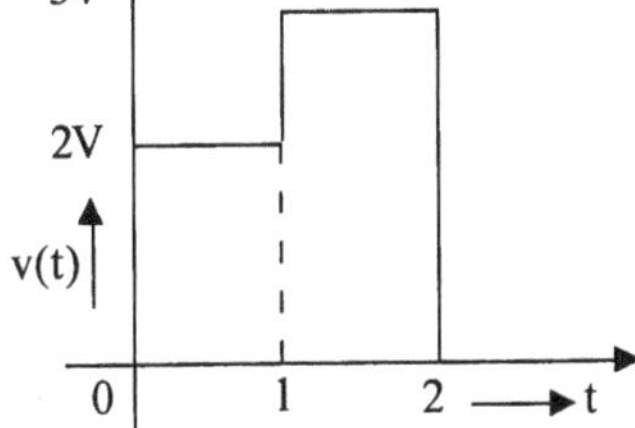

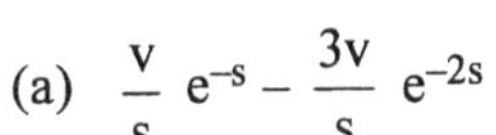

(a) $\frac{v}{s}e^{-s} - \frac{3v}{s}e^{-2s}$ (b) $\frac{2v}{s}e^{-s} - \frac{3v}{s}e^{-2s}$

(c) $\frac{2v}{s} + \frac{v}{s}e^{-s}$ (d) $\frac{2v}{s} + \frac{v}{s}e^{-s} - \frac{3v}{s}e^{-2s}$

9.30 For the network shown in the given figure, Z(O) = 3 Ω and Z (∞) = 2 Ω. The values of R_1 and R_2 will respectively be

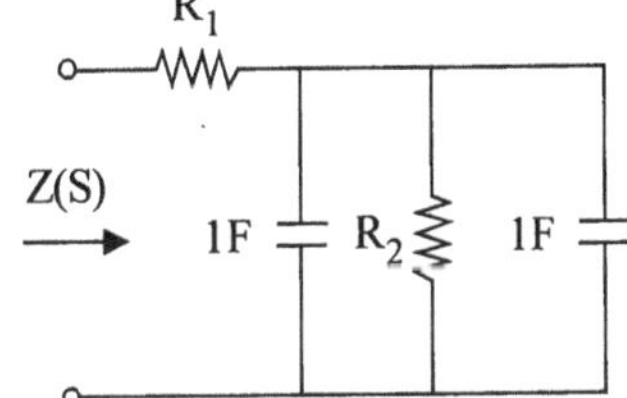

(a) 2 Ω, 1 Ω

(b) 1 Ω, 2 Ω

(c) 3 Ω, 2 Ω

(d) 2 Ω, 3 Ω

9.31 A signal is described by s(t) = r (t – a) – r (t – b), a < b, where r (t) is a unit ramp function starting at t = 0. The signal s (t) is represented as

(a)

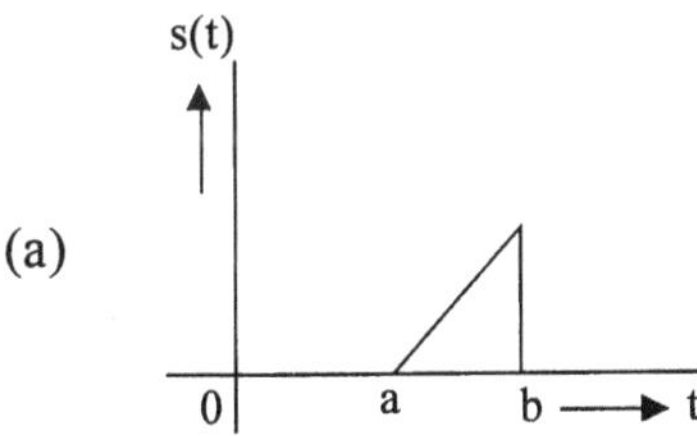

(b)

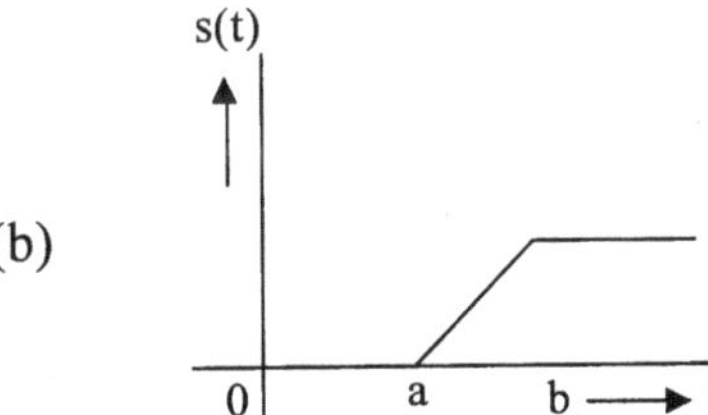

(c)

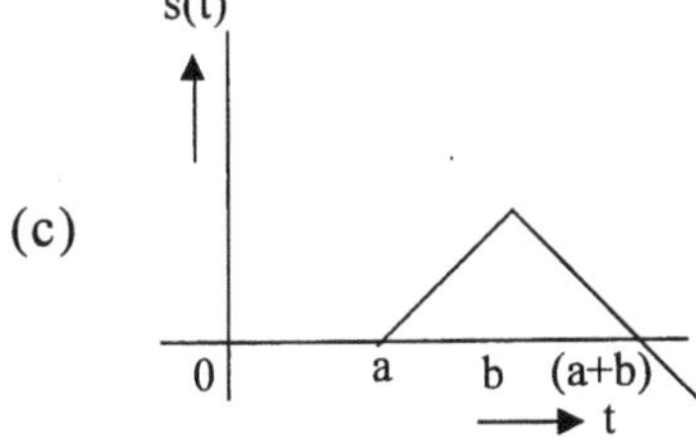

(d)

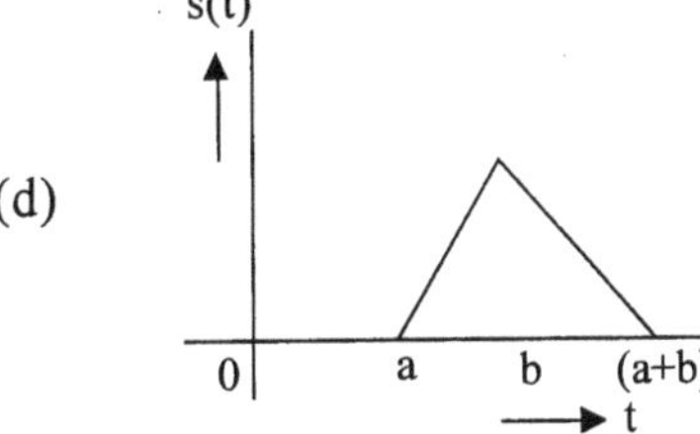

9.32 The open-circuit voltage ratio $\frac{V_2(s)}{V_1(s)}$ of the network shown in the given figure is

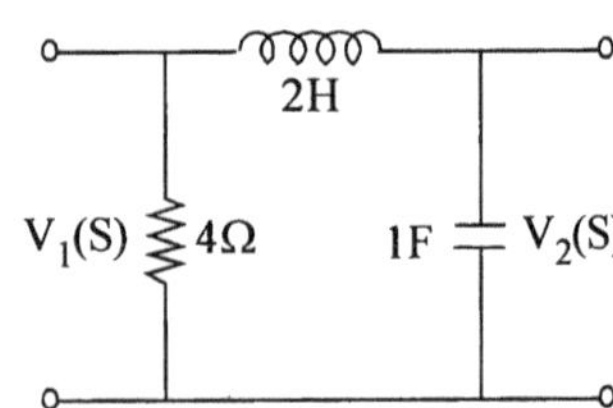

(a) $1 + 2s^2$

(b) $\frac{1}{1+2s^2}$

(c) $1+2s$

(d) $\frac{1}{1+2s}$

9.33 If $F(s) = \frac{1}{(s+1)(s^2+1)}$, then f (t) is

(a) $\int_0^t e^{-\tau} \sin(t-\tau)\, d\tau$ (b) $\int_0^t e^{-(t-\tau)} \sin(t-\tau)\, d\tau$

(c) $\int_0^t e^{-\tau} \cos(t-\tau)\, d\tau$ (d) $e^{-t} \sin t$

9.34 The value of G_{12} or $\left(\frac{V_2}{V_1}\right)$ for the circuit shown in the figure is

(a) $\frac{1}{4s^4+2s^2+1}$ (b) $\frac{s}{s^4+2s^2+1}$

(c) $\frac{1}{s^4+1}$ (d) $\frac{1}{16s^4+12s^2+1}$

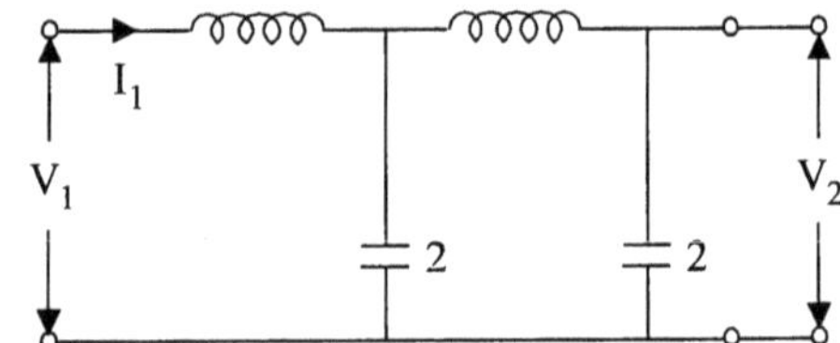

9.35 The Laplace transform of the waveform shown in the figure is

(a) $V(s) = \frac{1}{(1-e^{-as})}$

(b) $V(s) = \left(\frac{1-e^{-as}}{2s^2} - \frac{e^{-as}}{s}\right)$

(c) $V(s) = \left(\frac{1-e^{-as}}{2}\right)$

(d) $V(s) = \left(\frac{1-e^{-as}}{as^2} - \frac{e^{-as}}{s}\right)\frac{1}{(1-e^{-as})}$

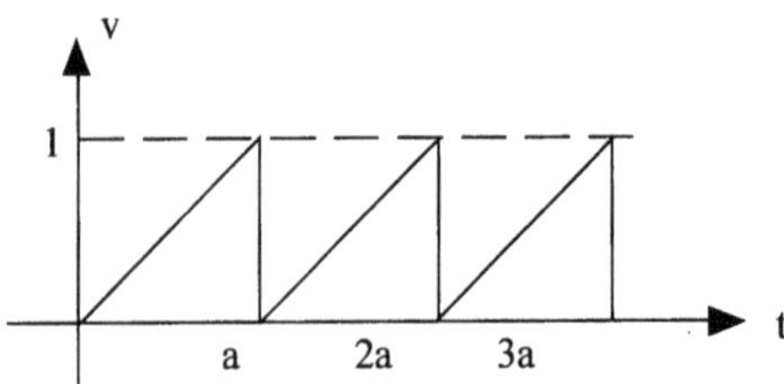

9.36 The impulse train shown in the figure represents the second derivative of a function f (t). The value of f (t) is

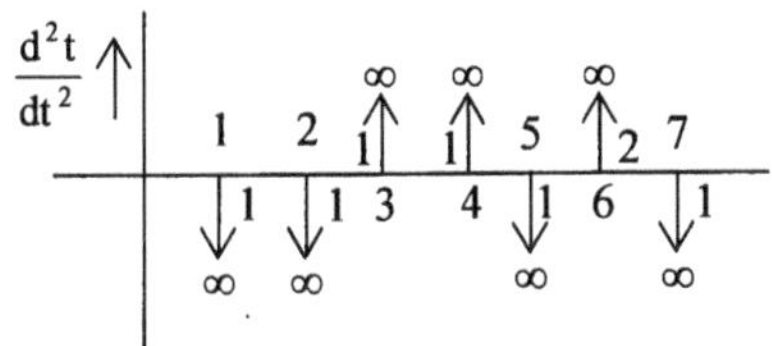

(a) $-(t-1)\,u(t-1) - (t-2)\,u(t-2) + (t-3)\,u(t-3) + (t-4)\,u(t-4) -$ $(t-5)\,u(t-5) + 2\,(t-6)\,u(t-6) - (t-7)\,u(t-7)$

(b) $-tu(t-1) - tu(t-2) - tu(t-3) - tu(t-4) + tu(t-5)$

(c) $tu(t-3) + tu(t-4) + 2\,tu(t-6)$

(d) $tu(t+1) + tu(t+2) + tu(t+3) + tu(t+4) + tu(t+5) + 2tu(t+6) + tu(t+7)$

9.37 The two networks given below are equivalent with respect to terminals 1 and 2 at all frequencies. Find C_A, L_B, L_C, C_C

1H 1F 1 2

2H C_A L_A 1/3 L_B L_C C_C 1 2

(a) 0.5, 0.33, 6, 0.166

(b) 0.5, 3, 5, 0.166

(c) 0.5, 3, 3, 2

(d) 0.5, 3, 6, 0.166

9.38 The driving point impedance function for the network shown in the figure has

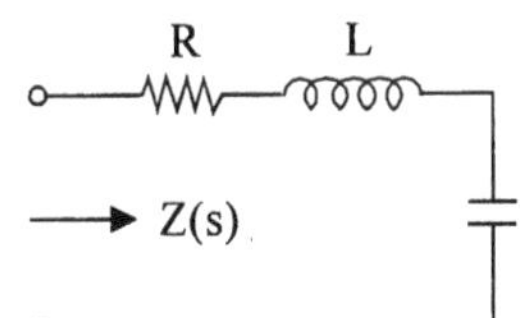

(a) poles at $s = 0$, $s = \infty$

(b) poles at $s = 0$ and zero at $s = \infty$

(c) no pole at $s = 0$

(d) no pole at $s = \infty$

9.39 The driving point admittance function of the network shown in the figure has a

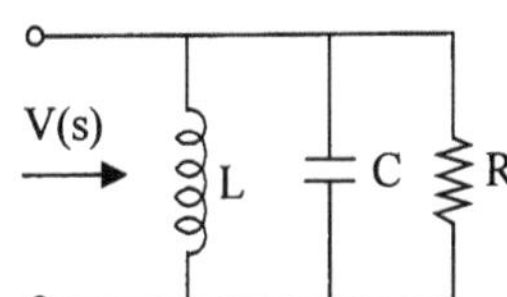

(a) pole at $s = 0$ and zero at $s = \infty$

(b) pole at $s = 0$ and pole at $s = \infty$

(c) pole at $s = \infty$ and zero at $s = 0$

(d) pole at $s = \infty$ and zero at $s = \infty$

9.40 The current transfer ratio $\frac{I_2(s)}{I_1(s)}$ for the circuit shown in the figure is,

(a) 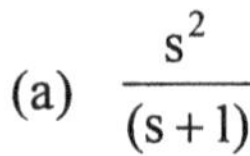 $\frac{s^2}{(s+1)}$ (b) 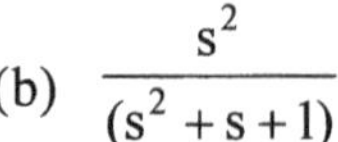 $\frac{s^2}{(s^2+s+1)}$

(c) $\frac{s^2}{(s^3+s^2+s+1)}$ (d) $\frac{s}{(s+1)}$

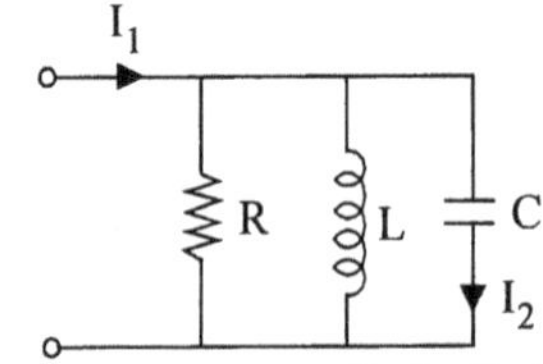

9.41 The transfer function $y_{12}(s) = \frac{I_2(s)}{V_1(s)}$ for the network shown in the figure, is

(a) $\frac{s^2}{s^2+s+1}$ (b) $\frac{s}{s+1}$

(c) $\frac{1}{s+1}$ (c) $\frac{s+1}{s^2+1}$

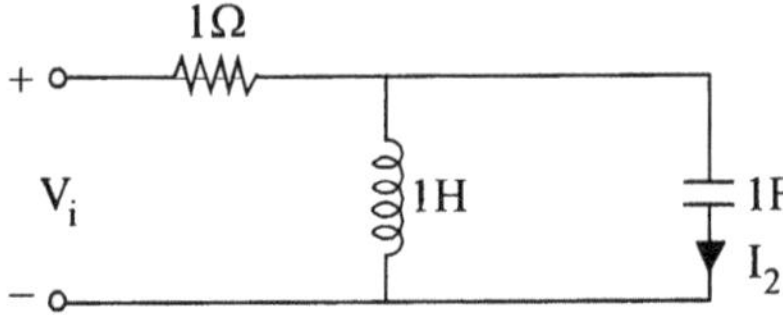

9.42 Given the Laplace transform, $V(s) = \int_0^{\infty} e^{-st}\, v(t)\, dt$. The inverse transform v(t) is

(a) $\int_{\sigma-j\infty}^{\sigma+j\infty} e^{st}\, V(s)\, ds$ (b) $\frac{1}{2\pi j}\int_{\sigma-j\infty}^{\sigma+j\infty} e^{st}\, V(s)\, ds$

(c) $\frac{1}{2\pi j}\int_{0}^{\infty} e^{st}\, V(s)\, ds$ (d) $\frac{1}{2\pi j}\int_{0}^{\infty} e^{-st}\, V(s)\, ds$

9.43 The transfer function of a low pass RC network is

(a) RCS (1 + RCS) (b) 1/(1 + RCS) (c) RC/(1 + RCS) (d) s/(1 + RCS)

9.44 Voltage transfer function of the two-port network given in the figure has

(a) a zero at the origin

(b) a zero at ∞

(c) no zero

(d) a zero at j1

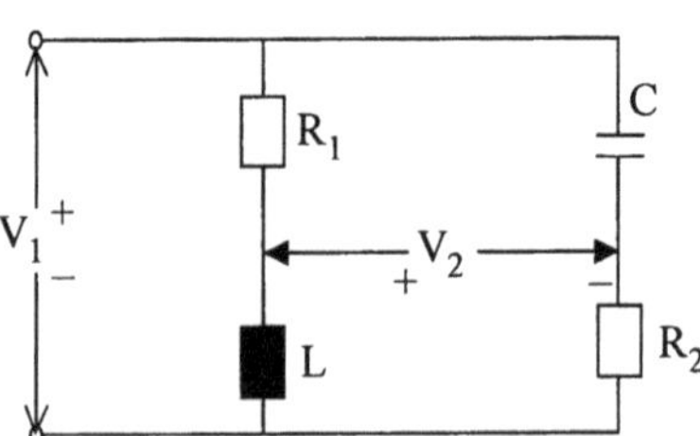

9.45 Consider the following statements :

1. Current through an inductor cannot change abruptly.
2. Voltage across the capacitor cannot change abruptly.
3. Initial value of a function f(t) is $\underset{s \to o}{\text{Lt}}\ s\ F\ (s)$
4. Final value of a function f(t) is $\underset{s \to \infty}{\text{Lt}}\ s\ F\ (s)$

Of these statements

(a) 3 and 2 are correct (b) 1 and 4 are correct
(c) 1 and 2 are correct (d) 2 and 3 are correct

9.46 In Laplace transform, the variable 's' equals $(\sigma + j\omega)$. Which of the following represent the true nature of σ ?

1. σ has a damping effect.
2. σ is responsible for convergence of integral $\int_0^{\infty} f(t) e^{-st}\, dt$
3. σ has a value less than zero

Select the correct answer using the codes given below

(a) 1, 2 and 3 (b) 1 and 2 (c) 2 and 3 (d) 1 and 3

9.47 Which one of the following is the correct Laplace transform of the signal in the given figure ?

(a) $\dfrac{1}{Ts^2}\ [1 - e^{-Ts}\ (1 + Ts)]$

(b) $\dfrac{1}{Ts^2}\ [e^{-Ts} - 1 + Ts]$

(c) $\dfrac{1}{Ts^2}\ [e^{-Ts} + 1 - Ts]$

(d) $\dfrac{1}{Ts^2}\ [1 - e^{-T}s + Ts]$

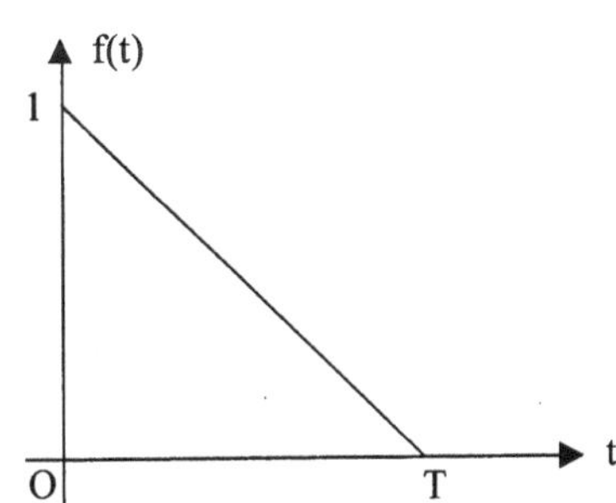

9.48 A network has response with time as shown in Fig. 1. Which one of the following diagrams represents the location of the poles of this network ?

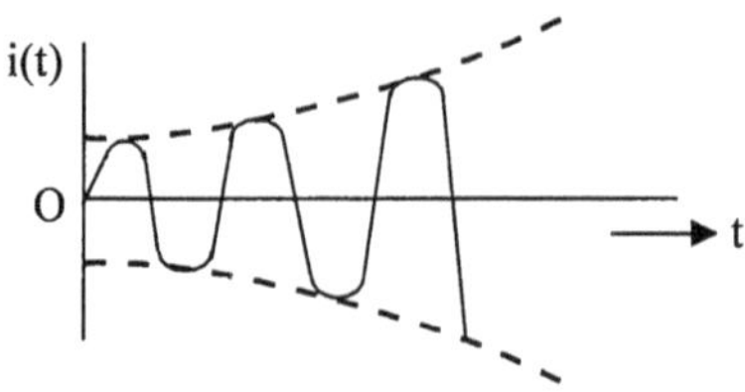

Fig. 1

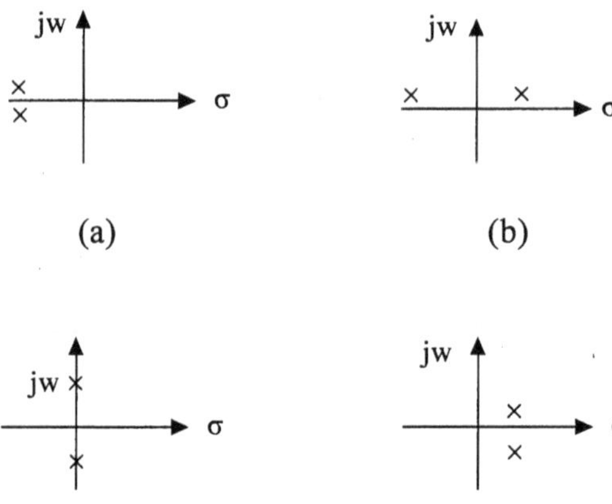

9.49 The following table gives some time functions and their Laplace transforms :

f (t)	**F (s)**
1. δ (t)	s
2. u (t)	1/s
3. tu (t)	$2/s^2$
4. t^2u (t)	$2/s^3$

Of these, the correctly matched pair is

(a) 2 and 4 (b) 1 and 4 (c) 3 and 4 (d) 1 and 2

9.50 The final value of $L^{-1}\dfrac{2s+1}{s^4+8s^3+16s^2+s}$ is

(a) infinity (b) 2 (c) 1 (d) zero

9.51 Which one of the following circuits has a driving-point impedance of

$$z(s) = \frac{2(s^2 + s + 1/2)}{(s^2 + s + 1)}$$

(a) 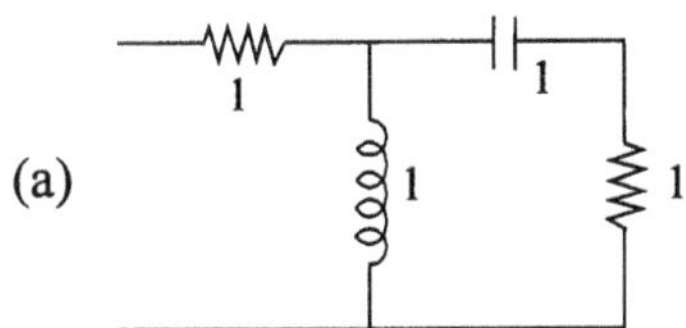

(b) 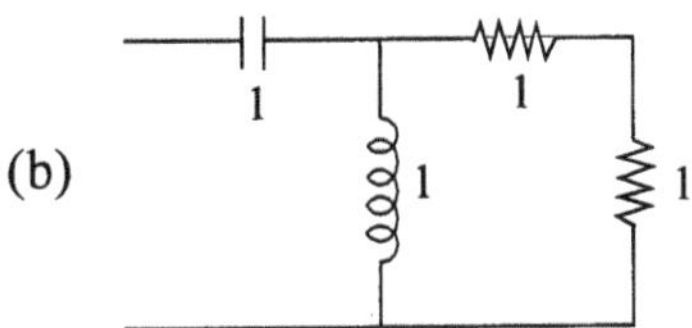

(c) 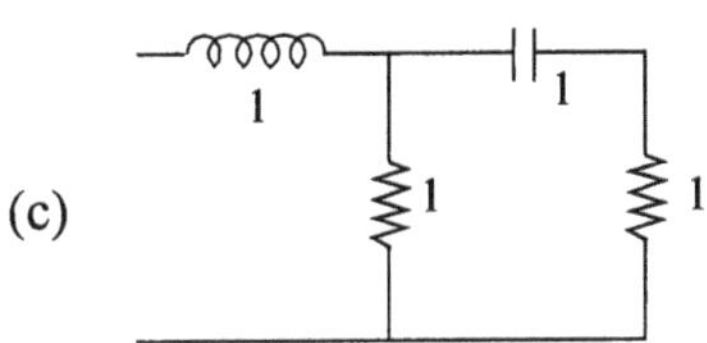

(d) 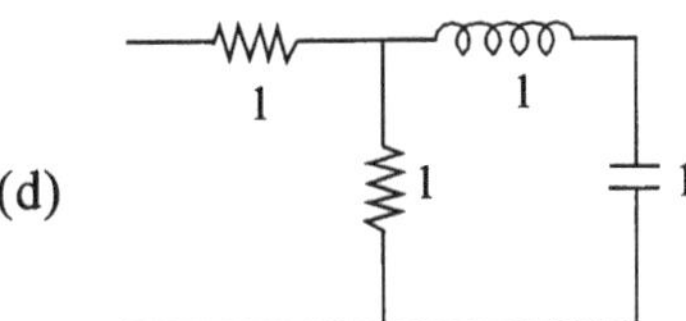

9.52 A system is represented by the transfer function $\frac{10}{(s+1)(s+2)}$. The d.c gain of the system is

(a) 1 (b) 2 (c) 5 (d) 10

9.53 The Laplace transformation method enables one to find the response of a network in

(a) the transient state only

(b) the steady state only

(c) both transient and steady state

(d) the transient state provided sinusoidal forcing functions do not exist.

9.54 Frequency response of the function T (s) = (s + 1)/(s + 2) exhibits a maximum phase at a frequency (in radian/sec.)

(a) 0 (b) $\frac{1}{\sqrt{2}}$ (c) $\sqrt{2}$ (d) ∞

9.55 Match List I with List II and select the correct answer using the codes given below the Lists :

List - I	List - II
A. 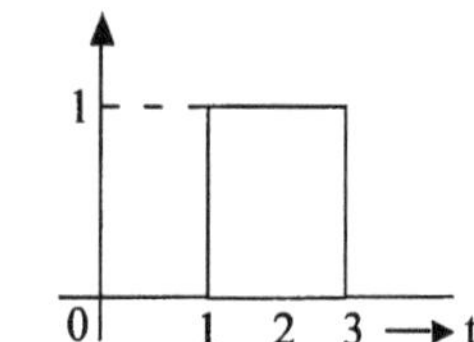	1. $v(t) = u(t + 1)$
B.	2. $v(t) = u(t) - 2u(t - 1) + 2u(t - 2) - 2u(t - 3) + ...$
C.	3. $v(t) = u(t - 1) - u(t - 3)$
D. 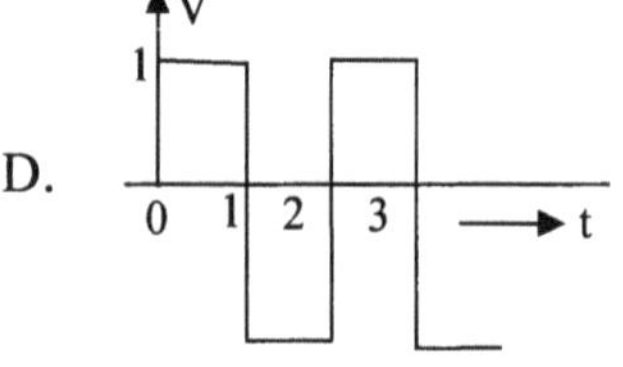	4. $\underset{a \to 0}{\text{Lt}}\ v(t) = \delta(t - 1)$

Codes :

	A	B	C	D
(a)	1	2	3	4
(b)	3	4	1	2
(c)	4	3	2	1
(d)	4	3	1	2

9.56 If $\delta(t)$ denotes a unit impulse, then the Laplace transform of $\frac{d^2\delta(t)}{dt^2}$ will be

(a) 1 (b) s^2 (c) s (d) s^{-2}

9.57 If x (t) and its first derivative are Laplace transformable and the Laplace transform of x (t) is X (s), then $\lim_{t \to 0} x(t)$

(a) $\lim_{s \to \infty} sX(s)$ (b) $\lim_{s \to 0} sX(s)$

(c) $\lim_{s \to \infty} \frac{X(s)}{s}$ (d) $\lim_{s \to 0} \frac{X(s)}{s}$

9.58 Given that $h(t) = 10\, e^{-10t}\, u(t)$, and $e(t) = \sin 10t\, u(t)$, the Laplace transform of the signal $f(t) = \int_0^t h(t-\tau)\, e(\tau)\, d\tau$ is given by

(a) $\frac{10}{(s+10)(s^2+100)}$ (b) $\frac{10(s+10)}{(s^2+100)}$

(c) $\frac{100}{(s+10)(s^2+100)}$ (d) $\frac{1}{(s+10)(s^2+100)}$

9.59 Of the following transfer function of second order linear time-invariant systems, the underdamped system is represented by

(a) $H(s) = \frac{1}{s^2+4s+4}$ (b) $H(s) = \frac{1}{s^2+5s+4}$

(c) $H(s) = \frac{1}{s^2+4.5s+4}$ (d) $H(s) = \frac{1}{s^2+3s+4}$

9.60 The impulse response of a single-pole system would approach a non-zero constant as $t \to \infty$ if and only if the pole is located in the s - plane

(a) on the negative real axis (b) at the origin

(c) on the positive real axis (d) on the imaginary axis

9.61 If x (t) and $\frac{dx(t)}{dt}$ are Laplace transformable and $\lim_{t\to\infty} x(t)$ exists, then $\lim_{t\to\infty} x(t)$ is equal to

(a) $\lim_{s\to\infty} sX(s)$ (b) $\lim_{s\to 0} sX(s)$

(c) $\lim_{s\to\infty} \frac{X(s)}{s}$ (d) $\lim_{s\to 0} \frac{X(s)}{s}$

9.62 If f_1 (t) and f_2 (t) are duration-limited signals such that

$$f_1(t) \neq 0 \quad \text{for } 1 < t < 3$$
$$= 0 \quad \text{elsewhere}$$
$$f_2(t) \neq 0 \quad \text{for } 5 < t < 7$$
$$= 0 \quad \text{elsewhere,}$$

then the convolution of f_1 (t) and f_2 (t) is zero everywhere except for

(a) $1 < t < 7$ (b) $3 < t < 5$

(c) $5 < t < 21$ (d) $6 < t < 10$

9.63 Voltage transfer function of a simple RC integrator has

(a) a finite zero and a pole at infinity (b) a finite zero and a pole at the origin

(c) a zero at the origin and a finite pole (d) a zero at infinity and a finite pole.

9.64 Driving-point impedance of the network shown in the figure is :

(a) $\dfrac{s^3 + 2s^2 + s + 1}{2s^2 + 1}$

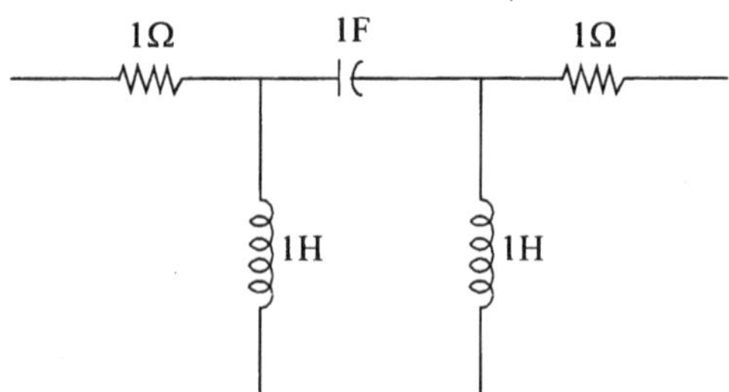

(b) $\dfrac{s^3 + 2s^2 + s + 1}{s^2 + 1}$

(c) 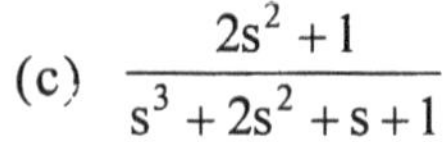 $\dfrac{2s^2 + 1}{s^3 + 2s^2 + s + 1}$

(d) 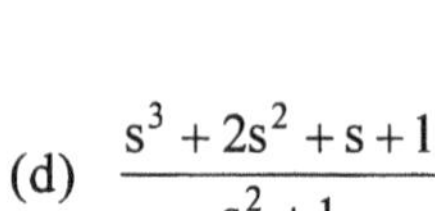 $\dfrac{s^3 + 2s^2 + s + 1}{s^2 + 1}$

9.65 If a function f(t) u(t) is shifted to right side by t_0, then the function can be expressed as

(a) $f(t - t_0)\, u(t)$ (b) $f(t)\, u(t - t_0)$

(c) $f(t - t_0)\, u(t - t_0)$ (d) $f(t + t_0)\, u(t + t_0)$

9.66 If Laplace transform of a function f (t) is $F(s) = \dfrac{s+2}{s+1}$, then f (t) is

(a) $e^{-2t} - e^{-t}$ (b) $\dfrac{d}{dt}(te^{-t})$ (c) $\delta(t) + e^{-t}$ (d) $\dfrac{d}{dt}(e^{-t})$

9.67 Match List I (System function) with list II (Impulse response) and select the correct answer using the codes given below the Lists :

List I | List II

A. 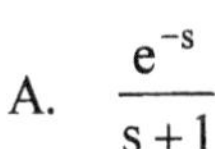 $\frac{e^{-s}}{s+1}$

1. 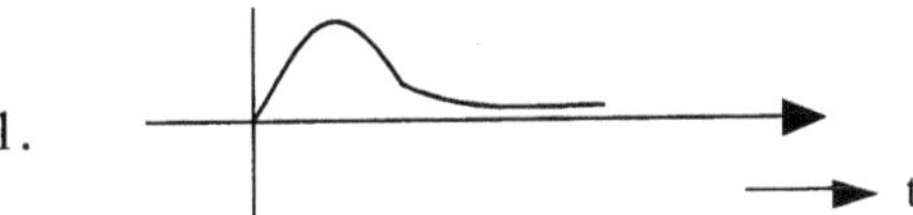

B. $\frac{1}{s^2+s+1}$

2. 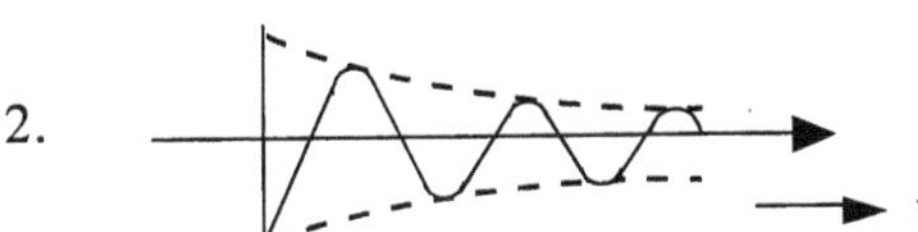

C. $\frac{1}{(s+1)^2}$

3. 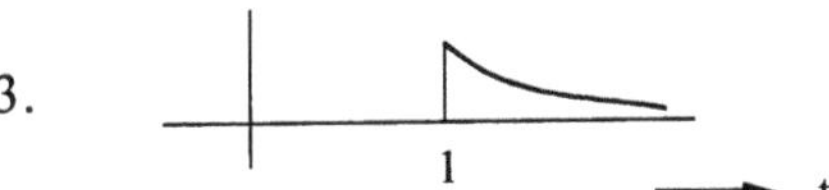

D. $\frac{1}{s^2+s}$

4. 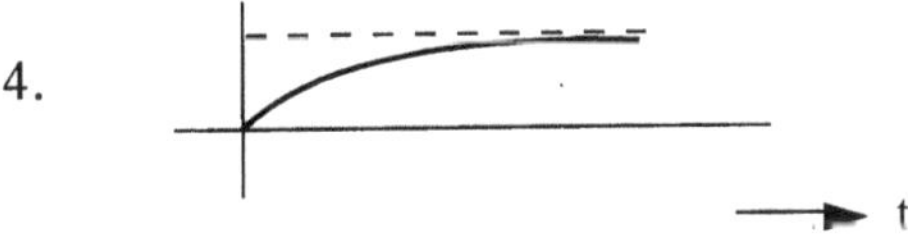

5. 

(a)	A	B	C	D
	3	4	1	2
(b)	A	B	C	D
	5	2	3	4
(c)	A	B	C	D
	3	2	1	4
(d)	A	B	C	D
	5	4	3	2

9.68 An input f(t) is applied to a linear network giving a response g(t). The function f(t) is delayed by 1 time unit. If the network is time-invariant, then the response will be

(a) g (t) (b) g(t – 1) (c) g(t + 1) (d) 1 – g(t)

9.69 A system function has a single zero and single pole. The constant multiplier 'K' is 1. For the given excitation sin t, the response is $\sqrt{2}$ with 45° lagging. The system has a pole and a zero respectively at

(a) zero and 1 (b) infinity and – 1 (c) – 1 and zero (d) zero and – 1

9.70 If the step response of an initially relaxed circuit is known, then the ramp response can be obtained by

(a) integrating the step response (b) differentiating the step response

(c) integrating the step response twice (c) differentiating the step response twice

9.71 An inductor with inductance L and initial current I_0 is shown as

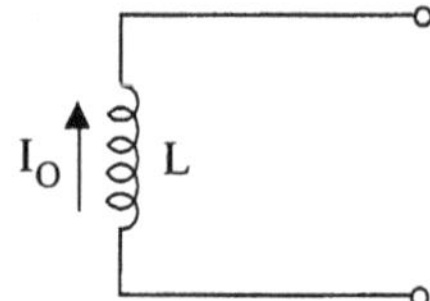

The correct admittance diagram for it is

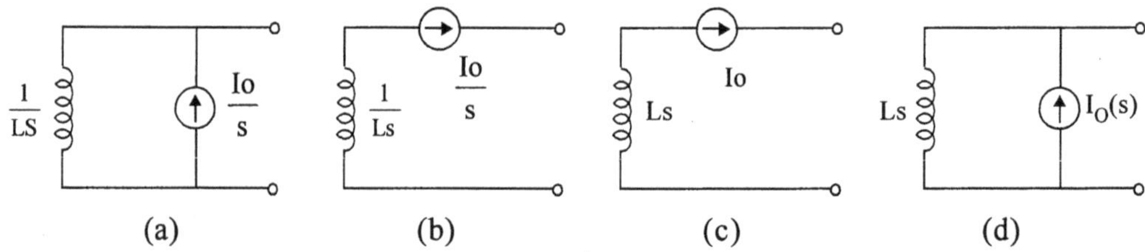

CHAPTER 10 – Two Port Networks

10.1 A two port network, shown in fig is described by the following equations

$I_1 = Y_{11} E_1 + Y_{12} E_2$

$I_2 = Y_{21} E_1 + Y_{22} E_2$

The admittance parameters, Y_{11} Y_{12} Y_{21} and Y_{22} for the network shown are

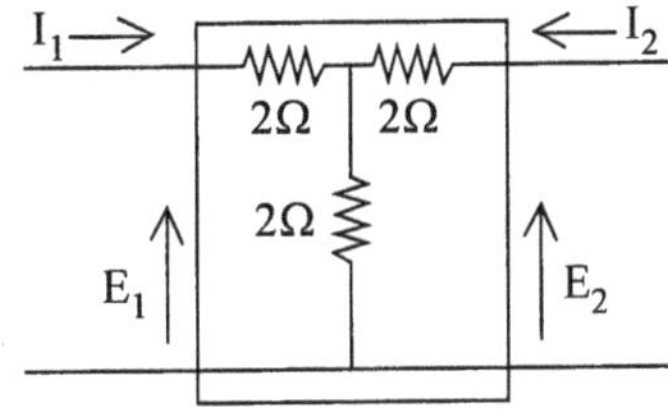

(a) 0.5 S, 1 S, 2 S and 1 S respectively

(b) $\frac{1}{3}$ S, $-\frac{1}{6}$ S, $-\frac{1}{6}$ and $\frac{1}{3}$ S respectively

(c) 0.5 S, 0.5 S, 1.5 S and 2 S respectively

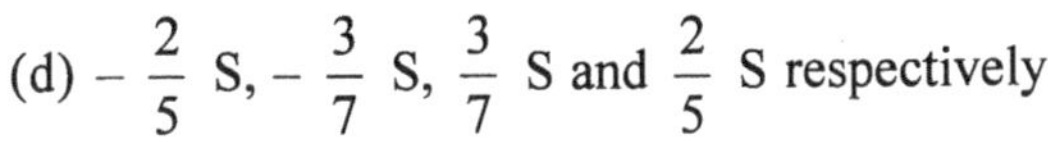

(d) $-\frac{2}{5}$ S, $-\frac{3}{7}$ S, $\frac{3}{7}$ S and $\frac{2}{5}$ S respectively

10.2 A passive 2-port network is in a steady-state. Compared to its input, the steady state output can never offer

(a) higher voltage (b) lower impedance (c) greater power (d) better regulation

10.3 A two-port device is defined by the following pairs of equations :

$I_1 = 2v_1 + v_2$ and $I_2 = v_1 + v_2$

Its impedance parameters (z_{11}, z_{12}, z_{21}, z_{22}) are given by

(a) (2, 1, 1, 1) (b) (1, –1, –1, 2) (c) (1, 1, 1, 2) (d) (2, –1, –1, 1)

10.4 When a number of 2-port networks are connected is cascade, the individual

(a) Z_{oc} matrices are added (b) Y_{sc}- matrices are added

(c) chain matrices are multiplied (d) h-matrices are multiplied

10.5 A two-port network is defined by the relations $I_1 = 2V_1 + V_2$, $I_2 = 2V_1 + 3V_2$

Then Z_{12} is

(a) – 2 Ω (b) – 1 Ω (c) –1/2 Ω (d) – 1/4 Ω

10.6 With the usual notations, a two-port resistive network satisfies the condition

$$A = D = \frac{3}{2} B = \frac{4}{3} C$$

The Z_{11} of the network is

(a) 5/3 (b) 4/3 (c) 2/3 (d) 1/3

10.7 In a two-port network, the condition for reciprocity in terms of 'h' - parameters is

(a) $h_{12} = h_{21}$ (a) $h_{11} = h_{22}$ (c) $h_{11} = -h_{22}$ (d) $h_{12} = -h_{21}$

10.8 The z and h parameters of the network shown in the given figure will be

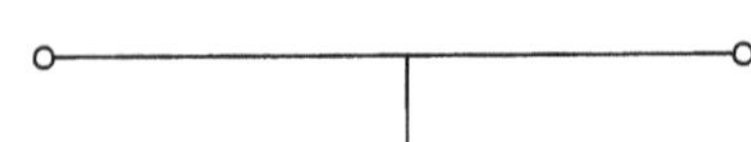

(a) $z = \begin{bmatrix} 0 & 1 \\ -1 & 1 \end{bmatrix}$; $h = \begin{bmatrix} 1 & 1 \\ 1 & 1 \end{bmatrix}$

(b) $z = \begin{bmatrix} 1 & 1 \\ 1 & 1 \end{bmatrix}$; $h = \begin{bmatrix} 0 & 1 \\ -1 & 1 \end{bmatrix}$

(c) $z = \begin{bmatrix} 1 & 0 \\ -1 & 1 \end{bmatrix}$; $h = \begin{bmatrix} -1 & 1 \\ 1 & -1 \end{bmatrix}$

(d) $z = \begin{bmatrix} 0 & 1 \\ 1 & 0 \end{bmatrix}$; $h = \begin{bmatrix} 0 & -1 \\ 2 & 1/2 \end{bmatrix}$

10.9 The Y_{21} parameter of the network shown in the given figure will be

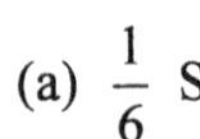

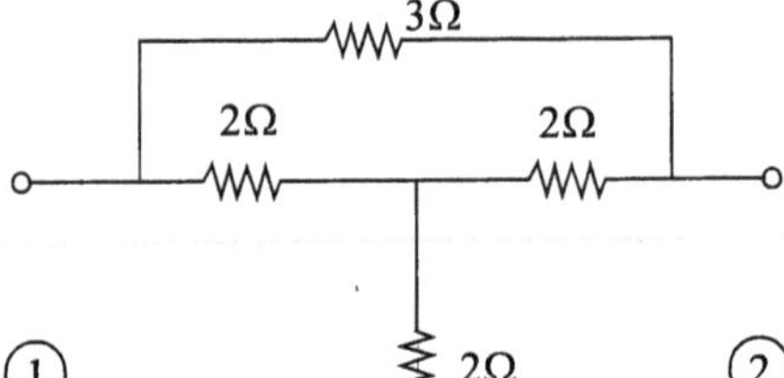

(a) $\frac{1}{6}$ S (b) $-\frac{1}{6}$ S

(c) $\frac{1}{2}$ S (d) $-\frac{1}{2}$ S

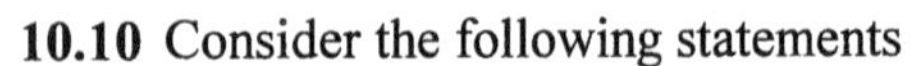

10.10 Consider the following statements :

1. The two-port network shown below does NOT have an impedance matrix representation.

2. The following two-port network does NOT have an admittance matrix representation.

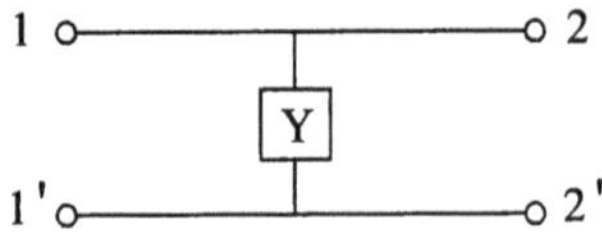

3. A two-port network is said to be reciprocal if it satisfies $z_{12} = z_{21}$ or an equivalent relationship.

Of these statements

(a) 1 and 2 are correct (b) 1, 2 and 3 are correct

(c) 1 and 3 are correct (d) none is correct

10.11 Match List-I with List-II for the two-port network shown in the given figure and select the correct answer using the codes given below the lists :

	List-I		**List-II**
A.	z_{11}	1.	R
B.	z_{12}	2.	R + L
C.	z_{21}	3.	R – Ls
D.	z_{22}	4.	R + Ls

Codes :

(a) A B C D — 1 2 1 4
(b) A B C D — 2 1 1 3
(c) A B C D — 1 1 1 4
(d) A B C D — 2 1 3 4

10.12 For a two-port network to be reciprocal, it is necessary that

(a) $z_{11} = z_{22}$ and $y_{21} = y_{12}$
(b) $z_{11} = z_{22}$ and AD – BC = 0
(c) $h_{21} = -h_{12}$ and AD – BC = 0
(d) $y_{21} = y_{12}$ and $h_{21} = -h_{12}$

10.13 Match List-I (Parameters) with List-II (Units) and select the correct answer using the codes given below the Lists :

	List-I		**List-II**
A.	h_{11}	1.	Dimensionless
B.	h_{12}	2.	Ohms
C.	h_{22}	3.	Siemens

Codes :

(a) A B C — 1 2 3
(b) A B C — 1 3 2
(c) A B C — 2 1 3
(d) A B C — 3 2 1

10.14 Two-port networks with transmission parameters A_1, B_1, C_1, D_1 and A_2, B_2, C_2, D_2 respectively are cascaded. The transmission parameter matrix of the cascaded network will be

(a) $\begin{bmatrix} A_1 & B_1 \\ C_1 & D_1 \end{bmatrix} + \begin{bmatrix} A_2 & B_2 \\ C_2 & D_2 \end{bmatrix}$

(b) $\begin{bmatrix} A_1 & B_1 \\ C_1 & D_1 \end{bmatrix} \begin{bmatrix} A_2 & B_2 \\ C_2 & D_2 \end{bmatrix}$

(c) $\begin{bmatrix} A_1A_2 & B_1B_2 \\ C_1C_2 & D_1D_2 \end{bmatrix}$

(d) $\begin{bmatrix} (A_1A_2 + C_1C_2) & (A_1A_2 - B_1D_2) \\ (C_1A_2 - D_1C_2) & (C_1C_2 + D_1D_2) \end{bmatrix}$

10.15 If the two-port network shown in the given figure has the constant $B = \frac{2s+1}{s^2}$, then z(s) will be

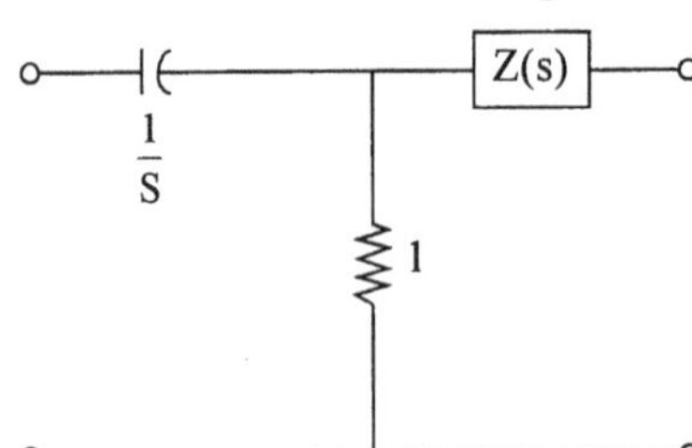

(a) s 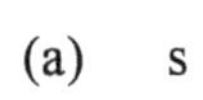 (b) $\frac{1}{s}$

(c) s + 1 (d) $1 + \frac{1}{s}$ 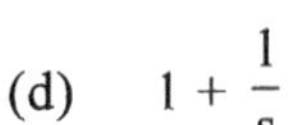

10.16 The short-circuit test of a 2-port network is shown in figure-I. The voltage across the terminals AA in the network shown in figure-II will be

(a) 20 V

(b) 10 V

(c) 5 V

(d) 2 V

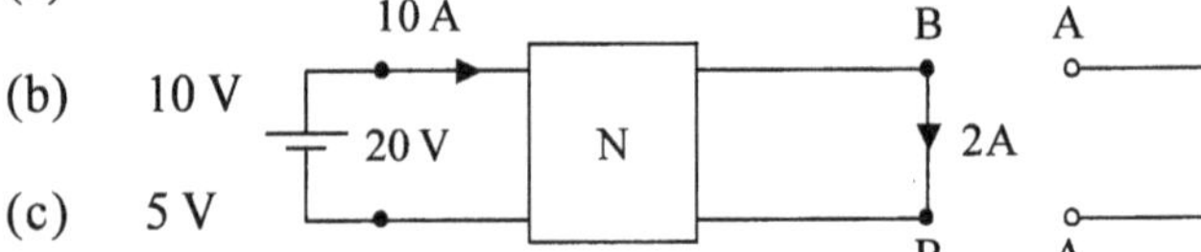

Figure - I

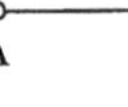
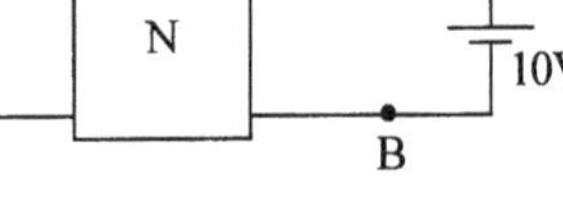

Figure - II

10.17 In a passive two-port network, the open-circuit impedance matrix is $\begin{bmatrix} 10\Omega & 2\Omega \\ 2\Omega & 5\Omega \end{bmatrix}$. If the input port is interchanged with the output port, then the open-circuit impedance matrix will be

(a) $\begin{bmatrix} 10\Omega & 2\Omega \\ 5\Omega & 2\Omega \end{bmatrix}$ (b) $\begin{bmatrix} 5\Omega & 2\Omega \\ 2\Omega & 10\Omega \end{bmatrix}$

(c) $\begin{bmatrix} 5\Omega & 10\Omega \\ 2\Omega & 2\Omega \end{bmatrix}$ (d) $\begin{bmatrix} 2\Omega & 5\Omega \\ 10\Omega & 2\Omega \end{bmatrix}$

10.18 For the given two port network, the Z_{12} parameter is

(a) $\frac{s+1}{s^2+s+1}$

(b) $\frac{s^2+s+1}{s+1}$

(c) $\frac{s}{s^2+s+1}$

(d) $\frac{s^2+1}{s^2+s+1}$

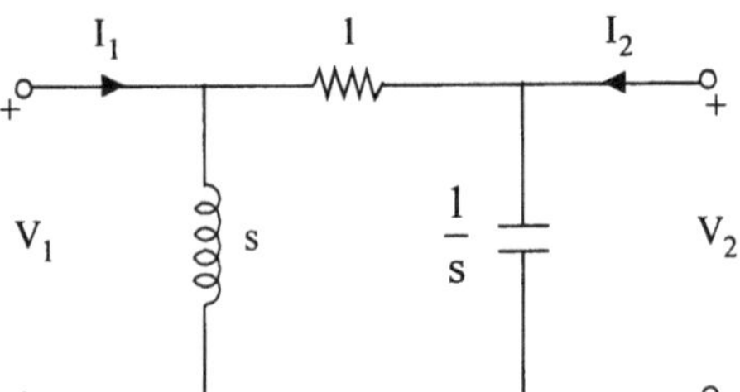

10.19 In the two-port network shown in the given figure, if G_{21} (s) is $\frac{1}{s^2+1}$, then Z(s) will be

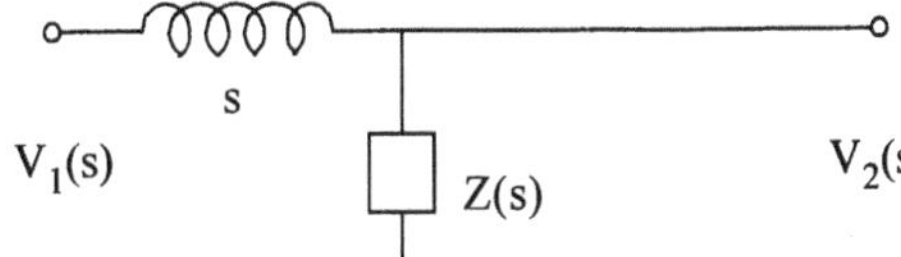

(a) s (b) s + 1

(c) $\frac{1}{s}$ (d) $s+\frac{1}{s}$

10.20 For a two-port symmetrical billateral network, if A = 3 and B = 1 Ω, the value of parameter C will be

(a) 4s (b) 6s (c) 8s (d) 16s

10.21 Consider the following two-port network configurations :

1. 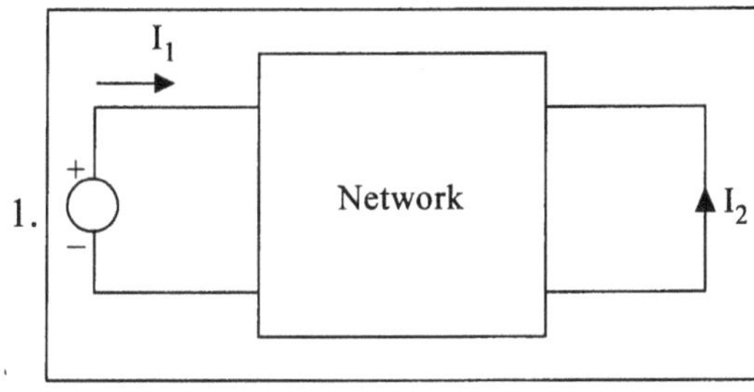

2. 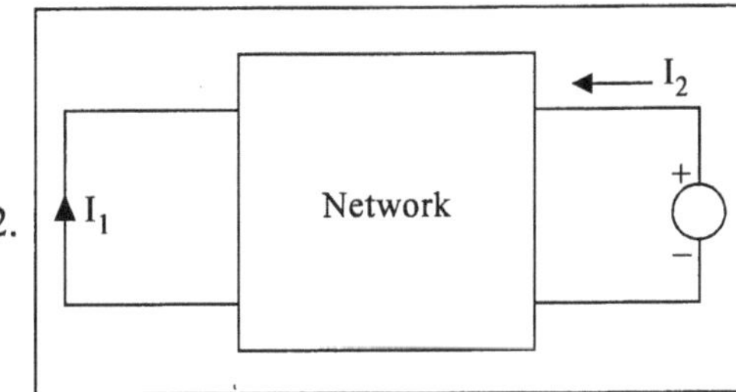

3. 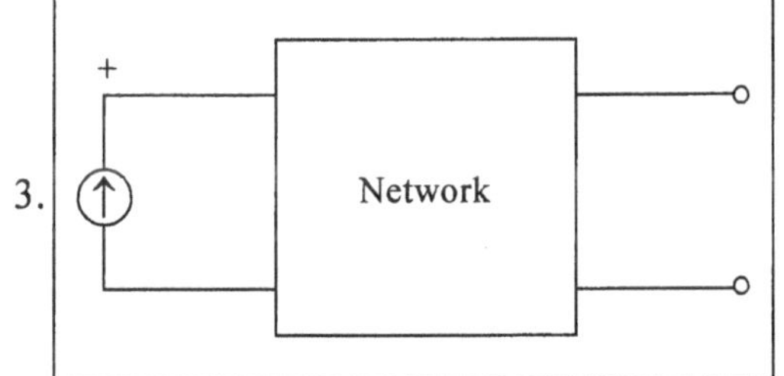

4. 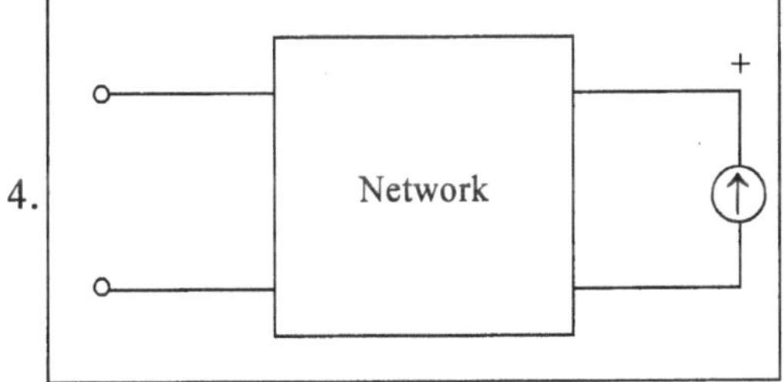

Which of these configurations relate the definition of short-circuit admittance parameters.

(a) 1 and 2 (b) 1 and 4 (c) 2 and 3 (4) 2 and 4

10.22 The h parameters h_{11} and h_{22} are related to z and y parameters as

(a) $h_{11} = z_{11}$ and $h_{22} = 1/z_{22}$ (b) $h_{11} = z_{11}$ and $h_{22} = y_{22}$

(c) $h_{11} = 1/y_{11}$ and $h_{22} = 1/z_{22}$ (d) $h_{11} = 1/y_{11}$ and $h_{22} = y_{22}$

10.23 The definiting equations of a two-part network in terms of the Laplace transform in matrix form are $\begin{bmatrix} V_1 \\ I_2 \end{bmatrix} = \begin{bmatrix} 0 & n \\ -n & 0 \end{bmatrix} \begin{bmatrix} I_1 \\ V_2 \end{bmatrix}$

The network so defined is called

(a) a gyrator (b) an ideal transformer

(d) T-network with resistive elements (d) T-network with resistive elements

10.24 For the given network, match List I with List II and select the correct answer using the codes given below the Lists :

List - I	**List - II**
A. z_{11}	1. $\frac{1}{3}$
B. z_{21}	2. $\frac{9}{8}$
C. y_{12}	3. $-\frac{3}{8}$
D. y_{22}	4. 1

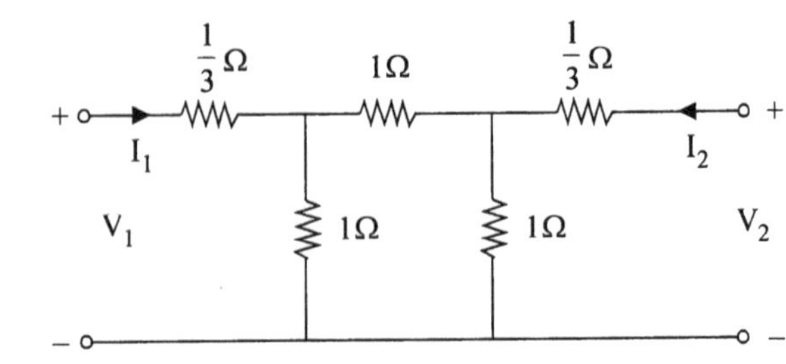

Codes :

(a) A B C D
 4 1 2 3

(b) A B C D
 1 4 3 2

(c) A B C D
 1 4 2 3

(d) A B C D
 4 1 3 2

10.25 The short-circuit admittance matrix of the network shown in the given figure is

(a) $\begin{pmatrix} \frac{1}{Z} & -\frac{1}{Z} \\ -\frac{1}{Z} & \frac{1}{Z} \end{pmatrix}$ (b) $\begin{pmatrix} \frac{1}{Z} & -\frac{1}{Z} \\ \frac{1}{Z} & \frac{1}{Z} \end{pmatrix}$

(c) $\begin{pmatrix} \frac{1}{Z} & \frac{1}{Z} \\ -1 & 1 \end{pmatrix}$ (d) $\begin{pmatrix} \frac{1}{Z} & 1 \\ 1 & \frac{1}{Z} \end{pmatrix}$

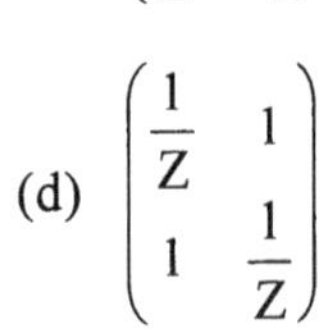

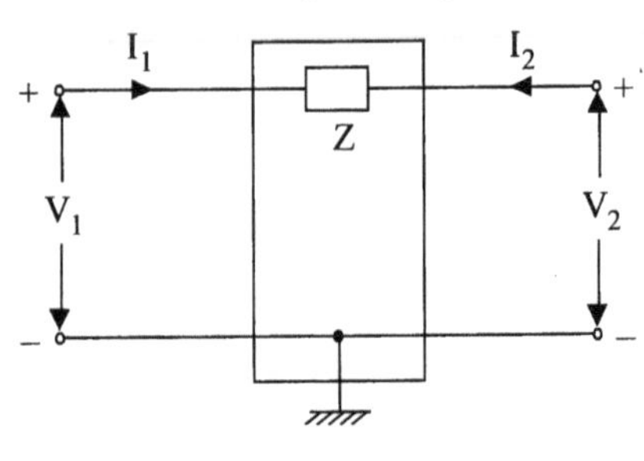

10.26 Consider the following statements :

For a reciprocal network,

1. A = D 2. $Z_{12} = Z_{21}$, 3. $h_{12} = -h_{21}$,

Of these statements

(a) 1, 2 and 3 are correct (b) 1 and 2 are correct

(c) 1 and 3 are correct (d) 2 and 3 are correct

10.27 Two-port network s α and β having ABCD parameters as

$A_\alpha = 4 = D_\alpha$ $A_\beta = 3 = D_\beta$

and

$B_\alpha = 5, C_\alpha = 3$ $B_\beta = 4$ and $C_\beta = 2$

are connected in cascade in the order of α, β. The equivalent 'A' parameter of the combination is

(a) 17 (b) 22 (c) 24 (d) 31

10.28 The voltage transfer ratio of two-port networks connected in cascade may be conveniently obtained from the

(a) product of the individual ABCD matrices of the two networks

(b) product of voltage transfer ratios of the two individual networks

(c) sum of the Z-matrices of the two networks

(d) sum of the h-matrices of the two networks

10.29 The Y parameters of a four-terminal block are $\begin{bmatrix} 4 & 2 \\ 1 & 1 \end{bmatrix}$

A single element of 1 ohm is connected across as shown in the given figure.

The new Y parameters will be

(a) $\begin{bmatrix} 5 & 1 \\ 0 & 2 \end{bmatrix}$ (b) $\begin{bmatrix} 4 & 3 \\ 2 & 2 \end{bmatrix}$

(c) $\begin{bmatrix} 3 & 2 \\ 1 & 1 \end{bmatrix}$ (d) $\begin{bmatrix} 4 & 2 \\ 1 & 1 \end{bmatrix}$

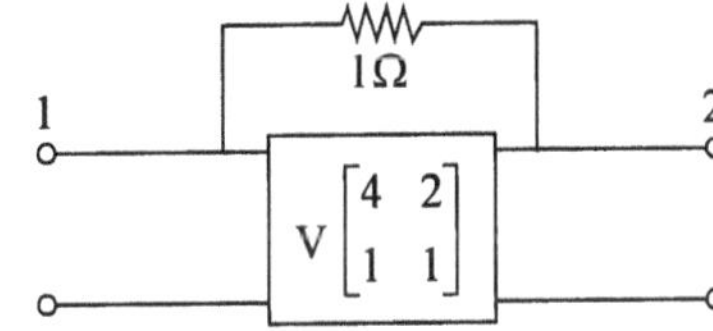

10.30 A two-port network is symmetrical if

(a) $z_{11}\, z_{22} - z_{11}\, z_{21} = 1$ (b) $AD - BC = 1$

(c) $h_{11}\, h_{22} - h_{12}\, h_{21} = 1$ (d) $y_{11}\, y_{22} - y_{12}\, y_{21} = 1$

10.31 Which one of the following pairs is correctly matched ?

(a) Symmetrical two-port network : $AD - BC = 1$

(b) Reciprocal two-port network : $z_{11} = z_{12}$

(c) Inverse hybrid parameter : A, B, C, D

(d) Hybrid parameter : $(V_1, I_2) = (I_1, V_2)$

10.32 An ideal transformer has turns ratio of 2 : 1. Considering high voltage side as port I and low voltage side as port 2, the transmission line parameters of transformer will be

(a) $\begin{bmatrix} 0 & -2 \\ 0.5 & 0 \end{bmatrix}$ (b) $\begin{bmatrix} -2 & 0 \\ 0 & -0.5 \end{bmatrix}$ (c) $\begin{bmatrix} 2 & 0 \\ 0 & 0.5 \end{bmatrix}$ (d) $\begin{bmatrix} 0.5 & 0 \\ 0 & 2 \end{bmatrix}$

10.33 A two-port network is represented by $V_1 = 24\,I_1 + 8\,I_2$ and $V_2 = 8\,I_1 + 32\,I_2$. Which one of the following networks is represented by these equations ?

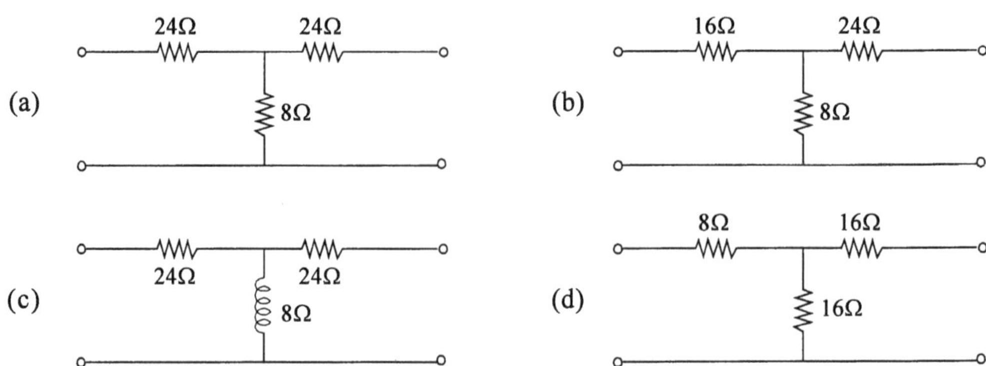

10.34 Ideal transformer cannot be described by

(a) h parameters (b) ABCD parameters

(c) G parameters (d) Z parameters

10.35 With reference to the equivalent circuit of a transistor shown in the given figure, match List I with List II and select the correct answer using codes given below the lists :

List I	**List II**
A. h_{11}	1. $(r_e + r_b) - \dfrac{\alpha r_c + r_b}{r_c + r_b}\, r_b$
B. h_{12}	2. $-\dfrac{r_b + \alpha r_c}{r_c + r_b}$
C. h_{21}	3. $\dfrac{1}{r_c + r_b}$
D. h_{22}	4. $\dfrac{r_b}{r_c + r_b}$

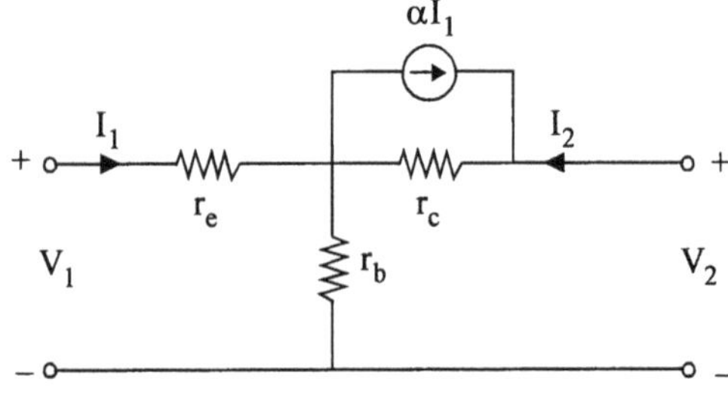

Codes :

	A	B	C	D
(a)	1	2	3	4
(b)	1	3	2	4
(c)	1	4	2	3
(d)	2	3	4	1

10.36 In a two-port network containing linear bilateral passive circuit elements, which one of the following conditions for z parameters would hold ?

(a) $z_{11} = z_{22}$ (b) $z_{12}\, z_{21} = z_{11}\, z_{22}$

(c) $z_{11}\, z_{22} = z_{22}\, z_{21}$ (d) $z_{12} = z_{21}$

10.37 The two-port network shown in the figure is characterised by the impedance parameters z_{11}, z_{12}, z_{21} and z_{22}. For the equivalent Thevenin's source looking to the left of port 2, the V_T and Z_T will be respectively.

(a) $V_T = \dfrac{Z_{11}}{Z_{11} + Z_g} V_g;\ Z_T = Z_{22} - Z_{12}$

(b) $V_T = \dfrac{Z_{12}}{Z_{11} + Z_g} V_g;\ Z_T = Z_{22} - Z_{12}$

(c) $V_T = \dfrac{Z_{21} V_g}{Z_{11} + Z_g};\ Z_T = Z_{22} - \dfrac{Z_{12} Z_{21}}{Z_{11} + Z_g}$

(d) $V_T = \dfrac{Z_{21} V_g}{Z_{11} + Z_g};\ Z_T = Z_{22} + \dfrac{Z_{12} Z_{21}}{Z_{11} + Z_g}$

10.38 Two sets of measurements on a linear passive two-port network are shown in the following figures :

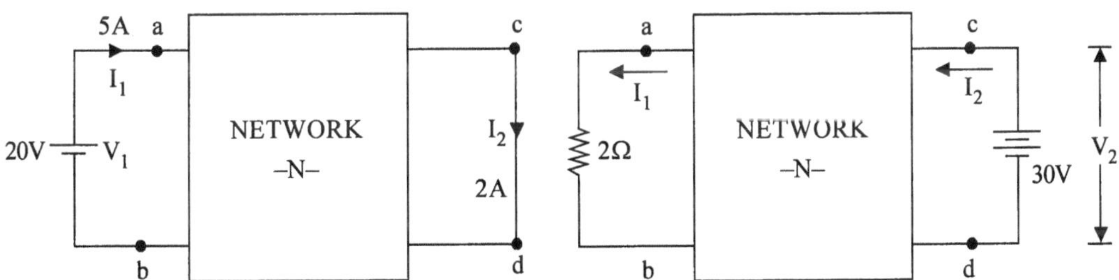

The current flowing through the 2-ohm resistor is

(a) 2 A (b) 1 A (c) 0.5 A (d) zero

10.39 In respect of the 2-port network shown in the figure, the admittance parameters are : $y_{11} = 8$ S, $y_{12} = y_{21} = -6$ S and $y_{22} = 6$ S.

The values of y_A, y_B and y_C (in units of Siemens) will be respectively

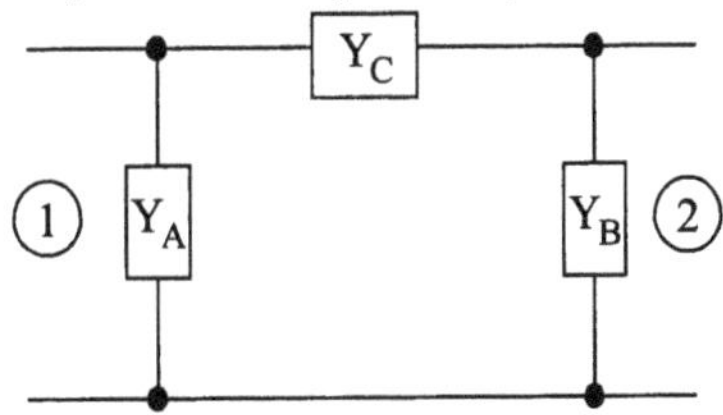

(a) 2, 6 and – 6

(b) 2, 6 and 0

(c) 2, 0 and 6

(d) 2, 6 and 8

10.40 The model of a transistor in the common emitter connection is shown in the following figure :

Match List-I (Parameters) with List-II (Values) and select the correct answer using the codes given below the Lists :

List I	**List II**
A. h_{22}	1. $r_b + r_c$
B. h_{11}	2. α_{cb}
C. h_{21}	3. $\dfrac{1}{r_e + r_d}$

Codes :

	A	B	C		A	B	C
(a)	3	1	2	(b)	1	3	2
(c)	2	3	1	(d)	3	2	1

10.41 Consider the following statements for a 2-port network :

1. $z_{11} = z_{22}$ 2. $h_{12} = h_{21}$ 3. $y_{12} = y_{21}$ 4. $BC - AD = -1$

The network is reciprocal if and only if

(a) 1 and 2 are correct (b) 2 and 3 are correct

(c) 3 and 4 are correct (d) 4 alone is correct

10.42 For a two-port reciprocal network, the output open-circuit voltage divided by the input current is equal to

(a) B (b) z_{12} (c) $\dfrac{1}{y_{12}}$ (d) h_{12}

10.43

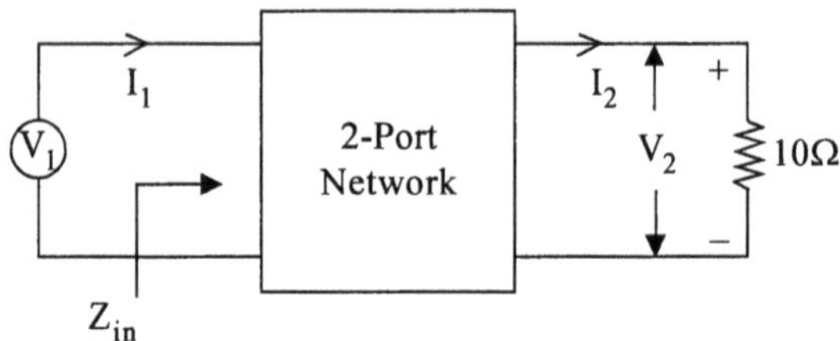

If the transmission parameters of the above network are A = C = 1, B = 2 and D = 3, then the value of Z_{in} is

(a) $\dfrac{12}{13}\ \Omega$ (b) $\dfrac{13}{12}\ \Omega$ (c) 3 Ω (d) 4 Ω

10.44 The impedance matrices of two-port networks are given by $\begin{bmatrix} 3 & 2 \\ 2 & 3 \end{bmatrix}$ and $\begin{bmatrix} 15 & 5 \\ 5 & 25 \end{bmatrix}$

If these two networks are connected in series, the impedance matrix of the resulting two-port network will be

(a) $\begin{bmatrix} 3 & 5 \\ 2 & 25 \end{bmatrix}$ (b) $\begin{bmatrix} 18 & 7 \\ 7 & 28 \end{bmatrix}$ (c) $\begin{bmatrix} 15 & 2 \\ 5 & 3 \end{bmatrix}$ (d) indeterminate

10.45 z-matrix for the network shown in the given figure is

(a) $\begin{bmatrix} 2s+1 & 2s \\ 2s & 2s+\frac{3}{s} \end{bmatrix}$ (b) $\begin{bmatrix} 2s+1 & -2s \\ -2s & 2s+\frac{3}{s} \end{bmatrix}$

(c) $\begin{bmatrix} 2s+1 & +2s \\ -2s & 2s+3/s \end{bmatrix}$ (d) $\begin{bmatrix} 2s+\frac{3}{2} & -2s \\ 2s & 2s+3/s \end{bmatrix}$

10.46 A T-network is shown in the given figure. Its Y_{sc} matrix will be (units in siemens)

(a) $\begin{bmatrix} \frac{10}{200} & \frac{5}{200} \\ \frac{5}{200} & \frac{10}{200} \end{bmatrix}$ (b) $\begin{bmatrix} \frac{10}{200} & \frac{-5}{200} \\ \frac{-5}{200} & \frac{10}{200} \end{bmatrix}$

(c) $\begin{bmatrix} \frac{15}{200} & \frac{-5}{200} \\ \frac{-5}{200} & \frac{15}{200} \end{bmatrix}$ (d) $\begin{bmatrix} \frac{15}{200} & \frac{5}{200} \\ \frac{5}{200} & \frac{15}{200} \end{bmatrix}$

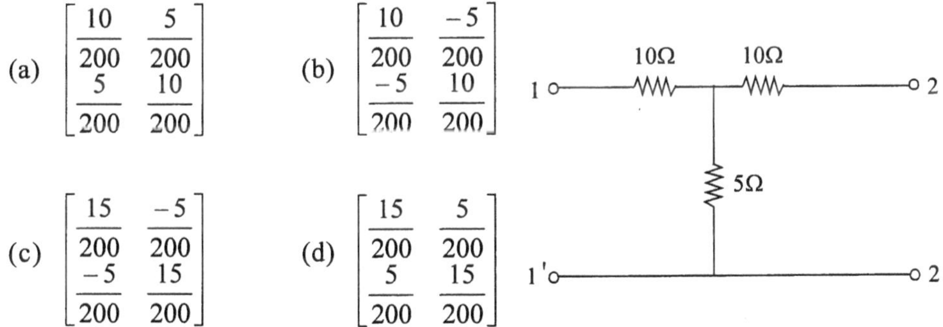

Chapter 11 - Network Filters and Attenuators

11.1 For the design of low pass prototype filter (see figure given) with load resistance R_L = 1 ohm and angular frequency ω = 1 rad/sec, the values of L and C would be

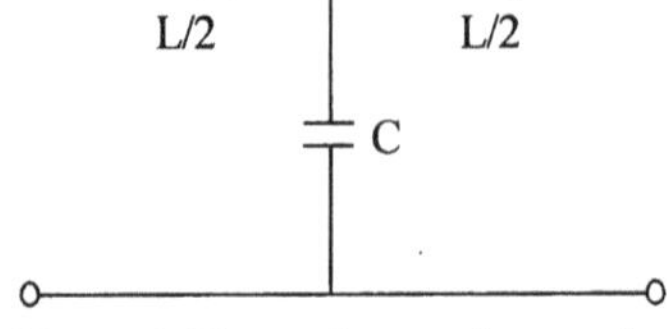

(a) 1 H and 1 F

(b) 1 H and 2 F

(c) 2 H and 1 F

(d) 2 H and 2 F

11.2 The passband of a typical filter network with Z_1 and Z_2 as the series and shunt arm impedances is characterised by

(a) $-1 < \dfrac{Z_1}{4Z_2} < 0$ (b) $-1 < \dfrac{Z_1}{4Z_2} < 1$

(c) $0 < \dfrac{Z_1}{4Z_2} < 1$ (d) none of these

11.3 In an m-drived low pass filter, the value of m is [f_c = cut-off frequency f_∞ = series resonant frequency (series arm)]

(a) $\sqrt{1-\left(\dfrac{f_\infty}{f_c}\right)^2}$ (b) $\sqrt{1-\left(\dfrac{f_c}{f_\infty}\right)^2}$ (c) $\sqrt{1+\left(\dfrac{f_\infty}{f_c}\right)^2}$ (d) $\sqrt{1+\left(\dfrac{f_c}{f_\infty}\right)^2}$

11.4 A four-terminal T-network is inserted between source and load resistance R as shown in the given figure. The resistance seen by source remain the same with or without the four terminal block when R is

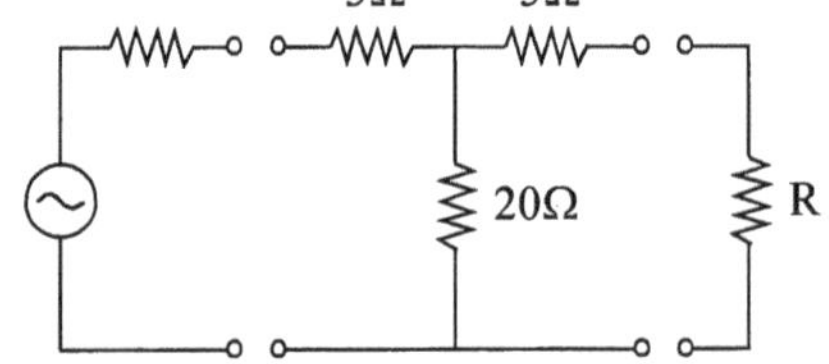

(a) 5 Ω

(b) 10 Ω

(c) 15 Ω

(d) 20 Ω

11.5 If X_A and X_B represent the series arm and the shunt arm reactance of a prototype filter, then in the stop band of the filter, signs of X_A and X_B are

(a) opposite and the characteristic impedance is resistive

(b) the same and the characteristic impedance is resistive

(c) opposite and the characteristic impedance is reacitve

(d) the same and the characteristic impedance is reactive

11.6 The pole-zero pattern shown in the given figure is for

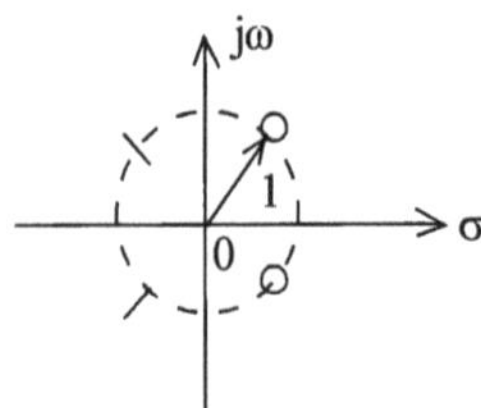

(a) a low-pass filter

(b) a high-pass filter

(c) an all-pass filter

(d) a band-pass filter

CHAPTER 12 - Fourier Series and Fourier Transforms

12.1 The amplitude and phase spectra for the few harmonics of a periodic signal of time period 1S are shown in the figure below.

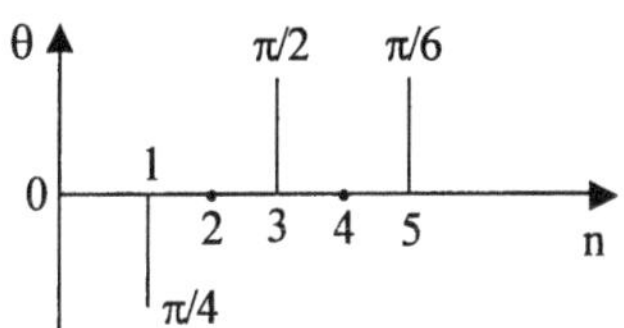

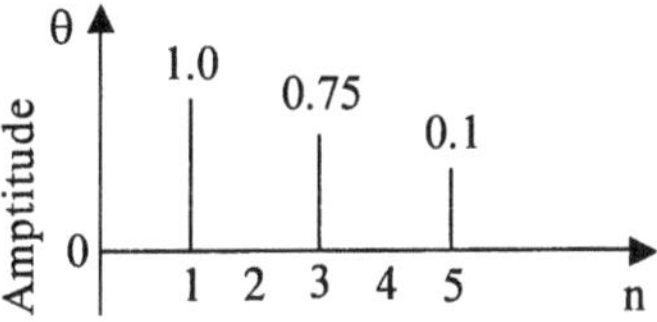

(a) $\cos\left(2\pi t - \frac{\pi}{4}\right) + 0.75\cos\left(6\pi t + \frac{\pi}{2}\right) + 0.1\cos\left(10\pi t + \frac{\pi}{6}\right)$.....

(b) $\cos\left(\frac{t}{2\pi} - \frac{\pi}{4}\right) + 0.75\cos\left(\frac{t}{6\pi} - \frac{\pi}{2}\right) + 0.1\cos\left(\frac{t}{10\pi} + \frac{\pi}{6}\right)$

(c) $\cos\left(2\pi t - \frac{\pi}{4}\right) + 0.75\cos\left(6\pi t + \frac{\pi}{2}\cdot\frac{1}{3}\right) + 0.1\cos\left(10\pi t + \frac{\pi}{6}\cdot\frac{1}{6}\right)$

(d) $-\cos\left(2\pi t - \frac{\pi}{4}\right) + 0.75\cos\left(6\pi t + \frac{\pi}{2}\right) + 0.1\cos\left(10\pi t + \frac{\pi}{6}\right)$

12.2 A periodic function of half-wave symmetry is necessarily

(a) an even function (b) an odd function

(c) neither odd nor even (d) both odd and even

12.3 About the Fourier series expansion of a periodic function it can be said that

(a) Even functions have only a constant and cosine terms in their FS expansion

(b) Odd function have only sine terms in their FS expansion

(c) Functions with half-wave symmetry contain only odd harmonics

(d) All the above three.

12.4 If from the function f(t) one forms the function, $\psi(t) = f(t) + f(-t)$, then $\psi(t)$ is

(a) even (b) odd (c) neither even nor odd (d) both even and odd.

12.5 When a voltage $V_0 \sin \omega_0 t$ is applied to the pure inductor, the ammeter shown in the figure reads I_0. If the voltage applied is :

$V_0 \sin \omega_0 t + 2V_0 \sin 2\omega_0 t + 3V_0 \sin 3\omega_0 t + 4V_0 \sin 4\omega_0 t$, the ammeter reading would be

(a) 0

(b) $10\, I_0$

(c) $\sqrt{4^3 + 3^3 + 2^2 + 1}\; I_0$

(d) $2\, I_0$

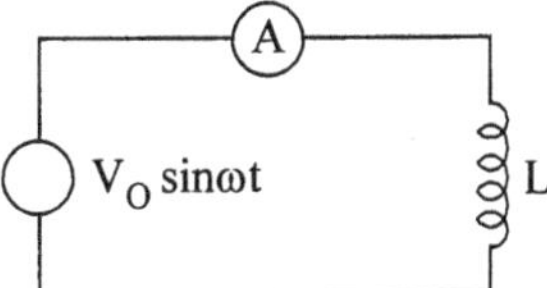

12.6 Which of the following periodic waveforms will have only odd harmonics of sinusoidal waveforms ?

1. 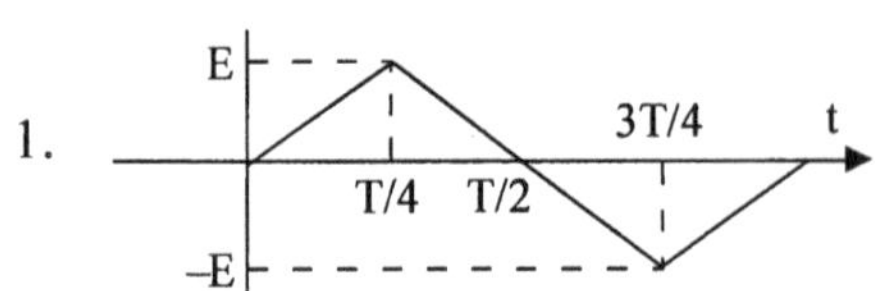

2. 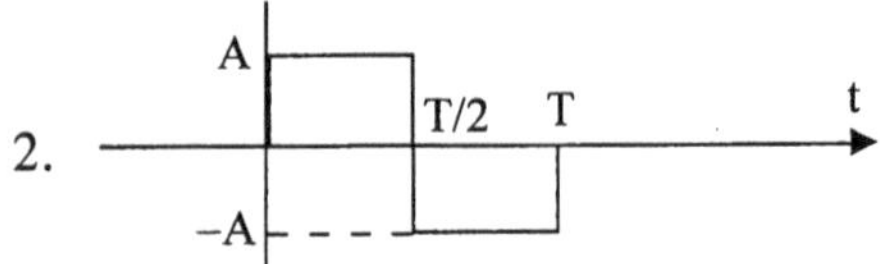

3. 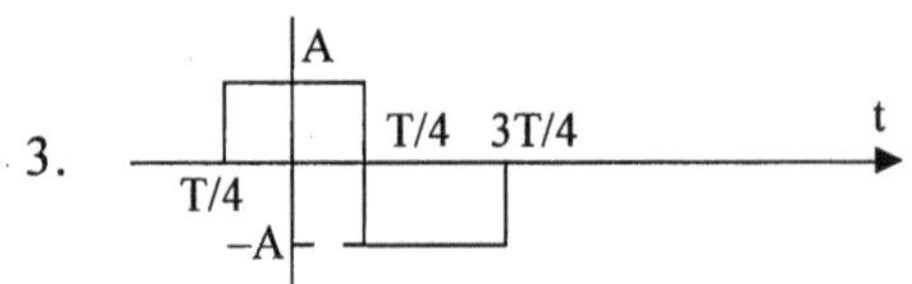

4. 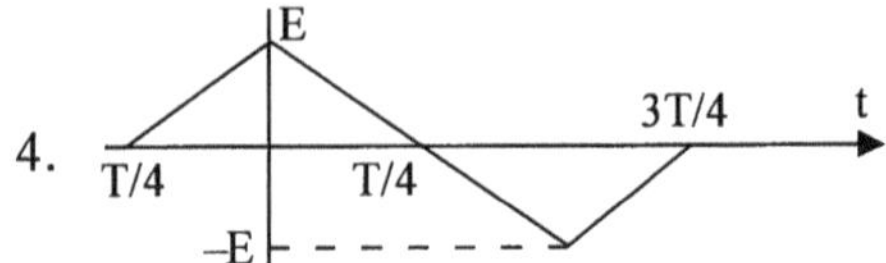

Select the correct answer using the codes given below :

Codes :

(a) 1 and 2 (b) 1 and 3 (c) 1 and 4 (d) 2 and 4

12.7 If f (t) = – f (– t) and f (t) satisfy the Dirichlet's conditions, then f (t) can be expanded in a Fourier series containing

(a) only sine terms (b) only cosine terms

(c) cosine terms and a constant term (d) sine terms and a constant term

12.8 The amplitude of the first odd harmonic of the square wave shown in the figure is equal to

(a) $\frac{4V}{3\pi}$ (b) $\frac{2V}{3\pi}$

(c) $\frac{V}{\pi}$ (d) 0

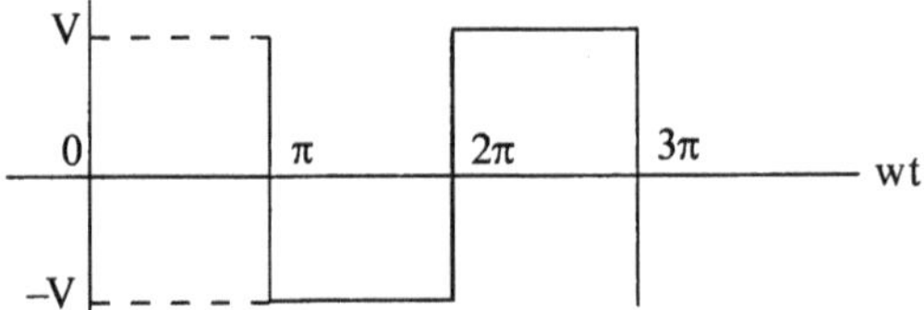

12.9 A periodic triangular wave is shown in the figure. Its Fourier components will consist only of

(a) all cosine terms

(b) all sine terms

(c) odd cosine terms

(d) odd sine terms

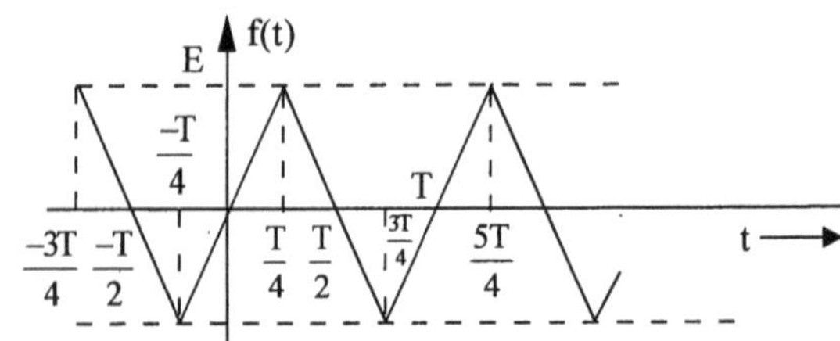

12.10 The Fourier series representation of a periodic current is

$$[2 + 6\sqrt{2}\ \cos \omega t + \sqrt{48}\ \sin 2\ \omega t]\ \text{A}.$$

The effective value of the current is

(a) $(2 + 6 + \sqrt{24})\,A$ (b) $8\,A$ (c) $6\,A$ (d) $2\,A$

12.11 Match List I (Properties) with List II (Characteristics of the trigonometric form) in regard to Fourier series of periodic f (t) and select the correct answer using the codes given below the Lists :

List I	List II
A. f (t) + f (–t) = 0	1. Even harmonics can exist
B. f (t) – f (–t) = 0	2. Odd harmonics can exist
C. f (t) + f (t – T/2) = 0	3. The dc and cosine terms can exist
D. f (t) – f (t – T/2) = 0	4. Sine terms can exist
	5. Cosine terms of even harmonics can exist

Codes :

	A	B	C	D
(a)	4	5	3	1
(b)	3	4	1	2
(c)	5	4	2	3
(d)	4	3	2	1

12.12 Consider the following statements regarding the fundamental component f_1 (t) of an arbitrary periodic signal f (t) :

It is possible for

1. the amplitude of f_1 (t) to exceed the peak value of f (t).
2. f_1 (t) to be identically zero for a non-zero f (t).
3. the effective value of f_1 (t) to exceed effective value of f (t).

Which of these statements is/are correct ?

(a) 1 alone (b) 1 and 2 (c) 2 and 3 (d) 1 and 3

12.13 Match List I with List II and select the correct answer using the codes given below the Lists :

List I	List II
A. $f(t) = -f(-t)$	1. Exponential form of Fourier series
B. $\sum_{n=-\infty}^{\infty} C_n e^{jn\omega_0 t}$	2. Fourier transform
C. $\int_{-\infty}^{\infty} f(t) e^{-j\omega t} dt$	3. Convolution integral
D. $\int_0^t f_1(\tau) f_2(t-\tau) d\tau$	4. z – transform
	5. Odd function wave symmetry

Codes :

(a)	A	B	C	D
	5	1	2	3
(b)	A	B	C	D
	2	1	5	3
(c)	A	B	C	D
	5	4	2	1
(d)	A	B	C	D
	4	5	1	2

12.14 For the expression of $f(\omega t)$ in Fourier series $a_0 + a_1 \cos \omega t + ... + b_1 \sin \omega t + ... + b_n \sin n \omega t + ...$ if $f(\omega t) = f(-\omega t)$

(a) $a_n = 0$ for all n including $n = 0$

(b) $b_n = 0$ for all n

(c) $a_0 = 0$

(d) $a_n = 0$ for all n except $n = 0$

12.15 The fourier transform of the function Sgn (t) defined in the figure is

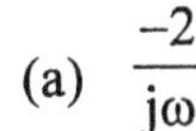 (a) $\frac{-2}{j\omega}$

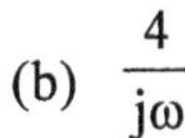 (b) $\frac{4}{j\omega}$

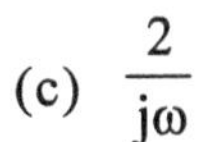 (c) $\frac{2}{j\omega}$

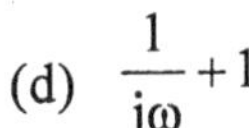 (d) $\frac{1}{j\omega}+1$

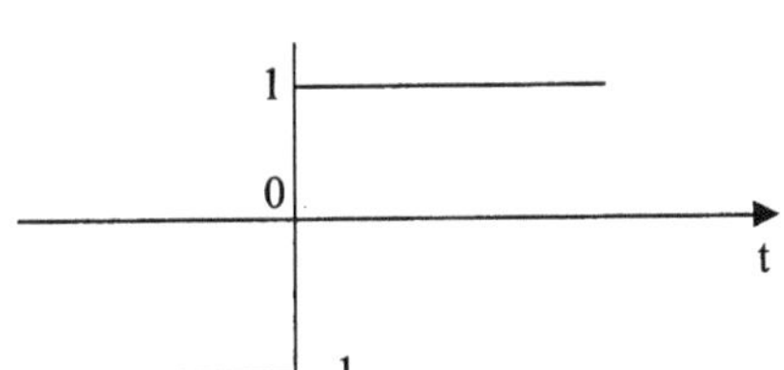

12.16 The inverse Fourier transform of $F(j\omega) = \int_{-\infty}^{\infty} \exp(-j\,\omega\,t)\, f(t)\, dt$ is

(a) $f(t) = \int_{-\infty}^{\infty} \exp(j\,\omega\,t)\, F(j\omega)\, d\omega$

(b) $f(t) = \frac{1}{2\pi} \int_{-\infty}^{\infty} \exp(j\,\omega\,t)\, F(j\omega)\, d\omega$

(c) $f(t) = \frac{1}{2\pi} \int_{-\infty}^{\infty} \exp(-j\,\omega\,t)\, F(j\omega)\, d\omega$

(d) $f(t) = \frac{1}{2\pi} \int_{-\infty}^{\infty} \exp(-j\,\omega\,t)\, F(-j\omega)\, d\omega$

12.17 Which one of the following is the Fourier transform of the signal given in Figure (B), if the Fourier transform of the signal in Figure (A) is given by $2\,\frac{\sin \omega T_1}{\omega}$?

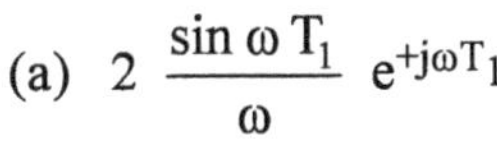 (a) $2\,\frac{\sin \omega T_1}{\omega}\, e^{+j\omega T_1}$

(b) $2\,\frac{\sin \omega T_1}{\omega}\, e^{-j\omega T_1}$

(c) $\frac{\sin \omega T_1}{\omega}\, e^{-j\omega T_1}$

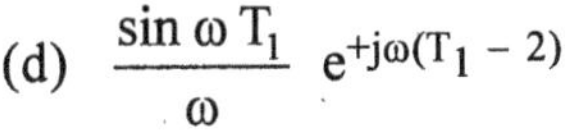 (d) $\frac{\sin \omega T_1}{\omega}\, e^{+j\omega (T_1 - 2)}$

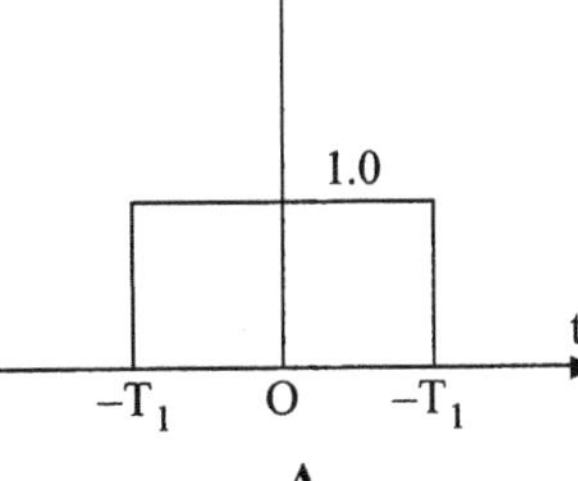

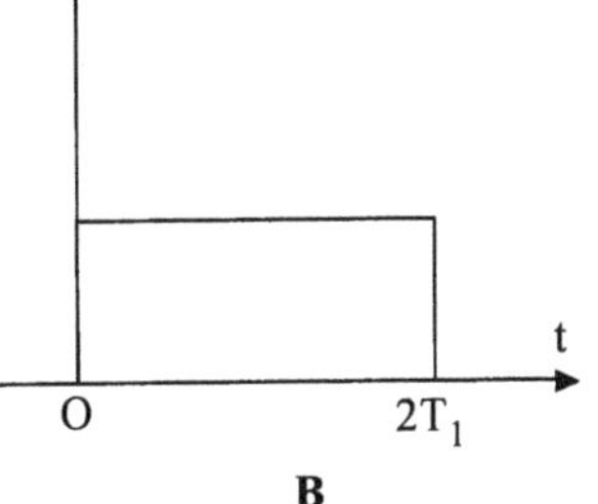

12.18 Fourier transform $F(j\omega)$ of an arbitrary signal has the property,

(a) $F(j\omega) = F(-j\omega)$ (b) $F(j\omega) = -F(-j\omega)$

(c) $F(j\omega) = F^*(-j\omega)$ (d) $F(j\omega) = -F^*(-j\omega)$

12.19 The inverse fourier transform of the function $F(\omega) = \dfrac{1}{j\omega} + \pi\delta(\omega)$ is

(a) $\sin \omega t$ (b) $\cos \omega t$ (c) sgn (t) (d) u (t)

12.20 Which one of the following is the correct Fourier transform of the unit step signal
u (t)= 1 for $t \geq 0$
= 0 for $t \leq 0$

(a) $\pi\,\delta(\omega)$ (b) $\dfrac{1}{j\omega}$ (c) $\dfrac{1}{j\omega} + \pi\delta(\omega)$ (d) $\dfrac{1}{j\omega} + 2\pi\delta(\omega)$

12.21 The Fourier transform of $v(t) = \cos \omega_0 t$ is given by

(a) $V(f) = \dfrac{1}{2}\,\delta(f - f_0)$ (b) $V(f) = \dfrac{1}{2}\,\delta(f + f_0)$

(c) $V(f) = \dfrac{1}{2}\,[\delta(f - f_0) - \delta(f + f_0)]$ (d) $V(f) = \dfrac{1}{2}\,[\delta(f - f_0) + \delta(f + f_0)]$

12.22 If $g(t) \leftrightarrow G(f)$ represents a Fourier transform pair, then according to the duality property of Fourier transforms,

(a) $G(t) \leftrightarrow g(f)$ (b) $G(t) \leftrightarrow g^*(f)$ (c) $G(t) \leftrightarrow g(-f)$ (d) $G(t) \leftrightarrow g^*(-f)$

12.23 A voltage signal v (t) has the following Fourier transform : $V(j\omega) = \begin{cases} e^{-j\omega d} & \text{for} |\omega| < 1 \\ 0 & \text{for} |\omega| > 1 \end{cases}$

The energy that would be dissipated in a 1 Ω resistor fed from v (t) is

(a) $\dfrac{2}{\pi}$ Joules (b) $\dfrac{2e^{-2d}}{\pi}$ Joules (c) $\dfrac{1}{\pi}$ Joules (d) $\dfrac{1}{2\pi}$ Joules

12.24 The Fourier transform of a double-sided exponential signal $x(t) = e^{-b|t|}$

(a) is $2b/(b^2 + \omega^2)$ (b) is $\dfrac{e^{-j\tan^{-1}\left(\frac{\omega}{b}\right)}}{(b^2 + \omega^2)}$

(c) does not exist (d) exists only when it is single sided

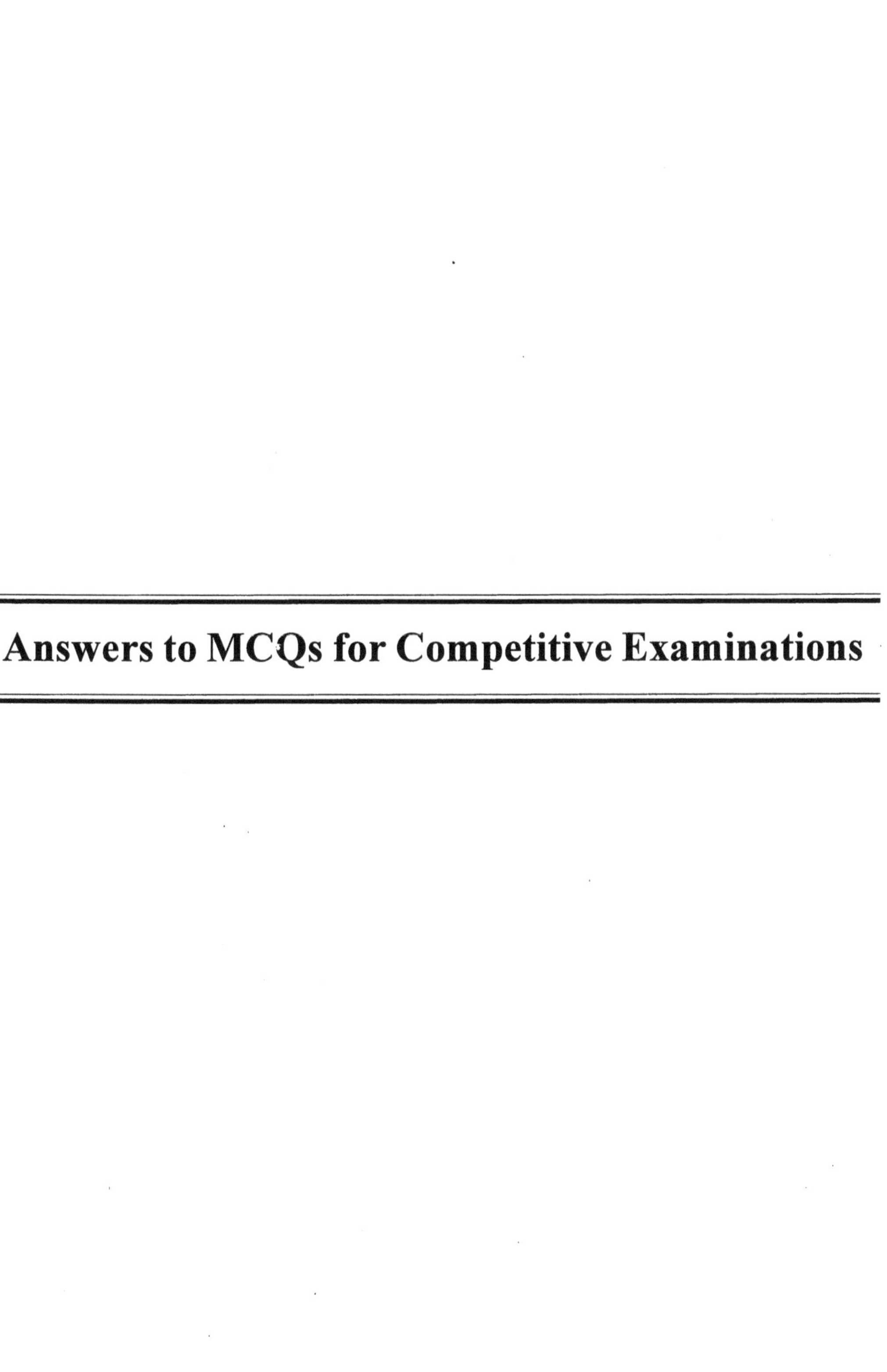

Answers to MCQs for Competitive Examinations

Chapter 1 Network Variables and Elements

1. (d)	2. (c)	3. (b)	4. (c)	5. (c)	6. (b)
7. (d)	8. (b)	9. (a)	10. (d)	11. (c)	12. (d)
13. (a)	14. (b)				

Chapter 2 Kirchhoff's Laws and Resistive Networks

1. (b)	2. (a)	3. (d)	4. (b)	5. (a)	6. (c)
7. (b)	8. (c)	9. (c)	10. (b)	11. (a)	12. (b)
13. (a)	14. (c)	15. (c)	16. (c)	17. (b)	18. (d)
19. (b)	20. (b)	21. (a)	22. (d)	23. (b)	24. (d)
25. (b)	26. (b)	27. (a)	28. (d)	29. (d)	30. (a)
31. (a)	32. (b)	33. (c)	34. (c)	35. (c)	36. (d)
37. (b)	38. (a)	39. (d)	40. (c)	41. (c)	42. (b)
43. (a)	44. (b)	45. (a)	46. (d)	47. (b)	48. (c)

Chapter 3 Single Phase Circuits

1. (b)	2. (d)	3. (a)	4. (a)	5. (d)	6. (a)
7. (a)	8. (a)	9. (c)	10. (a)	11. (c)	12. (b)
13. (d)	14. (b)	15. (d)	16. (c)	17. (d)	18. (b)
19. (c)	20. (d)	21. (b)	22. (b)	23. (b)	24. (b)
25. (d)	26. (a)	27. (a)	28. (a)	29. (a)	30. (a)
31. (d)	32. (c)	33. (b)	34. (a)	35. (d)	36. (a)
37. (b)	38. (a)	39. (a)	40. (b)	41. (a)	42. (c)
43. (b)	44. (d)	45. (d)	46. (c)	47. (c)	48. (a)
49. (a)	50. (a)	51. (d)	52. (b)	53. (d)	54. (d)
55. (d)	56. (b)	57. (a)	58. (a)	59. (d)	60. (d)

61. (c)	62. (a)	63. (b)	64. (a)	65. (a)	66. (c)
67. (d)	68. (c)	69. (c)	70. (c)	71. (c)	72. (d)
73. (a)	74. (c)	75. (c)	76. (a)	77. (c)	78. (a)
79. (b)	80. (a)	81. (c)	82. (c)	83. (c)	84. (d)
85. (c)	86. (a)	87. (b)	88. (c)	89. (c)	90. (b)
91. (b)	92. (d)	93. (d)	94. (b)	95. (a)	96. (c)
97. (b)	98. (d)	99. (d)	100. (b)	101. (d)	102. (d)
103. (c)	104. (b)	105. (a)	106. (b)	107. (c)	108. (d)
109. (c)					

Chapter 4 Three Phase Circuits

1. (c)	2. (b)	3. (b)	4. (c)	5. (d)	6. (a)
7. (c)	8. (a)	9. (c)	10. (d)	11. (a)	12. (d)
13. (d)	14. (b)				

Chapter 5 Network Topology

1. (a)	2. (a)	3. (c)	4. (d)	5. (b)	6. (b)
7. (b)	8. (b)	9. (a)	10. (b)	11. (d)	12. (b)
13. (a)	14. (c)	15. (c)	16. (b)	17. (d)	18. (d)
19. (a)	20. (a)	21. (d)	22. (b)		

Chapter 6 Network Theorems

1. (b)	2. (a)	3. (a)	4. (c)	5. (d)	6. (a)
7. (a)	8. (b)	9. (b)	10. (c)	11. (c)	12. (c)
13. (a)	14. (c)	15. (a)	16. (b)	17. (c)	18. (c)
19. (b)	20. (d)	21. (d)	22. (b)	23. (c)	24. (b)
25. (c)	26. (d)	27. (c)	28. (b)	29. (c)	30. (c)
31. (a)	32. (c)	33. (a)	34. (d)	35. (c)	36. (c)

37. (d)	38. (b)	39. (c)	40. (c)	41. (a)	42. (a)
43. (d)	44. (b)	45. (b)	46. (d)	47. (b)	48. (a)
49. (b)	50. (d)	51. (b)	52. (c)	53. (b)	54. (d)
55. (a)	56. (b)	57. (d)	58. (c)	59. (b)	

Chapter 7 Differential Equations and Initial in RLC Networks

1. (d)	2. (d)	3. (b)	4. (b)	5. (a)	6. (a)
7. (b)	8. (c)	9. (a)	10. (d)	11. (d)	12. (b)
13. (b)	14. (b)	15. (d)	16. (c)	17. (a)	18. (a)
19. (b)	20. (c)	21. (c)	22. (b)	23. (c)	24. (a)
25. (d)					

Chapter 8 Time Response of RLC Networks

1. (d)	2. (a)	3. (c)	4. (b)	5. (a)	6. (d)
7. (b)	8. (b)	9. (d)	10. (c)	11. (a)	12. (c)
13. (a)	14. (a)	15. (d)	16. (b)	17. (a)	18. (b)
19. (a)	20. (d)	21. (a)	22. (d)	23. (a)	24. (c)
25. (c)	26. (a)	27. (c)	28. (c)	29. (d)	30. (d)
31. (c)	32. (b)	33. (b)	34. (c)	35. (a)	36. (b)
37. (b)	38. (c)	39. (a)	40. (b)	41. (a)	42. (a)
43. (d)	44. (c)	45. (a)	46. (c)	47. (c)	48. (c)
49. (a)	50. (a)	51. (a)	52. (b)	53. (a)	54. (a)
55. (c)	56. (a)	57. (c)	58. (b)	59. (c)	60. (b)
61. (d)	62. (c)	63. (d)	64. (d)	65. (a)	66. (a)
67. (a)	68. (a)	69. (b)	70. (d)	71. (b)	72. (b)
73. (b)	74. (d)	75. (c)			

Chapter 9 Laplace Transform Method of Analysis of Networks

1. (a)	2. (a)	3. (d)	4. (a)	5. (c)	6. (c)
7. (d)	8. (b)	9. (b)	10. (d)	11. (d)	12. (d)
13. (b)	14. (a)	15. (a)	16. (a)	17. (c)	18. (d)
19. (a)	20. (c)	21. (d)	22. (a)	23. (b)	24. (a)
25. (d)	26. (b)	27. (c)	28. (c)	29. (d)	30. (a)
31. (b)	32. (b)	33. (a)	34. (d)	35. (d)	36. (a)
37. (d)	38. (a)	39. (b)	40. (b)	41. (a)	42. (b)
43. (b)	44. (a)	45. (c)	46. (b)	47. (b)	48. (d)
49. (a)	50. (c)	51. (a)	52. (c)	53. (c)	54. (c)
55. (b)	56. (b)	57. (a)	58. (c)	59. (d)	60. (b)
61. (b)	62. (d)	63. (d)	64. (a)	65. (c)	66. (c)
67. (c)	68. (b)	69. (d)	70. (a)	71. (a)	

Chapter 10 Two Port Networks

1. (b)	2. (c)	3. (b)	4. (c)	5. (d)	6. (b)
7. (d)	8. (b)	9. (d)	10. (b)	11. (c)	12. (d)
13. (c)	14. (b)	15. (b)	16. (d)	17. (b)	18. (c)
19. (c)	20. (c)	21. (a)	22. (c)	23. (b)	24. (d)
25. (a)	26. (d)	27. (b)	28. (a)	29. (a)	30. (c)
31. (d)	32. (c)	33. (b)	34. (d)	35. (c)	36. (d)
37. (c)	38. (a)	39. (c)	40. (a)	41. (d)	42. (b)
43. (a)	44. (b)	45. (a)	46. (c)		

CHAPTER 11 Network Filters and Attenuators

1. (d)	2. (a)	3. (b)	4. (c)	5. (d)	6. (c)

CHAPTER 12 Fourier Series and Fourier Transforms

1. (a)	2. (c)	3. (d)	4. (a)	5. (d)	6. (a)
7. (a)	8. (a)	9. (d)	10. (b)	11. (d)	12. (a)
13. (a)	14. (b)	15. (c)	16. (b)	17. (b)	18. (c)
19. (d)	20. (c)	21. (d)	22. (c)	23. (c)	24. (a)

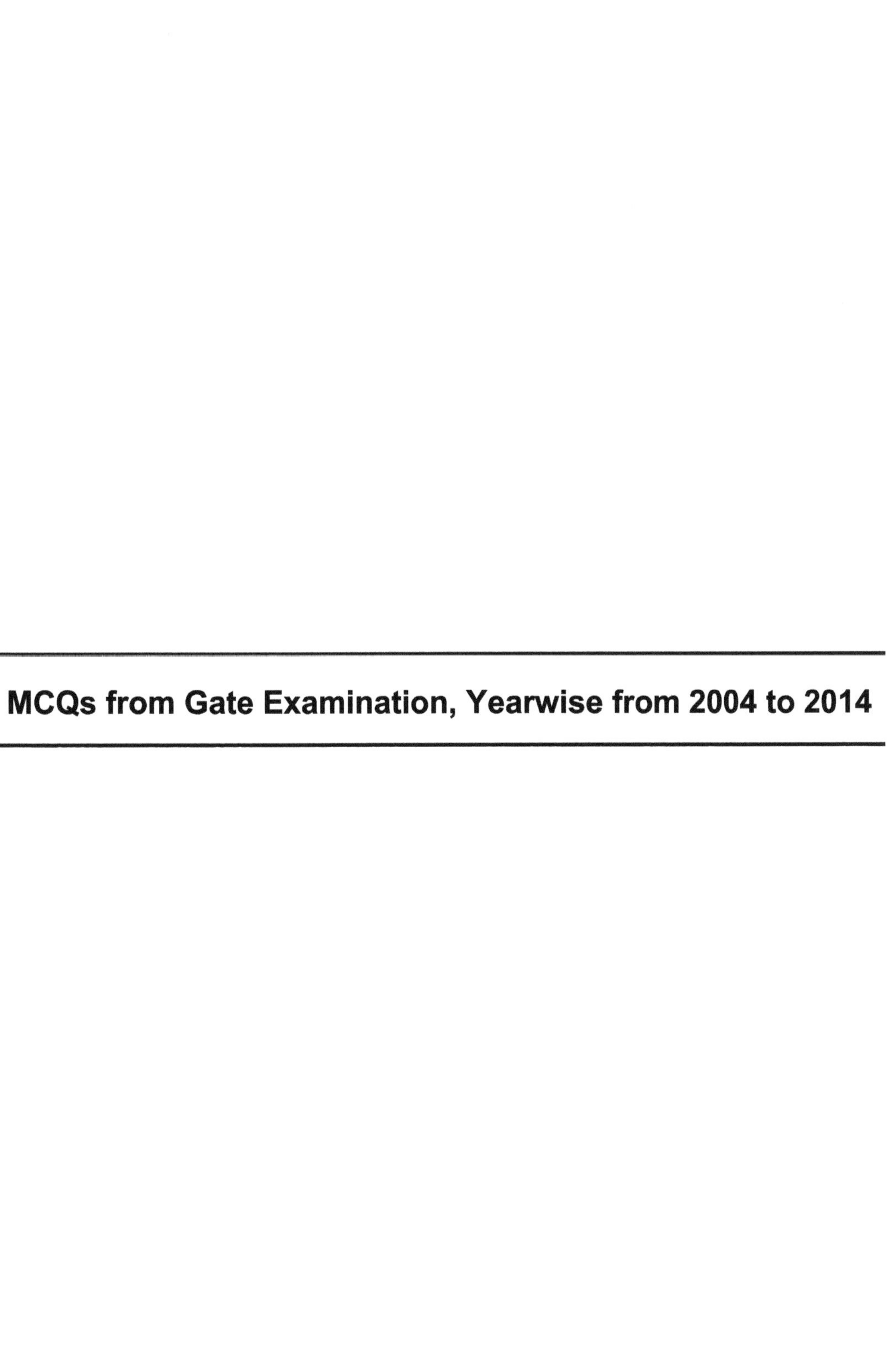

MCQs from Gate Examination, Yearwise from 2004 to 2014

2004 ECE

1. The equivalent inductance measured between the terminals 1 and 2 for the circuit shown in the figure is

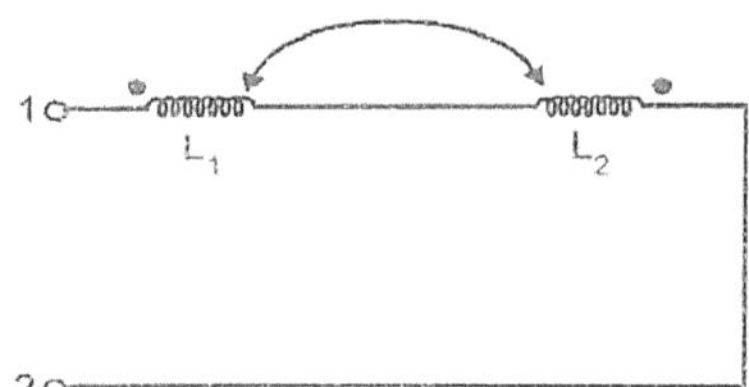

(a) $L_1 + L_2 + M$
(b) $L_1 + L_2 - M$
(c) $L_1 + L_2 + 2M$
(d) $L_1 + L_2 - 2M$

2. The circuit shown in the figure, with $R = \frac{1}{3}\ \Omega$, $L = \frac{1}{4}$ H, C = 3 F has input voltage v(t) = sin 2t. The resulting current i(t) is

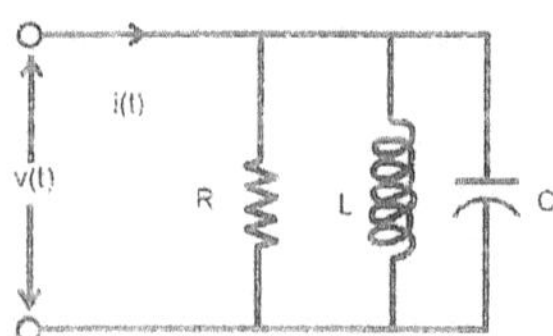

(a) 5 sin (2t + 53.1°)
(b) 5 sin (2t – 53.1°)
(c) 25 sin (2t + 53.1°)
(d) 25 sin (2t – 53.1°)

3. For the circuit shown in the figure, the time constant RC = 1 ms. The input voltage is $v(t) = \sqrt{2}\ \sin 10^3 t$. The output voltage $v_o(t)$ is equal to

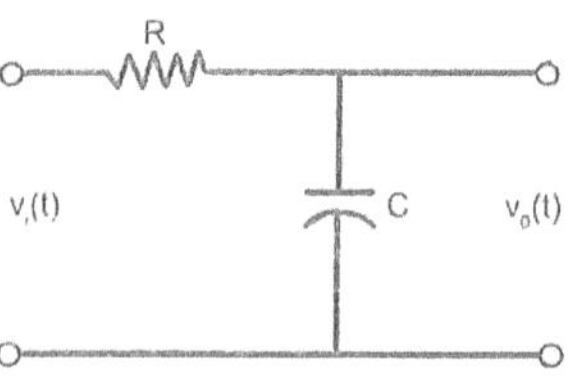

(a) $\sin (10^3 t - 45^\circ)$
(b) $\sin (10^3 t + 45^\circ)$
(c) $\sin (10^3 t - 53^\circ)$
(d) $\sin (10^3 t + 53^\circ)$

4. For the R-L circuit shown in the figure, the input voltage $v_i(t) = u(t)$. The current i(t) is

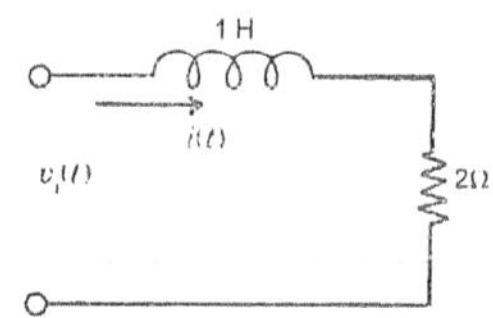

(a) 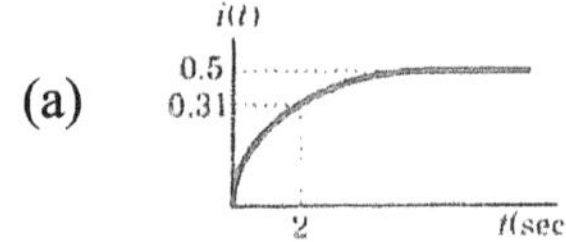

(b) 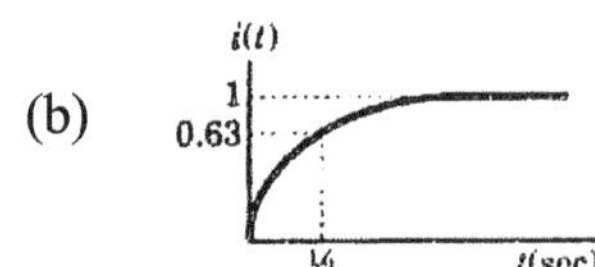

(c) 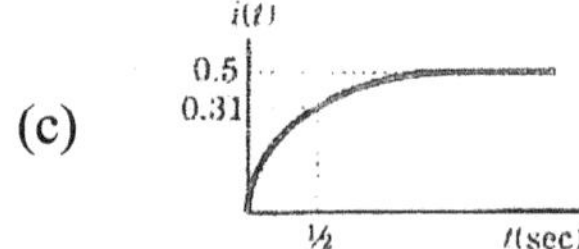

(d) 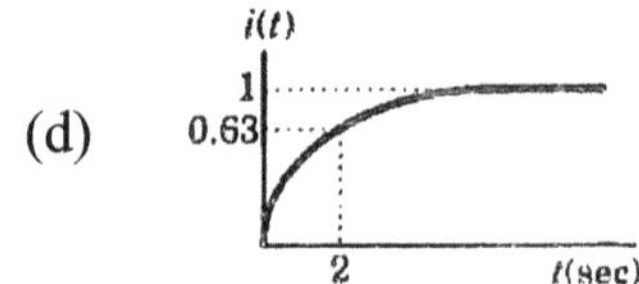

5. The Fourier transform of a conjugate symmetric function is always
   (a) Imaginary
   (b) Conjugate anti-symmetric
   (c) Real
   (d) Conjugate symmetric

6. For the lattice circuit shown in the figure, $Z_a = j2\Omega$ and $Z_b = 2\Omega$. The values of the open circuit impedance parameters $Z = \begin{bmatrix} Z_{11} & Z_{12} \\ Z_{21} & Z_{22} \end{bmatrix}$ are

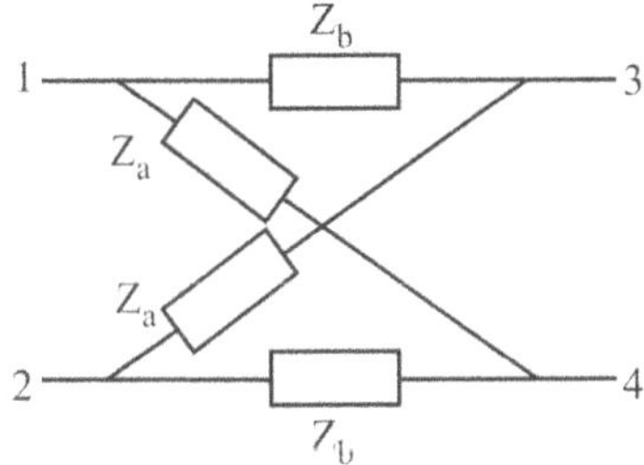

   (a) $\begin{bmatrix} 1-j & 1+j \\ 1+j & 1+j \end{bmatrix}$

   (b) $\begin{bmatrix} 1-j & 1+j \\ -1+j & 1-j \end{bmatrix}$

   (c) $\begin{bmatrix} 1+j & 1+j \\ 1-j & 1-j \end{bmatrix}$

   (d) $\begin{bmatrix} 1+j & -1+j \\ -1+j & 1+j \end{bmatrix}$

7. The circuit shown in the figure has initial current $i_L(0^-) = 1$ A through the inductor and an initial voltage $v_c(0^-) = -1$V across the capacitor. For input $v(t) = u(t)$, the Laplace transform of the current $i(t)$ for $t \geq 0$ is

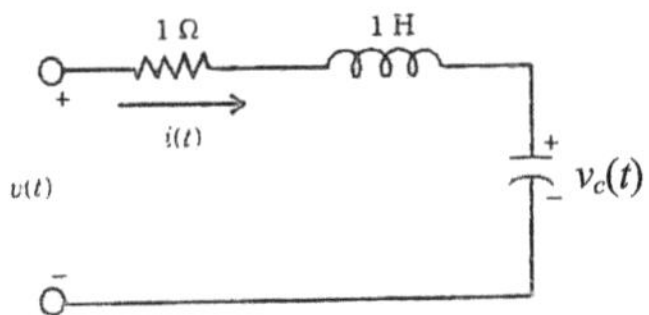

   (a) $\dfrac{s}{s^2+s+1}$ (b) $\dfrac{s+2}{s^2+s+1}$

   (c) $\dfrac{s-2}{s^2+s+1}$ (d) $\dfrac{s-2}{s^2+s+1}$

8. The transfer function $H(s) = \dfrac{V_o(s)}{V_i(s)}$ of an R-L-C circuit is given by $H(s) = \dfrac{10^6}{s^2 + 20s + 10^6}$. The quality factor (Q-factor) of this circuit is
   (a) 25 (b) 50
   (c) 100 (d) 5000

9. For the circuit shown in the figure, the initial conditions are zero. Its transfer function $H(s) = \dfrac{V_c(s)}{V_i(s)}$ is

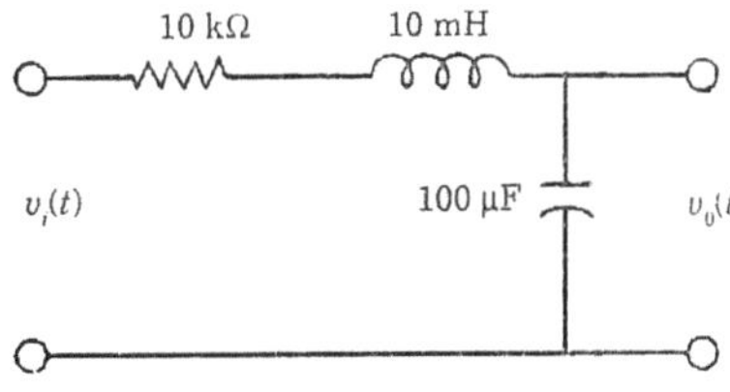

   (a) $\dfrac{1}{s^2+10^6 s+10^6}$

   (b) $\dfrac{10^6}{s^2+10^3 s+10^6}$

   (c) $\dfrac{10^3}{s^2+10^3 s+10^6}$

   (d) $\dfrac{10^6}{s^2+10^6 s+10^6}$

10. A system described by the following differential equation
$$\frac{d^2y}{dt^2}+3\frac{dy}{dt}+2y=x(t)$$ is initially at rest. For input x(t) = 2u(t), the output y(t) is
(a) $(1-2e^{-t}+e^{-2t})\,u(t)$
(b) $(1+2e^{-t}-2e^{-2t})\,u(t)$
(c) $(0.5+e^{-t}+1.5e^{-2t})\,u(t)$
(d) $(0.5+2e^{-t}+2e^{-2t})\,u(t)$

11. Consider the following statements S_1 and S_2
S_1: At the resonant frequency the impedance of a series R-L-C circuit is zero.
S_2: In a parallel G-L-C circuit, increasing the conductance G results in increase in its Q factor. Which one of the following is correct?
(a) S_1 is FALSE and S_2 is TRUE
(b) Both S_1 and S_2 are TRUE
(c) S_1 is TRUE and S_2 is FALSE
(d) Both S_1 and S_2 are FALSE

2004 EEE

12. The value of Z in the Fig. which is most appropriate to cause parallel resonance at 500 Hz is

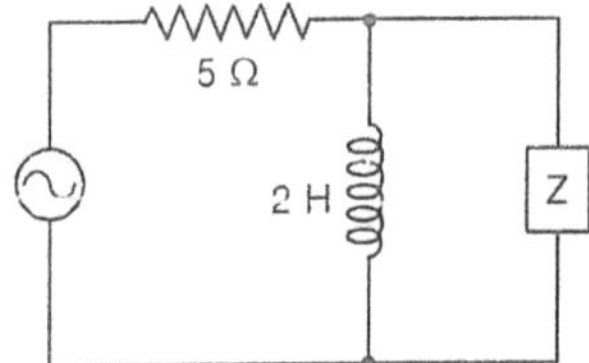

(a) 125.00 mH (b) 304.20 μF
(c) 2.0 μF (d) 0.05 μF

13. Total instantaneous power supplied by a 3-phase ac supply to a balanced R-L load is
(a) Zero (b) Constant
(c) Pulsating with zero average
(d) Pulsating with non-zero average

14. The rms value of the periodic waveform given in the Fig.

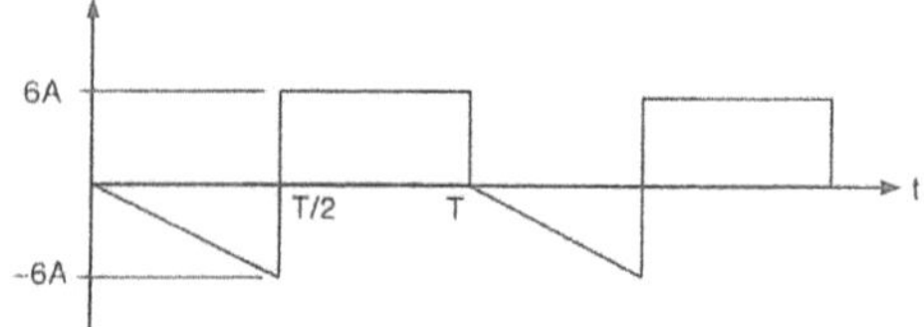

(a) $2\sqrt{6}$ A (b) $6\sqrt{2}$ A
(c) $\sqrt{4/3}$ A (d) 1.5 A

15. In the Fig. the value of the source voltage is

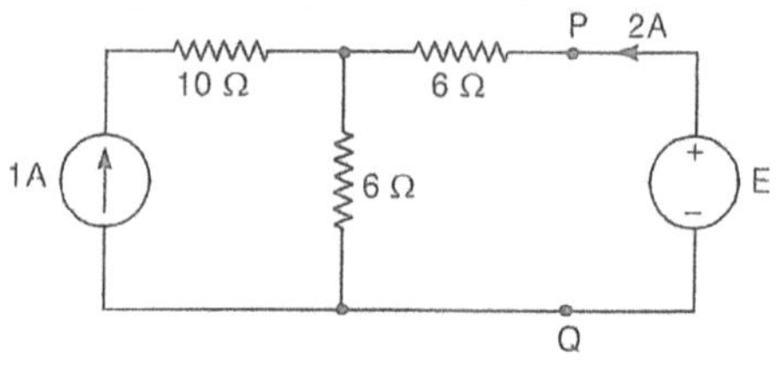

(a) 12 V (b) 24 V
(c) 30 V (d) 44 V

16. In the Fig. R_a, R_b and R_c are 20 Ω, 10 Ω and 10 Ω respectively. The resistances R_1, R_2 and R_3 in Ω of an equivalent star-connection are

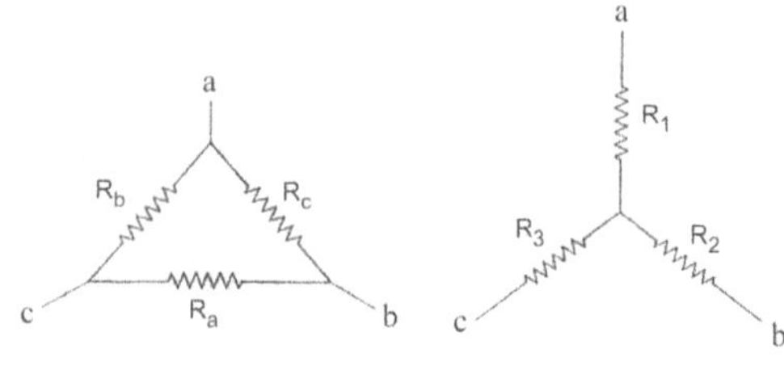

(a) 2.5, 5, 5 (b) 5, 2.5, 5
(c) 5, 5, 2.5 (d) 2.5, 5, 2.5

17. In the Fig. the value of resistance R in Ω is

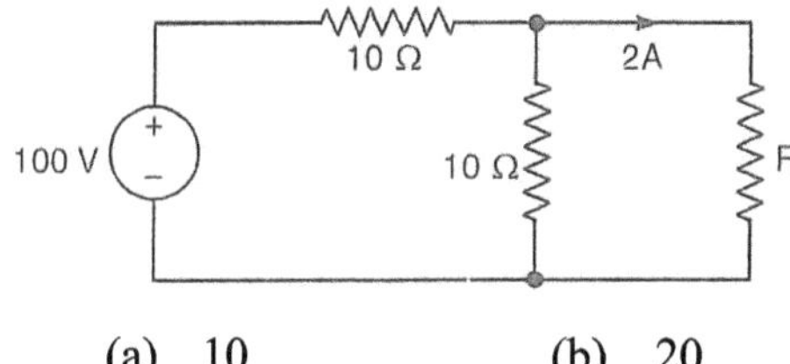

(a) 10 (b) 20
(c) 30 (d) 40

18. In the Fig. the capacitor initially has a charge of 10 Coulombs. The current in the circuit one second after the switch S is closed will be

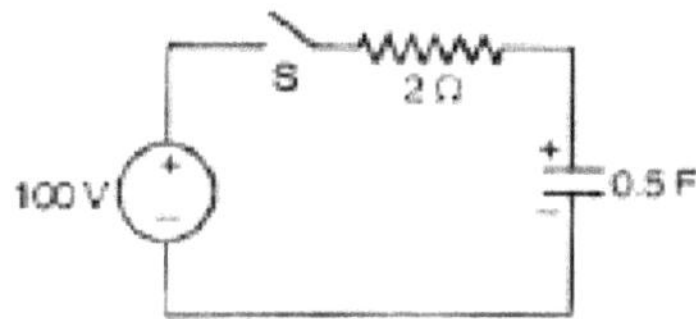

(a) 14.7 A (b) 18.5 A
(c) 40.0 A (d) 50.0 A

19. The rms value of the resultant current in a wire which carries a dc current of 10 A and a sinusoidal alternating current of peak value 20 A is

(a) 14.1 A (b) 17.3 A
(c) 22.4 A (d) 30.0 A

20. The Z matrix of a 2-port network is given by $\begin{bmatrix} 0.9 & 0.2 \\ 0.2 & 0.6 \end{bmatrix}$

The element Y_{22} of the corresponding Y matrix of the same network is given by

(a) 1.2 (b) 0.4
(c) – 0.4 (d) 1.8

2005 ECE

1. The condition on R, L and C such that the step response y(t) in the figure has no oscillations, is

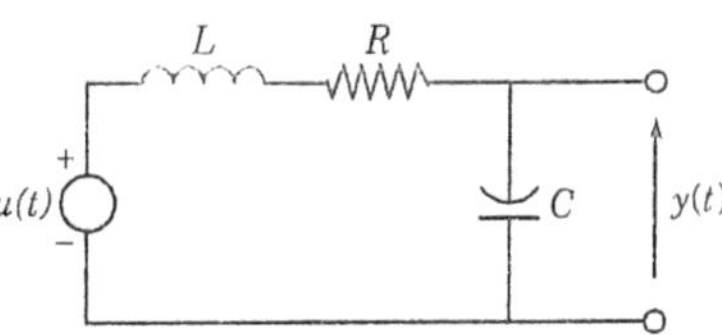

(a) $R \geq \frac{1}{2}\sqrt{\frac{L}{C}}$ (b) $R \geq \sqrt{\frac{L}{C}}$

(c) $R \geq 2\sqrt{\frac{L}{C}}$ (d) $R = \frac{1}{\sqrt{LC}}$

2. The ABCD parameters of an ideal n:1 transformer shown in the figure are $\begin{bmatrix} n & 0 \\ 0 & X \end{bmatrix}$. The value of X will be

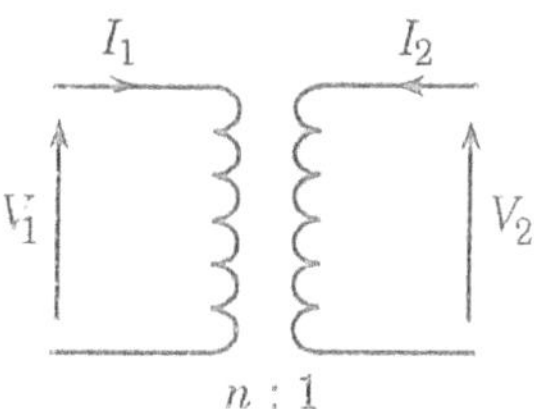

(a) n (b) $\frac{1}{n}$

(c) n^2 (d) $\frac{1}{n^2}$

3. In a series RLC circuit, $R = 2\ k\Omega$, L = 1 H, and $C = \frac{1}{400}\ \mu F$. The resonant frequency is

(a) 2×10^4 Hz (b) $\frac{1}{\pi} \times 10^4$ Hz

(c) 10^4 Hz (d) $2\pi \times 10^4$ Hz

4. The maximum power that can be transferred to the load resistor R_L from the voltage source in the figure is

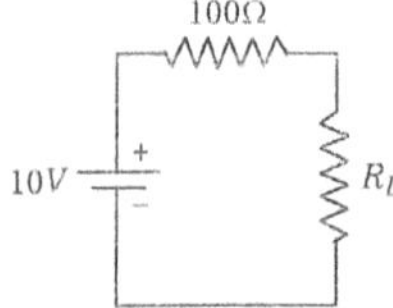

(a) 1 W (b) 10 W
(c) 0.25 W (d) 0.5 W

5. For the circuit shown in the figure, the instantaneous current $i_1(t)$ is

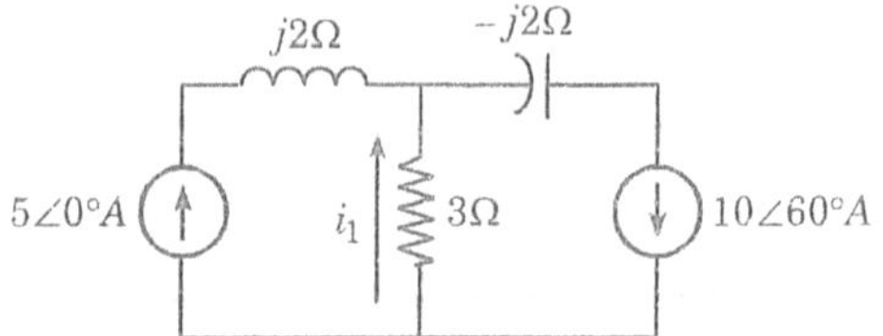

(a) $\frac{10\sqrt{3}}{2}\angle 90^\circ$ Amps

(b) $\frac{10\sqrt{3}}{2}\angle -90^\circ$ Amps

(c) $5\angle 60^\circ$ Amps

(d) $5\angle -60^\circ$ Amps

6. For the circuit shown in the Figure, Thevenin's voltage and Thevenin's equivalent resistance at terminals a – b is

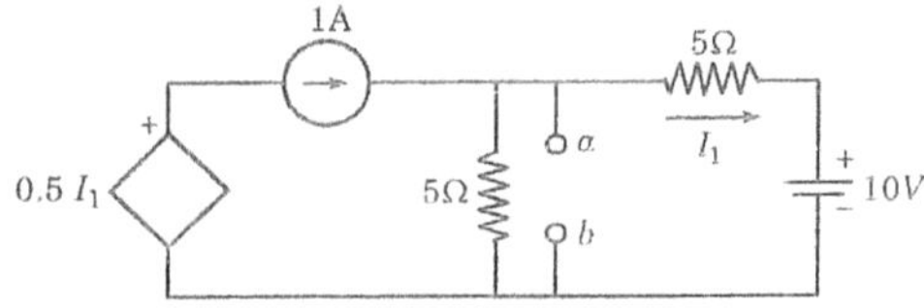

(a) 5 V and 2 Ω
(b) 7.5 V and 2.5 Ω
(c) 4 V and 2 Ω
(d) 3 V and 2.5 Ω

7. If $R_1 = R_2 = R_4 = R$ and $R_3 = 1.1R$ in the bridge circuit shown in the figure, then the reading in the ideal voltmeter connected between a and b is

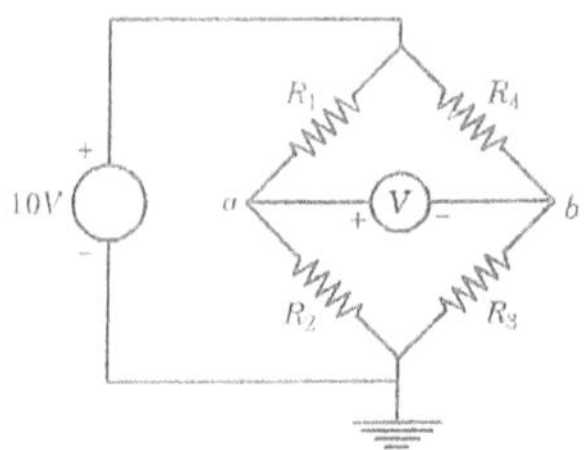

(a) 0.238 V (b) 0.138 V
(c) – 0.238 V (d) 1 V

8. The h parameters of the circuit shown in the figure are

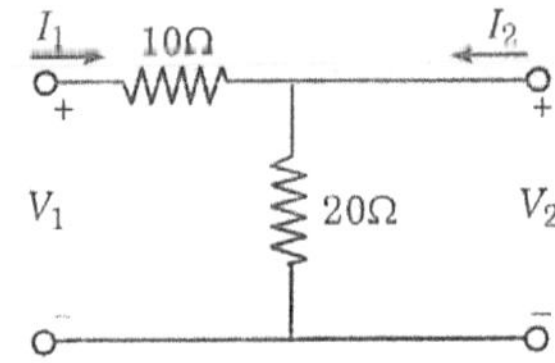

(a) $\begin{bmatrix} 0.1 & 0.1 \\ -0.1 & 0.3 \end{bmatrix}$ (b) $\begin{bmatrix} 10 & -1 \\ 1 & 0.05 \end{bmatrix}$

(c) $\begin{bmatrix} 30 & 20 \\ 20 & 20 \end{bmatrix}$ (d) $\begin{bmatrix} 10 & 1 \\ -1 & 0.05 \end{bmatrix}$

9. A square pulse of 3 volts amplitude is applied to C-R circuit shown in the figure. The capacitor is initially uncharged. The output voltage V_o at time t = 2 sec is

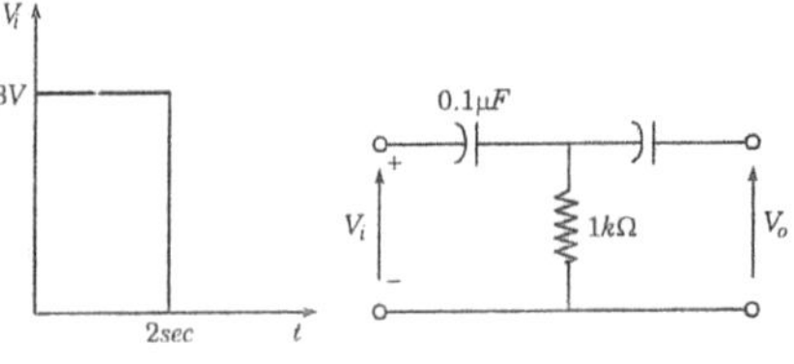

(a) 3V (b) – 3V
(c) 4V (d) – 4V

10. For a signal x(t) the Fourier transform is X(f). Then the inverse Fourier transform of X(3f + 2) is given by

(a) $\frac{1}{2}x\left(\frac{t}{2}\right)e^{-j3\pi t}$

(b) $\frac{1}{3}x\left(\frac{t}{3}\right)e^{-j\frac{4\pi t}{3}}$

(c) $3x(3t)e^{-j4\pi t}$

(d) $x(3t+2)$

2005 EEE

11. In the figure given below the value of R is

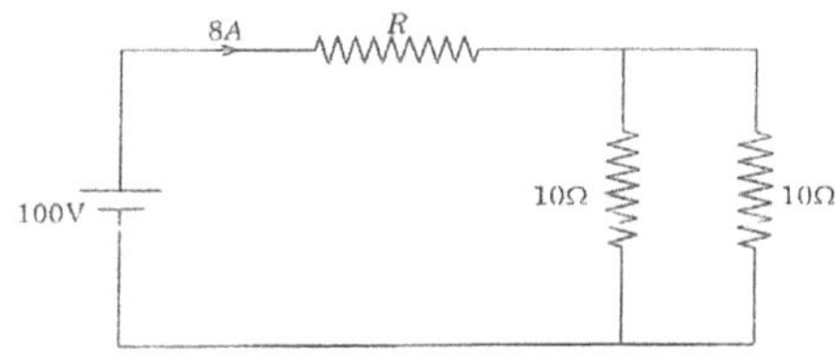

(a) 2.5 Ω (b) 5.0 Ω
(c) 7.5 Ω (d) 10.0 Ω

12. The RMS value of the voltage u(t) = 3 + 4 cos (3t) is

(a) $\sqrt{17}$ V
(b) 5V
(c) 7V
(d) $\left(3+2\sqrt{2}\right)$ V

13. For the two port network shown in the figure the Z-matrix is given by

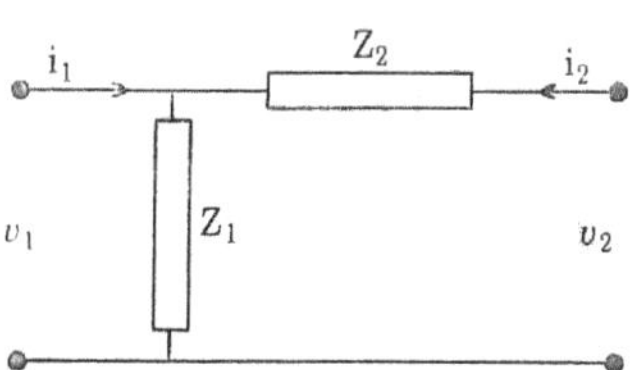

(a) $\begin{bmatrix} Z_1 & Z_1+Z_2 \\ Z_1+Z_2 & Z_2 \end{bmatrix}$

(b) $\begin{bmatrix} Z_1 & Z_1 \\ Z_1+Z_2 & Z_2 \end{bmatrix}$

(c) $\begin{bmatrix} Z_1 & Z_2 \\ Z_2 & Z_1+Z_2 \end{bmatrix}$

(d) $\begin{bmatrix} Z_1 & Z_1 \\ Z_1 & Z_1+Z_2 \end{bmatrix}$

14. In the figure given below, the initial capacitor voltage is zero. The switch is closed at t = 0. The final steady-state voltage across the capacitor is

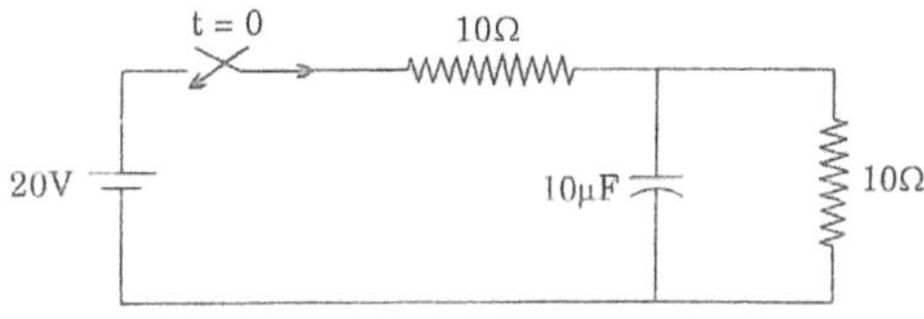

(a) 20 V (b) 10 V
(c) 5 V (d) 0 V

15. The RL circuit of the figure is fed from a constant magnitude, variable frequency sinusoidal voltage source v_{IN}. At 100 Hz, the R and L elements each have a voltage drop u_{RMS}. If the frequency of the source is changed to 50 Hz, then new voltage drop across R is

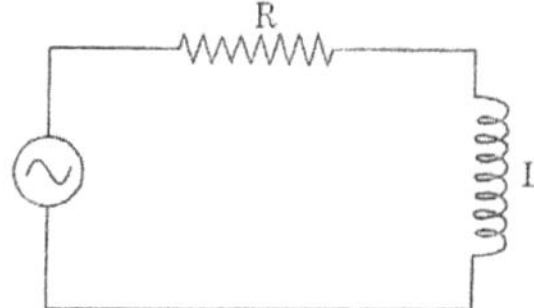

(a) $\sqrt{\frac{5}{8}}u_{RMS}$ (b) $\sqrt{\frac{2}{3}}u_{RMS}$
(c) $\sqrt{\frac{8}{5}}u_{RMS}$ (d) $\sqrt{\frac{3}{2}}u_{RMS}$

16. For the three-phase circuit shown in the figure the ratio of the currents $I_R : I_Y : I_B$ is given by

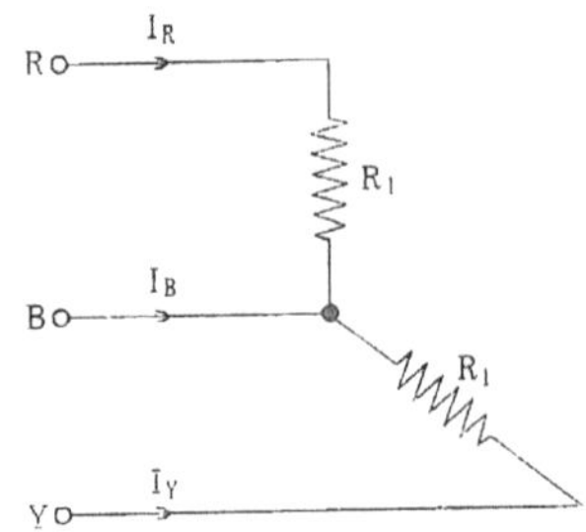

(a) $1:1:\sqrt{3}$ (b) 1:1:2

(c) 1:1:0 (d) $1:1:\sqrt{3/2}$

17. For the triangular wave form shown in the figure, the RMS value of the voltage is equal to

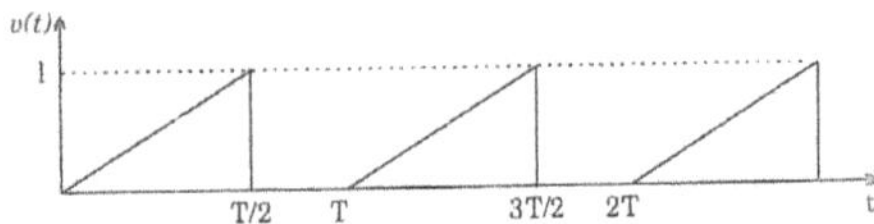

(a) $\sqrt{\frac{1}{6}}$ (b) $\sqrt{\frac{1}{3}}$

(c) $\frac{1}{3}$ (d) $\sqrt{\frac{2}{3}}$

18. The circuit shown in the figure is in steady state, when the switch is closed at t = 0. Assuming that the inductance is ideal, the current through the inductor at $t = 0^+$ equals

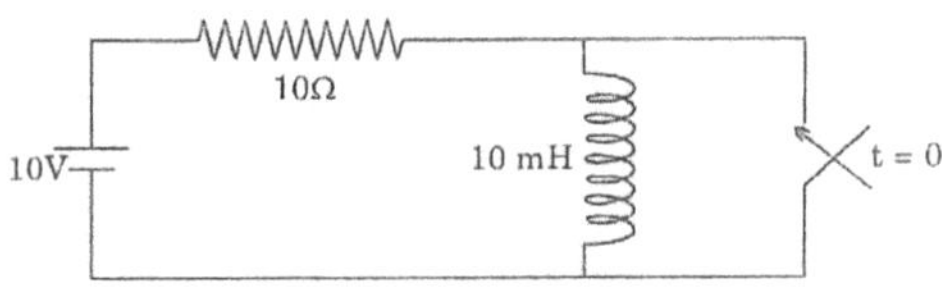

(a) 0 A (b) 0.5 A

(c) 1 A (d) 2 A

19. In the given figure, the Thevenin's equivalent pair (voltage, impedance), as seen at the terminals P-Q is given by

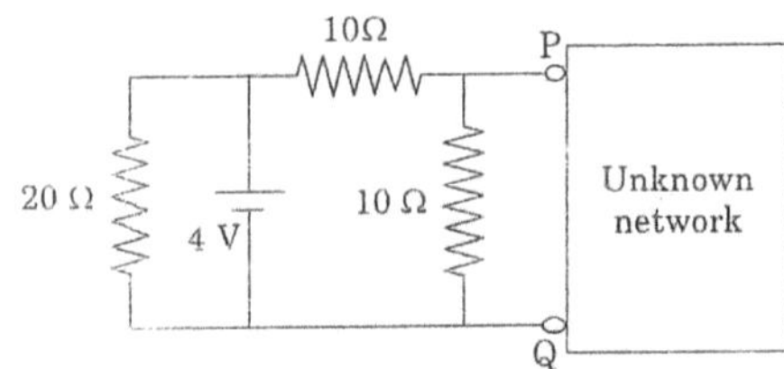

(a) $(2\text{ V}, 5\,\Omega)$ (b) $(2\text{ V}, 7.5\,\Omega)$

(c) $(4\text{ V}, 5\,\Omega)$ (d) $(4\text{ V}, 7.5\,\Omega)$

20. The Laplace transform of a function f(t) is $F(s) = \frac{5s^2 + 23s + 6}{s(s^2 + 2s + 2)}$. As $t \to \infty$, f(t) approaches

(a) 3 (b) 5

(c) $\frac{17}{2}$ (d) ∞

21. The Fourier series for the function $f(x) = \sin^2 x$ is

(a) sin x + sin 2x

(b) 1 – cos 2x

(c) sin 2x + cos 2x

(d) 0.5 – 0.5 cos 2x

22. Two wattmeters, which are connected to measure the total power on a three-phase system supplying a balanced load, read 10.5 kW and –2.5 kW, respectively. The total power and the power factor, respectively, are

(a) 13.0 kW, 0.334

(b) 13.0 kW, 0.684

(c) 8.0 kW, 0.52

(d) 8.0 kW, 0.334

Linked Answer Questions: Q. 23(a) in Q.23(b). Carry Two Marks Each

Statement for Linked Answer Questions 23(a) and 23(b)

A coil of inductance 10 H and resistance 40 Ω is connected as shown in the figure. After the switch S has been in contact with point 1 for a very long time, it is moved to point 2 at t = 0.

23a. If, at t = 0^+, the voltage across the coil is 120 V, the value of resistance R is

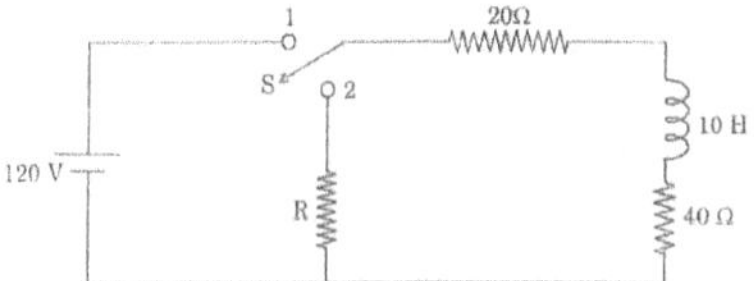

(a) 0 Ω (b) 20 Ω
(c) 40 Ω (d) 60 Ω

23b. For the value of obtained in (a), the time taken for 95% of the stored energy to be dissipated is close to

(a) 0.10 sec (b) 0.15 sec
(c) 0.50 sec (d) 1.0sec

2006 ECE

1. Consider the function *f*(*t*) having Laplace transform

$$F(s)=\frac{\omega_0}{s^2+\omega_0^2}\operatorname{Re}[s]>0$$

The final value of *f*(*t*) would be

(a) 0 (b) 1
(c) $-1\le f(\infty)\le 1$ (d) ∞

2. A two-port network is represented by ABCD parameters given by

$$\begin{bmatrix}V_1\\I_1\end{bmatrix}=\begin{bmatrix}A & B\\C & D\end{bmatrix}\begin{bmatrix}V_2\\-I_2\end{bmatrix}$$

If port-2 is terminated by R_L, the input impedance seen at port-I is given by

(a) $\frac{A+BR_L}{C+DR_L}$ (b) $\frac{AR_L+C}{BR_L+D}$

(c) $\frac{DR_L+A}{BR_L+C}$ (d) $\frac{B+AR_L}{D+CR_L}$

3. In the two port network shown in the figure below, z_{12} and z_{21} are, respectively.

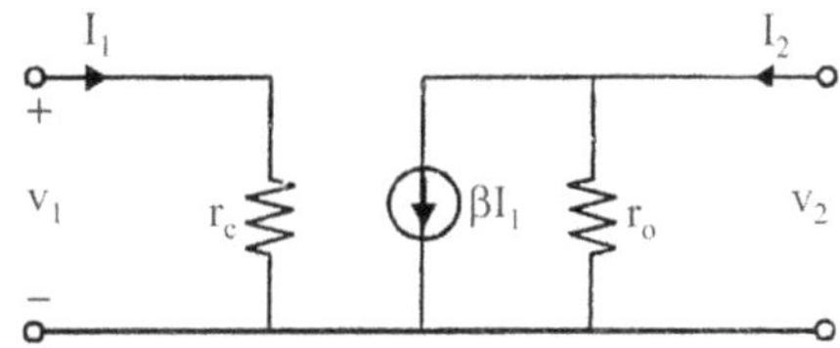

(a) r_e and βr_0 (b) 0 and $-\beta r_0$
(c) 0 and βr_0 (d) r_c and $-\beta r_0$

4. A 2 mH inductor with some initial current can be represented as shown below, where s is the Laplace Transform variable. The value of initial current is

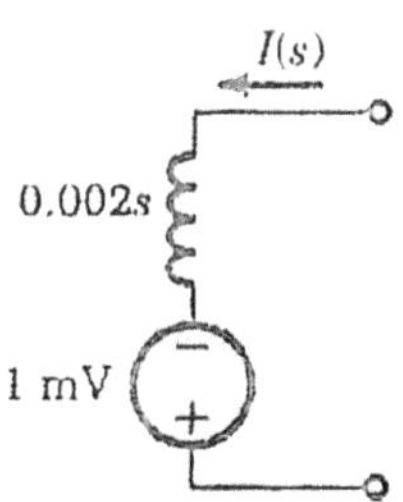

(a) 0.5 A (b) 2.0 A
(c) 1.0 A (d) 0.0A

5. In the figure shown, assume that all the capacitors are initially uncharged. If V_t (t) = 10u(t) Volts, then V_0(t) is given by

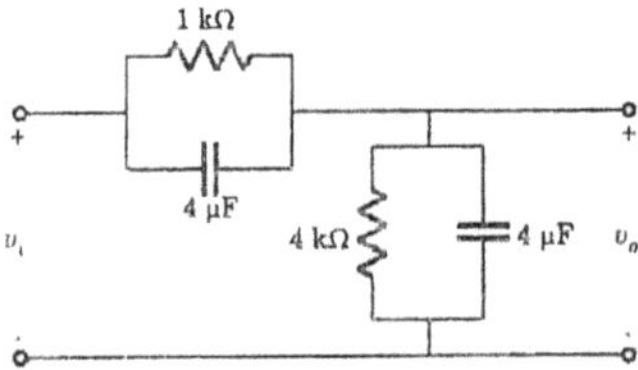

(a) $8e^{-0.004\,t}$ Volts
(b) $8\,(1 - e^{-0.004\,t})$ Volts
(c) $8u\,(t)$ Volts
(d) 8 Volts

6. The unit-step response of a system starting from rest is given by
$C(t) = 1 - e^{-2t}$ for $t \geq 0$
The transfer function of the system is
(a) $\frac{1}{1+2s}$ (b) $\frac{2}{2+s}$
(c) $\frac{1}{2+s}$ (d) $\frac{2s}{1+2s}$

7. The unit impulse response of a system is
$h(t) = e^{-t}, t \geq 0$
For this system, the steady-state value of the output for unit step input is equal to
(a) –1 (b) 0
(c) 1 (d) ∞

2006 EEE

8. $x(t)$ is a real valued function of a real variable with period T. Its trigonometric Fourier Series expansion contains no terms of frequency $\omega = 2\pi\,(2k)/T$; $k = 1, 2,\ldots$ Also, no sine terms are present. Then x(t) satisfies the equation
(a) $x(t) = -x\,(t - T)$
(b) $x(t) = x\,(T - t) = -x(-t)$
(c) $x(t) = x(T - t) = -x(t - T/2)$
(d) $x(t) = x(t - T) = x\,(t - T/2)$

9. In the figure the current source is $1\angle 0$ A, R = 1 Ω, the impedances are Z_C = jΩ, and Z_L = 2jΩ. The Thevenin equivalent looking into the circuit across X-Y is

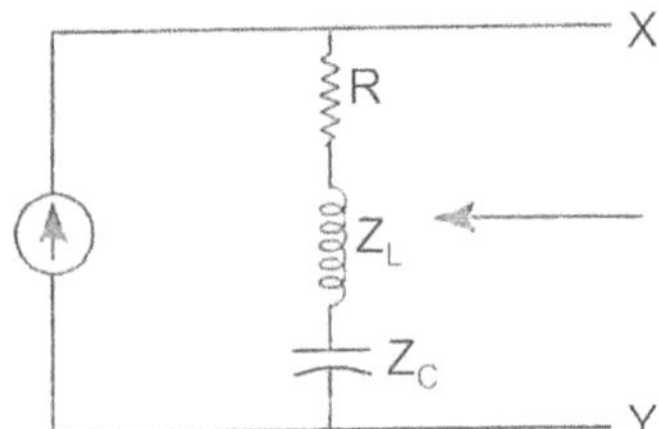

(a) $\sqrt{2}\angle 0\,\text{V}, (1+2j)\Omega$
(b) $2\angle 45^\circ\,\text{V}, (1-2j)\Omega$
(c) $2\angle 45^\circ$ V, $(1+j)\Omega$
(d) $\sqrt{2}\angle 45^\circ$ V, $(1+j)\Omega$

10. The parameters of the circuit shown in the figure are R_i = 1 MΩ, R_o = 10Ω, A = 10^6 V/V. If V_i = 1 μV, then output voltage, input impedance and output impedance respectively are

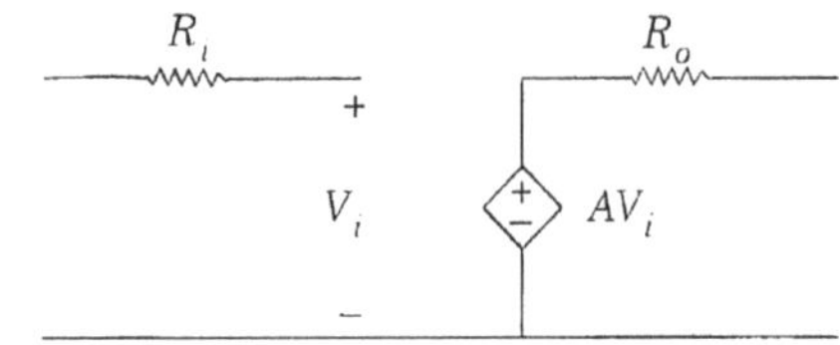

(a) 1V,∞,10Ω
(b) 1V,0,10Ω
(c) 1V,0,∞
(d) 10V,∞,10Ω

11. In the circuit shown in the figure, the current source I = 1 A, voltage source V = 5V, $R_1 = R_2 = R_3 = 1$ Ω, $L_1 = L_2 = L_3 = 1$ H, $C_1 = C_2 = 1$F.

The currents (in A) through R_3 and the voltage source V respectively will be

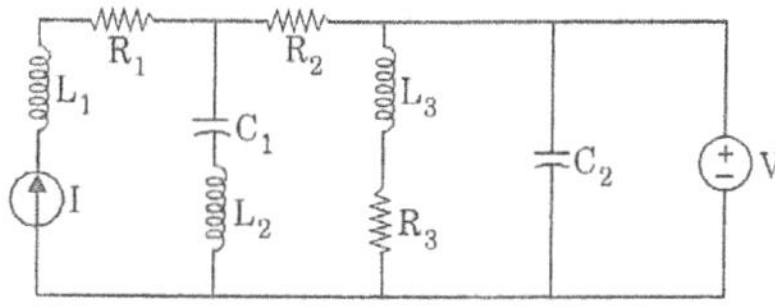

(a) 1, 4 (b) 5, 1
(c) 5, 2 (d) 5, 4

12. The circuit shown in the figure is energized by a sinusoidal voltage source V_1 at a frequency which causes resonance with a current of I.

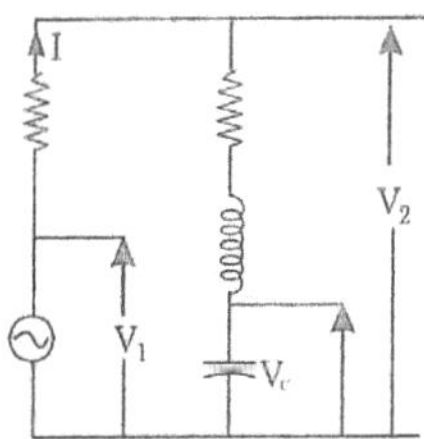

The phasor diagram which is applicable to this circuit

(a)

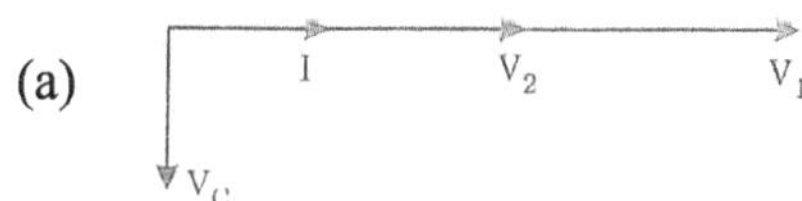

(b)

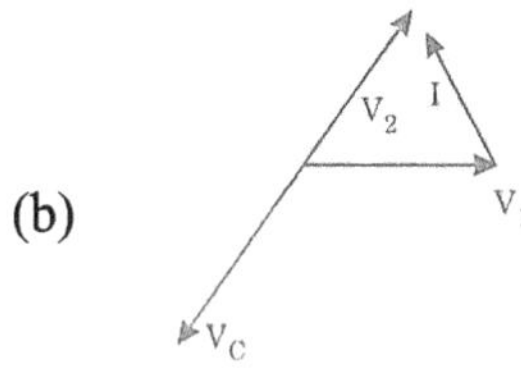

(c)

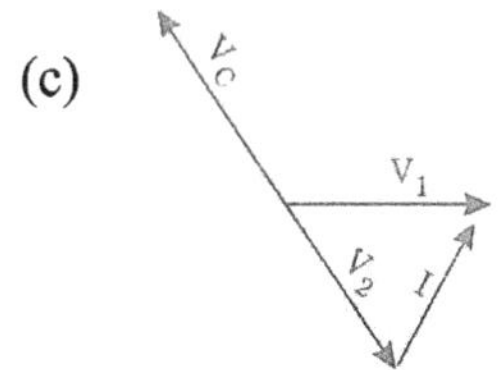

(d)

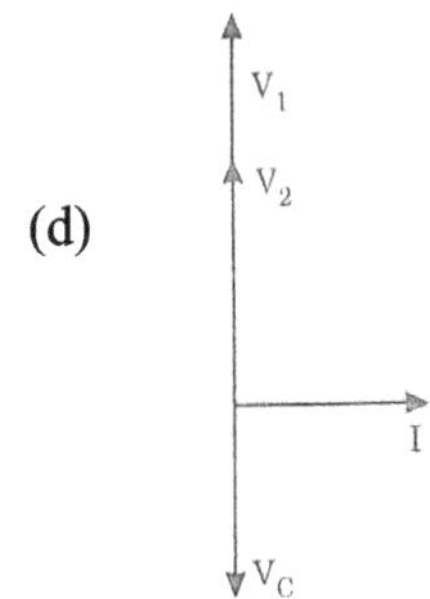

13. An ideal capacitor is charged to a voltage V_0 and connected at $t = 0$ across an ideal inductor L. (The circuit now consists of a capacitor and inductor alone). If we let $\omega_0 = \frac{1}{\sqrt{LC}}$, the voltage across the capacitor at time t > 0 is given by

(a) V_0 (b) $V_0 \cos(\omega_0 t)$
(c) $V_0 \sin(\omega_0 t)$ (d) $V_0 e^{-\omega_0 t} \cos(\omega_0 t)$

14. A 400V, 50 Hz, three phase balanced source supplies power to a star connected load whose rating is $12\sqrt{3}\,kVA$, 0.8 pf(lag). The rating (in kVAR) of the delta connected (capacitive) reactive power band necessary to bring the pf to unity is

(a) 28.78 (b) 21.60
(c) 16.60 (d) 12.47

2007 ECE

1. An independent voltage source in series with an impedance $Z_S = R_S + jX_s$ delivers a maximum average power to a load impedance Z_L when

(a) $Z_L = R_S + jX_S$ (b) $Z_L = R_S$
(c) $Z_L = jX_s$ (d) $Z_L = R_S - jX_S$

2. The RC circuit shown in the figure is

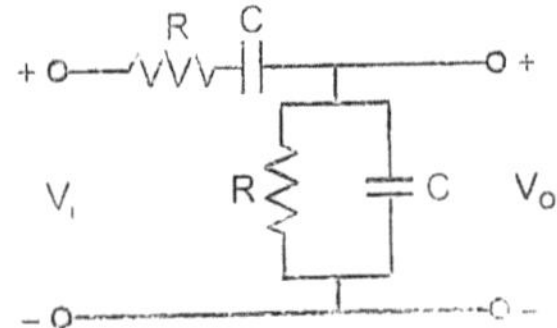

(a) a low-pass filter
(b) a high-pass filter
(c) a band-pass filter
(d) a band-reject filter

3. If the Laplace transform of a signal y(t) is $Y(s) = \dfrac{1}{s(s-1)}$, then its final value is

(a) –1 (b) 0
(c) 1 (d) Unbounded

4. Two series resonant filters are as shown in the figure. Let the 3-dB bandwidth of Filter 1 be B_1 and that of Filter 2 be B_2. The value of $\dfrac{B_1}{B_2}$ is

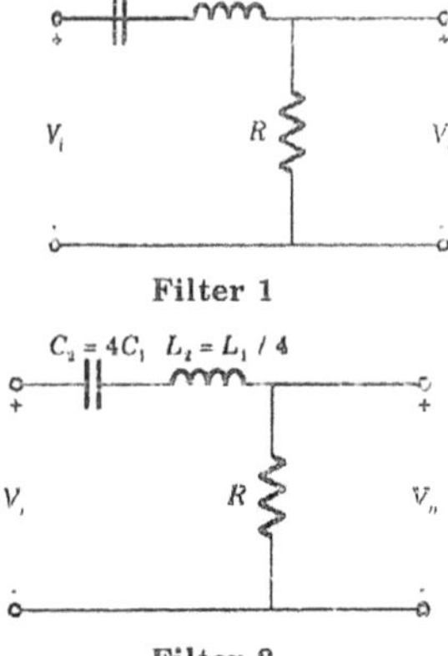

(a) 4 (b) 1
(c) $\dfrac{1}{2}$ (d) $\dfrac{1}{4}$

5. For the circuit shown in the figure, the Thevenin voltage and resistance looking into X-Y are

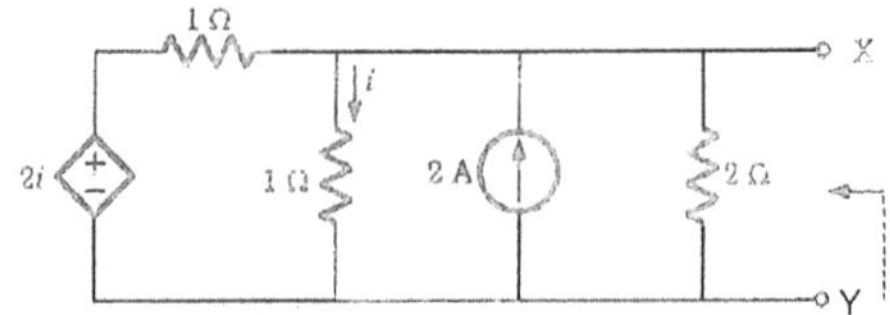

(a) $\dfrac{4}{3}$V, 2Ω (b) 4V, $\dfrac{2}{3}$Ω
(c) $\dfrac{4}{3}$V, $\dfrac{2}{3}$Ω (d) 4V, 2Ω

6. In the circuit shown, V_C is 0 volts at t = 0, sec. For t > 0, the capacitor current $i_C(t)$, where t is in seconds, is given by

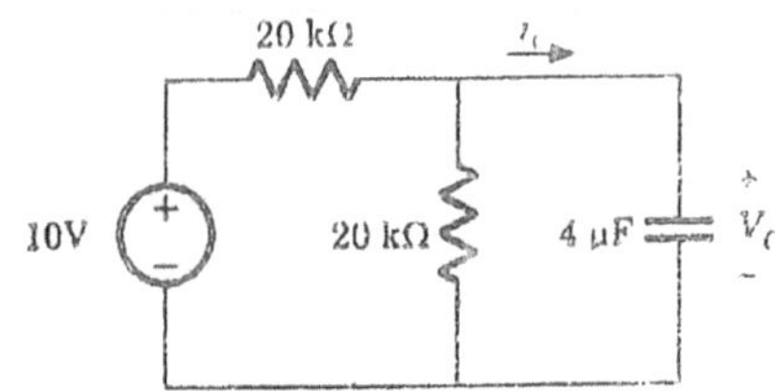

(a) 0.50 exp (– 25t) mA
(b) 0.25 exp(–25t) mA
(c) 0.50 exp (–12.5t) mA
(d) 0.25 exp (– 6.25t) mA

7. In the AC network shown in the figure, the phasor voltage V_{AB} (in Volts) is

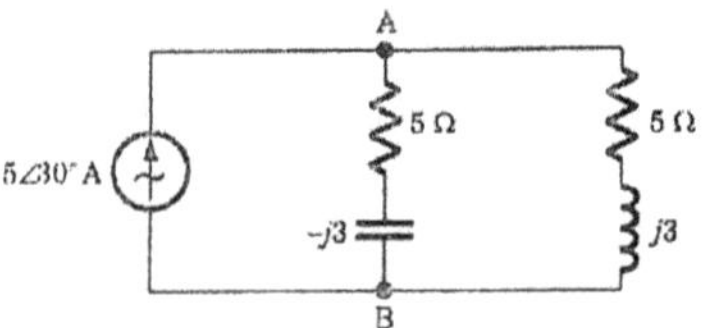

(a) 0 (b) 5∠30°
(c) 12.5∠30° (d) 17∠30°

8. The frequency response of a linear, time-invariant system is given by $H(f)=\frac{5}{1+j10\pi f}$. The step response of the system is

(a) $5(1-e^{-5t})u(t)$

(b) $5\left(1-e^{-\frac{t}{5}}\right)u(t)$

(c) $\frac{1}{5}\left(1-e^{-5t}\right)u(t)$

(d) $\frac{1}{5}\left(1-e^{-\frac{t}{5}}\right)u(t)$

2007 EEE

9. Let a signal $a_1 \sin(\omega_1 t+\phi_1)$ be applied to a stable linear time – invariant system. Let the corresponding steady state output be represented as $a_2 F(\omega_2 t+\phi_2)$. Then which of the following statements is true ?

(a) F is not ncccssarily a "sinc" or "cosine" function but must be periodic with $\omega_1=\omega_2$

(b) F must be a "sine" or "cosine" function with $a_1=a_2$

(c) F must be a "sine" function with $w_1=w_2$ and $\phi_1=\phi_2$

(d) F must be a "sine" or "cosine" function with $\omega_1=\omega_2$.

10. A signal x(t) is given by

$$x(t)=\begin{cases}1, -T/4<t\le 3T/4\\ -1, 3T/4<t\le 7T/4\\ -x(t+T)\end{cases}$$

which among the following gives the fundamental Fourier term of x(t) ?

(a) $\frac{4}{\pi}\cos\left(\frac{\pi t}{T}-\frac{\pi}{4}\right)$

(b) $\frac{4}{\pi}\cos\left(\frac{\pi t}{2T}+\frac{\pi}{4}\right)$

(c) $\frac{4}{\pi}\sin\left(\frac{\pi t}{T}-\frac{\pi}{4}\right)$

(d) $\frac{4}{\pi}\sin\left(\frac{\pi t}{2T}+\frac{\pi}{4}\right)$

11. A 230 V (phase), 50 Hz, three – phase, 4 – wire system has a phase sequence ABC. A unity power – factor load of 4 kW is connected between phase A and neutral N. it is desired to achieve zero neutral current through the use of a pure inductor and a pure capacitor in the other two phase. The values of inductor and capacitor are

(a) 72.95 mH in phase C and 139.02 μF in phase B

(b) 72.95 mH in phase B and 139.02 μF in phase C

(c) 42.12 mH in phase C and 240.79 μF in phase B

(d) 42.12 mH in phase B and 240.79 μF in phase C

12. If u(t), r(t) denote the unit step and unit ramp functions respectively and u(t) * r(t) their convolution, then the function u(t+1) * r(t–2) is given by

(a) (1/2) (t – 1) (t – 2)

(b) (1/2) (t – 1) (t – 2)

(c) $(1/2)(t-1)^2 u(t-1)$

(d) none of these

13. In the circuit shown in figure switch SW_1 is initially CLOSED and SW_2 is OPEN. The inductor L carriers a current of 10 A and the capacitor is charged to 10 V with polarities as indicated. SW_2 is initially CLOSED at t = 0 – and SW_1 is OPENED at t = 0. The current through C and the voltage across L at t = 0 + is

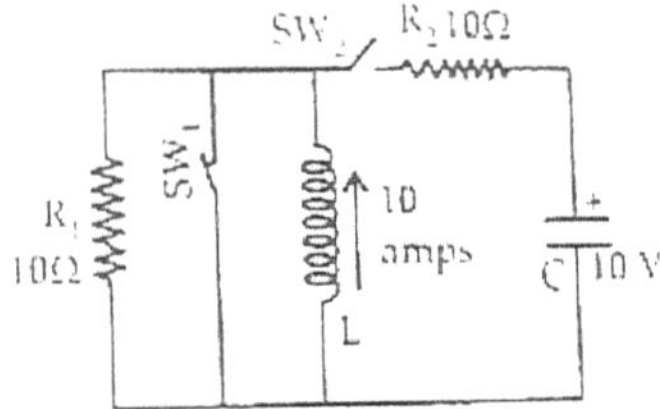

(a) 55 A, 4.5 V (b) 5.5 A, 45 V
(c) 45 A, 5.5 V (d) 4.5 A, 55 V

14. The R-L-C series circuit shown is supplied from a variable frequency voltage source. The admittance locus of the R-L-C network at terminals AB for increasing frequency ω is

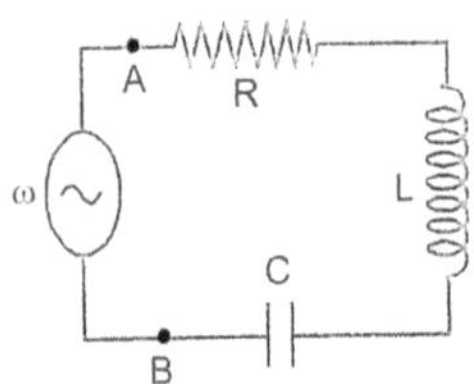

(a)

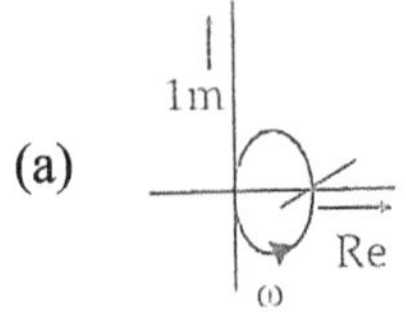

(b)

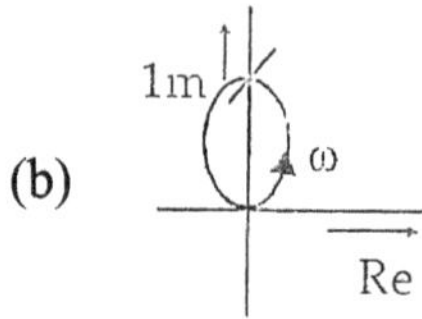

(c)

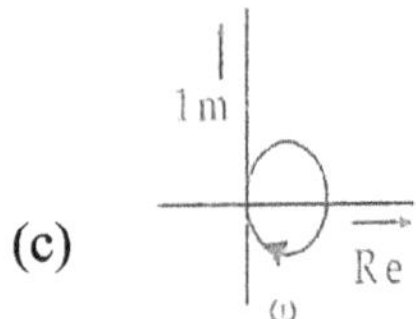

(d)

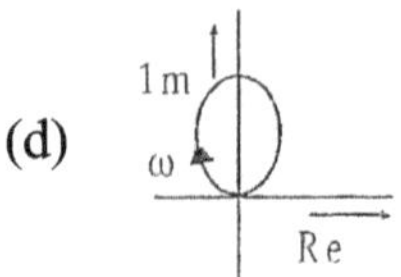

15. In the figure given below all phasors are with reference to the potential at point "O". The locus of voltage phasor V_{YX} as R is varied from zero to infinity is shown by

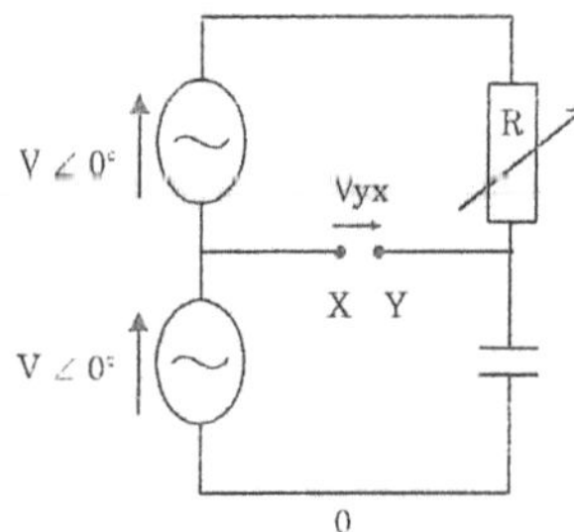

(a)

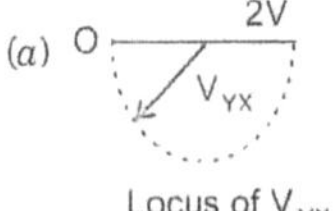

(b)

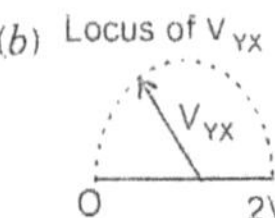

(c)

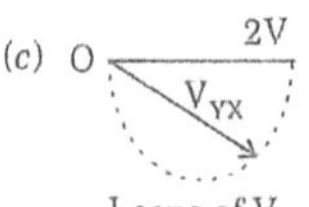

(d)

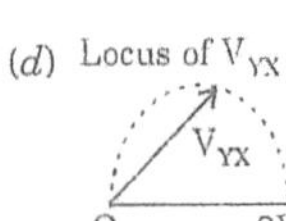

16. A 3 V dc supply with an internal resistance of 2 Ω supplies a passive non-linear resistance characterized by the relation VNL = I^2_{NL}. The power dissipated in the non-linear resistance is

(a) 1.0 W (b) 1.5 W
(c) 2.5 W (d) 3.0 W

Linked Answer Questions: Q. 17 to 18. Carry two marks each

Statement for Linked Answer Questions 17 and 18:

An inductor designed with 400 turns coil wound on an iron core of 16 cm^2 cross sectional area and with a cut of an air gap length of 1 mm. The coil is connected to a 230 V, 50 Hz ac supply. Neglect coil resistance, core loss, iron reluctance and leakage inductance. ($\mu_0 = 4\pi \times 10^{-7}$ H/m)

17. The current in the inductor is
    (a) 18.08 A (b) 9.04 A
    (c) 4.56 A (d) 2.28 A

18. The average force on the core to reduce the air gap will be
    (a) 832.29 N
    (b) 1666.22 N
    (c) 3332.47 N
    (d) 6664.84 N

2008 ECE

1. In the following graph, the number of trees (P) and the number of cut-sets (Q) are

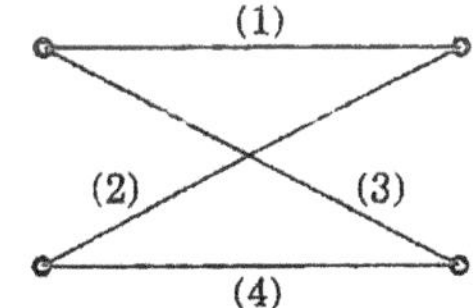

   (a) P = 2, Q = 2
   (b) P = 2, Q = 6
   (c) P = 4, Q = 6
   (d) P = 4, Q = 10

2. In the following circuit, the switch S is closed at t = 0. The rate of change of current $\frac{di}{dt}(0+)$ is given by

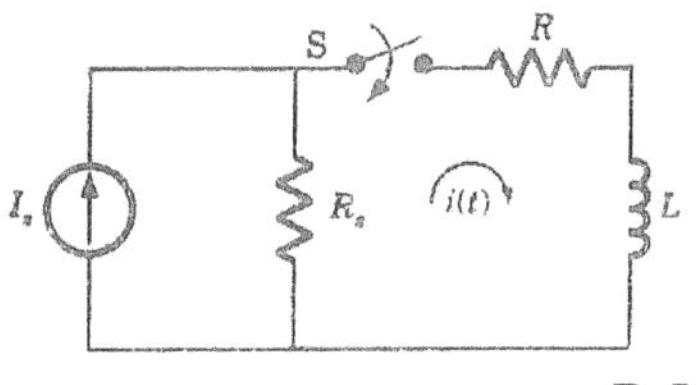

   (a) 0 (b) $\frac{R_s I_s}{L}$
   (c) $\frac{(R+R_s)I_s}{L}$ (d) ∞

3. The impulse response h(t) of a linear time-invariant continuous time system is described by h(t) = exp (αt) u (t) + exp (βt) u (–t), where u(t) denotes the unit step function, and α and β are real constants. This system is stable if
   (a) α is positive and β is positive
   (b) α is negative and β is negative
   (c) α is positive and β is negative
   (d) α is negative and β is positive

4. The Thevenin equivalent impedance Z_{th} between the nodes P and Q in the following circuit is

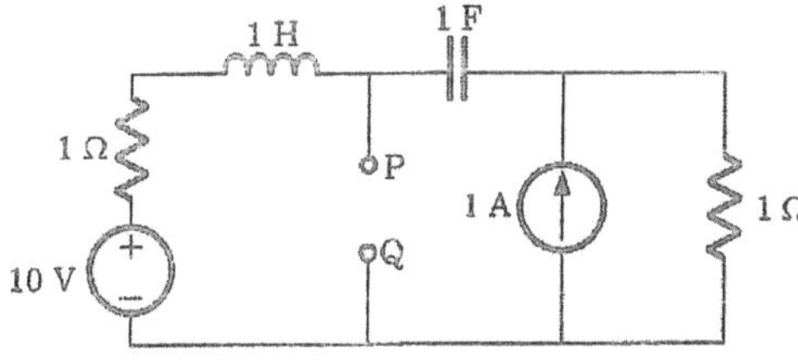

   (a) 1 (b) $1+s+\frac{1}{s}$
   (c) $2+s+\frac{1}{s}$ (d) $\frac{s^2+s+1}{s^2+2s+1}$

5. The driving point impedance of the following network is given by

$Z(s) = \dfrac{0.2\,s}{s^2 + 0.1\,s + 2}$. The component values are

(a) $L = 5H, R = 0.5\ \Omega, C = 0.1F$

(b) $L = 0.1\ H, R = 0.5\Omega, C = 5F$

(c) $L = 5H, R = 2\Omega, C = 0.1F$

(d) $L = 0.1H, R = 2\Omega, C = 5F$

6. Group I lists a set of four transfer functions. Group II gives a list of possible step responses y(t). Match the step responses with the corresponding transfer functions

Group I

$P = \dfrac{25}{s^2 + 25}$ $\quad Q = \dfrac{36}{s^2 + 20s + 36}$

$R = \dfrac{36}{s^2 + 12s + 36}$ $\quad S = \dfrac{49}{s^2 + 7s + 49}$

Group II

1.

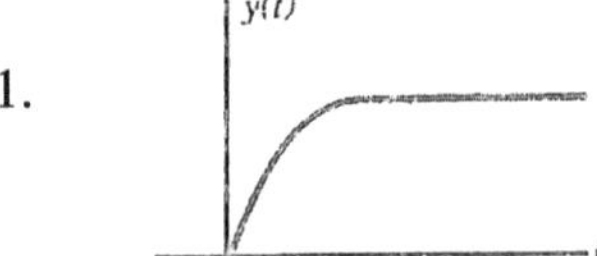

2.

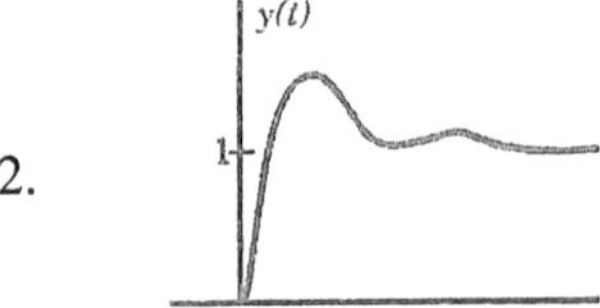

3.

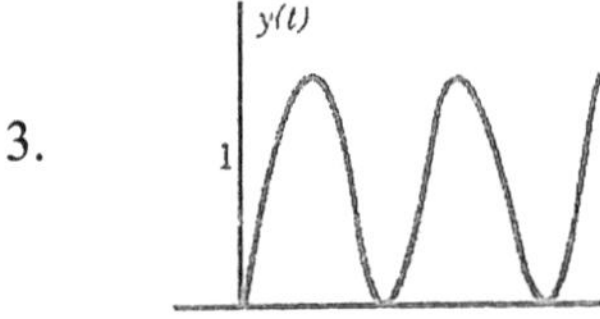

4.

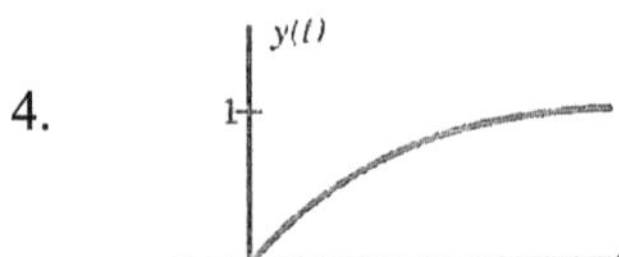

(a) P – 3, Q – 1, R –4, S – 2

(b) P – 3, Q – 2, R – 4, S – 1

(c) P – 2, Q – 1, R – 4, S – 3

(d) P – 3, Q – 4, R – 1, S – 2

Common Data for Questions 7 and 8

The following series RLC circuit with zero initial conditions is excited by a unit impulse function $\delta(t)$

1 H 1Ω

δ(t) 1 F $V_c(t)$

7. For $t > 0$, the output voltage $V_C(t)$ is

(a) $\dfrac{2}{\sqrt{3}}\left(e^{-\frac{1}{2}t} - e^{\frac{\sqrt{3}}{2}t}\right)$

(b) $\dfrac{2}{\sqrt{3}}t\,e^{-\frac{1}{2}t}$

(c) $\dfrac{2}{\sqrt{3}}e^{-\frac{1}{2}t}\cos\left(\dfrac{\sqrt{3}}{2}t\right)$

(d) $\dfrac{2}{\sqrt{3}}e^{-\frac{1}{2}t}\sin\left(\dfrac{\sqrt{3}}{2}t\right)$

8. For t > 0, the voltage across the resistor is

(a) $\dfrac{1}{\sqrt{3}}\left(e^{-\frac{\sqrt{3}}{2}} - e^{-\frac{1}{2}t}\right)$

(b) $e^{\frac{1}{2}t}\left[\cos\left(\dfrac{\sqrt{3}t}{2}\right) - \dfrac{1}{\sqrt{3}}\sin\left(\dfrac{\sqrt{3}t}{2}\right)\right]$

(c) $\dfrac{2}{\sqrt{3}}e^{-\frac{1}{2}t}\sin\left(\dfrac{\sqrt{3}t}{2}\right)$

(d) $\dfrac{2}{\sqrt{3}}e^{-\frac{1}{2}t}\cos\left(\dfrac{\sqrt{3}t}{2}\right)$

Linked Answer Questions: 9 to 10 carry two marks each.

Statement for Linked Answer Questions 9 and 10:

A two port network shown below is excited by external dc source. The voltages and the currents are measured with voltmeters V_1, V_2 and ammeters A_1, A_2 (all assumed to be ideal), as indicated. Under following switch conditions, the readings obtained are:

(i) S_1 – Open, S_2 – Closed $A_1 = 0$ A, $V_1 = 4.5$ V, $V_2 = 1.5$ V, $A_2 = 1$A

(ii) S_1 – Closed, S_2 – Open $A_1 = 4$ A, $V_1 = 6$V, V2 = 6 V A_2 = OA

9. The Z-parameter matrix for this network is

(a) $\begin{bmatrix} 1.5 & 1.5 \\ 4.5 & 1.5 \end{bmatrix}$ (b) $\begin{bmatrix} 1.5 & 4.5 \\ 1.5 & 4.5 \end{bmatrix}$

(c) $\begin{bmatrix} 1.5 & 4.5 \\ 1.5 & 1.5 \end{bmatrix}$ (d) $\begin{bmatrix} 4.5 & 1.5 \\ 1.5 & 4.5 \end{bmatrix}$

10. The h-parameter matrix for this network is

(a) $\begin{bmatrix} -3 & 3 \\ -1 & 0.67 \end{bmatrix}$ (b) $\begin{bmatrix} -3 & -1 \\ 3 & 0.67 \end{bmatrix}$

(c) $\begin{bmatrix} 3 & 3 \\ 1 & 0.67 \end{bmatrix}$ (d) $\begin{bmatrix} 3 & 1 \\ 3 & 0.67 \end{bmatrix}$

2008 EEE

11. The number of chords in the graph of the given circuit will be

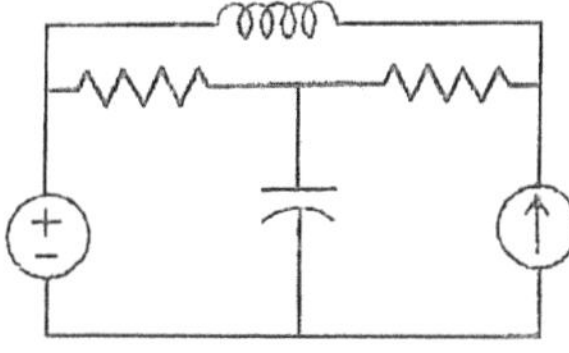

(a) 3 (b) 4

(c) 5 (d) 6

12. The Thevenin's equivalent of a circuit operating at $\omega = 5$ rad/s, has $V_\alpha = 3.71 \angle 15.9^\circ$ and $Z_o = 2.38 - j\,0.667\,\Omega$. At this frequency, the minimal realization of the Thevenin's impedance will have a

(a) Resistor and a capacitor and an inductor

(b) Resistor and a capacitor

(c) Resistor and an inductor

(d) Capacitor and an inductor

13. A signal $e^{-\alpha t}\sin(\omega t)$ is the input to a Linear Time Invariant System. Given K and ϕ are constants, the output of the system will be of the form $Ke^{-\beta t}\sin(\upsilon t + \phi)$ where

(a) β need not be equal to α but υ equal to ω

(b) υ need not be equal to ω but β equal to α

(c) β equal to α and υ equal to ω

(d) β need not be equal to α and υ need not be equal to ω

14. A function y(t) satisfies the following differential equation

$$\frac{dy(t)}{dt} + y(t) = \delta(t)$$

Where $\delta(t)$ is the delta function. Assuming zero initial condition, and denoting the unit step function by $u(t)$, $y(t)$ can be of the form

(a) e^1 (b) e^{-t}

(c) $e^t u(t)$ (d) $e^{-t} u(t)$

15. The time constant for the given circuit will be

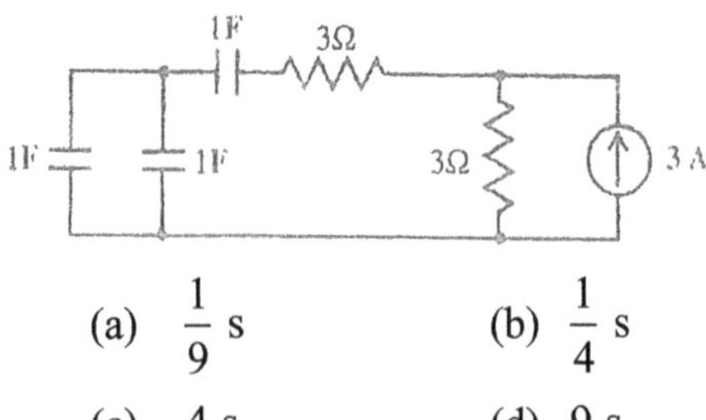

(a) $\frac{1}{9}$ s (b) $\frac{1}{4}$ s
(c) 4 s (d) 9 s

16. The resonant frequency for the given circuit will be

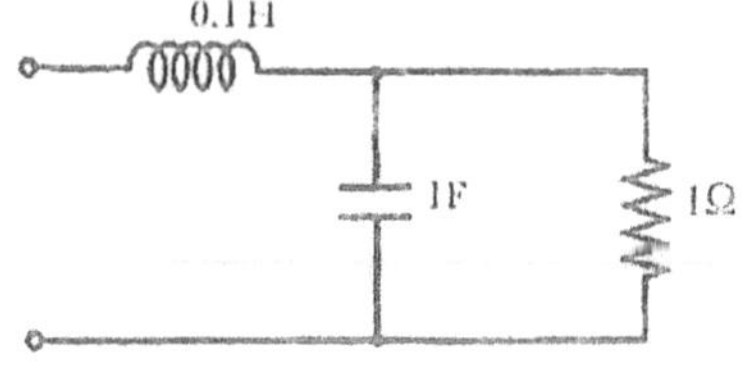

(a) 1 rad/s (b) 2 rad/s
(c) 3 rad/s (d) 4 rad/s

17. Assuming ideal elements in the circuit shown below, the voltage V_{ab} will be

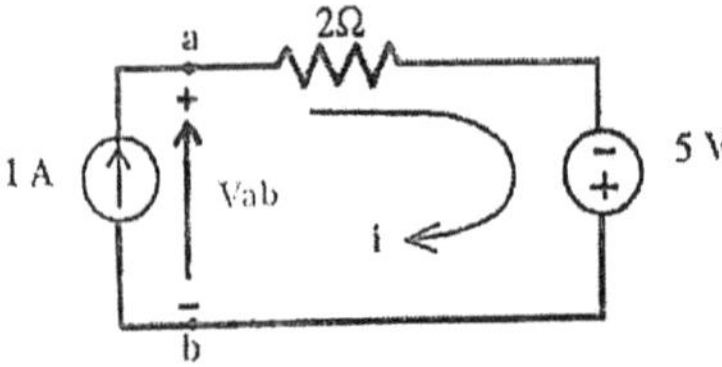

(a) – 3V (b) 0 V
(c) 3 V (d) 5 V

18. A coil of 300 turns is wound on a non-magnetic core having a mean circumference of 300 mm and a cross-sectional area of 300 mm^2. The inductance of the coil corresponding to a magnetizing current of 3A will be (Given that $\mu_0 = 4\pi \times 10^{-7}$ H/m)

(a) 37.68 μH (b) 113.04 μH
(c) 37.68 mH (d) 113.04 mH

19. In the circuit shown in the figure, the value of the current i will be given by

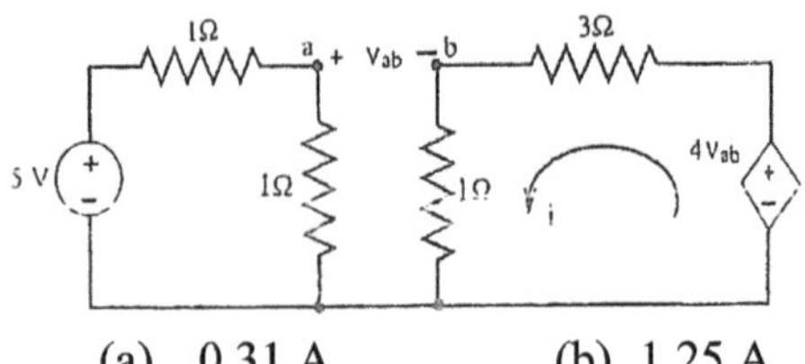

(a) 0.31 A (b) 1.25 A
(c) 1.75 A (d) 2.5 A

20. Let x(t) be a periodic signal with time period T. Let $y(t) = x(t - t_o) + x(t + t_o)$ for some t_o. The fourier Series coefficients of y(t) are denoted by b. If $b_k = 0$ for all odd k, the t_o can be equal to

(a) $\frac{T}{8}$ (b) $\frac{T}{4}$

(c) $\frac{T}{2}$ (d) 2T

Linked Answer Questions: Q. 21 to Q.22 carry two marks each.

Statement for Linked Answer Questions 21 and 22

The current i(t) sketched in the figure flows through an initially uncharged 0.3nF capacitor.

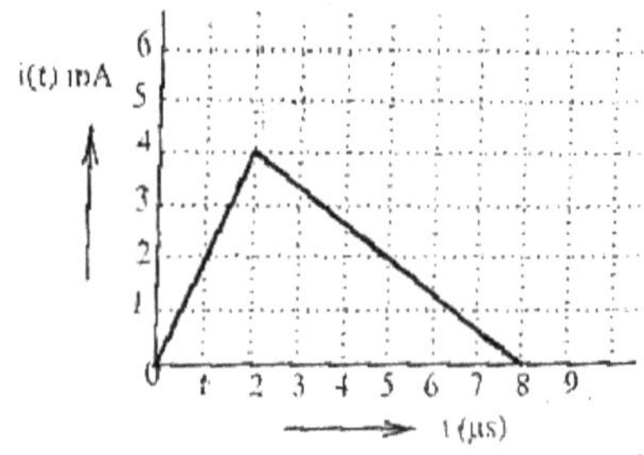

21. The charge stored in the capacitor at t = 5 μs, will be

(a) 8 nC (b) 10 nC
(c) 13 nC (d) 16 nC

22. The capacitor charged upto 5 μs, as per the current profile given in the figure, is connected across an inductor of 0.6 mH. Then the value of voltage across the capacitor after 1 μs will approximately be

(a) 18.8 V (b) 23.5 V
(c) – 23.5 V (d) – 30.6 V

2009 ECE

1. The Fourier series of a real periodic function has only

P. Cosine terms if it is even
Q. Sine terms if it is even
R. Cosine terms if it is odd
S. Sine terms if it is odd

Which of the above statements are correct?

(a) P and S (b) P and R
(c) Q and S (d) Q and R

2. A function is given $f(t) = \sin^2 t + \cos 2t$. Which of the following is true?

(a) f has frequency components at 0 and $1/2\pi$ Hz
(b) f has frequency components at 0 and $1/\pi$ Hz
(c) f has frequency components at $1/2\pi$ and $1/\pi$ Hz
(d) f has frequency components at 0, $1/2\pi$ and $1/\pi$ Hz

3. In the interconnection of ideal sources shown in the figure, it is known that the 60 V source is absorbing power

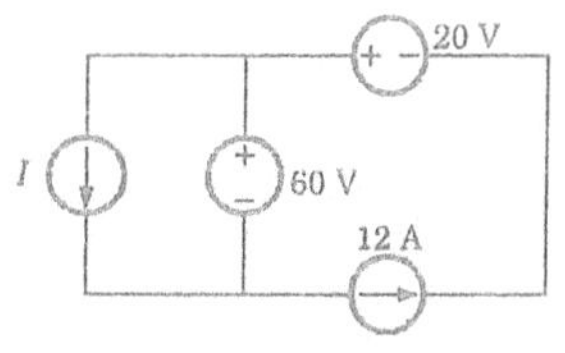

Which of the following can be the value of the current source?

(a) 10 A (b) 13 A
(c) 15 A (d) 18A

4. If the transfer function of the following network is $(V_0(s)/(V_1(s)) = 1/(2 + s\,CR)$

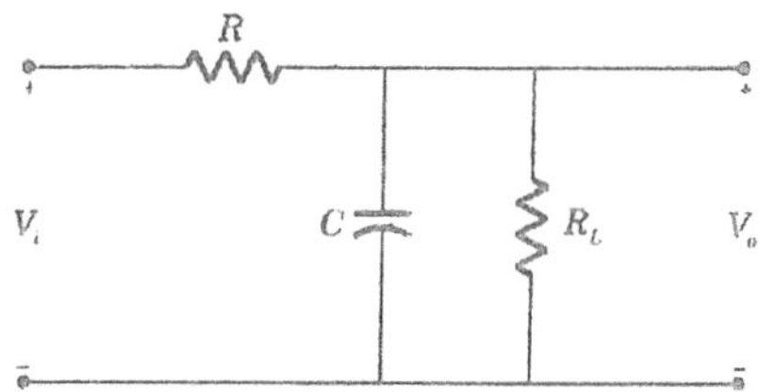

The value of the load resistance R_L is

(a) R/4 (b) R/2
(c) R (d) 2R

5. An AC source of RMS voltage 20 V with internal impedance $Z_S = (1 + 2j)\ \Omega$ feeds a load of impedance $Z_L = (7 + 4j)\ \Omega$ in the figure below. The reactive power consumed by the load is

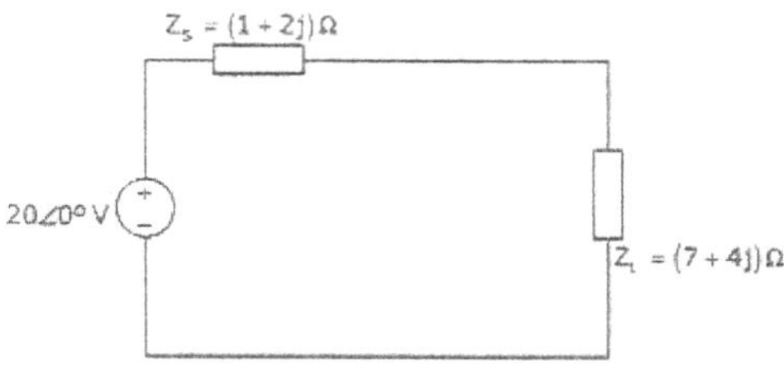

(a) 8 VAR (b) 16 VAR
(c) 28 VAR (d) 32 VAR

6. The switch in the circuit shown was on position a for a long time, and is moved to position b at time t = 0. The current i(t) for t > 0 is given by

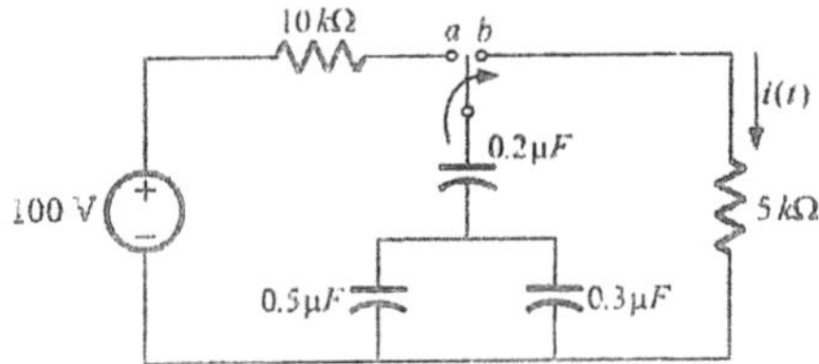

(a) $0.2e^{-125}tu(t)$ mA

(b) $20e^{-1250t}u(t)$ mA

(c) $0.2e^{-1250t}u(t)$ mA

(d) $20e^{-1000t}u(t)$ mA

7. In the circuit shown, what value of R_L maximizes the power delivered to R_L?

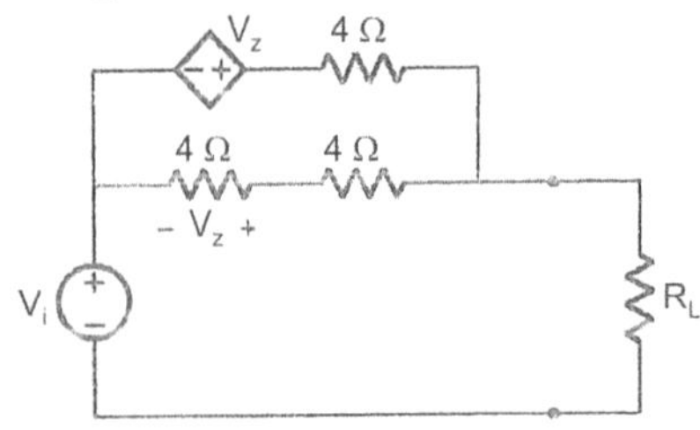

(a) 2. 4 Ω (b) (8/3) Ω
(c) 4 Ω (d) 6 Ω

8. The time domain behavior of an RL circuit is represented by

$L(di/dt) + Ri = V_0(1 + Be^{-Rt/L} \sin t)\, u(t)$

For an initial current of $i(0) = (V_0/R)$, the steady state value of the current is given by

(a) $i(t) \to (V_0/R)$

(b) $i(t) \to 2(V_0/R)$

(c) $i(t) \to (V_0/R)(1 + B)$

(d) $i(t) \to (2V_0/R)(1 + B)$

9. The unit step response of an under-damped second order system has steady state value of –2. Which one of the following transfer functions has these properties?

(a) $(-2.24)/(s^2 + 2.59s + 1.12)$

(b) $(-3.82)/(s^2 + 1.91s + 1.91)$

(c) $(-2.24)/(s^2 - 2.59s + 1.12)$

(d) $(-3.82)/(s^2 - 1.91s + 1.91)$

2009 EEE

10. The current through the 2 kΩ resistance in the circuit shown is

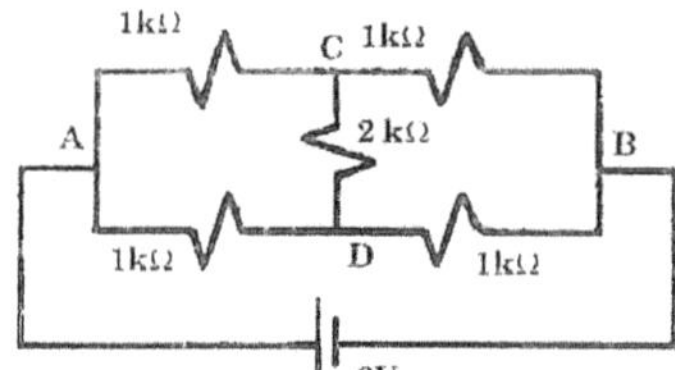

(a) 0 mA (b) 1 mA
(c) 2 mA (d) 6 mA

11. How many 200W/220V incandescent lamps connected is series would consume the same total power as a single 100W/220V incandescent lamp?

(a) not possible (b) 4
(c) 3 (d) 2

12. A Linear Time Invariant system with an impulse response h(t) produces output y(t) when input x(t) is applied. When the input $x(t - \tau)$ is applied to a system with impulse response $h(t - \tau)$, the output will be

(a) y(t) (b) y(2(t – τ))
(c) y(t – τ) (d) y(t – 2τ)

13. In the figure shown, all elements used are ideal. For time t < 0, S_1 remained closed and S_2 open. At t = 0, S_1 is opened and S_2 is closed. If the voltage V_{c2} across the capacitor C_2 at t = 0 is zero, the voltage across the capacitor combination at t = 0^+ will be

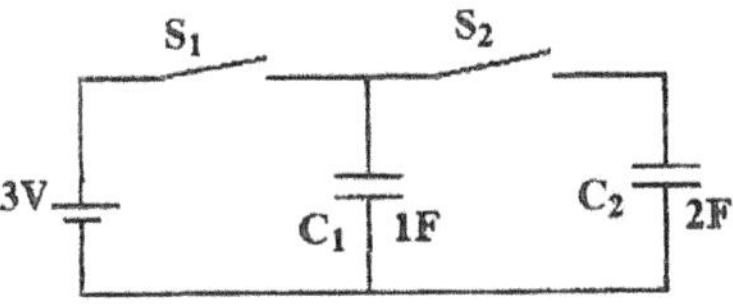

(a) 1 V (b) 2 V
(c) 1.5 V (d) 3 V

14. The equivalent capacitance of the input loop of the circuit shown is

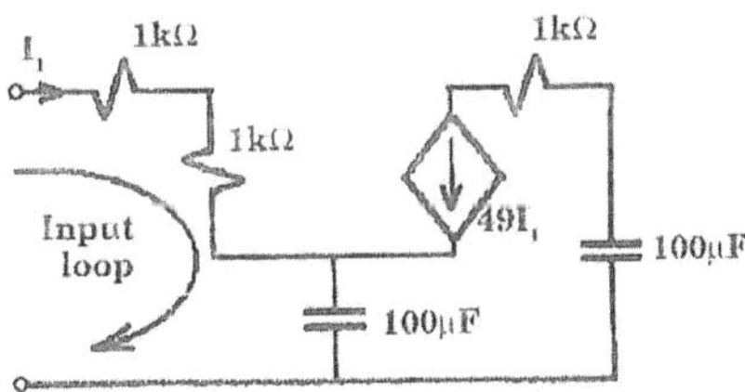

(a) 2 μF (b) 100 μF
(c) 200 μF (d) 4 μF

15. For the circuit shown, find out the current flowing through the 2 resistance. Also identify the changes to be made to double the current through the 2Ω resistance

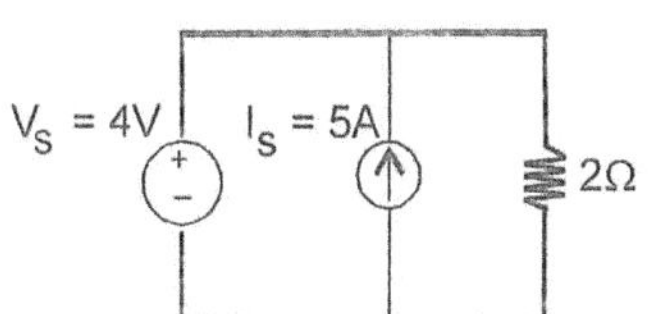

(a) (5A; Put V_S = 20 V)
(b) (2A; Put V_S = 8V)
(c) (5A; Put I_S = 10 A)
(d) (7A; Put I_S = 12A)

16. The figure shows a three-phase delta connected load supplied from a 400 V, 50 Hz, 3-phase balanced source. The pressure coil (PC) and current coil (CC) of a wattmeter are connected to the load as shown, with the coil polarities suitably selected to ensure a positive deflection. The wattmeter reading will be

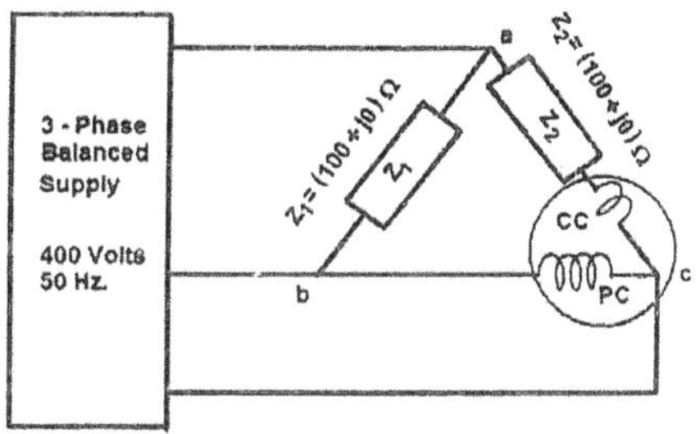

(a) 0 (b) 1600 Watt
(c) 800 Watt (d) 400 Watt

Common data for Q. No 17 and 18.

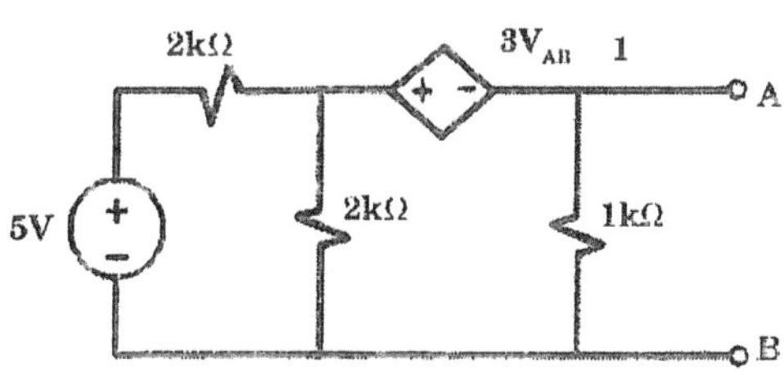

17. For the circuit given above, the Thevenin's resistance across the terminals A and B is

(a) 0.5 kΩ (b) 0.2 kΩ
(c) 1 kΩ (d) 0.11 kΩ

18. For the circuit given above, the Thevenin's voltage across the terminals A and B is

(a) 1.25V (b) 0.25V
(c) 1V (d) 0.5V

2010 ECE

1. The trigonometric Fourier series for the waveform f(t) shown below contains

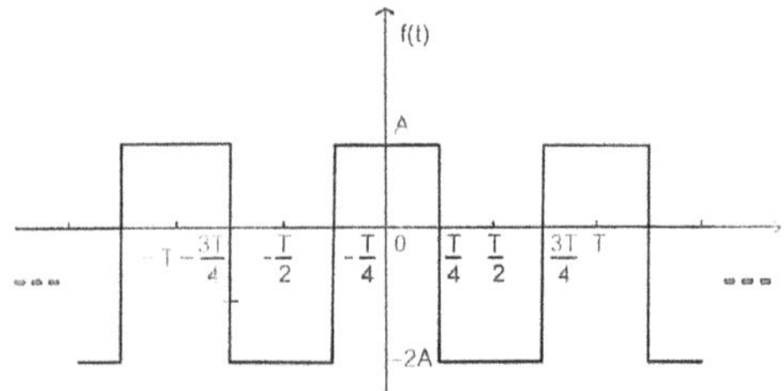

(a) only cosine terms and zero value for the dc component
(b) only cosine terms and a positive value for the dc component
(c) only cosine terms and a negative value for the dc component
(d) only sine terms and a negative for the dc component.

2. For the two-port network shown below, the short-circuit admittance parameter matrix is

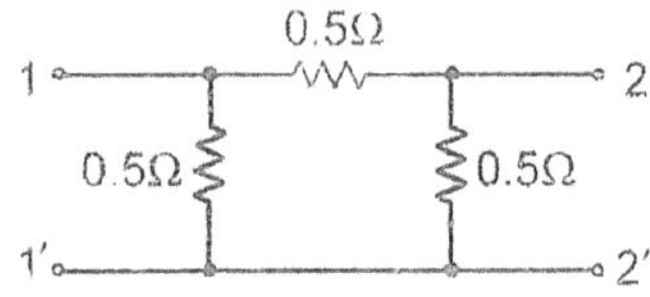

(a) $\begin{bmatrix} 4 & -2 \\ -2 & 4 \end{bmatrix}S$ (b) $\begin{bmatrix} 1 & -0.5 \\ -0.5 & 1 \end{bmatrix}S$

(c) $\begin{bmatrix} 1 & 0.5 \\ 0.5 & 1 \end{bmatrix}S$ (d) $\begin{bmatrix} 4 & 2 \\ 2 & 4 \end{bmatrix}S$

3. For parallel RLC circuit, which one of the following statements is NOT correct?

(a) The bandwidth of the circuit decreases if R is increased
(b) The bandwidth of the circuit remains same if L is increased
(c) At resonance, input impedance is a real quantity
(d) At resonance, the magnitude of input impedance attains its minimum value

4. Given

$$f(t) = L^{-1}\left[\frac{3s+1}{s^3+4s^2+(K-3)s}\right].$$

If $\lim_{t \to \infty} f(t) = 1$, then the value of K is

(a) 1 (b) 2
(c) 3 (d) 4

5. In the circuit shown, the switch S is open for a long time and is closed at t = 0. The current i(t) for $t \geq 0^+$ is

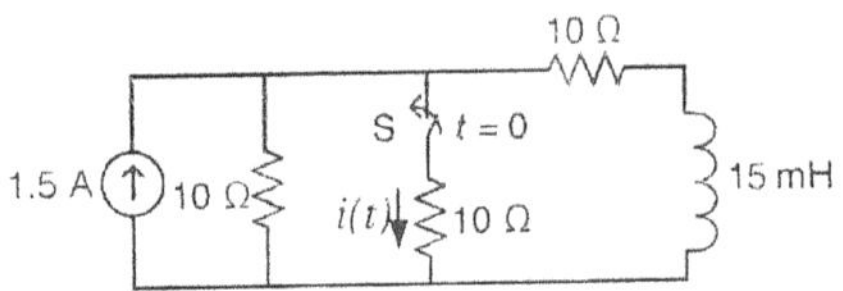

(a) $i(t) = 0.5 - 0.125e^{-1000t}A$
(b) $i(t) = 1.5 - 0.125e^{-1000t} A$
(c) $i(t) = 0.5 - 0.5 e^{-1000t} A$
(d) $i(t) = 0.375e^{-1000t} A$

6. The current I in the circuit shown is

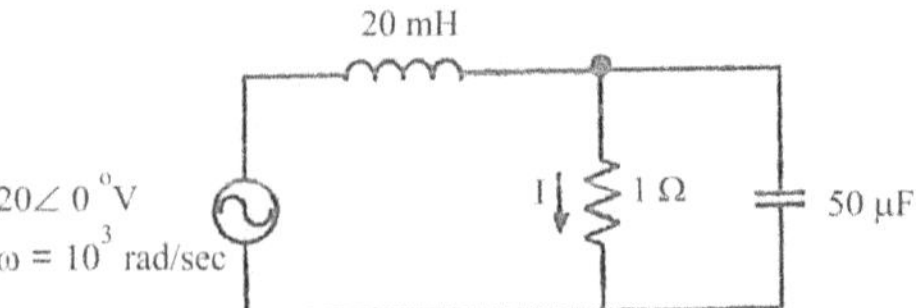

(a) –j1A (b) J1A
(c) 0A (d) 20A

7. In the circuit shown, the power supplied by the voltage source is

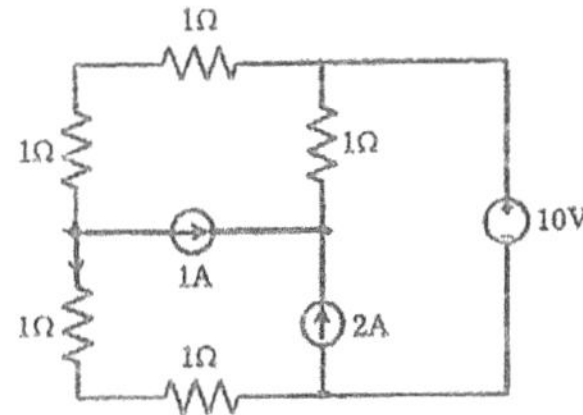

(a) 0 W (b) 5 W
(c) 10 W (d) 100 W

2010 EEE

8. The period of the signal
$x(t) = 8 \sin(0.8\pi t + (\pi/4))$ is
(a) 0.4π s (b) 0.8π s
(c) 1.25 s (d) 2.5 s

9. The switch in the circuit has been closed for a long time. It is opened at t = 0. At t = 0^+, the current through the 1 μF capacitor is

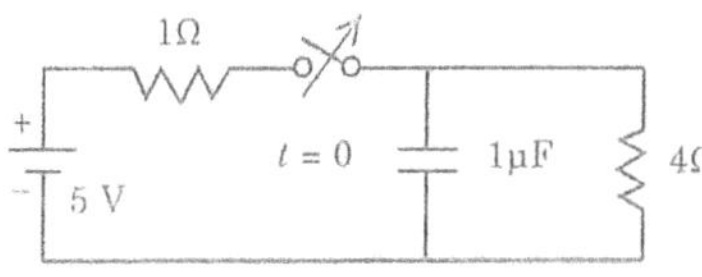

(a) 0 A (b) 1 A
(c) 1.25 A (d) 5 A

10. The second harmonic component of the periodic waveform given in the figure has an amplitude of

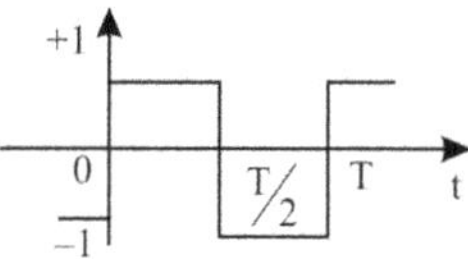

(a) 0 (b) 1
(c) $2/\pi$ (d) $\sqrt{5}$

11. A wattmeter is connected as shown in the figure. The wattmeter reads

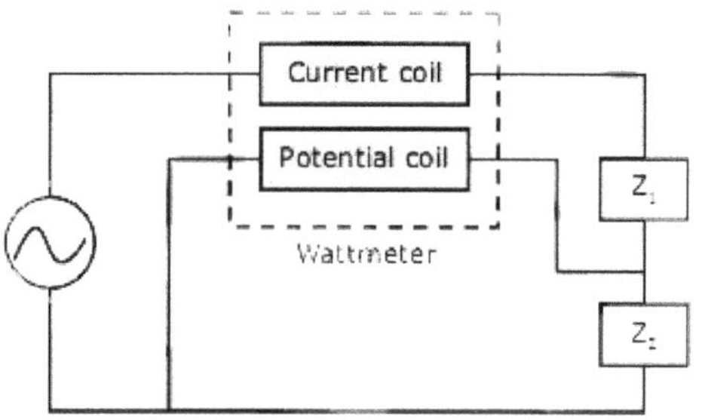

(a) Zero always
(b) Total power consumed by Z_1 and Z_2
(c) Power consumed by Z_1
(d) Power consumed by Z_2

12. If the 12 Ω resistor draws a current 1 A as shown in the figure, the value of resistance R is

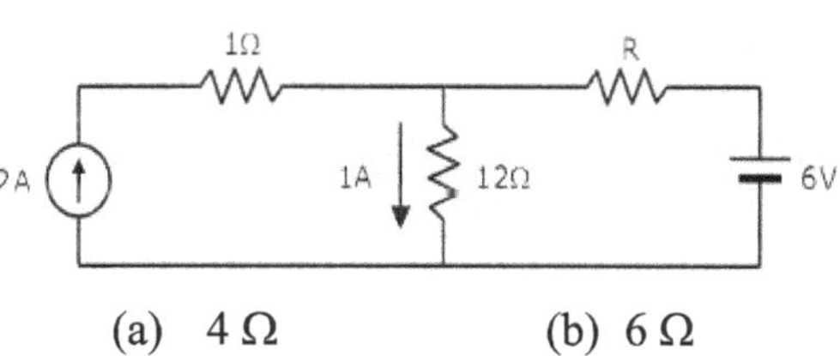

(a) 4 Ω (b) 6 Ω
(c) 8 Ω (d) 18 Ω

13. The two-port network P shown in the figure has ports 1 and 2, denoted by terminals (a, b) and (c, d), respectively. It has an impedance matrix Z with parameters denoted by z_{ij}. A 1 Ω resistor is connected in series with the network at port 1 as shown in the figure. The impedance matrix of the modified two-port network (shown as a dashed box) is.

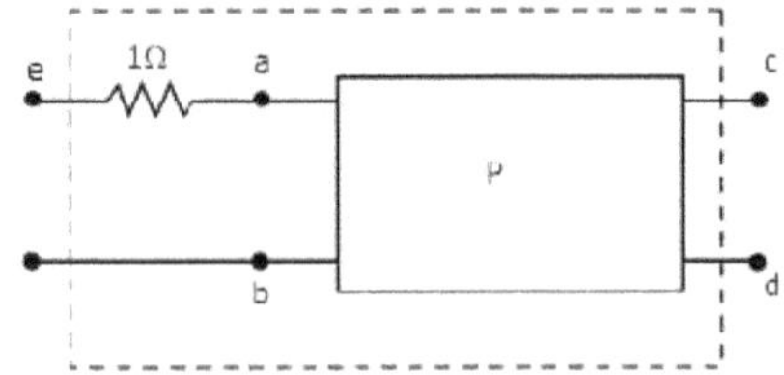

(a) $\begin{pmatrix} z_{11}+1 & z_{12}+1 \\ z_{21} & z_{22}+1 \end{pmatrix}$ (b) $\begin{pmatrix} z_{11}+1 & z_{12} \\ z_{21} & z_{22}+1 \end{pmatrix}$

(c) $\begin{pmatrix} z_{11}+1 & z_{12} \\ z_{21} & z_{22} \end{pmatrix}$ (d) $\begin{pmatrix} z_{11}+1 & z_{12} \\ z_{21}+1 & z_{22} \end{pmatrix}$

14. The initial current through the inductor is zero, while the initial capacitor voltage is 100 V in the figure. The switch is chosed at t = 0.

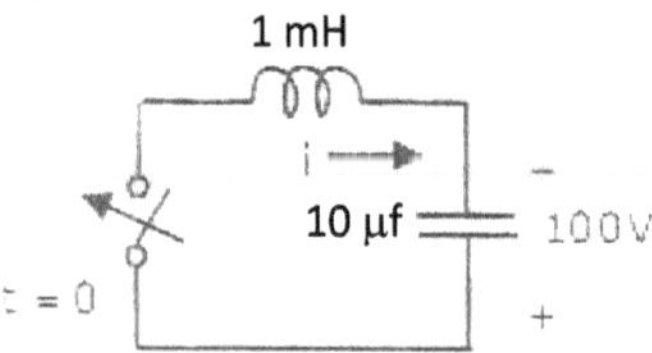

The current i through the circuit is:
(a) 5 cos (5 × 10^3t) A
(b) 5 sin (10^4t) A
(c) 10 cos (5 × 10^3t) A
(d) 10 sin (10^4t) A

Directions for question 15 to 16:
Common Data Questions

Given f(t) and g(t) as shown below:

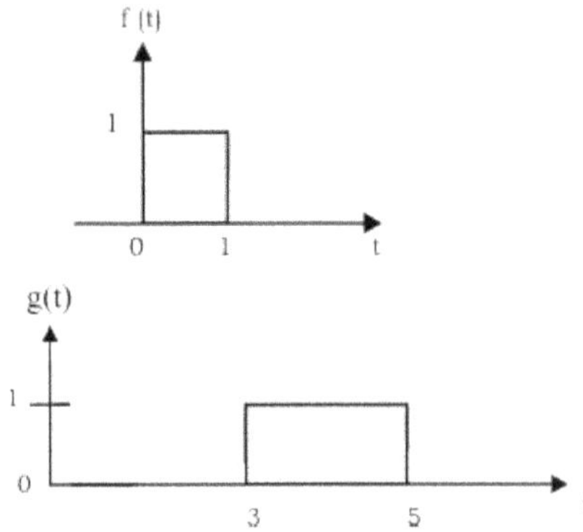

15. g(t) can be expressed as
(a) g(t) = f(2t – 3)
(b) g(t) = f((t/2) – 3)
(c) g(t) = f(2t – (3/2))
(d) g(t) = f((t/2) – (3/2))

16. The Laplace transform of g(t) is
(a) (1/s) $(e^{3s} - e^{5s})$
(b) (1/s) $(e^{-5s} - e^{-3s})$
(c) $(e^{-3s}/s)(1 - e^{-2s})$
(d) (1/s) $(e^{5s} - e^{3s})$

17. For the differential equation $\frac{d^2x}{dt^2} + 6\frac{dx}{dt} + 8x = 0$ with initial conditions x(0) = 1 and $\left.\frac{dx}{dt}\right|_{t-0} = 0$, the solution is
(a) $x(t) = 2e^{-6t} - e^{-2t}$
(b) $x(t) = 2e^{-2t} - e^{-4t}$
(c) $x(t) = -e^{-6t} + 2e^{-4t}$
(d) $x(t) = e^{-2t} + 2e^{-4t}$

2011 ECE

1. If the unit step response of a network is $(1 - e^{-\alpha t})$, then its unit impulse response is
(a) $\alpha e^{-\alpha t}$ (b) $\alpha^{-1}e^{-\alpha t}$
(c) $(1 - \alpha^{-1})e^{-\alpha t}$ (d) $(1 - \alpha)e^{-\alpha t}$

2. The circuit shown below is driven by a sinusoidal input $V_1 = V_p \cos(t/RC)$. The steady state output V_0 is

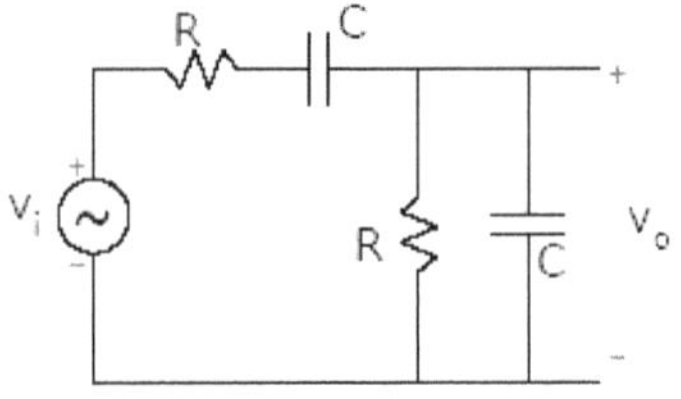

(a) $(V_p/3)\cos(t/RC)$
(b) $(V_p/3)\sin(t/RC)$
(c) $(V_p/2)\cos(t/RC)$
(d) $(V_p/2)\sin(t/RC)$

3. The differential equation $100\frac{d^2y}{dt^2}-20\frac{dy}{dt}+y=x(t)$ describes a system with an input x(t) and an output y(t). The system, which is initially relaxed, is excited by a unit step input. The output y(t) can be represented by the waveform.

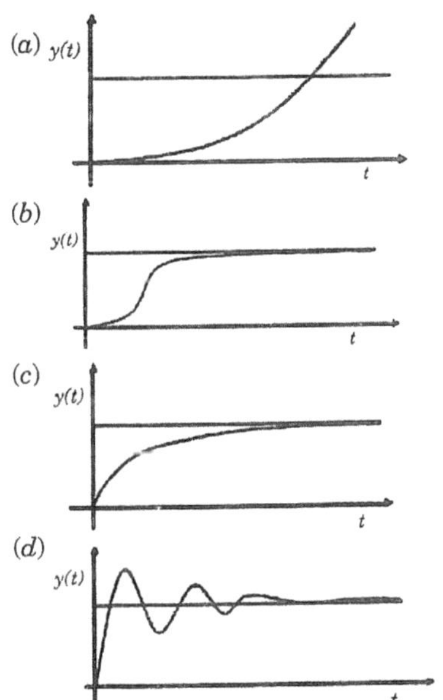

4. The trigonometric Fourier series of an even function does not have the
(a) dc term
(b) cosine terms
(c) sine terms
(d) odd harmonic terms

5. In the circuit shown below, the Norton equivalent current in amperes with respect to the terminals P and Q is

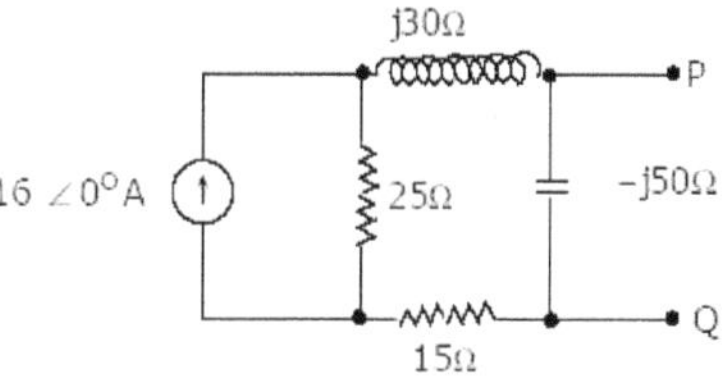

(a) $6.4-j4.8$
(b) $6.56-j7.87$
(c) $10+j0$
(d) $16+j0$

6. In the circuit shown below, the value of R_L such that the power transferred to RL is maximum is

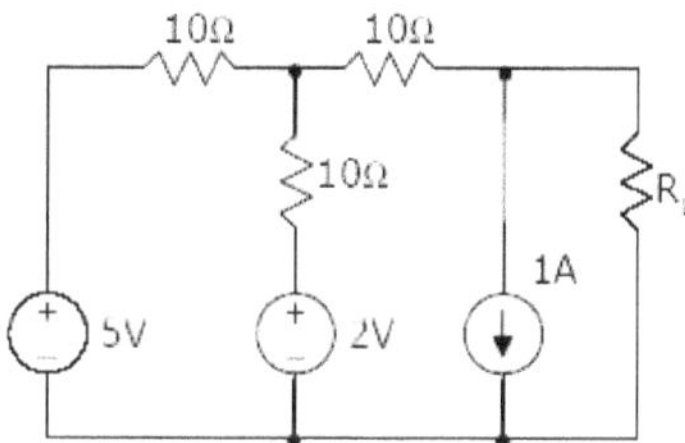

(a) 5 Ω (b) 10 Ω
(c) 15 Ω (d) 20 Ω

7. In the circuit shown below, the current I is equal to

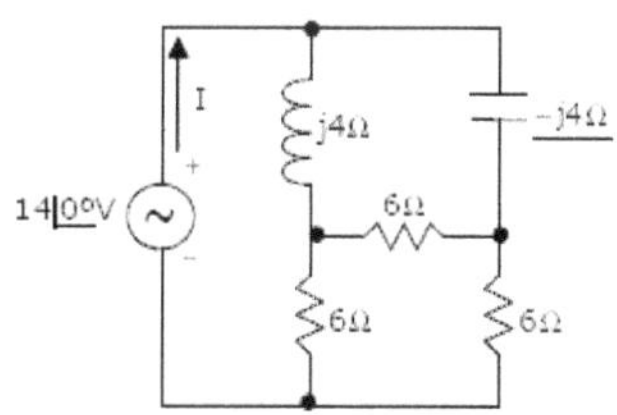

(a) $14\angle 0^\circ$ A (b) $2.0\angle 0^\circ$ A
(c) $2.8\angle 0^\circ$ A (d) $3.2\angle 0^\circ$ A

8. In the circuit shown below, the network N is described by the following Y matrix:

$Y = \begin{bmatrix} 0.1s & -0.01s \\ 0.01s & 0.1s \end{bmatrix}$, the voltage gain $\frac{V_2}{V_1}$ is

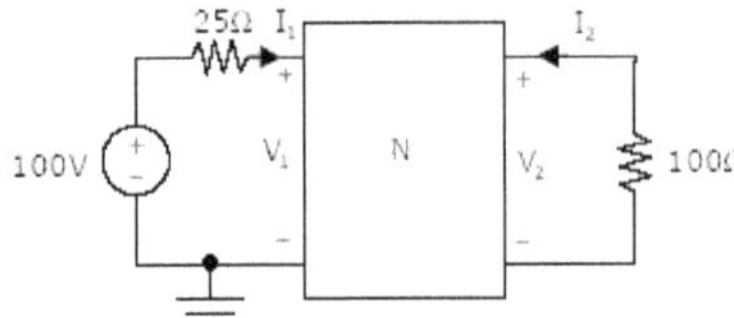

(a) 1/90 (b) –1/90
(c) –1/99 (d) –1/11

9. In the circuit shown below, the initial charge on the capacitor is 2.5 mC, with the voltage polarity as indicated. The switch is closed at time t = 0.

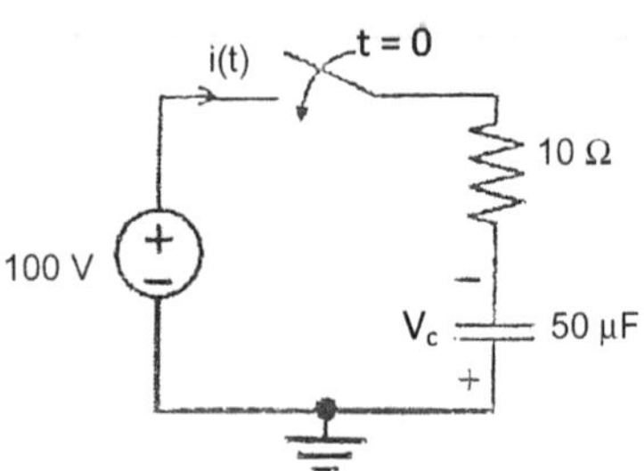

The current i(t) at a time t after the switch is closed is
(a) $i(t) = 15 \exp(-2 \times 10^3 t)$ A
(b) $i(t) = 5 \exp(-2 \times 10^3 t)$ A
(c) $i(t) = 10 \exp(-2 \times 10^3 t)$ A
(d) $i(t) = -5 \exp(-2 \times 10^3 t)$ A

10. An input $x(t) = \exp(-2t)\, u(t) + \delta(t-6)$ is applied to an LTI system with impulse response h(t) = u(t). The output is
(a) $[1 - \exp(-2t)]\, u(t) + u(t+6)$
(b) $[1 - \exp(-2t)]\, u(t) + u(t-6)$
(c) $0.5\,[1 - \exp(-2t)]\, u(t) + u(t+6)$
(d) $0.5\,[1 - \exp(-2t)]\, u(t) + u(t-6)$

2011 EEE

11. The r. m. s value of the current i(t) in the circuit shown below is

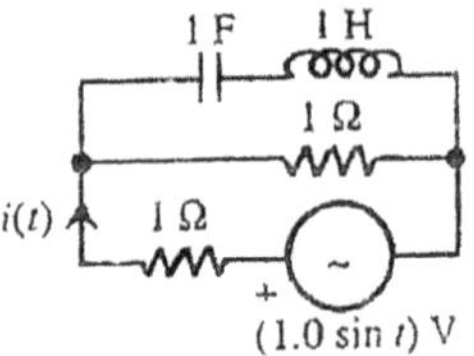

(a) $\frac{1}{2}$ A (b) $\frac{1}{\sqrt{2}}$ A
(c) 1 A (d) $\sqrt{2}$ A

12. Signal shown below will contain the following nonzero terms

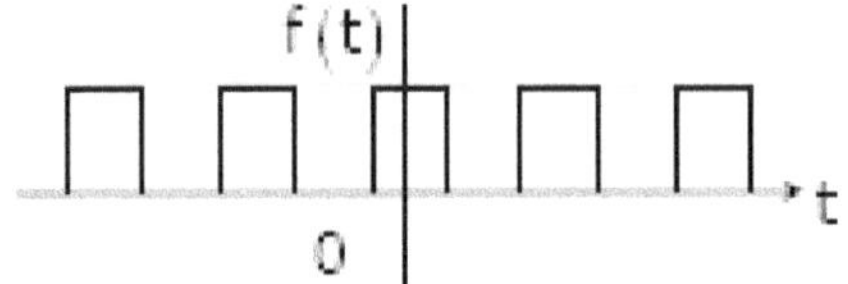

(a) a_0 and b_n, n = 1, 3, 5, …., ∞
(b) a_0 and a_n, n = 1, 2, 3, ….., ∞
(c) a_0 a_n and b_n, n = 1, 2, 3, …., ∞
(d) a_0 and a_n, n = 1, 3, 5,…., ∞

13. A low-pass filter with a cut-off frequency of 30 Hz is cascaded with a high-pass filter with a cut-off frequency of 20 Hz. The resultant system of filters will function as
(a) an all-pass filter
(b) an all-stop filter
(c) a band stop (band-reject) filter
(d) a band-pass filter

14. The voltage applied to a circuit is $100\sqrt{2}$ cos (100 πt) volts and the circuit draws a current of $10\sqrt{2}$ sin (100πt + π/4) amperes. Taking the

voltage as the reference phasor, the phasor representation of the current in amperes is

(a) $10\sqrt{2}\angle -\pi/4$

(b) $10\angle -\pi/4$

(c) $10\angle +\pi/4$

(d) $10\sqrt{2}\angle +\pi/4$

15. In the circuit given below, the value of R required for the transfer of maximum power to the load having a resistance of 3Ω is

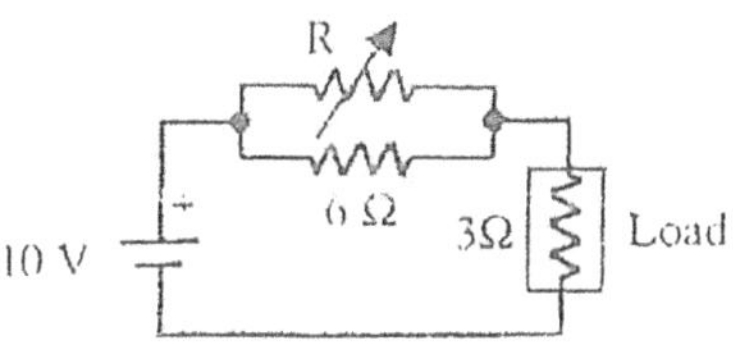

(a) zero (b) 3Ω

(c) 6Ω (d) infinity

16. Given two continuous time signals $x(t) = e^{-t}$ and $y(t) = e^{-2t}$ which exist for $t > 0$, the convolution $z(t) = x(t)*y(t)$ is

(a) $e^{-t} - e^{-2t}$ (b) e^{-3t}

(c) e^{t} (d) $e^{-t} + e^{-2t}$

17. The Laplace transform of the delayed version f(t – T) be F_2s, Let $F_1^*(s)$ be the complex conjugate of $F_1(s)$ with the Laplace variable set as $s = \alpha + jw$. If $G(s) = \dfrac{F_2(s).F_1^*(s)}{|F_1(s)|^2}$, then the inverse Laplace transform of G(s) is

(a) An ideal impulse $\delta(t)$

(b) An ideal delayed impulse $\delta(t-T)$

(c) An ideal step function u(t)

(d) An ideal delayed step function u(t – T)

18. The response h(t) of a linear time invariant system to an impulse $\delta(t)$, under initially relaxed condition is $h(t) = e^{-t} + e^{-2t}$. The response of this system for a unit step input u(t) is

(a) $u(t) + e^{-t} + e^{-2t}$

(b) $(e^{-t} + e^{-2t})\, u(t)$

(c) $(1.5 - e^{-t} - 0.5\, e^{-2t})\, u(t)$

(d) $e^{-1}\delta(t) + e^{-2t}u(t)$

Common Data Questions: 19 & 20

An RLC circuit with relevant data is given below.

$$V_s = 1\angle 0,\ I_s = \sqrt{2}\ \angle \pi/4,$$
$$I_{RL} = \sqrt{2}\ \angle -\pi/4$$

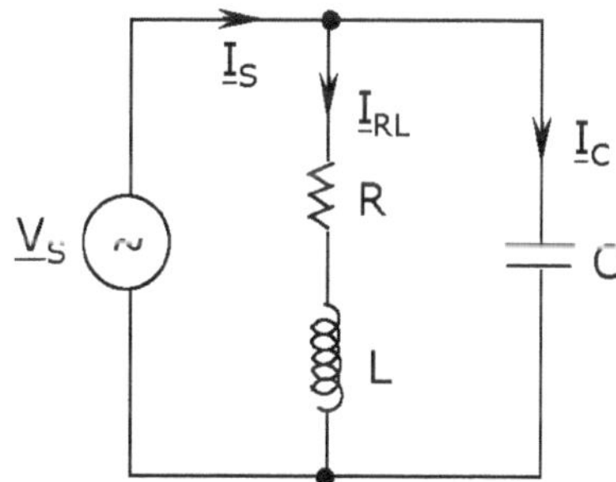

19. Power developed by the source is

(a) 0.5 W (b) 1 W

(c) $\sqrt{2}$ W (d) 2 W

20. The current I_c in the figure above is

(a) $-j2$A (b) $-j\dfrac{1}{\sqrt{2}}$ A

(c) $+j\dfrac{1}{\sqrt{2}}$ A (d) $+j2$A

2012 ECE &EEE

1. In the circuit shown below, the current through the inductor is

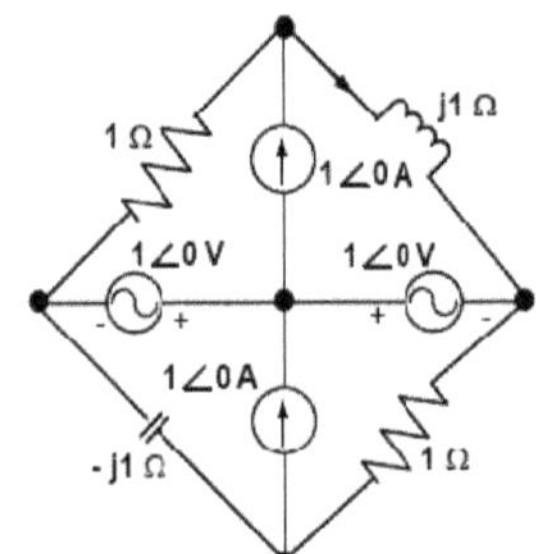

(a) $\frac{2}{1+j}$A (b) $\frac{-1}{1+j}$A

(c) $\frac{1}{1+j}$A (d) 0A

2. The impedance looking into nodes 1 and 2 in the given circuit is

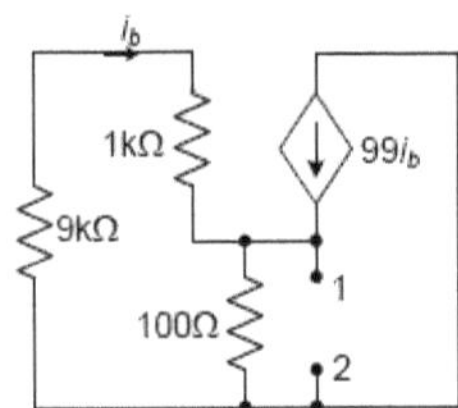

(a) 50 Ω (b) 100 Ω

(c) 5 kΩ (d) 10.1 kΩ

3. The unilateral Laplace transform of f(t) is $\frac{1}{s^2+s+1}$. The unilateral Laplace transform of t f (t) is

(a) $-\frac{s}{\left(s^2+s+1\right)^2}$ (b) $-\frac{2s+1}{\left(s^2+s+1\right)^2}$

(c) $\frac{s}{\left(s^2+s+1\right)^2}$ (d) $\frac{2s+1}{\left(s^2+s+1\right)^2}$

4. The i-v characteristics of the diode in the circuit given below are

$$i=\begin{cases}\frac{v-0.7}{500}\text{A}, & v\geq 0.7\text{ V}\\ 0\text{ A}, & v<0.7\text{ V}\end{cases}$$

The current in the circuit is

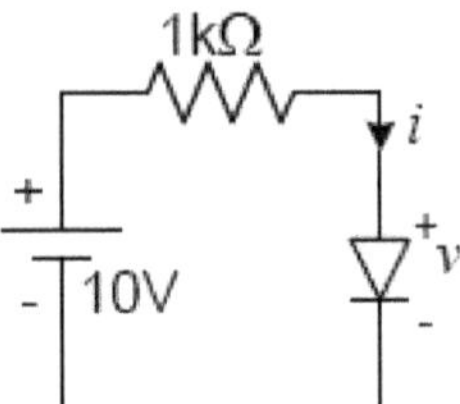

(a) 10 mA (b) 9.3 mA

(c) 6.67 mA (d) 6.2 mA

5. For the circuit shown in the figure, the voltage and current expressions are $v(t)=E_1\sin(\omega t)+E_3\sin(3\omega t)$ and $i(t)=I_1\sin(\omega t-\phi_1)+I_3\sin(3\omega t-\phi_3)+I_5\sin(5\omega t)$. The average power measured by the Wattmeter is

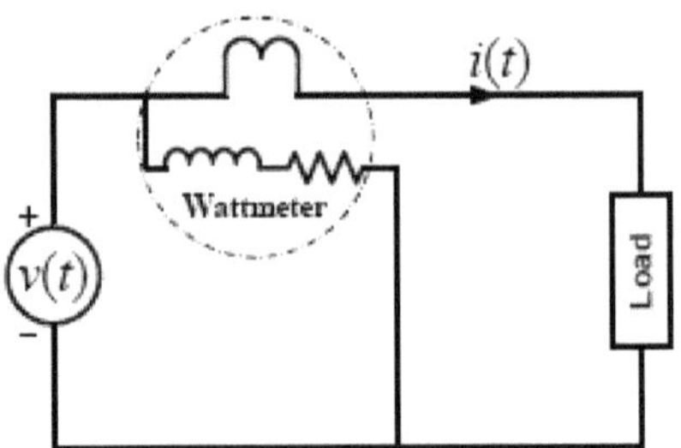

(a) $\frac{1}{2}E_1I_1\cos\phi_1$

(b) $\frac{1}{2}[E_1I_1\cos\phi_1+E_1I_3\cos\phi_3+E_1I_5]$

(c) $\frac{1}{2}[E_1I_1\cos\phi_1+E_3I_3\cos\phi_3]$

(d) $\frac{1}{2}[E_1I_1\cos\phi_1+E_3I_1\cos\phi_1]$

6. If $V_A - V_B = 6$ V, then $V_C - V_D$ is

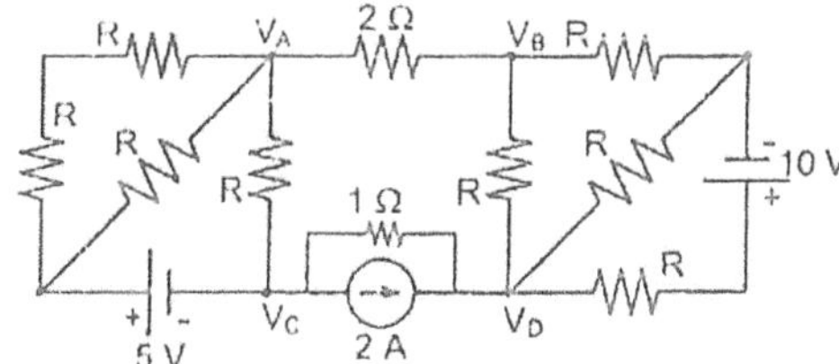

(a) –5 V (b) 2 V
(c) 3 V (d) 6 V

7. Consider the differential equation $\frac{d^2y(t)}{dt^2} + 2\frac{dy(t)}{dt} + y(t) = \delta(t)$ with $y(t)\big|_{t=0^-} = -2$ and $\frac{dy}{dt}\Big|_{t=0^-} = 0$.

The numerical value of $\frac{dy}{dt}\Big|_{t=0^+}$ is

(a) – 2 (b) – 1
(c) 0 (d) 1

8 Assuming both the voltage sources are in phase, the value of R for which maximum power is transferred from circuit A to circuit B is

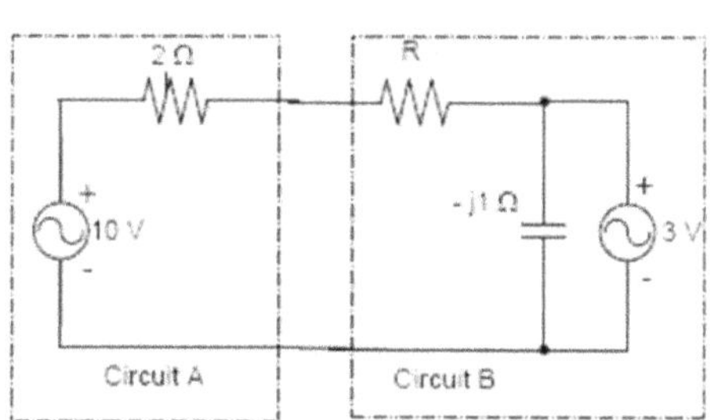

(a) 0.8 Ω (b) 1.4 Ω
(c) 2 Ω (d) 2.8 Ω

9. With 10 V dc connec ted at port A in the linear nonreciprocal two-port network shown below, the following were observed:

(i) 1 Ω connected at port B draws a current of 3 A

(ii) 2.5 Ω connected at port B draws a current of 2 A

With 10 V dc connected at port A, the current drawn by 7 Ω connected at port B is

(a) 3/7 A (b) 5/7 A
(c) 1 A (d) 9/7 A

Linked answer questions

Statement for linked answer questions 10 & 11

In the circuit shown, the three voltmeter readings are V_1 = 220 V, V_2 = 122 V, V_3 = 136 V.

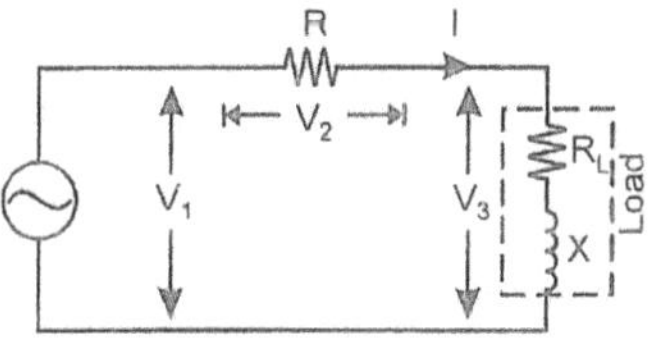

10. The power factor of the load is

(a) 0.45 (b) 0.50
(c) 0.55 (d) 0.60

11. If R_L = 5 Ω, the approximate power consumption in the load is

(a) 700 W (b) 750 W
(c) 800 W (d) 850 W

12. The Fourier transform of a signal h(t) is H(jω) = (2 cosω) (sin 2ω)/ω. The value of h(0) is

(a) 1/4 (b) 1/2
(c) 1 (d) 2

2013 EEE & ECE

1. The impulse response of a system is h(t) = t u(t). For an input u(t – 1), the output is

(a) $\frac{t^2}{2}u(t)$

(b) $\frac{t(t-1)}{2}u(t-1)$

(c) $\frac{(t-1)^2}{2}u(t-1)$

(d) $\frac{t^2-1}{2}u(t-1)$

2. Consider a delta connection of resistors and its equivalent star connection as shown below. If all elements of the delta connection are scaled by a factor k, k > 0, the elements of the corresponding star equivalent will be scaled by a factor of

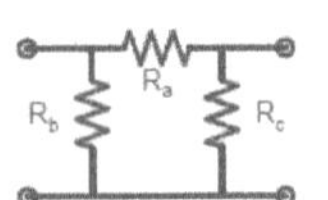

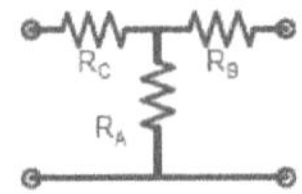

(a) k^2 (b) k

(c) 1/k (d) $\sqrt{k}$

3. For a periodic signal $v(t) = 30\sin 100t + 10\cos 300t + 6\sin(500t + \pi/4)$, the fundamental frequency in rad/s is

(a) 100 (b) 300

(c) 500 (d) 1500

4. Assuming zero initial condition, the response y (t) of the system given below to a unit step input u(t) is

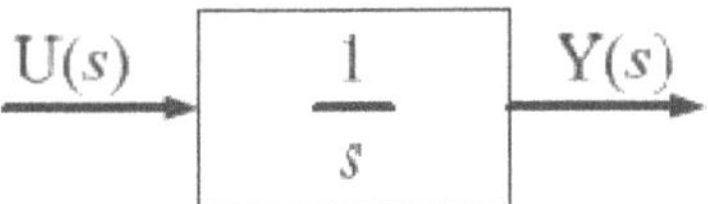

(a) u (t) (b) t u (t)

(c) $\frac{t^2}{2}u(t)$ (d) $e^{-t}u(t)$

5. The transfer function $\frac{V_2(s)}{V_1(s)}$ of the circuit shown below is

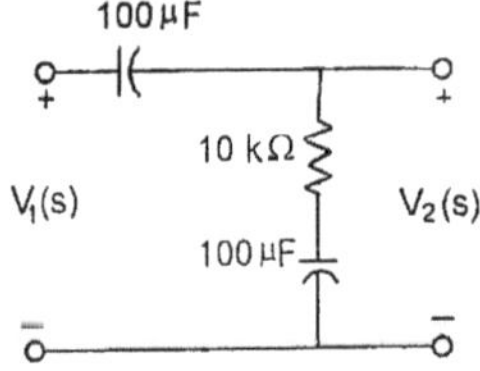

(a) $\frac{0.5s+1}{s+1}$ (b) $\frac{3s+6}{s+2}$

(c) $\frac{s+2}{s+1}$ (d) $\frac{s+1}{s+2}$

6. A source $v_s(t) = V\cos 100\pi t$ has an internal impedance of $(4 + j3)\ \Omega$. If a purely resistive load connected to this source has to extract the maximum power out of the source, its value in Ω should be

(a) 3 (b) 4

(c) 5 (d) 7

7. In the circuit shown below, if the source voltage $V_S = 100\angle 53.13°V$ then the Thevenin's equivalent voltage in Volts as seen by the load resistance R_L is

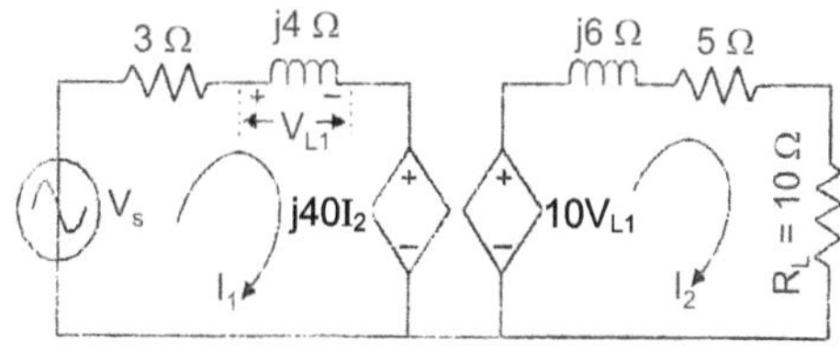

(a) $100\angle 90°$ (b) $800\angle 0°$
(c) $800\angle 90°$ (d) $100\angle 60°$

8. Two magnetically uncoupled inductive coils have Q factors q_1 and q_2 at the chosen operating frequency. Their respective resistances are R_1 and R_2. When connected in series, their effective Q factor at the same operating frequency is

(a) $q_1 + q_2$
(b) $(1/q_1) + (1/q_2)$
(c) $(q_1R_1 + q_2R_2)/(R_1 + R_2)$
(d) $(q_1R_2 + q_2R_1)/(R_1 + R_2)$

9. The impulse response of a continuous time system is given by $h(t) = \delta(t-1) + \delta(t-3)$. The value of the step response at $t = 2$ is

(a) 0 (b) 1
(c) 2 (d) 3

10. A system described by a linear, constant coefficient, ordinary, first order differential equation has an exact solution given by y(t) for t > 0, when the forcing function is x(t) and the initial condition is y(0). If one wishes to modify the system so that the solution becomes $-2y(t)$ for $t > 0$, we need to

(a) change the initial condition to $-y(0)$ and the forcing function to $2x(t)$
(b) change the initial condition to $2y(0)$ and the forcing function to $-x(t)$
(c) change the initial condition to $j\sqrt{2}y(0)$ and the forcing function to $j\sqrt{2}x(t)$
(d) change the initial condition to $-2y(0)$ and the forcing function to $-2x(t)$

Common data for Questions 11 & 12

Consider the following figure

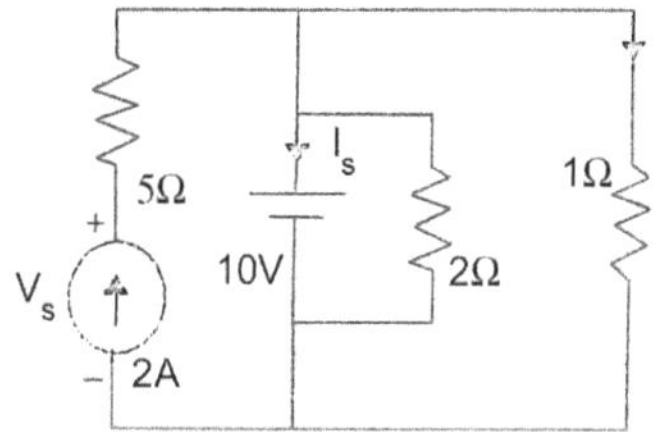

11. The current I_S in Amps in the voltage source, and voltage V_S in Volts across the current source respectively, are

(a) 13, – 20 (b) 8, – 10
(c) – 8, 20 (d) – 13, 20

12. The current in the 1 Ω resistor in Amps is

(a) 2 (b) 3.33
(c) 10 (d) 12

13. A system is described by the differential equation

$$\frac{d^2y}{dt^2} + 5\frac{dy}{dt} + 6y(t) = x(t)$$

Let x(t) be a rectangular pulse given by $x(t) = \begin{cases} 1 & 0 < t < 2 \\ 0 & \text{otherwise} \end{cases}$

Assuming that y(0) = and $\frac{dy}{dt} = 0$ at t = 0, the Laplace transform of y(t) is

(a) $\frac{e^{-2s}}{s(s+2)(s+3)}$

(b) $\frac{1-e^{-2s}}{s(s+2)(s+3)}$

(c) $\frac{e^{-2s}}{(s+2)(s+3)}$

(d) $\frac{1-e^{-2s}}{(s+2)(s+3)}$

2014 ECE

1. Consider the configuration shown in the figure which is a portion of a larger electrical network

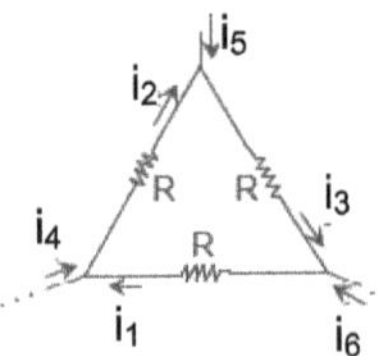

For R =1 Ω and currents i1 = 2 A, $i_4 = -1$ A, $i_5 = -4$ A, which one of the following is TRUE?

(a) $i_6 = 5$ A
(b) $i_3 = -4$ A
(c) Data is sufficient to conclude that the supposed currents are impossible
(d) Data is insufficient to identify the currents i_2, i_3, and i_6

2. A continuous, linear time-invariant filter has an impulse response $h(t)$ described by

$$h(t) = \begin{cases} 3 & \text{for } 0 \le t \le 3 \\ 0 & \text{otherwise} \end{cases}$$

When a constant input of value 5 is applied to this filter, the steady state output is _____.

3. A Y-network has resistances of 10 Ω each in two of its arms, while the third arm has a resistance of 11 Ω. In the equivalent Δ-network, the lowest value (in Ω) among the three resistances is _____.

4. A 230 V rms source supplies power to two loads connected in parallel. The first load draws 10 kW at 0.8 leading power factor and the second one draws 10 kVA at 0.8 lagging power factor. The complex power delivered by the source is

(a) (18 + j 1.5) k V A
(b) (18 − j 1.5) k VA
(c) (20 + j 1.5) k V A
(d) (20 − j 1.5) k VA

5. A periodic variable x is shown in the figure as a function of time. The root-mean-square (rms) value of x is _____.

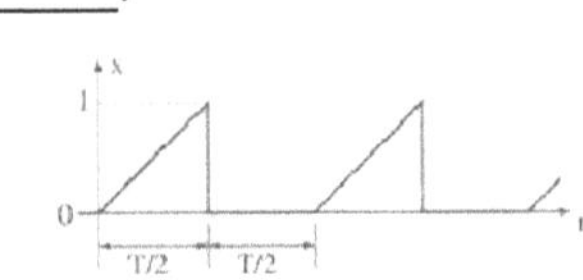

6. In the circuit shown in the figure, the value of capacitor C (in mF) needed to have critically damped response i(t) is ________.

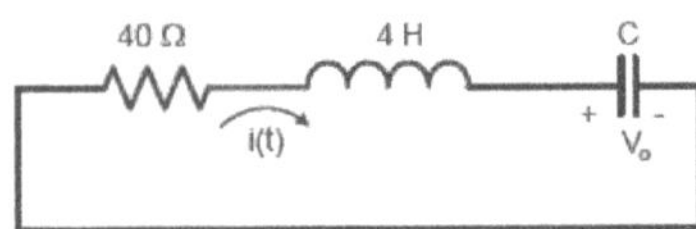

7. A system is described by the following differential equation,

where $u(t)$ is the input to the system and (t) is the output of the system.

$$\dot{y}(t)+5y(t)=u(t)$$

When y(0) = 1 and u(t) is a unit step function, $y(t)$ is

(a) $0.2+0.8e^{-5t}$

(b) $0.2-0.2e^{-5t}$

(c) $0.8+0.2e^{-5t}$

(d) $0.8-0.8e^{-5t}$

8. Norton's theorem states that a complex network connected to a load can be replaced with an equivalent impedance

(a) in series with a current source

(b) in parallel with a voltage source

(c) in series with a voltage source

(d) in parallel with a current source

9. In the figure shown, the ideal switch has been open for a long time. If it is closed at $t = 0$, then the magnitude of the current (in mA) through the 4 kΩ resistor at $t = 0^+$ is ________.

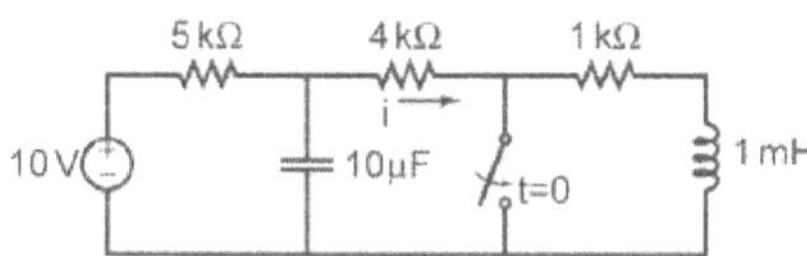

10. A series LCR circuit is operated at a frequency different from its resonant frequency. The operating frequency is such that the current leads the supply voltage. The magnitude of current is half the value at resonance. If the values of L, C and R are 1 H, 1 F and 1 Ω, respectively, the operating angular frequency (in rad/s) is ________.

11. In the h-parameter model of the 2-port network given in the figure shown, the value of h_{22} (in S) is ________.

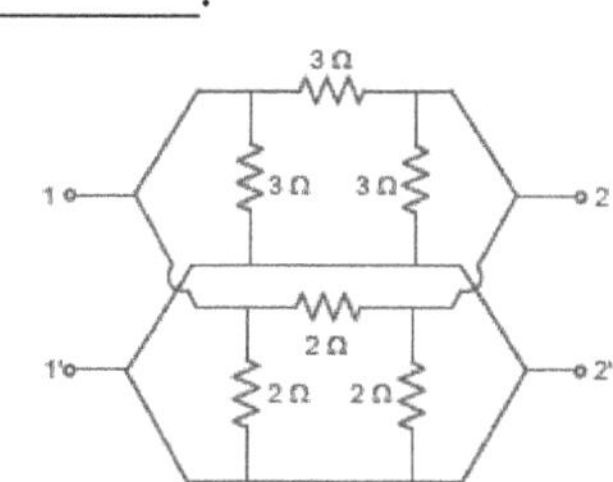

12. In the figure shown, the capacitor is initially uncharged. Which one of the following expressions describes the current I(t) (in mA) for $t > 0$?

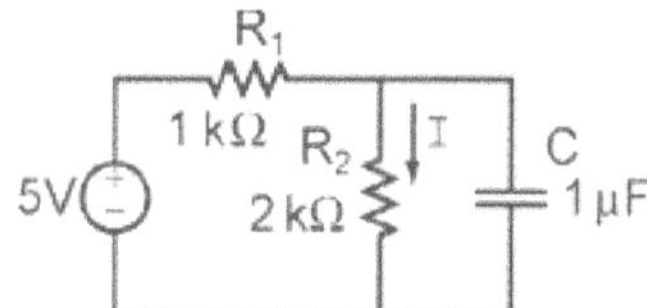

(a) $I(t)=\frac{5}{3}\left(1-e^{-t/\tau}\right), \tau=\frac{2}{3}\text{msec}$

(b) $I(t)=\frac{5}{2}\left(1-e^{-t/\tau}\right), \tau=\frac{2}{3}\text{msec}$

(c) $I(t)=\frac{5}{3}\left(1-e^{-t/\tau}\right), \tau=3\text{msec}$

(d) $I(t)=\frac{5}{2}\left(1-e^{-t/\tau}\right), \tau=3\text{msec}$

13. In the magnetically coupled circuit shown in the figure, 56 % of the total flux emanating from one coil links the other coil. The value of the mutual inductance (in H) is ________.

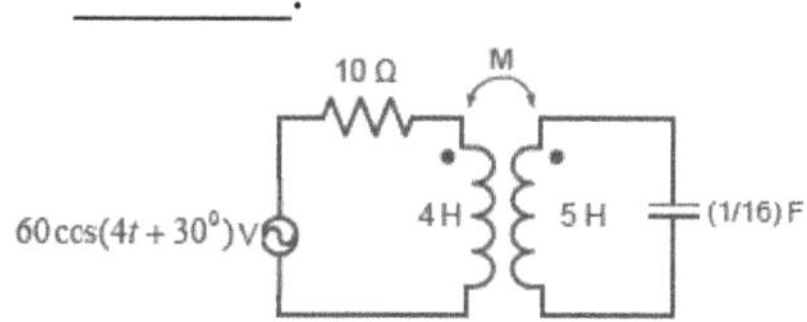

14. A series RC circuit is connected to a DC voltage source at time t = 0. The relation between the source voltage V_S, the resistance R, the capacitance C, and the current i(t) is given below:

$$V_s = Ri(t) + \frac{1}{c}\int_0^t i(u)\,du.$$

Which one of the following represent the current i(t) ?

(a) 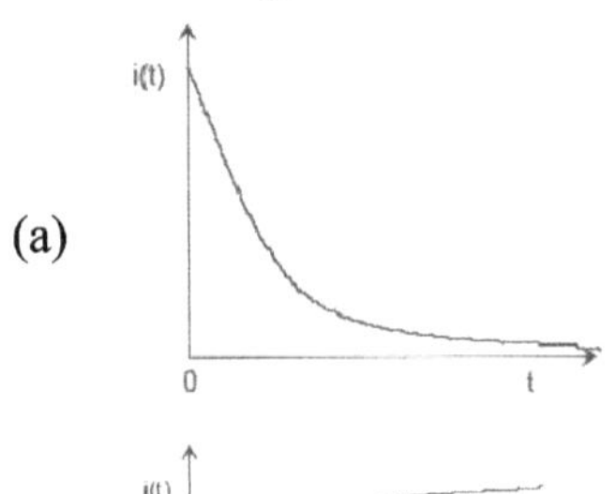

(b) 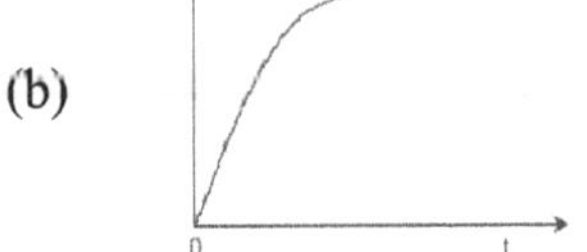

(c) 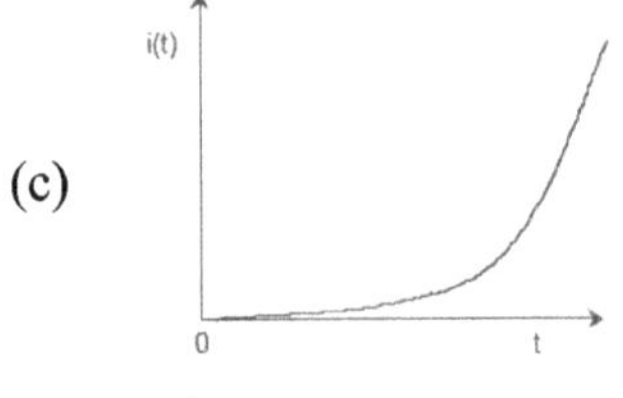

(d) 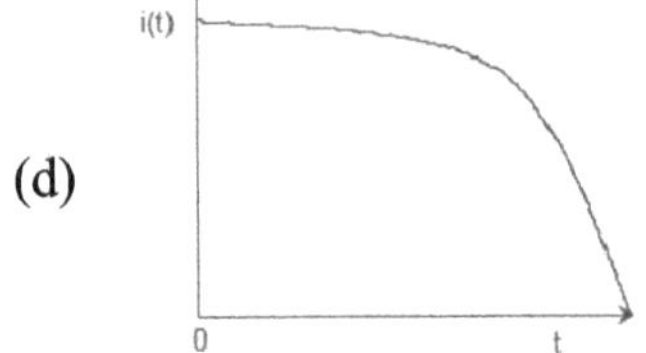

15. In the figure shown, the value of the current I (in Amperes) is ________.

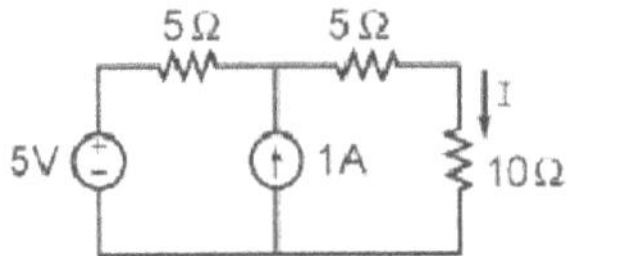

16. The input $-3e^{2t}u(t)$, where $u(t)$ is the unit step function, is applied to a system with transfer function $\frac{s-2}{s+3}$.

If the initial value of the output is −2, then the value of the output at steady state is ________.

17. Consider the building block called 'Network N' shown in the figure. Let C = 100 μF and R = 10 kΩ.

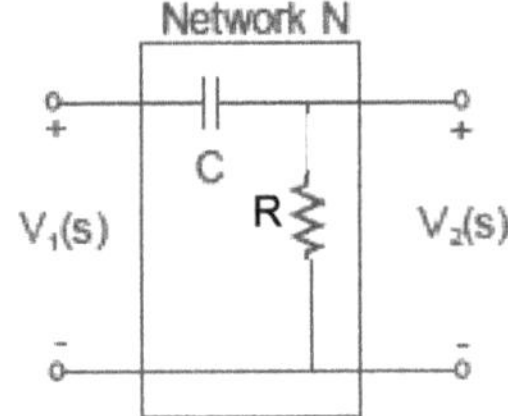

Two such blocks are connected in cascade, as shown in the figure.

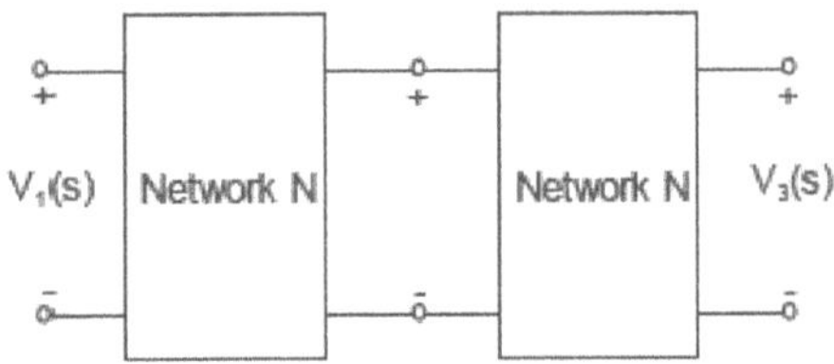

The transfer function $\frac{V_3(s)}{V_1(s)}$ of the cascaded network is

(a) $\frac{s}{1+s}$ (b) $\frac{s^2}{1+3s+s^2}$

(c) $\left(\frac{s}{1+s}\right)^2$ (d) $\frac{s}{2+s}$

18. In the circuit shown in the figure, the value of node voltage V_2 is

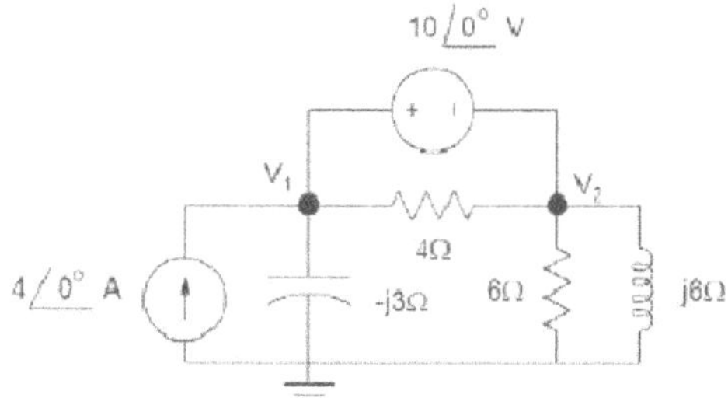

(a) 22 + j 2 V
(b) 2 + j 22 V
(c) 22 – j 2 V
(d) 2 – j 22 V

19. In the circuit shown in the figure, the angular frequency ω (in rad/s), at which the Norton equivalent impedance as seen from terminals b-b′ is purely resistive, is ________.

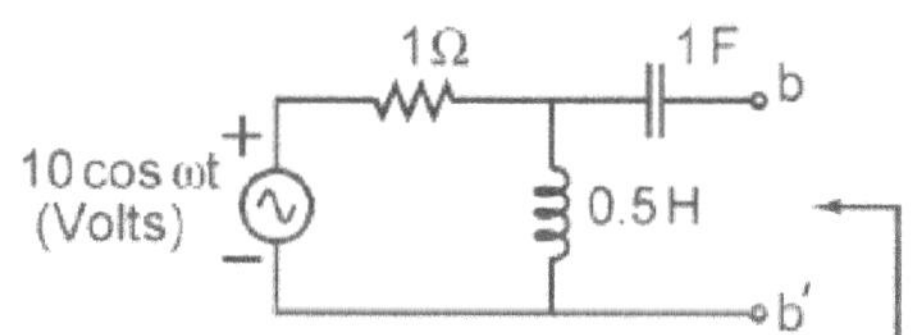

20. For the Y-network shown in the figure, the value of R_1 (in Ω) in the equivalent Δ-network is ________.

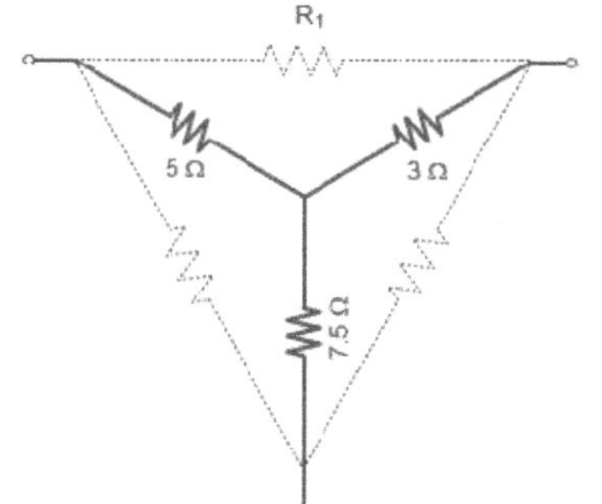

21. The circuit shown in the figure represents a

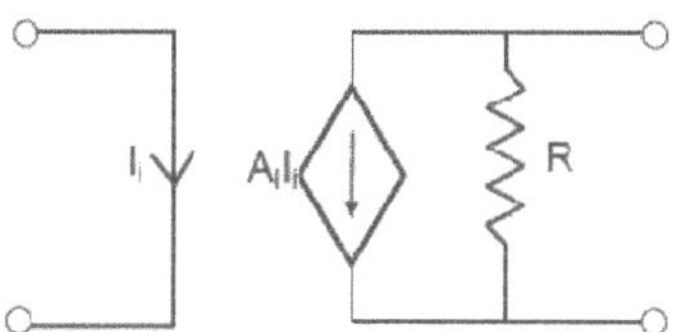

(a) voltage controlled voltage source
(b) voltage controlled current source
(c) current controlled current source
(d) current controlled voltage source

22. The magnitude of current (in mA) through the resistor R_2 in the figure shown is ________.

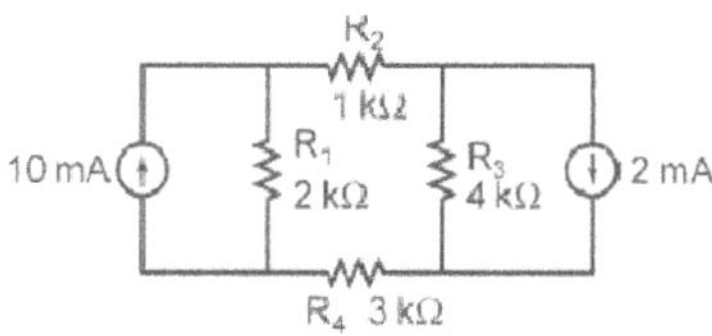

23. The unilateral Laplace transform of f(t) is $\frac{1}{s^2+s+1}$. Which one of the following is the unilateral Laplace transform of g(t) = t · f(t)?

(a) $\frac{-s}{\left(s^2+s+1\right)^2}$ (b) $\frac{-(2s+1)}{\left(s^2+s+1\right)^2}$

(c) $\frac{s}{\left(s^2+s+1\right)^2}$ (d) $\frac{2s+1}{\left(s^2+s+1\right)^2}$

24. The steady state output of the circuit shown in the figure is given by y(t) = A(ω) sin (ωt + ϕ(ω)). If the

amplitude $|A(\omega)| = 0.25,$ then the frequency ω is

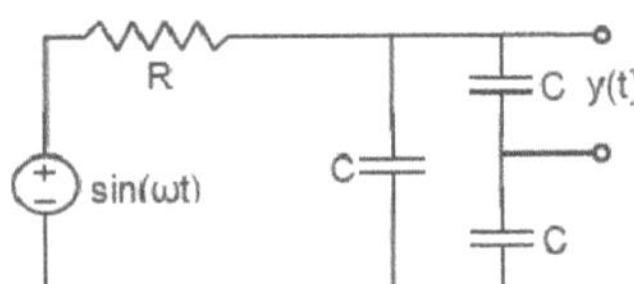

(a) $\frac{1}{\sqrt{3}RC}$ (b) $\frac{2}{\sqrt{3}RC}$

(c) $\frac{1}{RC}$ (d) $\frac{2}{RC}$

25. In the circuit shown in the figure, the value of v_0 (t) (in Volts) for $t \to \infty$ is ______.

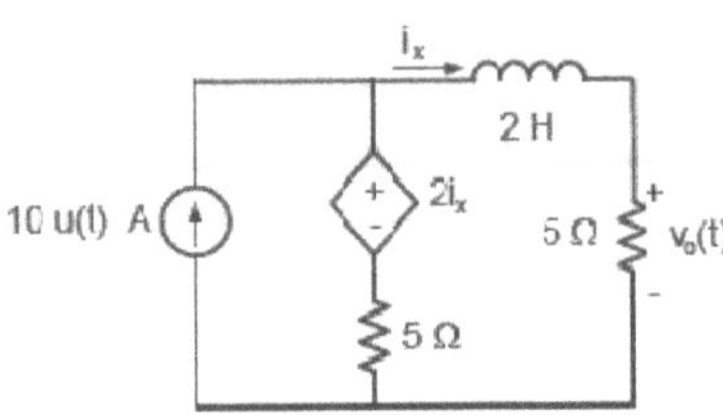

26. The equivalent resistance in the infinite ladder network shown in the figure, is R_e.

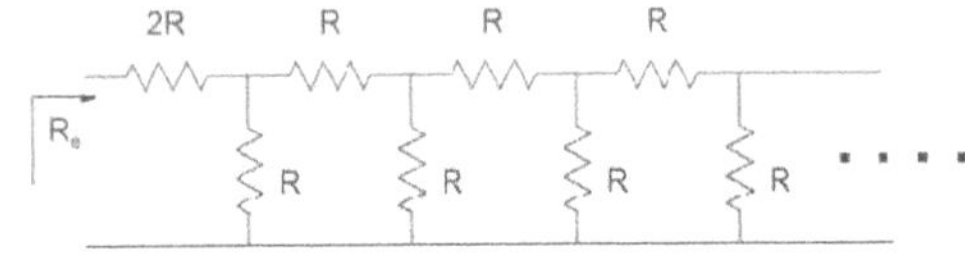

The value of R_e/R is ________

27. For the two-port network shown in the figure, the impedance (Z) matrix (in Ω) is

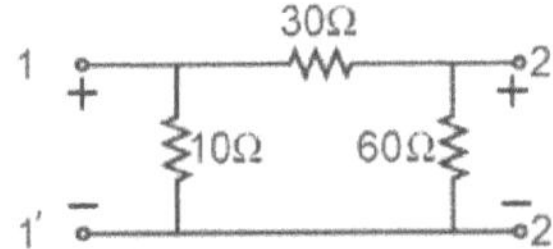

(a) $\begin{bmatrix} 6 & 24 \\ 42 & 9 \end{bmatrix}$ (b) $\begin{bmatrix} 9 & 8 \\ 8 & 24 \end{bmatrix}$

(c) $\begin{bmatrix} 9 & 6 \\ 6 & 24 \end{bmatrix}$ (d) $\begin{bmatrix} 42 & 6 \\ 6 & 60 \end{bmatrix}$

2014 EEE

28. A function f(t) is shown in the figure The Fourier transform F(ω) of f(t) is

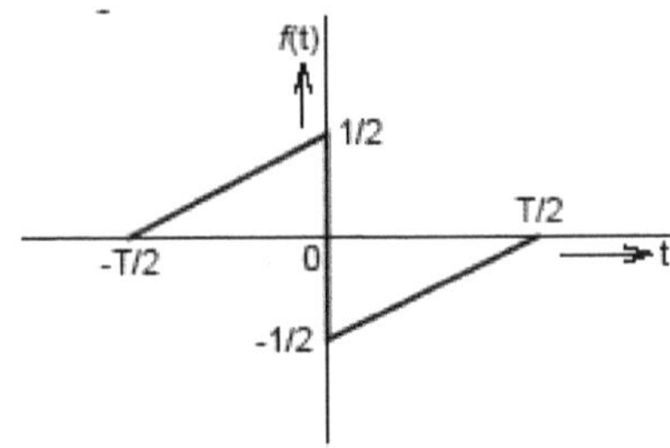

(a) real and even function of ω
(b) real and odd function of ω
(c) imaginary and odd function of ω
(d) imaginary and even function of ω

29. The line A to neutral voltage is $10\angle 15^\circ$ V for a balanced three phase star-connected load with phase sequence ABC. The voltage of line B with respect to line C is given by

(a) $10\sqrt{3}\angle 105^\circ$ V

(b) $10\angle 105^\circ$ V

(c) $10\sqrt{3}\angle -75^\circ$ V

(d) $-10\sqrt{3}\angle 90^\circ$ V

30. The driving point impedance Z(s) for the circuit shown below is

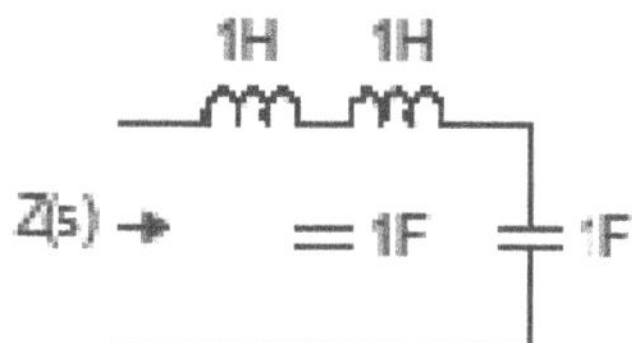

(a) $\dfrac{s^4+3s^2+1}{s^3+2s}$

(b) $\dfrac{s^4+2s^2+4}{s^2+2}$

(c) $\dfrac{s^2+1}{s^4+s^2+1}$

(d) $\dfrac{s^3+1}{s^4+s^2+1}$

31. A signal is represented by

$$x(t)=\begin{cases}1 & |t|<1\\ 0 & |t|>1\end{cases}$$

The Fourier transform of the convolved signal $y(t) = x(2t) * x\left(\dfrac{t}{2}\right)$ is

(a) $\dfrac{4}{\omega^2}\sin\left(\dfrac{\omega}{2}\right)\sin(2\omega)$

(b) $\dfrac{4}{\omega^2}\sin\left(\dfrac{\omega}{2}\right)$

(c) $\dfrac{4}{\omega^2}\sin(2\omega)$

(d) $\dfrac{4}{\omega^2}\sin^2\omega$

32. The Norton's equivalent source in amperes as seen into the terminals X and Y is _______.

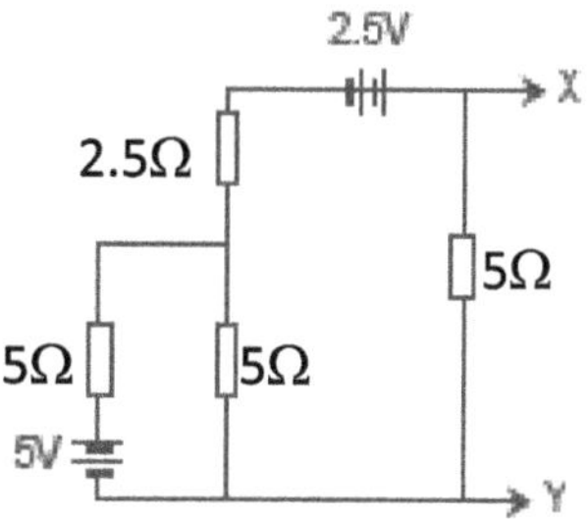

33. The power delivered by the current source, in the figure, is ________.

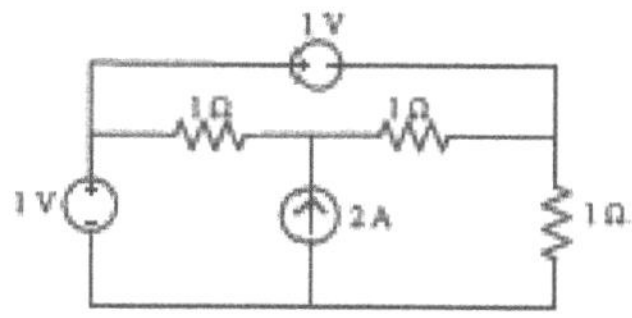

34. Assuming an ideal transformer, the Thevenin's equivalent voltage and impedance as seen from the terminals x and y for the circuit in figure are

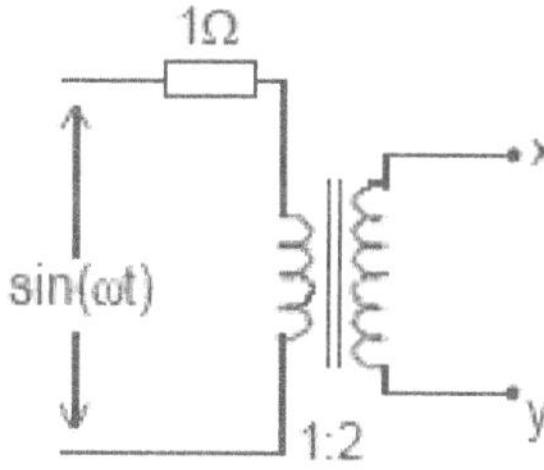

(a) $2\sin(\omega t), 4\Omega$

(b) $1\sin(\omega t), 1\Omega$

(c) $1\sin(\omega t), 2\Omega$

(d) $2\sin(\omega t), 0.5\Omega$

35. While measuring power of a three-phase balanced load by the two-wattmeter method, the readings are 100 W and 250 W. The power factor of the load is _________.

36. The voltage across the capacitor, as shown in the figure, is expressed as
$$V_c(t) = A_1 \sin(\omega_1 t - \theta_1) + A_2 \sin(\omega_2 t - \theta_2)$$

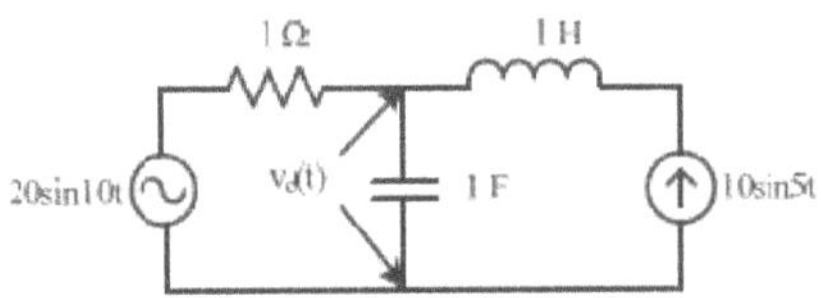

The values of A_1 and A_2 respectively, are

(a) 2.0 and 1.98
(b) 2.0 and 4.20
(c) 2.5 and 3.50
(d) 5.0 and 6.40

37. The total power dissipated in the circuit, shown in the figure, is 1 kW.

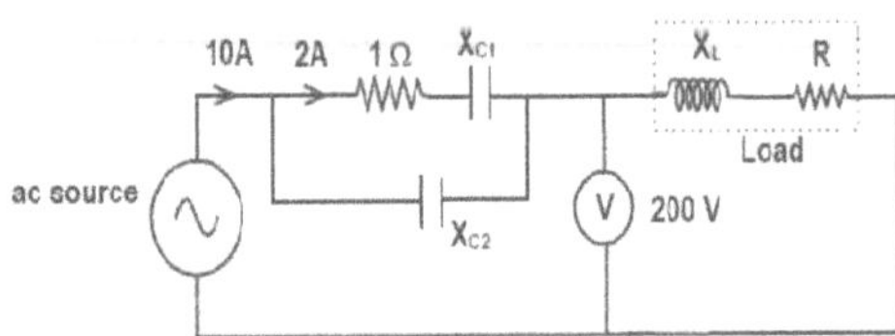

The voltmeter, across the load, reads 200 V. The value of X_L is ________.

38. Let $X(s) = \frac{3s+5}{s^2+10s+21}$ be the Laplace Transform of a signal x(t). Then, $x(0^+)$ is

(a) 0 (b) 3
(c) 5 (d) 21

39. The three circuit elements shown in the figure are part of an electric circuit. The total power absorbed by the three circuit elements in watts is ________.

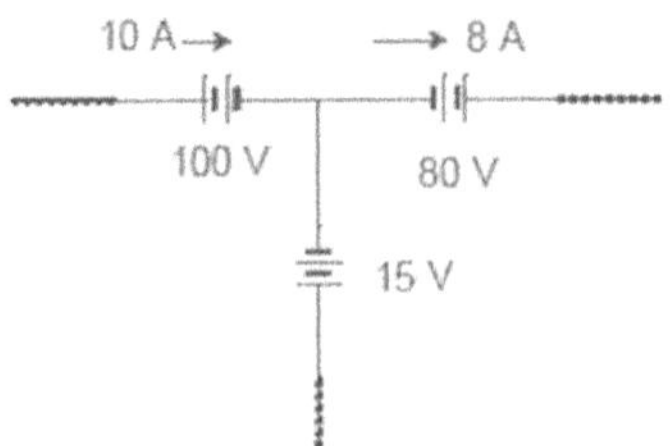

40. A combination of 1 μF capacitor with an initial voltage $v_c(0) = -2V$ in series with a 100 Ω resistor is connected to a 20 mA ideal dc current source by operating both switches at t = 0 s as shown. Which of the following graphs shown in the options approximates the voltage v_s across the current source over the next few seconds?

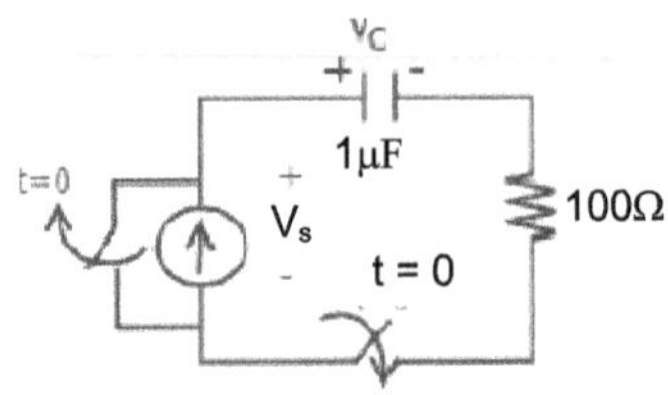

(a)

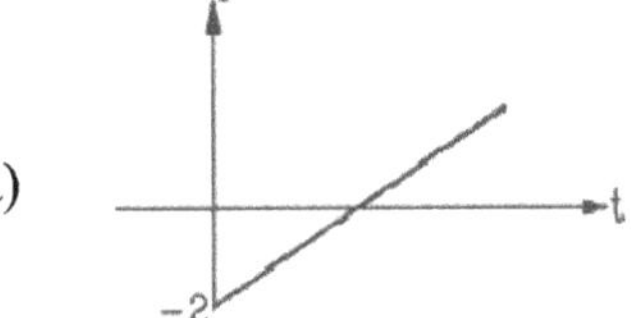

(b)

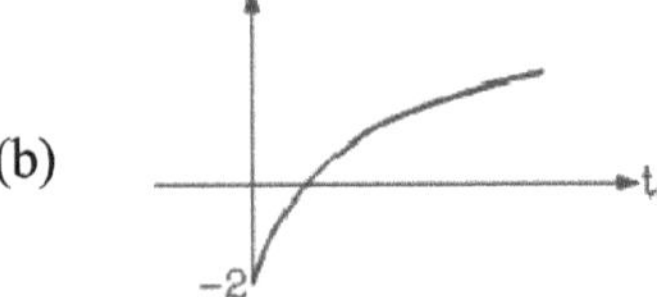

(c) 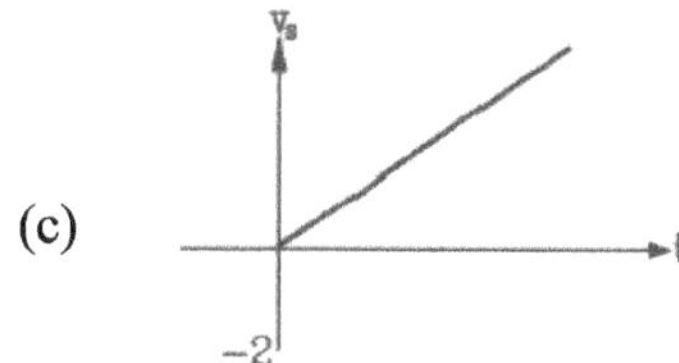

(d) 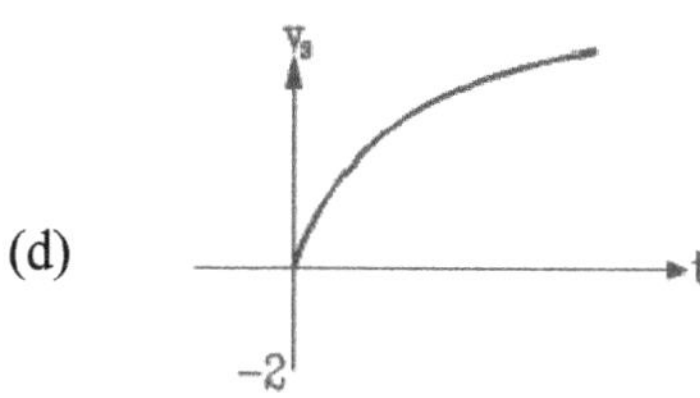

41. x(t) is nonzero only for $T_x < t < T'_x$ and similarly, y(t) is nonzero only for $T_y < t < T'_y$. Let z(t) be convolution on x(t) and y(t). Which one of the following statements is True?
   (a) z(t) can be nonzero over an unbounded interval.
   (b) z(t) is nonzero for $t < T_x + T_y$.
   (c) z(t) is zero outside of $T_x + T_y < t < T'_x + T'_y$.
   (d) z(t) is nonzero for $t > T'_x + T'_y$.

42. For a periodic square wave, which one of following statements is TRUE?
   (a) The Fourier series coefficients do not exist.
   (b) The Fourier series coefficients exist but the reconstruction converges at no point.
   (c) The Fourier series coefficients exist and the reconstruction converges at most points.
   (d) The Fourier series coefficients exist and the reconstruction converges at every point.

43. Power consumed by a balanced 3-phase, 3-wire load is measured by the two wattmeter method.
The first wattmeter reads twice that of the second. Then the load impedance angle in radians is
   (a) $\pi/12$ (b) $\pi/8$
   (c) $\pi/6$ (d) $\pi/3$

44. In the figure, the value of resistor R is (25 + I/2) ohms, where I is the current in amperes. The current I is ________ .

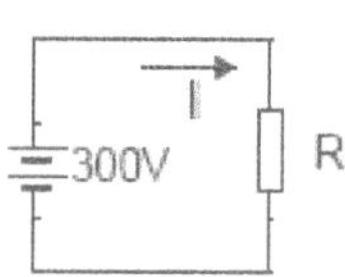

45. The function shown in the figure can be represented as

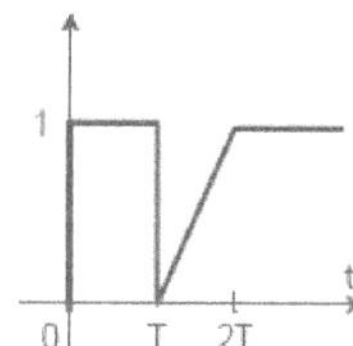

   (a) $u(t)-u(t-T)+\frac{(t-T)}{T}u(t-T)-\frac{(t-2T)}{T}u(t-2T)$
   (b) $u(t)+\frac{t}{T}u(t-T)-\frac{t}{T}u(t-2T)$
   (c) $u(t)-u(t-T)+\frac{(t-T)}{T}u(t)-\frac{(t-2T)}{T}u(t)$
   (d) $u(t)+\frac{(t-T)}{T}u(t-T)-2\frac{(t-2T)}{T}u(t-2T)$

46. The reading of the voltmeter (rms) in volts, for the circuit shown in the figure is _______.

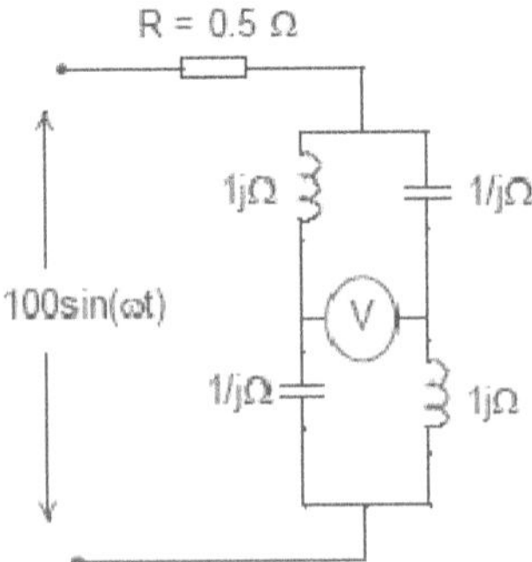

47. Let f(t) be a continuous time signal and let F(ω) be its Fourer Transform defined by

$$F(\omega) = \int_{-\infty}^{\infty} f(t)e^{-j\omega t}\,dt$$

Define g(t) by

$$g(t) = \int_{-\infty}^{\infty} F(u)e^{-jut}\,du$$

What is the relationship between f(t) and g(t)?

(a) g(t) would always be proportional to f(t).

(b) g(t) would be proportional to f(t) if f(t) is an even function.

(c) g(t) would be proportional to f(t) only if f(t) is a sinusoidal function.

(d) g(t) would never be proportional to f(t)

Answers

2004

1. d	2. a	3. a	4. c	5. c
6. d	7. b	8. b	9. d	10. a
11. d	12. d	13. b	14. a	15. c
16. a	17. d	18. a	19. b	20. d

2005

1. c	2. b	3. b	4. c	5. a
6. b	7. c	8. d	9. b	10. b
11. c	12. a	13. d	14. b	15. c
16. a	17. a	18. c	19. a	20. a
21. d	22. d	23a. c	23b. b	

2006

1. c	2. d	3. b	4. a	5. c
6. b	7. c	8. d	9. d	10. a
11. d	12. a	13. b	14. d	

2007

1. d	2. c	3. d	4. d	5. d
6. a	7. d	8. b	9. d	10. a
11. b	12. c	13. d	14. c	15. a
16. a	17. d	18. a		

2008

1. c	2. b	3. d	4. a	5. d
6. d	7. d	8. b	9. c	10. a
11. a	12. b	13. c	14. d	15. c
16. c	17. a	18. b	19. b	20. b
21. c	22. d			

2009

1. a	2. b	3. a	4. c	5. b
6. b	7. c	8. a	9. b	10. a
11. d	12. d	13. a	14. a	15. b
16. c	17. b	18. d		

2010

1. c	2. a	3. d	4. d	5. a
6. a	7. a	8. d	9. b	10. a
11. d	12. b	13. c	14. d	15. d
16. c	17. b			

2011

1. a	2. a	3. a	4. c	5. a
6. c	7. b	8. d	9. a	10. d
11. b	12. d	13. d	14. b	15. a
16. a	17. b	18. c	19. b	20. d

2012

1. c	2. a	3. d	4. d	5. c
6. a	7. d	8. a	9. c	10. a
11. b	12. c			

2013

1. c	2. b	3. a	4. b	5. d
6. c	7. c	8. c	9. b	10. d
11. d	12. c	13. b		

2014

1. a	2. 44 to 46	3. 29.08 to 28.10	4. b
5. 0.39 to 0.42	6. 9.99 to 10.01	7. a	8. d
9. 1.2 to 1.3	10. 0.45 to 0.47	11. 1.24 to 1.26	12. a
13. 2.49 to 2.52	14. a	15. 0.49 to 0.51	16. –0.01 to 0.01
17. b	18. d	19. 1.9 to 2.1	20. 9 to 11
21. c	22. 2.79 to 2.81	23. d	24. b
25. 31.24 to 31.26		26. 2.60 to 2.64	27. c
28. c	29. c	30. a	31. a
32. 1 to 1	33. 3 to 3	34. a	35. 0.78 to 0.82
36. a	37. 17.3 to 17.4	38. b	39. 330 to 330
40. c	41. c	42. c	43. c
44. 10 to 10	45. a	46. 140 to 142	47. b

www.ingramcontent.com/pod-product-compliance
Ingram Content Group UK Ltd.
Pitfield, Milton Keynes, MK11 3LW, UK
UKHW061535190726
13845UKWH00005B/112